ENVIRONMENTAL SCIENCE

earth as a living planet

CANADIAN EDITION

Daniel B. Botkin

Professor Emeritus
Department of Ecology, Evolution, and Marine Biology
University of California, Santa Barbara

President
The Center for the Study of the Environment
Santa Barbara, California

Edward A. Keller

Professor
Environmental Studies and Geological Sciences
University of California, Santa Barbara

Isobel W. Heathcote

Professor
Faculty of Environmental Sciences and School of Engineering
University of Guelph

John Wiley & Sons Canada, Ltd.

Library and Archives Canada Cataloguing in Publication

Botkin, Daniel B

Environmental science : earth as a living planet / Daniel B. Botkin, Edward A. Keller, Isobel W. Heathcote. -- Canadian ed.

Includes bibliographical references and index.

ISBN-13: 978-0-470-83688-0
ISBN-10: 0-470-83688-1

1. Environmental sciences--Textbooks. I. Keller, Edward A., 1942- II. Heathcote, Isobel W. III. Title.

GE105.B68 2005 363.7 C2005-904585-X

Production Credits

Acquisitions Editor: Michael Valerio
Publishing Services Director: Karen Bryan
Developmental Editor: Zoë Craig
Marketing Manager: Sean O'Reilly
New Media Editor: Elsa Passera
Cover Photography: Kristie Macor
Cover Design: Interrobang Graphic Design, Inc.
Interior Design: OrangeSprocket Communications
Printing and Binding: Tri-Graphic Printing Limited

Printed and bound in Canada
10 9 8 7 6 5 4 3 2 1

John Wiley & Sons Canada, Ltd.
6045 Freemont Blvd.
Mississauga, Ontario L5R 4J3
Visit our website at: www.wiley.ca

DOMTAR EarthChoice®

FSC
Mixed Sources
Product group from well-managed forests, controlled sources and recycled wood or fiber
Cert no. SW-COC-681
www.fsc.org
© 1996 Forest Stewardship Council

Printed in Canada by SW-COC-1352.
Item #PRO0304160

DEDICATIONS

For **Jane O'Brien** who, during her life, gave joy and spirit to my work.

Dan Botkin

For **Valery Rivera** who contributed so much to this book and is a fountain of inspiration in our work and lives.

Ed Keller

For **John H. McAndrews**, **George Evelyn Hutchinson**, and **James W. MacLaren**, whose wisdom, kindness, and passion for their disciplines have had a profound influence on my life and work.

Isobel Heathcote

ABOUT THE AUTHORS

Daniel B. Botkin is President of The Center for the Study of Environment and Professor Emeritus of Ecology, Evolution and Marine Biology, University of California, Santa Barbara. From 1978 to 1993, he was Professor of Biology and Environmental Studies at the University of California, Santa Barbara, serving as Chairman of the Environmental Studies Program from 1978 to 1985. For more than three decades, Professor Botkin has been active in the application of ecological science to environmental management. He is the winner of the Mitchell International Prize for Sustainable Development and the Fernow Prize for International Forestry, and he has been elected to the California Environmental Hall of Fame.

Trained in physics and biology, Professor Botkin is a leader in the application of advanced technology to the study of the environment. The originator of widely used forest gap-models, his research has involved endangered species, characteristics of natural wilderness areas, the study of the biosphere, and attempts to deal with global environmental problems. During his career, Professor Botkin has advised the World Bank about tropical forests, biological diversity, and sustainability; the Rockefeller Foundation about global environmental issues; the government of Taiwan about approaches to solving environmental problems; and the state of California on the environmental effects of water diversion on Mono Lake. He served as the primary advisor to the National Geographic Society for their centennial edition map on "The Endangered Earth." He recently directed a study for the states of Oregon and California concerning salmon and their forested habitats.

He has published many articles and books about environmental issues. His latest books are *Beyond the Stoney Mountains: Nature in the American West from Lewis and Clark to Today* (Oxford University Press), *Strange Encounters: Adventures of a Renegade Naturalist* (Penguin/Tarcher), *The Blue Planet* (Wiley), *Our Natural History: The Lessons of Lewis and Clark* (Putnam), *Discordant Harmonies: A New Ecology for the 21st Century* (Oxford University Press), and *Forest Dynamics: An Ecological Model* (Oxford University Press).

Professor Botkin was on the faculty of the Yale School of Forestry and Environmental Studies (1968–1974) and was a member of the staff of the Ecosystems Center at the Marine Biological Laboratory, Woods Hole, MA (1975–1977). He received a B.A. from the University of Rochester, an M.A. from the University of Wisconsin, and a Ph.D. from Rutgers University.

Edward A. Keller was chair of the Environmental Studies and Hydrologic Sciences Programs from 1993 to 1997 and is Professor of Geological Sciences at the University of California, Santa Barbara, where he teaches geomorphology, environmental geology, environmental science, river processes, and engineering geology. Prior to joining the faculty at Santa Barbara, he taught geomorphology, environmental studies, and earth science at the University of North Carolina, Charlotte. He was the 1982–1983 Hartley Visiting Professor at the University of Southampton and a Visiting Fellow in 2000 at Emmanuel College of Cambridge University, England.

Professor Keller has focused his research efforts in three areas: studies of Quaternary stratigraphy and tectonics as they relate to earthquakes, active folding, and mountain building processes; hydrologic process and wildfire in the chaparral environment of southern California; and physical habitat requirements for the endangered southern California steelhead trout. He is the recipient of various Water Resources Research Center grants to study fluvial processes and U.S. Geological Survey and Southern California Earthquake Center grants to study earthquake hazards.

Professor Keller has published numerous papers, and is the author of the textbooks *Environmental Geology, Introduction to Environmental Geology* and (with Nicholas Pinter) *Active Tectonics* (Prentice-Hall). He holds bachelors degrees in both geology and mathematics from California State University, Fresno; an M.S. in geology from the University of California; and a Ph.D. in geology from Purdue University.

Isobel W. Heathcote is Professor of Environmental Science and Environmental Engineering at the University of Guelph, where she is also Dean of the Faculty of Graduate Studies. Her work experience is diverse, encompassing employment in both the public and private sectors. From 1979 to 1985, she was employed by the Ontario Ministry of the Environment's Water Resources Branch, where she worked on water management issues in all the major lake and river systems in Ontario, with special emphasis on watershed management planning. As Supervisor, Great Lakes Investigations and Surveillance, she and her staff conducted water pollution studies throughout the Great Lakes, especially in the Areas of Concern, and contributed to the development of Remedial Action Plans for those areas.

From 1985 to 1991, Isobel directed and taught in the Environmental Studies and Environmental Sciences programs at the University of Toronto. She joined the University of Guelph in 1991. Her research interests centre on environmental management policy, integrated water management and watershed restoration, and public participation in environmental policy development. She has chaired and served on numerous environmental advisory committees, including providing advice to six Ontario Ministers of the Environment on the development of water management policy. Since 2001, Professor Heathcote has been the Canadian Co-Chair of the Canada-United States International Joint Commission's Science Advisory Board. She is also a former President of the Board of Directors of the Canadian Institute for Environmental Law and Policy, and a former member of the Board of Directors of the Canadian Pollution Prevention Centre. In addition to numerous papers, reports, and policy commentaries, she is the author of *Environmental Problem Solving: A Case Study Approach* (McGraw-Hill), and *Integrated Watershed Management: Principles and Practice* (Wiley). She holds a B.Sc. from the University of Toronto and an M.S. and Ph.D. in biology from Yale University.

PREFACE

Botkin and Keller's *Environmental Science: Earth as a Living Planet* is one of the classic texts in environmental science. Long recognized for its outstanding science content and explanatory graphics, the book is also known for its balance and clarity. This Canadian edition continues to offer the reader an introduction to the scientific foundations of environmental science and an appreciation of the wide-ranging sources of this highly topical discipline, all within the context of the Canadian environment and culture.

The Canadian edition has undergone extensive review by environmental instructors across Canada, and their feedback has been carefully and thoroughly incorporated into this edition. We believe the result is a thorough, relevant and engaging Canadian environmental science text.

GOALS OF THIS BOOK

The foremost goal of this text is to teach you, the reader, how to think about environmental issues using the natural sciences as a foundation, and how to be able to think critically about this crucial subject. Many disciplines influence the study of the environment, and this text serves to introduce this information in an analytical and interdisciplinary manner. As a result, *Environmental Science* provides an introduction to the entire spectrum of relationships between people and the environment. Based on the philosophy that several threads of inquiry are of particular importance to environmental science, we must do more than simply identify and discuss environmental problems and solutions. To be effective, we must know what science is, what it is not, and how to use it.

WHAT IS ENVIRONMENTAL SCIENCE AND WHY IS IT IMPORTANT?

Environmental science is a group of sciences that attempt to explain how life on Earth is sustained. We depend on our environment, and people can live only in an environment with certain kinds of characteristics and within certain ranges of availability of resources. Because modern science and technology give us the power to affect the environment, we have to understand how the environment works so that we can live within its constraints.

WHAT IS THE "SCIENCE" IN ENVIRONMENTAL SCIENCE?

Many sciences are important to environmental science. These include natural and physical sciences such as biology (especially ecology, the part of biology that deals with the relationships among living things and their environment), geology, hydrology, climatology, meteorology, oceanography, and soil science. It also includes several social sciences, especially economics and law.

WHAT IS DIFFERENT ABOUT ENVIRONMENTAL SCIENCE COMPARED TO OTHER SCIENCES?

It involves many sciences, but it also deals with many topics that have great emotional effects on people, and therefore are subject to political debate and strong feelings that often ignore scientific information. Being "mission-oriented," it is also aimed at solving real environmental problems, rather than just understanding how they arise.

ORGANIZATION

This text is divided into four parts. Part I provides a broad overview of the key themes in *Environmental Science*, the scientific method, thinking critically about the environment, and basic concepts of ecosystem structure and function. Part II focuses on how humans impact the natural environment through a range of activities including world food supply, agriculture, conserving and managing life on Earth, and energy production and use. Part III presents the water environment on Earth, in terms of water supply use and management, and water pollution treatment; the air environment, from global issues such as climate, global warming, and stratospheric ozone depletion to regional issues such as acid rain, to local issues including urban air pollution and indoor air pollution; and waste management. Finally, Part IV is concerned with relationships between environment and society. Topics include environmental economics, the urban environment, environmental impact and planning, and how we might achieve sustainability.

THEMES

Throughout the text six themes have been highlighted in boxes called A Closer Look in order to help the reader appreciate the inter-relationships in this discipline. They are:

Human Population

Underlying nearly all environmental problems is the rapidly increasing human population. Ultimately, we cannot expect to solve these other problems unless we can limit the total number of people on Earth to an amount the environment can sustain. We believe that education is important to solving the population problem. As people become more educated, and as the rate of literacy increases, population growth tends to decrease.

Sustainability

Sustainability is a term that has gained much popularity recently. Speaking generally, it means that a resource is used in such a way that it continues to be available. However, the term is used vaguely and it is something we are struggling to clarify. Some would define it as ensuring that future generations have equal opportunities in relation to the resources that our planet offers. Others would argue that sustainability refers to types of developments that are economically viable, do not harm the environment, and are socially just. We all agree that we must learn how to sustain our environmental resources so that they continue to provide benefits for people and other living things on our planet.

A Global Perspective

Until recently, we generally believed that human activity caused only local or, at most, regional environmental change. We now know that effects of human activity on Earth are of such an extent that we are involved in a series of unplanned planetary experiments. The main goal of the emerging science known as Earth System Science is to obtain basic understanding of how our planet works as a system. This understanding can then be applied to help solve global environmental problems. The emergence of Earth System Science has opened up a new area of inquiry for faculty and students. Understanding the relationships between biological and physical sciences requires interdisciplinary cooperation and education.

The Urban World

An ever-growing number of people are living in urban areas. Unfortunately, our urban centres have long been neglected and the quality of the urban environment has suffered. It is often here that we experience air pollution, waste disposal problems, social unrest, and other stresses of the environment. In the past we have centred our studies of the environment more on wilderness than the urban environment. In the future we must place greater focus on towns and cities as livable environments.

People and Nature

People seem to be always interested—amazed, fascinated, pleased, made curious—by our environment. Why is it suitable for us? How can we keep it that way? We know that people and our civilizations are having major effects on the environment, from a local one (the street where you live) to the entire planet (we have created a hole in the Earth's ozone layer that can affect ourselves and many forms of life).

Science and Values

Finding solutions to environmental problems involves more than simply gathering facts and understanding the scientific issues of a particular problem. It also has much to do with our systems of values and issues of social justice. To solve our environmental problems, we must understand what our values are and which potential solutions are socially just. Then, we can apply scientific knowledge about specific problems and find acceptable solutions.

FEATURES OF THE CANADIAN EDITION

Environmental Science, Canadian Edition contains extensive reviewer-based content revisions and new photographs that reflect the unique Canadian environment, culture, and political structure. New modules were also developed to provide relevancy, and to promote a critical appreciation of the applications of environmental methods and concepts.

BUILD YOUR ENVIRONMENTAL SKILLS

These modules encourage students to develop key skills for environmental science, including quantitative skills such as calculating exponential growth or biological productivity, and qualitative skills such as evaluating the quality of scientific writing and jurisdictional responsibility for Canadian environmental issues.

HOW GREEN IS YOUR CAMPUS?

Highlighting how environmental concepts can be applied locally, these modules challenge students to investigate the environmental practices of their academic institution. Topics focus on include energy efficiency, water conservation, and an institutional commitment to sustainability.

CRITICAL THINKING ISSUES

At the end of each chapter, an issue is presented as a method of encouraging critical thinking about the environment and to help students understand how these issues may be studied and evaluated. For example, Chapter 19 presents the environmental issue of how polluted waters can be restored. The Issue in Chapter 23 explores the potential for mining using microorganisms rather than toxic chemicals.

SUPPLEMENTARY MATERIALS

Student Web Site (www.wiley.com/canada/botkin)

A completely new, redesigned, content-rich Web site has been created to provide enrichment activities and resources for students. These features include review of Learning Objectives, Image Gallery, Web links, Critical Thinking Issues, Virtual Field Trips, Environmental Debates, and Web links to important data and research in the field of environmental studies. New "Ask the Expert" boxes present interviews with Canadian environmental practitioners on topics presented in the text. For example, in Ask the Expert 25.1, environmental studies graduate George Ferreira talks about how he uses video techniques to give remote First Nations communities a voice in the federal policy process.

EMBRACING ENVIRONMENTALLY SUSTAINABLE PRACTICES

In planning this publication, efforts were taken to choose paper that would be the most environmentally friendly while maintaining high standards of quality. As a result, this text is printed on Domtar EarthCote, part of the Domtar EarthChoice® family of papers. It is Forest Stewardship Council (FSC) certified and contains 30% post-consumer waste recycled fibre. The Domtar EarthChoice® family of products is endorsed by the Rainforest Alliance and is independently certified by their certification program SmartWood, in accordance with the global standards of the FSC. Printed materials can only bear the FSC logo if all members of the supply chain (forest, pulp mill, paper manufacturer, merchant and printer) have obtained FSC certification. John Wiley & Sons Canada is proud to be the first publisher to use this type of environmentally friendly paper in a higher education textbook.

SmartWood is the forestry certification program of the Rainforest Alliance, a global non-profit conservation organization. Through its certification programs, Rainforest Alliance works with participating companies, co-operatives and landowners to ensure that they meet rigorous standards that conserve biodiversity and provide sustainable livelihoods. Further information on the Rainforest Alliance can be found at www.rainforestalliance.org.

The Forest Stewardship Council is an international non-profit association that supports the sustainable management of forests by developing standards for forest certification. The FSC's membership comprises environmental and social groups, as well as progressive forestry and wood retail companies. Please go to www.fsccanada.org for further information.

DOMTAR EarthChoice®

ACKNOWLEDGEMENTS

Completion of this book was only possible due to the cooperation and work of many people. To all those who so freely offered their advice and encouragement in this endeavour, we offer our most sincere appreciation. We are especially grateful to colleagues and friends who provided technical advice and images for this book, including Mark Anderson, Edward and Zoë Belk, E. Ann Clark, Dave Chambers, Ward Chesworth, Elspeth Evans, Will Gorlitz, Carole Ann Lacroix, Jock McAndrews, Cecelia Paine, Danielle Fortin (University of Ottawa), and Dave Schultz (Grand River Conservation Authority).

Wiley's Acquisitions Editor, Michael Valerio, has been a guiding force throughout the project, providing invaluable support, encouragement, and many helpful suggestions. Wiley's Developmental Editor, Zoë Craig, brought a sure and experienced hand to a complex production process. We also extend thanks to Maureen Moyes for her careful editing of the manuscript, Elizabeth Chong for her proofreading of all of the pages, Edwin Durbin for creating the index, Marg Anne Morrison for her tireless photo research, and Bill McGrath, Rob Oviatt, and the rest of the staff at OrangeSprocket Communications for their outstanding design work. Special thanks also goes to William Mark Buhay, University of Winnipeg, for all his contributions related to the text. We are especially indebted to Laura Mousseau for her extensive and always cheerful research support.

Early reviewers provided valuable feedback on the feasibility of writing a Canadian edition of Botkin and Keller, and they helped greatly in its direction. They were: William Mark Buhay, University of Winnipeg; Tony Price, University of Toronto at Scarborough; Vivien Taylor, Trent University; and Susan Vajoczki, McMaster University. Of particular importance to the development of the book were the individuals who read the book chapter by chapter and provided valuable comments and constructive criticism. This was a particularly challenging job given the wide variety of topics covered in the text, and we believe that the book could not have been successfully completed without their assistance. These reviewers are offered our special gratitude:

William Mark Buhay, *University of Winnipeg*
John Buschek, *Carleton University*
Linda Campbell, *Queen's University*
Tim Elkin, *Camosun College*
Mark Hanson, *University of Manitoba*
Peter V. Hodson, *Queen's University*
Lawrence E. Licht, *York University*
Jeffrey M. Long, *University of Manitoba*
Marshall L. McCall, *York University*
Monica Mulrennan, *Concordia University*
Hilary Sandford, *Camosun College*
Susan Vajoczki, *McMaster University*
Michael C. Wilson, *Douglas College*
Carl Wolfe, *York University*
Bert Weichel, *University of Saskatchewan*

Isobel Heathcote
Guelph, Ontario July 2005

BRIEF CONTENTS

CONTENTS

chapter 11

chapter 12

chapter 13

chapter 24

Part IV Managing Sustainably

chapter 25

chapter 26

Part I

The Structure and Function of Ecosystems

Ground level in a southern Ontario mixed hardwood forest. The forest floor is an important interface between non-living and living parts of the forest ecosystem.

Photo courtesy Carole Ann Lacroix, Herbarium Curator, University of Guelph.

chapter 1

Old Growth Forest in Temagami, Ontario: Local and Global Connections

Just outside the town of Temagami, Ontario, lie some of Canada's oldest forests. Giant cedars, yellow birch, and pines pre-dating the arrival of the earliest European settlers cover the steep hills around Blueberry Lake. Although much of the surrounding area was logged early in the twentieth century, much of Blueberry Lake's watershed remains intact, in part because access to the area is difficult, so other forests were logged first. As those regions were cleared, however, pressure to cut Temagami's old-growth forests has grown. The area badly needs the jobs that logging would create. But others believe that the forest, if left intact, could become the focus of a growing tourism industry.

Longstanding tensions between tourist operations, foresters, First Nations, and environmental groups have escalated in recent years, as provincial controls on logging operations have tightened, and public awareness of the issues has increased. In 1989, hundreds of protesters blocked logging operations in Temagami, creating the single largest act of civil disobedience in Canada up to that time. The protests were successful in saving the Obabika old-growth pine forest, the largest stand of old growth red and eastern white pine remaining in the world. Other areas are now scheduled for logging, however, including forests to the north of Blueberry Lake. Critics argue that current lumber demand can be met from existing second-growth forest. Less than 1% of Ontario's old-growth white pine forests remains intact, and forests like those around Blueberry Lake are now extremely rare.[1,2,3]

At the heart of the Temagami dispute is our incomplete understanding about the value, role, and dynamics of old-growth forests, and the tension between economic development and ecosystem protection. Are government policies adequate to provide sufficient protection for our old growth forests? If our demand for wood and paper cannot be met from second-growth forests, should we allow the logging of ancient trees, or should we instead be trying to change our patterns of consumption and waste generation?

Figure 1.1 • **Old-growth forest, Temagami. Ontario.**

Key Themes in Environmental Sciences

LEARNING OBJECTIVES

Certain themes are basic to environmental science. After reading this chapter, you should understand:

- That people and the environment are intimately connected.
- Why rapid human population growth is *the* fundamental environmental issue.
- Why we must learn to sustain our environmental resources.
- How human beings affect the global environment.
- Why urban environments need attention.
- Why solutions to environmental problems involve making value judgements.

- *The case study of the Blueberry Lake forests in Temagami, Ontario illustrates the major themes of environmental science. People and nature are intimately connected, and changes in one lead to changes in the other. Environmental issues and their solutions involve values and attitudes as well as scientific understanding. The Temagami story also illustrates important questions that we all must face: Which individual actions contribute to environmental degradation? What actions can people, both as individuals and as groups, take to limit environmental damage?*

1.1 Major Themes of Environmental Science

In A. D. 1, the world's population was approximately 100 million, roughly three times the current Canadian population. In 1960, the global population was 3 billion people. Our population has more than doubled in the last 40 years, to 6.3 billion people today. If recent human population growth rates continue, our numbers will reach 10 billion before 2040. The problem is that the Earth has not grown any larger and its available resources have not increased. How can the Earth sustain all these people? What is the maximum number of people that could live on the Earth—not just for a short time, but sustained over a long time?

Estimates of how many people the planet can support range from 2.5 billion to 40 billion. Estimates vary so widely because it depends on quality of life people are willing to accept: the poorer that quality, the greater the number of people that can be squeezed onto the Earth's surface. How many people the Earth can sustain depends on science and values, and people and nature. The more people packed onto the Earth, the less room and resources there are for wild animals and plants, wilderness, recreation areas, and other aspects of nature, and the faster Earth's resources will be used. The answer also depends on the distribution of the human population on Earth, whether it is concentrated in groups, for instance in cities, or spread evenly across the land.

Although the environment is complex and environmental issues sometimes cover an unmanageable number of topics, three issues—the human population, urbanization, and sustainability within a global perspective—are central, and must be evaluated in light of the interrelation between people and nature. The answers ultimately depend on both science and the environment.

Therefore, this book approaches environmental science through six interrelated themes:

- *Human population growth*, including...
- *An urbanizing world*, both of which raise the question of...
- *Sustainability of our population and all of nature*, leading therefore to...
- *People and nature* within...
- *A global perspective*, and the solutions to these depend on a combination of...
- *Science and values*

You may ask, "If this is all there is to it, what is in the rest of this book?" [See A Closer Look 1.1] The answer lies with the old saying: "The devil is in the details." The solution to specific environmental problems requires a variety of knowledge, much of it changing as our understanding of environmental systems evolves. The six themes help us see the big picture and provide a valuable background. The opening case study illustrates linkages among the themes and the importance of details. Individual use of wood products would not cause a major environmental problem if the individual were the only user. It is the huge number of people who want to use wood and paper and who need ways to make a living that makes this local problem a worldwide one.

In this chapter, we introduce the six themes with brief examples, showing the linkages among them and touching on the importance of specific knowledge that will be the concern of the rest of the book. We start with human population growth.

1.2 Human Population Growth

The John Eli Miller Family

John Eli Miller was an ordinary man except for one thing—when he died in the mid-twentieth century, he was the head of one of the largest families in North America (Figure 1.2). He was survived by 5 children, 61 grandchildren, 338 great-grandchildren, and 6 great-great-grandchildren. Within his lifetime, John Miller witnessed a family population explosion. What was perhaps even more remarkable was that the explosion started with a family of just 7 children—not that unusual for the Victorian era.[4]

During most of John Miller's life, his family was not unusually large. It is just that he lived long enough to find out what simple multiplication can do, and he lived in a time when the death rate among infants, children, and young adults was very low compared with typical death rates during the history of most human populations. Of his 7 children 5 survived him; of 63 grandchildren, 61 survived him; and of 341 great-grandchildren (born to 55 married grandchildren—an average of slightly more than 6 children per parent)—338 survived him.

John Miller's family emphasizes a major factor in our modern population explosion. Modern technology, modern medicine, and the supply of food, clothing, and shelter have decreased death rates and increased the net rate of growth. As a result, the human population has increased greatly. At the same time, our per capita consumption of natural resources has skyrocketed, with serious implications for the environment. William Rees of the University of British Columbia has popularized the notion of **ecological footprint**—the area of land that would be necessary to produce the resources used, and assimilate the wastes produced by a given population. Canadians have a much larger ecological footprint than people in less developed countries, because our per capita consumption of resources, and our generation of wastes, is so much higher than in most non-industrialized

A CLOSER LOOK 1.1

A Little Environmental History

A brief historical explanation will help clarify what we seek to accomplish. Before 1960, few people had ever heard the word *ecology*, and the term *environment* meant little as a political or social issue. Then Rachel Carson's landmark book, *Silent Spring* (Houghton Mifflin, Boston, 1960, 1962) was published. At about the same time, several major environmental events occurred such as major oil spills and the highly publicized threats of extinction of many species, including whales, elephants, and songbirds. The environment became a popular issue.

As with any new social or political issue, at first relatively few people recognized its importance. Those who did found it necessary to stress the problems—to emphasize the negative—in order to bring public attention to environmental concerns. Adding to the limitations of the early approach to environmental issues was a lack of scientific knowledge and practical know-how. Environmental sciences were in their infancies. Some people even saw science as part of the problem.

The early days of modern environmentalism were dominated by confrontations between those labelled environmentalists and those labelled anti-environmentalists. Prominent environmental non-government organizations, including Greenpeace (founded in Vancouver around 1970) and Toronto's Pollution Probe trace their origins to this period. Stated in the simplest terms, environmentalists believed that the world was in peril. To them, economic and social development meant the destruction of the environment and ultimately the end of civilization, the extinction of many species, and perhaps threatening even the existence of human beings. Their solution was a new worldview that depended only secondarily on facts, understanding, and science. In contrast, once again stated in the simplest term, the anti-environmentalists believed that social and economic health and progress were necessary—whatever the environmental effects—if people and civilization were to prosper. From their perspective, environmentalists represented a dangerous and extreme view, with a focus on the environment to the detriment of people—a focus they thought would destroy the very basis of civilization and lead to the ruin of our modern way of life.

Today, the situation has changed considerably. Public opinion polls repeatedly show that people around the world rank the environment among the most important social and political issues. There is no longer a need to prove that environmental problems are serious.

Significant progress has been made in many areas of environmental science (although our scientific understanding of the environment still lags behind our need to know). Advances have also been made in the creation of legal frameworks for the management of the environment, thus providing a new basis for addressing environmental issues. The time is now ripe to seek truly lasting, more rational solutions to environmental problems.

Figure 1.2 • The population bomb starts with little sparks. *(a)* A simplified family tree of four generations of the John Eli Miller family. *(b)* The population explosion of the John Eli Miller family shown in graphic form.

(a)

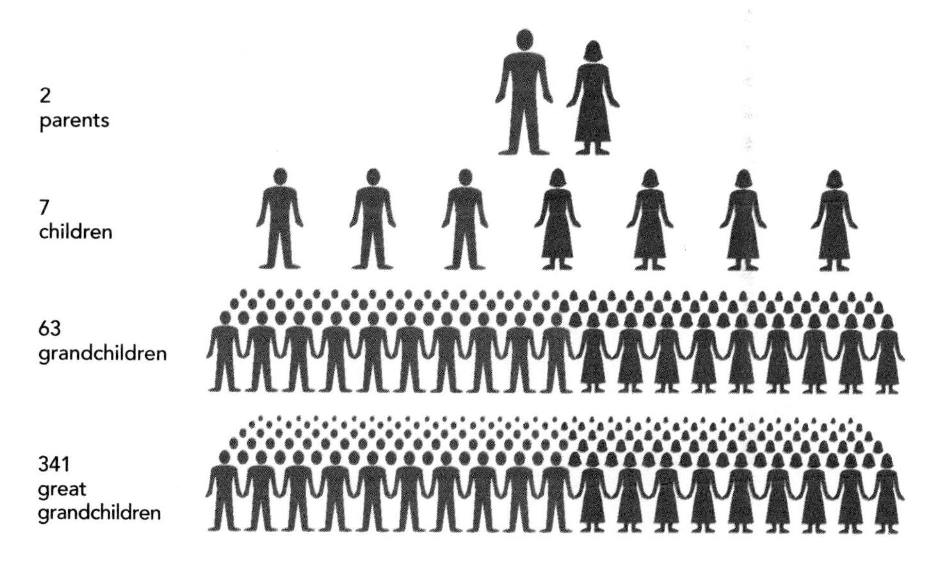

(b)

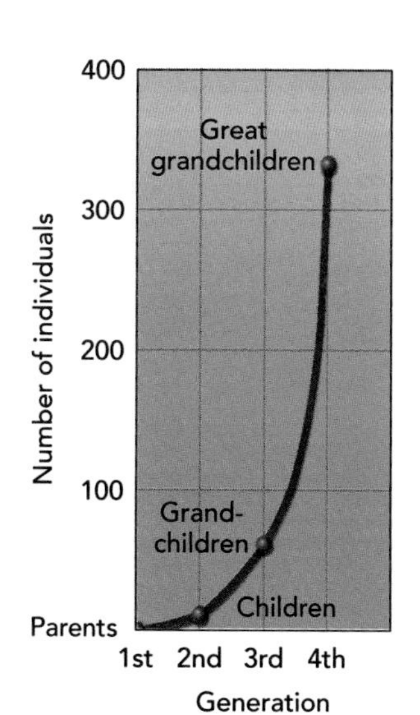

nations. By one estimate, Canada's average ecological footprint is 7.7 ha/capita. Although this is less than the United States at 10.3 ha/capita, it is well above the world average of 2.8 ha/capita. By contrast, India (0.8 ha/capita), Bangladesh (0.5 ha/capita), and China (1.2 ha/capita) have among the smallest ecological footprints in the world.[5,6,7]

Our Rapid Population Growth

The most dramatic increase in the history of the human population occurred in the last part of the twentieth century. As mentioned earlier, in merely the past 40 years, the human population of the world more than doubled, increasing from 2.5 billion to over 6.3 billion. Figure 1.3 illustrates the rapid explosion of the human population, sometimes referred to as the population bomb.[8] The figure breaks down the increases by region.

Human population growth is, in some important ways, *the* underlying issue of the environment. Much current environmental damage is directly or indirectly the result of the very large number of people on the Earth and our rate of increase. As you will see in Chapter 4, where we consider the human population in more detail, for most of human history the total population was small and the average long-term rate of increase was low relative to today's population growth rate.[9, 10]

Although it is customary to think of the population as increasing continuously without declines or fluctuations, the growth of the human population has not been a steady march. For example, great declines occurred during the time of the Black Death. [See A Closer Look 1.2.]

Figure 1.3 • Population change since 1950 projected to the year 2150 for major areas of the world, medium fertility scenario. The population of Africa will nearly quadruple. The only major area whose population is projected to drop over time is Europe—from 728 million to 595 million, a decline of 18% over 155 years. [*Source*: Modified from Population Division, Department of Economic and Social Affairs, United Nations Secretariat, *World Population Projections* to 2150 (New York: United Nations, 1998).]

Famines

Famine is one of the things that happen when a human population exceeds its environmental resources. Famines have occurred in recent decades in many regions of the world. In Africa, in the mid-1970s, following a drought in the Sahel region, 500,000 Africans starved to death and several million more were permanently affected by malnutrition.[12] Starvation in African nations gained worldwide attention 10 years later, in the 1980s.[13,14] In one year during that period, as many as 22 African nations suffered catastrophic food shortages and 150 million Africans faced starvation. Although there have not been such spectacularly acute famines in the last decade as occurred in the 1970s and 1980s, there is a continuing food crisis in southern Africa, particularly in Malawi, Zambia, and Zimbabwe.[15]

African famines had multiple interrelated causes. One, as suggested, is drought. Although drought is not new to Africa, the size of the population affected by drought is new. In addition, African deserts appear to be spreading, in part because of changing climate but also because of human activities. Poor farming practices have increased erosion, and deforestation may be helping to make the environment drier. The control and destruction of food has sometimes been used as a weapon in political disruptions (Figure 1.5).

These famines illustrate another key theme: people and nature. People affect the environment, and the environment affects people. The environment affects agriculture, and agriculture affects the environment.

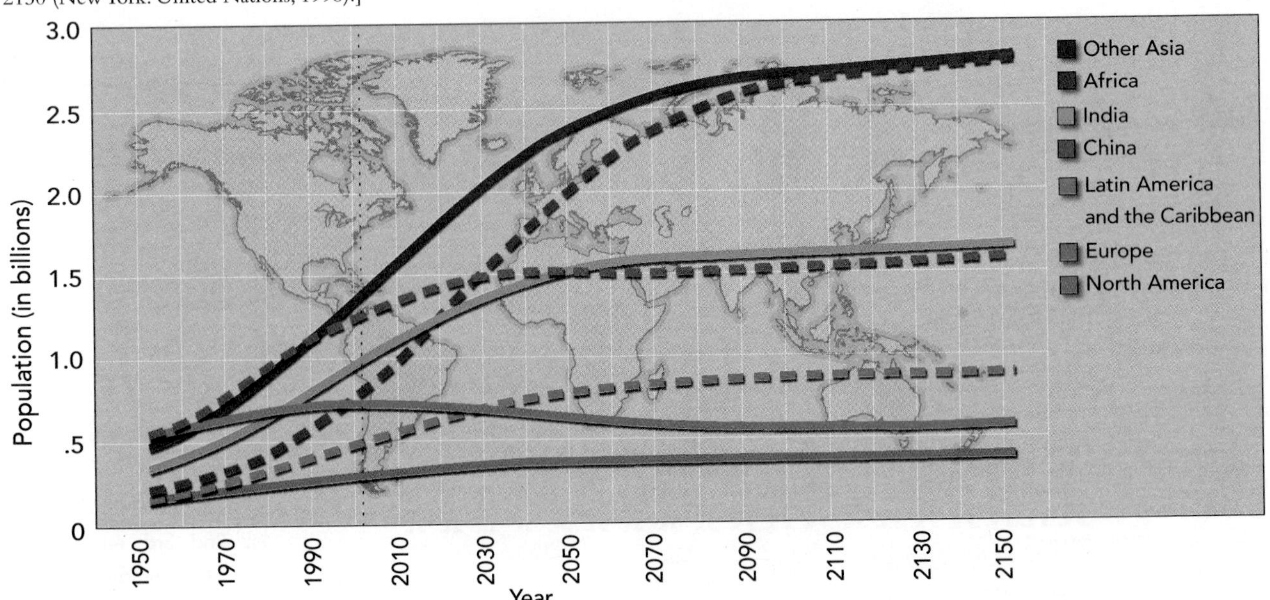

HUMAN POPULATION

A CLOSER LOOK 1.2

The Black Death

The epidemic disease bubonic plague, commonly known as the Black Death, spread throughout Europe during the fourteenth century. The most severe episodes occurred between 1347 and 1351, but there were many recurrences throughout the century (Figure 1.4).[11] The disease is caused by the bacteria *Yersinia pestis* and spread by fleas that live on rodents. It was first recorded in Western history as a major human problem in the seventh century in the Roman Empire and North Africa. The plague probably first appeared in India in the seventh century and spread rapidly north and west. Not until the fourteenth century did another major epidemic occur. The plague again spread rapidly, reaching Italy in 1348 and Spain, France, Scandinavia, and central Europe within two years. In England, one-fourth to one-third of the population died within a single decade, although mortality varied widely by region. Entire towns were abandoned, and the production of food for the remaining population was jeopardized.

The Black Death had many environmental and economic consequences. For example, the great reduction in the labour force led to an increase in wages and is believed to have been a contributing factor to a subsequent increase in the standard of living. Much agricultural land was abandoned because no one was available to work it.

As this example illustrates, human populations have not always increased continuously but have suffered setbacks and declines. The bubonic plague is one of the best-recorded and best-known setbacks in human history. It is likely, though, that other such episodes, resulting from changes in climate, declines in the food supply, or environmental catastrophes have happened many times in human history.

(a)

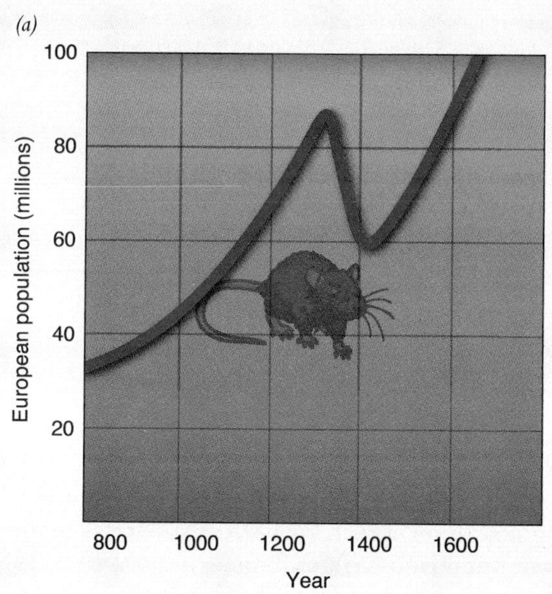

(b)

Figure 1.4 • Impact of the Black Death in Europe. *(a)* The change in the population of Europe during the time of the Black Death. *(b)* During the fourteenth century, the Black Death killed many people in Europe. The reduction in the human population had economic and environmental effects. This fourteenth-century illustration of two plague victims is from a miniature from the Toggenburg Bible. Note the swellings over the bodies, characteristic of the plague.

Human population growth in Africa has severely stretched the capacity of the land to provide sufficient food and has threatened its future productivity.

This situation involves yet another key theme: science and values. Scientific knowledge has led to an increase in agricultural production, a better understanding of population growth, and what is required to conserve natural resources. With this knowledge, we are forced to confront a choice: Which is more important, the survival of people alive today or conservation of the environment, on which future food production and human life depend?[16] Answering this question demands *value judgements* and the information and knowledge with which to make such judgements. Different cultures have different ways of examining such questions. Each culture has a unique perspective on the world, including notions of what is right and wrong (**ethical**), good and bad (**morals**), and what is important, valuable, and desirable (**values**). An individual's values are shaped by factors like gender, ethnic origin, religious preferences, age, socio-economic status, education, health, and occupation. These differences in personal experience lead us to different beliefs

(a)

(b)

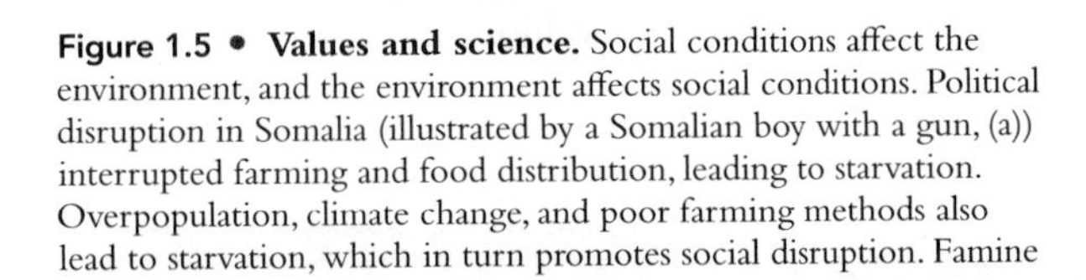

Figure 1.5 • Values and science. Social conditions affect the environment, and the environment affects social conditions. Political disruption in Somalia (illustrated by a Somalian boy with a gun, (a)) interrupted farming and food distribution, leading to starvation. Overpopulation, climate change, and poor farming methods also lead to starvation, which in turn promotes social disruption. Famine has been common in parts of Africa since the 1980s, as illustrated by gifts of food from aid agencies in southern Sudan *(b)*. Environmental degradation, especially in less developed countries, is often driven by political instability, poverty, famine, and war. Environmental issues cannot therefore be addressed without consideration of social, political, and economic influences.

about what is important, good, or desirable in life and in public policy, and affect our interpretation of environmental conditions and our responses to them. The growing awareness that scientific knowledge must be combined with social, political, and economic considerations began in the 1970s and is now an established part of environmental thought. Sometimes called **political ecology**, it reflects the need to incorporate multiple values in environmental awareness.

For example, we must determine whether we can continue to increase agricultural production without destroying the very environment on which agriculture and, indeed, the persistence of life on Earth depend. Although science provides one basis for a value judgement, cultural context can also have a strong influence on decisions about the importance of a problem and appropriate solutions for it. For this reason, Canada is increasingly employing collaborative, multi-interest frameworks for environmental management. This **co-management**, as it is called, allows the involvement of multiple value systems in environmental decision-making. Chapter 19 describes co-management in the Fraser River Basin of British Columbia. Now, traditional ecological knowledge is explicitly incorporated in many environmental assessment processes, alongside Western scientific analysis.

The human population is doubling every few decades, but human effects on the environment are growing even faster. Human beings cannot escape the laws of population growth, which are discussed in several chapters. The broad science and value question is: What will we do about the increase in our own species and its impact on our planet and our future?

1.3 An Urban World

In part because of rapid human population growth, and changes in technology, we are becoming an urban species, and our effects on the environment increasingly affect urban life (Figure 1.6). With economic development comes urbanization; people move from farms to cities and perhaps to suburbs. Cities and towns increase in size. Since cities are commonly located near rivers and along coastlines, urban sprawl often overtakes the good agricultural land of river floodplains as well as the coastal wetlands, which are important habitats for many rare and endangered species. As urban areas expand, wetlands are filled in, forests cut, and soils covered over with pavement and buildings.

In developed countries, about 75% of the population live in urban areas and 25% in rural areas; but in developing countries, only 40% of the people are city dwellers.[17] It is estimated that by 2025 almost two-thirds of the global population—5 billion people—will live in cities. Only a few urban areas had populations over 4 million in 1950. In 2004, Shanghai, China, was the world's largest city, surpassing Tokyo, Japan, which had held that position for many years. In 2015, both Shanghai and Tokyo will still be the world's largest cities, each with a projected population of close to 30 million. The number of **megacities**—urban areas with at least 8 million inhabitants—increased from two (New York City and London) in 1950 to 23 in

1995 (Figure 1.7). Most megacities—17—are in the developing world. Estimates suggest that by 2015 the world will have 36 megacities, 23 of them in Asia.[18,19] (With a current population of 4.2 million, Toronto, Ontario, is Canada's largest city and ranks fifth in North America, but it is still too small to qualify for megacity status.)

In the past, environmental organizations often focused on non-urban issues—wilderness, endangered species, and natural resources, including forests, fisheries, and wildlife. Although these will remain important issues, in the future we must place more emphasis on urban environments and on the effects of urban environments on the rest of the planet.

1.4 Sustainability and Carrying Capacity

The story of the Eli Miller family and the recent famines in Africa bring up one of the central environmental questions of our time: What is the maximum number of people the Earth can sustain? That is, what is the sustainable human carrying capacity of the Earth? Much of this book will deal with knowledge that helps answer this question. However, there is little doubt that we are using our renewable environmental resources faster than they can be replenished—that is, we are using these resources *unsustainably*. In general, we are using forests and fish faster than they can regrow and eliminating habitats of endangered species and wildlife faster than they can be replenished. We are extracting minerals, oil, and groundwater without sufficient concern for their limits or the need to recycle them. As a result, there is a present shortage of some resources and an expectation of more shortages in the future. Clearly, we must learn how to sustain our environmental resources so that they continue to provide benefits for people and other living things on our planet.

Figure 1.7 • **With a 2005 population of 6.8 million, Hong Kong is fast approaching megacity status.**

Defining Sustainability

Sustainability must be achieved, but we are unclear at present how to achieve it, in part because the word is used to mean different things. **Sustainability** refers to resources

Figure 1.6 • **An urban world and a global perspective.** When the world is viewed at night from space, the urban areas show up as bright lights. The number of urban areas reflects the urbanization of Earth.

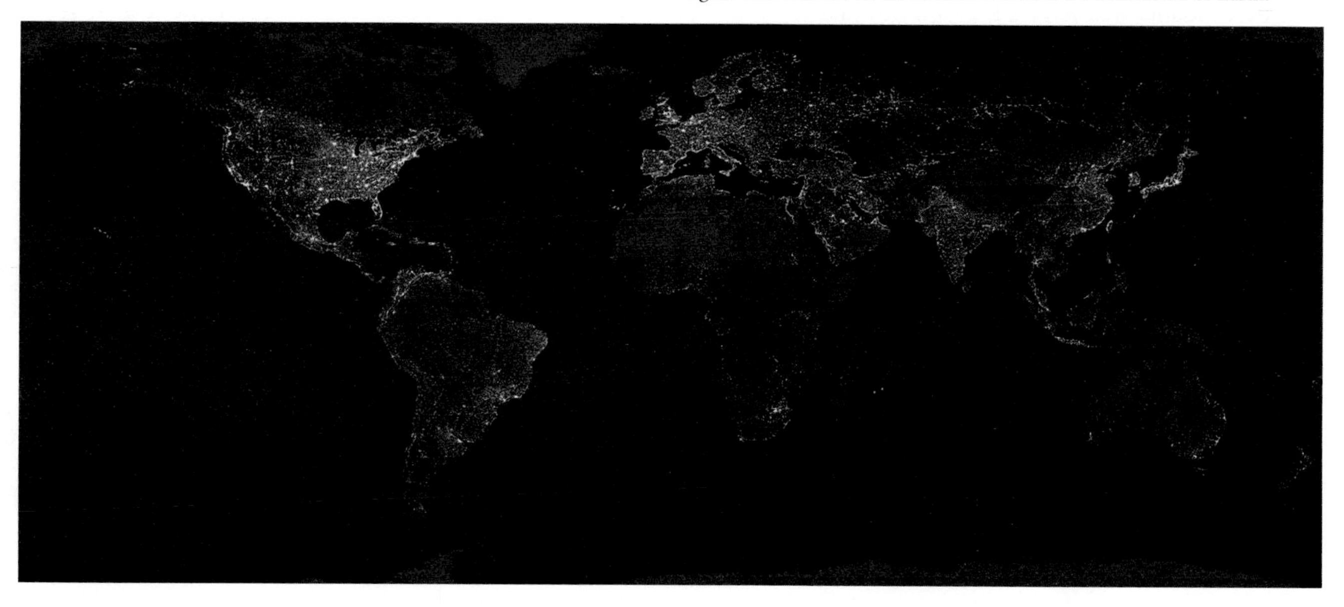

and their environment. In this book, sustainability has two scientific definitions: **sustainable resource harvest**, such as a sustainable supply of timber, meaning that the same quantity of that resource can be harvested each year (or other harvest interval) for an unlimited or specified length of time without decreasing the ability of that resource to continue to produce the same harvest level. (Sustainable harvest levels may vary from year to year, as the condition of a population responds to environmental conditions.) We also refer to a **sustainable ecosystem**, meaning an ecosystem that is still able to maintain its essential functions and properties even though we are harvesting one of its resources.

Other kinds of sustainability pertain to human societies. A sustainable economy is an economy that maintains its level of activity over time in spite of its uses of environmental resources. **Sustainable development** typically means that a society can continue to develop its economy and social institutions, and maintain its environment for an indefinite time.

Carrying Capacity of the Earth

Carrying capacity is a concept related to sustainability. It is usually defined as the maximum number of individuals of a species that can be sustained by an environment without decreasing the capacity of the environment to sustain that same amount in the future. When we ask, "What is the maximum number of people that the Earth can sustain?" we are asking about the Earth's carrying capacity—and we are also asking about sustainability.

As mentioned before, the desirable human carrying capacity depends in part on our values. Do we want those who follow us to live short lives in crowded surroundings without a chance to enjoy Earth's scenery and diversity of life? Alternatively, do we hope that our descendants will have a life of high quality and good health? Once we choose a goal for the quality of life, we can use scientific information to understand what the carrying capacity might be and how we might achieve it. [See A Closer Look 1.3.] We may choose to continue our wasteful resource consumption patterns, and maintain our large ecological footprint, or make changes to reduce our impact on the environment. Our choices will determine the local and global carrying capacity for the human population.

1.5 People and Nature

Today, we stand at the threshold of a major change in our approach to environmental issues. Two paths lie before us. One is to assume that environmental problems are the result of human actions and that the solution is simply to stop these actions based on the notion, popularized some 40 years ago, that people are separate from nature. This path has produced many advances such as reductions in air and water pollution, but also many failures. It has emphasized confrontation and emotionalism. It has been characterized by a lack of understanding of basic facts about the environment and how natural ecological systems function, as well as a willingness to base solutions on political ideologies and on ancient myths about nature. An example of one failure is presented in A Closer Look 1.3. In the late 1990s, fisheries managers placed strict controls on harvesting Pacific salmon, believing that stocks had been depleted by overfishing. Yet, despite those reductions, the salmon population remains severely depleted.

The second path is to begin with a scientific analysis of an environmental controversy and to move from confrontation to co-operative problem solving. It accepts the connection between people and nature. It offers the potential for long-lasting, successful solutions to environmental problems. One purpose of this book is to take the student down the second pathway.

People and nature are intimately integrated. Each affects the other. We depend on nature directly for many material resources such as wood, water, and oxygen in the air. We depend on nature indirectly through what are called "public service functions." For example, soil is necessary for plants and, therefore, for us (Figure 1.9); the atmosphere provides a climate in which we can

(a)

(b)

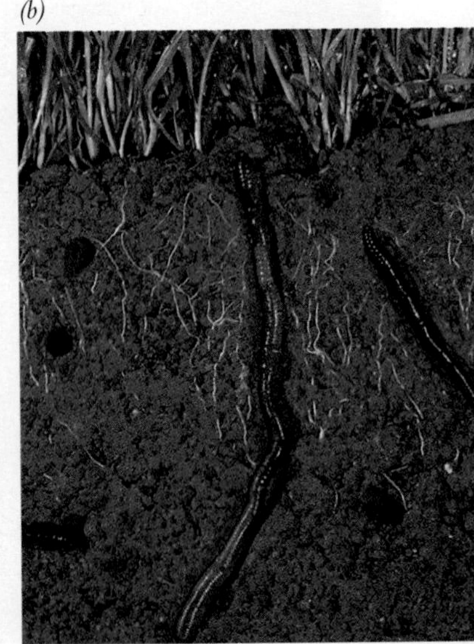

Figure 1.9 • *(a)* **Cross-section of a soil;** *(b)* earthworms are among the many soil animals important to maintain soil fertility and structure.

A CLOSER LOOK 1.3

Carrying Capacity of Pacific Salmon

One approach to determining the carrying capacity and the sustainable harvest of a resource is to examine historical records and see if previous catch levels were maintained. One problem of determining carrying capacity using this method is illustrated by the history of commercial salmon fishing in the Pacific Ocean off British Columbia. Figure 1.8 graphs the annual catch of salmon from 1900 (when commercial fishing began in earnest in the region) to 2000.[20] The catch increased rapidly from 1900 to 1925 to a peak of just over 80,000 metric tonnes, levelled off, then fluctuated between 60,000 and 80,000 metric tonnes until 1950, when it declined again. In 1990, after a brief recovery, the Pacific salmon catch began a dramatic decline, to hover just above 20,000 metric tones in 2000.

From this history, what, if any, levels of Pacific salmon are sustainable? The high catches between 1925 and 1950 are about the same, so a person in charge of managing the salmon fishery in 1920 might reasonably have concluded that the salmon had a sustainable catch of between 60,000 and 80,000 metric tonnes. However, a person who had only the information after 1995, when the catch represents the tail of a declining curve, would conclude that the approximate level of catch in the late 1990s was sustainable. This graph illustrates the difficulty in determining a truly sustainable harvest simply from historical records.

Figure 1.8 • The commercial catch of Pacific salmon by the BC commercial fishery from 1900 through 2002. [*Source*: Adapted from Brian E. Riddell and Art F. Tautz 2004, "State of Pacific salmon and their habitats: Canada and the United States," Chapter 7 in Patricia Gallauger and Laurie Wood (eds.). 2004. *Proceedings, The World Summit on Salmon*, June 10-13, 2003, Simon Fraser University Continuing Studies in Science. Vancouver, BC: Simon Fraser University.]

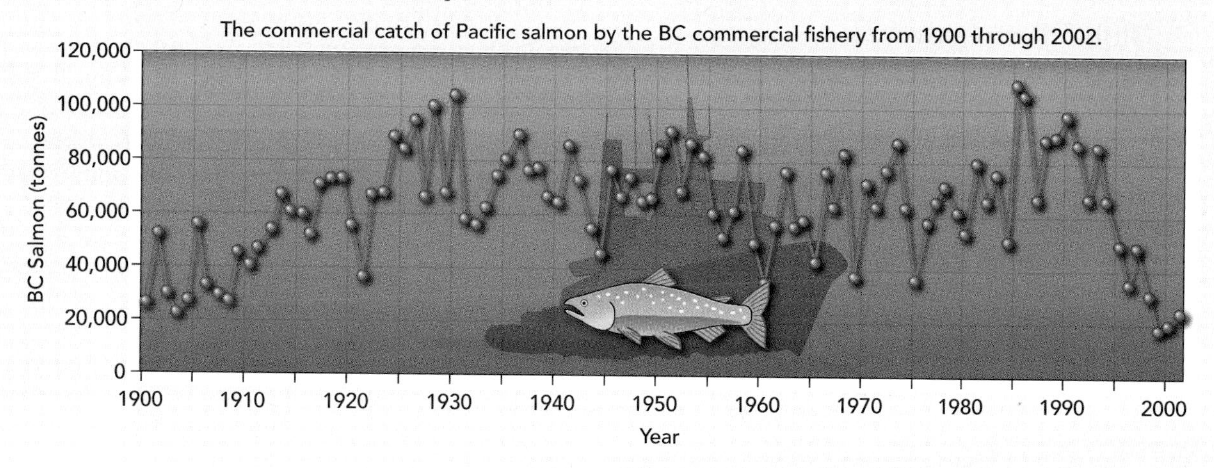

live; the ozone layer high in the atmosphere protects us from damaging ultraviolet radiation; trees absorb some air pollutants; wetlands can cleanse water. People also depend on nature for beauty and recreation—for the needs of our inner selves.

Simultaneously, we affect nature. For as long as people have had tools, including fire, they have changed nature, often in ways that we like, prefer, and have considered "natural." One argument is that it is natural for organisms to change their environment. Elephants topple trees, changing forests to grasslands, and people cut down trees and plant crops (Figure 1.10). Who is to say which is more natural? In fact, few organisms do **not** change their environment.

People have known this for a long time, but the idea that people might change nature to their advantage was unpopular in the last decades of the twentieth century. At that time, the word **environment** suggested

Figure 1.10 • Land cleared by African elephants, Travo National Park, Kenya [*Source*: D. B. Botkin.]

something separate—"out there"—so that people were perceived apart from nature. Today, environmental sciences are showing us how people and nature connect—and in what ways this is beneficial to both.

As the environment becomes increasingly recognized as important, we become more Earth-centered. We seek to spend more time in nature for recreation and spiritual activities. We accept that we have evolved on and with the Earth and are not separate from it. We understand the need to celebrate our union with nature as we work toward sustainability.

Most people recognize that we must seek sustainability not only of the environment, but also of our economic activities so that humanity and the environment can persist together. The dichotomy of the twentieth century is giving way to a new unity: the idea that a sustainable environment and a sustainable economy may be compatible—people and nature are intertwined, and a success for one involves success for the other.

1.6 A Global Perspective

One solution to famines such as those that occurred in recent decades in Africa is better food distribution throughout the world. Although the total amount of food produced worldwide each year is still sufficient to feed the world's population, famines occur because local food production is insufficient in some places, and because worldwide transportation of food is inadequate. Thus, local famines are a *global environmental problem.*

The recognition that, worldwide, civilization can change the environment at a global level is relatively recent. As discussed in detail in later chapters, scientists now believe that modern chemical emissions are changing the ozone layer high in the atmosphere. Scientists also believe that burning fossil fuels increases the concentration of greenhouse gases in the atmosphere, which may change Earth's climate. These atmospheric changes suggest that the actions of many groups of people at many locations affect the environment of the entire world.[21,22] Another idea explored in later chapters is that non-human life also affects the global environment and has changed it over the course of several billion years. These two new ideas have profoundly affected our approach to environmental issues.

At the heart of many environmental issues is the challenge of our growing human population. This burgeoning population must be fed, housed, and kept warm. With population growth comes increasing pressure on agriculture and resource use. Improvements in agricultural technology mean that we can now produce higher yields of key food crops in the same land area, but farmers everywhere must confront limitations imposed by depleted soils, soil erosion, and overgrazing. Although total global food production is probably sufficient to feed the human population, problems with getting the food where it is needed mean that some areas experience famine, while others have plenty to eat. Higher development status also brings higher rates of resource consumption and energy use. The most developed countries consume an astonishing 50 to 90 metric tons of resources per person per year, with associated high levels of waste production. In huge economies, like those of the United States and Western European nations, such consumption must inevitably result in massive environmental change. In less developed countries, resource and energy consumption and waste production rates are much lower. However, since those countries' populations are often also very large, it will be important to ensure that higher development status does not bring with it devastating resource exploitation and environmental degradation. The sheer size of the world population means that even small steps toward resource conservation and reuse can have important positive consequences for Earth's natural systems.

A growing world population, and higher per capita resource consumption, have resulted in widespread alteration of Earth's chemical and biological systems. Although industrial and manufacturing activities have added a range of chemical contaminants to the natural environment, the most pressing pollution issues may be those related to human and animal wastes. Chapter 5 describes how human activities have had a dramatic influence on **global biogeochemical cycles**, the chemical pathways that link the atmosphere, biosphere, hydrosphere, and lithosphere.

The impact of the human population is also apparent in global biological systems, which have been gravely altered by heavy fishing and logging pressures, and by the steadily expanding land area occupied by human habitation and farming systems. Natural forest systems in particular have been fragmented and reduced in size, to the point that they can no longer support viable populations of some plants and animals. More than half of Earth's forests are moderately or severely altered, and rates of deforestation are alarmingly high in some regions. In the Amazon River Basin of South America, forest clearance rates doubled in just two years during the mid-1990s. Increasing international trade, transportation, and even recreational travel have also created routes by which biological organisms can move more easily from one part of the world to another. As a result, invasions of non-native species are a growing problem in many environments. This influence, coupled with widespread damage to natural systems, has vastly increased the rate of biological extinctions, now 100 to 1,000 times the rate that existed 10,000 years ago.

Other important global trends are apparent. In the most developed countries, energy use has risen more than 70% since 1970—roughly 2% per year—and will continue to rise unchecked without major changes in energy consumption patterns. An important implication of this increase is the associated rise in air pollutants and the "greenhouse gases" responsible for climate change (described in more detail in Chapters 20 and 21), particularly if our heavy reliance on fossil fuels continues.

Comparison of these trends reveals the interlinkage of several important environmental indicators. It is also apparent that some phenomena are global in scale, for instance soil degradation, while others such as famine are more localized. These patterns provide important clues for environmental scientists. For example, large-scale phenomena like climate change require local action but also regional, national, and international coordination. Similarly, careful attention to high-risk areas—perhaps areas that have rapid population growth, valuable natural resources, and fragile biological systems—may be as or more important than a lower level of action across all systems. Clearly, actions taken with respect to one indicator may have implications for other indicators—therein lies a ray of hope. If we can understand the linkages between environmental and human phenomena, we can begin to develop a more protective means of producing food, using and reusing resources, and accommodating the growth of human settlements. Much good work has been accomplished to this end, and much of the remainder of this book will consider how we can better understand the roots of and find appropriate solutions for unwanted environmental change.

Awareness of the global interactions between life and the environment has led to the development of the **Gaia hypothesis**, originated by British chemist James Lovelock and American biologist Lynn Margulis. The Gaia hypothesis (discussed in Chapter 3) proposes that the Earth is a living, self-regulating organism, that the environment at a global level has been profoundly changed by life over the history of life on Earth and that these changes have tended to improve the chances for the continuation of life. Because life affects the environment at a global level, the environment of our planet is different from that of a lifeless one. Environmental change is a natural phenomenon and should be accommodated in our management strategies. However, our understanding of environmental processes is still incomplete. As Dan Botkin argues in his book, *Discordant Harmonies*, our challenge is to allow change to occur at natural rates and in natural forms. Effective environmental management will require rethinking our management policies to reflect the complex and dynamic nature of environmental systems.[23]

1.7 Science and Values

Deciding what to do about an environmental problem involves both values and science. We must choose what we want the environment to be. To make this choice, we must first know what is possible. That requires knowing and understanding the implications of—and uncertainties in—the scientific data. Once we know our options, we can select from among them. What we choose is determined by our values. An example of a value judgement regarding the world's human environmental problem is the choice between the desire of an individual to have many children and the need to find a way to limit the human population worldwide.

Once we have chosen a goal based on knowledge and values, we have to find a way to attain that goal. This step also requires knowledge. The more technologically advanced and powerful our civilization, the more knowledge is required. For example, current fishing methods make it possible to harvest very large numbers of Coho salmon from the Fraser River, British Columbia, and demand for salmon encourages us to harvest as many as possible. To determine whether Coho salmon are sustainable, we must know the current and historical salmon population. We must also understand the processes of birth and growth for this fish, its food requirements, habitat, life cycle, etc.—all the factors that ultimately determine the abundance of salmon in the Fraser River.

Consider, in contrast, the situation just over two centuries ago. When Alexander Mackenzie and later Simon Fraser first made expeditions to the Pacific coast, they found many small First Nations villages that depended, in large part, on the fish in the river for food (Figure 1.11). The human population was small, and the methods of fishing were simple. The maximum number of fish the people could catch probably posed no threat to the salmon. These people could fish without scien-

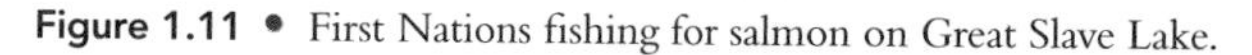

Figure 1.11 • First Nations fishing for salmon on Great Slave Lake.

CRITICAL THINKING ISSUE

How Can We Preserve the World's Coral Reefs?

Coral reefs are among the largest, oldest, most diverse, and most beautiful communities of plants and animals. Today, many coral reefs have been seriously damaged or are at risk. Scientists estimate that approximately 10% have already been destroyed, while another 30% are threatened. The major threats to reefs are direct or indirect results of human activities. Almost 60% of the world's reefs are threatened by activities including coastal development, destructive fishing practices, overexploitation of resources, and marine pollution.[25] [See Figure 1.12.]

The pieces of coral that most people know from souvenir and jewellery shops are limestone skeletons secreted by colonies of animals related to sea anemones and jellyfish. Like their relatives, these small individual coral animals, or polyps, use tentacles equipped with stinging cells to capture food. In addition, polyps obtain nourishment from photosynthetic algae that live in their cells. When polyps die, their skeletons remain while the next generation of individuals secrete new material. Thus, reefs grow slowly by accretion. Coral reefs that exist today are 5,000 to 10,000 years old. By taking the brunt of the force of waves, coral reefs protect coastlines from erosion, a function that has been estimated to have a value of over half a million dollars a year per square meter. In addition, reefs can provide humans with living resources (fish) and services (tourism, coastal protection) worth $375 billion per year.[26]

Coral reefs provide homes for a vast variety of plants and animals. Approximately 25% of all marine organisms, about 1 million species, are associated with coral reefs. Reef organisms are the source of many useful chemicals and medicines, and scientists are currently searching for others. The plant and animal species found around coral reefs are linked in intricate ways, so that removing only one or two key elements may cause a catastrophic collapse. For example, overfishing in the waters off the Cook Islands in the South Pacific in the 1980s removed most of the reef's parrot fish and sea urchins, both of which feed on algae. Soon algae overgrew the reef, and the entire community of reef life collapsed.

Coral reefs have long been the main source of protein for tropical people, who today number approximately 1 billion. Because of modern transportation and preservation methods, fish and other food organisms from coral reefs are now eaten by many other people as well. In fact, reef fish constitute about 15% of the entire worldwide catch. Unfortunately, because many of the world's reefs are being overfished, some species are now rare and endangered. In some regions, fishing for food and for aquarium fish involves use of dynamite to stun fish, or cyanide to poison them temporarily. Both methods can kill or damage other organisms, and dynamite can destroy the reef material itself. Fishing is not the only threat to coral reefs, however. The limestone material that forms the bulk of a reef is sometimes mined for use as a construction material. Millions of tourists from around the world who flock to reef areas to fish, swim, dive, and enjoy their beauty pose an additional threat. However, perhaps the greatest threat to coral reefs comes from increasing population in the tropics. Population densities as high as 500 people per square kilometre are found in parts of tropical Asia and the Caribbean. Half a billion people live within one kilometre of a coral reef.[27] In many areas, raw or inadequately treated sewage, and runoff from land development forestation, add to the burden of sediment and pollution.

Critical Thinking Questions

1. How does the current state of the world's coral reefs illustrate each of the six key themes of this book?
2. What are the utilitarian, ecological, aesthetic, and moral justifications for preserving coral reefs?
3. What things can you do in your everyday life to contribute to the preservation of coral reefs?

(a)

(b)

Figure 1.12 • *(a)* A pristine coral reef ecosystem, *(b)* coral reef bleaching in the Caribbean.

tific understanding of numbers and processes. (This example does not suggest that prescientific societies lacked an appreciation for the idea of sustainability. On the contrary, many so-called primitive societies held strong beliefs about the limits of harvests.)

Placing a Value on the Environment

How do we place a value on any aspect of our environment? How do we choose between two different concerns? The value of the environment is based on eight justifications: aesthetic, creative, recreational, inspirational, moral, cultural, ecological, and utilitarian (materialistic).

The **utilitarian justification** sees some aspect of the environment as valuable because it benefits individuals economically or is directly used for or necessary to human survival. For example, mangrove swamps provide shrimp that are the basis of the livelihood of fishers in many countries. The **ecological justification** is that an ecosystem is necessary for the survival of some species, or that the system itself provides some benefit. For example, mangrove swamps provide habitat for marine fish, and although we do not eat mangrove trees, we may eat the fish that depend on them. Therefore, conservation of the mangrove is important ecologically. In addition, mangroves are habitat for many non-commercial species, some endangered. As another example, burning coal and oil adds greenhouse gases to the atmosphere, which may lead to a change in climate that could affect the entire earth. Such ecological reasons form a basis for the conservation of nature that is essentially enlightened self-interest.

Aesthetic justification has to do with our appreciation of the beauty of nature. For example, many people find wilderness scenery beautiful and would rather live in a world with wilderness than without it. One way we enjoy nature's beauty is to seek recreation in the outdoors. The aesthetic and recreational justifications are gaining a legal basis. The Saguenay region of Québec acknowledges that whales have an important role related to recreation: People observe and photograph the whales and enjoy viewing them in a natural setting (**recreational justification**). Many examples illustrate the importance of the aesthetic values of the environment. When people grieve following the death of a loved one, they typically seek out places with grass, trees, and flowers, and thus we decorate our graveyards. Conservation of nature can be based on its benefits to the human spirit (**inspirational justification**)—to benefit what is sometimes called our inner selves. Nature is an aid to human creativity (the **creative justification**). Artists and poets, among others, find a source of their creativity in their contact with nature. This is a widespread reason that people like nature, but it is rarely used in formal environmental arguments. Although popular discussions of environmental issues might make aesthetic, recreational, and inspirational elements seem superficial as justifications for the conservation of nature, in fact, beauty in their surroundings is of profound importance to people. Frederick Law Olmsted, the great landscape planner, argued that plantings of vegetation provide medical, psychological, and social benefits, and are essential to city life.

Moral justification has to do with the belief that various aspects of the environment have a right to exist and that it is our moral obligation to allow them to continue or help them to persist. Moral arguments have been extended to many non-human organisms, to entire ecosystems, and even to inanimate objects. For example, the historian Roderick Nash has written an article entitled "Do Rocks Have Rights?" that discusses such moral justification.[24] The United Nations General Assembly World Charter for Nature, signed in 1982, states that species have a moral right to exist.

The analysis of environmental values is the focus of a discipline known as environmental ethics. Environmental ethics is a challenging field. The resolution of one environmental problem, for instance the establishment of a wilderness preserve to protect fragile habitat, can create new problems such as reduced income for residents. Which interests should prevail? Another concern of environmental ethics is our obligation to future generations: Do we have a moral obligation to leave the environment in good condition for our descendants, or are we at liberty to use environmental resources to the point of depletion within our own lifetimes?

Summary

- Six threads, or themes, run through this text: the urgency of the population issue, the importance of urban environments, the need for sustainable use and management of resources, the importance of a global perspective, people and nature, and the role of science and values in the decisions we face.
- People and nature are intertwined. Each affects the other.
- The human population grew at a rate unprecedented in history in the twentieth century. Population growth is the underlying environmental problem.
- When the impact of technology is combined with the impact of population, the impact on the environment is multiplied, as reflected in the larger ecological footprints of industrialized nations compared to the global average.
- In an increasingly urban world, we must focus much of our attention on the environments of cities and on the effects of cities on the rest of the environment.
- Determining the Earth's carrying capacity for people and levels of sustainable harvests of resources is difficult but crucial if we are to plan effectively to meet our needs in the future. Estimates of the carrying capacity of the Earth for people range from 2.5 to 40 billion. The difference has to do with the quality of life projected for people—the poorer the quality of life, the more people can be packed onto the Earth.
- Awareness of how people at a local level affect the environment globally gives credence to the Gaia hypothesis. Future generations will need a global perspective on environmental issues.
- Placing a value on various aspects of the environment requires knowledge and understanding of the science but also depends on our judgements concerning the uses and aesthetics of the environment and on our moral commitments to other living things and to future generations.

STUDY QUESTIONS

1. Refer to Figure 1.8. How would you respond to the statement "the catch of salmon in the Fraser River is sustainable at 1960s levels"?
2. In what ways do the effects on the environment of a resident of a large city differ from the effects of someone living on a farm? In what ways are the effects similar?
3. Programs have been established to supply food from Western nations to starving people in Africa. Some people argue that such food programs, which may have short-term benefits, actually increase the threat of starvation in the future. What are the pros and cons of international food relief programs?
4. What are the values involved in deciding whether to create an international food relief program? What are five kinds of information required to determine the long-term effects of such programs?
5. Consider the following environmental problems. Describe how each is both a local issue and a part of larger global environmental systems.
   a. The growth of the human population.
   b. The furbish lousewort is a small flowering plant found in a few locations on the New Brunswick-Maine border. It is so rare that it has been seen by few people and is considered endangered.
   c. The blue whale, listed as an endangered species under the *Canadian Species at Risk Act.*
   d. A car that has air-conditioning.
   e. Seriously polluted harbours and coastlines in major ocean ports.
6. Give some examples of how cultural values affect the perception of an environmental problem and its potential solutions.
7. Is it possible that sometime in the future all the land on Earth will become one big city? If not, why not? To what extent does the answer depend on the following:
   a. global environmental considerations
   b. scientific information
   c. values

FURTHER READING

Botkin, D. B. *No Man's Garden: Thoreau and a New Vision for Civilization and Nature*. Washington, D.C.: Island Press, 2000. Discusses many of the central themes of this textbook, with special emphasis on values and science and on an urban world. Henry David Thoreau's life and works illustrate approaches that can be beneficial in our dealing with modern environmental issues.

Botkin, D. B. *Discordant Harmonies: A New Ecology for the 21st Century*. New York: Oxford University Press, 1990. An analysis of the myths that lie behind attempts to solve environmental issues.

Kent, M. M. *World Population: Fundamentals of Growth*. Washington, D.C.: Population Reference Bureau, 1990. Facts and data about the growing human population.

Kessler, E., ed. "Population, Natural Resources and Development." *AMBIO*: (1992) 21(1). A special issue of the journal AMBIO, addressing many problems concerning human population growth and its economic and environmental implications.

Leopold, A. *A Sand County Almanac*. New York: Oxford University Press, 1949. Perhaps, along with Rachel Carson's *Silent Spring*, one of the most influential books of the post–World War II and pre–Vietnam War era about the value of the environment. Leopold defines and explains the land ethic and writes poetically about the aesthetics of nature.

Lutz, W. *The Future of World Population*. Washington, D.C.: Population Reference Bureau, 1994. Summary of current information on population trends and future scenarios of fertility, mortality, and migration.

Nash, R. F. *The Rights of Nature: A History of Environmental Ethics*. Madison: University of Wisconsin Press, 1988. An introduction to environmental ethics.

chapter 2

The Experimental Lakes Area: Applying Science to Solve Environmental Problems

What causes algae blooms? Today, almost any aquatic scientist could give you a straightforward answer. Indeed, Chapter 19 of this book discusses the mechanisms of this environmental nuisance. Forty years ago, however, the picture was much less clear. Blue-green algae blooms were common in temperate lakes and rivers. Lake Erie's condition was so grave that it was considered "dead," its shoreline clotted with rotting algae throughout the late summer, and its waters green with excessive plant growth. What could be done? Scientific information was needed to answer the key questions: What causes algae blooms? And what can be done to control them?

Clearly, it would take more than experiments with test tubes and beakers to understand these ecosystem-scale problems. Large-scale studies of real lakes were needed, but few suitable facilities were available. In 1967, the governments of Canada and Ontario recognized that need, establishing the Experimental Lakes Area (ELA) near Kenora in northwestern Ontario. Comprising 58 small lakes ranging from 1 to 84 ha, the ELA is a natural laboratory relatively isolated from the industrial activity and rapid population growth typical of more southerly regions. For almost 40 years, the facility has operated under a Canada–Ontario agreement that limits activity in the area to ecological research activities.

Early research in the ELA used artificial loading of nutrients in experimental lakes to answer the key questions about algae blooms. The answers: In freshwater ecosystems, algae growth is limited by the availability of phosphorus, the nutrient that is usually in lowest supply in those systems. Controlling phosphorus should reduce nuisance algae growth in most lakes and rivers.[1] Science provided knowledge about what would happen and what management approaches were possible to restore a "dead" lake. This knowledge was combined with values about people and nature to choose policies and actions.[2] Those strategies included the establishment of phosphorus reduction targets in the Canada–U.S. Great Lakes Water Quality Agreement, new laws to reduce phosphorus discharges from sewage treatment plants, and limits on the amount of phosphorus in detergents and other cleaning products.

Figure 2.1 • The Experimental Lakes Area.

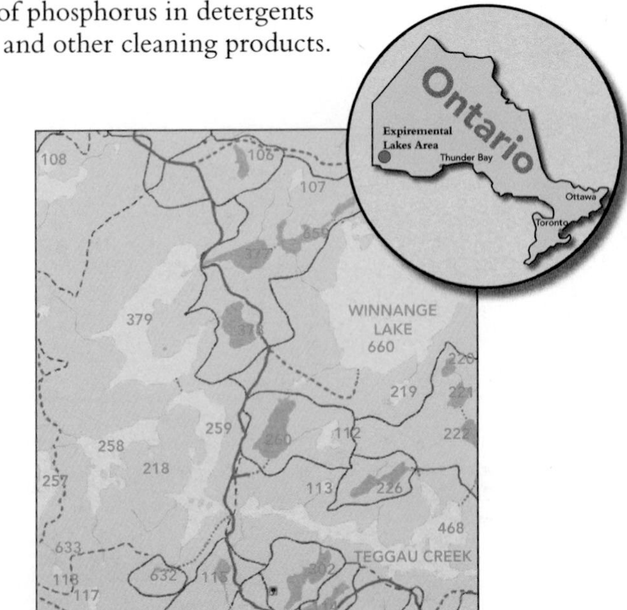

Science as a Way of Knowing: Critical Thinking about the Environment

LEARNING OBJECTIVES

Science is a process of refining our understanding of nature through continual questioning and active investigation. It is more than a collection of facts to be memorized. After reading this chapter, you should understand that:

- Thinking about environmental issues requires thinking scientifically.
- Scientific knowledge is acquired through observations of the natural world that can be tested through additional observations and experiments and can be disproved.
- Scientific understanding is not fixed, but changes over time as new data, observations, theories, and tests become available.
- Scientific reasoning combines induction and deduction-different but complementary ways of thinking.
- Every measurement involves some degree of approximation—that is, uncertainty—and that a measurement without a statement about its degree of uncertainty is meaningless.
- Scientific reasoning combines induction and deduction-different but complementary ways of thinking.
- Scientific discovery involves a number of processes, including the scientific method, and that science and scientists are too diverse to be described by just one method.
- Technology, the application of scientific knowledge, is not science but that science and technology interact, stimulating growth in each other.
- That decision-making about environmental issues involves society, politics, culture, economics, and values, as well as scientific information.

Since the early research on phosphorus, the ELA has been used to answer other ecosystem-scale questions, including those related to acid precipitation and lake acidification, and the changes in flow and water chemistry associated with the creation of reservoirs. In 1999, five lakes in the ELA became part of an international network of Long-term Ecosystem Research sites. To date, more than a thousand scientific papers and monographs have been produced as a result of ELA research.

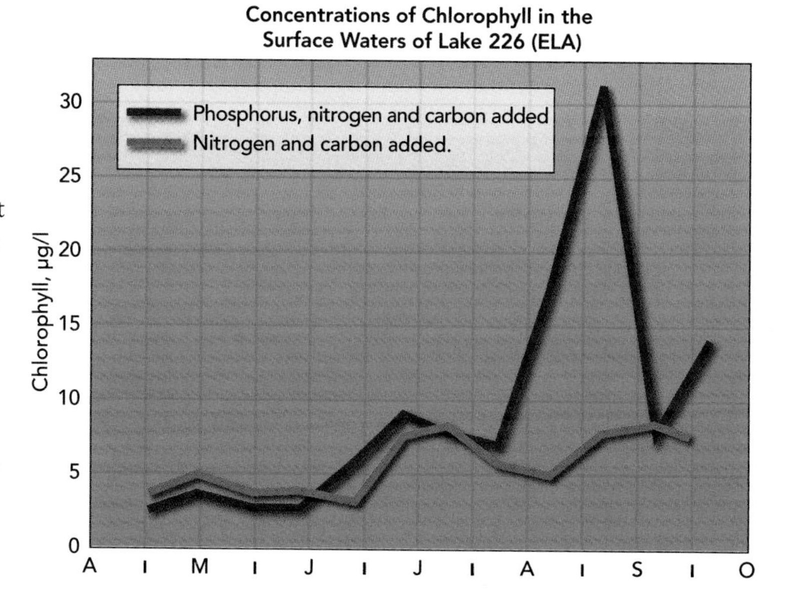

Figure 2.2 • Concentrations of chlorophyll in the surface waters of Lake 226 (ELA) in response to artifical loading of carbon, nitrogen, and phosphorus in 1973. (*Source*: D.W. Schindler and E.J. Fee. 1974. Experimental Lakes Area: Whole-Lake experiments in eutrophication. *J. Fish. Res. Board Canada*. 31:937-953.)

2.1 Understanding What Science Is (and What It Isn't)

Science as Process

The complexity of environmental sciences suggests several questions:

- How can we understand ecological phenomena?
- How can we take a scientific approach to the great, ancient questions about nature around us, the source of great wonder at the complexity of life and the marvellous adaptations of living things?
- How can we seek answers to practical questions about the effect of people on nature and the actions people should take in solving environmental problems?

To begin to tackle the problem of a scientific understanding of life and its environment, it is useful to review the fundamentals of the scientific method and consider in what ways ecology can fit that mould and in what ways it may need to develop new approaches.

Science has a long history in Western civilization (See A Closer Look 2.1). Science is a process, a way of knowing. It results in conclusions, generalizations, and sometimes scientific theories and scientific laws. These comprise a body of beliefs. Often people confuse the process of science with a fixed set of beliefs—the results. But science does not lead so much to a fixed set of beliefs as to a set of beliefs that, at the present time, allow us to explain all known observations about a kind of phenomenon and make predictions about that kind of phenomenon.[3]

Science is a process of discovery—a continuing process whose essence is change in ideas. Often, the fact that scientific ideas change seems frustrating. Why can't scientists agree on what is the best diet for people? Why is a chemical considered dangerous in the environment for a while and then determined not to be? Why do scientists in one decade believe that fire in nature is an undesirable disturbance and in a later decade decide that it is important and natural? Is there going to be global warming or not? Can't scientists just find out the truth for each of these questions, once and for all, and agree on it?

Rather than looking to science for answers to such questions, it is more accurate to think of science as a continuing adventure of making increasingly better approximations about how the world works. Sometimes changes in ideas are small, and the major context remains the same. Sometimes a science undergoes a fundamental revolution in ideas.

Science is one way of looking at the world. It begins with observations about the natural world such as: How much phosphorus does it take to cause an algae bloom? What algae species are present in a typical lake? What happens to the algal community when lake chemistry changes? From these observations, scientists formulate hypotheses that can be tested. For example, one hypothesis could be that that chlorophyll production (reflecting increased photosynthesis) will increase in the presence of excess phosphorus. The Western scientific tradition does not deal with things that do not have the potential to be tested by observation such as the ultimate purpose of life or the existence of a supernatural being. Science also does not deal with questions that involve values such as standards of beauty or issues of good and evil—for example, whether the scenery in the Experimental Lakes Area is beautiful. Both science and values are important, as the case study of the Experimental Lakes Area illustrates; that is why this connection is one of the key themes of this book. The criterion by which we decide whether a statement is in the realm of science is:

Whether it is possible, at least in principle, to disprove the statement.

Disprovability

It is generally agreed today that the essence of the scientific method is **disprovability** (see Figure 2.3). A statement can be said to be scientific if someone can state a method by which it could be disproved. Thus, if you can think of a test that could disprove a statement, then that statement can be said to be scientific. If you cannot think of such a test, then the statement is said to be non-scientific. For example, consider the crop circles discussed in A Closer Look 2.2. One website states that some people believe the crop circles are a "spiritual nudge," which is designed to awaken us to our larger context and milieu, which is none other than our "collective earth soul." Whether or not this is true, it does not seem to be a statement open to disproof. The statement that "the scenery at the Experimental Lakes Area

Figure 2.3 • Ways of thinking about the world with and without science.

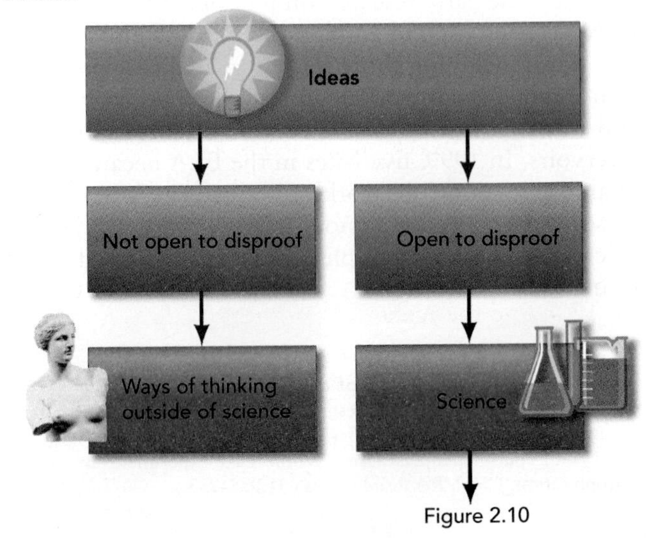

SCIENCE AND VALUES

A CLOSER LOOK 2.1

A Brief History of Science

Thinking scientifically about the environment is as old as science itself. Science had its beginnings in the ancient civilizations of Babylonia and Egypt. There, observations of the environment were carried out primarily for practical reasons, such as planting crops, or for religious reasons, such as using the positions of the planets and stars to predict human events. These ancient practices differed from modern science in that they did not distinguish between science and technology or between science and religion. As we will see in this chapter, science is a way of knowing. Technology is the means people use to obtain physical needs and benefits. In our scientific age, technology has become the application of scientific knowledge to achieve some practical goal.

Because of their interest in ideas, the ancient Greeks developed a more theoretical approach to science. Knowledge for its own sake, rather than for practical purposes, became the primary goal. At the same time, the Greeks' philosophical approach began to move science away from religion and toward philosophy.

Modern science is usually considered to have its roots in the end of the sixteenth and the beginning of the seventeenth centuries, with the development of the scientific method by Gilbert (magnets), Galileo (physics of motion), and Harvey (circulation of blood). Unlike earlier classical scientists—who asked "Why"? in the sense of "For what purpose?"—they made important discoveries by asking "How"? in the sense of "How does it work?" Galileo also pioneered the use of numerical observations and mathematical models. The scientific method, which quickly proved very successful in advancing knowledge, was first described explicitly by Francis Bacon in 1620. Although not a practical scientist himself, Bacon recognized the importance of the scientific method, and his writings did much to promote scientific research.[3]

Our cultural heritage, then, gives us several ways of thinking about the environment, including the kind of thinking we do in everyday life and the kind of thinking scientists do (Table 2.1). There are many similarities between thinking in modern, everyday life and in science. This means that thinking scientifically is something all of us can do. At the same time, there are crucial differences. Ignoring these differences can lead to invalid conclusions, in general, and to serious errors in decisions about the environment.

Table 2.1 • Knowledge in Everyday Life Compared with Knowledge in Science

Parameter	*Everyday Life*	*Science*
Goal	To lead a satisfying life (implicit)	To know, predict, and explain (explicit)
Requirements	Context-specific knowledge; no complex series of inferences; can tolerate ambiguities and lack of precision	General knowledge; complex, logical sequences of inferences; must be precise and unambiguous
Resolution of questions	Through discussion, compromise, consensus	Through observation, experimentation, logic
Understanding	Acquired spontaneously through interacting with world and people; criteria not well defined	Pursued deliberately; criteria clearly specified
Validity	Assumed, no strong need to check; based on observations, common sense, tradition, authorities, experts, social mores, faith	Must be checked; based on replications, converging evidence, formal proofs, statistics, logic
Organization of knowledge	Network of concepts acquired through experience; local, not integrated	Organized, coherent, hierarchical, logical; global, integrated
Acquisition of knowledge	Perception, patterns, qualitative; subjective	Plus formal rules, procedures, symbols,statistics, mental models; objective
Quality control	Informal correction of errors	Strict requirements for eliminating errors and making sources of error explicit

Source: Based on F. Reif and J. H. Larkin, "Cognition in Scientific and Everyday Domains: Comparison and Learning Implications," *Journal of Research in Science Teaching* 28(9), pp. 733–760. Copyright © 1991 by National Association for Research in Science Teaching. Reprinted by permission of John Wiley & Sons.

is pretty" also is not disprovable. On the other hand, the statement that "more than 50% of the people who visit the ELA find the scenery beautiful" could be tested by public opinion surveys and can be treated as a scientific statement—actually as a hypothesis (discussed later).

What, then, is truth in the context of science? In his book *Lectures on Physics*, the famous Nobel Prize–winning physicist Richard Feynman said, "The test of all knowledge is experiment. Experiment is the sole judge of scientific Truth." By this, he means that a large and persuasive body of evidence brings us closer to the truth. After millennia of daily observation, for example, we can say that it is true that the sun will rise tomorrow. Albert Einstein and Stephen Hawking have been influential in developing a new theory of gravity, incorporating an explanation of what happens inside black holes. Their

A CLOSER LOOK 2.2

The Case of the Mysterious Crop Circles

For 13 years, circular patterns appeared "mysteriously" in grain fields in southern England (Figure 2.4). Proposed explanations included aliens, electromagnetic forces, whirlwinds, and pranksters. The mystery generated a journal and a research organization headed by a scientist, as well as a number of books, magazines, and clubs devoted solely to crop circles. Scientists from Great Britain and Japan brought in scientific equipment to study the strange patterns. Then, in September 1991, two men confessed to having created the circles by entering the fields along paths made by tractors (to disguise their footprints) and dragging planks through the fields. When they made their confession, they demonstrated their technique to reporters and some crop-circle experts.[4, 5]

In spite of their confession, some people continue to believe that the crop circles have some alternative causes. Crop-circle organizations continue to exist and now have websites. For example, a report published on the World Wide Web in 2003 stated that "strange orange lightning" was seen one evening and that crop circles appeared the next day.[4, 6]

How is it that so many people, including some scientists, took, and still take, crop circles seen in England seriously? The answer is that they misunderstood the scientific method and engaged in fallacious reasoning—and that at some level some people want to believe in a mysterious cause for the circles and choose to ignore standard scientific analyses and methods. The failure of some people to think critically about crop circles causes no harm, but the same type of thinking applied to other, more serious environmental issues, can have serious consequences. There are two kinds of failures regarding the scientific method illustrated by crop circles: first, that the scientific method is used incorrectly; and second, that scientific information is rejected by choice. Both failures occur in environmental problems.

Figure 2.4 • **English crop circles.** A crop circles close-up at the Vale Pewsey in southern England in July 1990.

theory is now widely accepted as "mostly true." If a theory—in Stephen Jay Gould's words, a network of ideas—can satisfactorily explain both existing and new observations, over a sustained period, it is generally held to be approaching the truth.

There are many ways of looking at the world, such as the religious, aesthetic, and moral views; they are not science, however, because their assertions are not open to disproof in the scientific sense. They are based ultimately on faith, beliefs, and cultural and personal choices. The distinction between a scientific statement and a non-scientific statement is not a value judgement—the distinction is not meant to imply that science is the only "good" kind of knowledge. The distinction is simply a philosophical one about kinds of knowledge and logic. To say that these other ways of looking at the world are not science is not to denigrate them. Each way of viewing the world gives us a different way of perceiving and of making sense of our world, and each is valuable to us.

Foundations of Science

Science makes certain assumptions about the natural world. To understand what science is, it is important to be aware of these assumptions.

- Events in the natural world follow patterns that can be understood through careful observation and scientific analysis, which we will describe later.
- These basic patterns and the rules that describe them are the same throughout the universe.
- Science is based on a type of reasoning known as induction; it begins with specific observations about the natural world and extends to generalizations.
- Generalizations can be subjected to tests that may disprove them. If such a test cannot be devised, then a generalization cannot be treated as a scientific statement.
- Although new evidence can disprove existing scientific theories, science can never provide absolute proof of the truth of its theories.

The Nature of Scientific Proof

One source of serious misunderstanding about science is the use of the word *proof*, which most students encounter in mathematics, particularly in geometry. Proof in mathematics and logic involves reasoning from initial definitions and assumptions. If a conclusion follows logically from these premises, the conclusion is said to be proven. This process is known as **deductive reasoning**.

An example of deductive reasoning is the following syllogism, or series of logically connected statements.

Premise: A straight line is the shortest distance between two points.
Premise: The line from A to B is the shortest distance between

points A and B.
Conclusion: Therefore, the line from A to B is a straight line.

Note that the conclusion in this syllogism follows directly from the premises.

Deductive proof does not require that the premises be true, only that the reasoning be foolproof. Logically valid but untrue statements can result from false premises, as in the following example:

Premise: Humans are the only toolmaking organisms.
Premise: The New Caledonian crow makes tools.
Conclusion: Therefore, the New Caledonian crow is a human being.

In this case, the concluding statement must be true if both the preceding statements are true. However, we know that the conclusion is not only false but ridiculous. If the second statement is true (which it is), then the first cannot be true. The conclusion that a bird, the New Caledonian crow, which makes hooked tools from twigs and wire to retrieve food (Figure 2.5), is human follows logically in the syllogism but defies common sense.

The rules of deductive logic govern only the process of moving from premises to conclusion. Science, in contrast, requires not only *logical reasoning* but also *correct premises.* Because science is based on observations, its conclusions are only as true as the premises from which they are deduced.

Returning to the example of the New Caledonian crow, to be scientific the three statements should be expressed conditionally (with reservation):

If humans are the only toolmaking organisms
and
the New Caledonian crow is a toolmaker,
then
the New Caledonian crow is a human being.

When we formulate generalizations based on a number of observations, we are engaging in **inductive reasoning**. To illustrate, consider the eared grebe. The grebe's "ears" are a fan of golden feathers that occur behind the eyes of males during the breeding season. Let us define birds with these golden feather fans as eared grebes (Figure 2.6). If we always observe that the breeding male grebes have this feather fan, we may make the inductive statement, "All male eared grebes have golden feathers during the breeding season." What we really mean is, "All of the male eared grebes we have seen in the breeding season have golden feathers." We never know when our very next observation will turn up a bird that is like a male eared grebe in all ways except that it lacks these feathers in the breeding season. This is not impossible, somewhere, due to a mutation.

Proof in inductive reasoning is therefore very different from proof in deductive reasoning. When we say something is proven in induction, what we really mean is that it has a very high degree of likelihood, or probability. **Probability** is a way of expressing our certainty (or uncertainty)—our estimation of how good our observations are, how confident we are of our predictions.

Figure 2.5 • The New Caledonian crow has been observed making hook-shaped tools from twigs and wire, to retrieve bits of food. [*Source:* Weir, A. A. S., J. Chappell, and A Kacelnik. 2002. Shaping of hooks in New Caledonian crows. *Science* 297: 981.]

When we have a fairly high degree of confidence in our conclusions in science, we often forget to state the degree of certainty or uncertainty. Instead of saying, "It is very likely that...," we say, "It has been proven that..." Unfortunately, many people interpret this as a deductive statement, meaning the conclusion is absolutely true, which has led to much misunderstanding about science. Although science begins with observations and, therefore, inductive reasoning, deductive reasoning is useful in helping scientists analyse whether conclusions based on inductions are logically valid. *Scientific reasoning combines induction and deduction—different but complementary ways of thinking.*[7]

Figure 2.6 • Male eared grebe.

What we have just described is the classic scientific method. Scientific advances, however, often begin with instances of insight—leaps of imagination, which are then subjected to the stepwise inductive process. Some scientists have made major advances by being in the right place at the right time and knowing the right things at the right time. For example, penicillin was discovered "by accident" in 1928 when Sir Alexander Fleming was studying the pus-producing bacteria *Staphylococcus aureus*. A culture of these bacteria was accidentally contaminated by the green fungus *Penicillium notatum*. Fleming noticed that the bacteria did not grow in those areas of a culture where the fungus grew. He isolated the mould, grew it in a fluid medium, and found that it produced a substance that killed many of the bacteria that caused diseases. Eventually, this discovery led other scientists to develop an injectable agent to treat diseases. *Penicillium notatum* is a common mould found on stale bread. No doubt many others had seen the mould, perhaps even noticing that other strange growths on bread did not overlap with Penicillium notatum. It took Fleming's knowledge and observational ability for this piece of "luck" to occur.

2.2 Measurements and Uncertainty

A Word about Numbers in Science

We communicate scientific information in several ways. One is the written word for conveying synthesis, analysis, and conclusions. When we add numbers to our analysis, we add another dimension of understanding that goes beyond qualitative understanding and synthesis of a problem. Using numbers and statistical analysis allows us to visualize relationships in graphs and make predictions. It also allows us to analyse the strength of a relationship and in some cases discover a new relationship. This is the main reason we have added mathematical Build Your Environmental Skills boxes to this book.

People in general put more faith in the accuracy of measurements than do scientists. Scientists realize that every measurement is only an approximation. Measurements are limited; they depend on the instruments used and the people who use the instruments. Measurement uncertainties are inevitable; they can be reduced, but they can never be completely eliminated. Because all measurements have some degree of uncertainty, a measurement is meaningless unless it is accompanied by an estimate of its uncertainty.

Consider the loss of the *Challenger* space shuttle in 1986, the first major space shuttle accident, which appeared to be the result of the failure of rubber O-rings that were supposed to seal sections of rockets together. Imagine a simplified scenario in which an engineer is given a rubber O-ring used to seal fuel gases in a space shuttle. She is asked to determine the flexibility of the O-rings under different temperature conditions to help answer the questions "At what temperature do the O-rings become brittle and subject to failure?" and "At what temperature(s) is it unsafe to fire the shuttle?"

After doing some tests, the engineer says that the rubber becomes brittle at –1°C (30°F). Is it safe to launch the shuttle at 0°C (32°F)? At this point, you do not have enough information to answer the question. You assume that the temperature data may have some degree of uncertainty, but you have no idea how large it is. Is the uncertainty ±5°C, ±2°C, or ±0.5°C? To make a reasonably safe and economically sound decision about whether to launch the shuttle, you must know the uncertainty of the measurement.

Dealing with Uncertainties

There are two sources of uncertainty. One is the real variability of nature. The other is the fact that every measurement has some error. Various kinds of errors can occur in experimentation. They include *measurement uncertainties*, random error (errors due to natural variability in the process being studied), errors in experimental design, equipment malfunction, the mis-recording of data, and other errors. Errors that occur consistently such as those resulting from incorrectly calibrated instruments are *systematic errors*.

Scientists traditionally include a discussion of experimental errors when they report results. Often, error analysis leads to greater understanding and sometimes even to important discoveries. For example, the existence of the eighth planet in our solar system, Neptune, was discovered when scientists investigated apparent inconsistencies—observed systematic "errors"—in the orbit of the seventh planet, Uranus. Measurement uncertainties can be reduced by improving the instruments used and by requiring a standard procedure in making measurements. They can also be reduced by using carefully designed experiments and appropriate statistical procedures. However, uncertainties are inherent in every measurement and can never be completely eliminated.

As difficult as it is for us to live with uncertainty, that is the nature of nature, as well as the nature of measurement and of science. We need to use our understanding of the uncertainties in measurements to read reports of scientific studies critically, whether they appear in science journals or in popular magazines and newspapers. (See A Closer Look 2.3.)

Accuracy and Precision

Certain measurements have been made very carefully by many people over a long period, and accepted values have been determined. *Accuracy* is the extent to which a measurement agrees with the true or accepted value. A dart thrown at a target is accurate if it hits the bull's eye. *Precision* is the degree of exactness with which a quantity is measured. A very precise measurement may not be accurate. If three darts thrown at a target all hit close together, but far from the bull's eye, they would be precise but not accurate. As another example, if the accepted value for the temperature of boiling water at sea level is 100.000°C and you measure it as 99.875°C, your measurement is just as precise as the accepted value—that is, it is measured to the nearest 1/1000 degree—but it is still slightly inaccurate. The accuracy of observations is checked by comparison with observations or predictions made by scientists.

Although it is important to make measurements as precisely as possible, it is equally important not to report measurements with more precision than they warrant. Doing so conveys a misleading sense of both precision and accuracy. A measurement based on a ruler divided into tenths of a centimetre should be precise to the nearest hundredth of a centimetre, but no more. A small bone that appears to end exactly at the 2.1-cm mark should be reported as 2.10 cm long, not as 2.100 cm. The latter would imply that the right-hand zero was read directly, when in fact it was estimated.

2.3 Observations, Facts, Inferences, and Hypotheses

It is important to distinguish between observations and inferences, which are ideas based on observations. **Observations**, the basis of science, may be made through any of the five senses or by instruments that measure beyond what we can sense. An **inference** is a generalization that arises from a set of observations. When what is observed about a particular thing is agreed on by all, or almost all, it is often called a **fact**.

We might observe that a substance is a white, crystalline material with a sweet taste. We might *infer* from these observations alone that the substance is sugar. Before this inference can be accepted as fact, however, it must be subjected to further tests. Confusing observations with inferences and accepting untested inferences as facts are kinds of sloppy thinking often described by the phrase "thinking makes it so."

When scientists wish to test an inference, they convert it into a statement that can be disproved. This type of statement is known as a **hypothesis**. A hypothesis continues to be accepted until it is disproved. If a hypothesis has not been disproved, it still has not been proved true in the deductive sense; it has only been found to be probably true until and unless evidence to the contrary is found.

Typically hypotheses take the form of if-then statements. For example, a scientist is trying to understand how growth of a plant will change with the amount of light it receives. He measures the rate of photosynthesis at a variety of light intensities (Figure 2.7). The rate of photosynthesis is called the **dependent variable** because it is affected by, and in this sense *depends on*, the amount of light, which is called the **independent variable**. (A variable is simply a quantity or condition that may vary.) The independent variable is also sometimes called a **manipulated variable**, because the scientist deliberately changes, or manipulates, it. The dependent variable is then referred to as a **responding variable**—one that responds to changes in the manipulated variable.

As trees grow, many **variables** exist. Some, like the position of the North Star, can be assumed to be irrelevant. Others, like the duration of daylight, are potentially relevant. In testing a hypothesis, a scientist tries to keep all relevant variables constant, except for the independent and dependent variables. This practice is known as *controlling variables*. In a *controlled experiment*, the experiment is compared to a standard, or control—an exact duplicate of the experiment except for the condition of the one variable being tested (the independent variable). Any difference in outcome (dependent variable) between the experiment and the control can be attributed to the effect of the independent variable.

An important aspect of science, but one often overlooked in descriptions of the scientific method, is the need to define or describe variables in exact terms that can be understood by all scientists. The least ambiguous

Figure 2.7 • Dependent and Independent variable. Photosynthesis as affected by light. In this diagram, photosynthesis is represented by carbon dioxide (CO_2) uptake. Light is the independent ariable (CO_2), uptake is the dependent variable.

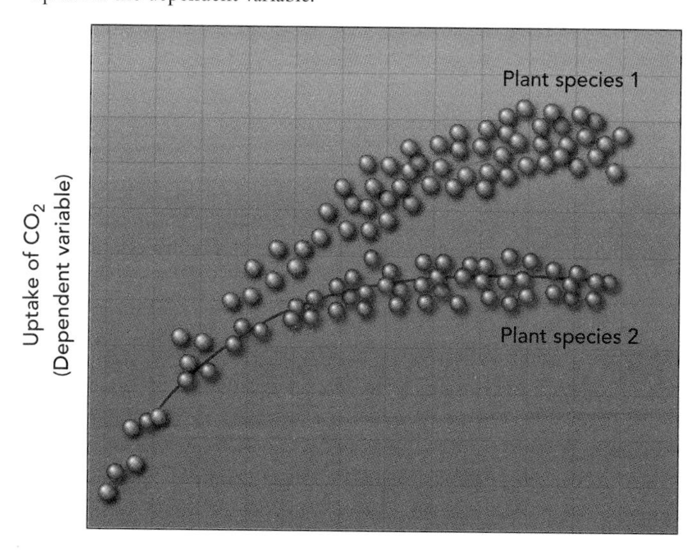

A CLOSER LOOK 2.3

Measurement of Carbon Stored in Vegetation

A number of people have suggested that a partial solution to global climate change—one way of meeting our obligations under the Kyoto Protocol (see Chapter 20)—might be a massive worldwide program of tree planting. Trees take carbon dioxide (an important greenhouse gas) out of the air in the process of photosynthesis. Because they live a long time, trees can store carbon for decades, even centuries. But how much carbon can be stored? Many books and reports published during the past 20 years contained numbers representing the total stored carbon in the vegetation of Earth, but all were presented without any estimate of error (Table 2.2). Without an estimate of that uncertainty, the figures are meaningless, yet important environmental decisions have been based on them.

Recent studies have reduced the error by replacing guesses and extrapolations with scientific sampling techniques, similar to the procedures used to predict the outcomes of elections. Even these improved data would be meaningless, however, without an estimate of error. The new figures show that the earlier estimates were three to four times too large, grossly overestimating the storage of carbon in vegetation and, therefore, the contribution tree planting could make in offsetting global warming.

Table 2.2 • Estimates of Above-Ground Biomass in the North American Boreal Forest

Source	*Biomass*[a] *(kg/m²)*	*Carbon*[b] *(kg/m²)*	*Total Biomass*[c] *(10⁹ metric tons)*	*Total Carbon*[c] *(10⁹ metric tons)*
This study[d]	4.2 ± 1.0	1.9 ± 0.4	22 ± 5	9.7 ± 2
Previous extimates[e]				
Estimate 1	17.5	7.9	90	40
Estimate 2	15.4	6.9	79	35
Estimate 3	14.8	6.7	76	34
Estimate 4	12.4	5.6	64	29
Estimate 5	5.9	2.7	30	13.8

Source: D. B. Botkin and L. Simpson, "The First Statistically Valid Estimate of Biomass for a Large Region," *Biogeochemistry 9* (1990): 161–274. Reprinted by permission of Kluwer Academic, Dordrecht, The Netherlands.

[a] Values in this column are for total above-ground biomass. Data from previous studies giving total biomass have been adjusted using the assumption that 23% of the total biomass is in below-ground roots. Most references use this percentage; Leith and Whittaker use 17%. We have chosen to use the larger value to give a more conservative comparison.

[b] Carbon is assumed to be 45% of total biomass following R. H. Whittaker, *Communities and Ecosystems* (New York: Macmillan, 1974).

[c] Assuming our estimate of the geographic extent of the North American boreal forest: 5,126,427 km² (324,166 mi²).

[d] Based on a statistically valid survey; above-ground woodplants only.

[e] Lacking estimates of error: Sources of previous estimates by number (1) G. J. Ajtay, P. Ketner, and P. Duvigneaud, "Terrestrial Primary Production and Phytomass," in B. Bolin, E. T. Degens, S. Kempe, and P. Ketner, eds., *The Global Carbon Cycle* (New York: Wiley, 1979), pp. 129–182. (2) R. H. Whittaker and G. E. Likens, "Carbon in the Biota," in G. M. Woodwell and E. V. Pecam, eds., *Carbon and the Biosphere* (Springfield, Va.: National Technical Information Center, 1973), pp. 281–300. (3) J. S. Olson, H. A. Pfuderer, and Y. H. Chan, *Changes in the Global Carbon Cycle and the Biosphere*, ORNL/EIS-109 (Oak Ridge, Tenn.: Oak Ridge National Laboratory, 1978). (4) J. S. Olson, I. A. Watts, and L. I. Allison, *Carbon in Live Vegetation of Major World Ecosystems*, ORNL-5862 (Oak Ridge, Tenn.: Oak Ridge National Laboratory, 1983). (5) G. M. Bonnor, *Inventory of Forest Biomass in Canada* (Petawawa, Ontario: Canadian Forest Service, Petawawa National Forest Institute, 1985).

way to define or describe a variable is in terms of what one would have to do to duplicate the variable's measurement. Such definitions are called operational definitions. Before carrying out an experiment, both the independent and dependent variables must be defined operationally. Operational definitions allow other scientists to repeat experiments exactly and to check on the results reported.

While performing an experiment, a scientist must record the values of the input (independent variable) and the output (dependent variable). These values are referred to as data (singular:datum). They may be numerical, quantitative data or nonnumerical, qualitative data. In our example, qualitative data would be the species of a tree; quantitative data would be the mass in grams or the diameter in centimetres.

Knowledge in an area of science grows as more hypotheses are supported. Because hypotheses in science are continually being tested and evaluated by other scientists, science has a built-in self-correcting feedback system. Scientists use accumulated knowledge to develop explanations that are consistent with currently accepted hypotheses. Sometimes an explanation is presented as a model. A model is "a deliberately simplified construct of nature."[8] It may be a physical working model, a pictorial model, a set of mathematical equations, or a computer simulation. For example, the Canadian Hydraulics Centre of the National Research Council is Canada's largest coastal and hydraulic engineering laboratory, and one of the world's most sophisticated facilities for simulation of ocean waves and other physical phenomena. The Centre also develops and tests mathematical equations and computer simulations that attempt to explain some aspects of water movement, as a basis of comparison with large-scale physical models and field measurements. Models require scientists to make assumptions about the way the world works. Flawed assumptions usually lead to incorrect conclusions. Consider the case of a scientist constructing a model to predict the growth of bacteria in food. Lacking information to the contrary, the scientist might assume that the rate of bacterial growth depends only on food temperature. Modelled predictions might therefore suggest that if the food is kept cool, no bacterial growth will occur. If the initial assumption is false, and growth in fact depends on other factors such as starting bacterial population, size of container, or acidity of the food, the scientist's conclusions could seriously over- or under-estimate the actual bacterial growth that would occur.

Models can be considered a form of deductive reasoning, in that they begin with a theory or premise, and proceed to conclusions about specific cases. For example, the ECOL computer model was developed in the late 1970s, as a means of simulating the growth of algae in the Grand River, Ontario. The work of Schindler and others, described in this chapter's opening case, had suggested that high phosphorus concentrations caused excess algae growth. The model was constructed with a simple premise: that phosphorus was the only factor that caused aquatic plants to grow. All other influences, such as temperature, light, and flow velocity were intentionally ignored. Early simulation results were disappointing. The model could not duplicate observed algal growth patterns. Clearly, something was missing. In the late 1990s, the model was extensively revised based on the most recent available research, which indicated that both temperature and light were important factors in nuisance algae blooms. Figure 2.8 shows the results of this simulation. The graph shows observed and predicted dissolved oxygen levels in the stream, which change as the plants produce oxygen in the process of photosynthesis (described in more detail in Chapter 5). The updated model successfully predicts the lowest dissolved oxygen levels, which are important for fish health. It is less accurate in predicting peak oxygen concentrations. We can deduce from these results that the phosphorus, heat, and light influences currently reflected in the model are either incomplete or faulty. As this example shows, however, model predictions are not facts or data; they are guesses or deductions based on an underlying model. Therefore, it is common to see model results reported with their likely predictive error.

As new knowledge accumulates, the assumptions underlying models may have to be revised or replaced, with the goal of creating models more consistent with observations.[8] Computer simulation of the atmosphere has become important in scientific analysis of the possibility of global warming. Computer simulation is increasingly important in understanding biological systems such as forest growth, and physical systems such as the tsunami that devastated Southeast Asia on December 26, 2004 (Figure 2.9).

A body of ideas that offers broad, fundamental explanations of many observations is called a **theory**. A scientific **law** is simply the description of a repeatable phenomenon. For example, the law of gravity states that each body in the universe attracts every other body in the universe. In daily life, we see this law in action when we toss an object into the air and see it fall to the ground. As long as the object is heavier than air, it will fall every time. The law of gravity describes what will happen,

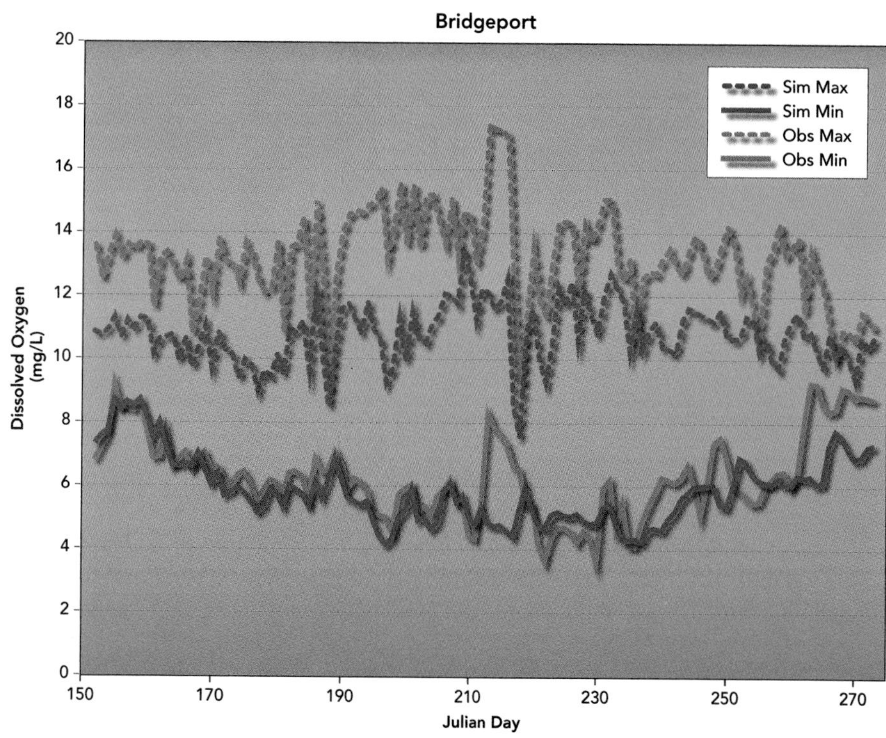

Figure 2.8 • Simulated oxygen concentrations based on estimated aquatic plant growth in the Grand River, Ontario. "Sim" means simulated values; "Obs" are observed conditions. Note that the model does not predict maximum concentrations as well as minimum levels, suggesting that the theory on which the model is based is incomplete or otherwise incorrect. (*Source*: Grand River Conservation Authority.)

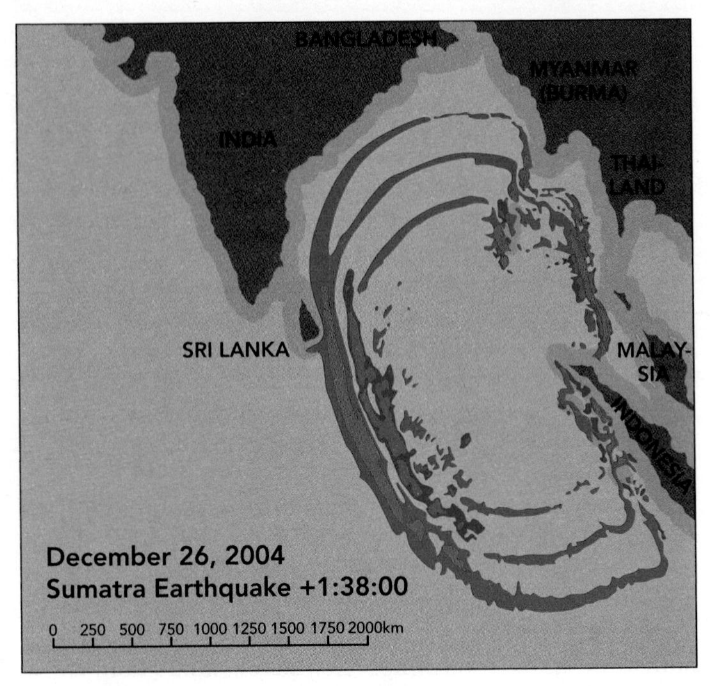

Figure 2.9 • A computer simulation of the tsunami that devastated south Asia on December 26, 2004. [*Source*: University of California Santa Cruz.]

but does not explain why. Various theories have been proposed to explain the law of gravity, and even today there is debate on this question.

The ideas discussed in this section are usually referred to as the scientific method and can be presented as a series of steps:

1. Make observations and develop a question about the observations.
2. Develop a tentative answer to the question—a hypothesis.
3. Design a controlled experiment to test the hypothesis (implies identifying and defining independent and dependent variables).
4. Collect data in an organized form, such as a table.
5. Interpret the data through graphic or other means.
6. Draw a conclusion from the data.
7. Compare the conclusion with the hypothesis and determine whether the results support or disprove the hypothesis.
8. If the hypothesis is shown to be consistent with observations in some limited experiments, conduct additional experiments to test it further. If the hypothesis is rejected, make additional observations and construct a new hypothesis (Figure 2.10).

An example will illustrate these steps. Chapter 14 describes the habitat destruction in the Hudson Bay lowlands caused by intensive foraging by snow geese. Scientists have been studying this problem and trying to determine what habitat protection and restoration mechanisms will be most effective. In one study, scientists wanted to test the **hypothesis** that if the geese were excluded from an area, native vegetation would recover. They then designed a **controlled experiment** to test this hypothesis. For each native species they chose three experimental sites carefully matched for size, elevation, and vegetative condition. On each site, they erected a wire enclosure one metre square. Forty-two samples of native grasses were then transplanted into each enclosure, using a consistent planting scheme. One-fifth of the plots were left untreated; the rest were fertilized or treated with peat mulch, or both. The plants were monitored for two years, and scored as "living," "dying," or "dead" at regular intervals. The resulting data were **collected and analysed** using statistical methods. Based on the results, the scientists concluded that plants in the treated plots grew better than those left in natural condition. **Comparing** these conclusions with their hypothesis, the scientists found that their results **disproved** their hypothesis. Therefore, they constructed a **new hypothesis**: successful habitat restoration would require both exclusion of the geese, and the addition of fertilizer and mulch to the degraded soils.[9] The scientists went on to conduct additional field studies in this area, carefully matching sites and controlling conditions so that any differences could be attributed to experimental treatment. Their findings proved robust and repeatable, allowing them eventually to develop a theory (model) for vegetation change in the Hudson Bay lowlands.[10]

2.4 A Word about Creativity and Critical Thinking

Creativity in science, as in other areas of knowledge, has to do with original thoughts (those never stated before). Many people have the ability to be creative, but it helps to be intensely curious about how something in nature works. Sometimes creativity comes as an inspiration or sudden idea—such as the possible answer to an old question. Suppose you are trying to find out why a particular species of tree is dying in a particular region. You may have a creative hunch that it's due to human-caused acid precipitation (rain or snow that is acidic due to air pollution; see Chapter 21). This is a hypothesis that requires testing, and this is where critical thinking in applying the scientific method enters the picture. To think critically, you think of ways to test a hypothesis and use your creative skills to form alternative hypotheses that will also be tested. You synthesize what is known about the question you are asking, then gather and analyse data in an attempt to disprove your hypotheses. You examine all sides of the problem you are studying, evaluate each in terms of the available evidence—physical, chemical, and biological data—and reach a tentative conclusion.

You may also think creatively and critically about a particular environmental problem without gathering your own data and testing hypotheses. In this case, you

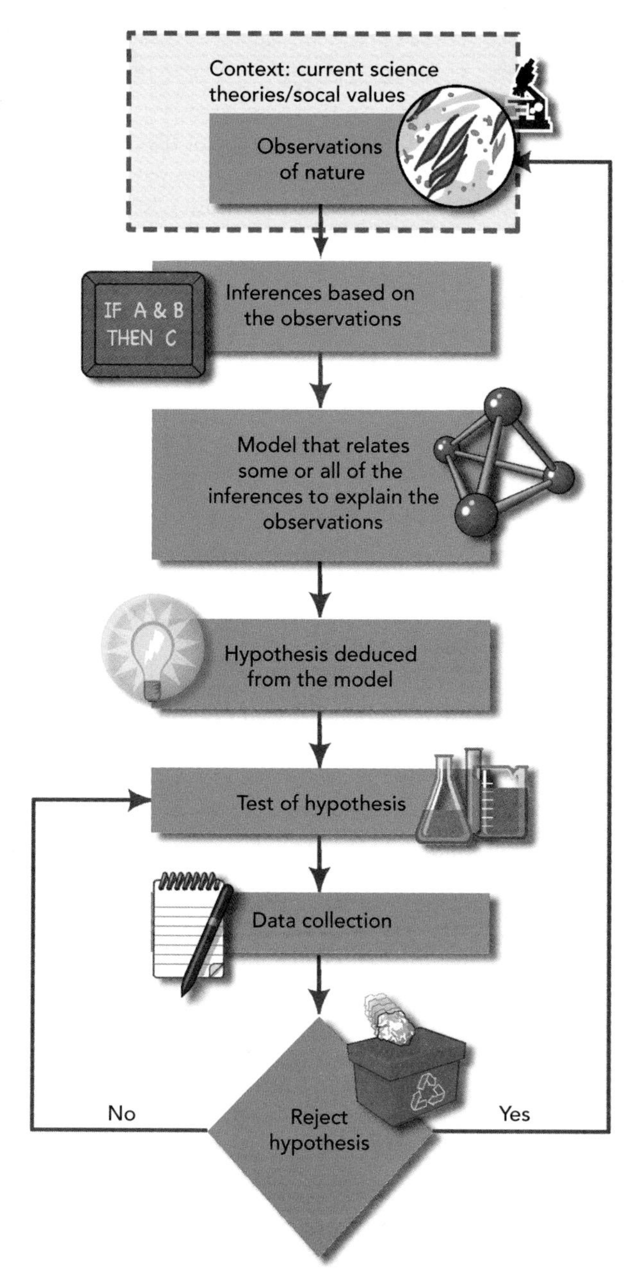

Figure 2.10 • The scientific method: Scientific investigation as a feedback process. (*Source:* Modified from C. M. Pease and J. J. Bull, *Bioscience* 42[April 1992]:293–298.)

obtain and synthesize available data, analyse the data in terms of its uncertainties, examine all sides of the problem to the extent possible, and draw a tentative conclusion. Some scientists have made important creative discoveries by re-examining available data gathered for other purposes. For example, gathering data on the abundance of tiny single-celled marine organisms (*Foraminifera*) and the sulphur particles they produce and release into the atmosphere led to the discovery that these small particles of sulphur compounds serve as nuclei around which water vapour condenses and droplets form. They are important in making clouds that produce precipitation in the ocean and on land when the clouds move over land. This discovery was important to our understanding about the global linkages among ocean water, organisms, chemical compounds, clouds, and rain.

2.5 Misunderstandings about Science

The scientific method as we have described it so far does not take into account differences among the various disciplines of science. The logic of research is not the same in physics, for example, as in biology. In fact, within biology itself, research on evolution, for example, differs in significant ways from research on ecology. It is much more realistic to speak of the *methods of science* than of the scientific method.

Theory in Science and Language

A common misunderstanding about science arises from confusion between the use of the word *theory* in science and its use in everyday language. A **scientific theory** is a grand scheme that relates and explains many observations and is supported by a great deal of evidence. In contrast, in everyday usage, a theory can be a guess, a hypothesis, a prediction, a notion, or a belief. We often hear the phrase "It's just a theory." That may make sense in everyday language, but not in the language of science. In fact, theories have tremendous prestige and are considered the greatest achievements of science.[11]

Further misunderstanding arises when scientists use the word *theory* in several different senses.[7] For example, we may encounter references to a currently accepted, widely supported theory, such as the theory of evolution by natural selection; a discarded theory, such as the theory of inheritance of acquired characteristics; a new theory, such as the theory of evolution of multicellular organisms by symbiosis; and a model dealing with a specific or narrow area of science, such as the theory of enzyme action.[11]

One of the most important misunderstandings about the scientific method pertains to the relationship between research and theory. Although theory is usually presented as growing out of research, in fact theories also guide research. The very observations a scientist makes often occur in the context of existing theories. At times, discrepancies between observations and accepted theories become so great that a scientific revolution occurs; old theories are discarded and replaced with new or significantly revised theories.[12]

Science and Technology

Another misunderstanding about science occurs when science is confused with technology. As noted earlier, science is a search for understanding of the natural world, whereas technology is the application of scientific knowledge in an attempt to benefit human beings. Science often leads to technological developments, just as new technologies lead to scientific discoveries. The telescope began as a technological device such as an aid to sailors, but when Galileo used it to study the heavens, it became a source of new scientific knowledge. That knowledge stimulated the technology of telescope making, leading to the production of better telescopes, which in turn led to further advances in astronomy.

Science is limited by the technology available. Before the invention of the electron microscope, scientists were limited to magnifications of 1,000 times and to studying objects about the size of one-tenth of a micrometre. (A micrometre is 1/1,000,000 of a metre, or 1/1,000 of a millimetre.) The electron microscope enabled scientists to view objects far smaller by magnifying more than 100,000 times. The electron microscope, a basis for new science, was also the product of science. Without prior scientific knowledge about electron beams and how to focus them, the electron microscope could not have been developed.

Most of us do not come in direct contact with science in our daily lives; instead, we come in contact with the products of science—technological devices such as cars, toasters, and microwave ovens. Thus, people tend to confuse the products of science with science itself. As you study science, it will help if you keep in mind the distinction between science and technology.

Science and Objectivity

One myth about science is the myth of objectivity, or value-free science—the idea that scientists are capable of complete objectivity independent of their personal values and the culture in which they live and that science deals only with objective facts. Objectivity is certainly a goal of scientists, but it is unrealistic to think that they can be totally free of influence by their social environments and personal values. A more realistic view is to admit that scientists do have biases and to try to identify these biases rather than ignore them. In some ways, this situation is similar to that of measurement error. It is inescapable, and we can best deal with it by recognizing it and estimating its effects.

To find examples of how personal and social values affect science, we have only to look at recent controversies about environmental issues, such as whether or not to adopt more stringent automobile emission standards. Genetic engineering, nuclear power, and the preservation of threatened or endangered species involve conflicts among science, technology, and society.

The idea that science is not entirely value-free should not be taken to mean that fuzzy thinking is acceptable in science. It is still important to think critically and logically about science and related social issues. Without the high standards of evidence held up as the norm for science, we run the danger of accepting unfounded ideas about the world. When we confuse what we would like to believe with what we have the evidence to believe, we have a weak basis for making critical environmental decisions—decisions that could have far-reaching and serious consequences.

Science, Pseudoscience, and Frontier Science

Some ideas presented as scientific are in fact not scientific, because they are inherently untestable, lack empirical support, or are based on faulty reasoning or poor scientific methodology. Such ideas are referred to as **pseudoscientific** (*pseudo* means "false").

Pseudoscientific ideas arise from various sources, as shown in Figure 2.11. With more research, however, some of these models may move into the realm of accepted science and new ideas will take their place at the advancing frontier of knowledge.[13] Research may not support other hypotheses at the frontier, and these will be discarded by scientists. People may support pseudoscience because they don't understand the basis on which sound science is built, or because they are intentionally trying to deceive others for financial or political gain.

Some people continue to believe in discarded scientific ideas, or pseudoscience. For example, although scientists long ago discarded the notion that the movements of the heavenly bodies could affect the personalities and fates of humans, a substantial number of people believe in or are not sure about astrology (in one study, 25% believed and 22% were not sure).[14] Interestingly,

Figure 2.11 • Beyond the fringe? A diagrammatic view of different kinds of knowledge and ideas.

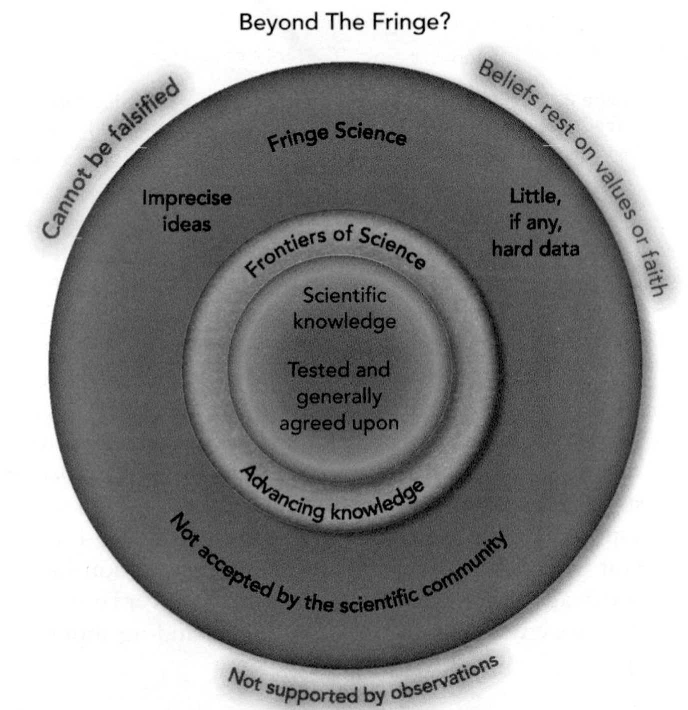

many of the things that astrologers learned about the movements of the stars and planets were so accurate that they became the basis for astronomy, the scientific study of the heavens. Parts of astrology moved into accepted science, parts into pseudoscience.

Although scientists have no trouble distinguishing between accepted science and pseudoscience, they do have trouble identifying ideas at the frontier that will become accepted and those that will be relegated to pseudoscience. That trouble arises because science is a process of continual investigation (See Fig. 2.11). This ambiguity at the frontiers of science leads many people to accept some frontier science before it has been completely verified and to confuse pseudoscience with frontier science. (See the discussion of the Gaia hypothesis in Chapter 3.)

Whereas pseudoscience is not based on the scientific method, **junk science** is scientific research that is intentionally biased to support a particular agenda. Junk science can include falsification of data, selective presentation or interpretation of experimental results, or active opposition to research that runs counter to the intended agenda. Junk science uses scientific tools to sway public debate, often related to environmental or human health, and often in the mass media.

2.6 Environmental Questions and the Scientific Method

Environmental sciences deal with especially complex systems and include a relatively new set of sciences. Therefore, the process of scientific study has not always neatly followed the formal scientific method discussed earlier in this chapter. Often, observations are not used to develop formal hypotheses. Controlled laboratory experiments have been the exception rather than the rule. Much environmental research has been limited to field observations of processes and events that have been difficult to subject to controlled experiments.

Environmental research presents several obstacles to following the classic scientific method. The long time frame of many ecological processes relative to human lifetimes, professional lifetimes, and lengths of research grants poses problems for establishing statements that can in practice be subject to disproof. What do we do if a theoretical disproof through direct observation would take a century or more? Other obstacles include difficulties in setting up adequate **experimental controls** in field studies, in developing laboratory experiments of sufficient complexity, and in developing theory and models for complex systems. Throughout this text, we present differences between the "standard" scientific method and the actual approach that has been used in environmental sciences.

An Example: The Vancouver Island Marmot

An environmental question that has been difficult to investigate using the traditional scientific method is the problem of the Vancouver Island Marmot, one of Canada's most endangered species (Figure 2.12). In the early 1990s, the total population of this species, which is endemic to Vancouver Island, had declined to fewer than 150 individuals, all located in less than 200 sq km of the island. The marmot had never been abundant since European settlement, but its numbers had dropped by more than 60% in just 10 years. Several suggestions were made about what to do to save this species from extinction.

One idea was to remove marmots from the wild and attempt to breed them in captivity. Then, once the numbers had increased, some of the marmots could be released back into the wild. Another suggestion was to move animals currently living in degraded habitat to areas with more favourable conditions. Over much of the marmots' natural range, widespread logging had reduced suitable habitat during the twentieth century, and reduced the availability of key food species, the deep soils and talus slopes marmots need for hibernation, and the boulders they use for "lookouts."

In 1996, following careful screening of sites and existing populations, a family group of six marmots was selected for reintroduction to ideal habitat. The actual transfer of animals went smoothly, with none showing signs of agitation or disrupted feeding. In the weeks following the reintroduction, however, some animals were observed moving over a wider area than would normally

Figure 2.12 • **The Vancouver Island Marmot.**

be the case, while others simply disappeared, possibly because of predation. In the fall of 1996, the surviving animals began hibernation in a normal fashion, but failed to emerge in the spring and were later found dead in their burrows. The dead animals had good body fat levels and should have been well equipped to survive hibernation. Few clues to their deaths were apparent. More than a year after this heartbreaking failure, researchers identified high concentrations of two bacteria strains, *Yersinia frederiksenii* and *Carnobacterium divergens*, in the dead animals. These bacteria are common in marmots, but only rarely cause death. Other possible explanations of the marmots' mortality include excessive stress following the reintroduction, pre-existing disease in the family group, or the presence of some other unidentified bacteria or virus. The high body fat levels in the animals suggest that stress was not a primary cause, but it may have been a contributing factor. Although traditional scientific methods were used in counting and assessing the health of individual animals, in the end, the researchers were unable to understand the reasons for the animals' disappearance. They could only acknowledge the need to base reintroductions upon larger numbers of animals, pursue captive breeding efforts, and try to fill the many gaps in our knowledge about this endangered species.[15]

To summarize this discussion, you will find that environmental sciences—young sciences dealing with complex phenomena—are filled with unanswered questions, and that many of these questions do not seem open to the traditional scientific method. They require new approaches that retain the essential key of the scientific method: the ability to disprove a statement. The large number of unanswered questions and the difficulty in answering them may seem discouraging. But they can also be viewed as an adventure in which doubt and uncertainty are a wonderful challenge. Richard Feynman expressed this idea when he said, "You cannot understand science and its relation to anything else unless you understand and appreciate [science as] the great adventure of our time. You do not live in your time unless you understand that this is a tremendous adventure and a wild and exciting thing."[16] We can approach environmental science in this spirit and seek a set of scientific methods that will allow us to answer its fascinating questions.

Some Alternatives to Direct Experimentation

So how have environmental scientists tried to answer these difficult questions? Several approaches have been taken, including use of historical records and observations of modern catastrophes and disturbances.

Historical Evidence

Ecologists have made use of both human and ecological historical records. A classic example is a study of the forest history at Crawford Lake, Ontario. Crawford Lake is unusual because it is a meromictic system—a lake that is very deep relative to its surface area, so that its waters never fully mix and its sediments are largely undisturbed. As sediment and biological materials settle in the lake, they are preserved in clearly defined layers called varves. Like tree rings, varves can be counted to estimate the age of lake sediment at different depths.

For more than 30 years, Dr. John McAndrews and his students have studied the sediments of Crawford Lake and the microfossils they contain, especially pollen from ancient forests. McAndrews and his colleagues argued that pollen analysis could be used to reconstruct the forest history around Crawford Lake. The unusual varve structure would allow very precise dating of each ecological stage, while pollen content would reflect forest composition over time. If the original forest was mainly maple and beech, then conservation efforts could encourage those species and discourage those like birch, cedar, and pine that currently dominate the system.

The pollen record from Crawford Lake did indeed show an early maple-beech forest (Figure 2.13). Those findings were confirmed by macrofossil evidence, like twigs, leaves and seeds preserved in lake sediment. However, McAndrews found much more of interest in the varved sediments. Beginning in about 1360 AD, he observed a striking increase in grass and corn pollen, coupled with a decline in beech and maple. About the same time, oak and pine apparently began to increase in the forests around Crawford Lake. These changes persisted until about 1660 AD, when grass, corn, and now ragweed increased dramatically, with a steep drop in oak and pine. Based on these findings, McAndrews concluded that the Crawford Lake sediments reveal three distinct stages. The earliest is a relatively stable mixed hardwood forest, dominated by maple and beech. The second, lasting about 300 years, is a period of Iroquois agriculture. The third, commencing in about 1750, reflects extensive European settlement, land clearance, and farming in the area.

McAndrews' findings prompted detailed archaeological investigation of the area, revealing the remains of six long-houses and evidence of a substantial Iroquois settlement. From the accounts of French explorers and missionaries, we know that these kinds of villages were occupied for 10 to 20 years, during which their occupants cut and burned forests and planted corn using manual methods. With time, soil fertility and crop yields declined and the fields were abandoned, to be recolonized by poplar, oak, and white pine. With the coming of European settlers, and especially the mechanization of farming, came a more dramatic and permanent change in the Crawford Lake forest ecosystem.

Four kinds of historical records—archeological excavations, pollen analysis, plant fossils, and written

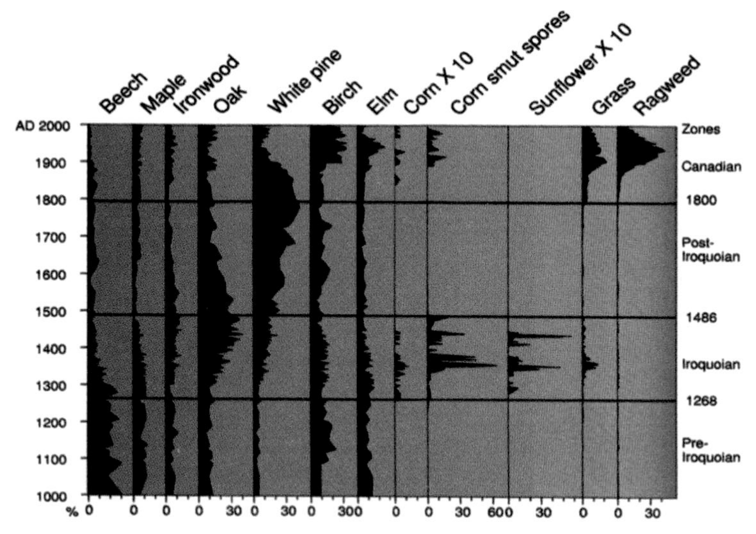

Figure 2.13 • Vegetation history at Crawford Lake, as reflected in the pollen record. The pollen percentages are based on 200 tree pollen grains. Since 1867, the chronology is based on varves but before then, it is based on 23 carbon dates. [*Source*: Ekdahl, E. J., J. L. Teranes, T. P. Guilderson, C. L. Turton, J. H. McAndrews, C. A. Wittkop and E. C. Stoermer. 2004. A prehistorical record of cultural eutrophication from Crawford Lake, Canada. Geology 32: 745-748.]

records—provided important evidence to disprove two hypotheses: that the early Crawford Lake forest was dominated by pine and oak, and that First Nations agriculture had only a negligible influence on the forest ecosystem. Thus, the use of historical information meets the primary requirement of the scientific method—the ability to disprove a statement. Historical evidence is a major source of data that can be used to test scientific hypotheses in ecology.

Modern Catastrophes and Disturbances as Experiments

Sometimes a large-scale catastrophe provides a kind of modern ecological experiment. The volcanic eruption of Mount St. Helens in 1980 supplied such an experiment, destroying vegetation and wildlife over a wide area. The recovery of plants, animals, and ecosystems following this explosion gave scientists insights into the dynamics of ecological systems and provided some surprises. The main surprise was the rapidity with which vegetation recovered and wildlife returned to parts of the mountain. In other ways, the recovery followed expected patterns in ecological succession (see Chapter 9). A more recent example of a catastrophe that has become the subject of ecological research was the 1988 wildfire in Yellowstone National Park (Figure 2.14).

It is important to point out that the greater the quantity and the better the quality of ecological data prior to such a catastrophe, the more that can be learned from the response of ecological systems to the event. This calls for careful monitoring of the environment.

2.7 Science and Decision Making

Like the scientific method, the process of making decisions is sometimes presented as a series of steps:

1. Formulate a clear statement of the issue to be decided.
2. Gather the scientific information related to the issue.
3. List all alternative courses of action.
4. Predict the positive and negative consequences of each course of action and the likelihood that each consequence will occur.
5. Weigh the alternatives and choose the best solution.

Such a procedure is a good guide to rational decision-making, but it assumes a simplicity not often found in real-world issues. It is difficult to anticipate all the consequences of a course of action, and unintended consequences are at the root of many environmental problems. Often the scientific information is incomplete and even controversial. For example, the insecticide DDT causes eggshells of birds that feed on insects to be so thin that unhatched birds die. When DDT first came into use, this consequence was not predicted. Only when populations of species such as the bald eagle became seriously endangered did people become aware of it.

In the face of incomplete information, scientific controversies, conflicting interests, and emotionalism, how are environmental decisions made? We need to begin with the scientific evidence from all relevant sources and with estimates of the uncertainties in each. Where scientists disagree about the interpretation of data, it may be possible to develop a consensus or a series of predictions based on the different interpretations. The impacts of the scenarios need to be identified and the risks associated with each analysed in comparison with the benefits. Avoiding emotionalism and resisting slogans and propaganda are essential to developing sound approaches to environmental issues. Ultimately, however, environmental decisions are policy decisions negotiated through the political process. Policymakers are rarely professional scientists; generally they are political leaders and ordinary citizens. Therefore, the scientific education of those in government and business, as well as of all citizens, is crucial.

2.8 Learning about Science

Science is an open-ended process of finding out about the natural world. In contrast, science lectures and texts are usually summaries of the answers arrived at through this process, and science homework and tests are exercises in finding the right answer.

Perhaps as a result of this teaching approach, students often perceive science as a body of facts to be memorized and view lectures and texts as authoritative sources of absolute truths about the world. In contrast, scientists view scientific knowledge as currently accepted truth, always subject to change as new observations and interpretations are made. Students tend not to question the material in textbooks and lectures, whereas the essence of science is questioning and critically examining accepted truths.

Viewing science as a collection of facts may lead students to see science as difficult to understand and remember. In contrast, scientists emphasize the use of facts to form coherent pictures of the world that have the power to explain many phenomena. These pictures (models, theories) are so powerful that scientists find them easy to remember and expect that students will also.

Learning about science requires solving problems. Here, too, the attitudes of students and scientists differ. Whereas students may look to formulas, typical problems, and algorithms to help solve problems, scientists look to general principles, critical thinking, and creativity.

We encourage you to learn environmental science in an active mode, to be critical of what you hear, what you see, and what you read—including this textbook. We encourage you to try to understand what science is—its assumptions, methods, and limitations—and to apply this understanding to your study of environmental science. Most of all, we hope that you will see not isolated facts to be learned by rote but the connections among facts that reflect interdependence and unity in the environment.

2.9 Science and Media Coverage

Most media reports on scientific issues deal with new discoveries, fringe science, and pseudoscience. [See A Closer Look 2.4.] It is important to analyse the claims in such reports to decide in which category they should be placed. Critical reading and listening require thinking carefully about whether a statement is based on observations and data, an objective interpretation of data, an interpretation based on the experience of an expert in the area, or a subjective opinion.

The need for critical examination applies to statements made by scientists as well as by other sources. On the one hand, scientists' training in analyzing data may make them more qualified than the general public to decide certain complex issues; on the other, as discussed before, scientists' own values, interests, and cultural backgrounds may influence their interpretations of data and bias their pronouncements. The situation is complicated by the fact that some scientists become "media scientists," whose opinions are sought on a wide variety of topics, some of which they have not studied as scientists. The opinions of experts do need to be heard, but appealing to scientists as authorities outside their area of expertise is contrary to the anti-authoritarian nature of science.

Figure 2.14 • Forested land in Yellowstone National Park *(a)* before and *(b)* after the 1988 fire. The fire was a catastrophic event to which forest species have adapted.

(a)

(b)

A CLOSER LOOK 2.4

Evaluating Media Coverage

The following questions will help you evaluate a report on science:[17]

1. Where is the report? Is it in a scientific journal, a newspaper, a popular science magazine, or a tabloid?
2. Is the report based on observations of actual occurrences? Were those observations made by more than one person? More than a few?
3. Are the sources of the report identified specifically? Are the sources specific, named scientists, scientific journals, or scientific organizations?
4. Does the report provide evidence that the claims are supported by other members of the scientific community?
5. Does the evidence for the claims seem sufficient? Is there contradictory evidence that might offset the evidence given for them?
6. Do the claims follow logically from the evidence? Do the claims violate reason (for example, a 99-year-old woman gives birth to baby)? Is there a simpler explanation for the observations?
7. Is there a clear basis for suspecting bias on the part of the source or the writer of the report?
8. Can you describe the line of reasoning that leads from the evidence to the claims?

Summary

- Science is one path to critical thinking about the natural world. Its goal is to gain an understanding of how nature works. Decisions on environmental issues must begin with an examination of the relevant scientific evidence. However, environmental decisions also require careful analysis of economic, social, and political consequences. Solutions will reflect religious, aesthetic, and ethical values as well.
- Science begins with careful observations of the natural world, from which scientists formulate hypotheses. Whenever possible, scientists test hypotheses with controlled experiments.
- Although the scientific method is often taught as a prescribed series of steps, it is better to think of it as a general guide to scientific thinking, with many variations.
- Scientific knowledge is acquired through inductive reasoning, in which general conclusions are based on specific observations. Conclusions arrived at through induction can never be proved with certainty. Because of the inductive nature of science, it is possible to disprove hypotheses, but it is not possible to prove them with 100% certainty.
- Measurements are approximations that may be more or less exact, depending on the measuring instruments and the people who use the instruments. A measurement is meaningful when it is accompanied by an estimate of the degree of uncertainty, or error.
- Accuracy in measurement is the extent to which the measurement agrees with an accepted value. Precision is the degree of exactness with which a measurement is made. A precise measurement may not be accurate. The estimate of uncertainty provides information on the precision of a measurement.
- A specific statement that describes a repeatable phenomenon is called a law. The law of gravity describes the behaviour of bodies in the universe, but does not explain that behaviour.
- A general statement that relates and explains a great many hypotheses is called a theory. Theories are the greatest achievements of science.
- Critical thinking can help us distinguish science from pseudoscience. It can also help us recognize possible bias on the part of scientists and in the media. Critical thinking involves questioning and synthesizing what is learned in order to achieve knowledge, rather than merely acquire information.

STUDY QUESTIONS

1. Which of the following are scientific statements and which are not? What is the basis for your decision in each case?
   a. The amount of carbon dioxide in the atmosphere is increasing.
   b. Vancouver Island Marmots are ugly.
   c. Vancouver Island Marmots are endangered.
   d. There are 150 Vancouver Island Marmots.
   e. Crop circles are a sign from Earth to us that we should act better.
   f. Crop circles can be made by people.
   g. Excess phosphorus causes algae blooms in temperate freshwater lakes.

CRITICAL THINKING ISSUE

How Do We Decide What to Believe about Environmental Issues?

When you read about an environmental issue in a newspaper or magazine, how do you decide whether to accept the claims made in the article? Are they based on scientific evidence, and are they logical?

Scientific evidence is based on observations, but media accounts often rely mainly on inferences (interpretations) rather than evidence. Distinguishing inferences from evidence is an important first step in evaluating articles critically.

Second, it is important to consider the source of a statement. Is the source a reputable scientific organization or publication? Does the source have a vested interest that might bias the claims? When sources are not named, it is impossible to judge the reliability of claims.

If a claim is based on scientific evidence presented logically from a reliable, unbiased source, it is appropriate to accept the claim tentatively, pending further information. Practice your critical evaluation skills by reading the article in the box and answering the critical thinking questions.

Critical Thinking Questions

1. What is the major claim made in the article?
2. What evidence does the author present to support the claim?
3. Is the evidence based on observations, and is the source of the evidence reputable and unbiased?
4. Is the argument for the claim, whether based on evidence or not, logical?
5. Would you accept or reject the claim?
6. Even if the claim were well supported by evidence based on good authority, why would your acceptance be only tentative?

Biologist Sounds Battle Cry Against Unwelcome Amphibian

Last Updated Mon, 21 Feb 2005 11:26:32 EST
CBC News

VANCOUVER—A conservation biologist is urging an all-out assault on the American bullfrog in British Columbia this spring, saying the amphibians are a threat to aquatic ecosystems. Stan Orchard says the bullfrog has a varied diet. Snakes, turtles, and birds have been prey. Even cats aren't safe from the invaders, which can grow to the size of a dinner plate.

The American bullfrog was imported as a delicacy for the dinner table, but now they're doing the dining in increasing numbers. They'll eat just about anything that will fit into their mouths. At greatest risk are other species of frogs and small ducks. The bullfrogs are blamed for wiping out the Cranberry Lake duck population.

"There's no question about the ecological impact of bullfrogs on the region," Orchard says. "It's going to change water chemistry in some areas, and it's certainly going to remove some species from the local ecosystems." For years, local biologists have studied the population boom; now Orchard says it's time to act. "We don't have the luxury of time to really get numbers on the population. What we'll be doing is what's called the removal technique – you simply keep removing them and removing them until your numbers drop off to zero."

Once the ice thaws, Orchard hopes to get to work, zapping the frogs with an electrical charge to stun them and make trapping easier. Local governments have chipped in cash, but so far, Orchard says it's not nearly enough to win the war.

2. What is the logical conclusion of each of the following syllogisms? Which conclusions correspond to observed reality?
   a. All men are mortal.
      Socrates is a man.
      Therefore ____________________
   b. All sheep are black.
      Mary's lamb is white.
      Therefore ____________________
   c. All Amazons are women.
      No men are Amazons.
      Therefore ____________________
   d. All elephants are animals.
      All animals are living beings.
      Therefore ____________________
3. Which of the following statements are supported by deductive reasoning and which by inductive reasoning?
   a. The sun will rise tomorrow.
   b. The square of the hypotenuse of a right triangle is equal to the sum of the squares of the other two sides.
   c. Only male deer have antlers.
   d. If $A = B$ and $B = C$, then $A = C$.
   e. The net force acting on a body equals its mass times its acceleration.
4. The accepted value for the number of inches in a centimetre is 0.3937. Two students mark off a centimetre on a piece of paper and then measure the distance using a ruler (in inches). Student A finds the distance equal to 0.3827 in., and student B finds it equal to 0.39

in. Which measurement is more accurate? Which is more precise? If student B measured the distance as 0.3900 in., what would be your answer?

5. a. A teacher gives five students each a metal bar and asks them to measure the length. The measurements obtained are 5.03, 4.99, 5.02, 4.96, and 5.00 cm. How can you explain the variability in the measurements? Are these systematic or random errors?
   b. The next day, the teacher gives the students the same bars but tells them that the bars have contracted because they have been in the refrigerator. In fact, the temperature difference would be too small to have any measurable effect on the length of the bars. The students' measurements, in the same order as in part a, are 5.01, 4.95, 5.00, 4.90, and 4.95 cm. Why are the students' measurements different from those of the day before? What does this illustrate about science?
6. Identify the independent and dependent variables in each of the following:
   a. Change in the rate of breathing in response to exercise.
   b. The effect of study time on grades.
   c. The likelihood that people exposed to smoke from other people's cigarettes will contract lung cancer.
7. a. Identify a technological advance that resulted from a scientific discovery.
   b. Identify a scientific discovery that resulted from a technological advance.
   c. Identify a technological device you have used today. What scientific discoveries were necessary before the device could be developed?
8. What is fallacious about each of the following conclusions?
   a. A fortune cookie contains the statement "A happy event will occur in your life." Four months later, you find a $100 bill. You conclude that the fortune was correct.
   b. A person claims that aliens visited Earth in prehistoric times and influenced the cultural development of humans. As evidence, the person points to ideas among many groups of people about beings who came from the sky and performed amazing feats.
   c. A person observes that light-coloured animals almost always live on light-coloured surfaces, whereas dark forms of the same species live on dark surfaces. The person concludes that the light surface causes the light color of the animals.
   d. A person knows three people who have had fewer colds since they began taking vitamin C on a regular basis. The person concludes that vitamin C prevents colds.
9. Find a newspaper article on a controversial topic. Identify some loaded words in the article—that is, words that convey an emotional reaction or a value judgment. Why do you think the authors have used this kind of language?
10. Identify some social, economic, aesthetic, and ethical issues involved in a current environmental controversy.

FURTHER READING

American Association for the Advancement of Science (AAAS). *Science for All Americans.* Washington, D.C.: AAAS, 1989. This report focuses on the knowledge, skills, and attitudes a student needs to be scientifically literate.

Botkin, D. B. *No Man's Garden: Thoreau and a New Vision for Civilization and Nature.* Washington, D.C.: Island Press, 2001. The author discusses how science can be applied to the study of nature and to problems associated with people and nature. He also discusses science and values.

Grinnell, F. *The Scientific Attitude.* New York: Guilford, 1992. Examples from biomedical research are used to illustrate the processes of science (observing, hypothesizing, experimenting) and how scientists interact with each other and with society.

Kuhn, Thomas S. *The Structure of Scientific Revolutions.* Chicago: University of Chicago Press, 1996. A modern classic in the discussion of the scientific method, especially regarding major transitions in new sciences like environmental sciences.

McCain, G., and E. M. Segal. *The Game of Science.* Monterey, Calif.: Brooks/Cole, 1982. The authors present a lively look into the subculture of science.

Sagan, C. *The Demon-Haunted World.* New York: Random House, 1995. The author argues that irrational thinking and superstition threaten democratic institutions and discusses the importance of scientific thinking to our global civilization.

chapter 3

Amboseli National Park: Disturbance, Change, and Environment Connectivity

Figure 3.1 • African Elephant.
Elephant damage to trees is considered a factor in loss of woodland habitat in Amboseli National Park. However, elephants probably play a relatively minor role compared with oscillations in climate and groundwater conditions.

Environmental change is often caused by a complex web of interactions, and the most obvious answer to the question of what caused a particular change may not be the right answer. Amboseli National Park is a case in point. In the short span of a few decades, this park, located in southern Kenya at the foot of Mount Kilimanjaro, underwent a significant environmental change. An understanding of physical, biological, and human-use factors—and how these factors are linked—was needed to explain what happened.

The park is centered on an ancient lake bed, remnants of which include the seasonally flooded Lake Amboseli and some swampland. Mount Kilimanjaro is a well-known volcano, composed of alternating layers of volcanic rock and ash deposits. Rainfall that reaches the slopes of Mount Kilimanjaro infiltrates the volcanic material (becomes groundwater) and moves slowly down the slopes to saturate the ancient lake bed, eventually emerging at springs in the swampy, seasonally flooded land. The groundwater becomes very saline (salty) as it percolates through the lake bed, since salt stored in the lake bed sediments dissolves easily when the sediments are wet.

Before the mid-1950s, the dominant vegetation in the park was fever tree woodlands, which provided habitat for mammalian species such as kudu, baboons, vervet monkeys, leopards, and impalas. Starting in the 1950s, and accelerating in the 1960s, these woodlands disappeared and were replaced by short grass and brush, which provided habitat for typical plains animals such as zebras and wildebeest. The annual amounts of rainfall increased during these decades.

Loss of the woodland habitat was initially blamed on overgrazing of cattle by the Masai people and damage to the trees from elephants (Figure 3.1). However, investigators found that most dead trees were in an area that had been free of cattle since 1961 before the major decline in the woodland environment. Furthermore, some of the woodlands that suffered the least decline had the highest density of people and cattle. These observations suggested that overgrazing by cattle was not responsible for loss of the trees. Elephant damage was also thought to be a major factor because elephants had stripped bark from more than 83% of the trees in some areas and had pushed over some younger, smaller trees. However, researchers concluded that elephants played only a secondary role in changing the habitat. As the density of fever trees and other woodland plants decreased, the incidence of damage caused by elephants increased. In other words, elephant damage interacted with some other, primary factor in changing the habitat.

Environmental scientists eventually rejected these hypotheses as the main causes of the environmental change, demonstrating that changes in rainfall and soils were the primary culprits, rather than people or elephants.[1, 2] Research on rainfall, groundwater history, and soils suggested that the Amboseli National Park area

The Big Picture: Systems of Change

LEARNING OBJECTIVES

Changes in systems may occur naturally or may be induced by humans. Many complex and far-reaching interactions can result. After reading this chapter, you should understand:

- Why solutions to many environmental problems involve the study of systems and rates of change.
- How positive and negative feedback operate in a system.
- The implications of exponential growth and doubling time.
- That natural disturbances and changes in systems such as forests, rivers, and coral reefs are important to their continued existence.
- What an ecosystem is and why sustained life on Earth is a characteristic of ecosystems.
- What the Gaia hypothesis is and how life on Earth has affected Earth itself.
- What the principle of uniformitarianism is and how it can be used to anticipate future changes.
- Why the principle of environmental connectivity is important in studying environmental problems.
- Why environmental problems are difficult to solve.
- How human activities amplify the effects of natural disasters.

is very sensitive to changing amounts of rainfall. During dry periods, the salty groundwater sinks lower in the earth, and soil near the surface has a relatively low salt content. The fever trees grow well in the nonsalty soil.

Figure 3.2 • **Masai people grazing cattle in Amboseli National Park, Kenya.** Grazing activities were prematurely blamed for loss of fever tree woodlands.

During wet periods, the groundwater rises close to the surface, bringing with it salt, which invades the root zones of trees and kills them. The groundwater level rose as much as 3.5 m in response to unusually wet years in the 1960s. Soil analysis confirmed that the tree stands that suffered the most damage were those associated with highly saline soils.[1, 2] As the trees died, they were replaced by salt-tolerant grasses and low brush.[1,2]

Based on the historic record from early European explorers, accounts from Masai herders, and fluctuating lake levels in other East African lakes, the scientists concluded that cycles of greater and lesser rainfall change hydrology and soil conditions, which in turn change the plant and the animal life of the area.[1]

- *The Amboseli story illustrates how environmental science attempts to work out sequences of events that follow a particular change. At Amboseli, rainfall cycles change hydrology and soil conditions, which in turn change the vegetation and animals of the area. In this chapter, we examine environmental systems and changes that occur as a result of natural and human-induced processes.*

3.1 Systems and Feedback

As the opening case illustrates, environmental problems often have multiple causes. Solving one problem may create another. In environmental science at every level, we must deal with a variety of systems that range from simple to complex. We must understand systems and how different parts of systems interact with one another. We begin our discussion of systems by defining what a system is and how it can operate.

System Defined

A system is a set of components or parts that function together to act as a whole. A single organism (such as your body) is a system, as are a sewage treatment plant, a city (Figure 3.3), a river (Figure 3.4), and your residence room. On a much different scale, the entire earth is a system.

Systems may be open or closed. An open system is not generally contained within boundaries, and some energy or material (solid, liquid, or gas) moves into or out of the system. The ocean is an open system with regard to water, because water moves into the ocean from the atmosphere and out of the ocean to the atmosphere. Conversely, in a closed system, no such movements take place. Earth is a closed system (for all practical purposes) with regard to material.

Systems respond to inputs and have outputs. Your body, for example, is a complex system. If you are hiking and see a grizzly bear, the sight of the bear is an input. Your body reacts to that input: The adrenaline level in your blood goes up, your heart rate increases, and the hair on your head and arms may rise. Your response—perhaps moving slowly away from the bear—is an output. We call these responses feedback.

Feedback

Feedback occurs when the output of the system also serves as an input and leads to further changes in the system. A good example of feedback is human temperature regulation. If you go out in the sun and get hot, the increase in temperature affects your sensory perceptions (input). If you stay in the sun, your body responds physiologically: Your skin pores open, and you are cooled by evaporating water (you sweat). The cooling is output, and it is also input to your sensory perceptions. You may also respond behaviorally: Because you feel hot (input), you walk into the shade and your temperature returns to normal.

In our example, an increase in temperature is followed by a response that leads to a decrease in temperature. This is an example of negative feedback, in which an increase in output leads to a later decrease. Negative feedback is self-regulating, or stabilizing; it usually keeps a system in a relatively constant condition.

Figure 3.3 • Sydney, Australia is an example of a large, complex urban system that includes air, water, and land resources. Urban systems are particularly important in environmental science because more and more people are living in urban areas.

Positive feedback occurs when an increase in output leads to a further increase in the output. A fire starting in a forest provides an example of positive feedback. The wood may be slightly damp at the beginning and so may not burn well. Once a fire starts, wood near the flame dries out and begins to burn, which in turn dries out a greater quantity of wood and leads to a larger fire. The larger the fire, the more wood is dried and the more rapidly the fire increases. Positive feedback, sometimes called "a vicious cycle," is destabilizing.

Environmental damage can occur when human use of the environment leads to positive feedback. Off-road

Figure 3.4 • Reid Brook, Labrador is a system that includes water, sediment, vegetation, and animals such as fish and insects that function together as a whole.

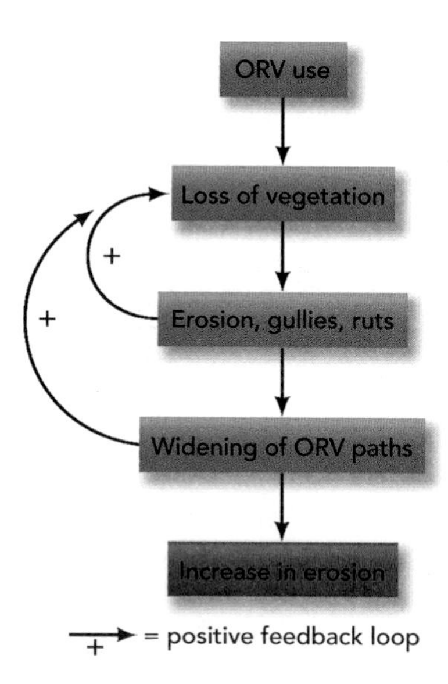

Figure 3.5 • How use of off-road vehicles (ORVs) produces positive feedback that increases erosion.

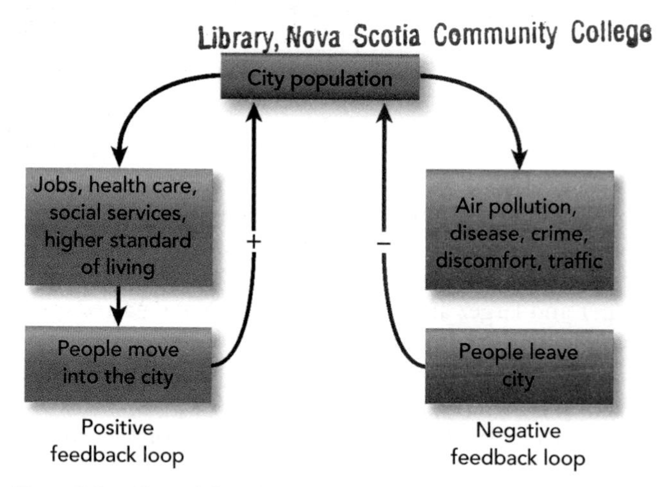

Figure 3.6 • Potential positive and negative feedback loops for changes of human population in large cities. The left side of the figure shows that as jobs, health care, and a higher standard of living increase, so do migration and city population. Conversely, the right side of the figure shows that increases in air pollution, disease, crime, discomforts, or traffic tend to reduce the city pollution. [*Source*: Modified from M. Maruyama. "The Second Cybernetics: Deviation-Amplifying Mutual Causal Processes," *American Scientist* 51 (1963):164–670. Reprinted by permission of *American Scientist*, magazine of Sigma Xi, The Scientific Research Society.)

vehicle use—including bicycles (Figure 3.5)—may cause positive feedback with respect to soil erosion. These vehicles' churning tires, designed to grip the earth, erode the soil and uproot plants, increasing the rate of erosion. As more soil is exposed to erosion by running water, ruts, and gullies are carved. Drivers then avoid the ruts and gullies, driving on adjacent sections that are not as eroded, thus widening paths and further increasing erosion. The gullies themselves cause an increase in erosion because they concentrate runoff and have steep side slopes. Once formed, gullies tend to grow in length, width, and depth, causing additional erosion (Figure 3.5). Eventually, an area of intensive off-road vehicle use may become a wasteland of eroded paths and gullies. Positive feedback has made the situation get worse and worse.

Some situations can be examined in terms of both positive and negative feedback loops. Changes in human population in large cities present an example, as illustrated in Figure 3.6. Positive feedback, which increases the population in cities, may occur when people perceive greater opportunities in cities and hope for a higher standard of living. Negative feedback may result from air and water pollution, disease, crime, and discomfort, if these factors encourage some people to migrate from the cities to rural areas.

Practicing your critical thinking skills, you may ask, "Is negative feedback generally desirable, and is positive feedback generally undesirable?" Reflecting on this question, we can see that although negative feedback is self-regulating, it may in some instances not be desirable. The period of time over which the positive or negative feedback occurs is the important factor.

For example, suppose we are interested in restoring the ecology of the Canadian prairies through the reintroduction of swift foxes. We will expect positive feedback in the population for a time as the number of swift foxes grows. In this case, positive feedback for a period of time is desirable because it produces a change we want. We might also envision a system in a state that is stable but undesirable. An example is a polluted stream in an urban environment. Urban runoff and its associated pollutants such as oil and other chemicals from streets entering the stream's system may, through negative feedback mechanisms, reach a stable state between the water and the pollutants in the stream. However, this system would be considered by most people to be undesirable. A channel restoration project might be implemented to control pollutants by collecting and treating them before they enter the stream. As a result, the stream might reach a new state that is ecologically more desirable.

We can see that whether we view positive or negative feedback as desirable depends on the system and potential changes. Nevertheless, some of the major environmental problems we face today result from positive feedback mechanisms that are out of control. These include resource use and growth of human population, among others.

Interestingly, throughout most of human history, strong negative feedback cycles resulted in a very low growth in human population. Disease and limited capacity to produce food kept growth low. However, in the past hundred years, modern medicine, sanitation, and agricultural practices turned negative feedback into positive feedback, and rapid increase in human population occurred.

3.2 Exponential Growth

A particularly important example of positive feedback occurs with exponential growth. Simply stated, growth is exponential when it occurs at a constant *rate* per time period (rather than a constant *amount*). For instance, suppose you have $1,000 in the bank, and it grows at 10%

per year. The first year, $100 in interest is added to your account. The second year, you earn more because you earn 10% on the new total amount, $1,100. The greater the amount, the greater the interest earned, so the money (or the population, or some other quantity) increases by larger and larger amounts. When we plot data in which exponential growth is occurring, the curve we obtain is said to be "J" shaped. It looks like a skateboard ramp, starting out nearly flat and then rising steeply. (The actual shape depends on the scale of the units of the curve.) Figure 3.7 shows a typical exponential growth curve.

Calculating exponential growth involves two related factors: the rate of growth measured as a percentage and the doubling time in years. The doubling time is the time necessary for the quantity being measured to double. A useful rule is that the doubling time is approximately equal to 70 divided by the annual percentage growth rate. Build Your Environmental Skills 3.1 describes exponential growth calculations and explains why 70 divided by the annual growth rate is the doubling time. If you understand algebra and natural logarithms, you should find the example interesting. However, it is the general principles that are most important.

One significant principle of exponential growth is that it is incompatible with the concept of sustainability, which is a long-term process (decades to hundreds of years or more). In fact, the term *sustainable growth* is an oxymoron—a self-contradiction. Even at modest growth rates, the number of whatever is growing will eventually reach extraordinary levels that are impossible to maintain.[3,4]

Exponential growth has interesting (and sometimes alarming) consequences, as illustrated in a fictional story by Albert Bartlett.[3] Imagine a hypothetical strain of bacteria in which each bacterium divides into two every 60 seconds (the doubling time is 1 minute). Assume that one bacterium is put in a bottle at 11:00 a.m. The bottle (its world) is full at 12:00 noon. When was the bottle half full? The answer is 11:59 a.m. If you were a bacterium in the bottle, at what time would you realize that you were running out of space? There is no single answer to this question. Consider, though, that at 11:58 the bottle was 75% empty, and at 11:57 it was 88% empty. Now assume that at 11:58 some farsighted bacteria realized that the population was running out of space and started looking around for new bottles. Let's suppose that they were able to find and move into three more bottles. How much time did they buy? Two additional minutes. They will run out of space at 12:02 p.m. If they had found 16 additional bottles, how much more time would they have?

The story of bacteria in bottles, while obviously hypothetical, illustrates the power of exponential growth. We look at exponential growth and doubling time again in Chapter 4, when we consider the growth of the human population. Here, we simply note that many systems in nature display exponential growth for some period of time, so it is important that we be able to recognize it. In particular, it is important to recognize exponential growth in a positive feedback cycle, as accompanying changes may be very difficult to control or stop.

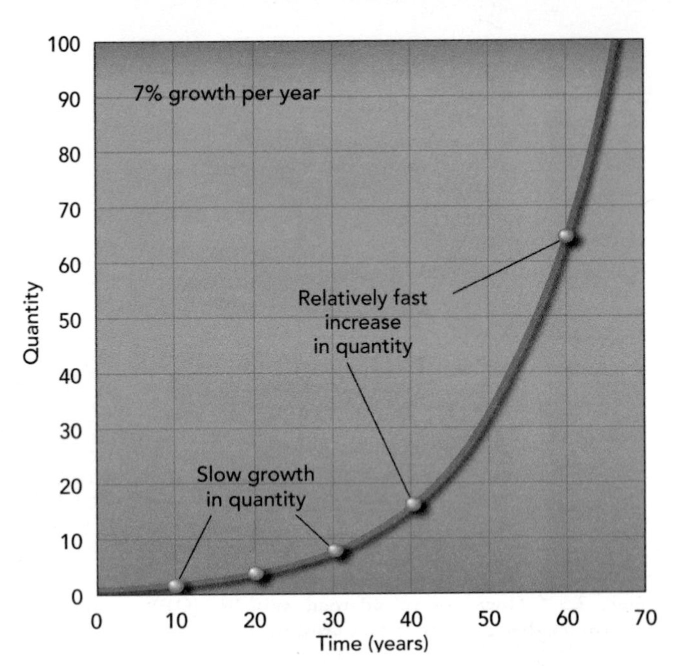

Figure 3.7 • **Exponential growth.** Idealized curve illustrating exponential growth. Growth rate is constant at 7%, and time necessary to double the quantity is constant at 10 years. Notice that growth is slow at first and much faster after several doubling times. For example, the quantity changes from 2 to 4 (absolute increase of 2) for the doubling from 10 to 20 years. It increases from 32 to 64 (absolute increase of 32) during the doubling from 50 to 60 years.

3.3 Environmental Connectivity

Our discussion of positive and negative feedback sets the stage for another fundamental concept in environmental science: environmental connectivity. Simply stated, environmental connectivity means that it is impossible to change only one thing; everything affects everything else. Of course, this is something of an overstatement. The concept is not absolutely true; the extinction of a species of snails in North America, for instance, is hardly likely to change the flow characteristics of the Amazon River. Many aspects of the natural environment are closely linked, however. Changes in one part of a system often have secondary and tertiary effects within the system and effects on adjacent systems. Earth and its ecosystems are complex entities in which any action may have several or many effects. The case study of Amboseli National Park is a good example of the principle of environmental connectivity. We can find many other examples in both human-made and natural settings.

An Urban Example

Consider the changes occurring in Canadian cities such as Toronto and Winnipeg during a major shift in land use from forests or agricultural land to urban development. Clearing the land for urban use increases runoff and the

Build Your Environmental Skills 3.1

Exponential Growth

Section 3.2 describes the idea of exponential growth, one of the most important concepts in science and in everyday life. In exponential growth, a quantity increases at a fixed rate per unit of time. The mathematical equation used to describe this is:

$$N = N_0\, e^{kt}$$

where N = the future value
$N0$ = the present (starting value)
e = the base of natural logarithms, a constant 2.71828
k = the rate of growth, expressed as a fraction of 1 (e.g., for 2% growth, k = 0.002)
t = the number of time steps (for instance, years) over which growth is to be calculated.

Use this simple equation to calculate:

- The world population in 2020, given a starting population of 6.3 million and an annual growth rate of 1.36% (k = 0.0136). You may also want to calculate the expected population at some intermediate dates (2008, 2010, 2013, 2015, 2018, and 2020), and graph the results.
- The world population in 2020 using the same starting population but a growth rate of 2%.
- The world population in 2020 using the same starting population but a growth rate of 1%.
- The expected population in your home town, using its current population and one or more of the three growth rates suggested above.

We can also use the exponential growth equation to estimate the doubling time of Earth's population. Set $N = 2N_0$ and solve for the value of t (let's call it *Td*, to represent doubling time) by taking the natural logarithm of both sides:

$$\ln 2 = kT_d, \text{ so } T_d = \ln 2/k = 0.693/k$$

Multiplying both numerator and denominator by 100, we find that doubling time is approximately 70 divided by the growth rate expressed as a percent (e.g., 10%). This result is our general rule—that the doubling time is approximately 70 divided by the growth rate. For example, if the growth rate is 10% per year, then T^d = 7 years.

amount of sediment eroded from the land (soil erosion). Increased runoff from streets and sediment eroded from bare ground during construction affect the form and shape of the river channel. The river carries more sediment, and some of it is deposited on the bottom of the channel, reducing channel depth and increasing flood hazard.

Eventually, as more land is paved, the amount of sediment eroded from the land decreases, runoff increases further, and the streams readjust to a lower sediment load (the amount of sediment carried by the stream) and more runoff. The readjustment is a form of negative feedback inherent in streams and rivers.

Urbanization is also likely to pollute the streams or otherwise change water quality. The increased fine sediment makes the water muddy, and chemicals from street and yard runoff pollute the stream. These changes affect the biological systems in the stream and adjacent banks. Thus, land-use conversion can set off a series of changes in the environment, and each change is likely to trigger additional changes.

A Forest Example

The interaction among forests, streams, and fish in British Columbia rivers provides another example of environmental connectivity. In that region's forests, large pieces of woody debris such as tree trunks and roots are necessary to form and maintain nearly all the pool environments in small streams (Figure 3.8). Large branches fall naturally into streams and partially block their flow, producing pools of deeper water. These pools provide much of the rearing habitat for young salmon, which spend part of their lives in the streams before migrating to the ocean.

It was formerly common practice to remove woody debris from streams because it was thought to block the migration of adult salmon attempting to return to spawning beds in the streams. We now know that this practice degrades the fish habitat. Stream restoration projects now often place large woody debris into channels to improve the fish habitat. The role of large woody debris in stream processes and salmon habitat illustrates the value of studying relations between physical and biological systems to help provide for sustainable fish populations. Such studies are at the heart of environmental science.

3.4 Uniformitarianism

Earth and its life forms have changed many times, but the processes necessary to sustain life and the environment for life have occurred throughout much of Earth's history. The principle that physical and biological processes presently forming and modifying Earth can help explain the geologic and evolutionary history of Earth is known as uniformitarianism. The principle may be more simply stated as "the present is the key to the past." For example, if a deposit of gravel and sand found at the top of a mountain is similar to stream gravel found today in an adjacent valley, we may infer by uniformitarianism

Figure 3.8 • Stream processes are significantly modified by fallen trees. The large tree trunk in the central portion of the photograph produced a small pool, which is a good fish habitat.

that a stream once flowed in a valley where the mountaintop is now. Uniformitarianism can also be used as a key to the future. If a process like stream gravel formation is operating today, it is likely to be operating in the same fashion in the future.

Uniformitarianism was first suggested in 1785 by the Scottish scientist and geologist James Hutton. Charles Darwin was impressed by the concept of uniformitarianism, and the concept pervades his ideas on biological evolution. Today, uniformitarianism is considered one of the fundamental principles of the biological and earth sciences.

Uniformitarianism does not demand or even suggest that the magnitude and frequency of natural processes remain constant. Obviously, some processes do not extend back through all of geologic time. For example, the early Earth atmosphere did not contain free oxygen. However, for the past several billion years, the continents, oceans, and atmosphere have been similar to those of today. We assume that physical and biological processes that form and modify the Earth's surface have not changed significantly over this period. To be useful from an environmental standpoint, the principle of uniformitarianism has to be more than a key to the past. We must turn it around and say that a study of past and present processes is the key to the future. That is, we can assume that in the future the same physical and biological processes will operate, although the rates will vary as the environment is influenced by natural change and human activity. Geologically short-lived landforms such as beaches (Figure 3.9) and lakes will continue to appear and disappear in response to storms, fires, volcanic eruptions, and earthquakes. Extinctions of animals and plants will continue in spite of, as well as because of, human activity. We want to improve our ability to predict what the future may bring, and uniformitarianism can assist in this task.

3.5 Changes and Equilibrium in Systems

Uniformitarianism, then, suggests that changes in natural systems may be predictable. We turn next to an examination of how systems may change. This will involve looking at the relation of system inputs and outputs.

Where the input into a system is equal to the output (Figure 3.10*a*), there is no net change in the size of the reservoir (the amount of whatever is being measured), and the system is said to be in a **steady state**. The steady state is a dynamic equilibrium, because material or energy is entering and leaving the system in equal amounts. The opposing processes occur at equal rates. An approximate steady state may occur on a global scale, such as in the balance between incoming solar radiation and outgoing radiation from Earth, or on the smaller scale of a university, where first-year students begin their studies and others graduate at about the same rate.

When the input into the system is less than the output (Figure 3.10*b*) the size of the reservoir declines. For example, if a resource, such as groundwater, is consumed faster than it can be naturally or humanly replaced, it may be used up. Conversely, in a system where input exceeds output (Figure 3.10*c*), the reservoir will increase. Examples are the buildup of heavy metals in lakes and the pollution of groundwater. Climate change (Chapter 20) is another example. The global reservoir of greenhouse gases is steadily increasing, as a result of emissions from a variety

Figure 3.9 • This beach on the island of Bora Bora, French Polynesia, is an example of a geologically short-lived landform, vulnerable to rapid change from storms and other natural processes.

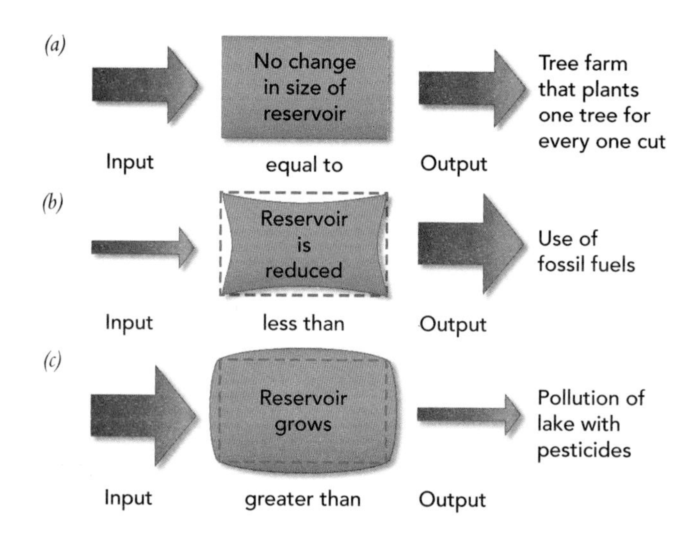

Figure 3.10 • Major ways in which a reservoir, or stock, of some material can change. (*Source*: Modified from P. R. Ehrlich, A. H. Ehrlich, and J. P. Holvren, *Ecoscience: Population, Resources, Environment*, 3rd ed. [San Francisco: W. H. Freeman, 1977].) Row *(a)* represents steady state conditions, rows *(b)* and *(c)* are examples of negative and positive changes in storage.)

of human activities. Even though less developed countries have much smaller emissions—by some estimates, 20 to 30 times less than Canada or the United States—we all share in the global problem of climate change.

By using rates of change or input–output analysis of systems, we can derive an average residence time for objects or material moving through a system. The **average residence time** is the time it takes for a given part of the total reservoir of a particular material to be cycled through the system. To compute the average residence time when the size of the reservoir and the rate of transfer are constant, we divide the total size of the reservoir by the average rate of transfer through that reservoir. For example, suppose the university mentioned above has 10,000 students. Each year 2,500 first year students start and 2,500 fourth year students graduate. The average residence time for students is 10,000 divided by 2,500, or four years.

Average residence time has important implications for environmental systems. A system such as a small lake with an inlet and an outlet and a high transfer rate of water has a short residence time for water. (See Build Your Environmental Skills 3.2.) On the one hand, that makes the lake especially vulnerable to change if, for example, a pollutant is introduced. On the other hand, the pollutant soon leaves the lake. Large systems with a slow rate of transfer of water, such as oceans, have a long residence time and are much less vulnerable to quick change. However, once polluted, large systems with slow transfer rates are difficult to clean up.

Let us look more closely at system inputs and outputs. Inputs to systems may be thought of as causes and outputs or responses as effects. For example, we may add a nitrogen fertilizer to an orchard of apple trees. Adding the fertilizer is an input (or cause), and the output (or effect) is the number of apples the tree produces.

If the relationship between a cause (input) and effect (output) is strictly proportional for all values, then we call this relationship *linear.* Input and output coupled with feedback in a system may result in relationships between cause and effect that are *nonlinear,* and there may be *delays* in the response.[4] Some relationships in systems are linear over a particular range of input and then become nonlinear. For example, if you apply 0.25 kg of fertilizer per apple tree and the yield increases by 5%, and you then apply 0.50 kg per tree and the yield increases by 10%, and you then apply 0.75 kg per tree and the yield increases by 15%, the relationship is linear over these values of input of fertilizer. But what if you apply 50 kg per tree in the hope of increasing the yield by 1,000%? You would probably damage or kill the tree and the yield would be zero!

Over the entire range from 0.25 to 50 kg, then, the relationship between cause and effect changes. You might also note delays in response. When you add fertilizer, for example, it takes time for it to enter the soil and be used by the tree. Many responses to environmental inputs (including human population change; pollution of the land, water, and air; and use of resources) are nonlinear and may involve delays that must be recognized if we are to understand and solve environmental problems.

With an understanding of input and output, we have a framework for interpreting some of the changes that may affect systems. An idea that has been used and defended in the study of our natural environment is that natural systems that have not been affected by human activity tend toward some sort of steady state, or dynamic equilibrium. Sometimes this is called the *balance of nature.* Certainly, negative feedback operates in many natural systems and may tend to hold a system at equilibrium. Nevertheless, we need to ask how often the equilibrium model really applies.

If we examine natural systems in detail and perform our evaluation over a variety of time frames, it is evident that a steady state, or dynamic equilibrium, is seldom obtained or maintained for very long. Rather, systems are characterized not only by human-induced disturbances but also by natural disturbances (sometimes called natural disasters, such as flood and wildfires). Thus, changes over time can be expected. In fact, studies of such diverse systems as forests, rivers, and coral reefs suggest that disturbances due to natural events such as storms, floods, and fires are necessary for the maintenance of those systems.

The environmental lesson is that systems change naturally. If we are going to manage systems for the betterment of the environment, we need to gain a better understanding of the following:[5,6]

Build Your Environmental Skills 3.2

Average Residence Time (ART)

The *average residence time (ART)* for a chemical element or compound is an important concept in evaluating many environmental problems. ART is defined as the ratio of the size of a reservoir or pool of some material—say, the amount of water in a reservoir—to the rate of transfer through the reservoir. The equation is

$$ART = S/F$$

where S is the size of the reservoir and F is the rate of transfer.

Figure 3.11 shows a map of Big Lake, a reservoir of water impounded by a dam. The lake has three rivers that feed a combined 10 m^3/sec of water into the lake, and the outlet structure releases an equal 10 m^3/sec. We assume evaporation of water from the lake is negligible in this simplified example. The lake is contaminated with a water pollutant, MTBE (methyl tertiary–butyl ether), a toxic gasoline additive that has reached the lake through urban runoff, gasoline spills, and gasoline-fueled boats on the lake. MTBE readily dissolves in water and so travels with it. If we are able to control the sources of MTBE to Big Lake, we can calculate the time necessary to flush the lake of the toxin.

- Start by calculating the ART of water in the lake using the above formula.
- Now determine the amount of MTBE in the lake. *Hint*: remember to convert micrograms (μg) to grams, and then kilograms and finally tonnes. Also convert seconds to hours and then to years.
- Calculate the pre-control rate at which MTBE is entering the lake.
- Finally, calculate the ART of MTBE in the lake.

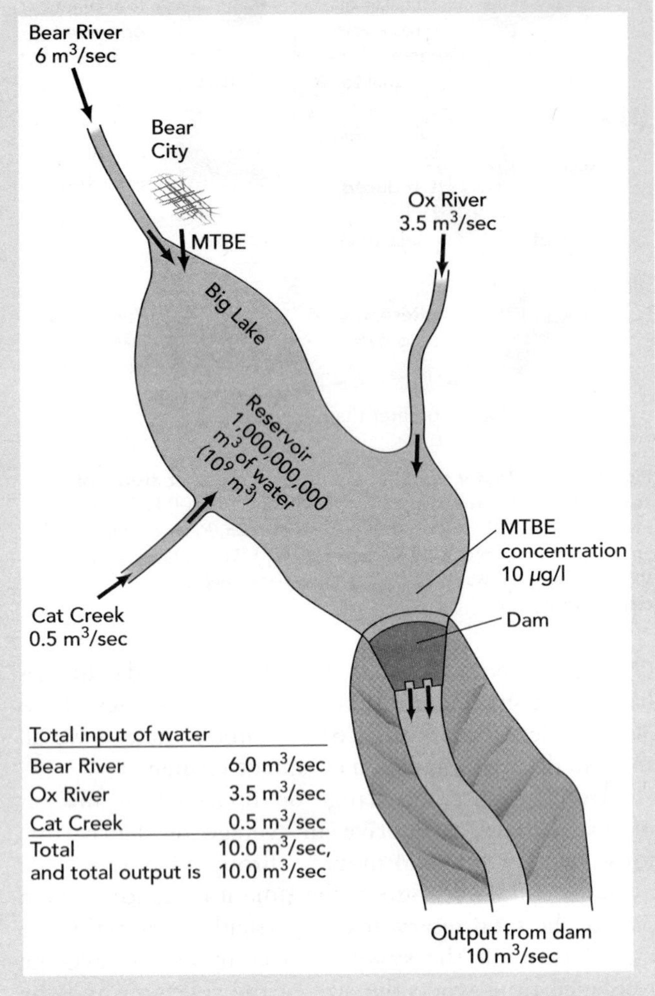

Figure 3.11 • Idealized diagram of a lake system with MTBE contamination.

For these calculations, use multiplication factors in the Appendix at the end of this book.

- the types of disturbances and changes that are likely to occur
- the time periods over which changes occur
- the importance of each change to the long-term productivity of the system.

These concepts are at the heart of understanding the principles of environmental connectivity and sustainability.

3.6 Earth and Life

Next, we turn next from a general discussion of systems to a more direct focus on Earth as a living planet. Earth formed approximately 4.6 billion years ago when a cloud of interstellar gas known as a solar nebula collapsed, creating protostars and planetary systems. Life on Earth began approximately 1.6 billion years later (3 billion years ago) and since that time has profoundly affected the planet. Since the emergence of life, many kinds of organisms have evolved, flourished, and died, leaving only their fossils to record their place in history.

Several million years ago, there were evolutionary beginnings for the eventual dominance of humans on Earth. Eventually, the fossil record tells us, we, too, will disappear. The brief moment of humanity in Earth history may not be particularly significant. However, to those living now and to future generations, how we affect our environment is important.

Human activities increase and decrease the magnitude and frequency of some natural Earth processes. For example, rivers may rise and flood the surrounding countryside regardless of human activities, but the magnitude and frequency of flooding may be greatly increased or decreased by human activity. In order to predict the long-range effects of such processes as flooding, we must be able to determine how our future activities will change the rates of physical processes.

From a biological and geological point of view, we know that the ultimate fate of every species is extinction. However, humans have accelerated this fate for many species. As the human population has increased, a parallel increase in the extinction of species has occurred. These extinctions are closely related to land-use change—to agricultural and urban uses that change the ecological conditions of an area. Some species are domesticated or cultivated, and their numbers grow; others are removed as pests.

3.7 Earth as a Living System

Earth as a planet has been profoundly altered by the life that inhabits it. Earth's air, oceans, soils, and sedimentary rocks are very different from what they would be on a lifeless planet. In many ways, life helps control the makeup of the air, oceans, and sediments.

Life interacts with its environment on many levels. A single bacterium in the soil interacts with the air, water, and particles of soil around it within the space of a fraction of a cubic centimeter. A forest extending hundreds of square kilometers interacts with large volumes of air, water, and soil. All of the oceans, all of the lower atmosphere, and all of the near-surface part of the solid Earth are affected by life.

A general term, **biota**, is used to refer to all living things (animals and plants, including microorganisms) within a given area—from an aquarium to a continent to Earth as a whole. The region of Earth where life exists is known as the **biosphere**. It extends from the depths of the oceans to the summits of mountains. The biosphere includes all life as well as the lower atmosphere and the oceans, rivers, lakes, soils, and solid sediments that actively interchange materials with life. All living things require energy and materials. In the biosphere, energy is received from the sun and the interior of Earth and is used and given off as materials are recycled.

To understand what is required to sustain life, consider the following question: How small a part of the biosphere could be isolated from the rest and still sustain life? Suppose you put parts of the biosphere into a glass container and sealed it. What minimum set of contents would sustain life? If you placed a single green plant in the container along with air, water, and some soil, the plant could make sugars from water and from carbon dioxide in the air. It could also make many organic compounds, including proteins and woody tissue, from sugars and from inorganic compounds in the soil. But no green plant can decompose its own products and recycle the materials. Eventually, your green plant would die.

We know of no single organism, population, or species that both produces all its own food and completely recycles all its own metabolic products. For life to persist, there must be several species within an environment that includes fluid media—air and water—to transport materials and energy. Such an environment is an ecosystem, our next important topic of discussion.

3.8 Ecosystems

An **ecosystem** is a community of organisms and its local nonliving environment in which matter (chemical elements) cycles and energy flows. Chapter 5 describes the laws of the thermodynamics governing these. It is a fundamental principle that *sustained life on Earth is a characteristic of ecosystems*, not of individual organisms or populations or single species.

The Nature of Ecosystems

The term *ecosystem* is applied to areas of all sizes, from the smallest puddle of water to a large forest or the entire global biosphere. Ecosystems differ greatly in composition—that is, in the number and kinds of species, the kinds and relative proportions of non-biological constituents, and the degree of variation in time and space. Sometimes the borders of an ecosystem are well defined, as in the transition from the ocean to a rocky coast or from a pond to the surrounding woods. Sometimes the borders are vague, as in the subtle gradation of forest to prairie in western Canada or from grasslands to savannas or forests in East Africa. What is common to all ecosystems is not physical structure—size, shape, variations of borders—but the existence of the processes we have mentioned—the flow of energy and the cycling of chemical elements.

Ecosystems can be natural or artificial (Figure 3.12). A pond constructed as part of a waste treatment plant is an artificial ecosystem. Ecosystems can be natural or managed, and the management can vary over a wide range of actions. Agriculture can be thought of as partial management of certain kinds of ecosystems.

Natural ecosystems carry out many public service functions for us. Wastewater from houses and industries is often converted to drinkable water by passage through natural ecosystems such as soils. Pollutants, such as those in the (smoke from industrial plants or in the exhaust from automobiles), are often trapped on leaves or converted to harmless compounds by forests.

The Gaia Hypothesis

Our discussion of Earth as a system—life in its environment, the biosphere, and ecosystems—leads us to the question of how much life on Earth has affected our planet. In recent years, the **Gaia hypothesis**—named for Gaia, the Greek

(a)

(b)

Figure 3.12 • Natural and artificial ecosystems. *(a)* Natural ponds. *(b)* An artificial pond in a sewage treatment plant.

goddess Mother Earth—has become a hotly debated subject.[7] The hypothesis states that life manipulates the environment for the maintenance of life. For example, some scientists believe that algae floating near the surface of the ocean influence rainfall at sea and the carbon dioxide content of the atmosphere, thereby significantly affecting the global climate. It follows, then, that the planet Earth is capable of physiological self-regulation.

According to James Lovelock, a British scientist who has been developing the Gaia hypothesis since the early 1970s, the idea of a living Earth is probably as old as humanity.[7] James Hutton, whose theory of uniformitarianism was discussed earlier, stated in 1785 that he believed Earth to be a superorganism and compared the cycling of nutrients from soils and rocks in streams and rivers to the circulation of blood in an animal.[7] In this metaphor, the rivers are the arteries and veins, the forests are the lungs, and the oceans are the heart of Earth.

The Gaia hypothesis is really a series of hypotheses. The first is that life, since its inception, has greatly affected the planetary environment. Few scientists would disagree. Certainly, there is some evidence that life has had such an effect on Earth's climate. The second hypothesis asserts that life has altered Earth's environment in ways that have allowed life to persist. The clearing of land for agriculture is an example here. A popularized extension of the Gaia hypothesis is that life *deliberately* (consciously) controls the global environment. Few scientists accept this idea.

The extended Gaia hypothesis may have merit in the future, however. Humans have become conscious of our effects on the planet, some of which influence future changes in the global environment. Thus, the concept that humans can consciously make a difference in the future of our planet is not as extreme a view as would once have been thought. The future status of the human environment may depend, in part, on actions we take now and in coming years. This aspect of the Gaia hypothesis exemplifies the key theme of thinking globally, which was introduced in Chapter 1.

The decisions we make in managing our global environment depend on our values as well as our understanding of how Earth works (another key theme from Chapter 1). With this in mind, we explore in greater depth how human processes are linked to environmental change.

3.9 Why Solving Environmental Problems Is Often Difficult

Global environmental systems—whether the entire planet or the hydrosphere (the water portion), lithosphere (rocks, minerals, and soils), biosphere (living organisms), or human population—are open systems characterized by poorly defined boundaries and the transfer of material and energy. They are inherently difficult to work with. Global environmental problems are particularly difficult because of the following three characteristics:[4]

- exponential growth and the positive feedback that accompanies growth
- lag times between stimuli and responses of systems that can be decades or longer, and
- consequences of events that have the potential to result in irreversible changes on the human time scale.

Figure 3.13 illustrates these concepts graphically.

Exponential growth, long lag time, and the possibility of irreversible consequences have special implications for environmental problems and finding solutions to those problems. We now recognize the potential dangers of exponential growth and realize that when these are coupled to long lag times and irreversible consequences, we must pay special attention to finding solutions. Thus, again we see the importance of the principle of environmental connectivity, which states that one activity or change often leads to a sequence of changes, some of which may be difficult to recognize. Recognition of lag time and the irreversible consequences associated with exponential growth is paramount in addressing environmental problems.

Most changes brought on by human activity involve rather slow processes with cumulative effects. For example, our present global warming began with the Industrial Revolution, when people started burning massive amounts of fossil fuels. Levels of carbon dioxide in the atmosphere added by human activities have nearly doubled since the Industrial Revolution and are causing human-enhanced global warming. Certainly, we are aware that climate may change naturally and that such change may be rapid (see Chapter 20). However, present global warming is significant and results in part from human activity.

The decline of fisheries is another process that generally takes place over a relatively long time. In such cases, recognizing that irreversible damage may have been done often involves crossing a threshold that is difficult to identify in time to avoid problems. More commonly, we recognize that a threshold has been exceeded after we begin to identify consequences of a particular activity. For example, we have painfully come to the conclusion that the crash of the salmon population in British Columbia and the Pacific Northwest is the combined result of overfishing, building dams, and land-use practices such as timber harvesting. We did not recognize the crash until it occurred and we reflected on past activities. This example further supports the idea that environmental science problems are often complex, with linkages among various parts of ocean, river, land, and forest systems.

Change can also be chaotic. This occurs when some small change is amplified, resulting in complex and perhaps periodic activity or behavior. Chaos theory is a mathematical description of change that has been used to describe the behavior of a variety of systems, including population fluctuations, and changes in circulation and patterns of air current in the atmosphere. An often-cited hypothetical example is a scenario in which the beating of the wings of a hummingbird in Brazil causes, through a series of amplified activities, a hurricane in Miami. Although this example seems unlikely to occur in nature, many surprises confront us when we consider natural Earth systems. For example, changes in the temperature of the Pacific Ocean cause far-ranging changes in storms, floods, and other natural hazards on a global basis (see Chapter 20)—and are even thought to exert a powerful control on salmon survival in British Columbia.

As stated, one of our goals in understanding the role of human processes in environmental change is to help manage our global environment. To accomplish this, we need to be able to predict changes before they occur. But as the examples above demonstrate, prediction presents great challenges. Although some changes are anticipated,

Figure 3.13 • The concept of overshoot, illustrating the influence of exponential growth, lag time, and collapse on carrying capacity. Carrying capacity starts out relatively high, but as exponential growth increases the population beyond the carrying capacity, overshoot occurs and population collapses. If environmental damage occurs as a result of overuse and damage to resources on which the carrying capacity depends, then the carrying capacity also crashes, as shown here. [*Source*: Modified after D.H. Meadows and others 1992.]

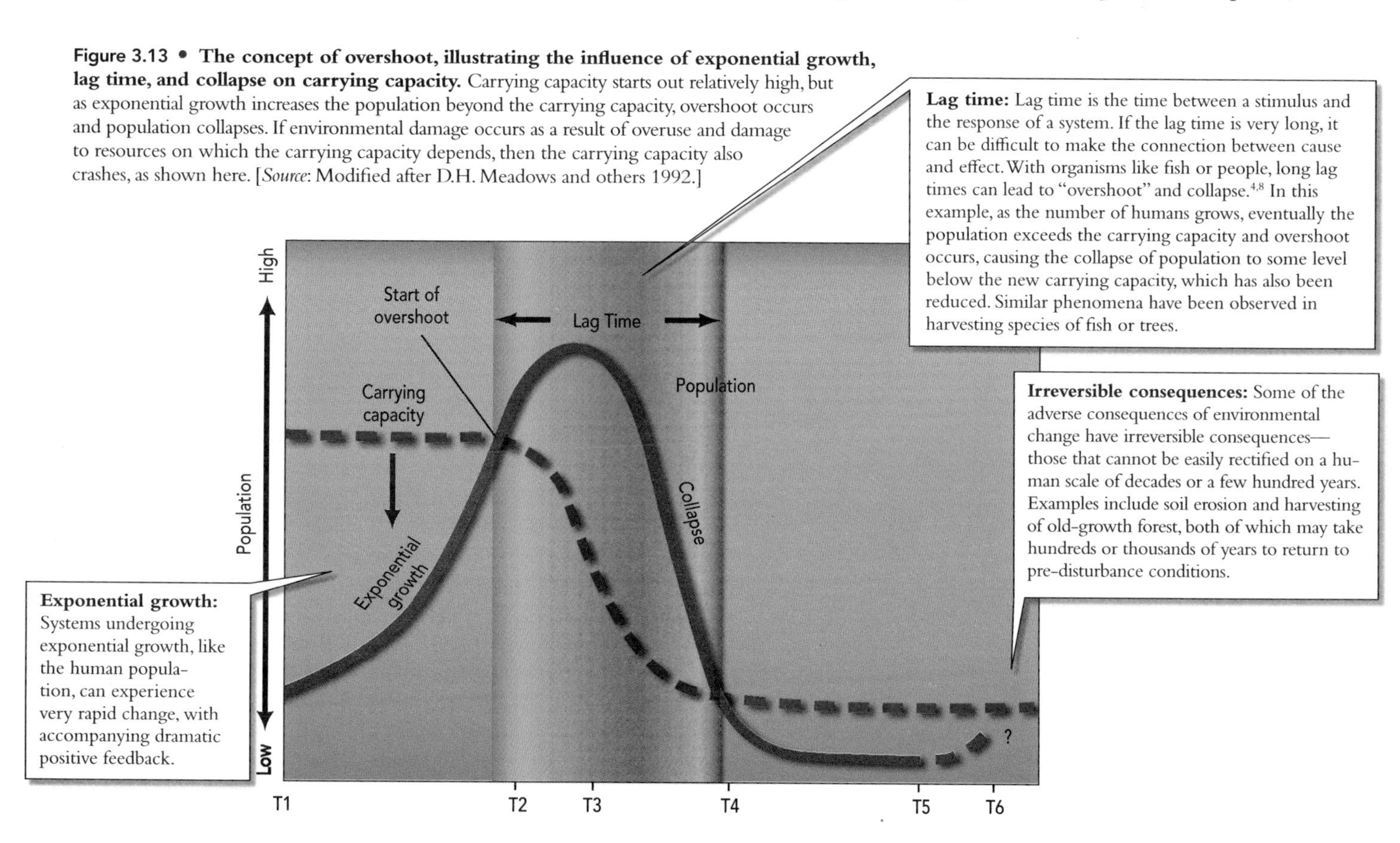

CRITICAL THINKING ISSUE

Is the Gaia Hypothesis Science?

According to the Gaia hypothesis, Earth and all living things form a single system, with interdependent parts, communication among these parts, and the ability to self-regulate. Are the Gaia hypothesis and its component hypotheses science, fringe science, or pseudoscience? Is the Gaia hypothesis anything more than an attractive metaphor? Does it have religious overtones? Answering these questions is more difficult than answering similar questions about, say, crop circles, as described in Chapter 2. Analyzing the Gaia hypothesis forces us to deal with some of our most fundamental ideas about science and life.

Critical Thinking Questions

1. What are the main hypotheses included in the Gaia hypothesis?
2. What kind of evidence would support each hypothesis?
3. Which of the hypotheses can be tested?
4. Is each hypothesis a science, fringe science, or pseudoscience?
5. What are the strengths and weaknesses of the Gaia hypothesis?

others come as a surprise. As we learn to apply the principles of environmental connectivity and uniformitarianism more skillfully, we will be better able to anticipate changes that would otherwise have been surprises.

As the chapter's opening case illustrates, environmental problems often have multiple, interlinked causes. Many are international or even global in scale, yet we do not have political systems that cross national boundaries. Governments must therefore work together to achieve environmental improvement, despite very different regulatory systems, political styles, and economic status. As we mentioned in Chapter 1, different cultures have different perspectives on the identification and resolution of environmental problems. Successful management often requires inputs of knowledge other than science, including traditional aboriginal ecological knowledge, review of written records (as was the case in the Crawford Lake example discussed in Chapter 2), and aesthetic, spiritual, or economic considerations.

Recognizing these challenges, the 1992 United Nations Conference on Environment and Development adopted a resolution on the **Precautionary Principle**: that in order to protect the environment, a precautionary approach should be taken. Where there are threats of serious or irreversible damage to the environment, lack of full scientific certainty should not be used as a reason for postponing action. Although the principle has proven difficult to implement in environmental policy, it has prompted constructive and widespread debate on how best to solve environmental problems and prevent further degradation.[9]

Summary

- A system is a set of components or parts that function together as a whole. Environmental studies deal with complex systems at every level, and solutions to environmental problems often involve understanding systems and rates of change.
- Systems respond to inputs and have outputs. Feedback is a special kind of system response. Positive feedback is destabilizing, whereas negative feedback tends to stabilize or encourage more constant conditions in a system.
- Relationships between the input (cause) and output (effect) of systems may be nonlinear and may involve delays.
- The principle of environmental connectivity, simply stated, holds that everything affects everything else. It emphasizes linkages among parts of systems.
- The principle of uniformitarianism can help predict future environmental conditions on the basis of the past and the present.
- A particularly important aspect of positive feedback is exponential growth, in which the increase per time period is a constant fraction or percentage of the current amount. Exponential growth involves two factors: the rate of growth and the doubling time.
- Changes in systems can be studied through input–output analysis. The average residence time is the average time it takes for the total reservoir of a particular material to be cycled through the system.
- Life on Earth began about 3 billion years ago and since that time has profoundly changed our planet. Sustained life on Earth is a characteristic not of individual organisms or populations but of ecosystems—

local communities of interacting populations and their non-biological environments.

- The general term *biota* refers to all living things, and the region of Earth where life exists is known as the *biosphere*.
- The Gaia hypothesis states that life on Earth, through a complex system of positive and negative feedback, regulates the planetary environment to help sustain life.
- Exponential growth, long lag times, and the possibility of irreversible change combine to make solving environmental problems difficult.
- The Precautionary Principle states that where there are threats of serious or irreversible damage to the environment, lack of full scientific certainty should not be used as a reason for postponing action.

STUDY QUESTIONS

1. How does the Amboseli National Park case history exemplify the principle of environmental connectivity?
2. What is the difference between positive and negative feedback in systems? Provide an example of each.
3. What is the main point concerning exponential growth? Is exponential growth good or bad?
4. Why is the idea of equilibrium in systems somewhat misleading in regard to environmental questions? Is the establishment of a balance of nature ever possible?
5. Why is the concept of the ecosystem so important in the study of environmental science? Should we be worried about disturbing ecosystems? Under what circumstances should we worry or not worry?
6. Is the Gaia hypothesis a true statement of how nature works, or is it simply a metaphor? Explain.
7. How might you use the principle of uniformitarianism to help evaluate environmental problems? Is it possible to use this principle to help evaluate the potential consequences of too many people on Earth?
8. Why does overshoot occur, and what could be done to anticipate and avoid it?

FURTHER READING

Government of Canada. Canada's National Implementation Strategy on Climate Change. Ottawa: Queen's Printer, 2000. This report outlines what Canada is doing to address the challenges of global change and what it can do in the future.

Bunyard, P., ed. 1996. *Gaia in Action: Science of the Living Earth*. Edinburgh: Floris Books. Investigations into implications of the Gaia hypothesis.

Ehrlich, P. R., A. H. Ehrlich, and J. P. Holdren. *Ecoscience*. San Francisco: Freeman, 1970. Although this is an older book, Chapter 2 provides a good overview of the physical world and how systems and changes may affect our environment.

Lovelock, J. *The Ages of Gaia: A Biography of Our Living Earth*. New York: Norton, 1995. This small book explains the Gaia hypothesis, presenting the case that life very much affects our planet and in fact may regulate it for the benefit of life.

chapter 4

Death in Bangladesh

Bangladesh has 138 million people, and in 2002, an annual population growth rate of 2.2%. Although this is not a particularly high rate—it is only slightly higher than the global average of 1.8%—it translates to an additional 2.76 million people per year.[1] In the 1990s, more than 100,000 people died in Bangladesh when storms pushed ocean waters onshore, flooding the low-lying coastal lands that make up most of this nation (Figure 4.1).[2] This disaster was made even more tragic by the fact that the average rate of population growth in Bangladesh replaced those 100,000 people in just two weeks. The growth curve of the population showed barely a ripple (Figure 4.2).

Bangladesh is one of the poorest nations in the world, and this poverty affects human survival. According to World Bank standards for poverty, over half the population (53%) qualifies as "poor" and 36% as "very poor." Nine out of 10 children are malnourished, and 50% of children are stunted from lack of proper nourishment. Six hundred children die each day from malnutrition. The average number of food calories available per person is only 88% of that required for good health. Less than half the population (47%) living in urban areas has access to safe drinking water, and less than a fifth has access to adequate modern sanitation. The average life expectancy is about 59 years. With inadequate resources for each individual and a relatively rapid growth rate among nations, Bangladesh struggles to maintain even its existing poor standard of living.

Some argue that the low-lying coastal areas of Bangladesh are fundamentally uninhabitable at high population densities for extended periods and are inhabited now only because of the population overcrowding on the Indian subcontinent. For Bangladesh, solving major environmental problems, conserving biological diversity, or optimizing production of fisheries and vegetation is difficult when people barely have sufficient resources to survive and the rising population erases any advances.

Figure 4.1 • Flood victims gather to collect survival assistance following one of the catastrophic floods in Bangladesh that have claimed hundreds of thousands of lives. Over crowding in Bangladesh increases the number of deaths from these floods. [Photo courtesy of World Food Programme.]

The Human Population and the Environment

LEARNING OBJECTIVES

The current human population represents something unprecedented in the history of the world. Never before has one species had such a great impact on the environment in such a short time and continued to increase at such a rapid rate. These qualities make the human population *the* underlying environmental issue. After reading this chapter, you should understand:

- Ultimately, there can be no long-term solutions to environmental problems unless the human population stops increasing.
- Two major questions about the human population involve what controls its rate of growth and how many people Earth can sustain.
- The rapid increase in the human population has occurred with little or no change in the maximum lifetime of an individual.
- Modern medical practices and improvements in sanitation, control of disease-spreading organisms, and supplies of human necessities have decreased death rates and accelerated the net rate of human population growth.
- Countries with a high standard of living have moved more quickly to a lower birth rate than countries with a low standard of living.
- Although we cannot predict with absolute certainty what the future human carrying capacity of Earth will be, understanding human population dynamics is helpful in making useful forecasts.

- *Progress is being made in Bangladesh in family planning and reducing the growth rate. However, as this case study illustrates, long-term solutions to environmental issues in that nation, and elsewhere, are difficult to find as long as the population continues its momentous rise. If the human population continues to grow rapidly, it will ultimately overwhelm the environment. That is why human population growth is a major theme of this textbook.*

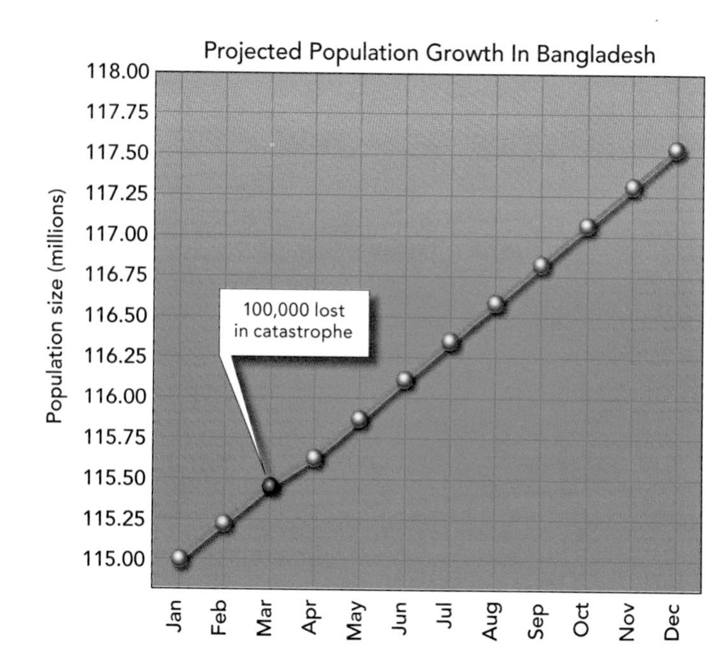

Figure 4.2 • **Projected population growth in Bangladesh** for the months shortly before and after a loss of 100,000 people in a single catastrophe. The population is so large that growth continues with little slowdown and rapidly recovers the number lost.

4.1 How Populations Change Over Time: Basic Concepts of Population Dynamics

Basic Concepts

The rapid regrowth of Bangladesh's population following devastating storms vividly illustrates the human population's great capacity for growth and suggests the problems that this growth poses for the environment.

The central question: What would make it possible for the human population to cease to grow? This question can be answered only if we understand the basics of a population. A **population** is a group of individuals of the same species living in the same area or interbreeding and sharing genetic information. A **species** is a group individuals that are capable of interbreeding. A species is made up of populations. Five key properties of any population are: abundance, which is the size of a population—now, in the past, and in the future—**birth rates, death rates, growth rates**, and age structure.

Populations change over time and over space. The general study of population changes is called **population dynamics**. How rapidly a population changes depends on the growth rate, which is the difference between the birth rate and the death rate. (See Table 4.1 for other useful terms and Build Your Environmental Skills 4.1.)

Table 4.1 • Human Population Terms

Crude birth rate: number of births per 1,000 individuals per year; "crude" because population age structure is not taken into account
Crude death rate: number of deaths per 1,000 individuals per year
Crude growth rate: net number added per 1,000 individuals per year; also equal to crude birth rate minus crude death rate
Fertility: pregnancy or the capacity to become pregnant or to have children
General fertility rate: number of live births expected in a year per 1,000 women aged 15 to 49 years, considered the childbearing years
Age-specific rate: number of births expected per year among a fertility-specific age group of women in a population
Total fertility rate (TFR): average number of children expected to be born to a woman throughout her childbearing years
Cause-specific death rate: number of deaths from one cause per 100,000 total deaths
Incidence rate: number of people contracting a disease during a time period, usually measured per 100 people
Prevalence rate: number of people afflicted by a disease at a particular time
Case fatality rate: percentage of people who die once they contract a disease
Morbidity: general term meaning the occurrence of disease and illness in a population
Rate of natural increase (RNI): birth rate minus death rate, implying annual rate of population growth not including migration
Doubling time: number of years it takes for a population to double, assuming a constant rate of natural increase
Infant mortality rate: annual number of deaths of infants under age 1 per 1,000 live births
Life expectancy at birth: average number of years a newborn infant can expect to live under current mortality levels
GNP per capita: gross national product (GNP), which includes the value of all domestic and foreign output, per person

Source: C. Haub and D. Cornelius, *World Population Data Sheet* (Washington, D.C.: Population Reference Bureau, 1998).

Age Structure

An important factor in population growth is the population age structure, the proportion of the population in each age class. The age structure of a population affects current and future birth rates, death rates, and growth rates; has an impact on the environment; and implications for current and future social and economic status.

We can picture an age structure as if it were a pile of blocks, one for each age group, where the size of each block represents the number of people in that group. Although age structures can take many shapes, four general types are most important to our discussion: a pyramid, a column, an inverted pyramid (top-heavy), and a column with a bulge. The pyramid age structure occurs in a population with many young people and a high death rate at each age—and therefore a short average lifetime. A column shape occurs where the birth rate and death rate are low and a high percentage of the population is elderly. A bulge occurs if some event in the past caused a high birth rate or death rate for some age group but not others.

Age structure varies considerably by nation (Figure 4.3). Developing countries, with high birth and death rates, usually have many young people and few older people in the population, creating a pyramid-shaped age structure. Kenya's age structure is typical of this pattern. In developing countries today, about 34% of the population is under 15 years of age. Such an age structure indicates that the population will grow very rapidly in the future, when the young reach marriage and reproductive ages. It suggests that the future for such a nation requires more jobs for the young, and it has many other social implications that go beyond the scope of this book.

In contrast, industrialized nations have slower population growth and a more stable age structure. The diagrams for Canada and the United States show this more rectangular pattern. Australia and Italy are examples of transitional age structures. Italy's slightly top-heavy pyramid shows a nation with declining growth. Australia's pyramid is rectangular for age groups under 50, showing stable growth there, but has much smaller representation of people over 50. Elderly people comprise a small percentage (3%) of Kenya's population, but a much larger

Build Your Environmental Skills 4.1

Forecasting Population Change

Chances are that the population of your home town is either growing or shrinking—with significant implications for local schools, health care, and municipal systems. Populations change in size through births, deaths, immigration, and emigration. We can write a formula to represent population change:

$$P_2 = P_1 + (B - D) + (I - E)$$

where P_1 = the number of individuals in a population at time 1

P_2 = the number of individuals in that population at some later time 2

B = the number of births in the period from time 1 to time 2

D = the number of deaths from time 1 to time 2

I = the number entering as immigrants

E = the number leaving as emigrants

Ignoring for the moment immigration and emigration, how rapidly a population changes depends on the growth rate, which is the difference between the birth rate and the death rate (see Table 4.1 for other useful terms). The human population growth rate is usually expressed in the rate per 1,000, called the crude rate, rather than the more familiar percentage, which is the rate per 100. The crude growth rate (g) is just the net population change—the birth rate (b) minus the death rate (d); it is expressed in numbers of individuals per unit time:

$$g = b - d$$

Recall from Chapter 3 that doubling time—the time it takes a population to reach twice its present size—can be estimated by the formula

$$T = 70/\text{annual growth rate}$$

where T is the doubling time and the annual growth rate is expressed as a percentage. For example, a population growing 2% per year would double in approximately 35 years.

We can also express population growth as a percentage change, relative to the starting population, by dividing the crude growth rate by the total number in the population:

$$g = (B - D)/N \quad \text{or} \quad g = G/N$$

It is important to be consistent in using the population at the beginning, middle, or end of the period. Usually, the number at the beginning or the middle is used.

Using these formulas, calculate the birth, death, and growth rate for Australia in 2002, using the following data:

- Total population in mid-2002 = 19,700,000 people
- Total births from mid-2002 to mid-2003 = 394,000 births
- Total deaths from mid-2002 to mid-2003 = 137,900 deaths

Was Australia's 2002-2003 growth rate close to the global average of 1.3?

Figure 4.3 • Age structure of developing, transitional, and developed countries. (*Sources*: Modified from Statistics Canada (**www.statcan.ca**); Council of Europe, Recent Demographic Developments in Europe 1997, Table 1–1; United Nations, The Sex and Age Distribution of the World Populations—The 1996 Revision, 500–1.) Developing countries like Kenya typically display pyramid-shaped population age pyramids, reflecting a high proportion of young people in the population. Industrialized nations like Canada have more columnar pyramids, because of their more stable age structure. Countries like Italy with declining populations have a high proportion of older people in the population, and therefore a somewhat top-heavy pyramid.

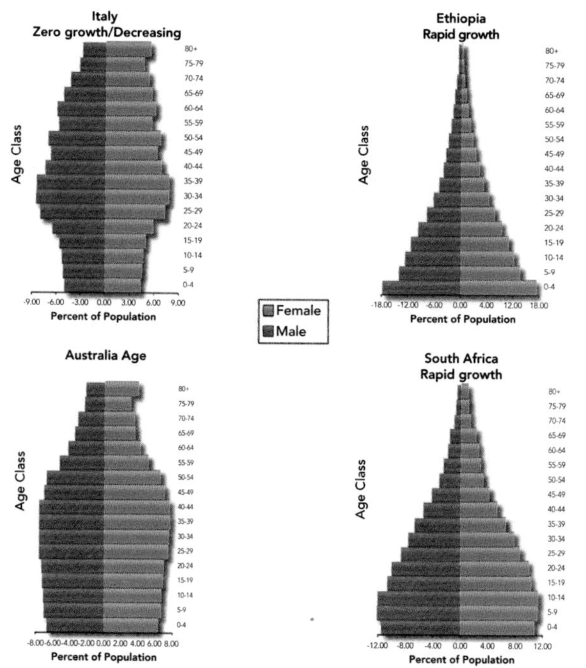

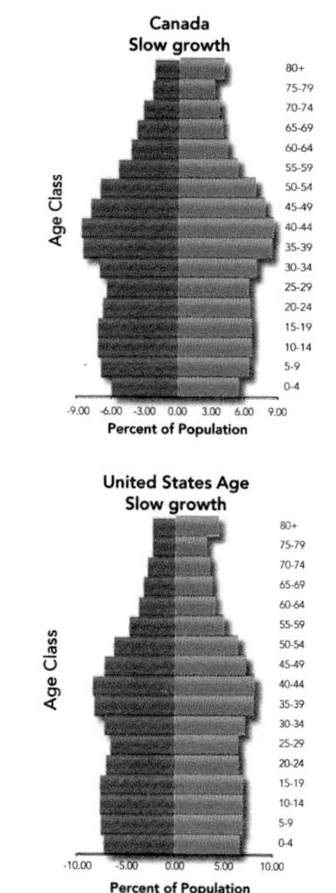

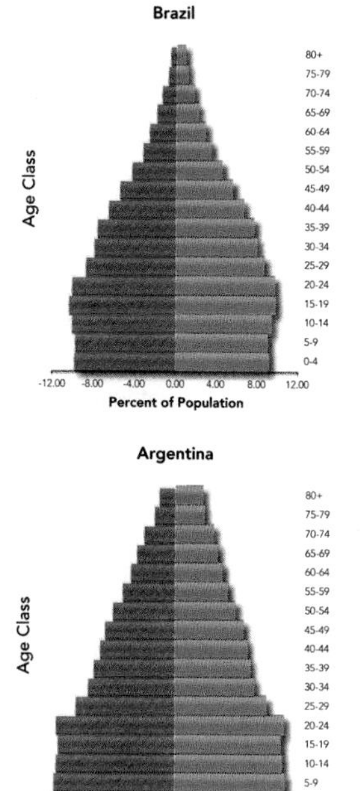

percentage of the populations of Canada and Italy (13% and 17%, respectively).[1]

Age structure provides insight into a population's history, its current status, and its likely future. For example, the baby boom that occurred after World War II in Canada and the United States (a great increase in births from 1946 through 1964) forms a pulse in the population that can be seen as a bulge in the age structure, especially of those aged 40 to 50 in 2000 (see Figure 4.3). A secondary, smaller bulge results from offspring of the first baby boom, which can be seen as a slight increase in 5- to 15-year-olds. This second peak shows that the baby boom pulse is moving through the age structure. Each baby boom increases demand for social and economic resources; for example, schools were crowded when the baby boomers were of primary and secondary school age.

One economic implication of age structure involves care of the elderly. In pre-industrial and non-industrial societies, average lifetimes are short, children care for their parents, and it benefits parents to have many children. In modern technological societies, family size is smaller, and care for the elderly is distributed throughout the society through taxes, so that those who work provide funds to care for those who cannot. Parents can benefit when their children are well educated and have high-paying jobs. Rather than relying on a large family in which each child has fewer resources, parents tend to have fewer children and invest more in each. This makes *zero population growth* possible. However, a shift from a youthful age structure (like Kenya's) to an elderly age structure (like Italy's) means that a smaller percentage of the population works—and fewer tax monies are available for elder care.

A population heavily weighted toward the elderly poses problems for a nation. The easiest way to increase tax income is to increase the percentage of young people and promote rapid population growth. Thus, short-term economic pressures at a national level can lead to political policies supporting rapid population growth, which is not in the long-term interest of the nation.

4.2 Kinds of Population Growth

Exponential Growth

As discussed in Chapter 3, a population experiencing exponential growth increases by a constant percentage per unit of time. The growth rate of the human population increased but varied during the first part of the twentieth century, peaking in 1965–1970 at 2.1% because of improvements in health care, medicine, and food production. The human population has actually increased at a rate faster than the rate of exponential growth. This increase in the human growth rate has stopped, however, and the growth rate is declining globally. As mentioned, it is now approximately 1.3%.[1]

A Brief History of Human Population Growth

The history of the human population (see Figure 4.4) can be viewed as consisting of four major periods:

1. In the early period of hunters and gatherers, the world's total human population was probably less than a few million.
2. A second period, beginning with the rise of agriculture, allowed a much greater density of people and the first major increase in the human population.

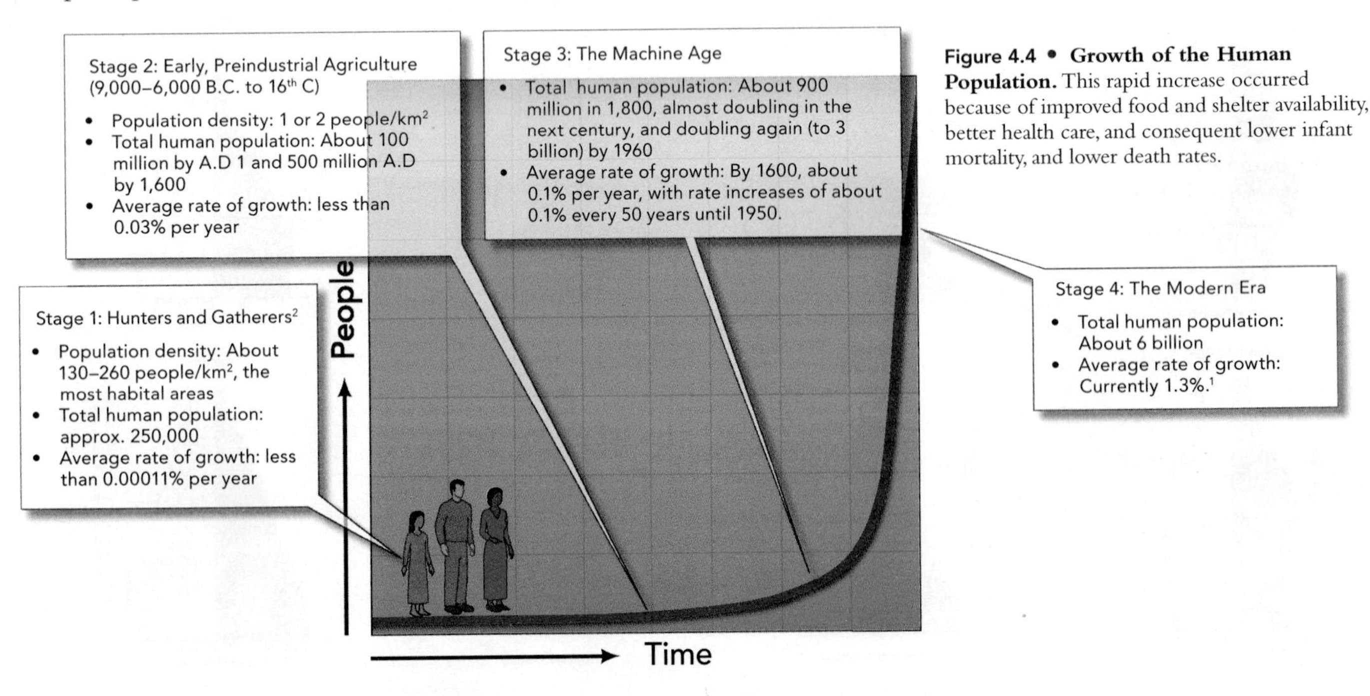

Figure 4.4 • Growth of the Human Population. This rapid increase occurred because of improved food and shelter availability, better health care, and consequent lower infant mortality, and lower death rates.

3. The Industrial Revolution, with improvements in health care and the supply of food, led to a rapid increase in the human population.
4. Today, the rate of population growth has slowed in wealthy, industrialized nations but continues to increase rapidly in many poorer, less developed nations (Figures 4.5 and 4.6).

It is also interesting to look at the total cumulative population over the history of humans, which is explored in A Closer Look 4.1, "How Many People Have Lived on Earth?"

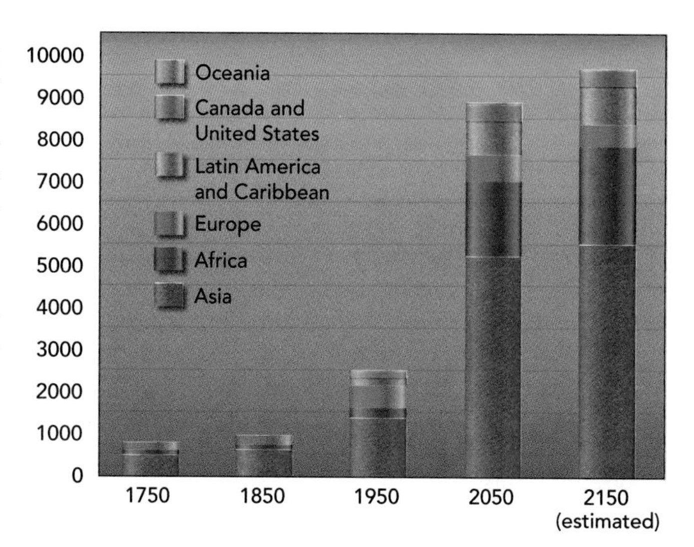

Figure 4.5 • World human population since 1750 A.D. showing the contribution of major world regions. Most growth in the twenty-first century will take place in developing nations. Note the increasing contribution from Africa and Asia after 1950. (*Source*: United Nations.)

4.3 Present Human Population Rates of Growth

Currently, the world population is approximately 6.4 billion people, with an annual growth rate of approximately 1.3% (see Figures 4.5 and 4.6). At this rate, 84 million people are added to Earth's population in a single year, a number almost three times the population of Canada![1] Human population trends vary greatly among the major regions of the world (see Figure 4.5) and among countries. In developing countries, the average population growth rate in 2002 was 1.6%,[1] whereas in the developed nations—those in Western Europe and North America, for example—the growth rate was less than 1% and, in some cases much lower, even negative.[1] The growth rate of the population of the United States has declined, while that of Canada continues to rise (Figure 4.7), mainly as a result of immigration.

Thus, there may be a correlation between poverty and population growth. The poorer a nation, the more likely the population growth rate will be high; the wealthier a nation—and the higher the average per-capita income—the lower the population growth rate. Since a high growth rate works against an increase in per capita income, poor nations are in danger of a positive feedback (see Chapter 3). The more people, the greater the growth rate; the greater the growth rate, the more people.

Figure 4.6 • Growth of the world population. *(a)* Total world population illustrating the logistic growth curve. *(b)* Number of people added to the world population by decade. Between 1980 and 1990, 82 million people were added each year, mainly in developing countries.

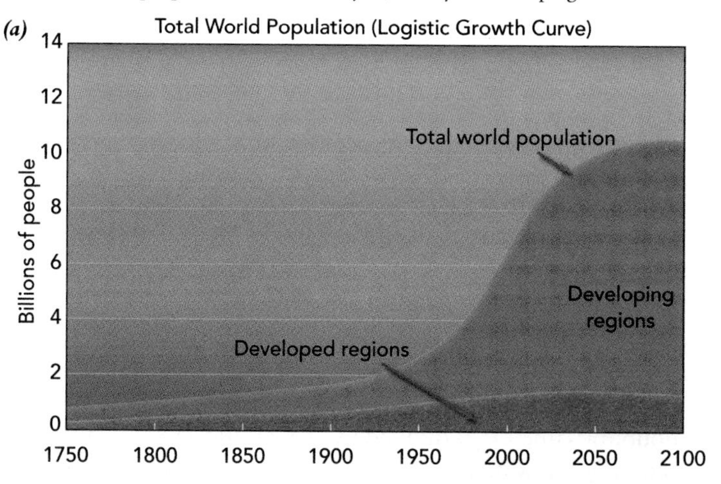

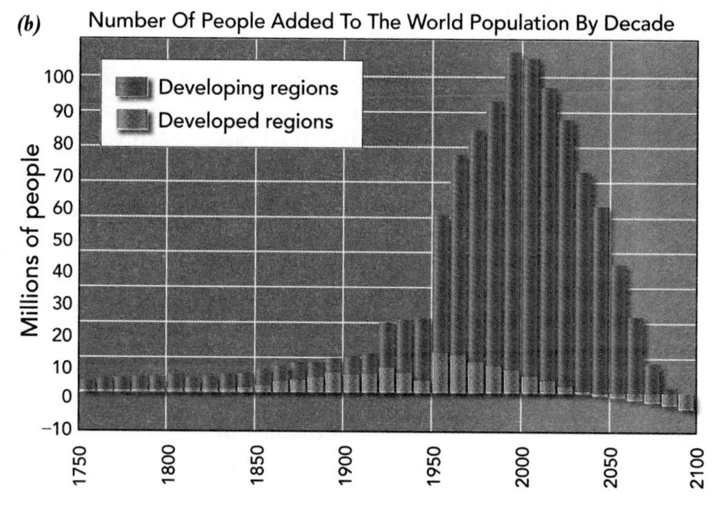

4.4 Projecting Future Population Growth

With human population growth a central issue, it is important that we develop methods to forecast what will happen to our population in the future. One of the simplest approaches is to calculate the doubling time.

A CLOSER LOOK 4.1

How Many People Have Lived on Earth?

How many people have lived on Earth? Of course, before written history, there were no censuses. There is evidence of periodic censuses in ancient Babylon (about 3,800 B.C.), Egypt (beginning about 2,500 B.C.) and China (550 B.C.). The first estimates of population in Western civilization were attempted in the Roman era. During the Middle Ages and the Renaissance, scholars occasionally estimated the number of people. The first modern census was taken in 1655 in the Canadian colonies by the French and British.[3] The first series of regular censuses taken by a country began in Sweden in 1750, and the United States has taken a census every decade since 1790. Most countries began much later. The first Russian census, for example, was taken in 1870. Even today, many countries do not take censuses or do not do so regularly. The population of China has only recently begun to be known with any accuracy. By studying modern primitive peoples and applying principles of ecology, however, we can gain a rough idea of the total number of people who may have lived on Earth. Summing all the values, including those since the beginning of written history, about 50 billion people are estimated to have lived on Earth.[4] If this is so, then, surprisingly, the 6 billion people alive today represent more than 10% of all of the people who have ever lived.

Exponential Growth and Doubling Time

Recall from Chapter 3 that doubling time, a concept used frequently in discussing human population growth, is the time required for a population to double in size (see Build Your Environmental Skills 4.1). This concept was illustrated by Albert Bartlett's fictional story about the growth of a hypothetical strain of bacteria in the laboratory. If the bottle were full at 12:00 p.m., it was half full at 11:59 a.m.! Doubling time is clearly a powerful concept in population studies.

The standard way to calculate doubling time is to assume that the population is growing exponentially (has a constant growth rate). We can then estimate doubling time by dividing 70 by the annual growth rate stated as a percentage.

The doubling time based on exponential growth is very sensitive to growth rate; that is, it changes quickly as the growth rate changes (Figure 4.8). A few examples demonstrate this sensitivity. The Canadian population grew approximately 1.0% per year during the 1980s, a doubling time of 70 divided by 1.0, or 70 years. During the same period, the Ivory Coast's population grew at 3.8% per year; 70 divided by 3.8 is approximately 18. So, while the doubling time for the Canadian population was 70 years, the doubling time for the Ivory Coast's population was a little more than 18 years. The populations of Belgium, Denmark, Portugal, and Poland are growing at an annual rate of about 0.1%—their doubling time is 700 years. The world's most populous country, China, has grown at an annual rate of 0.7%; its doubling time is 100 years. In the latter part of the twentieth century, the world population growth rate was 1.3%, which would lead to a doubling time of 53 years.[1]

No population can sustain an exponential rate of growth indefinitely. Eventually, the population will run out of food and space. A population of 100 increasing at 5% per year, for example, would grow to 1 billion in less than 325 years. If the human population had increased at this rate since the beginning of recorded history, it would now exceed all the known matter available in the universe.

The Logistic Growth Curve

An exponentially growing population theoretically increases forever, but on Earth, which is limited in size, this is not possible. If a population cannot increase forever, what changes in the population can we expect over time? One suggestion made about the human population is that it would follow a smooth S-shaped curve known as the **logistic growth curve** (Figure 4.6). The population would increase exponentially only temporarily. After

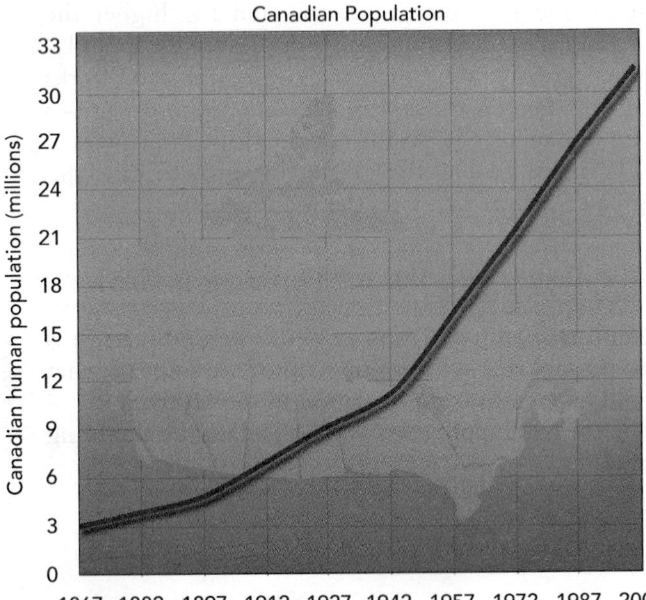

Figure 4.7 • **Canadian population 1867 to 2002.** Note that the rate of growth of the Canadian population does not appear to be slowing.

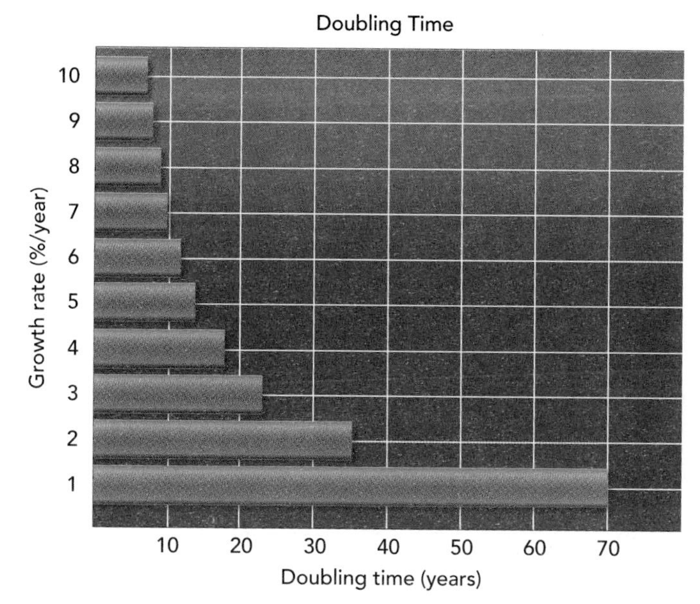

Figure 4.8 • Doubling time. Doubling time changes rapidly with the population growth rate. Because the world's population is increasing at a rate between 1% and 2%, we expect it to double within the next 35 to 70 years.

that, the rate of growth would gradually decline (i.e., the population would increase more slowly) until an upper population limit, called the **logistic carrying capacity**, was reached (Figure 4.6 and Figure 4.9). Once the logistic carrying capacity had been reached, the population would remain at that number.

The logistic growth curve was first suggested in 1838 by a European scientist, P. F. Verhulst, as a theory for the growth of animal populations. It has been applied widely to the growth of many animal populations, including those important in wildlife management, endangered species, and in fisheries (see Chapter 14), and the human population. Unfortunately, there is little evidence that human populations—or any animal populations, for that matter—actually follow this growth curve.

The logistic curve involves assumptions that are unrealistic for humans and for other mammals. These assumptions include a constant environment, a constant carrying capacity, and a homogeneous population (one in which all individuals are identical in their effects on each other). The logistic curve is especially unlikely if death rates continue to decrease from improvements in health care, medicine, and food supplies. Once a human population has benefitted from these improvements, it must pass through what has become known as the demographic transition to achieve **zero population growth**, which leads to a stabilized population. The demographic transition is discussed later in this chapter.

Forecasting Human Population Growth Using the Logistic Curve

Although unrealistic, the logistic curve has been the method used for most long-term forecasts to determine the population size of specific countries. This S-shaped curve first rises steeply upward and then changes slope, curving toward the horizontal carrying capacity (Figure 4.9). The point at which the curve changes is the inflection point. Until a population has reached the inflection point, we cannot project the final logistic size.

Unfortunately for those who want to make this calculation, the human population has not yet made the bend around the inflection point. Typically, forecasters have dealt with this problem by assuming that today's population is just reaching the inflection point. This standard practice inevitably leads to a great underestimate of the maximum population. For example, an early projection of the upper limit of the U.S. population, made in the 1930s, assumed that the inflection point had already occurred. That assumption resulted in an estimate that the final population of the United States would be approximately 200 million. That number was exceeded years ago; the U.S. population has passed 295 million.[5]

The World Bank, an international organization that makes loans and provides technical assistance to developing countries, has made a series of projections based on current birth rates and death rates and assumptions about how these rates will change. These projections form the basis for the logistic curves presented in Figure 4.9. The projections assume that (1) mortality will fall everywhere and level off when female life expectancy reaches 82 years; (2) fertility will reach replacement levels everywhere between 2005 and 2060; (3) there will be no major worldwide catastrophe. This approach projects an equilibrium world population of 10.1–12.5 billion.[6] Developed countries would experience population growth from 1.2 billion today to 1.9 billion, but populations in developing countries would increase from 4.5 billion to 9.6 billion. Bangladesh (an area roughly twice the size of New Brunswick) would reach 257 million; Nigeria, 453 million; and India, 1.86 billion. In these projections, the developing countries contribute 95% of the increase.[6] These findings have important implications for the management of both population and environment, but they have not been without controversy. Some people argue that focusing attention on controlling population growth in developing countries (whose ecological footprint is very low) diverts attention from the wasteful consumption practices of affluent developed countries like Canada. Finding solutions for environmental problems will require both population controls and, for developed countries, reduction in our per capita resource use and waste generation.

4.5 The Demographic Transition

The **demographic transition** is a three-stage pattern of change in birth rates and death rates that occurred during the process of industrial and economic development of Western nations. It leads to a decline in population growth.

A decline in the death rate is the first stage of the demographic transition (Figure 4.10). In a non-industrial country, birth rates and death rates are high, and the growth rate is low.[6] With industrialization, health and sanitation improve, and the death rate drops rapidly. The birth rate remains high, however, and the population enters stage II, a period with a high growth rate. Most European nations passed through this period in the eighteenth and nineteenth centuries.

As education and the standard of living increase, and as family-planning methods become more widely used, the population reaches stage III. The birth rate drops toward the death rate, and the growth rate therefore decreases, eventually to a low or zero growth rate. However, the birth rate declines only if families believe there is a direct connection between future economic well-being and funds spent on the education and care of their young. Such families have few children and put all their resources toward the children's education and well-being.

Historically, parents preferred to have large families. Without other means of support, parents can depend on children for a kind of "social security" in their old age, and children help with many kinds of hunting, gathering, and low-technology farming. Unless a change in attitude occurs among parents—they see more benefits from a few well-educated children than from many poorer children—nations face a problem in making the transition from stage II to stage III (see Figure 4.10c).

Some developed countries are approaching stage III, but it is an open question whether other developing nations will make the transition before a serious population crash occurs. *The key point here is that the demographic transition will take place only if parents come to believe that having a small family is to their benefit. Here we see again the connection between science and values.* Scientific analysis can show the value of small families, but this knowledge must become part of cultural values to have an effect.

4.6 Population and Technology

The danger that the human population poses to the environment is the result of two factors: the number of people and the impact of each person on the environment. When there were few people on Earth and technology was limited, human impact was local. In that situation, overuse of a local resource had few or no large or long-lasting effects. The fundamental problem now is that there are so many people and our technologies are so powerful that our effects on the environment are global and important. With more people, and higher levels of industrialization, there are higher rates of resource extraction and waste production, higher emissions of air and water pollutants, and a larger ecological footprint. In rapidly urbanizing areas, with congested transportation routes and the proximity of major industrial operations to residential development, these impacts may be particularly apparent.

The simplest way to characterize the total impact of the human population on the environment is as follows: The total environmental effect is the average impact of an individual multiplied by the total number of individuals,[7] or

$$T = P \times I$$

where P is the population size—the number of people—and I is the average environmental impact per person. The impact per person varies widely. The average impact of a person who lives in Canada is much greater than the impact of a person who lives in a low-technology society. But even in a poor low-technology nation like Bangladesh, the sheer number of people leads to large-scale environmental effects. Almost 200 years ago, Thomas Malthus foresaw the human population

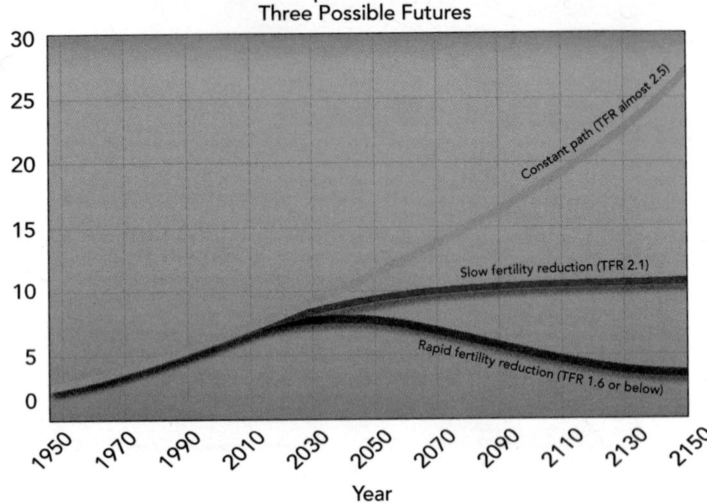

Figure 4.9 • Exponential and logistic growth curves. Three possible paths of future world population growth, as projected by the United Nations. The constant path assumes that the current growth rate will continue unchanged, resulting in an exponential increase. The slow fertility reduction path assumes that the world's fertility will decline to reach replacement level by the year 2050 and that the world's population will stabilize at about 11 billion by the twenty-second century. The rapid fertility reduction path assumes that the world's fertility will go into decline in the twenty-first century, peaking at 7.7 billion in 2050 and dropping to 3.6 billion by 2150. These are theoretical curves. (*Source*: United Nations Population Division, 1998.)

HUMAN POPULATION

A CLOSER LOOK 4.2

The Prophecy of Malthus

Almost 200 years ago, the English economist Thomas Malthus eloquently stated the human population problem. He based his argument on three simple premises:[8]

- Food is necessary for people to survive.
- "Passion between the sexes is necessary and will remain nearly in its present state"—so children will continue to be born.
- The power of population growth is "indefinitely greater than the power of Earth to produce subsistence."

Malthus reasoned that it would be impossible to maintain a rapidly multiplying human population on a finite resource base. His projections of the ultimate fate of humankind were dire, as dismal a picture as that painted by the most extreme pessimists of today. The power of population growth is so great, he wrote, that "premature death must in some shape or other visit the human race. The vices of mankind are active and able ministers of depopulation, but should they fail, sickly seasons, epidemics, pestilence and plague, advance in terrific array, and sweep off their thousands and ten thousands." Should these fail, "gigantic famine stalks in the rear, and with one mighty blow, levels the population with the food of the world." Some would argue that recent famines in Africa and North Korea are evidence that Malthus' predictions were correct.

Critics of Malthus continue to point out that these patterns are local, not global, and that Malthus' predictions have not yet been substantiated. Whenever things have looked bleak, technology has provided a way out, allowing us to live at greater densities. These critics have argued that our technologies will continue to save us from a Malthusian fate and that therefore we need not worry about the growth of the human population. They point for example to the dramatic increases in food production over the past century, enabled by new technologies, new crop varieties, and new approaches to agriculture.

Who is correct? Ultimately, in a finite world, Malthus must be correct about the final outcome of unchecked growth. He may have been wrong about the timing; he did not anticipate the capability of technological changes to delay the inevitable. But although some people believe that Earth can support many more people than it does now, in the long run there must be an upper limit. The basic issue that confronts us is this: How can we achieve a constant world population, or at least halt the increase in the population, in a way that is most beneficial to most people? This is undoubtedly one of the most important questions that has ever faced humanity.

---

problem (see A Closer Look 4.2). Modern technology increases the use of resources and enables us to affect the environment in many new ways, compared with hunters and gatherers or people who farmed with simple wooden and stone tools. For example, before the invention of chlorofluorocarbons (CFCs), used as propellants in spray cans and as coolants in refrigerators and air conditioners, we were not causing depletion of the ozone layer in the upper atmosphere. Similarly, before we started driving automobiles, there was much less demand for steel, little demand for oil, and much less air pollution. These linkages between people and the global environment illustrate the global theme and the people and nature theme of this book.

The population-times-technology equation reveals a great irony involving two standard goals of international aid: improving the standard of living and slowing the overall human population growth. Improving the standard of living increases the total environmental impact, countering the environmental benefits of a decline in population growth.

4.7 The Human Population, the Quality of Life, and the Human Carrying Capacity of Earth

What is the human carrying capacity of Earth—that is, how many people can live on Earth at the same time? The answer depends on what quality of life people desire and are willing to accept.

Estimates of the human carrying capacity of Earth have typically involved two methods. One method is to extrapolate from past growth. This approach, as discussed earlier, assumes that the population will follow an S-shaped logistic growth curve, so that it will gradually level off (see Figure 4.9).

The second method can be referred to as the "packing problem" approach. This method simply considers how many people might be packed onto Earth, not taking into sufficient account the need for lands and oceans to provide food, water, energy, construction materials, and scenic beauty, and the need to maintain biological diversity. It could be called the "standing-room-only approach." This has led to very high estimates of the total number of people that might occupy Earth—as many as 50 billion.

More recently, a philosophical movement has developed at the other extreme, known as "deep ecology." This philosophy makes sustaining the biosphere the primary moral imperative for people. Its proponents argue that the whole Earth is necessary to sustain life. Therefore, everything else must be sacrificed to the goal of sustaining the biosphere. People are considered active agents of destruction of the biosphere, therefore, the total number of people should be greatly reduced. Based on this rationale, estimates for the desirable number of people vary greatly, from a few million up.

Between the packing-problem approach and deep-ecology approach are numerous options. Setting goals between these extremes is possible, but each of these goals is a value judgement, again reminding us of the theme of science and values. What constitutes a desirable quality of life is a value judgement. What kind of life is possible is affected by technology, which in turn is affected by science. Scientific understanding tells us what is required to meet each quality-of-life level.

The options vary according to the quality of life for the average person. If everyone in the world were to live at the same level as those in Canada, with our high resource use, then the carrying capacity would be comparatively low. Conversely, if everyone were to live at the level of those in Bangladesh, with all of its risks plus its poverty and its heavy drain on biological diversity and scenic beauty, the carrying capacity would be much higher.

In summary, the acceptable carrying capacity is not simply a scientific issue; it is an issue combining science and values, one theme of this book. Science plays two roles. First, by leading to new knowledge that leads to new technology, there is possibly a greater impact per individual on Earth's resources and a higher population density. Second, scientific methods can be used to forecast a probable carrying capacity once a goal for the average quality of life, in terms of human values, is chosen. In this second use, science can tell us the implications of our value judgements, but it cannot provide those value judgements.

Potential Effects of Medical Advances on Demographic Transition

Although the demographic transition is traditionally defined as consisting of three stages, advances in treating chronic health problems such as heart disease can lead a stage III country to a second decline in the death rate. This could cause a second transitional phase of population growth (stage IV), in which the birth rate would remain the same while the death rate fell. A second stable phase of low or zero growth (stage V) would be achieved only when the birth rate declined to match the decline in the death rate. Thus, there is danger of a new spurt of growth even in industrialized nations that have passed through the standard demographic transition.

Recent medical advances in the understanding of aging and the potential of new biotechnology to increase both the average longevity and the maximum lifetime of human beings have major implications for human population growth. As these medical advances take place, then the death rate will drop and the growth rate will increase even more. Thus, a prospect that is positive from each individual's point of view—a longer, healthier, and more active life—could have negative effects on the environment. Ultimately, we will need to decide among the following choices: stop medical research dealing with chronic diseases of old age and attempts to increase people's maximum lifetime; reduce the birth rate; or chose neither and wait for Malthus's projections to come true—for famine, environmental catastrophes, and epidemic diseases to cause large and sporadic episodes of human death. The first choice seems inhumane, but the second is highly controversial. This is one of the most important issues concerning science and values and people and nature.

Human Death Rates and the Rise of Industrial Societies

More consideration is needed of the first stage in the demographic transition. We can get an idea of the first stage by comparing a modern industrialized country, such as Switzerland, which has a crude death rate of 9 per 1,000, with a developing nation, such as Sierra Leone, which has a crude death rate of 25.[1] Modern medicine has greatly reduced death rates from disease in countries such as Switzerland, particularly with respect to death from acute or epidemic diseases.

An acute or epidemic disease appears rapidly in the population, affects a comparatively large percentage of it, and then declines or almost disappears for a while, only to reappear later. Epidemic diseases typically are rare but have occasional outbreaks during which a large proportion of the population is infected. Recall the example of the bubonic plague, discussed in Chapter 1, which killed a third of the population of Europe in multiple outbreaks. Influenza, measles, mumps, and cholera are other examples of epidemic diseases. The influenza epidemic of 1918-1919 killed almost 22 million people in every continent but Antarctica. A billion people—half the global population of the time—were affected by the disease. A chronic disease, in contrast, is always present in a population, typically occurring in a relatively small but relatively constant percentage of the population. Heart disease, cancer, and stroke are examples.

The great decrease in the percentage of deaths due to acute or epidemic diseases can be seen in a comparison of causes of deaths in Ecuador in 1987 and Canada in the early 1920s and 1997 (Figure 4.11).[9] In Ecuador, a developing nation, acute diseases and those listed as "all others" accounted for about 60% of mortality in 1987. In

Canada, in 1997, these accounted for only about 12% of mortality. Chronic diseases account for about 70% of mortality in modern Canada. In contrast, these accounted for only about 45% of the deaths in Canada in 1921-1925 and about 33% in Ecuador in 1987. In 1987, then, Ecuador resembled the Canada of the 1920s more than it resembled modern Canada.

Although outbreaks of the well-known traditional epidemic diseases have declined greatly during the past century in industrialized nations, there is now concern that the incidence of these diseases may increase due to several factors. One is that as the human population grows, people live in new habitats, where previously unknown diseases occur. Another is that strains of disease organisms have developed resistance to antibiotics and other modern methods of control.

A broader view of why diseases are likely to increase comes from an ecological and evolutionary perspective (which is explained in later chapters). Stated simply, 6.3 billion people on Earth is a great resource and opportunity for other species: We offer a huge and easily accessible host. Thinking that other species will not take advantage of the opportunity humans present is naïve. From this perspective, the future holds more diseases, rather than fewer. This is a new perspective. In the mid-twentieth century, it was easy to believe that modern medicine would eventually cure all diseases and that most people would live the maximum human life span.

The sudden occurrence of a new disease in February 2003, severe acute respiratory syndrome (SARS), demonstrated that modern transportation and the world's huge human population could lead to the rapid spread of epidemic diseases. Jet airliners daily carry vast numbers of people and goods around the world. SARS began in China; its source is unclear, but it may have spread from some wild animal to human beings. China has become much more open to foreign travellers, with more than 90 million people visiting in a recent year.[10] As a result, it has become easier for diseases originating in China to reach other parts of the world. By late spring 2003, SARS had spread to two dozen countries. Quick action led by the World Health Organization (WHO) contained the disease, which as of this writing appears well under control.[11]

West Nile virus is another example of how rapidly and widely diseases spread today. Before 1999, West Nile virus occurred in Africa, West Asia, and the Middle East, but not in the New World.

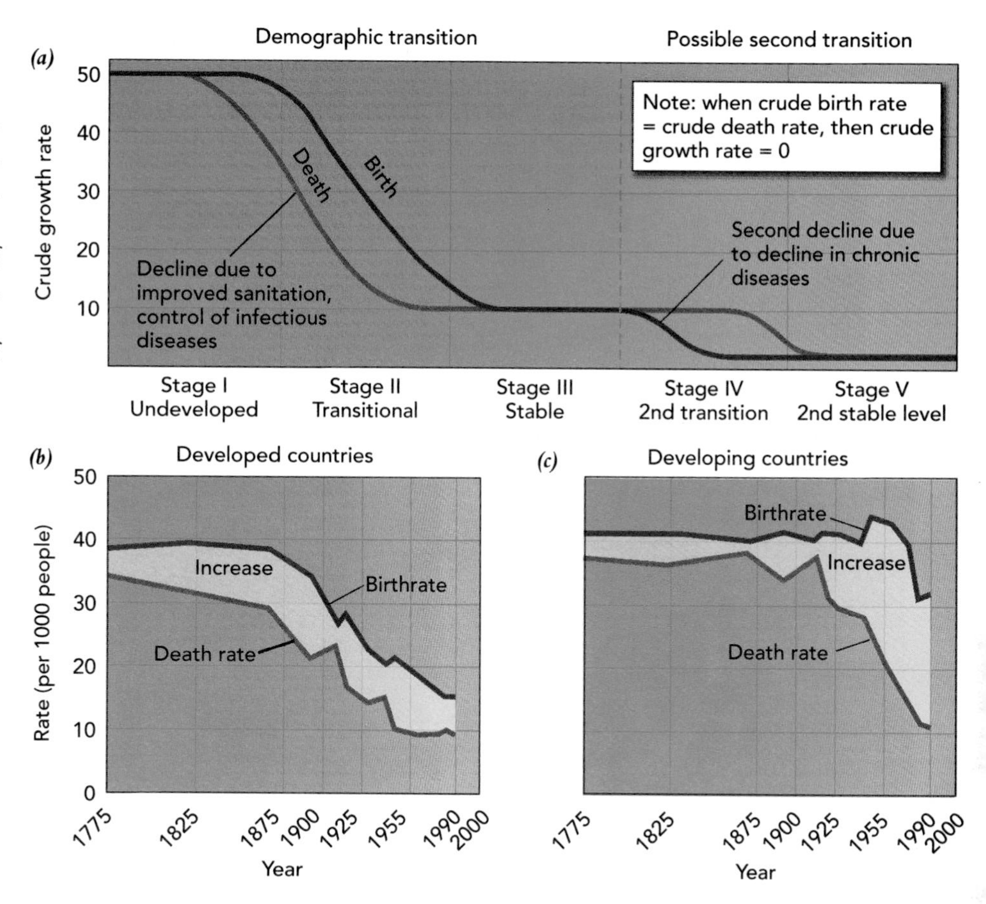

Figure 4.10 • **The demographic transition:** *(a)* theoretical, including possible fourth and fifth stages that might take place in the future; *(b)* as has taken place for developed countries since 1775; and *(c)* as may be occurring in developing nations since 1775. (*Source*: Modified from M. M. Kent and K. A. Crews, *World Population: Fundamentals of Growth* [Washington, D.C.: Population Reference Bureau, 1990]. Copyright 1990 by the Population Reference Bureau, Inc. Reprinted by permission.)

Related to encephalitis, West Nile virus is spread by mosquitoes, which bite and thus infect birds and people. It reached the Western Hemisphere through infected birds. A growing number of bird species, many of them native to Canada, are now known to be affected by the virus. They include familiar roadside birds like crows and ravens, and common visitors to bird feeders like blue jays and black-capped chickadees. Fortunately, although the disease can be fatal, in most people it appears to last only a few days and rarely causes severe symptoms.[12] In Canada, there were only four confirmed cases of West Nile Virus in the summer of 2004, down from a peak of 1,388 in 2003. Of those 1,388 cases, 848 of them originated in Saskatchewan, which had no cases in the previous year. In that same year, a total of 14 people died from West Nile-related illness in Canada, 10 in Saskatchewan, two in Manitoba, and two in Ontario. In some of these cases, West Nile was considered a contributing factor but not the primary cause of death. In 2002, there were 409 cases

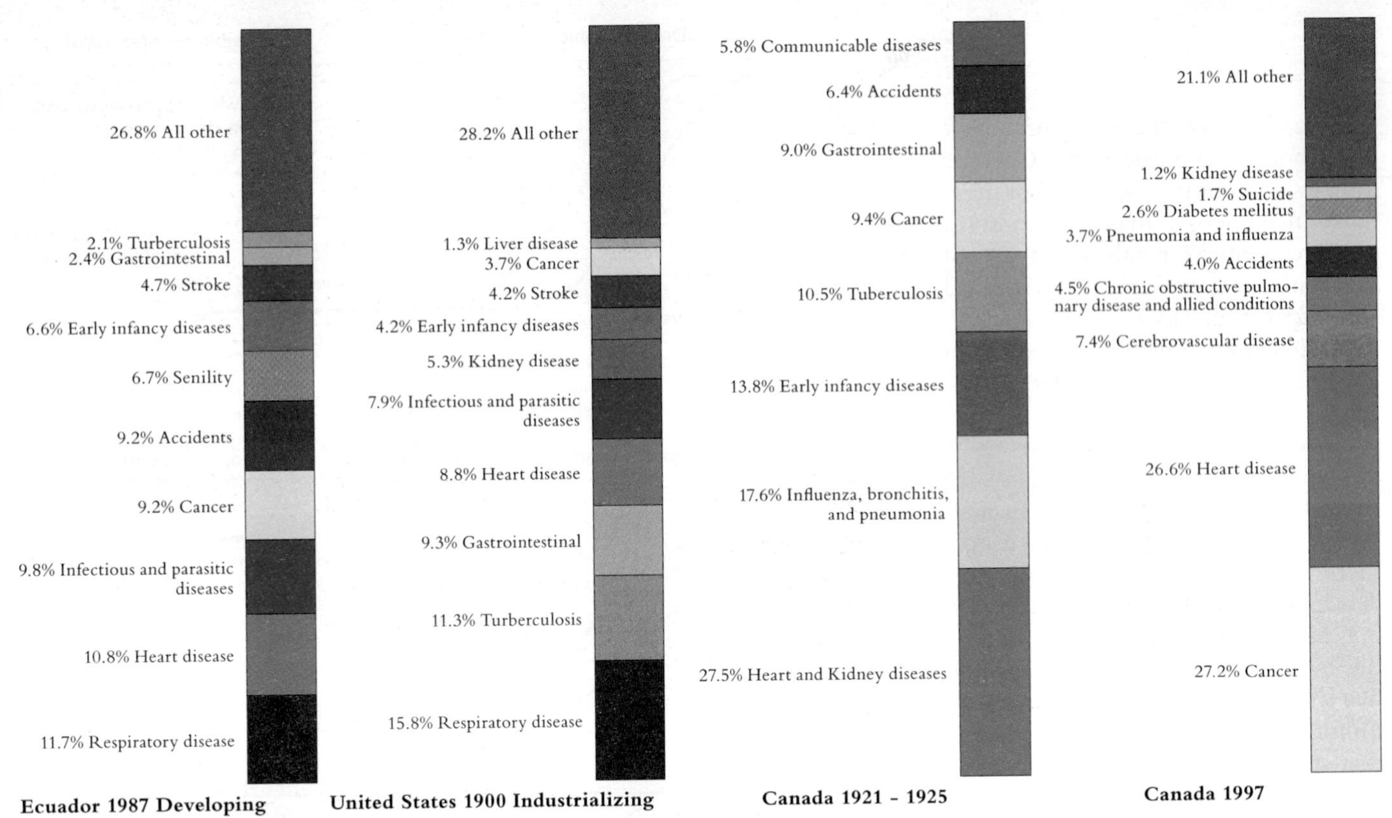

Figure 4.11 • Causes of mortality in developing, industrializing, and industrialized nations. (*Source*: Ecuador 1987 data from M. M. Kent and K. A. Crews, *World Population: Fundamentals of Growth* [Washington, D.C.: Population Reference Bureau, 1990]. Copyright 1990 by the Population Reference Bureau, Inc. Reprinted by permission. National Vital Statistics Report 48 [11], July 24, 2000. Canadian 1921–1925 and 1997 data from Statistics Canada (www.statcan.ca.)

of the disease in Canada, 308 of them in Ontario; 18 of those people died.

Increasing density in populations of poor people also makes the spread of diseases more likely. In China more than 550 million people are infected with the tuberculosis bacteria (TB), and 120,000 to 205,000 die each year from this disease. TB is one disease causing concern among health experts because of its persistence and capacity to increase worldwide.[13] AIDS is a factor in this, because individuals with AIDS lack resistance to diseases. Research has indicated that up to 70% of TB patients are infected with HIV; TB appears to be one of the first opportunistic diseases to strike HIV-infected people.[13] Some 95% of the 33.6 million people with full-blown AIDS or infected with HIV (the virus that causes AIDS) live in the developing world. Of these 33.6 million, 70% live in sub-Saharan Africa, which is home to 10% of the world's population.[14] However, because birth rates in African countries continue to be high, populations will continue to grow.[15]

Natural disasters, such as the Southeast Asia tsunami on December 26, 2004, can also have a powerful impact on local and regional population density. As many as 300,000 people are thought to have died in the event, although estimates are still uncertain (500 bodies a day were still being discovered in February 2005). Millions of others were displaced, changing local population structures and associated demands on food, water, and sanitation resources. Earthquakes, landslides, volcanoes, and floods have also had a dramatic impact on local population in countries throughout the world.

Finally, as we discussed in Chapter 1, famine has had a dramatic impact on population growth in some regions of the world. Over half a billion people in South and Southeast Asia are at risk of death from starvation; roughly a quarter of a billion people in Africa face the same challenge. Famine killed one in five people in Ireland in the mid-nineteenth century, and famines also occurred in Iceland in 1783, when a volcanic eruption caused a crop failure, in China between 1946 and 1948, and in India in 1943. Every year, starvation kills about 18 million people. Children, especially those under five years of age, are most at risk. Even Argentina, once one of the world's most prosperous countries, has experienced rising deaths from starvation. Although the country produces sufficient food to feed its population, recent economic crises have doubled food costs and caused the collapse of the social security program, making it difficult for all but the most affluent families to buy food.

Longevity and its Effect on Population Growth

The **maximum lifetime** is the genetically determined maximum possible age to which an individual of a species can live. **Life expectancy** is the average number of years an individual can expect to live given the individual's present age. (The term is often applied, without qualification, to mean the life expectancy of a newborn.)

A surprising aspect of the second and third periods in the history of human population is that population growth occurred with little or no change in the maximum lifetime. What changed were birth rates, death rates, population growth rates, age structure, and average life expectancy.

Immigration, Emigration, and Population Growth

The December 2004 Southeast Asia tsunami is one example of how the movement of people can affect population density. National rates of immigration (entering the country) and emigration (leaving the country) can also have an impact on population growth. Indeed, Canada's immigration policies are one of the factors in its continued population growth, compared to the United States, where population growth has leveled off in the last decade. Immigration to Canada exceeded 1.5 million people per year around 1900, dropping to less than 200,000 per year in mid-century. Since 1945, immigration rates have climbed steadily because of welcoming immigration policies, and are beginning to approach 1.5 million people a year. Emigrations from Canada also peaked around 1900, at 1.2 million people a year (so net population growth was actually small), but today emigration rates are low, about 350,000 people per year. This high net immigration accounts for Canada's steadily growing population, despite our relatively low birth rates. There are, of course, regional differences in population growth and immigration/emigration. For example, the Northwest Territories experienced a huge population increase during the gold rush years of the 1870s, but that population had just as quickly declined by 1900. Today, the population of the Northwest Territories and Nunavut is relatively stable. By contrast, Ontario's population growth rate has been declining slowly since 1850, and its overall population growth is slower than other regions of the county. In the 2001 Canadian census, Alberta had the highest population growth rate, with most growth concentrated in Calgary and Edmonton. There is a small but a noticeable trend of movement away from rural areas into cities; rural populations in Canada declined 0.4% between 1996 and 2001. Over the same period, Canada's population overall increased only 4%, with most of that growth attributable to immigration.[9] It is estimated that by 2070, Canada's population will approach 70 million people.

4.8 Limiting Factors

Basic Concepts

On our finite planet, human populations will eventually be limited by some factor or combination of factors. We can group limiting factors into those that affect a population during the year in which they become limiting (short-term factors), those whose effects are apparent after one year but before 10 years (intermediate-term factors), and those whose effects are not apparent for 10 years (long-term factors). Some factors fit into more than one category, having, say, both short-term and intermediate-term effects.

An important short-term factor is the disruption of food distribution in a country, commonly caused by drought, or a shortage of energy for transporting food. The outbreak of a new disease or a new strain of a previously controlled disease is another important short-term factor.

Intermediate-term factors include desertification (see Chapter 11); dispersal of certain pollutants such as toxic metals into waters and fisheries; disruption in the supply of non-renewable resources, such as rare metals used in making steel alloys for transportation machinery; and a decrease in the supply of firewood or other fuels for heating and cooking.

Long-term factors include soil erosion, a decline in groundwater supplies, and climate change. Changes in resources available per person suggest that we may have already exceeded the long-term human carrying capacity of Earth. For example, wood production peaked at 0.67 m^3/person in 1967, fish production at 5.5 kg/person in 1970, beef at 11.81 kg/person in 1977, mutton at 1.92 kg/person in 1972, wool at 0.86 kg/person in 1960, and cereal crops at 342 kg/person in 1977.[16] Before these peaks were reached, per-capita production of each resource had grown rapidly.

Life Expectancy

Life expectancy, as stated earlier, is the estimated average number of years a person of a specific age can expect to live. Technically, it is an age-specific number: Each age class within a population has its own life expectancy. For general comparison, however, the life expectancy at birth is used. Life expectancy differs by nation and by sex, age, and other factors. The life expectancy in a hunter-gatherer society is short. For example, among the Kung Bushmen of Botswana, life expectancy at birth is 39 years.[4, 17]

As mentioned earlier, the human population's great rate of increase has occurred even though life expectancy has not increased. In fact, studies of the age at death carved on tombstones tell us that the chances of a 75 year-old living to age 90 were greater in ancient Rome than they are today in modern society (Figure 4.12). These studies also suggest that death rates were much higher among young people in Rome than in twentieth-century England. In ancient Rome, the life expectancy of a one-year-old was about 22 years, whereas in twentieth-century England it was about 50 years. Life expectancy in twentieth-century England was greater than in ancient Rome for all ages until about age 55, after which the life expectancy appears to have been higher for ancient Romans than for a twentieth-century Briton. This would suggest that many hazards of modern life may be concentrated more on the aged. Pollution-induced diseases are one factor in this change.

4.9 How Can We Achieve Zero Population Growth?

We have surveyed several aspects of population dynamics. We now return to our earlier question: How can we achieve zero population growth?

Age of First Childbearing

The simplest and one of the most effective means of slowing population growth is to delay the age of first childbearing by women.[18] As more women enter the workforce and as education levels and standards of living increase, this delay occurs naturally. Social pressures that lead to deferred marriage and childbearing can also be effective.

Countries with high growth rates have early marriages. In South Asia and in sub-Saharan Africa, about 50% of women marry between the ages of 15 and 19. In Bangladesh, women marry at age 16 on average, whereas in Sri Lanka the average age for marriage is 25. The World Bank estimates that if Bangladesh adopted Sri Lanka's marriage pattern, families could average 2.2 fewer children.[19] Increases in the marriage age could account for 40–50% of the drop in fertility required to achieve zero population growth for many countries.

Age at marriage has risen in some countries, especially in Asia. For example, in Korea, the average marriage age went from 17 in 1925 to 24 in 1975. China passed laws fixing minimum marriage ages, first at 18 for women and 20 for men in 1950, then at 20 for women and 22 for men in 1980.[6] As a result of this law and a one-child policy introduced in 1978, between 1972 and 1985, China's birth rate dropped from 32 to 18 per thousand people, and the average fertility rate went from 5.7 to 2.1 children.

Birth Control: Biological and Societal

Another simple means of decreasing birth rates is breast-feeding, which can delay resumption of ovulation.[20] This is the primary birth-control method for women in a number of countries. In the mid-1970s, the practice of breast-feeding provided more protection against conception in developing countries than did family-planning programs, according to the World Bank.[6]

Nevertheless, much emphasis is placed on the need for family planning.[21] Traditional methods range from abstinence to induction of sterility with natural agents. Modern methods include the birth-control pill, which prevents ovulation through control of hormone levels; surgical techniques for permanent sterility; and mechanical devices. Contraceptive devices are used widely in many parts of the world, especially in East Asia, where data show that 78% of women use them. In Africa, only 18% of women use them; in Central and South America, the numbers are 53% and 62%, respectively.[1]

Abortion is also widespread. Although usually medically safe, abortion is one of the most controversial methods from a moral perspective. Ironically, it is one of the most important birth-control methods in terms of its effects on birth rates—approximately 46 million abortions are performed each year.[22]

National Programs to Reduce Birth Rates

Reducing birth rates requires a change in attitude, knowledge of the means of birth control, and the ability to afford these means. As we have seen, a change in attitude can occur simply with an increase in the standard of living. In many countries, however, it has been necessary to providing formal family-planning programs to explain the problems arising from rapid population growth and describe the benefits to individuals of reduced population growth. These programs also provide information about birth-control methods and provide access to these methods.[23] The choice of population-control methods is an issue that involves social, moral, and religious beliefs, which vary from country to country.

The first country to adopt an official population policy was India in 1952. Few developing countries had official family-planning programs before 1965. Since 1965, many similar programs have been introduced, and the World Bank has lent $4.2 billion to more than 80 countries to support "reproductive" health projects.[20, 24] Although most countries now have some kind of family-planning program, effectiveness varies greatly. A wide range of approaches have been used, from simply providing more information, to promoting and providing means for birth control, to offering rewards and

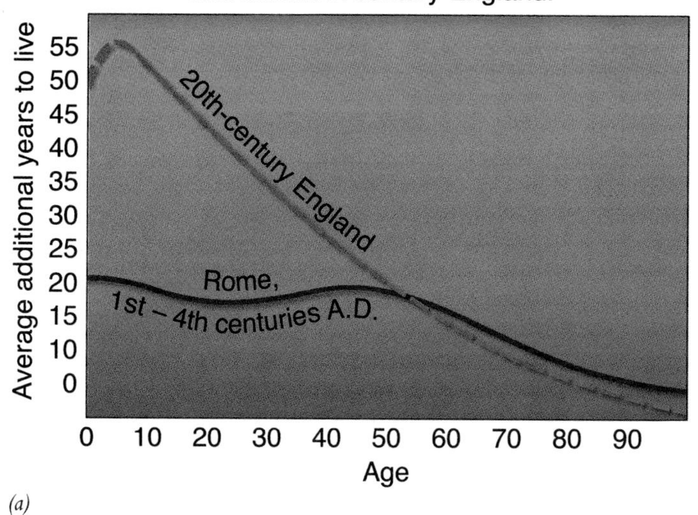

(a)

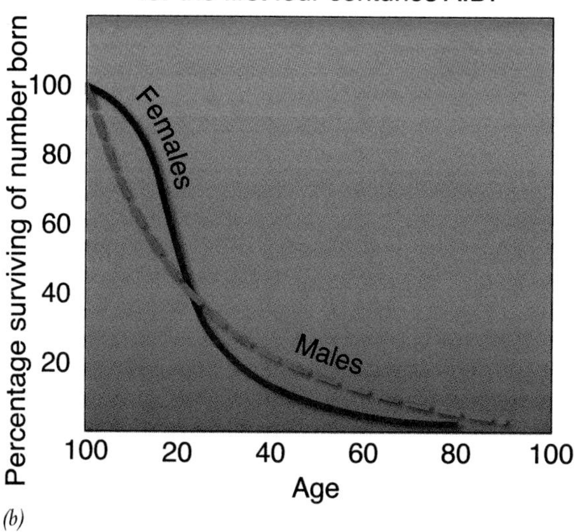

(b)

Figure 4.12 • Life expectancy and survivorship in ancient Rome and twentieth century England. *(a)* Life expectancy in ancient Rome and twentieth-century England. This graph shows the average additional number of years one could expect to live having reached a given age. For example, a 10-year-old in England could expect to live about 55 more years; a 10-year-old in Rome could expect to live about 20 more years. Among the young, life expectancy was greater in twentieth-century England than in ancient Rome. However, the graphs cross at about age 60. An 80-year-old Roman could expect to live longer than an 80-year-old Briton. The graph for Romans is reconstructed from ages given on tombstones. *(b)* Approximate survivorship curve for Rome for the first four centuries a.d. The percent surviving decreases rapidly in the early years, reflecting the high mortality rates for children in ancient Rome. Females had a slightly higher survivorship rate until age 20, after which males had a slightly higher rate. (*Source*: Modified from G. E. Hutchinson, An Introduction to Population Ecology [New Haven, Conn.: Yale University Press, 1978]. Copyright 1978 by Yale University Press. Used by permission.)

imposing penalties. Penalties usually take the form of taxes. Ghana, Malaysia, Pakistan, Singapore, and the Philippines have used a combination of methods, including limits on tax allowances for children and maternity benefits. Tanzania has restricted paid maternity leave for women to a frequency of once in three years. Singapore does not take family size into account in allocating government-built housing, so larger families are more crowded. Singapore also gives higher priority in school admission to children from smaller families.[6] Some countries, including Bangladesh, India, and Sri Lanka, have paid people to be voluntarily sterilized. In Sri Lanka, this practice has applied only to families with two children, and a voluntary statement of consent is signed.

China has one of the oldest and most effective family-planning programs. In 1978, China adopted an official policy to reduce its human population growth from 1.2% in that year to zero by the year 2000. In 2000, China's growth rate had slowed to 1.0% a year rather than to zero; the program has done much to curb the country's rapid population growth. The Chinese program places emphasis on single-child families. The government has used education, a network of family planning that provides information and means for birth control, and a system of rewards and penalties. Women have been given paid leave for abortions and for sterilization operations. Families with a single child have received benefits, including financial subsidies in some areas. In some parts of China, families that have a second child have had to return the bonuses received for the first. Other rewards and penalties vary from province to province.[6] The program has raised questions related to several of this book's theme, including science and values, and people and nature. In the past, some Chinese practices were mandatory, raising the question: How can freedom of choice be exercised at the same time that emphasis is being placed on reducing the birth rate? The Government of China responded to these concerns with a new law in 2002 that prevented the use of coercion to force families to control family size.

This focus on family planning raises an important question about the role of women and education in global population control. Women have been called the "key to development" because of their central influence in childbearing decisions and infant survivorship. Empowerment of women through community groups has been a powerful force in raising development status. Even a single charismatic individual in a community can begin a dialogue, drawing other women into the discussion through social activities or simple conversation. Through sharing similar experiences, women find opportunities for support, advice, and education. With knowledge comes confidence and better decisions about appropriate family size and how to achieve it. Literacy and the status of women are therefore two key determinants in development status and management of population growth.

Summary

- The human population is the underlying environmental issue, because most current environmental damage results from the very high number of people on Earth and their influence on the environment.
- Throughout most of our history, the human population and its average growth rate were small. The growth of the human population can be divided into four major phases. Although the population has increased in each phase, the current situation is unprecedented.
- Countries with declining birth rates experienced a demographic transition marked by a decrease in death rates followed by a decrease in birth rates. Many developing nations experienced a great decrease in their death rates but still have a very high birth rate. It remains an open question whether some of these nations will be able to achieve a lower birth rate before reaching disastrously high population levels.
- The maximum population Earth can sustain and how large a population will ultimately be attained by human beings are controversial questions. Standard estimates suggest that the human population will reach 10–16 billion before stabilizing.
- How the human population might stabilize or be stabilized raises questions concerning science and values, and people and nature.
- Considerable lags in the responses of human populations to changes in birth rates and death rates occur because of the age structure of the population. A population that achieves replacement-level fertility will continue to grow for several generations.
- An effective method to lower a population's growth rate is to delay the age of first childbearing. This also involves relatively few societal and value issues.
- Literacy and the status of women are key determinants of population size and development status.

STUDY QUESTIONS

1. What are the principal reasons that the human population has grown so rapidly in the twentieth century?
2. Given a world population in 1990 of 5.3 billion with a growth rate of 1.8%, what is the doubling time?
3. Why is it important to consider the age structure of a human population?
4. Three characteristics of a population are birth rate, growth rate, and death rate. How has each been affected by (a) modern medicine, (b) modern agriculture, and (c) modern industry?
5. Strictly from a biological point of view, why is it difficult for a human population to achieve a constant size?
6. What environmental factors are likely to increase the chances of an outbreak of an epidemic disease?
7. Why is it so difficult to predict the growth of Earth's human population?
8. To which of the following can we attribute the great increase in human population since 1800: changes in human (a) birth rates, (b) death rates, (c) longevity, or (d) death rates among the very old? Explain.
9. What is the demographic transition? When would one expect replacement-level fertility to be achieved—before, during, or after the demographic transition?
10. Based on the history of human populations in various countries, how would you expect the following to change as per capita income increased: (a) birth rates, (b) death rates, (c) average family size, and (d) age structure of the population? Explain.
13. Consider the data presented in Figure 4.3 and the calculations you have completed in Build Your Environmental Skills 4.1. With no interventions or changes in the population dynamics of these countries, what would the age structure diagram look like in 20 years or in 50 years? What implications would that structure have for social systems like education, health care, and national pension plans?

FURTHER READING

Brown, L. R., G. Gardner, and B. Halueil. *Beyond Malthus: Nineteen Dimensions of the Population Challenge.* New York: W. W. Norton, 1999. A discussion of recent changes in human population trends and their implications.

Cohen, J. E. *How Many People Can the Earth Support?* New York: Norton, 1995. A detailed discussion of world population growth, Earth's human carrying capacity, and factors affecting both.

How Many People Can Earth Support?

In mid-1992 the world population reached 5.5 billion. Today it exceeds 6 billion. Estimates of how many people the planet can support range from 2.5 billion to 40 billion. Why do the estimates vary so widely? What is a realistic carrying capacity? What factors need to be considered to answer this question?

Food Supply

World grain production has apparently leveled off since reaching its highest levels in the mid-1980s. The production of grain from 1984 to 1994 remained at approximately 1.7 billion tonnes, up from 631 million tonnes in 1950. The remarkable increase in productivity after 1950 resulted from the development of high-yielding varieties, use of chemical fertilizers, application of pesticides, and doubling of cropland area. If the present harvest was distributed evenly and everyone ate a vegetarian diet, it could support 6 billion people. As the world population has continued to grow, the per capita allotment of grain has been falling since 1984, when it stood at 346 kg per capita. By 1994 it had fallen to 311 kg per capita.

Land and Soil Resources

Almost all the usable agricultural land, approximately 1.5 billion ha, is already being cultivated. An increase of 13% in agricultural lands is possible but would be costly. Land area devoted to raising crops has dropped since 1950 to 1.7 ha per capita and will likely continue to drop, to approximately 1 ha per capita by 2025 if present population projections hold. More soil is lost each year to erosion (about 26 billion tonnes) than is formed.

Water Resources

Water suitable for drinking and irrigation amounts to only a small proportion (less than 3%) of the water on Earth. Underground reservoirs are being depleted on the order of metres per year but are being replaced in centimetres or even fractions of centimetres per year. Per capita water consumption varies from 350–1,000 L a day in the developed countries to 2–5 L a day in the driest rural areas, where people may obtain water directly from streams or primitive wells.

Net Primary Production

Human and domestic animals use about 4% of the net primary production of the world's land area and 2% of that of the oceans. (Primary production is discussed in Chapter 9.) Net primary productivity on the planet has dropped about 13% since the 1950s.

Population Density

Population density varies greatly, from 3,076 people/km^2 on the tiny island of Malta to 66 people/km^2 in Africa as a whole. Bangladesh has 2,261 people/km^2; the Netherlands, 1,002/km^2; and Japan, 869/km^2.

Technology

Carrying capacity is not merely a matter of numbers of people. It also involves the impact they have on the world's resources—most critically on energy resources. Multiplying population by per capita energy consumption by the population gives a relative measure of the impact people have on the environment—their ecological footprint. By that measure, each Canadian has the impact of 35 people in India or 140 people in Bangladesh.

Critical Thinking Questions

1. Can agricultural technology improve food production so that it keeps pace with, population growth, and thus increases Earth's carrying capacity above 6 billion. What evidence supports this point of view? What evidence refutes it?
2. The land area of the world is approximately 150 million km^2. What would the world population be if the average density were 400 people/km^2? Is this a sound basis for determining carrying capacity? Explain.
3. What factors, in addition to the six listed in this box, do you think should be considered in determining the planet's carrying capacity?

Kent, M. M., and K. A. Crews. *World Population: Fundamentals of Growth*. Washington, D.C.: Population Reference Bureau, 1990. A clear and concise introduction to population concepts and trends.

Kessler, E., ed. "Population, Natural Resources and Development," *AMBIO* (1992): 2:1(1). A special issue of the journal focusing on human population growth and its economic and environmental implications.

McFalls, J. A. "Population: A Lively Introduction," *Population Bulletin*, 46 (1991): 2:1–43. A good introduction to human population dynamics.

World Bank. *Population Change and Economic Development*. Oxford: Oxford University Press, 1985. An excellent compilation of facts about human populations and methods for calculating population trends.

chapter 5

Canada's Melting Glaciers

Humans may unknowingly alter the natural cycling of chemicals and thus change the environment in unacceptable ways. Science can provide potential solutions, but our value systems dictate whether we are willing to pursue these solutions. The following story concerning Canada's melting glaciers shows that an understanding of biogeochemical cycles and mass balance can help us address adverse environmental effects.

For at least 150 years, scientists have documented the steady shrinking of Canada's great ice sheets, but there is now growing concern about the potential for climate change (discussed in Chapter 20 of this book) to speed that melting. Worldwide, more than 97% of Earth's water is saltwater; less than 3% is fresh. Of that amount, more than three-quarters is stored in ice caps and glaciers. (Most of the rest is groundwater; less than 0.01% of the world's water is found in lakes and rivers; see Chapter 18.) If predictions about rising concentrations of greenhouse gases in the atmosphere are correct, Arctic temperatures will rise and polar ice caps and glaciers will shrink. As glaciers melt, vast quantities of fresh water will enter the oceans. Sea levels can be expected to rise, but more subtle impacts on the health and composition of marine ecosystems are also predicteded. It has even been suggested that large additions of fresh water to northern oceans could change patterns of ocean circulation in the North Atlantic and elsewhere.

Canadian scientists working at John Evans Glacier, Ellesmere Island, Nunavut, have been developing tools to predict changes in Arctic glaciers. Researchers use a combination of detailed field monitoring and mathematical simulations of glacier behaviour to evaluate the magnitude and significance of the changes in glacier volume that might occur with climate change. Central to their work is a mass balance model, a representation that accounts for all the inputs (such as snow and rain) and outputs (such as evaporation and runoff) in a system. Coupled with simulations of glacier chemistry and movement developed by the University of British Columbia's Ice Sheet Modelling Program, the mass balance model will help scientists understand instabilities like those recently observed in the Laurentide Ice Sheet near Baffin Island, and the iceberg surges they can cause. Canada's new National Glaciology Program, which brings together expertise from Environment Canada and Natural Resources Canada, is geared to helping us to anticipate these massive changes and develop appropriate strategies for adapting to them.[1]

The John Evans Glacier, Nunavut.

The Biogeochemical Cycles

LEARNING OBJECTIVES

Life is composed of many chemical elements, which exist in specific amounts, concentrations, forms, and ratios to one another. If these conditions are not met, then life is limited. The study of chemical availability and biogeochemical cycles is important to the solution of many environmental problems. After reading this chapter, you should understand:

- What are the major biogeochemical cycles.
- What are the major factors and processes that control biogeochemical cycles.
- Why some chemical elements cycle quickly and some slowly.
- How each major component of Earth's global system (the atmosphere, water, solid surfaces, and life) is involved and linked with biogeochemical cycles.
- How the biogeochemical cycles most important to life, especially the carbon cycle, generally operate.
- How humans affect biogeochemical cycles.

- *The story of Canada's shrinking glaciers illustrates the importance of understanding how materials cycle in ecosystems. The story confirms several key themes from Chapter 1. As the global human population has increased, so have emissions of greenhouse gases and the climate changes they cause. The scientific study of the glaciers and the growing understanding of their place in the hydrologic cycle reflect the value the people of Canada place on their landscape and freshwater resources. Recognizing the need for early planning and predictive tools, glacier scientists developed a model that helped them understand the changes in Canada's glaciers. Their work provides an important foundation for the development of appropriate strategies to manage future climate change.*

Figure 5.1 • Images of Canada's glaciers

(a) Photograph of an iceberg calving off a glacier.

(b) Simplified mass balance diagram for an Arctic glacier.

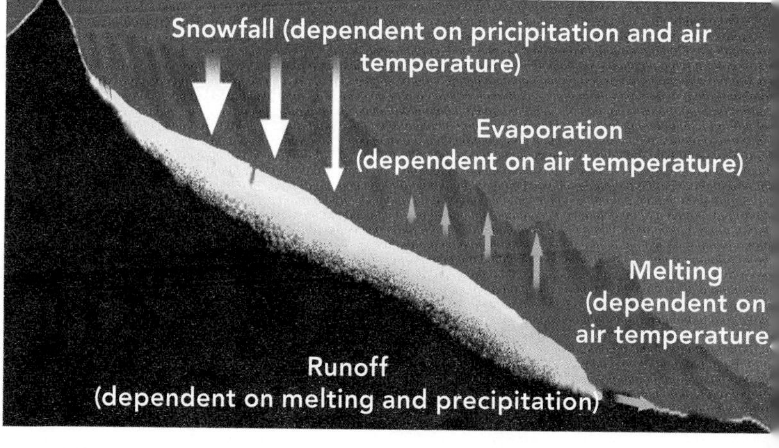

5.1 How Chemicals Cycle

Earth is a particularly good planet for life from a chemical point of view. The atmosphere contains plenty of oxygen, which humans and other aerobic organisms need to breathe, and water needed to sustain life. In many places, soils are fertile, containing the chemical elements necessary for plants to grow; and Earth's bedrock contains valuable metals and fuels. Of course, some parts of the Earth's surface are not perfect for life—deserts with little water, chemical deserts (such as the middle regions of the oceans) where the necessary nutrients necessary are not abundant, and certain soils in which some chemical elements required for life are lacking or others toxic to life are present.[2] Scientific questions reflecting our values for a quality environment are: What kinds of chemical processes benefit or harm the environment, ourselves, and other life forms? How can we manage chemicals in the environment to improve and sustain ecosystems from the local to the global level? To answer these questions, we need to know how chemical elements cycle, and this is our starting point.

Biogeochemical Cycles

The term chemical refers here both to an element, such as carbon (C) or phosphorus (P), and to a compound, such as water (H_2O). (See A Closer Look 5.1 for a discussion of matter and energy.) A **biogeochemical cycle** is the complete path a chemical takes through the four major components, or reservoirs, of Earth's system: atmosphere, hydrosphere (oceans, rivers, lakes, groundwater, and glaciers), lithosphere (rocks and soils), and biosphere (plants and animals). A biogeochemical cycle is *chemical* because it is chemicals that are cycled, *bio-* because the cycle involves life, and *geo-* because a cycle may include atmosphere, water, rocks, and soils.

For example, consider an atom of carbon (C) in carbon dioxide (CO_2) emitted from burning coal (which is made up of fossilized plants hundreds of millions of years old). The carbon atom is released into the atmosphere and then taken up by a plant and incorporated into a seed. A mouses eats the seed. A coyote eats the mouse, and the carbon atom is expelled following digestion as scat onto the soil. Decomposition of the scat allows our carbon atom to enter the atmosphere again. It may also enter another organism, such as an insect, which uses the scat as a resource.

Chemical Reactions

It is important to acknowledge in our discussion of how chemical cycles work that the emphasis is on chemistry. Many chemical reactions occur within and between the living and non-living portions of ecosystems. A **chemical reaction** is a process in which new chemicals are formed from elements and compounds that undergo a chemical change. For example, a simple reaction between rainwater (H_2O) and carbon dioxide (CO_2) in the atmosphere produces weak carbonic acid (H_2CO_3):

$$H_2O + CO_2 \rightarrow H_2CO_3$$

This weak acid reacts with earth materials, such as rock and soil, to release chemicals into the environment. The released chemicals include calcium, sodium, magnesium, and sulphur, with smaller amounts of heavy metals such as lead, mercury, and arsenic. The chemicals appear in various forms such as compounds and ions in solution.

Many other chemical reactions determine whether chemicals are available to life. For example, photosynthesis is a series of chemical reactions by which living green plants, with sunlight as an energy source, convert carbon dioxide (CO_2) and water (H_2O) to sugar ($C_6H_{12}O_6$) and oxygen (O_2). The general chemical reaction for photosynthesis is

$$6CO_2 + 6H_2O \xrightarrow{\text{sunlight}} C_6H_{12}O_6 + 6O_2$$

Photosynthesis produces oxygen as a by-product, and that is why we have free oxygen in our atmosphere. We return to the topic of photosynthesis later in this chapter (A Closer Look 5.3).

After considering the two chemical reactions and applying critical thinking, you may recognize that both reactions combine water and carbon dioxide, but the products are very different: carbonic acid in one combination and a sugar in the other. How can this be so? The answer lies in an important difference between the simple reaction in the atmosphere to produce carbonic acid, and the production of sugar and oxygen in the series of reactions of photosynthesis. Green plants use the energy from the sun, which they absorbed through the chemical chlorophyll. Thus, active solar energy is converted to a stored chemical energy in sugar.

Perhaps the simplest way to think of a biogeochemical cycle is as a "box-and-arrow" diagram, which shows where a chemical is stored and the pathways along which it is transferred from one storage place to another. (See Figure 5.3 and A Closer Look 5.2.) We can consider biogeochemical cycles on any spatial scale of interest to us, from a single ecosystem to the whole Earth. Considering such a cycle from a global perspective is often useful. The problem of potential global warming, for example, calls for an understanding of the cycling of carbon transferred into and out of Earth's atmosphere. Sometimes it is useful to consider a cycle at a more local level, such as in the John Evans Glacier case study. The key that unifies all these cycles is the involvement of the four principal components of the Earth system: lithosphere, atmosphere, hydrosphere, and biosphere. By their nature, chemicals in these four major components have different average times of storage and exist in various physical forms and chemical availability. Overall, the average residence time of chemicals is long in rocks, short in the atmosphere, and intermediate in the hydrosphere and biosphere.

5.2 Environmental Questions and Biogeochemical Cycles

With a general idea of how chemicals cycle, we next consider some of the environmental questions that the science of biogeochemical cycles can help answer. These questions include the following:

Atmospheric Questions

- What determines the concentrations of elements and compounds in the atmosphere?
- Where the atmosphere is polluted as the result of human activities, how might we alter a biogeochemical cycle to lower the pollution?

Hydrologic Questions

- What determines whether a body of water will be biologically productive?
- When a body of water becomes polluted, how can we alter biogeochemical cycles to reduce the pollution and its effects?

Biological Questions

- What factors, including chemicals necessary for life, place limits on the abundance and growth of organisms and their ecosystems?
- What toxic chemicals might be present that adversely affect the abundance and growth of organisms and their ecosystems?
- How can people improve the production of a desired biological resource?
- What are the sources of chemicals required for life, and how might we make these more readily available?
- What problems occur when a chemical is too abundant, as in the case of artificial loading of phosphorus in the Experimental Lakes Area (Chapter 2)?

Geologic Questions

- What physical and chemical processes control the movement and storage of chemical elements in the environment?
- How are chemical elements transferred from solid earth to the water, atmosphere, and life forms?
- How does the long-term storage of chemical elements (for thousands of years or longer) in rocks and soils affect ecosystems on a local to global scale?

5.3 Biogeochemical Cycles and Life: Limiting Factors

All living things are made up of chemical elements, but of the 103 known elements, only 24 are required for biological processes. These 24 are divided into the **macronutrients**, elements required in large amounts by all life, and **micronutrients**, elements required either in small amounts by all life or in moderate amounts by some forms of life and not at all by others.

The macronutrients, in turn, include the "big six," the elements that form the fundamental building blocks of life. These are carbon, hydrogen, nitrogen, oxygen, phosphorus, and sulphur. Each element plays a special role in organisms. Carbon is the basic building block of organic compounds. Along with oxygen and hydrogen, carbon forms carbohydrates. Nitrogen, along with these other three, makes proteins. Phosphorus is the "energy element"; it occurs in the compounds called adenosine triphosphate (ATP) and adenosine diphosphate (ADP), important in the transfer and use of energy within cells.

In addition to the "big six," other macronutrients also play important roles. Calcium, for example, is the structure element, occurring in bones of vertebrates, shells of shellfish, and wood-forming cell walls of vegetation. Sodium and potassium are important to nerve signal transmission. Many metals required by living things are necessary for specific enzymes. (An enzyme is a complex organic compound that acts as a catalyst—it causes or speeds up chemical reactions such as digestion.)

For any life form to persist, chemical elements must be available at the right times, in the right amounts, in the right chemical form, and in the right concentrations relative to each other. When this does not happen, a chemical can become a **limiting factor**, preventing the growth of an individual, population, or species, or even causing its local extinction. Limiting factors were discussed in Chapter 4 and are discussed in Chapter 11. Chemical elements may also be toxic to some life forms and ecosystems. Mercury, for example, is toxic even in low concentrations. Copper and other elements are required in low concentrations for life processes but are toxic when present in high concentrations.

5.4 General Concepts Central to Biogeochemical Cycles

Although there are as many biogeochemical cycles as there are chemicals, certain general concepts hold true for these cycles.

- Some chemical elements cycle quickly and are readily regenerated for biological activity. Oxygen and nitrogen are among these. Typically, these elements have a

A CLOSER LOOK 5.1

Matter and Energy

The universe as we know it consists of two entities: energy and matter. Energy is the ability to do work. Matter is the material that makes up our physical and biological environments (you are composed of matter). The first law of thermodynamics—also known as the law of conservation of energy or the first energy law—states that energy cannot be created or destroyed but can change from one form to another. This law stipulates that the total amount of energy in the universe does not change.

Our sun produces energy through nuclear reactions at high temperatures and pressures that change mass (a measure of the amount of matter) into energy. At first glance, this may seem to violate the law of conservation of energy. However, this is not the case, because energy and matter are interchangeable. Albert Einstein first described the equivalence of energy and mass in his famous equation $E = mc^2$, where E is energy, m is mass, and c is the velocity of light in a vacuum, such as outer space (approximately 300,000 km/s). Because the velocity of light squared is a very large number, even a small amount of change in mass produces very large amounts of energy.[3]

Energy, then, may be thought of as an abstract mathematical quantity that is always conserved. This means that it is impossible to get something for nothing when dealing with energy; it is impossible to extract more energy from any system than the amount of energy that originally entered the system. In fact, the second law of thermodynamics states that you cannot "break even." When energy is changed from one form to another, it always moves from a more useful form to a less useful one. Thus, as energy moves through any system and is changed from one form to another, energy is conserved, but it becomes less useful. We return to the two energy laws when we discuss energy in ecosystems (Chapter 9) and in modern society (Chapter 16).

Next, we discuss a brief introduction to the basic chemistry of matter, which will help you understand biogeochemical cycles. An atom is the smallest part of a chemical element that can take part in a chemical reaction with another atom. An element is a chemical substance composed of identical atoms that cannot be separated by ordinary chemical processes into different substances. Each element is given a symbol. For example, the symbol for the element carbon is C, and that for phosphorus is P.

A model of an atom (Figure 5.2) shows three subatomic particles: neutrons, protons, and electrons. The atom is visualized as a central nucleus composed of neutrons with no electrical charge and protons with positive charge. A cloud of electrons, each with a negative charge, revolves about the nucleus. The number of protons in the nucleus is unique for each element and is the atomic number for that element. For example, hydrogen, H, has one proton in its nucleus, and its atomic number is 1. Uranium has 92 protons in its nucleus, and its atomic number is 92. A list of known elements with their atomic numbers, called the Periodic Table, is shown in the Appendix of this text .

Electrons in our model of the atom are arranged in shells (representing energy levels), and the electrons closest to the nucleus are bound tighter to the atom than those in the outer shells. Electrons have negligible mass compared with neutrons or protons; therefore, nearly the entire mass of an atom is in the nucleus.

The sum of the number of neutrons and protons in the nucleus of an atom is known as the atomic weight. Atoms of the same element always have the same atomic number (the same number of protons in the nucleus), but they can have different numbers of neutrons and, therefore, different atomic weights. Two atoms of the same element with different numbers of neutrons in their nuclei and different atomic weights are known as isotopes of that element. For example, two isotopes of oxygen are ^{16}O) and ^{18}O, where 16 and 18 are the atomic weights. Both isotopes have an atomic number of 8, but ^{18}O has two more neutrons than ^{16}O. Such information is proving very useful in studying how the earth works. For

Figure 5.2 • The structure of an atom. *(a)* Conceptual diagram showing the basic structure of an atom as a nucleus surrounded by a cloud of electrons. *(b)* Conceptualized model of an atom of carbon with six protons and six neutrons in the nucleus and six orbiting electrons in two energy shells. *(c)* Three-dimensional view of *(b)*. Size of nucleus relative to size of electron shells is greatly exaggerated. [*Source:* After F. Press, and R. Siever *Understanding Earth* (New York: Freeman, 1994).]

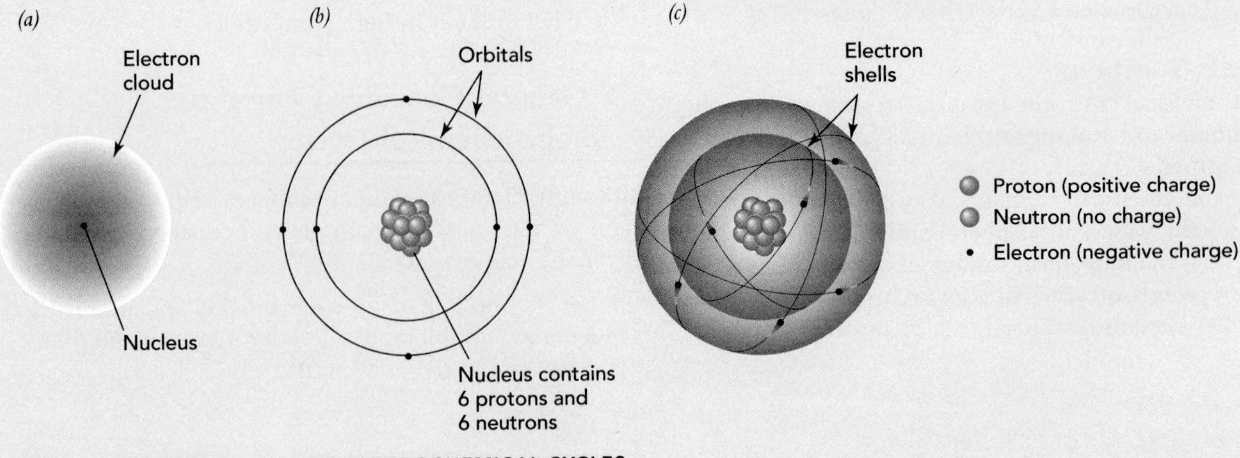

example, the understanding of oxygen isotopes has resulted in a better understanding of how the global climate has changed. This topic is beyond the scope of our present discussion but can be found in many basic textbooks on oceanography.

An atom is chemically balanced in terms of electric charge when the number of protons in the nucleus is equal to the number of electrons. However, an atom may lose or gain electrons, changing the balance in the electrical charge. An atom that has lost or gained electrons is called an ion. An atom that has lost one or more electrons has a net positive charge and is called a cation. For example, the potassium ion K^+ has lost one electron, and the calcium ion Ca^{2+} has lost two electrons. An atom that has gained electrons has a net negative charge and is called an anion. For example, O^{2-} is an anion of oxygen that has gained two electrons.

A compound is a chemical substance composed of two or more atoms of the same or different elements. The smallest unit of a compound is a molecule. For example, each molecule of water, H_2O, contains two atoms of hydrogen and one atom of oxygen held together by chemical bonds. Minerals that form rocks are compounds, as are most chemical substances found in a solid, liquid, or gaseous state in the environment.

The atoms that constitute a compound are held together by chemical bonding. The four main types of chemical bonds are covalent, ionic, Van der Waals, and metallic. It is important to recognize that when talking about chemical bonding of compounds, we are dealing with a complex subject. Although some compounds have a particular type of bond, many other compounds have more than one type of bond. With this caveat in mind, let us define each type of bond. Covalent bonds result when atoms share electrons. This sharing takes place in the region between the atoms, and the strength of the bond is related to the number of pairs of electrons that are shared. Some important environmental compounds are held together solely by covalent bonds. These include carbon dioxide (CO_2) and water (H_2O). Covalent bonds are stronger than ionic bonds, which form as a result of attraction between cations and anions. An example of an environmentally important compound with ionic bonds is table salt (mineral halite), or sodium chloride ($NaC1$). Compounds with ionic bonds such as sodium chloride tend to be soluble in water and thus dissolve easily, as salt does. Van der Waals bonds are weak forces of attraction between molecules that are not bound to each other. Such bonding is much weaker than either covalent or ionic bonding. For example, the mineral graphite (which is the "lead" in pencils) is black and consists of sheets of carbon atoms that easily part or break from one another because the bonds are of the weak Van der Waals type. Finally, metallic bonds are those in which electrons are shared, as with covalent bonds. However, they differ because in metallic bonding the electrons are shared by all atoms of the solid rather than by specific atoms. As a result, the electrons can flow. For example, the mineral and element gold is an excellent conductor of electricity and pounds easily into thin sheets because the electrons have the freedom of movement that is characteristic of metallic bonding.

In summary, in the study of the environment, we are concerned with energy and matter (chemicals) that moves in and between the major components of the Earth system. An example is the element carbon, which moves through the atmosphere, hydrosphere, lithosphere, and biosphere in a large variety of compounds. These include carbon dioxide (CO_2) and methane (CH_4), which are gases in the atmosphere; sugar ($C_6H_{12}O_6$) in plants and animals; and complex hydrocarbons (compounds of hydrogen and carbon) in coal and oil deposits.

gas phase and are present in Earth's atmosphere, and/or they are easily dissolved in water and are carried by the hydrologic cycle.

- Other chemical elements are easily tied up in relatively immobile forms and are returned slowly, by geological processes, to where they can be reused by life. Typically, these elements lack a gas phase and are not found in significant concentrations in the atmosphere. They also are relatively insoluble in water and are more likely to end up as deep-ocean sediment. Phosphorus is an example of this kind of chemical.
- Since life evolved, it has greatly altered biogeochemical cycles, and this alteration has changed our planet in many ways, such as in the development of the fertile soils on which agriculture depends.
- The continuation of processes that control biogeochemical cycles is essential to the long-term maintenance of life on Earth.
- Through modern technology, we have begun to transfer chemical elements among air, water, and soil at rates and in amounts comparable to natural processes. These transfers can benefit society, as when they improve crop production, but they can also pose environmental dangers, as illustrated by the opening case study. To live in the manner we currently and historically have known, we must recognize the full range of consequences of altering biogeochemical cycles.

Discussion of biogeochemical cycles beyond the general concepts listed above requires an understanding of geologic and hydrologic cycles. Of particular importance are geologic processes linked to cycling of chemicals in the biosphere.

A CLOSER LOOK 5.2

A Biogeochemical Cycle

The simplest way to visualize a biogeochemical cycle is as a box-and-arrow diagram, where the boxes represent places where a chemical is stored (called storage compartments) and the arrows represent pathways of transfer. The rate of transfer or flux is the amount of chemical per unit time that enters or leaves a storage compartment. When a chemical enters a storage compartment from another compartment, we say that the receiving compartment is a sink. For example, the forests of the world (which are a storage compartment for carbon) may function as a sink for carbon from the atmosphere, sequestering it in wood, leaves, and roots. The amount of carbon transferred from the atmosphere to the forest on a global basis is the flux, which can be measured in units, such as billions of tonnes of carbon per year.

A biogeochemical cycle is generally drawn for a single chemical element, but sometimes it is drawn for a compound—for example, water (H_2O). Shown are the basic elements of a biogeochemical cycle for water, which represents three parts of the hydrological cycle. Water is stored temporarily in a lake (compartment B). It enters the lake from the atmosphere (compartment A) as precipitation and from the land around the lake as runoff (compartment C). It leaves the lake through evaporation to the atmosphere or runoff to a surface stream or infiltration to subsurface flows.

In each compartment, we can identify an average length of time that an atom is stored before it is transferred. This is called the residence time.

As an example, consider a salt lake with no transfer out except by evaporation. Assume that the lake contains 3,000,000 m^3 of water and the evaporation is 3,000 m^3/day. Surface runoff into the lake is also 3,000 m^3/day, so the volume of water in the lake remains constant. We can calculate the average residence time of the water in the lake as the volume of the lake divided by the evaporation rate, or 3,000,000 m^3 divided by 3,000 m^3/day, which is 1,000 days (or 2.7 years).

Another crucial aspect of a biogeochemical cycle is the set of factors or processes that control the flow from one compartment to another. To understand a biogeochemical cycle, these factors and processes should be quantified and understood. For example, understanding how air temperature and wind velocity vary across a lake is important in understanding the rate of evaporation of water from the lake.

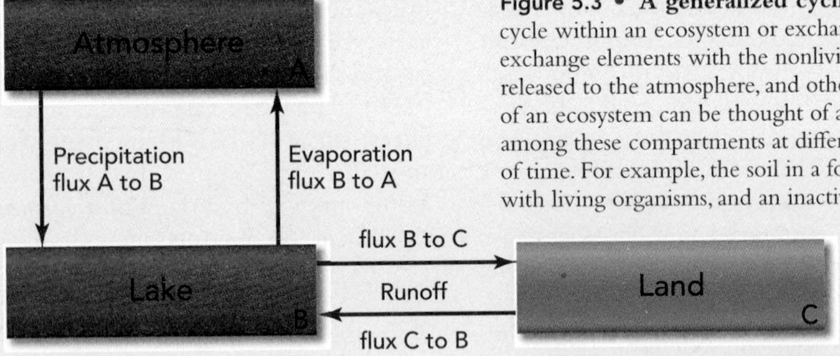

Figure 5.3 • A generalized cycling of a chemical in an ecosystem. Chemical elements cycle within an ecosystem or exchange between an ecosystem and the biosphere. Organisms exchange elements with the nonliving environment; some elements are taken up from and released to the atmosphere, and others are exchanged with water and soil or sediments. The parts of an ecosystem can be thought of as storage compartments for chemicals. The chemicals move among these compartments at different rates and remain within them for different average lengths of time. For example, the soil in a forest has an active part, which rapidly exchanges elements with living organisms, and an inactive part, which exchanges elements slowly (as shown in the lower part of the diagram). Generally, life benefits if needed chemicals are kept within the ecosystem and are not lost through geologic processes, such as erosion, that remove them from the ecosystem.

Annual rates of transfer from the storage compartments in the hydrologic cycle are shown in Figure 5.8. These rates of transfer define a global water budget. If we sum the arrows going up in the figure, and then sum the arrows going down, we find that the two sums are the same, 577,000 km^3/year. Similarly, precipitation on land (119,000 km^3/year) is balanced by evaporation from land plus surface and subsurface runoff.

Especially important from an environmental perspective is that rates of transfer on land are small relative to what's happening in the ocean. For example, most of the water that evaporates from the ocean falls again as precipitation into the ocean. On land, most of the water that falls as precipitation comes from evaporation of water from land. This means that regional land-use changes, such as the building of large dams and reservoirs, can change the amount of water evaporated into the atmosphere and change the location and amount of precipitation on land—water we depend on to raise our crops and supply water for our urban environments. One of the most important implications of urbanization is that by paving and roofing large areas, we reduce the ability of rainwater to infiltrate into soils, and from there into groundwater (aquifers). Therefore, urbanization can have a significant effect on groundwater recharge potential. Furthermore, as we pave over large areas of land in cities, storm water runs off more quickly and in greater volume, thereby increasing flood hazards. Bringing water into semi-arid cities by pumping groundwater or transporting water from distant mountains through aqueducts may increase evaporation, thereby increasing humidity and precipitation in a region.

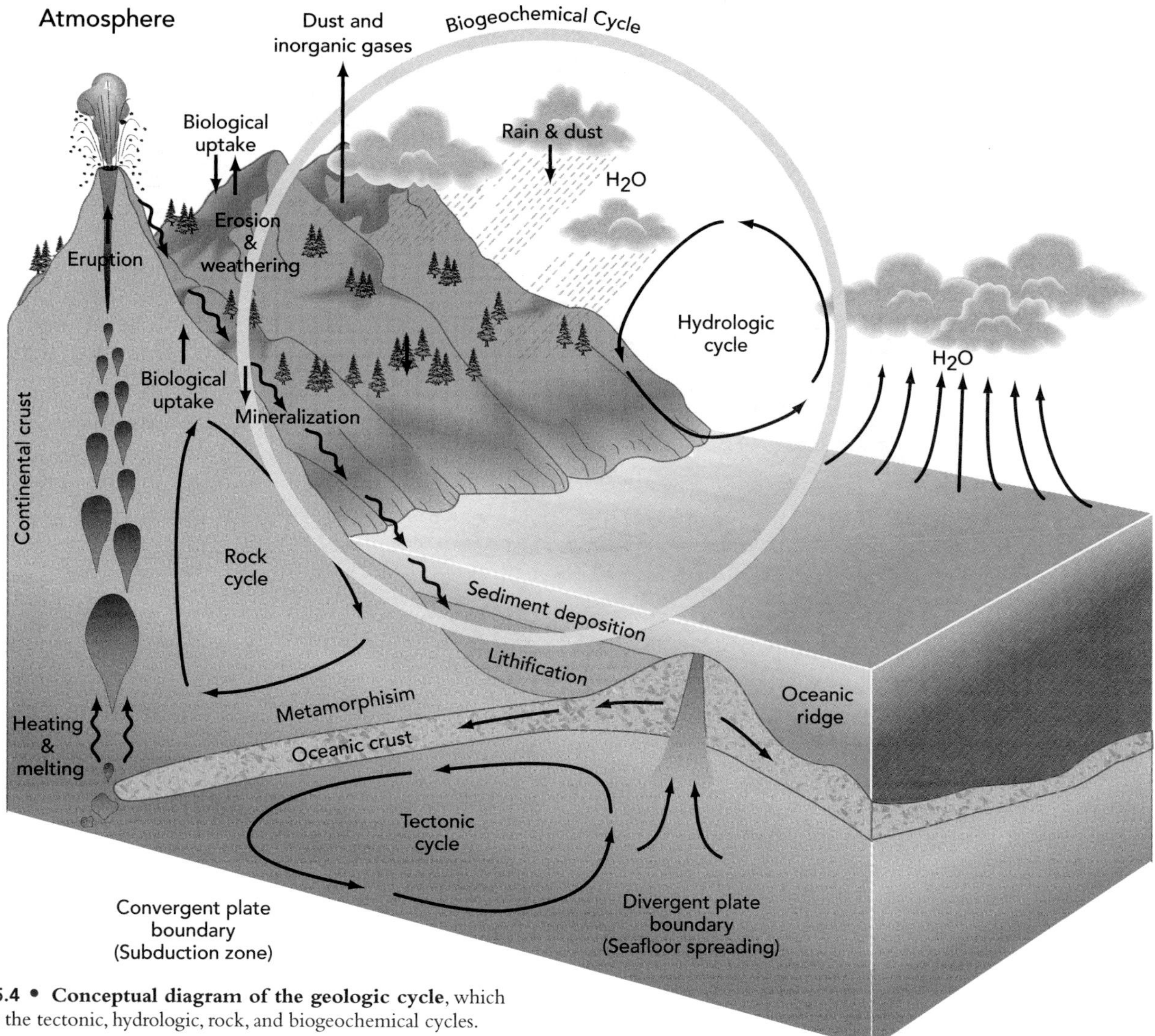

Figure 5.4 • **Conceptual diagram of the geologic cycle**, which includes the tectonic, hydrologic, rock, and biogeochemical cycles.

plates are geologically active areas, and most volcanic activity and earthquakes occur there. Earthquakes occur when the brittle upper lithosphere fractures along faults. Movement of several meters between plates can occur within a few seconds or minutes, in contrast to the slow, deeper plate movement described above.

Three types of plate boundaries occur: divergent, convergent, and transform faults.

- A divergent plate boundary occurs at a spreading ocean ridge, where plates are moving away from one another and new lithosphere is produced. This process, known as seafloor spreading, produces ocean basins.
- A convergent plate boundary occurs when plates collide. When a plate composed of relatively heavy ocean basin rocks dives, or subducts, beneath the leading edge of a plate composed of lighter continental rocks, a subduction zone is present. Such a convergence may produce linear coastal mountain ranges, such as the Andes in South America. When two plates, each composed of lighter continental rocks, collide, a continental mountain range such as the Himalayas in Asia may form.[4, 5]
- A transform fault boundary occurs where one plate slides past another. An example is the San Andreas Fault in California, which is the boundary between the North American and Pacific plates. The Pacific plate is moving north relative to the North American plate at about 5 cm/year. As a result, Los Angeles is moving slowly toward San Francisco about 500 km north. If this motion continues, in about 10 million years, San Francisco will be a suburb of Los Angeles.

The Hydrologic Cycle

The **hydrologic cycle** (Figure 5.6) is the transfer of water from the oceans to the atmosphere to the land and back to the oceans. The processes involved include evaporation of water from the oceans; precipitation on land; evaporation from land; and runoff from streams, rivers, and subsurface groundwater. The hydrologic cycle is driven by solar

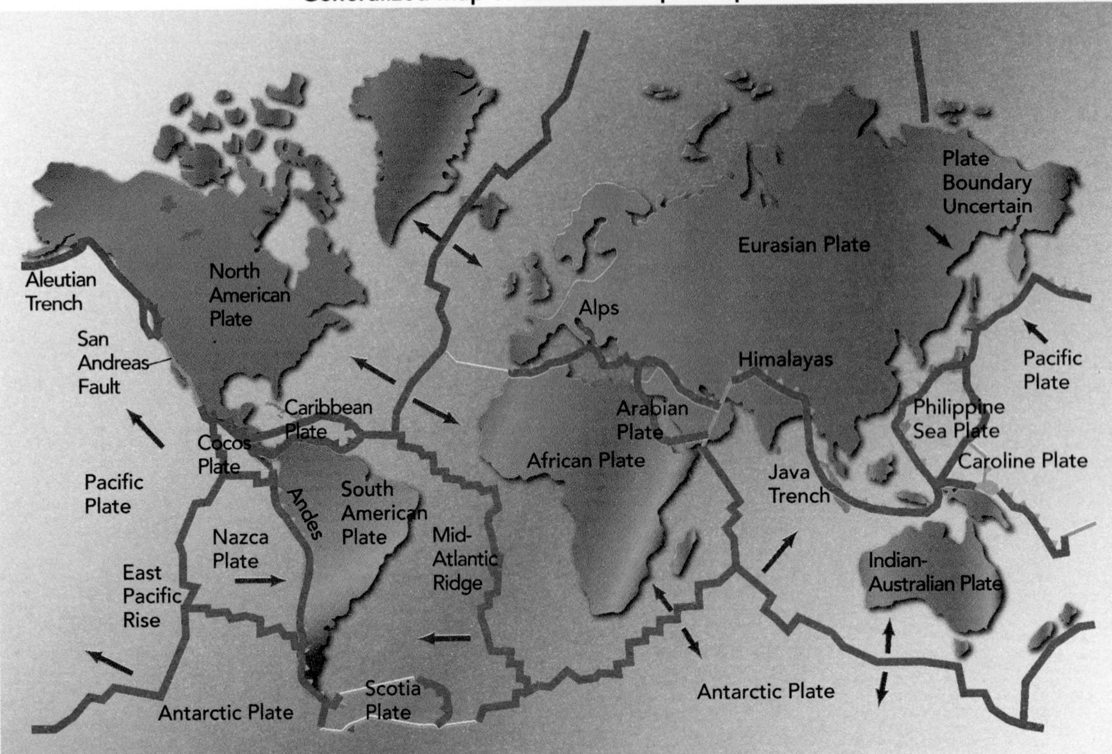

Figure 5.5 • Generalized map of Earth's lithospheric plates. Divergent plate boundaries are shown as heavy lines (for example, the Mid-Atlantic ridge). Convergent boundaries are shown as barbed lines (for example, the Aleutian trench). Transform fault boundaries are shown as thinner yellow lines (for example, the San Andreas Fault). Arrows indicate directions of relative plate motions. [*Source*: Modified from B. C. Birchfiel, R. J. Foster, E. A. Keller, W. N. Melhorn, D. G. Brookins, L. W. Mintz, and H. V. Thurman, *Physical Geology: The Structures and Processes of the Earth* (Columbus, Ohio: Merrill, 1982).]

energy, which evaporates water from oceans, freshwater bodies, soils, and vegetation. Of the total 1.3 billion km^3 of water on Earth, about 97% is in oceans and about 2% is in glaciers and ice caps. The rest is in freshwater on land and in the atmosphere. Although it represents only a small fraction of the water on earth, the water on land is important in moving chemicals, sculpting landscape, weathering rocks, transporting sediments, and providing our water resources. The water in the atmosphere—only 0.001% of the total on Earth—cycles quickly to produce rain and runoff for our water resources.

Annual rates of transfer from the storage compartments in the hydrologic cycle are shown in Figure 5.8. These rates of transfer define a global water budget. If we sum the arrows going up in the figure, and then sum the arrows going down, we find that the two sums are the same, 577,000 km^3/year. Similarly, precipitation on land (119,000 km^3/year) is balanced by evaporation from land plus surface and subsurface runoff.

Especially important from an environmental perspective is that rates of transfer on land are small relative to what's happening in the ocean. For example, most of the water that evaporates from the ocean falls again as precipitation into the ocean. On land, most of the water that falls as precipitation comes from evaporation of water from land. This means that regional land-use changes, such as the building of large dams and reservoirs, can change the amount of water evaporated into the atmosphere and change the location and amount of precipitation on land—water we depend on to raise our crops and supply water for our urban environments. One of the most important implications of urbanization is that by paving and roofing large areas, we reduce the ability of rainwater to infiltrate into soils, and from there into groundwater (aquifers). Therefore, urbanization can have a significant effect on groundwater recharge potential. Furthermore, as we pave over large areas of land in cities, storm water runs off more quickly and in greater volume, thereby increasing flood hazards. Bringing water into semi-arid cities by pumping groundwater or transporting water from distant mountains through aqueducts may increase evaporation, thereby increasing humidity and precipitation in a region.

Figure 5.6 shows that approximately 60% of water that falls on land each year from precipitation evaporates into the atmosphere. A smaller component (about 40%) returns to the ocean as surface and subsurface runoff. This small annual transfer of water supplies resources for rivers, groundwater, and urban and agricultural lands. Unfortunately, distribution of water on land is far from uniform. As a result, water shortages occur in some areas. As human population increases, water

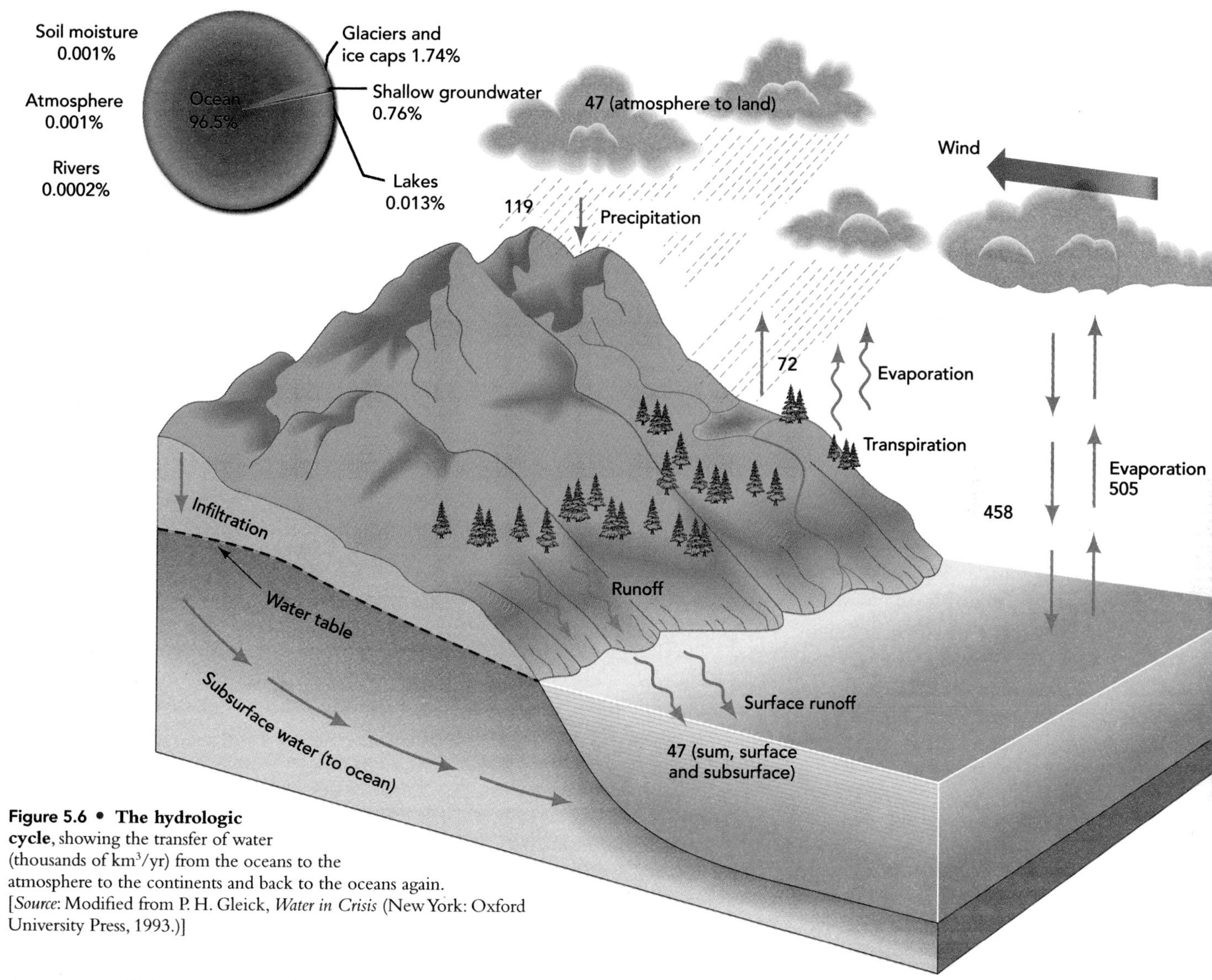

Figure 5.6 • The hydrologic cycle, showing the transfer of water (thousands of km^3/yr) from the oceans to the atmosphere to the continents and back to the oceans again. [*Source*: Modified from P. H. Gleick, *Water in Crisis* (New York: Oxford University Press, 1993.)]

shortages will become more frequent in arid and semiarid regions, where water is naturally scarce.

At the regional and local level, the fundamental hydrologic unit of the landscape is the **drainage basin** (also called a watershed or catchment). A drainage basin is the area that contributes surface runoff to a particular stream or river. The term drainage basin is usually used in evaluating the hydrology of an area, such as the stream flow or runoff from hill slopes. Drainage basins vary greatly in size, from less than a hectare to millions of square kilometers. A drainage basin is usually named for its main stream or river, such as the St. Lawrence or Mackenzie River drainage basin.

The Rock Cycle

The **rock cycle** consists of numerous processes that produce rocks and soils. The rock cycle depends on the tectonic cycle for energy and on the hydrologic cycle for water. As shown in Figure 5.7 rock is classified as igneous, sedimentary, or metamorphic.

These types of rock are involved in a worldwide recycling process. Internal heat from the tectonic cycle produces igneous rocks from molten material near the surface, such as lava from volcanoes. These new rocks weather when exposed at the surface. The freezing of water in cracks of rocks creates a physical weathering process. The water expands when it freezes, breaking the rocks apart. Physical weathering makes smaller particles of rocks from bigger ones, producing sediment such as gravel, sand, and silt. Chemical weathering occurs when the weak acids in water dissolve chemicals from rocks. The sediments and dissolved chemicals are then transported by water, wind, or ice (glaciers).

Weathered materials accumulate in depositional basins such as the oceans. Sediments in these basins are compacted by overlying sediments and converted to sedimentary rocks by compaction and the cementing together of particles. After sedimentary rocks are buried to sufficient depths (usually tens to hundreds of kilometers), they may be altered by heat, pressure, or chemically active fluids and transformed to metamorphic rocks. Deeply buried rocks may be transported to the surface by an uplift associated with plate tectonics and then subjected to weathering.

You can see on Figure 5.7 that life processes play an important role in the rock cycle through the formation of organic carbon-containing rocks like coal (described in Chapter 17) and limestone, which is mostly a calcium carbonate (the material of seashells and bones).

Plate tectonic processes of uplift and subsidence of rocks, along with erosion, produce the Earth's varied topography. The spectacular Marble Canyon of Kootenay National Park in British Columbia (Figure 5.8), sculpted from mostly sedimentary rocks, is one example. Another is the beautiful Tower Karst in China (Figure 5.9); these resistant blocks of limestone have survived chemical weathering and erosion that removed the surrounding rocks. Impressive karst caves occur in several regions of Canada, including central Manitoba between Lakes Winnipeg and Manitoba.

Our discussion of geologic cycles has emphasized tectonic, hydrologic, and rock-forming processes. We can now begin to integrate biogeochemical processes into the picture.

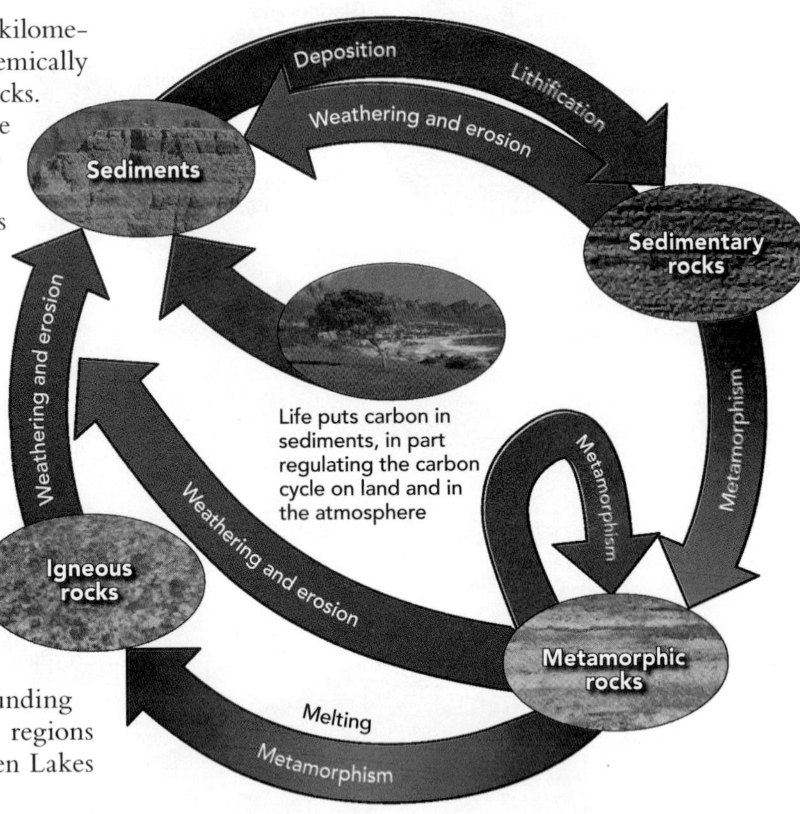

Figure 5.7 • The rock cycle and major paths of material transfer as modified by life.

5.6 Biogeochemical Cycling in Ecosystems

When we ask questions about what chemicals might limit the abundance of a specific individual organism, population, or species, we look for the answer first at the ecosystem level. As explained in Chapter 3, an ecosystem is a community of different species and their nonliving environment in which energy flows and chemicals cycle. The boundaries that we choose for this investigation may be somewhat arbitrary, selected for convenience of measurement and analysis. On land, we often evaluate biogeochemical cycles for a fundamental element of the landscape, usually a drainage basin. Freshwater bodies—lakes, ponds, and bogs—are also convenient landforms for analysis of ecosystems and biogeochemical cycling.

Chemical cycling in an ecosystem begins with inputs from outside. On land, chemical inputs to an ecosystem come from the atmosphere via rain, wind-transported dust (called dry fallout or dry deposition), and volcanic ash from eruptions and from adjoining land via stream flow from flooding and groundwater from springs. Ocean and freshwater ecosystems have the same atmospheric and land inputs (including large, nearshore submarine springs). Ocean ecosystems, in

Figure 5.8 • In response to slow tectonic uplift of the region, Tokumm Creek has eroded through the sedimentary rocks of Marble Canyon in Kootenay Park, British Columbia, to produce spectacular scenery, as shown in this photograph.

addition, have inputs from ocean currents and submarine vents (hot springs) at divergent plate boundaries.

Chemicals cycle internally within an ecosystem through air, water, rocks, soil, and food chains by way of physical transport and chemical reactions. With the death of individual organisms, decomposition through chemical reactions returns chemical elements to other parts of the ecosystem. In addition, living organisms release some chemical elements directly into an ecosystem. Defecation by animals and dropping of fruit by plants are examples.

An ecosystem can lose chemical elements to other ecosystems. For example, rivers transport chemicals from the land to the sea. An ecosystem that has little loss of chemical elements can function in its current condition for longer periods than a "leaky" ecosystem that loses chemical elements rapidly. All ecosystems, however, lose chemicals to some extent. Therefore, all ecosystems require some external inputs of chemicals.

Ecosystem Cycles of a Metal and a Nonmetal

Within an ecosystem, different chemical elements may have very different pathways, as illustrated in Figure 5.10 for calcium and in Figure 5.11 for sulphur. The calcium cycle is typical of a metallic element, and the sulphur cycle is typical of a nonmetallic element.

An important difference between these cycles is that calcium, like most metals, does not form a gas on the Earth's surface. Therefore, calcium is not present as a gas in the atmosphere. In contrast, sulphur forms several gases, including sulphur dioxide (a major air pollutant and component of acid rain; see Chapter 21) and hydrogen sulfide (swamp or rotten egg gas, usually produced biologically).

Figure 5.9 • **This landscape in the People's Republic of China features Tower Karst, steep-sided hills or pinnacles composed of limestone.** The rock has been slowly dissolving through chemical weathering. The pinnacles and hills are remnants of the weathering and erosion processes.

Because sulphur has gas forms, it can be returned to an ecosystem more rapidly than can calcium. Annual input of sulphur from the atmosphere to a forest ecosystem has been measured to be 10 times that of calcium. Therefore, calcium and other elements without a gas phase are more likely to become limiting factors.

Humans have influenced the cycles of various metals, including iron, copper, nickel, zinc, and tin, through mining activities (Chapter 23). These influences typically involve extracting ore (rock containing the desired metal), crushing, grinding, and washing the rock to concentrate the ore, and refining the metal through smelting (heat) or electrochemical processes. With mining and refining of metals, humans increase the transfer of metal from minerals to dissolved form in water, and from water into sediment and (depending on the metal) biological tissue. They also increase the rate of mineral dust (normally a small component) entering the atmosphere. For example, human influences on the global iron cycle include higher rates of dust emissions to the atmosphere and oceans (where iron can be a limiting factor in the growth of organisms), higher inputs of dissolved iron to rivers, lakes and oceans from runoff, contributing to increases in marine plant growth. Realizing that this is a one-directional flow of material is important, at least within geologic time. Iron from the Canadian shield is transported to the deep ocean, where it becomes part of biological tissue and ultimately marine sediments; it is not returned to the Canadian north.

Chemical Cycling and the Balance of Nature

To sustain life indefinitely within an ecosystem, energy must be continuously added, and the store of essential chemicals must not decline. We have already discussed the common belief that, without human interference, life would be sustained indefinitely in a steady state, or "balance of nature." We also discussed the belief that life tends to function to preserve an environment beneficial to itself. Both beliefs presume that chemical elements necessary for life exist in a dynamic steady state within an ecosystem.[6] These beliefs can be rephrased as a scientific hypothesis: Without human disturbance, the net storage of chemical elements within an ecosystem will remain constant over time. However, as noted, inevitably, some fraction of the chemical elements stored in an ecosystem is lost and must be replaced. Are ecosystems ever in a dynamic steady state in regard to chemical elements? Studies indicate that they are not, because rates of input and output of chemicals do not balance, and concentrations of some chemicals decrease over time.

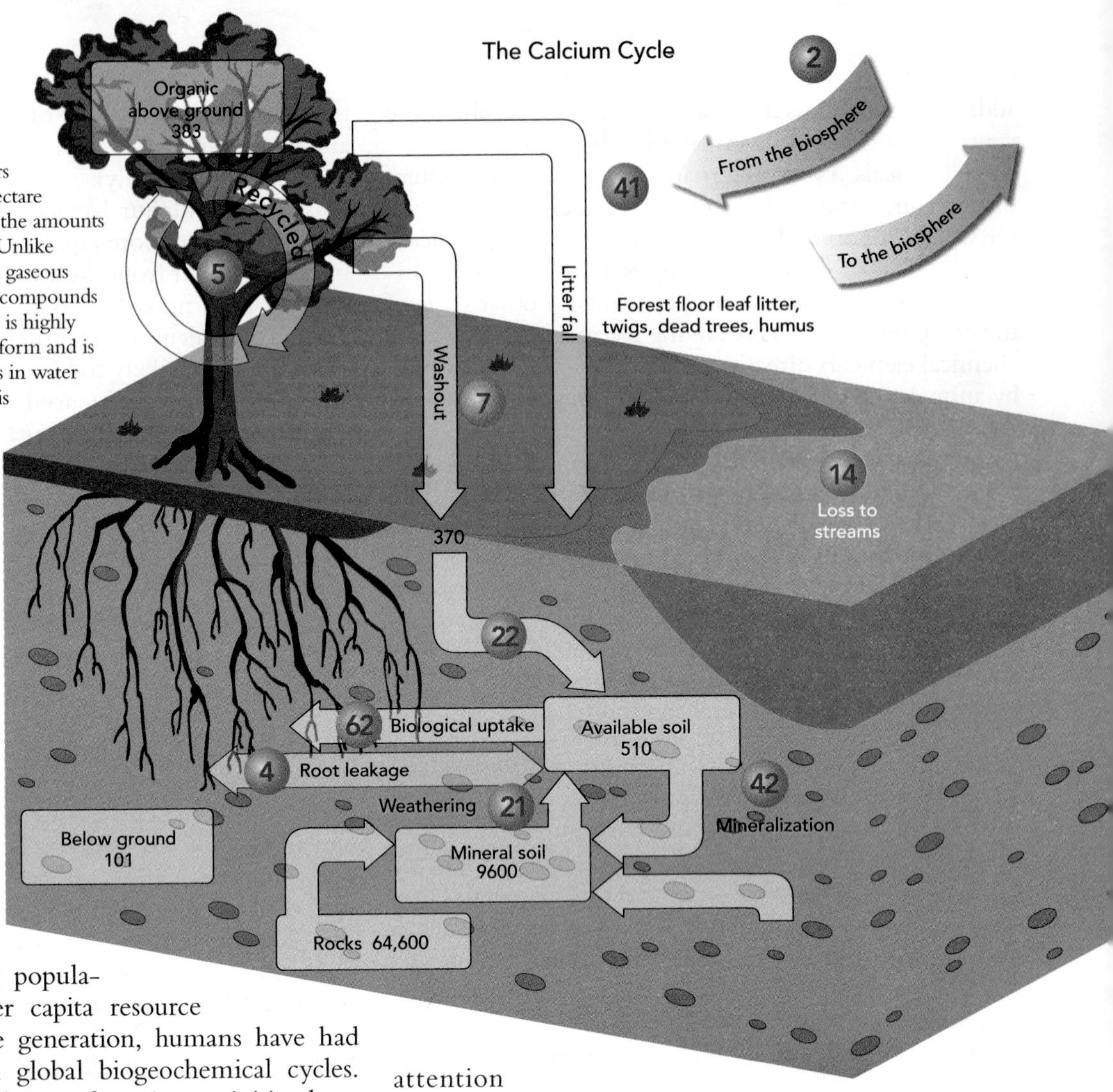

Figure 5.10 • Annual calcium cycle in a forest ecosystem. The circled numbers are flux rates in kilograms per hectare per year. The other numbers are the amounts stored in kilograms per hectare. Unlike sulphur, calcium does not have a gaseous phase, although it does occur in compounds as part of dust particles. Calcium is highly soluble in water in its inorganic form and is readily lost from land ecosystems in water transport. The information in this diagram was obtained from Hubbard Brook Ecosystem. [*Source*: Modified from G. E. Likens, F. H. Bormann, R. S. Pierce, J. S. Eaton, and N. M. Johnson, *The Biogeochemistry of a Forested Ecosystem*, 2nd ed. (New York: Springer-Verlag, 1995).]

With our growing population and increasing per capita resource consumption and waste generation, humans have had widespread impacts on global biogeochemical cycles. Although industrial and manufacturing activities have added a range of chemical contaminants to the natural environment, the most pressing pollution issues may be those related to human and animal wastes. We can now observe marked changes, particularly in the carbon, nitrogen, and phosphorus cycles, as a result of human activities. These points are discussed in more detail in the following section.

5.7 Some Major Global Chemical Cycles

Earlier in this chapter, we asked what chemical elements limit the abundance of life. We pointed out that the chemical elements required by life are divided into two major groups: macronutrients, which are required by all forms of life in large amounts, and micronutrients, which are either required by all forms of life in small amounts or required by only certain life forms. In this section, we consider the global cycles of four macronutrients—carbon, nitrogen, phosphorus, and sulphur. We focus on these in part because they are four of the "big six"—the elements that are the basic building blocks of life. Each is also related to important environmental problems that have attracted attention in the past and will continue to do so in the future. The global oxygen and hydrogen cycles, while central to biological systems, are strongly linked to other global biogeochemical cycles, particularly the carbon cycle. In the following discussion, we will focus on carbon, nitrogen, phosphorus, and sulphur—the elements that are the principal targets of environmental management activities.

The Carbon Cycle

Carbon is the element that anchors all organic substances, from coal and oil to DNA (deoxyribonucleic acid), the compound that carries genetic information. Although of central importance to life, carbon contributes only 0.032% of the mass of the crust, ranking far behind oxygen (45.2%), silicon (29.5%), aluminum (8.0%), iron (5.8%), calcium (5.1%), and magnesium (2.8%.).[7, 8]

The major pathways and storage reservoirs of the **carbon cycle** are shown in Figure 5.12. Notice that carbon has a gaseous phase as part of its cycle. It occurs in the Earth's atmosphere as carbon dioxide (CO_2) and methane (CH_4), both greenhouse gases (see Chapter 20). Carbon

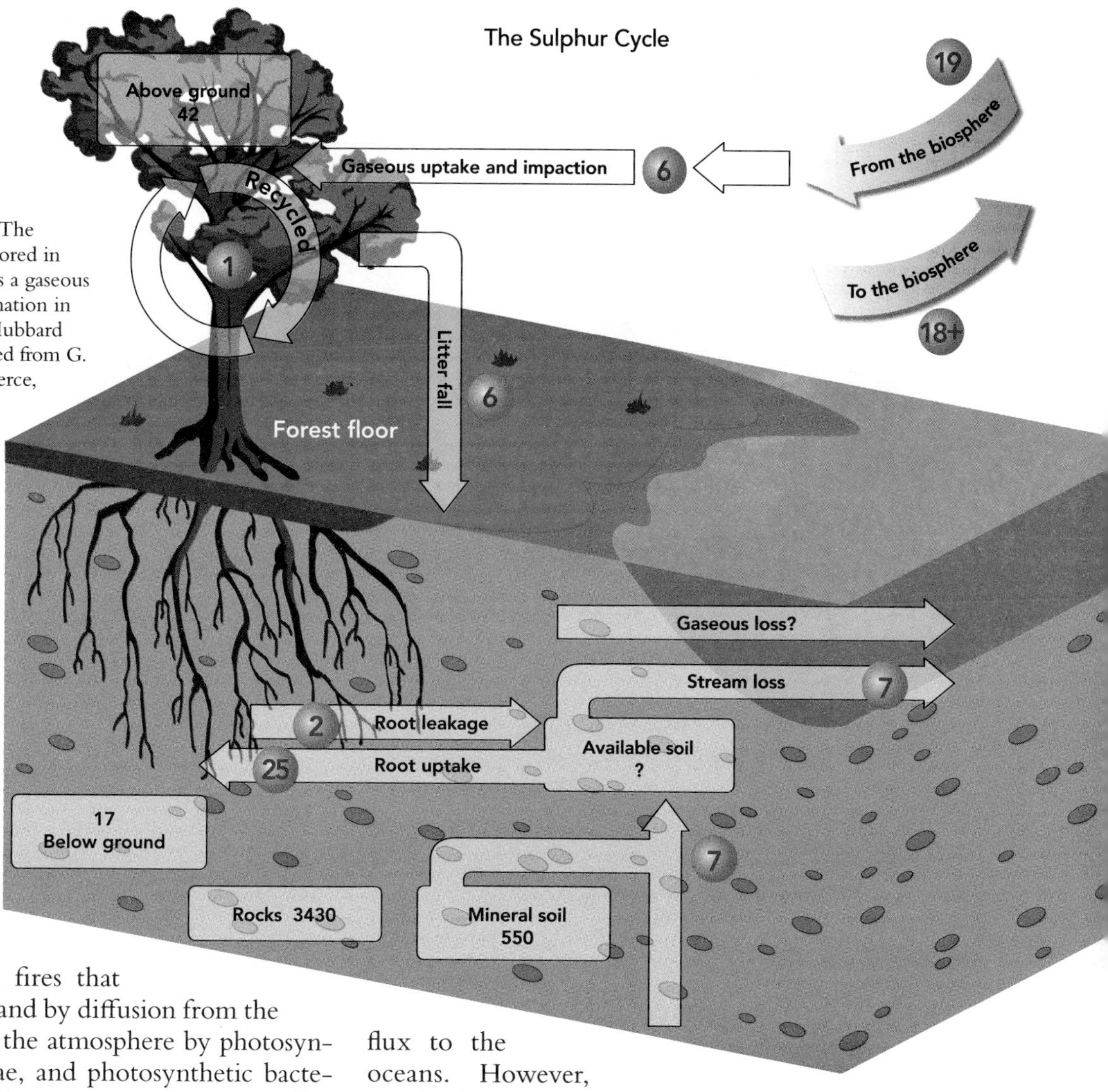

Figure 5.11 • Annual sulphur cycle in a forest ecosystem. The circled numbers are the flux rates in kilograms per hectare per year. The other numbers are the amounts stored in kilograms per hectare. Sulphur has a gaseous phase as H_2S and SO_2. The information in this diagram was obtained from Hubbard Brook Ecosystem. [*Source*: Modified from G. E. Likens, F. H. Bormann, R. S. Pierce, J. S. Eaton, and N. M. Johnson, *The Biogeochemistry of a Forested Ecosystem*, 2nd ed. (New York: Springer-Verlag, 1995).]

enters the atmosphere through the respiration of living things, through fires that burn organic compounds, and by diffusion from the ocean. It is removed from the atmosphere by photosynthesis of green plants, algae, and photosynthetic bacteria, and stored in various long-term repositories called carbon sinks. (See A Closer Look 5.3.) Over the past 3 billion years of Earth's history, the rate at which biological processes have removed carbon dioxide from the atmosphere has exceeded the rate of addition. Therefore, Earth's atmosphere has much less carbon dioxide than would occur on a lifeless Earth.

Carbon occurs in the ocean in several inorganic forms, including dissolved carbon dioxide as well as carbonate (CO_3^{2}) and bicarbonate (HCO_3^-). It also occurs in organic compounds of marine organisms and their products, such as seashells ($CaCO_3$). Carbon enters the ocean from the atmosphere by simple diffusion of carbon dioxide. The carbon dioxide then dissolves and is converted to carbonate and bicarbonate. Marine algae and photosynthetic bacteria obtain the carbon dioxide they use from the water in one of these three forms. Carbon is transferred from the land to the ocean in rivers and streams as dissolved carbon, including organic compounds, and as organic particulates (fine particles of organic matter). Winds also transport small organic particulates from the land to the ocean. Transfer via rivers and streams makes up a relatively small fraction of the total global carbon flux to the oceans. However, on the local and regional scale, input of carbon from rivers to nearshore areas such as deltas and salt marshes, which are often highly biologically productive, is important.

Carbon enters the biota through photosynthesis and is returned to the atmosphere or waters by respiration or by fire. When an organism dies, most of its organic material decomposes to inorganic compounds, including carbon dioxide. Some carbon may be buried where there is not sufficient oxygen to make this conversion possible or where the temperatures are too cold for decomposition. In these locations, organic matter is stored. Over years, decades, and centuries, storage of carbon occurs in wetlands, including parts of floodplains, lake basins, bogs, swamps, deep-sea sediments, and near-polar regions. Over longer periods (thousands to several million years), some carbon may be buried with sediments that become sedimentary rocks. This carbon is transformed into fossil fuels such as natural gas, oil, and coal. Nearly all of the carbon stored in the lithosphere exists as sedimentary rocks. Most of this is in the form of carbonates such as limestone, much of which has a direct biological origin. Over 8,000 plant

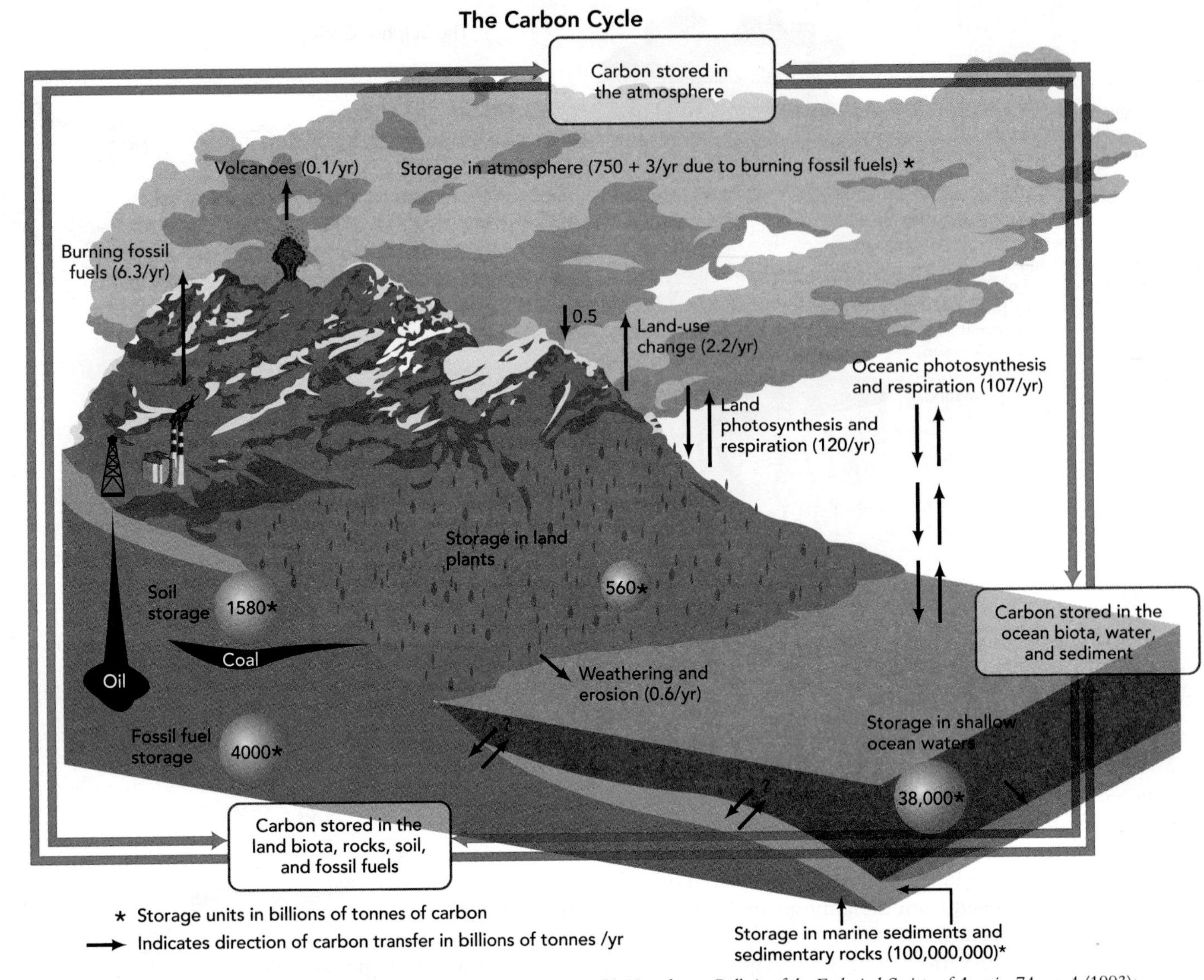

Figure 5.12 • Generalized global carbon cycle. [*Source*: Modified from G. Lambert, *La Recherche* 18 (1987): 782–783, with some data from R. Houghton, *Bulletin of the Ecological Society of America* 74, no. 4 (1993): 355–356, and R. Houghton, *Tellus* 55B, no. 2 (2003): 378–390.]

species have developed adaptations that allow them to use a supplementary method of carbon dioxide uptake using a four-carbon molecule rather than the normal three-carbon molecule. These plants, called C4 plants, tend to be those that grow in habitats with high daytime temperatures and intense sunlight. They include plants like crabgrass, corn, and sugar cane. Other plants, called CAM plants, employ both C3 and C4 cycles, using C3 processes in the daytime and C4 processes at night. Many arid climate plants like cacti, bromeliads like the pineapple, and sedums, are CAM plants. C3 processes are more efficient at lower oxygen concentrations and temperatures (below 27°C); C4 processes have the advantage over 32°C. Because plants in a given habitat and region tend to have similar carbon-cycle adaptations, C3, C4, and CAM systems have implications for plant adaptation to changes in the carbon cycle, and affect how carbon is stored and cycled within a region.

The cycling of carbon dioxide between land organisms and the atmosphere is a large flux. Approximately 15% of the total carbon in the atmosphere is taken up by photosynthesis and released by respiration on land annually. Thus, as noted, aerobic life has a large effect on the chemistry of the atmosphere.

The Missing Carbon Sink

Because carbon forms two of the most important greenhouse gases—carbon dioxide and methane—much research has been devoted to understanding the carbon cycle. However, at a global level, some key issues remain unanswered. For example, monitoring atmospheric carbon dioxide levels over the past several decades suggests that of the approximately 8.5 units released per year by human activities into the atmosphere, approximately 3.2 units remain there. It is estimated that about 2.4 units diffuse into the ocean. This leaves 2.9 units unaccounted for.[9, 10] Several hundred or more million tonnes of carbon

A CLOSER LOOK 5.3

Photosynthesis and Respiration

Recall that the general chemical reaction for photosynthesis is

$$6CO_2 + 6H_2O \xrightarrow{\text{sunlight}} C_6H_{12}O6 + 6O_2$$

Carbon dioxide enters biological cycles through photosynthesis, the process whereby the cells of living organisms (such as plants) convert energy from sunlight in a series of chemical reactions to chemical energy. In the process, carbon dioxide and water are combined to form organic compounds such as simple sugars and starch, with oxygen as a by-product (Figure 5.13). Carbon leaves living biota through respiration, the process by which organic compounds are broken down to release gaseous carbon dioxide. For example, animals (including people) take in air, which has a relatively high concentration of oxygen. Oxygen is absorbed by blood in the lungs. The chemical reaction for respiration is the reverse of that for photosynthesis. Through respiration, carbon dioxide is released into the atmosphere. Figure 5.14 illustrates the role of photosynthesis and respiration, along with other processes, in the carbon cycle of a lake.

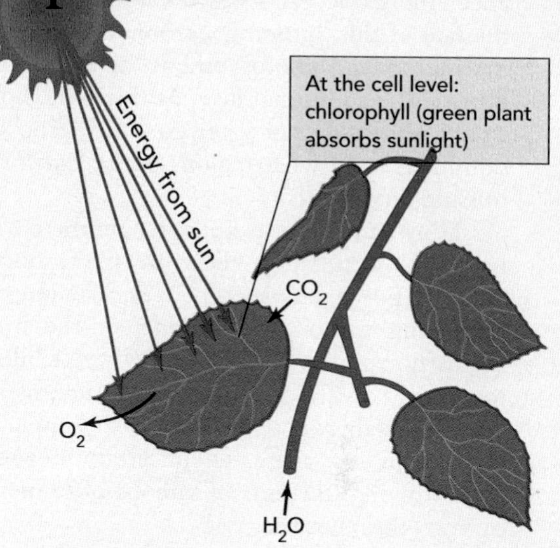

Figure 5.13 • **The process of photosynthesis in a green plant**

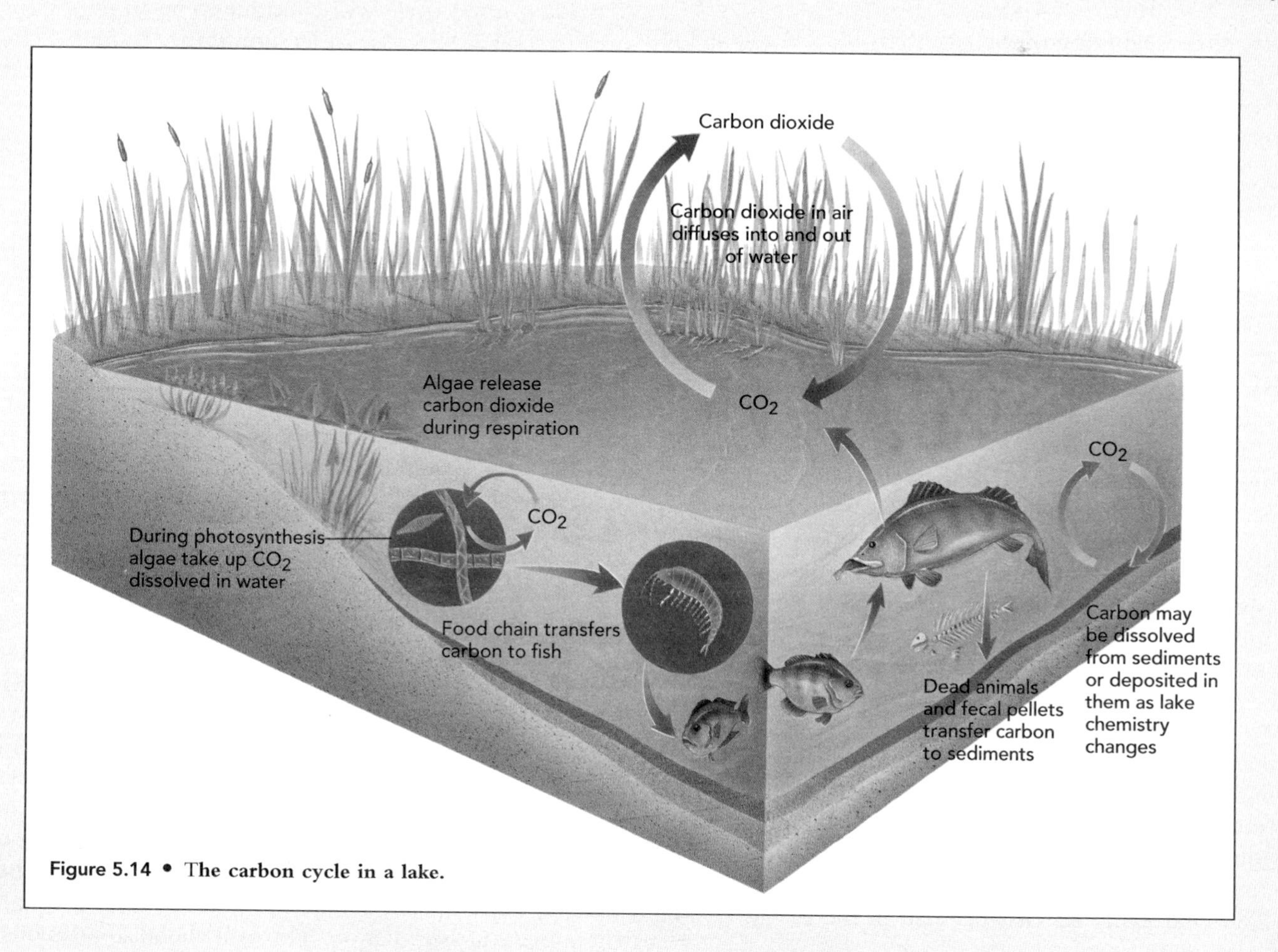

Figure 5.14 • **The carbon cycle in a lake.**

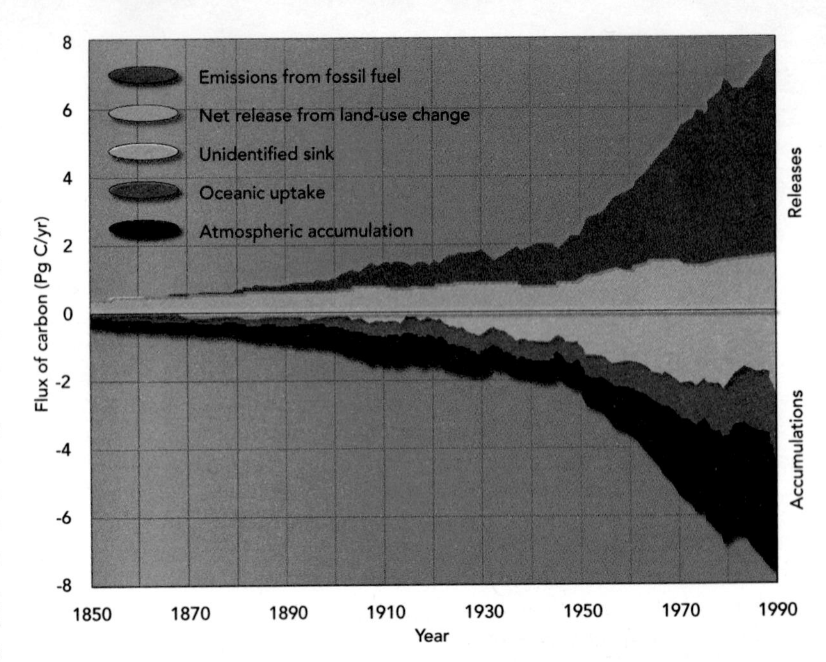

Figure 5.15 • Global flux of carbon, 1850–1990, showing estimated missing carbon sink. Modified after Woods Hole Research Center, Global Carbon Cycle, "The Missing Carbon Sink," 2000.

are released each year from fossil fuel combustion and end up somewhere unknown to science. Inorganic processes do not account for the fate of this "missing carbon sink." Either marine or land photosynthesis, or both, must provide the additional flux. At this time, however, scientists do not agree on which processes dominate or in what regions of the Earth this missing flux occurs.

Why can scientists not agree on these basic issues? Unfortunately, there are many uncertainties. For example, when uncertainties in measuring carbon are considered, the missing carbon sink has a flux of 2.9 ± 1.1 billion tonnes/yr, or 2.9 ± 1.1 Pg/yr (petagrams/yr). The uncertainty is about 40% of the suspected flux. Including estimated uncertainties yields the following, in units of billions of tonnes of carbon (GtC) per year.[11–13]

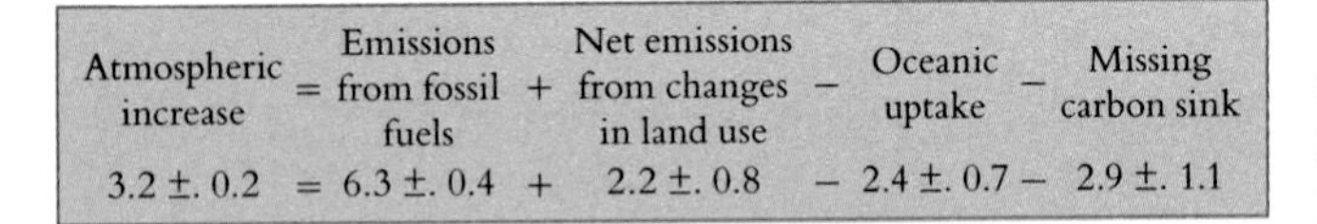

Atmospheric increase	=	Emissions from fossil fuels	+	Net emissions from changes in land use	−	Oceanic uptake	−	Missing carbon sink
3.2 ±. 0.2	=	6.3 ±. 0.4	+	2.2 ±. 0.8	−	2.4 ±. 0.7	−	2.9 ±. 1.1

Since the beginning of the Industrial Revolution in the mid-nineteenth century, the unidentified sink has apparently steadily increased in size (Figure 5.15). It is believed that about 0.7 ± 0.8 billion tonnes/yr could be stored in terrestrial ecosystems, including forests and, to a lesser extent, soils. Of particular importance are those forests in the northern hemisphere that are recovering from timber harvesting during the past two centuries and are growing fast today. Fast growing forests remove carbon from the atmosphere at an increasing rate as biomass is added. However, fires and melting frozen ground in the boreal forests also release carbon, and it is not certain whether the forests comprise a net sink or a net source of carbon to the atmosphere.[14] This is reflected by the large uncertainty in the terrestrial flux of 0.7 ± 0.8 GtC per year. The uncertainty of ± 0.8 units is greater than the estimated flux of 0.7 units! Two major uncertainties in estimating carbon fluxes are rates of land-use change, especially those cleared for agriculture harvested, or burned.[12, 13] The amount of carbon in ecosystem storage compartments affected by human use. Uncertainties will be reduced in the future if we are more successful in measuring and monitoring land-use change (deforestation, burning, and clearing) and estimating the flux of carbon in the ecosystems and the atmosphere. Due to changes in land-use and the burning of fossil fuels, the missing carbon sink is not static but changes on annual and decadal time scales. It was about 2.4 ± 1.1 GtC per year in the 1980s and 2.9 ± 1.1 GtC per year in the 1990s.[12,13] The missing carbon problem illustrates the complexity of biogeochemical cycles, especially ones in which the biota play an important role.

The carbon cycle will continue to be an important area of research because of its significance to global climate investigations, especially to climate change, and the future of Canada's glaciers.[15,16,17,18]

The Nitrogen Cycle

Nitrogen is essential to life because it is necessary for the manufacture of proteins and DNA. Free nitrogen (N_2 uncombined with any other element) makes up approximately 78% of Earth's atmosphere. However, many organisms cannot use this nitrogen directly. Some, such as animals, require nitrogen in an organic compound. Others, including plants, algae, and bacteria, can take up nitrogen either as the nitrate ion (NO_3^-), the nitrite ion (NO_2^-), or the ammonium ion (NH_4^+). Since nitrogen is a relatively unreactive element, few processes convert molecular nitrogen to one of these compounds. Lightning oxidizes nitrogen, producing nitric oxide. In nature, essentially all other conversions of molecular nitrogen to biologically useful forms are conducted by bacteria.

The **nitrogen cycle** is one of the most important and most complex of the global cycles (Figure 5.16). The process of converting inorganic, molecular nitrogen in the atmosphere to ammonia or nitrate is called **nitrogen fixation**. Ammonia is converted by nitrifying bacteria to nitrate and nitrite (nitrification). Nitrite is an intermediate stage between ammonia (which contains no oxygen) and nitrate. Once in these forms, nitrogen can be used on land by plants and in the oceans by algae through the process of assimilation. Through chemical reactions,

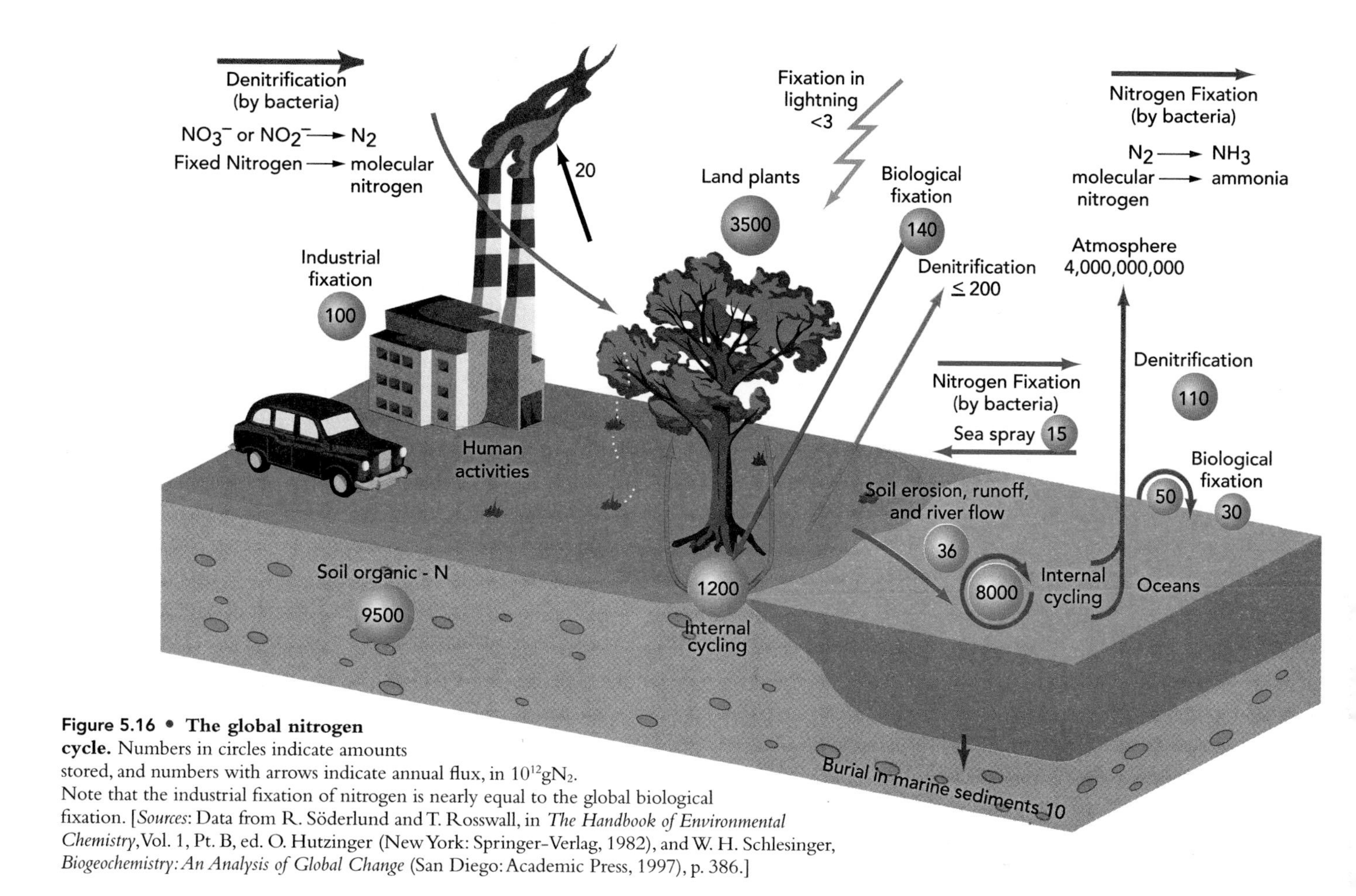

Figure 5.16 • The global nitrogen cycle. Numbers in circles indicate amounts stored, and numbers with arrows indicate annual flux, in $10^{12}gN_2$. Note that the industrial fixation of nitrogen is nearly equal to the global biological fixation. [*Sources:* Data from R. Söderlund and T. Rosswall, in *The Handbook of Environmental Chemistry*, Vol. 1, Pt. B, ed. O. Hutzinger (New York: Springer-Verlag, 1982), and W. H. Schlesinger, *Biogeochemistry: An Analysis of Global Change* (San Diego: Academic Press, 1997), p. 386.]

bacteria, plants, and algae then convert these inorganic nitrogen compounds into organic ones, and the nitrogen becomes available to ecological food chains. When organisms die, other bacteria convert the organic compounds containing nitrogen back to ammonia, nitrate, or molecular nitrogen, which enters the atmosphere. The conversion of nitrates and nitrites to ammonia by bacterial action is called **ammonification**. The process of releasing fixed nitrogen back to molecular nitrogen is called **denitrification**.

Nearly all organisms depend on nitrogen-converting bacteria. Some organisms have evolved symbiotic relationships with these bacteria. For example, the roots of the pea family have nodules that provide a habitat for the bacteria. The bacteria obtain organic compounds for food from the plants, and the plants obtain usable nitrogen. Such plants can grow in otherwise nitrogen-poor environments. When these plants die, they contribute nitrogen-rich organic matter to the soil, thereby improving the soil's fertility. Alder trees, too, have nitrogen-fixing bacteria as symbionts in their roots. (Symbionts are organisms in symbiotic relationships.) These trees grow along streams, and their nitrogen-rich leaves fall into the streams and increase the supply of the element in a biologically usable form to freshwater organisms.

Nitrogen-fixing bacteria are also symbionts in the stomachs of some animals, particularly ruminants (the cud-chewing animals). These animals, which include buffalo, cows, deer, moose, and giraffes, have a specialized four-chambered stomach. The bacteria provide as much as half of the total nitrogen needed by the animals, with the rest provided by protein in the green plants the animals eat.

In terms of availability for life, nitrogen falls somewhere between carbon and phosphorus. Like carbon, nitrogen has a gaseous phase and is a major component of Earth's atmosphere. Unlike carbon, however, it is not very reactive, and its conversion depends heavily on biological activity. Thus, the nitrogen cycle is not only essential to life but also is primarily driven by life.

In the early part of the twentieth century, scientists discovered that industrial processes could convert molecular nitrogen into compounds usable by plants. This greatly increased the availability of nitrogen in fertilizers. Today, most commercial nitrogen fertilizer is produced through industrial fixation of nitrogen. The amount of industrial fixed nitrogen is about 50% of the amount fixed in the biosphere. Nitrogen in agricultural runoff is a potential source of water pollution.

Nitrogen combines with oxygen in high-temperature atmospheres. As a result, many modern industrial combustion processes produce oxides of nitrogen. These processes include the burning of fossil fuels in gasoline and diesel engines. Thus, oxides of nitrogen, which are air pollutants, are indirect results of modern industrial activity and modern technology. Nitrogen oxides play a significant role in urban smog (see Chapter 21). Human influences on the nitrogen cycle occur very rapidly, and their variability can be monitored from year to year. By contrast, the phosphorus cycle, described below, is a "slow" cycle, with many influences occurring over geologic time.

The Phosphorus Cycle

Phosphorus, one of the "big six" elements required in large quantities by all forms of life, is often a limiting nutrient for plant and algal growth because, despite its relative abundance in the lithosphere, it is both physically and chemically inaccessible. However, if phosphorus is too abundant, it can cause environmental problems, as illustrated by the Experimental Lakes Area studies described in Chapter 2.

Unlike carbon and nitrogen, phosphorus does not have a gaseous phase on the Earth (Figure 5.17). Thus, the **phosphorus cycle** is significantly different from the carbon and nitrogen cycles. The rate of transfer of phosphorus in the Earth's system is slow compared with that of carbon or nitrogen. Phosphorus exists in the atmosphere only in small particles of dust. In addition, phosphorus tends to form compounds that are relatively insoluble in water. Consequently, phosphorus is not readily weathered chemically. It does occur commonly in an oxidized state as phosphate, which combines with calcium, potassium, magnesium, or iron to form minerals.

Phosphorus enters the biota through uptake as phosphate by plants, algae, and some bacteria. In a relatively stable ecosystem, much of the phosphorus that is taken up by vegetation is returned to the soil when the plants die. Nevertheless, some phosphorus is inevitably lost to ecosystems. It is transported by rivers to the oceans, either in a water-soluble form or as suspended particles.

An important way in which phosphorus returns from the ocean to the land involves ocean-feeding birds, such as the Chilean pelican. Areas of rising ocean currents, called upwellings, bring nutrients like phosphorus from deep waters to the surface. Plankton (and many other species) thrive in areas with upwellings, and provide food for small fish like anchovies. Fish-eating birds like the pelican in turn feed on small fish, and nest on offshore islands, where they are protected from predators. Over time, their nesting sites become covered with their phosphorus rich excrement, called guano. The birds nest by the thousands, and deposits of guano accumulate over centuries. In relatively dry climates, guano hardens into a rocklike mass that may be up to 40 m thick. The guano results from a combination of biological and non-biological processes. Without the plankton, fish, and birds, the phosphorus would have remained in the ocean. Without the upwellings, the phosphorus would not have been available.

Guano deposits were once major sources of phosphorus for fertilizers. In the mid-1800s, as much as 9 million tonnes per year of guano deposits were shipped to London from islands near Peru. Today, most phosphorus fertilizers come from mining of phosphate-rich sedimentary

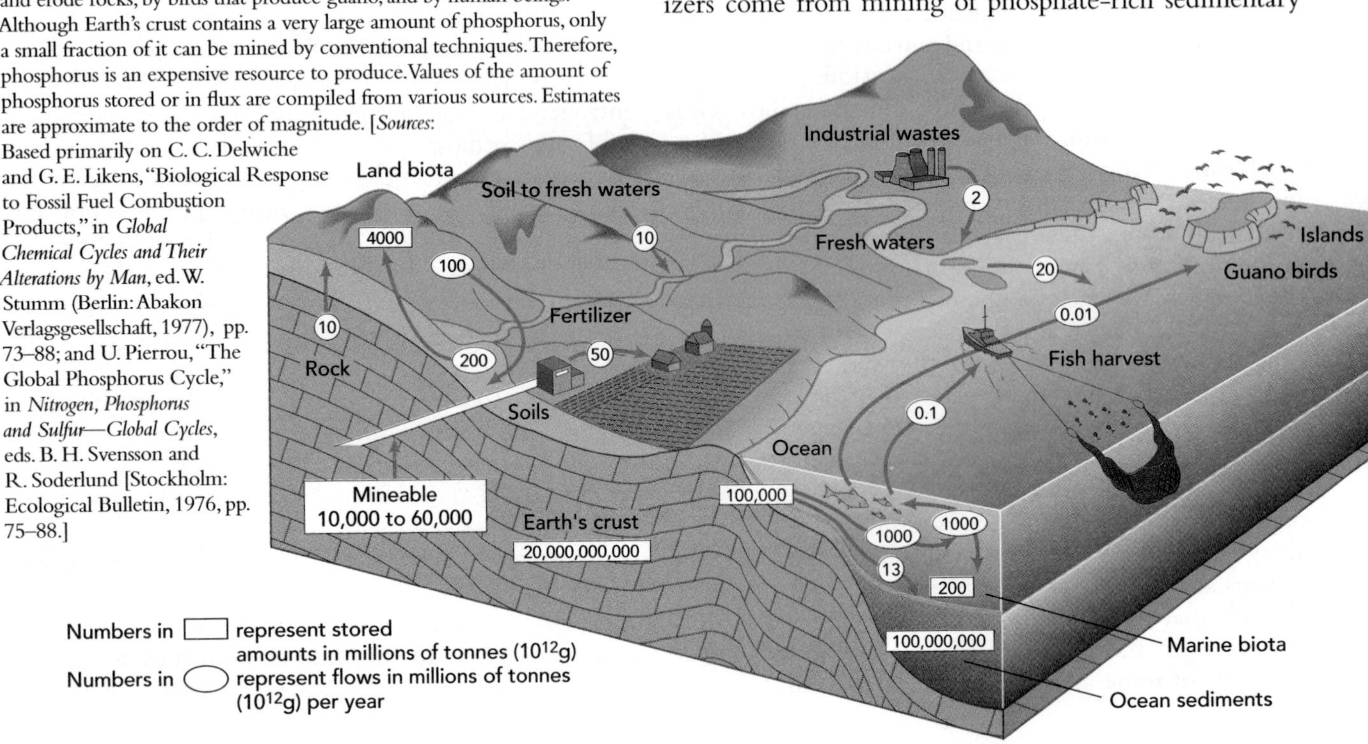

Figure 5.17 • The global phosphorus cycle. Phosphorus is recycled to soil and land biota through geologic processes that uplift the land and erode rocks, by birds that produce guano, and by human beings. Although Earth's crust contains a very large amount of phosphorus, only a small fraction of it can be mined by conventional techniques. Therefore, phosphorus is an expensive resource to produce. Values of the amount of phosphorus stored or in flux are compiled from various sources. Estimates are approximate to the order of magnitude. [*Sources:* Based primarily on C. C. Delwiche and G. E. Likens, "Biological Response to Fossil Fuel Combustion Products," in *Global Chemical Cycles and Their Alterations by Man*, ed. W. Stumm (Berlin: Abakon Verlagsgesellschaft, 1977), pp. 73–88; and U. Pierrou, "The Global Phosphorus Cycle," in *Nitrogen, Phosphorus and Sulfur—Global Cycles*, eds. B. H. Svensson and R. Soderlund [Stockholm: Ecological Bulletin, 1976, pp. 75–88.]

rocks containing fossils of marine animals. The richest phosphate mine in the world is Bone Valley, 40 km east of Tampa, Florida. Between 10 and 15 million years ago, Bone Valley was the bottom of a shallow sea where marine invertebrates lived and died.[19] Through very slow tectonic processes, Bone Valley was slowly uplifted, and in the 1880s and 1890s phosphate ore was discovered there. Today, Bone Valley provides more than one-third of the world's phosphate production.

Canada is not a major producer of phosphate and has only a single open-pit phosphate mine—northern Ontario's Kapuskasing Phosphate Project. By contrast, the United States, with total reserves of 2.2 billion tonnes (roughly 16% of the world total), is the world's largest producer of phosphate. As with other mining activities, phosphate mining may have negative effects on the land and ecosystems. For example, in some phosphorus mines, huge pits and waste ponds have scarred the landscape, damaging biologic and hydrologic resources. Balancing the need for phosphorus with the adverse environmental impacts of mining is a major environmental issue.

Human impacts on the phosphorus cycle centre on the use of mined and manufactured phosphates as fertilizer and cleaning agents. This phosphate then enters lakes and streams in runoff and in municipal wastewaters. Despite laws to control the amount of phosphate in detergents and similar products (see Chapter 2), phosphorus enrichment remains a major cause of algae blooms in temperate freshwater systems. Chapter 19 describes this process, called eutrophication, in more detail. The opening case in Chapter 2 discussed important Canadian research that proved the link between phosphorus pollution and nuisance aquatic plant growth.

The Sulphur Cycle

Sulphur, a non-metallic nutrient important in many biological processes, is another member of the "big six" elements, and one whose cyle has been altered by human activities. It exists in many chemical forms, including gases such as hydrogen sulphide (H_2S) and sulphur dioxide (SO_2), as sulphuric acid (H_2SO_4), and as various sulphur-containing salts. Figure 5.11 illustrates the cycling of sulphur in a typical forest ecosystem.

Like phosphorus, much of the world's sulphur is found in rocks and soils, or in ocean sediments. Unlike phosphorus, sulphur has an important atmospheric cycle. It enters the atmosphere through natural sources such as volcanic eruptions, growth and decay of biological organisms, and evaporation from water bodies. Human sources of sulphur to the atmosphere mainly arise from the burning of sulphur-containing fossil fuels. Acid precipitation, discussed in Chapter 21, is caused by industrial emissions of acid-forming sulphur and nitrogen compounds to the air. Human emissions of sulphur are thought to be roughly equivalent to natural sources in terms of total mass. The magnitude of these sources, and recent efforts to control them, are discussed in more detail in Chapter 21.

Sulphur has an important role in biological systems, especially as a component of proteins and enzymes. Primary producers take up sulphur dissolved in water, for instance as sulphates (SO_4^{-2}), and transfer that sulphur to the consumers that eat them. As biological material decays, sulphur is released to the atmosphere, where it is available for biological uptake or other chemical reactions. Oxygen-poor environments such as deep lake or ocean waters provide conditions that allow the release of hydrogen sulphide gas to the air, and the creation of inorganic, insoluble sulphides like iron sulphide (FeS_2—iron pyrite, or "fool's gold"). These sulphides are deposited as solids, returning the sulphur they contain to rock and soil.

Humans influence the cycling of many other materials in the environment, including metals and anthropogenic chemicals. These issues are discussed in later chapters of this book, especially Chapters 19-24.

CRITICAL THINKING ISSUE

How Are Human Activities Affecting the Nitrogen Cycle?

Scientists estimate that nitrogen deposition to Earth's surface will double in the next 25 years. What is causing this increase, and how will it affect the nitrogen cycle?

The natural rate of nitrogen fixation on land is estimated to be 140 teragrams (Tg) of nitrogen a year (1 teragram = 1 million tonnes). Human activities, such as use of fertilizers, draining of wetlands, clearing of land for agriculture, and burning of fossil fuels, are causing additional nitrogen to enter the environment. Currently, human activities are responsible for more than half of the fixed nitrogen that is deposited on land. Before the twentieth century, fixed nitrogen was recycled by bacteria with no net accumulation. Since 1900, however, the use of commercial fertilizers has increased exponentially (see Figure 5.18). Nitrates and ammonia from burning fossil fuels have increased about 20% in the last decade or so. These inputs have overwhelmed the denitrifying part of the nitrogen cycle and the ability of plants to use fixed nitrogen.

Nitrate ions in the presence of soil or water may form nitric acid. With other acids in the soil, nitric acid can leach out chemicals important to plant growth, such as magnesium and potassium. When these chemicals are depleted, more toxic ones, such as aluminum, may be released, damaging plant roots. Acidification of soil by nitrate ions is also harmful to organisms. When toxic chemicals wash into streams, they can kill fish. Excess nitrates can cause also cause algae to overgrow. High levels of nitrates in drinking water from streams or groundwater contaminated by fertilizers are a human health hazard.

The nitrogen and carbon cycles are linked because nitrogen is a component of chlorophyll, the molecule plants use in photosynthesis. Because nitrogen is a limiting factor on land, it has been predicted that increasing levels of global nitrogen may increase plant growth.

Recent studies have suggested that a beneficial effect from increased nitrogen would be short-lived, however. As plants use additional nitrogen, some other factor will become limiting. When that occurs, plant growth will slow, and so will the uptake of carbon dioxide. More research is needed to understand the interactions between carbon and nitrogen cycles and predict long-term effects of human activities.

Critical Thinking Questions

1. Compare the rate of human contributions to nitrogen fixation with the natural rate.
2. How does the change in fertilizer use relate to the change in world population? Why?
3. Develop a diagram to illustrate the links between the nitrogen and carbon cycles.
4. Make a list of ways in which human activities could be modified to reduce human contributions to the nitrogen cycle.

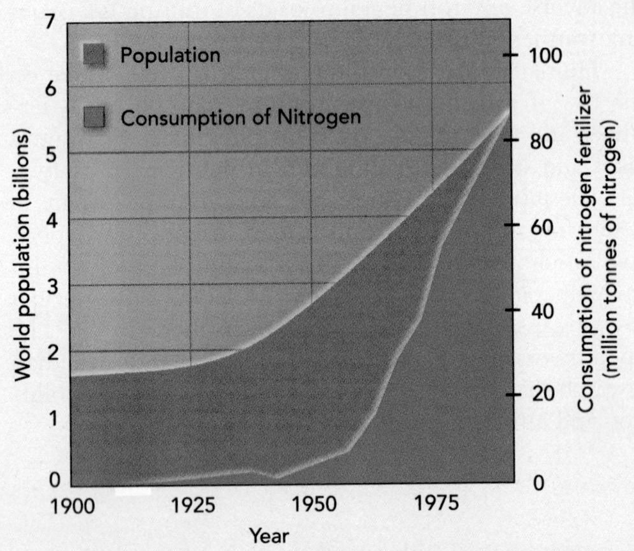

Figure 5.18 • Impact of the human population on nitrogen consumption.

Summary

- Biogeochemical cycles are the major way in which elements important to Earth processes and life are moved through the atmosphere, hydrosphere, lithosphere, and biosphere.
- Biogeochemical cycles can be described as a series of reservoirs, or storage compartments, and pathways, or fluxes, between reservoirs.
- In general, some chemical elements cycle quickly and are readily regenerated for biological activity. Elements whose biogeochemical cycles include a gaseous phase in the atmosphere tend to cycle more rapidly.
- Modern technology has begun to alter and transfer chemical elements in biogeochemical cycles at rates and amounts comparable to those of natural processes. Some of these activities are beneficial to society, but others pose dangers.
- To be better prepared to manage our environment, we must recognize both positive and negative consequences of activities that transfer chemical elements and deal with them appropriately.

- Biogeochemical cycles tend to be complex, and Earth's biota has greatly altered the cycling of chemicals through the air, water, and soil. Continuation of these processes is essential to the long-term maintenance of life on Earth.
- Every living thing, plant or animal, requires a number of chemical elements. These chemicals must be available at the appropriate time and in the appropriate form and amount.
- Chemicals can be reused and recycled, but in any real ecosystem some elements are lost over time and must be replenished if life in the ecosystem is to persist.
- Change and disturbance of natural ecosystems are the norm. A steady state, in which the net storage of chemicals in an ecosystem does not change with time, cannot be maintained.
- There are many uncertainties in measuring either the amount of a chemical in storage or the rate of transfer between reservoirs. For example, the global carbon cycle includes a large sink that science has not yet been able to locate.

STUDY QUESTIONS

1. Why is an understanding of biogeochemical cycles important in environmental science? Explain your answer with two examples.
2. What are some of the general rules that govern biogeochemical cycles, especially the transfer of material?
3. Identify the major aspects of the carbon cycle and the environmental concerns associated with it.
4. What are differences in the biogeochemical cycles for phosphorus and nitrogen, and why are the differences important in environmental science?
5. List five ways in which humans impact biogeochemical cycles. Which of these activities do you think has the greatest impact on the most cycles? Which do you think should be the focus of environmental management activities? Explain your answer.

FURTHER READING

Berner, R. A., and E. K. Berner. *Global Environment: Water, Air, and Geochemical Cycles.* Upper Saddle River, N.J.: Prentice-Hall, 1996. A good discussion of environmental geochemical cycles, focusing on Earth's air and water systems.

Kasting, J. F., O. B. Toon, and J. B. Pollack. "How Climate Evolved on the Terrestrial Planets," *Scientific American*, 258 (1988):2:90–97. This paper provides a good discussion of the carbonate–silicate cycle and why it is important in environmental science.

Lerman, A. "Weathering and Erosional Controls of Geologic Cycles," *Chemical Geology*, (1990) 84:13–14. Natural transfer of elements from the continents to the oceans is largely accomplished by erosion of the land and transport of dissolved material in rivers.

Post, W. M., T. Peng, W. R. Emanual, A. W. King, V. H. Dale, and D. L. DeAngelis. "The Global Carbon Cycle," *American Scientist* 78 (1990):310–326. The authors describe the natural balance of carbon dioxide in the atmosphere and review why the global climate hangs in the balance.

Schlesinger, W. H. *Biogeochemistry: An Analysis of Global Change*, 2nd ed. San Diego: Academic Press, 1997. This book provides a comprehensive and up-to-date overview of the chemical reactions on land, in the oceans, and in the atmosphere of Earth.

chapter 6

The Acorn Connection

As young children, most people learn that acorns grow into oak trees. In fact, though, most acorns do not grow into trees, but become food for mice, chipmunks, squirrels, and deer.

In the woodlands of eastern Canada, where oak trees abound, large crops of acorns (Figure 6.1*a*) are produced every three to four years. Acorns are rich in proteins and fats and an excellent source of nutrition. A steady supply of acorns would be an excellent food base for the woodland animals.

Acorn production is affected by the amount of light and rain, temperature patterns over the year, and the quality of the soil. Scientists reason that if oaks produced the same number of acorns each year, the animal population that feeds on them would grow so large that very few acorns would survive.

In reality, the number of acorns produced varies from year to year, with "mast" years—years of high production—occurring occasionally. In the years between bumper crops of acorns, mice populations decline. With the next bumper crop, there are more acorns than can be eaten by acorn consumers; so many acorns survive to become oaks. Also, because of the abundance of food, the mice populations increase.

White-footed mice (Figure 6.1*b*), feeders on acorns, also carry tick larvae (Figure 6.1*c*). As the ticks feed on the blood of the mice, they pick up the microorganisms responsible for Lyme disease. Mice populations are highest during the summer following a bumper crop of acorns, and so are tick larvae.

In later stages of their life cycle, ticks attach to other animals, including deer (Figure 6.1*d*). As deer brush against plants, ticks are deposited. The ticks can be picked up by people brushing against the plants as they walk past. If an infected tick bites a person, the person may contract Lyme disease. Following the regrowth of forests originally cut for agriculture in the nineteenth century, second-growth forest area has increased and deer populations have soared. Lyme disease has become the most common tick-borne disease in northern temperate regions,[1,2] although its incidence in Canada is still far lower than in the United States.

Let us return to the mice. Besides feeding on acorns (and other grains), mice feed on insects, including larvae of the gypsy moth. Gypsy moth larvae (Figure 6.1*e*) feed on leaves of trees and are particularly fond of oak leaves. Studies suggest that in years when mice populations are low—the years between bumper crops of acorns—gypsy moth populations can increase dramatically. During these periodic outbreaks, gypsy moth larvae can virtually denude an area, stripping the leaves from the trees. Oaks that have lost most or all of their leaves may not produce bumper crops of acorns.

Once the leaves are off the trees, more light reaches the ground, and seedlings of many plants that could not do well in deep forest shade begin to grow. As a result, other species of trees may gain a foothold in the forest and change its species profile. Of course, the next generation of gypsy moth larvae finds little to eat, and the population of gypsy moths begins to decline again.

Abundant acorns draw deer into the woods, where they browse on small plants and tree seedlings. Ticks drop off the deer and lay eggs in the leaf litter. When the eggs hatch, the larvae attach to mice, and the cycle of Lyme disease continues. Deer do not eat ferns, however, and in areas where deer populations are dense, many ferns but few wildflowers and tree seedlings are found.

Predators are also affected by the periodic nature of acorn crops. For example, birds that feed on gypsy moth larvae lose a food source when moth populations are low. When moth populations are high, however, bird nests are more exposed to predation because trees lose so many leaves.[2]

- *The acorn connection illustrates many basic characteristics of ecosystems and ecological communities. First, all of the living parts of the oak forest community depend on the nonliving parts of the ecosystem for their survival: water, soil, air, and the light that provides energy for photosynthesis. Second, the members of the ecological community affect the nonliving parts of the ecosystem. When gypsy moths denude an area, for example, more sunlight can reach the forest floor. Third, the living organisms in the ecosystem are connected in complex relationships that make it difficult to change one thing without changing many others. Fourth, the relationships among the members of the ecological community are dynamic and constantly changing. Fifth, the implication for human management of ecosystems is that any management practice involves trade-offs. In this case, managing the forest to protect people against Lyme disease only results in the potential for more gypsy moth damage.*[1]

Ecosystem Structure and Function

LEARNING OBJECTIVES

Life on Earth is sustained by ecosystems, which vary greatly but have certain attributes in common. After reading this chapter, you should understand:

- Why the ecosystem is the basic system that supports life and allows it to persist.
- What are food chains, food webs, and trophic levels.
- What the concept of ecosystem management involves.
- How conservation and management of the environment might be improved through ecosystem management.
- What a community-level effect is.

Figure 6.1 • **The acorn connection.** The tick that carries Lyme disease *(b)* feeds on both the white-footed mouse *(a)* and the white-tailed deer *(c)*. Oak leaves *(e)* are an important food for the deer and for gypsy moths *(d)*, while oak acorns are important food for the mouse. But the mouse also eats the moths. The more mice, the fewer gypsy moths, but the more ticks.

6.1 The Ecosystem: Sustaining Life on Earth

We associate life with individual organisms that are alive. However, sustaining life on Earth requires more than individuals or even single populations or species. Life is sustained by the interactions of many organisms functioning together, in ecosystems, interacting through their physical and chemical environments. Sustained life on Earth, then, is a characteristic of ecosystems, not of individual organisms or populations.[3] To understand important environmental issues, such as conserving endangered species, sustaining renewable resources, and minimizing the effects of toxic substances, we must understand certain basic principles about occurring ecosystems.

Basic Characteristics of Ecosystems

Ecosystems have several fundamental characteristics.

- **Structure.** An ecosystem is made up of two major parts: nonliving and living. The nonliving part is the physical-chemical environment, including the local atmosphere, water, and mineral soil (on land) or other substrate (in water). The living part, called the ecological community, is the set of species interacting within the ecosystem.
- **Processes.** Two basic kinds of processes must occur in an ecosystem: a cycling of chemical elements and a flow of energy.
- **Change.** An ecosystem changes over time and can undergo development through a process called succession, which is discussed in Chapter 9.

The processes that occur in an ecosystem are necessary for the life of the ecological community, but no member of that community can carry out these processes alone. That is why we have said that sustained life on Earth, rather than individuals or populations, is a characteristic of ecosystems.

We can see this by looking at cycling in an ecosystem. As mentioned in Chapter 5, chemical cycling is complex. Each chemical element required for growth and reproduction must be made available to each organism at the right time, in the right amount, and in the right ratio relative to other elements. These chemical elements must also be recycled—converted to a reusable form. Wastes are converted into food, which is converted into wastes, which is converted again into food, with the cycling going on indefinitely, if the ecosystem is to remain viable.

For complete recycling of chemical elements to take place, several species must interact. In the presence of light, green plants, algae, and photosynthetic bacteria produce sugar from carbon dioxide and water. From sugar and inorganic compounds, they make many other organic compounds, including proteins and woody tissue. Nevertheless, no green plant can decompose woody tissue back to its original inorganic compounds. Other forms of life—primarily bacteria and fungi—can decompose organic matter; but they cannot produce their own food. Instead, they obtain energy and chemical nutrition from the dead tissues on which they feed.

LEVELS OF ECOLOGICAL ORGANIZATION	
Organism	A living being, such as an individual plant or animal.
Species	A group of organisms capable of interbreeding.
Population	A group of individuals of the same species living in a specific place at a specific time, or interbreeding and sharing genetic information.
Community	The set of interacting populations that makes up the living part of an ecosystems.
Ecosystem	A community of organisms and its abiotic environment. An ecosystem is the minimum system that includes and sustains life. It must include at least an autotroph, a decomposer, a liquid medium, a source and sink of energy, and all the chemical elements required by the autotroph and the decomposer.
Landscape	A set of ecosystems connected across a geographic area.
Biome	A kind of ecosystem. The rain forest is an example of a biome; rain forests occur in many parts of the world but are not all connected with each other.
Biosphere	The part of a planet where life exists. On Earth it extends from the depths of the oceans to the summit of mountains, but most life exists within a few meters of the surface.

Table 6.1 • **Levels of Ecological Organization.**

Theoretically, at its simplest, the ecological community in an ecosystem consists of at least one species that produces its own food from inorganic compounds in its environment and another species that decomposes the wastes of the first species, plus a fluid medium (air, water, or both). We turn next to a more detailed discussion of ecological communities—in particular, of food chains in ecological communities.

Populations, Communities, and Ecosystems

Thus far, we have spoken of species, but what really concerns us is the population of a given species that lives in the ecosystem of interest. A population is a group of individuals of the same species living in the same geographical area. The concept of population is important for two reasons. First, each population experiences a unique combination of biotic and abiotic conditions, and thus may respond to those conditions differently than another population of the same species. Think of human populations, for example. A population of humans in sub-Saharan Africa, for example, faces a very different set of challenges than a population living in Paris or New York. The second reason that consider populations is important relates to their genetic composition, the combination of genes, or hereditary material, encoded in their cells. Chapter 7 discusses these concepts in more detail. Table 6.1 describes major levels of ecological organization.

We have identified an ecological community as the set of interacting species that makes up the living part of an ecosystem. In practice, the term ecological community is defined by ecologists in two ways. One method is to define the community as a set of interacting species found in the same place and functioning together to make possible the persistence of life. That is essentially the definition we used earlier. A problem with this definition is that it is often difficult in practice to know the entire set of interacting species. Ecologists therefore may use a pragmatic or operational definition, in which the community consists of all the species found in an area, whether or not they are known to interact. Animals in different cages in a zoo could be called a community according to this definition.

One way in which individuals in a community interact is by feeding on one another. Energy, chemical elements, and some compounds are transferred from creature to creature along food chains, the linkage of who feeds on whom. A **food chain** is a simplistic representation of complex feeding relationships, typically showing a single species at each level. It is more realistic to think in terms of complex linkages involving multiple species and their interrelationships. These linkages are called **food webs**.

Ecologists group the organisms in a food web into trophic levels. A **trophic level** consists of all those organisms in a food web that are the same number of feeding levels away from the original source of energy. The original source of energy in most ecosystems is the sun. In other cases, it is the energy in certain inorganic compounds.

Green plants, algae, and certain bacteria produce sugars through the process of photosynthesis, using only the energy of the sun and carbon dioxide (CO^2) from the air, so they are grouped into the first trophic level. Organisms in the first trophic level, which make their own food and inorganic chemicals and a source of energy, are called **autotrophs**. Photosynthesis requires the presence of chlorophyll, a specialized molecule present in all green plants. Some bacteria, termed chemoautotrophs, lack chlorophyll but are able to produce food without sunlight from sulphur compounds and carbon dioxide, through the process of chemosynthesis.

Organisms that must rely on others for their energy supply are called **heterotrophs**. There may be several levels of heterotrophs in an ecosystem. Herbivores, organisms that feed on plants, algae, or photosynthetic bacteria, are members of the second trophic level. Carnivores, or meat-eaters, that feed directly on herbivores make up the third trophic level. Carnivores that feed on third-level carnivores are in the fourth trophic level, and so on.

Most ecosystems have two complementary food chains, one based on herbivores and one on detritus-feeders. Grazing food chains, such as temperate forests and grasslands are familiar to us. Less visible, but far more important in terms of energy flow, are detritus food chains. At the base of a detritus food chain are the organic carbon molecules derived from dead plant and animal matter. These molecules are the food source for the primary consumers, the bacteria and fungi (microdecomposers). Feeding on the microdecomposers, the secondary consumers in a detritus food chain are macrodecomposer organisms like earthworms, nematodes and insect larvae. At the top of the food chain are micropredators such as spiders. No ecosystem could function properly without detritus food chains. By one estimate, 80% of the nutrients needed by a forest for growth are derived from detritus (the remainder comes from rock weathering and rainfall). This is an important clue for the management of forests and agricultural lands. Ecosystems need these nutrients for successful growth. If plant debris such as branches, stalks, and leaves are removed from an ecosystem through human activity, the future recovery of that system will be slowed unless an external source of nutrients (such as fertilizer) is provided.

Food chains and food webs are often quite complicated and thus difficult to analyze. A detailed look at one of the simplest food chains is provided in A Closer Look 6.1. Next, we look briefly at several more complicated food chains.

A Terrestrial Food Chain

An example of terrestrial food chains and trophic levels is shown in Figure 6.4. This north temperate woodland food web existed in North America before European settlement and included human beings. The first trophic level, autotrophs, includes grasses, herbs, and trees. The second trophic level, herbivores, includes mice, an insect called the pine borer, and other animals (such as deer) not shown here. The third trophic level, carnivores, includes foxes and wolves, hawks and other predatory birds, spiders, and predatory insects. People are omnivores (eaters of both plants and animals) and feed on several trophic levels. In

A CLOSER LOOK 6.1

Hot Spring Ecosystems in Banff National Park

Perhaps the simplest natural ecosystem is a hot spring such as those found in Banff National Park, Alberta.[4] Few organisms can live in these hot springs because the environment is so severe. Maximum water temperature in Banff's Upper Hot Springs can approach 66°C. Some organisms that can live in hot springs are especially adapted for those harsh conditions.

Each of Canada's Rocky Mountain hot springs has its own ecosystem, characterized by a unique combination of minerals, gases, and temperature. Radium Hot Springs, in Kootenay National Park, British Columbia, is a fast-flowing, odourless spring high in silica, magnesium, sulphate, fluoride, calcium, bicarbonate, and the traces of radon that give the spring its name. Its pH, at 7.05, is close to neutral. In Jasper National Park, Alberta, Miette Hot Springs flow more slowly than those in Kootenay National Park, but have much higher concentrations of sulphate, calcium, and potassium, a much stronger sulphur smell, and a slightly acidic pH. Banff Upper Hot Springs is the most alkaline, has a high level of dissolved salts, and a very strong sulphur smell and taste.

The Rocky Mountain hot springs are formed from precipitation infiltrating into the high western slopes of Mount Rundle (Figure 6.2*b*). As the water seeps downward through the mountain's sedimentary rocks, it accumulates dissolved minerals and grows hotter from pressure and the heat of the Earth's crust. Where there is a crack or fault in the rock, such as the Sulphur Mountain Thrust Fault, the water can flow up to the surface and emerge as a spring.

Banff's hot springs support an unremarkable little snail that is found nowhere else in the world. First described in 1926, the Banff Springs Snail (*Physella johnsoni*) is an endangered species that is known to inhabit only four thermal springs on Sulphur Mountain in Banff National Park. Parks Canada studies have demonstrated that the populations of this species fluctuate widely, with lowest numbers in spring. The tiny snails—most are less than half a centimeter in length—live in only a small portion of each spring, feeding and laying their eggs on floating mats of photosynthetic bacteria and algae (Figure 6.2*a*). Their distribution is limited to waters with temperatures between 30° and 36°C.[5]

First trophic level. At the base of the hot springs' food web are the autotrophs, photosynthetic bacteria and algae. In the hot springs, as in most communities, the source of energy is sunlight (Figure 6.3)

Second trophic level. The Banff Springs Snail, a herbivore, functions at the second trophic level. Populations of the snail fluctuate widely, possibly in response to the abundance and distribution of its microbial food sources.

Third trophic level. Above the snails, at the third trophic level, are the carnivores that feed on the snails and their eggs. These include a number of duck species, including mallards, green-winged teal, and Barrow's goldeneyes.

Fourth trophic level. The fourth and final trophic level consists of decomposers, which in the hot springs are primarily bacteria. These organisms feed off the wastes and dead organisms of higher trophic levels

The entire hot springs community of organisms—photosynthetic bacteria and algae, snails, carnivores, and decomposers—is maintained by two factors: (1) sunlight, which provides an input of usable energy for the organisms; and (2) a constant flow of hot water, which provides a continual new supply of chemical elements required for life and a habitat in which the bacteria, algae and snails can persist.

Even though this is one of the simplest ecological communities in terms of the numbers of species, a fair number of species are found. At least a dozen species in all are important in this ecosystem. The ecological community they form has been sustained for long periods in these unusual habitats.

Another interesting aspect of the hot springs ecosystem is the snail's close reliance on biotic and abiotic conditions within the spring. The dramatic population fluctuations make the species particularly vulnerable when numbers are at their lowest, from March to June.

(a)

(b)

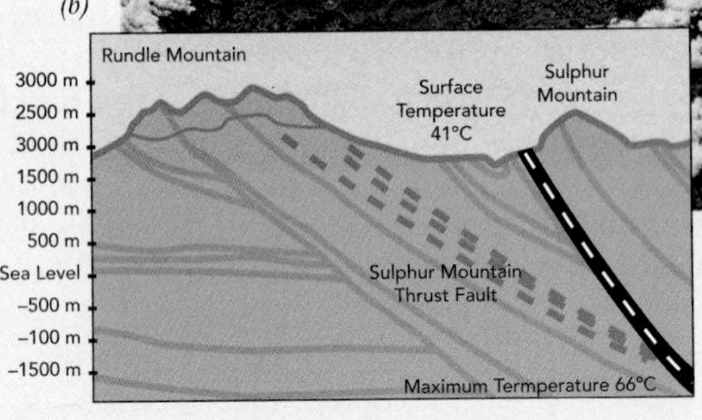

Figure 6.2 • (*a*) **Banff Upper Hot Springs** (*b*) **Diagram of Sulphur Mountain Fault.**

Particularly critical are the floating algae mats on which the snails feed and lay their eggs. Disturbance of those mats can kill the snails or their eggs. Insect repellents, sunscreen, and cosmetics can also introduce unwanted chemical changes in the simple ecosystem. These kinds of influences may have been responsible for the snail's disappearance over the past decade from four of the nine springs in which it was previously found.

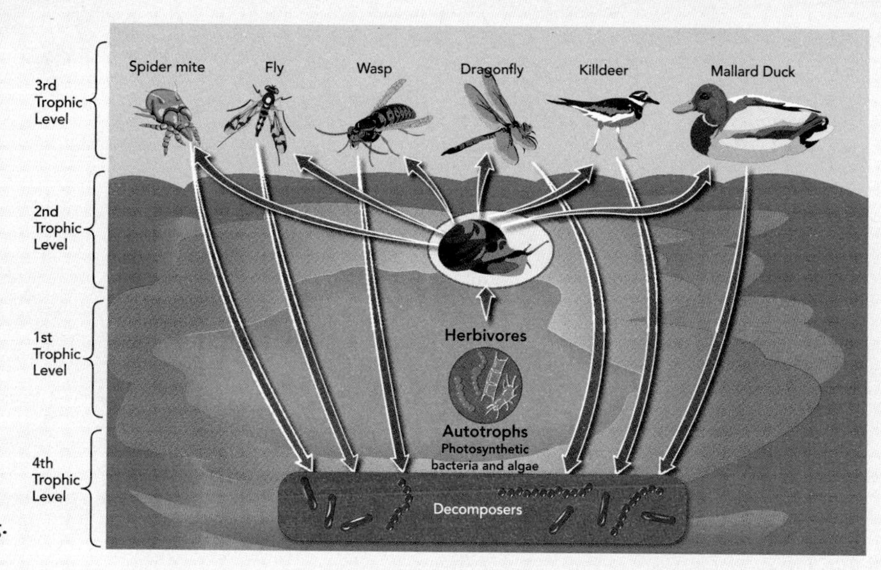

Figure 6.3 • Food web of a Banff hot spring.

Figure 6.4, people would be included in the fourth trophic level, the highest level in which they would take part. Decomposers, such as bacteria and fungi, feed on wastes and dead organisms of all trophic levels. Decomposers are also shown here on the fourth level.

An Oceanic Food Chain

Marine food webs differ in some respects from terrestrial food webs. For example, they tend to involve fewer species of primary producers and more species of predators than they do in the terrestrial ecosystem just considered. In a typical oceanic ecosystem (Figure 6.5), microscopic single-cell planktonic algae and planktonic photosynthetic bacteria are in the first trophic level. Small invertebrates called zooplankton and some fish feed on the algae and photosynthetic bacteria, forming the second trophic level. Other fish and invertebrates feed on these herbivores and form the third trophic level. The great baleen whales filter seawater for food, feeding primarily on small herbivorous zooplankton (mostly crustaceans), and thus the baleen whales are also in the third level. Some fish and marine mammals, such as killer whales, feed on predatory fish and form higher trophic levels.

The Food Web of the Harp Seal

In the abstract, a diagram of a food web and its trophic levels seems simple and neat; but in reality, food webs are complex, because most creatures feed on several trophic levels. For example, consider the food web of the harp seal (Figure 6.6). The harp seal is shown at the fifth level.[6] It feeds on flatfish (fourth level), which feed on sand lances (third level), which feed on euphausiids (second level), which feed on phytoplankton (level 1). But the harp seal actually feeds at several trophic levels, from the second through the fourth, and it feeds on predators of some of its prey and thus is a competitor with some of its own food.[6] A species that feeds on several trophic levels typically is classified as belonging to the trophic level above the highest level from which it feeds. Thus, we place the harp seal on the fifth trophic level. As we have seen, the number and health of living organisms are determined by a range of abiotic and biotic factors. When abiotic conditions are not ideal, the organism may fail to grow or reproduce successfully or, in extreme cases, may die. This preferred set of conditions is called a **range of tolerance**. In the center of this range is a zone of preferred conditions, in which the organism will thrive. Shelford's Law of Tolerance states that for an organism to succeed in a given environment, each ecosystem condition must remain within the tolerance range of that organism. If any condition exceeds maximum or minimum tolerance levels, the organism will fail to thrive and will eventually die or leave the ecosystem. Although the distribution of many bird species depends on climatic factors, the ranges of others depend on the presence of particular vegetation (which is itself determined by climate). Still others depend on both temperature and vegetation. This suggests that if our climate is warming, temperature-dependent species will gradually move north, while those controlled by both temperature and vegetation cannot move until the vegetation moves. If vegetation changes slowly, as with forest trees, some species may be at risk if they are unable to find both acceptable minimum temperature conditions and an acceptable habitat (the combination of abiotic factors necessary for their survival) in the same location.

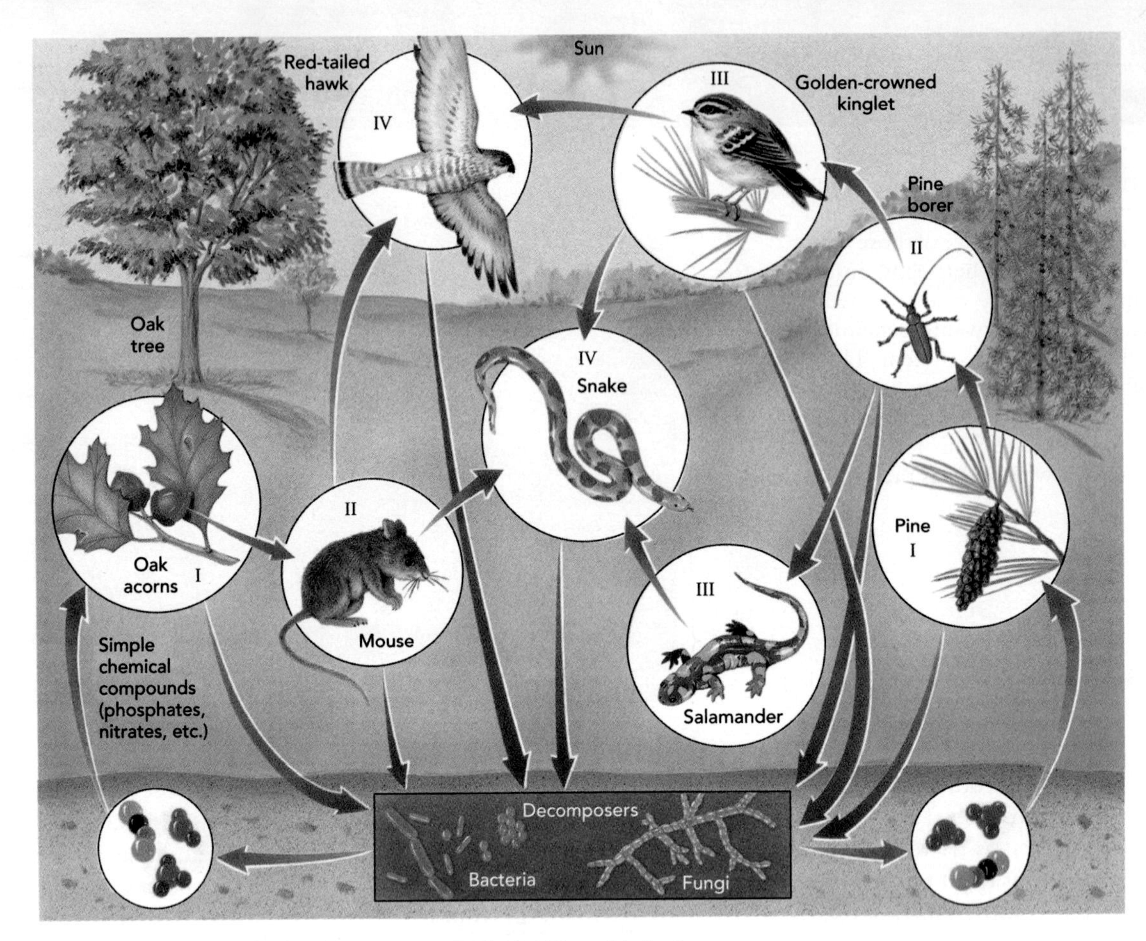

Figure 6.4 • A typical terrestrial food web. Roman numerals identify trophic levels.

There is usually a single limiting factor that determines the size, spatial range and condition of a population. In Chapter 2, we examined research into the factors that limit the growth of nuisance aquatic algae. For an Ontario cornfield, the limiting factor may be soil nutrient content if there is an abundance of rain. In a very dry year, however, moisture itself may become the limiting factor. For fish like trout and salmon, temperature and dissolved oxygen are often limiting factors. (The two are interrelated, in that colder water can hold more oxygen.)

Indicator species are those with known tolerance levels, whose presence or absence can provide insight into abiotic ecosystem conditions. For many years, aquatic invertebrate species have been used as indicators of water quality in lakes and streams. They are good indicators because they spend all of their lives in the water, have limited mobility, are easy to collect, and differ in their tolerance to water pollution. Mayfly, dragonfly, and damselfly larvae are very sensitive to low oxygen levels, for instance, and are therefore good indicators of adequate oxygenation. On the other hand, the presence of pollution-tolerant species like midge larvae and leeches can signal more polluted conditions.

Species vary widely in their tolerance of abiotic conditions. Some, like the carp, are able to survive in a wide range of conditions; others are acutely sensitive to small changes in habitat conditions. A species is not necessarily broad or narrow in all its preferences. Trout and salmon, for example, are acutely sensitive to changes in temperature and dissolved oxygen concentration, but can tolerate a relatively wide range of some toxins (copper and zinc are two examples). Generally speaking, younger individuals and those that are actively reproducing are more susceptible to changes in abiotic conditions than are older, non-reproductive individuals.

6.2 The Community Effect

Species can interact directly through food chains, as we have just seen. They also interact directly through symbiosis and competition, discussed in the next chapter. But a species can also affect other species indirectly, by affecting a third, a fourth, or many other species that, in turn, affect the second species. In addition, a species can affect the nonliving environment, which then affects a group of species in the community. Changes in that group affect another group. Such indirect and more complicated interactions are referred to as community-level interactions.

Interactions of populations at the community level are illustrated by the woodland caribou of northern Canada. In fact, the community-level interactions of the

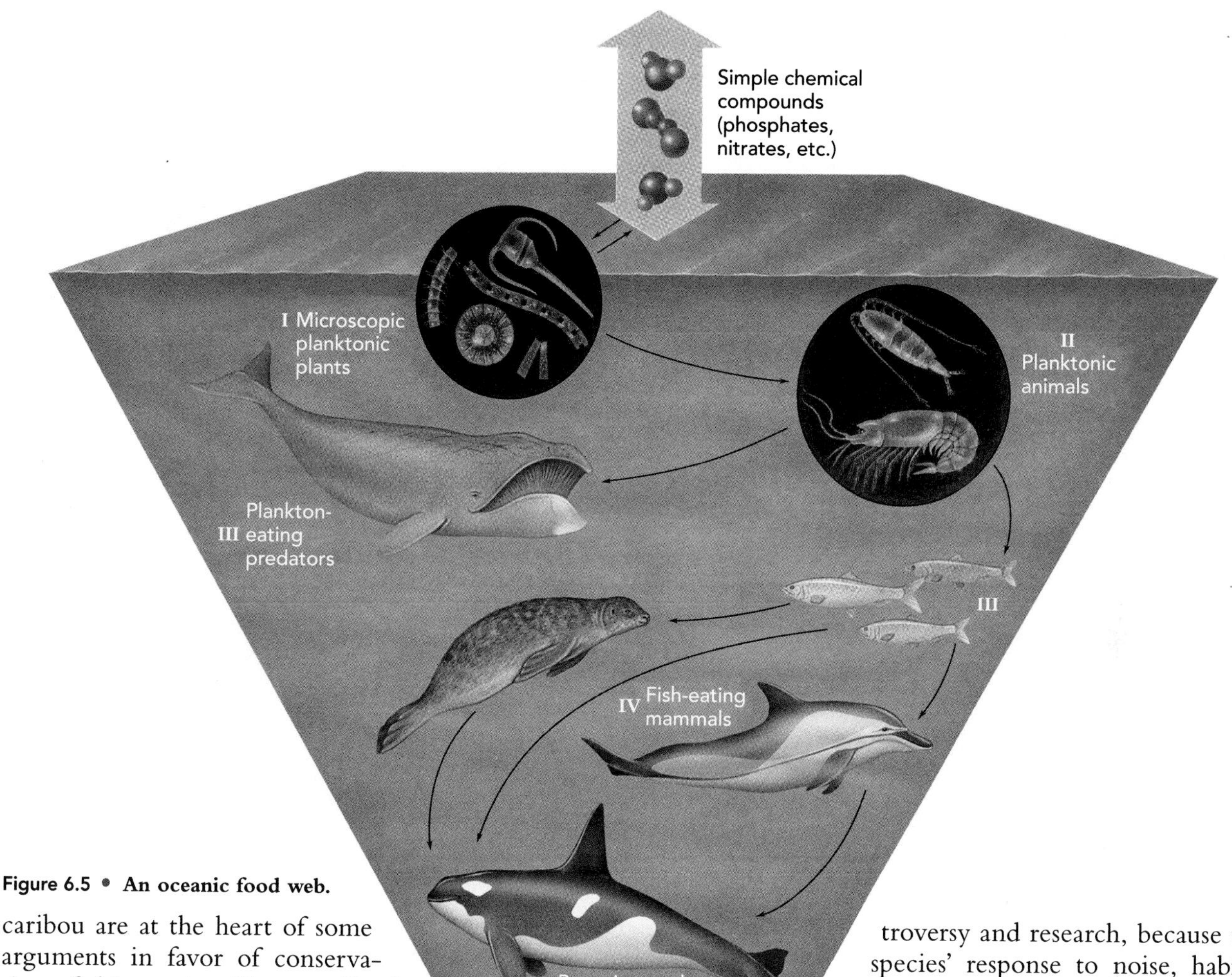

Figure 6.5 • An oceanic food web.

caribou are at the heart of some arguments in favor of conservation of this species. The woodland caribou is the southernmost subspecies of caribou in Canada, and has the widest range: from northern British Columbia and the Yukon through the Northwest Territories, Nunavut, and eastward through most of northern Canada. The Canadian population totals about 200,000, with roughly half of those animals located in Newfoundland and Labrador. Woodland caribou prefer to feed on lichens, and their metabolism is adapted for this purpose.[7, 8, 9] The caribou are in turn prey for large carnivores such as wolves and bears. Smaller predators such as lynx, wolverine, and coyote are able to bring down smaller caribou and calves.

Caribou have three features that make them a good indicator of habitat disturbance. First, through their unique lichen metabolism ability and role in predator-prey systems, they are central to the health of northern ecosystems. Second, they are found in low densities and reproduce slowly. And third, they are sensitive to habitat degradation.[10]

The caribou's preferred habitat is deep boreal forest, a forest type that is declining rapidly in northern Canada as a result of logging, mining, agriculture and habitat fragmentation due to pipeline and road building.[11] The caribou's protection has been a focus of controversy and research, because the species' response to noise, habitat alteration, and other factors are still poorly understood. Hunters argue that if harvesting levels are kept low, the population will not be affected. On the other hand, conservationists argue that it is which animals, not only how many, that are hunted that affects herd health. There is some evidence that killing the lead animal in a herd will frighten the remainder and divert it from suitable feeding and reproductive habitat. Conservationists say that there are still too few caribou for the population to be maintained at a sufficient level. If the caribou herd declines, lichen populations will increase while large carnivores will decline. This example shows that such effects can occur through food chains and can alter the distribution and abundance of individual species.

A species such as the woodland caribou that has a large effect on its community or ecosystem is called a **keystone species**, or a key species.[12] Its removal or a change in its role within the ecosystem changes the basic nature of the community.

Community-level effects demonstrate the reality behind the concept of an ecological community; they show us that certain processes can take place only because of a set of species interacting together. These effects also suggest that an ecological community is more than the

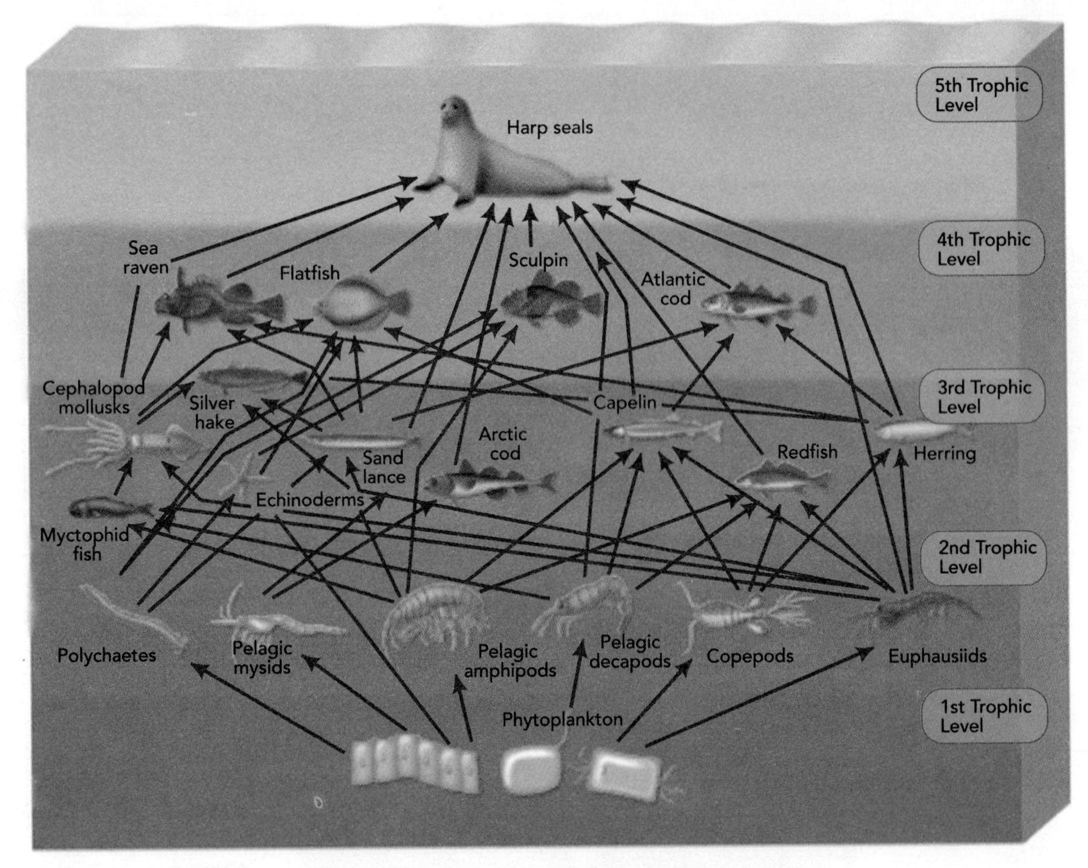

Figure 6.6 • **Food web of the harp seal** showing how complex a real food web can be.

sum of its parts—a perception called the holistic view. (Refer to the discussion of environmental connectivity in Chapter 3.)

In spite of their importance, community-level interactions are often difficult to recognize. One difficulty is knowing when and how species interact. Community-level interactions are not always as clear as those involving the woodland caribou. Even there, considerable scientific research was required to understand the interactions. Adding to the complexity, the set of species that make up an ecological community is not fixed completely but varies within the same kind of ecosystem from time to time and place to place. This brings us to the question of how to identify ecosystems.

6.3 How Do You Know When You Have Found an Ecosystem?

An ecosystem is the minimal entity that has the properties required to sustain life. This implies that an ecosystem is real and important and therefore that we should be able to find one easily. However, ecosystems vary greatly in structural complexity and in the clarity of their boundaries. Ecosystems differ in size, from the smallest puddle of water to a large forest. Ecosystems and their communities differ in composition, from a few species in the small space of a hot spring to many species interacting over a large area of the ocean. Furthermore, ecosystems differ in the kinds and relative proportions of their non-biological constituents and in their degree of variation in time and space.

Sometimes the borders of the ecosystem are well defined, such as the border between a lake

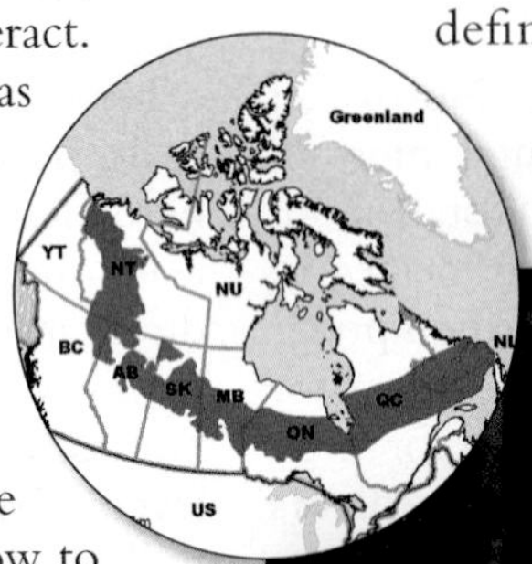

Figure 6.7 • **The Woodland Caribou and its distribution in Canada.**

and the surrounding countryside. But sometimes the transition from one ecosystem to another is gradual, as in the transition from prairie to deciduous forest to boreal forest in Riding Mountain National Park, Manitoba (Figure 6.8*b*), in the subtle gradations from grasslands to savannas in East Africa, and from boreal forest to tundra in the far north, where the trees thin out gradually (Figure 6.9).

A commonly used practical delineation of the boundary of an ecosystem on land is the watershed. A watershed is defined as follows: Within a watershed, any drop of rain that reaches the ground flows out in the same stream. Topography (the lay of the land) determines the watershed. When a watershed is used to define the boundaries of an ecosystem, the ecosystem is unified in terms of chemical cycling. Some classic experimental studies of ecosystems have been conducted on forested watersheds in Canadian Forest Service and United States Forest Service experimental areas. One of the most famous of these is the Hubbard Brook experimental forest in New Hampshire (Figure 6.10).

What all ecosystems have in common is not a particular physical size or shape but the processes we have mentioned: the flow of energy and the cycling of chemical elements. Ecological communities change over time, and it is the interactions among the species—a dynamic set of processes—that are the key to the community concept.

Some authors have referred to ecology as the "study of complexity"[13] and this is indeed a useful concept. Ecosystems are spatially complex mixtures of biotic and abiotic components. They are dynamic, naturally changing systems, not fixed in time. In one sense, they are functional units, but they are also connected to external systems through water, air, and the migration of organisms into and out of the system. A healthy ecosystem exhibits **resilience**, the ability to withstand disturbance because of the presence of complex, interactive processes and structures. In theory, the more diverse and complex an ecosystem, the more resilient it will be. C. S. Holling has described ecological resilience as the amount of disturbance that an ecosystem can absorb before it changes its structure by changing the variables and processes that control its behaviour. These ideas are consistent with Dan Botkin's "discordant harmonies". Managed within a reasonable range of variation, a forest will remain a forest, and a marsh will remain a wetland ecosystem. Changed beyond the point of resilience, the forest will become a grassland; the marsh will become a grassland.

6.4 Ecosystem Management

Ecosystems can be natural or artificial or a combination of both. An artificial pond that is a part of a waste-treatment plant is an example of an artificial ecosystem. Ecosystems can also be managed, and the management can include a large range of actions. Agriculture can be thought of as partial management of certain kinds of ecosystems (see Chapters 11 and 12), as can forests managed for timber production (see Chapter 13). Wildlife preserves are

Figure 6.8 • Ecosystem transitions. *(a)* Sometimes the transition from one ecosystem to another is sharp and distinct, as in the transition from lake to forest at Lake Moraine in Banff National Park, Alberta. *(b)* Sometimes the transition is gradual and fuzzy, as in the transitions among vegetation types and their associated ecosystems from prairie ecosystem to deciduous forest to boreal forest in Riding Mountain National Park, Manitoba.

(a)

(b)

examples of partially managed ecosystems (see Chapters 13 and 14).

Sometimes, when we manage or domesticate individuals or populations, we separate them from their ecosystems. We also do this to ourselves (see Chapter 3). When we do this, we must replace the ecosystem functions of energy flow and chemical cycling with our own actions. This is what happens in a zoo, where we must provide food and remove the wastes for individuals separated from their natural environments.

The ecosystem concept, then, lies at the heart of the management of natural resources. When we try to conserve species or manage natural resources so that they are sustainable, we must focus on their ecosystem and make sure that it continues to function. If it doesn't, we must replace or supplement ecosystem functions with our own actions. Ecosystem management, however, involves more than compensating for changes we make in ecosystems. It means managing and conserving life on Earth by considering chemical cycling, energy flow, community-level interactions, and the natural changes that take place within ecosystems. It means, as Dan Botkin has suggested in his book *Discordant Harmonies*, managing ecosystems within natural ranges of variation, in time scales appropriate for each system's processes and species.

Because of their complexity, ecosystems are difficult to describe. Measurements of the number of species or the sizes of individual populations reflect the structure of an ecosystem, but not its function. They fail to reveal the relationships that may exist between species or trophic levels, and they do not describe how efficiently or inefficiently the system functions. Measurements of the flow of energy (discussed in more detail in Chapter 9), or the rate at which biological tissue is produced, or the total mass of tissue present (biomass) reflect the system's function, but not its structure. Each approach provides only part of the picture; measures of both structure and function are necessary. These concepts will be explored in later chapters of this book.

Figure 6.9 • **Where do ecosystems begin and end?** Often, the transition between boreal forest and tundra is subtle and occurs over a large area. Here we see thinly scattered, small trees interspersed among tundra vegetation, within the transition in Kluane Park, Yukon.

Figure 6.10 • **A watershed ecosystem.** The V-shaped logged area in this picture is the famous Hubbard Brook ecosystem study. Here, a watershed defines the ecosystem, and the V shape is an entire watershed cut as part of the experiment.

CRITICAL THINKING ISSUE

How Are the Borders of an Ecosystem Defined?

The borders between ecosystems may be well defined or gradual. Those considered well defined include freshwater streams. Such ecosystems are often studied separately from surrounding ecosystems by researchers with different training and using different methods. Research on streams in coastal British Columbia in which salmon spawn has raised questions about the practice of studying aquatic and terrestrial ecosystems separately.

Salmon are anadromous fish—fish that come from the ocean to spawn in freshwater streams. In British Columbia, enormous numbers of salmon spawn in thousands of streams. Although salmon are born in freshwater, they migrate to the ocean, where most of their growth occurs. After they return to their home streams, they spawn and die. In one sense, therefore, salmon are a means of transporting resources from the ocean to freshwater. Because of their large numbers, salmon have the potential to make significant contributions to organic and mineral content of streams.

Salmon have a high lipid content compared with many other fish and are thus a good energy source for the animals that prey on them. In addition, their decay adds nitrogen, phosphorus, carbon, and other inorganic elements to freshwater. One study of a lake in western Alaska has shown that 24 million fish add 170 tons of phosphorus to the lake each year—an amount equal to or greater than recommended rates for applying fertilizers to trees. When the fish die, their carcasses decay and provide nourishment for algae, fungi, and bacteria. Invertebrates feed on these and on decaying bits of fish. Other fish feed on the invertebrates. Finally, bears and other carnivores eat salmon, both live and dead, during their upstream migration. In that way, nutrients derived from salmon pass into the soil and vegetation surrounding the streams.

Spawning fish have higher proportions of heavy isotopes of nitrogen and carbon (^{15}N and ^{13}C). These can be used to trace the relative contributions of anadromous fish to the nitrogen and carbon content of organisms in the food web. One such study showed that spawning salmon contributed 10.9% of the nitrogen found in invertebrate predators and 17.5% in the foliage of riparian plants. While it is not surprising to find aquatic invertebrates, which feed on salmon eggs and juveniles, with large amounts of nitrogen derived from salmon, researchers were surprised at the high levels in stream-side vegetation. When terrestrial mammals and birds feed on salmon, their feces and any uneaten salmon carcasses decay and add nutrients to the soil, where they can be taken up through the roots of plants. In coastal British Columbia, many species of mammals and birds feed on salmon. Salmon migrations attract large numbers of predators to streams and lakes. Salmon and other anadromous fish thus appear to link the ocean, freshwater, and land to an extent that is only beginning to be appreciated.

Critical Thinking Questions

1. Given the intricate connections between the aquatic and terrestrial ecosystems along salmon streams, how would you define the boundaries of the ecosystems?
2. When more adult salmon reach the spawning grounds than are needed to maintain the population, some are considered excess. How might the research described here affect that view?
3. Some biologists have called salmon a keystone species. Given what you know about keystone species, how would you argue for or against this designation?
4. In recent years, the numbers of anadromous fish along the Pacific Coast of Canada have declined precipitously because of overfishing and habitat destruction. What effects would you predict this might have on the ecology of freshwater streams and their adjoining land areas?
5. What types of management decisions about fish, wildlife, and forests would follow from recognizing the connection between aquatic and terrestrial ecosystems?

Summary

- An ecosystem is the simplest entity that can sustain life. At its most basic, an ecosystem consists of several species and a fluid medium (air, water, or both). The ecosystem must sustain two processes—the cycling of chemical elements and the flow of energy.
- The living part of an ecosystem is the ecological community, a set of species connected by food webs and trophic levels. A food web or chain describes who feeds on whom. A trophic level consists of all the organisms that are the same number of feeding steps from the initial source of energy.
- Community-level effects result from indirect interactions among species, such as those that occur when wolves hunt woodland caribou.
- Ecosystems are real and important, but it is often difficult to define the limits of a system or to pinpoint all the interactions that take place.
- Ecosystem management is considered key to the successful conservation of life on Earth.

STUDY QUESTIONS

1. What is the difference between an ecosystem and an ecological community?
2. Based on the discussion in this chapter, would you expect a highly polluted ecosystem to have many species or few species? Explain the reasons for your answer.
3. Is our human species a keystone species? Why or why not?
4. Which of the following are ecosystems? Which are ecological communities? Which are neither? Why?
   a. Québec City
   b. A 1,000-ha farm in Saskatchewan
   c. A sewage-treatment plant
   d. The St. Croix River
   e. Lake Ontario

FURTHER READING

Bilby, R. E., B. R. Fransen, and P. A. Bisson. "Incorporation of Nitrogen and Carbon from Spawning Coho Salmon into the Trophic System of Small Streams: Evidence from Stable Isotopes," *Canadian Journal of Fisheries and Aquatic Sciences* 53 (1996): 164–173. A well-documented example of a community-level effect.

Bormann, F. H., and G. E. Likens. *Pattern and Process in a Forested Ecosystem.* New York: Springer-Verlag, 1979. A synthetic view of the northern hardwood ecosystem, including its structure, function, development, and relationship to disturbance.

Botkin. Daniel B. 1990. *Discordant Harmonies: A New Ecology for the 21st Century.* New York: Oxford University Press.

Brown, D. E., F. Reichenbacher, and S. E. Franson. *A Classification of North American Biotic Communities/ Maps.* Salt Lake City: University of Utah Press, 1998. A description of a hierarchical classification system for biotic communities occurring from the Arctic Circle through Central America.

Diamond, J., and T. J. Case, eds. *Community Ecology.* New York: Harper & Row, 1986. An examination of community ecology from several perspectives, including spatial and temporal scales, equilibrium theory, and community structuring mechanisms.

Pahl-Wostl, C. *The Dynamic Nature of Ecosystems: Chaos and Order Entwined.* New York: Wiley, 1995. A book arguing that the trade-off between the irregular, chaotic dynamics at the level of populations and the spatiotemporal organization at the level of the system as a whole shapes ecological systems.

Strong, D. R., D. Simberloff, L. G. Abele, and A. B. Thistle, eds. *Ecological Communities: Conceptual Issues and the Evidence.* Princeton, N.J.: Princeton University Press, 1984.. A collection of papers that illuminates species interactions and the ecological effects of these interactions.

Willson, M. R., and K. C. Halupka. "Anadromous Fish as Keystone Species in Vertebrate Communities." *Conservation Biology* 9 (1995) 3:489–497. An article describing community-level effects of salmon.

chapter 7

Biodiversity in the Bay of Fundy

The Bay of Fundy is one of Canada's most diverse and productive ecosystems. Part of its fame derives from its exceptionally high tides and "tidal bore" —the wall of water that forms when an incoming tide advances up a constricted opening like a river mouth. In fact, the Bay's tides are the basis of much of its diversity. Like other estuarine systems, where the fresh water of rivers mixes with the salt water of the ocean, the Bay of Fundy supports a mixture of marine and freshwater species. Unlike other estuaries, however, the Bay of Fundy's very high tides play a role in increasing the diversity of species. At the mouth of the Bay, tidal action causes upwelling of deeper, nutrient-rich waters, encouraging the growth of a wide range of plankton species, and the numerous fish, seabird and mammal species that feed on them. Farther up the Bay, vast tidal flats and salt marshes provide excellent habitat for migratory waterfowl, bottom-feeding fish, and many shellfish and other invertebrates.

The Bay's many habitat types provide abundant food supply and nursery areas for fish and shorebirds, while tributary streams and adjacent terrestrial habitats offer additional opportunities for breeding, spawning, foraging, and the raising of young. Each habitat not only has its own biological diversity, but also affects the diversity of life over a much larger region.

In recent years, scientists have observed declining numbers and altered behaviour in the rare Northern Right Whale *Eubalaena glacialis* and three important Fundy bird species: phalaropes, petrels, and puffins. The fish community is also showing a subtle shift, with favoured food species like cod, halibut, and haddock declining, and dogfish (a small shark) and skate increasing. In some parts of the Bay, sediment quality has also changed. Chignecto Bay, one of two narrow arms at the head of the Bay of Fundy, receives flows from several rivers, including the Petitcodiac and Missaguash. Perhaps because of upstream land clearance in the watersheds of those rivers, sediments in Chignecto Bay have become noticeably finer over the last several decades. By contrast, sediments in nearby Minas Basin, a more open channel, appear to have become coarser over time. These changes had an important impact on the distribution and foraging behaviour of shorebirds such as the semipalmated sandpiper, for which the area is an important feeding ground.[1]

Many explanations for these changes have been proposed. The region has experienced considerable alteration over the past 150 years, especially from logging, pulp and paper production, and commercial fishing. More recently, many tributary streams have been dammed and causeways have been constructed to improve traffic flow, causing alterations in water flow and sediment deposition patterns.

Some scientists have suggested that these human influences had the result of simplifying the Bay of Fundy ecosystem. There is evidence of a decline in the size and number of large fish and mammals, and changes in the Bay's intricate food web. Some of these changes have been attributed to habitat degradation, such as the changes in Chignecto Bay's sediments, and chemical contamination. In some areas, pollution-tolerant algae species are increasing, with associated declines in more sensitive species. Overall, the Bay remains an unusually productive and diverse ecosystem. If measures can be implemented to reduce pollution, protect sensitive habitat, and avoid excessive harvesting, this rich resource can be conserved for future generations.

- *The case study of the Bay of Fundy shows that the conditions of local habitats and ecosystems can affect other, much larger ones. Conservation of biological diversity in the Bay of Fundy is representative of the complexity involved in conserving biological diversity throughout most of the world, and it illustrates two of the major themes of this book: science and values, and people and nature. How to preserve the region's biological diversity—and why this matters—are the subjects of this chapter.*

Biological Diversity

LEARNING OBJECTIVES

People have long wondered how the amazing diversity of living things on Earth occurred. This diversity developed through biological evolution and is affected by interactions between species and by the environment. After reading this chapter, you should understand:

- How mutation, natural selection, migration, and genetic drift lead to the evolution of new species.
- Why people value biological diversity.
- How people can affect biological diversity.
- How biological diversity may affect biological production, energy flow, chemical cycling, and other ecosystem processes.
- What environmental major problems are associated with biological diversity.
- Why so many species have evolved and persist.
- The concepts of the ecological niche and habitat.

Figure 7.1 • The Bay of Fundy.

7.1 What Is Biological Diversity?

Biological diversity refers to the variety of life forms commonly expressed as the number of species in an area, or the number of genetic types in an area. However, discussions about biological diversity are complicated by the fact that people mean various things when they talk about it. They may mean conservation of a single rare species, of a variety of habitats, of the number of genetic varieties, of the number of species, or of the relative abundance of species. These concepts are interrelated, but each has a distinct meaning.

Newspapers and television frequently cover the problem of disappearing species around the world and the need to conserve these species. Before we can intelligently discuss the issues involved in conserving the diversity of life, we must understand how this diversity came to be. First, this chapter addresses the principles of biological evolution. We then turn to biological diversity itself: its various meanings, how interactions among species increase or decrease diversity, and how the environment affects diversity.

Figure 7.2 • **Inuit images of nature from Canada's arctic.**

7.2 Biological Evolution

The long asked question about biological diversity is: How did it all come about? Before modern science, the diversity of life and the adaptations of living things to their environment seemed too amazing to have happened by chance. The only possible explanation seemed to be that this diversity was created by God (or gods). The great Roman philosopher and writer Cicero put it succinctly: "Who cannot wonder at this harmony of things, at this symphony of nature which seems to will the well-being of the world?" He concluded that "everything in the world is marvelously ordered by divine providence and wisdom for the safety and protection of us all".[2]

With the rise of modern science, however, other explanations became possible. In the nineteenth century, Charles Darwin found an explanation which became known as **biological evolution**. (His insights, based in part on a visit to the Galápagos Islands and his observation of the species there, are described in more detail in Chapter 8.) Biological evolution refers to the change in inherited characteristics of a population from generation to generation. It can result in new species—populations that can no longer reproduce with members of the original species. *Along with self-reproduction, biological evolution is one of the features that distinguishes life from everything else in the universe.*

According to the theory of biological evolution, new species arise as a result of competition for resources and the difference among individuals in their adaptations to environmental conditions. Since the environment continually changes, which individuals are best adapted changes too. As Charles Darwin wrote "can it be doubted, from the struggle each individual has to obtain subsistence, that any minute variation in structure, habits, or instincts, adapting that individual better to the new [environmental] conditions, would tell upon its vigour and health? In the struggle it would have a better chance of surviving; and those of its offspring that inherited the variation, be it ever so slight, would also have a better chance." Sounds plausible, but how does this evolution occur? *Four processes lead to evolution: mutation, natural selection, migration, and genetic drift.*

Mutation

Genes, contained in chromosomes within cells, are inherited—passed from one generation to the next. A *genotype* is the genetic makeup of an individual or group. Genes are made up of a complex chemical compound called deoxyribonucleic acid (DNA). DNA in turn is made up of chemical building blocks that form a code, a kind of alphabet of information. The DNA alphabet consists of four letters (specific nitrogen-containing compounds, called bases), which are combined in pairs: (A) adenine, (C) cytosine, (G), guanine, and (T) thymine. How these letters are combined in long strands determines the "message" interpreted by a cell to produce specific compounds.

Sets of the four base pairs form a **gene**, which is a single piece of genetic information. The number of base pairs that make up a gene varies. To make matters more

(a)

(b)

(c)

Figure 7.3 • Mutation. *(a)* a normal fruit fly and a fruit fly with an antennae mutation. *(b)* Trandescantia is a small flowering plant used in the study of effects of mutagens. The color of stamen hairs in the flower (pink versus clear) is the result of a single gene, and changes when that gene is mutated by radiation or certain chemicals, such as ethylene chloride.

complex, some base pairs found in DNA are nonfunctional—they are not active and do not determine any chemicals that are produced by the cell. Furthermore, some genes affect the activity of others, turning those other genes on or off. And creatures such as humans have genes that limit the number of times a cell can divide—and therefore determine maximum longevity.

When a cell divides, the DNA is reproduced and each new cell gets a copy. Sometimes an error in reproduction changes the DNA and therefore changes inherited characteristics. Sometimes an external agent comes in contact with DNA and alters it. Radiation, such as X-rays and gamma rays, can break the DNA apart or change its chemical structure. Certain chemicals can also change DNA, as can viruses. When DNA changes in any of these ways, then it is said to have undergone **mutation**.

In some cases, a cell or offspring with a mutation cannot survive. In other cases, the mutation simply adds variability to the inherited characteristics. But in still other cases, individuals with mutations are so different from their parents that they cannot reproduce with normal offspring of their species, so a new species has been created (Figure 7.3).

Natural Selection

When there is variation within a species, some individuals may be better suited to the environment than others. (Change is not always for the better. Mutation can result in a new species whether or not that species is better adapted than its parental species to the environment.) Organisms whose biological characteristics make them better able to survive and reproduce in their environment leave more offspring than others. Their descendants form a larger proportion of the next generation. In this way, these individuals are more "fit" for the environment; this process of increasing the proportion of offspring is called **natural selection**. Which inherited characteristics lead to more offspring depends on the specific characteristics of an environment, and as the environment changes over time, the characteristics "fit" will also change. In summary, natural selection has four primary characteristics:

1. Inheritance of traits from one generation to the next and some variation in these traits—that is, genetic variability.
2. Environmental variability.
3. Differential reproduction that varies with the environment.
4. Influence of the environment on survival and reproduction.

Natural selection is illustrated in A Closer Look 7.1, which describes how the mosquitoes that carry malaria develop resistance to DDT and how the species of microorganism that causes malaria develops a resistance to quinine, a treatment for the disease.

When natural selection takes place over a long time, a number of characteristics can change. The accumulation of these changes may be so great that the present generation can no longer reproduce with individuals that have the original DNA structure, resulting in a new species. A **species** is a group of individuals that can (and at least occasionally do) reproduce with each other.

Geographic Isolation and Migration

Sometimes two populations of the same species become geographically isolated from each other for a long time. During that time, the two populations may change so much that they can no longer reproduce together even when they are brought back into contact. In this case, two new species have evolved from the original species. This can happen even if the genetic changes are not more fit but simply different enough to prevent reproduction.

Ironically, the loss of geographic isolation can also lead to a new species. This can happen when one population of a species migrates into a habitat previously occupied by another population of that species, thereby changing gene frequency in that habitat. For example, this change

A CLOSER LOOK 7.1

Natural Selection: Mosquitoes and the Malaria Parasite

Malaria poses a great threat to 2.4 billion people—over one-third of the world's population—living in more than 90 countries, most of them located in the tropics. In the United States, Florida has recently experienced a small but serious malaria outbreak. Worldwide, an estimated 300 to 400 million people are infected each year, 1.1 million of whom die.[3] In Africa alone, more than 3,000 children die daily from this disease.[4]

Once thought to be caused by filth or bad air (hence the name *malaria*, from the Latin for "bad air"), malaria is actually caused by parasitic microbes (four species of the protozoa *Plasmodium*). These microbes affect and are carried by *Anopheles* mosquitoes, which then transfer the protozoa to people.

One solution to the malaria problem, then, would be the eradication of *Anopheles* mosquitoes. By the end of World War II, scientists had discovered that the pesticide DDT was extremely effective against *Anopheles* mosquitoes. They had also found chloroquine highly effective in killing *Plasmodium* parasites. (Chloroquine is an artificial derivative of quinine, a chemical from the bark of the quinine tree that was an early treatment for malaria.)

In 1957, the World Health Organization (WHO) began a $6 billion campaign to rid the world of malaria using a combination of DDT and chloroquine. At first, the strategy seemed successful. By the mid-1960s, malaria was nearly gone or had been eliminated from 80% of the target areas.

However, success was short-lived. The mosquitoes began to develop a resistance to DDT, and the protozoa became resistant to chloroquine. In many tropical areas, the incidence of malaria worsened. For example, as a result of the WHO program, the number of cases in Sri Lanka had dropped from 1 million to only 17 by 1963. But by 1975, 600,000 cases had been reported, and the actual number is believed to be four times higher. Resistance among the mosquitoes to DDT became widespread, and resistance of the protozoa to chloroquine was found in 80% of the 92 countries where malaria was a major killer.[3, 5]

The mosquitoes and the protozoa developed this resistance through natural selection. When they were exposed to DTT and chloroquine, the susceptible individuals died. The most resistant organisms survived and passed their resistant genes to their offspring. Since the susceptible individuals died, they left few or no offspring, and any offspring they left were susceptible.

Thus, a change in the environment—the human introduction of DDT and chloroquine—caused a particular genotype to become dominant in the populations. A practical lesson from this experience is that if we set out to eliminate a disease-causing species, we must attack it completely at the outset and destroy all the individuals before natural selection leads to resistance. But sometimes this may be an impossible task, in part because of the natural genetic variation in the target species.

Since the drug chloroquine is generally ineffective now, new drugs have been developed to prevent malaria. However, these second- and third-line drugs will eventually become unsuccessful, too, as a result of the same process of biological evolution by natural selection. This process is speeded up by the ability of the *Plasmodium* to mutate rapidly. In South Africa, for example, the protozoa became resistant to mefloquine immediately after the drug became available as a treatment.

An alternative is to develop a vaccine against the *Plasmodium* protozoa. Biotechnology has made it possible to map the structure of these malaria-causing organisms. Scientists are currently mapping the genetic structure of *P. falciparum*, the most deadly of the protozoa, and expect to finish within several years. With this information, they expect to create a vaccine containing a variety of the species that is benign in human beings but that produces an immune reaction.[6]

In addition, scientists are mapping the genetic structure of *Anopheles gambiae*, the carrier mosquito. This project could provide insight about genes that could prevent development of the malaria parasite within the mosquito. In addition, it could identify genes associated with insecticide resistance and provide clues to developing a new pesticide.[6]

The development of resistance to DDT by mosquitoes and to chloroquine by *Plasmodium* is an example of biological evolution in action today. With the aid of biotechnology, scientists are working to understand the specific chemical structure of the inheritance of characteristics.

Figure 7.4 • **The *Anopheles* mosquito.**

in gene frequency can result from the migration of seeds of flowering plants blown by wind or carried in the fur of mammals—if the seed lands in a new habitat, the environment may be different enough to favor genotypes not as favored by natural selection in the parents' habitat. Natural selection, in combination with geographic isolation and subsequent migration, can thus lead to new dominant genotypes and eventually to new species.

Migration has been an important evolutionary process over geologic time (a period long enough for geologic changes to take place). For example, during intervals between recent ice ages and at the end of the last ice age, Alaska and Siberia were connected by a land bridge that permitted the migration of plants and animals. When the land bridge was closed off by rising seawater, populations that had migrated to the New World were cut off, and new species evolved. Similarly, marsupials that migrated to Australia were cut off from other continents after geographical shifts.

The word, *evolution* in the term *biological evolution* has a special meaning. Outside biology, *evolution* is used broadly to mean the history and development of something. Within biology, however, the term has a more specialized meaning. For example, geologists talk about the evolution of Earth, which simply means Earth's history and the geologic changes that occurred over that history. Book reviewers talk about the evolution of the plot of a novel, meaning how the story unfolds. Biological evolution is a one-way process. Once a species is extinct, it is gone forever. You can run a machine, such as a mechanical clock, forward and backward. But when a new species evolves, it cannot evolve backward into its parents.

Genetic Drift

Another process that can lead to evolution is **genetic drift**—changes in the frequency of a gene in a population due not to mutation, selection, or migration but simply to chance. Chance may determine which individuals become isolated in a small group from a larger population and thus which genetic characteristics are most common in that isolated population. The individuals may not be better adapted to the environment; in fact, they may be more poorly adapted or neutrally adapted. Genetic drift can occur in any small population and may also present problems when a small group is by chance isolated from the main population. For example, Rocky Mountain bighorn sheep (*Ovis canadensis canadensis*) live in grasslands and shrublands along the western flanks of British Columbia's southern Rocky Mountains. In the summer, these sheep feed high up in the mountains, where it is cooler and wetter and there is more vegetation. Before high-density European settlement of the region, the sheep could move freely and sometimes migrated from one mountain to another by descending into the valleys and crossing them in the winter. In this way, large numbers of sheep interbred.

With the development of cattle ranches and other human activities, many populations of bighorn sheep could no longer migrate among the mountains by crossing the valleys. These sheep became isolated in very small groups—commonly, a dozen or so—so chance may play a large role in what inherited characteristics remain in the population.

Genetic drift can improve the adaptation of these populations. Suppose, for example, that you begin a study of mosquitoes that carry the malaria parasite (see A Closer Look 7.1). You want to study their resistance to DDT, and so you go out and collect a sample of the mosquitoes. After you grow the population and obtain a number of generations of offspring, you begin your research by exposing them to DDT—and none of them die. By chance, the ones you collected were all resistant to DDT, as are their offspring. Your experiment ends up with a population completely resistant to DDT, unlike the wild population, which has a few individuals who are resistant.

But genetic drift is generally considered a serious problem when populations become very small, as happens with endangered species. It can be a problem for a rare or endangered species for two reasons: (1) characteristics that are less adapted to existing environmental conditions may dominate, making survival of the species less likely; and (2) the small size of the population reduces genetic variability and hence the ability of the population to adapt to future changes in the environment.

Evolution as a Game

Biological evolution is so different from other processes that it is worthwhile to spend some extra time exploring this idea. There are no simple rules that species must follow to win or to stay in the game of life. Sometimes when we try to manage species, we assume that evolution will follow simple rules. But species play tricks on us; they adapt or fail to adapt over time in ways that we did not anticipate. Such unexpected outcomes result from our failure to understand fully how species have evolved in relation to their ecological situations. Nevertheless we continue to hope and plan as if life and its environment will follow simple rules. This is true even for the most recent work in genetic engineering.

Complexity is a feature of evolution. Species have evolved many intricate and amazing adaptations that have allowed them to persist. It is essential to realize that these adaptations have evolved not in isolation but in the context of relationships to other organisms and to the environment. The environment sets up a situation within which evolution, by natural selection, takes place. The great ecologist G. E. Hutchinson referred to this interaction in the title of one of his books, *The Ecological Theater and the Evolutionary Play*. Here, the ecological situation—the condition of the environment and other species—is the scenery and theatre within which natural selection occurs, and natural selection results in a story of evolution played out in that theatre—over the history of life on Earth.[7]

7.3 Basic Concepts of Biological Diversity

Now that we have explored biological evolution, we can turn to biological diversity. To develop workable policies for conserving biological diversity, we must be clear about the meaning of the term. Biological diversity involves the following concepts:

- *Genetic diversity*: the total number of genetic characteristics of a specific species, subspecies, or group of species. In terms of genetic engineering and our new understanding of DNA, this could mean the total base-pair sequences in DNA; the total number of genes, active or not; or the total number of active genes.
- *Habitat diversity*: the different kinds of habitats in a given unit area.
- *Species diversity*, which, in turn, has three qualities:
  species richness—the total number of species;
  species evenness—the relative abundance of species;
  and
  species dominance—the most abundant species.

To understand the differences among species richness, species evenness, and species dominance, imagine two ecological communities, each with 10 species and 100 individuals, as illustrated in Figure 7.5. In the first community (Figure 7.5*a*), 82 individuals belong to a single species, and the remaining 9 species are represented by 2 individuals each. In the second community (Figure 7.5*b*), all the species are equally abundant; each has 10 individuals. Which community is more diverse?

At first, one might think that the two communities have the same species diversity because they have the same number of species. However, if you walked through both communities, the second would appear more diverse. In the first community, most of the time you would see individuals only of the dominant species (in the case shown in Figure 7.5*a*, corn plants); you probably would not see many of the other species at all. In the second community, even a casual visitor would see many of the species in a short time. The first community would appear to have relatively little diversity until it was subjected to careful study.

As this example suggests, merely counting the number of species is not enough to describe biological diversity. Species diversity has to do with the relative chance of seeing species as much as it has to do with the actual number present. Ecologists refer to the total number of species in an area as *species richness*, the relative abundance of species as *species evenness*, and the most abundant species as *dominant*.

7.4 The Evolution of Life on Earth

For the mosquitoes and their malaria parasite (see A Closer Look 7.1), evolution occurred rapidly. In contrast, during most of Earth's history, evolution seems to have proceeded on average much more slowly. How do we know about the history of evolution? In part, from the study of fossils. The earliest known fossils, 3.5 billion years old, are micro-organisms that appear to be ancestral forms of bacteria and what some microbiolo-

Figure 7.5 • Species richness versus species evenness. The corn field and the mixed hardwood forest may have the same number of species (species richness) but very different species abundance (species evenness) but very different relative species abundance (evenness).

gists now called Archaea.[8] For the next 2 billion years, only such microbial forms lived on Earth.

Amazingly, these organisms greatly changed the global environment, especially by altering the chemistry of the atmosphere. A major way this change came about was from photosynthesis, a capability that evolved during those 2 billion years. As with all photosynthetic organisms, these early ones removed carbon dioxide from the atmosphere and released large amounts of oxygen into it (illustrating our ongoing assertion that life has always changed the environment on a global scale). This led to a high concentration of oxygen in the atmosphere (familiar to us today), setting the ecological stage for the evolution of new forms of life. That free oxygen allowed the evolution of respiration, which paved the way for oxygen-breathing organisms, including humans.

The earliest fossils of multicellular organisms appear in approximately 600-million-year-old rocks in southern Australia. These had shells, gills, filters, efficient guts, and a circulatory system, and in these ways they were relatively advanced—they must have had ancestors that do not appear in known fossils in which these organs and systems evolved. Among these were jellyfish-like animals, trilobites (Figure 7.6), mollusks (clams), echinoderms (such as sea urchins), and sea snails.

During the first major period of multicellular life, called the *Cambrian period*, which lasted until about 500 millions years ago, living things remained in the oceans. Almost 100 million years later in the Silurian period, plants evolved to live on land. One of the world's most remarkable Middle Cambrian deposits is the Burgess Shale, located in Yoho National Park, in the Rocky Mountains near Field, British Columbia. The Burgess Shale is dated to about 540 million years ago and has been the subject of intense study since it was discovered in 1909. It is remarkable because of the large number of fossil invertebrates preserved in the shale, and the excellent state of their preservation. (See Figure 7.7.)

Figure 7.6 • Photograph of Fossil Trilobite.

For larger, multicellular organisms, life on land required some major innovations, including the following:

- Structural support needed, because while aquatic organisms are buoyed up by the water, on land gravity becomes a real force with which to contend.
- An internal aquatic environment, with a plumbing system giving it access to all parts of the organism and devices for conserving the water against losses to the surrounding atmosphere.
- Means for exchanging gases with air instead of with water.
- A moist environment for the reproductive system, essential for all sexually reproducing organisms.

The first fish to venture onto land, an obscure group called the crossopterygians, did so in the Devonian period (about 400 million years ago). These gave rise to the amphibians. The crossopterygians had several features that served to make the transition possible. Their lobe-like fins, for example, were preadapted as limbs because the lobes contain (much foreshortened) the elements of a quadruped limb, complete with small bones to form the extremity. They also had internal nostrils characteristic of air-breathing animals. Being fish, the crossopterygians already had a serviceable vascular system that was adequate for making a start on land. Fossils of amphibians occur even later, in the Devonian period about 360 million years ago. Water conservation, however, never became a strong point with amphibians: they retain permeable skins to this day, which is one reason they have never become independent of the aquatic environment.

The earliest land plants were seedless, and the earliest of these could reproduce only in water, so they were limited to wet habitats. These plants reached their peak in dominance of the land in the Carboniferous period (see appendices for dates for these periods). Seed plants evolved during the Devonian, starting with conifers with naked seeds (plants called *gymnosperms*, which means "naked seed"). The last frontiers for plants—so far, at least—were dry steppes, savannas, and prairies. These were not colonized until the Tertiary period, about 55 million years ago, when grasses evolved (Figure 7.8).

Despite their limitations, the amphibians ruled the land for many millions of years during the Devonian period. They had one difficulty that limited their expansion into many niches: They never met the reproductive requirement for life on land. In most species, the female amphibian lays her eggs in the water, the male fertilizes them there after a courtship ritual, and the young hatch as

Figure 7.7 • **The Burgess Shale.**

tadpoles. Like the seedless plants, the amphibians—with one foot on the land, so to speak— have remained tied to the water for breeding. Some became quite large (2–3 m long). One branch evolved to become reptiles; the rest that survive are frogs, toads, newts, salamanders, and limbless water "snakes" that seem to have decided that, after all, they prefer a fish's life.

The reptiles freed themselves from the water by evolving an egg that could be incubated outside of the water and by getting themselves a watertight skin. These two "inventions" gave them the versatility to occupy terrestrial niches that the amphibians had missed because of their bondage to the water. The amniotic egg did for reptilian diversity what jaws did for diversity in fishes. Originating in the Carboniferous coal swamps (about 375 million years ago), by the Jurassic period (some 185 million years later) the reptiles had moved onto the land, up into the air, and back to the water (as veritable sea monsters). This resulted in the production of the two orders of dinosaurs (the largest quadrupeds ever to walk the Earth) and gave rise to two new vertebrate classes—mammals and birds.

Mammals are in many ways better equipped as quadrupeds to occupy terrestrial niches than were the great reptiles. It is difficult to pick out a single mammalian "invention" comparable to the jaws of fish or the reptilian egg, for the mammal is a fine-tuned quadruped, adapted to a faster and more versatile life than the reptiles could ever have led (notwithstanding the current vogue of popularity they are enjoying). The mammalian "invention" is perhaps just that: a set of interdependent improvements managed by a more capable brain and supported by a faster metabolism. The placental uterus is sometimes regarded as the key to mammalian success; but it's really only a piece of equipment mandated by the delicate intricacy of the fetus that lives in it, especially its brain.

Thus, life evolved on Earth, bringing us, in the most general way, to the present, where we confront problems about the great diversity of life. The mechanism of biological evolution, the rate at which species evolved and became extinct, and the kinds of environments in

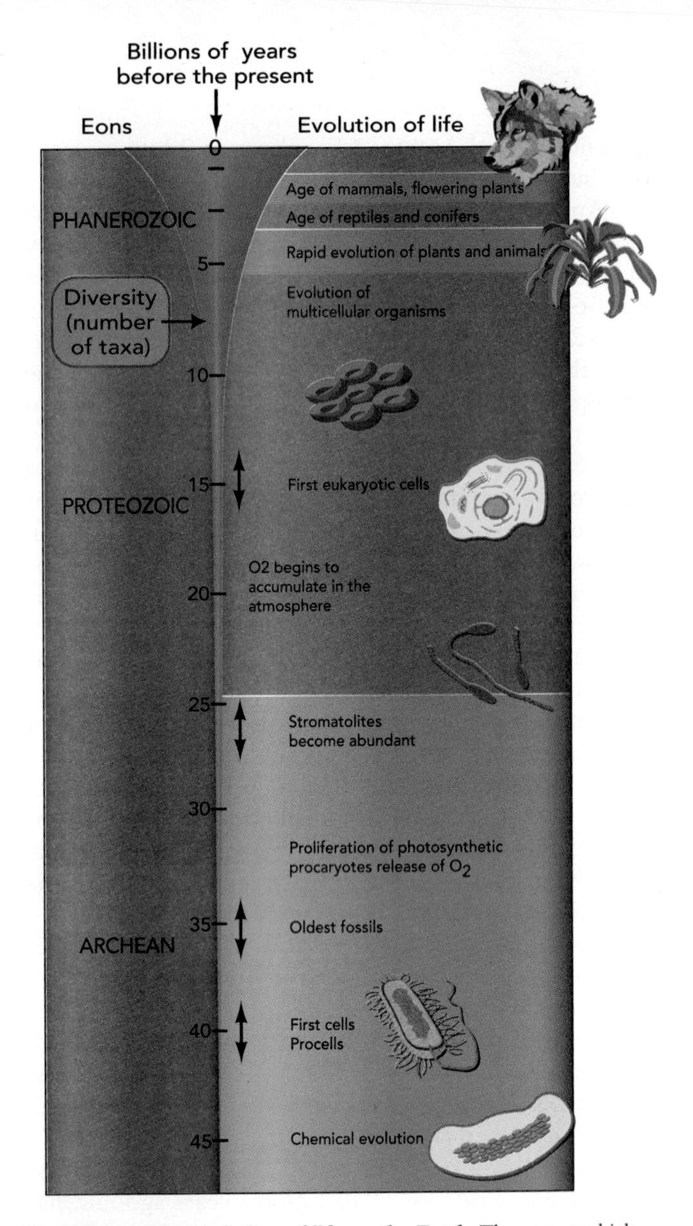

Figure 7.8 • **The evolution of life on the Earth.** The rates at which new organisms appear and of biological diversity both increase with time. A simplified representation of global diversity through geologic time.

which species evolve provide essential background to understanding today's biological diversity issues.

During the history of life on Earth, evolution generally proceeded comparatively slowly, as did the extinction of species. But major catastrophes, including the crashing of asteroids onto Earth, rapidly changed the environment at a global level, extinguished many species in a comparatively short time, and opened up niches to which new species then evolved (Figure 7.9). Environmental change at many scales of time and space is a characteristic of our planet. Species have evolved within this environment and adapted to it. As a result, many species require certain kinds and rates of change. When we slow down or speed up environmental change, we impose novel risks on species.

7.5 The Number of Species on Earth

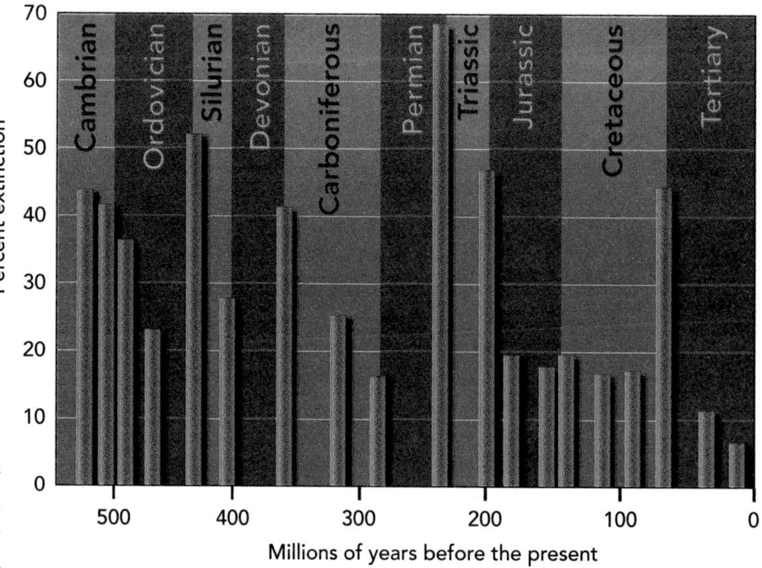

Figure 7.9 • **The extraordinary frequency of great extinction events that have occurred during the Phanerozoic Eon.** The percentage of extinction was determined from the disappearance of genera of well-skeletonized animals.

Many species have come and gone on Earth. But how many exist today? No one knows the exact number because new species are discovered all the time, especially in little-explored areas such as tropical rain forests. For example, since 1992, five new mammals have been discovered in Laos and two more have been "rediscovered." The rediscovered animals were known from a few specimens but had not been seen for a long time. The new species include (1) the spindle-horned oryx (which is not only a new species but also represents a previously unknown genus); (2) the small black muntjak; (3) the giant muntjak (the muntjak, also known as "barking deer," is a small deer; the giant muntjak is so called because it has large antlers); (4) the striped hare (whose nearest relative lives in Sumatra); and (5) a new species of civet cat. The rediscovered species are a wild pig, previously known only from a skull found by a hunter in the 1890s, and Roosevelt's muntjak, previously known from one specimen obtained in the 1920s.

That such a small country with a long history of human occupancy would have so many new mammal species suggests how little we still know about the total biological diversity on Earth. Some 1.4 million species have been named, but biologists estimate that the total number is probably considerably higher, from 3 million to as many as 100 million (Figure 7.10).

Many people think in terms of two major kinds of life: animals and plants. Scientists, however, group living things on the basis of evolutionary relationships—a biological genealogy. In the recent past, scientists classified life into five kingdoms: animals, plants, fungi, protists, and bacteria. New evidence from the fossil record and studies in molecular biology suggest that it may be more appropriate to describe life as existing in three major domains, one called Eukaryota or Eukarya, which includes animals, plants, fungi, and protists (mostly single-celled organisms); Bacteria; and Archaea.[8,9] Eukarya have a nucleus and other organelles; Bacteria and Archaea do not. Archaea used to be classified among Bacteria, but they have substantial molecular differences that suggest ancient divergence in heritage (Figure 7.11).

Figure 7.10 • **Comparison of species estimates** for selected taxa. With about 15,000 new species described every year, a changing classification system, and no central data registry, it is not surprising that species estimates differ among publications and years. (*Sources*: Modified from Wilson, E. O., ed. 1988. *Biodiversity*, (Washington, D.C.: National Academy Press), Lynn Margulis, Karlene V. Schwartz (contributor), and Stephen Jay Gould, 1988. *Five Kingdoms: An Illustrated Guide to the Phyla of Life on Earth*.)

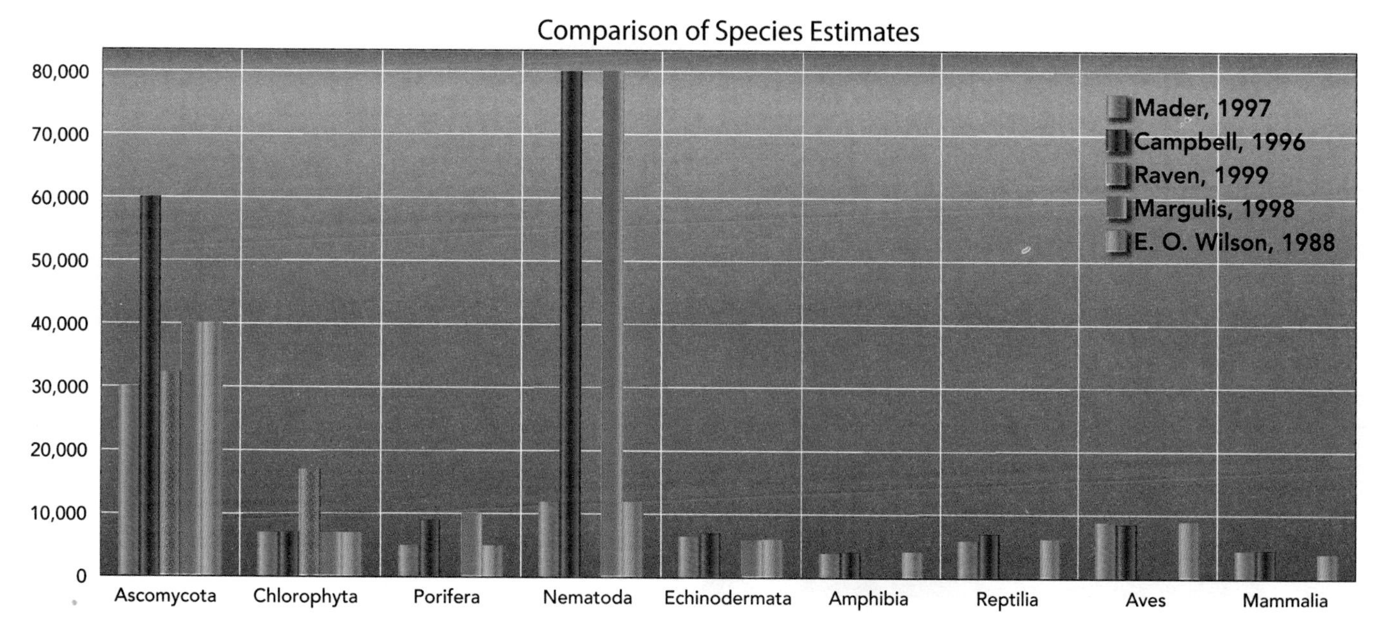

Insects and plants make up most of the known species (see Figure 7.10). Many of the insects are tropical beetles in rain forests. Mammals, the group to which people belong, include a comparatively small number of species, about 4,000.

7.6 Why Are There So Many Species?

Why are there so many species? Since species compete with one another for resources, wouldn't the losers drop out, leaving only a few winners? How did so many different species survive? Part of the answer lies in the different ways in which organisms interact, and the other part lies with the idea of the ecological niche.

Interactions between Species

Fundamentally, species interact in three ways: competition, in which the outcome is negative for both groups; symbiosis, which benefits both participants; and predation–parasitism, in which the outcome benefits one and is detrimental to the other. Each type of interaction affects evolution, the persistence of species, and the overall diversity of life. In the next chapter, we will examine the case of purple loosestrife, an alien invasive plant that has "out-competed" many native wetland species in Canada. Symbiosis and predator-prey relationships are discussed in more detail later in this chapter.

The Competitive Exclusion Principle

The **competitive exclusion principle** states that two species that have *exactly* the same requirements cannot coexist in *exactly* the same habitat. Garrett Hardin expressed the idea most succinctly: "Complete competitors cannot coexist."[10] This principle would seem to result in only a few species surviving on Earth.

The Eastern grey squirrel and the British red squirrel illustrate the competitive exclusion principle. The grey squirrel (*Sciurus carolinensis*) was introduced into Great Britain because some people thought it was attractive and would be a pleasant addition to the landscape. Thus, its introduction was not accidental but intentional. In fact, about a dozen attempts were made, the first perhaps as early as 1830. By the 1920s, the grey squirrel was well established in Great Britain, and in the 1940s and 1950s its numbers expanded greatly.

Today, the grey squirrel is a problem; it competes with the native red squirrel (*Sciurus vulgaris*) and is winning. The two species have almost exactly the same habitat requirements. At present, the red squirrel population is approximately 160,000. Although red squirrels used to be found in deciduous woodlands throughout the lowlands of central and southern Britain, they now are common only in Cambria, Northumberland, and Scotland, with scattered populations in East Anglia, Wales, Isle of Wight, and on islands in Poole Harbor, Dorset.[11] If present trends continue, the red squirrel may disappear from the British mainland in the next 20 years.

One reason for the shift in the balance of these species may be that the red squirrels' main food source in the winter is hazelnuts, while grey squirrels prefer acorns. Thus, red squirrels have a competitive advantage in areas with hazelnuts, and grey squirrels have the advantage in oak forests. When grey squirrels were introduced, oaks were the dominant mature trees in Great Britain; about 40% of the timber planted was oaks.

The introduction of the grey squirrel into Great Britain illustrates one of the major causes of modern threats to biological diversity: the introduction by peo-

Figure 7.11 • The structure of cells. Photomicrographs of *(a)* a eukaryote cell and *(b)* a bacterial (prokaryote) cell. From these images you can see that the eukaryotic cell has a much more complex structure, including many organelles.

(a)

(b)

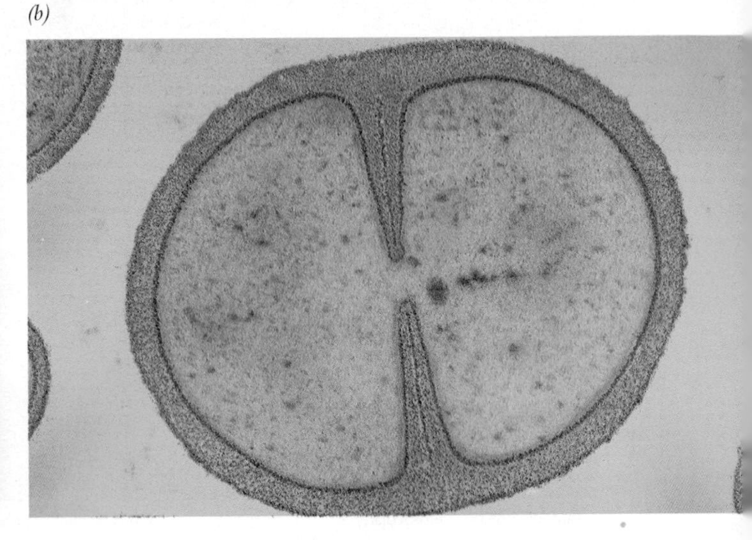

ple of species into new habitats. Introduced competitors frequently threaten native species. We will look at this issue again in Chapter 8.

So, according to the competitive exclusion principle, complete competitors cannot coexist; one will always exclude the other. Therefore, we might expect that over a very long period this would reduce the number of species over a large area. Taking this idea to its logical extreme, we could imagine Earth with very few species—perhaps one green plant on the land, one herbivore to eat it, one carnivore, and one decomposer. If we added four species for the ocean and four for freshwater, we would have only twelve species on our planet.

Being a little more realistic, we could take into account adaptations to major differences in climate and other environmental aspects. Perhaps we could specify 100 environmental categories: cold and dry, cold and wet, warm and dry, warm and wet, and so forth. Even so, we would expect that within each environmental category, competitive exclusion would result in the survival of only a few species. Allowing four species per major environmental category would result in only 400 species. Yet about 1.4 million species have been named, and scientists speculate that many more millions may exist—so many that we do not have even a good estimate. How can they all coexist?

7.7 Niches: How Species Coexist

For species to coexist, each must have—or develop—a unique role in the ecosystem. This role is akin to the species' job or function, and is called its ecological niche. Whereas the species' (or individual organism's) habitat is the place that it lives—its home—the ecological niche includes not only the species' preferred abiotic and biotic living conditions, but also considerations such as feeding and reproductive behavior. The timing of these behaviors can, for instance, be an important factor defining a species' niche. The ecological niche is a much more comprehensive concept than habitat, and encompasses not only place but the functional role or "lifestyle" of the species.

In 1961, the ecologist George Evelyn Hutchinson wrote a paper entitled "The Paradox of the Plankton", asking how so many aquatic invertebrate (plankton) species could coexist in the ocean. Surely, he suggested, many of those species were competing for the same light and nutrients, and surely some species were particularly efficient at harvesting those resources. So why didn't those species gradually come to dominate the system, and the others die out? The answer, he found, was that the ocean is not as well mixed or homogeneous as it might appear. Deeper waters are colder but oxygen-poorer than surface waters, and wind blowing over the surface pushes surface debris—and nutrients—into "windrows." The result is a system that appears uniform but is in fact a complex structure of microenvironments, each supporting a range of specialized species.

The niche concept is illustrated most easily by experiments done with small, common insects—flour beetles (*Tribolium*), which, as their name suggests, live on wheat flour. Flour beetles make good experimental subjects because they require only small containers of wheat flour to live and are easy to grow. In a classic experiment, flour beetles of two species were placed in small containers of flour, which were then maintained at various temperature and moisture levels—some are cool and wet, others warm and dry. Periodically, the beetles in each container were counted. Eventually, one species was found to succeed—to thrive and grow—in the container, while the other species became extinct. One species does better when it is cold and wet, the other when it is warm and dry (Figure 7.12). Curiously, when conditions are in between, sometimes one species persists and sometimes the other, seemingly randomly, but invariably, becomes extinct. The competitive exclusion principle holds for these beetles, but both species can survive in a complex environment—one that has cold and wet habitats as well as warm and dry habitats. In no location, however, do the species coexist.

The little beetles provide us with the key to the coexistence of many species. Species that require the same resources can coexist by utilizing those resources under different environmental conditions. So it is habitat complexity that allows complete competitors—and not-so-complete competitors—to coexist because they avoid competing with each other.[12]

In the 1950s, the ecologist Robert MacArthur conducted a study of warblers—small songbirds—in a coniferous forest. The five species he examined all fed on spruce budworm from spruce or balsam fir trees, and appeared very similar not only in feeding habits, but also in body form and general lifestyle. MacArthur was able to demonstrate that each species had a unique feeding niche, for instance higher or lower on the tree, or closer or farther from the trunk. In this way, several similar species were able to coexist successfully.[13]

Professions and Places: The Ecological Niche and the Habitat

The flour beetles are said to have the same ecologically functional niche, which means they have the same *profession*—eating flour. But they have different habitats. Where a species lives is its *habitat*, but what it does for a living (its profession) is its ecological niche. Suppose you have a neighbor who is a bus driver. Where your neighbor lives and works is your town—that's his habitat. What your neighbor does is drive a bus—that's his

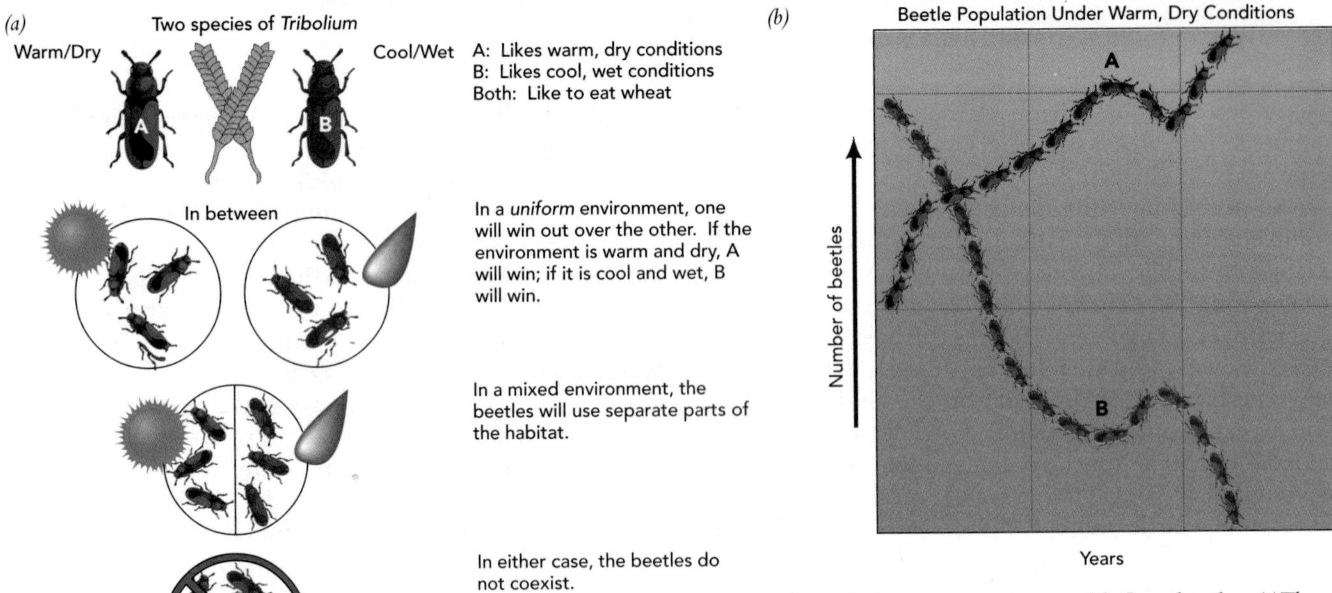

Figure 7.12 • A classical experiment with flour beetles. *(a)* The general process illustrating competitive exclusion in these species; *(b)* Results of a typical experiment under warm, dry conditions.

niche. Similarly, if someone says, "Here comes a wolf," you think not only of a creature that inhabits the northern forests (its habitat) but also of a predator that feeds on large mammals (its niche).

Understanding the niche of a species is useful in assessing the impact of development or of changes in land use. Will the change remove an essential requirement for some species' niche? A new highway that makes car travel easier might eliminate your neighbor's bus route (an essential part of his habitat) and thereby eliminate his profession (his niche). Other things could also eliminate his niche. Suppose a new school were built so that all the children could walk to school. Then a bus driver would not be needed; his niche no longer existed in your town. In the same way, cutting a forest may drive away prey and eliminate the niche of the wolf.

Measuring Niches

The ecological niche is a useful idea, but, as scientists, we want to be able to measure it—to make it quantitative. How can we do that? How do you measure a profession? One answer is to describe the niche as the set of all environmental conditions under which a species can persist and carry out its life functions. This measured niche is known as the Hutchinsonian niche, after G. E. Hutchinson, who first suggested it.[14] It is illustrated by the distribution of two species of flatworm, a tiny worm that lives on the bottom of freshwater streams. A study was made of two species of these small worms in Great Britain, where it was found that some streams contained one species, some the other, and still others both.[15]

The stream waters are cold at their source in the mountains and become progressively warmer as they flow downstream. Each species of flatworm occurs within a specific range of water temperatures. In streams where species A occurs alone, it is found from 6° to 17°C (Figure 7.13*a*). Where species B occurs alone, it is found from 6° to 23°C (Figure 7.13*b*). When they occur in the same stream, their temperature ranges are much narrower. Species A lives in the upstream sections, where the temperature ranges from 6° to 14°C, and species B lives in the warmer downstream areas, where temperatures range from 14° to 23°C (Figure 7.13*c*).

The temperature range in which species A occurs when it has no competition from B is called its *fundamental temperature niche.* The set of conditions under which it persists in the presence of B is called its *realized temperature niche.* The flatworms show that species divide up their habitats so that they use resources from different parts of it.

Of course, temperature is only one aspect of the environment. Flatworms also have requirements in terms of the acidity of the water and other factors. We could create graphs for each of these factors, showing the range within which A and B occurred. The collection of all those graphs would constitute the complete Hutchinsonian description of the niche of a species.

A Practical Implication: From the discussion of the competitive exclusion principle and the ecological niche, we learn something important about the conservation of species. If we want to conserve a species in its native habitat, we must make sure that all the requirements of its niche are present. Conservation of endangered species is more than a matter of putting many individuals of that species into an area; all the life requirements for that species must also be present—we have to conserve not only a population, but its habitat and its niche.

Symbiosis

Our discussion up to this point might leave the impression that species interact mainly through competition—by interfering with one another. But **symbiosis** is also important. This term, derived from a Greek word meaning "living together," describes a relationship between two organisms that is beneficial to both and enhances each organism's chances of persisting. Each partner in symbiosis is called a **symbiont**.

Symbiosis is widespread and common; most animals and plants have symbiotic relationships with other species. We humans have symbionts—microbiologists tell us that about 10% of a person's body weight is actually the weight of symbiotic micro-organisms in the intestines. The resident bacteria help our digestion; we provide a habitat that supplies all their needs; both we and they benefit. We become aware of this intestinal community when it changes—for example, when we travel to a foreign country and ingest new strains of bacteria. Then we suffer a well-known traveller's malady, gastrointestinal upset.

Another important kind of symbiotic interaction occurs between certain mammals and bacteria. A reindeer on the northern tundra may appear to be alone but carries with it many companions. Like domestic cattle, the reindeer is a ruminant, with a four-chambered stomach (Figure 7.14) teeming with microbes (a billion per cubic centimetre). In this partially closed environment, the respiration of micro-organisms uses up the oxygen ingested by the reindeer while eating. Other micro-organisms digest cellulose, take nitrogen from the air in the stomach, and make proteins. The bacterial species that digest the parts of the vegetation that the reindeer cannot digest itself (in particular, the cellulose and lignins of cell walls in woody tissue) require a peculiar environment: they can survive only in an environment without oxygen. One of the few places on Earth's surface where such an environment exists is the inside of a ruminant's stomach.[16] The bacteria and the reindeer are symbionts, each providing what the other needs; and neither could survive without the other. They are therefore called **obligate symbionts**.

A Practical Implication: We can see that symbiosis promotes biological diversity and that if we want to save a species from extinction, we must save not only its habitat and niche but also its symbionts. This suggests another important point that will become more and more evident in

Figure 7.13 • The occurrence of freshwater flatworms in cold mountain streams in Great Britain. *(a)* The presence of species A in relation to temperature in streams where it occurs alone. *(b)* The presence of species B in relation to temperature in streams where it occurs alone. *(c)* The temperature range of both species in streams where they occur together. Inspect the three graphs; what is the effect of each species on the other?

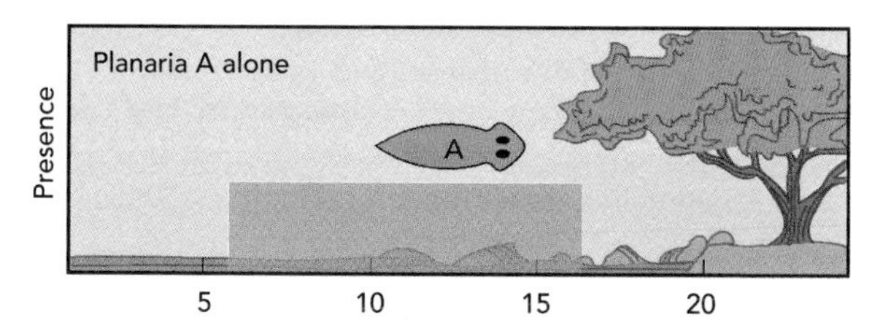

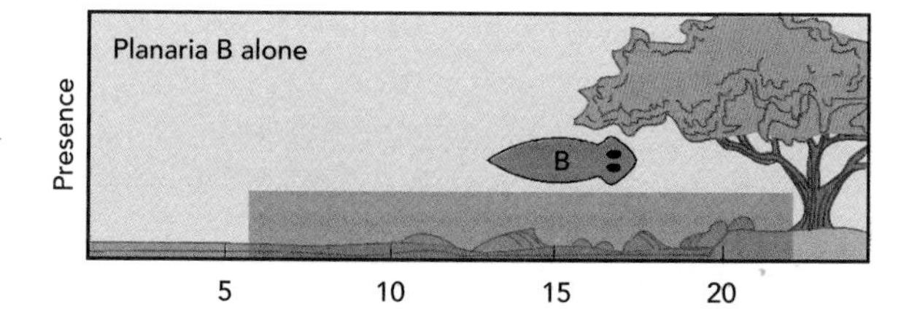

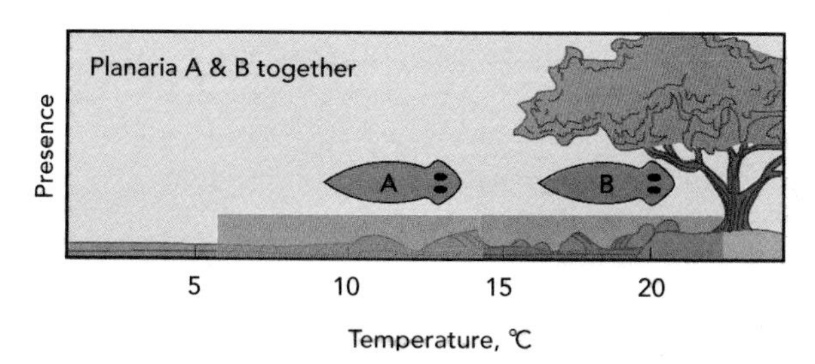

Figure 7.14 • The stomach of a reindeer illustrates complex symbiotic relationships. For example, in the rumen, bacteria digest woody tissue the reindeer could not otherwise digest. The result is food for the reindeer and food and a home for the bacteria, which could not survive in the local environment outside.

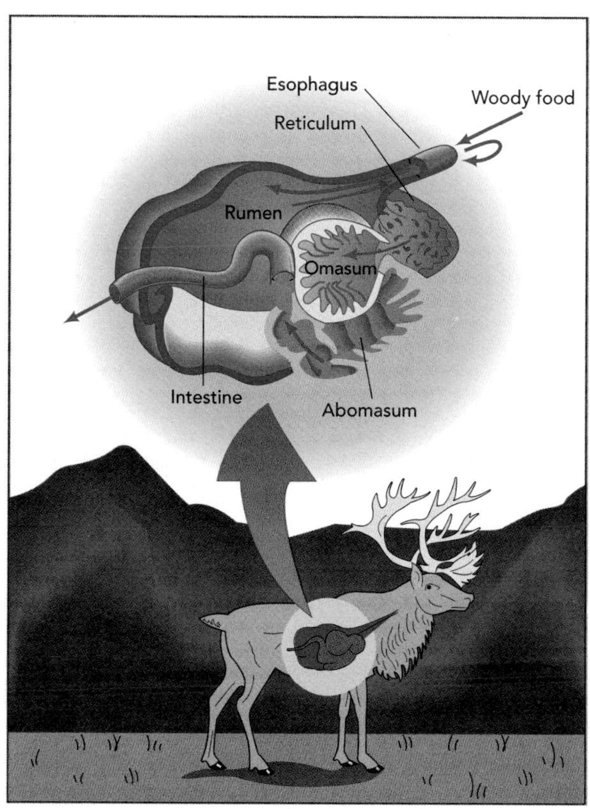

later chapters: The attempt to save a single species almost invariably leads us to conserve a group of species, not just a single species or a particular physical habitat.

Predation and Parasitism

Predation–parasitism is the third way in which species interact. **Predation** occurs when an organism (a predator) feeds on other live organisms (prey), usually of another species. **Parasitism** occurs when one organism (the parasite) lives on, in, or within another (the host) and depends on it for existence but makes no useful contribution to it and may harm it.

Predation can increase the diversity of prey species. Think again about the competitive exclusion principle. Suppose two species are competing in the same habitat and have the same requirements. One will win out. But if a predator feeds on the more abundant species, it can keep that prey species from overwhelming the other. Both might persist whereas, without the predator, only one would. For example, some studies have shown that a moderately grazed pasture has more species of plants than an ungrazed one. The same seems to be true for natural grasslands and savannas. Without grazers and browsers, then, African grasslands and savannas might have fewer species of plants.

7.8 Environmental Factors that Influence Diversity

Species are not uniformly distributed over Earth's surface; diversity varies greatly from place to place. For instance, suppose you were to go outside and count all the species in a field or any open space near where you are reading this book (that would be a good way to begin to learn for yourself about biodiversity). The number of species you found would depend on where you are. If you live in northern Alaska or Canada, Scandinavia, or Siberia, you would probably find a significantly smaller number of species than if you live in the tropical areas of Brazil, Indonesia, or central Africa. Variation in diversity is partially a question of latitude—in general, greater diversity occurs at lower latitudes. Diversity also varies within local areas. If you count species in the relatively sparse environment of an abandoned city lot, for example, you will find quite a different number than if you counted species in an old, long-undisturbed forest.

The large-scale geographic pattern in the distribution of species, called *biogeography*, is the topic of the next chapter. For now, we will look at some of the environmental factors that influence diversity locally. Table 7.1 summarizes several of these factors.

Which species and ecosystems occur on the land change with soil type and topography: slope, aspect (the direction the slope faces), elevation, and nearness to a drainage basin. These factors influence the number and kinds of plants. The kinds of plants, in turn, influence the number and kinds of animals. Some of the possible interrelationships are illustrated in Figure 7.15.[17]

Such a change in species can be seen with changes in elevation in mountainous areas like these near Mt. Logan, Canada's highest mountain, in the Saint Elias Range of the Yukon Territory(Figure 7.16). Although such patterns are most easily seen in vegetation, they occur for all organisms. See, for example, the pattern of distribution of African mammals on Mount Kilimanjaro (Figure 7.17).

Some habitats harbour few species because they are stressful to life, as a comparison of vegetation in two areas of Africa illustrates. In eastern and southern Africa, well-drained, sandy soils support diverse vegetation, including many species of *Acacia* and *Combretum* trees as well as many grasses. In contrast, woodlands on the very heavy clay soils of wet areas near rivers, such as the Sengwa River in Zimbabwe, are composed almost exclusively of a single species called Mopane. Very heavy clay soils store water and prevent most oxygen from reaching roots. As a result, only tree species with very shallow roots survive.

Moderate environmental disturbance can also increase diversity. For example, fire is a common disturbance in many forests and grasslands. Occasional light

Table 7.1 • Some Major Factors That Increase and Decrease Biological Diversity.

A. Factors that tend to increase diversity

1. A physically diverse habitat.
2. Moderate amounts of disturbance (such as fire or storm in a forest or a sudden flow of water from a storm into a pond).
3. A small variation in environmental conditions (temperature, precipitation, nutrient supply, etc.).
4. High diversity at one trophic level increases the diversity at another trophic level. (Many kinds of trees provide habitats for many kinds of birds and insects.)
5. An environment highly modified by life (e.g., a rich organic soil).
6. Middle stages of succession.
7. Evolution.

B. Factors that tend to decrease diversity

1. Environmental stress.
2. Extreme environments (conditions near the limit of what living things can withstand).
3. A severe limitation in the supply of an essential resource.
4. Extreme amounts of disturbance.
5. Recent introduction of exotic species (species from other areas).
6. Geographic isolation (being on a real or ecological island).

fires produce a mosaic of recently burned and unburned areas. These patches favor different kinds of species and increase overall diversity.

Of course, people also affect diversity. In general, urbanization, industrialization, and agriculture decrease diversity, reducing the number of habitats and simplifying habitats. (See, for example, the effects of agriculture on habitats, discussed in Chapter 11.) In addition, we intentionally favour specific species and manipulate populations for our own purposes, as when a person plants a lawn or when a farmer plants a single crop over a large area.

Most people don't think of cities as having any beneficial effects on biological diversity. Indeed, the development of cities tends to reduce biological diversity. This is, in part, because cities have typically been located at good sites for travel, such as along rivers or near oceans, where biological diversity is often high. However, in recent years we have begun to realize that cities can contribute in important ways to the conservation of biological diversity. We discuss this topic in Chapter 26.

The number of species per unit area varies over time as well as space. The changes occur over many time periods, from very short (a year), to moderate (years, decades), to even longer (centuries or more in a forest), and finally to very long geologic periods. As an example of the latter, diversity of animals in the early Paleozoic era (beginning about 500 million years ago) was much

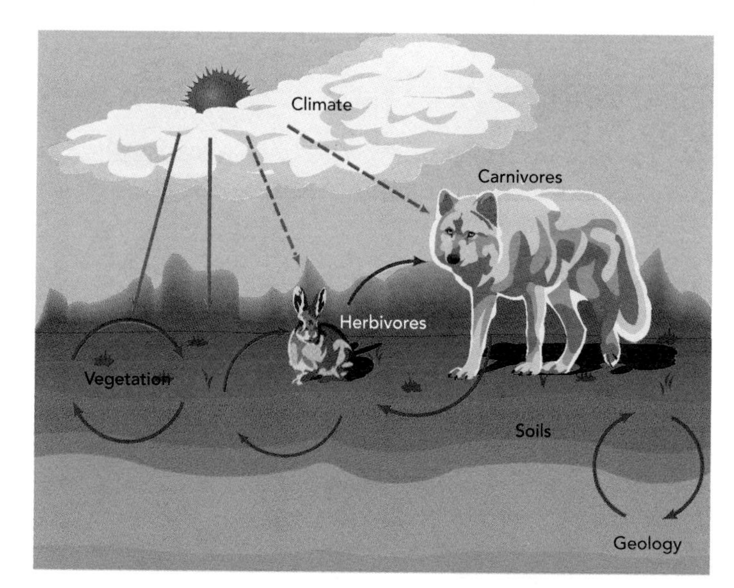

Figure 7.15 • **Interrelationships among climate, geology, soil, vegetation, and animals.** What lives where depends on many factors. Climate, geologic features (bedrock type, topographic features), and soils influence vegetation. Vegetation in turn influences soils and the kinds of animals that will be present. Animals affect the vegetation. Arrows represent a causal relationship; the direction is from cause to effect. A dashed arrow indicates a relatively weak influence, and a solid arrow a relatively strong influence.

Figure 7.16 • **Ecological gradient.** Change in the relative abundance of a species over an area or a distance is referred to as an *ecological gradient.* Such a change can be seen with changes in elevation in mountainous areas. The altitudinal zones of vegetation in Jasper National Park, Alberta.

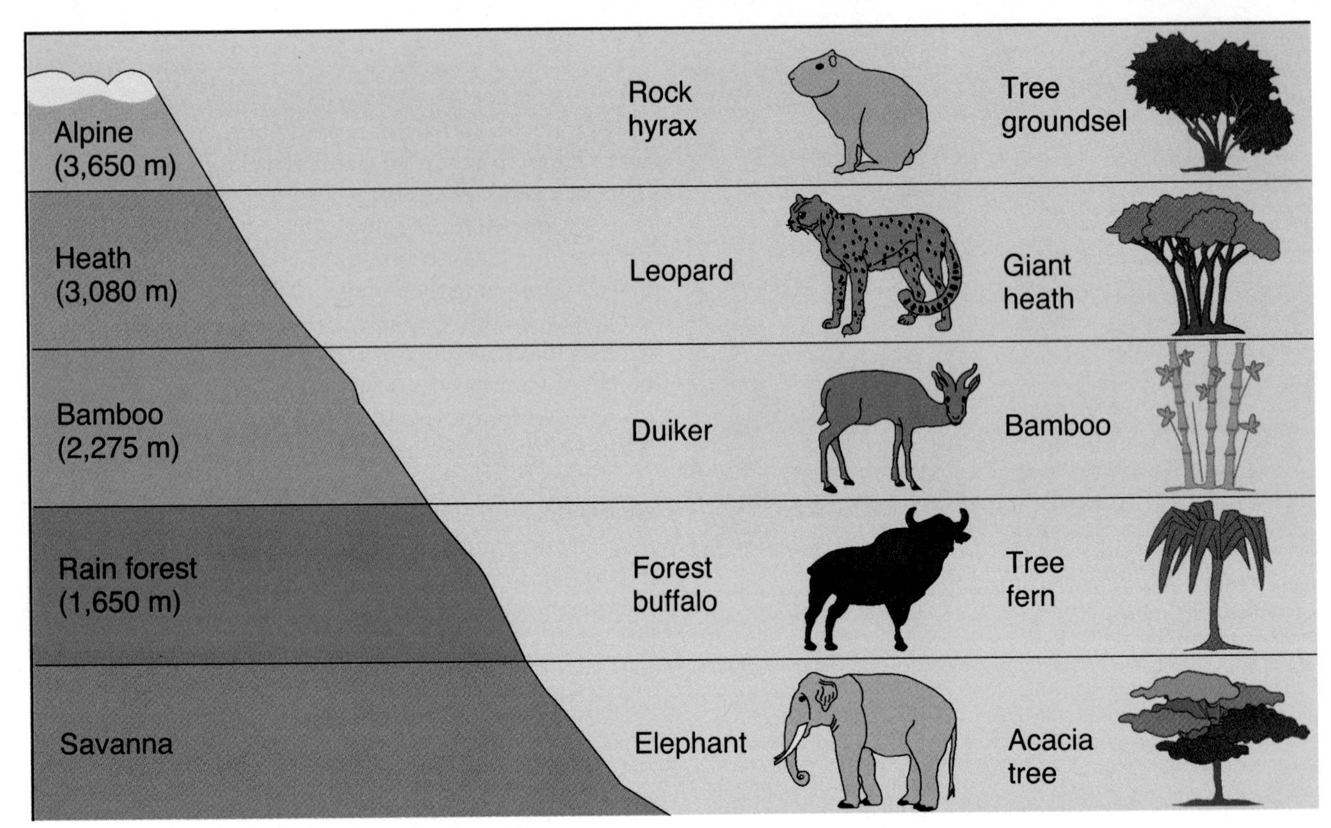

Figure 7.17 • Changes in the distribution of animals with elevation on a typical mountain in Kenya. (*Source*: From C. B. Cox, I. N. Healey, and P. D. Moore, *Biogeography* [New York: Halsted, 1973]. Copyright of Blackwell Scientific Publishing, Ltd.)

lower than in the late Mesozoic (beginning about 130 million years ago) and the Cenozoic (beginning about 75 million years ago) eras.[15]

7.9 Genetic Engineering and Some New Issues about Biological Diversity

Our understanding of evolution today owes a lot to the modern science of molecular biology and the practice of genetic engineering, which are creating a revolution in how we think about and deal with species. Currently, scientists have essentially the complete DNA code for five species: the fruit fly (*Drosophila*), a nematode worm called *C. elegans* (a very small worm that lives in water); yeast; a small weed plant, thale cress (*Arabidopsis thaliana*); and humans. Scientists focused on these species either because they are of great interest to us (as with humans) or because they are relatively easy to study—having either few base pairs (the nematode worm) or having genetic characteristics that were well known (the fruit fly).

Environmental Issues as Information Issues

The amount of information contained in DNA is enormous. Thale cress's DNA consists of 125 million base pairs, and this is comparatively few. Important crop plants have even larger numbers of base pairs. Rice has 430 million and wheat more than 16 billion! The base pairs in thale cress appear to make up approximately 25,000 genes, but many are duplicates; so there are probably about 15,000 unique genes that determine what thale cress will be like. This is about the same number of genes in the nematode worm and the fruit fly. In contrast, the number of human genes is estimated to be between 30,000 and 130,000.

A Practical Implication: Scientists can now manipulate DNA, which affects inherited characteristics of crops, bacteria, and other organisms, giving them new combinations of characteristics not found before and demonstrating that characteristics are inherited and can be altered, as predicted by evolutionary theory. These new capabilities pose novel problems and hold new promise for biological diversity. On the one hand, we may be able to help rare, endangered species by increasing their genetic variability or by overcoming some of their less adaptive genetic characteristics that result from genetic drift. On the other hand, we may inadvertently create superpests, predators, or competitors of endangered species. Such new organisms may be to our benefit, but we must be careful not to release new strains into the environment that can reproduce rapidly and become unexpected pests. Genetic engineering poses new challenges for the environment, as we will discuss in later chapters. (We discuss some environmental implications of genetic engineering in Chapters 11 and 13.)

CRITICAL THINKING ISSUE

Why Preserve Biodiversity?

Preserving the diversity of life on Earth has become an accepted goal for many people. But when that goal comes into conflict with other goals, such as economic development, the question becomes: How much diversity and at what cost? In 1980, the International Union for Conservation of Nature and Natural Resources (IUCN) summarized the reasons for conserving diversity. Read the statements reproduced below and answer the questions that follow.

An Ethical Basis for Preserving Biodiversity

1. The world is an interdependent whole made up of natural and human communities. The well-being and health of any one part depends on the well-being and health of the other parts.
2. Humanity is part of nature, and humans are subject to the same immutable ecological laws as are all other species on the planet.
3. All life depends on the uninterrupted functioning of natural systems that ensures the supply of energy and nutrients, so ecological responsibility among all people is necessary for the survival, security, equity, and dignity of the world's communities.
4. Human culture must be built on a profound respect for nature, a sense of being at one with nature, and a recognition that human affairs must proceed in harmony and in balance with nature.
5. The ecological limits within which we must work are not limits to human endeavor; instead, they give direction and guidance as to how human affairs can sustain environmental stability and diversity.
6. All species have an inherent right to exist. The ecological processes that support the integrity of the biosphere and its diverse species, landscapes, and habitats are to be maintained. Similarly, the full range of human cultural adaptations to local environments is to be enabled to prosper.
7. Sustainability is the basic principle of all social and economic development.
8. Personal and social values should be chosen to accentuate the richness of flora, fauna, and human experience. This moral foundation will enable the many utilitarian values of nature—for food, health, science, technology, industry, and recreation—to be equitably distributed and sustained for future generations.
9. The well-being of future generations is a social responsibility of the present generation. Therefore, the present generation should limit its consumption of nonrenewable resources to the level that is necessary to meet the basic needs of society and ensure that renewable resources are nurtured for their sustainable productivity.
10. All persons must be empowered to exercise responsibility for their own lives and for the life of Earth. They must, therefore, have full access to educational opportunities, political enfranchisement, and sustaining livelihoods.
11. Diversity in ethical and cultural outlooks toward nature and human life is to be encouraged by promoting relationships that respect and enhance the diversity of life, irrespective of the political, economic, or religious ideology in a society.

Critical Thinking Questions

1. Which of the statements are scientific—that is, which are based on repeatable observations, can be tested, and are supported by evidence?
2. Using the table that follows, identify which statement from the argument for preserving biodiversity is implied by each type of value. Write the number of the statement in the right-hand column. A statement may fit into more than one category.

Type of Value	Source of Value of Living Organisms	Statement Number
Ethical	The fact that they are alive	
Aesthetic	Their beauty and the rewards humans derive from their beauty	
Economic	The direct and indirect ways in which they benefit humans	
Ecological	Their contributions to the health of the ecosystem	
Intellectual	Their contributions to knowledge	
Emotive	The sense of awe and wonder they inspire in humans	
Religious	Having been created by a supernatural being or force	
Recreational	Sport, tourism, and other recreations	

3. How would you rank the types of values given?
4. Which type or types of values could be used to support preservation of smallpox viruses? All 1,200 (or more) species of beetles in the tropical rain forest? The northern spotted owl? Dolphins? Sharks?

Summary

- Biological evolution—the change in inherited characteristics of a population from generation to generation—is responsible for the development of the many species of life on Earth. Four processes that lead to evolution are mutation, natural selection, migration, and genetic drift.

- Biological diversity involves three concepts: genetic diversity (the total number of genetic characteristics), habitat diversity (the diversity of habitats in a given unit area), and species diversity. Species diversity, in turn, involves three ideas: species richness (the total number of species), species evenness (the relative abundance of species), and species dominance (the most abundant species).
- About 1.4 million species have been identified and named. Insects and plants make up most of these species. With further explorations, especially in tropical areas, the number of identified species, especially of invertebrates and plants, will increase.
- Species engage in three basic kinds of interactions: competition, symbiosis, and predation–parasitism. Each type of interaction affects evolution, the persistence of species, and the overall diversity of life. It is important to understand that organisms have evolved together so that predator, parasite, prey, competitor, and symbiont have adjusted to one another. Human interventions frequently upset these adjustments.
- The competitive exclusion principle states that two species that have exactly the same requirements cannot coexist in exactly the same habitat; one must win. The reason that more species do not die out from competition is that they have developed a particular niche and thus avoid competition.
- The number of species in a given habitat is affected by many factors, including latitude, elevation, topography, the severity of the environment, and the diversity of the habitat. Predation and moderate disturbances, such as fire, can actually increase the diversity of species. The number of species also varies over time. Of course, people affect diversity as well.

STUDY QUESTIONS

1. Why do introduced species often become pests?
2. On which of the following planets would you expect a greater diversity of species?
   a. A planet with intense tectonic activities
   b. A tectonically dead planet (Remember that tectonics refers to the geologic processes that involve the movement of tectonic plates and continents, processes that lead to mountain building and so forth.)
3. You are going to conduct a survey of national parks. What relationship would you expect to find between the number of species of trees and the size of the parks?
4. A city park manager has run out of money to buy new plants. How can the park labor force alone be used to increase the diversity of (a) trees and (b) birds in the parks?
5. A plague of locusts visits a farm field. Soon after, many kinds of birds arrive to feed on the locusts. What changes occur in animal dominance and diversity? Begin with the time before the locusts arrive and end with the time after the birds have been present for several days.
6. What is the difference between dominance and diversity in (a) a zoo and (b) a natural wildlife preserve?
7. What is the difference between habitat and niche?
8. There are more than 600 species of trees in Costa Rica, most of which are in the tropical rain forests. What might account for the coexistence of so many species with similar resource needs?
9. Which of the following can lead to populations that are less adapted to the environment than were their ancestors?
   a. Natural selection
   b. Migration
   c. Mutation
   d. Genetic drift

FURTHER READING

Leveque, C., and J. Mounolou. *Biodiversity*. New York: John Wiley, 2003.

Charlesworth, B., and C. Charlesworth. *Evolution: A Very Short Introduction*. Oxford: Oxford University Press, 2003.

Darwin, C. A. T*he Origin of Species by Means of Natural Selection, or the Preservation of Proved Races in the Struggle for Life*. London: Murray, 1859. Reprinted variously. A book that marked a revolution in the study and understanding of biotic existence.

Dawkins, R. *Climbing Mount Improbable*. New York: Viking, 1996. A discussion of some implications of modern discoveries in genetics and evolution.

Margulis, L., and D. Sagan. *What Is Life?* New York: Simon & Schuster, 1995. A beautifully illustrated and well-written introduction to the major forms of life on Earth and the effect of life's diversity on the global environment.

Pimentel, D. "Conserving Biological Diversity in Agricultural/Forestry Systems," *BioScience* 42 (1992):354–362. A journal article in which the author argues that most biological diversity exists in human-managed ecosystems and identifies means to conserve that diversity.

Wilson, E. O., ed. *The Diversity of Life*. New York: Norton, 1992. A book outlining the story of evolution of life on Earth, how species became diverse, and the scope of the current threat to that diversity.

chapter 8

Too Much of a Good Thing: The Invasion of the Purple Loosestrife

Purple loosestrife was introduced into North America from Europe in the early nineteenth century—deliberately, as a herbal remedy for diarrhea, dysentery, bleeding ulcers, and sores—and accidentally when ships from Europe released ballast water contaminated with loosestrife seeds in North American waters (Figure 8.1).[1] Although a beautiful and useful plant, purple loosestrife spread rapidly, grew densely, and eliminated many of the native plants that provided food and cover for wildlife in wetlands of Canada and the United States.[2] Without natural enemies and highly tolerant of variations in moisture and temperature, purple loosestrife soon migrated and grew along the New England coast. Canals and waterways constructed in the 1880s helped it spread inland in New York and the St. Lawrence River Valley. By the mid-1990s, it grew throughout southeastern Canada, and the eastern and midwestern United States, and now it has started to appear in the western United States and southwestern Canada.

One purple loosestrife plant can produce more than 2.5 million seeds a year.[3] The seeds survive for a time and can spread in the water and in mud that clings to wildlife, livestock, and people. When disturbances clear the land and expose the soil to the sun, as many as 20,000 seedlings may germinate in 1 square metre. Few native plants can compete with this productivity.

Mature loosestrife reaches 2 m and overtops many native wetland herbs and shrubs, shading them—especially their seedlings—so they cannot grow. Some of these native plants are essential for some native animals. Purple loosestrife has endangered many species across Canada, including swamp rose mallow, a species of special concern under the *Species at Risk Act*, and wetland habitat for huge flocks of migratory waterfowl in Manitoba. It is less palatable to livestock than are native grasses and sedges, so it has decreased livestock productivity and led to agricultural losses.

The plant is difficult to eliminate—if cut down, the roots persist. Even when burned or sprayed with herbicides, stems sprout from those roots. What to do? Removing every bit of the plant eradicates it, but that is very time-consuming. Instead, scientists are trying to find biological controls. Their effort focuses on three European insects that are natural parasites of loosestrife. The most promising is being released in hopes of eliminating 75–80% of purple loosestrife in Canada and the United States (Figure 8.1*b*).

The story of purple loosestrife illustrates the need to be careful about introducing exotic species. Such introductions are a major environmental problem; examples range from the introduction into South America of Africanized "killer" bees, which then migrated north to the United States, to the introduction of cane toads into the wet tropics of Australia.

- *The story of the purple loosestrife illustrates features of biogeography—the geographic distribution of life—and how this distribution came about, as well as some of the environmental problems that result when we alter this geography. The geography of life is influenced by plate tectonics, climate, and ocean dynamics. Over the long time that life has existed on Earth, groups of species have become isolated by geographic barriers. While the ability to invade and succeed in new habitats is a necessary quality of life, the introduction of species into new areas often creates problems for existing ecosystems. This chapter explores the basic concepts of biogeography, introduces the major biomes—the kinds of ecosystems—found on Earth; and discusses environmental problems that arise because of disruption of biogeographic patterns.*

Biogeography

LEARNING OBJECTIVES

If we are to conserve biological diversity, we must understand the large-scale, global patterns of biogeography. After reading this chapter, you should understand:

- How large-scale global patterns and the environment affect biological diversity.
- How climate, bedrock, soils, and the geography of life are related to one another.
- What biotic provinces and biomes are and how they differ.
- How plate tectonics affects biogeography.
- What island biogeography is, and what it implies for the general geography of life, especially the geography of biological diversity.
- What the major patterns in the distribution of biomes on Earth are and the major characteristics of each of the 17 biomes found on Earth.
- How people affect the geography of life.
- How the introduction of exotic species into new habitats typically affects the new habitats.

Fiure 8.1 • *(a)* **Purple loosestrife**, a European plant introduced into North America, has caused many problems for the biological diversity of native plants. *(b)* An insect parasite of purple loosestrife, from its native habitat, may be useful in controlling this plant in North America.

(a)

(b)

8.1 Why Were Introductions of New Species into Europe So Popular Long Ago?

Introduction of species is one of the most troublesome problems with life on Earth. The irony is that some introductions, such as crops and decorative plants, have been of great benefit. In 1749 Linnaeus, one of the first scientific botanists and the originator of the modern taxonomy of organisms, sent a colleague, Peter Kalm, to North America to collect plants to decorate the gardens of Europe. Western Europe's climate was similar to the climate in parts of China and eastern North America, but those two areas had a much greater variety of plant species. For this reason, people who wanted beautiful gardens in Europe during the eighteenth century sought to augment native vegetation with flowering trees, shrubs, and herbs from the New World. But why were there fewer species in Europe? And if introductions of exotic species are such a problem today, why were these introductions into Europe from North America not a problem then? This puzzle is explained by theories of **biogeography**—large-scale, global patterns—beginning with the concepts of the biotic province and the biome.

8.2 Wallace's Realms: Biotic Provinces

In 1876, the great British biologist Alfred Russell Wallace (co-discoverer of the theory of biological evolution with Charles Darwin) suggested that the world could be divided into six biogeographic regions on the basis of fundamental features of the animals found in those areas.[4] Wallace referred to these regions as **realms** and named them Nearctic (North America), Neotropical (Central and South America), Palaearctic (Europe, northern Asia, and northern Africa), Ethiopian (central and southern Africa), Oriental (the Indian subcontinent and Malaysia), and Australian. These have become known as *Wallace's realms* (Figure 8.2). The recognition of these worldwide patterns in animal species was the first step in understanding biogeography.

All living organisms are classified into groups called **taxa**, usually on the basis of their evolutionary relationships or similarity of characteristics (Linnaeus played a crucial role in working all this out.) The hierarchy of these groups (from largest and most inclusive to smallest and least inclusive) begins with a domain or kingdom. The plant kingdom is made up of divisions. The animal kingdom is made up of phyla (singular: *phylum*). A phylum or division is, in turn, made up of classes, which are made up of orders, which are made up of families, which are made up of genera (singular: *genus*), which are made up of species.

In each major biogeographic area (Wallace's realm), certain families of animals are dominant, and animals of these families fill the ecological niches (see Chapter 7.) Animals filling a particular ecological niche in one similar niches in the other realms. For example, bison and pronghorn antelope are among the large mammalian herbivores in North America. Rodents such as the capybara fill the same niches in South America, and kangaroos fill them in Australia. In central and southern Africa, many species, including Cape buffalo and antelopes, fill these niches.

This is the basic concept of Wallace's realms, and it is still considered valid and has been extended to all life forms,[5] including plants (Figure 8.3)[6,7] and invertebrates. These realms are now referred to as **biotic provinces**.[8] A biotic province is a region inhabited by a characteristic set of taxa (species, families, orders), bounded by barriers that prevent the spread of those distinctive kinds of

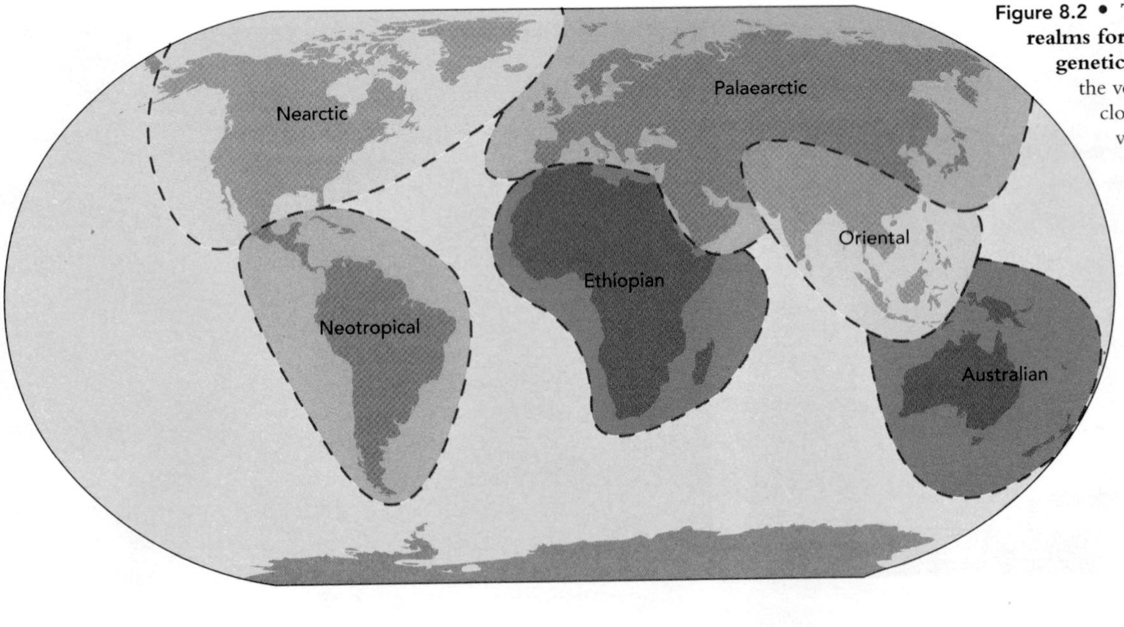

Figure 8.2 • **The main biogeographic realms for animals are based on genetic factors.** Within each realm, the vertebrates are, in general, more closely related to each other than to vertebrates filling similar niches in other realms.

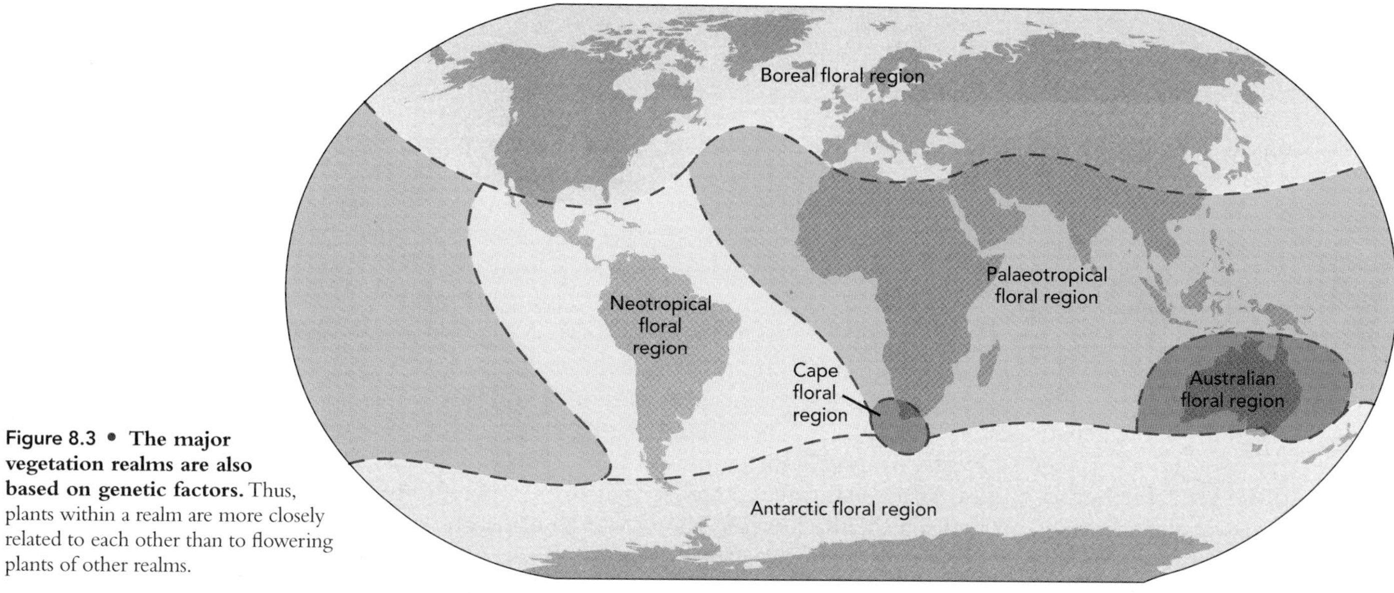

Figure 8.3 • The major vegetation realms are also based on genetic factors. Thus, plants within a realm are more closely related to each other than to flowering plants of other realms.

life to other regions and the immigration of foreign species; it is more a geographical concept than a biological one.[9] So in a biotic province, organisms share a common genetic heritage but may live in a variety of environments as long as they are genetically isolated from other regions.

Wallace did not have the benefit of our modern understanding of geologic processes to explain how distinct biological groups could have evolved in isolation from one another. The modern explanation is that continental drift, caused by plate tectonics, has periodically joined and separated the continents (see the discussion in Chapter 5).[10,11] Connection of continents allowed genetic mixing; separation imposed geographic isolation. Continental continuity and land bridges (which can occur during very low sea levels, like the Bering land bridge connecting Siberia and Alaska during the last glacial period) enabled organisms to enter new habitats. Continental separation led to genetic isolation and the evolution of new species. Within a realm, therefore, species are more likely to be related and to have evolved and adapted in the same place for a long time. But when people bring home a species from far away, they are likely to be introducing a species that is unrelated, or only distantly related, to native species. And this new and unrelated species has not evolved and adapted in the presence of the home species, so ecological and evolutionary adjustments are yet to take place. Sometimes the introduction brings in a superior competitor. That is what happened with purple loosestrife.

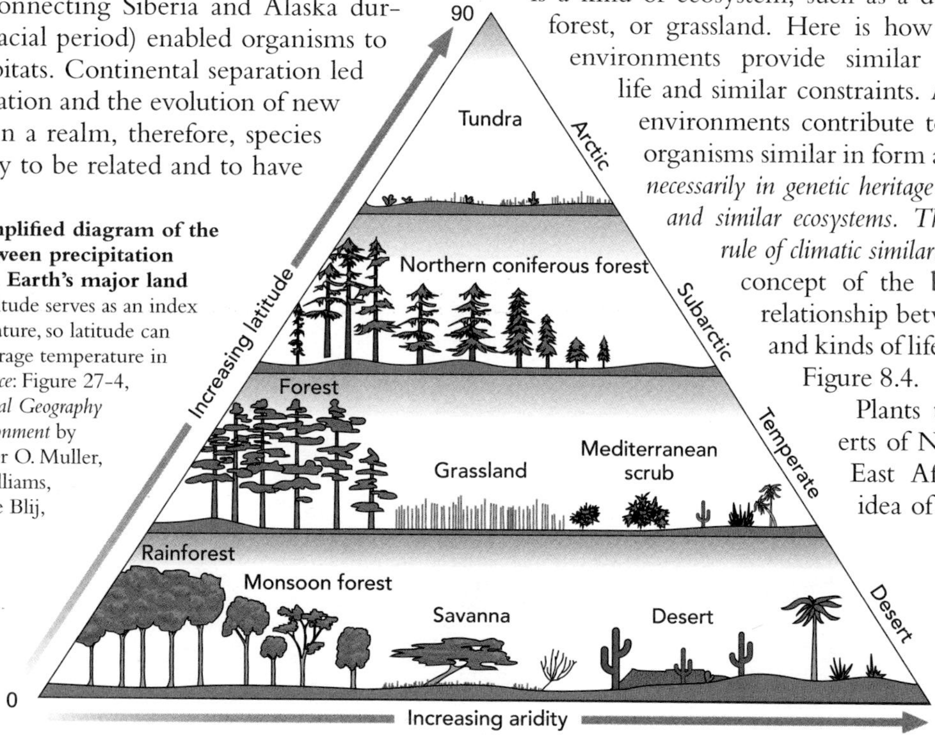

Figure 8.4 • Simplified diagram of the relationship between precipitation and latitude and Earth's major land biomes. Here, latitude serves as an index of average temperature, so latitude can be replaced by average temperature in this diagram. [*Source*: Figure 27-4, p. 293, from *Physical Geography of the Global Environment* by Harm de Blij, Peter O. Muller, and Richard S. Williams, edited by Harm de Blij, copyright 2004 by Oxford University Press, Inc. Used by permission of Oxford University Press, Inc.]

8.3 Biomes: The Relationship between Climate and Life

The biome is another major biogeographic pattern. It is a kind of ecosystem, such as a desert, tropical rain forest, or grassland. Here is how it works: Similar environments provide similar opportunities for life and similar constraints. As a result, similar environments contribute to the evolution of organisms similar in form and *function (but not necessarily in genetic heritage or internal makeup) and similar ecosystems. This is known as the rule of climatic similarity* and leads to the concept of the **biome**. The close relationship between environment and kinds of life forms is shown in Figure 8.4.

Plants that grow in deserts of North America and East Africa illustrate the idea of a biome. Some of these plants look similar but belong to very different biological families. Geographically isolated for 180 million years, they have been subjected to similar climates, which imposed similar stresses and opened up similar ecological opportunities. On both continents, desert plants evolved to adapt to these stresses and potentials, and have come to look alike and prevail in like habitats.

The ancestral differences between these look-alike plants can be found in their flowers, fruits, and seeds, which change the least over time and thus provide the best clues to the genetic history of a species. The Joshua tree and saguaro cactus of North America look similar to the giant Euphorbia of East Africa. All three are tall, have succulent green stems that replace the leaves as the major sites of photosynthesis, and have spiny projections, but these plants are not closely related. The Joshua tree is a member of the agave family, the saguaro is a member of the cactus family, and the Euphorbia is a member of the spurge family. Their similar shapes result from evolution in similar desert climates, a process known as convergent evolution (see Figure 8.5). The Euphorbia and the Joshua tree are in the same biome but in different biotic provinces. They function similarly and have the same niche, but they are not closely related.

So here is the difference between a biotic province and a biome: A biotic province is based on who is related to whom. A biome is based on niches and habitats. Species within a biotic province, in general, are more closely related to each other than to species in other provinces. In two different biotic provinces, the same ecological niche will be filled with species that perform a specific function and may look very similar to each other but have quite different genetic ancestries. In this way, a biotic province is an evolutionary unit.

The strong relationship between climate and life suggests that if we know the climate of an area, we can make a pretty good prediction about what biome will be found there, what its approximate biomass (amount of living matter) and production will be, and what the shapes and forms of dominant organisms will be.[12,13] The general relationship between biome type and the two most important climatic factors—rainfall and temperature—is diagrammed in Figure 8.6.

Another important process that influences life's geography is **divergent evolution**. In this process, a population is divided, usually by geographic barriers. Once separated into two populations, each evolves

(a)

(b)

(c)

Figure 8.5 • Convergent evolution. Given sufficient time and similar climates in different areas, species similar in shape and form will tend to occur. The Joshua tree *(a)* and saguaro cactus *(b)* of North America look similar to the giant Euphorbia *(c)* of East Africa. All three are tall, have green succulent stems that replace the leaves as the major sites of photosynthesis, and have spiny projections. However, these plants are not closely related; the Joshua tree is a member of the agave family, the saguaro is a member of the cactus family, and the Euphorbia is a member of the spurge family. Their similar shapes result from evolution under similar desert climates, a process known as **convergent evolution**.

separately, but the two groups retain some characteristics in common. It is now believed that the ostrich (native to Africa), the rhea (native to South America), and the emu (native to Australia) have a common ancestor but evolved separately (Figure 8.7). In open savannas and grasslands, a large bird that can run quickly but feed efficiently on small seeds and insects has certain advantages over other organisms seeking the same food. Thus, these species maintained the same characteristics in widely separated areas. Both convergent and divergent evolution increase biological diversity.

People make use of convergent evolution when they move decorative and useful plants around the world. Cities that lie in similar climates in different parts of the world now share many of the same decorative plants. Bougainvillea, a spectacularly bright flowering shrub originally native to Southeast Asia, decorates cities as distant from each other as Los Angeles and the capital of Zimbabwe. In Toronto, Norway maple from Europe and the Tree-of-Heaven and gingko tree from China grow along with such native species as white ash, sugar maple, and beech. People intentionally introduced the Asian and European trees.

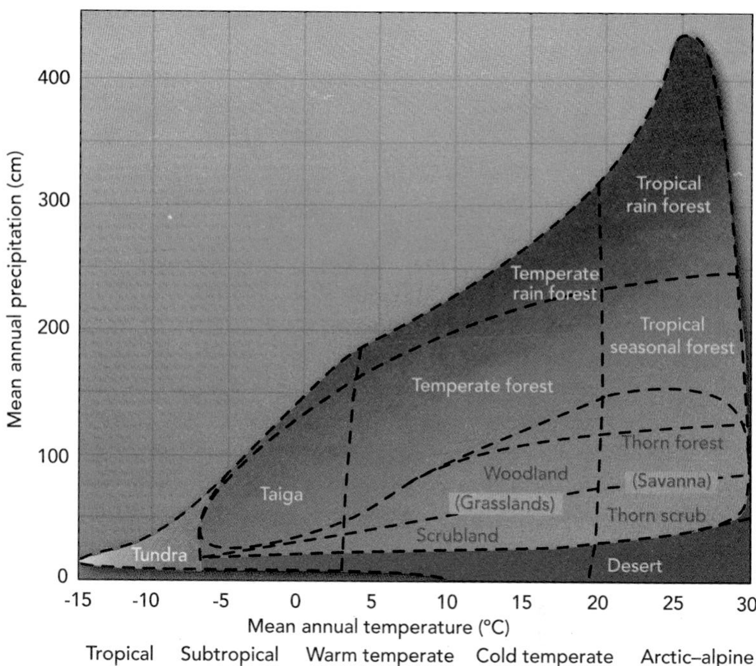

Figure 8.6 • Another view of the relationship between climate and vegetation (compare Figure 8.4). A pattern of the types of vegetation in relation to precipitation and temperature. Boundaries between types are approximate. Note that deserts occur over a wide range of annual temperatures, from 30°–5°C, as long as rainfall is less than about 50 cm/year. The warmer the climate, the more rainfall is required to change from desert to another biome. [*Source*: Adapted from R. H. Whittaker, *Communities and Ecosystems*, 2nd ed. (New York: Macmillan, 1975).]

Figure 8.7 • Divergent evolution. These three large, flightless birds evolved from a common ancestor but are found in widely separated regions: *(a)* the ostrich in Africa, *(b)* the rhea in South America, and *(c)* the emu in Australia.

(a)

(b)

(c)

The Distribution of Life on Earth

Earth has 17 major biomes: tundra, taiga (boreal forests), temperate deciduous forests, temperate rain forests, temperate woodlands, temperate shrublands, temperate grasslands, tropical rain forests, tropical seasonal forests and savannas, deserts, wetlands, freshwaters, intertidal areas, open oceans, benthos, upwellings, and hydrothermal vents.

The first rule of moving species around the planet is: *It is less likely to be harmful if you move a species within its biotic province.* The second rule is: *Moving a species into the same biome from a different biotic province is likely to be harmful.* The third rule is: *Local moves are less likely to be harmful than global moves (from one continent to another).* This does not mean that we should stop all introductions of species, but it does mean that such introductions have to be done very carefully, especially introductions from one part of a biome to another, across continents.

To know when such moves are likely to cause problems, we need to understand Earth's biomes. The next part of this chapter describes Earth's major biomes and their locations. Figure 8.8 shows the major land biomes. Subtypes occur within these major types. Biomes are usually named for the dominant vegetation (for example, coniferous forests, grasslands); for the dominant shape and form, or physiognomy, of the dominant organisms (forest, shrubland); or for the dominant climatic conditions (cold desert, warm desert).

In case you think the biome concept is abstract, look at the satellite image of Earth from space (Figure 8.9). The biomes show up and give Earth's surface a distinct pattern, a pattern that corresponds with a world map of average summer temperature (Figure 8.10)—showing that the importance of climate to biomes, discussed earlier, is also a reality. For example, July average temperatures above 30°C (Figure 8.10), coupled with low precipitation and other factors, encourage the formation of deserts (Figures 8.9 and 8.10) in the Americas and in Africa.

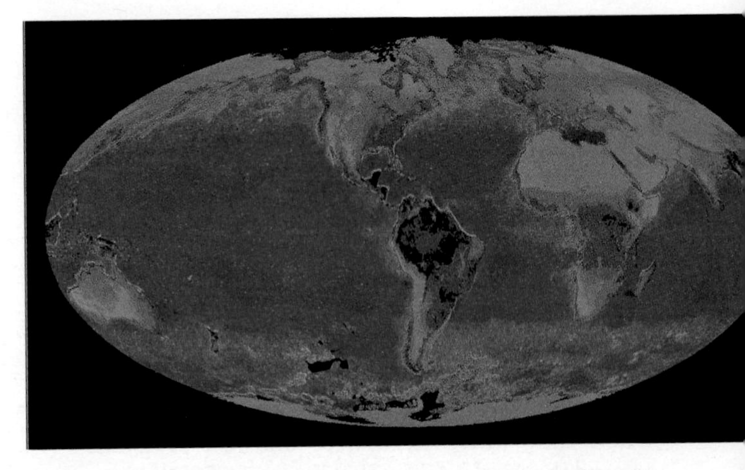

Figure 8.9 • NASA Landsat imagery of the vegetation of Earth from space.

Boreal forests occur where July average temperatures are below 20°C. Precipitation patterns, combined with temperature patterns, lead to an even closer correspondence between climate and biomes, as revealed by vegetation. Vegetation is the form of life most visible from space, but other forms of life have similar geographic relationships.

Climate also correlates with biological *productivity*. Warm, wet climates favor high vegetation production—as long as other factors, such as availability of chemical elements that plants require, are not limiting. Cold or dry climates do not generally allow such high rates of vegetation growth.

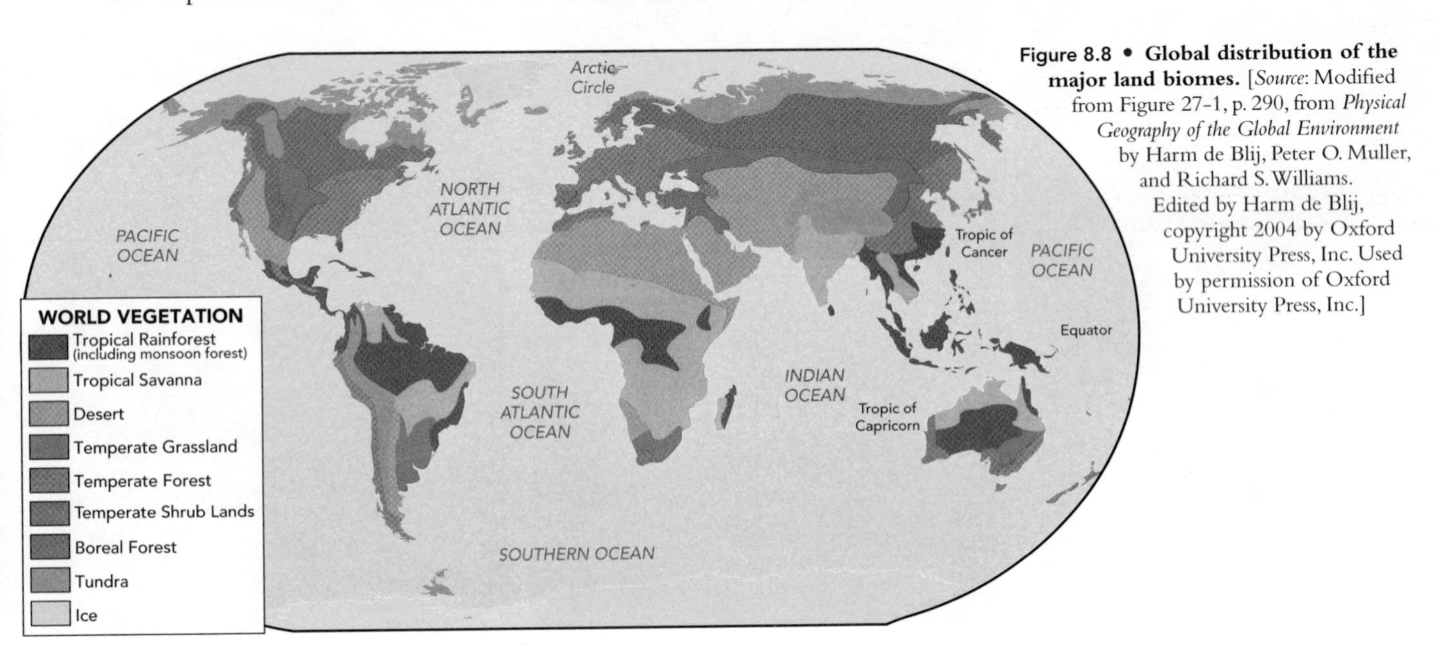

Figure 8.8 • Global distribution of the major land biomes. [*Source*: Modified from Figure 27-1, p. 290, from *Physical Geography of the Global Environment* by Harm de Blij, Peter O. Muller, and Richard S. Williams. Edited by Harm de Blij, copyright 2004 by Oxford University Press, Inc. Used by permission of Oxford University Press, Inc.]

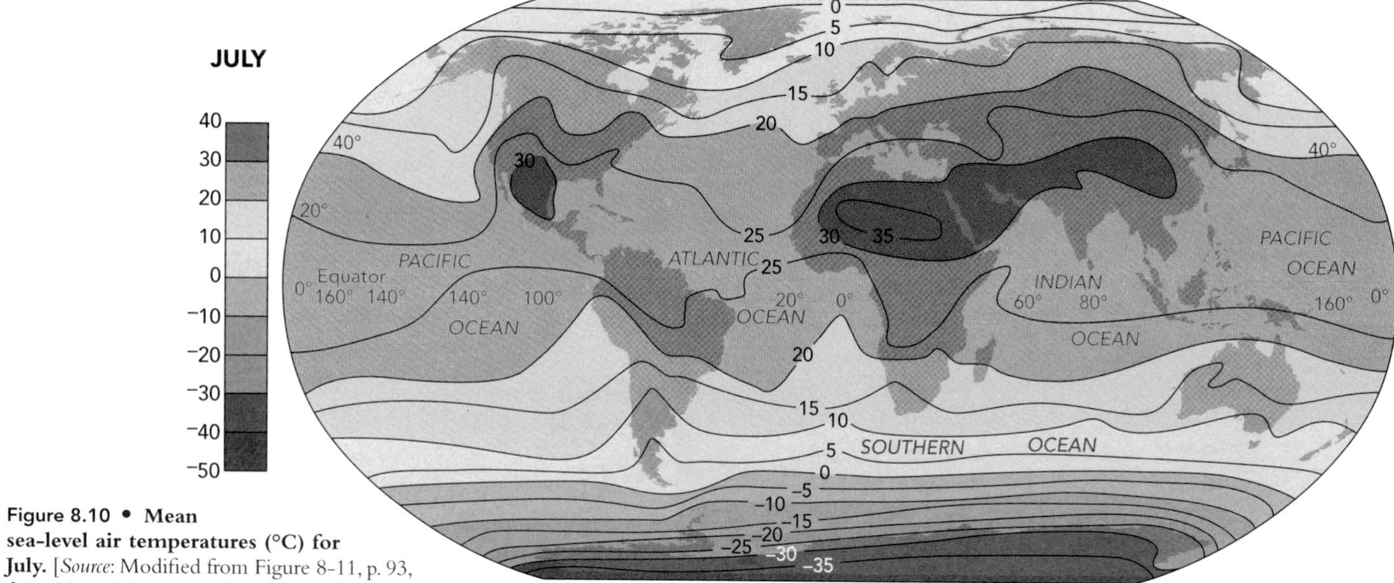

Figure 8.10 • Mean sea-level air temperatures (°C) for July. [*Source*: Modified from Figure 8-11, p. 93, from *Physical Geography of the Global Environment* by Harm de Blij, Peter O. Muller, and Richard S. Williams. Edited by Harm de Blij, copyright 2004 by Oxford University Press, Inc. Used by permission of Oxford University Press, Inc.]

Biological diversity varies among biomes. A geographic pattern that has long intrigued and puzzled biologists is the general decline in biological diversity with increasing latitude. Mexico has 23,000 plant species. Brazil, Colombia, Peru, Madagascar, Mexico, India, China, Indonesia, and Australia contain about 60% of the world's named plant species. In contrast, boreal forests have relatively low biological diversity—while hundreds of tree species occur in a tropical rain forest, a typical boreal forest has fewer than ten and often no more than five tree species. (In the far north, some of this pattern may be explained by the short time since the last glacial maximum roughly 15,000 years ago, because some boreal species are still expanding their ranges northward in the wake of glacial retreat.) The Serengeti Plains of east Africa have dozens of species of large mammals, while in the boreal forest you find fewer than 10—for example, moose, deer, bear (one or a few species), and mountain lions.

What explains this pattern? Ecologists continue to debate the cause of latitudinal patterns in biological diversity. Our discussion so far suggests one theory—simply that the more favorable the temperature and precipitation for life, the more diverse it will be. But the same pattern seems to hold in the oceans, so high diversity cannot simply be linked to rainfall. Another theory is that the greater the *variability* of climate, the lower the diversity. According to this theory, temperature and rainfall in tropical rain forests not only remain within boundaries good for life, but also remain relatively constant throughout the year. Meanwhile, high latitudes, such as around Fairbanks, Alaska, have warm summers—the daily temperatures can be about 20°C—but in the winter the temperatures are very low. The basis of this theory lies with the niche concept (see Chapter 7). Part of the explanation seems to be that where environment varies greatly, each species must be a generalist—adapted to a wide range of environmental conditions—so there are fewer available niches. Where environment is relatively constant, species can become specialists, with a narrow range of tolerance, and therefore can divide the environment into many niches (see the discussion of the Hutchinsonian niche in Chapter 7).

But most likely the pattern is the result of several factors, as there are exceptions to the general rule that relatively constant climates have relatively high diversity. For example, in Israel, areas long disturbed by the grazing of goats and sheep and by human settlements have high vegetation diversity. The cause of geographic patterns in biological diversity is a topic that has fascinated people for thousands of years, and a complete discussion goes beyond what is possible in this book. The important thing to remember is that here is an ancient question still not completely answered, and ready for more scientific insights in the future.

Earth's 17 Major Biomes

In the next section, we take a closer look at the Earth's 17 major biomes.

Tundra

Tundras are treeless plains that occur in the harsh climates of low rainfall and low average temperatures (Figure 8.11). The dominant vegetation includes grasses and their relatives (sedges), mosses, lichens, flowering dwarf shrubs, and mat-forming plants. As the environment becomes harsher, dwarf shrubs and grasslike plants give way to mosses and lichens and finally to bare rock surfaces with occasional lichens.

There are two kinds of tundra: arctic, which occurs at high latitudes, and alpine, which occurs at high

Figure 8.11 • Tundra biome. Shown here is tundra in the Canadian arctic.

elevations. The vegetation of both is similar, but the dominant kinds of animals are different. Arctic tundras typically have some large mammals, such as the caribou of North America, as well as important small mammals, birds, and insects. In alpine tundras, the dominant animals are small rodents and insects. This is partly because alpine tundras occupy comparatively small, isolated areas, whereas arctic tundras cover the large territories required for populations of large mammals. These patterns are consistent with the theory of island biogeography, discussed in Section 8.5.

Parts of tundra have *permafrost*—permanently frozen ground—which is extremely fragile. When the vegetative cover is removed from permafrost area, for instance in road building, the land surface begins to warm and the permafrost melts. Meltwater pools and becomes a heat reservoir that encourages further melting. Permafrost areas damaged in this way may be permanently changed or take a very long time to recover. One of the possible implications of global climate change (Chapter 20) is more widespread melting of permafrost, and permanent alteration of the tundra biome.

Taiga, or Boreal Forests

The taiga, or boreal forest, biome includes the forests of the cold climates of high latitudes and high altitudes. Taiga forests are dominated by conifers, especially spruces, firs, larches, and some pines. Aspens and birches are important flowering trees (Figure 8.12). Typically, boreal forests form dense stands of relatively small trees, usually less than 30 m tall. Boreal forests cover very large areas. Their biological diversity is low—only about 20 major species occur in North American boreal forests—but they contain some of the most commercially valuable trees, such as white pine, various species of spruce, and cedar.

As mentioned in Section 8.2, the northern areas of North America and Eurasia were connected by a land bridge during the last glacial period, allowing the animals and vegetation of the boreal forest to spread widely. Thus, the boreal forests of North America and Eurasia share both genetic heritage (biotic provinces) and similar climates. The similar climate has led to dominant life forms similar in shape and form (biome characteristics). Moose, for example, are found on both continents, as are small flowering plants called *Saxifraga flagellaris*. The dominant animals of boreal forests include a few large mammals (moose, deer, wolves, and bears), small carnivores (foxes), small rodents (squirrels and rabbits), many insects, and migratory birds, especially waterfowl and carnivorous land birds, such as owls and eagles.

Disturbances—particularly fires, storms, and outbreaks of insects—are common in the boreal forests. For example, the whole of Ontario's Quetico Provincial Park burns over (through numerous small fires) an average of once every 80 years, and individual forest stands are rarely more than 60 years old.

Boreal forests contain some of Earth's largest remaining roadless areas and are appreciated for conservation of wilderness, their natural resources and for various spiritual and recreational uses. The fur-bearing animals of boreal and tundra biomes have long been of commercial value, although that value has declined greatly in Western nations because of concerns about the animals' welfare.

Temperate Deciduous Forests

Temperate forests occur in climates somewhat warmer than those of the boreal forest. These forests grow throughout North America, Eurasia, and Japan and have many genera in common but different species. Dominant vegetation includes tall deciduous trees; common species are maples, beeches, oaks, hickories, and chestnuts, typically taller than trees of the boreal forest. These forests are important economically for their hardwood trees, which are used for furniture, among other things. Temperate deciduous forests are among the biomes most changed by human beings because they occur in regions long dominated by civilization, including much of China, Japan, Western Europe, the United States, and the urbanized parts of Canada.

Large mammals, like deer, in this biome depend on young forests. Because the deep shade of temperate forests allows little vegetation growth near the ground, dominant animals tend to be small mammals that live in trees (e.g., squirrels) and those (e.g., mice) that feed on soil organisms and small plants. Other typical mammals include raccoons, porcupines, and skunks. Birds and insects are abundant.

There are few remaining uncut or unplowed stands of temperate forest, and some of them are important nature preserves. As with the boreal forest biome, fire is a natural, recurring feature, although the frequency of fire in many temperate forests is typically lower than in

Figure 8.12 • The boreal forest. A stand of boreal forest near Whitehorse, Yukon.

the boreal forests. In this biome, hunting-and-gathering cultures appear to have contributed to the frequency of fire in many regions.

Temperate Rain Forests

Temperate rain forests occur where temperatures are moderate and precipitation exceeds 250 cm/year. Such rain forests are rare but spectacular. They are important commercially, for timber crops, and culturally, as the home of several First Nations. The dominant trees are evergreen conifers, in contrast to the temperate deciduous forests, where flowering deciduous trees dominate. An intriguing question is why evergreens dominate the temperate rain forests but not the deciduous temperate forests. The best explanation appears to be that winters are wet and relatively mild in the temperate rain forests, so evergreen trees can carry out photosynthesis and grow when the temperature warms above freezing in the winter, while the deciduous plants cannot. In contrast, in the temperate deciduous forests, winter temperatures remain below freezing, and the conifers are at a disadvantage—they must pay the metabolic costs of maintaining green needles without receiving a metabolic benefit.[14]

The temperate rain forests are the giant forests (Figure 8.13). In the Northern Hemisphere, they include the redwood forests of California and Oregon, where the tallest trees in the world exist, as well as forests in British Columbia and the state of Washington, dominated by such large trees as Douglas firs and western cedars. The trees grow taller than 70 m and are long-lived; Douglas firs live more than 400 years, redwoods several thousand.

Temperate rain forests also occur in the Southern Hemisphere. The best known of these are the forests of western New Zealand. The trees here are typically smaller than those in North America.

Temperate rain forests have comparatively low biological diversity, because they create deep shade in which few other plants can grow and so provide food for herbivores. Yet, the rarity of these magnificent ancient forests, and their associated species such as the spotted owl and the marbled murrelet, have made them a focus of biological conservation efforts.[15]

Figure 8.13 • Temperate rain forest biome, British Columbia.

Temperate Woodlands

Temperate woodlands occur where the temperature patterns are like those of deciduous forests but the climate is slightly drier. In North America, such areas occur from Eastern Canada south to Georgia and the Caribbean islands. Temperate woodlands are dominated by small trees, such as piñon pines and evergreen oaks. The stands tend to be open, with wide spaces between trees, allowing considerable light to reach the ground. These generally pleasant areas are often used for recreation.

Fire is a common disturbance, and many species are adapted to it and require it. Temperate woodlands are often economically valuable because the pines typical of these areas are fast growing and produce wood for timber, pulp, and paper. Forest plantations are common in this biome. The pleasant combination of trees and grasses, along with the abundance of some large mammals such as mule deer in Ponderosa pine, has also made these a focus for biological conservation, once again bringing science and values to the fore.

Temperate Shrublands

In still drier climates, temperate shrublands (sometimes called **chaparral** or scrub forest) occur. These miniature woodlands are dominated by dense stands of shrubs that rarely exceed a few meters in height. Chaparral occurs in Mediterranean climates—climates with low rainfall that is concentrated in the cool season. It is found along the California coast and in Chile, South Africa, and the Mediterranean region. While only about 5% of Earth's land area is in this biome, it is among the most attractive to people because of its moderate, sunny climate. Therefore, it has been highly modified around the world by human action, and relatively few examples of native chaparral remain. Ancient Greek and Roman civilizations were located in this biome.

Typically, chaparral vegetation is distinctively aromatic; an example is sage. Some scientists believe the aromatic compounds produced by the plants are a kind of chemical warfare—that these compounds are toxic to competing plants and give the producers of them a competitive advantage. Although this explanation seems plausible, establishing that it is true has been difficult, and its demonstration still eludes scientists. There are few large mammals; reptiles and small mammals are more characteristic of this biome. The animals and plants have little economic value at present, but the biome is important for watersheds and erosion control.

The vegetation is adapted to fire; many species regenerate rapidly, and some actually promote fire by producing abundant fuel in the form of dead twigs and branches. As a result, stands are rarely more than 50 years old. When intense precipitation follows fire, erosion can be exceptionally severe until renewed vegetation again protects the slopes.

Temperate Grasslands

Temperate grasslands occur in regions too dry for forests and too moist for deserts. Dominant plant species are grasses, forbs (flowering plants that are not grasses, trees, or shrubs) and other flowering plants, many of which are perennials with extensively developed roots. Temperate grasslands cover large areas of Earth—or did before many were converted to agriculture. They include the great North American prairies (Figure 8.14), the steppes of Eurasia; the plains of eastern and southern Africa; and the pampas of South America. In Canada, remnants of original mixed prairie grasslands are preserved in Grasslands National Park, Saskatchewan, and Riding Mountain National Park, Manitoba.

The soils often have a deep organic layer, formed by the decaying roots of the grasses, and the decaying stems and leaves of the prairie plants. The result is some of the best soils in the world for agriculture. By volume, most of the food of the world comes from this biome—all of the small grains and most of the large hoofed herbivores, like cattle and bison, that provide meat for people.

Figure 8.14 • **Temperate grassland biome.** The great prairie of western Canada and the United States was one of the world's largest areas of temperate grasslands before European settlement of North America.

In North America, unploughed prairie is rare. People have made a considerable effort in recent years to restore prairies, a labour-intensive task. Prior to European settlement, the North American prairie extended from southeastern Manitoba northwest to central Alberta, and south along the eastern fringe of the Rocky Mountains to Oklahoma and Texas. Other grasslands occurred in the high plains of northern California and in eastern Oregon and Washington.

Grasslands are home to the highest abundance and greatest diversity of large mammals: the wild horses, asses, and antelopes of Eurasia; the once-huge herds of bison that roamed the North American prairies; the kangaroos of Australia; and the antelope and other large herbivores of Africa. Fossil evidence suggests that grasslands and grazing mammals evolved together, beginning about 60 million years ago. Grassland plants are adapted to certain kinds of grazing, and animals are important in the transport of seeds. The animals, of course, require edible grasses and forbs, so there is a kind of symbiosis among the animals and plants of this biome, which is essential for its continuation.

Fire is a natural, recurring feature. If fire and grazing are eliminated, the land tends to become desert shrubland in the drier regions and open woodlands in the wetter areas. In many grasslands, hunting-and-gathering cultures appear to have contributed to the fire frequency and, in this way, increased the area and persistence of this biome.

Tropical Rain Forests

Tropical rain forests occur where the average temperature is high and relatively constant throughout the year and

Figure 8.15 • Tropical rain forest biome. The vegetation in Puerto Rico's El Yunque rain forest illustrates montane tropical rain forests.

Figure 8.16 • Desert biome. These desert lands are in the Mojave desert, California.

where rainfall is high and relatively frequent throughout the year. Such conditions occur in northern South America, Central America, western Africa, northeastern Australia, Indonesia, the Philippines, Borneo, Hawaii, and parts of Malaysia. Tropical rain forests have long been home to hunting-and-gathering cultures, but relatively few civilizations have been able to persist in this biome.

Tropical rain forests are famous for their diversity of vegetation. Hundreds of species of trees may be found within a few square kilometers. Typically, some trees are very tall, and some, such as palm trees, remain relatively small. Some plants, like bromeliads and some ferns, grow on trees (Figure 8.15). Approximately two-thirds of the 300,000 known species of flowering plants occur in the tropics, mainly in tropical rain forests. Many species of animals occur as well. The mammals tend to live in trees, but some are ground dwellers. Insects and other invertebrates are abundant and show a high diversity. Rain forests occur in some of the most remote regions of Earth and remain unexplored; many undiscovered species are believed to exist there.

Except for dead organic matter at the surface, soils in this biome tend to be very low in nutrients. Most chemical elements (nutrients) are held in the living vegetation, which has evolved to survive in this environment; otherwise, rainfall would rapidly remove many chemical elements necessary for life.

Tropical Seasonal Forests and Savannas

Tropical seasonal forests occur at low latitudes, where the average temperature is high and relatively constant throughout the year and rainfall is abundant but very seasonal. Such forests are found in India and Southeast Asia, Africa, and South and Central America. In areas of even lower rainfall, tropical savannas—grasslands with scattered trees—are found. These include the savannas of Africa, which, along with the grasslands, have the greatest abundance of large mammals remaining anywhere in the world. The number of plant species is high as well.

Disturbances, including fires and the impact of herbivory on the vegetation, are common but may be necessary to maintain these areas as savannas; otherwise, they would revert to woodlands in wetter areas or to shrublands in drier areas. In still drier climates, savannas are replaced by shrublands, characterized by small shrubs, a generally low abundance of vegetation, and a low density of vertebrate animals.

Deserts

Deserts occur in the driest regions where vegetation can survive, typically where the rainfall is less than 50 cm/year. Most deserts—such as the Sahara of North Africa (Figure 8.16) and the deserts of the southwestern United States, Mexico, and Australia—occur at low latitudes. However, cold deserts occur in the basin and range areas of Utah and Nevada and in parts of western Asia.

Most deserts have a considerable amount of specialized vegetation, as well as specialized vertebrate and invertebrate animals. Soils often have little or no organic matter but abundant nutrients and need only water to become very productive. Common disturbances are occasional fires; occasional cold weather; and sudden, infrequent, and intense rains that cause flooding.

Relatively few large mammals live in deserts. The dominant animals of warm deserts are non-mammalian vertebrates (snakes and reptiles). Mammals are usually small, like the kangaroo mice of North American deserts. Many desert animals are adapted for burrowing and night activity, to escape the heat of the sun.

Figure 8.17 • **Wetland biome.** Wetlands include areas of open standing water and areas of herbs, grasses, shrubs, and trees that can withstand persistent or frequent flooding.

Wetlands

Wetlands include freshwater swamps, marshes, and bogs and saltwater marshes. All have standing water: The water table is at the surface, and the ground is saturated with water (Figure 8.17). Standing water creates a special soil environment with little oxygen, so decay takes place slowly, and only plants with specialized roots can survive. Bogs—wetlands with rainwater or stream input but no surface water outlet—are characterized by floating mats of vegetation. Swamps and marshes are wetlands with surface inlets and outlets.

Dominant plants are small, ranging from small trees—such as the mangroves of warm coastal wetlands and the black spruces and larches of the North—to shrubs, sedges, and mosses. Small changes in elevation make a great difference. On slight rises, roots can obtain oxygen, and small trees can grow; in lower areas are patches of open water with algae and mosses.

Although wetlands occupy a relatively small portion of Earth's land area, they are important in the biosphere. In the oxygenless soils, bacteria survive that cannot live in high oxygen atmospheres. These bacteria carry out chemical processes, such as the production of methane and hydrogen sulfide, that have important effects in the biosphere. In addition, over geologic time, wetland environments produced the vegetation that today is coal.

Saltwater marshes, such as those in the Bay of Fundy (Chapter 7), are important breeding areas for many oceanic animals and are home to many invertebrates. Dominant animals include crabs and other shellfish such as clams. Saltwater marshes are an important economic resource. Dominant animals in freshwater wetlands include many species of insects, birds, and amphibians; few mammals are exclusive inhabitants of this biome. The larger swamps of warm regions are famous for large reptiles and snakes, as well as for a relatively high diversity of mammals where topographic variation includes small upland areas.

Although, because of the high water table, people tend not to live within wetlands, this biome can produce many edible plants; plants useful for making things such as baskets and other containers; and animals, including fish, that provide food. Wetlands are often used for recreation and have been a favorite biome for many naturalists and conservationists.

Freshwaters

Freshwater lakes, ponds, rivers, and streams make up a very small portion of Earth's surface but are critical to the water supply for homes, industry, recreation, and agriculture and play important ecological roles (Figure 8.18). Rivers and streams are also important in the biosphere as major transporters of materials from land to ocean.

Dominant freshwater plants are floating algae, referred to as *phytoplankton*. Flowering plants such as water lilies are rooted along shores and in shallow areas. Animal life is often abundant. Open waters have numerous small invertebrate animals (collectively called *zooplankton*), both herbivores and carnivores, and many species of finfish and shellfish.

Estuaries—areas at the mouths of rivers, where river water mixes with ocean water—are rich in nutrients. They usually support an abundance of fish and are important breeding sites for many commercially significant fish. Furthermore, many species that spend a greater portion of their life cycle in other biomes depend on freshwaters for reproduction or for food, as well as for drinking water. Freshwaters, then, are among the most important biomes for life's diversity.

Freshwaters are also among the areas most altered by human activities, especially by modern technology. Waterpower was one of the first sources of non-biological energy, and rivers have long provided major transportation routes for people. Impoundment of streams and rivers (for example, by building dams and reservoirs), along

Figure 8.18 • **Freshwater biome.** The Muskoka River, Ontario.

Figure 8.19 • Intertidal biome. The intertidal zone at the Twelve Apostles, on the south coast of Victoria, Australia.

with channelization to make transportation on them easier, has led to major changes in many freshwaters.

In recent years, the importance of freshwaters and their surrounding riparian and wetland areas has been recognized, and a major change has occurred in Western civilization's attitude toward rivers and streams. A generation ago, waterpower was considered one of the "cleanest" and most environmentally friendly forms of energy production. Today, there is growing interest in protecting free-flowing rivers and streams, whose complex channels, backwaters, meanders, floodplains, and seasonal ponds are important to fish, water birds, and other wildlife. Much conservation effort is now being expended to restore freshwaters. For example, water quality improvement is a major focus of restoration efforts in the Great Lakes and their connecting channels under the Canada–United States Great Lakes Water Quality Agreement. Through that Agreement, the International Joint Commission, established under the 1909 Boundary Waters Treaty between the two nations, oversees a binational program of Remedial Action Plans aimed at restoration of 42 freshwater "areas of concern" in the Great Lakes system.

Meanwhile, however, in developing nations, waterpower remains one of the least expensive sources of energy, and the development of major dams continues, as with the recently completed Three Gorges Dam in China. We can expect freshwaters to be a major focus of environmental conflicts in the next decades.[16]

Intertidal Areas

The intertidal biome is made up of areas exposed alternately to air during low tide and ocean waters during high tide (Figure 8.19). Constant movement of waters transports nutrients into and out of these areas, which are usually rich in life and important to people as a direct source of food and as a spawning and breeding ground for many important foods. As a result, they are major economic resources. Large algae are found here, from giant kelp of temperate and cold waters to algae of coral reefs in the tropics. Birds and attached shellfish are usually abundant and are economically important. Nearshore areas are often important breeding grounds for many species of fish and shellfish, often also of economic significance. They are also major resting sites for migratory waterfowl. For example, the intertidal mudflats of the Fraser River estuary in southwestern British Columbia are important stopovers on the Pacific Flyway. Millions of birds from three continents congregate in these locations to feed on abundant invertebrates, replenishing their energy on the way to and from wintering grounds in California, western Mexico, and Central and South America.

The nearshore part of the oceanic environment is most susceptible to pollution from land sources. It is often heavily polluted by human activities, because major cities and civilizations tend to develop at the mouths of major rivers and along productive intertidal coastlines. In addition, as a major recreational area, it is subject to considerable alteration by people. Some of the oldest environmental laws concern the rights to use resources of this biome, and today major legal conflicts continue about access to intertidal areas and harvesting of biological resources.

Disturbances are common in the intertidal biome. Indeed, some of the most extreme variations in environmental conditions occur in this biome, among which are daily changes in sea level with the tides, seasonal changes in tidal minima and maxima, and ocean storms. Adaptation to these disturbances is essential to survival within the intertidal area. Consider, for example, a barnacle or mussel that twice daily experiences a change from a cool or cold saline water environment to direct exposure to bright sunlight and the highly oxygenated atmosphere.

Open Ocean

Called the *pelagic region*, the open ocean biome includes open waters in all of the oceans. These vast areas tend to be low in nitrogen and phosphorus—chemical deserts with low productivity and low diversity of algae. Many species of large animals occur but at low density. Linking deep regions of the ocean to intertidal zones are shallow plateaus called continental shelves. These "banks," like the famed Grand Banks and Georges Bank off Newfoundland, are among the most productive regions of the ocean. Georges Bank is more than 100 m higher than the sea floor of the Gulf of Maine, just north of it, and was in fact exposed land during the last glacial advance. The role of these continental shelves in Canada's Atlantic fishery is discussed in Chapter 14.

Benthos

The bottom portion of oceans is called the *benthos* (deeps). The primary input of food is dead organic matter that falls from above. The waters are too dark for photosynthesis, so no plants grow there.

Upwellings

Deep-ocean waters are cold and dark, and life is scarce. However, these waters are rich in nutrients because of numerous creatures that die in surface waters and sink. (See Chapter 9 for a discussion of energy flow in ocean ecosystems.) Upward flows, or upwellings, of deep-ocean waters bring nutrients to the surface, allowing abundant growth of algae, and animals that depend on algae. Upwellings occur off the west coast of North America, the Bay of Fundy (see Chapter 7), South America, West Africa, and near the Arctic and Antarctic ice sheets. In some areas, deeper waters are brought to the surface by winds that push coastal waters away from shore. These fertile upwelling zones are among the most important regions for the production of commercial fish.

Hydrothermal Vents

Hydrothermal vents, a recently discovered biome, occur in the deep ocean, where plate tectonic processes create vents of hot water with a high concentration of sulfur compounds. These sulfur compounds provide an energy basis for chemosynthetic bacteria, which support giant clams, worms, and other unusual life forms (see Chapter 8). Water pressure is high, and temperatures range from the boiling point in waters of vents to the frigid (about 4°C) waters of the deep ocean.

8.4 Geographic Patterns of Life within a Continent

So far, the discussion has focused on continental similarities and differences among species and on biological diversity. The same concepts—convergent evolution, divergent evolution, common ancestry, and similar environments—led to geographic patterns within a continent. The theory of continental drift provides us with a moving picture show of huge landmasses moving ponderously over Earth's surface, periodically isolating and remixing groups of organisms and leading to an increase in the diversity of species. If each continent were a uniform plot of land in a uniform climate, there would be fewer potential ecological niches (see Chapter 7), and biological diversity within a continent would be low.

Plate tectonics results in continents with complex topography, including mountain ranges and alterations in drainage patterns and the paths of rivers. These can be barriers to the migration of species, leading to geographic isolation within a continent. When the Rocky Mountains began to form 90 million years ago, they created a barrier to various kinds of non-mountain life forms. As a result of this—as well as climatic changes, some of which stemmed partially from the mountain building—California's vegetation is quite distinct from that found at similar elevations and latitudes east of the Rocky Mountains.

Patterns of life on a continent are also affected by the proximity of a habitat to an ocean or other large body of water, ocean currents near shore, position relative to mountain ranges, latitude, and longitude. Figure 8.20 shows the pattern of life from west to east across North America (see A Closer Look 8.1). Modern DNA analysis is revolutionizing our understanding of biogeography. For examples, recent studies have used these techniques to re-evaluate assumptions about the impact of human hunting on large mammal populations during the last glaciation. Their results have demonstrated that widespread Pleistocene extinctions of large mammals were likely the result of declining genetic diversity associated with climate cooling about 37,000 years ago, at the beginning of the last glacial advance. The impact of human hunters was probably insignificant until at least 15,000 years later.[18]

8.5 Island Biogeography

The many jokes and stories about people becoming castaways on an island have a basis in facts about the biogeography of islands. Islands have fewer species than continents, and the smaller the island, the fewer the species, on average. Also, the farther away an island is from a continent, the fewer species it will have. These two observations form the basis for the *theory of island biogeography.*

Darwin's visit to the Galápagos Islands gave him his most powerful insight into biological evolution.[18] There he found many species of finches that were related to a single species found elsewhere. On the Galápagos, each species was adapted to a different niche.[19] Although all the finches are of similar size, body shape, and colour, each is uniquely adapted to a particular habitat and food source. Vegetarian and ground-dwelling forms have crushing bills, for example, while finches that inhabit trees have grasping bills. Those that probe for food in cactus or tree bark have long, tapered bills suited to that task. Some forms feed on insects, while others feed on seeds. One, known colloquially as the "vampire finch," feeds on the blood of larger birds by pecking at their backs. Finches elsewhere in the world are generalized feeders. Darwin speculated that finches isolated from other species on the continents eventually separated into a number of groups, each adapted to a more specialized role. This process is called **adaptive radiation**.

Adaptive radiation has also been observed in other systems. On the Hawaiian Islands, a finchlike ancestor evolved into several species, including fruit and seed eaters, insect eaters, and nectar eaters, each with a beak adapted for its specific food.[20] Lake Malawi, Africa, supports over 500 species of cichlid fishes

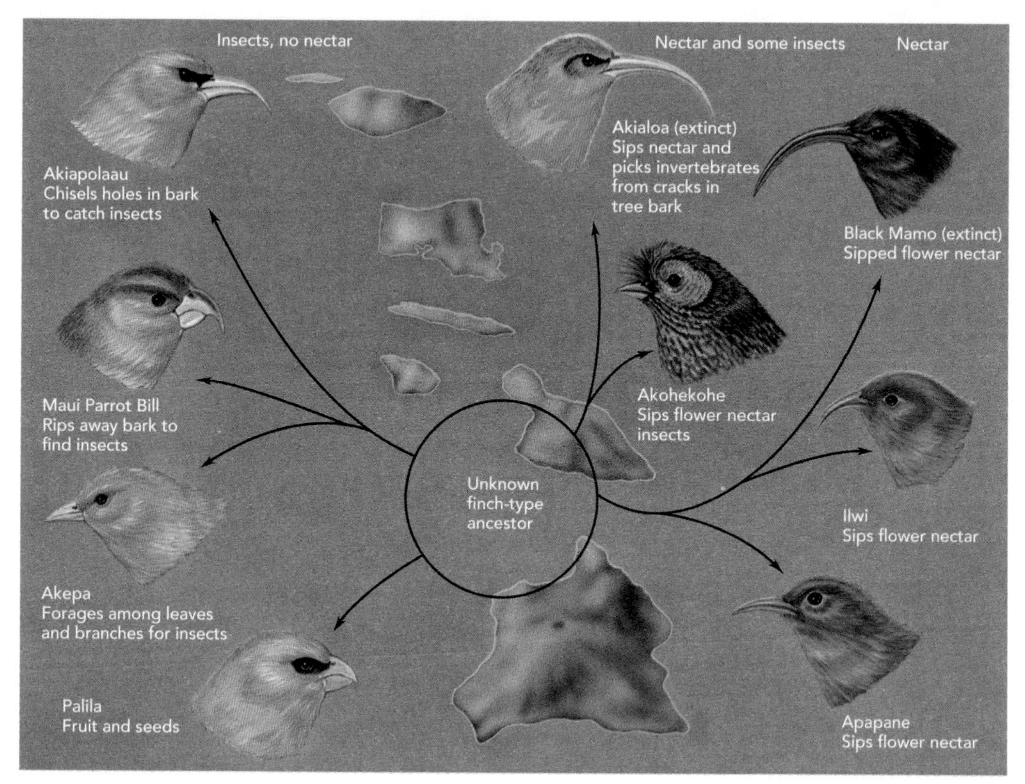

Figure 8.20 • Evolutionary divergence among honeycreepers in Hawaii. Sixteen species of birds, each with a beak specialized for its food, evolved from a single ancestor. Eight of the species are shown here. The species evolved to fit ecological niches that, on the North American continent, had previously been filled by other species not closely related to the ancestor. [*Source*: From C. B. Cox, I. N. Healey, and P. D. Moore, *Biogeography* (New York: Halsted, 1973). Copyright of Blackwell Scientific Publishing, Ltd.]

(Figure 8.20), all believed to have arisen from a common origin.

We can make several generalizations about species diversity on islands:

- The two sources of new species on an island are migration from the mainland and evolution of new species in place (as with Darwin's finches in the Galápagos).
- Islands have fewer species than continents.
- The smaller the island, the fewer the species, as can be seen in the number of reptiles and amphibians in various West Indian islands.
- The farther the island from a mainland (continent), the fewer the species (Figure 8.21).[21]

Why these generalizations? Small islands tend to have fewer habitat types. Some habitats on a small island may be too small to support a population large enough to have a good chance of long-term survival. A small population might be easily extinguished by a storm, flood, or other catastrophe or disturbance. Every species is subject to risk of extinction by predation, disease (parasitism), competition, climatic change, or habitat alteration. Generally, a smaller population is at greater risk of extinction. On a smaller island it is difficult for the smaller population of a particular species to be supported.

The farther the island is from the mainland, the harder it will be for an organism to travel the distance.

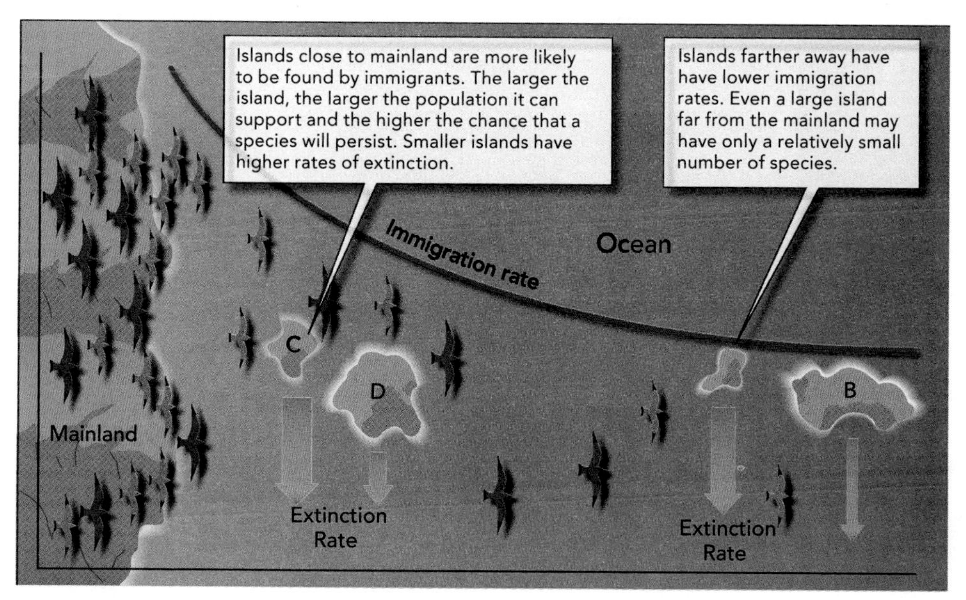

Figure 8.21 • Idealized relation of an island's size, distance from the mainland, and number of species. The average number of species on the island depends on the rate of immigration and the rate of extinction. Thus, a small island near the mainland may have the same number of species as a large island far from the mainland. The thickness of the arrow represents the magnitude of the rate. [*Source*: Modified from R. H. MacArthur and E. O. Wilson, *The Theory of Island Biogeography* (Princeton, NJ: Princeton University Press, 1967).]

A CLOSER LOOK 8.1

A Biogeographical Cross Section of North America

A generalized cross section of North America shows the relationships among weather patterns, topography, and biota (Figure 8.22). Off the West Coast in the Pacific basin occur the pelagic ecosystems, where sufficient light for photosynthesis penetrates the waters. This region is populated by small, mainly single-cell algae. Other oceanic zones with too little light for photosynthesis are populated by animals that feed on dead organisms that sink from above. Near the shore, particularly in areas of upwelling, are abundant algae, fish, birds, shellfish, and marine mammals. Where the tides and waves alternately cover and uncover the shore, a long, thin line of intertidal ecosystems is found, dominated by kelp and other large algae that are attached to the ocean bottom; by shellfish, such as mussels, barnacles, abalone, crabs, and other invertebrates; and by shorebirds, such as sandpipers.

Weather systems move generally from west to east in Canada. As air masses are forced over the coastal mountains and Rocky Mountains, they are cooled, and the moisture condenses to form clouds and rain. The West Coast is an area of moderate temperature because water has a high capacity to store heat and the Pacific Ocean moderates the land's temperature. The annual precipitation increases with elevation on the western slopes of the mountains. In general, the colder, wetter heights of the mountains support coniferous forests.

Along the coast of British Columbia, cool temperatures year-round lead to heavy rains near the shore, producing an unusual temperate-climate rain forest. Canada possesses the largest remaining intact temperate rain forest in the world—roughly a quarter of the global total. The best-known Canadian examples occur on Vancouver Island, the central British Columbia coast, and the Queen Charlotte Islands.

The eastern slopes of the coastal ranges form a so-called rain shadow. First, the air that passes over these eastern slopes gives up most of its moisture to the mountains; as a result, it is dry as it passes to the east. In addition, as air sinks to lower elevations, it is warmed, and can hold more moisture. This dry air tends to take up moisture from the ground, producing the dry prairie climate and the deserts of the

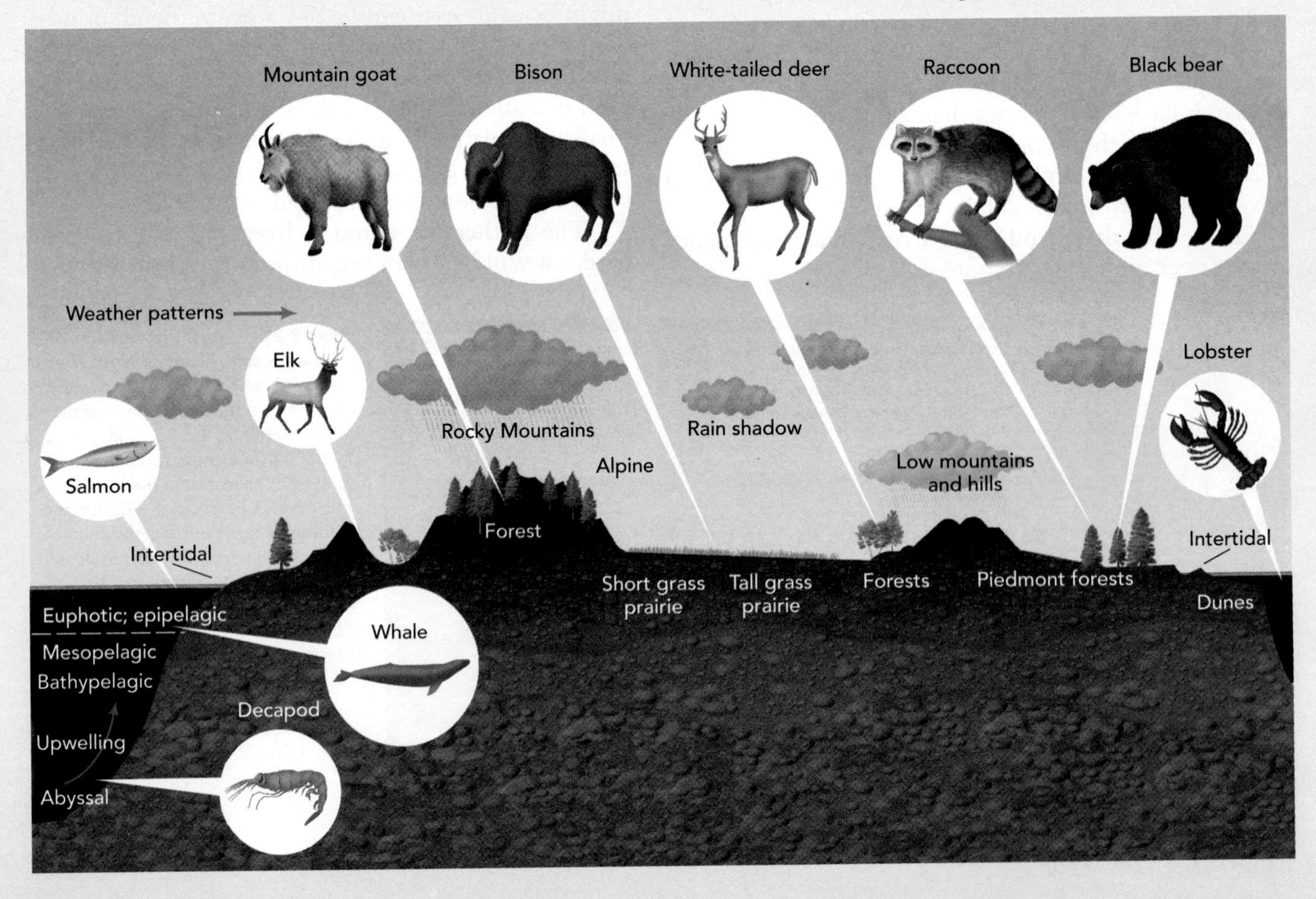

Figure 8.22 • Generalized cross section of North America showing weather, landforms, and the geography of life. The weather patterns move from west to east.

American Southwest. Whereas annual rainfall in coastal British Columbia can approach 6,000 mm/year, east of the Rocky Mountains it falls to about 275 mm/year, but then increases steadily eastward: 450 mm/year at Winnipeg, Manitoba; 700 mm/year near Montréal, Québec; and 1200 mm/year near St. John's, Newfoundland.[22]

The biomes reflect these changes in rainfall. In southwestern Alberta are shortgrass prairies, which become mixed-grass prairies (a mixture of shortgrass and tall-grass prairies) and then tall-grass prairies in the Red River Valley of Manitoba as we continue eastward. Rainfall reaches levels sufficient to support forests farther east, near the Manitoba-Ontario border, where the annual rainfall reaches 50-70 cm. From there to the East Coast, the eastern deciduous forest (dominated by trees that lose their leaves during the winter) and the boreal forest of eastern North America predominate.

The patterns described for Canada occur worldwide. One sees changes with elevation from warm, dry-adapted woodlands to moist, cool-adapted woodlands in Spain, where beech and birch, characteristic of middle and northern Europe (Germany, Scandinavia), are found at high elevations and alpine tundra is found at the summits. Similar patterns occur in Venezuela, where a change in elevation from sea level to 5,000 m at the summits of the Andes is equivalent to a latitudinal change from the Amazon basin to the southern tip of the South American continent. The seasonality of rainfall, as well as the total amount, often determines which ecosystems occur in an area.

Two other general concepts of biogeography, illustrated by both the latitudinal patterns from the Arctic to the tropics and the altitudinal patterns from mountaintops to valleys, are that (1) the number of species declines as the environment becomes more stressful; and (2) on land, the height of vegetation decreases as the environment becomes more stressful (Figure 8.23). These concepts apply to most stresses, including those that people impose by adding pollutants to the environment, decreasing the fertility of soils or otherwise impoverishing habitats, and increasing the rate of environmental disturbance. From these concepts, we can predict that highly polluted and disturbed landscapes and seascapes will have few species and that, on the land, the dominant species of plants will have small stature.

Figure 8.23 • Environmental stress and biogeography. Certain general patterns can be found as an environment becomes more stressful. This diagram shows the effects of increasing water stress. Where rainfall is plentiful, vegetation is abundant, and there are forests of tall trees of many species. As rainfall lessens, the size of the plants decreases to small trees, then shrubs and grasses, then scattered plants. The total biomass decreases and, in general, the number of species decreases. Similar changes accompany increases in other kinds of stress, including the stress of certain pollutants. [*Source*: Adapted from R. H. Whittaker, *Communities and Ecosystems*, 2nd ed. (New York: Macmillan, 1975).]

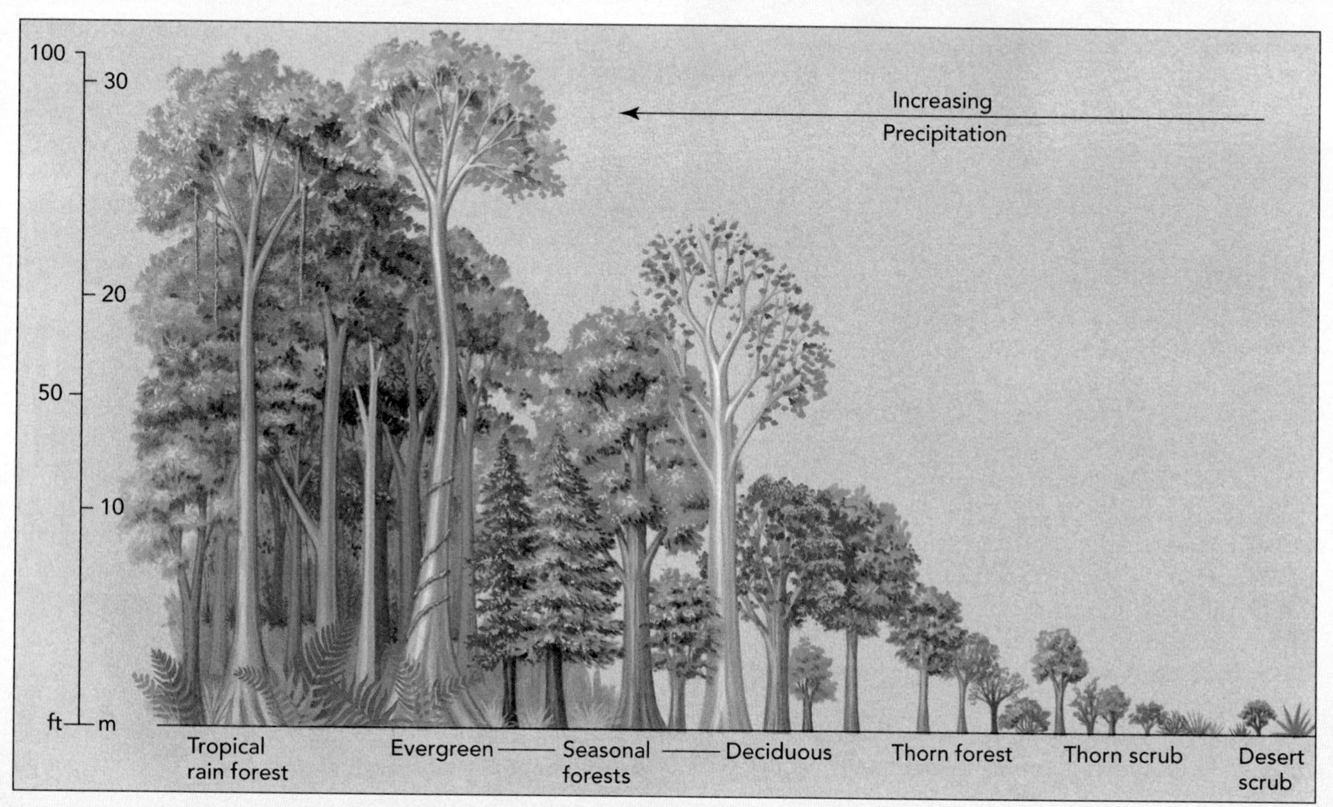

CRITICAL THINKING ISSUE

Escape of an Exotic Species

About 20 years ago, a delicate seaweed species named *Caulerpa taxifolia* (Figure 8.24) was brought from its native habitat in the Pacific Ocean to a zoo in Germany, where it was cultivated and used to embellish saltwater aquarium exhibits, a seemingly harmless action. The seaweed was such a success that samples were sent to other institutions, including the Oceanographic Museum in Monaco.

Within about five years of its introduction there, an unfortunate accident took place. The seaweed was inadvertently flushed into the Mediterranean when exhibit tanks were cleaned. This accident might seem innocuous, but considering it so would ignore the tremendous power of species to act as biological invaders.

Once freed in the Mediterranean, *Caulerpa* quickly changed its growth pattern and adapted to its new habitat along the southern coast of France. This may have occurred through a mutation, through hybridization with native seaweeds, or because its genetic code contained information that allowed for considerable plasticity. Whatever the exact genetic explanation, today *Caulerpa* grows about six times larger in the Mediterranean than it does in its native Pacific Ocean. It also tolerates colder temperatures, surviving in waters as chilly as 10°C, compared with a limit of 21°C in its home waters.

Figure 8.24 • ***Caulerpa taxifolia*.** This delicate seaweed brought from the Pacific Ocean has become a pest in the Mediterranean.

Over the past two or three years, *Caulerpa* has spread to the Adriatic, and it now appears to threaten the entire Mediterranean with its ability to choke out competing algae. It grows on rocks, sand, and mud, unlike most seaweed species, which do best on only one kind of substrate. It grows so profusely that it blankets competing native seaweeds, excluding them. It appears to be toxic to local animals that feed on seaweed, such as sea urchins, which will not touch it.

Controlling this invader has proved very difficult. When mechanically removed, it soon recovers to form even denser stands. Scientists are seeking a biological control method and are currently studying an exotic snail for release.

Thus, a seemingly harmless algae that gracefully decorated indoor aquariums has become an invading monster, affecting algae and animal life in the Mediterranean Sea, with consequences for commerce, recreation, and aesthetics. The problem caused by *Caulerpa taxifolia* is the result of the geography of life and the alteration of this geography by human action. The question that faces us today is: How can we control *Caulerpa taxifolia*?

Critical Thinking Questions

1. Suppose you were in charge of a program to control or eliminate this pest from the Mediterranean. Would you try to eliminate *Caulerpa* completely or just reduce its abundance?
2. Develop a plan for the control of this species of algae, based on material in this and previous chapters. What kinds of biogeographic exploration might help you develop a plan to control this species?
3. If nothing is done, what would be a likely long-term outcome?

In addition, a smaller island is a smaller "target," less likely to be found by individuals of any species.

A final generalization about island biogeography is that over a long time, an island tends to maintain a rather constant number of species, which is the result of the rate at which species are added minus the rate at which they become extinct. For any island, the number of species of a particular life form can be predicted from the island's size and distance from the mainland.

The concepts of island biogeography apply not only to real islands in an ocean but also to ecological islands. An **ecological island** is a comparatively small habitat separated from a major habitat of the same kind. For example, a pond in the Ontario woods is an ecological island relative to the Great Lakes that border Ontario. A small stand of trees within a prairie is a forest island. A city park is also an ecological island. Is a city park large enough to support a population of a particular species? To know whether it is, we can apply the concepts of island biogeography.

8.6 Biogeography and People

We have seen that biogeography affects biological diversity. Changes in biological diversity in turn affect people and the living resources that we depend on. These effects extend from individuals to civilizations. For example, the last ice ages had dramatic effects on plants and animals and thus on human beings. Europe and Great Britain have fewer native species of trees than other temperate regions of the world. Only 30 tree species are native to Great Britain (that is, they were present prior to human settlement), although hundreds of species grow there today.

Why are there so few native species in Europe and Great Britain? It is because of the combined effects of climatic change and the geography of European mountain ranges. In Europe, major mountain ranges run east–west, whereas in North America and Asia, the major ranges run north–south. During the past 2 million years, Earth has experienced several episodes of continental glaciation, when glaciers several kilometers thick expanded from the Arctic over the landscape. At the same time, glaciers formed in the mountains and expanded downward. Trees in Europe were caught between the ice from the North and the ice from the mountains and had few refuges; many species became extinct. In contrast, in North America and Asia, as the ice advanced, tree seeds could spread southward, where they became established and produced new plants. Thus, the tree species "migrated" southward and survived each episode of glaciation.[23]

Since the rise of modern civilization, these ancient events have had many practical consequences. As we mentioned before, soon after Europeans discovered North America, they began to import **exotic species** (a species introduced into a new geographic area) of trees and shrubs to Europe and Great Britain, where these were used to decorate gardens, homes, and parks and formed the basis of much of the commercial forestry in the region. For example, in the famous gardens of the Alhambra in Granada, Spain, Monterey cypress from North America are grown as hedges and cut in elaborate shapes. Douglas fir and Monterey pine are important commercial timber trees in Great Britain and Europe. These are only two examples of how knowledge of biogeography—being able to predict what will grow where based on climatic similarity—has been used for both aesthetic and economic benefits.

As mentioned in Chapter 7, people have altered biodiversity mainly by (1) direct hunting, which can cause extinction or serious decline in a species; (2) directly disrupting habitats; and (3) introducing exotic species into new habitats. The last is particularly relevant to this chapter.

Human introduction of exotics has had mixed results. On the one hand, the major foods of the world come from only a few species, and these species have been introduced widely by people. Without such introductions, people could not live in great numbers in most locations (see Chapter 12). The use of exotic species has also beautified landscapes in many cultures, although the value placed on particular species varies considerably from culture to culture. In addition, pets such as cats and dogs are routinely introduced into new habitats by people.

On the other hand, as this chapter's opening case study and Critical Thinking Issue show, the introduction of exotics into new habitats has often had disastrous ecological consequences. For example, European explorers and settlers brought many infectious diseases to the New World. These introductions had devastating consequences for indigenous peoples who had no immunity to diseases such as cholera, smallpox, typhoid, influenza, scarlet fever, diphtheria, and venereal diseases. One estimate suggests that Canadian First Nations populations were reduced by as much as 90% in hardest-hit areas, over a 200-year period, by exposure to these diseases.[24] Similarly, the introduction of rabbits into Australia for food and sport has had many unintended consequences, including widespread habitat alteration and associated impacts on native species.

This leads to some general rules:

- Unless there is a very good reason to introduce a species into a new habitat, don't do it.
- If you are going to introduce a species into a new habitat, do it very carefully. Check first about the natural pests and parasites of the species you plan to introduce. Are some of them essential to keep the species from reaching undesirable abundances? Are some of them likely to cause problems themselves?

As the technology of travel increases, the inadvertent introduction of exotics becomes more likely. We must increase our awareness of the problems associated with these introductions, be on our guard to know when introductions happen, and prevent those that are clearly undesirable or whose desirability is unknown.

Summary

- To conserve biological diversity, we must understand the large-scale global patterns of life. This is known as biogeography.
- Geographic isolation leads to the evolution of new species. Wallace's realms, or biotic provinces, are major geographic divisions (generally continents) based on fundamental features of the species found in them. Species filling specific niches within one realm are of different stock from those filling the same niches in other realms.
- The rule of climatic similarity holds that similar environments contribute to the evolution of biota and biological communities similar in external form and function but not in genetic heritage or internal makeup. Areas of climatic similarity with similar biota are biomes. A biome is a kind of ecosystem; examples are desert, grasslands, and rain forest.
- Convergent evolution occurs when two genetically dissimilar species that inhabit geographically separate parts of a biome develop along similar lines and have similar external form and function. Divergent evolution occurs when several species evolve from a common ancestral species but develop separately because of geographic isolation.
- Earth has 17 major biomes, each with its own characteristic dominant shapes and forms of life. Biomes vary in their importance to people, and some are of great importance. Most biomes have been heavily altered by human actions. Understanding the major characteristics of these biomes is important to the conservation and sustainable use of their resources.
- The study of island life has led to a theory of island biogeography that includes several important concepts. One is that islands have fewer species than mainlands because of their smaller size and distance from the mainland. Another is that the smaller an island and the farther it is from the mainland, the fewer species the island will contain.
- Ecological islands—habitats separated from the main part of a biome—show the same diversity characteristics as physical islands. The smaller the ecological island and the greater its distance from its "mainland," the fewer species it can support.
- People have long introduced exotic species into new habitats, sometimes creating benefits, often causing new problems. From the study of biogeography, certain general rules can be set down concerning the introduction of exotic species. The primary rule is this: Unless there is a very clear and good reason to introduce an exotic species into a new habitat, don't do it; and take precautions to prevent such introductions from occurring inadvertently.

STUDY QUESTIONS

1. What is a geologic barrier, and why is this concept important in the geography of living things?
2. What are the major factors that determine which species live in a particular location on a continent?
3. What are the consequences of geographic isolation?
4. In Jules Verne's classic novel *The Mysterious Island*, a group of people find themselves on an isolated volcanic island inhabited by kangaroos and large rodents closely related to the agoutis of South America. Why is this situation unrealistic? What would make this co-occurrence possible?
5. What do we learn from biogeography that helps us understand why there are so many species on Earth? Make a list of the major factors that contribute to high biodiversity.
6. What are three ways in which people have altered the distribution of living things?
7. From the perspective of biogeography, why do people attach so much importance to the conservation of tropical rain forests?
8. What ideas from the theory of island biogeography might explain why there are large mammals in arctic tundra but not in alpine tundra?
9. If you were to travel to Mars, which is dry and has wide daily variations in temperature, and were to search for life, what kind of biome would you search for first?
10. Suppose you were going to build a spacecraft for long-term voyages and you planned to use an ecological life-support system—that is, use ecosystems to provide food; to recycle oxygen, carbon dioxide, and water; and to treat wastes. What biomes do you believe would be most important to take along? Would you create a "new" biome?

FURTHER READING

Dice, Lee R. "The Canadian biotic province with special reference to the mammals." *Ecology* 19(4) (1938): 503-514.

Elton, C. S. *The Ecology of Invasions by Animals and Plants.* New York: Oxford University Press, 2000. (Reprinted with new introduction by Daniel Simberloff.) A classic work by one of the major ecologists of the twentieth century.

Levin, S. A., ed. *Encyclopedia of Biodiversity.* San Diego: Academic Press, 2000. A new, broad-based review of biodiversity.

Noss, R. F., and A. Y. Cooperrider. *Saving Nature's Legacy: Protecting and Restoring Biodiversity.* Washington, D.C.: Island Press, 1994.

Reid, W. V., and K. R. Miller. *Keeping Options Alive: The Scientific Basis for Conserving Biodiversity.* Washington, D.C.: World Resources Institute, 1989.

Rosenzweig, M. L. *Species Diversity in Space and Time.* New York: Cambridge University Press, 2003.

Strahler, A. and A. Strahler. *Physical Geography,* 3rd ed. Canadian Version. Toronto: Wiley, 2005. A book long recognized as the leading introduction to geography, including an introduction to biogeography.

Wilson, E. O., ed. *Biodiversity.* Washington, D.C.: National Academy Press, 1988. A major overview by some leading scientists.

Wilson, E. O. *Conserving Earth's 1999 Biodiversity.* Washington, D.C.: Island Press, 1999. A CD-ROM that includes text of some of Wilson's writing. He is recognized as one of the leading authorities on biodiversity.

World Resources Institute. *National Biodiversity Planning: Guidelines Based on Early Experiences around the World.* Washington, D.C.: World Resources Institute, 1996. An "illustrative biodiversity-planning process" based on the real-world experiences of 17 countries or regions—Australia, Canada, Chile, China, Costa Rica, Egypt, Germany, Indonesia, Kenya, Mexico, the Netherlands, Norway, the Philippines, Poland, the South Pacific, the United Kingdom, and Vietnam—already developing national strategies, plans, and programs.

chapter 9

Harvesting Forests in Iran and England

Life can be wondrously productive, but in many cases the harvesting of wild living resources, from marine mammals to forests, has exceeded Earth's bounty. People may be in the process of destroying many resources by harvesting them faster than they can regrow, and have been doing so for a long time.

The landscape of Western Iran is dry and barren, a study in brown and beige. Vegetation is sparse; the few camelthorn bushes are heavily grazed by herds of goats. For thousands of years, village houses have been built of mud, because wood is scarce and expensive. A casual observer would draw the logical conclusion that this barren landscape is the product of a hot, dry climate, inhospitable for plants and animals. Yet in the fifth century BC, the Greek historian and geographer Herodotus of Halicarnassus wrote that he and his companions "...walked all day without seeing the sun," so thick were the forests of the Zagros Mountains of Western Iran. Figure 9.1 shows that landscape today. Where did all the trees go? A modest decrease in winter precipitation, coupled with centuries of overgrazing and overharvesting of wood for fuel, has created a parched and treeless landscape that will never regain its former productivity.

In contrast, in medieval England, some small forested areas were harvested carefully and slowly. For example, in 1356, a survey of the estates of Bishop Ely stated that a "certain wood called Heylewode" was 32.4 ha in size and that every year the "underwood" (shrubs and young, small trees) was harvested in 4.5 of the 32.4 ha, leaving mature trees everywhere and young trees on the remaining 27.9 ha. This practice was continued, "without causing waste or destruction"; in modern terminology, the practice was sustainable. Similar forests were managed in much the same way in other locations in England from the fourteenth century until World War I with little decline in the woodlands.[1] Today, we are seeing projects designed to redevelop these methods (Figure 9.2).

These two contrasting examples suggest that at certain levels of harvesting, forests can be sustained, but once these levels are exceeded, forests will decline and may not recover. If they do recover, the recovery may take an exceedingly long time.

Figure 9.1 • Unsustained Forest Production. This area of Western Iran was once heavily forested. Sometimes, forests that are cut do not regenerate. Many areas of the world have experienced overgrazing and over-harvesting of trees for fuel wood. Over many years, such practices, along with climatic change, prevented regeneration.

Biological Productivity and Energy Flow

LEARNING OBJECTIVES

To conserve and manage our biological resources wisely, we must understand the basic concepts of energy, energy flow in ecosystems, and biological production. After reading this chapter, you should understand:

- That energy flow determines the upper limit on the production of biological resources.
- How the first and second laws of thermodynamics affect energy and production.
- That energy flow is one way through the ecosystem.
- That a basic quality of life is its ability to create order from energy on a local scale.
- What determines the efficiency of biological production.

Figure 9.2 • **Sustained Timber Production.** Some English woodlands have been managed since medieval times for sustainable forestry.

- *Throughout the history of civilization, people have cut trees faster than trees have regrown. Beginning with the earliest civilizations, forests were cleared to make way for agriculture and to provide fuel and structural materials, the basis of early civilization.[2] The practice of clearing the land and eliminating forests has continued into our own time. Clearly, the amount of timber harvested cannot always exceed forever the amount of new timber that grows between harvests. Eventually, the supply of timber will run out. With modern pressures for the increased use of wood and other biological resources, more than ever people need to understand there are limits to the growth of timber—and to all biological production.*

9.1 How Much Can We Grow?

Determining how much organic matter can be produced in any time period is important to many environmental topics, especially those that concern biological resources. How many bushels of wheat can a farmer produce in a field in a year? What is the upper limit of food that can be produced to feed the Earth's population? What is the limit on the number of whales in the ocean? What is the maximum production that we can expect of forests like those in Iran—or Canada?

Many factors can limit growth, but the ultimate limit on production of organic matter is energy flow. To estimate the actual production and the maximum possible production of organic matter of any kind, it is necessary to understand basic concepts of energy, and energy flow in ecosystems.

9.2 Biological Production

The total amount of organic matter on Earth or in any ecosystem or area is called its **biomass**. Biomass is usually measured as the amount per unit surface area of Earth (e.g., as grams per square meter [g/m^2] or metric tons per hectare [MT/ha]).

Biomass is increased through biological production (growth). Change in biomass over a given period is called *net production*. Biological production is the capture of usable energy from the environment to produce organic compounds in which that energy is stored. In photosynthesis, the environmental energy is from visible light. This light is transferred to energy in the chemical bonds of the organic compounds. This capture is often referred to as energy "fixation," and it is often said that the organism has "fixed" energy. Three measures are used for biological production: biomass, energy stored, and carbon stored. (Carbon is important because of its central role in photosynthesis, which is discussed in the following section. See also Chapter 5, where the carbon cycle is discussed in more detail.) We can think of these measures as the currencies of production. (General relationships for calculating production are given in Build Your Environmental Skills 9.1.)

Two Kinds of Biological Production

There are two kinds of biological production. Some organisms make their own organic matter from a source of energy and inorganic compounds. These organisms, introduced in the discussion of trophic levels in Chapter 6, are called autotrophs (meaning self-nourishing). The autotrophs include green plants (those containing chlorophyll), such as herbs, shrubs, and trees; algae, which are usually found in water but occasionally grow on land; and certain kinds of bacteria that grow in water.

The production carried out by autotrophs is called **primary production**. Most autotrophs make sugar from sunlight, carbon dioxide, and water in a process called **photosynthesis**, which releases free oxygen. Some autotrophic bacteria can derive energy from inorganic sulfur compounds; these bacteria are referred to as **chemoautotrophs**, or chemotrophs. Such bacteria have been discovered in deep-ocean vents, where they provide the basis for a strange ecological community. Chemoautotrophs are also found in marsh mud, where there is no free oxygen. In chemosynthesis, the energy in hydrogen sulfide (H_2S) is used by certain bacteria to make simple organic compounds. The reactions differ among species and depend on characteristics of the environment (Figure 9.3).

Other kinds of life cannot make their own organic compounds from inorganic ones and must feed on other living things. These are called **heterotrophs**. All animals, including human beings, are heterotrophs, as are fungi, many kinds of bacteria, and many other small life forms. Production by heterotrophs is called **secondary production** because it depends on the production of autotrophic organisms. This dependence is the basis of the food web described in Chapter 6. (The associated energy flow is illustrated in Figure 9.5.)

Once an organism has obtained new organic matter, it can use the energy in that organic matter to do things: to move, make new kinds of compounds, grow, reproduce, or store it for future uses. The use of energy from organic matter by most heterotrophic and autotrophic organisms is accomplished through **respiration**. In respiration, an organic compound is combined with oxygen to release energy and produce carbon dioxide and water. The process is similar to the burning of organic compounds but takes place within cells at much lower temperatures through enzyme-mediated reactions. *Respiration is the*

Figure 9.3 • Deep-sea vent chemosynthetic bacteria.

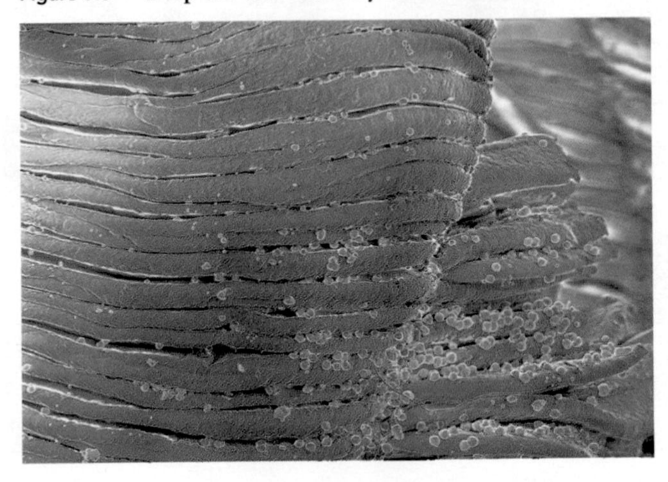

Build Your Environmental Skills 9.1

Calculating Production, Biomass, and Energy Flow

As we have seen, net production is simply gross production less respiration, can therefore also be calculated as

$$NP = GP - R$$

We also know that net production is the amount of biomass remaining after an organism has used what it needs for fuel. Therefore, we can write a general relation between net production (NP) and biomass (B):

$$B_2 = B_1 + NP$$

where:

B_2 = the biomass at the end of the time period
B_1 = the amount of biomass at the beginning of the time period, and
NP = the change in biomass during the time period (Figure 9.4).

Thus,

$$NP = B_2 - B_1$$

The three currencies of energy flow are biomass, energy content, and carbon content. The average of the energy in vegetation is approximately 21 kilojoules per gram (kJ/g)★. Energy content of organic matter varies. Ignoring bone and shells, woody tissue contains the least energy per gram, about 17 kJ/g; fat contains the most, about 38 kJ/g; and muscle contains approximately 21–25 kJ/g. Leaves and shoots of green plants have 21–23 kJ/g; roots have about 19 kJ/g.[3]

Why do we care about these relationships? Ecosystem production provides many clues to ecosystem health, and is a critical measure for planning resource extraction. An example will illustrate this.

In the Zhujiang River Delta, Guangdong Province, southern China, fish and mulberry leaves for silkworm production are produced simultaneously in an agro-forestry operation. Mulberry trees produce leaves that are eaten by silkworms. The silkworm excrement is fed to carp in an adjacent pond, and nutrient-rich pond mud is then returned to the land to fertilize the mulberry trees. Is this an efficient system? Production and biomass calculations can provide insight.

Data from Shunde County, Guangdong, China, show that the mean yield of mulberry leaves for that region was 28.5 tonnes/ha/yr with a maximum yield of 75 tonnes/ha/yr. There are about 125 mulberry trees per hectare in the plantation. Assume that the initial biomass of silkworms is approximately 1.5 kg/ha/year (roughly 150,000 silkworms), and that those 150,000 silkworms and their cocoons have a final biomass 2,025−2,250 kg/ha/yr. Assume also that the silkworms produce about 35 tonnes/ha/year of excrement.[4]

- What is the NP of the silkworms over a typical year in Shunde County?
- What is the approximate energy transfer efficiency between the primary producers of this system (the mulberry trees) and the silkworms? Is this an efficient system?
- If the transfer efficiency of the detritus-feeding carp is 10%, approximately how many tonnes of carp can be produced annually in this system?
- If individual carp have an average mass of 1 kg at the end of one year, approximately how many carp can be produced from the 1.5 kg of silkworm larvae?
- Draw a pyramid of numbers and a pyramid of biomass for this system (see the Critical Thinking Issue later in this chapter). How do the two pyramids compare?
- Compare the production of this system with the data provided in Table 9.1 (A Closer Look 9.2). Is the mulberry-silkworm-carp system more or less efficient than the trophic types presented in Table 9.1?

★ The kilojoule (1 kJ = 1,000 J = 0.24 kcal) is the International System (SI) unit preferred in scientific notation for energy and work. It replaces the calorie or kilocalorie of earlier studies of energy flow. The kilocalorie is the amount of energy required to heat a kilogram of water 1 degree Celsius (from 15.5° to 16.5°C). (The calorie, which is one-thousandth of a kilocalorie, is the amount of energy required to heat a gram of water the same 1 degree Celsius.) Note that the calorie referred to in diet books is actually the kilocalorie, though for convenience it is referred to in popular literature as a calorie. To keep all this straight, just remember that almost nobody uses the "little" calorie, regardless of what they call it. To compare, an average apple contains about 419 kJ or 100 kcal. The calorie is typically used in studies of diets; the joule is used in physics and engineering.

Figure 9.4 • Net production: *(a)* A carp-mulberry-silkworm agroforestry operation in China. *(b)* Mature silkworm larvae and their cocoons. You can think of the mature crop shown in (b) as B_2 and the starting mass of silkworm larvae as B_1. The difference between mulberry leaves and mature silkworm larvae illustrates net secondary production (*NSP*).

(a)

(b)

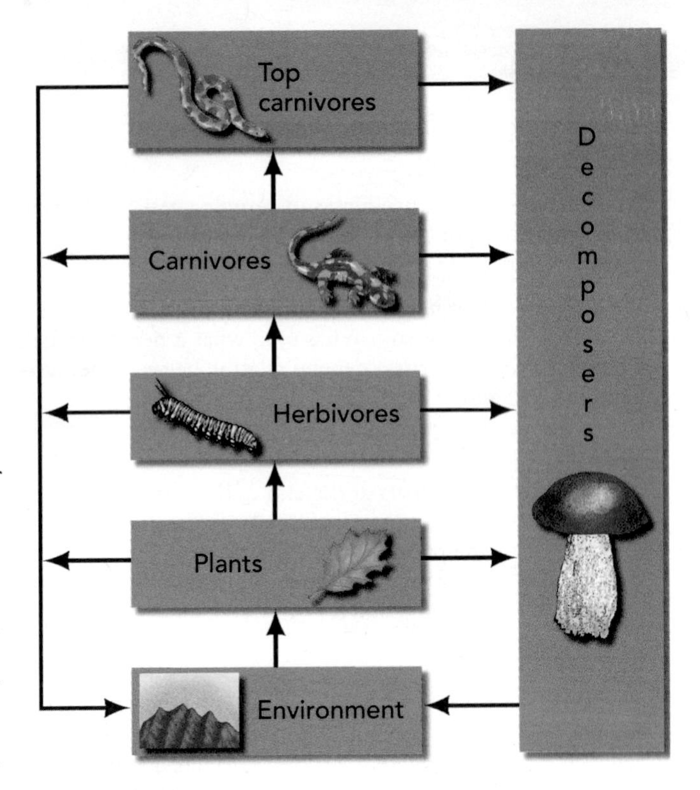

Figure 9.5 • Energy Pathways Through an Ecosystem. Usable energy flows from the external environment (the sun) to the plants, then to the herbivores, carnivores, and top carnivores. Death at each level transfers energy to decomposers. Energy lost as heat is returned to the external environment.

use of biomass to release energy that can be used to do work. Respiration returns to the environment the carbon dioxide that had been removed by photosynthesis.

Gross and Net Production

The production of biomass and its use as a source of energy by autotrophs includes three steps:

1. An organism produces organic matter within its body (gross production).
2. It uses some of this new organic matter as a fuel (respiration).
3. It stores some of the newly produced organic matter for future use (net production).

The basis of gross and net production lies in the processes of photosynthesis and respiration. Recall from Chapter 5 that photosynthesis—the process by which autotrophs make sugar from sunlight, carbon dioxide, and water—is defined as

$$6CO_2 + 6H_2O = C_6H_{12}O_6 + 6O_2$$

When organisms use energy through respiration, the process is simply the reverse of photosynthesis and can be most simply expressed as

$$C_6H_{12}O_6 + 6O_2 = 6CO_2 + 6H_2O + \text{Energy}$$

The first step in production of biomass, production of organic matter before use, is called **gross production**. The amount left after utilization is called net production.

Net production = Gross production − Respiration.

The difference between gross and net production is like the difference between a person's gross and net income. Your gross income is the amount you are paid. Your net income is what you have left after taxes and other fixed costs. Respiration is like the necessary expenses that are required in order for you to do your work.

The gross production of a tree—or any other plant—is the total amount of sugar it produces by photosynthesis before any is used. Within living cells in a green plant, some of the sugar is oxidized in respiration. Energy is used to convert sugars to other carbohydrates, those carbohydrates to amino acids, amino acids to proteins and new leaf tissue. Energy is also used to transport material within the plant to roots, stems, flowers, and fruits. Some energy is lost as heat in the transfer. Some energy is used to make other organic compounds in other parts of the plant: cell walls, proteins, etc. Some energy is stored in these other parts of the plant for later use. For woody plants such as trees, this stored energy, or biomass, includes new wood laid down in the trunk, new buds that will develop into leaves and flowers the next year, and new roots.

Net production for autotrophs is given as

$$NPP = GPP - R_a$$

where NPP = net primary production
GPP = gross primary production, and
R_a = respiration of autotrophs.

Organisms that do not make their own food are called secondary producers. Secondary production of a population is given as:

$$NSP = B_2 - B_1$$

where NSP = net secondary production
B_2 = biomass at time 2, and
B_1 = biomass at time 1

The change in biomass is the result of the addition of mass of living individuals, including the addition of newborns and immigrants, and loss through death and emigration.

We can express gross production on an ecosystem basis:

$$GEP = GPP$$

where GEP = gross ecosystem production, and
GPP = gross primary production.

Similarly, we can express respiration on an ecosystem basis, as the sum of respiration of autotrophs and that of heterotrophs:

$$R_e = R_a + R_h$$

where R_e = net ecosystem respiration
R_a = respiration of autotrophs, and
R_h = respiration of heterotrophs.

Figure 9.6 • Making Energy Visible. Top: A birch forest in a temperate hardwood forest as we see it, using normal photographic film *(a)* and the same forest photographed with infrared film *(b)*. Red colour means warmer temperatures; the leaves are warmer than the surroundings because they are heated by sunlight. Bottom: A nearby rocky outcrop as we see it, using normal photographic film *(c)* and the same rocky outcrop photographed with infrared film *(d)*. Blue means that a surface is cool. The rocks appear deep blue, indicating that they are much cooler than the surrounding trees.

These ecosystem equalities allow us to calculate net ecosystem production as:

$$NEP = GEP - R_e$$

where NEP = net ecosystem production
GEP = gross ecosystem production, and
R_e = net ecosystem respiration

9.3 Energy Flow

Energy is a difficult and abstract concept. When we buy electricity, what are we buying? We cannot see it or feel it, even though we have to pay for it.[5]

At first glance, energy flow seems simple enough: We take energy in and use it, just like a machine. But if we dig a little deeper into this subject, we discover a philosophical importance: We learn what distinguishes life and life-containing systems from the rest of the universe.

Although most energy is invisible, infrared film shows differences between warm and cold objects, which indicate factors about how energy flow affects life. With infrared film, warm objects appear red, and cool objects blue. Figure 9.6 shows birch trees in a temperate hardwood forest as normally viewed using standard film, and with infrared film, which shows tree leaves bright red, indicating that they have been warmed by the sun and are absorbing and reflecting energy, whereas the white birch bark remains cooler. The ability of tree leaves to absorb energy is essential; it is this source of energy that ultimately supports all life in a forest. Energy flows through life, and *energy flow* is a key concept.

All life requires energy. Energy is the ability to do work, to move matter. As anyone who has dieted knows, our weight is a delicate balance between the energy we take in through our food and the energy we use. What we do not use and do not pass on, we store. Our use of energy, and whether we gain or lose weight, follows the laws of physics. This is not only true for people, it is also true of all living things, of all ecological communities and ecosystems, and of the entire biosphere.

Ecosystem energy flow is the movement of energy through an ecosystem from the external environment through a series of organisms and back to the external environment. It is one of the fundamental processes common to all ecosystems.

Energy enters an ecosystem by two pathways. The first is energy fixed by organisms, as discussed. In the second pathway, heat energy is transferred by the air or water currents or by convection through soils and sediments and warms living things. For instance, when a warm air mass passes over a forest, heat energy is transferred from the air to the land and to the organisms.

9.4 The Ultimate Limit on the Abundance of Life

What ultimately limits the amount of organic matter that can be produced? What limits the maximum rate of that production? How closely do ecosystems, species, populations, and individuals approach this limit? Are any of these near to being as productive as possible? The answer lies in the laws of thermodynamics.

The Laws of Thermodynamics

The *law of conservation of energy* states that in any physical or chemical change, energy is neither created nor destroyed but merely changed from one form to another. The law of conservation of energy is also called the *first law of thermodynamics* (discussed in A Closer Look 5.1).

If the total amount of energy is always conserved—if it remains constant—then why can we not just recycle energy inside our bodies? Similarly, why cannot energy be recycled in ecosystems and in the biosphere?

Let us imagine how that might work with frogs and mosquitoes, for example. Frogs eat insects, including mosquitoes. Mosquitoes suck blood from vertebrates, including frogs. Consider an imaginary closed ecosystem consisting of water, air, a rock for frogs to sit on, frogs, and mosquitoes. In this system, the frogs get their energy from eating the mosquitoes, and the mosquitoes get their energy from biting the frogs (Figure 9.7). Such a closed system would be a biological perpetual-motion machine: It would continue indefinitely without an input of any new material or energy. This is impossible. Why? The general answer is found in the *second law of thermodynamics*, which addresses how energy changes in form.

Energy always changes from a more useful, more highly organized form to a less useful, more disorganized form. That is, energy cannot be completely recycled to its original state of organized, high-quality usefulness. For this reason, the mosquito–frog system will eventually stop when not enough useful energy is left. (There is also a more mundane reason: Only female mosquitoes require blood and then only in order to reproduce. Mosquitoes are otherwise herbivorous.)

From the discussion presented in A Closer Look 9.1, we reach a new understanding of a basic quality of life.[3] *It is the ability to create order on a local scale that distinguishes life from its nonliving environment.* This ability requires obtaining energy in a usable form, and that is why we eat. This principle is true for every ecological level: individual, population, community, ecosystem, and biosphere. Energy must continually be added to an ecological system in a usable form. Energy is inevitably degraded into heat, and this heat must be released from the system. If it is not released, the temperature of the system will increase indefinitely. *The net flow of energy through an ecosystem, then, is a one-way flow.*

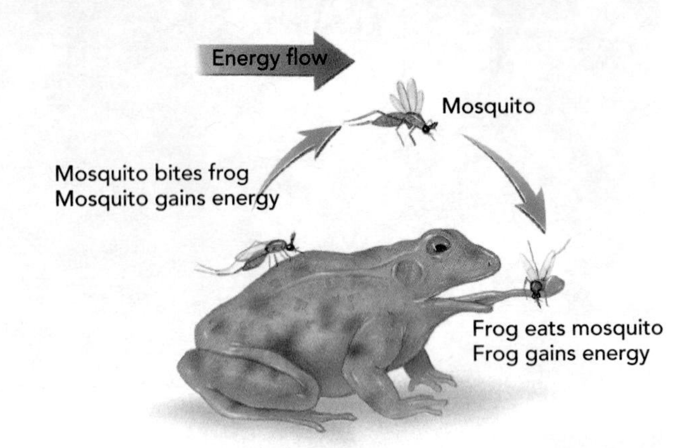

Figure 9.7 • **An impossible ecosystem.**

Based on what we have said about the energy flow through an ecosystem, we can see that an ecosystem must lie between a source of usable energy and a sink for degraded (heat) energy. The ecosystem is said to be an *intermediate system* between the energy source and the energy sink. The energy source, ecosystem, and energy sink together form a **thermodynamic system**. The ecosystem can undergo an increase in order, called a *local increase*, as long as the entire system undergoes a decrease in order, called a *global decrease. To put all this simply, creating local order involves the production of organic matter. Producing organic matter requires energy, and organic matter is where energy is stored.*

Energy Efficiency and Transfer Efficiency

How efficiently do living things use energy? This is an important question for the management and conservation of all biological resources. We might expect biological resources to be efficient in their use of energy—to produce a maximum amount of biomass from a given amount of energy.

No system can be 100% efficient. As energy flows through a food web it is degraded, and less and less is usable. Generally, the more energy an organism gets, the more it has for its own use. However, organisms differ in how efficiently they use the energy they obtain. A more efficient organism has an advantage over a less efficient one.

Efficiency can be defined for both artificial and natural systems: machines, individual organisms, populations, trophic levels, ecosystems, and the biosphere.[6] *Energy efficiency* is defined as the ratio of output to input, and it is usually further defined as the amount of useful work obtained from some amount of available energy. *Efficiency* has different meanings to different users. From the point of view of a farmer, an efficient corn crop is one that

A CLOSER LOOK 9.1

The Second Law of Thermodynamics

To understand why we cannot recycle energy, imagine a closed system (a system that receives no input after the initial input) containing a pile of coal, a tank of water, air, a steam engine, and a carpenter (Figure 9.8). Suppose the engine runs a lathe that makes furniture. The carpenter lights a fire to boil the water, creating steam to run the engine. As the engine runs, the heat from the fire gradually warms the entire system.

When all the coal is completely burned, the carpenter will not be able to boil any more water, and the engine will stop. The average temperature of the system is now higher than the starting temperature. The energy that was in the coal is dispersed throughout the entire system, much of it as heat in the air.

Why can't the carpenter recover all that energy, recompact it, put it under the boiler, and run the engine? The answer is found in the second law of thermodynamics, which essentially tells us that with each energy transfer, energy becomes less organized and less able to do useful work. Physicists have discovered that no real use of energy can ever be 100% efficient. Whenever useful work is done, some energy is inevitably converted to heat. Collecting all the energy dispersed in this closed system would require more energy than could be recovered.

Our imaginary system begins in a highly organized state, with energy compacted in the coal. It ends in a less organized state, with the energy dispersed throughout the system as heat. The energy has been degraded, and the system is said to have undergone a decrease in order. The measure of the decrease in order (the disorganization of energy) is called entropy. The carpenter did produce some furniture, converting a pile of lumber into nicely ordered tables and chairs. The system had a local increase of order (the furniture) at the cost of a general increase in disorder (the state of the entire system). All energy of all systems tends to flow toward states of increasing entropy.

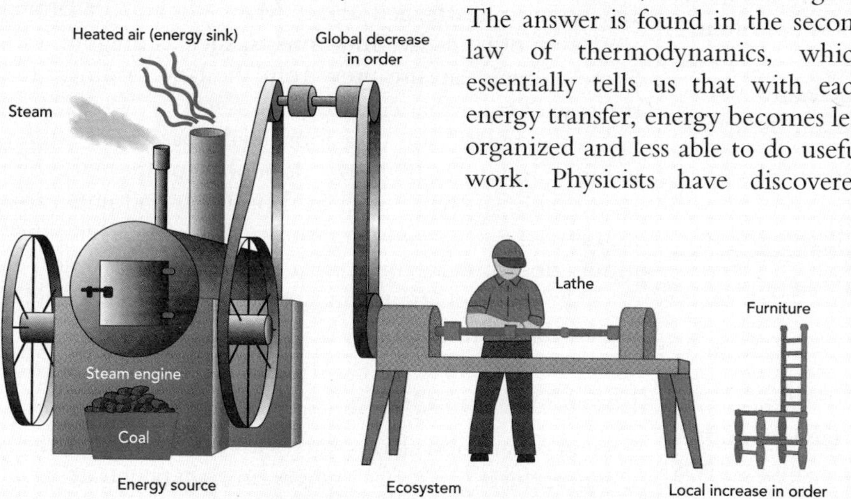

Figure 9.8 • A system closed to the flow of energy.

converts a great deal of solar energy to sugar and uses little of that sugar to produce stems, roots, and leaves. In other words, the most efficient crop is the one that has the most harvestable energy left at the end of the season. A truck driver views an efficient truck just the opposite. For the driver, an efficient truck uses as much energy as possible from its fuel and stores as little energy as possible (in its exhaust). When we view organisms as food, we define efficiency as the farmer does, in terms of energy storage (net production from available energy). When we are energy users, we define efficiency as the truck driver does, in terms of how much useful work we accomplish with the available energy.

A common ecological measure of energy efficiency is called food-chain efficiency, or **trophic-level efficiency**, which is the ratio of production of one trophic level to the production of the next lower trophic level. (See Chapter 6.) This efficiency is never very high. Green plants convert only 1–3% of the energy received from the sun during the year to new plant tissue. The efficiency with which herbivores convert the potentially available plant energy into herbivorous energy is usually less than 1%, as is the efficiency with which carnivores convert herbivores into carnivorous energy. It is frequently written in popular literature that the transfer is 10%—for example, that 10% of the energy in corn can be converted into energy in a cow. However, this is a managed ecological efficiency rather than the natural, trophic-level efficiency. In natural ecosystems, the organisms in one trophic level tend to take in much less energy than the potential maximum amount available to them, and they use more energy than they store for the next trophic level.

Consider an example. At La Mauricie National Park, Québec, wolves feed on moose in a natural wilderness. Studies in other systems have shown that a pack of 18 wolves kill an average of one moose approximately every 2.5 days,[7] resulting in a trophic-level efficiency of wolves of about 0.01%. Wolves use most of the energy they take in from eating moose, especially in the search for prey.[8] From the wolves' point of view, wolves are efficient, but from the viewpoint of a predator who wants to feed on wolves, they appear inefficient.

The rule of thumb for ecological trophic energy efficiency is that more than 90% (usually much more) of all energy transferred between trophic levels is lost as heat. Less than 10% (approximately 1% in natural ecosystems) is fixed as new tissue. In highly managed ecosystems, such as ranches, the efficiency may be greater. But even in such systems, it takes an average of 3.5 kg of vegetable matter to produce 0.5 kg of edible meat. Cattle are among the least efficient producers, requiring around 7.5 kg of vegetable matter to produce 0.5 kg of edible meat. Chickens are much more efficient, using approximately 1.6 kg of vegetable matter to produce 0.5 kg of eggs or meat. Much attention has been paid to the idea that humans should eat at a lower trophic level in order to use resources more efficiently. (See The Critical Thinking Issue, "Should People Eat Lower on the Food Chain?")

Many other kinds of energetic efficiencies are widely used in ecological studies. Some of these are described in A Closer Look 9.2.[9]

9.5 Some Examples of Energy Flow

We conclude the chapter with several examples of energy flow in ecosystems.

Energy Flow in an Old Field Food Chain

In an old field in Ontario, meadow mice feed on grasses and herbs, and long-tailed weasels feed on mice (this is just one of many food chains in this old field).[11] The first step in the flow of energy is the fixation of light energy by photosynthesis in the leaves of grasses, herbs, and shrubs. In this step, the energy in light is transferred to and stored in sugar or carbohydrates in the plants. Recall that this energy—the energy that is stored by autotrophs before any is used—is called gross primary production. Some of the energy is used immediately by the leaves to keep their own life processes going. As explained earlier, the amount stored by the autotrophs after using what is needed is net primary production, reflected in ecosystem biomass.

As we have seen, only a small fraction of the energy available to each trophic level is *used for net production of new tissue. A large fraction of the energy available to each trophic level is used in respiration.* In the old field in Ontario, about 15% of the vegetation's gross production is used in respiration; 68% of the energy taken up by the mice is used in respiration; and 93% of the energy taken up by the long-tailed weasel is used in respiration. Only a part of the energy flow in the old field moves through the food chain of vegetation–mice–weasels. One reason is that mice eat only the plant seeds. Most of the energy remains in the vegetation until it is transferred from dead vegetation to other animals, fungi, and bacteria by what is called the decomposer food chain.

Energy Flow in a Stream or River

In most ecosystems, the original fixing of energy occurs within the ecosystem; however, some freshwater streams are an exception. The amount of organic matter produced by algae living in a stream is small relative to the amount of organic matter that falls into the stream from dead leaves and twigs of vegetation on the land.[12] *Detritivores* (organisms that feed on dead organic material) are common in streams and feed mainly on this deposited vegetation. Some animals are shredders that tear up leaves; others feed on the smaller pieces.

Other grazing animals move along rock surfaces and scrape off attached algae. Many stream predators are larvae of land-dwelling insects, such as dragonflies. Some animals capture prey from the land or air, as in the case of trout that catch flying insects.

An extreme case of a food chain based on external food input occurs in the floodplain of the Amazon River basin, in which fish feed on fruits and nuts carried into the streams during the rainy season. Here, the production of herbivorous fish exceeds what would be possible from aquatic primary production alone, yielding an abundant food supply for people in the region.[11]

Energy Flow in Ocean Ecosystems

Several ocean food chains start with phytoplankton that live near the ocean surface where sunlight penetrates and oxygen levels are comparatively high. One of these food chains continues near the ocean surface where a variety of animals feed on those algae—these animals include tiny floating invertebrates called zooplankton, and some of Earth's huge whales, the baleen whales that sweep up the zooplankton as they swim. In turn, these animals are fed on by other animals that live near the surface or spend much of their time in the upper ocean.

A second, curious food chain exists primarily deep in the ocean. It begins at the surface with the production of phytoplankton and includes surface feeding on those algae by small invertebrate animals. Wastes released by these animals, including fecal material and dead individuals, sink and descend to the ocean depths. There, deep-living animals feed on this "marine snow" of organic material, produced far away, high above them where sunlight shines. It is a curious and comparatively little known kind of life, so different from our land habitat, and a subject of much research and discovery.

SUSTAINABILITY

A CLOSER LOOK 9.2

Ecological Efficiencies

Table 9.1 gives some values for growth efficiency, or gross production efficiency (*P/C*), which is the ratio of the material produced (*P* is the net production) by an organism or population to the material consumed (*C*). The amount consumed is normally much less than the maximum amount available. Estimates show, for example, that less than 1% to about 20% of the leaves available in forests and woodlands are consumed annually by leaf-eating insects.[10]

Table 9.1 also gives examples of net growth efficiency, or net production efficiency (*P/A*). This is the ratio of the material produced (*P*) to the material assimilated (*A*), which is less than the material consumed because some food taken in is discharged as waste and never used by an organism. Table 9.1 shows that ecological efficiencies vary widely among animals. Generally speaking, vertebrates are less efficient producers than invertebrates or microorganisms.

Table 9.1 • Ecological Efficiencies for Animal Populations

	Ecological Efficiencies (%)	
Trophic Types	*Net Production Efficiency (P/A)*	*Gross Production Efficiency (P/C)*
Terrestrial Animals		
Microorganisms[a]	~40	
Invertebrates		
Herbivores	20–40	8–27
Carnivores	10–37	~34
Saprophages	17–40	5–8
Vertebrates		
Herbivores[a]	2–10	
Carnivores[a]	2–10	
Fishes[a,b]	—	1–7

Sources: T. Penczak, *Comp. Biochem. Physiology* 101 (1992): 791–798; D. E. Reichle, "The Role of Soil Invertebrates in Nutrient Cycling," in U. Lohm and T. Persson, eds., *Ecol. Bull. (Stockholm)* 25 (1977): 145–156; and M. Schaefer, "Secondary Production and Decomposition," in E. Rohrig and B. Ulrich, eds., *Temperate Deciduous Forests*, vol. 7 of *Ecosystems of the World* (Amsterdam: Elsevier, 1991).

[a] Data are based on characteristic values for trophic levels and populations.

[b] Populations in a tropical river.

Chemosynthetic Energy Flow in the Ocean

Earlier, we mentioned organisms that make their own food from energy in sulfur compounds. This process creates a curious class of food chains in the depths of the oceans that support previously unknown life forms. The basis of the food chains is *chemosynthesis*, in which the source of energy is not sunlight but hot, inorganic sulfur compounds emitted from vents in the ocean floor.

Sulfur-laden water is emitted from hot water vents at depths of 2,500–2,700 m in areas where flowing lava causes seafloor spreading. A rich biological community exists in and around the vents, including large white clams up to 20 cm in diameter, brown mussels, and white crabs. Clams and mussels filter chemoautotrophic bacteria and particles of dead organic matter from the water. Some vent communities contain limpets, pink fish, tube worms, and octopuses. Among the most curious creatures found in vents are giant worms, some 3.9 m long (Figure 9.9).[13]

Although large areas of the ocean have low productivity, combined, they account for a major portion of total energy fixed. Highly productive areas of the oceans are found in upwelling zones, which occur when deep-ocean waters, rich in nutrients from dead organic material, flow upward, allowing abundant growth of algae and photosynthetic bacteria. Herbivores and carnivores move organic nutrients through the food chain. Although only 1/1,000 of the oceans' surface has natural upwellings, these zones account for more than 44% of the fish eaten by the world's human population.[14]

Figure 9.9 • **Among the Earth's most curious creatures are giant worms that live in deep sea vents and feed on chemosynthetic organisms.**

CRITICAL THINKING ISSUE

Should People Eat Lower on the Food Chain?

The energy content of a food chain is often represented by an *energy pyramid*, such as the one shown here in Figure 9.10a for a hypothetical, idealized food chain. In an energy pyramid, each level of the food chain is represented by a rectangle whose area is more or less proportional to the energy content for that level. For the sake of simplicity, the food chain shown here assumes that each link in the chain has one and only one source of food.

Assume that if a 75-kg person ate frogs (and some people do!), he would need 10 a day, or 3,000 a year (approximately 300 kg). If each frog ate 10 grasshoppers a day, the 3,000 frogs would require 9,000,000 grasshoppers a year to supply their energy needs, or approximately 9,000 kg of grasshoppers. A horde of grasshoppers of that size would require 333,000 kg of wheat to sustain them for a year.

As the pyramid illustrates, energy content decreases at each higher level of the food chain. The result is that the amount of energy at the top of a pyramid is related to the number of layers the pyramid has. For example, if people fed on grasshoppers rather than frogs, each person could probably get by on 100 grasshoppers a day. The 9,000,000 grasshoppers could support 300 people for a year, rather than only one. If, instead of grasshoppers, people ate wheat, then 333,000 kg of wheat could support 666 people for a year.

This argument is often extended to suggest that people should become herbivores (vegetarians, in human parlance) and eat directly from the lowest level of all food chains, the autotrophs. Consider, however, that humans can eat only parts of some plants. Herbivores can eat some parts of plants that humans cannot eat and some plants that humans cannot eat at all. When people eat these herbivores, more of the energy stored in plants becomes available for human consumption. The most dramatic example of this is in aquatic food chains. Because people cannot digest most kinds of algae, which are the base of most aquatic food chains, they depend on eating fish that eat algae and fish that eat other fish. So if people were to become entirely herbivorous, they would be excluded from many food chains. In addition, there are major areas of Earth where crop production damages the land but grazing by herbivores does not. In those cases, conservation of soil and biological diversity lead to arguments that support the use of grazing animals for human food. This creates an environmental issue: How low on the food chain should people eat?

1 person

3,000 frogs

9,000,000 grasshoppers

333,000 kg of wheat

(a)

(b)

Figure 9.10 • The energy pyramid. *(a)* The total bulk (mass) of all the organisms making up each level of a food chain is less than that of the previous level. *(b)* Eating lower on the food chain.

Critical Thinking Questions

1. Why does the energy content decrease at each higher level of a food chain? What happens to the energy that is lost at each level?
2. The pyramid diagram uses mass as an indirect measure of the energy value for each level of the pyramid. Why is it appropriate to use mass to represent energy content?
3. Using the average of 21 kJ of energy to equal 1 g of completely dried vegetation and assuming that wheat is 80% water, what is the energy content of the 333,000 kg of wheat shown in the pyramid?
4. Make a list of the environmental arguments for and against an entirely vegetarian diet for people. What might be the consequences for Canadian agriculture if everyone in the country began to eat lower on the food chain?
5. How low do you eat on the food chain? Would you be willing to eat lower? Explain.

Summary

- The study of energy flow is important in determining limits on food supply and on the production of all biological resources, such as wood and fiber.
- In every ecosystem, energy flow provides a foundation for life and thus imposes a limit on the abundance and richness of life. The amount of energy available to each trophic level in a food chain depends not only on the strength of the energy source but also on the efficiency with which the energy is transferred along the food chain.
- Energy is fixed by autotrophs—organisms that make their own food from energy and small inorganic compounds. The initial energy comes from two sources: light (mainly sunlight) and small sulphur compounds. Plants, algae, and some bacteria are autotrophs.
- Only autotrophs can make their own food; all other organisms are heterotrophs, which must feed on other organisms.
- Biological production is the production of new organic matter, which we measure as change in biomass, change in stored energy, or change in stored carbon. Another way to think about biological production is that it is the change in biomass over time.
- Gross production is production measured before any utilization. Net production is the amount stored (not used) at the end of some time period. Respiration uses stored energy, so net production equals gross production minus respiration.
- The laws of thermodynamics connect life to order in the universe. The second law of thermodynamics tells us that order always decreases when any real process occurs in the universe. However, life is more ordered than its environment. The ability to create order is the essence of what we get from our food.
- Energy efficiency is the ratio of output to input, or the amount of useful work obtained from some amount of available energy. Trophic-level efficiency is the ratio of production of one trophic level to the production of the next lower trophic level. This efficiency is never very high, often only about 1%.

STUDY QUESTIONS

1. What is the difference between gross production and net production? Primary and secondary production?
2. What's the meaning of the statement "Any living or life-containing system is always more ordered than its nonliving environment"?
3. Keep track of the food you eat during one day and make a food chain linking yourself with the sources of those foods. Determine the biomass (grams) and energy (kilocalories) you have eaten. Using an average of 5 kcal/g, then using the information on food packaging or assuming that your net production is 10% efficient in terms of the energy intake, how much additional energy might you have stored during the day? What is your weight gain from the food you have eaten?
4. Referring to question 3, what amount of vegetation did you eat during one day? If vegetation was 1% efficient in converting sunlight to organic matter stored as net production, how much sunlight was required to provide the vegetation you took in during the day?

FURTHER READING

Although in most cases we try to provide the most up-to-date references, with this subject we believe some of the easiest-to-read and most important references are among the classical earlier works.

Blum, H. F. *Time's Arrow and Evolution*. New York: Harper & Row, 1962. A very readable book discussing how life is connected to the laws of thermodynamics.

Gates, D. M. *Biophysical Ecology*. New York: Springer-Verlag, 1980. A discussion about how energy in the environment affects life.

Morowitz, H. J. *Energy Flow in Biology*. Woodbridge, Conn.: Oxbow, 1979. The most thorough discussion available about the connection between energy and life at all levels.

Morowitz, H. J. "The Six Million Dollar Man." *In The Wine of Life and Other Essays on Societies, Energy, and Living Things*. New York: Bantam, 1981. A fun essay about the second law of thermodynamics and life.

Schrödinger, E. (ed. Roger Penrose). *What Is Life?: With Mind and Matter and Autobiographical Sketches (Canto)*. Cambridge: Cambridge University Press, 1992. A classic, easy to read statement about how the use of energy differentiates life from other phenomena in the universe.

Sherman, K. *Large Marine Ecosystems: Patterns, Processes, and Yields*. Portland, Ore.: Book News, 1990. A book discussing possible impacts of global change on ocean productivity.

Part II

Human Activities and the Environment

Rush hour traffic in downtown Bangkok, Thailand. Bangkok has some of the world's worst traffic congestion, costing the country almost a billion dollars every year in lost productivity, reduced quality of life, and human health impacts.

Photo courtesy Isobel Heathcote.

chapter 10

Restoring an Oak Savanna at Pinery Provincial Park, Ontario

Every year, thousands of campers visit Pinery Provincial Park, south of Grand Bend, Ontario, never knowing that they pitch their tents in one of the world's last remaining fragments of oak savanna. In its natural condition, oak savanna is a transition community between tallgrass prairie and temperate forest. The name is misleading. Although a number of oak species are present, oak savanna is a diverse, open habitat with scattered trees intermixed with low-growing plants. The Pinery's oak savanna supports at least 800 plant species and 300 bird species.

Oak savannas burn naturally, and the species that live in them have evolved with and adapted to fire over millions of years. Natural fires occur frequently and as a result are light, clearing out woody debris and many young trees but not burning through the bark of most of the large, thick-barked, seed-producing trees. Nor are the fires intense enough to destroy much of the organic matter in the soil. Even young seedlings can regenerate following a fire, and reach sapling height faster than unburned plants.

A century ago, the woodlands were open, with trees scattered among the grasses. Most of the trees were large and mature. Throughout most of the twentieth century, fires were considered bad for forests, and the general practice was to suppress and fight fires. The oak savanna at the Pinery was protected from fires for over half a century, and during that time, the forest changed. Originally, an open landscape of oaks and grasslands on sandy soils, the Pinery was purchased for parkland in 1957. In the early 1960s, more than 3 million red pine trees were planted that soon dominated the landscape and fundamentally altered its ecological dynamics.

A restoration project was launched at the Pinery with the goal of returning the forest to the way it was before European settlement. This could not be done simply by reintroducing fire into the forest, which was

Figure 10.1 • Oak savanna in Pinery Provincial Park, Ontario. *(a)* After a period of fire suppression. The forest became much denser and the threat of seriously damaging wildfires was great. *(b)* In a restoration project, stands were thinned and controlled burns were conducted.

Ecological Restoration and Succession

LEARNING OBJECTIVES

Restoration ecology is a new field. In this chapter, we explore the concepts of restoration ecology, with a special emphasis on how ecosystems restore themselves through the process of ecological succession. After reading this chapter, you should understand:

- What ecological restoration means.
- What kinds of goals are possible for ecological restoration.
- What basic approaches, methods, and limits apply to restoration.
- How an ecosystem restores itself through ecological succession after a disturbance.
- What role disturbances play in the persistence of ecosystems.
- How physical forces and biological processes affect the land.
- Why ecosystems do not remain in a steady state.

now not only dense but also had a large accumulation of organic matter in the soil and on the soil surface. The kind of fire that would occur under these conditions could be highly destructive, hot enough to kill even the mature, seed-bearing trees and burning through the soil as well, destroying its organic matter. Such a fire could either permanently eliminate oak from the burned area or so damage the trees and the soil that recovery would take a very long time.

Before fire could be reintroduced, the forest had to be returned to a lower fuel condition. Pine trees were systematically removed to reproduce conditions typical of the nineteenth century, with oak trees clumped around large grassy openings. The researchers also raked organic matter from the forest floor, and then conducted controlled burns in selected areas.

When these modified stands were burned, the fire intensity was low, with flames averaging only about 15 cm high. Figure 10.1*a* shows part of the Pinery forest before the restoration project. The same area is shown on Figure 10.1*b* after the pines are removed and controlled burning conducted. The fires did not kill the mature oak trees and left some younger trees to replace the older ones. Since then, grass has returned to the openings. Now the forest can again follow its historical pattern of frequent small fires. The forest has been restored, but the case study of the Pinery illustrates that ecological restoration can be complex and can require great care and considerable effort.[1]

- *Wherever people have lived, they have changed their environment. In North America, this was true of the original people to arrive on the continent, just as it has been since European settlement. While some of the changes people make are considered desirable, others are considered degradation. For lands and waters that have been degraded, the question is: How can we restore them? In recent years a new field called restoration ecology has developed within the science of ecology. Its goal is to return damaged ecosystems to some set of conditions considered functional, sustainable, and "natural". Whether restoration can always be successful is still an open question. For some ecosystems and species, success appears achievable, but at the Pinery, success has required great effort.*

10.1 Restore to What?

The idea of ecological restoration raises a curious question: Restore to what?

The Balance of Nature

Until the second half of the twentieth century, the predominant belief in Western civilization was that any natural area—a forest, a prairie, an intertidal zone—left undisturbed by people achieved a single condition that would persist indefinitely. This condition, as mentioned in Chapter 3, is known as the balance of nature. The major tenets of a belief in the balance of nature are as follows:

1. Undisturbed, nature achieves a permanency of form and structure that persists indefinitely.
2. If it is disturbed and the disturbing force is removed, nature returns to exactly the same permanent state.
3. In this permanent state of nature, there is a "great chain of being" with a place for each creature (a habitat and a niche) and each creature in its appropriate place.

These ideas had their roots in Greek and Roman philosophies about nature, but they have played an important role in modern environmentalism as well. In the early twentieth century, ecologists formalized the belief in the balance of nature. They said that succession proceeded to a fixed, classic condition, which they called a climax state and defined as a steady-state stage that would persist indefinitely and have maximum organic matter, maximum storage of chemical elements, and maximum biological diversity. At that time, people thought that wildfires were always detrimental to wildlife, vegetation, and natural ecosystems. *Bambi*, a 1942 Walt Disney movie, expressed this belief, depicting a fire that brought death to friendly animals. In Canada and the United States, Smokey the Bear is a well-known symbol employed for decades to warn visitors to national forests to be careful with fire and avoid setting wildfires. The message is that wildfires are always harmful to wildlife and ecosystems.

All of this suggests a belief that the balance of nature does in fact exist. If that were true, the answer to the question "Restore to what?" would be simple: restore to the original, natural, permanent condition. The method of restoration would be simple, too: get out of the way and let nature take its course. Since the second half of the twentieth century, though, ecologists have learned that nature is not constant and that forests, prairies—all ecosystems—undergo change. Moreover, since change has been a part of natural ecological systems for millions of years, many species have adapted to change. Indeed, many require specific kinds of change in order to survive.

Dealing with change—natural and human-induced—poses questions of human values as well as science. This is illustrated by wildfires in forests, grasslands, and shrublands, which can be extremely destructive to human life and property. Scientific understanding tells us that fires are natural and that some species require them. Whether we choose to allow fires to burn, or light fires ourselves, is a matter of values. In 2003, extremely dry conditions created one of British Columbia's worst fire seasons and costliest natural disasters in over 100 years. Despite fire-fighting costs of over half a billion dollars, almost 2,500 fires burned that year, destroying hundreds of homes and displacing tens of thousands of people. Other natural hazards like volcanoes, earthquakes, landslides, and floods have also affected millions of people over the last century. Even ice can take a toll. The great ice storm of January 1998 extended from central Ontario, east to the Bay of Fundy in New Brunswick, prevented more than 2.6 million people from traveling to work, damaged one-third of crop lands in southern Ontario and Québec, and toppled or killed millions of trees. Total costs of that storm are estimated at over $1 billion.

Restoration ecology depends on science to discover what used to be, when it existed, what is now possible, and how different goals can be achieved. The selection of goals for restoration is a matter of human values.

Quetico Provincial Park and the Boundary Waters Canoe Area Wilderness: An Example of the Naturalness of Change

One of the best-documented examples of natural disturbance is the role of fire in the northern woods of North America. At 475,782 ha, Quetico Provincial Park is Ontario's second largest wilderness park and, with its neighbour and partner, the Superior National Forest, Minnesota, forms the largest international wilderness recreational area in the world. Protected since 1913 by provincial legislation, the park exemplifies nature relatively undisturbed by people. The area is no longer open to logging or other direct disturbances by people. In the early days of European exploration and settlement of North America, French voyageurs travelled through this region hunting and trading for furs. In some places, logging and farming were common in the nineteenth and early twentieth centuries, but for the most part the land has been relatively untouched. Despite the lack of human influence, the forests show a persistent history of fire. Fires occur somewhere in this forest almost every year, and, on average, the entire area burns once every 80 years. Fires cover areas large enough to be visible by satellite remote sensing (Figure 10.2).

When fires occur in Quetico Provincial Park at natural rates and natural intensities, they have some

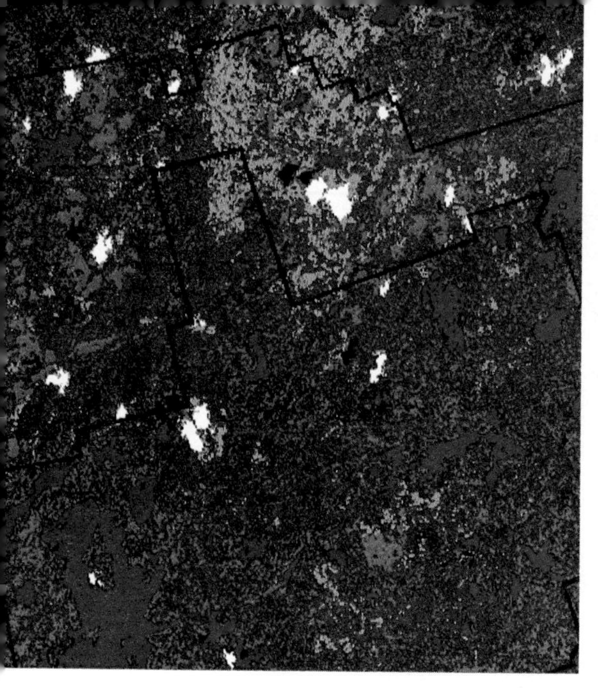

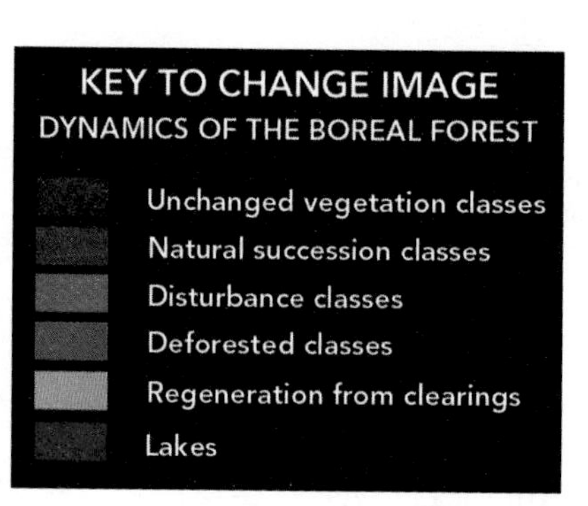

Figure 10.2 • Forest fires can be natural or caused by people. This figure shows the change in a large area of the Superior National Forest in Minnesota between 1973 and 1983, as observed by the Landsat satellite. The black boundaries show a central corridor where logging is permitted, surrounded by the Boundary Waters Canoe Area at the top and bottom. This area is protected from all uses except certain types of recreation. The bright yellow shows areas that were clear of trees in 1973 but had regenerated to young forest by 1983. Most of this change is due to regrowth following a large fire that burned both inside and outside the wilderness. Red areas were forested in 1973 but cleared in 1983. Most of these are outside the wilderness, and some of these are due to logging (red) and some to fire or storms. Greens show areas that were forested both years.

beneficial effects. For example, trees in unburned forests appear more susceptible to insect outbreaks and disease. Thus, recent ecological research suggests that wilderness depends on change and that succession and disturbance are a continual processes. The landscape is dynamic.[2]

Goals of Restoration: What Is "Natural"?

With the examples of the Pinery and Quetico Provincial Park forests, we can now return to the question "Restore to what?" If an ecosystem passes naturally through many different states, and all of them are "natural," and if change itself—including certain kinds of wildfire—is natural, then what can it mean to "restore" nature? How can restoration that involves such things as wildfires occur without undue damage to human life and property?

Can we restore an ecological system to any one of its past states and claim that this is natural and successful restoration? A frequently accepted answer is that restoration means restoring an ecosystem to its historical range of variation and to an ability to sustain itself and its crucial functions, including the cycling of chemical elements (see Chapter 5), the flow of energy (Chapter 9), and the maintenance of the biological diversity that existed previously (Chapters 7 and 8). According to this interpretation, ecological restoration means restoring processes and a set of conditions known to have existed for that ecosystem. From this viewpoint, we examine altered ecosystems and populations that have declined, to try to determine what is needed to restore balance.

A variety of answers have been put forward to answer the question: What is restoration? At the extreme are those who argue that all human impacts on nature are "unnatural" and therefore undesirable, and that the only true goal of restoration is to bring nature back to a condition that existed before human influence. The anthropologist Paul S. Martin takes this position. He proposes that the only truly "natural" time is before any significant human influence occurs. Specifically, he suggests that restoration to conditions of 10,000 B.C.—before farming—should be our goal. He even suggests introducing the African elephant into North America to replace the mastodon, whose extinction, he argues, was the result of hunting by First Nations peoples. As we saw in Chapter 2, First Nations agricultural communities had a distinct and probably permanent impact on the landscape of southern Ontario.

Overall, new ways of thinking about restoration leave open choice, which is a matter of science and values. Science tells us what nature has been and what it can be; our values determine what we want nature to be. There is no single perfect condition. However, for some goals in Table 10.1, specific conditions are especially desirable. It is possible to restore an ecosystem so that most of the time it supports conditions that people find desirable.

10.2 What Needs to Be Restored?

Ecosystems of all kinds have undergone degradation and need restoration. However, some ecosystems have undergone especially widespread loss and degradation, and are therefore the focus of current attention. In addition to forests, these include wetlands, both freshwater (Figure 10.3) and coastal; grasslands, especially the North American prairie; streams and rivers and the riparian zones alongside them; lakes; and habitats of threatened and endangered species. Also included are areas that people want to restore for aesthetic and moral reasons, showing again that restoration involves values. In this section, we discuss restoration of wetlands, rivers, streams, and prairies.

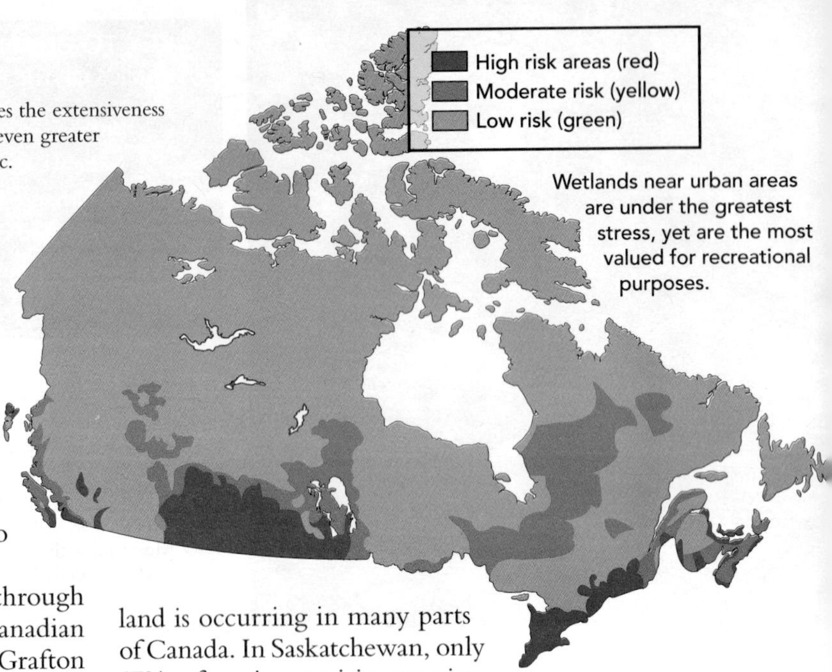

Figure 10.3 • **Canadian wetlands at risk.** This map illustrates the extensiveness of threatened wetland losses. In some regions, the risk of loss is even greater because of proximity to urban areas and heavy recreational traffic.

Wetlands, Rivers, and Streams

Large areas of both freshwater and coastal wetlands have been greatly altered during the past 200 years, especially in North America. Over the past 200 years, more than half the wetlands in Canada—as much as three-quarters in some regions—have disappeared because of human activities, mainly agriculture and urban or industrial development (see Chapter 18). The shorelines of many lakes and rivers have also changed, as humans fill and build on them.

In some cases, these changes can be reversed through targeted restoration efforts. An example of a Canadian restoration project is the removal of a dam at Grafton Lake, Nova Scotia. Located in Kejimkujik National Park, Grafton Lake was built by damming Grafton Brook in 1938, to support a salmon hatchery. Three species at risk, the Blandings turtle, piping plover, and water pennywort, occur in the area. Uncertain of the potential impact, researchers have been wary of removing the Grafton Lake dam. Recent studies have confirmed that the dam is a physical barrier to the natural migration patterns of turtle and trout. Removing the dam to restore natural lake hydrology was necessary to restore biological integrity and address conservation concerns. The task provided a unique opportunity to document ecosystem change as the lake was restored to its natural condition.

Prairie Restoration

As noted in Chapter 8, prairies once occupied more land in Canada and the United States than any other kind of ecosystem. Today, only a few small remnants of prairie remain. (Figure 8.14)

There are two kinds of prairie restoration. In a few places, original prairie exists that has never been ploughed. Here, the soil structure is intact, and restoration is simpler. One of the best known of these areas is the CFB Suffield National Wildlife Area, one of the largest blocks of unploughed grassland remaining in the Canadian prairies. A complex landscape of grasslands, sand hills, rivers and wetlands, CFB Suffield provides a refuge for more than 1,100 known species, 14 of which are listed as species at risk in Canada. The area has been zoned "Out of Bounds" to all military training and defence research activities since 1971, and thus protected from many of the human activities that have altered other ecosystems.

Restoration is more complicated in areas where the land was ploughed. Nevertheless, prairie restoration has gained considerable attention in recent decades, and restoration of the prairie on previously ploughed and farmed land is occurring in many parts of Canada. In Saskatchewan, only 17% of native prairie remains. The Saskatchewan Watershed Authority is working with local landowners and conservation groups on local stewardship programs. School groups like those at Winston Knoll Collegiate and Luther College, both in Regina, have been particularly active in these projects.

A peculiarity about the history of prairies is that although most prairie land was converted to agriculture, this was not done along roads and railroads, so that long, narrow strips of unploughed native prairie remain on these rights-of-way.

In Saskatchewan, the prairie once covered more than 80% of the province—24 million ha. About 96% of the Saskatchewan prairie land has been converted to other uses, primarily agriculture. The remaining native prairie is highly fragmented and occurs in hundreds of small parcels of land, few of which are over 1,000 ha in area.

Table 10.1 • Some Possible Restoration Goals

Goal	*Approach*
1. Pre-Industrial	Maintain ecosystems as they were in 1500 a.d.
2. Presettlement (e.g., of North America)	Maintain ecosystems as they were about 1492 a.d.
3. Preagriculture	Maintain ecosystems as they were about 5,000 b.c.
4. Before any significant impact of human beings	Maintain ecosystems as they were about 10,000 b.c.
5. Maximum production	Independent of a specific time
6. Maximum diversity	Independent of a specific time
7. Maximum biomass	Independent of old growth
8. Preserve a specific endangered species	Whatever stage it is adapted to
9. Historic range of variation	Create the future like the known past

(a) (b)

Figure 10.4 • Primary succession. *(a)* Forests developing on new lava flows in Hawaii and, *(b)* at the edge of a retreating glacier.

These small parcels of prairie (some of them along railway rights-of-way) provide some of the last habitats for native plants. Restoration of prairies elsewhere in Saskatchewan employs stewardship programs and public seed harvesting, to use these habitats as seed sources for other regions.[3]

10.3 When Nature Restores Itself: The Process of Ecological Succession

We have been discussing people's attempts to restore altered ecosystems. Often, the change has been caused by people, but natural areas are subject to natural disturbances as well. Storms and fires, for example, have always been a part of the environment.[4] Recovery of disturbed ecosystems can also occur naturally, through a process called ecological succession. This natural recovery can occur if the alteration is not too great. Sometimes, though, recovery takes longer than people would like.

We can classify ecological succession as either primary or secondary. Primary succession is the initial establishment and development of an ecosystem where one did not exist previously. Secondary succession is re-establishment of an ecosystem following disturbances. In secondary succession, there are remnants of a previous biological community, including such things as organic matter and seeds. Forests that develop on new lava flows (Figure 10.4*a*) and at the edges of retreating glaciers (Figure 10.4*b*) are examples of primary succession. Forests that develop on abandoned pastures or following hurricanes, floods, or fires are examples of secondary succession (see A Closer Look 10.1).

Succession is one of the most important ecological processes, and the patterns of succession have many management implications. We see examples of succession all around us. When a house lot is abandoned in a city, weeds begin to grow. After a few years, shrubs and trees can be found; secondary succession is taking place. A farmer weeding a crop and a homeowner weeding a lawn are both fighting against the natural processes of secondary succession.

Patterns in Succession

Succession occurs in most kinds of ecosystems, and when it occurs, it follows certain general patterns. Below we consider succession in three classic cases involving forests: (1) on dry sand dunes along the shores of the Atlantic Ocean, (2) in a northern freshwater bog, and (3) in an abandoned farm field.

Dune Succession

Sand dunes are continually being formed along sandy shores and then breached and destroyed by storms. In coastal regions such as Malpeque Bay, Prince Edward Island, soon after a dune is formed, dune grass invades. This grass has special adaptations to the unstable dune. Just under the surface, it puts out runners with sharp ends (if you step on one, it will hurt). The dune grass rapidly forms a complex network of underground runners, crisscrossing almost like a coarsely woven mat. Above the ground, the green stems carry out photosynthesis, and the grasses grow.

Once the dune grass is established, its runners stabilize the sand, and seeds of other plants are less easily buried too deep or blown away. The seeds germinate and grow, and a more diverse ecological community begins to develop. The plants of this early stage tend to be small, wind-resistant, grow well in bright light, and withstand the harshness of the environment—high temperatures in the summer, low temperatures in the winter, and intense storms.

Slowly, larger plants, such as eastern red cedar and white pine, are able to grow on the dunes. Eventually, a forest develops, which may include tree species such as beech and maple. Such a forest can persist for many years, but at some time a severe storm breaches even these heavily vegetated dunes, and the process begins again (Figure 10.6).

A CLOSER LOOK 10.1

An Example of Forest Secondary Succession

Within a few years after a field is abandoned, seeds of many kinds sprout, some of short-lived weedy plants and some of trees (Figure 10.5*a*). After a few years, certain species, generally referred to as pioneer species, become established. In southern Ontario, Eastern white cedar, white pine, white birch, and yellow birch are such species (Figure 10.5*b*). These trees are fast growing in bright light and have widely distributed seeds. For example, seeds of white cedar are eaten by birds and dispersed; birches have very light seeds that are widely dispersed by the wind. After several decades, forests of the pioneer species are well established, forming a dense stand of trees.

Once the initial forest is established, other species begin to grow and become important. Typical dominants in eastern Canada are sugar maple and beech. These later successional trees are slower growing than the pioneer species, and they have other characteristics that make them well adapted to the later stages of succession. Ecologists refer to these species as shade-tolerant: they grow relatively well in the deep shade of the redeveloping forest.

After three or four decades, most of the short-lived species have matured, borne fruit, and died. Since they cannot grow in the shade of a forest that has been re-established, they do not regenerate. For example, after five or six decades an Ontario forest is a rich mixture of birches, maples, beeches, and other species (Figure 10.5*c*). The trees vary in size, but the trees that dominate now are generally taller than those dominating at earlier stages. After one or two centuries, such a forest will be composed mainly of shade-tolerant species.

Figure 10.5 • A series of photographs of succession on an abandoned pasture in southern Ontario: *(a)* a second-year field; *(b)* young white cedar trees, grasses, and other plants some years after a field was abandoned; *(c)* mature forests developed along old stone farm walls.

(a)

(b)

(c)

Bog Succession

A bog is an open body of water with surface inlets—usually small streams—but no surface outlet. As a result, the waters of a bog are quiet, flowing slowly if at all. Many bogs that exist today originated as lakes that filled depressions in the land, which in turn were created by glaciers during the Pleistocene ice age.

Succession in a northern bog, such as Burns Bog in Delta, British Columbia (Figure 10.7), begins when sedges (grasslike herbs) put out floating runners (Figure 10.8*a, b*). These runners form a complex matlike network, similar to that formed by dune grass. The stems of the sedge grow on the runners and carry out photosynthesis. Wind blows particles onto the mat, and soil, of a kind, develops. Seeds of other plants land on the mat and do not sink into the water. They can germinate. The floating mat becomes thicker, and small shrubs and trees, adapted to wet environments, grow. In the north, these include species of the blueberry and willow families.

The bog also fills in from the bottom as streams carry fine particles of clay into it (Figure 10.8*b, c*). At the shoreward end, the floating mat and the bottom sediments meet, forming a solid surface. Farther out, a quaking bog occurs. You can walk on this quaking bog mat; and if you jump up and down, all the plants around you bounce and shake. The mat is really floating. Eventually, as the bog fills in from the top and the bottom, trees that can withstand wetter conditions—such as northern cedar, black spruce, and balsam fir—grow. The formerly open water bog becomes a wetland forest or swamp.

Old-Field Succession

In eastern Canada, a great deal of land was cleared and farmed in the eighteenth and nineteenth centuries. Today, much of this farmland has been abandoned and allowed

Figure 10.6 • Dune succession on the shores of Prince Edward Island. Dune grass shoots appear scattered on the slope, where they emerge from underground runners.

Figure 10.7 • Burns Bog, a famous bog in Delta, British Columbia.

to grow back to forest (see Figure 10.5). The first plants to enter the abandoned farmlands are those like annual (living only one year) herbs and grasses, which are adapted to the harsh and highly variable conditions of a clearing—a wide range of temperatures and precipitation. As these plants become established, other, larger perennial plants, like raspberry and small shrubs, enter. Eventually, large trees grow, such as sugar maple, beech, yellow birch, and white pine, forming a dense forest.

General Patterns of Succession

Reviewing these three examples of ecological succession involving forests, you can see common elements in them, even though the environments are very different. Common elements of succession in such cases include the following:

1. An initial kind of fast-growing, opportunistic vegetation specially adapted to the unstable conditions. These plants are typically small in stature, with adaptations that help stabilize the physical environment.
2. A second stage with plants still of small stature, rapidly growing, with seeds that spread rapidly.
3. A third stage in which larger plants, including trees, enter and begin to dominate the site.
4. A fourth stage in which a mature forest develops.

Although we list four stages, it is common practice to combine the first two and to speak of early, middle, and late successional stages. These general patterns of succession can be found in most ecosystems, although the species differ. The stages of succession are described here in terms of vegetation, but similarly adapted animals and other life forms are associated with each stage. We discuss other general properties of the process of succession later in this chapter.

Species characteristic of the early stages of succession are called pioneers, or early-successional species.

Figure 10.8 • Diagram of bog succession. Open water *(a)* is transformed through formation of a floating mat of sedge and deposition of sediments *(b)* into wetland forest *(c)*.

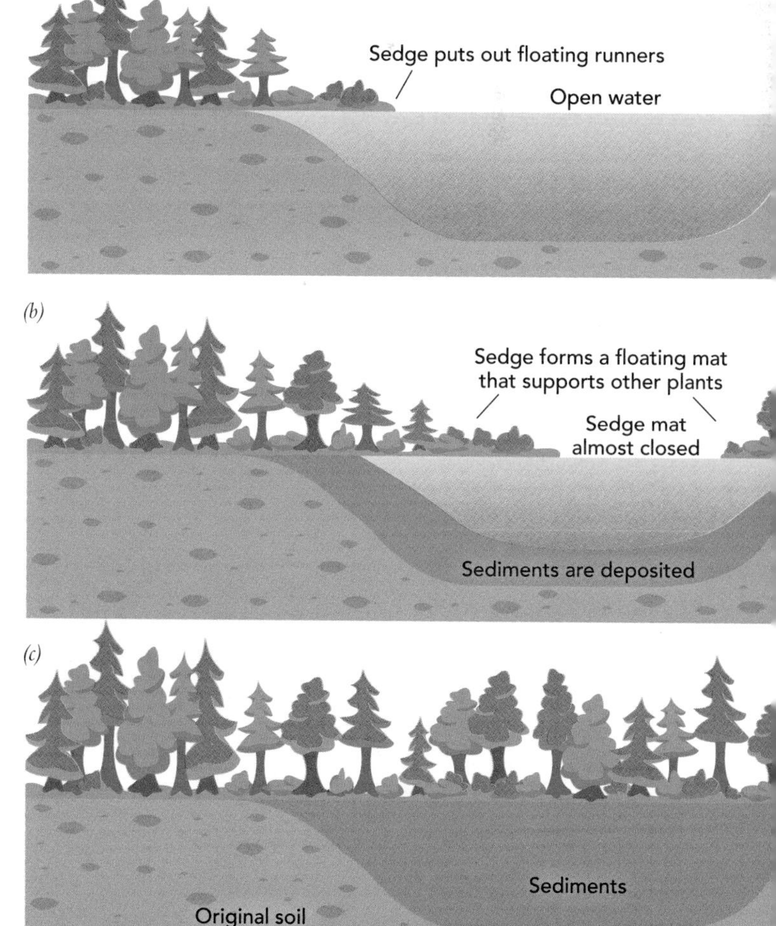

They have evolved and are adapted to the environmental conditions in early stages of succession. Plant species that dominate late stages of succession, called late-successional species, tend to be slower growing and longer-lived. These species have evolved and are adapted to environmental conditions in the late stages. For example, they grow well in shade and have seeds that, while not as widely dispersing, can persist a rather long time.

In early stages of succession, biomass and biological diversity increase (Figure 10.9). In middle stages of succession, we find trees of many species and many sizes. Gross and net production (see Chapter 8) change during succession as well: Gross production increases, while net production decreases. Chemical cycling also changes: The organic material in the soil increases, as does the amount of chemical elements stored in the soils and trees.[5] We look more closely at chemical cycling changes in the next section.

10.4 Succession and Chemical Cycling

An important effect of succession is a change in the storage of chemical elements necessary for life. On land, the storage of chemical elements (including nitrogen, phosphorus, potassium, and calcium, essential for plant growth and function) generally increases during the progression from the earliest stages of succession to middle succession. There are two reasons for this.

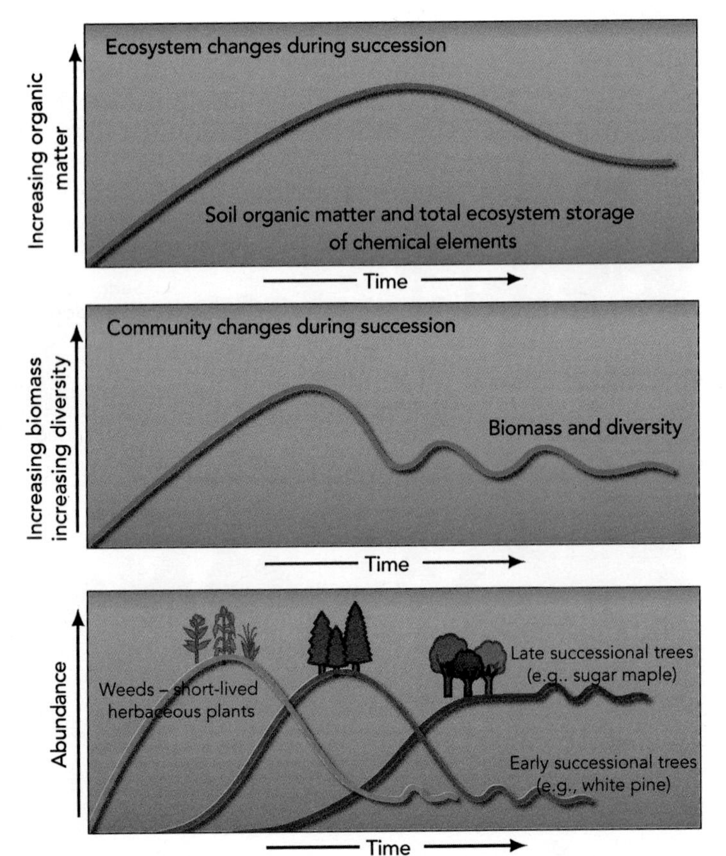

First, organic matter stores chemical elements; as long as there is an increase in organic matter within the ecosystem, there will be an increase in the storage of chemical elements. This is true for live and dead organic matter. Additionally, many plants have root nodules containing bacteria that can assimilate atmospheric nitrogen, which is then used by the plant in the process known as *nitrogen fixation* (Chapter 5).

The second reason is indirect: The presence of live and dead organic matter helps retard erosion. Both organic and inorganic soil can be lost to erosion because of the effects of wind and water. Vegetation tends to reduce or prevent such losses and therefore causes an increase in total stored material.

Organic matter in soil not only stores chemical elements; it also functions as an ion-exchange column, holding metallic ions that would otherwise be transported in dissolved form in groundwater and lost to the ecosystem. Generally, the greater the volume of soil and the greater the percentage of organic matter in the soil, the more chemical elements will be retained. Another important factor is a soil's storage capacity for each element, which varies with the average size of the soil particles. Soils composed mainly of large, coarse particles, like sand, have a smaller total surface area and can store a smaller quantity of chemical elements. Clay, which is composed of the smallest particles, stores the greatest quantity of chemical elements.

Soils contain greater quantities of chemical elements than do live organisms. However, much of what is stored in a soil may be relatively unavailable, or may only become available slowly because the elements are tied up in complex compounds that decay slowly. In contrast, the elements stored in living tissues are readily available to other organisms through food chains.

Rates of cycling and average storage times are system characteristics (as discussed in Chapter 3). The increase in chemical elements that occurs in the early and middle stages of succession does not continue indefinitely. If an ecosystem persists for a very long period with no disturbance, it will experience a slow but definite loss of stored chemical elements. Thus, the ecosystem will slowly degrade and become depauperate—literally, impoverished—and thus less able to support rapid growth, high biomass density, and high biological diversity (Figure 10.10).[6] The changes in chemical cycling during disturbance and successional recovery are discussed in A Closer Look 10.2.

Figure 10.9 • Graphs showing changes in biomass and diversity with succession.

10.5 Species Change in Succession: Do Early Successional Species Prepare the Way for Later Ones?

To restore an ecosystem, understanding what causes one species to replace another during the succession process is important. If we understand these causes and effects, we can use them for more effective ecosystem restoration. Earlier and later species in succession may interact in three ways: through (1) facilitation, (2) interference, or (3) life history differences. If they do not interact, the result is called *chronic patchiness* (Table 10.2).[7, 8]

Facilitation

In dune and bog succession, the first plant species—dune grass and floating sedge—prepare the way for other species to grow. This is called facilitation (early-successional species facilitate the establishment of later successional species; Figure 10.10*a*).

Facilitation has been found to take place in tropical rain forests.[9] Early-successional species speed the reappearance of the microclimatic conditions that occur in a mature forest. Because of the rapid growth of early-successional plants, the temperature, relative humidity, and light intensity at the soil surface in tropical forests can reach levels similar to those of a mature rain forest after only 14 years.[5] Once these conditions are established, species that are adapted to deep forest shade can germinate and persist. As another example, recent studies have shown that restoration of tallgrass prairie, which is native to parts of Manitoba and Ontario, can be impeded by the presence of weed species, but facilitated by enrichment of soils with carbon and certain kinds of native fungi.[11, 12]

Knowing the role of facilitation can be useful in the restoration of ecosystems. Plants that facilitate the presence of others should be planted first. On sandy areas, for example, dune grasses can help hold the soil before attempts are made to plant larger shrubs or trees.

Interference

Facilitation does not always occur. Sometimes, certain early-successional species interfere with the entrance of other species (Figure 10.10*b*). For example, in some old fields of Québec and Ontario, among the early-successional species are grasses that form dense cover, including the prairie grass little bluestem. The living and dead stems of this grass form a mat so dense that other plant seeds that fall onto it cannot reach the ground and do not germinate. The grass interferes with the entrance of other plant species—a process called, naturally enough, interference.

Interference does not last forever, however. Eventually, some breaks occur in the grass mat—perhaps from surface water erosion, the death of a patch of grass from disease, or removal by fire. Breaks in the grass mat allow seeds of trees such as cedar to reach the ground. Cedar is also adapted to early succession because birds who feed on it spread its seeds are spread rapidly and widely and because this species grows well in the bright light and otherwise harsh conditions of early succession. Once started, cedar soon grows taller than the grasses, shading them so much that they cannot grow. More ground is open, and grasses are eventually replaced.

In parts of Asia, interference occurs where bamboo grows, and, in tropical areas, where a smaller grass,

(a)

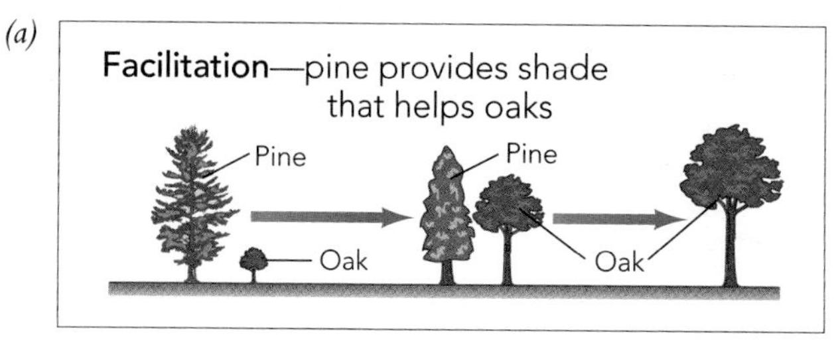

(b)

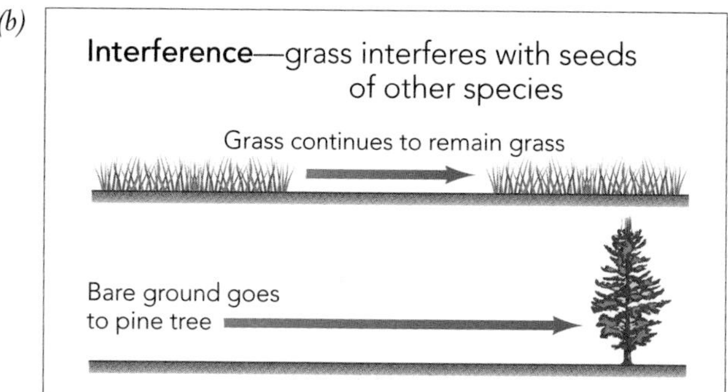

(c)

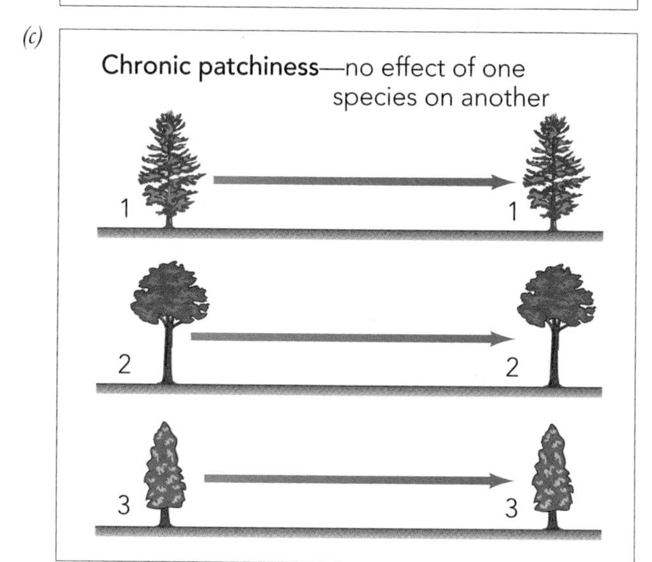

Figure 10.10 • Patterns of interaction among species in ecological succession. *(a)* Facilitation. As Henry David Thoreau observed in Massachusetts more than 100 years ago, pines provide shade and act as "nurse trees" for oaks. Pines do well in openings. If there were no pines, few or no oaks would survive. Thus the pines facilitate the entrance of oaks. *(b)* Interference. Some grasses that grow in open areas form dense mats that prevent seeds of trees from reaching the soil and germinating. The grasses interfere with the addition of trees. *(c)* Chronic patchiness. Earlier-entering species neither help nor interfere with other species; instead, as in a desert, the physical environment dominates.

A CLOSER LOOK 10.2

Changes in Chemical Cycling During a Disturbance

When an ecosystem is disturbed by fire, storms, or human activities, changes occur in chemical cycling. For example, when a forest is burned, complex organic compounds, such as wood, are converted to smaller inorganic compounds, including carbon dioxide, nitrogen oxides, and sulfur oxides. Some of the inorganic compounds from the wood are lost to the ecosystem during the fire as vapours that escape into the atmosphere and are distributed widely or as particles of ash that are blown away, but some of the ash falls directly onto the soil. These compounds are highly soluble in water and readily available for vegetation uptake. Therefore, immediately after a fire, the availability of chemical elements increases. This is true even if the ecosystem as a whole has undergone a net loss in total stored chemical elements.

If sufficient live vegetation remains after a fire, then the sudden, temporary increase, or pulse, of newly available elements is taken up rapidly, especially if it is followed by a moderate amount of rainfall (enough for good vegetation growth, but not so much as to cause excessive erosion). The pulse of inorganic nutrients can then lead to a pulse in the growth of vegetation and an increase in the amount of stored chemical elements in the vegetation. In turn, this boosts the supply of nutritious food for herbivores, which can subsequently undergo a population increase. The pulse in chemical elements in the soil can therefore have effects that extend throughout the food chain.

Other disturbances on the land produce effects similar to those of fire. For example, severe storms, such as hurricanes and tornadoes, knock down and kill vegetation. The vegetation decays, increasing the concentration of chemical elements in the soil, which are then available for vegetation growth. Storms also have another effect in forests: When trees are uprooted, chemical elements that were near the bottom of the root zone are brought to the surface, where they are more readily available. Knowledge of the changes in chemical cycling and the availability of chemical elements from the soil during succession can be useful to us in restoring ecosystems. We know that nutrients must be available within the rooting depth of the vegetation. The soil must have sufficient organic matter to hold on to nutrients. Restoration will be more difficult where the soil has lost its organic matter and has been leached. A leached soil has lost nutrients as water drains through it, especially acidic water, dissolving and carrying away chemical elements. Thus, heavily leached soils, such as those subjected to acid rain and acid mine drainage (Chapter 19), pose special challenges to the process of land restoration.

Imperata, grows. Like little bluestem in Canada, these grasses form stands so dense that seeds of later-successional species cannot reach the ground, germinate, or obtain enough light, water, and nutrients to survive. *Imperata* either replaces itself or is replaced by bamboo, which then replaces itself.[9] Once established, *Imperata* and bamboo appear able to persist for a long time. Once again, when and if breaks occur in the cover of these grasses, other species can germinate and grow, and a forest eventually develops.

Life History Differences

In other cases, species do not affect one another so much. Rather, differences in the life histories of the species allow some to arrive first and grow quickly, while others arrive later and grow more slowly. An example of a life history difference is seed dispersal. The seeds of early-successional species are readily transported by wind or animals and reach a clearing sooner and grow faster than seeds of late-successional species. In many forested areas of eastern North America, for example, birds eat the fruit of cherries and red cedar, and their droppings contain the seeds, which spread widely.

In contrast, late-successional species often can have life histories that bring them into an ecosystem much later. For example, although sugar maple can grow in open areas, it is slow to colonize them because its seeds take longer to travel and its seedlings can tolerate shade. Beech produces large nuts that store a lot of food for

Table 10.2 • Patterns of Interaction among Earlier and Later Species in Succession

1. ***Facilitation.*** One species can prepare the way for the next (and may even be necessary for the occurrence of the next).
2. ***Interference.*** For a time, early successional species can prevent the entrance of later successional species.
3. ***Life history differences.*** One species may not affect the time of entrance of another; two species may appear at different times during succession because of differences in transport, germination, growth, and longevity of seeds.
4. ***Chronic patchiness.*** Succession never occurs, and the species that enters first remains until the next disturbance.

Source: J. H. Connell and R. O. Slatyer, "Mechanisms of Succession in Natural Communities and Their Role in Community Stability and Organization," *American Naturalist III* (1977): 1119–1144; S. T. A. Pickett, S. L. Collins, and J. J. Armesto, "Models, Mechanisms, and Pathways of Succession," *Botanical Review* 53 (1987): 335–371.

CRITICAL THINKING ISSUE

How Can We Evaluate Constructed Ecosystems?

What happens when restoring altered ecosystems is not an option? In such cases, artificial ecosystems may be constructed to replace the damaged ones.

The Artificial Reef Society of British Columbia (ARSBC) has long promoted the sinking of decommissioned military ships to create artificial reefs. In 1992, the ARSBC bought the decommissioned destroyer HMCS Chaudière from the federal government for $1 plus tax, and sank it off Kunechin Point in Sechelt Inlet, British Columbia. Marine life has flourished on the 111 m, 2,900 tonne ship since that time, but the project remains controversial. Critics, including the Georgia Strait Alliance and the Fraser River Coalition, note that although ships like the Chaudière may attract fish and create exciting scuba diving opportunities, they also have the potential to contaminate sensitive ecosystems. Paints (often containing the anti-fouling substance tributyl tin), fuels, oils, solvents, asbestos and even simple rust can have unanticipated impacts on marine waters, sediments, and biota. While artificial reefs may attract marine organisms, they do not necessarily increase ecosystem diversity or productivity. Unless projects are carefully designed, they may fail to meet the needs of target species and attract others instead. As a result, their environmental impacts can outweigh their benefits.

Environment Canada regulates the sinking of decommissioned ships under the Disposal at Sea regulations of the *Canadian Environmental Protection Act*, and all proposed sunken-ship reef projects are reviewed under the *Canadian Environmental Assessment Act*. Responsibility for monitoring is less clear. There is little baseline data for the Chaudière site before the sinking. Most current monitoring is conducted by volunteer divers, either under Fisheries and Oceans Canada's Reefkeeper program or through organizations like the Georgia Strait Alliance. The quality of this information is not always consistent. Volunteer divers vary in their ability to identify individual organisms or even to make an accurate count of the species present. In one paired-diver survey, one diver saw considerably more species than any other, even including the professional biologist leading the dive! In addition to errors in counting the number and type of species, divers may become confused by background conditions such as light and algal growth. Certain species are particularly difficult for "beginner" volunteer divers to identify. Diver training and debriefing at the end of each dive is important to reduce discrepancies between divers, and to learn how each diver collects data.

How should we evaluate the success or failure of the Chaudière sinking, relative to the restoration goals suggested in Table 10.1? As with many other environmental issues, perspectives on the Chaudière differ depending on personal values and priorities. For some people, it is enough to have new habitat created, even if that habitat does not meet the needs of native species. For other people, protection of native ecosystems is a more important goal. Evaluation of the Chaudière experience is currently hampered by a lack of data before and after the sinking. More systematic sampling is required, with each reef surveyed by at least one "expert" research diver, supported by volunteer divers who provide backup data. Good information about the ecological changes caused by the Chaudière sinking could provide a basis for more subjective discussions of desired ecosystem outcomes.

Critical Thinking Questions

1. Do you consider the Chaudière artificial reef project an environmental success or failure? Explain your answer.
2. Propose a monitoring scheme for the Chaudière that would produce more systematic, reliable data. Who should pay for your proposed program?
3. How do you think one can decide whether a constructed ecosystem is an adequate replacement for a natural ecosystem?
4. The term adaptive management refers to the use of scientific research in ecosystem management. In what ways could adaptive management have been used in the Chaudière project? What lessons from the project could be used to improve similar projects in the future?

a newly germinated seedling. This helps the seedling establish itself in the deep shade of a forest until it is able to feed itself through photosynthesis. But these seeds are heavy and are moved relatively short distances by seed-eating animals.

Chronic Patchiness

A fourth possibility is that species just do not interact and that succession, as described, does not take place. This is called chronic patchiness (Figure 10.10*c*), and it is the case in some deserts. For example, in the warm deserts of California, Arizona, and Mexico, the major shrub species grow in patches, which often consist of mature individuals with few seedlings. These patches persist for long periods until there is a disturbance.[10] Similarly, in highly polluted environments, a sequence of species replacement may not occur.

What kinds of changes occur during succession depends on the complex interplay between life and its environment. Life tends to build up, or aggrade, whereas non-biological processes in the environment tend to erode and degrade. In harsh environments, where energy or chemical elements required for life are limited and disturbances are frequent, the physical, degrading environment dominates, and succession does not occur.

10.6 Applying Ecological Knowledge to Restore Heavily Altered Lands and Ecosystems

An example of how ecological succession can aid in the restoration of heavily altered lands is the ongoing effort to undo mining damage in Great Britain, where mining caused widespread destruction to land. In Great Britain, where some mines have been in use since medieval times, approximately 55,000 ha are damaged by mining. Recently, programs were initiated to remove toxic pollutants from the mines and mine tailings, to restore these lands to useful biological production, and to restore the visual attractiveness of the landscape.[11] Similar restoration activities are underway in many parts of Canada, including Sudbury, Ontario, where mining and smelting activities contributed to environmental degradation for over 100 years.[14, 15, 16]

One area altered by a long history of mining lies within the British Peak District National Park, where lead has been mined since the Middle Ages and waste tailings are as much as 5 m deep. The first attempts to restore this area used a modern agricultural approach: heavy application of fertilizers and planting of fast-growing agricultural grasses to revegetate the site rapidly. These grasses quickly green on the good soil of a level farm field, and the hope was that, with fertilizer, they would do the same in this situation. However, after a short period of growth, the grasses died. On the poor soil, leached of its nutrients and lacking organic matter, erosion continued, and water runoff soon leached the added fertilizers away. As a result, the areas were soon barren again.

Figure 10.11 • An old lead-mining area in Great Britain, now undergoing restoration. Restoration involves planting of early-successional native grasses adapted to low-nutrient soils with little physical structure.

When the agricultural approach failed, an ecological approach was tried, using knowledge about ecological succession. Instead of planting fast-growing but vulnerable agricultural grasses, ecologists planted slow-growing native grasses known to be adapted to minerally deficient soils and the harsh conditions that exist in cleared areas. In choosing these plants, the ecologists relied on their observations of what vegetation first appeared in areas of Great Britain that had undergone succession naturally.[11] The result of the ecological approach has been successful restoration of once-degraded lands (Figure 10.11).

Heavily altered landscapes are found in many places. Restoration similar to that in Great Britain is done in North America to reclaim lands damaged by mining and smelting. In such cases, restoration often begins during the mining process rather than afterward. In many areas of North America, similar methods are used to restore "brownfield" sites—those formerly occupied by industries or waste disposal sites.

Summary

- Restoration of altered ecosystems is a major new emphasis in environmental sciences and is developing into a new field. Restoration involves a combination of human activities and natural processes of ecological succession.
- Disturbance, change, and variation in the environment are natural processes, and ecological systems and species have evolved, and continue to evolve, in response to these changes.
- When ecosystems are disturbed, they undergo a process of recovery known as ecological succession, the establishment and development of an ecosystem. Knowledge of succession is important in the restoration of altered lands.
- During succession, there is usually a clear, repeatable pattern of changes in species. Some species, called early-successional species, are adapted to the first stages, when the environment is harsh and variable but when necessary resources may be available in abundance. This contrasts with late stages in succession, when biological effects have modified the environment and reduced some of the variability but also have tied up some resources. Typically, early-successional species are fast growing and short-lived, whereas late-successional species are slow growing and long-lived.
- Biomass, production, diversity, and chemical cycling change during succession. Biomass and diversity peak

in mid-succession, increasing at first to a maximum, then declining and varying over time.

- Changes in the kinds of species found during succession can be due to facilitation, interference, or simply life history differences. In facilitation, one species prepares the way for others. In interference, an early-successional species prevents the entrance of later-successional ones. Life history characteristics of late-successional species sometimes slow their entrance into an area.

STUDY QUESTIONS

1. Farming has been described as managing land to keep it in an early stage of succession. What does this mean, and how is it achieved?
2. Redwood trees reproduce successfully only after disturbances (including fire and floods), yet individual redwood trees may live more than 1,000 years. Is redwood an early- or late-successional species?
3. Why could it be said that succession does not take place in a desert shrubland (an area where rainfall is very low and the only plants are certain drought-adapted shrubs)?
4. Develop a plan to restore an abandoned field in your town to natural vegetation for use as a park. The following materials are available: bales of hay; artificial fertilizer; and seeds of annual flowers, grasses, shrubs, and trees.
5. Oil has leaked for many years from the gasoline tanks of a gas station. Some of the oil has oozed to the surface. As a result, the gas station has been abandoned and revegetation has begun to occur. What effects would you expect this oil to have on the process of succession?
6. You are put in charge of a project to grow yellow birch, a species common in midsuccession, as a commercial crop. (Yellow birch is used to make dowels and hardwood plywood panels.) Considering the process of succession, how would you go about setting up plantations of yellow birch? How often would you log these plantations? Would the logging be complete or selective—that is, would you remove all the trees at one time or only some at any one time? Explain your answer.
7. Early in the twentieth century, a large meteorite collided with the earth in Siberia and destroyed boreal forests over a large area. How might restoration and succession following this large-scale disturbance differ from restoration and succession after a fire burned a few hectares in the same forest? (See Chapter 8 for information about boreal forests.)

FURTHER READING

Berger, J. J. *Environmental Restoration: Science and Strategies for Restoring the Earth.* Washington, D.C.: Island Press, 1990. An informed and lively overview of the beginning of the restoration movement that includes scientific and technical papers given at the first national conference on restoration, held in 1933.

Botkin, D. B. *Discordant Harmonies: A New Ecology for the 21st Century.* New York: Oxford University Press, 1992.

Botkin, D. B. *No Man's Garden: Thoreau and a New Vision for Civilization and Nature.* Washington, D.C.: Island Press, 2001.

Cairns, J., Jr., ed. *Rehabilitating Damaged Ecosystems,* 2nd ed. Lewis Publishers, 1995. Discussions of natural and human-assisted restoration of various ecosystem types after either natural or human-caused disturbance.

Foster, D. R., and J. F. O'Keefe. *New England Forests through Time: Insights from the Harvard Forest Dioramas.* Cambridge, Mass.: Harvard University Press, 2000. A beautifully illustrated short book that discusses secondary succession of forests in New England, using as the centerpiece a famous set of miniatures at the Harvard Forest.

Parks Canada. *2001 State of Protected Heritage Areas. 2001 Report.* Ottawa: Queen's Printer. A comprehensive report on research and field monitoring aimed at managing, protecting, and preserving Canada's system of national parks. This is the first Parks Canada report following the enactment of the *Canada National Parks Act* in 2000.

Petts, G., and P. Calow, eds.. *River Restoration.* London: Blackwell Science, 1996. An overview of international efforts to restore rivers, designed for undergraduates as well as other audiences. The book begins with a general introduction to rivers and includes chapters on the control of weeds, the conservation and restoration of fish, and the relationships between disturbance and recovery.

chapter 11

Food for China

China, with one-fifth of the world's population, is the most populous country on Earth. Between 1980 and 1995, China's population grew by 200 million people—about six times the present population of Canada—to reach 1.2 billion. Although its growth rate is expected to slow somewhat in the coming decades, population experts predict that there will be 1.4 billion Chinese by 2025 (Figure 11.1*a*).[1] Farming is, of course, an ancient practice in China, but can China's food production continue to keep pace with its growing population? If not, China may need to import more grain than other countries can spare. Increased demand by China for world grain supplies could increase food prices dramatically and precipitate famines in other areas of the world (Figure 11.1*b*).

To gain some idea of the potential impact of China on the world's food supply, consider this: to supply two more beers a year to each person in China would require all the grain produced in Norway. If the average Chinese ate as much fish as the average citizen of Japan, China would consume the entire world's fish catch. All the chickens necessary to reach China's goal of 200 eggs per person per year would require all the grain exported by Canada—the world's second largest grain exporter.[2]

Between 1959 and 1961, 30 million Chinese starved to death during a famine brought on by a state policy of modernization that forced millions of farmers to work on large construction projects. In the wake of famine, China attempted to slow population growth through a "one couple, one child" policy (see Chapter 4) and to increase food production through extensive agricultural reform. Between 1980 and 1995, the total fertility rate in China fell from 4.8 children per woman to fewer than 2 children. In the same period, China achieved a remarkable increase in annual grain production from the 200 kg per person needed to maintain a minimal level of physical activity to 300 kg.

Can China continue to produce enough food for its people? The country has relatively little cropland—approximately 100 million hectares (ha), or 0.08 ha per capita. By comparison, India has 0.19 hectares per capita—170 million ha of land with which to feed its population of more than 1 billion.[1] Industrialization, including the recently completed vast hydroelectric Three Gorges Dam, is eliminating some farmland. China has been losing agricultural land to roads, railroads, and manufacturing plants at the rate of 1.6% per year.[3] At the same time, with half its cropland under irrigation, China is facing a severe water crisis due to seasonal variations in rainfall, periodic floods, and diversion of water to non-agricultural uses. With increased prosperity, Chinese are eating higher up the food chain—consuming more meat, which in turn requires more grain production per person (Figure 11.2). Whether China's agricultural resources can sustain its population in coming decades will be a critical test of whether the world can find a

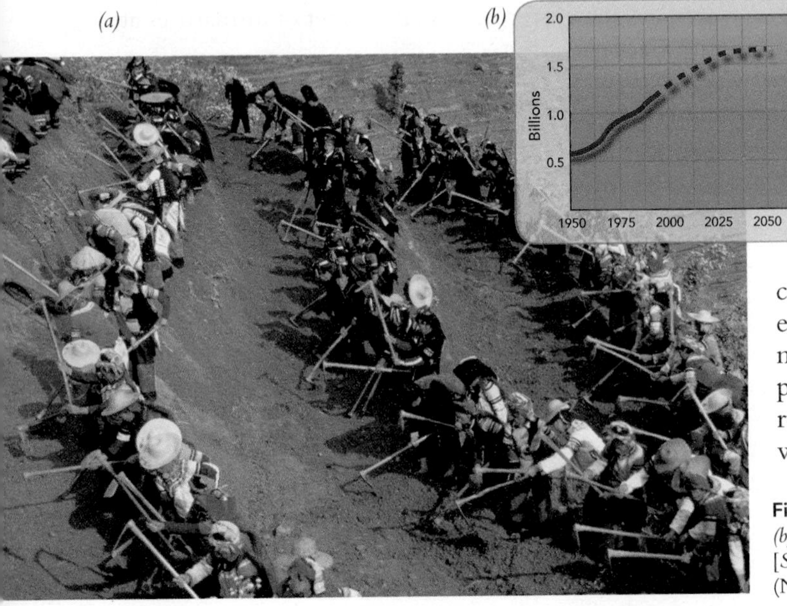

Figure 11.1 • *(a)* **Photograph of traditional farming in China.** *(b)* **Population of China, 1950–1994, with projections to 2050.** [*Source*: L. R. Brown, *Who Will Feed China? Wakeup Call for a Small Planet* (New York: Norton, 1995), p. 36.]

Agriculture, Food Production, and the Environment

LEARNING OBJECTIVES

The big question about farming and the environment is: Can we produce enough food to feed Earth's growing human population, and do this sustainably? The major challenges are to increase the productivity of the land, hectare by hectare; to distribute food adequately around the world; to decrease the negative environmental effects of agriculture; and to avoid creating new environmental problems as agriculture advances. After reading this chapter, you should understand:

- What it means to take an ecological perspective on agriculture.
- How agroecosystems differ from natural ecosystems.
- How the food supply depends on the environment.
- What role limiting factors play in determining crop yield.
- How the concept of sustainability applies to agriculture.
- How the growing human population, the loss of fertile soils, and the lack of water for irrigation can affect future food shortages worldwide.
- The relative importance of food distribution and food production.
- The potential benefits and environmental effects of genetic engineering of crops.

sustainable balance among food production, consumer demand, and population growth.[4, 5, 6, 7, 8, 9]

- *The needs of China, then, illustrate several of the themes of this book. We see once again that the human population is a fundamental, underlying problem that needs to be viewed from a global perspective. The desires and demands of the people of China, including an improvement in the material quality of their lives, will put pressure on world food production. This leads to questions about the sustainability of agriculture, both within a nation and worldwide, and ultimately leads to decisions that involve values and science.*

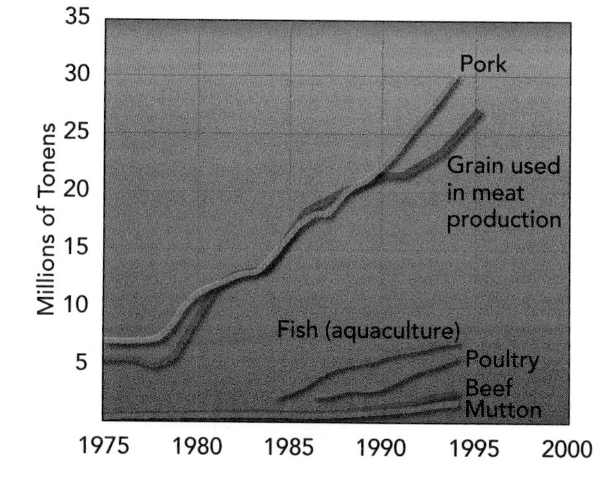

Figure 11.2 • Meat in China. Meat consumption in China, 1975–1994. [*Source*: L. R. Brown, *Who Will Feed China? Wakeup Call for a Small Planet* (New York: Norton, 1995), p. 47 and 51.]

11.1 Can We Feed the World?

Can we produce enough food to feed Earth's growing human population? Can we grow crops sustainably, so that both crop production and agricultural ecosystems remain viable? Can we produce this food without seriously damaging other ecosystems that receive agricultural wastes? These are basic environmental questions about agriculture. To answer them, we must first understand how crops grow and how productive they can be—the topic of this chapter. Then we have to consider the environmental effects of agriculture—the subject of the following chapter.

Of all human activities, agriculture has arguably been proven the most sustainable, simply because people have farmed the Nile Valley, the fertile crescent of the Middle East, rice fields in China, and elsewhere for thousands of years. Few, if any, other human activities have been maintained in the same place for equally long times.

Even so, great concern remains about the sustainability of most agriculture. Where crops have been sustainably produced, farming has changed local ecosystems. Perhaps the most notable exception is the Nile Valley. Before the building of the Aswan High Dam, annual floods deposited new soil every year and farming did not displace this process (Figure 11.3). In most places, however, farming degrades soil; fertilizers and pesticides affect soil, water, and downstream ecosystems; and many other changes occur that are described in this and the next chapter.

The history of agriculture is a series of human attempts to overcome environmental limitations and problems. Each new solution has created new environmental problems, which, in turn, have required their own solutions. Thus, in seeking to improve agricultural systems, people should expect some undesirable side effects and be ready to cope with them.

Figure 11.3 • **Traditional farming along the Nile River, Egypt.**

A surprisingly large percentage of the world's land area is in agriculture: approximately 11% of the world's total land area excluding Antarctica—an area about the size of South and North America combined—enough to make agriculture a human-induced biome (Table 11.1 and Figure 11.4). The percentage of land in agriculture varies considerably among continents, from 30% of the land in Europe to 6% in Australia. In Canada, roughly 11% of the land can support some form of agriculture, but only about 6% is suitable for crop production. The most productive farmland—only about 2% of the country's total land area—is found in the southern prairies of Manitoba, Saskatchewan, and Alberta, and the southern portions of Ontario and Québec.

A major problem in the future is when the human population doubles, agricultural production must double

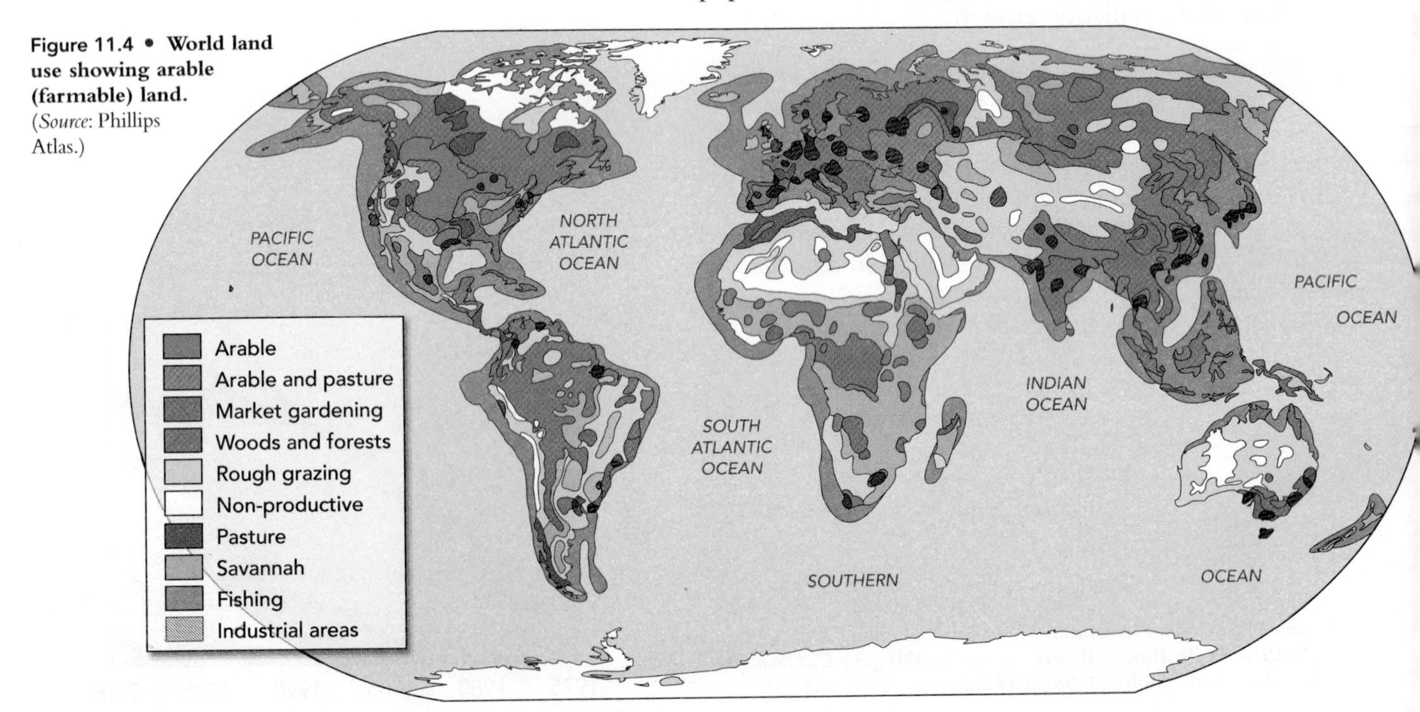

Figure 11.4 • **World land use showing arable (farmable) land.** (*Source*: Phillips Atlas.)

Table 11.1 • Land, People, and Agriculture

Location	Total Land Area (mi²)	Total Land Area (km²)	Human Population	People per km²	Crop Land Area (km²)	Crop Land per Person (ha/person)	Crop Land as % of Total Land
Asia	17,388,686	45,036,697	3,600,000,000	79.93	4,827,520	0.13	11%
Africa	11,715,721	30,343,717	800,000,000	26.36	2,737,040	0.34	9%
N. America	9,529,129	24,680,444	300,000,000	12.16	2,247,000	0.75	9%
S. America	6,878,572	17,815,501	500,000,000	28.07	1,595,940	0.32	9%
Europe	4,065,945	10,530,798	800,000,000	75.97	3,112,050	0.39	30%
Australia	3,405,792	8,821,001	30,000,000	3.40	521,480	1.74	6%
World	52,983,845	137,228,159	6,030,000,000	43.94	15,041,030	0.25	11%

Notes: World land area excludes Antarctica; S. America includes Central America and the Caribbean; Europe includes former USSR states; Australia includes Oceania; Africa cropland includes Middle East. An additional 300 million ha is in pasture and rangeland.
Source: For cropland and populations, *Information Please Almanac*, 1998, p. 487; for total land area, *Random House Atlas of the World*, 2000.

just to meet the present level of per capita food consumption; and for some people, the current food supply is inadequate. If there is not an increase in production per unit area, an additional area equal to the entire New World must become farmland. Where will we find it? Think about biogeography (Chapter 8). Which biomes should be greatly altered? Land best suited for agriculture has already been put to this use. Existing good farmland is already under pressure for conversion to cities, towns, and suburbs. As the human population increases, cities and suburbs will continue to grow (see Chapter 26) and more of the best agricultural land will be converted to other uses. This suggests that future generations will depend not on better farmland and better farming conditions, but on increasingly marginal land. It is a huge challenge.

The world food supply is also greatly influenced by social disruptions and social attitudes, which affect the environment and, in turn, affect agriculture. In Africa, social disruptions since 1960 have included 12 wars, 70 coups, and 13 assassinations. Such social instability makes sustained agricultural yields difficult.[10] So does variation in weather, the traditional bane of farmers.[11, 12]

The key to food production in the future seems to be increased production per unit area, probably on worsening land, and with the risk of increasing environmental damage. Can this be done? Some agricultural scientists and agricultural corporations believe that production per unit area will continue to increase, partially through advances in genetically modified crops (GMCs). This new methodology, however, raises some important potential environmental problems, discussed in Chapter 12. Furthermore, increased production in the past has depended on increased use of water and fertilizers (Figure 11.5). Water is a limiting factor in many parts of the world and will become a limiting factor in more areas in the future (Chapters 18 and 19).

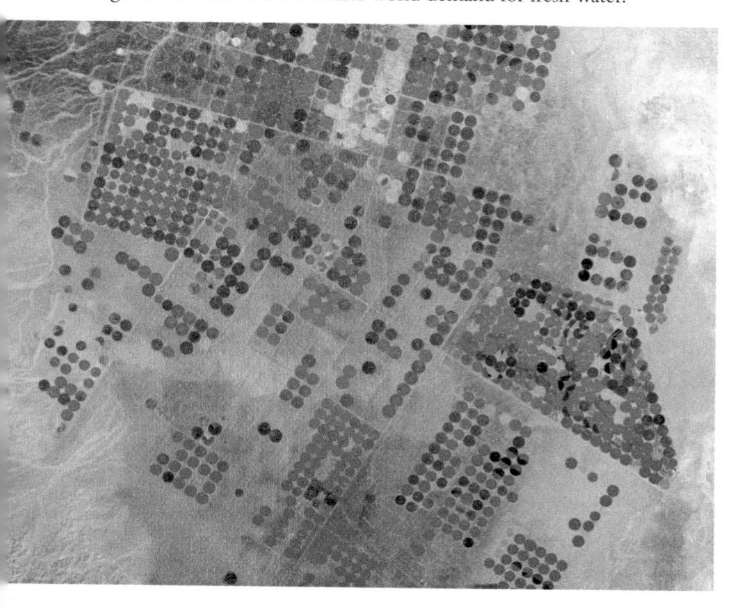

Figure 11.5 • Irrigated farmland in Saudi Arabia as seen from space. Although productive where such agriculture was previously impossible, such irrigation in desert areas increases world demand for fresh water.

11.2 How We Starve

People "starve" in two ways: undernourishment and malnourishment. Undernourishment results from a lack of sufficient calories in available food, so that one has little or no ability to move or work and eventually dies from the lack of energy. Malnourishment results from a lack of specific chemical components of food, such as proteins, vitamins, or other essential chemical elements. Both are global problems.

Widespread undernourishment manifests itself as famines that are obvious, dramatic, and fast-acting. Malnourishment is long term and insidious. Although people may not die outright, they are less productive than normal and can suffer permanent impairment and even brain damage. Among the major problems of undernourishment are marasmus, progressive emaciation caused by a lack of protein and calories; kwashiorkor, a lack of sufficient protein in the diet, which leads to a failure of neural development in infants and, therefore, to learning disabilities (Figure 11.6); and chronic hunger, which occurs when people have enough food to stay alive but not enough to lead satisfactory and productive lives. This means that world food production must provide adequate nutritional quality, not just total quantity. The supply of

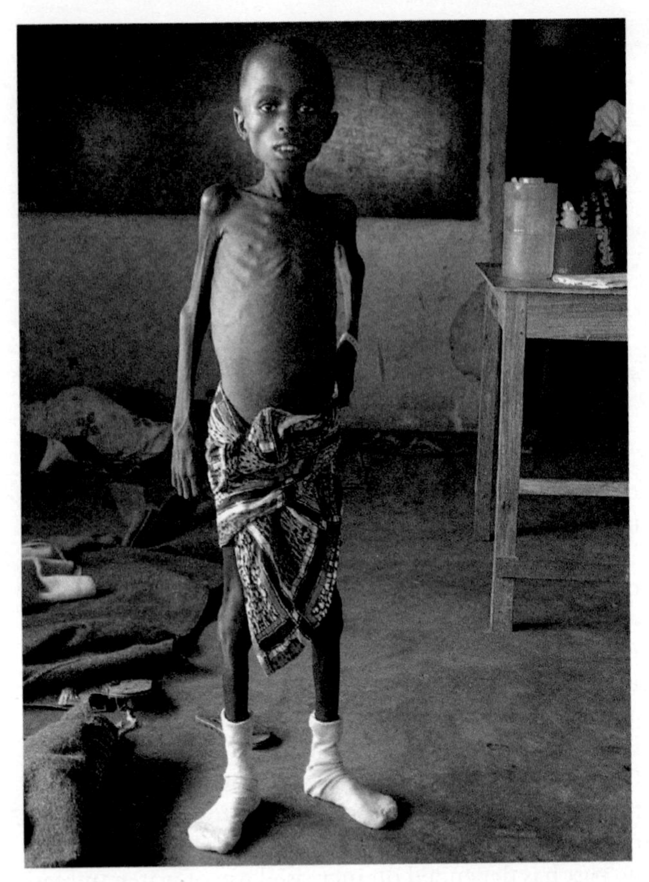

Figure 11.6 • Photograph of a child suffering from kwashiorkor.

protein has been the major nutritional-quality problem. Animals provide the easiest protein food source for people, but depending on animals for protein raises several questions of values, including ecological ones (Is it better to eat lower on the food chain?), environmental ones (Do domestic animals erode soil faster than crops?), and ethical ones (Is it morally right to eat animals?). How people answer these questions affects approaches to agriculture and thereby the environmental effects of agriculture. Again, the theme of science and values arises.

Since the end of World War II, rarely has a year passed without a famine somewhere in the world (Figure 11.7).[13] Food emergencies affected 34 countries worldwide at the end of the twentieth century. Varying weather patterns in Africa, Latin America, and Asia, as well as an inadequate international trade in food, contributed to these emergencies.[14] Africa remains the continent with the most acute food shortages, due to adverse weather and civil strife.[15] Examples include famines in Brazil (1979–1984), Ethiopia (1984–1985), Somalia (1991–1993), and the 1998 crisis in Sudan. The distribution problem is clearly illustrated by these recent famines. Food distribution fails because poor people cannot buy food and pay for its delivery; because transportation is lacking or too expensive; or because food is withheld for political or military reasons. Although there is considerable international trade in food, most of the trade is among rich nations.

Another common solution is food aid among nations, where one nation provides food to another or gives or lends money to purchase food. In the 1950s and 1960s, only a few industrialized countries provided food aid, using stocks of surplus food. A peak in international food aid occurred in the 1960s, when a total of 13.2 million tonnes per year of food was given. A world food crisis in the early 1970s raised awareness of the need for greater attention to food supply and stability. But during the 1980s, donor commitments totaled only 7.5 million tonnes. Then a record level of 15 million tonnes of food aid in 1992–1993 met less than 50% of the minimum caloric needs of the people fed. If food aid alone is to bring the world's malnourished people to a desired nutritional status, an estimated 55 million tons will be required by the year 2010—more than six times the amount available in 1995.[16]

When a group of people starve, the world feels sorrow for them. Humanitarian gestures are important, but such efforts cannot solve the world's food problem. Food aid is a short-term answer. In the long run, when food distribution is the primary problem the best solution is to increase local production. Ironically, food aid can work against increased availability of locally grown food. Free food undercuts local farmers; they cannot compete with it. Availability of food grown locally also avoids disruptions in distribution and the need to transport food over long distances. The only complete solution to famine is to develop long-term sustainable agriculture locally. The old saying "Give a man a fish and feed him for a day; teach him to fish and feed him for life" is true. We must develop and teach agricultural techniques that can be maintained over long periods without using up resources.

11.3 What We Eat and What We Grow

Crops

Of Earth's half-million plant species, only about 3,000 have been used as agricultural crops and only 150 species have been cultivated on a large scale. In Canada and the United States, 200 species are grown as crops. Most of the world's food is provided by only 14 crop species. In approximate order of importance, these are wheat, rice, maize, potatoes, sweet potatoes, manioc, sugarcane, sugar beet, common beans, soybeans, barley, sorghum, coconuts, and bananas (Figure 11.8). Of these, six provide more than 80% of the total calories consumed by people either directly or indirectly (Figure 11.9).[17]

Some crops, called forage, are grown as food for domestic animals. These include alfalfa, sorghum, and various species of grasses grown as hay. Alfalfa is the most important forage crop in Canada and the United

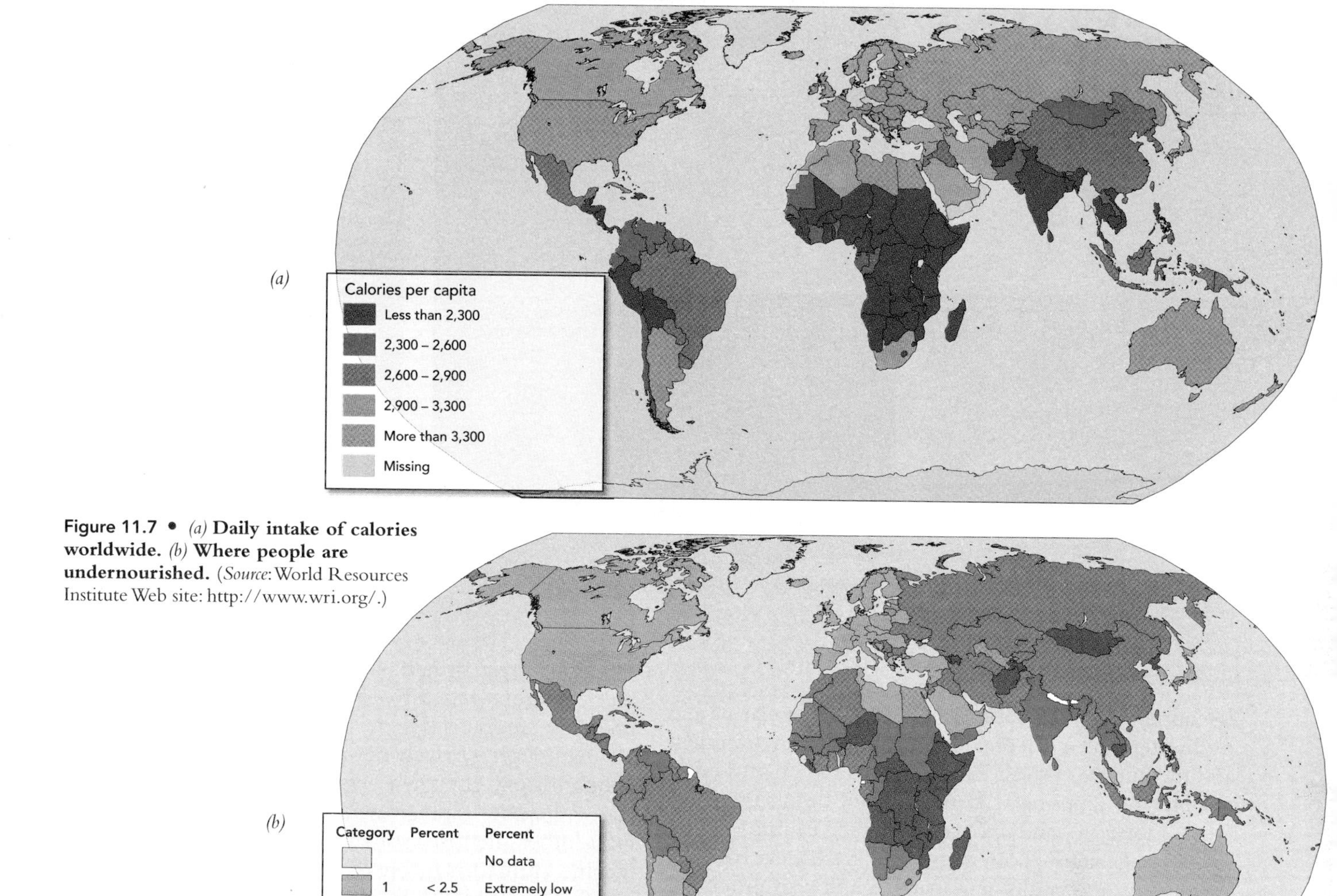

Figure 11.7 • *(a)* **Daily intake of calories worldwide.** *(b)* **Where people are undernourished.** (*Source:* World Resources Institute Web site: http://www.wri.org/.)

States, which together have almost 20 million ha planted in alfalfa—over one-half the world's total.

Worldwide, people keep 14 billion chickens, 1.3 billion cattle, more than 1 billion sheep, more than a billion ducks, almost a billion pigs, 700 million goats, more than 160 million water buffalo, and about 18 million camels (Figure 11.10).[18] These important food sources have a major impact on the land, as discussed in Chapter 12. Interestingly, the number of cattle in the world has not increased in the past 10 years, the number of sheep has declined slightly (nearly 100,000), and the number of goats increased from 600 million in 1992 to 750 million in 2002. The production of beef, however, increased from 53 million tonnes (*t*) in 1992 to 57 million *t* in 2002. During the same period, the production of meat from chickens increased greatly, from 40 million *t* to 60 million *t*, and meat from pigs increased from 72 million *t* to more than 90 million *t*.

Most cattle live on rangeland or pasture. **Rangeland** provides food for grazing and browsing animals without ploughing and planting; **pasture** is ploughed, planted, and harvested to provide forage for animals. More than 34 million km^2 are in permanent pasture worldwide—an area larger than the combined sizes of Canada, the United States, Mexico, Brazil, Argentina, and Chile.[18]

There is a large world trade in small grains. Only Canada, the United States, Australia, and New Zealand are major exporters; the rest of the world's nations are net importers. In 2002, world small-grain production was slightly more than 2 billion tonnes, slightly below a record crop of 2.1 billion tonnes in 2001.[18] World small-grain production was 0.8 billion tonnes in 1961, reached 1 billion in 1966, then doubled to 2 billion in 1996, a remarkable increase in 30 years. But production has remained relatively flat since then. The question we must ask, and cannot answer at this time, is whether this means that the world's carrying capacity for small grains has been reached or simply that the demand is not growing (Figure 11.11).

Figure 11.8 • **Some of the world's major crops**, including *(a)* wheat, *(b)* rice, and *(c)* soybeans. See text for a discussion of the relative importance of these three crops.

Aquaculture

In contrast to food obtained on the land, most marine and freshwater food is still obtained by hunting. Environment and the hunt for fish in Earth's fisheries are discussed in Chapter 14. Hunting wild fish has not been sustainable (see Chapter 14), and **aquaculture**, the farming of food in aquatic habitats—both marine and freshwater is growing rapidly, and could be a major solution to providing nutritional quality. Popular aquacultural products include carp, tilapia, oysters, and shrimp, but in many nations other species are farm-raised and culturally important, such as salmon, trout and mussels in Canada, yellowtail (important in Japan); crayfish (United States); eels and minnows (China); catfish (southern and midwestern United States); plaice, sole, and the Southeast Asian milkfish (Great Britain); and sturgeon (Ukraine). A few species—trout and carp—have been subject to genetic breeding programs (Figure 11.12).[19]

Although relatively new in Canada, aquaculture has a long history elsewhere, especially in China, where at least 50 species are grown, including finfish, shrimp, crab, other shellfish, sea turtles, and sea cucumbers (a marine animal).[20] In the Szechwan area of China, fish are farmed in more than 100,000 ha of flooded rice fields. This is an ancient practice that can be traced back to a treatise on fish culture written by Fan Li in 475 B.C.[19] Build Your Environmental Skills 9.1 described how carp are raised in an integrated agroforestry-aquaculture system in southern China.

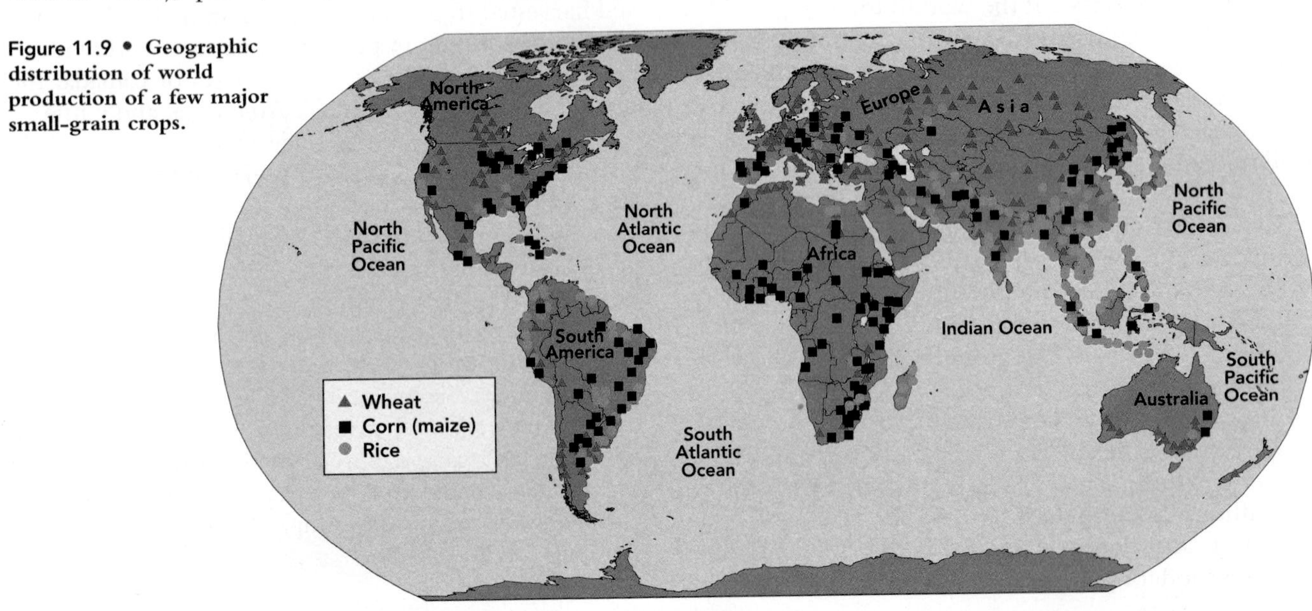

Figure 11.9 • **Geographic distribution of world production of a few major small-grain crops.**

Aquaculture can be extremely productive on a per area basis, especially because flowing water brings food into the pond or enclosure from outside. Although the area of Earth that can support freshwater aquaculture is small, we can expect this kind of aquaculture to increase in the future and become a more important source of protein. In China and other Asian countries, farmers often grow several species of fish in the same pond, exploiting their different ecological niches. Ponds developed mainly for carp, a bottom-feeding fish, also contain minnows, which feed at the surface on leaves added to the pond.

Sometimes fishponds use otherwise wasted resources, such as fertilized water from treated sewage; fishponds exist in natural hot springs (Idaho) and contain warmed water used in cooling electric power plants (Great Britain).[19]

Mariculture, the farming of ocean fish, although producing a small part of the total marine fish catch, has grown rapidly in past decades and will likely continue to do so. Mariculture of abalone and oysters, whose natural production is limited, is increasing. In the United States and Canada, for example, researchers are working to learn how to attract the young, swimming stages of these shellfish to areas where they can be conveniently grown and harvested.

Oysters and mussels are grown on rafts that are lowered into the ocean, a common practice in the Atlantic Ocean in Portugal and in the Mediterranean in such nations as France. As filter feeders, these animals obtain food from water that moves past them in currents. Because a small raft is exposed to a large volume of water, and thus a large volume of food, rafts can be extremely productive. Mussels grown on rafts in bays of Galicia, Spain, produce 300 tonnes per hectare, whereas public harvesting grounds of wild shellfish in Canada and the United States yield only about 10 kg/ha.[19] Oysters and mussels are also grown on rafts and artificial pilings in the intertidal zone in coastal British Columbia and Washington state (Figure 11.13).

Aquaculture operations have often been controversial in Canada, because of the potential impact on water quality— for instance nutrient enrichment and addition of the pharmaceutical chemicals used in fish and shellfish rearing—and aesthetics. In inland regions and some coastal areas, fish are grown in on-land tanks, whose wastewaters can be carefully controlled and treated to reduce pollution.

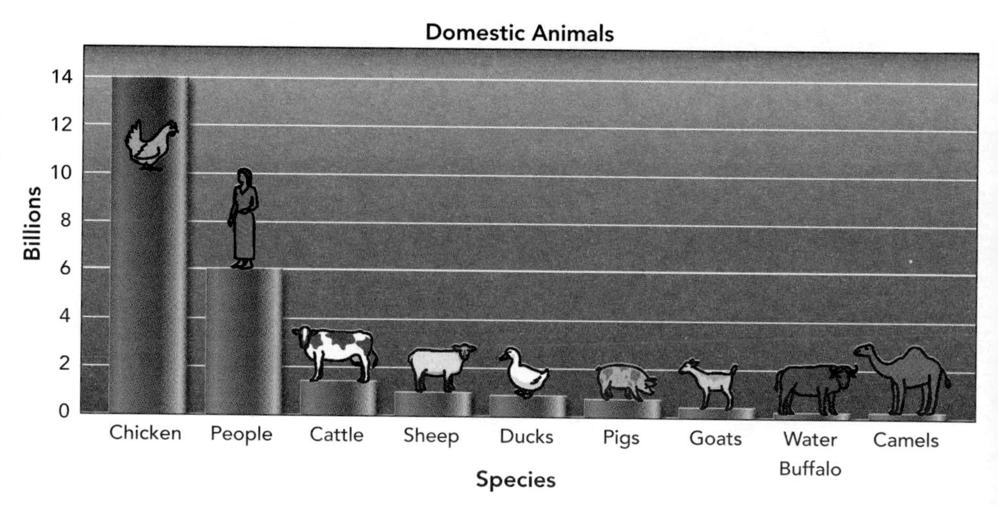

Figure 11.10 • **Billions of domestic animals compared to numbers of human beings.** There are more than three chickens per person in the world, and about one cow for every five people.

11.4 An Ecological Perspective on Agriculture

Farming creates novel ecological conditions (Figure 11.14). These **agroecosystems** differ from natural ecosystems in six ways.

- In farming we try to stop ecological succession and keep the agroecosystem in an early successional state (see Chapter 10). Most crops are early successional species, which means that they do best when sunlight, water, and chemical nutrients in the soil are abundant. Under natural conditions, later successional plants would eventually replace crop species. Any attempt to prevent natural successional processes from occurring requires time and effort on our part. Most crops

Figure 11.11 • **World small-grains production since 1983.** (*Source*: FAO statistics FAOSTATS web site.)

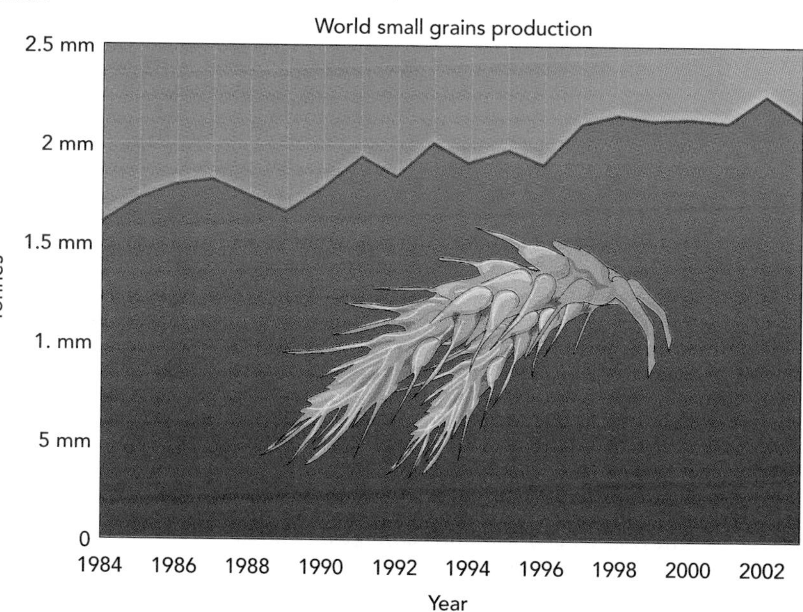

Figure 11.12 • Aerial photograph of fish farms at Lake San Isabel on Luzon Island in Philippines.

Figure 11.13 • An oyster farm in British Columbia. Oysters are grown on artificial rafts in the intertidal zone.

are planted on cleared land, which is then kept clear of other vegetation. In contrast, when a clearing is created by a natural disturbance, such as a fire or a storm, the vegetation returns—first early successional species and then later ones.

- Most agriculture involves **monoculture**—the planting of a single species, or even a single strain or subspecies, such as a single hybrid of corn. Monoculture can make the entire crop vulnerable to attack by a single disease or a single change in environmental conditions. Repeated planting of a single species can reduce the soil content of certain essential elements, thereby reducing overall soil fertility. This can be counteracted to a certain extent by artificial fertilizers and by the ancient practice of **crop rotation**. In crop rotation, different crops are planted in turn in the same field, with the field occasionally left fallow. A fallow field is allowed to grow with a cover crop (sometimes planted, sometimes whatever germinates) that is not harvested for at least one season. Often the vegetation that grows on the fallow field is ploughed under to add to soil fertility.
- Crops are planted in neat rows, which makes life easy for pests, because the crop plants have no place to hide. In natural ecosystems, many plant species grow mixed together in complex patterns, so it is harder for pests to find their favorite victims.
- Farming greatly simplifies biological diversity and food chains. Most pest-control methods reduce the abundance and diversity of natural predators and make the agroecosystem more susceptible to undesirable changes.
- Ploughing is unlike any natural soil disturbance—nothing in nature repeatedly and regularly turns over the soil to a specific depth. Ploughing exposes the soil to erosion and damages its physical structure, leading to a decline in organic matter and a loss of chemical elements to erosion. This is illustrated by studies in southwestern Nova Scotia, of previously ploughed land that was abandoned for agriculture. Instead of returning to its original mixed woodlands, the ploughed land is colonized by pure stands of white pine or undergoes succession to mixed forest much more slowly than in nontilled land.
- The newest difference is genetic modification of crops—a novel situation.

11.5 Limiting Factors

High-quality agricultural soil has all the chemical elements required for plant growth and a physical structure that lets both air and water move freely through the soil, yet it retains water well. The best agricultural soils have a high organic content and a mixture of sediment particle sizes. Small particles, especially fine clays, help to retain moisture and chemical elements; larger particles, including sand and pebbles, help the flow of water. But different crops require different soils. Lowland rice grows in flooded ponds and requires a heavy, water-saturated soil, while watermelons grow best in very sandy soil.

Soils rarely have everything a crop needs. Canada's glacial history has created a variety of soils of varying nutrient and organic content. In any given region, the question for a farmer is: What needs to be added or done to make a soil more productive for a crop? The traditional answer is that, at any time, just one factor is limiting. If that factor can be improved, the soil will be more productive; if that single factor is not improved, nothing else will make a difference. The idea that some single factor determines the growth and therefore the presence of a species is known as *Liebig's law of the minimum*, after Justus von Liebig, a nineteenth-century agriculturalist who is credited with first stating this idea. He knew that crops required a number of nutrients in the soil and that adding these as fertilizers could increase crop yields.

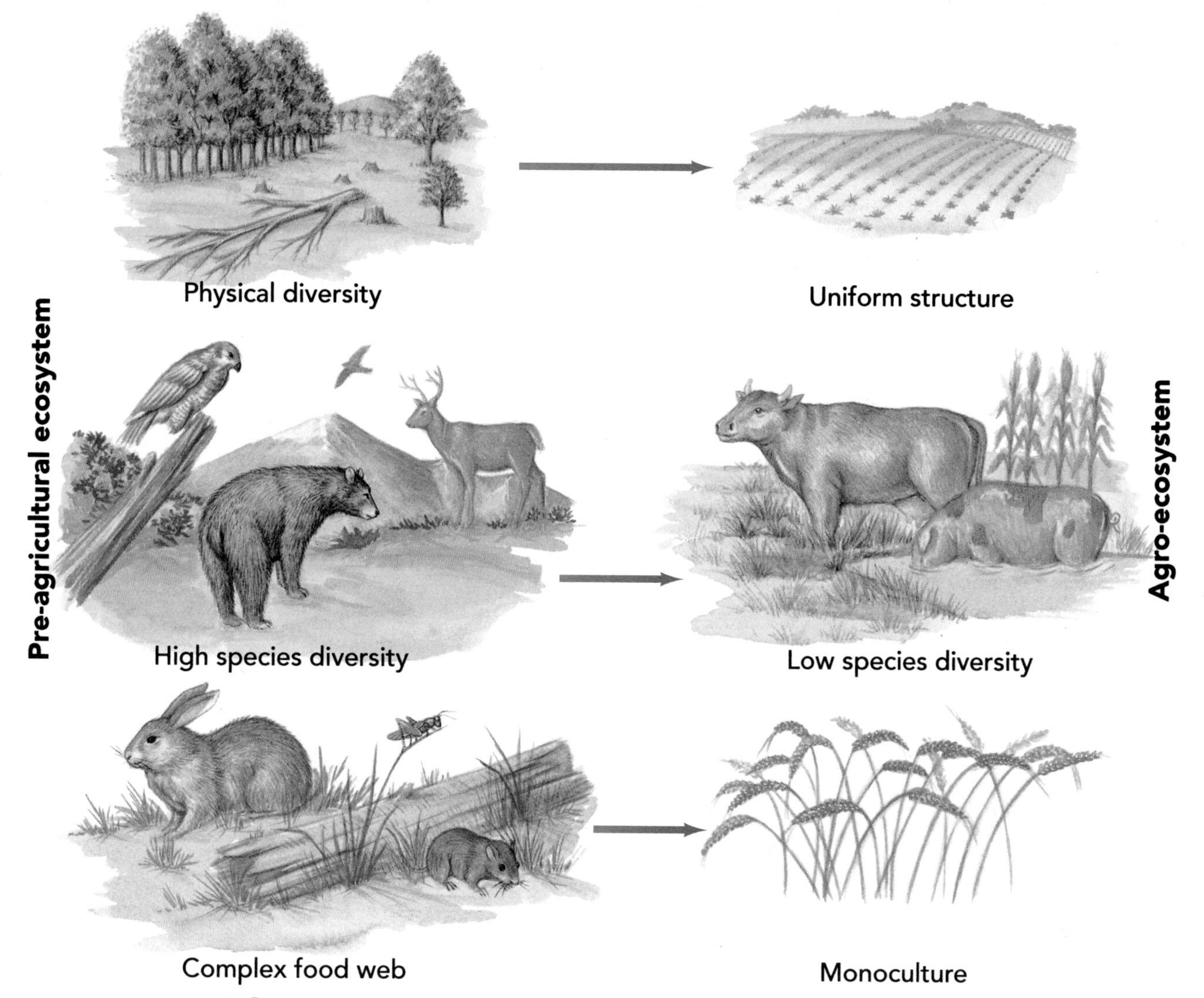

Figure 11.14 • How farming changes an ecosystem. It converts complex ecosystems of high species diversity and structural diversity to a monoculture of uniform structure. The soil is greatly modified. See text for additional information about agricultural effects on ecosystems.

However, the factor that caused an increase varied from time to time and place to place. A general statement of Liebig's law is: the growth of a plant is affected by one limiting factor at a time—the one whose availability is the least in comparison to the needs of a plant.

But the reality can be much more complicated. Crops require about 20 chemical elements. These must be available in the right amounts, at the right times, and in the right proportions to each other. It is customary to divide these life-important chemical elements into two groups, macronutrients and micronutrients. A macronutrient is a chemical element required by all living things in relatively large amounts. Macronutrients are sulfur, phosphorus, magnesium, calcium, potassium, nitrogen, oxygen, carbon, and hydrogen. A micronutrient is a chemical element required in small amounts—either in extremely small amounts by all forms of life or in moderate to small amounts for some forms of life. Micronutrients are often rarer metals, such as molybdenum, copper, zinc, manganese, and iron. (Macronutrients and micronutrients are also discussed in Chapter 4.)

If Liebig were always right, then environmental factors would always act one by one to limit the distribution of living things. However, there can be exceptions to this rule. For example, nitrogen is a necessary part of every protein, and proteins are essential building blocks of cells. Enzymes, which make many cell reactions possible, contain nitrogen. A plant given little nitrogen and phosphorus might not make enough of the enzymes involved in taking up and using phosphorus. Increasing nitrogen to the plant might increase the plant's uptake and use of phosphorus. If this were the case, the two elements would have a synergistic effect. In a synergistic effect, a change in availability of one resource affects the response of an organism to another resource.

So far we have discussed effects of chemical elements when they are scarce. But it is possible to have too much of a good thing—most chemical elements become toxic when they are present in high concentrations. As a simple example, plants die when they have too little water but also when they are flooded, unless they have specific adaptations to living in water. So it is with chemical elements required for life. In this section, we have discussed the problems of too little; we discuss the problems of too much in Chapter 15.

The older a soil, the more likely it is to lack trace elements, because as a soil ages its chemical elements may be leached by water from the upper layers to deeper layers (see Chapter 10). When crucial chemical elements are leached below the reach of crop roots, the soil becomes infertile. Striking cases of soil-nutrient limitations have been found in Australia, which has some of the oldest soils in the world—on land that has been above sea level for many millions of years, during which time severe leaching has taken place. Sometimes trace elements are required in extremely small amounts. For example, in certain Australian soils it is estimated that adding 30 grams of molybdenum to a field increases the yield of grass by 1 tonne per year.

The idea of a limiting growth factor, originally used in reference to crop plants, has been extended by ecologists to include all life requirements for all species in all habitats.

11.6 The Future of Agriculture

We can identify three major technological approaches to agriculture. One is modern mechanized agriculture, where production is based on highly mechanized technology that has a high demand for resources—including land, water, and fuel—and makes little use of biologically based technologies. Another approach is resource-based agriculture, which is based on biological technology and conservation of land, water, and energy. An offshoot of the second is organic food production, where crops are grown without artificial chemicals (including pesticides), where genetic engineering of crops is not used, and where ecological control methods are employed. The third is bioengineering.

In mechanized agriculture, production is determined by economic demand and limited by that demand, not by resources. In resource-based agriculture, production is limited by environmental sustainability and the availability of resources and economic demand usually exceeds production (Figure 11.15).

The history of agriculture can be summarized most simply as four stages:

- Resource-based agriculture and what we now call organic agriculture were introduced about 10,000 years ago.
- A shift to mechanized, demand-based agriculture occurred during the Industrial Revolution of the eighteenth and nineteenth centuries.
- A return to resource-based agriculture began in the twentieth century, using new technologies.
- Today there is a growing interest in organic agriculture as well as a potentially large-scale use of genetically engineered crops. (See A Closer Look 11.2.)

What can be done to help crop production keep pace with human population growth? Since there are so many plant species, perhaps some yet unused ones could provide new sources of food and grow in environments little used for agriculture. Those interested in the conservation of biological diversity urge a search for new crops because that this is one utilitarian justification for the conservation of species. Another suggestion is that some of these new crops may be easier on the environment and more likely to allow sustainable agriculture. However, it may be that over the long history of human existence, those edible species have been found, and the number is small. Research is underway to seek new crops or plants that have been eaten locally but whose potential for widespread, intense cultivation has not been tested. Among current candidates are guayule, crambe, pigeon peas, and grain amaranth.[24]

11.7 Increasing the Yield per Hectare

In the nineteenth century, advances in selective breeding and mechanization of agriculture allowed much higher agricultural productivity than had been the case in previous eras. During the twentieth century, rapid strides were made in increasing crop production per unit area. Some of these advances, which involved the development of new hybrids, became known as the green revolution. Several other methods of increasing the food supply may hold some promise as well, although always with limitations. Future increases in agricultural production will likely come mainly from the development of high-yield strains of crops.

The Green Revolution

The green revolution is the name attached to post–World War II programs that have led to the development of new strains of crops with higher yields, better resistance to disease, or better ability to grow under poor conditions. One advancement of the green revolution was the development of superstrains of rice at the International Rice Research Institute in the Philippines (see Figure 11.17). Although hybridization of rice vastly increased rice production per hectare, the new strains required greater use of fertilizers and as much as four to seven times the water. In some cases, they produced a rice that was not considered desirable to eat. Another development in the green revolution was strains of maize with improved disease resistance at the International Maize and Wheat Improvement Center in Mexico.

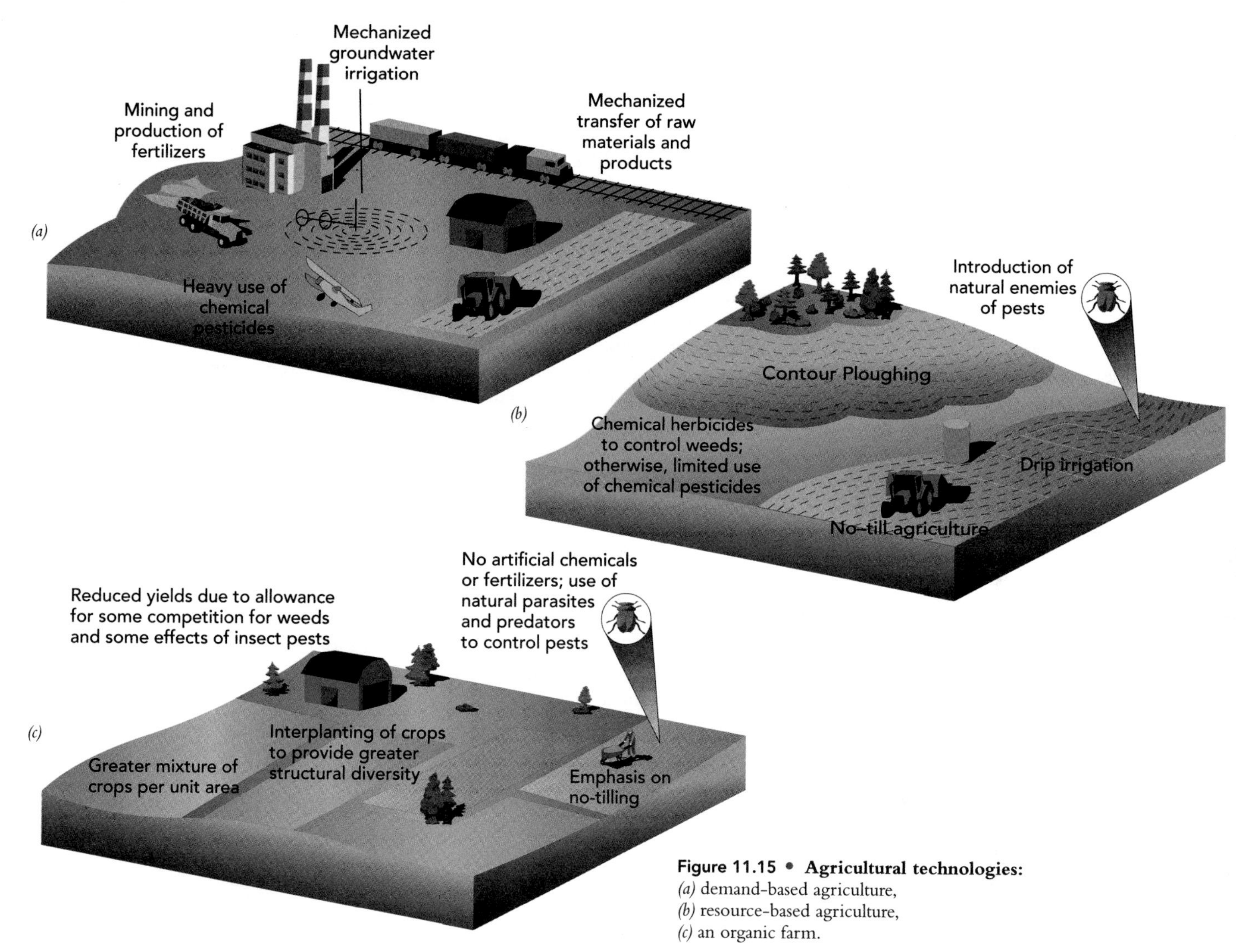

Figure 11.15 • Agricultural technologies: *(a)* demand-based agriculture, *(b)* resource-based agriculture, *(c)* an organic farm.

Improved Irrigation

Better irrigation techniques could improve crop yield and reduce overall water use. Drip irrigation—from tubes that drip water slowly—greatly reduces the loss of water from evaporation and increases yield. However, it is expensive and most likely to be used in developed nations or nations with a large surplus of hard currency—in other words, in few of the countries where hunger is most severe.

Some people suggest that in the future we will rely increasingly on artificial agriculture, such as hydro ponics, which is the growing of plants in a fertilized water solution on a completely artificial substrate in an artificial environment, such as a greenhouse. This approach is extremely expensive and unlikely to be effective where hunger is the greatest.

11.8 Organic Farming

Future farming must be easier on the environment than past farming if agriculture is to be widely sustainable. Organic farming is an example of one solution. Organic farming is typically considered to have three qualities: It is more like natural ecosystems than monocultures; it minimizes negative environmental impacts; and the food that results from it does not contain artificial compounds. Organic farming has been one of the fastest-growing sectors in Canadian agriculture, although it still occupies a small fraction of Canadian farmland and contributes only a small amount of agriculture income. In 2003, there were 3,317 certified organic producers in Canada, representing 1.3% of all farms in the country. With 1,049 certified producers, Saskatchewan has almost half the organic farms in Canada. Québec, Ontario, and British Columbia are also important organic producers, reporting 793, 487, and 420 organic farms, respectively, in 2003. There are also regional differences in organic crops.[25, 26] About half the Québec organic producers are maple syrup operations; in Saskatchewan, almost 20% of organic growers produce wild rice. Roughly a quarter of Canada's organic farms are small, with less than $10,000 per year in gross sales. Almost 46% of Canadian organic farms are considered "large"—that is, having more than $50,000 in gross sales. By contrast, *average* income for Canada's traditional potato, cattle, dairy, swine, and poultry farms approaches $100,000 per year. To be certified

A CLOSER LOOK 11.1

Traditional Farming Methods

In industrialized countries of temperate zones, there is a long history of ploughing to clear land for agriculture; but in less industrialized, tropical areas, there is a history of agricultural methods that depend on clearing the vegetation without ploughing. Where the loss of nutrients from the soil occurs rapidly following clearing, as in some tropical rain forests, the traditional practice is to cut the forest in small patches but not cut it completely (Figure 11.16). Some shrubs and herbaceous plants are left. Several crops are planted together among existing vegetation. Crops are harvested for a few years. Then the land is allowed to grow back to a forest. The natural process of secondary succession—the redevelopment of the ecosystem—is allowed to occur. In fact, these farming practices promote this redevelopment and increase the conservation of chemical elements in the ecosystem. After the forest has grown back, the process is repeated.

This kind of agriculture has many names. It is sometimes called cultivation with forest or bush fallow. In Latin America, it is called milpa agriculture; in Great Britain, swidden agriculture; in western Africa, fang agriculture. In this type of agriculture, a mixture of crops is utilized, including root, stem, and fruit crops. For example, in western Africa, fang agriculture includes yams (a root crop) plus maize; in Southeast Asia, root crops are grown with rice and millet or with rice and maize.[21]

In theory, this method could be sustainable if human population density remained low. Erosional losses are minimized, and the soil eventually recovers its fertility. Uncut vegetation provides future seed sources. When human population pressures are low, a longer time occurs between the periods of use of any site. This is known as a long rotation period. Under high population pressures, such as occur in many places today, the rotation period is much shorter, and the land may not be able to recover sufficiently from previous usage. In such cases, production is not sustainable.

For many years, agricultural experts from developed nations viewed this method of agriculture as a poor process with low, short-term productivity, used only by primitive peoples. Now it is understood that this kind of agriculture is well suited to high-rainfall lands where soils readily become impoverished when the land is completely cleared. The mixture of crops allows different species to contribute to soil fertility in different ways. Some perennial plants slow physical erosion; native legumes add nitrogen to the soil; and so forth.

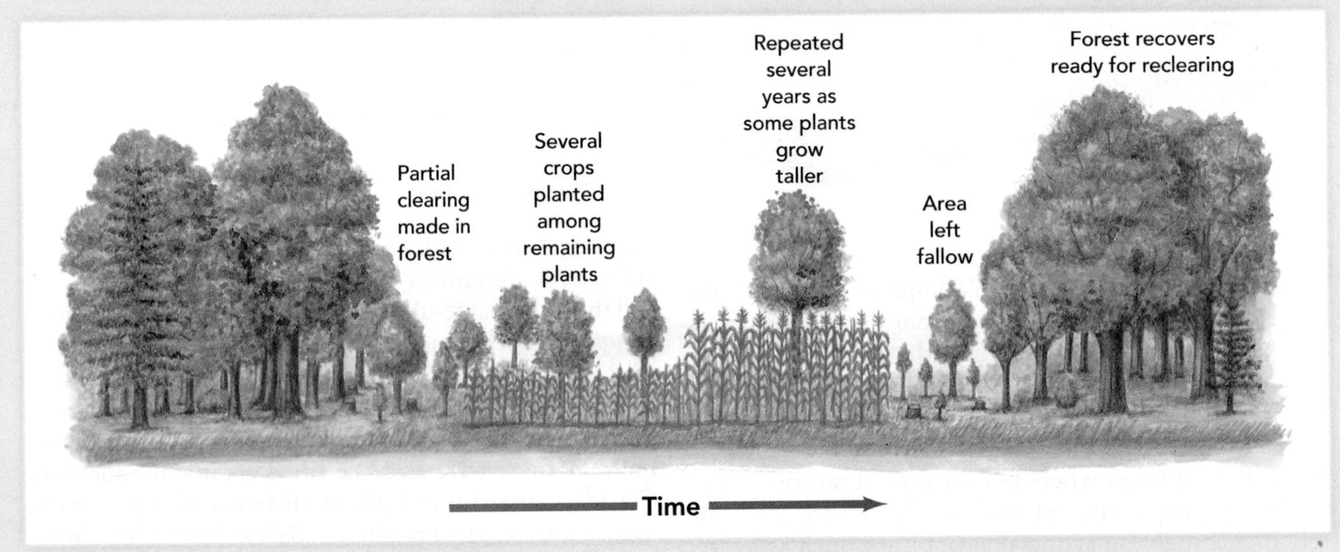

Figure 11.16 • Bush fallow, also called milpa, fang, or swidden agriculture. Over time, secondary succession will take place on land partially cleared by slash and burn.

organic, a farm must comply with the Canadian National Standard for Organic Agriculture, which was approved by the Standards Council of Canada in 1999.

11.9 Alternatives to Monoculture

An important trade-off is implicit in the choice to plant a single hybrid. Each year, seed companies make use of climate forecasts for the growing season and knowledge of the most likely strains of diseases and insect pests in the area to be planted. They then develop hybrids resistant to those strains and well adapted to the forecast weather. If the predictions are correct, crop production can be very high. If the predictions are incorrect, crop production in the entire area can be very low.

An alternative is to plant a mixture of crops and/or a broad range of genotypes at a particular time and place. This approach is typical of the pre-industrial agriculture

A CLOSER LOOK 11.2

Potential Future Advances in Agriculture

New Genetic Strains and Hybrids

From the beginning, farming has affected the genetics of domesticated animals and plants, as people selected strains that were easy to grow and harvest (Figure 11.17). The act of farming makes certain species abundant where they were rare; thus, farming selectively favours certain genotypes (populations with certain genetic characteristics). Features that make a species or a genotype a weaker competitor under natural conditions sometimes make it more desirable as a crop. For example, wild relatives of wheat lose their seeds when the seeds are ripened and gently shaken by wind or animals. If you try to cut wild wheat stems and take them home, most of the seeds will be shaken off and few will be left when you arrive. This is an adaptation that helps spread the seeds.

Figure 11.17 • Experimental rice plots at the International Rice Research Institute, Philippines, showing visual crop variation based on the use of fertilizers.

Some individuals of wild wheat have a mutation that makes the seeds remain on the stalk. In the wild, these mutants leave fewer offspring and do not persist in the population—they are less fit genetically (see Chapter 7). Early farmers selected these mutants because they were easier to collect, transport, and use. In this way, people changed selective pressures on wheat and hastened the evolution of useful wheat strains that could not survive on their own. The result has been a sort of symbiotic relationship between people and these otherwise less competitive forms of wheat.

People also domesticated wheat by moving it to habitats to which it was not originally adapted, then developing new strains that could persist in these environments and also changing the environmental conditions of the new habitat to better suit the wheat.[22] Corn (maize) went through a similar process of domestication. Likewise, domestic animals have been bred to make them more docile and better producers of the meat and dairy products that we prefer. Modern agriculture has carried this process a step further with the frequent intentional development of hybrids of different genotypes, bred to overcome new strains of disease and changes in climate.

New Crops

Developing new crops by domesticating species that are currently wild offers considerable potential. Although new crops are unlikely to replace current crop species as major food sources in this century, there is great interest in new crops to increase production in marginal areas and to increase the production of non-food products such as oils. The development of new crops has been a continuing process in the history of agriculture. As people spread around the world, new crops were discovered and transported from one area to another. The process of introduction and increase in the production of crops continues.[23]

Among the likely candidates for new crops are amaranth for seeds and leaves; Leucaena, a legume useful for animal feed; and triticale, a synthetic hybrid of wheat and rye. A promising source of new crops is the desert; none of the 14 major crops are plants of arid or semiarid regions, yet there are vast areas of desert and semi-desert. There are about 450 million hectares of arid land in North America, two-thirds of it moderately desertified and the remainder severely desertified. In Africa, Australia, and South America, the areas are even greater. Several species of plants can be grown commercially under arid conditions, allowing us to use a biome for agriculture that has been little used in this way in the past. Examples are guayule (a source of rubber), jojoba (for oil), bladderpod (for oil from seeds), and gumweed (for resin). Jojoba, a native shrub of the American Sonoran Desert, produces an extremely fine oil, remarkably resistant to bacterial degradation, which is useful in cosmetics and as a fine lubricant. Jojoba is now grown commercially in Australia, Egypt, Ghana, Iran, Israel, Jordan, Mexico, Saudi Arabia, and the United States.[21]

still found in many developing countries, and organic farmers promote it today. It gives lower average yearly production but reduces the risk of very low production years. Monoculture trades off long-term stability for the opportunity to have very high production in one year. Which approach to choose is a question of values and therefore part of the theme of science and values.

Regions of Earth differ greatly in their capacity for crop production. Factors that influence which crops are produced in which areas include tradition, access to technology and supplies, and local politics. In Canada alone, provinces differ greatly in their crop production (Figure 11.18). Saskatchewan produces more than half of Canadian Western Red Spring (CWRS) wheat, the principal class of wheat grown in Canada. Albert and Manitoba are second, with 29% and 20% of total CWRS production. By contrast, Ontario leads the country in soybean production. With a 2002 total production of 1.91 million tonnes, Ontario dwarfed Québec (315,000 tonnes) and Manitoba (108,900 tonnes), the next largest producers.[27, 28]

In some parts of Canada, there is a movement to reintroduce flavourful older grain varieties such as Red Fife wheat, which were once an important part of Canada's agricultural economy but are now almost extinct. Flour and baked goods made from these heritage strains are popular products in specialty bakeries and restaurants.

In the prairie provinces (where shortgrass prairie originally grew), rangelands and irrigated farmlands are common. There, irrigated cropland is mainly planted in small grains (barley, oats, wheat, etc.). Farther east in Ontario and Québec are areas where corn and soybeans are most important. Those two provinces also produce more than three-quarters of the country's dairy cattle. About 72% of Canada's beef production occurs in Western Canada. In the warmest regions of the country, particularly British Columbia and southern Ontario, major crops are vegetables and fruits, including grapes for eating and wine production.

11.10 Eating Lower on the Food Chain

Some people believe that it is ecologically unsound to use domestic animals as food on the grounds that eating each step farther up a food chain leaves much less food to eat per hectare. This argument is as follows (you will remember this from the Critical Thinking Issue in Chapter 9): No organism is 100% efficient. Only a fraction of the energy in food taken in is converted to new organic matter. Crop plants may convert 1–10% of sunlight to edible food, and cows may convert only 1–10% of hay and grain to meat. Thus, the same area could produce 10–100 times as much vegetation per year as meat (see the Critical Thinking Issue in Chapter 9). This holds true for the best agricultural lands, which have deep, fertile soils on level ground.

However, as with so many issues, a simple generalization does not apply to all cases. Land too poor for crops that people can eat can be excellent rangeland, with grasses and woody plants that domestic livestock can eat (Figure 11.19 and Figure 11.20). These lands occur on steeper slopes, with thinner soils or with less rainfall. Thus, from the point of view of sustainable agriculture, there is value in rangeland or pasture. The wisest approach to sustainable agriculture involves a combination of different kinds of land use: using the best agricultural lands for crops, using poorer

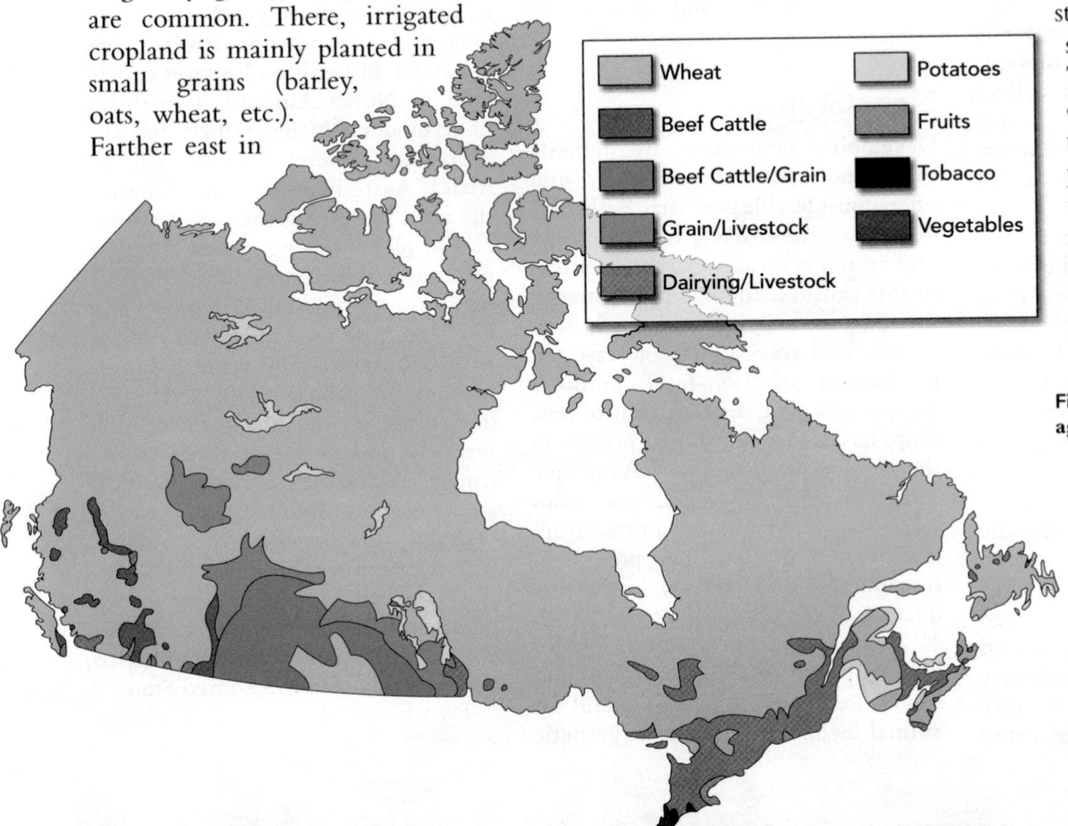

Figure 11.18 • Major types of agricultural production in Canada.

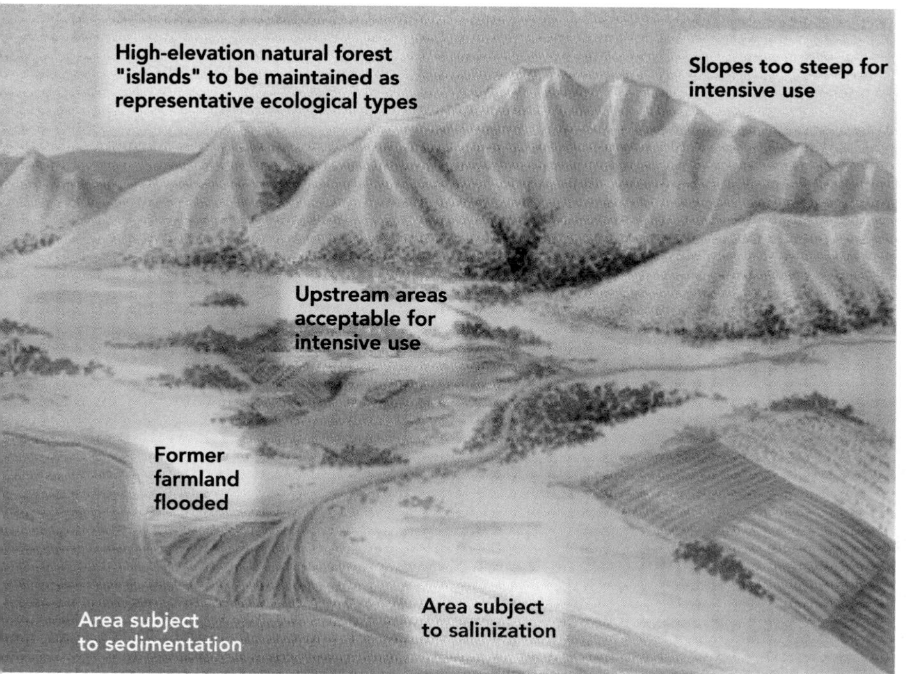

Figure 11.19 • Physical and ecological considerations in watershed development—such as slope, elevation, floodplain, and river delta location—limit land available for agriculture.

lands for pastures and rangelands, and avoiding the use of the best lands for grain production for animal feed.

Another problem with the argument that we should eat lower on the food chain is that food is more than just calories, and animals are important sources of proteins and minerals. Animals provide the major source of protein in human diets—56 million tonnes of edible protein per year worldwide. In Canada and the United States, 75% of the protein, 33% of the energy, and most of the calcium and phosphorus in human nutrition come from animal products.

A third factor to keep in mind is that domestic animals are often used for other purposes; such as ploughing, carrying loads, and transportation; and they are sources of wool and leather as well as food. In addition, their excrement is an important fertilizer and, in some areas of the world, an important—sometimes the only—fuel for fire. In these ways, the use of animals as part of food production represents an increase in efficiency.

Some people maintain vegetarian diets because of specific dietary problems. Others do not eat meat for moral, ethical, or religious reasons. Thus, the decision about when and whether to use animals as part of food production is an issue of both science and values.

11.11 Genetically Modified Food: Biotechnology, Farming, and Environment

The discovery that DNA is the universal carrier of genetic information has led to the development of a completely new technology known as genetic engineering. Now that the chemistry of inheritance is understood, scientists have been able to develop methods to transfer specific genetic characteristics from one individual to another, from one population to another, and from one species to another. This has led to the genetic modification of crops, with major implications for agriculture. The development and use of genetically modified crops (GMCs) has given rise to new environmental controversies as well as a promise of increased agricultural production.

Genetic engineering in agriculture involves several different practices, which we can group as follows: faster and more efficient ways to develop new hybrids; the introduction of the "terminator gene" (discussed in the following chapter); and the transfer of genetic properties from widely divergent kinds of life. These three practices have quite different potentials and problems. Of the three, hybridization is the least novel—crop hybridization has become a standard methodology in modern agriculture; biotechnology adds a new way to create hybrids, but it is not an entirely new process. In contrast, the second and third practices were never previously possible.

There is considerable interest in the potential for genetic engineering to develop strains of crops with entirely new characteristics. One focus of this research is the development of new crops that have the same symbiotic relationship found in legumes (members of the pea family) so that they "fix" nitrogen (convert atmospheric

Figure 11.20 • Land that is unsuitable for crop production on a sustainable basis can be used for other purposes.

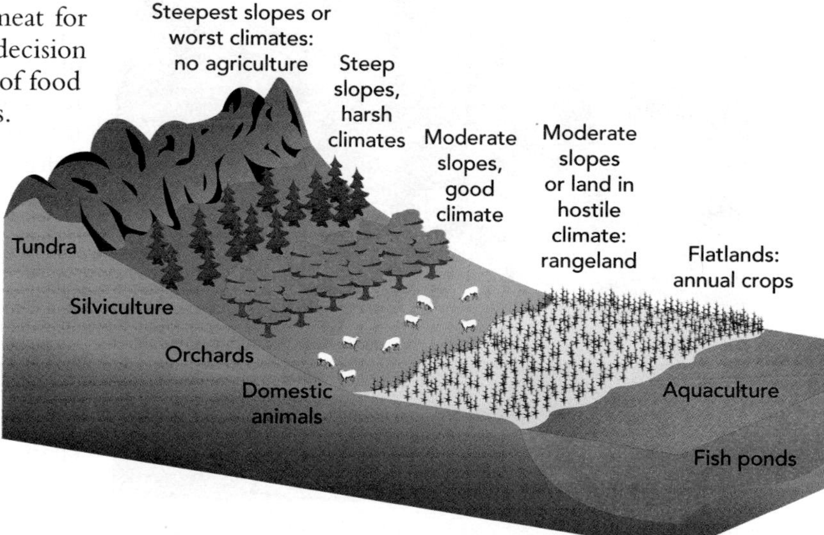

gaseous nitrogen to a form that can be used by green plants). Recall that bacteria grow in the root nodules of legumes. The bacteria live on substances produced by the legumes; in turn, the bacteria fix nitrogen. Legumes are often rotated with other crops so that the soil is enriched in nitrogen. It may be possible to develop new strains of corn and other crops that, along with new strains of bacteria, can form a symbiotic nitrogen-fixing relationship. Such an accomplishment would increase the production of these crops and reduce the need for fertilizers.

Another goal of agricultural genetic engineering is the development of strains with improved tolerance to drought, cold, heat, and toxic chemical elements. For example, one effort is aimed at developing wheat that is resistant to high levels of aluminum, an element that has negative effects on many plants.[29] Another goal is to create crops that produce their own pesticides. This is discussed in the next chapter. Although genetic modifications have proved to have great benefit, they involve both limitations and environmental concerns. These, too, are discussed in the next chapter.

The world land area planted with GMCs has grown rapidly from the first plantings in 1996 to about 58 million hectares in 2002 (Figure 11.21), over two-thirds of which is in the United States. Among the major genetically modified crops are corn, cotton, soybeans, canola, squash, and papaya.[30] It is difficult for the average consumer to avoid food and non-food agricultural products from GMCs, given that three-quarters of the soybeans, almost a third of the corn, and more than half of the plants that produce canola oil grown in Canada and the United States are genetically modified.[30] Currently, it is not possible to separate GMC products from non-GMC products when they reach the retail market.

Figure 11.21 • Genetically modified cotton.

The jury is out as to whether the benefits will outweigh undesirable effects discussed in the next chapter. What is important for environmental science is that the use of genetically modified crops is underway before the environmental effects are fully understood. As with many new technologies of the industrial age, application has preceded environmental investigation and understanding. The challenge for environmental science is to gain an understanding of environmental effects of GMCs quickly.

11.12 Climate Change and Agriculture

Climate change is more likely to decrease yield than to increase it, because areas with the best soils in the world also happen to have climates well suited to agriculture. Thus, most climatic changes are likely to make things worse. If global warming takes place as forecast by models of global climate, major disruptions to agriculture will result.[31] The best climates for agriculture may shift from their present locations to locations farther north. For example, the climate of the Midwestern United States corn belt, which has been so good for crop production, could move northward, favouring Canada. This could have a real impact on the amount of grain produced if soils in Canada prove not to be as suitable for grain production as those in the U.S. Midwest. Global warming could also lead to an increase in evapotranspiration (loss of water from the soil both by evaporation and by transpiration from plants) in many midlatitude areas[32]; supplying water for irrigation would become an even greater problem than it is today. Those concerned about the future of agriculture must take the possibility of global warming into account in their planning.[33] Plants require water, and much modern agriculture involves irrigation. Sources of irrigation water include groundwater; water diversion from nearby streams, rivers, and natural lakes; and artificial reservoirs. But large-scale irrigation projects cause environmental problems (discussed in Chapter 12). Construction of reservoirs changes the local environment. Some habitats disappear. Stream patterns change, and erosion rates increase in the watershed of the reservoir (see Chapter 18).

Summary

- The basic environmental questions about agriculture are: Can we produce enough food to feed Earth's growing human population? Can we grow crops sustainably, so that both crop production and agricultural ecosystems maintain their viability? Can we produce this food without seriously damaging other ecosystems that receive the wastes of agriculture?
- Agriculture changes the environment; the more intense the agriculture, the greater the changes.
- From an ecological perspective, agriculture is an attempt to keep an ecosystem in an early successional stage.
- The history of agriculture can be viewed as a series of attempts to overcome environmental limitations and problems. Each new solution has created new environmental problems, which have in turn required their own solutions.
- Farming greatly simplifies ecosystems, creating short and simple food chains, resulting in large areas planted in a single species or genetic strain grown in regular rows, reducing species diversity, and reducing the organic content and overall fertility of soils.
- These simplifications open farmed land to pests, both predators and parasites.
- Biotechnology makes genetic manipulation possible, creating entirely new ways to produce new crops and new strains within a crop. This holds a potential for increased production and for the production of new products from crops and domesticated animals, but it also poses novel environmental threats and raises many issues regarding science and values.
- At present, world food production is adequate in quantity and quality (nutritional value) to feed the current human population. Our modern food problem is the result of two factors: the great increase in human populations, which outstrips local food production in many areas, and inadequate world food distribution.
- Today, inadequate distribution is the most important cause of starvation. However, in the future, if the human population continues to increase, there will be a limit on the ability of Earth to produce sufficient food.
- If the human population continues to grow at something like its present rate, it will double within the next hundred years, meaning that world food production will have to double just to provide the same amount and quality of food per person available today.
- As living standards rise in nations such as China, the demand for higher-quality food will increase. As a result, per capita food demand will grow faster than the human population.
- In the past, we have increased world food production by (1) increasing the area devoted to food production and (2) increasing the production per unit area. The first method dominated most of human history; but during the twentieth century, major increases were due largely to increasing yields per unit area.
- Some of the best farmland in the world is being withdrawn from food production and put to other uses, including urbanization and suburbanization.
- Future global environmental problems could decrease world food production. These problems include possible effects of global warming (Chapter 20) and possible transportation of pests from one part of the world to another (Chapter 8).
- Many agricultural experts believe that production per unit area can increase through the use of genetically engineered crops. However, there are concerns that these crops may cause major, novel environmental problems, and the jury is out as to whether such crops will be a net benefit or a net disbenefit.

STUDY QUESTIONS

1. What will be the best way to feed the world in the next 10 years? The next 100 years?
2. A city garbage dump is filled; it is suggested that the area be turned into a farm. What factors in the dump might make it a good area to farm, and what might make it a poor area to farm?
3. Most crops are characteristic of what stages in ecological succession? How might we use our knowledge of succession to make agriculture sustainable?
4. Ranching wild animals—that is, keeping them fenced but never tamed—has been suggested as a way to increase food production in Africa, where wildlife is abundant. Based on this chapter, what are the environmental advantages and disadvantages of such game ranching?
5. Explain what is meant by the following statement: The world food problem is one of distribution, not of production. What are the major solutions to this world food problem?
6. You are sent into the Amazon rain forest to look for new crop species. In what kinds of habitats would you look? What kinds of plants would you look for?

CRITICAL THINKING ISSUE

Will There Be Enough Water to Produce Food for a Growing Population?

Between 2000 and 2025, scientists estimate that the world population will increase from 6.1 billion to 7.8 billion,[34] approximately double what it was in 1974. To keep pace with the growing population, the United Nations Food and Agricultural Organization predicts that food production will have to double by 2025, and so will the amount of water consumed by food crops. Will the supply of freshwater be able to meet this increased demand, or will water supply limit global food production?

Growing crops consume water through transpiration (loss of water from leaves as part of the photosynthetic process) and evaporation from plant and soil surfaces. The volume of water consumed by crops worldwide—including rainwater and irrigated water—is estimated at 3,200 billion m^3 per year. An almost equal amount of water is used by other plants in and near agricultural fields; thus, it takes 7,500 billion m^3 per year of water to supply crop ecosystems around the world. (See Table 11.2.) Grazing and pasture land account for another 5,800 billion m^3 and evaporation from irrigated water another 500 billion m^3 for a total of 13,800 billion m^3 of water per year for food production, or 20% of the water evaporated and transpired worldwide. By 2025, people will appropriate almost half of all the water available to grow food for human use. Where will the additional water come from?

Although the amount of rainwater cannot be increased, it can be more efficiently used through farming methods such as terracing, mulching, and contouring. Forty percent of the global food harvest now comes from irrigated land, and some scientists estimate that the volume of irrigated water available to crops will have to triple by 2025—a volume equal to that of 24 Nile Rivers or almost 300 Fraser Rivers.[35] A significant saving of water can, therefore, come from more efficient irrigation methods, such as improved sprinkler systems, drip irrigation, night irrigation, and surge flow.

Additional water could be diverted from other uses to irrigation. But this may not be as easy as it sounds because of competing needs for water, especially drinking and household water. The additional billions of people to the world population in the next decades will also need water. Already, humans use 54% of the world's runoff. Increasing this to more than 70%, as will be required to feed the growing population, may result in loss of freshwater ecosystems, decline in world fisheries, and extinction of aquatic species.

Irrigation projects are costly to construct, and fewer sites than in the past are now available that are acceptable ecologically and socially. As a result, irrigation has grown slower than population in recent years. If this trend continues, the irrigation gap will become even larger. Irrigation also causes salinization of agricultural areas, making them less suitable for growing crops and requiring more water to flush the salts from the soil. In many places, groundwater and aquifers are being used faster than they are replaced—a process that is unsustainable in the long-run. Many rivers are already so heavily used that they release little or no water to the ocean. These include the Ganges and most other rivers in India, the Huang He (Yellow River) in China, the Chao Phraya in Thailand, the Amu Dar'ya and Syr Dar'ya in the Aral Sea basin, and the Nile and Colorado Rivers.

Table 11.2 • Estimated Water Requirements of Food and Forage Crops

Crop	*Liters/kg*
Potatoes	500
Wheat	900
Alfalfa	900
Sorghum	1,110
Corn	1,400
Rice	1,912
Soybeans	2,000
Broiler chicken	3,500*
Beef	100,000*

* Includes water used to raise feed and forage.

Source: D. Pimentel et al., "Water Resources: Agriculture, the Environment, and Society." *Bioscience* 4, no. 2 (February 1997): 100.

Two hundred years ago, Thomas Malthus put forth the proposition that population grows more rapidly than the ability of the soil to grow food and that at some time the human population will outstrip the food supply (see A Closer Look 5.1). Malthus might be surprised to know that, by applying science and technology to agriculture, food production has so far kept pace with population growth. For example, between 1950 and 1995, world population increased by 122% while grain productivity increased 141%. Since 1995, however, grain production has slowed down (see Figure 11.22), and the question remains as to whether Malthus will be proven right in the twenty-first century. Will science and technology be able to solve the problem of water supply for growing food for people, or will water prove a limiting factor in agricultural production?

Critical Thinking Questions

1. How might changes in diet in developed countries affect water availability?
2. How might global warming affect estimates of the amount of water needed to grow crops in the twenty-first century?
3. Withdrawing water from aquifers faster than the replacement rate is sometimes referred to as "mining water." Why do you think this term is used?
4. Many countries in warm areas of the world are unable to raise enough food, such as wheat, to supply their populations. Consequently, they import wheat and other grains. How is this equivalent to importing water?

5. Malthusians are those who believe that eventually, unless population growth is checked, there will not be enough food for the world's people. Anti-Malthusians believe that technology will save the human race from a Malthusian fate. Analyse the issue of water supply for agriculture from both points of view.

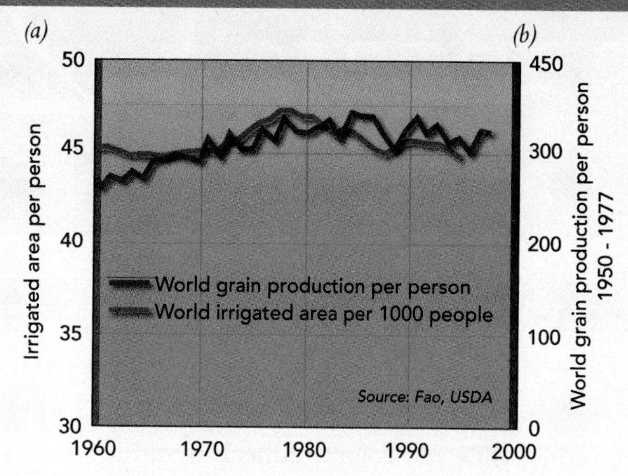

Figure 11.22 • *(a)* **World irrigated area per thousand people, 1961–1995.** [*Source*: L. R. Brown, M. Renner, and C. Flavin, *Vital Signs*: 1998 (New York: Norton, 1998), p. 47.] *(b)* **World grain production per person, 1960–1977.** [*Source*: L. R. Brown, M. Renner, and C. Flavin, *Vital Signs*: 1998 (New York: Norton, 1998), p. 29.]

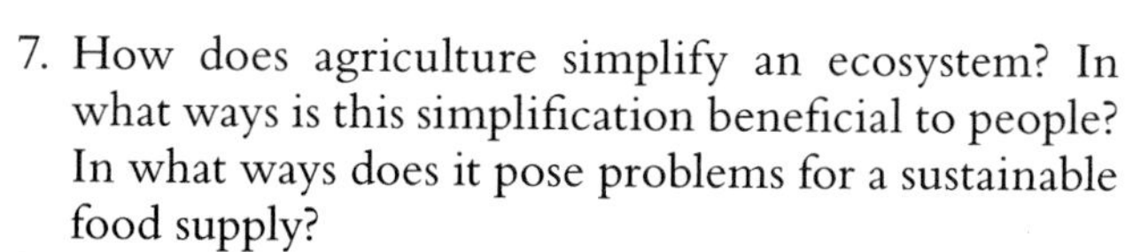

7. How does agriculture simplify an ecosystem? In what ways is this simplification beneficial to people? In what ways does it pose problems for a sustainable food supply?
8. A vegetable garden is planted in a vacant city lot. Peas and beans grow well, but tomatoes and lettuce do poorly. What is the likely problem? How could it be corrected?
9. A second vegetable garden is planted in another vacant lot. Nothing grows well. Outside the city, in otherwise similar environments, vegetables grow vigorously. What might explain the difference?
10. Should organic farming be allowed to include genetically modified organisms? Why or why not?

FURTHER READING

Achebe, C., G. Hyden, C. Magadza, and A. P. Okeyo, eds. *Beyond Hunger in Africa: Conventional Wisdom and an Alternative Vision*. Portsmouth, N.H.: Heinemann, 1990.

Borgstrom, G. *The Hungry Planet: The Modern World at the Edge of Famine*. New York: Macmillan, 1965. The major book by one of the leaders of agricultural change.

Fisher, G., K. Frohberg, M. L. Parry, and C. Rosenzweig. "Climate Change and World Food Supply, Demand, and Trade: Who Benefits, Who Loses," *Global Environmental Change 4* (1994) (1):7–23. Uses computer simulation of climate change to examine implications for agriculture.

Gliessman, S. *Agroecology: Ecological Processes in Sustainable Agriculture*. Sleeping Bear or Ann Arbor Press, 1998.

MacRae, R. J., S.B. Hill, J. Henning, and A. J. Bentley. "Policies, programs and regulations to support the transition to sustainable agriculture in Canada," *American Journal of Alternative Agriculture* (1990) 5 (2): 76-92.

Muir, W. H. and R. D. Howard. "Possible ecological risks of transgenic organism release when transgenes affect mating success: sexual selection and the Trojan gene hypothesis," *Proc. Nat. Acad. Sci.* (1999) 96: 13853-13856.

Wittwer, S., Y. Yee, H. Sun, and L. Wong. *Feeding a Billion: Agriculture in China.* Lansing: Michigan State University Press, 1987. A look at modern and traditional agricultural technologies of China and a discussion of how China feeds almost one-quarter of the world's population on 7% of the world's arable land.

chapter 12

Clean-Water Farms

Brothers Alex and Randall Affleck are fifth generation farmers near Bedeque Bay, Prince Edward Island, growing potatoes to augment income from the family farm's small dairy herd. In the ocean nearby, shellfishers harvest oysters, but their haul is increasingly contaminated by bacteria from manure spread on neighbouring farms like the Afflecks'. Nitrates from manure also contribute to algae blooms. As the algae die, decaying plant matter increases silt in the water and chokes the oysters.

For the Afflecks and the local fishers, a solution was critical. Runoff from the Affleck farm not only causes problems for the shellfish industry, it also represents a costly loss of topsoil. What is the solution? Like thousands of other farmers across Canada, the Afflecks turned to the Environmental Farm Plan (EFP), a voluntary program funded by Agriculture and Agri-Food Canada that helps farmers identify and fix pollution sources. Established in the early 1990s, the EFP program is now underway in all Canadian provinces. About 10% of Canadian farms have participated in some form of environmental farm planning. Ontario's EFP program is one of the longest-running; Alberta has just begun to test the program. In some regions, including Ontario, Alberta, Newfoundland, and Prince Edward Island, farmers attend an information workshop, but (like the Afflecks) conduct their own assessment using a workbook. In other regions such as Nova Scotia, an EFP program coordinator and an agricultural engineer inspect the farm property with the farmer, and help with the risk assessment. In Québec, farmers can join an advisory club called the Club Conseil, where they can meet to discuss concerns with agronomists and other agricultural experts.

The EFP process helped the Afflecks identify contaminated runoff as a key concern on their farm. Inexpensive measures, suggested in the workbook, led to major improvements. A marsh uphill of Bedeque Bay is now fenced, to keep cows out and prevent manure-contaminated waters from running into the Bay. Sloping fields are now tilled across the slope rather than downhill, creating horizontal ridges to trap runoff and eroded sediment on the land. The potato crop is rotated with hay and grain, to reduce fertilizer use (and associated costs) and increase soil productivity. The Afflecks also plant winter wheat and rye grass as cover crops, to guard against soil erosion, while newly planted hedgerows provide additional protection against wind erosion.[1]

Why are farmers buying into the EFP program? For the Afflecks, the payoff was increased environmental awareness and confidence that they are moving toward more sustainable farming practices. There are also real financial benefits. The program can help farmers access funds for new structures and practices, and can provide the farmer-to-farmer technical support necessary to maintain a viable farming operation while contributing to environmental stewardship.

Effects of Agriculture on the Environment

LEARNING OBJECTIVES

Agriculture changes the environment in many ways, both locally and globally. After reading this chapter, you should understand:

- How agriculture can lead to soil erosion, the severity of the problem, the methods available to reduce erosion, and how these methods have reduced soil erosion in Canada.
- How farming can deplete soil fertility and why agriculture often requires the use of fertilizers.
- Why some lands are best used for grazing and how overgrazing can damage land.
- What causes desertification.
- How farming creates conditions that promotes pest species, the importance of controlling pests (including weeds), and the problems associated with chemical pesticides.
- How alternative agricultural methods—including integrated pest management, no-till agriculture, mixed cropping, and other methods of soil conservation—can provide major environmental benefits.
- That genetic modification of crops could improve food production and benefit the environment but perhaps also could create new environmental problems.

- *This case study shows that practices that are environmentally benign can be economically advantageous. The environmental advantages of such alternative approaches to agriculture and the environmental effects of various forms of agriculture are the subjects of this chapter.*

Figure 12.1 • **A potato farm near the ocean, Prince Edward Island.**

12.1 How Agriculture Changes the Environment

Agriculture is both one of humanity's and civilization's greatest triumphs and the source of some of its greatest environmental problems. Agriculture has an ancient lineage, going back thousands of years, and it has always changed the local environment. Nothing in nature resembles a plough or does what a plough does, so when ploughing was invented, the environmental effects of agriculture increased.

Environmental effects of agriculture expanded greatly with the scientific–industrial revolution. Major environmental problems that result from agriculture include soil erosion; sediment transport and deposition downstream; on-site pollution from overuse and secondary effects of fertilizers and pesticides; off-site pollution of other ecosystems, soils, water, and air; deforestation; desertification; degradation of water aquifers; salinization; accumulation of toxic metals; accumulation of toxic organic compounds; and loss of biodiversity.

12.2 The Plough Puzzle

Here is a curious puzzle about agriculture and the plough. There are big differences between the soils of an unploughed forest and soils of previously forested land ploughed and used for crops for several thousand years—in Italy, for example, iron ploughs were pulled by oxen many centuries ago.[2] These differences were observed and written about by an originator of the modern study of the environment, George Perkins Marsh. Born in Vermont in the nineteenth century, Marsh became the American ambassador to Italy and Egypt. While in Italy, he was so struck by the differences between the Vermont forest soils and soils farmed for thousands of years on the Italian peninsula that he made this a major theme in his landmark book, *Man and Nature*. The farmlands he observed in Italy had once been forests. While the soil in Vermont was rich in organic matter and had definite layers, the soil of Italian farmland had little organic matter and lacked definite layers (Figure 12.4*a, b*).

There would be an expectation that farming in such soil would eventually become unsustainable, but much of the farmland in Italy and France has been in continuous use since pre-Roman times and is still highly productive. How can this be? What has been the long-term effect on the environment? The answers lie within this chapter.

12.3 Our Eroding Soils

The 1930s Dust Bowl increases the puzzle about the plough. Soils are keys to sustainable farming, and are easily damaged by farming (see A Closer Look 12.1). When land is cleared of its natural vegetation, such as forest or grassland, the soil begins to lose its fertility. Some of this occurs by physical erosion. According to Statistics Canada, urban uses have consumed 12,000 square kilometres of land since 1971, half of it high-quality farmland. In Ontario alone, more than 18% of Class 1 (best) farmland has been converted to urban development. In the United States, about 1 million ha are lost each year to urbanization and soil erosion.

Soil erosion became an international issue in North America in the 1930s, when intense ploughing, combined with a major drought, loosened the soil over large areas. The soil blew away, creating dust storms that buried automobiles and houses, destroyed many farms, and impoverished or displaced many people.

The land that became the Dust Bowl had been part of North America's great prairies, where grasses rooted deep, creating a heavily organic soil a metre or more down, protecting the soil from water and wind. When the plough turned over those roots, the soil was exposed directly to sun, rain, and wind, which further loosened the soil. It was a great tragedy of that time and a lesson people thought would be remembered forever. Nevertheless, soil continues to erode.

The introduction of heavy earthmoving machinery after World War II added to the problem by further compacting the soil and damaging the soil structure so important for crop production.

As the Dust Bowl made clear, when a forest or prairie is cleared for agriculture, the soil changes. The original soil developed over a long period; it is typically rich in organic matter, rich in chemical nutrients, and provides a physical structure conducive to plant growth. When the original vegetation is cleared and the land is planted in crops, most of whose organic matter is harvested and removed, there is less input of dead organic matter from the vegetation to the soil, which is exposed to sunlight that warms it and speeds the rate of decomposition of its organic matter. For these reasons, the amount of organic matter declines, and the soil's physical structure becomes less conducive to plant growth.

The rate of loss of fertility is sometimes measured as the time required for the soil to lose one-half of its original store of the chemical elements necessary for crops. How long this takes depends partly on the climate. It happens much faster in warmer and wetter areas, such as tropical rain forests, than in colder or dryer areas, such as those where the natural vegetation is a temperate-zone grassland or forest.[3]

(a)

(b)

Figure 12.2 • The Dust Bowl. Poor agricultural practices and a major drought created the Dust Bowl, which lasted about 10 years during the 1930s. Heavily ploughed lands lacking vegetative cover blew away easily in the dry winds, creating dust storms *(a)* and burying houses and *(b)* trucks.

Traditionally, farmers combated the decline in soil fertility by using organic fertilizers such as animal manure. These have the advantage of improving the soil's chemical and physical characteristics. Organic fertilizers can have drawbacks, especially under intense agriculture on poor soils—in these situations, they do not provide enough of the chemical elements needed to replace what is lost. The development of industrially produced fertilizers, commonly called "chemical" or "artificial" fertilizers, was a major factor in the great increases in crop production in the twentieth century. Among the most important advances were industrial processes to convert molecular nitrogen gas in the atmosphere to nitrate that can be used directly by plants. Phosphorus is mined, usually from a fossil source that was biological in origin such as deposits of bird guano (excrement) on islands used for nesting (Figure 12.3). Nitrogen, phosphorus, and other elements are combined in proportions that are appropriate for specific crops in specific locations.

Figure 12.3 • A guano island mining of phosphorus.

Since the end of World War II, farming has seriously damaged more than 1 billion ha of land (about 10.5% of the world's best soil), equal in area to China and India. Overgrazing, deforestation, and destructive crop practices have so seriously damaged approximately 9 million ha that recovery will be difficult; restoration of the rest will require serious action.[4] Until a few years ago, more than 3 billion tonnes of topsoil was lost from North American farms every year. Prince Edward Island has lost half its natural topsoil since 1900.[5] But things are getting better. Although soil loss in Canada continues, measurements suggest the losses have slowed considerably thanks to improvements in ploughing and the use of no-till agriculture (discussed later in this chapter). The area of Canadian cropland at risk of erosion and soil loss declined by 10% overall between 1981 and 1996. One U.S. study showed that, on average, soil erosion declined from 17 metric tonnes per hectare per year (t/ha/yr) to about 13 t/ha/yr.[6] Ideally, soil loss should not exceed 6.7 t/ha/year, the rate at which soil is naturally replenished.

A recent study examined the potential for soil loss reduction in northwestern New Brunswick potato farms. The results showed that mean annual soil loss could be reduced by 98% using simple hay mulching. The mulching also decreased the loss of nitrogen by up to 24%, while the yield of potatoes either remained the same or increased slightly.[7, 8]

A CLOSER LOOK 12.1

Soils

To most people soil is something we walk on and do not think much about —it is just "dirt." But soils are a key to life on the land, affecting life and affected by it. Upon closer examination, soils are quite remarkable. You will not find anything like Earth soil on Mars or Venus or the moon. Why not? Because water and life have greatly altered the land surface. Geologically, soils are earth materials modified over time by physical, chemical, and biological processes into a series of layers called soil horizons. Each kind of soil has its own chemical composition. Soils develop over time—sometimes a very long time, perhaps thousands of years. If you dig carefully into a soil so that you leave a nice, clean vertical side, you will see the soil's layers—that is, if the soil has not been disturbed by a plough or other human activity and if the soil developed in a climate that promoted certain processes. In a northern forest, a soil is dark at the top, then has a white powdery layer, pale as ash, then a brightly coloured layer, which is usually much deeper than the white one and is typically orangish. Below that is a soil whose colour is close to that of the bedrock (which geologists call "the parent material" for obvious reasons) (Figure 12.4*c*). We call the layers horizons.

Overall, water flows down through the soil. With a pH of about 5.5, rainwater is naturally slightly acid because it has some carbon dioxide from the air dissolved in it, and this forms carbonic acid, a mild acid. It does not taste sour or acid, but it is acid enough to leach metals from the soil. As a result, minerals such as iron, calcium, and magnesium are leached from the upper horizons (A and E) and may be deposited in a lower horizon (B) (Figure 12.4*c*). The upper horizons are usually full of life and are viewed by ecologists as complex ecosystems, or ecosystem units (horizons O and A). Decomposition is the name of the game as fungi, bacteria, and small animals live on what plants and animals on the surface produce and deposit. Actual chemical decomposition of organic compounds from the surface is done by bacteria and fungi, the great chemical factories of the biosphere. Soil animals, such as ear-

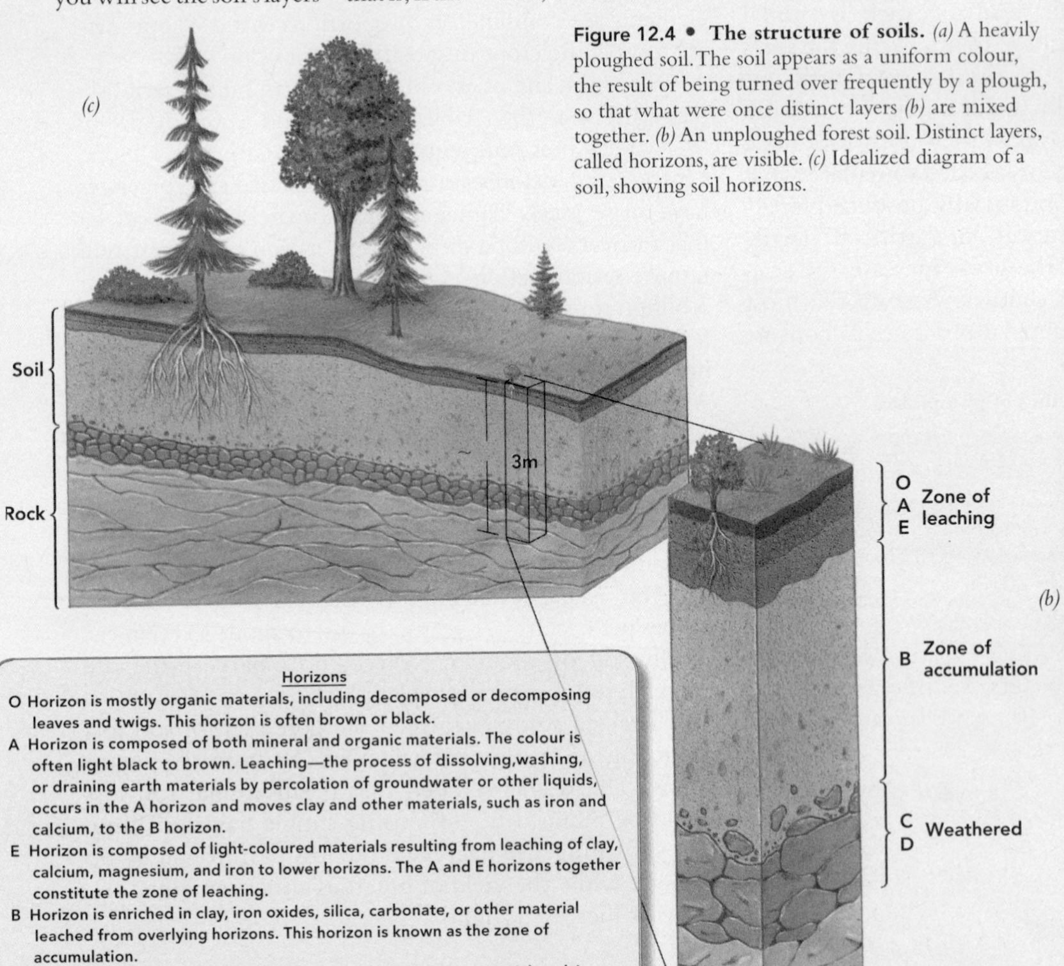

(a) (b)

Figure 12.4 • The structure of soils. *(a)* A heavily ploughed soil. The soil appears as a uniform colour, the result of being turned over frequently by a plough, so that what were once distinct layers *(b)* are mixed together. *(b)* An unploughed forest soil. Distinct layers, called horizons, are visible. *(c)* Idealized diagram of a soil, showing soil horizons.

thworms, eat leaves, twigs, and other remains and break them into smaller pieces that are easier for the fungi and bacteria to process. The animals affect the rate of chemical reactions in the soil. There are also predators on soil animals, so there is a soil ecological food chain.

What a soil is like is determined by climate, parent material (bedrock), topography, biological activity, and time. The soil horizons shown in Figure 12.4*c* are not necessarily all present in any one soil. Very young soils may have only an upper A horizon over a C horizon, whereas mature soils may have nearly all the horizons shown.

Soil fertility is the capacity of a soil to supply nutrients necessary for plant growth. Soils that have formed on geologically young materials—for example, glacial deposits in southern Canada that form productive land for corn cultivation—are nutrient-rich. Soils in semi-arid regions are often nutrient-rich and need only water to become very productive for agriculture, such as in the Great Central Valley of California; soils in humid areas and tropics may be heavily leached and relatively nutrient-poor due to the high rainfall. In such soils, nutrients may be cycled through the organic-rich upper horizons; and if forest cover is removed, reforestation may be very difficult (see Chapter 13). Soils that accumulate certain clay minerals in semi-arid regions may swell when they get wet and shrink as they dry out, cracking roads, walls, buildings, and other structures. In the north, permafrost can also thaw and refreeze under certain conditions, for instance when fragile soils are disturbed allowing water to collect on the surface, acting as a heat sink. Expansion and contraction of soils in Canada and the United States cause billions of dollars' worth of property damage each year.

Soils with small clay particles (less than 0.004 mm in diameter) retain water well and retard the movement of water because the spaces between the particles are very small. Soils with coarser grains (greater than 0.06 mm in diameter), such as sand or gravel, have relatively large spaces between grains, so water moves quickly through them. Soils with a mixture of clay and sand can retain water well enough for plant growth but also drain well. Soils with a high percentage of organic matter also retain water and chemical nutrients for plant growth. It is an advantage to have good drainage, so a coarse-grained soil is a good place to build your house. If you are going to farm, you will do best in a loam soil that has a mixture of particle sizes. Thus, the type of soil particles present is important in determining where to build a house and where to farm and in siting facilities such as landfills, where retention of pollutants on site is an objective (see Chapter 24).

Coarse-grained soils, especially those composed primarily of sand (0.06–2.0 mm in diameter), are particularly susceptible to water and wind erosion. Soils composed of coarser (heavier) particles or finer particles that are usually more cohesive (held together by clay minerals) are more resistant to erosion.

It is difficult to think of a human use of the near-surface land environment that does not involve consideration of the soils present. As a result, the study of soils continues to be an important part of environmental sciences.

12.4 Where Eroded Soil Goes: Sediments Also Cause Environmental Problems

Soil eroded from one location has to go somewhere else. Much of it travels down streams and rivers and is deposited at their mouths. The Fraser, Mackenzie, Peace, and St. Lawrence rivers together carry an annual load of more than 170 million tonnes, much of it from agricultural lands. That is more than 5,600 kg of sediment for each person in Canada. Much of this sediment is deposited in reservoirs, rivers, and lakes. Eventually, these sediments fill otherwise productive waters, destroying some fisheries. In tropical waters, sediments entering the ocean can destroy coral reefs near a shore.

Sedimentation has chemical environmental effects as well. Nitrates, ammonia, and other fertilizers carried by sediments enrich the waters downstream. This enrichment, called eutrophication (Chapter 19), promotes the growth of algae. It is a straightforward process: fertilizers that were meant to increase the growth of crops affect algae in the water, but people generally do not want water enriched with algae because dead algae are decomposed by bacteria that, in turn, remove oxygen from the water. As a result, fish can no longer live in the water. The water becomes thick with a greenish-brown mat, unpleasant for recreation and a poor base for drinking water. Sediments also can transport toxic chemical pesticides. Since the 1930s, agriculture-induced sedimentation has decreased with the decrease in the rate of soil erosion. Even so, taking into account the costs of dredging and the decline in the useful life of reservoirs, sediment damage costs Canada millions of dollars every year.

Making Soils Sustainable

Soil forms continuously. In ideal farming, the amount of soil lost would never exceed the amount of new soil produced. Production is slow—on good lands, a layer of soil 1 mm deep, thinner than a piece of paper, forms at a rate ranging from one per decade to one in 40 years. Sustainability of soils can be aided by fall ploughing; multiculture (planting several crops intermixed in the same field), terracing, crop rotation, contour ploughing, and no-till agriculture. At this point in our discussion, we have reached a partial answer to the question: How could farming be sustained for thousands of years, while the soil has been degraded? We must recognize a distinction between the sustainability of a product (in this case crops) and the sustainability of the ecosystem. In agriculture, crop production can be sustained, but the ecosystem may not be. If the ecosystem is not sustained, then people must provide additional input of energy and chemical elements to replace what is lost.

Contour Ploughing

Ploughed furrows make paths for water to flow, and if the furrows go downhill, then the water moves rapidly along them, increasing the erosion rate. In **contour ploughing**, the land is ploughed perpendicular to the slopes, and as horizontally as possible (Figure 12.5*a*).

Contour ploughing has been the single most effective way to reduce soil erosion. This was demonstrated by an experiment on sloping land planted in potatoes. Part of the land was ploughed in uphill and downhill rows, and part was contour-ploughed. The uphill and downhill section lost 32 metric tonnes/ha of topsoil; the contour-ploughed section lost only 0.22 metric tonnes/ha. It would take almost 150 years for the contour-ploughed land to erode as much as the traditionally ploughed land eroded in a single year! Besides drastically reducing soil erosion, contour ploughing uses less fuel and time. Even so, today contour ploughing is used on only about 10% of the agricultural land in Canada and the United States.

No-Till Agriculture

An even more efficient way to slow erosion is to avoid ploughing all together. **No-till agriculture** (also called conservation tillage) involves not ploughing the land, using herbicides and integrated pest management (discussed later in this chapter) to keep down weeds, and allowing some weeds to grow. Stems and roots that are not part of the commercial crop are left in the fields and allowed to decay in place (Figure 12.5*b*). In contrast to standard modern approaches, the goal in no-till agriculture is to suppress and control weeds but not to eliminate them at the expense of soil conservation. Worldwide, no-till agriculture is increasing. Paraguay leads the world with 55% of its farmland in no-till. The United States, with 17.5% in no-till, lags behind many other nations. Canada reached 30% in 2001, up from 24% a decade earlier (Figure 12.6); Argentina has 45%, Brazil 39%.

12.5 Controlling Pests

From an ecological point of view, pests are undesirable competitors, parasites, or predators. The major agricultural pests are insects that feed mainly on the live parts of plants, especially leaves and stems; nematodes (small worms), which live mainly in the soil and feed on roots and other plant tissues; bacterial and viral diseases; weeds (flowering plants that compete with the crops); and vertebrates (mainly rodents and birds) that feed on grain or fruit. Even today, with modern technology, the total losses from all pests are huge; in the United States, pests account for an estimated loss of one-third of the potential harvest and about one-tenth of the harvested crop. Preharvest losses are due to competition from weeds, diseases, and herbivores; postharvest losses are largely due to herbivores.[9]

People usually think that the major agricultural pests are insects, but in fact, weeds are the major problem. Farming produces special environmental and ecological conditions that tend to promote weeds. Remember that the process of farming is an attempt to hold back the natural processes of ecological succession, prevent

Figure 12.5 • **Alternative agricultural ploughing and tilling methods:** *(a)* contour strip crops; *(b)* no-till soybean crop planted in wheat stubble.

(a)

(b)

migrating organisms from entering an area, and prevent natural interactions (including competition, predation, and parasitism) between populations of different species.

Because a farm is maintained in a very early stage of ecological succession and is enriched by fertilizers and water, it is a good place not only for crops but also for other early-successional plants. These non-crop and, therefore, undesirable plants are weeds. A weed is just a plant in a place we do not want it to be. Recall that early-successional plants tend to be fast-growing and to have seeds that are easily blown by the wind or spread by animals. These plants spread and grow rapidly in the inviting habitat of open, early-successional croplands.

There are about 30,000 species of weeds, and in any year a typical farm field is infested with between 10 and 50 of them. Weeds compete with crops for all resources: light, water, nutrients, and space to grow. With more weeds, there are less crops. Some weeds can have a devastating effect on crops. For example, the production of soybeans is reduced by 60% if a weed called cocklebur grows three individuals per square metre (one individual per square foot).[10] Weeds have been shown to reduce crop yields in Canada by up to 35%. Across North America, agricultural losses from weeds are roughly tens of billions of dollars a year. In Canada, an additional $1.4 billion is spent annually for chemical weed control, amounting to almost 60% of all pesticide sales.

12.6 The History of Pesticides

Before the Industrial Revolution, farmers could do little to prevent pests except remove them when they appeared or use farming methods that decreased their density. For example, slash-and-burn agriculture (also known as swidden agriculture) allows succession to take place. The greater diversity of plants and the long time between the use of each plot reduces the density of pests (see Chapter 11). Pre-industrial farmers also planted aromatic herbs and other vegetation that repel insects.

With the beginning of modern science-based agriculture, people began to search for chemicals that would reduce the abundance of pests. More important, they searched for a "magic bullet"—a chemical (referred to as a *narrow-spectrum pesticide*)—that would have a single target, just one pest, and not affect anything else. But this proved elusive. Living things share many chemical reactions (see Chapters 4 and 6), so a chemical that is toxic to one species is likely to be toxic to another. The story of the scientific search for pesticides is the search for a better and better magic bullet. The earliest pesticides were simple inorganic compounds that were widely toxic. One of the earliest was arsenic, a chemical element toxic to all life, including people. It was effective in killing pests, but it killed beneficial organisms as well and was very dangerous to use.

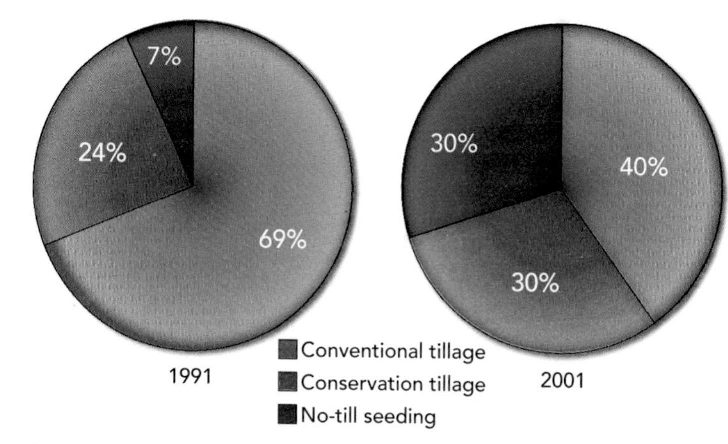

Figure 12.6 • To plough or not to plough: Canada's choices. In Canada, the percentage of no-till agriculture increased greatly in a decade. (*Source:* Statistics Canada's Internet Site, http://www.statcan.ca, 2003.)

A second stage in the development of pesticides began in the 1930s and involved petroleum-based sprays and natural plant chemicals. Many plants produce chemicals as a defense against disease and herbivores, and these chemicals are effective pesticides. Nicotine, from the tobacco plant, is the primary agent in some insecticides still in wide use today. However, although natural plant pesticides are comparatively safe, they were not as effective as desired.

The third stage in pesticide development was the development of artificial organic compounds. Some, like DDT, are broad-spectrum, but more effective than natural plant chemicals. These chemicals are effective and have been important to agriculture, but unexpected environmental effects keep cropping up, and the magic bullet remains elusive. For example, aldrin and dieldrin have been widely used to control termites as well as pests on corn, potatoes, and fruits. Dieldrin is about 50 times as toxic to people as DDT. These chemicals are designed to remain in the soil and typically do so for years, but they are readily drained from tropical rain forest soils. As a result, they have spread widely and are found in organisms in arctic waters. The chemicals accumulate in people. A study of breast-fed babies in Australia showed that every day 88% took in an amount that exceeded the World Health Organization standard for a daily intake.[11]

As a result, a fourth stage in the development of pesticides began, which returned to biological and ecological knowledge. This was the beginning of modern **biological control**, the use of biological predators and parasites to control pests. One of the most effective of these is a bacterium named *Bacillus thuringiensis* (BT), a disease that affects caterpillars and the larvae of other insect pests. Spores of BT are sold commercially (you can buy them at your local garden store and use this method for your home garden). BT has been one of the most important ways to control gypsy moth epidemics, an introduced moth whose larvae periodically strip most of the leaves from large areas of forests in Eastern Canada and the northeastern United States. BT has proved safe and effective—safe because it causes disease only in specific insects and is harmless to people and other mammals, and as a natural biological "product," its presence and its decay is non-polluting.

Another group of effective biological control agents are small wasps that are parasites of caterpillars. The wasps lay their eggs on the caterpillars; the larval wasps then feed on the caterpillars, killing them. These wasps tend to have very specific relationships (one species of wasp will be a parasite of one species of pest), and so they are both effective and narrow-spectrum (Figure 12.7).

And in the list of biological control species we cannot forget ladybugs, which are predators of many pests. You can buy these, too, at many garden stores and release them in your garden.

Another technique to control insects involves the use of sex pheromones, chemicals released by most species of adult insects (usually the female) to attract members of the opposite sex. In some species, pheromones are effective up to 4.3 km away. These chemicals have been identified, synthesized, and used as bait in insect traps, in insect surveys, or simply to confuse the mating patterns of the insects involved.

12.7 Integrated Pest Management

While biological control works well, it has not solved all problems with agricultural pests. As a result, a fifth stage developed, known as integrated pest management (IPM). IPM uses a combination of methods, including biological control, certain chemical pesticides, and some methods of planting crops. A key idea underlying IPM is that the goal can be control rather than complete elimination of a pest. This is justified for several reasons. Economically, it becomes increasingly expensive to eliminate a greater percentage of pests, while the value of that elimination, in terms of crops to sell, becomes even less. This suggests that it make economic sense to leave some pests and eliminate only enough to provide benefit. In addition, allowing some of a pest population to remain, small but controlled, does less damage to ecosystems, soils, water, and air. *Some like to think of IPM as an ecosystem approach to pest management, because it uses the characteristics of ecological communities and ecosystems* as discussed in Chapters 4, 6, and 10.

Another characteristic of IPM is the attempt to move away from monoculture of a single strain growing in perfectly regular rows. Studies show that just the physical complexity of a habitat can slow the spread of parasites. In effect, a pest like a caterpillar or mite is trying to find its way through a maze. If the maze consists of regular rows of nothing but what the pest likes to eat, the maze problem is easily solved by the dumbest of animals. But if there are several species, even two or three, arranged in a more complex pattern, the pests have a hard time finding their prey.

No-till or low-till agriculture is another feature of IPM, because this helps natural enemies of some pests to build up in the soil (ploughing destroys the habitats of these pest enemies).

Control of the oriental fruit moth, which attacks numerous fruit crops, is an example of IPM biological control. The moth is a prey of a species of wasp, *Macrocentrus ancylivorus*,[12] and introducing the wasp into fields helped control the moth. Interestingly, in peach fields the wasp was more effective when strawberry fields were nearby. The strawberry fields provided an alternative habitat for the wasp, especially important for overwintering (Figure 12.8).[10] As this example shows, spatial complexity and biological diversity also become part of the IPM strategy.

Although artificial pesticides are used, they are used along with the other techniques, so the application of

Figure 12.7 • Integrated pest management: the biological control of pests.

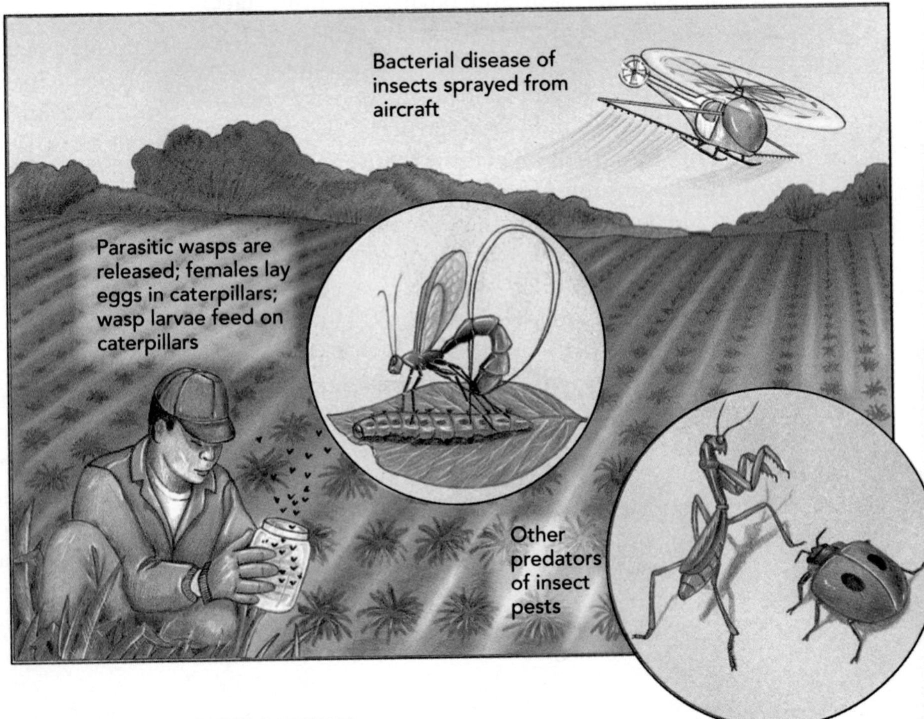

these pesticides can be sparing and specific.[13] A recent study concluded that IPM could reduce the use of pesticides by as much as 75% while reducing preharvest pest-caused losses by 50%. This would also greatly reduce the costs to farmers for pest control.[9] Figure 12.9 shows the oriental fruit moth larvae, a pest of peaches, apples, and other fruit.

Current agricultural practices in Canada and the United States involve a combination of approaches, but in most cases they are more restricted than an IPM strategy. Biological control methods are used to a comparatively small extent. They are the primary tactics for controlling vertebrate pests (mice, voles, and birds) that feed on lettuce, tomatoes, and strawberries but are not major techniques for grains, cotton, potatoes, apples, or melons. Chemicals are the principal control methods for insect pests. For weeds, the principal controls are methods of land culture. The use of genetically resistant stock is important for disease control in wheat, corn, cotton, and some vegetable crops, such as lettuce and tomatoes.

Figure 12.8 • **The wasp that preys on the oriental fruit moth and that has been used in biological control.**

Monitoring Pesticides in the Environment

The Organisation for Economic Cooperation and Development (OECD) ranks Canada 22nd out of 28 nations in pesticide use, at just under 1 kg per capita per year (1994 data). Only six countries, Australia, Italy, France, Belgium, the United States, and Portugal, use more pesticides per capita. The use of pesticides has grown. Statistics Canada data indicate that pesticide use in Canada quadrupled between 1970 and 1995. Worldwide, about 80% of the pesticides in use are applied in developing countries.

Agricultural scientists work continuously to improve pesticides. Once applied, these chemicals may decompose in place or may be blown by the wind or transported by surface and subsurface waters, meanwhile continuing to decompose. Sometimes the initial breakdown products (the first, still complex chemicals produced from the original pesticides) are also toxic, as happens with DDT (see A Closer Look 12.2). Eventually, the toxic compounds are decomposed to their original inorganic or simple, non-toxic organic compounds. However, for some chemicals, this can take a very long time.

Where do all these pesticides go? How long do they last in the environment, both on the site where they were deposited and downstream and downwind? What is the concentration of these in our waters? To establish useful standards for pesticide levels in the environment, and to understand the environmental effects of pesticides, it is necessary to monitor the concentrations. Public health standards and environmental effects standards were established for some but not all of these compounds. The collection of water quality data, including pesticide concentrations in surface waters, is a provincial responsibility. The frequency and extent of this monitoring vary considerably from region to region. In Ontario, for example,

Figure 12.9 • **The oriental fruit moth larvae, a pest of fruit crops, is controlled by a parasite wasp that attacks the larvae.** *(a)* The larvae. *(b)* Apples damaged by the moth.

(a)

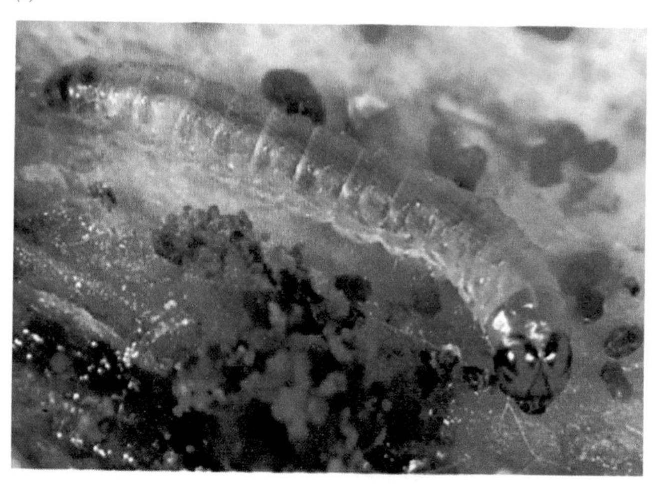

(b)

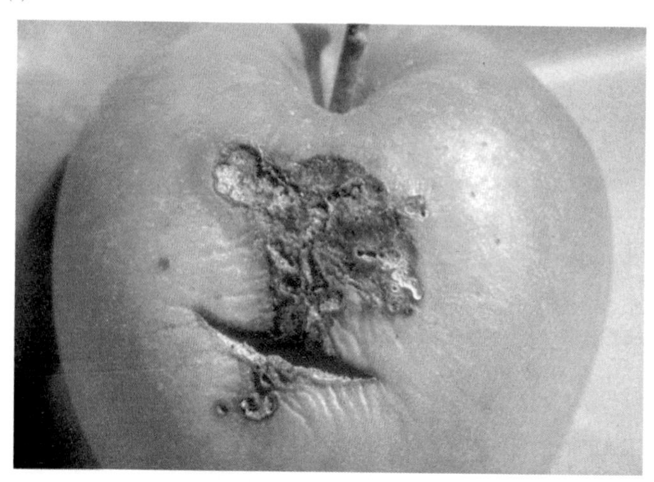

A CLOSER LOOK 12.2

DDT

The real revolution in chemical pesticides—the development of more sophisticated pesticides—began with the end of World War II and the discovery of DDT and other chlorinated hydrocarbons, including aldrin and dieldrin. When DDT was first developed in the 1940s, it seemed to be the long-sought magic bullet, with no short-term effects on people and deadly only to insects. At the time, scientists believed that a chemical could not be readily transported from its original site of application unless it was water-soluble. DDT was not very soluble in water and therefore did not appear to pose an environmental hazard. DDT was used very widely until three things were discovered.

- It has long-term effects on desirable species. Most spectacularly, it decreases the thickness of eggshells as they develop within birds.
- It is stored in oils and fats and is transferred up food chains as one animal eats another; it becomes concentrated as it is passed up food chains, so that the higher an organism is on a food chain, the higher its concentration of DDT. This process is known as food-chain concentration or biomagnification (discussed in detail in Chapter 15).
- The storage of DDT in fats and oils allows the chemical to be transferred biologically even though it is not very soluble in water.

In birds, DDT and the products of its chemical breakdown (known as DDD and DDE) thin eggshells so that they break easily, reducing the success of reproduction. This was found to be especially severe in birds that are high on the food chain—predators that feed on other predators, such as the bald eagle, osprey, and pelican, which feed on fish that may be predators of other fish.

As a result, DDT was banned in most developed nations—banned in Canada in 1969 and in the United States in 1971. Since then, a dramatic recovery has occurred in the affected bird populations. The brown pelican of the Florida and California coasts, which had become rare and endangered and whose reproduction had been restricted to offshore islands where DDT had not been used, became common again. The bald eagle became abundant again in the north woods, where it can be seen in many regions including Quetico Provincial Park, Banff National Park, and coastal British Columbia. However, DDT is still being produced in the United States for use in the developing and less developed nations, especially as a control for malaria-spreading mosquitoes.

The use of DDT has had some benefits. It was primarily responsible for eliminating malaria and yellow fever as major diseases, reducing the incidence of malaria in the United States from an average of 250,000 cases a year prior to the spraying program to fewer than 10 per year in 1950. Even for these uses, DDT's effectiveness has declined over the years because many insect species have developed a resistance to it. Nevertheless, DDT continues to be used because it is cheap and sufficiently effective and because people have become accustomed to using it. About 35,000 metric tonnes of DDT are produced annually in at least five countries, and it is legally imported and used in dozens, including Mexico.

Although people in developed nations believe they are free from the effects of DDT, in fact this chemical is transported back to industrial nations in agricultural products from nations still using the chemical. Also, migrating birds that spend part of the year in malarial regions are still subject to DDT. Thus, despite being banned in the developed nations, DDT remains an important world issue in pest control. (The problem of developing nations' use of pesticides banned in other nations is an issue not only for DDT but also for other chemicals.)

With the banning of DDT in developed nations, other chemicals became more prominent, chemicals that were less persistent in the environment. Among the next generation of insecticides were organophosphates—phosphorus-containing chemicals that affect the nervous system. These chemicals are more specific and decay rapidly in the soil. Therefore, they do not have the same persistence as DDT. But they are toxic to people and must be handled very carefully by those who apply them.

Chemical pesticides have created a revolution in agriculture. However, in addition to the negative environmental effects of chemicals such as DDT, they have other major drawbacks. One problem is secondary pest outbreaks, which occur after extended use (and possibly because of extended use) of a pesticide. Secondary pest outbreaks can come about in two ways: (1) reducing one target species reduces competition with a second species, which then flourishes and becomes a pest, or (2) the pest develops resistance to the pesticides through evolution and natural selection, which favor those who have a greater immunity to the chemical.[14] Resistance has developed with many pesticides. For example, Dasanit (fensulfothion), an organophosphate, first introduced in 1970 to control maggots that attack onions, was originally successful but is now so ineffective that it is no longer used for that crop.

there is no systematic monitoring of pesticides or other organic contaminants in surface waters or treated drinking waters, although some special studies have provided insight into those conditions. Available data suggest that during heavy spring runoff, concentrations of some herbicides might be reaching or exceeding established public health standards. This research is just beginning, and it is difficult to reach definitive conclusions whether present concentrations are causing harm in public water supplies or to wildlife, fish, algae in fresh waters, or vegetation. Advances in knowledge give us much more information, on a more regular basis, about how much of many artificial compounds are in our waters, but we are still unclear about their environmental effects. A wider and better program to monitor pesticides in water and soil is important to provide a sound scientific basis for dealing with pesticides.

12.8 Genetically Modified Crops

Remember from Chapter 11 that the genetic modification of organisms currently uses three methods: (1) faster and more efficient development of new hybrids, (2) introduction of the "terminator gene," and (3) transfer of genetic properties from widely divergent kinds of life.

Each of these methods poses different potential environmental problems. It is important to remember a general rule of environmental actions: if actions we take are similar in kind and frequency to natural changes, then the effects on the environment are likely to be benign. This is because species have had a long time to evolve and adapt to these changes. In contrast, changes that are novel—that do not occur in nature—are more likely to have negative or undesirable environmental effects, both direct and indirect. We can apply this rule to the three categories of genetically engineered crops.

New Hybrids

The development of hybrids within a species is a natural phenomenon (Chapter 7), and the development of hybrids of major crops, especially of small grains, has been a major factor in the great increase in productivity of twentieth-century agriculture. Strictly from an environmental perspective, genetic engineering to develop hybrids within a species is likely to be as benign as the development of agricultural hybrids has been with conventional methods.

There is an important caveat, however. Some people are concerned that the great efficiency of genetic modification methods may produce "superhybrids" that are so productive they can grow where they are not wanted and become pests. Another concern is that some new hybrid characteristics could be transferred by interbreeding with closely related weeds (Figure 12.10). This could inadvertently create a "superweed" whose growth, persistence, and resistance to pesticides would make it difficult to control. Another environmental concern is that new hybrids might be developed that could grow on ever more marginal lands. The development of crops on such marginal lands might increase erosion and sedimentation and lead to decreased biological diversity in specific biomes. Still another potential problem is that "superhybrids" might require much more fertilizers, pesticides, and water. This could lead to greater pollution and the need for more irrigation.

On the other hand, genetic engineering could lead to hybrids that require less fertilizer, pesticide, and water. For example, presently only legumes (peas and their relatives) have symbiotic relationships with bacteria and fungi that allow them to fix nitrogen. Attempts are underway to transfer this capability to other crops so that more kinds of crops would enrich the soil with nitrogen and require much less external application of nitrogen fertilizer.

The Terminator Gene

The terminator gene makes seeds from a crop sterile. This is done for environmental and economic reasons. In theory, it prevents genetically modified crops from spreading. It also protects the market for the corporation that developed it: farmers cannot get around purchasing seeds by using some of their crops' hybrid seeds the next year. But this poses social and political problems. Farmers in less developed nations and governments of nations that lack genetic-engineering capabilities are concerned that the terminator gene will allow major multinational corporations to control the world food supply. Concerned observers believe that farmers in poor nations must be able to grow next year's crops from their own seeds because they cannot afford to buy new seeds every year. This is not directly an environmental problem, but it can become an environmental problem indirectly by affecting total world food production, which then affects the human population and how land is used in areas that have been in agriculture.

Transfer of Genes from One Major Form of Life to Another

Most environmental concerns deal with the third kind of genetic modification of crops: the transfer of genes from one major kind of life to another. This is a novel effect and more likely to have negative and undesirable impacts. In several cases, this type of genetic modification has led to unforeseen and undesirable environmental effects. Perhaps the best known involves potatoes and corn (Figure 12.11), caterpillars that eat these crops, a disease of caterpillars that controls these pests, and an endangered species, monarch butterflies. Here is what happened.

As discussed earlier, the bacterium *Bacillus thuringiensis* is a successful pesticide, causing a disease in many caterpillars. With the development of biotechnology, agricultural scientists studied the bacteria and discovered the toxic chemical and the gene that caused its production within the bacteria. This gene was then transferred to potatoes and corn so that the biologically engineered plants produced their own pesticide. At first, this was believed to be a constructive step in pest control, because it was no longer necessary to spray

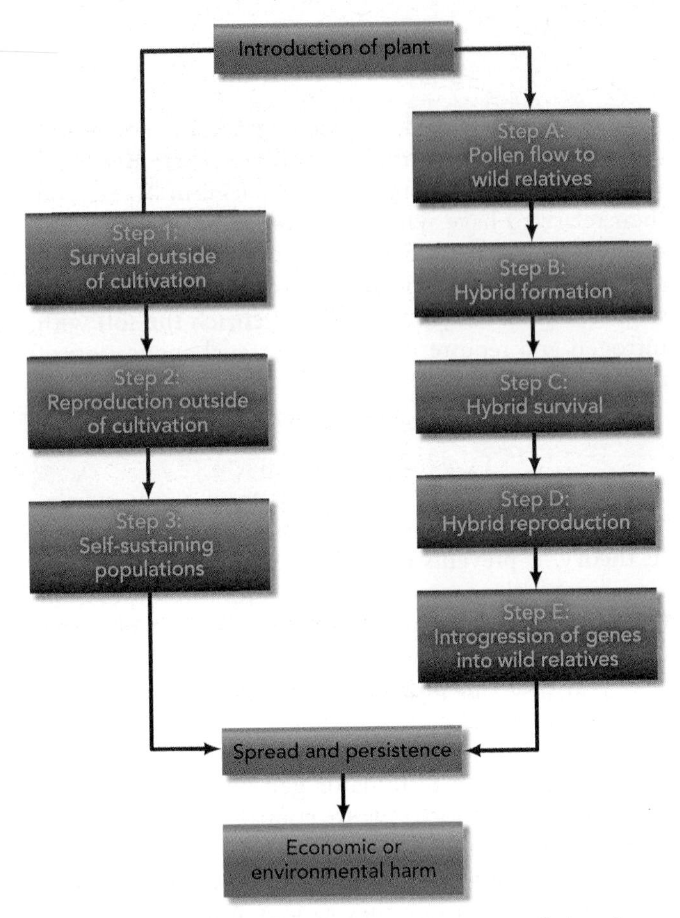

Figure 12.10 • **Ways in which the genetic characteristics of a modified crop might spread.**

a pesticide. However, the genetically engineered potatoes and corn produced the toxic BT substance in every cell—not just in the leaves that the caterpillars ate, but also in the potatoes and corn sold as food, in the flowers, and in the pollen.

Pollen released from fields of these genetically modified crops spread widely, and people began to worry that the toxic pollen could be ingested by monarch butterflies as they migrated through areas where the crops were grown. The monarch butterfly is a protected species under the federal *Species at Risk Act* (see Chapter 14). In fact, experiments showed that the butterflies inadvertently ingested the toxic pollen and died. (This pollen is not their food; they feed on milkweed, but they ingested pollen that fell on the milkweed.) Although this effect has not yet been observed outside experimental plots, it is a cause for concern. The indirect, complex effect just described was not originally expected, but in hindsight it is easy to see that such complex, indirect effects are typical of ecosystems and ecological interactions. It is an example of gene transfer from one kind of life (bacteria) to a different type of life (green plants). This is a novel effect and one that should be carefully monitored.

A strain of rice was developed that produces beta-carotene, important in human nutrition. The rice may have added nutritional benefits, particularly valuable for the poor of the world who depend on rice as a primary food. The gene that enables rice to make beta-carotene comes from daffodils, but the modification actually required the introduction of four specific genes and would likely be impossible without genetic-engineering techniques. That is, genes were transferred between plants that would not exchange genes in nature. Again, the rule of natural change suggests that we should monitor such actions carefully.

Although the genetically modified rice appears to have beneficial effects, the government of India has refused to allow it to be grown in that country. There is worldwide concern about the political, social, and environmental effects of genetic modification of crops. This is a story in process, one that will change rapidly in the next few years. You can check on these fast-moving events on the textbook's website.

12.9 Grazing on Rangelands: An Environment Benefit or Problem?

Almost half Earth's land area is used as rangeland, and about 30% of Earth's land is *arid* rangeland, land that is easily damaged by grazing, especially during drought (Figure 12.12). In Canada, more than 99% of rangeland is in the Prairie Provinces. Much of the world's rangeland is in poor condition from overgrazing. In Canada, rangeland conditions have improved since the 1930s, especially in upland areas. However, land near streams, and the streams themselves, continue to be heavily affected by grazing.

Grazing cattle trample stream banks and release their waste into stream water. Maintaining a high-quality stream environment requires that cattle be fenced behind a buffer zone.

Agriculture and Agri-Food Canada actively manages 915,000 hectares of rangeland in 87 "community pastures" in western Canada. The agency's goal is to restore this land to "good" condition, with 50–75% of its biomass natural vegetation. Currently, about half the land is at this level, although studies have shown that modest improvements in range condition, for instance by reducing grazing during drought conditions, can improve carrying capacity by as much as 25%.[15]

Traditional and Industrialized Use of Grazing and Rangelands

Traditional herding practices and industrialized production of domestic animals have different effects on the environment. In modern industrialized agriculture, cattle are initially raised on open range and then transported to feedlots, where they are fattened for market. Feedlots have become widely known in recent years as sources of local pollution. The penned cattle are often

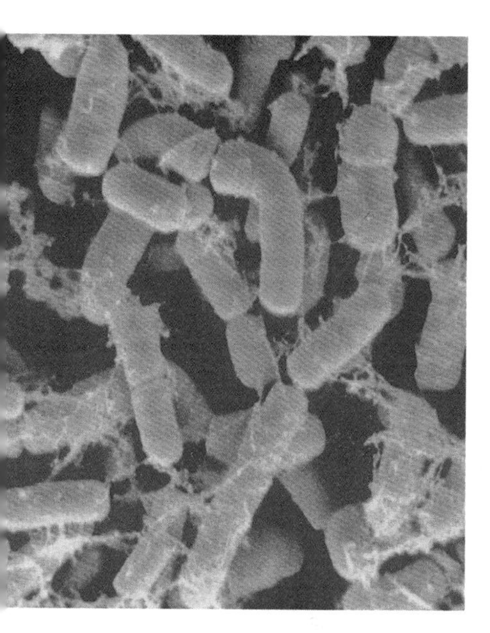

(a) *Bacillus thuringiensis* bacteria (a natural pesticide). The gene that caused the pesticide (BT) was placed into corn through genetic engineering.

(b) BT corn is identical in appearance to ordinary strains of corn, shown here. BT corn contains its own pesticide in every cell of the plant.

(c) Pollen from the BT corn is also toxic and when it lands on milkweed, monarch butterflies that eat the milkweed may die.

Figure 12.11 • **The flow of the BT toxin from bacteria** *(a)* **to corn through genetic engineering** *(b)* **and the ecological transfer of the toxic substances to monarch butterflies** *(c)*.

crowded and are fed grain or forage that is transported to the feedlot. Manure builds up in large mounds. When it rains, the manure pollutes local streams. Feedlots are popular with meat producers because they are economical for rapid production of good-quality meat. However, large feedlots require intense use of resources and have negative environmental effects.

Traditional herding practices, by comparison, chiefly affect the environment through overgrazing. Goats are especially damaging to vegetation, but all domestic herbivores can destroy rangeland. The effect of domestic herbivores on the land varies greatly with their density relative to rainfall and soil fertility. At low to moderate densities, the animals may actually aid growth of above-ground vegetation by fertilizing soil with their manure and stimulating plant growth by clipping off plants ends in grazing, just as pruning stimulates plant growth. But at high densities, the vegetation is eaten faster than it can grow; some species are lost, and the growth of others is greatly reduced.

The Biogeography of Agricultural Animals

People have distributed cattle, sheep, goats, and horses, and other domestic animals, around the world and then promoted the growth of these animals to densities that have changed the landscape. Pre-industrial people made such introductions. For example, Polynesian settlers brought pigs and other domesticated animals to Hawaii and other Pacific islands. Since the age of exploration by Western civilization, starting in the fifteenth century, domestic animals have been introduced into Australia, New Zealand, and the Americas. Horses, cows, sheep, and goats were brought to North America after the sixteenth century. The spread of cattle brought new animal diseases and new weeds, which arrived on the animals' hooves and in their manure. Introductions of domestic animals into new habitats have many environmental effects. Two important effects are:

(1) native vegetation, not adapted to the introduced grazers, may be greatly reduced and threatened with extinction; and
(2) animals introduced to a new habitat may compete with native herbivores, reducing their numbers to

Figure 12.12 • **Traditional sheep grazing, a practice that has occurred for thousands of years and affects almost half of Earth's land.**

a point at which they, too, may be threatened with extinction.

A recent important issue in cattle production is the opening of tropical forest areas and their conversion to rangeland (e.g., Brazilian Amazon Basin). In a typical situation, the forest is cleared by burning and crops are planted for about four years. After that time, the soil has lost so much fertility that crops can no longer be grown economically. Ranchers then purchase the land, already cleared, and run cattle bred to survive in the hot, humid conditions. After about another four years, the land can no longer support even grazing and is abandoned. In such areas, grazing has greatly impaired the land's capability for many uses, including forest growth.[16] Clearly, this is an unsustainable approach to agriculture and therefore undesirable.

The spread of domestic herbivores around the world is one major way people have changed the environment through agriculture. As the human population increases, and as income and expectations rise, the demand for meat increases. As a result, we can expect greater demand for rangeland and pastureland in the next decades. A major challenge in agriculture will be to develop ways to make the production of domestic animals sustainable.

Carrying Capacity of Grazing Lands

Carrying capacity is the maximum number of a species per unit area that can persist without decreasing the ability of that population or its ecosystem to maintain that density in the future. The carrying capacity of land for cattle varies with rainfall, topography, soil type, and soil fertility.

When the carrying capacity is exceeded, the land is overgrazed. **Overgrazing** slows the growth of the vegetation, reduces the diversity of plant species, leads to dominance by plant species that are relatively undesirable to the cattle, hastens the loss of soil by erosion as the plant cover is reduced, and subjects the land to further damage from the cattle's trampling on it (Figure 12.13). The damaged land can no longer support the same density of cattle.

In areas with moderate to high rainfall evenly distributed throughout the year, cattle can be maintained at high densities; but in arid and semi-arid regions, the density drops greatly. In Canada, the carrying capacity for cows drops from about 240 animal units/100 ha in Atlantic Canada, Québec, and Ontario, to about 3/100 ha in the dry central prairies of Manitoba and Saskatchewan. Farther west, as rainfall increases, carrying capacity also increases, to a maximum of about 80 animal units/100 ha in southern British Columbia. (Figure 12.14).

12.10 Desertification: Regional Effects and Global Impact

Deserts occur naturally where there is too little water for substantial plant growth. Because the plants that do grow are too sparsely spaced and unproductive to create a soil rich in organic matter, desert soils are mainly inorganic, coarse, and typically sandy (see the discussion of succession and soils in Chapter 10). When rain does come, it is often heavy, and erosion is severe. The principal climatic condition that leads to desert is low or undependable precipitation. The warmer the climate, the greater the rainfall required to convert an area from desert to non-desert, such as grassland. But even in cooler climates or at higher altitudes, deserts may form if precipitation is too low to support more than sparse plant life. The crucial factor is the amount of water in the soil available for plants to use. Factors that destroy the ability of a soil to store water can create a desert.

Earth has five natural warm desert regions, all of which lie primarily between latitudes 15° and 30° north and south of the equator. These include the deserts of the southwestern United States and Mexico; Pacific Coast deserts of Chile and southern Ecuador; the Kalahari Desert of southern Africa; the Australian deserts that cover most of that continent; and the greatest desert region of all—the desert that extends from the Atlantic Coast of North Africa (the Sahara) eastward to deserts of Arabia, Iran, Russia, Pakistan, India, and China.[17] Only Europe lacks a major warm desert; it lies north of the desert latitudinal band.

Based on climate, about one-third of Earth's land area should be desert, but estimates suggest that 43% of the land is desert. This additional desert area is believed

Figure 12.13 • **Soil erosion caused by overgrazing and other land-use practices.**

Figure 12.14 • **Carrying capacity of pasture and rangeland in Canada, in average number of cattle per square kilometre.** [*Source:* Statistics Canada, 2001 Census of Agriculture.]

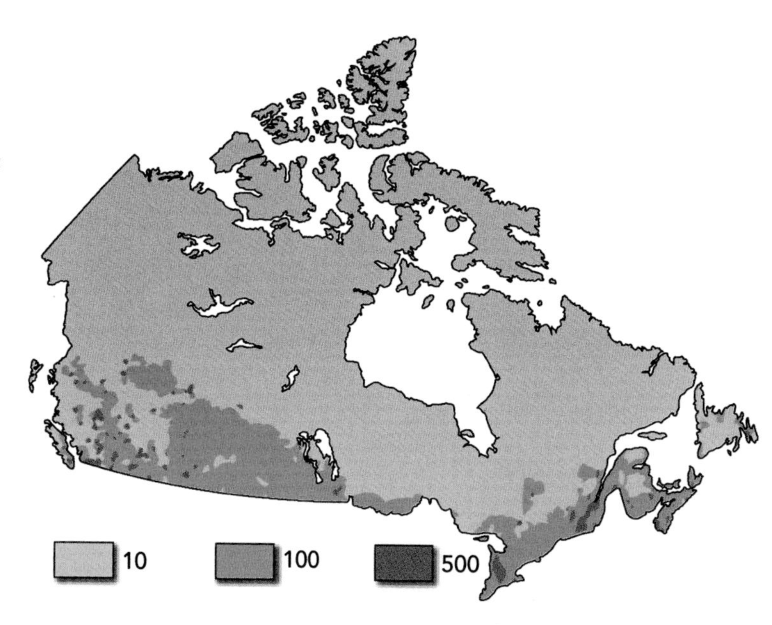

to be a result of human activities.[18] **Desertification** is the deterioration of land in arid, semi-arid, and dry subhumid areas due to changes in climate and human activities.[19]

Desertification is a serious global problem. It affects one-sixth of the world's population (about 1 billion people), 25% of the world's total land area, and 70% of all drylands (3.6 billion ha). Land degradation caused by people has altered 73% of drier rangelands (3.3 billion ha) and the soil fertility and structure of 47% of dryland areas with marginal rainfall for crops. Land degradation also affects 30% of dryland areas with high population density and agricultural potential.

A large part of desertification occurs in the poorest countries. These regions include Asia, Africa, and South America. Worldwide, 6 million hectares of land per year are lost to this process, with an estimated economic loss of $40 billion. And the cost of recovery of these lands could reach $10 billion per year.[20]

What Causes Deserts?

Some areas of Earth are marginal lands; even light grazing and crop production can turn them into deserts. In semi-arid regions, rainfall is just barely enough to enable the land to produce more vegetation than a desert, and even light grazing is a problem. The leading human causes of desertification are bad farming practices, such as failure to use contour ploughing or simply too much farming (Figure 12.15); overgrazing (Figure 12.13); the conversion of rangelands to croplands in marginal areas where rainfall is not sufficient to support crops over the long term; and poor forestry practices, including cutting all the trees in an area marginal for tree growth.

In northern China, areas that were once grasslands were overgrazed, and then some of these rangelands were converted to croplands. Both practices led to the conversion of the land to desert. Between 1949 and 1980, some 65,000 km^2 (an area larger than Denmark) became desert, and an additional 160,000 km^2 are in danger of becoming desert. As a result of desertification, the frequency of sandstorms increased from about three days per year in the early 1950s to an average of 17 days per year in the next decade, to more than 25 days per year by the early 1980s.[17] In 2001, Environment Canada reported that fine sand particles in dust clouds over British Columbia's lower mainland were actually sand from the Gobi Desert in Central Asia. High winds

Figure 12.15 • **Soil erosion.** *(a)* Gully soil erosion on cleared and ploughed farmland in South Australia. *(b)* Agricultural runoff carrying heavy sediment load.

(a)

(b)

in China and Mongolia had lifted the particles and carried them over 15,000 km to Canada!

Desertlike areas can be created anywhere by poisoning the soil. Poisoning can result from the application of persistent pesticides or other toxic organic chemicals from industrial processes that lead to improper disposal of toxic chemicals, and from airborne pollutant acidification, excessive manure in feedlots, and oil or chemical spills. All of these can poison soil, forcing abandonment or reduced agricultural use of lands. Worldwide, chemicals account for about 12% of all soil degradation. Ironically, irrigation in arid areas can also lead to desertification. When irrigation water evaporates, a residue of salts is left behind. Although these salts may have been in very low concentrations in the irrigation water, over time the salts can build up in the soil to the point at which they become toxic. This effect can sometimes be reversed if irrigation is increased greatly; the larger volume of water then redissolves the salts and carries them with it as it percolates down into the water table.[21]

Preventing Desertification

The first step in preventing desertification is detection of initial symptoms. The major symptoms of desertification are the following:

- Lowering of the water table (wells have to be dug deeper and deeper).
- Increase in the salt content of the soil
- Reduced surface water (streams and ponds dry up).
- Increased soil erosion (the dry soil, losing its organic matter, begins to be blown and washed away in heavy rains).
- Loss of native vegetation (not having adapted to desert conditions, native vegetation can no longer survive).[21]

Preventing desertification begins with monitoring these factors. Monitoring aquifers and soils is important in marginal agricultural lands. When we observe undesirable changes, we can try to control the activities producing these changes. Proper methods of soil conservation, forest management, and irrigation can help prevent the spread of deserts. (See Chapters 11 and 18 for a background discussion of soil and farming and irrigation practices.) In addition to the practices discussed earlier, good soil conservation includes the use of windbreaks (narrow lines of trees that help slow the wind) to prevent wind erosion of the soil. A landscape with trees is a landscape with a good chance of avoiding desertification. Practices that lead to deforestation in marginal areas should be avoided. Reforestation, including the planting of windbreaks, should be encouraged.

12.11 Does Farming Change the Biosphere?

People have long recognized local and regional impacts of agriculture, but it is a recent idea that farming might affect Earth's entire life-support system. This possibility came to people's attention in the twentieth century, first with events like the Dust Bowl of the 1930s, discussed earlier in this chapter, which led some to speculate that such disasters could become worldwide.[22] The idea gained supporters in the late twentieth century as satellites and astronauts gave us views of Earth from space and the idea of global ecology began to develop. How might farming change the biosphere? First, agriculture changes land cover, resulting in changes in the reflection of light by the land surface, the evaporation of water, the roughness of the surface, and the rate of exchange of chemical compounds (such as carbon dioxide) produced and removed by living things. Each of these changes can have regional and global climatic effects.

Second, modern agriculture increases carbon dioxide (CO_2) in two ways. As a major user of fossil fuels, it increases carbon dioxide in the atmosphere, adding to the buildup of greenhouse gases (discussed in detail in Chapter 20). In addition, clearing land for agriculture speeds the decomposition of organic matter in the soil, transferring the carbon stored in the organic matter into carbon dioxide, which also increases the CO_2 concentration in the atmosphere.

Agriculture can also affect climate through fire. Fires associated with clearing land for agriculture, especially in tropical countries, may have significant effects on the climate because they add small particulates to the atmosphere. Another global effect of agriculture results from the artificial production of nitrogen compounds for use in fertilizer, which may be leading to significant changes in global biogeochemical cycles (see Chapter 5).

Finally, agriculture affects species diversity. The loss of competing ecosystems (because of agricultural land use) reduces biological diversity and increases the number of endangered species.

Summary

- The Industrial Revolution and the rise of agricultural sciences have led to a revolution in agriculture, with many benefits and some serious drawbacks, including increased soil loss, erosion, and resulting downstream sedimentation, and associated water pollution.
- Modern fertilizers have greatly increased the yield per unit area. A wide variety of new pesticides have reduced, but not eliminated, the loss of crops to weeds, diseases, and herbivores.
- Most twentieth-century agriculture relied on machinery and the use of abundant energy, with relatively little attention paid to the loss of soils, the limits of groundwater, and the negative effects of chemical pesticides.
- Overgrazing has caused severe damage to lands. It is important to properly manage livestock, including using appropriate lands for grazing and keeping livestock at a sustainable density.
- Desertification is a serious problem that can be caused by poor farming practices and by the conversion of marginal grazing lands to croplands. Additional desertification can be avoided by improving farming practices, planting trees as windbreaks, and monitoring land for symptoms of desertification.
- Two revolutions are occurring in agriculture, one ecological and the other genetic. The ecological approach to agriculture will emphasize integrated pest management, ecosystems and biomes, taking into account the complexity of these systems. Soil conservation through no-till agriculture and contour ploughing, and water conservation. The genetic revolution is already the subject of controversy, offering both benefits and environmental dangers. Dangers will result if genetic modification is used without considering the ecosystem, landscape, biome, and global context in which it is done.

STUDY QUESTIONS

1. How can an insect pest species become resistant to a pesticide?
2. How can farming lead to the spread of deserts? What might be done to stop this desertification?
3. It has been said that farming can never be sustainable. What does this statement mean? Do you agree or disagree? List your reasons.
4. Design an integrated pest-management scheme for use in a small vegetable garden in a city lot behind a house. How would this scheme differ from IPM used on a large farm?
5. Under what conditions might grazing cattle be sustainable when growing wheat is not? Under what conditions might a herd of bison provide a sustainable supply of meat when cattle might not?
6. Consider the genetically modified rice that produces vitamin A. Make a table of the potential benefits and potential environmental problems that such a crop could produce.
7. Should genetically modified crops be considered acceptable for "organic" farming?

FURTHER READING

Acton, D. F. and L. J. Gregorich (eds.) *The Health of Our Soils: Toward Sustainable Agriculture in Canada.* Centre for Land and Biological Resources Research, Research Branch, Publication 1906/E. Ottawa: Agriculture and Agri-Food Canada, 1995.

Agriculture and Agri-Food Canada. *Profile of Production Trends and Environmental Issues in Canada's Agriculture and Agri-Food Sector.* Catalogue No. A22-166/2-1997E. Ottawa: Agriculture and Agri-Food Canada, 1997.

Colburn, T., D. Dumanoski, and J. P. Myers. *Our Stolen Future: Are We Threatening Our Fertility, Intelligence, and Survival?—A Scientific Detective Story.* New York: Plume, 1997.

Grainger, A. *Desertification: How People Make Deserts, How People Can Stop and Why They Don't,* 2nd ed. London: Earthscan Books, 1982.

Mitchell, B. and D. Shrubsole. *Canadian Water Management: Visions for Sustainability.* Cambridge, Ontario: Canadian Water Resources Association, 1994.

Rissler, J., and M. Mellon. *The Ecological Risks of Engineered Crops.* Cambridge, Mass.: MIT Press, 1996.

Toy, Terrence J., George R. Foster, and Kenneth G. Renard. *Soil Erosion: Processes, Prediction, Measurement, and Control.* New York: Wiley, 2002.

chapter 13

Conserving Carolinian Forests in Southern Ontario

One of Canada's rarest ecosystems is found in southernmost Ontario, in a gentle climate afforded by the closeness of lakes Ontario, Erie, and Huron. The broad-leaved Carolinian forest is one that properly belongs, as its name suggests, much farther south along the Atlantic coast. In Ontario, southern species like magnolia, pawpaw, and tuliptree, unfamiliar to the rest of Canada, are at the far northern limit of their range. Other Canadian rarities like chestnut, mockernut and pignut hickories, chinquapin, chestnut, scarlet, black, and pin oaks, black gum, blue ash, Kentucky coffee tree, redbud, red mulberry, and sassafras round out the flora. In all, some 2,200 plant species, including 70 species of trees, make this one of Canada's most diverse ecosystems, with a varied landscape of forests, grasslands, wetlands, and streams, which support numerous bird, reptile, amphibian, and insect species. At least 124 species in the Carolinian forest are considered vulnerable, threatened, or endangered provincially or nationally; 400 are considered rare. Successful conservation efforts must address a range of habitat types. In southern Ontario, this is a major challenge. Since European settlement, more than 70% of the province's natural wetlands have been drained or filled, and old growth forest exists only in scattered pockets.

How then do we conserve and save what is left of Canada's Carolinian ecosystem? In 1984, a conservation group called Carolinian Canada identified 38 "critical natural areas" (totalling 40,800 ha) and set them aside in a system of conservation areas and preserves. Today, we recognize that these habitat "islands" may not be adequate for conservation purposes. Conserving and restoring these systems require actions at many spatial scales, consistent with the variety of spatial scales that create and maintain their forests, grasslands, and wetlands. As we saw in Table 7.1, physical diversity and small variations in environmental conditions can increase biodiversity. In Carolinian ecosystems, the complex geographical patterns within wetlands and between them and upland forested and grassland areas are important for conservation. Moreover, many Carolinian species, such as the hooded warbler and king rail, need large areas of habitat for nesting and foraging. When the landscape is fragmented by development, or interrupted by major barriers such as highways, these species can suffer. Because of present land use pressures, much of the land currently available for Carolinian conservation and restoration is in relatively small parcels—approximately 10 to 100 ha—that lie within a complex landscape matrix of farmlands.

One solution is to link existing conservation areas into a larger, connected network (Figure 13.2) that allows free movement of plants and animals between larger habitat areas. This new approach is called "landscape perspective." In Carolinian Canada's Big Picture program[1], conservation experts used available data and state-of-the-art information technology to identify core natural areas, other significant natural areas, and potential habitat corridors to link the natural areas together. Tools such as land purchase, landowner stewardship, and education programs were used to acquire and protect the most critical Carolinian habitat fragments and create corridors to link them. For example, the

Figure 13.1 • Carolinian forest in the Catfish Creek Slope and Floodplain Forest, Southwestern Ontario.

Forests, Parks, and Landscapes

LEARNING OBJECTIVES

Forests and parks are among our most valued resources. Their conservation and management require that we understand landscapes—groups of interconnected ecosystems. This is a larger view that includes populations, species, and ecosystems. After reading this chapter, you should understand:

- What ecological services are provided by various kinds of landscapes.
- The landscape context for conservation and management of forests and parks.
- The basic principles of forest management, including its historical context.
- The roles that parks and nature preserves play in the conservation of wilderness.

entire habitat in the 233-ha Catfish Creek Slope and Floodplain Forest, on the central north shore of Lake Erie, is privately owned.[2] Through Natural Heritage Stewardship Agreements with the Catfish Creek Conservation Authority, landowners have made binding commitments to conserve the site and its many rare and threatened species in their natural state.

- *Those who seek to manage and conserve the Carolinian ecosystem of southern Ontario need to understand ecological processes on a variety of spatial scales. This chapter discusses modern approaches to conserving and managing landscapes for the extraction of useful products and the conservation of ecosystems and biological diversity.*

Figure 13.2 • Distribution of Carolinian habitat in Canada. The red dots represent Carolinian Canada's designated "signature" Carolinian habitat sites.

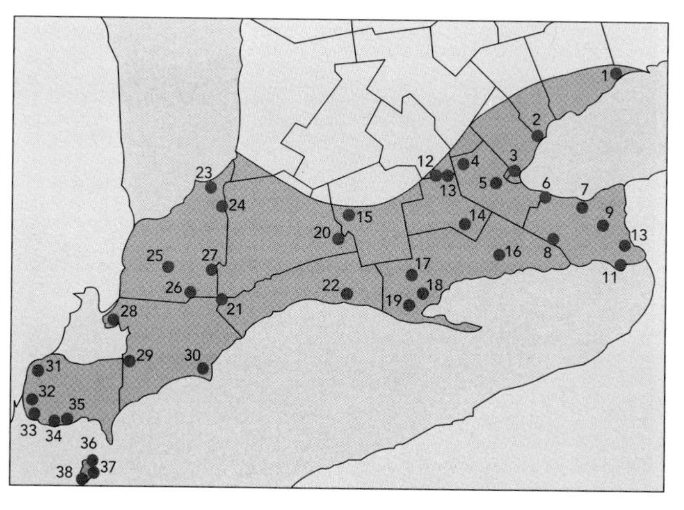

Site Names:

1. Rouge River Valley
2. Iroquois Shoreline Woods
3. Sassafras Woods
4. Beverly Swamp
5. Dundas Valley
6. Grimsby-Winona Escarpment and Beamer Valley
7. Jordan Escarpment Valley
8. Caistor-Canborough Slough Forest
9. Fonthill Sandhill Valley
10. Willoughby Clay Plain
11. Point Abino Peninsula Sandland Forest
12. Sudden Bog
13. Grand River Valley Forests and Spotiswood Lakes
14. Six Nations Forests
15. Embro Upland Forest
16. Oriskany Sandstone and Woodlands
17. Delhi Big Creek Valley
18. St.Williams Dwarf Oak Forest
19. Big Creek Valley - South Walsingham Sand Ridges
20. Dorchester Swamp
21. Skunk's Misery
22. Catfish Creek Slope and Floodplain Forest
23. Port Franks Wetlands and Forested Dunes
24. Ausable River Valley
25. Plum Creek Upland Woodlots
26. Shetland Kentucky Coffeetree Woods
27. Sydenham River Corridor
28. Walpole Island
29. Lake St. Clair Marshes
30. Sinclair's Bush
31. Ojibway Prairie Remnants
32. Canard River Kentucky Coffee-tree Woods
33. Big Creek Marsh
34. Oxley Poison Sumac Marsh
35. Cedar Creek
36. Middle Point Woods
37. Stone Road Alvar
38. Middle Island

13.1 The Landscape Concept

As the case study of the Carolinian ecosystem illustrates, populations, species, and ecosystems are connected across landscapes. Each ecosystem affects others and is affected by others. Many species require complex habitats involving more than one type of ecosystem to complete their life cycle. Thus, life is connected spatially on surprisingly large regional scales. While in some cases it is possible to conserve a small area without considering its surrounding landscape, in most cases management and conservation of natural living resources require a larger view. A recent advance in environmental sciences is the acceptance of this larger view—the **landscape perspective** (Figure 13.3).

In the early part of the twentieth century, when most activities concerning natural-resource management were just beginning to develop as science-based professions, the perspective was quite different. When the Canadian Forest Service was founded in 1899, the goal of natural-resource management was to maximize production of a single species or population in a single place. The focus was usually on a local landscape—a "stand"—an informal term used by foresters for a specific forest area. Management was designed primarily at the stand level.

Today, management and conservation of various habitat types require knowledge on several spatial scales, from the local (the forester's stand) to the regional. The need for both breadth and depth of knowledge makes conservation and management of landscapes more challenging, but it also makes successful management and conservation more likely.

Several questions remain unanswered, however. One question is what percentage of a landscape to include in parks and preserves and what percentage to set aside for growing and harvesting natural resources. The percentages will vary with the kind of habitat, the conditions required by individual species, and the values of the people participating in the decision making. Another important question involves the topology of the landscape: What shapes work best for parks, preserves, and areas open to harvest, and how are these different land uses best connected?

This chapter discusses the landscape perspective of forests and parks, because forests provide economically and socially valuable resources and are the subject of major environmental controversies. However, the general principles apply to all biomes. The chapter also presents basic principles of forest management and conservation. Only a few examples are presented, but, again, the principles discussed apply widely. The chapter discusses parks and nature preserves as well as the spatial connections among these areas.

13.2 Contemporary Conflicts over Forest Land and Forest Resources

In recent decades, forest conservation has become an international *cause célèbre*, especially conservation of remaining old-growth forests such as the giant trees of the rain forests in British Columbia and the Pacific Northwest and of tropical rain forests (Figure 13.4). Forestry, however, has a long history as a profession. The professional growing of trees is called **silviculture** (from silvus, Latin for "forest," and culture, as in "agriculture").

Silviculture has a very long history, but forestry as we know it today developed into a science-based activity, and what we today consider a profession in the late nineteenth and early twentieth centuries. The first modern Canadian professional forestry school was established at the University of New Brunswick in 1908, spurred by growing concerns about the depletion of Canada's living resources. Today, many of these schools have broadened their programs and become centers for environmental science.

Current conflicts about forests centre on the following questions:

- What is the appropriate balance between resource extraction and conservation of biodiversity, including endangered species, in a forest?
- Can a forest serve elements of both functions simultaneously and in the same place?

Figure 13.3 • **The landscape idea.** As this view of Reid Brook, Labrador shows, a real landscape often includes different ecosystems adjacent to one another. Different colours and different shapes indicate different habitats and ecosystems. In the foreground is the stream and its grassy floodplain. The stream flows into a lake ecosystem in the middle distance. Beyond is upland boreal forest indicated by the conical shapes and pointed tops of spruce. In the far distance are mountains with alpine vegetation.

Figure 13.4 • Temperate rain forest on Vancouver Island.

- Can a forest be managed sustainably for either use? If so, how?
- Can a forest be managed for either resources or conservation and provide recreation and landscape beauty as well as meet the spiritual needs of people? Does this matter anyway?
- What role do forests play in our global environment such as their effects on the climate?

Table 13.1 lists more specific issues in forestry.

Forests have always been important to people, and forests and society have always been closely linked. Since the earliest human cultures, wood has been one of the major building materials and the main fuel source. Forests provided materials for the first boats and first wagons. Today, nearly half the people in the world depend on wood for cooking, and in developing nations wood remains the primary heating fuel.[3]

At the same time, people have appreciated forests for spiritual and aesthetic reasons. There is a long history of sacred forest groves. When Julius Caesar was trying to conquer the Gauls in what is now southern France, he found the enemy difficult to defeat on the battlefield. He burned the society's sacred groves to demoralize them—demonstrating the spiritual importance of forests to the Gauls (and serving as an early example of psychological warfare). In the Pacific Northwest, the great forests of Douglas fir provided many necessities of life to First Nations peoples from housing to boats, and were, as they are today, important spiritually.

Forests also benefit people and the environment indirectly through **public service functions**. Forests sequester carbon, and thus counteract the influence of greenhouse gases on global climate (Chapter 20). At regional and global levels, forests may be significant influences on the climate. Forests reduce erosion and moderate the availability of water (Figure 13.5), improving the water supply from major watersheds to cities. Forests are habitats for endangered species and other wildlife. They are important for recreation, hiking, hunting, and bird and wildlife viewing, and for their simple aesthetic value, as beautiful places to be.

In the early twentieth century, a major goal of silviculture was to maximize the yield in the harvest of a single resource. The ecosystem was a minor concern, as were non-target, non-commercial species and associated wildlife. (The notion of sustainable forestry does have its roots in that time, however. For instance, Bernard Fernow, an American forester who became the head of the University of Toronto's School of Forestry, recognized the need for the forest industry and national governments to ensure that forests could be managed sustainably, and for research and education to support that need.[4]) Today, most forestry schools take a much broader view, considering timber harvest and ecosystem sustainability as part of the entire range of goals for managing forests.

Table 13.1 • Major Forestry Issues

- *Sustainability*: How can we achieve sustainable forestry? (This is the fundamental question.)
- *Clear-cutting*: Is clear-cutting ever good? Should it ever be allowed?
- *Old-growth forests*: Should all old-growth forests be preserved, or should some logging be allowed in some old-growth stands?
- *Plantations*: Are plantations intrinsically bad because they involve intentional manipulation of land to grow trees, or are they one of the keys to achieving the goals of biological conservation of forests?
- *Stream-protection zones*: Should all streams everywhere have a wooded buffer zone in which no logging and no other harvesting or destructive activities are permitted? If so, how wide should that buffer zone be?
- *National forests*: Is the purpose of national forests to provide the nation with a reliable source of timber, or is it to conserve living resources? What is the role of recreation in national forests.
- *Forest fires*: Are forests fires almost all bad, are they occasionally beneficial to a forest, or are they frequently important and essential to forests?
- *Certification*: Should our society certify forestry practices as sustainable? If so, how should this be done and who should do it?
- *Scale of management*: What is the appropriate spatial and temporal scale for management of forests?
- *People's role*: Should people play any active role in the management of forests? If so, what kinds of activities should they undertake, how often, and where?

Figure 13.5 • A forested watershed, showing the effects of trees in evaporating water, slowing erosion, and providing wildlife habitat.

13.3 Tree Niches

To understand the role of trees in a forested ecosystem, we must understand something about how trees grow (Figure 13.6). Each species of tree has its own niche (see Chapter 6) and is adapted to specific environmental conditions. For example, in boreal forests (see Chapter 8), a determinant of a tree niche is the water content of the soil. White birch grows well in dry soils; balsam fir grows well in well-watered sites; and black spruce grows well in bogs (Figure 13.7).

Another way in which niches are determined is through forest tolerance of shade. Some trees that grow only in the bright sun of open areas are found in clearings. Other species can grow in shade; seedlings and saplings of these species are found within the deep shade of an old forest. Sugar maple and beech are typical of these shade-tolerant trees. Birch and cherry, examples of trees that require bright sunlight, are "shade-intolerant."

Most of the big trees of western Canada and the United States require open, bright conditions and certain kinds of disturbances in order to germinate and survive the early stages of their lives. These include coastal redwood, which out-competes other species only if both fires and floods occasionally occur; Douglas fir, which begins its growth in openings; and the giant sequoia, whose seeds will germinate only on bare, mineral soil. We discussed these requirements in somewhat different terms in Chapter 10. Some trees are adapted to early stages in succession, when sites are open and there is bright sunlight. Others are adapted to later stages in succession, when there is a high density of trees.

Understanding the specific requirements of individual tree species helps us to determine where they will grow, the best place to plant them as a commercial crop, and where they might best contribute to biological conservation or to the beauty of a landscape. Clearly, there is no single best set of conditions for a forest. Many kinds of forests adapted to many kinds of conditions. The same is true for all landscapes—grasslands, deserts, rivers, and lakes.

13.4 A Forester's View of a Forest

Traditionally, foresters have managed trees at the local level of a stand. As noted earlier, a **stand** is an informal term that foresters use to refer to a group of trees. Stands can be small (half a hectare) to medium-size (several hundred hectares). Foresters classify stands based on tree composition. The two major kinds of commercial stands are even-aged stands, where all live trees began growth from seeds and roots germinating the same year, and uneven-aged stands, which have at least three distinct age classes. In even-aged stands, trees are approximately the same height but differ in girth and vigour.

A forest that has never been cut is called an **old-growth forest**. A forest that has been cut and has

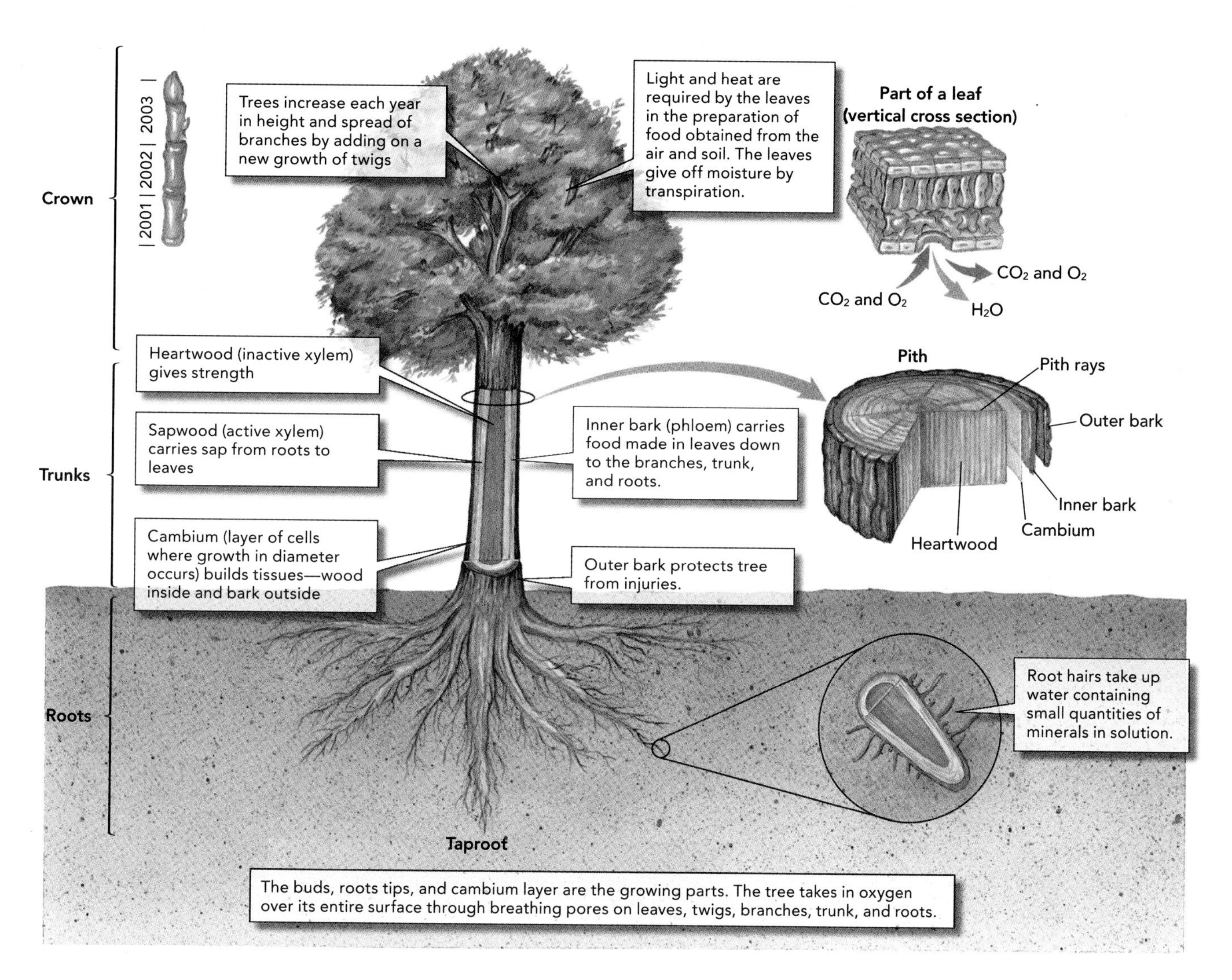

Figure 13.6 • **How a tree grows.** [*Source*: Modified from C. H. Stoddard, *Essentials of Forestry Practice*, 3rd ed. (New York: Wiley, 1978).]

regrown is called a **second-growth forest**. Although the term old-growth forest has gained popularity in several well-publicized disputes about forests, it is not a scientific term and does not yet have an agreed-on, precise meaning. In popular usage, it often refers to virgin forest. Another important management term is **rotation time**. As noted earlier, this is the time between cuts of a stand.

Foresters and forest ecologists group the trees in a forest into the **dominants** (tallest, most common, and most vigorous), **codominants** (fairly common, sharing the canopy or top part of the forest), **intermediate** (forming a layer of growth below dominants), and **suppressed** (growing in the understory).

Productivity of a forest varies according to soil fertility, water supply, and local climate. Foresters classify sites by **site quality**, which is the maximum timber crop the site can produce in a given time. Site quality can decrease with poor management. For example, too-frequent burning of forests decreases the potential for tree growth by lowering soil fertility. Traditionally, foresters developed site indexes for types of forestlands and derived yield tables to estimate future production.

Figure 13.7 • **Some characteristics of tree niches.** Tree species have evolved to be adapted to different kinds of environments. In northern boreal forests, white birch grows on dry sites (and early successional sites); balsam fir grows in wetter soils, up to wetlands; and black spruce grows in the wetter sites associated with northern bogs.

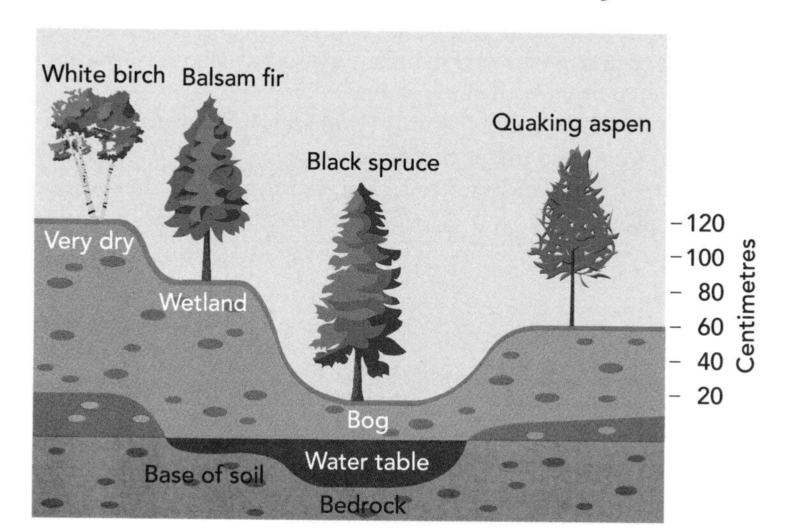

Today, forecasts of forest production frequently rely on computer simulation.

Although forests are complex and difficult to manage, one advantage they have over many other ecosystems is that trees provide easily obtained and helpful information. For example, the age and growth rate of trees can be measured from tree rings (see Figure 13.6). In temperate and boreal forests, trees produce one growth ring per year. This information can be used in forest management.

13.5 Approaches to Forest Management

Managing forests can involve removing poorly formed and unproductive trees (or selected other trees) to permit larger trees to grow more rapidly, planting genetically controlled seedlings, and fertilizing the soil. Forest geneticists breed new strains of trees just as agricultural geneticists breed new strains of corn, wheat, tomatoes, and other crop plants. New "supertrees" are allegedly able to maintain a high rate of growth and increase the total production of forests.

Another aspect of silviculture is the control of diseases and pests. There has been relatively little success in controlling diseases in forests. Tree diseases are primarily fungal. Often, as with Dutch elm disease, an insect spreads the fungus from tree to tree. Insect outbreaks occur infrequently, but when they do occur, they can have large-scale results. Some insect problems are due to the introduction of exotic species. For example, the gypsy moth, introduced intentionally into New England around the turn of the twentieth century as a source of silk, escaped and has spread through many eastern states. Other insect outbreaks have recurred naturally for a long time. For example, a nineteenth-century New Hampshire gazetteer referred to a "plague of loathesome [sic] worms" that removed all the leaves from large areas of forest.

Insects affect trees by defoliating them; by eating the buds at the tops of the trees and destroying the main trunk, causing forked growth; by eating fruits; and by serving as carriers of diseases. Insecticides are sometimes used to combat these pests.

The United Nations Food and Agriculture Organization recognizes two broad categories of current forest management practices: clear-cutting (sometimes called clearfelling) and selective harvesting. These practices are discussed in more detail in the following sections.

Clear-cutting

Clear-cutting (Figure 13.8) is the cutting of all trees in a stand at the same time. In North America, most clear-cutting operations are highly mechanized, and most harvesting is done by specialized contractors, despite whether the stand is privately or publicly owned. The scale of many major forestry operations in Canada, especially in the boreal forest, is such that huge, heavy machinery is usually required. While such methods are efficient, they can cause serious disturbance and compaction of the forest floor. By contrast, clear-cutting in Central and Eastern Europe often employs manual methods or light harvesting equipment, typically limited to a few square hectares. These methods are much more expensive and can be difficult, especially when large trees must be felled, but they are much more protective of forest soils.

Scientists have tested the effects of clear-cutting.[5-8] In one experiment, a consortium of the Alberta Environmental Centre, the Canadian Forest Service, and Alberta Land and Forest Services studied the differences between clear-cut hardwood forests and those affected by fire. They concluded that current clear-cutting practices reduce the structural complexity of young forests, and thus change the composition and abundance of biota and the nature of ecological processes. The consortium concluded that Alberta's current forest management practices are inadequate to ensure the ecological integrity—the ability to maintain natural processes and biodiversity over time—of Alberta's boreal mixed forests.[7] In another study, at the U.S. Forest Service Hubbard Brook Experimental Forest in New Hampshire, an entire watershed was clear-cut (see Figure 6.10) and herbicides were applied to prevent regrowth for two years.[6] The results were dramatic. Erosion increased and the pattern of water runoff changed substantially. The exposed soil decayed rapidly and the concentrations of nitrates in the stream water exceeded public health standards.

Clear-cutting also changes chemical cycling in forests and causes the soil to lose chemical elements necessary for life. When a forest is clear-cut, trees are no longer available to take up nutrients. Open to the sun

Figure 13.8 • A clear-cut forest in western British Columbia.

Figure 13.9 • Effects of clear-cutting on forest chemical cycling. Chemical cycling *(a)* in an old-growth forest and *(b)* after clear-cutting. *(c)* Increase in nitrate concentration in streams following logging and the burning of slash (leaves, branches, and other tree debris). [*Sources*: *(a)* and *(b)* adapted from R. L. Fredriksen, "Comparative Chemical Water Quality—Natural and Disturbed Streams Following Logging and Slash Burning," in *Forest Land Use and Stream Environment*. Corvallis: Oregon State University, 1971), pp. 125–137.]

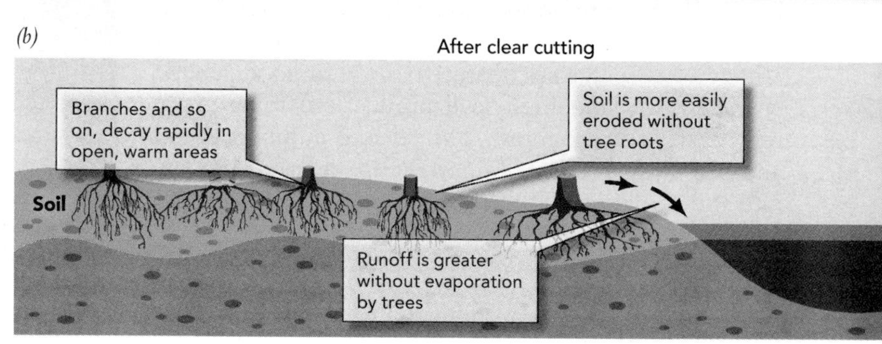

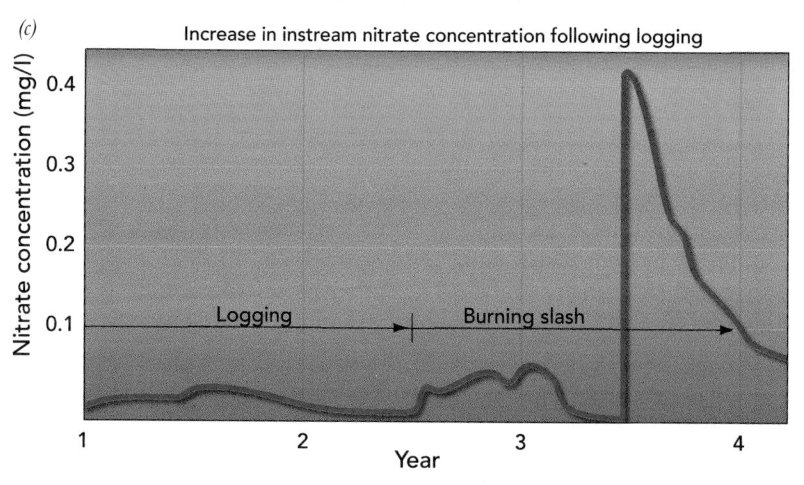

and rain, the ground becomes warmer. This accelerates decay, with chemical elements, such as nitrogen, converted more rapidly to forms that are water-soluble and readily lost in runoff during rains (Figure 13.9).[8] The greater the damage to the forest floor, for instance from the movement of heavy machinery, the greater the potential for erosion and loss of nutrients from the site.

The Hubbard Brook experiments showed that clear-cutting can be a particularly poor practice on steep slopes in areas of moderate to heavy rainfall. The worst effects of clear-cutting resulted from the logging of vast areas of North America during the nineteenth and early twentieth centuries. Clear-cutting on such a large scale is neither necessary nor desirable for the best timber production. However, where the ground is level or only slightly sloped, where rainfall is moderate, and where the desirable species require open areas for growth, clear-cutting on an appropriate spatial scale may be a useful way to regenerate desirable species. The key here is that clear-cutting is neither all good nor all bad for timber production or forest ecosystems. The use of clear-cutting must be evaluated on a case-by-case basis, taking into account the size and shape of cuts, the environment, and the available species of trees.

Selective Harvesting

The term selective harvesting refers to a group of alternatives to clear-cutting, including selective cutting, strip-cutting, shelterwood-cutting, and seed-tree cutting. In **selective cutting**, individual trees are marked and cut. Sometimes smaller, poorly formed trees are selectively removed; this practice is called **thinning**. At other times, trees of specific species and sizes are removed. For example, in Costa Rica, some forestry companies use a combination of these methods and cut only some of the largest mahogany trees, leaving other, less valuable trees to help maintain the ecosystem and permitting some of the large mahogany trees to continue to provide seeds for future generations. Selective cutting and thinning are common in Central and Eastern Europe, where clear-cutting is limited to small areas. These alternatives are gaining prominence in North America, especially in forested ecosystems that require careful management. A traditional type of selective cutting in Central Europe is the Plenterwald (selection forestry) system, which creates and manages mixed hardwood forests of uneven age.[9] In the Plenterwald system, mature trees are cut one at a time, with smaller trees removed to maintain the desired forest composition, encourage particular shade and light conditions, or control disease.

In **strip-cutting**, narrow rows of forest are cut, leaving wooded corridors. Strip-cutting offers several advantages. The uncut strips protect regenerating trees from wind and direct sunlight, and these remaining trees provide seeds. In addition, strip-cutting can minimize the negative aesthetic effects of logging by leaving buffer zones and allowing the corridors of forest that remain to be used for recreation and as wildlife habitats. The size of the cuts, relative to remaining forest corridors, is important in determining the recovery rate of the cut area. **Shelterwood-cutting** is the practice of cutting dead and less desirable trees first and later cutting mature trees. As a result, there are always young trees left in the forest to provide some shelter for wildlife and light shade for seedlings. **Seed-tree cutting** removes all but a few seed trees (mature trees with good genetic characteristics and high seed production) to promote regeneration of the forest. The relative numbers of cut and standing trees are important in determining the rate and composition of the regrowth.

Plantation Forestry

Sometimes foresters grow trees in a **plantation**, which is a stand of a single species planted in straight rows (Figure 13.10). Typically, the land is fertilized, sometimes by helicopter, and modern machines make harvesting rapid—some remove the entire tree, root and all. Plantation forestry is thus much like modern agriculture. Intensive management like this is common in Europe and parts of Canada. In some countries such as Chile and New Zealand, large forest plantations have replaced natural forest in many areas, and now supply most of those countries' demand for timber.[9] Like other monocultures, however, forest plantations are more vulnerable to the effects of insects and disease, and may be less productive than more diverse systems over the long term.

Other forests such as those in much of northern and eastern Canada are managed less actively. In these, seeds from existing trees are allowed to regenerate naturally. Ecological succession proceeds without management intervention (see Chapter 10). Which approach is best depends on the type of forest, the environment, and the characteristics of the commercially valuable species. Which is better for trees, ecosystems, and people—plantations, or forests allowed to grow without human interference until it is time to harvest, or something in between? Which is sustainable? Which is better for biodiversity, for landscape beauty, etc.?

Forest plantations offer an important alternative solution to the pressure on natural forests. If plantations were used where forest production is high, then a comparatively small percentage of the world's forestland could provide all the world's timber. For example, high-yield forests

Figure 13.10 • **A modern Canadian forest plantation.**

produce 15–20 m^3/ha/yr. According to one estimate, if plantations were put on timberland that could produce at least 10 m^3/ha/yr, then 10% of the world's forestland could provide enough timber for the world's timber trade.[10] This could reduce pressure on old-growth forests, on forests important for biological conservation, and on forestlands important for recreation. However, critics of plantation forestry argue that although plantations grow trees, they do not grow forests, and cannot replace the diverse ecological services that forests provide.

13.6 Sustainable Forestry

A major goal today is to have **sustainable forestry**. Stated in the most general terms, a sustainable forest is one from which a resource can be harvested at a rate that does not decrease the ability of the forest ecosystem to continue to provide that same rate of harvest indefinitely. In reality, the situation is more complicated.

What Is Forest Sustainability?

There are two basic kinds of ecological sustainability: sustainability of the harvest of a specific resource that grows within an ecosystem; and sustainability of the entire ecosystem—and therefore of many species, habitats, and environmental conditions. For forests, this translates into sustainability of the harvest of timber and sustainability of the forest as an ecosystem. Although sustainability has long been discussed in forestry, there is inadequate scientific data to demonstrate that sustainability of either kind has ever been achieved in forests, except in a few unusual cases.[4]

Certification of Forest Practices

If the data do not indicate whether a particular set of practices has led to sustainable forestry, how can we achieve it? The current general approach is to compare the actual practices of specific corporations or government agencies with practices believed to be consistent with sustainability. This has become a formal process called **certification of forestry**, and there are organizations whose main function is to certify forest practices. The catch here is that nobody actually knows what exactly is required to ensure sustainability, or whether the recommended practices will actually turn out to be sustainable. Since trees take a long time to grow, and a series of harvests is necessary to prove sustainability, the proof lies in the future. Despite this limitation, certification of forestry is becoming common. As practiced today, it is as much an art or craft as it is a science.

Worldwide concern about the need for forest sustainability has led to international efforts to ban imports of wood produced from purportedly non-sustainable forest practices and to the development of international programs for certification of forest practices. Some European nations have banned the import of certain tropical woods, and some environmental organizations have organized demonstrations in support of such bans.

There is a gradual movement away from calling certified forest practices "sustainable" and to the alternative terms "well-managed forests" or "improved management."[11,12] A small industry has developed, comprising companies that review the management of forests and provide certification. However, standardized criteria have not been established. The search for such acceptable uniform criteria has led to an ongoing series of international meetings, including major ones in Montreal and Helsinki. Since 1993, the Canadian Sustainable Forestry Certification Coalition, a group of 22 forest industry organizations, has worked with the Canadian Standards Association to develop and promote Sustainable Forest Management (SFM) standards. Certified forest management is rapidly gaining acceptance in Canada. The total certified area within Canada tripled between 2002 and 2004, from about 28 million ha to 86 million ha.

Certification programs began because environmental organizations were concerned that forests were being cut in an unsustainable fashion. The certification process has had an unintended consequence: In some cases it has evolved into a means for producers of wood and wood products (private corporations and nations that depend on wood exports) to reach new markets, keep existing ones, or increase the value of the product (because people will pay more for a "certified" wood product).

Some scientists have begun to call for a new forestry that includes a variety of practices that they believe would increase the likelihood of sustainability. Most basic is accepting the dynamic characteristics of forests—that to remain sustainable over the long term, a forest may have to change in the short term. Some of the broader, science-based concerns are presented in terms of the need for ecosystem management and a landscape context. Scientists point out that any application of a certification program should be regarded as an experiment requiring a precautionary approach. Therefore, any new programs that claim to provide sustainable practices must include, for comparison, control areas where no cutting is done and must also include adequate scientific monitoring of the status of the forest ecosystem.

13.7 A Global Perspective on Forests

The functioning of an individual tree provides the basic understanding about how forests can affect the entire biosphere. Vegetation of any kind can affect the atmosphere in four ways (Figure 13.11):

Figure 13.11 • Four ways that a forest can affect the atmosphere: (1) some solar radiation is absorbed by trees and some is reflected, changing the local energy budget, compared to a non-forest environment; (2) evaporation and transpiration from trees, together called evapotranspiration, transfer water to the atmosphere; (3) carbon dioxide is removed from the air and oxygen is released to the atmosphere by photosynthesis from trees (carbon dioxide is a greenhouse gas associated with climate change, and reducing the gas cools the temperature of the atmosphere, see Chapter 20); and (4) near-surface wind is reduced because the trees produce a roughness near the ground that slows the wind.

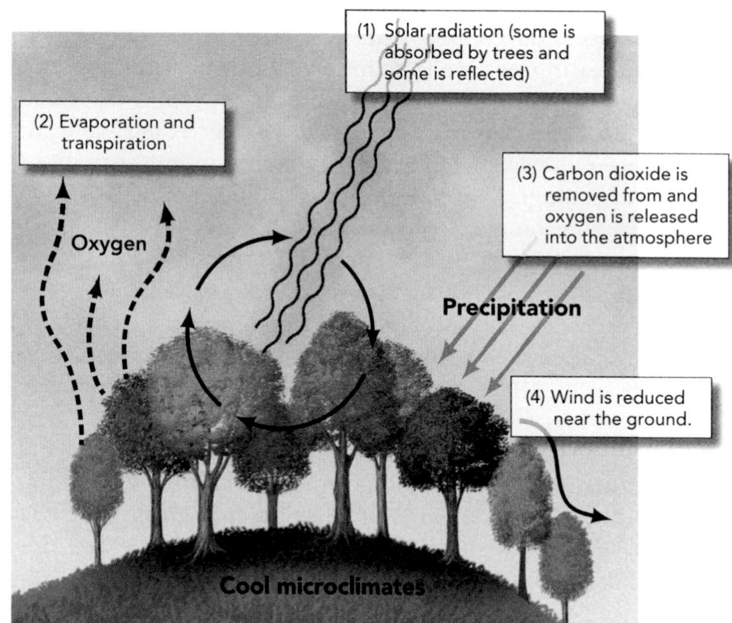

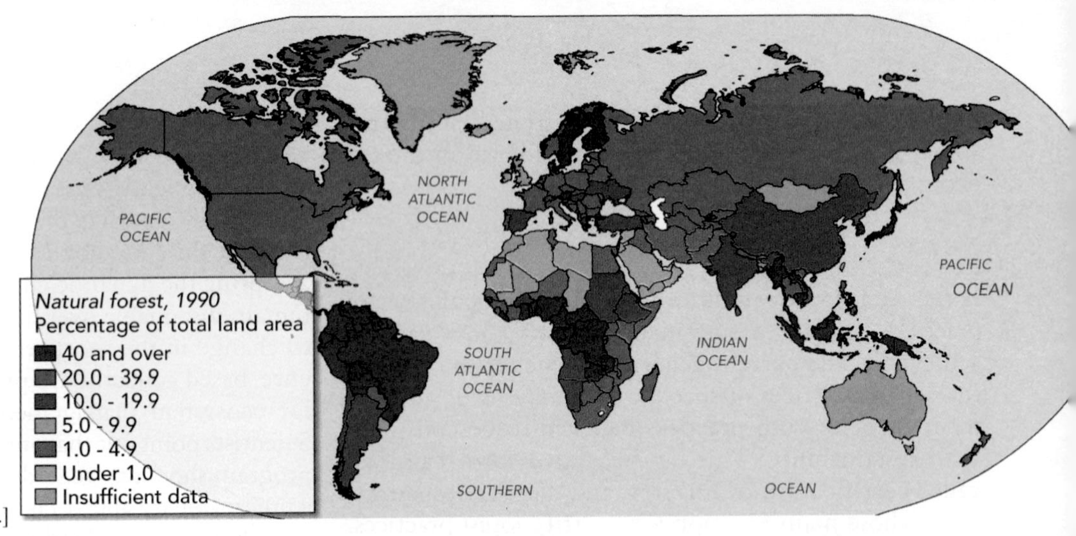

Figure 13.12 • The world's forested area, by nation. [*Source: State of the World's Forests 2001* (Rome: U.N. Food and Agriculture Organization), available at http://www.fao.org/docrep/U8480E56.jpg.]

1. By changing the color of the surface and therefore the amount of sunlight reflected and absorbed.
2. By increasing the amount of water transpired and evaporated from the surface to the atmosphere.
3. By sequestering carbon, and thus changing the rate at which greenhouse gases are released from Earth's surface into the atmosphere (Chapter 20).
4. By changing "surface roughness," which affects wind speed at the surface, and thus microclimate close to ground level.

In general, vegetation makes the surface darker, so it absorbs more sunlight and reflects less, warming the Earth. The contrast is especially strong between the dark needles of conifers and winter snow in northern forests and between the dark green of shrublands and the yellowish soils of many semi-arid climates.

Vegetation in general and forests in particular tend to evaporate more water than bare surfaces. This is because the total surface area of the many leaves is many times larger than the area of the soil surface. Is this increased evaporation good or bad? That depends on one's goals. Increasing evaporation means that less water runs off the surface. This reduces erosion. Although increased evaporation also means that less water is available for our own water supply and for streams, in most situations the ecological and environmental benefits of increased evaporation outweigh the disadvantages.

Vegetation also takes up and stores carbon through the process of photosynthesis. In addition to creating new biomass for ecosystem function and resource extraction, this process sequesters carbon that would otherwise enter the atmosphere, contributing to the effect of greenhousse gases and the challenge of climate change.

At the local level, vegetation can cool streams and undergrowth by shading, slow prevailing winds, and otherwise moderate a local or "micro" climate. If you have ever sought shade under a tree on a hot day, this effect will be familiar to you.

World Forest Area, Global Production, and Consumption of Forest Resources

At the beginning of the twenty-first century, the world contained approximately 3.87 billion ha of forested area covering approximately 26.6% of Earth's surface (Figure 13.12).[13] This works out to about 0.6 ha per person. The forest area is up from 3.45 billion ha estimated in 1990, but down from 4 billion ha in 1980. Countries differ greatly in their forest resources, depending on the potential of their land and climate for tree growth and on their history of land use and deforestation. Ten nations have two-thirds of the world's forests. In descending order, these are the Russian Federation, Brazil, Canada, the United States, China, Australia, the Democratic Republic of the Congo, Indonesia, Angola, and Peru.

Developed countries account for 70% of the world's total production and consumption of industrial wood products; developing countries produce and consume about 90% of wood used as firewood. North America is the world's dominant supplier of timber for construction, pulp, and paper, which together make up approximately 90% of the world timber trade (the rest consists of hardwoods used for furniture such as teak, mahogany, oak, and maple).[3] Total global production/consumption is about 1.5 billion m3 annually—roughly the same volume as half a million Olympic-size swimming pools.

Canada has approximately 401.5 million ha of treed land.[14] Of this, 97 million ha are treed wetlands and land with scattered trees, such as wooded prairie, leaving 309.5 million ha of true forest. Of the total treed area, 93% is owned by the Crown; only 7% is privately owned. More than three-quarters of Canada's forest is composed of coniferous species, mainly spruce, fir, and pine. The remaining forest is broadleaved deciduous or mixed forest, with species like oak, maple, birch and ash. Of Canada's true forest, only 144.6 million ha are considered accessible for commercial forest activities. Most of Canada's timber production is in Québec (86.4 million ha, equating to $10.7 billion in export value), closely followed by British Columbia (64.1 million ha; $12.6 billion in export value) and Ontario (68.3 million ha; $10.7 billion in export value), although all areas of Canada possess some exportable timber or other wood products.[15]

In recent years, the world trade in timber has not grown substantially. Thus, the amount traded annually (about 1.5 billion m^3, as mentioned earlier) is a reasonable estimate of the total present world demand for the 6 billion people on Earth, at their present standards of living. The fundamental questions are whether and how Earth's forests can continue to produce at least this amount of timber for an indefinite period, and whether and how they can produce even more as the world's human population continues to grow and as standards of living rise worldwide. All of this has to happen while forests continue to perform their other functions, which include ecosystem goods and services, biological conservation functions, and functions involving the aesthetic and spiritual needs of people. In terms of the themes of this book, the question is: How can forest production be sustainable, even increase, while meeting the needs of people and nature? The answer involves science and values.

Again, a useful way to approach the problems and goals associated with forests is to begin by understanding the life and needs of an individual tree. From that, we can expand outward.

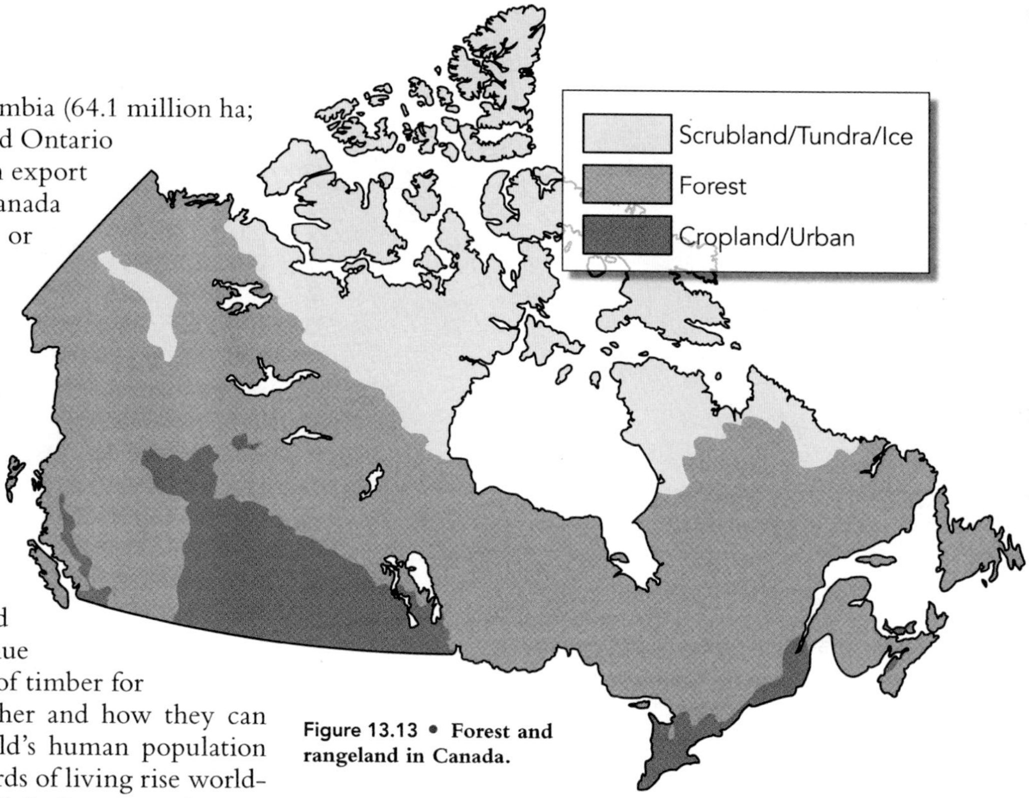

Scrubland/Tundra/Ice
Forest
Cropland/Urban

Figure 13.13 • **Forest and rangeland in Canada.**

13.8 Deforestation: A Global Dilemma

Another global effect of forestry is that cutting forests in one country affects other countries. For example, deforestation is estimated to have increased erosion and caused the displacement of 562 million ha of soil worldwide, at an estimated rate of 5 to 6 million ha/yr.[16] Nepal, one of the most mountainous countries in the world, lost more than half its forest cover between 1950 and 1980, mainly due to excessive harvesting of fuelwood and livestock fodder. Such cutting destabilizes soil, increasing the frequency of landslides, amount of runoff, and sediment load in streams. Many Nepalese streams feed rivers that flow into India (Figure 13.14). Recent heavy flooding in India's Ganges Valley has caused $1 billion a year in property damage and is blamed on the loss of large forested watersheds in other countries.[17] The loss of forest cover in Nepal continues at a rate of about 100,000 ha per year. Reforestation efforts replace less than 15,000 ha per year. If present trends continue, little forestland will remain in Nepal, thus permanently exacerbating India's flooding problems.[18]

Determining the worldwide net rate of change in forest resources is difficult. Some experts argue that there is a worldwide net increase in forests because large areas in the temperate zone, such as the eastern and midwestern United States, were cleared in the nineteenth and early twentieth centuries and are now regenerating. Most experts disagree. Because few forests are successfully managed to achieve sustainability, it seems likely that the world's forests are undergoing a net decline, perhaps rapidly. However, until recently there has been little information on which to base an accurate evaluation. Since forests cover large, often remote areas that are seldom visited or studied, it has been difficult to assess the total amount of forest area. Only recently have satellite remote sensing technologies made is possible to obtain accurate estimates of the distribution and abundance of forests, and these suggest that past methods overestimated forest biomass by 100% to 400%.[19]

Recognizing the limits of existing information, it is nonetheless useful to review standard information about forests. It is estimated that forests covered one-quarter of Earth's land area in 1950 but only one-fifth in 1980.[20] Modern technologies, including remote sensing and geographic information systems, now make it possible to assess the total world land area made up of forests and will enable us to track future changes.

(a)

(b)

Figure 13.14 • The role of forests in controlling erosion. *(a)* Planting pine trees on the steep slopes in Nepal to replace forests that have been cut. The dark green in the background is yet uncut forest, and the contrast between foreground and background suggests the intensity of clearing that is taking place. *(b)* The Ganges River in northern India carries a heavy load of sediment, as shown by the sediments deposited within and along the flowing water and by the color of the water itself.

History of Deforestation

As we saw in Chapter 9, forests were cut in the Near East, Greece, and the Roman Empire before the modern era. Removal of forests continued northward in Europe as civilization advanced. Fossil records suggest that prehistoric farmers in Denmark cleared forests so much that early-successional weeds occupied large areas. In medieval times, Great Britain's forests were cut and many forested areas were eliminated. With colonization of the New World, much of North America was cleared.[21]

Today, deforestation is largely concentrated in the developing world, which lost approximately 200 million ha from 1980 to 1995.[22] Many of these forests are in the tropics, mountain regions, or high latitudes, places difficult to exploit before the advent of modern transportation and machines.[23] The problem is especially severe in the tropics because of high human population growth. For example, fire-related deforestation rose sharply in Indonesia in 1997. Estimates of forest area destroyed in 1997 range from 150,000 to 300,000 ha.[22] On the other hand, the rate of deforestation in tropical rain forests appears to be slowing, based on a United Nations analysis of 300 satellite images[24]—and that satellite images provide a new way to detect deforestation (Figure 13.15).

Other areas where tropical deforestation is a major concern include the Amazon basin, western and central Africa, and Southeast Asia and the Pacific regions, including Borneo, the Philippines, and Malaysia. According to the World Resources Institute, Asia cleared 30%, Africa 18%, Latin America 18%, and the world as a whole 20% of tropical forests between 1960 and 1990. It is important to note that estimates are, if anything, underestimates, because organizations must rely primarily on official statements from nations that do not necessarily include losses from illegal cutting or other unstated cutting.[23]

Causes of Deforestation

The most common reason that people cut forests is to use or sell timber for lumber, paper products, or fuel. Logging by large timber companies and local cutting by villagers are both major causes of deforestation. Another important cause is the clearing of forestland for agriculture, including conversion to cropland and pasture. This is a principal cause of deforestation in Nepal and Brazil, and it was one of the major reasons for clearing forests in

Figure 13.15 • A satellite image showing clearings in the tropical rain forests in the Amazon in Brazil. The image is in false infrared. Rivers are black, and the bright red is the leaves of the living rain forest. The straight lines of other colors, mostly light blue to gray, are clearing sites of deforestation by people extending from roads. Much of the clearing is for agriculture. The distance across the image is about 100 km.

A CLOSER LOOK 13.1

Community Forestry

In many parts of the world, people cut nearby forests to meet the needs of small communities. This is particularly true in developing nations, where firewood remains a major fuel and constitutes an important part of energy used. In the past, most government forestry departments concentrated their efforts on government-owned forestland or merely policed their countries' forests. Now, many realize that this approach must change.

Some countries have placed new emphasis on community forestry, in which professional foresters help villagers develop woodlots with the goal of achieving some kind of sustainable local harvest to meet local needs. The United Nations Food and Agricultural Organization (FAO) and the World Bank are supporting these programs. For example, in Malawi, Africa, the World Bank and FAO sponsor a reforestation project in which almost 40% of the households have planted trees. In South Korea, villagers have been reforesting the country at the rate of 40,000 ha per year.

In community forestry, good management practices include limiting access; cutting the slower-growing and poorer-burning species first to promote the growth of better firewood species; making use of plantations; and supplementing firewood with more easily renewable fuels. However, some of these practices conflict with traditional local activities or are difficult to implement for other reasons.

Such community efforts are impressive, but in total, they have only a small effect on the worldwide shortage of firewood. It is not clear whether developing nations can implement successful management policies in time to prevent serious damage to their forests and the land. If alternative fuels for developing nations cannot be found, the effects will be severe, not only for the land but also for the people.

eastern Canada and New England during the first settlement by Europeans.

The World Firewood Shortage

In many parts of the world, wood is a major energy source. Some 63% of all wood produced in the world, or 2.1 million m^3, is used for firewood. Firewood provides 2% of total commercial energy in developed countries, but it provides 15% of the energy in developing countries and is the major source of energy for most countries of sub-Saharan Africa, Central America, and continental Southeast Asia.[25]

As the human population grows, use of firewood increases. In this situation, management is essential, including management of woodland stands to improve growth. However, well-planned management of firewood stands has been the exception rather than the rule. Some successful community-based projects are discussed in A Closer Look 13.1.

Indirect Deforestation

A more subtle cause of the loss of forests is indirect deforestation—the death of trees from pollution or disease. Acid rain and other pollutants may be killing trees in many areas in and near industrial countries. In Germany, there is talk of Waldsterben ("forest death"). The German government has estimated that one-third of the country's forests have suffered damage: death of standing trees, yellowing of needles, or poorly formed shoots. The causes are unclear but appear to involve a number of factors, including acid rain, ozone, and other air pollutants that tend to weaken trees and increase their susceptibility to disease. This problem extends throughout central Europe and is especially acute in Poland, the Czech Republic, and Slovakia. In eastern Canada and the New England area of the United States, similar, curious damage is affecting red spruce.

If global warming occurs as projected by global models of climate, indirect forest damage could occur over vast regions, with major die-offs in many areas and major shifts in the areas of potential growth for each species of tree.[26] The combination of temperature and rainfall required for various tree species could be altered by global warming, and some species may not continue to grow in their current locations. Such changes would dwarf damage from other causes, affecting the habitat for many endangered species.[27]

Some suggest that global warming would merely change the location of forests, not their total area or production. However, even if a climate conducive to forest growth were to move to new locations, trees would have to reach these areas. This process would take a long time because changes in the geographic distribution of trees depend primarily on seeds blown by the wind or carried by animals. In addition, for production to remain as high as it is now, climates that meet the needs of forest trees would have to occur where the soils also meet these needs. This combination of climate and soils occurs widely now but might become scarcer with large-scale climate change. Climate change will likely also affect the distribution of insects and disease, reducing forest productivity over a larger area.

13.9 Parks, Nature, Preserves, and Wilderness

Landscapes are often protected from harvest and other potentially destructive uses by establishing parks, nature preserves, and legally designated wilderness areas. In addition, some privately held lands have been set aside for biological conservation. Organizations such as Carolinian Canada, the Nature Conservancy of Canada, the Nature Trust of New Brunswick, and the Land Trust Alliance of British Columbia purchase lands privately and maintain them as nature preserves. Whether government or private conservation areas succeed better in reaching the goals listed in Table 13.2 is a matter of considerable controversy.

Parks, natural areas, and wilderness provide benefits within their boundaries and can also serve as migratory corridors between other natural areas. Originally, parks were established for specific purposes related to the land within the park boundaries (see A Closer Look 13.2). In the future, the design of large landscapes to serve a combination of land uses—including parks, preserves, and wilderness—needs to become more important and a greater focus for discussion.

Parks and Preserves as Islands

The International Union for the Conservation of Nature (IUCN) defines "park" as a protected area managed mainly for ecosystem protection and recreation; the concept of park implies the presence of people to some degree. A nature preserve, although people may use it, has as its primary purpose the protection and conservation of some resource, typically a biological one. Every park or preserve is an ecological island of one kind of landscape surrounded by a different kind, or several different kinds, of landscape.

Ecological and physical islands have special ecological qualities (discussed in detail in Chapter 8), and concepts of island biogeography are relevant to the design and management of parks. Specifically, park and preserve planners know that the size of the park and the diversity of habitats determine the number of species that can be maintained there. In addition, the farther the park is from other parks or sources of species, fewer species are found. Even the shape of a park can determine what species can survive within it (Figure 13.16).

Parks have definite boundaries. These boundaries are usually arbitrary from an ecological viewpoint and have been established for political, economic, or historical reasons unrelated to the natural ecosystem. In fact, many urban parks have been developed on former landfills (see Chapter 24), or abandoned industrial

Table 13.2 • Goals of Parks, Natural Preserves, and Wilderness Areas

Parks are as old as civilization. The goals of park and nature-preserve management can be summarized as follows:

1. Preservation of unique geological and scenic wonders of nature, such as Niagara Falls and the Grand Canyon
2. Preservation of nature without human interference (preserving wilderness for its own sake)
3. Preservation of nature in a condition thought to be representative of some prior time (e.g., Canada prior to European settlement)
4. Wildlife conservation, including conservation of the required habitat and ecosystem of the wildlife
5. Conservation of specific endangered species and habitats
6. Conservation of the total biological diversity of a region
7. Maintenance of wildlife for hunting
8. Maintenance of uniquely or unusually beautiful landscapes for aesthetic reasons
9. Maintenance of representative natural areas for an entire country
10. Maintenance for outdoor recreation, including a range of activities from viewing scenery to wilderness recreation (hiking, cross-country skiing, rock climbing) and tourism (car and bus tours, swimming, downhill skiing, camping)
11. Maintenance of areas set aside for scientific research, both as a basis for park management and for the pursuit of answers to fundamental scientific questions
12. Provision of corridors and connections between separated natural areas

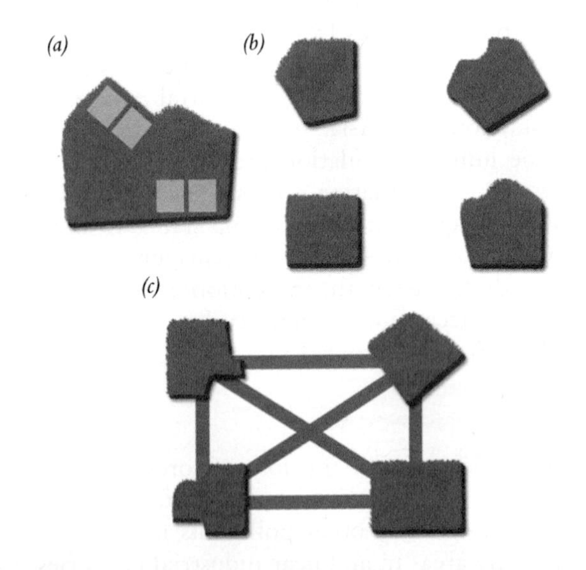

Figure 13.16 • Park shapes and island biogeography. A large park *(a)* can maintain more species, but several small parks *(b)* provide a kind of insurance against catastrophe. For example, if a storm struck one park and killed all individuals of one species in it, other populations of that species could survive in the other parks. A combination (c) can provide the benefits of both a single large park and several small ones. Here, the small parts are connected by migration corridors that allow occasional migration among them; the total area is equal to that of *(a)*, and the distribution over the landscape provides greater insurance.

A CLOSER LOOK 13.2

A Brief History of Parks

The French word *parc* once referred to an enclosed area for keeping wildlife to be hunted. In Europe, such areas were set aside for the nobility and excluded the public. One example is the area now known as Depone National Park on the southern coast of Spain. Originally a country home of nobles, today it is one of the most important natural areas of Europe, used by 80% of birds migrating between Europe and Africa.

The first major public park of the modern era was Victoria Park in Great Britain, authorized in 1842. The concept of a national park, whose purposes include protection of nature as well as public access, originated in North America in the nineteenth century. In the twentieth century, the purpose of national parks was broadened to emphasize biological conservation, and this idea was applied worldwide.[28]

The first designated national park in the world was Yosemite National Park in California, which was made a park by an act signed by President Lincoln in 1864. The term national park, however, was first used with the establishment of Yellowstone in 1872. Canada's first national park, Banff National Park, was established in 1885.

Although the purpose of the earliest national parks in the United States was to preserve the unique, awesome landscapes of the country, a purpose the historian Alfred Runte refers to as "monumentalism," the same was not true in Canada. The costs of Canada's new railway were immense, and parks like Banff, with its hot springs, were developed as tourist destinations whose revenues could be used to offset the costs of the railway. With time, however, Canadians and Americans came to see their national parks as a contribution to civilization equivalent to the architectural treasures of the Old World and sought to preserve them as a matter of national pride.[29]

In the second half of the twentieth century, the emphasis of park management became more ecological, with parks established both for scientific research and to maintain examples of representative natural areas. For instance, Zimbabwe established Sengwa National Park (now called Matusadona National Park) solely for scientific research; there are no tourist areas, and tourists are not generally allowed. Its purpose is the study of natural ecosystems with as little human interference as possible so that the principles of wildlife and wilderness management can be better formulated and understood. Other national parks in the countries of eastern and southern Africa—including those of Kenya, Uganda, Tanzania, Zimbabwe, and South Africa—have been established primarily for viewing wildlife and for biological conservation.

In recent years, the number of national parks throughout the world has increased rapidly. The law establishing national parks in France was first enacted in 1960. Taiwan had no national parks prior to 1980 but now has six.[30] In Canada, national and provincial agencies have added more than 25 million ha in national, provincial, and territorial parks since 1992, for a current total of 68 million ha. The work is not yet done, however. Canada's national park system is only two-thirds complete, with 14 of 39 natural regions not yet represented. Land for national parks within five of these 14 regions have been purchased or are under negotiation.[31] Furthermore, aquatic parks are relatively recent additions to the Canadian suite of protected spaces and likely still under-represent these important ecosystems.

The conservation of representative natural areas of a country is an increasingly common goal of national parks. For example, the goal of New Zealand's national park planning is to include at least one area representative of each major ecosystem of the nation, from seacoast to mountain peak.

sites—areas that would otherwise be considered wastelands, useless for any other purpose.

Even where parks or preserves have been set aside for the conservation of some species, the boundaries are usually arbitrary, which has caused problems. For example, Lake Manyara National Park in Tanzania, famous for its elephants, was originally established with boundaries that were poorly matched with the habitat requirements of the resident elephant population. Before this park was established, elephants would spend part of the year feeding along a steep incline above the lake. At other times of the year, they would migrate down to the valley floor, depending on the availability of food and water. These annual migrations were necessary for the elephants to obtain food of sufficient nutritional quality throughout the year.

When the park was established, there were farms along its northern border. These farms crossed the traditional pathways of the elephants, creating two negative effects. First, elephants came into direct conflict with farmers. Elephants crashed through farm fences, eating corn and other crops and causing general disruption. Second, whenever the farmers succeeded in keeping elephants out, the animals were cut off from reaching their feeding ground near the lake. When it became clear that the park boundaries were arbitrary and inappropriate, adjustments were made to extend the boundaries to include the traditional migratory routes. This eased the conflicts between elephants and farmers.

Parks necessarily isolate populations. If the habitat they provide is too small for the needs of a given population, the population may become genetically isolated and impoverished. If parks are to function as biological preserves, they must be adequate in size and habitat diversity to maintain a population large enough to avoid the serious genetic difficulties that can develop in small populations. An alternative, if necessary, is for a manager to move individuals of one species from one park to another to maintain genetic diversity. Canada's national recovery plan for the Wood Bison includes provisions for movement of individuals from one area to another, to improve herd health and increase genetic diversity. As another example, the Yellowstone-to-Yukon initiative, a network of over 270 organizations, emphasizes the development of multiple-park management institutions that would allow the linkage of parks in a binational habitat corridor. Similar approaches are planned for an Algonquin-to-Adirondacks (A2A) corridor, and for a Heart of the Continent area centred on Quetico Provincial Park.

Conflicts in Managing Parks

The idea of a national, provincial or city park is well accepted in Canada, but conflicts arise over what kinds of activities and the intensity of activities should be allowed in parks. Many of the recent conflicts relating to national parks have concerned the use of motor vehicles. Algonquin Provincial Park in northern Ontario, established in 1893—comparatively early compared with many other parks—occupies land that was once used by a variety of logging and recreational vehicles and provided livelihoods for hunting and fishing guides and other tourism companies. These people felt that restricting motor-vehicle use would destroy their livelihoods. In the United States, travel into Yellowstone National Park by snowmobiles in the winter has become popular, but this has led to air and noise pollution and marred the experience of the park's beauty for many visitors.[29]

Interactions between people and wildlife can also be a problem. While many people like to visit parks to see wildlife, some wildlife, such as grizzly bears in Banff National Park, can be dangerous. There has been conflict in the past between conserving the grizzly and making the park as open as possible for recreation. Another challenge is the management of fire in parks. Many ecosystems, such as the oak savannah of Ontario's Pinery Provincial Park (Chapter 10), are adapted to the presence of fire. Should fires in those areas be controlled to protect human life and property? Or should they be allowed to burn, as was the case in Yellowstone National Park's extensive fire in 1988. British Columbia's 2003 fires, which were caused by human actions, were a graphic reminder of the costs of fire we cannot or choose not to control.

How Much Land Should Be in Parks?

Another important controversy in managing parks is what percentage of a landscape should be in parks or nature preserves, especially with regard to the goals of biological diversity. One project in Canada and the United States, The Wildlands Project, argues that large areas are necessary to conserve ecosystems, so that even large parks, such as Wood Buffalo National Park on the Alberta–Northwest Territories border, and Yellowstone National Park in Wyoming, need to be connected by conservation corridors. The rationale is, in part, the landscape concept discussed at the beginning of this chapter. At the same time, however, the growing human population and the desire for a higher standard of living put pressure on existing parks in many parts of the world.

Nations differ widely in the percentage of their total area set aside as national parks. In their 1987 report, *Our Common Future*, the World Commission on Environment and Development (often called the Brundtland Commission after its chair, former Norwegian Prime Minister Gro Harlem Brundtland) recommended tripling the global area of protected land. That recommendation has been widely interpreted as a target of 12% protected land in each country, based on the area of protected land that existed in 1987. (Greenpeace and other non-government organizations have subsequently challenged the 12% target as far too low for adequate protection of biodiversity.)

Many countries, including Canada, have now adopted a goal of 12% protected land in formal policy. Costa Rica, a small country with high biological diversity, has more than 12% of its land in national parks.[20] Kenya, a larger nation that also has numerous biological resources, has 7.6% of its land in national parks.[31] In France, an industrialized nation in which humans have altered the landscape for several thousand years, only 0.7% of the land is in the nation's six national parks. However, France has 38 regional parks that encompass 11% (5.9 million ha) of the nation's area.[32]

The total amount of protected natural area in Canada is more than 90 million ha, approximately 9.6% of the total Canadian land area.[33] This places Canada well below the OECD average of 12.6% protected area. However, the percentage of land in parks, preserves, and other conservation areas differs greatly among the Canadian provinces and territories. For example, Nunavut has protected areas, whereas New Brunswick and Alberta allocate much less of their land area to parks and still less to designated wilderness.[34] Furthermore, it can be argued that undue emphasis on the amount of land under protection detracts from focus on the level and nature of required protection measures, regional differences in biodiversity, and linkages between the two.

Conserving Wilderness

What is wilderness? To some, it is an area undisturbed by people. To others, it is simply a natural area. The Alberta Wilderness Association defines wilderness as an area "characterized by the dominance of natural processes, the presence of the full complement of plant and animal communities characteristic of the region, and the absence of human constraints on nature." This definition is consistent with that of the IUCN, which defines wilderness as "a large area of unmodified or slightly modified land, and/or sea, retaining its natural character and influence, without permanent or significant habitation."

The conservation of wilderness is an idea introduced in the second half of the twentieth century and one that is likely to become more important as the human population increases and the effects of human society become more pervasive throughout the world (Figure 13.17).

Figure 13.17 • A feeling of wilderness. Quttinirpaaq National Park, Nunavut, designated in 1999 and now covering 37,775 sq. km. As the photograph suggests, this vast area gives a visitor a sense of wilderness as a place where a person is only a visitor and human beings seem to have no impact.

In 2000, a new *Canada National Parks Act* consolidated a patchwork of legal provisions into a new, streamlined Act. The new law makes it easier to establish new parks, control commercial development in existing parks, and conserve nationally and internationally significant heritage resources. It also establishes larger fines and penalties for illegal hunting and trafficking in wildlife and other natural resources. New regulations under the Act provide for the declaration of wilderness areas within national parks, and zoning of other park areas depending on intended use. The Act does not explicitly manage wilderness, but does make it easier to manage areas considered wilderness within existing parks. For example, Fundy National Park, on the shore of the Bay of Fundy (Chapter 7) has established four park zones under the new legislation. Zone I (Special Preservation) offers the highest level of protection in the park system, and is intended to conserve unique, rare, sensitive or otherwise special natural features. Motorized vehicles are prohibited in Zone I, and visitor traffic is discouraged. In Fundy National Park, the Point Wolfe Coastal Cliffs and the Goos River Coastal Cliffs are designated Zone I, because they contain the two known New Brunswick sites of the bird's-eye primrose, and provide excellent nesting sites for the peregrine falcon, a species-at-risk. More than 65% of Fundy National Park is designated as Zone II (Wilderness). Zone II is intended to capture extensive areas representing the best natural features of the region. Visitors are allowed, but motorized traffic is prohibited. Within these designated wilderness areas, only basic infrastructure such as trails and backcountry campsites is provided. Zone III (Natural Environment) areas are those intended for a moderate level of visitor use, including swimming, hiking, and non-motorized boating. Most visitor use occurs in Zone IV (Outdoor Recreation). Zone IV areas include both day use and overnight camping opportunities, and require more extensive infrastructure such as roads and permanent structures such as washrooms and golf courses.

Countries with a significant amount of wilderness include New Zealand, Canada, Sweden, Norway, Finland, Russia, and Australia; some countries of eastern and southern Africa; many countries of South America, including parts of the Brazilian and Peruvian Amazon Basin; the mountainous high-altitude areas of Chile and Argentina; some of the remaining interior tropical forests of Southeast Asia; and the Pacific Rim countries (parts of Borneo, the Philippines, Papua New Guinea, and Indonesia). In addition, wilderness can be found in the polar regions, including Antarctica, Greenland, and Iceland. In all, Canada possesses 20% of the world's remaining wilderness. According to the World Wildlife Fund, Canada has one-quarter of the world's temperate rain forest, one-third of the world's boreal forest, and virtually all of the remaining old-growth red and white pine. However, it would be a mistake to think that Canada's national and provincial park systems adequately represent our country's natural regions. For example, the World Wildlife Fund identified 75 natural regions in Québec, noting that none of these are adequately represented by protected areas, 15 have some degree of representation, and 60 (two-thirds of which are in the northern part of the province) are not represented in Québec's park system.[35] Some Canadian ecosystems are particularly under-represented in our national and provincial park system. They include grassland ecosystems, such as the fescue prairie preserved in Saskatchewan's Prince Albert National Park, and marine and freshwater parks, despite the importance of those ecosystems to Canada and internationally.

Many countries have no wilderness left to preserve. In the Danish language, even the word for wilderness

CRITICAL THINKING ISSUE

How Does Fragmentation of Tropical Forests Contribute to Habitat Destruction?

Although tropical rain forests occupy only about 7% of the world's land area, they provide habitat for at least half of the world's plant and animal species. Approximately 100 million people live in rain forests or depend on them for their livelihood. Tropical plants provide products such as chocolate, nuts, fruits, gums, coffee, wood, rubber, pesticides, fibres, and dyes. Drugs used to treat high blood pressure, Hodgkin's disease, leukemia, multiple sclerosis, and Parkinson's disease have been made from tropical plants, and medical scientists believe many more are yet to be discovered.

Most of the interest in tropical rain forests has focused on Brazil, whose forests are believed to have more species than any other habitat. Estimates of destruction in the Brazilian rain forest range from 6% to 12%, but numerous studies have shown that deforested area alone does not adequately measure habitat destruction, because surrounding habitats are also affected. For example, the more fragmented a forest is, the more edges there are, the greater the impact on the living organisms. Such edge effects vary depending on the species, the characteristics of the land surrounding the forest fragment, and the distance between fragments. For example, a forest surrounded by farmland is more deeply affected than one surrounded by abandoned land in which secondary growth presents a more gradual transition between forest and deforested areas. Some insects, small mammals, and many birds find only 80 m a barrier to movement from one fragment to another. Corridors between forested areas also help to offset the negative effects of deforestation on plants and animals of the forest.

Critical Thinking Questions

1. Assuming an edge effect of 1 km, what is the approximate area affected by deforestation of 100 km^2 in the form of a square, 10 km on each side? If the 100-km^2 area were in the form of 10 rectangles, each 10 km long and 1 km wide, separated from each other by a distance of 5 km, how large an area would be affected?
2. What environmental factors at the edge of a fragment would differ from those in the centre? How might the differences affect plants and animals at the edge?
3. Why is a simple rule of thumb, such as assuming an edge effect of 1 km, too simplistic as a model of the effects of deforestation? Given that it is too simplistic, what advantages are there in using the rule?
4. Forest fragments are sometimes compared with islands. What are some ways in which this is an appropriate comparison? What are some ways in which it is not?

has disappeared, although that word was important in the ancestral languages of the Danes.[35] A national park in Switzerland lies in view of the Alps—scenery that inspired awe in English romantic poets of the early nineteenth century. The park is in an area that has been heavily exploited for such activities as mining and foundries since the Middle Ages. All Swiss forests are planted.[36]

Henry David Thoreau distinguished between "wilderness" and "wildness." He thought of wilderness as a physical place and wildness as a state of mind. During his travels through the Maine woods in the 1840s, he concluded that wilderness was an interesting place to visit but not to live in. He preferred long walks through the woods and near swamps around his home in Concord, Massachusetts, where he was able to experience a feeling of wildness. Thus, Thoreau raised a fundamental question: Can one experience true wildness only in a huge area set aside as a wilderness and untouched by human actions, or can wildness be experienced in small, heavily modified, and naturalistic landscapes, such as those around Thoreau's home?

As Thoreau suggested, small, local, naturalistic parks may have more value as places of solitude and beauty than some more traditional wilderness areas. In Japan, for instance, there are roadless recreation areas, but they are filled with people. One two-day hiking circuit leads to a high-altitude marsh where people can stay in small cabins. Trash is removed from the area by helicopter. People taking this hike experience a sense of wildness.

In some ways, the answer to the question raised by Thoreau is highly personal. We must discover for ourselves what kind of natural or naturalistic area meets our spiritual, aesthetic, and emotional needs. This is yet another area in which one of our key themes, science and values, is evident.

Conflicts in Managing Wilderness

The definition of wilderness has given rise to several controversies. Those interested in developing the natural resources of an area, including mineral ores and timber, have argued that wilderness is defined too narrowly, and that proposed protections will restrict resource exploitation unnecessarily when, they say, there is plenty of wilderness elsewhere. Those who wish to conserve additional wild areas have argued that wilderness should be broadly defined, and that mining and logging are inconsistent with wilderness protection.

The notion of managing wilderness may seem a paradox—a true wilderness would need no management. However, with the world's high population, even wilderness must be carefully defined and set aside. We can view the goal of managing wilderness in two ways: in terms of the wilderness itself and in terms of people. Current thinking suggests that the goal of wilderness management is to preserve nature, while recognizing the role of humans, for instance First Nations, in using and managing those resources.

Legally designated wilderness can be seen as one extreme in a spectrum of environments to manage. At the other extreme are heavily modified systems like urban parks and reclaimed industrial lands. You can think of many stages in between these on this spectrum.[37, 38] Wilderness management should involve as little direct action as possible, so as to minimize human influence. Ironically, one of the necessities is to control human access so that a visitor has little, if any, sense that other people are present. Access to wilderness by automobiles, light aircraft and helicopters, snowmobiles, and all-terrain off-road vehicles must also be restricted.

Consider Yoho National Park in Golden, British Columbia, consisting of more than 1300 sq km and part of UNESCO's Canadian Rocky Mountain Parks World Heritage Site. Mountain bikes, automobiles, and a public bus are permitted in the park. Can each visitor really have a wilderness experience there, or is the human carrying capacity exceeded? The answer will be a subjective judgment. If all visitors see only their own companions and believe they are alone, then the actual number of visitors does not matter for each visitor's wilderness experience. On the other hand, if every visitor finds the solitude ruined by strangers, then the management has failed, no matter how few people have visited the area.

Wilderness designation and management must also take into account adjacent land uses. A wilderness next to a garbage dump or a smoke-emitting power plant is a contradiction in terms. Whether a wilderness can be adjacent to a high-intensity campground or near a city is a subtler question that must be resolved by citizens.

Today, those involved in wilderness management recognize that wild areas change over time and that these changes should be allowed to occur as long as they are consistent with naturally occurring types, rates, and scales of change. This is different from earlier views that nature undisturbed was unchanging and should be managed so that it did not change. In addition, it is generally argued that in choosing what activities can be allowed in a wilderness, emphasis should be placed on activities that depend on wilderness (the experience of solitude or the observation of shy and elusive wildlife) rather than on activities that can be carried out elsewhere (such as downhill skiing).

A source of conflict is that wilderness areas frequently contain economically important resources, including timber, mineral ores, and sources of energy. There has been heated debate about whether wilderness areas should be open to the extraction of oil and mineral ores. In Canada, there is widespread agreement that such activities should not occur in national parks, but they are underway in a number of provincial parks and elsewhere in the world. In 2005, the United States Congress approved a plan to allow oil exploration in the Arctic National Wildlife Refuge (ANWR) on the North Slope of Alaska, a pristine wilderness area that provides habitat for several protected species (Figure 13.18). The oil industry has long argued in favour of drilling for oil in the ANWR, but the idea was unpopular for decades among many members of the public and the U.S. government, and no drilling was permitted. When George W. Bush, who favoured drilling in the ANWR, was elected president in 2000, the controversy

Figure 13.18 • The National Arctic Wildlife Refuge, North Slope, Alaska, is valued for its scenery, wildlife, and oil.

Figure 13.19 • Those in favour of ANWR argue that new technology can reduce the impact of developing oil fields in the Arctic: Wells are located in a central area and use directional drilling; roads are constructed of ice in the winter, melting to become invisible in the summer; pipelines are elevated to allow animals, in this case caribou, to pass through the area, and oil-field and drilling wastes are injected deep underground. See text for arguments against drilling.

surrounding drilling emerged. In March 2005, the U.S. Senate approved the drilling project, although no significant activity is expected for at least seven years.

Another controversy involves the need to study wilderness versus the desire to leave the wilderness undisturbed. Those in favour of scientific research in the wilderness argue that it is necessary for the conservation of wilderness. Those opposed—a minority in the scientific community—argue that scientific research creates an unnecessary disturbance in wilderness and contradicts the purpose of wilderness protection. One solution is to establish separate research preserves.

Summary

- In the past, land management for the harvest of resources and the conservation of nature was mostly local, with each parcel of land considered independently. Today, a landscape perspective has developed, and lands used for harvesting resources are seen as part of a matrix that includes lands set aside for the conservation of biological diversity and for landscape beauty.
- Forests are among civilization's most important renewable resources. Forest management seeks a sustainable harvest and sustainable ecosystems.
- There are few examples of successful sustainable forestry. As a result, a practice has developed called "certification of sustainable forestry." Certification involves determining which methods appear most consistent with sustainability and then comparing the management of a specific forest with those standards.
- The continued use of firewood as an important fuel in developing nations is a major threat to forests, given the rapid population growth of those areas. It is doubtful that developing nations can implement successful management programs in time to prevent serious damage to their forests and severe effects on their people.
- Clear-cutting is a major source of controversy in forestry. Some tree species require clearing to reproduce and grow, but the scope and method of cutting must be examined carefully in terms of the needs of the species and the type of forest ecosystem.
- Properly managed plantations can relieve pressure on forests.
- Managing parks for biological conservation is a relatively new idea that began in the nineteenth century. The manager of a park must be concerned with its shape and size. Parks that are too small or the wrong shape may have too small a population of the species for which the park was established and thus may not be able to sustain the species.
- A special extreme in conservation of natural areas is the management of wilderness. In Canada, the 2000 *Canada National Parks Act* provided a legal basis for such conservation. Many of Canada's wilderness areas are located in northern lands, where they are under the control and co-management of First Nations peoples.
- Parks, nature preserves, wilderness areas, and actively harvested forests affect one another. The geographic pattern of these areas on a landscape, including corridors and connections among different types, is part of the contemporary approach to biological conservation and the harvest of forest resources.

STUDY QUESTIONS

1. What environmental conflicts might arise when a forest is managed for the multiple uses of *(a)* commercial timber, *(b)* wildlife conservation, and *(c)* a watershed for a reservoir? In what ways could management for one use benefit another?
2. What arguments could you offer for and against the statement, "Clear-cutting is natural and necessary for forest management"?
3. Can a wilderness park be managed to supply water to a city? Explain your answer.
4. A park is being planned in rugged mountains with high rainfall. What are the environmental considerations if the purpose of the park is to preserve a rare species of deer? What if the purpose is recreation, including hiking and hunting?
5. What are the environmental effects of decreasing the rotation time in forests from an average of 60 years to 10 years? Compare these effects for *(a)* a woodland in a dry climate on a sandy soil and *(b)* a rain forest.
6. In a small but heavily forested nation, two plans are put forward for forest harvests. In Plan A, all the forests to be harvested are in the eastern part of the nation, while all the forests of the west are set aside as wilderness areas, parks, and nature preserves. In Plan

B, small areas of forests to be harvested are distributed throughout the country, in many cases adjacent to parks, preserves, and wilderness areas. Which plan would you choose? Note that in Plan B, wilderness areas would be smaller than in Plan A.

7. The smallest biosphere reserve in Canada is Mont Saint-Hilaire, east of Montréal, Québec. It covers 5,500 ha. Do you think this can meet the meaning of wilderness, and the intent of the wilderness provisions under the Canada National Parks Act? Justify your answer.

FURTHER READING

Aird, P. L. "Conservation for the sustainable development of forests worldwide: a compendium of concepts and terms." *The Forestry Chronicle* 70(6) (1994), pp. 666-674.

Barnes, B. V., D. R. Aak, S. R. Denton, and S. H. Spurr. *Forest Ecology*, 4th ed. New York: Wiley, 1998.

Bolgiano, Chris. *Living in the Appalachian Forest: True Tales of Sustainable Forestry.* Stackpole Books, 2002. Positives and negatives about attempts to do sustainable forestry in America's most diverse forests.

Food and Agricultural Organization. *State of the World's Forests.* Washington, D.C.: United Nations, 1999.

Grossman, D. H., D. Faber-Langendoen, A. S. Weakley, M. Anderson, P. Bourgeoron, R. Crawford, K. Goodin, S. Landaal, K. Metzler, K. D. Patterson, M. Pyne, M. Reid, and M. L. Hunter, Jr. *Wildlife, Forests and Forestry: Principles of Managing Forests for Biological Diversity.* Englewood Cliffs, N.J.: Prentice-Hall, 1990.

Hunter, Malcolm L. Jr. *Wildlife, Forests and Forestry – Principles of Managing Forests for Biological Diversity.* Englewood Cliffs, NJ: Prentice Hall, 1990.

Jenkins, Michael B. *The Business of Sustainable Forestry.* Washington, D.C.: Island Press, 1999.

Kimmins, J. P. *Forest Ecology: A Foundation for Sustainable Management.* New York: Prentice-Hall, 1996. A textbook that applies recent developments in ecology to the practical problems of managing forests.

Kohm, K. A., and J. F. Franklin. *Creating a Forestry for the 21st Century: The Science of Ecosystem Management.* Washington, D.C.: Island Press, 1996. A book that applies new understanding of ecological processes, including the role of disturbances, to management of forests. The book views forest management within an economic and societal context.

Lavender, D. P., R. Parish, C. M. Johnson, G. Montgomery, A. Vyse, R. A. Willis, and D. Winston (eds.). *Regenerating British Columbia's Forests.* Vancouver, BC: UBC Press, 1990..

Noss, R. "Protecting Natural Areas in Fragmented Landscapes." Natural Areas Journal 7 (1987), pp. 2–7.

Oliver, C. D., and B. C. Larson. *Forest Stand Dynamics.* New York: Wiley, 1996.

Perlin, J. *A Forest Journey: The Role of Wood in the Development of Civilization.* New York: Norton, 1989. A fascinating presentation of the story of wood, forests, and people, covering a period of 5,000 years and spanning five continents.

Sharma, N., ed. *Contemporary Issues in Forest Management: Policy Implications.* Washington, D.C.: The World Bank, 1992. A series of articles that analyze the reasons for conserving and sustaining forests.

Spurr, S. H., and B. V. Barnes. *Forest Ecology.* New York: Ronald Press, 1992. The most recent edition of a classic textbook that provides an ecological basis for the management of forestland.

chapter 14

A Story of Three Geese

Three closely related species—the Aleutian goose (Figure 14.1), the snow goose, and the Hawaiian goose (the state bird of Hawaii, also called the nene)—had different histories in the twentieth century. It is a sort of *Goldilocks and The Three Bears* story: One species (the snow goose) became too abundant, one (the nene) not abundant enough, and one (the Aleutian goose) about right.

The three birds share a common heritage with the Canada goose—the goose that has become so common in much of North America—and all also feed on grasses. The habitats of the nene and the other geese, however, are distant from each other and very different in climate. The Aleutian goose nests in the high-latitude Aleutian Islands surrounded by cold waters.[1] The snow goose also nests in the Far North, along coastal marshes—many along Hudson Bay.[2] In contrast to the other two species, which spend only part of the year in warm climates, the nene lives year-round on lava slopes in the balmy climes of Hawaii (Figure 14.1).[3]

Threats to the continued existence of the Aleutian goose and the nene were the result of human actions—people introduced non-native predators and exotic grasses that were not palatable to the geese or were much less nutritious for them. Meanwhile, European settlement of North America brought improved conditions for snow geese. By the mid-1900s, degradation and draining of the Gulf Coast marshes pushed many snow geese inland, where they learned to feed on the stubble and seeds of wheat, corn, and rice.[2]

This agricultural food supply improved the snow geese's winter survival rates and springtime weight gain, which in turn increased the number of goslings born. In 25 years, the mid-continent population increased from about 2 million to 5 million birds.[2] As a result, snow geese overgrazed their shoreline nesting grounds, transforming them into mudflats. Between 1973 and 1996, bare shoreline spread inland. About one-third of the snow geese's 2,000 km habitat along Hudson Bay has been damaged and another third destroyed. Bare soils are warmer, have higher evaporation rates, and are more susceptible to erosion.

Despite the loss of forage, snow geese keep coming back to the same area, and they keep reproducing, using energy gained during migration. The goslings are not so lucky. In some habitats, snow goose families walk as much as 500 km to find food, and only 10% of the goslings survive. Adults commonly live and reproduce into their teens, so the population keeps growing. However, a lack of suitable habitat could make the aging snow goose population collapse.

Meanwhile, by the end of the 1960s, the Aleutian goose population had declined to 800 individuals. The decline began early in the twentieth century, when arctic foxes were introduced for fur farming on most of the goose's nesting islands. A recovery program removed the foxes from more than half of the geese's current range and 25–50% of its historical range. Many introduced grasses were also removed, so that the native food grasses, more palatable and more nutritious, are abundant again. The Aleutian goose now numbers more than 15,000, primarily inhabiting three of the islands of its former range (Figure 14.2).

Figure 14.1 • The nene goose or Hawaiian goose (*Branta sandwicensis*). Only about 1,000 of these nonmigratory and nonaquatic birds remain in the wild today. Inset: range of the Hawaiian goose.

Wildlife, Fisheries, and Endangered Species

LEARNING OBJECTIVES

Wildlife, fish, and endangered species are among the most popular environmental issues. People love to see wildlife; fishing is an important commercial and recreational activity, and provides protein for many people and since the nineteenth century, endangered species have drawn public attention. You would think that by now we would be doing a good job of conserving and managing these kinds of life. This chapter tells you how we are doing. After reading this chapter, you should understand:

- Why people want to conserve wildlife, fish, and endangered species.
- The importance of habitat, ecosystems, and landscape in the conservation of endangered species.
- Current causes of extinction.
- Steps we can take to achieve sustainability of wildlife, fisheries, and endangered species.
- The concepts of maximum sustainable yield, the logistic growth curve, carrying capacity, optimum sustainable yield, and minimum viable populations.

During the same period, the nene of Hawaii declined as a result of human introductions of mongooses, rats, feral dogs, and pet cats, and also because of hunting (outlawed in 1911) and the introduction of exotic grasses. Only a few small flocks remained by the 1950s except in captivity.[3] Nenes breed successfully in captivity, but have done poorly when released into their native habitat. In contrast to the recovery program for the Aleutian goose, the Hawaiian program has had only limited success in removing introduced predators, and exotic grasses continue to dominate much of the birds' habitat, so palatable and nutritious food is much less abundant and harder to find. At present, only a thousand or so nenes exist outside of captivity.

Figure 14.2 • **The Aleutian goose, *Branta canadensis leucopareia*,** is one of five subspecies of the Canada goose. Inset: range of the Aleutian goose (yellow areas are where the goose occurs today). (*Source*: Alaska Department of Fish and Game.)

- *The comparison among the three geese—overpopulation of the snow goose, successful recovery of the Aleutian goose, and the lack of recovery of their near relative, the nene—provides some insight into how to help endangered species. In modern times, many species faced extinction because of human actions, such as the introduction of non-native species and habitat alteration. Restoring the habitat is essential to the recovery of a species. In most cases, it is better to have a small population in a healthy habitat than to have a large population in a badly degraded habitat. In this chapter, we discuss the conservation and management of endangered species, building on the concepts introduced in the previous chapter with reference to forests, parks, and landscapes.*

14.1 Introduction

Wildlife, fisheries, and endangered species are considered together in this chapter because they have a common history of exploitation, management, and conservation, and because our attempts to manage and conserve them follow the same approaches. Although any form of life, from bacteria and fungi to flowering plants and animals, can become endangered, concern about endangered species has tended to focus on wildlife. We will maintain that focus, but we ask you to be sure to remember that the general principles apply to all forms of life.

The opening case study about the Aleutian and Hawaiian geese raises an interesting philosophical question: When we say that we want to save a species, what is it that we really want to save? There are four possible answers:

1. A wild creature in a wild habitat that is a symbol of wilderness.
2. A wild creature in a managed habitat, so that the species can feed and reproduce with little interference and so that we can see it in a naturalistic habitat. (The recovery of the Aleutian goose fits this goal.)
3. A population in a zoo, so that the genetic characteristics are maintained in live individuals.
4. Genetic material only—frozen cells containing DNA from a species for future scientific research.

Which of these goals people choose involves not only science but also values. People have different reasons for wishing to save endangered species—utilitarian, ecological, cultural, recreational, spiritual, inspirational, aesthetic, and moral (see A Closer Look 14.1). Policies and actions differ widely depending on which goal is chosen.

14.2 Traditional Single-Species Wildlife Management

Attempts to apply science to the conservation and management of wildlife and fisheries, and therefore to endangered species, began around the turn of the twentieth century and tended to view each species as a single population in isolation. Assumptions included the following:

- The population could be represented by a single number, its total size.
- Undisturbed by human activities, a population would grow to a fixed size, called the "carrying capacity."
- Environment, except for human-induced changes, is constant.

This perception of wildlife and fisheries was formalized in the S-shaped logistic growth equation, discussed in Chapter 4 (Figure 14.3). Two goals resulted from these ideas and the logistic equation: for a species that we intend to harvest, the goal was maximum sustainable yield (MSY); for a species that we wish to conserve, the goal was to have that species reach, and remain at, its carrying capacity. **Maximum sustainable yield** was defined as the population size that yielded maximum production (measured either as a net increase in the number of individuals in the population or as a net change in biomass) that would allow this population to be sustained indefinitely without decreasing its ability to provide the same level of production. More simply, the population was viewed as a factory that could keep churning out exactly the same quantity of a product year after year.

This approach failed. None of the assumptions are true, as you have been learning in this book. Populations cannot be represented only by a single number—their total abundance—and do not remain at a fixed carrying capacity. Nor is the environment constant.

Today a broader view is developing. It acknowledges that a population exists in a changing environment (including human-induced changes), that populations interact, and that it is necessary to include an ecosystem and landscape context for conservation and management (as we learned in Chapter 13). With this new understanding, the goal for a species to be harvested is to sustain a harvestable population in a sustainable ecosystem. The goal for a threatened or endangered species is sometimes stated as a **minimum viable population**, which is the estimated smallest population that can maintain itself and its genetic variability indefinitely. Other times the goal is stated as the *carrying capacity*, and still other times as the *optimum sustainable population.* (These terms will be explained later in this chapter.)

Figure 14.3 • The basic logistic growth curve.

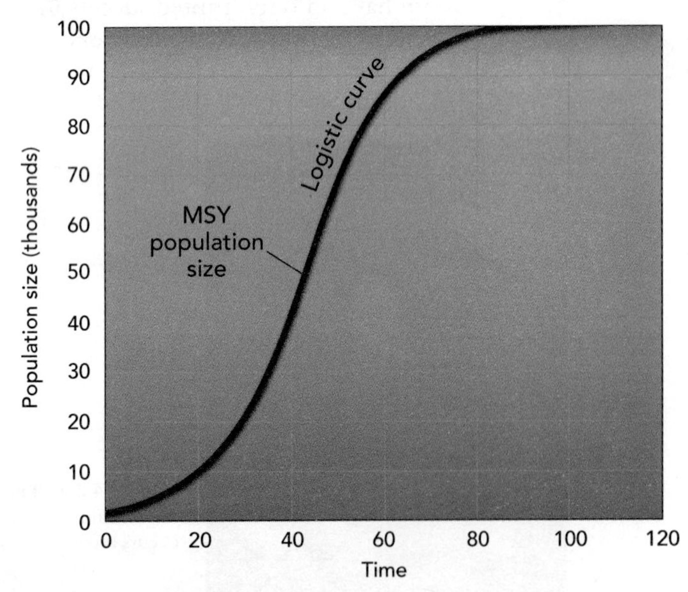

A CLOSER LOOK 14.1

Reasons for the Conservation of Endangered Species (and All Life on Earth)

Some important reasons for conserving endangered species can be classified as utilitarian, ecological, aesthetic, moral, and cultural.

Utilitarian Justification

Utilitarian justification is based on the consideration that many wild species might be useful to us, so it is unwise to destroy them before we have a chance to test their uses. Many of the arguments for conserving endangered species, and for biological diversity in general, have focused on the utilitarian justification.

One such argument is the need to conserve wild strains of grains and other crops. Disease organisms that attack crops evolve continually, and as new disease strains develop, crops become vulnerable. Crops such as wheat and corn depend on the continued introduction of fresh genetic characteristics from wild strains to create new, disease-resistant genetic hybrids. We might also find new crops among the many species of plants. Many horticultural crops and products have come from tropical rain forests.[4] For example, of 275 species found in one hectare in a Peruvian tropical forest, 72 yielded products with direct economic value.

Another utilitarian justification for biological conservation is that many important chemical compounds come from wild organisms. Digitalis, an important drug in treating certain heart ailments, comes from purple foxglove. Aspirin is a derivative of willow bark. Other important medicines derived from plants include cancer-fighting chemicals from rosy periwinkle and the Pacific yew tree, steroids from Mexican yams, anti-hypertensive drugs from serpent wood, and antibiotics from tropical fungi.[5] Some 25% of prescriptions dispensed today contain ingredients extracted from vascular plants.[6] And these represent only a small fraction of the estimated 270,000 existing plant species.[5] Other plants and organisms may produce useful medical compounds that are as yet unknown.[7]

Some species are also used directly in medical research. For example, the armadillo, one of only two animal species known to contract leprosy, is important to studies seeking a cure for that disease.[8] Other animals, such as the horseshoe crabs and barnacles, are important because of physiologically active compounds they make. Still others may have similar uses as yet unknown to us.

Another utilitarian justification is that many species help to control pollution control. Plants, fungi, and bacteria remove toxic substances from air, water, and soils. Carbon dioxide and sulfur dioxide are removed by vegetation, carbon monoxide is reduced and oxidized by soil fungi and bacteria, and nitric oxide is incorporated into the biological nitrogen cycle. Because species vary in their capabilities, diversity of species provides the best range of pollution control.

Tourism provides yet another utilitarian justification. Ecotourism is a growing source of income for many developing countries. Ecotourists value nature, including its endangered species, for aesthetic or spiritual reasons, but the result can be utilitarian.

Ecological Justification

When we reason that organisms are necessary to maintain the functions of ecosystems and the biosphere, we are using an ecological justification for their conservation of these organisms. When bees pollinate flowers, for example, they provide a benefit to us that would be costly to replace with human labour. Trees remove certain pollutants from the air; and some soil bacteria fix nitrogen, converting it from molecular nitrogen in the atmosphere to nitrate and ammonia that can be taken up by other living things. That some such functions involve the entire biosphere reminds us of the global perspective on conserving nature and specific species.

Aesthetic Justification

An aesthetic justification asserts that biological diversity enhances the quality of our lives, providing some of the most beautiful and appealing aspects of our existence. Biological diversity is an important quality of landscape beauty. Many organisms—birds, large land mammals, and flowering plants, as well as many insects and ocean animals—have long been appreciated for their beauty. Whatever other reasons Pleistocene people had for creating paintings in caves in France and Spain, their paintings of wildlife, done about 14,000 years ago, are beautiful. The paintings include species that have since become extinct, such as mastodons. Poetry, novels, plays, paintings, and sculpture often celebrate the beauty of nature. It is a very human quality to appreciate nature's beauty and is a strong reason for the conservation of endangered species.

Moral Justification

Moral justification is based on the belief that species have a moral right to exist, independent of our need for them. Consequently, the argument follows that in our role as global stewards, we should promote the continued existence of species and conserve biological diversity. This right to exist was stated in the UN General Assembly World Charter for Nature, 1982. Thus, a moral justification for the conservation of endangered species is part of the intent of the law.

Moral justification has deep roots within human culture, religion, and society. Moral justification also has economic ramifications. As more and more citizens of the world assert the validity of moral justification, more actions that have economic effects are taken to defend a moral position.

Arnee Naess, a proponent of deep ecology, one of its principle philosophers, explains: "The right of all the forms [of life] to live is a universal right which cannot be quantified. No single species of living being has more of this particular right to live and unfold than any other species."[9]

Cultural Justification

Certain species, some threatened or endangered, are of great importance to many indigenous peoples, who rely on these species for food, shelter, tools, fuel, materials for clothing, and medicine. For poor indigenous people who depend on forests, there may be no reasonable replacement except continual outside assistance, which development projects are supposed to eliminate.

More about the Logistic Growth Curve

The concept of maximum sustainable population was made explicit in the logistic growth curve, first proposed in 1838. Recall from Chapter 4 that this is an S-shaped curve representing the growth of a population over time [see Figures 14.3 and 14.4(*a*) and (*b*)]. The logistic growth curve includes the following ideas and assumptions:

- A population that is small in relation to its resources grows at a nearly exponential rate.
- Competition among individuals in the population slows the growth rate.
- The greater the number of individuals, the greater the competition and the slower the rate of growth.
- Eventually, a point is reached, called the "logistic carrying capacity," at which the number of individuals is just sufficient for the available resources.
- At this level, the number of births in a unit of time equals the number of deaths, and the population is constant.
- A population can be described simply by its total number.
- Therefore, all individuals are equal.
- The environment is assumed to be constant.

An important result of all of this is that a *logistic population is stable in terms of its carrying capacity*—it will return to that number after a disturbance. If the population grows beyond its carrying capacity, deaths exceed births, and the population declines back to the carrying capacity. If the population falls below the carrying capacity, births exceed deaths, and the population increases. Only if the population is exactly at the carrying capacity do births exactly equal deaths, and then the population does not change.

Carrying capacity is an important term in wildlife management. It has three definitions. The first is the carrying capacity as defined by the logistic growth curve. We will refer to this as the logistic carrying capacity. The second definition contains the same idea but is not dependent on that specific equation. It states that the carrying capacity is an abundance at which a population can sustain itself without any detrimental effects that would decrease the ability of that species to maintain that abundance. The third, more recent definition is usually referred to as the optimum sustainable population and is defined as the maximum population that can be sustained indefinitely without decreasing the ability of that species *or its habitat or ecosystem* to sustain that population level for some specified time.

Another key concept from the logistic growth curve is the population size that provides the maximum sustainable yield. *In the logistic curve, the greatest production*

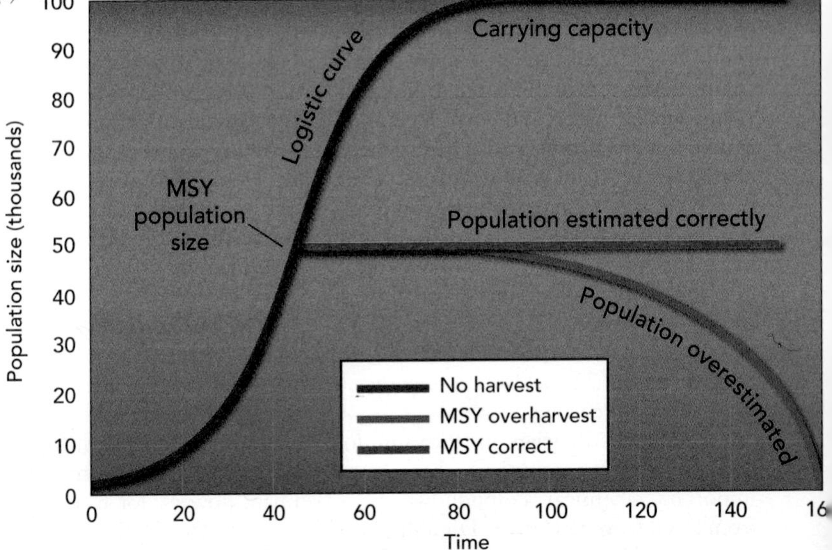

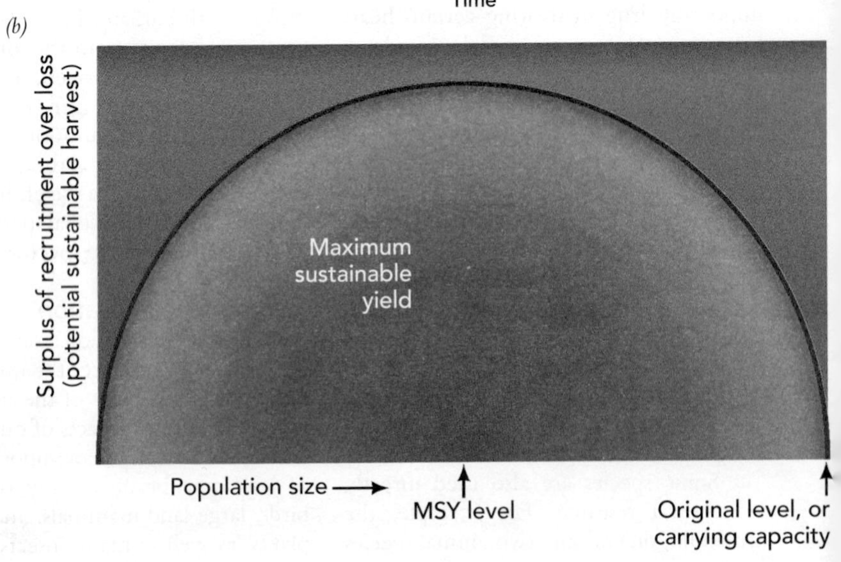

Figure 14.4 • Logistic growth. *(a)* The logistic growth curve, showing the carrying capacity and the maximum sustainable yield (MSY) population (where the population size is one-half the carrying capacity). The figure shows what happens to a population when we assume it is at MSY and it is not. Suppose a population grows according to the logistic curve from a small number to a carrying capacity of 100,000 with an annual growth rate of 5%. The correct maximum sustainable yield would be 50,000. When the population reaches exactly that amount and we harvest at exactly the calculated maximum sustainable yield, the population continues to be constant. But if we make a mistake in estimating the size of the population (for example, if we believe that it is 60,000 when it is only 50,000), then the harvest will always be too large, and we will drive the population to extinction. *(b)* Another view of a logistic population. The growth in the population here is graphed against population size. The growth peaks when the population is exactly at one-half the carrying capacity. This is a mathematical consequence of the equation for the curve. It is rarely, if at all, observed in nature.

occurs when the population is exactly one-half of the carrying capacity (see Figure 14.4). This is nifty, because it makes everything seem simple—all you have to do is figure out the carrying capacity and keep the population at one-half of it. But what seems simple can easily become troublesome. Even if the basic assumptions of the logistic curve were true, which they are not, the slightest overestimate of carrying capacity, and therefore MSY, would lead to overharvesting, a decline in production, and a decline in the abundance of the species. *If a population is harvested as if it were actually at one-half its carrying capacity, then unless a logistic population is actually maintained at exactly that number, its growth will decline.* Since it is almost impossible to maintain a wild population at some exact number, the approach is doomed from the start.

Despite its limitations, the logistic growth curve was used for all wildlife and fisheries (including endangered species) throughout much of the twentieth century.

An Example of Problems with the Logistic Curve

Suppose you are in charge of managing a deer herd for recreational hunting. Your goal is to maintain the population at its MSY level, which as noted in Figure 14.4, occurs at exactly one-half of the carrying capacity. At this abundance, the population increases by the greatest number during any period.

To accomplish this goal, first you have to determine the logistic carrying capacity. You are immediately in trouble because, first, only in a few cases has the carrying capacity ever been determined by legitimate scientific methods (see Chapter 2) and, second, we now know that the carrying capacity varies with changes in the environment. The procedure in the past has been to estimate the carrying capacity by *non-scientific* means and then attempt to maintain the population at one-half that level. This method requires accurate counts each year. It also requires that the environment not vary or, if it does, that it vary in a way that does not affect the population. Since these conditions cannot be met, the logistic curve has to fail as a basis for managing the deer herd.

An interesting example of the staying power of the logistic growth curve can be found in Canada's Marine Mammal Regulations under the federal *Fisheries Act*. The goal of these regulations is to "set in place a contemporary management framework designed to address the non-consumptive use and protection of marine mammals in Canada." This seems consistent with the modern approach, and so the act seems to be off to a good start. However, much of the commentary surrounding proposed revisions to the regulations suggests that a secondary goal is to maintain an "optimum sustainable population" of marine mammals. What is this? The wording of the act allows two interpretations. One is the logistic carrying capacity, and the other is the MSY population level of the logistic growth curve. Therefore, the Act takes us back to square one, the logistic curve.

14.3 Stories Told by the Grizzly Bear and the Bison: Wildlife Management Questions Requiring New Approaches

As discussed in Chapter 2, studies in the environmental sciences sometimes do not seem suited to use of the standard scientific method. This is also true of some aspects of wildlife management and conservation. Several examples illustrate the needs and problems.

The Grizzly Bear

A classic example of wildlife management is the North American grizzly bear. An endangered species, the grizzly has been the subject of efforts by the Canadian Wildlife Service and the U.S. Fish and Wildlife Service to meet the requirements of the Canadian *Species at Risk Act* and the U.S. *Endangered Species Act*, which include restoring the population of the grizzly.

The grizzly became endangered from hunting and habitat destruction. It is arguably the most dangerous North American mammal, famous for its unprovoked attacks on people, and has been eliminated from much of its range for that reason. Males weigh as much as 270 kg, females as much as 160 kg. When they rear up on their hind legs, they are almost 3m tall—no wonder they are frightening (Figure 14.5). Although they are frightening, and perhaps because of it, grizzlies intrigue

Figure 14.5 • Grizzly bear. Records of bear sightings by Lewis and Clark have been used to estimate their population at the beginning of the nineteenth century.

people, and watching grizzlies from a safe distance has become a popular recreation.

At first glance, the goal of restoring the grizzly seems simple enough. Then the question arises: Restore to what? One answer is to its abundance at the time of the European discovery and settlement of North America. It turns out that there is very little historical information about the abundance of the grizzly at that time, so it is not easy to determine the number of grizzlies (or a density of grizzlies per unit area) to select as a target "restored" population.

Adding to the difficulty is the lack of a good estimate of the grizzly's present abundance. Unless we know the present abundance, we do not know how far we will have to take the species to "restore" it to some hypothetical past abundance. The grizzly is also difficult to study. It is large and dangerous and tends to be reclusive. The U.S. Fish and Wildlife Service attempted to count the grizzlies in Yellowstone National Park by installing automatic flash cameras that were set off when the grizzlies took the bait. This seemed a good idea, but the grizzlies did not like the flash cameras and destroyed them. At this time, there is not a good scientific estimate of the present number of grizzlies.

How do we arrive at an estimate of a population that existed at a time when nobody thought of counting its members? Where possible, we make use of historical records, as discussed in Chapter 2. In 1691, Henry Kelsey reached the Canadian prairies and described "a great sort of Bear, which is Bigger than any white Bear is Neither White nor Black But silver hair'd like our English Rabbit." Writing in the late 1700s, the explorer David Thompson describes several encounters with grizzly bears in western Canada, where at one time the species was more common than the black bear. By the 1800s, the grizzly population had begun to decline, as a growing human population, equipped with better weapons, hunted them for fur and trophies. In Alberta, grizzly bears persisted in reduced numbers into the late twentieth century. By the early 1970s, hunting limitations slowed the decimation of the bear population, but by then numbers were so small that recovery proved a major challenge. The detailed records of grizzly sightings left by the Lewis and Clark expedition (1804-1806) provide an especially detailed account. On that expedition, they saw 37 grizzly bears over a distance of approximately 1,600 kilometres.[10]

Lewis and Clark saw grizzlies from near what is now Pierre, South Dakota, to what is today Missoula, Montana. A northern and southern geographic limit to the grizzly's range can be obtained from other explorers. Assuming Lewis and Clark could see a half-mile to each side of their line of travel on average, the density of the bears was approximately 3.7 per 100 square miles. If we estimate that the bear's geographic range was an area of 320,000 square miles in the mountain and western plains states, we arrive at a total population of 320,000 $\times$ 0.37, or about 12,000 bears.

Suppose we phrase this as a hypothesis: "The number of grizzly bears in 1805 in what is now the United States was 12,000." Is this a statement open to disproof? Not without time travel. Therefore, it is not a scientific statement; it can only be taken as an educated guess or, more formally, an assumption or premise. Still, it has some basis in historical documents, and it is better than no information, since we have few alternatives to determine what used to be. We can use this assumption to create a plan to restore the grizzly to that abundance. But is this the best approach?

Another approach is to determine the minimum viable population of the grizzly bears. Forget completely about historical references and use present knowledge of population dynamics and genetics, along with food requirements and potential production of that food. Using this approach, we could estimate how many bears appear to be a "safe" number—that is, a number that carries small risk of extinction and loss of genetic diversity. More precisely, we could phrase this statement as "How many bears are necessary so that the probability that the grizzly will become extinct in the next 10 years (or some other period considered reasonable for planning) is less than 1% (or another acceptable percentage)?" With appropriate studies, this approach could have a scientific basis. Consider a statement of this kind phrased as a hypothesis: "A population of 1,000 bears (or some other number) results in a 99% chance that at least one mature male and one mature female will be alive 10 years from today." We can disprove this statement, but only by waiting for the next 10 years to go by. Although it is a scientific statement, it is a difficult one to deal with in planning for the present.

The American Bison

Another classic case of wildlife management, or mismanagement, is the demise of the bison, or buffalo. Once numbering in the tens of millions, the bison was brought close to extinction in the nineteenth century for two reasons. Bison were hunted because coats made of bison hides had become fashionable in Europe. Bison were also killed as part of warfare against the Plains peoples (Figure 14.6*a*). Colonel R. I. Dodge was quoted in 1867 as saying, "Kill every buffalo you can. Every buffalo dead is an Indian gone."[11] In Canada, as many as 200,000 bison were killed every year between the 1830s and the 1860s.

Unlike the grizzly bear, the bison has recovered, in large part because ranchers have begun to find them a profitable animal to raise and sell for meat and other

(a)

(b)

Figure 14.6 • The bison, past and present. *(a)* Painting of a buffalo hunt by George Catlin in 1832–1833 at the mouth of the Yellowstone River. *(b)* A commercial bison ranch. In recent years, interest in growing bison on ranches has increased greatly. In part, the goal is to restore bison to a reasonable percentage of its numbers before the mid-nineteenth century. In part, bison are ranched because people like them. In addition, there is a growing market for bison meat and other products, including cloth made from bison hair.

products (Figure 14.6*b*). Informal estimates, including herds on private and public ranges, suggest there are more than 200,000 and perhaps as many as 350,000 today. In Canada, there are two subspecies, the plains bison, which once ranged across the North American prairies, and the smaller wood bison, whose range was centred in Northern Alberta and extended north into the Northwest Territories. About 600 plains bison are located at Elk Island National Park, Alberta, with smaller numbers in Prince Albert National Park, Saskatchewan; Riding Mountain National Park, Manitoba; and Waterton Lakes National Park, Alberta. Many more are in private collections and commercial ranches throughout Canada and the United States. The wood bison was always less numerous, probably never more than 200,000 individuals even before European settlement. Today, there are about 3,000 wood bison in Canada, mostly in free-roaming herds, the largest of which are in Wood Buffalo National Park, on the Alberta-Northwest Territories border, and the Mackenzie Bison Sanctuary near Fort Providence, Northwest Territories.

How many bison were there before European settlement of the American West? And how low did their numbers drop? Historical records provide insight. In 1865, General Mitchell, in response to Indian attacks in the fall of 1864, set fires to drive away the Indians and the buffalo, killing vast numbers of animals.[11] The speed with which bison were almost eliminated was surprising—even to many of those involved in hunting them.

Many early writers tell of immense herds of bison, but few counted them. One exception was General Isaac I. Stevens, who, on July 10, 1853, was surveying for the transcontinental railway in North Dakota. He and his men climbed a high hill and saw "for a great distance ahead every square mile" having "a herd of buffalo upon it." He wrote that "their number was variously estimated by the members of the party—some as high as half a million. I do not think it any exaggeration to set it down at 200,000."[11] He suggested that just one herd had about the same number of bison as exist in total today!

One of the better attempts to estimate the number of buffalo in a herd was made by Colonel R. I. Dodge, who took a wagon from Fort Zarah to Fort Larned on the Arkansas River in May 1871, a distance of about 50 km. For at least 30 of those kilometres, he found himself in a "dark blanket" of buffalo. Dodge estimated that the mass of animals he saw in one day totalled 480,000. At one point, he and his men travelled to the top of a hill from which he estimated that he could see 10 to 15 km, and from that high point there appeared to be a single solid mass of buffalo extending over 40 km. At 25 animals per hectare, not a particularly high density, the herd would have numbered 2.7–8.0 million animals.[11]

In the fall of 1868, "a train traveled one hundred twenty miles between Ellsworth and Sheridan through a continuous, browsing herd, packed so thick that the engineer had to stop several times, mostly because the buffaloes would scarcely get off the tracks for the whistle and the belching smoke."[11] That spring, a train had been delayed for eight hours while a single herd passed "in one steady, unending stream." We can use accounts of such experiences to set bounds on the possible number of animals seen. At the highest extreme, we can assume that the train bisected a circular herd with a diameter of just over 1900 km. Such a herd would cover about 30 square kilometers. If we suppose that people exaggerated the density of the buffalo, and there were only 25 per hectare—a moderate density for a herd—this single herd would have numbered 70 million animals!

Some might say that this estimate is probably too high, because the herd would more likely have formed a broad, meandering, migrating line, rather than a circle. The impression remains the same—there were huge numbers of buffalo in the American West even as late as 1868, numbering in the tens of millions and probably 50 million or more. Ominously, that same year, the Kansas Pacific Railroad advertised a "Grand Railway Excursion and Buffalo Hunt."[11] Some say that many hunters believed the buffalo could never be brought to extinction because there were so many. This belief was common about all of America's living resources throughout the nineteenth century.

We tend to think that environmentalism is a social and political movement of the twentieth century; but to the contrary, after the American Civil War there were said to have been angry protests in every legislature over the slaughter of the buffalo. In 1871, the U.S. Biological Survey sent George Grinnell to survey the herds along the Platte River. He estimated that only 500,000 buffalo remained there and that at the then-current rate of killing, the animals would not last long. As late as the spring of 1883, a herd of an estimated 75,000 crossed the Yellowstone River near Miles City, Montana, but fewer than 5,000 reached the Canadian border.[11] By the end of that year—only 15 years after the Kansas Pacific train was delayed for eight hours by a huge herd of buffalo—only a thousand or so buffalo could be found—256 in captivity and about 835 roaming the plains. A short time later, there were only 50 buffalo wild on the plains.

Today, more and more ranchers are finding ways to maintain bison, and the market for bison meat and other bison products is growing, along with an increasing interest in re-establishing bison herds for aesthetic, spiritual, and moral reasons.

The history of the bison raises once again the question of what we mean by "restore" a population. Even with our crude estimates of original abundances, the numbers would have varied from year to year. So we would have to "restore" bison, not to a single number independent of the ability of its habitat to support the population, but to some range of abundances. How do we approach that problem and obtain an estimate of the range?

14.4 Improved Approaches to Wildlife Management

The World Wildlife Fund Canada, the Ecological Society of America, the Smithsonian Institution, and the International Union for the Conservation of Nature (IUCN) have proposed four principles of wildlife conservation:

- A safety factor in terms of population size, to allow for limitations of knowledge and the imperfections of procedures. An interest in harvesting a population should not allow the population to be depleted to some theoretical minimum size.
- Concern with the entire community of organisms and all the renewable resources, so that policies developed for one species are not wasteful of other resources.
- Maintenance of the ecosystem of which the wildlife are a part, minimizing risk of irreversible change and long-term adverse effects as a result of use.
- Continual monitoring, analysis, and assessment. The application of science and the pursuit of knowledge about the wildlife of interest and its ecosystem should be maintained and the results made available to the public.

These principles broaden the scope of wildlife management from a narrow focus on a single species to inclusion of the ecological community and ecosystem. They call for a safety net in terms of population size, meaning that no population should be held at exactly the MSY level or reduced to some theoretical minimum abundance. These new principles provide a starting point for an improved approach to wildlife management.

Time Series and Historical Range of Variation

As the history of the buffalo illustrates, we would like to have an estimate of population over a number of years. This set of estimates is called a **time series** and could provide us with a measure of the **historical range of variation**—the known range of abundances of a population or species over some past time interval. Such records exist for few species. One is the American whooping crane (Figure 14.7), North America's tallest bird, standing about 1.6m tall. Because this species became so rare and because it migrated as a single flock, people began counting the total population in the late 1930s. At that time, they saw only 14 whooping cranes. Not only was the total number counted, but also the number born that year. The difference between these two numbers gives the number dying each year as well. And from this time series, we can estimate the probability of extinction. The first estimate of the probability of extinction based on the historical range of variation, made in the early 1970s, was a surprise. Although the birds were few, the probability of extinction was less than one in a billion. How could this number be so low? Use of the historical range of variation carries with it the assumption that causes of variation in the future will be those and only those that occurred during the historical period. For the whooping cranes, one catastrophe—such as a long, unprecedented drought on the wintering grounds—could cause a population decline not observed in the past.

Figure 14.7 • **The whooping crane** *(a)* is one of many species that appear always to have been rare. Rarity does not necessarily lead to extinction but a rare species, especially one that has undergone a rapid and large decrease in abundance, needs careful attention and assessment as to threatened or endangered status. *(b)* Migration route; and *(c)* change in population from 1940.

Even with this limitation, this method provides invaluable information. Unfortunately, at present, mathematical estimates of the probability of extinction have been done for just a handful of species. The good news is that the whooping cranes have continued to increase, and a record number of 189 reached the wintering grounds at Aranas National Wildlife Refuge in Texas during December 2003.[3] In addition, with help from a variety of governmental and non-governmental organizations, eastern populations of whooping cranes have been established through breeding programs. A joint Canada–United States recovery team manages the Whooping Crane Recovery Strategy for the two countries. The program includes captive breeding and reintroduction of hand-reared birds into the wild. In 2003, there were approximately 420 wild and captive birds, including 185 wild birds of the original flock that breeds in Wood Buffalo National Park.[12, 13]

Age Structure as Useful Information

An additional key to successful wildlife management is monitoring the population's age structure (see Chapter 4), which can provide many different kinds of information. For example, populations of the Steller sea lion (along with harbour and northern fur seals and some sea birds) have been decreasing over the past several decades. The causes of this decline are unknown. Possibilities include increased incidence of parasites, higher predation rates, inadequate food, environmental toxins, and meteorological changes. A recent study compared the age structure of Steller sea lion samples taken in the 1970s and the 1980s. Distinct differences were apparent in the two samples, with significantly fewer young sea lions in the populations of the 1980s. The main predators of Steller sea lions are killer whales, which may select pups and young seals over mature animals. Pup mortality also seems to be influenced by factors such as storms, because although young seals can swim, they may not be able to climb out of the water in storm conditions. The shift in population structure toward older animals, coupled with declining population size, suggests that the Steller sea lions are only marginally viable and thus require special protection. Similar patterns have been observed in populations of other marine mammals in Canada, such as the right whale population of the Bay of Fundy.

Harvests as an Estimate of Numbers

Another method of estimating animal populations is to use the number harvested. Records of the number of buffalo killed were neither organized nor all that well kept, but they were sufficient to give us some idea of the number taken. In 1870, about 2 million buffalo were killed. In 1872, one company in Dodge City, Kansas, handled 200,000 hides. Estimates based on the sum of reports from such companies, together with guesses at how many animals were likely taken by small operators and not reported, suggest that about 1.5 million hides were shipped in 1872 and again in 1873.[11] In those years, buffalo hunting was the main economic activity in Kansas. Aboriginal people were also killing large numbers of buffalo for their own use and for trade. Estimates range to 3.5 million buffalo killed per year during the 1870s.[11] The bison numbered at least in the low millions.

Another way in which harvest counts are used to estimate previous animal abundance is called the **catch per unit effort**. This method assumes that the same effort is exerted by all hunters/harvesters per unit of time, as long as they have the same technology. So if you know the total time spent in hunting/harvesting and you know the catch per unit of effort, you can estimate the total population. The method leads to rather crude estimates with a large observational error; but where there is no other source of information, it can offer unique insights.

An interesting application of this method is the reconstruction of the harvest of the bowhead whale (Figure 14.8*a*) and, from that, an estimate of the total bowhead population. Taken traditionally by Inuit people, the bowhead was the object of European whaling (Figure 14.8*b*) from 1820 until the beginning of World War I. (See A Closer Look 14.2 for a general discussion of marine mammals.) Every ship's voyage was recorded, so we know essentially 100% of all ships that went out to catch bowheads. In addition, on each ship a daily log was kept, with records including the number of whales caught, their size in terms of barrels of oil, the sea conditions, the visibility, and the ice conditions. Of these logbooks, 20% still exist, and their entries have been computerized. Using some crude statistical techniques, it was possible to estimate the abundance of the bowhead in 1820, which was 20,000 plus or minus 10,000. Indeed, it was possible to estimate the total catch of whales and the catch for each year—and therefore the entire history of the hunting of this species.

In summary, new approaches to wildlife conservation and management include (1) historical range of abundance; (2) estimation of the probability of extinction based on historical range of abundance; (3) use of age-structure information; and (4) better use of harvests as sources of information. These, along with an understanding of the ecosystem and landscape context for populations, are improving our ability to conserve wildlife.

14.5 Fisheries

Fish are important to our diets, providing about 16% of the world's protein, and they are especially important protein sources in developing countries.[21] Fish provide 6.6% of food in North America (where people are less interested in fish than are people in most other areas), 8% in Latin America, 9.7% in Western Europe, 21% in Africa, 22% in central Asia, and 28% in the Far East.

The total global fish harvest continues to climb because of increases in the number of boats, improvements in technology, and especially increases in aquaculture production. Fishing is an international trade, but a few countries dominate. Japan, China, Russia, Chile, and the United States are among the major fisheries nations.[22] And commercial fisheries are concentrated in relatively few areas of the world's oceans (Figure 14.10). Continental shelves, which make up only 10% of the oceans, provide more than 90% of the fishery harvest. Fish are abundant where their food is abundant and, ultimately, where there is high production of algae at the base of the food chain. Algae are most abundant in areas with relatively high concentrations of the chemical elements necessary for life, particularly nitrogen and phosphorus. These areas occur most commonly along the continental shelf, particularly in regions of wind-induced upwellings and sometimes quite close to shore.

The world's total fish harvest has increased greatly since the middle of the twentieth century. The total harvest was 35 million metric tonnes (MT) in 1960. It more than doubled in just 20 years (an annual increase of

(*a*)

(*b*)

Figure 14.8 • (*a*) A bowhead whale. (*b*) A nineteenth-century whaling ship.

A CLOSER LOOK 14.2

Conservation of Whales and Other Marine Mammals

Fossil records show that all marine mammals were originally inhabitants of the land. During the last 80 million years, several separate groups of mammals returned to the oceans and underwent adaptations to marine life. Each group of marine mammals shows a different degree of transition to ocean life. Understandably, the adaptation is greatest for those that began the transition longest ago. Some marine mammals—such as dolphins, porpoises, and great whales—complete their entire life cycle in the oceans and have organs and limbs that are highly adapted to life in the water. They cannot move on the land. Others, such as seals and sea lions, spend part of their time on shore.

Whales

Whales fit into two major categories: baleen and toothed [Figure 14.9 *(a)* and *(b)*]. The sperm whale is the only great whale that is toothed; the rest of the toothed group are smaller whales, dolphins, and porpoises. The other great whales, in the baleen group, have highly modified teeth that look like giant combs and act as water filters. Baleen whales feed by filtering ocean plankton.

Drawings of whales have been dated as early as 2200 BC.[14] Inuit peoples used whales for food and clothing as long ago as 1500 b.c. In the ninth century AD, whaling by Norwegians was reported by travelers whose accounts were written down in the court of the English king Alfred.

The earliest whale hunters killed these huge mammals from the shore or from small boats near shore, but gradually whale hunters ventured farther out. In the eleventh and twelfth centuries, Basques hunted the Atlantic right whale from open boats in the Bay of Biscay, off the western coast of France. Whales were brought ashore for processing, and the boats returned to land once the search for a whale was finished.

Eventually, whaling became pelagic: whalers took to the open ocean and searched for whales from ships that remained at sea for long periods. The whales were brought on board and processed there. North American fleets developed in the eighteenth century; by the nineteenth century, the United States dominated the industry, providing most of the ships and even more of the crews for whaling.[15,16]

Whales provided many nineteenth-century products. Whale oil was used for cooking, lubrication, and lamps. Whales also provided the main ingredients for the base of perfumes. The elongated teeth (whalebone, or baleen) that enable baleen whales to strain the ocean waters for food are flexible and springy and were used for corset stays and other products before the invention of inexpensive steel springs.

Conservation of whales has been a concern among conservationists for many years. Attempts to control whaling began with the League of Nations in 1924. The first agreement, the Convention for the Regulation of Whaling, was signed by 21 countries in 1931. In 1946, a conference in Washington, D.C., initiated the International Whaling Commission (IWC), and in 1982, the IWC established a moratorium on commercial whaling. Currently, 12 of approximately 80 species of whales are protected.[14]

In the past, each marine mammal population was treated as if it were isolated, had a constant supply of food, and was subject only to the effects of human harvesting. That is, it was assumed that its growth followed the logistic curve. We now realize that management policies for marine mammals must be expanded to include ecosystem concepts and the understanding that populations interact in complex ways.

The goal of marine mammal management is to prevent extinction and maintain large population sizes rather than to maximize production. For this reason, marine mammal protection programs are geared to establishing an optimum sustainable population (OSP)—the largest population that can be sustained indefinitely without deleterious effects on the population or its

Figure 14.9 • *(a)* **A Sperm whale.** *(b)* **A Blue whale.**

(a)

(b)

Table 14.1 • Recent Whale Population Estimates

Area/Region	*Blue*	*Bowhead*	*Humpback*	*Fin*	*Right*	*Grey*	*Sperm*	*Sei*	*Minke*
Western Arctic		8,000					—		
Pacific						23,109			
California–Mexico	2,134		482						
Central Pacific			1,407						
North Pacific				UNK	UNK		UNK	UNK	
California–Washington			597	575			756		526
East Tropical Pacific	1,400						22,700		
Atlantic									
Western North	308		5,543	2,700	295		337	870	2,650
Northern Gulf of Mexico							213		
Antarctica	610		15,000	18,000		UNK	290,000	17,000	700,000
IWC Estimates for South 30°S									

Sources: National Marine Fisheries Service, 1994; J. Barlow et al., *U.S. Pacific Marine Mammal Stock Assessments*, NOAA Technical Memorandum NMFS (Washington, D.C.: U.S. Department of Commerce, NOAA, National Marine Fisheries Service, 1995); R. J. Small and D. P. DeMaster, *Alaska Marine Mammal Stock Assessments 1995*, NOAA Technical Memorandum NMFS-AFSC-57 (Washington, D.C.: U.S. Department of Commerce, NOAA, National Marine Fisheries Service, 1995). Antarctic estimates are from the International Whaling Commission.

environment—rather than a maximum or optimum sustainable yield.

The IWC has played a major role in reducing (almost eliminating) the commercial harvest of whales. Although an informal commission operating on a consensus basis, the IWC has become a powerful force for conservation and an important forum for discussing international conservation and establishing maximum and optimum sustainable yields. Since the IWC was formed, no species has become extinct, the total take of whales has decreased, and harvesting of species considered endangered has ceased, although endangered species protected from hunting have had a mixed history (see Table 14.1).[17,18] Global climate change, pollution, and ozone depletion now pose greater risks to whale populations than does whaling.[19] The IWC is itself not without controversy, however, especially with respect to the setting of a "research" hunt quota for Japan. And although it is one of the IWC's founding nations, Canada has relinquished its membership on the IWC because of concerns about controls on aboriginal whale hunts.

Dolphins and Other Small Cetaceans

Among the many species of small "whales," or cetaceans, are dolphins and porpoises, more than 40 species of which have been hunted commercially or have been killed inadvertently by other fishing efforts.[20] A classic case is the inadvertent catch of the spinner, spotted, and common dolphins of the eastern Pacific. Because these carnivorous, fish-eating mammals often feed with yellow fin tuna, a major commercial fish, more than 100,000 of these dolphins have been netted and killed inadvertently in recent years.[20]

about 3.6%), to 72 million MT in 1980 and almost doubled again by 2001, reaching 130 million MT. Much of this increase is from aquaculture, which also more than doubled between 1992 and 2001, from about 15 million MT to more than 37 million MT. Aquaculture presently provides more than one-fifth of all fish harvested, up from 15% in 1992.[23] Scientists estimate that there are 27,000 species of fish and shellfish in the oceans. People catch many of these species for food, but only a few kinds provide most of the food—anchovies, herrings, and sardines provide almost 20% (Table 14.2).

Although the total marine fisheries catch has increased during the past half-century, the effort required to catch a fish has increased as well. More fishing boats with improved gear sail the oceans (Figure 14.11). That is why the total catch can increase while the total population of a fish species declines.

The Decline of Fish Populations

Evidence that the fish populations were declining came from the catch per unit effort. A unit of effort varies with the kind of fish sought. For marine fish caught with lines and hooks, the catch rate generally fell from 6–12 fish caught per 100 hooks—the success typical of a previously unexploited fish population—to 0.5–2.0 fish per 100 hooks just 10 years later (Figure 14.11*b*). These observations suggest that fishing depletes fish quickly—

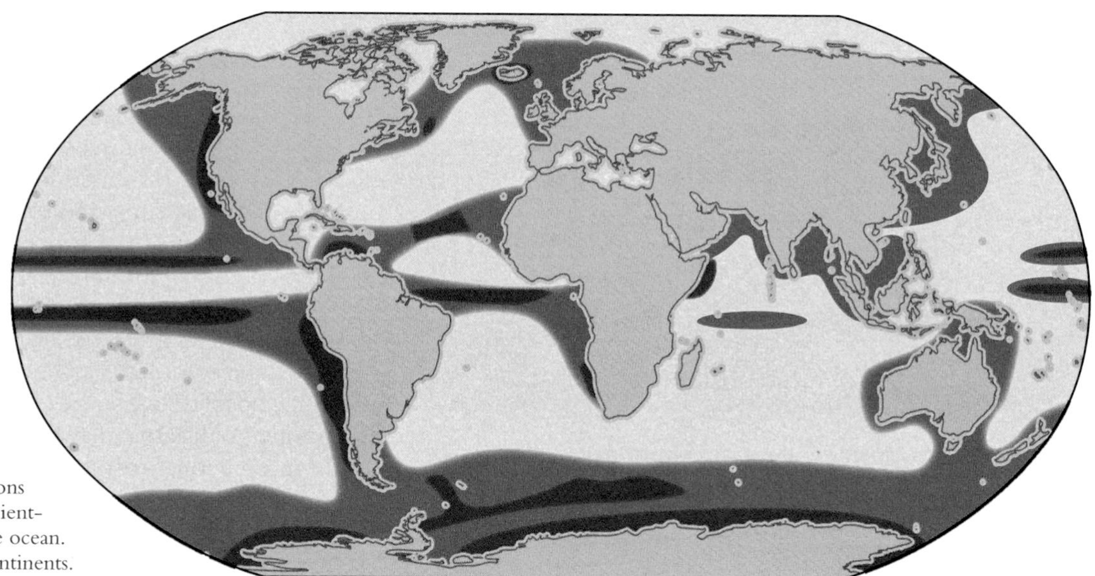

Figure 14.10 • The world's major fisheries. Red areas are major fisheries; the darker the red, the greater the harvest and the more important the fishery. Most major fisheries occur in areas of ocean upwellings—locations where currents rise, bringing nutrient-rich waters from the depths of the ocean. Upwellings tend to occur near continents.

about an 80% decline in 15 years (Figure 14.12). Many of the fish that people eat are predators, and on fishing grounds, the biomass of large predatory fish appears to be only about 10% of pre-industrial levels. These changes indicate that the biomass of most major commercial fish has declined greatly, so that we are mining, not sustaining, these living resources.

Species suffering these declines include cod, flatfishes, tuna, swordfish, sharks, skates, and rays (Table 14.3).[24,25] The North Atlantic, whose George's Bank and Grand Banks have for centuries provided some of the world's largest fish harvests, is suffering. The Atlantic cod catch was 3.7 million MT in 1957, peaked at 7.1 million MT in 1974, then declined to 4.3 million MT in 2000, climbing slightly to 4.7 million MT in 2001.[26] European scientists recently called for a total ban on cod fishing in the North Atlantic, and the European Union came close to accepting this, stopping just short of a total ban and instead establishing a 65% cut in the allowed catch for North Sea cod for 2004 and 2005.[27] Fishing, some of it illegal, continued largely unabated into the early 1990s. In 1992, the Canadian government imposed a moratorium on fishing for Atlantic cod and in 1994 major restrictions on the Atlantic fishery, including new powers to board and arrest foreign vessels. Today, the recovery of the Atlantic cod population is uncertain, and the United Nations Food and Agriculture Organization predicts that stocks will remain very low relative to historic levels.

Table 14.2 • World Fisheries Catch

Kind	*Harvest (Millions of metric tons)*	*Percent of World Total*	*Accumulated Percentage*
Herring, sardines, and anchovies	25.0	19.23%	19.23%
Carp and relatives	15.0	11.54%	30.77%
Cod, hake, and haddock	8.6	6.62%	37.38%
Tuna and their relatives	6.0	4.62%	42.00%
Oysters	4.2	3.23%	45.23%
Shrimp	4.0	3.08%	48.31%
Squid and octopus	3.7	2.85%	51.15%
Other mollusks	3.7	2.85%	54.00%
Clams and relatives	3.0	2.31%	56.31%
Tilapia	2.3	1.77%	58.08%
Scallops	1.8	1.38%	59.46%
Mussels and relatives	1.6	1.23%	60.69%
Subtotal	78.9	60.69%	
TOTAL ALL SPECIES	130.0	100.00%	

Source: NOAA World Fisheries.

Scallops in the western Pacific show a typical harvest pattern, starting very low in 1964 at 200 MT, increasing rapidly to 10,000 MT in 1975, declining by more than a half by the 1980s, increasing to about 10,000 MT in 1992, and then declining to 3,000 MT in 2000.[20] Catch of tuna and their relatives peaked in the early 1990s at about 730,000 MT and fell to 680,000 MT in 2000, a decline of 14% (Figure 14.12).

Food webs are also complex adding to the difficulty of managing fisheries. Typical of marine food webs, the food chain of the Atlantic bluefish shows links to a number of other species, each requiring its own habitat and depending on processes that have a variety of scales of space and time (Figure 14.13). In addition to this complexity, marine systems are influenced by many events on surrounding lands—runoff from farms that is highly polluted with fertilizers and pesticides; introductions of exotic species; and direct alteration of habitats from fishing and the development of shoreline homes. Add to these the varied temperature and salinity of nearshore waters—freshwater inlets from rivers and streams, seawater from the Atlantic, and brackish water resulting from the mixture of these.

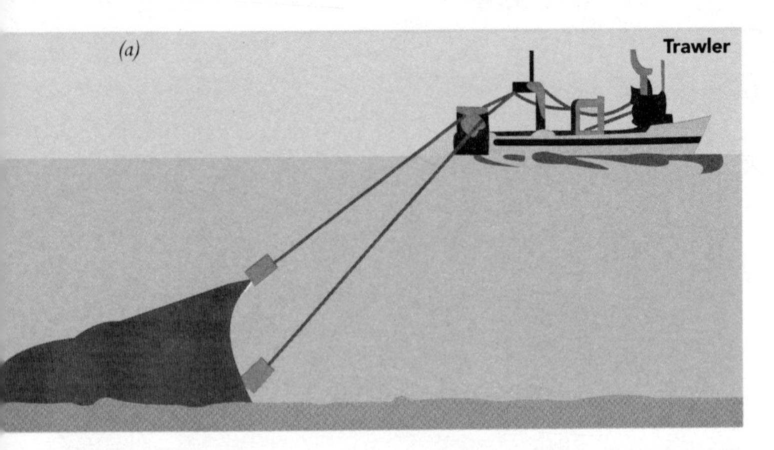

Figure 14.11 • Some modern commercial fishing methods. *(a)* Trawler; *(b)* long-liner; *(c)* gillnetter.

Just determining which of these factors, if any, are responsible for a major change in the abundance of any fish species is difficult enough, let alone finding a solution that is economically feasible and maintains traditional levels of fisheries employment. The Atlantic cod fisheries are at the limit of what environmental sciences can deal with at this time. Scientific theory remains inadequate, as do observations, especially of fish abundance. In some cases, this has allowed politicians to dismiss sound scientific advice on fisheries management as uncertain, and allow unsustainable fishing practices to continue. Our incomplete knowledge of the Atlantic fisheries is sufficient reason to manage with caution: the Precautionary Principle that was discussed in Chapter 3.

Ironically, this crisis has arisen for one of the living resources most subjected to science-based management. How could this have happened? First, management has been based, by and large, on the logistic growth curve, whose problems were discussed earlier. Second, fisheries are an open resource, subject to the problems of "the tragedy of the common," a phrase coined by Garrett Hardin. In an open resource, often in international waters, the numbers of fish that may be harvested can be limited only by international treaties, which are not tightly binding. Open resources offer ample opportunity for unregulated or illegal harvest, or harvest contrary to agreements.

Exploitation of a new fishery usually occurs before scientific assessment, so the fish are depleted by the time any reliable information about them is available. Furthermore, some fishing gear is destructive to the habitat. Ground trawling equipment destroys the ocean floor, ruining habitat for both the target fish and its food. Long-line fishing causes the death of sea turtles and other non-target surface animals. Large tuna nets have killed many dolphins that are hunting the tuna.

In addition to highlighting the need for better management methods, the harvest of large predators raises questions about ocean ecological communities, especially whether these large predators play an important role in controlling the abundance of other species.

Human beings began as hunter-gatherers, and some hunter-gatherer cultures still exist. Wildlife on the land used to be a major source of food for hunter-gatherers. It is now a minor source of food for people of developed nations, although still a major food source for some indigenous peoples, such as the Inuit. In contrast, even developed nations are still primarily hunter-gatherers in the harvesting of fish (see the discussion of aquaculture in Chapter 11).

Can Fishing Ever Be Sustainable?

Suppose you went into fishing as a business and expected a reasonable increase in that business in the first 20 years. As we said earlier, when the world's ocean catch of fish doubled in the 20 years between 1960 and 1980 that was an average annual increase of 3.7%. From a business point of view, even assuming all fish caught are sold, that is not a rapid improvement in sales; but it is a heavy burden on a living resource. There is a general lesson in this: few wild biological resources can sustain a harvest at a level that meets even low requirements for a growing business. Most wild biological resources are not a good business over the long run. People discovered this fact with the harvest of the bison, discussed earlier, and it is true for whales as well (see A Closer Look 14.2). There have been

Table 14.3 • Problems of Some Major Fisheries

Anchovy: Reached peak in 1970 (10 million metric tons), then declined.

Atlantic herring: Exploitation so high that recruitment was decreased.

Arcto-Norwegian cod: High fishing level followed by four years of poor recruitment.

Downs' stock of herring in the North Sea: Managers failed to grasp stock and recruitment problems.

North Atlantic haddock: Catch averaged 50,000 MT for many years; increased to 155,000 in 1965 and 127,000 in 1966, then fell to 12,000 in 1971–1974. In 1973, the International Commission for the Northwest Atlantic Fisheries (ICNAF) established a quota of 6,000 MT. Apparently, haddock could sustain a 50,000 MT catch; but when this was tripled, the population declined to a point where only a smaller catch could be sustained.

Atlantic menhaden: Catch peaked at 712,000 MT in 1956 but declined to 161,400 by 1969. Fisheries experts believe the drop was due to overfishing.

Salmon: Loss has occurred worldwide wherever salmon and their relatives once used fresh waters to spawn and rear. This has been the result of construction of dams on rivers, channelization of rivers, pollution, overharvesting, and habitat alteration of many kinds.

Pacific sardines: Declined catastrophically from the 1950s through the 1970s.

Source: D. Cushing, *Fisheries Resources of the Sea and Their Management* (London: Oxford University Press, 1975).

a few exceptions, such as the several hundred years of fur trading by the Hudson's Bay Company in northern Canada. *Past experience suggests that economically beneficial sustainability is unlikely for most wild populations.*

With that in mind, we can turn to farming fish—aquaculture, discussed in Chapters 11 and 12. This has been an important source of food in China for centuries and is an increasingly important food source worldwide. But aquaculture can create its own environmental problems. One of the best-known environmental issues involves Atlantic salmon, the major species of salmon grown as a crop. In 2000, Atlantic salmon were put on the endangered species list. One of the explanations proposed for the problems of this species is extensive mariculture of Atlantic salmon in such places as British Columbia and the coast of Maine. Observers have identified two concerns. First, water pollution from salmon excrement and excess food provided for the salmon damage the habitat of wild salmon. Second, breeding between native salmon and non-native farmed salmon that escape from farming enclosures may lead to genetic strains that are less fit for a particular stream.

In summary, fish are an important food and world harvests of fish are large, but the fish populations on which the harvests depend are generally declining, easily exploited, and difficult to restore. There is a desperate need for new approaches to forecasting acceptable harvests and establishing workable international agreements to limit catch. This is a major environmental challenge, needing solutions within the next decade. The greatest hope of developing sustainable fisheries may lie in managing small-boat fishing operations, whose practices are more easily modified than those of large industrial fishers with extensive and costly infrastructure. Small-boat fishing provides an important livelihood for many local and indigenous Canadian communities.

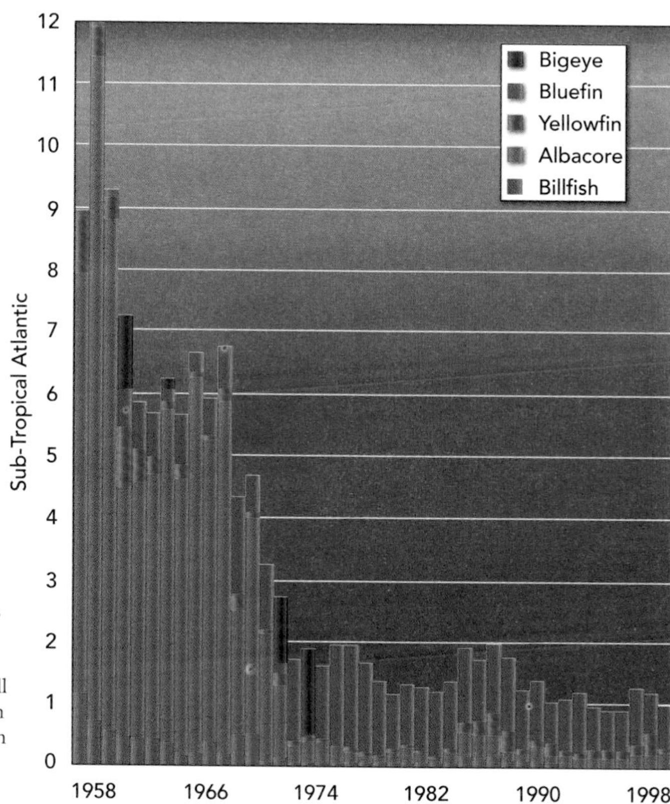

Figure 14.12 • Tuna Catch Decline. The catch per unit of effort—represented here as the number of fish caught per 100 hooks—for tuna and their relatives in the subtropical Atlantic Ocean. The vertical axis shows the number of fish caught per 100 hooks. The catch per unit of effort was 12 in 1958, when heavy modern industrial fishing for tuna began, and had declined rapidly by 1974, to about 2. This pattern occurred worldwide in all the major fishing grounds for these species. [*Source:* Modified from Ransom A. Meyers and Boris Worm, "Rapid Worldwide Depletion of Predatory Fish Communities," *Nature* (May 15, 2003).]

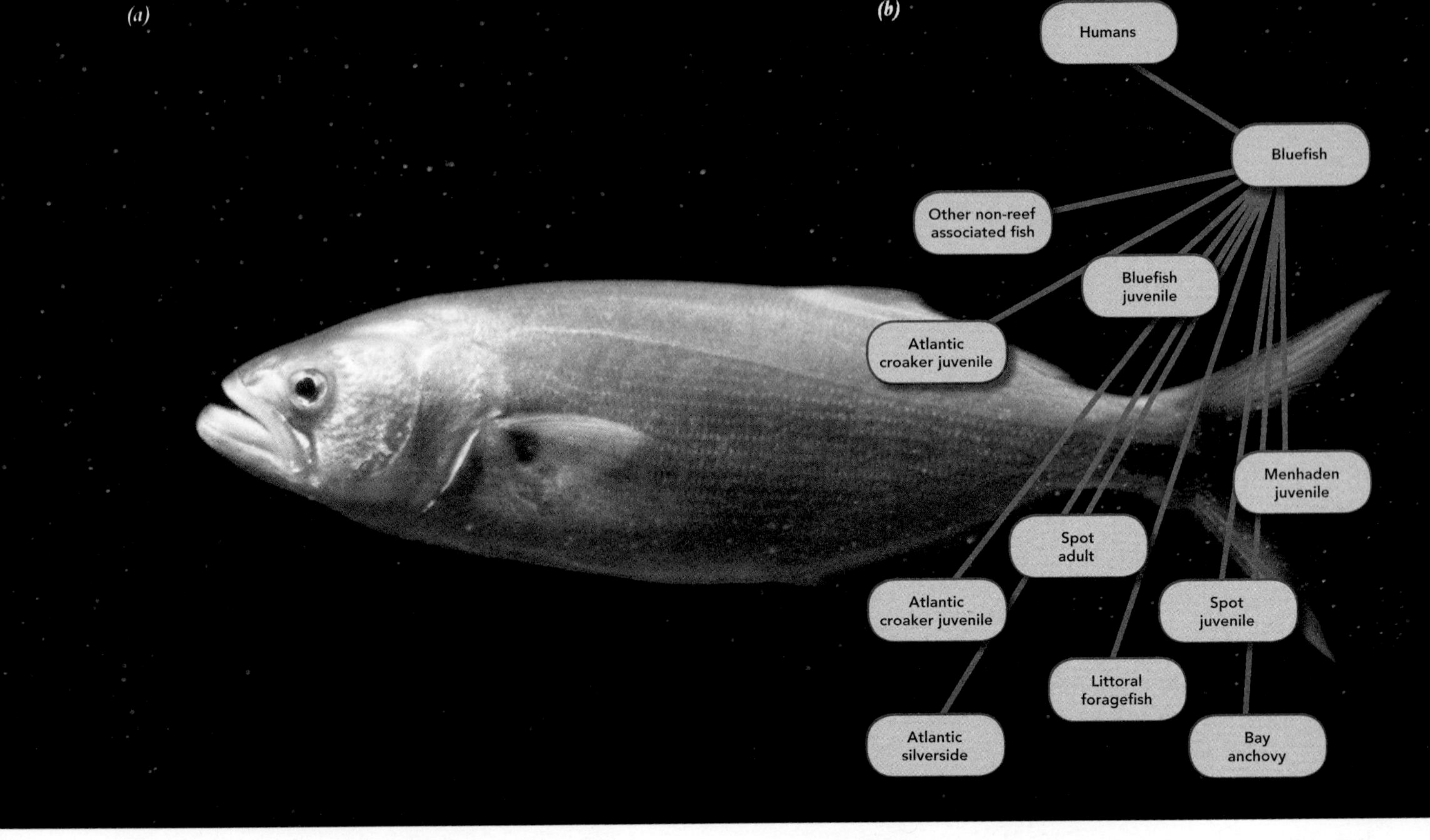

Figure 14.13 • *(a)* **Bluefish**; *(b)* food chain of the bluefish in Chesapeake Bay. [*Source:* Chesapeake Bay Foundation.]

14.6 The Current Status of Endangered Species

The human population has had a dramatic impact on the populations of many wildlife and fish species, some of which are now endangered. As our awareness of this problem grows, public interest in rare and endangered species, especially large mammals and birds, has expanded, so it is time for us to turn our attention to that topic. First some facts. The number of species of animals listed as threatened or endangered increased from about 1,700 in 1988 to 3,800 in 1996 and 5,400 in 2000 (Table 14.4).[10] The IUCN *Red Book of Threatened Animals* reports that about 25% of all known species of mammals are at risk of extinction, as well as 11% of known birds, 20% of known reptiles, 25% of amphibians, and 34% of fish, primarily freshwater fish.

The IUCN Red List of *Threatened Plants* estimates that 33,798 species of vascular plants (the familiar kind of plants—trees, grasses, shrubs, flowering herbs), or 12.5% of those known, have recently become extinct or endangered.[3] Of the approximately 100,000 species of trees, the IUCN lists more than 8,700, or approximately 9%, as globally threatened (see Table 14.4). As of May 2004, the Committee on the Status of Endangered Wildlife in Canada (COSEWIC) has listed 169 Canadian species of plants and animals as endangered, 114 species as threatened, and 140 as of special concern.

What does it mean to call a species "endangered"? The term can have a strictly biological meaning, or it can have a legal meaning. The words endangered and threatened are defined in the Canadian *Species at Risk Act* of 2003. The Act defines an *endangered species* as "a wildlife species that is facing imminent extirpation or extinction." The term "wildlife" is further defined as "a species, subspecies, variety or geographically or genetically distinct population of animal, plant, or other organism, other than a bacterium or virus, that is wild by nature and (a) is native to Canada; or (b) has extended its range into Canada without human intervention and has been present in Canada for at least 50 years." The term *"threatened species,"* according to the Act, means "a wildlife species that is likely to become an endangered species if nothing is done to reverse the factors leading to its extirpation or extinction."

14.7 How a Species Becomes Endangered and Extinct

Extinction is the rule of nature (see the discussion of biological evolution in Chapter 7). **Local extinction** (extirpation) occurs when a species disappears from a part of its range but persists elsewhere. **Global extinction** occurs when a species can no longer be found anywhere. Although extinction is the ultimate fate of all species,

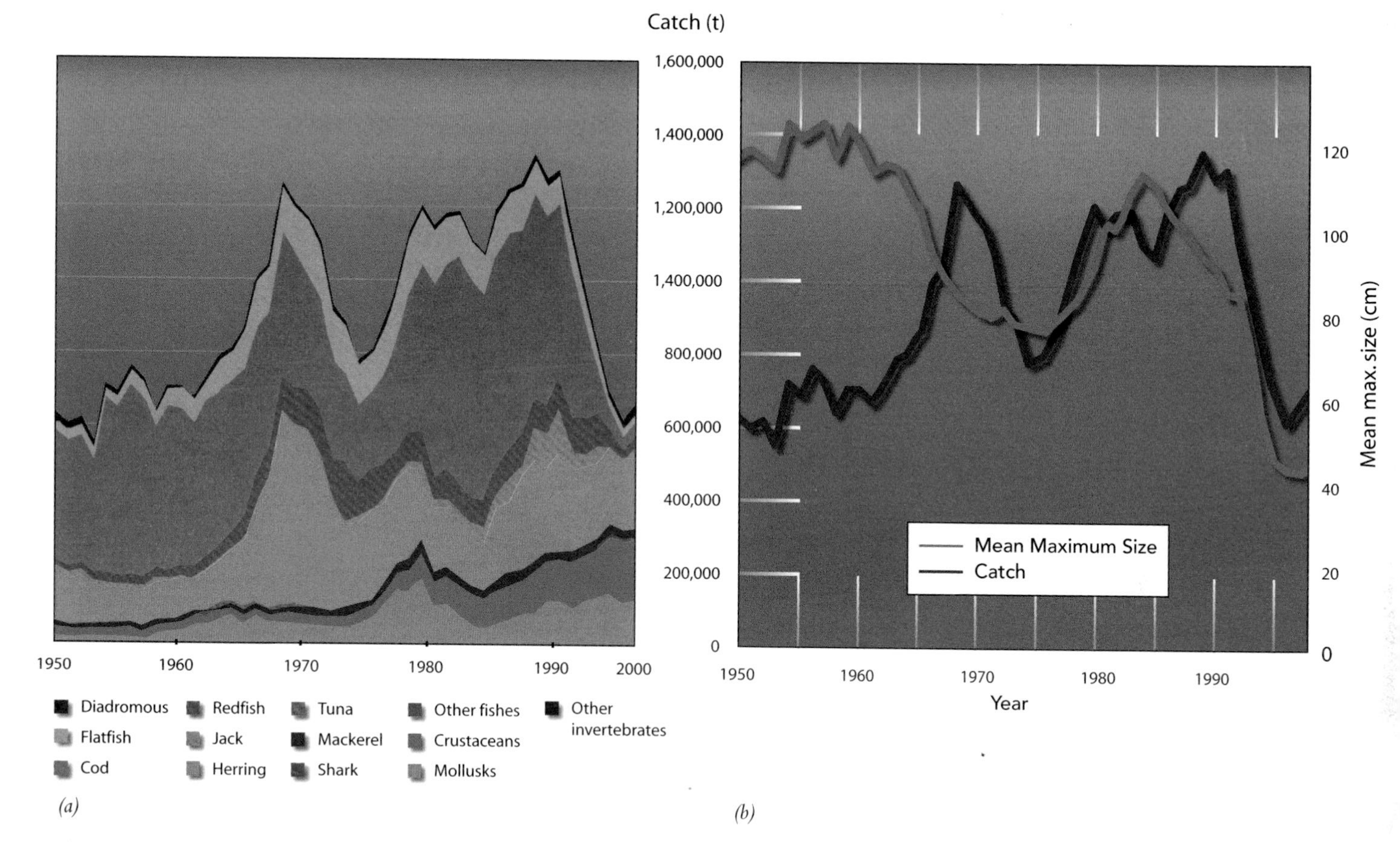

Figure 14.14 • Canada's Northwest Atlantic fishery. *(a)* Time series of catch composition for Canada, Northwest Atlantic fishery; *(b)* Time series of total catch and average maximum size of species catch for Canada, Northwest Atlantic. [*Source:* UN Food and Agriculture Organization.]

the rate of extinctions has varied greatly over geologic time and has accelerated since the Industrial Revolution. From 580 million years ago until the beginning of the Industrial Revolution, about one species per year, on average, became extinct. Over much of the history of life on Earth, the rate of evolution of new species equaled or slightly exceeded the rate of extinction. The average longevity of a species has been about 10 million years.[16] However, as discussed in Chapter 7, the fossil record suggests that there have been periods of catastrophic losses of species and other periods of rapid evolution of new species [see Figures 7.9 and 14.15(*a–c*)], which some refer to as "punctuated extinctions." About 250 million years ago, a mass extinction occurred in which approximately 53% of marine animal species disappeared; about 65 million years ago, most of the dinosaurs became extinct. Interspersed with the episodes of mass extinctions, there seem to have been periods of hundreds of thousands of years with comparatively low rates of extinction.

An intriguing example of punctuated extinctions occurred about 10,000 years ago, at the end of the last great continental glaciation. At that time, massive extinctions of large birds and mammals occurred: 33 genera of large mammals—those weighing 50 kg or more—became extinct, whereas only 13 genera had become extinct in the preceding 1 or 2 million years (Figure 14.16). Smaller mammals were not so affected, nor were marine mammals. As early as 1876, Alfred Wallace, the English biogeographer, noted, "we live in a zoologically impoverished world, from which all of the hugest, and fiercest, and strangest forms have recently disappeared."[28] It has been suggested that these sudden extinctions coincided with the arrival, on different continents at different times, of Stone Age people and therefore may have been caused by hunting.[29] Causes of extinction are summarized in A Closer Look 14.3.

Table 14.4 • Worldwide Status of Endangered and Threatened Animals

Animal Group	*Critically Endangered*	*Endangered*	*Vulnerable (threatened)*	*Total*
Mammals	180	340	610	1130
Birds	182	321	680	1183
Reptiles	56	79	161	296
Amphibians	25	38	83	146
Fish	152	126	431	709
Crustaceans	56	72	280	408
Insects	45	118	392	555
Molluscs	170	209	467	846

Source: Data from International Union for the Conservation of Nature (IUCN) Red List 2000.

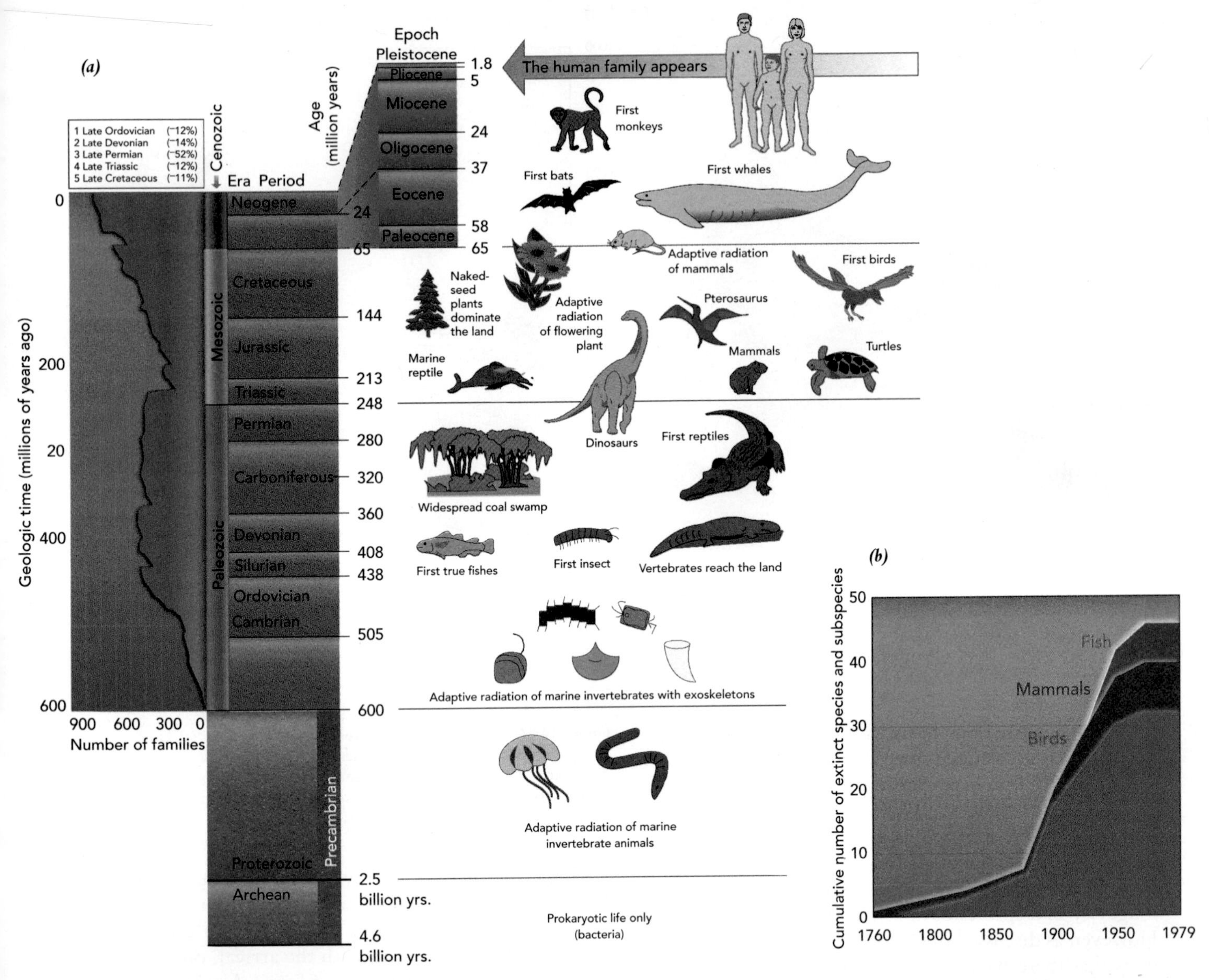

Figure 14.15 • *(a)* **Brief diagrammatic history of evolution and extinction of life on Earth.** There have been periods of rapid evolution of new species and episodes of catastrophic losses of species. Two major catastrophes were the Permian loss, which included 52% of marine animals, as well as land plants and animals, and the Cretaceous loss of dinosaurs. The graph on the left shows the number of families of marine animals in the fossil records. Note the long periods of overall increase in the number of families punctuated by brief periods of major declines. *(b)* Extinct vertebrate species and subspecies, 1760–1979. The number of species becoming extinct increases rapidly after 1860. Note that most of the increase is due to the extinction of birds. (*Sources:* [a] Modified from D. M. Raup, "Diversity Crisis in the Geological Past," in E. O. Wilson, ed., *Biodiversity* [Washington, D.C.: National Academy Press, 1988], p. 53; derived from S. M. Stanley, Earth and Life through Time [New York: W. H. Freeman, 1986]. Reprinted with permission; Modified from Council on Environmental Quality; additional data from B. Groombridge, England IUCN, 1993. [b] Modified from D. M. Raup and J. J. Sepkoski, Jr., "Mass Extinctions in the Marine Fossil Record," *Science* 215 [1982]:1501–1502.].)

14.8 How People Cause Extinctions and Affect Biological Diversity

People have become an important cause of extinction and an important factor in causing species to become threatened and endangered. Among the ways we cause extinction are:

- By intentional hunting or harvesting (for commercial purposes, for sport, or to control a species that is considered a pest).
- By disrupting or eliminating habitats.
- By introducing exotic species, including new parasites, predators, or competitors of a native species.
- By polluting the natural environment.

People have caused extinctions over a long time, not just in recent years. The earliest people probably caused extinctions through hunting. This practice continues, especially for specific animal products considered valuable such as elephant ivory and rhinoceros horns. When people learned to use fire, they began to change habitats over large areas. The development of agriculture and the rise of civilization led to rapid deforestation and other habitat changes. Later, as people explored new areas, introductions of exotic species increased as a cause of extinction (see Chapter 8), especially after Columbus's voyage to the New World, Magellan's circumnavigation of the globe, and the resulting spread of European civilization and technology. The introduction of thousands of novel chemicals into the environment made pollution an increasing cause of extinction in the twentieth century, and pollution control has proved to be a successful way to help species.[21]

The IUCN estimates that 75% of the extinctions of birds and mammals since 1600 have been caused by human beings. Hunting is estimated to have caused 42% of the extinctions of birds and 33% of the extinctions of mammals. The current extinction rate among most groups of mammals is estimated to be 1,000 times greater than the extinction rate at the end of the Pleistocene epoch.[30]

Figure 14.16 • Artist's rereconstruction of an extinct saber-toothed cat with prey. The cat is an example of one of the many large mammals that became extinct about 10,000 years ago.

The Good News: Species Whose Status Has Improved

There is some good news about endangered species as a result of human activities. A number of previously endangered species have recovered, like the Aleutian goose of the opening case study. Among the recovered species are the following:

- The elephant seal, which had dwindled to about a dozen animals around 1900 and now numbers in the hundreds of thousands.
- The sea otter, reduced in the nineteenth century to several hundred and now numbering approximately 10,000.
- Many species of birds endangered because the insecticide DDT caused thinning of eggshells and failure of reproduction. With the elimination of DDT in Canada and the United States, many bird species recovered, including the bald eagle, brown pelican, white pelican, osprey, and peregrine falcon.
- The blue whale, thought to have been reduced to about 400 when whaling was still actively pursued by a number of nations. Today, 400 blue whales are sighted annually in the Santa Barbara channel along the California coast. Sightings off British Columbia remain rare, however.[17]
- The grey whale, which was hunted to near-extinction but has recovered and is abundant along the California coast and British Columbia's Gulf Islands in its annual migration to Alaska.

Can a Species Be Too Abundant? If So, What Do We Do?

Sometimes we succeed too well in increasing the numbers of a species. Case in point: marine mammals are protected under federal legislation in both Canada and the United States, which has led to improvement in the status of many marine mammals. In southern regions, seas lions are now so abundant that they are a local problem. For example, in San Francisco Harbor and Santa Barbara Harbor, sea lions haul out and sun themselves on boats and pollute the water with their excrement near shore. In one case, so many sea lions hauled out on a sailboat in Santa Barbara Harbor that they sank the boat, and some of the animals were trapped and drowned. (For more information on the history and status of marine mammals, see A Closer Look 14.2.)

Mountain lions also have become locally overabundant. In the 1990s, California voters passed an initiative that protected the endangered mountain lion but contained no provisions for management of the lion should it become abundant, except in cases where it threatened human life and property. Few people thought the mountain lion could ever recover enough to become a problem, but in several cases in recent years, mountain lions have attacked and even killed people. These attacks become more frequent as both the mountain lion and human populations grow and people build houses in what was mountain lion habitat.

In Canada, the annual harp seal hunt—the largest marine mammal hunt in the world—has been the focus of international controversy for decades. The seal population has tripled in size since 1970, and now numbers about 5 million animals. In the absence of controls, the seals will quickly exceed the carrying capacity of the natural environment. Wildlife managers therefore faced a difficult choice: allow the population to self-regulate, dying off naturally from disease or starvation or through

A CLOSER LOOK 14.3

Causes of Extinction

Causes of extinction are usually grouped into four risk categories: population risk, environmental risk, natural catastrophe, and genetic risk. Risk here means the chance that a species or population will become extinct owing to one of these causes.

Population Risk

Random variations in population rates (in birth rates and death rates) can cause a species in low abundance to become extinct. This is termed *population risk.* For example, blue whales swim over vast areas of ocean. Because whaling once reduced their total population to only several hundred individuals, the success of individual blue whales in finding mates probably varied from year to year. If in one year most whales were unsuccessful in finding mates, then births could be dangerously low. Such random variation in populations, typical among many species, can occur without any change in the environment. It is a risk especially to species that consist of only a single population in one habitat. Mathematical models of population growth can help calculate the population risk and determine the minimum viable population size.[4]

Environmental Risk

Population size can be affected by changes in the environment that occur from day to day, month to month, and year to year, even though the changes are not severe enough to be considered environmental catastrophes. Environmental risk involves variation in the physical or biological environment, including variations in predator, prey, symbiotic, or competitor species. In some cases, species are so rare and isolated that such normal variations can lead to their extinction.

For example, Paul and Anne Ehrlich described the local extinction of a population of butterflies in the Colorado mountains.[32] These butterflies lay their eggs in the unopened buds of a single species of lupine (a member of the legume family), and the hatched caterpillars feed on the flowers. One year, however, a very late snow and freeze killed all the lupine buds, leaving the caterpillars without food and causing local extinction of the butterflies. Had this been the only population of that butterfly, the entire species would have become extinct.

Natural Catastrophe

A sudden change in the environment, not the result of human action, is a natural catastrophe. Fires, major storms, earthquakes, and floods are natural catastrophes on land; changes in currents and upwellings are ocean catastrophes. For example, the explosion of a volcano on the island of Krakatoa in Indonesia in 1883 caused one of recent history's worst natural catastrophes. Most of the island was blown to bits, bringing about local extinction of most life forms there.

Genetic Risk

Detrimental change in genetic characteristics, not caused by external environmental changes, is called *genetic risk.*[33] Genetic changes can occur in small populations from reduced genetic variation, genetic drift, and mutation (see Chapter 7). In a small population, only some of the possible inherited characteristics will be found. The species is vulnerable to extinction because it lacks variety or because a mutation can become fixed in the population.

Consider the small population of the whooping crane. In 1940, numbers had dropped to only 22 individuals. The modern population all comes from that small group. It stands to reason that this small number was likely to have less genetic variability than the much larger population that existed several centuries ago. This increased the cranes' vulnerability. Suppose that the last 22 cranes, by chance, had inherited characteristics that made them less able to withstand lack of water. If left in the wild, these cranes would have been more vulnerable to extinction than a larger, more genetically variable population.

reduced fertility, or allow humans to act as predators through hunting. Non-governmental organizations like the Sea Shepherd Conservation Society advocate the former approach, while Fisheries and Oceans Canada (DFO) continues to endorse an annual hunt. DFO's position is undoubtedly influenced by the contribution of the hunt to the economy of Atlantic Canada—over $20 million in 2003 alone. The seal hunt remains a highly emotional environmental issue, both in Canada and internationally, and will likely remain so. In 2005, DFO completed a three-year seal management plan, developed in consultation with stakeholders, which calls for a continuation of the hunt for the foreseeable future.

14.9 The Kirtland's Warbler and Environmental Change

Many endangered species are adapted to natural environmental change and require it. When human actions eliminate that change, a species can become threatened with extinction. This happened with the Kirtland's warbler, which nests in jack pine forests (Figure 14.17). In 1951, the Kirtland's warbler became the first songbird in the United States to be subject to a complete census, and about 400 nesting males were found. Concern about the species grew in the 1960s and increased when only 201 nesting males were found in the third census, in 1971.[31] Conservationists and scientists tried to understand what was causing the decline, which threatened the species with extinction. Kirtland's warblers were probably nesting in parts of southern Ontario during the late 1800s and early 1900s, but the species is now considered extirpated in Canada.

Kirtland's warblers are known to nest only in jack pine woodlands that are between 6 and 21 years old. At these ages, the trees, 1.5 – 7 metres tall, retain dead branches at ground level. Jack pine is a "fire species," which persists only where there are periodic forest fires. Cones of the jack pine open only after they have been heated by fire. The trees are intolerant of shade, able to grow only when their leaves can reach into full sunlight; so even if seeds were to germinate under mature trees, the seedlings could not grow in the shade and would die. Jack pine produces an abundance of dead branches, which some people view as an evolutionary adaptation to promote fires, essential to the survival of the species.

Kirtland's warblers thus require change at rather short intervals—forest fires approximately every 20–30 years, which was about the frequency of fires in jack pine woods in pre-settlement times.[31] At the time of the first European settlement of North America, jack pine may have covered a large area in what is now Michigan. Even as recently as the 1950s, the pine was estimated to cover nearly 200,000 hectares in that state. Small, and poorly formed, jack pine was considered a trash species by commercial loggers and was left alone; but many large fires followed the logging operations when large amounts of slash—branches, twigs, and other economically undesirable parts of the trees—were left in the woods. Elsewhere, fires were set to clear jack pine areas and promote the growth of blueberries. Some experts think that the population of Kirtland's warblers peaked in the late nineteenth century as a result of these fires. After 1927, fire suppression became the practice, and people were encouraged to replace jack pine with economically more useful species. One result was that the areas conducive to the nesting of the warbler shrank.[31]

Although it may seem obvious today that the warbler requires forest fires, this was not always understood. In 1926, one expert wrote, "fire might be the worst enemy of the bird."[34] Only with the introduction of controlled burning after vigorous advocacy by conservationists and ornithologists was habitat for the warbler maintained. The Kirtland's Warbler Recovery Plan, published by the U.S. Department of the Interior and the Fish and Wildlife Service in 1976 and updated in 1985, calls for the creation of 15,375 hectares of new habitat for the warbler for which "prescribed fire will be the primary tool used to regenerate non-merchantable jack pine stands on poor sites."[35,36] In Canada, the species is protected under the federal *Species at Risk* Act. There is considerable jack pine habitat in southern Ontario that would be suitable for the Kirtland's warbler, if a self-sustaining population can be established.

14.10 Ecological Islands and Endangered Species

The history of the Kirtland's warbler illustrates that a species may inhabit "ecological islands," which the isolated jack pine stands of the right age range are for that bird. Recall from our discussion in Chapter 8 that an ecological island is an area that is biologically isolated, so a species living there cannot mix (or only rarely mixes) with any other population of the same species (Figure 14.18). Mountaintops and isolated ponds are ecological islands. Real geographic islands are also ecological islands. Insights gained from studies of the biogeography

Figure 14.17 • *(a)* **A Kirtlands warbler** and *(b)* its jack pine habitat.

(a)

(b)

of islands have important implications for the conservation of endangered species and for the design of parks and preserves for biological conservation.

Almost every park is a biological island for some species. A small city park between buildings may be an island for trees and squirrels. At the other extreme, even a large national park is an ecological island. For example, the Masai Mara Game Reserve in the Serengeti Plain, which stretches from Tanzania to Kenya in East Africa, and other great wildlife parks of eastern and southern Africa are becoming islands of natural landscape surrounded by human settlements. Lions and other great cats exist in these parks as isolated populations, no longer able to roam completely freely and to mix over large areas. Other examples are islands of uncut forests left by logging operations and oceanic islands, where intense fishing has isolated parts of fish populations.

How large must an ecological island be to ensure the survival of a species? The size varies with the species, but can be estimated. Some islands that seem large to us are too small for species we wish to preserve. For example, a preserve was set aside in India in an attempt to reintroduce the Indian lion into an area where it had been eliminated by hunting and changing patterns of land use. In 1957, a male and two females were introduced into a 95-km^2 preserve in the Chakia forest, known as the Chandraprabha Sanctuary. The introduction was carried out carefully and the population was counted annually. There were four lions in 1958, five in 1960, seven in 1962, and eleven in 1965, after which they disappeared and were never seen again.

Why did they go? Although 95 km^2 seems large to us, male Indian lions have territories of 130 km^2. Within such a territory, females and young also live. A population that could persist for a long time would need a number of such territories, so an adequate preserve would require 640–1,300 km^2. Various other reasons were suggested for the disappearance of the lions, including poisoning and shooting by villagers; but regardless of the immediate cause, a much larger area than was set aside was required for long-term persistence of the lions.

14.11 Using Spatial Relationships to Conserve Endangered Species

The white-headed woodpecker (Figure 14.19) is an endangered species of western Canada. The woodpecker makes its nests in old dead or dying pines, and its foods include beetles and pine seeds. To conserve this species of woodpecker, these pines must be preserved. But the old pines are home to the beetles, which are pests to the trees and damage them for commercial logging. This presents an intriguing problem: How can we maintain the

(a)

(b)

(c)

Figure 14.18 • Ecological islands: *(a)* Stanley Park in Vancouver; *(b)* a mountaintop in Alberta; *(c)* an African wildlife park.

Figure 14.19 • **Endangered white-headed woodpecker.**

woodpecker and its food (the beetles) and also maintain productive forests?

The classic twentieth-century way to view the relationship among the pines, the beetles, and the woodpecker would be to show a food chain (see Chapter 6). But this alone does not solve the problem for us. A newer approach is to consider the habitat requirements of the beetle and the woodpecker. Their requirements are somewhat different. But if we overlay a map of one's habitat requirements over a map of the other's, the co-occurrence of habitats can be compared. Beginning with such maps, it becomes possible to design a landscape that would allow the maintenance of all three—pines, beetles, and birds.

Summary

- Modern approaches to management and conservation of wildlife use a broad perspective that considers interactions among species as well as the ecosystem and landscape contexts.
- To successfully manage wildlife for harvest, conservation, and protection from extinction, we need certain information. Measures of total abundance and of births and deaths over a long period will be helpful. We also need to know the habitat, viewed spatially and in quantitative terms. Age structure and other characteristics of a population can help in management, forecasting, and conservation. However, it is often difficult to obtain these data, especially for historical information.
- A common goal of wildlife conservation today is to "restore" the abundance of a species to some previous number, usually a number thought to have existed prior to the influence of modern technological civilization. Information about long-ago abundances is rarely available. Sometimes numbers can be estimated indirectly—for example, by using the Lewis and Clark journals to reconstruct the 1805 population of grizzly bears or using logbooks from ships hunting the bowhead whale. Adding to the complexity, the abundance of wildlife changes over time even in natural systems uninfluenced by modern technological civilization. And historical information often cannot be subjected to formal tests of disproof and therefore does not by itself qualify as scientific. Adequate information exists for relatively few species.
- Another approach is to seek a minimum viable population, a carrying capacity, or an optimal sustainable population or harvest based on data that can be obtained and tested today. This approach abandons the goal of restoring a species to some hypothetical past abundance.
- The good news is that many species once endangered have been successfully restored to an abundance that suggests they are unlikely to become extinct. Success is achieved when the habitat is restored to conditions required by a species. The conservation and management of wildlife presents great challenges but also offers great rewards of long-standing and deep meaning to people.

Figure 14.20 • *(a)* **The Eastern Grey wolf;** *(b)* Canadian distribution of the Grey Wolf

CRITICAL THINKING ISSUE

Should Wolves be Re-established in Eastern North America?

When European settlers first came to eastern Canada and the northeastern United States, the land was, like much of the rest of North America, inhabited by grey wolves. The settlers feared the wolves and their potential impact on precious livestock, and quickly set about eradicating them. By the 1870s, wolves were rare in eastern Canada; by 1880, they were extinct in New Brunswick. Twenty years later, they had disappeared from Nova Scotia, and by the beginning of the First World War, they had vanished from Newfoundland. By 1960, wolves had been exterminated in all of the lower 48 United States except for northern Minnesota. The last official sighting of a wolf in the Adirondack Mountains of the U.S. Northeast was in the 1890s. Today, there are approximately 2,000 grey wolves in eastern North America, mainly in Québec and northern Ontario, especially Algonquin Park. The population of these wolves appears to have stabilized over the past decade, although the species is still listed as of special concern under the federal *Species at Risk Act*. The challenge that now confronts wildlife managers is whether the species should be reintroduced in areas where it was formerly native.

The idea of wolf reintroduction is not new. The U.S. Fish and Wildlife Service has had a grey wolf recovery plan in place since the early 1970s, and small populations have been successfully reintroduced in Minnesota and Yellowstone National Park. In Canada, there have been proposals to bring the wolf back to New Brunswick. Although a survey seemed to show some support for the plan, many residents and organizations vigorously opposed reintroduction. Concerns expressed focused primarily on the potential dangers to humans, livestock, and pets and the possible impact on the deer population.

Wolves prey primarily on moose, deer, and beaver. All three species are abundant in New Brunswick. However, determining whether there are sufficient prey base to support a population of wolves is complicated by the fact that coyotes have moved into New Brunswick and occupy the niche once filled by wolves. Whether wolves would add to the deer kill or replace coyotes, with no net impact on the deer population, is difficult to predict.

Much of the northern part of New Brunswick is suitable for wolf habitat, because of its low human population density and few roads. Almost 50,000 sq. km. of potential wolf habitat have already been mapped in adjacent Maine and Québec, and could be made contiguous with the proposed New Brunswick tracts. As we saw in Chapter 13's Carolinian case study, such connections can be critical for species like the grey wolf that have large home ranges. Ideally, wolf population density should not exceed four wolves per 100 sq. km., and some estimates suggest that lower densities are better. Although sufficient habitat apparently exists for a wolf reintroduction program, human communities are scattered throughout much of northern New Brunswick, and many residents are concerned that wolves would threaten their pets and livestock, and interfere with the activities of back-country hikers and hunters. Ultimately, it may be the humans who need to change. Evidence from Europe indicates that wolves can learn to tolerate relatively high human densities; the question is whether the humans can tolerate the wolves.

Critical Thinking Questions

1. Who should make decisions about wildlife management, such as returning wolves to New Brunswick—scientists, government officials, or the public?
2. Some people advocate leaving the decision to the wolves—that is, waiting for them to disperse from Québec and Maine into New Brunswick. Study a map of the northeastern United States and southeastern Canada (Figure 14.20). What do you think is the likelihood of natural recolonization of New Brunswick by wolves?
3. Do you think that wolves should be reintroduced to northern New Brunswick? If you lived in the area, would that affect your opinion? How would removal of the wolf from the Species at Risk Act "special concern" list affect your opinion?
4. Some biologists recently concluded that wolves in Yellowstone and the Great Lakes region belong to a different subspecies, the Rocky Mountain timber wolf, from those that formerly lived in the northeastern United States, the eastern timber wolf. This means that although healthy populations occur in Québec and parts of Ontario, the eastern timber wolf is still extinct in the lower 48 United States. Would this affect your opinion about reintroducing wolves into New Brunswick?

STUDY QUESTIONS

1. Why are we so unsuccessful in making rats an endangered species?
2. What have been the major causes of extinction *(a)* in recent times and *(b)* before people existed on Earth?
3. As mentioned in the text, wildlife management agencies have suggested three key indicators of the status of the grizzly bear: (1) sufficient reproduction to offset the existing levels of human-caused mortality, (2) adequate distribution of breeding animals throughout the area, and (3) a limit on total

human-caused mortality. Are these indicators sufficient to assure the recovery of this species? If not, what would you suggest instead?
4. What are some arguments for and against the use of historical records that are not necessarily scientific to "restore" an abundance?
5. Five justifications for preserving endangered species were discussed in this chapter. Which of them apply to the following? (You can decide that none apply.)
   a. The black rhinoceros of Africa
   b. The Furbish lousewort, a rare small flowering plant of New Brunswick, seen by few people
   c. An unnamed and newly discovered beetle from the Amazon rain forest
   d. Smallpox
   e. Wild strains of potatoes in Peru
   f. The North American bald eagle
6. Locate an ecological island close to where you live and visit it. Which species are most vulnerable to local extinction?
7. Visit a local zoo or botanical garden. What activities are conducted there to promote biological conservation? Are these activities likely to be of much benefit?
8. What are the primary reasons that a single-species approach to conservation and management of wildlife did not succeed?
9. Using information available in libraries, determine the minimum area required for a minimum viable population of the following:
   a. Domestic cats
   b. Cheetahs
   c. American bison
   d. Swallowtail butterflies
10. Both a ranch and a preserve will be established for the North American bison. The goal of the ranch owner is to show that bison can be a better source of meat than introduced cattle and at the same time have a less detrimental effect on the land. The goal of the preserve is to maximize the abundance of the bison. How will the plans for the ranch and preserve differ, and how will they be similar?

FURTHER READING

Binkley, Clark S. and Richard S. Miller. "Recovery of the Whooping Crane." *Biological Conservation* 45(1988): 11-20.

Botkin, D. B. *No Man's Garden: Thoreau and a New Vision for Civilization and Nature.* Washington, D.C.: Island Press, 2001. A work that discusses deep ecology and its implications for biological conservation, as well as reasons for the conservation of nature, both scientific and beyond science.

Cadieux, C. L. *Wildlife Extinction.* Washington, D.C.: Stone Wall Press, 1991. An engaging overview of the issues surrounding endangered wildlife. Case studies are presented, along with lists of endangered wildlife and key players in their protection.

Caughley, G. and A. R. E. Sinclair. *Wildlife Ecology and Management.* London: Blackwell Scientific, 1994. A valuable textbook based on new ideas of wildlife management.

DiSilvestro, R. L. *The Endangered Kingdom: The Struggle to Save America's Wildlife.* New York: Wiley, 1991.

Ehrlich, P. R. and E. O. Wilson, eds. *Biodiversity.* Washington, D.C.: National Academy Press, 1988. An excellent compilation of papers providing an overview of biological diversity and the threats that can lead to extinction.

Ferry, Luc. *The New Ecological Order.* Chicago: University of Chicago Press, 1995. Translated by Carol Volk from the original French. An important book in environmental philosophy. It includes a fascinating discussion of deep ecology and its implications for biological conservation, as well as views of nature from medieval times to the present.

McCullough, D. R. *Metapopulations and Wildlife Conservation.* Washington, D.C.: Island Press, 1996. Introduction to a new geographically based approach to wildlife conservation, including the idea that isolated patches, such as parks and preserves, can help stabilize wildlife populations.

Pauly, D., J. Maclean, and J. L. Maclean. *The State of Fisheries and Ecosystems in the North Atlantic Ocean.* Washington, D.C.: Island Press, 2002.

Talbot, L. "Living Resource Conservation: An International Overview." Washington, DC: Marine Mammal Commission, 1996. A description of how the conservation of wild living resources is actually being done today, based on interviews and conferences involving more than 400 scientists and policy makers.

chapter 15

Demasculinization and Feminization of Frogs in the Environment

The story of wild leopard frogs from a variety of areas sounds something like a science fiction horror story. In affected areas between 10% and 92% of male frogs exhibit gonadal abnormalities, including retarded development and hermaphroditism, meaning they have both male and female reproductive organs. Other frogs have vocal sacs with retarded growth. Since male frogs use vocal sacks to attract female frogs, these frogs are less likely to mate. What is apparently causing some of the changes in the male frogs is exposure to atrazine, which is the most widely used agricultural herbicide in North America today. Particularly high frequency of sex reversal (92% of male frogs) has been observed in Wyoming along the North Platte River. The region is not near any large agricultural activity, and the use of atrazine there is not particularly significant. Hermaphrodite frogs are common in this area because the North Platte River flows from areas in Colorado where atrazine is commonly used. The amount of atrazine released into the environment is estimated to be approximately 7.3 million kg per year. The chemical degrades very slowly in the environment. Because of its continual application every year, the Mississippi River, which drains about 40% of the lower United States, discharges approximately 0.5 million kg of atrazine per year to the Gulf of Mexico. Atrazine becomes easily attached to dust particles and has been found in rain, fog, and snow, and even in groundwater and surface water in regions where it is not used.

Wild leopard frogs in America have been affected by human-made chemicals in the environment

Health Canada's interim guideline for this compound and its metabolites is 5 parts per billion (ppb); the U.S. Environmental Protection Agency (EPA) allows up to 3 ppb of atrazine in drinking water. At those concentrations, atrazine definitely affects frogs that swim in the water. Other studies around the world have confirmed this. For example, in Switzerland, where atrazine is banned, it commonly occurs with a concentration of about 1 ppb, and that is sufficient to change some male frogs into females. Available evidence suggests that feminization occurs even at much lower concentrations.

Of particular interest and importance is the process that causes the changes in the leopard frogs. We begin the discussion with the endocrine system, which is composed of glands that internally secrete hormones directly into the bloodstream. Endocrine hormones are carried by the blood to parts of the body where they regulate and control functions of growth and sexual development. Testosterone and estrogen are examples of hormones. Testosterone in male frogs is in part responsible for development of male characteristics. The atrazine is believed to switch on a gene that turns testosterone into estrogen (a female sex hormone). It is the hormones, not the genes that actually regulate the development and structure of reproductive organisms. Frogs are particularly vulnerable during their early development, before and as they metamorphose from tadpoles to adult frogs. This change occurs in the spring, when atrazine levels are often at a maximum level in surface water. Apparently, a single exposure to the chemical may affect the frog's development. Thus, the herbicide is known as a *hormone disrupter.* In a more general sense, substances that interact with the hormone systems of an organism, whether or not they are linked to disease or abnormalities, are known as *hormonally active agents* (HAAs). These HAAs have the ability to trick the organism's body into believing that the chemicals have a role to play in the functional development of the frog. HAAs interact with the mechanisms for regulating growth and development, thus

Environmental Health, Pollution, and Toxicology

LEARNING OBJECTIVES

Serious environmental health problems and diseases may arise from toxic elements in water, air, soil, and even the rocks on which we build our homes. After reading this chapter, you should understand:

- How the terms toxic, pollution, contamination, carcinogen, synergism, and biomagnification are used in environmental health.
- What the classification and characteristics are of major groups of pollutants in environmental toxicology.
- Why there is controversy and concern about synthetic organic compounds such as dioxin.
- Whether we should be concerned about exposure to human-produced electromagnetic fields.
- What the dose-response concept is and how it relates to LD-50, TD-50, ED-50, ecological gradients, and tolerance.
- How the process of biomagnification works and why it is important in toxicology.
- Why the threshold effects of environmental toxins are important.
- What the process of risk assessment in toxicology is and why such proces are often difficult and controversial.
- What the precautionary principle is and why it is important.

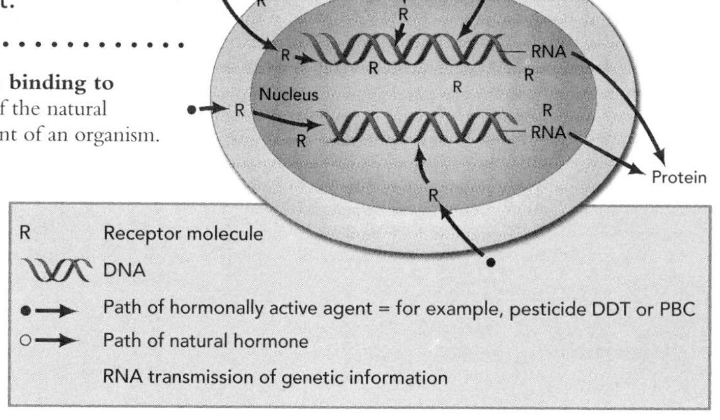

Figure 15.1 • Conceptual diagram of hormonally active agents (HAAs) binding to receptors on the surface of and inside a cell. HAAs can obstruct the role of the natural hormones that produce proteins that in turn regulate the growth and development of an organism.

disrupting normal growth functions. Figure 15.1 illustrates what happens when HAAs—in particular, hormone disrupters (such as pesticides and herbicides)—are introduced into the system. Natural hormones produced by the body send chemical messages to cells, where receptors for the hormone molecules are found on the outside and inside of cells. These natural hormones then transmit instructions to the cell's DNA, eventually directing development and growth. We now know that chemicals such as some pesticides and herbicides can also bind to the receptor molecules and either mimic or obstruct the role of the natural hormones. Thus, hormonal disrupters may also be known as HAAs.[1-4]

- *The story of sex reversal in wild leopard frogs dramatizes the importance of carefully evaluating the role of human-made chemicals in the environment. Studies to evaluate past or impending extinction of organisms often focus on global processes such as climate change, but the story of leopard frogs leads us down another path, one associated with human use of the natural environment. It also raises a number of more disturbing questions: Are we participating in an unplanned experiment on how human-made chemicals such as herbicides and pesticides might transform the bodies of other living beings, perhaps even people? Perhaps we will look back on this moment of understanding as a new beginning in meaningful studies that will answer some of these important questions.*

15.1 Some Basics

Disease is often due to an imbalance resulting from a poor adjustment between the individual and the environment.[5] However, disease seldom has a one-cause–one-effect relationship with the environment. Rather, the incidence of a disease depends on several factors, including physical environment, biological environment, and lifestyle. Linkages between these factors are often related to other factors, such as local customs and the level of industrialization. More primitive societies that live directly off the local environment are usually plagued by different environmental health problems than is our urban society. Industrial societies have nearly eliminated such diseases as cholera, dysentery, and typhoid.

People are often surprised to learn that the water we drink, the air we breathe, the soil in which we grow crops, and the rocks on which we build our homes and places of work may affect our chances of experiencing serious environmental health problems and diseases[5] (although, as suggested, direct causative relationships between the environment and disease are difficult to establish). At the same time, the environmental factors that contribute to disease—soil, rocks, water, and air—can also influence our chances of living a longer, more productive life.

Many people believe that soil, water, or air in a so-called natural state must be good and that if human activities have changed or modified them, they have become contaminated, polluted, and therefore bad.[6] This is by no means the entire story; many natural processes, including dust storms, floods, and volcanic processes, can introduce materials harmful to humans and other living things into the soil, water, and air.

Figure 15.2 • Lake Nyos, Cameroon, Africa. *(a)* In 1986, the lake released carbon dioxide that moved down the slopes of the hills to settle in low places, asphyxiating animals and people. *(b)* Animals asphyxiated by carbon dioxide.

(a)

(b)

A tragic example occurred on the night of August 21, 1986, when there was a massive natural release of carbon dioxide (CO_2) gas from Lake Nyos in Cameroon, Africa. The carbon dioxide was probably initially released from volcanic vents at the bottom of the lake and accumulated there with time. Pressure of the overlying lake water normally kept the dissolved gas at the bottom of the lake. However, the water was evidently agitated by a slide or small earthquake, and the bottom water moved upward. When the CO_2 gas reached the surface of the lake, it was released quickly into the atmosphere. The CO_2 gas, which is heavier than air, flowed downhill from the lake and settled in nearby villages, killing many animals and more than 1,800 people by asphyxiation (Figure 15.2). It was estimated that a similar event could recur within about 20 years, assuming that carbon dioxide continued to be released at the bottom of the lake.[7] Fortunately, an international aid project (scheduled to be completed early in the twenty-first century) is inserting pipes into the bottom of Lake Nyos. The gas-rich water is pumped to the surface where it is safely discharged into the atmosphere. In 2001, a warning system was installed, and one degassing pipe released a little more CO_2 than was seeping naturally into the lake. Recent data suggest that the single pipe now there barely keeps ahead of CO_2 that continues to enter the bottom, so the lake's 500,000 tons of built-up gas have only dropped 6%. At this rate, it could take 30 to 50 years to make Lake Nyos safe. In the meantime, there could be another eruption.[8]

Terminology

What do we mean when we use the terms *pollution, contamination, toxic,* and *carcinogen*? A polluted environment is one that is impure, dirty, or otherwise unclean. The term **pollution** refers to the occurrence of an unwanted change in the environment caused by the introduction of harmful materials or the production of harmful conditions (heat, cold, sound). **Contamination** has a meaning similar to that of pollution and implies making something unfit for a particular use through the introduction of undesirable materials—for example, the contamination of water by hazardous waste. The term toxic refers to materials (pollutants) that are poisonous to people and other living things. **Toxicology** is the science that studies chemicals that are known to be or could be toxic, and toxicologists are scientists in this field. A **carcinogen** is a particular kind of toxin that increases the risk of cancer. Carcinogens are among the most feared and regulated toxins in our society.

An important concept in considering pollution problems is **synergism**, the interaction of different substances resulting in a total effect greater than the sum of the effects of the separate substances. For example, both sulphur dioxide and coal dust particulates are air pollutants. Either one taken separately may cause adverse health effects, but when they combine, as when sulphur dioxide (SO_2) adheres to the coal dust, the dust with SO_2 is inhaled deeper than sulphur dioxide alone, causing greater damage to lungs. Another aspect of synergistic effects is that the body may be more sensitive to a toxin if it is simultaneously subjected to other toxins.

Pollutants are commonly introduced into the environment by way of **point sources**, such as smokestacks (see A Closer Look 15.1), pipes discharging into waterways (Figure 15.3), or accidental spills. **Area sources**, also called non-point sources, are more diffused over the land and include urban runoff and mobile sources such as automobile exhaust. Area sources are difficult to isolate and correct because the problem is often widely dispersed over a region, as in agricultural runoff that contains pesticides (Chapter 19).

Measuring the Amount of Pollution

How the amount or concentration of a particular pollutant or toxin present in the environment is reported varies widely. The amount of treated wastewater entering Toronto Harbour is a big number reported in millions of litres per day. Emission of nitrogen and sulphur oxides into the air is also a big number reported in millions of tonnes per year. Small amounts of pollutants or toxins in the environment, such as pesticides, are reported in units such as parts per million (ppm) or parts per billion (ppb). It is important to keep in mind that the concentration in ppm or ppb may be by either volume, mass, or weight. In some toxicology studies, the units used are milligrams of toxin per kilogram of body mass (1 mg/kg is equal to 1 ppm). Concentration may also be recorded as a percent. For example, 100 ppm (100 mg/kg) is equal to 0.01%. (How many ppm is equal to 1%?)

Figure 15.3 • This urban stream flows into the Pacific Ocean at a coastal park. The stream water often carries high counts of fecal coliform bacteria. As a result, the stream is a point source of pollution for the beach, which is sometimes closed to swimming following runoff events.

When dealing with water pollution, units of concentration for a pollutant may be milligrams per litre (mg/L) or micrograms per litre (μg/L). A milligram is one-thousandth of a gram, and a microgram is one-millionth of a gram. For water pollutants that do not cause significant change in density of water (1 g/cm^3), a concentration of pollution of 1 mg/L is approximately equivalent to 1 ppm. Air pollutants are commonly measured in units such as micrograms of pollutant per cubic meter of air (μg/m^3).

Units such as ppm, ppb, or μg/m^3 reflect very small concentrations. For example, if you were to use 3 g (one-tenth of an ounce) of salt to season popcorn in order to have salt at a concentration of 1 ppm by weight of the popcorn, you would have to pop approximately three metric tonnes of kernels!

15.2 Categories of Pollutants

A partial classification of pollutants by arbitrary categories is presented below. Examples of other pollutants are discussed in other parts of the book.

Infectious Agents

Infectious diseases, spread from the interactions between individuals and food, water, air, or soil, constitute some of the oldest health problems that humans face. Today, infectious diseases have the potential to pose rapid local to global threats by spreading in hours through airplane travellers. Terrorist activity may also spread diseases. Inhalation anthrax, caused by a bacterium, sent in a powdered form in envelopes through the mail in 2001, killed several people. New diseases are emerging, and previous ones may emerge again. Although we have cured many

A CLOSER LOOK 15.1

Sudbury Smelters: A Point Source

A famous example of a point source of pollution is provided by the smelters that refine nickel and copper ores at Sudbury, Ontario. Sudbury contains one of the world's major nickel and copper ore deposits. A number of mines, smelters, and refineries lie within a small area. The smelter stacks at one time released large amounts of particulates containing nickel, copper, and other toxic metals into the atmosphere. In addition, because the area rocks contain a high percentage of sulphur, the emissions included large amounts of sulphur dioxide (SO_2). During its peak output in the 1960s, this complex was the largest single source of sulphur dioxide emissions in North America.

In 1969, regulations were mandated to improve local air quality, forcing a reduction in emissions. Concentrations of sulphur dioxide were reduced locally by more than 50% after 1972. However, attempts to minimize the pollution problem in the immediate vicinity of the smelting operation by increasing smokestack height spread the problem as wind carried the pollutants greater distances. In order to better control emissions from Sudbury, the Ontario government set standards to reduce emissions to less than 365,000 tons per year by 1994 (about 14% of earlier emissions of 2,560,000 tonnes per year). The goal was achieved by reducing production from the smelters and by treating the emissions to reduce pollution.[9]

As a result of years of pollution, nickel has been found to contaminate soils 50 km from the stacks. The forests that once surrounded Sudbury were devastated by decades of acid rain (produced from SO_2 emissions) and the deposition of particulates containing metals. An area of approximately 250 km^2 was nearly devoid of vegetation, and damage to forests in the region has been visible over an area of approximately 3,500 km^2 (see Figure 15.4). Secondary effects, in addition to loss of vegetation, include soil erosion and drastic changes in soil chemistry resulting from the influx of the metals.

Reductions in emissions from Sudbury have allowed surrounding areas to slowly begin to recover from these effects. Species of trees once eradicated from some areas have begun to grow again. Recent restoration efforts have included planting of over 7 million trees, and 75 species of herbs, moss, and lichens—all of which have contributed to the increase of biodiversity. Lakes that were damaged due to acid precipitation in the area are rebounding and now support populations of plankton and fish.[9] The case of the Sudbury smelters thus provides a positive example of pollution reduction, emphasizing the key theme of thinking globally but acting locally to reduce air pollution. It also illustrates the theme of science and values: Scientists and engineers can design pollution abatement equipment, but spending the money to purchase the equipment reflects what value we place on clean air.

(a)

(b)

Figure 15.4 • Lake St. Charles, Sudbury, Ontario. *(a)* Prior to restoration. Note high stacks (smelters) in the background and lack of vegetation in the foreground resulting from air pollution (acid and metal deposition). *(b)* Recent photo showing regrowth and restoration.

diseases, we have no known reliable vaccines for others, such as HIV, hantavirus, and dengue fever.

Diseases that can be controlled by manipulating the environment, such as by improving sanitation or treating water, are classified as environmental health concerns. Although there is great concern about the toxins and carcinogens produced in industrial society today, the greatest mortality in developing countries is caused by environmentally transmitted infectious disease. In North America, thousands of cases of water-borne illness and food poisoning occur each year. These diseases can be spread by people; mosquitoes or fleas; or contact with contaminated food, water, or soil. They can also be transmitted through ventilation systems in buildings.

Some examples of environmentally transmitted infectious diseases are:

- Legionellosis, or Legionnaires' disease, which often occurs where air-conditioning systems have been contaminated by disease-causing organisms.

- Giardiasis, a protozoan infection of the small intestine spread via food, water, or person-to-person contact.
- Salmonella, a food-poisoning bacterial infection spread via water or food.
- Malaria, a protozoan infection transmitted by mosquitoes.
- Lyme Borreliosis, or Lyme disease, transmitted by ticks.
- Cryptosporidosis, a protozoan infection transmitted via water or person-to-person contact (see Chapter 19).[10]
- Anthrax, spread by terrorist activity.

Occasionally, there is media focus on epidemics in developing nations. An example is the highly contagious Ebola virus in Africa, which causes external and internal bleeding resulting in death of 80% of those infected. Many consider such epidemics as problems only in developing nations. This belief may provide a false sense of security! Monkeys and bats spread Ebola, but the origin of the virus in the tropical forest remains unknown. Developed countries, where outbreaks may occur in the future, will need to learn from the developing countries' experiences. To accomplish this and avoid potential global tragedies, more funds are needed to study infectious diseases in developing countries.

Toxic Metals

The major **metals** that pose health hazards to people and ecosystems are those with relatively high atomic weight (see Chapter 5), including mercury, lead, cadmium, nickel, gold, platinum, silver, bismuth, arsenic, selenium, vanadium, chromium, and thallium. Each of these elements may be found in soil or water not contaminated by humans. Each metal has uses in our modern industrial society, and each is also a by-product of the mining, refining, and use of other elements. Metals often have direct physiological toxic effects. Some are stored or incorporated in living tissue, sometimes permanently. A little arsenic each day may eventually result in a fatal dose, the subject of more than one murder mystery.

The content of metals in our bodies is referred to as the *body burden*. The body burden of toxic heavy elements for an average human body (70 kg) is about 8 mg for antimony, 13 mg for mercury, 18 mg for arsenic, 30 mg for cadmium, and 150 mg for lead. Lead (for which we apparently have no biological need) has an average body burden of about twice that of the others combined, reflecting our heavy use of this potentially toxic metal.

Mercury, thallium, cadmium, and lead are very toxic to humans. They have long been mined and used, and their toxic properties are well known. Mercury, for example, is the "Mad Hatter" element. At one time, mercuric salts were used in making felt hats stiff; because mercury damages the brain, hatters were known to act peculiarly in Victorian England. Thus, the Mad Hatter in Lewis Carroll's *Alice in Wonderland* had real antecedents in history.

In the past, the term "heavy metals" was used to refer to metals toxic to humans. This term has no clear basis in science, however. Not all "heavy" metals—for instance, those with densities greater than 4 grams per cubic centimetre—have highly toxic properties. There is, in fact, no clear relationship between the density of a metal and its behaviour in living organisms. For example, silver, with an atomic weight of 107.8682, is less toxic to humans and other organisms than either arsenic (74.92160) or zinc (65.409). In practice, different regulatory agencies define "heavy metals" differently. Current practice is now to use "toxic metals" or simply "metals" to refer to this group.

Toxic Pathways

Chemical elements released from rocks or human processes can become concentrated in humans (see Chapter 5) through many pathways (Figure 15.5). These pathways may involve what is known as **bioaccumulation** of a substance in the tissue of a single organism, and **biomagnification**, the increase in concentration of a substance in living tissue at higher trophic levels in a food web. For example, cadmium, which influences the risk of heart disease, may enter the environment via ash from burning coal. The cadmium in coal exists in very low concentrations (less than 0.05 ppm). After coal is burned in a power plant, the ash is collected in a solid form and disposed of in a landfill. The landfill is covered with soil and revegetated. The low concentration of cadmium in the ash and soil is taken into the plants as they grow. But the concentration of cadmium in the plants is three to five times greater than the concentration in the ash. As the cadmium moves through the food chain, it becomes more and more concentrated. By the time it is incorporated in the tissue of people and other carnivores, the concentration is approximately 50 to 60 times the original concentration in the coal.

Mercury in aquatic ecosystems offers another example of biomagnification. Mercury is a potentially serious pollutant of aquatic ecosystems such as ponds, lakes, rivers, and oceans. Natural sources of mercury in the environment include volcanic eruptions and erosion of natural mercury deposits, but we are most concerned with human input of mercury into the environment through processes such as burning coal in power plants, incinerating waste, and processing metals such as gold. Rates of input of mercury into the environment through human processes are poorly understood. However, it is believed that human activities have doubled or tripled the amount of mercury in the atmosphere, and it is increasing at about 1.5% per year.[11]

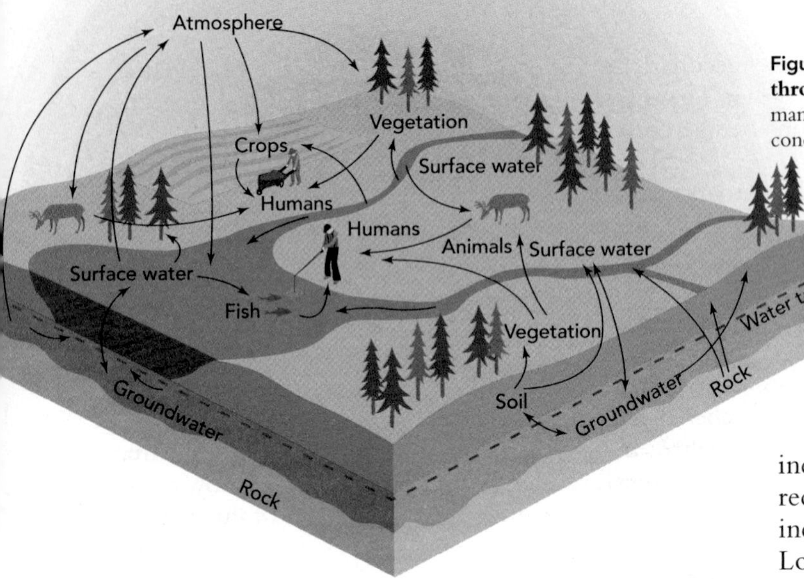

Figure 15.5 • Potential complex pathways for toxic materials through the living and non-living environment. Note the many arrows into humans and other animals, sometimes in increasing concentrations, as they move through the food chain.

A major source of mercury in many aquatic ecosystems is deposition from the atmosphere through precipitation. Most of the deposition is of inorganic mercury (Hg^{++}, ionic mercury). Once this mercury is in surface water, it enters into complex biogeochemical cycles and a process known as methylation may occur. Methylation changes inorganic mercury to methyl mercury (CH_3Hg^+) through bacterial activity.

Methyl mercury is much more harmful (toxic) than is inorganic mercury, in part because it crosses the blood-brain barrier more quickly and is eliminated more slowly from animals' systems. As the methyl mercury works its way through food chains, biomagnification occurs, so that higher concentrations of methyl mercury are found farther up the food chain. Thus, big fish that eat little fish contain a higher concentration of mercury than do smaller fish and the aquatic insects that these fish feed upon.

Selected aspects of the mercury cycle in aquatic ecosystems are shown in Figure 15.6. The figure emphasizes the input side of the cycle, from deposition of inorganic mercury through formation of methyl mercury, biomagnification, and sedimentation of mercury at the bottom of a pond. On the output side of the cycle, the mercury that enters fish may be taken up by animals that eat the fish; and sediment may release mercury by a variety of processes, including resuspension in the water, where eventually the mercury enters the food chain or is released into the atmosphere through volatilization (conversion of liquid mercury to a vapour form).

Biomagnification also occurs in the ocean. Large fish such as tuna and swordfish have elevated mercury concentrations and limiting consumption of these fish is recommended. Pregnant women are advised not to eat them at all.

During the twentieth century, several significant incidents of methyl mercury poisoning have been recorded. One, in Minamata Bay, Japan, involved the industrial release of methyl mercury (see A Closer Look 15.2). Another, in Iran, involved a methyl mercury fungicide used to treat wheat seeds. In each of these cases, hundreds of people were killed and thousands suffered permanent health impacts.[11]

Organic Compounds

Organic compounds are compounds of carbon produced naturally by living organisms or synthetically by human industrial processes. It is difficult to generalize about the environmental and health effects of artificially produced organic compounds because there are so many of them, they have so many uses, and they can produce so many different kinds of effects.

Synthetic organic compounds are used in industrial processes, pest control, pharmaceuticals, and food additives. People have produced over 20 million synthetic chemicals, and new ones are appearing at a rate of

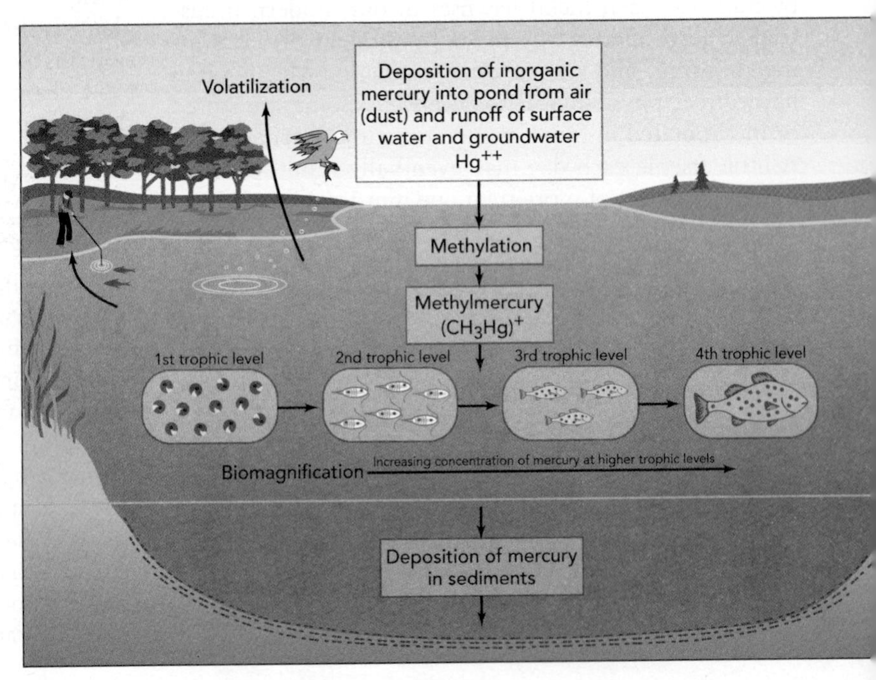

Figure 15.6 • Conceptual diagram showing selected pathways for movement of mercury into and through an aquatic ecosystem. [*Source*: Modified from G. L. Waldbott, *Health Effects of Environmental Pollutants*, 2nd ed. (Saint Louis: C.V. Mosby, 1978).]

Table 15.1 • Selected Common Persistent Organic Pollutants (POPs)	
Chemical	*Typical of Use*
Aldrin[a]	Insecticide
Atrazine	Herbicide
DDT[a]	Insecticide
Dieldrin[a]	Insecticide
Endrin[b]	Insecticide
Mirex[a]	Insecticide
PCBs[a]	Liquid insulators in electric transformers
Dioxins	By-product of herbicide production

Source: Data in part from Anne Platt McGinn, "Phasing Out Persistent Organic Pollutants," in Lester R. Brown et al., *State of the World 2000* (New York: Norton, 2000).

[a] Banned in Canada, the U.S. and many other countries.
[b] Restricted or banned in many countries.

about 1 million per year! Most are not produced commercially, but up to 100,000 chemicals are now being, or in the past have been, used. Once used and dispersed in the environment, they may produce a hazard for decades to hundreds of years.

Persistent Organic Pollutants

Some synthetic compounds are called "**persistent organic pollutants**" or **POPs**. Many were first produced decades ago when their harm to the environment was not known, and they are now banned or restricted (see Table 15.1 and A Closer Look 15.3).

POPs have several properties that define them:[12]

- They have a carbon-based molecular structure, often containing highly reactive chlorine.
- Most are manufactured by humans—they are synthetic chemicals.
- They are persistent in the environment. That is, they do not easily break down in the environment.
- They are polluting and toxic.
- They are soluble in fat and likely to accumulate in living tissue.
- They occur in forms that allow them to be transported by wind, water, and sediments for long distances.

For example, consider the marine sediments off Fortune, Newfoundland, which are contaminated with polychlorinated biphenyls (PCBs). PCBs are heat-stable oils originally used as an insulator in electric transformers.[12] The source of these materials is not clear, but likely includes PCB contaminated fuels and oils from the area's heavy shipping traffic. Other possibilities include deteriorating electrical transformers and contaminated dust transported by wind around the world. The dust and fluids containing PCBs were deposited in the marine nearshore zone, and in freshwater ponds, lakes, or rivers, where it entered the food chain. First, it entered algae along with nutrients it combined with. Insects ate the algae, which were eaten by shrimp and fish. In each stage up the food web, the concentration of PCBs increased. When people eat fish, the PCBs are concentrated in their fatty tissues and breast milk.

Concern about the toxic properties of chlorinated organic compounds prompted the development of replacements incorporating bromine or fluorine in place of chlorine. However, these materials also proved to be toxic and persistent and, like chlorinated compounds, they are now found in fish and marine mammal tissue worldwide.

Hormonally Active Agents (HAAs)

HAAs are also POPs. The opening case study discussed the feminization of frogs resulting from exposure to the herbicide atrazine, and you may wish to review that case study in context of the continued discussion here. There is an increasing body of scientific evidence that points to certain chemicals in the environment, known as **hormonally active agent (HAAs)**, as having potential to cause developmental and reproductive abnormalities in animals and humans. This evidence comes from studies of wildlife in the field and from laboratory studies of human diseases such as breast, prostate, and ovarian cancer, as well as abnormal testicular development and thyroid-related abnormalities.[2]

In partnership with Health Canada, provincial governments, industry, and universities, Environment Canada manages the Toxic Substances Research Initiative, which includes a variety of research on HAAs. Their research has revealed a variety of wildlife impacts of HAAs in Canada, including feminization of fish near municipal sewage treatment plant discharges, impaired reproductive function and development in fish exposed to pulp and paper mill effluent, and abnormal reproduction in snails exposed to antifouling substances in marine paints. The major disorders that have been studied in wildlife have centered on abnormalities including thinning of eggshells of birds, decline in populations of various animals and birds, reduced viability of offspring, and changes in sexual behaviour.[1]

With respect to human diseases, there has been much research on linkages between HAAs and breast cancer through exploring relationships between environmental estrogens and cancer. Other studies are ongoing to understand relationships between PCBs and neurological damage, reflected in poor performance on standard intelligence tests. Finally, there is concern that exposure of people to phthalates found in chlorinated-based plastics is also causing problems. The consumption of phthalates in North America is considerable, with the highest exposure in women of childbearing age. The products being tested as the source of contamination include perfumes and other cosmetics such as nail polishes and hairsprays.[1]

A CLOSER LOOK 15.2

Mercury, Minamata, and the English-Wabigoon River

In the Japanese coastal town of Minamata, on the island of Kyushu, a strange illness began to occur in the middle of the twentieth century. It was first recognized in birds that lost their coordination and fell to the ground or flew into buildings, and in cats that went mad, running in circles and foaming at the mouth.[13] The affliction, known by local fishermen as the "disease of the dancing cats," subsequently affected people, particularly families of fishermen. The first symptoms were subtle: fatigue, irritability, headaches, numbness in arms and legs, and difficulty in swallowing. More severe symptoms involved the sensory organs; vision was blurred and the visual field was restricted. Afflicted people became hard of hearing and lost muscular coordination. Some complained of a metallic taste in their mouths; their gums became inflamed, and they suffered from diarrhea. Eventually, 43 people died and 111 were severely disabled; in addition, 19 babies were born with congenital defects. Those affected lived in a small area, and much of the protein in their diet came from fish from Minamata Bay.

A vinyl chloride factory on the bay used mercury in an inorganic form in its production processes. The mercury was released in waste that was discharged into the bay. Mercury forms few organic compounds, and it was believed that the mercury, although poisonous, would not get into food chains. But the inorganic mercury released by the factory was converted by bacterial activity in the bay into methyl mercury, an organic compound that turned out to be much more harmful. Unlike inorganic mercury, methyl mercury readily passes through cell membranes. It is transported by the red blood cells throughout the body, and it enters and damages brain cells.[13] Fish absorb methyl mercury from water 100 times faster than they absorb inorganic mercury. (This was not known before the epidemic in Japan.) Once absorbed, methyl mercury is retained two to five times longer than is inorganic mercury.

In the late 1960s, scientists in Canada and other nations began to examine other major water systems for evidence of mercury pollution. In the English-Wabigoon River system, flowing westward from northwestern Ontario into Manitoba, they discovered fish contaminated with some of the highest concentrations of mercury in the world. The source was traced to a chlor-alkali process in a Dryden, Ontario, pulp and paper mill. Although that process was closed down in 1973, mercury emissions from the plant continue to this day, and high levels of methyl mercury persist in over 100 km of the river system. Members of two First Nations bands, the Grassy Narrows and Wabaseemoong bands, worked as fishing guides on the river, and fish was a regular part of both bands' diets. Many individuals from these bands had high blood levels of mercury, and some exhibited symptoms of mercury poisoning. The bands also experienced severe economic and cultural hardship following the closure of the river's fishery.

Harmful effects of methyl mercury depend on a variety of factors, including the amount and route of intake, the duration of exposure, and the species affected. The effects of the mercury are delayed from three weeks to two months from the time of ingestion. If mercury intake ceases, some symptoms may gradually disappear, but others are difficult to reverse.[14]

The mercury episodes at Minamata and the Wabigoon River illustrate four major factors that must be considered in evaluating and treating toxic environmental pollutants.

1. *Individuals vary in their response to exposure to the same dose, or amount, of a pollutant.* Not everyone responded in the same way; there were variations even among those most heavily exposed. Because we cannot predict exactly how any single individual will respond, we need to find a way to state an expected response of a particular percentage of individuals in a population.
2. *Pollutants may have a threshold*—that is, a level below which the effects are not observable and above which the effects become apparent. Symptoms appeared in individuals with concentrations of 500 ppb of mercury in their bodies; no measurable symptoms appeared in individuals with significantly lower concentrations.
3. *Some effects are reversible.* Some people recovered when the mercury-filled fish and shellfish was eliminated from their diet.
4. *The chemical form of a pollutant, its activity, and its potential to cause health problems may be changed markedly by ecological and biological processes.* In the case of mercury, its chemical form and concentration changed as the mercury moved through the food webs.

In summary, there is good scientific evidence that some chemical agents in sufficient concentrations will affect human reproduction through endocrine and hormonal disruption. The endocrine system is of primary importance because it is one of the two main systems (along with the nervous system) that regulate and control growth, development, and reproduction. In humans, the endocrine system is composed of a group of hormone-secreting glands, including the thyroid, pancreas, pituitary, ovaries (women), and testes (men). The hormones are transported in the bloodstream to virtually all parts of the body, where they act as chemical messengers to control growth and development of the body.[2]

Build Your Environmental Skills 15.1

Types of Scientific Writing

As A Closer Look 15.3 shows, scientists often disagree about the conclusions to be drawn from a body of evidence. It is important to learn to distinguish between different types of scientific writing, and separate sound science from opinion.

Scholarly journals are usually the most reliable. They contain reports on original experiments and other studies, and are usually written in technical language. The famous journals *Science* and *Nature* are good examples. Most articles in scholarly journals have been reviewed by other prominent researchers working in the same area—the "peers" of the authors. For that reason, scholarly journals are sometimes referred to as peer-reviewed.

Secondary sources summarize the results of primary research, and are usually geared to a more general audience. Although they contain technical information, they are written without jargon and can be understood by people without detailed knowledge of the subject matter. Secondary sources include review articles that summarize a body of work in a given field. Many, but not all, secondary sources are peer reviewed.

Government publications often include the results of original research and in that sense are primary sources, However, few government publications are peer reviewed, and as a result, they may not be as reliable a source as scholarly journals.

Popular magazines like *Canadian Geographic* are intended for a general rather than a technical audience. They may include summaries of recent research as well as opinion and news items. Articles in popular magazines are not peer reviewed and do not necessarily reflect sound scientific thought.

Technical books and encyclopedias may be of different kinds. Textbooks such as this one are typically reviewed and approved by instructors in a wide range of universities. Technical reference books, however, may not undergo such a rigorous review. If you are using books as a source, it is wise to refer to several sources and compare your findings. Encyclopedias provide a concise overview of key topics and are a useful place to start when beginning a research project.

The World Wide Web has revolutionized research by making billions of information sources available at the click of a mouse button. These sources vary widely in their content and reliability, however, and it can be hard to determine whether the content of a site is current or credible. It is wise to check information obtained from the web against primary sources.

Give credit where credit is due! At the heart of science is the expectation of academic integrity—not claiming others' work as your own. The World Wide Web in particular has made it easy to copy text and paste it into research papers. *This practice is never acceptable.* In any research report, it is important to explain your thinking in your own words. If you want to quote another author directly, make sure to use quotation marks and always cite your sources in your text and in the list of references at the end of your research paper.

Some Guides to Good Scientific Writing:

Day, Robert A. *How to Write and Publish a Scientific Paper.* Phoenix: Oryx Press, 1998.

Katz, Michael J. *Elements of the Scientific Paper: a Step-by-Step Guide for Students and Professionals.* New Haven: Yale University Press, 1985.

Brusaw, Charles T., Gerald J. Alfred and Water E. Oliu. *Handbook of Technical Writing.* New York: St. Martin's Press, 1976.

Health Canada's Science Advisory Board and the National Academy of Sciences have reviewed the available scientific evidence concerning HAAs and recommended that there should be continued monitoring of wildlife and human populations for abnormal development and reproduction. Furthermore, where wildlife species are known to be experiencing declines in population associated with abnormal deformities, experiments should be designed to study the phenomena with respect to chemical contamination. With respect to humans, the recommendation is for additional studies that will document the presence or absence of associations between HAAs and human cancers. When associations are discovered, the causality must also be investigated in terms of potential latency, relationships between exposure and disease, and indicators of susceptibility to disease of certain groups of people by age and sex.[1]

Radiation

Nuclear radiation is introduced here as a category of pollution. It is discussed in detail in Chapter 17 in conjunction with nuclear energy. We are concerned about nuclear radiation because excessive exposure is linked to serious health problems, including cancer. (See Chapter 21 for a discussion of radon gas as an indoor air pollutant.)

Thermal Pollution

Thermal pollution, also called heat pollution, occurs when heat released into water or air produces undesirable effects. Heat pollution can occur as a sudden, acute event or as a long-term, chronic release. Sudden heat releases may result from natural events, such as brush or forest fires and volcanic eruptions, or from human-induced events, such as agricultural burning. The major sources of chronic heat pollution are electric power plants that produce electricity in steam generators.

A CLOSER LOOK 15.3

Dioxin: The Big Unknown

Dioxin, a persistent organic pollutant or POP, may be one of the most toxic of the human-made chemicals in the environment. The history of the scientific study of dioxin and its regulation illustrate once again the interplay of science and values. Although science is not entirely certain of the toxicity of dioxin to humans and ecosystems, society has made a number of value judgments involving regulation of the substance. Controversy has surrounded these judgments and will surely continue to do so.

Dioxin is a colourless crystal made up of oxygen, hydrogen, carbon, and chlorine. It is classified as an organic compound because it contains carbon. About 75 types of dioxin (and dioxin-like compounds) are known; they are distinguished from one another by the arrangement and number of chlorine atoms in the molecule.

Dioxin is not normally manufactured intentionally but is a by-product resulting from chemical reactions, including the combustion of compounds containing chlorine in the production of herbicides.[15] Dioxins and dioxin-like compounds (specifically, chlorinated dibenzo-p-dioxin, or CDD, and chlorinated dibenzofurans, or CDF) are emitted into the air through such processes as incineration of municipal waste (the major source), incineration of medical waste, burning gasoline and diesel fuels in vehicles, burning wood as a fuel, and refining metals such as copper. The good news is that releases of CDDs and CDFs decreased about 75% from 1987 to 1995. However, our knowledge of the biogeochemical cycling of dioxins is still incomplete. In too many cases, the amounts of dioxins emitted are based more on expert opinion rather than high-quality data or even limited data.[16] As a result of scientific uncertainty the controversy concerning dioxin is sure to continue.

Although dioxin is known to be extremely toxic to mammals, its actions in the human body are not well known. What is known is that sufficient exposure to dioxin (usually from meat or milk containing the chemical) produces a skin condition (a form of acne) that may be accompanied by loss of weight, liver disorders, and nerve damage.[17]

The herbicide "Agent Orange," used as a defoliant in the Vietnam War, was a mixture of two herbicides, 2,4-D and 2,4,5-T, mixed with kerosene or diesel fuel and stored in orange containers. An unintended by-product of the manufacture of these materials was dioxin, which has been linked to health problems in the Vietnam veterans who used or were exposed to Agent Orange.

More recently, a dramatic case of dioxin poisoning was that of Ukrainian president Viktor Yushchenko. In September 2004, Yushchenko became ill with severe pancreatitis and a serious skin rash characterized by large raised blisters and skin discolouration. Laboratory testing confirmed that his symptoms were caused by extremely high concentrations of dioxin, probably administered orally. Yushchenko has accused a political rival of poisoning him at a dinner for senior Ukrainian officials.

In animals, research has shown that even small amounts of dioxin can cause birth defects and even death in sensitive species. However, the concentration necessary to cause human health hazards is still controversial. Studies suggest that workers exposed to high concentrations of dioxin for longer than a year have an increased risk of dying of cancer.[18]

Although Health Canada has yet to take a firm position on the subject, the United States EPA recently reclassified dioxin from a "probable" to a "known" human carcinogen. For most of the exposed people, such as those eating a diet high in animal fat, the EPA puts the risk of developing cancer between 1 in 1,000 and 1 in 100. This estimate represents the highest possible risk for the most exposed individuals. For most people the risk will likely be much lower, or near zero.[19] Some scientists argue that EPA's suggested acceptable intake level for dioxin (0.006 pg per kilogram of body weight per day; 1 pg = 10^{-12} g) is too low, and ought to be 100–1,000 times higher, or approximately 1–10 pg/day.[18] Others assert that lack of data precludes establishment of a specific threshold concentration of dioxin at which health hazards begin.[20] As indicated by these uncertainties, toxicity of dioxin will remain unclear until additional studies better delineate the potential hazard. (Build Your Environmental Skills 15.1 discusses how to distinguish between different types of scientific writing, and assess the merits of scientific arguments.)

Dioxin is a stable, long-lived chemical that is accumulating in the environment. Analysis of sediments taken from the bottom of Lake Superior suggests that the rate of deposition of dioxin increased eightfold from 1940 to 1970. However, since then, rates have slowly declined.[21, 22] As yet, we have not been able to determine a safe, reliable, and economically feasible way to clean up areas contaminated by dioxin. Many old waste disposal sites are contaminated by dioxin; it may also be found in soil and streams several kilometres around the sites (see Figure 15.7).

As noted, the controversy concerning the toxicity of dioxin is not over.[23,24] Some environmental scientists argue that the regulation of dioxin must be tougher, whereas the industries producing the chemical argue that the dangers of exposure are overestimated.

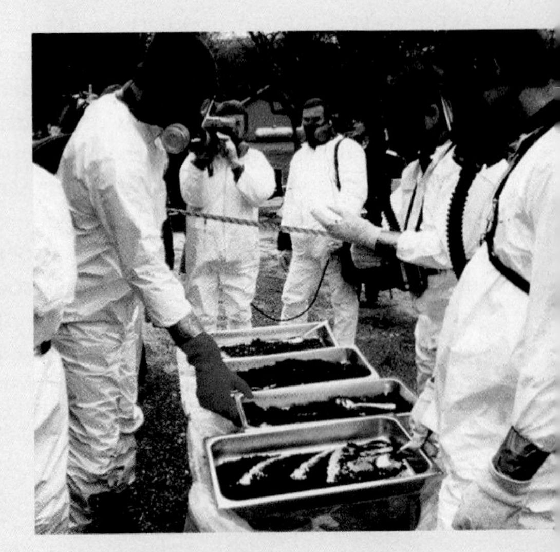

Figure 15.7 • **Soil samples from Times Beach, Missouri, thought to be contaminated by dioxin.**

The release of large amounts of heated water into a river changes the average water temperature and the concentration of dissolved oxygen (warm water holds less oxygen than cooler water), thereby changing the river's species composition (see the discussion of eutrophication in Chapter 19). Every species has a range of temperature within which it can survive and an optimal temperature for living. For some species of fish, the range is small, and even a small change in water temperature is a problem. Lake fish move away when the water temperature rises more than about 1.5°C above normal; river fish can withstand a rise of about 3°C.[25] Heating river water can change its natural conditions and disturb the ecosystem in several ways. Fish spawning cycles may be disrupted, and the fish may have a heightened susceptibility to disease; warmer water causes physical stress in some fish, and they may be easier for predators to catch; and warmer water may change the type and abundance of food available for fish at various times of the year.

There are several solutions to chronic thermal discharge into bodies of water. The heat can be released into the air by cooling towers (Figure 15.8), or the heated water can be temporarily stored in artificial lagoons until it is cooled to normal temperatures. Some attempts have been made to use the heated water to grow organisms of commercial value that require warmer water temperatures. Waste heat from a power plant can also be captured and used for a variety of purposes, such as warming buildings (see Chapter 16 for a discussion of cogeneration).

Figure 15.8 • Two types of cooling towers. *(a)* Wet cooling tower. Air circulates through tower; hot water drips down and evaporates, cooling the water. *(b)* Dry cooling tower. Heat from the water is transferred directly to the air, which rises and escapes the tower. *(c)* Cooling towers emitting steam at Didcot power plant, Oxfordshire, England. Red and white lines are vehicle lights resulting from long exposure time (photograph taken at dusk).

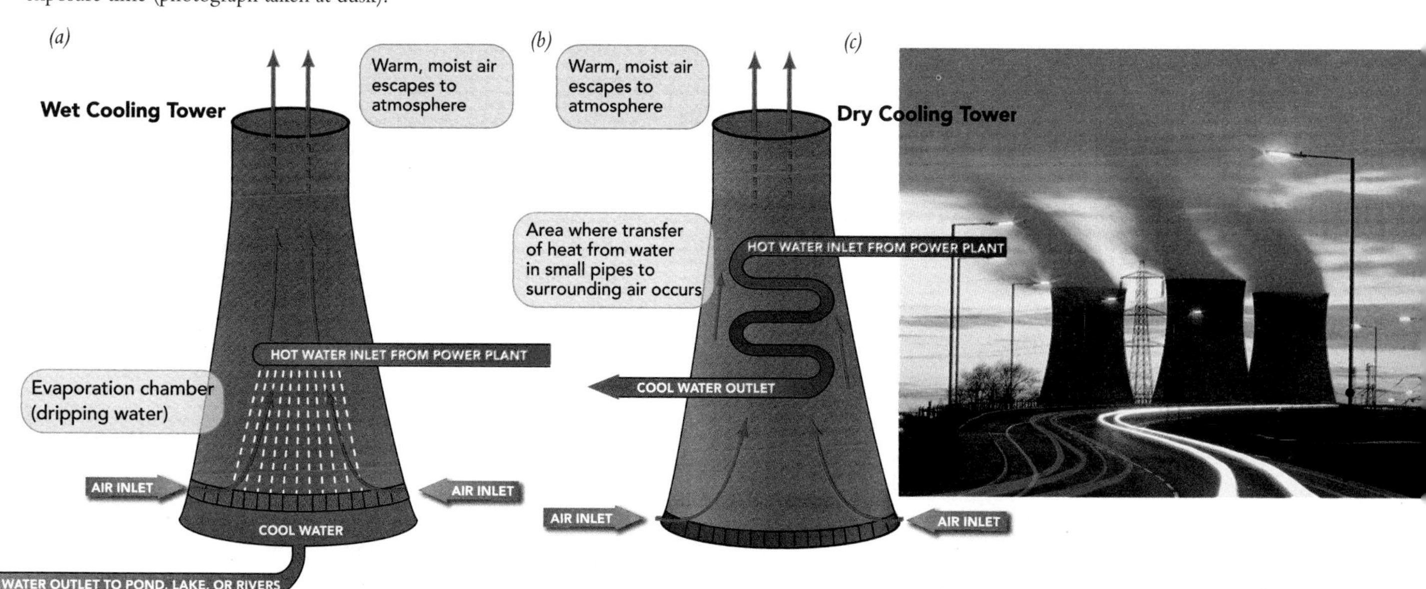

Particulates

Particulates are small particles of dust (including soot and asbestos fibres, discussed below) released into the atmosphere by many natural processes and human activities. Modern farming and combustion of oil and coal add considerable amounts of particulates to the atmosphere, as do dust storms, fires (Figure 15.9), and volcanic eruptions. The 1991 eruptions of Mount Pinatubo in the Philippines were the largest volcanic eruptions of the twentieth century, explosively delivering huge amounts of volcanic ash, sulphur dioxide, and other volcanic material and gases into the atmosphere to elevations up to 30 km. Eruptions can have a significant impact on the global environment and are linked to global climate change and stratospheric ozone depletion (see Chapters 20 and 22). In addition, many chemical toxins, such as metals, enter the biosphere as particulates. Sometimes, non-toxic particulates link with toxic substances, creating a synergetic threat (see discussion of particulates in Chapter 21).

Asbestos

Asbestos is a term for several minerals that take the form of small, elongated particles, or fibres. Industrial use of asbestos has contributed to fire prevention and has provided protection from the overheating of materials. Asbestos is also used as insulation for a variety of purposes. Unfortunately, however, excessive contact with asbestos has led to asbestosis (a lung disease caused by the inhalation of asbestos) and to cancer in some industrial workers. Experiments with animals have demonstrated that asbestos can cause tumours to develop if the fibres are embedded in lung tissue.[26] The hazard related to certain types of asbestos under certain conditions is thought to be so serious that extraordinary steps have been taken to reduce the presence of asbestos or to ban

Figure 15.9 • Smoke from open fires is an important source of air pollution in many cities. Shown here: Agra, India.

it outright. The expensive process of asbestos removal from old buildings (particularly schools) in the United States is one of those steps.

There are several types of asbestos, and they are not equally hazardous. The form most commonly utilized in Canada and the United States is white asbestos, which comes from the mineral chrysolite. It has been used as an insulation material around pipes, in floor and ceiling tiles, and for brake linings of automobiles and other vehicles. Approximately 95% of the asbestos that is now in place in the United States is of the chrysolite type. Most of this asbestos was mined in Canada, and environmental health studies of Canadian miners show that exposure to chrysolite asbestos is not particularly harmful. Studies involving another type of asbestos, known as crocidolite asbestos (blue asbestos), suggest that exposure to this mineral can be very hazardous and evidently does cause lung disease. Several other types of asbestos have also been shown to be harmful.[26]

A great deal of fear has been associated with non-occupational exposure to chrysolite asbestos. Tremendous amounts of money have been expended to remove it from homes, schools, public buildings, and other sites in spite of the fact that there has been no asbestos-related disease recorded among those exposed to chrysolite in non-occupational circumstances. It is now thought that much of the removal was unnecessary and that chrysolite asbestos does not pose a significant health hazard. Additional research into health risks from other varieties of asbestos is necessary to better understand the potential problem and to outline strategies to eliminate potential health problems.

Electromagnetic Fields

Electromagnetic fields (EMFs) are part of everyday urban life. Electric motors, electric transmission lines for utilities, and electrical appliances—such as toasters, electric blankets, and computers—all produce magnetic fields. There is currently a controversy over whether these fields produce a health risk.

Early on, investigators did not believe that magnetic fields were harmful, because fields drops off quickly with distance from the source and the strength of the fields that most people come into contact with are relatively weak. For example, the magnetic fields generated by power transmission lines or by a computer terminal are normally only about 1% of Earth's magnetic field; directly below power lines the electric field induced in the body is about what the body naturally produces within cells.

Several early studies, however, concluded that children exposed to EMFs from power lines have an increased risk of contracting leukemia, lymphomas, and nervous system cancers.[27] Investigators concluded that children so exposed are about one and a half to three times as likely to develop cancer as children with very low exposure to EMFs, but the results were questioned because of perceived problems with the research design (problems of sampling, tracking children, and estimating exposure to EMF).

A later study analysed over 1,200 children, approximately half of whom suffered from acute leukemia. It was necessary to estimate residential exposure to magnetic fields generated by nearby power lines in the children's present and former homes. Results of that study, which is the largest such investigation to date, concluded that there is no association between childhood leukemia and measured exposure to magnetic fields.[27,28]

Another study compared exposure of electric utility workers to magnetic fields with incidents of brain cancer and leukemia. That study revealed a weak association between exposure to magnetic fields and both brain cancer and leukemia. However, the associations are not strong and were not statistically significant.[29]

Saying that data are not statistically significant is another way of stating that the relationship between exposure and disease cannot be reasonably established, given the database that was analysed. It does not mean that additional data in a future study will not find a statistically significant relationship. Statistics can predict the strength of the relationship between variables such as exposure to a toxin and the incidence of a disease, but statistics cannot prove a cause-and-effect relationship between them.

In summary, in spite of the many studies that have been completed to evaluate relationships between disease and exposure to magnetic fields in our modern urban environment, the jury is still out. There seems to be some indication that magnetic fields may cause problems, but so far, the risks are relatively small and difficult to quantify.

Noise Pollution

Noise pollution is unwanted sound. Sound is a form of energy that travels as waves. We hear sound because our ears respond to sound waves through vibrations of the eardrum. The sensation of loudness is related to the intensity of the energy carried by the sound waves and is measured in units of decibels (dB). The threshold for human hearing is 0 dB; the average sound level in the interior of a home is about 45 dB; the sound of an automobile, about 70 dB; and the sound of a jet aircraft taking off, about 120 dB (see Table 15.2). A 10-fold increase in the strength of a particular sound adds 10 dB units on the scale. An increase of 100 times adds 20 units.[12] The decibel scale is logarithmic; it increases exponentially as a power of 10. For example, 50 dB is 10 times louder than 40 dB and 100 times louder than 30 dB.

Environmental effects of noise depend not only on the total energy but also on the sound's pitch, frequency, and time pattern and the length of exposure to the sound. Very loud noises (more than 140 dB) cause pain, and high levels can cause permanent hearing loss. Human ears can take sound up to about 60 dB without damage or hearing loss. Any sound above 80 dB is potentially dangerous. The noise of a lawn mower or motorcycle will begin to damage hearing after about eight hours of exposure. In recent years, there has been concern for teenagers (and older people, for that matter) who have suffered some permanent loss of hearing following extended exposure to amplified rock music (110 dB). At a noise level of 110 dB, damage to hearing can occur after an exposure time of only half an hour. Loud sounds at the workplace are another hazard. Levels of noise that are below the hearing-loss level may still interfere with human communication and may cause irritability. Noise in the range of 50 to 60 dB is sufficient to interfere with sleep, producing a feeling of fatigue upon awakening.

Table 15.2 • Examples of Sound Levels

Sound Source	*Intensity of Sound (dB)*	*Human Perception*
Threshold of hearing	0	
Rustling of leaf	10	Very quiet
Faint whisper	20	Very quiet
Average home	45	Quiet
Light traffic (30 m away)	55	Quiet
Normal conversation	65	Quiet
Chain saw (15 m away)	80	Moderately loud
Jet aircraft flyover at 300m	100	Very loud
Rock music concert	110	Very loud
Thunderclap (close)	120	Uncomfortably loud
Jet aircraft takeoff at 100m	125	Uncomfortably loud
	140	Threshold of pain
Rocket engine (close)	180	Traumatic injury

Voluntary Exposure

Voluntary exposure to toxins and potentially harmful chemicals is sometimes referred to as exposure to personal pollutants. The most common of these are tobacco, alcohol, and other drugs. Use and abuse of these substances have led to a variety of human ills, including death and chronic disease, criminal activity such as reckless driving and manslaughter, loss of careers, street crime, and the straining of human relations at all levels.

Scientific evidence has demonstrated that use of tobacco, in all of its forms, is both habit forming and dangerous to human health. Tobacco contains a variety of components that are toxic, carcinogenic, radioactive, and addictive. It has been estimated that 30% of all cancers in Canada and the United States are tied to smoking-related disorders. According to the American Cancer Society, cigarette smoking is responsible for approximately 80% of lung cancers; second-hand smoke is also a hazard, as are tobacco products such as chewing tobacco. Although the number of people who smoke in Canada and the United States, as a percentage of adults, has decreased in recent years, young people and people in developing countries are still becoming addicted to cigarettes and cigars.

Many people in our society use alcohol at social gatherings and celebrations. Approximately 70% of all Canadian adults drink some alcohol, and moderate use of alcohol is legal and accepted in our society. However, when abused, alcohol causes very serious problems. Approximately one-half of all deaths in automobile accidents are related to alcohol use by drivers. Furthermore, significant numbers of violent crimes and other criminal activities are committed by people under the influence of alcohol. Some people believe that alcohol is the most abused drug in our society today. Young people unfamiliar with the potential toxicity of alcohol have died from alcohol overdose (for example, by drinking 21 drinks on their 21st birthdays). Chronic alcoholism has many toxic consequences, including liver and heart disease.

A variety of illegal drugs are commonly used throughout the world. These drugs have various effects on their users, but the end result is often the degradation of the mind and/or body. Illegal drugs are particularly dangerous, because their strength, composition, and other chemical characteristics are seldom subject to quality control. Of particular concern in recent years has been the development of synthetic (designer) drugs that are addictive and capable of causing significant health problems and even death—in some cases—in first-time users.

15.3 General Effects of Pollutants

Almost every part of the human body is affected by one pollutant or another, as shown in Figure 15.10*a*. For example, lead and mercury (remember the Mad Hatter) affect the brain; arsenic, the skin; carbon monoxide, the heart; and fluoride, the bones. Wildlife is affected as well. Sites of effects of major pollutants in wildlife are shown in Figure 15.10*b*; effects of pollutants on wildlife populations are listed in Table 15.3.

The lists of potential toxins and affected body sites for humans and other animals in Figure 15.10 may be somewhat misleading. For example, chlorinated hydrocarbons, such as dioxin, are stored in the fat cells of animals, but they cause damage not only to fat cells but to the entire organism through disease, damaged skin, and birth defects. Similarly, a toxin that affects the brain, such as mercury, causes a wide variety of problems and symptoms, as discussed in A Closer Look 15.2. The value of Figure 15.10 is in helping us to understand in general the adverse effects of excess exposure to chemicals.

Concept of Dose and Response

Five centuries ago, the physician and alchemist Paracelsus wrote, "everything is poisonous, yet nothing is poisonous." By this, he meant essentially that a substance in too great an amount can be dangerous yet in an extremely small amount can be relatively harmless. Every chemical element has a spectrum of possible effects on a particular organism. For example, selenium is required in small amounts by living things but may be toxic or increase the probability of cancer in cattle and wildlife when it is present in high concentrations in the soil. Copper, chromium, and manganese are other chemical elements required in small amounts by animals but toxic in higher amounts.

It was recognized many years ago that the effect of a certain chemical on an individual depends on the dose. This concept is termed dose response. Dose dependency can be represented by a generalized dose-response curve such as that shown in Figure 15.11.

When various concentrations of a chemical present in a biological system are plotted against the effects on the organism, two things are apparent. First, relatively large concentrations are toxic and even lethal (points *D*, *E*, and *F* in Figure 15.12). Second, trace concentrations may be beneficial for life (between points *A* and *D*); and the dose–response curve forms a plateau of optimal concentration and maximum benefit between two points (*B* and *C*). Points *A*, *B*, *C*, *D*, *E*, and *F* in Figure 15.11 are important thresholds in the dose–response curve. Unfortunately, the amounts at which points *E* and *F* occur are known only for a few substances, for a few organisms, including people; and the very important point *D* is all but unknown. Doses that are beneficial, harmful, or lethal may differ widely for different organisms and are difficult to characterize.

Fluorine provides a good example of the general dose–response concept. Fluorine forms fluoride compounds that prevent tooth decay and promote the development of a healthy bone structure. Relationships between the concentration of fluoride (in a compound of fluorine, such as sodium fluoride, NaF) and health show a specific dose–response curve (Figure 15.12). The plateau for an optimal concentration of fluoride (point *B* to point *C*) to reduce dental caries (cavities) is from about 1 ppm to just less than 5 ppm. Levels greater than 1.5 ppm do not significantly decrease tooth decay but do increase the occurrence of discoloration of teeth. Concentrations of 4 to 6 ppm reduce the prevalence of osteoporosis, a disease characterized by loss of bone mass; and toxic effects are noticed between 6 and 7 ppm (point *D* in Figure 15.12).

Dose–Response Curve (LD-50, ED-50, and TD-50)

Individuals differ in their response to chemicals, and it is difficult to predict the dose that will cause a response in a particular individual. For this reason, it is practical to predict what percentage of a population will respond to a specific dose of a chemical.

For example, the dose at which 50% of the population die is called the lethal dose 50, or LD-50. The LD-50 is a crude approximation of a chemical's toxicity. It is a gruesome index that does not adequately convey the sophistication of modern toxicology and is of little use in setting a standard for toxicity. However, the LD-50 determination is required for new synthetic chemicals as a way of estimating their toxic potential. Table 15.4 lists, as examples, LD-50 values in rodents for selected chemicals.

The ED-50 (effective dose 50%) is the dose that causes an effect in 50% of the population of observed

Table 15.3 • Effects of Pollutants on Wildlife

Effect on Population	*Examples of Pollutants*
Changes in abundance	Arsenic, asbestos, cadmium, fluoride, hydrogen sulphide, nitrogen oxides, particulates, sulphur oxides, vanadium, POPs[a]
Changes in distribution	Fluoride, particulates, sulphur oxides, POPs
Changes in birth rates	Arsenic, lead, POPs
Changes in death rates	Arsenic, asbestos, beryllium, boron, cadmium, fluoride, hydrogen sulfide, lead, particulates, selenium, sulphur oxides, POPs
Changes in growth rates	Boron, fluoride, hydrochloric acid, lead, nitrogen oxides, sulphur oxides, POPs

[a] Pesticides, PCBs, hormonally active agents, dioxin, and DDT are examples (see Table 15.1).

Source: J. R. Newman, *Effects of Air Emissions on Wildlife*, U.S. Fish and Wildlife Service, 1980. Biological Services Program, National Power Plant Team, FWS/OBS-80/40, U.S. Fish and Wildlife Service, Washington, D.C.

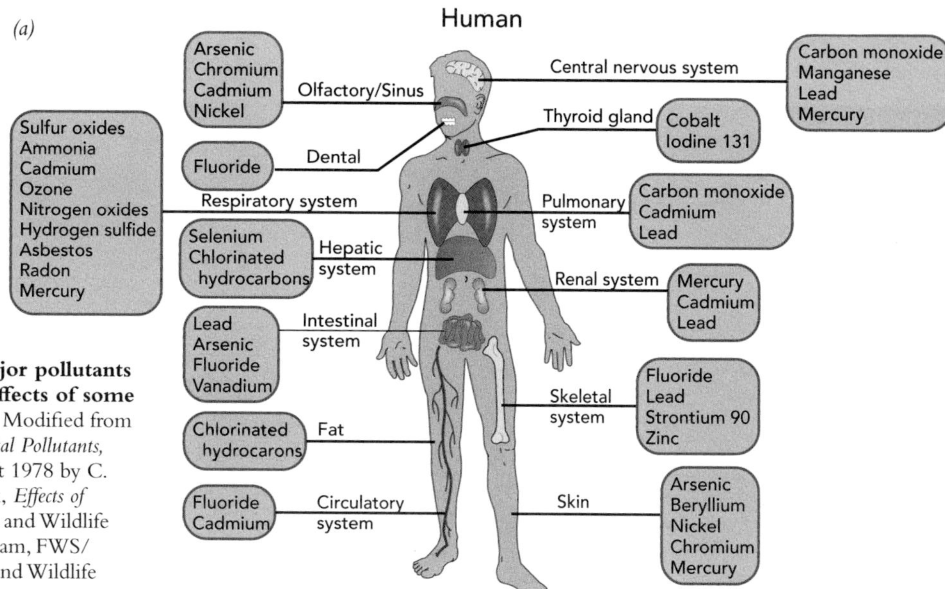

Figure 15.10 • *(a)* **Effects of some major pollutants in human beings.** *(b)* **Known sites of effects of some major pollutants in wildlife.** [*Source*: *(a)* Modified from G. L. Waldbott, *Health Effects of Environmental Pollutants,* 2nd ed. (St. Louis: Mosby, 1978). Copyright 1978 by C. V. Mosby. *(b)* Modified from J. R. Newman, *Effects of Air Emissions on Wildlife Resources*, U.S. Fish and Wildlife Services Program, National Power Plant Team, FWS/OBS-80/40 (Washington, D.C.: U.S. Fish and Wildlife Service, 1980).]

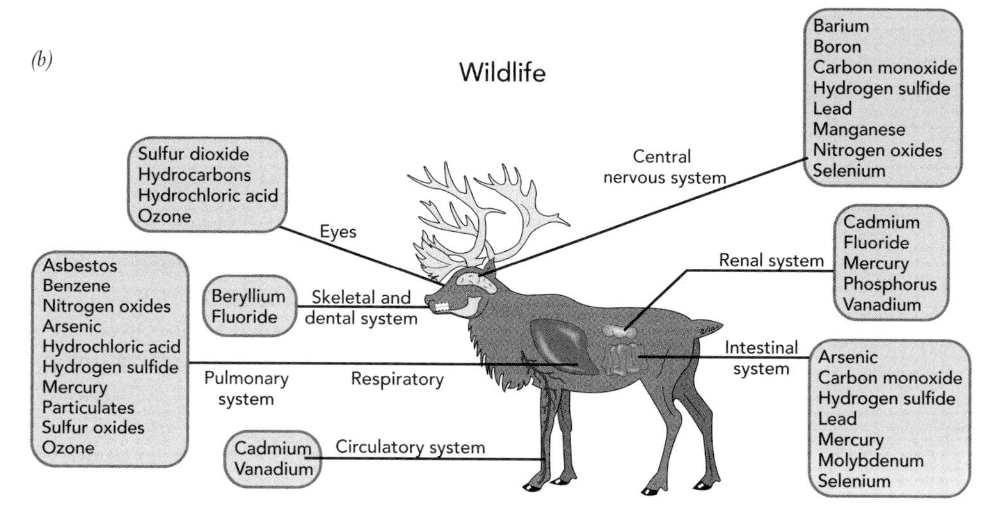

Figure 15.11 • **Generalized dose–response curve.** Low concentrations of a chemical may be harmful to life (below point *A*). As the concentration of the chemical increases from *A* to *B,* the benefit to life increases. The maximum concentration that is beneficial to life lies within the benefit plateau *(B–C)*. Concentrations greater than this plateau provide less and less benefits *(C–D)* and will harm life *(D–F)* as toxic concentrations are reached. Increased concentrations above the toxic level may result in death.

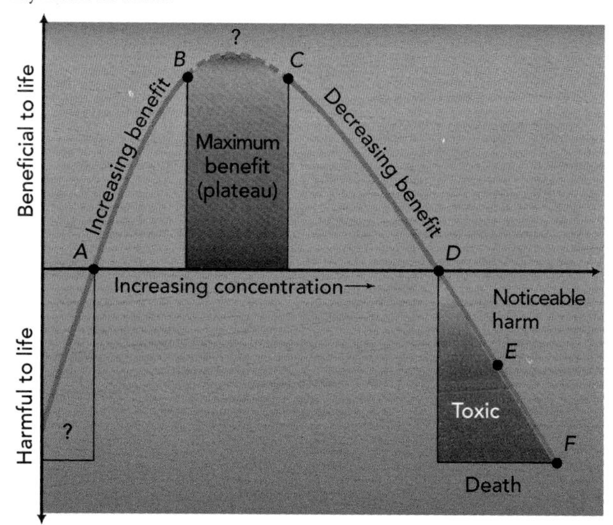

Figure 15.12 • **General dose–response curve for fluoride** showing the relationship between fluoride concentration and physiological benefit.

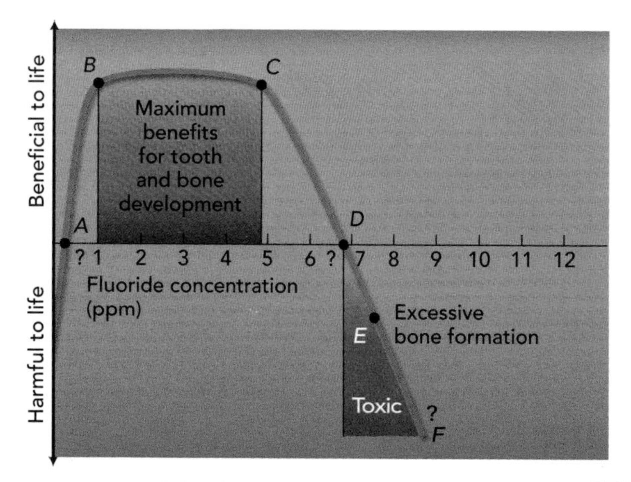

subjects. For example, the ED-50 of aspirin would be the dose that relieves headaches in 50% of the people.[30]

The TD-50 (toxic dose 50%) is defined as the dose that is toxic to 50% of the population. TD-50 is often used to indicate responses such as reduced enzyme activity, decreased reproductive success, or onset of specific symptoms, such as loss of hearing, nausea, or slurred speech.

For a particular chemical, there may be a whole family of dose–response curves, as illustrated in Figure 15.13. Which dose is of interest depends on what is being evaluated. For example, for insecticides we may wish to know the dose that will kill 100% of the insects exposed; therefore the LD-95 (the dose that kills 95% of the insects) may be the minimum acceptable level. However, when considering human health and the exposure to a particular toxin, we often want to know the LD-0—the maximum dose that does not cause any deaths.[30] For potentially toxic compounds such as insecticides, which may form a residue on food or food additives, we want to ensure that the expected levels of human exposure will have no known toxic effects. From an environmental perspective, this is important because of concerns about increased risk of cancer associated with exposure to toxic agents.[30]

For drugs used to treat a particular disease, the efficiency of the drug as a treatment is of paramount importance. In addition to knowing what the therapeutic value (ED-50) is, it is also important to know the relative safety of the drug. For example, there may be an overlap between the therapeutic dose (ED) and the toxic dose (TD) (see Figure 15.13). That is, the dose that causes a positive therapeutic response in some individuals might be toxic to others. A quantitative measure of the relative safety of a particular drug is the therapeutic index, defined as the ratio of the LD-50 to the ED-50. The greater the therapeutic index, the safer the drug is believed to be.[31] In other words, a drug with a large difference between the lethal and therapeutic dose is safer than one with a smaller difference.

Table 15.4 • Approximate LD-50 Values (for rodents) for Selected Agents

Agent	*LD50(mg/kg)*[a]
Sodium chloride (table salt)	4,000
Ferrous sulphate (to treat anemia)	1,520
2,4-D (a weed killer)	368
DDT (an insecticide)	135
Caffeine (in coffee)	127
Nicotine (in tobacco)	24
Strychnine sulphate (used to kill certain pests)	3
Botulinum toxin (in spoiled food)	0.00001

[a] Milligrams per kilogram of body mass (termed mass weight, although it really isn't a weight) administered by mouth to rodents. Rodents are commonly used in such evaluations, in part because they are mammals (as we are), are small, have a short life expectancy, and their biology is well known.

Source: H. B. Schiefer, D. C. Irvine, and S. C. Buzik, *Understanding Toxicology* (New York: CRC Press, 1997).

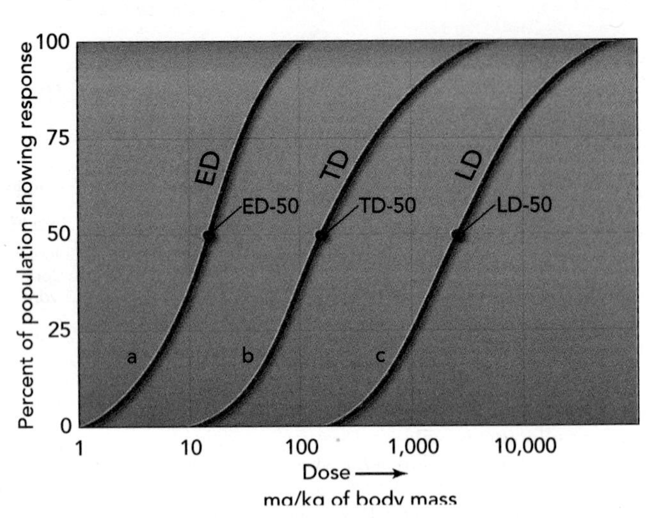

Figure 15.13 • Idealized diagram illustrating a family of dose–response curves for a specific drug: ED (effective dose), TD (toxic dose), and LD (lethal dose). Notice the overlap for some parts of the curves. For example, at ED-50, a few percent of the people exposed to that dose will suffer a toxic response, but none will die. At TD-50 dose, about 1% of the people exposed to that dose will die.

Threshold Effects

Recall from A Closer Look 15.2 that a **threshold** is a level below which no effect occurs and above which effects begin to occur. If a threshold dose of a chemical exists, then a concentration of that chemical in the environment below the threshold is safe. If there is no threshold dose, then even the smallest amount of the chemical has some negative toxic effect (Figure 15.14).

Whether or not there is a threshold effect for environmental toxins is an important environmental issue. For example, Ontario's *Environmental Protection Act* originally stated goal was to reduce pollutant discharges to zero. The goal implies there is no such thing as a threshold effect, since no level of toxin is to be legally permitted. However, it is unrealistic to believe zero discharge of a water pollutant can be achieved or to believe that we can reduce to zero the concentration of chemicals shown to be carcinogenic.

A problem in evaluating thresholds for toxic pollutants is that it is difficult to account for synergistic effects. Little is known about if or how thresholds might change if an organism is exposed to more than one toxin at the same time or to a combination of toxins and other chemicals, some of which are beneficial. Exposures of people to chemicals in the environment are complex, and we are only beginning to understand and conduct research on the possible interactions and consequences of multiple exposures.

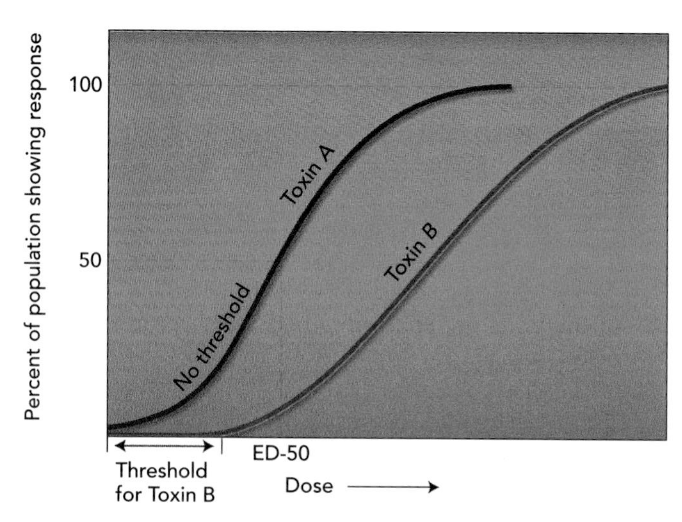

Figure 15.14 • In this hypothetical toxic dose–response curve, toxin A has no threshold; even the smallest amount has some measurable effect on the population. The TD-50 for toxin A is the dose required to produce a response in 50% of the population. Toxin B has a threshold (flat part of curve) where response is constant as dose increases. After the threshold dose is exceeded, the response increases.

Ecological Gradients

Dose-response effects differ among species. For example, the kinds of vegetation that can live nearest to a toxic source are often small plants with relatively short lifetimes (grasses, sedges, and weedy species usually regarded as pests) that are adapted to harsh and highly variable environments. Farther from the toxic source, trees may be able to survive. Changes in vegetation with distance from a toxic source define the ecological gradient.

Ecological gradients may be found around smelters and other industrial plants that discharge pollutants into the atmosphere from smokestacks. For example, ecological gradient patterns can be observed in the area around the smelters of Sudbury, Ontario, discussed earlier in this chapter (A Closer Look 15.1). Near the smelters, an area that was once forest is now a patchwork of bare rock and soil occupied by small plants.

Tolerance

The ability to resist or withstand stress resulting from exposure to a pollutant or harmful condition is referred to as tolerance. Tolerance can develop for some pollutants in some populations, but not for all pollutants in all populations.

Tolerance may result from behavioural, physiological, or genetic adaptation. *Behavioural tolerance* results from changes in behaviour. For example, mice learn to avoid traps.

Physiological tolerance results when an individual's body adjusts to tolerate a higher level of pollutant. For example, in a recent series of studies, students were exposed to ozone (O_3), an air pollutant often present in large cities (Chapter 21). The students at first experienced symptoms that included irritation of eyes and throat and shortness of breath. However, after a few days, their bodies adapted to the ozone, and they reported that they believed they were no longer breathing ozone-contaminated air, even though the concentration of O_3 stayed the same. This phenomenon explains why some people who regularly breathe polluted air report that they do not notice the pollution. Of course, it does not mean that the ozone is doing no damage; it is, especially in people with existing respiratory problems. There are many mechanisms for physiological tolerance, including *detoxification,* in which the toxic chemical is converted to a non-toxic form, and the internal transport of the toxin to a part of the body where it is not harmful, such as fat cells.

Genetic tolerance, or adaptation, results when some individuals in a population are naturally more resistant to a toxin than others. They are less damaged by exposure and more successful in breeding. Resistant individuals pass on the resistance to future generations, who are also more successful at breeding. Adaptation has been observed among some insect pests following exposure to particular chemical pesticides. For example, certain strains of malaria-causing mosquitoes are now resistant to DDT (see the discussion in Chapter 12); and some organisms that cause deadly infectious diseases have become resistant to common antibiotic drugs, such as penicillin.

Acute and Chronic Effects

Pollutants can have acute and chronic effects. An *acute effect* is one that occurs soon after exposure, usually to large amounts of a pollutant. A *chronic effect* takes place over a long period, often as a result of exposure to low levels of a pollutant. For example, a person exposed all at once to a high dose of radiation may be killed by radiation sickness soon after exposure (an acute effect). However, that same total dose, received slowly in small amounts over an entire lifetime, may instead cause mutations and lead to disease or affect the person's DNA and offspring (a chronic effect).

15.4 Risk Assessment

Risk assessment can be defined as the process of determining potential adverse environmental health effects to people exposed to pollutants and potentially toxic materials (recall the discussion of measurements and methods of science in Chapter 2). Such an assessment generally includes four steps:[32]

1. *Identification of the hazard.* Identification consists of testing materials to determine whether exposure is likely to cause environmental health problems. One method is to investigate people previously exposed

to a hazard. For example, to understand the toxicity of radiation produced from radon gas, researchers studied workers in uranium mines. Another method is to perform experiments to test effects on animals, such as mice, rats, or monkeys. This method has drawn increasing criticism from groups of people who believe such experiments are unethical. Another approach is to try to understand how a particular chemical works at the molecular level on cells. For example, research has been done to determine how dioxin interacts with living cells to produce an adverse response. After quantifying the response, scientists can develop mathematical models to predict or estimate dioxin's risk.[18] This relatively new approach might also be applicable to other potential toxins that work at the cellular level.

2. *Dose-response assessment.* The next step involves identifying relationships between the dose of a chemical (therapeutic drug, pollutant, or toxin) and the health effects to people. Some studies involve administering fairly high doses of a chemical to animals. The effects, such as illness or symptoms (rash, tumour development) are recorded for varying doses, and the results are used to predict the response in people. This is difficult, and the results are controversial for several reasons:
   - The dose that results in a particular response may be very small and subject to measurement errors.
   - There may be arguments over whether thresholds are present or absent.
   - Experiments on animals such as rats, mice, or monkeys may not be directly applicable to humans.
   - The assessment may rely on probability and statistical analysis. Although statistically significant results from experiments or observations are accepted as evidence to support an argument, statistics cannot establish that the substance tested caused the observed response.
3. *Exposure assessment.* Exposure assessment evaluates the intensity, duration, and frequency of human exposure to a particular chemical pollutant or toxin. The hazard to society is directly proportional to the total population exposed. The hazard to an individual is generally greater closer to the source of exposure. Like dose-response assessment, exposure assessment is difficult, and the results are often controversial, in part because of difficulties in measuring the concentration of a toxin present in doses as small as parts per million, billion, or even trillion. Some questions that exposure assessment attempts to answer are:
   - How many people were exposed to concentrations of a toxin thought to be dangerous?
   - How large an area was contaminated by the toxin?
   - What are the ecological gradients for exposure to the toxin?
   - How long were people exposed to a particular toxin?
4. *Risk characterization.* During this final step, the goal is to delineate health risk in terms of the magnitude of the potential environmental health problem that might result from exposure to a particular pollutant or toxin. To do this, it is necessary to identify the hazard, complete the dose-response assessment, and evaluate the exposure assessment as outlined above. This step involves all the uncertainties of the prior steps, and results are again likely to be controversial.

In summary, risk assessment is difficult, costly, and controversial. Each chemical is different, and there is no one method of determining responses of humans for specific EDs or TDs. Toxicologists use the scientific method of hypothesis testing with experiments (see Chapter 2) to generate predictions of how specific doses of a chemical may affect humans. Warning labels listing potential side effects of using a specific medication are required by law, and these warnings result from toxicology studies to determine a drug's safety.

Finally, risk assessment requires making scientific judgments and formulating actions to help minimize environmental health problems related to human exposure to pollutants and toxins. The process of risk management integrates the assessment of risk with technical, legal, political, social, and economic issues.[18] Scientific arguments concerning the toxicity of a particular material are often open to debate. For example, there is debate concerning whether the risk from dioxin is linear. That is, do effects start at minimum levels of exposure to dioxin and gradually increase, or is there a threshold exposure beyond which environmental health problems occur? (See A Closer Look 15.3).[18,23] It is the task of people in appropriate government agencies assigned to manage risk to make judgments and decisions based on the risk assessment and then to take appropriate actions to minimize the hazard resulting from exposure to toxins.

15.5 Precautionary Principle

Science has the role of trying to understand physical and biological processes associated with environmental problems such as global warming, depletion of ozone in the upper atmosphere, loss of biodiversity and endangered species, and management of forestry and fisheries resources among others. However, in a sense, all science is preliminary and it is difficult to prove relationships between physical and biological processes linked to human actions that result in degradation of the environment from a local to a global scale. For this

CRITICAL THINKING ISSUE

Is Lead in the Urban Environment Contributing to Antisocial Behaviour?

Lead is one of the most common toxic (harmful or poisonous) metals in our inner-city environments, and it may be linked to delinquent behaviour in children. Lead is found in all parts of the urban environment (air, soil, older pipes, and some paint, for example) and in biological systems, including people (Figure 15.15). There is no apparent biological need for lead, but it is sufficiently concentrated in the blood and bones of children living in inner cities to cause health and behaviour problems. In some populations, over 20% of the children have blood concentrations of lead that are higher than those believed safe.

Lead affects nearly every system of the body. Acute lead toxicity may be characterized by a variety of symptoms, including anemia, mental retardation, palsy, coma, seizures, apathy, poor coordination, subtle loss of recently acquired skills, and bizarre behaviour. Lead toxicity is particularly a problem for young children, who apparently are more susceptible to lead poisoning than are adults. Following acute toxic response to lead, some children manifest aggressive, difficult-to-manage behaviour.

The occurrence of lead toxicity or lead poisoning has cultural, political, and sociological implications. Over 2,000 years ago, the Roman Empire produced and used tremendous amounts of lead for a period of several hundred years. Production rates were as high as 55,000 metric tonnes per year. Romans used lead for storage and drinking vessels, in pipes, and as a base for cosmetics and medicines. It has been argued that lead poisoning among the upper class in Rome was partly responsible for Rome's decline.

Lead poisoning has also been suggested as causing the demise of the 1845 expedition led by Sir John Franklin to find a northwest passage to Asia. The expedition's two ships, the HMS *Erebus* and HMS *Terror*, were some of the first to be equipped with canned provisions, sealed with lead solder. All 129 members of the party perished. Forensic investigations of their remains, discovered near Baffin Bay, revealed high levels of lead, suggesting that lead poisoning may have caused many deaths and—more subtly—impaired the ability of expedition leaders to make sound decisions to protect their crew.

Figure 15.15 • A vacant lot in Toronto, Ontario. Lead is one of the most common toxic metals in urban environments, and is often found in inner city soils like these.

Lead poisoning probably results in widespread stillbirths, deformities, and brain damage. Studies analysing the lead content of bones of ancient Romans tend to support this hypothesis. More recently, the occurrence of lead has been studied in glacial ice cores from Greenland. Glaciers have an annual growth layer of ice. Older layers are buried by younger layers, allowing us to identify the age of each layer. Researchers drill glaciers, taking continuous samples of the layers that look like long solid rods of glacial ice; these are called cores. Measurements of the concentration of lead from cores show that during the Roman period, from approximately 500 b.c. to a.d. 300, lead concentrations in the glacial ice are about four times higher than before and after this period. This suggests that the mining and smelting of lead in the Roman Empire added small particles of lead to the atmosphere that eventually settled out in the glaciers of Greenland.

Lead toxicity, then, seems to have been a problem for a long time. Now, an emerging, interesting, and potentially significant hypothesis is that, in children, lead concentrations below the levels known to cause physical damage may be associated with an increased potential for antisocial, delinquent behaviour. This is a testable hypothesis. (See Chapter 2 for a discussion of hypotheses). If the hypothesis is correct, then some of our urban crime may be traced to environmental pollution!

A recent study in children aged 7 to 11 years old measured the amount of lead in bones and compared it with data concerning behaviour over a four-year period. The study, which reviewed the behaviour of 216 delinquent and 201 nondelinquent youths based on reports from parents, teachers, and the children themselves, concluded that an above-average concentration of lead in children's bones was associated with an increased risk for attention-deficit disorder, aggressive behaviour, and delinquency. The study took into account factors such as maternal intelligence, socioeconomic status, and quality of child rearing.

Critical Thinking Questions

1. What is the main point of the discussion about lead in the bones of children and behaviour?
2. What are the main assumptions of the argument? Are they reasonable?
3. What other hypotheses might be proposed to explain the behaviour?

reason, in 1992, the Rio Earth Summit on sustainable development listed as one of its principles what we now define as the **precautionary principle**. The argument supporting the principle is that when there exists a threat of potential serious and/or irreversible environmental damage, scientific certainty is not required to take a precautionary approach in an attempt to prevent potential environmental degradation. The precautionary principle thus contributes to critical thinking on a variety of environmental concerns, including the use of chemicals such as pesticides, herbicides, and drugs; fossil fuels; nuclear energy; land conversion; and management of wildlife, fisheries, and forest resources.[33]

The precautionary principle recognizes that scientific proof is not possible in most instances and that management practices are needed to reduce or eliminate environmental degradation believed to result from a particular human action. In other words, in spite of the fact that full scientific certainty is not available, we should still take cost-effective precautions to solve environmental problems. One of the difficulties in applying the precautionary principle is the decision concerning how much scientific evidence is needed before action on a particular environmental problem should be suggested. This is a significant question that is being debated. The principle recognizes the need to evaluate existing scientific evidence and the importance of continued scientific work and testing. In other words, the issue being evaluated has some preliminary data with provisional conclusions, but final analysis awaits additional or more reliable scientific data and analysis. For example, when considering environmental health issues related to the use of a particular chemical, there may be an abundance of scientific data—but with gaps, inconsistencies, and other scientific uncertainties. Those in favour of continuing a particular activity, such as use of pesticides or herbicides, may argue that there is not sufficient proof to warrant restricting the use of specific chemicals. Others would argue that absolute proof of safety is necessary before a new chemical is used. The precautionary principle, applied to this case, would be that lack of full scientific certainty concerning the use of pesticides and herbicides should not be used as a reason for not taking or for postponing cost-effective measures to reduce or prevent environmental degradation or health problems. This raises another important question: What constitutes a cost-effective measure? Certainly an examination of the benefits and costs of taking a particular action compared to taking no action should be done, but other economic analysis may also be appropriate.[34]

There will always be arguments over what constitutes sufficient scientific knowledge for decision making. Nevertheless, the precautionary principle, even though it may be difficult to apply, is becoming a common part of the process of environmental analysis when applied to environmental protection and environmental health issues. It requires us to apply the *principles of environmental unity* and predict potential consequences of activities before they occur. Therefore, the precautionary principle has the potential to become a proactive, rather than reactive, tool in reducing or eliminating environmental degradation resulting from human activity.

Summary

- Pollution produces an impure, dirty, or otherwise unclean state. Contamination means making something unfit for a particular use through the introduction of undesirable materials. Toxic materials are poisonous to people and other living things; toxicology is the study of toxic materials. A concept important in studying pollution problems is synergism, whereby actions of different substances produce a combined effect greater than the sum of the effects of the individual substances.
- How we measure the amount of a particular pollutant introduced into the environment or the concentration of that pollutant varies widely, depending on the substance. Common units for expressing the concentration of water pollutants are parts per million (ppm) and parts per billion (ppb). Air pollutants are commonly measured in units such as micrograms of pollutant per cubic meter of air ($\mu g/m^3$).
- Categories of environmental pollutants include toxic chemical elements (particularly metals), organic compounds, radiation, heat, particulates, electromagnetic fields, and noise.
- Organic compounds of carbon are produced by living organisms or synthetically by humans. Artificially produced organic compounds may have physiological, genetic, or ecological effects when introduced into the environment. Organic compounds vary with respect to their potential hazards: some are more readily degraded in the environment than others; some are more likely to undergo biomagnification; and some are extremely toxic even at very low concentrations. Organic compounds that cause serious concern include persistent organic pollutants such as pesticides, dioxin, PCBs, and hormonally active agents.
- The effect of a chemical or toxic material on an individual depends on the dose. It is also important to

determine tolerances of individuals as well as acute and chronic effects of pollutants and toxins.
- Risk assessment involves hazard identification, assessment of dose response, assessment of exposure, and risk characterization.
- The precautionary principle is emerging as an important concept in environmental protection.

STUDY QUESTIONS

1. Do you think the hypothesis that some crime is caused in part by environmental pollution is valid? Why? Why not? How might the hypothesis be further tested? What are the social ramifications of the tests?
2. What kinds of life forms would most likely survive in a highly polluted world? What would be their general ecological characteristics?
3. Some environmentalists argue that there is no such thing as a threshold for pollution effects. What is meant by this statement? How would you determine whether it was true for a specific chemical and a specific species?
4. What is biomagnification, and why is it important in toxicology?
5. You are lost in Transylvania while trying to locate Dracula's castle. Your only clue is that the soil around the castle has an unusually high concentration of the heavy metal arsenic. You wander in a dense fog, able to see only the ground a few metres in front of you. What changes in vegetation warn you that you are nearing the castle?
6. Distinguish between acute and chronic effects of pollutants.
7. Design an experiment to test whether tomatoes or cucumbers are more sensitive to lead pollution.
8. Why is it difficult to establish standards for acceptable levels of pollution? In giving your answer, consider physical, climatological, biological, social, and ethical reasons.
9. A new highway is built through a pine forest. Driving along the highway, you notice that the pines nearest the road have turned brown and are dying. You stop at a rest area and walk into the woods. One hundred metres away from the highway, the trees seem undamaged. How could you make a crude dose-response curve from direct observations of the pine forest? What else would be necessary to devise a dose-response curve from direct observation of the forest? What else would be necessary to devise a dose-response curve that could be used in planning the route of another highway?
10. Do you think that the precautionary principle is necessary to protect the environment? When? Why not?

FURTHER READING

Amdur, M., J. Doull and C. D. Klaasen, eds. *Casarett & Doull's Toxicology: The Basic Science of Poisons*, 4th ed. Tarrytown, N.Y.: Pergamon, 1991. A comprehensive and advanced work on toxicology.

Carson, R. *Silent Spring*. Boston: Houghton Mifflin, 1962. A classic book on problems associated with toxins in the environment.

Schiefer, H. B., D. G. Irvine, and S. C. Buzik. *Understanding Toxicology: Chemicals, Their Benefits and Risks*. Boca Raton, Fla.: CRC Press, 1997. A concise introduction to toxicology as it pertains to everyday life, including information about pesticides, industrial chemicals, hazardous waste, and air pollution.

Travis, C. C. and H. A. Hattemer-Frey. "Human Exposure to Dioxin." *The Science of the Total Environment* 104 (1997):97–127. An extensive technical review of dioxin accumulation and exposure.

chapter 16

The August 14, 2003, Blackout

The most serious blackout (failure of electric power) in North American history occurred on August 14, 2003. Most of southern and northeastern Ontario, and parts of eight U.S. states (approximately 24,000 square kilometres) suddenly lost power at about 4:00 p.m. More than 50 million people were affected, some trapped in elevators or electric trains underground. People streamed into city streets not certain whether the power failure was due to a terrorist attack or not. Power was restored within 48 hours to most places, but the event was an energy shock that demonstrates our dependence on aging power distribution systems and centralized electric power generation. Terrorists had nothing to do with the blackout, but the event caused harm, anxiety, and financial loss to millions of people.

Immediately following the episode, Natural Resources Canada and the U.S. Department of Energy established a joint task force to analyse the causes of the blackout and make recommendations to prevent a recurrence. Their conclusion was that the August 14, 2003, blackout was entirely preventable. In fact, it had many of the same causes as several other large-scale blackouts in earlier decades. Clearly, we had not learned enough from those earlier mistakes.

In its March 2004 final report, the task force concluded that the blackout had seven main causes:[1]

- Failure to react to changing voltage conditions, so as to maintain a stable output of electricity;
- Failure to operate power generating equipment within safe limits;
- Insufficient tree trimming, allowing storm-blown trees to damage power lines;
- Inadequate operator training;
- Failure to identify emergency conditions promptly, and communicate those conditions to neighbouring jurisdictions;
- Insufficient ability to monitor the interlinked regional system over its full range; and
- Inadequate system-wide planning.

Surprising though it seems, the task force found that the system was operating reliably just an hour before the blackout. There seemed to be enough voltage available, enough generators to create it, and enough transmission lines to carry it. Operators did not observe any unusual system frequencies or other irregularities.

In fact, this situation was misleading. Parts of the system were highly vulnerable, and operating on the very edge of reliability standards. As a result, even a small disruption somewhere in the system could—and

Blackout, August 14, 2003. The blackout affected more than 50 million people in over 24,000 square kilometres of southern Canada and the adjacent United States

Before

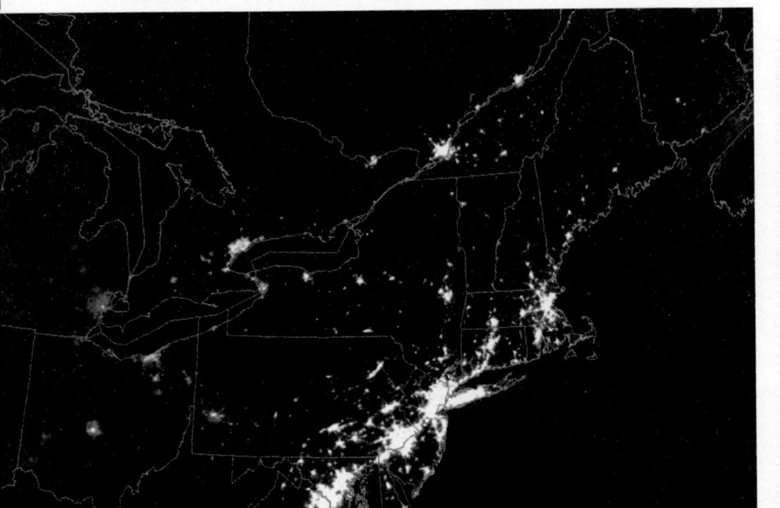

After

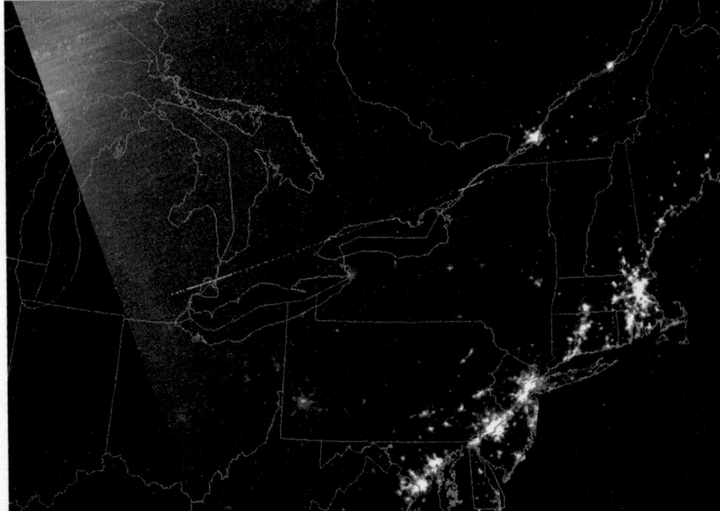

Energy: Some Basics

LEARNING OBJECTIVES

Understanding the basics of what energy is, as well as the sources and uses of energy, is essential for effective energy planning. After reading this chapter, you should understand:

- That energy is neither created nor destroyed but is transformed from one kind to another.
- Why in all transformations energy tends to go from a more usable to a less usable form.
- What energy efficiency is and why it is always less than 100%.
- That people in industrialized countries consume a disproportionately large share of the world's total energy, and how energy efficiency and conservation can help make better use of global energy resources.
- Why some energy planners propose a hard-path approach to energy provision and others a soft-path approach, and why both of these approaches have positive and negative points.
- Why integrated energy planning is an important goal for sustainable energy use.
- What elements are needed to develop integrated energy planning.

did—have major implications. Nuclear power plants (Chapter 17) were largely unaffected by the blackout. However, they and alternative power sources such as solar and wind energy were simply inadequate to provide the power lost in fossil-fuelled parts of the system.

The August 2003 blackout was a dramatic reminder of how much we rely on centralized power generation systems, and our heavy dependence on fossil fuels. The key to real energy planning requires a diversity of energy sources with a better mix of the fossil fuels and alternative sources that must eventually replace them. What is apparent is that in the first decades of the twenty-first century we are going to be continually plagued by dramatic price changes in energy with accompanying shortages. Using the remaining fossil fuels, particularly the cleaner fuels such as natural gas, should represent a transitional phase to more sustainable sources. What is really necessary is a major program to develop sources such as wind and solar energy much more vigorously than has been done in the past or apparently will be done in the next few years. If we are unable to make the transition from fossil fuels to alternative energy sources, then we will face an energy crisis that is unsurpassed in our history.

- *This case history demonstrates that Canada and the United States face serious energy problems. With this in mind, this chapter explores some of the basic principles associated with what energy is, how much energy we consume, and how we might manage energy for the future.*

***Written with assistance from Mel S. Manalis.**

16.1 The Concept of Energy

The concept of **energy** is somewhat abstract; you cannot see it or feel it, even though you have to pay for it, both biologically (see Chapter 9) and in fuel costs. To understand energy, it is easiest to begin with the idea of a force. We all have had the experience of exerting force by pushing or pulling. The strength of a force can be measured by how much it accelerates an object.

What if your car stalls while you are going up a hill and you get out to push it to the side of the road (Figure 16.1)? Because you are standing on a slope, you must push the car a short distance uphill as you move it off the road. In doing so, you apply a force against gravity, which would otherwise cause the car to roll downhill. If the brake is on, the brakes, tires, and bearings might heat up from friction. The longer the distance over which you exert force in pushing the car, the greater the change in the car's position and the greater the amount of heat from friction in the brakes, tires, and bearings.

In physicists' terms, exerting the force over the distance moved is work. That is, **work** *is the product of a force times a distance. Conversely, energy is the ability to do work*. If you push hard but the car does not move, you have exerted a force but you have not done any work on the car (according to the definition), even if you feel very tired and sweaty.

In pushing the stalled car, you have moved it against gravity and caused some of its parts (brakes, tires, bearings) to be heated. You have also displaced the air in front of the car, as it moves forward. These effects have something in common: they are forms of energy. You have converted chemical energy in your body to the energy of motion of the car and the air in front of it (kinetic energy). When the car is higher on the hill, the potential energy of the car has been increased, and friction produces heat energy.

Energy can be, and often is, converted or transformed from one kind to another, but the total energy is always conserved. The principle that energy cannot be created or destroyed but is always conserved is known as the **first law of thermodynamics**. (Recall the discussion of matter and energy in Chapter 5 and the detailed discussion of energy flow in the biosphere in Chapter 9.) Thermodynamics is the science that keeps track of energy as it undergoes various transformations from one type to another. We use the first law to keep track of the quantity of energy.[2]

To illustrate the conservation and conversion of energy, think about a tire swing over a creek (Figure 16.2). When the tire swing is held in its highest position, it is not moving. It does, however, contain stored energy owing to its position. We refer to the stored energy as *potential energy*. Other examples of potential energy are the gravitational energy in water behind a dam; the chemical energy in coal, fuel oil, and gasoline, as well as in the fat in your body; and nuclear energy, which is related to the forces binding the nuclei of atoms.[3]

The tire swing, when released from its highest position, moves downward. At the bottom (straight down), the speed of the tire swing is greatest, and no potential energy remains. At this point, all the swing's energy is the energy of motion, called *kinetic energy*. As the tire swings back and forth, the energy continuously changes between the two forms, potential and kinetic. However, with each swing, the tire slows down a little more and goes a little less high because of friction created by the movement of the tire and rope through air and friction at the pivot where the rope is tied to the tree. The friction slows the swing, generating *heat energy*, which is energy from random motion of atoms and molecules. Eventually, all the energy is converted to heat and emitted to the environment, and the swing stops.[3]

The example of the swing illustrates the tendency of energy to dissipate and end up as heat. Indeed, physicists have found that it is possible to change all the gravitational energy in a tire swing (a type of pendulum) to heat. However, it is impossible to change all the heat energy thus generated back into potential energy.

Energy is conserved in the tire swing. All the initial gravitational potential energy has been transformed by way of friction to heat energy when the tire swing finally stops. If the same amount of energy, in the form of heat, were returned to the tire swing, would you expect the swing to

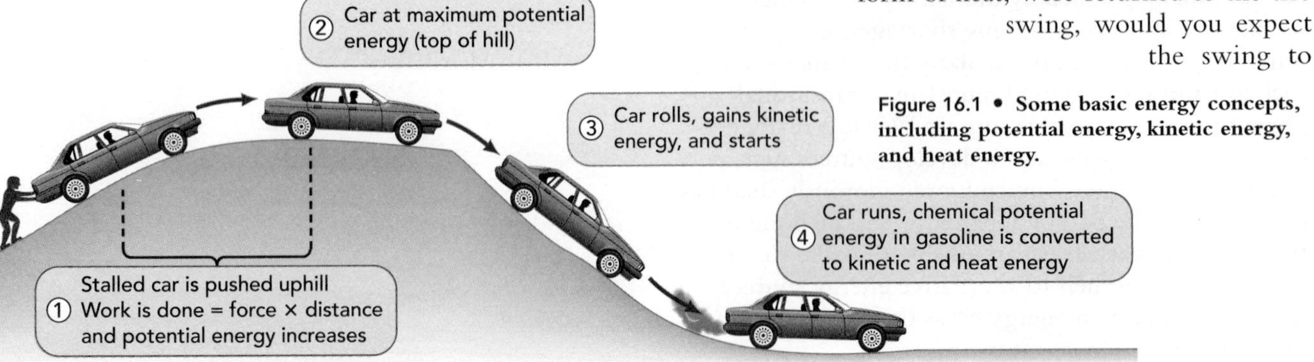

Figure 16.1 • **Some basic energy concepts, including potential energy, kinetic energy, and heat energy.**

start again? The answer is no. What, then, is used up? It is not energy, because energy is always conserved. What is used up is the energy *quality*—or the availability of the energy to perform work. The higher the quality of the energy, the more easily it can be converted to work; the lower the energy quality, the more difficult it is to convert to work.

This example illustrates another fundamental property of energy: Energy always tends to go from a more usable (higher-quality) form to a less usable (lower-quality) form. This is the **second law of thermodynamics**, and it means that, when you use energy, you lower its quality.

Let us return to the example of the stalled car, which you have now pushed to the side of the road. Pushing the car uphill a short distance increased its potential energy. This can be converted to kinetic energy by letting it roll back downhill. You restart the car. As the car idles, the potential chemical energy (from the gasoline) is converted to waste heat energy and other energy forms, including electricity to charge the battery and play the radio.

Why can we not collect the wasted heat and use it to run the engine? According to the second law of thermodynamics, once energy is degraded to low-quality heat, it can never regain its original availability or energy grade. When we refer to low-grade heat energy, we mean that relatively little of it is available to do useful work. High-grade energy, such as that of gasoline, coal, or natural gas, has high potential to do useful work. The biosphere continuously receives high-grade energy from the sun and radiates low-grade heat to the depths of space.[2,3] There is a way to capture waste heat, rather than simply release it to the environment. Section 16.5 describes the process of cogeneration, in which heat from combustion is used to create steam that is used to generate additional electricity. In a sense, we already use cogeneration in automobiles, in that we use waste heat from the engine to heat the passenger compartment. In 2004, Honda was awarded the German Prize for Natural Gas Industry Innovation, for its domestic cogeneration systems, which produce electricity and heated water for homes.

Figure 16.2 • **Diagram of a tire swing, illustrating the relation between potential and kinetic energy.**

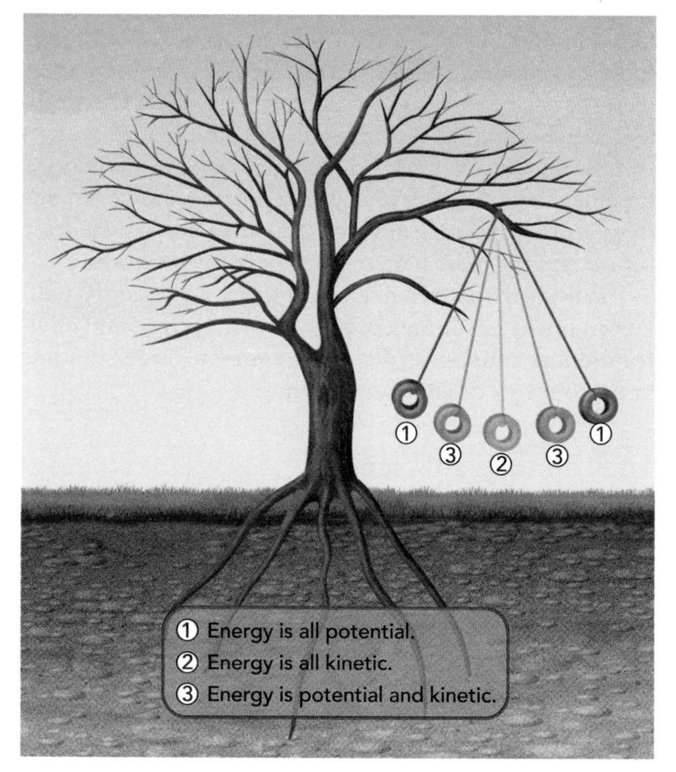

16.2 Past and Present Energy Crises

Energy crises are nothing new. People have encountered energy problems for thousands of years, as far back as the early Greek and Roman cultures.

Energy Crises in Ancient Greece and Rome

The climate in coastal areas of Greece 2,500 years ago was characterized by warm summers and cool winters, much as it is today. To warm their homes in winter, the Greeks used small, charcoal-burning heaters that were not very efficient. Since charcoal is produced from burning wood, wood was their primary source of energy, as it is today for half the world's people.

By the fifth century B.C., fuel shortages had become common, and much of the forest in many parts of Greece was depleted of firewood. As local supplies diminished, it became necessary to import wood from farther away. Olive groves became sources of fuel, which was made into charcoal for burning, reducing a valuable resource. By the fourth century B.C., the city of Athens had banned the use of olive wood for fuel.

At about this time, the Greeks began to build their houses facing the south, designing them so that the low winter sun entered the houses, providing heat, and the higher summer sun was partially blocked, cooling the houses. Recent excavations of ancient Greek cities suggest that large areas were planned so that individual homes could make maximum use of this solar energy. The Greeks' use of solar energy in heating homes was a logical answer to their energy problem.[4]

The use of wood in ancient Rome is somewhat analogous to the use of oil and gas in Canada today. Wealthy Roman citizens about 2,000 years ago had central heating in their large homes, burning as much as 125 kg of wood every hour. Not surprisingly, local wood supplies were quickly exhausted, and the Romans had to import wood from outlying areas. Eventually, wood had to be imported from as far away as 1,600 km.[4]

The Romans turned to solar energy for the same reasons as the Greeks but with much greater application and success. The Romans used glass windows to increase the effectiveness of solar heating, developed greenhouses to raise food during the winter, and oriented large public bathhouses (some of which accommodated up to 2,000 people) to use passive solar energy (Figure 16.3). The Romans believed that sunlight in bathhouses was healthy, and it also saved greatly on fuel costs. The use of solar energy in ancient Rome was widespread and resulted in laws to protect a person's right to solar energy. In some areas, it was illegal for one person to construct a building that shaded another's.[4]

The ancient Greeks and Romans experienced an energy crisis in their urban environments. In turning to solar energy, they moved toward what today we call sustainability. We are on that same path as fossil fuels become scarce.

Figure 16.3 • Roman bathhouse (lower level) in the town of Bath, England. The orientation of the bathhouse and the placement of windows are designed to maximize the benefits of passive solar energy.

Energy Today and Tomorrow

The energy situation facing the world today is in some ways similar to that faced by the early Greeks and Romans. The use of wood in Canada and the United States peaked in the 1880s, when the use of coal became widespread. The use of coal began to decline after 1920, with the availability of oil and gas. Today, we are facing the global peak of oil production, which is expected by about 2020. At the same time, rising oil prices have prompted a return to coal as a fuel source in thermal power generation. Since 1990, Canada's coal consumption has increased by 16% (40% in the Atlantic provinces), nearly all of it used for electricity generation.[5] Fossil fuel resources, which took millions of years to form, may be essentially exhausted in just a few hundred years.

The decisions we make today will affect energy use for generations. Should we choose complex, centralized energy production methods, or use simpler and widely dispersed energy production methods, or use a combination of the two? Which sources of energy should be emphasized? Which uses of energy should be emphasized for increased efficiency? How can we rely on current energy sources and provide for developing a sustainable energy policy? There are no easy answers to these questions.

Use of fossil fuels, especially oil, has resulted in improvements in sanitation, medicine, and agriculture. These improvements made possible the global human population increase that was discussed in other chapters. However, burning fossil fuels imposes growing environmental costs that are causing concerns ranging from urban pollution to a change in global climate.

In the future, there will be uncertainty when it comes to energy availability and cost. The sources and patterns of energy utilization will undoubtedly change. Energy supply and cost are uncertain as a result of growing demand and insufficient supply. Supplies will continue to be regulated, and there exists a growing potential for the disruption of supplies. Oil embargoes could cause significant economic impact; and a war or revolution in a petroleum-producing country would cause exports of petroleum to be reduced significantly.

Energy availability also has important implications for peace and conflict. In the days following the August 2003 blackout discussed at the beginning of this chapter, the U.S. publicly blamed Canada for the failure, even though these accusations later proved unfounded. Although there were flaws in both countries' power systems, the problem actually started in Ohio and spread, through a series of failures, into Canada. The negative tone of political statements about the event was worrying, however, and suggests that more serious consequences could arise from future energy crises here or elsewhere in the world.

It is clear that we need to rethink our energy policies in terms of sources, supply, consumption, and environmental concerns. We can begin by understanding basic concepts of energy efficiency.

16.3 Energy Efficiency

Two fundamental types of energy efficiencies are derived from the first and second laws of thermodynamics: the first-law efficiency and the second-law efficiency. The **first-law efficiency** deals with the amount of energy without any consideration of the quality or availability of the energy. It is calculated as the ratio of the actual amount of energy delivered where it is needed to the amount of energy supplied to meet that need. Expressions for efficiencies are given as fractions; multiplying the fraction by 100 converts it to a percentage. For example, consider a furnace system that keeps a home at a desired temperature of 18°C when the outside temperature is 0°C. The furnace, which burns natural gas, delivers one unit of heat energy to the house for every 1.5 units of energy extracted from burning the fuel. That means it has a first-law efficiency of 1 divided by 1.5, or 67% (see Table 16.1 and A Closer Look 16.1 for other examples).[2]

First-law efficiencies are misleading, because a high value suggests (often incorrectly) that little can be done to save energy through additional improvements in efficiency. This problem is addressed by the use of the **second-law efficiency**. Second-law efficiency refers to how well matched the energy end use is with the quality of the energy source. For our home-heating example, the second-law efficiency would compare the minimum energy necessary to heat the home with the energy actually used by the gas furnace. If we calculated the second-law efficiency (which is beyond the scope of this discussion), the result might be 5%—much lower than the first-law efficiency of 67%.[4] (We will see why later.) Table 16.1 also lists some second-law efficiencies for common uses of energy.

Values of second-law efficiency are important because low values indicate where improvements in energy technology and planning may save significant amounts of high-quality energy. Second-law efficiency tells us whether the energy quality is appropriate to the task. For example, you could use a welder's acetylene blowtorch to light a candle, but a match is much more efficient (and safer as well).

We are now in a position to understand why the second-law efficiency is so low (5%) for the house-heating example discussed earlier. This low efficiency implies that the furnace is consuming too much high-quality energy in carrying out the task of heating the house. In other words, the task of heating the house requires heat at a relatively low temperature, near 18°C, not heat with temperatures in excess of 1,000°C, such as is generated inside the gas furnace. Lower-quality energy, such as solar energy, could do the task and yield a higher second-law efficiency, because there is a better match between the required energy quality and the house-heating end use. Through better energy planning, such as matching the quality of energy supplies to the end use, higher second-law efficiencies can be achieved, resulting in substantial savings of high-quality energy.

Examination of Table 16.1 indicates that electricity-generating plants have nearly the same first-law and second-law efficiencies. These generating plants are examples of *heat engines*. A heat engine produces work from heat. Most of the electricity generated in the world today comes from heat engines that use nuclear fuel, coal, gas, or other fuels. Our own bodies are examples of heat engines, operating with a capacity (power) of about 100 watts (W) and fuelled indirectly by solar energy. (See A Closer Look 16.1 for an explanation of watts and other units of energy.) The internal combustion engine (used in automobiles) and the steam engine are additional examples of heat engines. A great deal of the world's energy is used in heat engines, with profound environmental effects such as thermal pollution, urban smog, acid rain, and global warming.

The maximum possible efficiency of a heat engine, known as *thermal efficiency*, was discovered by the French engineer Sadi Carnot in 1824, before the first law of thermodynamics was formulated.[6] Modern heat engines have thermal efficiencies that range between 60% and 80% of their ideal Carnot efficiencies. Modern 1,000-megawatt (MW) electrical generating plants have thermal efficiencies ranging between 30% and 40%; that means at least 60% to 70% of the energy input to the plant is rejected as waste heat. For example, assume that the electric power output from a large generating plant is 1 unit of power (typically 1,000 MW). Producing that 1 unit of power requires 3 units of input (such as burning coal) at the power plant, and the entire process produces 2 units of waste heat, for a thermal efficiency of 33%. The significant number here is the waste heat, two units, which amounts to twice the actual electric power produced.

Electricity may be produced by large power plants that burn coal or natural gas, by plants that use nuclear fuel, or by smaller producers, such as geothermal, solar, or wind sources (see Chapter 17). Once produced, the electricity is fed into the grid, which is the network

Table 16.1 • Examples of First- and Second-Law Efficiencies

Energy (End Use)	*First-Law Efficiency (%)*	*Waste Heat (%)*	*Second-Law Efficiency (%)*
Incandescent lightbulb	5	95	
Fluorescent light	20	80	
Automobile**	20–25	75–80	10
Power plants (electric); fossil fuel and nuclear**	30–40	60–70	30
Burning fossil fuels (used directly for heat)	65	35	
Water heating*			2
Space heating and cooling*			6
All energy (Canada and U.S.)*	50	50	10–15

*High potential for energy savings
**Moderate potential for energy savings

A CLOSER LOOK 16.1

A Little Environmental History

When we buy electricity by the kilowatt-hour, what are we buying? We say we are buying energy, but what does that mean? Before we go deeper into the concepts of energy and its uses, we need to define some basic units.

The fundamental energy unit in the metric system is the joule; 1 joule is defined as a force of 1 newton* applied over a distance of one metre. To work with large quantities such as the amount of energy used in Canada in a given year, we use the unit exajoule, which is equivalent to 10^{18} (a billion billion) joules. To put these big numbers in perspective, the United States today consumes approximately 100 exajoules of energy per year, and Canada consumes about 13% of that amount. World consumption is about 425 exajoules annually.

In many instances, we are particularly interested in the rate of energy use, or power, which is energy divided by time. In the metric system, power may be expressed as joules per second, or watts, W (1 joule per second is equal to 1 watt). When larger power units are required, we can use multipliers, such as kilo (thousand), mega (million), and giga (billion). For example, the rate of production of electrical energy in a modern nuclear power plant is typically 1,000 megawatts (MW) or 1 gigawatt (GW).

Sometimes it is useful to use a hybrid energy unit, such as the watt-hour, Wh (remember energy is power multiplied by time). Electrical energy is usually expressed and sold in kilowatt-hours (kWh, or 1,000 Wh). This unit of energy is 1,000 W applied for 1 hour (3,600 seconds), the equivalent energy of 3,600,000 J (3.6 MJ).

The average estimated electrical energy in kilowatt-hours used by various household appliances over a period of a year is shown in Table 16.2. The total annual energy used is the power rating of the appliance multiplied by the time the appliance was actually used. The appliances that use most of the electrical energy are water heaters, refrigerators, clothes dryers, and washing machines. A list of common household appliances and the amounts of energy they consume is useful in identifying those appliances that might help save energy through conservation or improved efficiency.

***A newton (N) is the force necessary to produce an acceleration of 1 m per sec per sec (m/s²) to a mass of 1 kg.**

Table 16.2 • Average Estimated Electrical Energy Use per Year for Typical Household Appliances

Appliance	*Power (W)*	*Average Hours Used per Year*	*Approximate Energy Used (kWh/yr)*
Clock	2	8,760	17
Clothes dryer	4,600	228	1,049
Hair dryer	1,000	60	60
Lightbulb	100	1,080	108
Compact fluorescent[a]	18	1,080	19
Television	350	1,440	504
Water heater (150 L)	4,500	1,044	4,698
Energy-efficient model[a]	2,800	1,044	2,900
Toaster	1,150	48	552
Washing machine	700	144	1,008
Refrigerator	360	6,000	2,160
Energy-efficient model[a]	180	6,000	1,100

Source: Data from U.S. Department of Energy and D. G. Kaufman and C. M. Franz, *Biosphere 2000: Protecting Our Global Environment* (New York: HarperCollins, 1993).

[a] Newer, energy-efficient model.

of power lines, or the distribution system. Eventually, it reaches residential, commercial and industrial consumers, where it lights, heats, and drives motors and other machinery used by society. As electricity moves through the grid, there are losses. The wires that transport electricity (power lines) have a natural resistance to electrical flow, known as *electrical resistivity.* This resistance converts some of the electric energy in the transmission lines to heat energy, which is radiated into the environment surrounding the lines. Energy is also lost from power lines as electromagnetic radiation. (Chapter 15)

16.4 Energy Sources and Consumption

People living in industrialized countries make up a relatively small percentage of the world's population, but they consume a disproportionate share of the total energy produced in the world. For example, the United States, with only 5% of the world's population, uses approximately 25% of the total energy consumed in the world. This is a reflection of its huge ecological footprint—20 times higher than many developing countries, and four times the world average—as discussed in Chapter 1. There is a direct relationship between a country's standard of living (as measured by gross national product) and energy consumption per capita. After the peak in oil production, expected in 2020–2050, supplies of oil and gasoline will be reduced and will be more expensive.

Before then, use of these fuels may be curtailed to reduce potential global climate change. As a result, within the next 30 years, both developed and developing countries will need to find innovative ways to obtain energy.

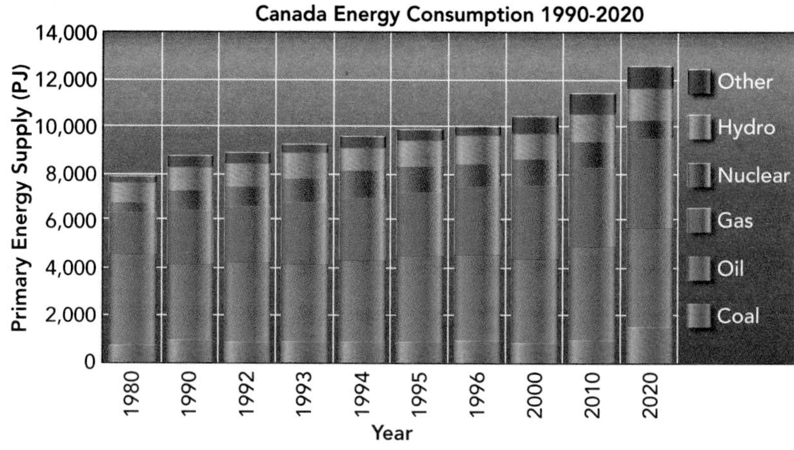

Figure 16.4 • **Total energy consumption for Canada, 1990–2020 (estimated).** [*Source:* Energy Council of Canada]

Fossil Fuels and Alternative Energy Sources

Today, approximately 75% of the energy consumed in Canada is produced by petroleum, natural gas, and coal. Because of their organic origin, these are called fossil fuels. They are produced from plant and animal material and are forms of stored solar energy that are part of our geologic resource base. They are essentially non-renewable. Other sources of energy—which include geothermal, nuclear, hydropower, and solar power, among others—are referred to as alternative energy sources. The term *alternative* designates these as sources that might replace fossil fuels in the future. Some of the alternative sources, such as solar and wind energy, are not depleted by consumption and are known as *renewable energy.*

The shift to alternative energy sources may be gradual, as fossil fuels continue to be utilized, or it could be accelerated as a result of rising fuel costs, diminishing fossil fuel resources, or concern over the potential environmental effects of burning fossil fuels. Regardless of which path we take, one thing is certain: fossil fuel resources are finite. It has taken millions of years for them to form, yet fossil fuels will be consumed in only a few hundred years of human history. Using even the most optimistic predictions, the fossil fuel epoch that started with the Industrial Revolution will represent only about 500 years of human history. Therefore, although fossil fuels have been extremely significant in the development of modern civilization, their use will be a short-lived event in the span of human history.

Energy Consumption Today

Actual and predicted energy consumption in Canada from 1990 to 2020 is shown in Figure 16.4. The figure illustrates our present dependence on the three major fossil fuels (coal, natural gas, and petroleum). From approximately 1950 through the mid-1970s, energy consumption increased tremendously, from approximately 30 exajoules to 70 exajoules (energy units are defined in A Closer Look 16.1). Consumption rates are slowing, however. Since about 1980, energy consumption has increased by only about 20 exajoules. This situation is encouraging because it suggests that policies to improve energy conservation through efficiency improvements (such as requiring new automobiles to be more fuel efficient and buildings to be better insulated) have been at least partially successful.

What is not shown in the figure, however, is the tremendous energy loss. For example, only about half the energy consumption in Canada and the United States was effectively used. The largest energy losses are associated with the production of electricity and with transportation.

We have already seen that energy is lost both in the generation of electricity at power plants and in the transmission of the electricity through the grid system. In the transportation sector, major energy losses occur in burning fossil fuels in car, bus, and truck engines, which produce waste heat that is lost to the environment. (In cities, this vehicular heat loss contributes to the formation of urban "heat islands," as discussed in Chapter 26.)

Another way to examine energy use is to look at the generalized energy flow of Canada for a particular year and by end use (Figure 16.5). The data in Table 16.3 show that although Canada is generally a net exporter of energy, it is a major consumer of oil. By contrast, the U.S. imports considerably more oil than it produces, with that consumption of energy fairly evenly distributed across residential/commercial, industrial, and transportation uses. It is clear that Canada and the United States remain vulnerable to changing world conditions affecting the production and delivery of crude oil. Evaluation of the full spectrum of potential energy sources is necessary to ensure the availability of sufficient energy in the future while sustaining environmental quality. Energy consumption patterns have also come under scrutiny.

16.5 Energy Conservation, Increased Efficiency, and Cogeneration

There is a movement to change patterns of energy consumption through measures such as conservation, increased energy efficiency, and cogeneration. Energy **conservation** refers simply to getting by with less demand for energy. In a pragmatic sense, this has to do with adjusting our energy needs and uses to minimize the amount of high-quality

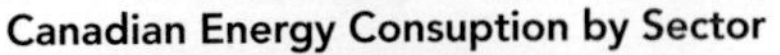

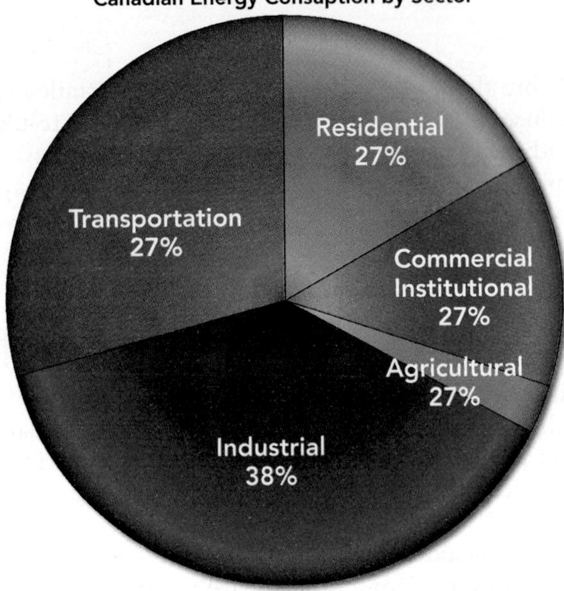

Figure 16.5 • Energy consumption in Canada by sector (approximate). [*Source*: Natural Resources Canada. *Improving Energy Performance in Canada: Report to Parliament Under the Energy Efficiency Act. 2002–2003 Report*. Ottawa: Natural Resources Canada, 2004.]

energy necessary to accomplish a given task. Increased **energy efficiency** involves designing equipment to yield more energy output from a given amount of input energy (first-law efficiency) or better matches between energy source and end use (second-law efficiency). **Cogeneration** includes a number of processes designed to capture and use waste heat rather than simply to release it into the atmosphere, water, or other parts of the environment as a thermal pollutant. An example of cogeneration is the *natural gas combined cycle power plant* that produces electricity in two ways: gas cycle and steam cycle. In the gas cycle, the natural gas fuel is burned in a gas turbine to produce electricity. In the steam cycle, hot exhaust from the gas turbine is used to create steam that is fed into a steam generator to produce additional electricity. The combined cycles capture waste heat from the gas cycle, nearly doubling the efficiency of the power plant from about 30% to 50–60%. Energy conservation is particularly attractive because it provides more than a one-to-one savings. Remember that it takes 3 units of fuel such as coal to produce 1 unit of power such as electricity (two-thirds is waste heat). Therefore, not using (conserving) 1 unit of power saves 3 units of fuel!

These three concepts—energy conservation, energy efficiency, and cogeneration—are all interlinked. For example, when electricity is produced at large, coal-burning power stations, large amounts of heat may be emitted into the atmosphere. Cogeneration, by using that waste heat, can increase the overall efficiency of a typical power plant from 33% to as much as 75%, effectively reducing losses from 67% to 25%. Cogeneration also involves generating electricity as a by-product from industrial processes that produce steam as part of their regular operations. Optimistic energy forecasters estimate that eventually we may meet approximately one-half the electrical power needs of industry through cogeneration. Cogeneration currently contributes about 6% to Canada's energy needs, but it has been estimated that up to 30% of the country's power capacity could be provided from that source.[7]

The average first-law efficiency of only 50% (Table 16.1) illustrates that large amounts of energy are currently lost in producing electricity and in transporting people and goods. Innovations in how we produce energy for a particular use can help prevent this loss, raising second-law efficiencies. Of particular importance will be energy uses with applications below 100°C, because a large portion of total energy consumption (for uses below 300°C) in Canada and the U.S. is for space heating and water heating. Figure 16.6 shows how some new buildings are incorporating energy efficient designs to reduce this reliance on external energy sources.

In considering where to focus our efforts to develop more energy efficiency, we need to look at the total energy-use picture. In Canada and the United States, space heating and cooling homes and offices, water heating, industrial processes (to produce steam), and automobiles account for nearly 60% of the total energy use. By comparison, transportation by train, bus, and airplane account for only about 5%. Therefore, the areas we should target for development of more energy efficiency are building design, industrial energy use, and automobile design.[7] We should note, however, that debate continues as to how much efficiency improvements and conservation can reduce future energy demands and the need for increased production from traditional sources, such as fossil fuel.

Table 16.3 • Generalized (approximate) Annual Energy Flow for Canada in 2002

Energy Source	*Energy Production*[a]	+	*Net Imports (Imports – Exports)*	=	*Energy Consumed*
Coal	1.43		−0.11		1.32
Natural gas	7.26		−3.96		3.30
Oil	5.2		−1.27		3.93
Hydropower and nuclear	1.51		−0.07		1.43
Other	0.63		−0.29		0.33
Total	16.02		−5.71	=	10.32

Source: Statistics Canada
[a] Exajoules (10^{18}J).

Building Design

A spectrum of possibilities exists for increasing energy efficiency and conservation in residential buildings. For new homes, the answer is to design and construct homes that minimize the energy consumption necessary to ensure comfortable living.[8] For example, building design can take advantage of passive solar potential, as did the early Greeks and Romans and the Native American cliff dwellers. Passive solar energy systems collect solar heat without using moving parts. Windows and overhanging structures can be positioned so that the overhangs shade the windows from incoming summer solar energy, thereby keeping the house cool while allowing winter sun to penetrate the windows and warm the house (Figure 16.6).

The potential for energy savings through architectural design for older buildings is extremely limited. The position of the building on the site is already established, and reconstruction and modifications are often not cost-effective. The best approach to energy conservation for these buildings is insulation, caulking, weather-stripping, installing window coverings and storm windows, and regular maintenance.

Buildings constructed to conserve energy are more likely to develop indoor air pollution problems, as pollutants emitted in buildings become concentrated due to reduced ventilation. Indoor air pollution is emerging as one of our most serious environmental problems. Potential problems can be reduced by better designs for air circulation systems that purify indoor air and bring in fresh, clean air (see Chapter 21). Construction that incorporates environmental principles is more expensive owing to higher fees for architects and engineers as well as higher initial construction costs. Nevertheless, moving toward improved design of homes and residential buildings to conserve energy remains an important endeavour.

Figure 16.6 • **Interior of Mountain Equipment Co-op retail store, Montréal.** Note the use of heat-retaining concrete and passive solar heating to reduce the use of energy for space heating.

Industrial Energy

According to the Organisation for Economic Cooperation and Development, Canada ranks an embarrassing 27th out of 29 OECD nations in terms of energy use per capita, and 26th out of 29 for total energy use. This again relates to our huge ecological footprint—the area of land necessary to produce the resources used, and assimilate the wastes produced, by a given population. Only the United States, Germany, and Japan rank higher in energy use per capita. Worse, OECD reports that total Canadian energy use grew 20.3% between 1980 and 1997, faster than the OECD average of 18%.[5] In other words, our ecological footprint is expanding, when it should be shrinking. The news is not all bad, however. Natural Resources Canada reports that many Canadian industrial sectors have greatly reduced energy consumption over the past decade. Most sectors report reductions of at least 10%, and some, including electric and electronic manufacturing, as high as 67%, despite significant growth in production.[7]

Today, Canadian industry consumes about 39% of the energy produced. The reason we have had higher productivity with less energy use is that more industries are using cogeneration and more energy-efficient machinery, such as motors and pumps designed to use less energy.[5, 9]

Automobile Design

Steady improvements have been made in the development of fuel-efficient automobiles during the last 30 years. In the early 1970s, the average North American automobile burned approximately 17.9 litres of gas for every 100 kilometres travelled. By 1996, the consumption had dropped to an average of 8.1 L/100 km for highway driving and as low as 5.1 L/100km for some automobiles. Fuel consumption rates did not improve much from 1996 to 1999. Today, about one fourth of the vehicles sold are SUVs and light trucks with fuel consumption of 12.5 to 25 L/100 km. A loophole in regulation, which classifies SUVs as light trucks, permits these vehicles to have poorer fuel consumption than conventional automobiles (see Chapter 17, Opening Case History). Today, however, the fuel consumption of some hybrid (gasoline-electric) vehicles is as low as

2.8 L/100 km on the highway and 4.2 L/100 km in the city (Figure 16.7). This improvement has several causes: increased efficiency and resulting conservation of fuel; smaller cars with engines constructed of lighter materials[5]; and the combination of a fuel-burning engine with an electric motor.[10] Of course, there is a price to pay for this change. Smaller cars may be more prone to damage on impact; and while cars have gotten smaller, trucks have tended to stay the same or increased in size. As a result, the number of serious accidents between cars and trucks has increased.

Figure 16.7 • **The Toyota Prius** is an example of new energy-efficient gas-electric hybrid automobiles. Canadian David Suzuki was one of the first people in North America to buy and promote the use of hydrid automobiles.

Values, Choices, and Energy Conservation

A potentially effective method of conserving energy is to change our behaviour by using less energy. This involves our values and the choices we make to act at a local level to address global environmental problems, such as human-induced warming caused by the burning of fossil fuels. For example, we make choices as to how far we commute to school or work and what method of transport we use to get there. Some people commute more than an hour by car to get to work, while others ride a bike, walk, or take a bus or train. Other ways of modifying behaviour to conserve energy include the following:

- Using carpools to travel to and from work or school
- Purchasing a hybrid car (gasoline-electric)
- Turning off lights when leaving rooms
- Taking shorter showers (conserves hot water)
- Putting on a sweater and turning down the thermostat during winter.

How Green is Your Campus? asks you to think about what your institution is doing to conserve energy and make sound energy choices.

16.6 Energy Policy

Canadian energy policy is driven by three main factors. The first is the unequal distribution of energy resources and users across the country. Whereas most consumers live in Ontario and Québec, most resources are located in Atlantic Canada (coal and offshore oil), Saskatchewan, Alberta, and British Columbia (coal, oil, and gas). A second major factor is the huge United States market for Canadian energy imports and exports, which drives the price of energy resources bought and sold in this country. The third and possibly most important factor affecting Canadian energy policy derives from Canada's structure as a confederation of provincial powers, not a strongly centralized federal state like the United States. Section 109 of Canada's *Constitution Act, 1982*, gives the provinces considerable power over their natural resources, and the right to impose taxes on those resources. The federal government is granted powers over matters crossing provincial borders, such as transportation, and the right to make laws for the "peace, order, and good government of Canada."

As we will see in later chapters of this book, this division of powers has often caused confusion in the management of Canada's environment. In the case of energy, nuclear power (Chapter 17) is considered to be a matter of national concern, and is solely regulated by the federal government. For all other energy sources, federal and provincial responsibilities overlap. As a result, Canadian energy policy is fragmented in a system of separate provincial laws overlapping with several federal statutes. Provincial governments control the generation and distribution of electricity (with the exception of nuclear activities controlled under federal law), and the limits that are placed on energy production and pricing.

Decisions about the management of a particular resource usually require negotiation between the federal and provincial government. Nowhere has this been more obvious than in Newfoundland Premier Danny Williams' demand that the Canadian federal government grant his province full royalty rights for offshore energy. After a long and acrimonious battle, in January 2005, the federal government agreed to give Newfoundland what it wanted, shielding the provincial revenues (expected to exceed $2 billion by 2012) from future clawbacks. Nova Scotia was granted a similar arrangement.

In 1997, Canada spent $2.4 billion on research and development (R&D); about $207 million of that went to energy research. Today, federal funding for energy R&D remains stable at about 7% of the national total. Much of this funding went to a variety of energy options, including traditional and emerging fossil fuel approaches, nuclear energy, and solar, biomass, and geothermal sources.

How Green is Your Campus?

Energy Efficiency

Universities and colleges across Canada are, like everyone else, facing the challenges of rising energy costs and ageing infrastructure. Many, like McGill University, have instituted energy conservation and awareness programs. Some, like the University of Waterloo, through its WATgreen Greening the Campus initiative, have formal sustainability programs. Many are modelled after the South Carolina Sustainable Universities Initiative, which began in 1998 as a collaboration among three South Carolina universities and now encompasses dozens of institutions in the U.S. and Canada.

Look around as you travel across campus, and see if you can identify any of the following initiatives:

- Special low-rate or preferred-location parking for carpool drivers.
- Signs reminding users to turn off light switches, or automatic switches that switch off lights in empty rooms.
- Room temperatures no higher than 20°C.
- A "cool pool"—swimming pools heated to no more than 27°C.
- Other evidence of energy conservation actions and awareness.
- A campus-wide master plan for energy efficiency (*hint*: search your institution's website for this kind of information).

An energy-efficient campus requires commitment of time and resources from the institution's senior administration. Is this commitment apparent from your institution's website, on-campus signs, promotional materials, or other communications? What factors might have helped or hindered the introduction of energy efficiency initiatives on your campus?

Climate change initiatives have provided an important incentive for Canadian energy R&D, and have forced a re-examination of energy use and technologies in all sectors of society (Figure 16.8).

Hard Path versus Soft Path

Energy policy today is at a crossroads. One road leads to what is termed the **hard path**, which involves finding greater amounts of fossil fuels and building larger power plants. Taking the hard path means continuing the past emphasis on the quantity of energy we use. In this respect, the hard path is more comfortable. It requires no new thinking; no realignment of political, economic, or social conditions; and little anticipation of coming reductions in production of oil.

Figure 16.8 • **Natural Resources Canada's CANMET laboratories in Ottawa** are conducting research on the kinetics of industrial combustion processes, to develop more fuel-efficient technologies.

People heavily invested in the continued use of fossil fuels and nuclear energy favour the hard path. They argue that much environmental degradation around the world has been caused by people who have been forced to utilize local resources, such as wood, for energy. As a result, their lands have suffered from loss of plant and animal life and from soil erosion (as discussed in Chapters 10 and 11). Their argument is that the way to solve these environmental problems is to provide cheap, high-quality energy, such as fossil fuels or nuclear energy.

In countries such as Canada and the United States, with sizable energy resources of coal, natural gas, and petroleum, people supporting the hard path argue that we should exploit those resources while finding ways to reduce the environmental impact of their use. According to these hard-path proponents, we should let the energy industry develop the available energy resources, and let industry, free from government regulations, provide a steady supply of energy with less total environmental damage.

The second road of energy policy is called the **soft path**.[11] Amory Lovins, the scientist who has defined and championed this soft path, states that it involves energy alternatives that emphasize energy quality, are renewable, are flexible, and are environmentally more benign than those of the hard path. As defined by Lovins, these alternatives have several characteristics:

- They rely heavily on renewable energy resources, such as sunlight, wind, and biomass (wood and other plant material).
- They are diverse and are tailored for maximum effectiveness under specific circumstances.
- They are flexible, accessible, and understandable to many people.

- They are matched in energy quality, geographic distribution, and scale to end-use needs, increasing second-law efficiency.

Lovins points out that people are not particularly interested in having a certain amount of oil, gas, or electricity delivered to their homes; rather, they are interested in having comfortable homes, adequate lighting, food on the table, and energy for transportation.[11] According to Lovins, only about 5% of end uses require high-grade energy such as electricity. Nevertheless, a lot of electricity is used to heat homes and water. Lovins shows that there is an imbalance in using nuclear reactions at extremely high temperatures and in burning fossil fuels at high temperatures simply to meet needs where the necessary temperature increase may be only a few 10s of degrees. Such large discrepancies are thought wasteful and a misallocation of high-quality energy.

Energy for Tomorrow

The availability of energy supplies and the future demand for energy are difficult to predict because the technical, economic, political, and social assumptions underlying predictions are constantly changing. In addition, seasonal and regional variations in energy consumption must also be considered. For example, in areas with cold winters and hot, humid summers, energy consumption peaks during the winter months, with a secondary peak in the summer (the former resulting from heating and the latter from air conditioning). Regional variations in energy sources and consumption are significant. For example, the transportation sector uses about one-fourth of the energy consumed. However, in western regions, where people often commute long distances to work, about one-half of the energy is used for transportation, more than double the national average. Energy sources also vary by region. For example, in Ontario and Québec, most energy is generated in hydroelectric systems, but in the prairie provinces, power plants are more likely to burn oil or natural gas to produce electricity.[12]

Future changes in population densities as well as intensive conservation measures will probably change existing patterns of energy use. This might involve a shift to more reliance on alternative (particularly renewable) energy sources.

To bring about an environmental scenario that would also stabilize the climate in terms of global warming, use of energy from fossil fuels would need to be cut by about 50%. One reason such a large reduction in the use of fossil fuels may not come quickly is that some policy-makers are still not convinced that significant global warming will actually occur as a result of burning fossil fuels. (Global climate change is discussed at length in Chapter 20.)

Over the past 30 years, energy scenarios have consistently overestimated the energy demands of the future. Low-energy scenarios for the future generally assume a moderate decrease in energy consumption accompanied by a shift from dependence on fossil fuels to the softer technologies (renewable, alternative energy resources). Reductions in energy use need not be associated with a lower quality of life. As noted, most OECD countries, including many nations with standards of living as high as or higher than Canada's, consume significantly less energy per person. What is needed is increased energy conservation and more efficient use of energy, including the following:[7]

- More energy-efficient land-use planning that maximizes the accessibility of services and minimizes the need for transportation.
- Agricultural practices and personal choices that emphasize eating more locally grown foods, to reduce energy use for transporting crops, and eating more foods such as vegetables, beans, and grains. These foods require less total energy to produce than beef, chicken, and pork when crops are grown to feed these animals.
- Industrial guidelines for factories that promote energy conservation and minimize production of waste.

All projections of specific sources of energy use in the future must be considered speculative. Perhaps most speculative of all is the idea that we really can obtain most of our energy needs from alternative renewable energy sources in the next several decades. From an

Figure 16.9 • Two possible future energy scenarios for Canada. [*Source*: National Energy Board. *Canada's Energy Future: Scenarios for Supply and Demand to 2025.* Ottawa: National Energy Board, 2003.].

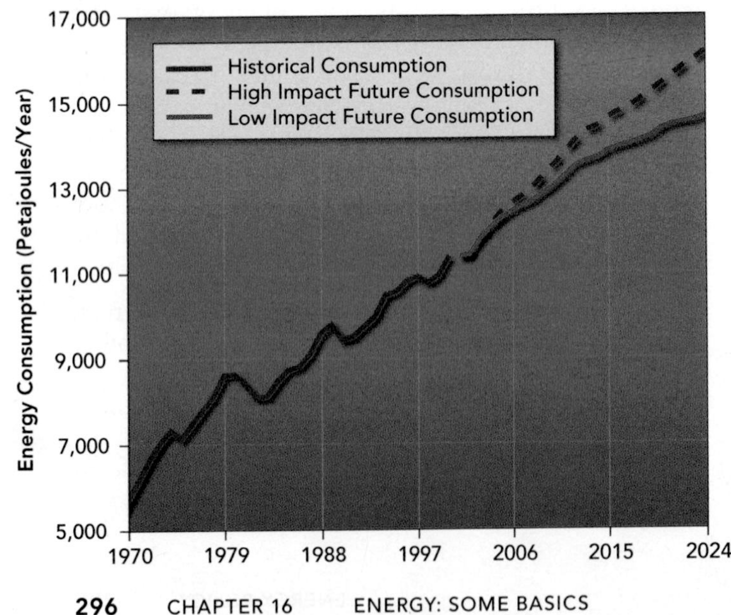

A CLOSER LOOK 16.2

Micropower

It is likely that sustainable energy management will include the emerging concept of micropower—smaller, distributed systems for production of electricity. Such systems are not new. The inventor Thomas Edison anticipated that electricity-generating systems would be dispersed; and by the late 1890s, many small electrical companies were marketing and building power plants, often located in the basements of businesses and factories. These early plants evidently used cogeneration principles, since waste heat was reused for heating buildings.[16] Imagine if we had followed this early model: homes would have their own power systems, electric power lines wouldn't snake through our neighbourhoods, and we could replace older, less efficient systems as we do refrigerators.

Instead, in the twentieth century Canadian and United States power plants grew larger. Industrializing countries, by the 1930s, had set up utility systems based on large-scale central power plants, as diagrammed in Figure 16.10*a*. Today, though, we are again evaluating the merits of distributive power systems, as shown in Figure 16.10*b*.

Large, centralized power systems are consistent with the hard path, while the distributive power system is more aligned with the soft path. Micropower devices rely heavily on renewable energy sources such as wind and sunlight, which feed into the electric grid system shown in Figure 16.10*b*. Examples of micropower include small residential wind turbines, private hydropower installations using small dams and turbines to generate electricity, and even solar-powered ovens. Use of micropower systems in the future is being encouraged because they are reliable and are associated with less environmental damage than are large fossil-fuel-burning power plants.[16]

Uses for micropower are emerging in both developed and developing countries. In countries that lack a centralized power-generating capacity, small-scale electrical power generation from solar and wind energy has become the most economical option. In nations with a high degree of industrialization, micropower may emerge as a potential replacement for aging electric power plants. For micropower to be a significant factor in energy production, a shift in policies and regulations to allow micropower devices to be more competitive with centralized generation of electrical power is required. Irrespective of the obstacles that micropower devices face, distributive power systems will probably play an important role in our goal of achieving integrated, sustainable energy management for the future.

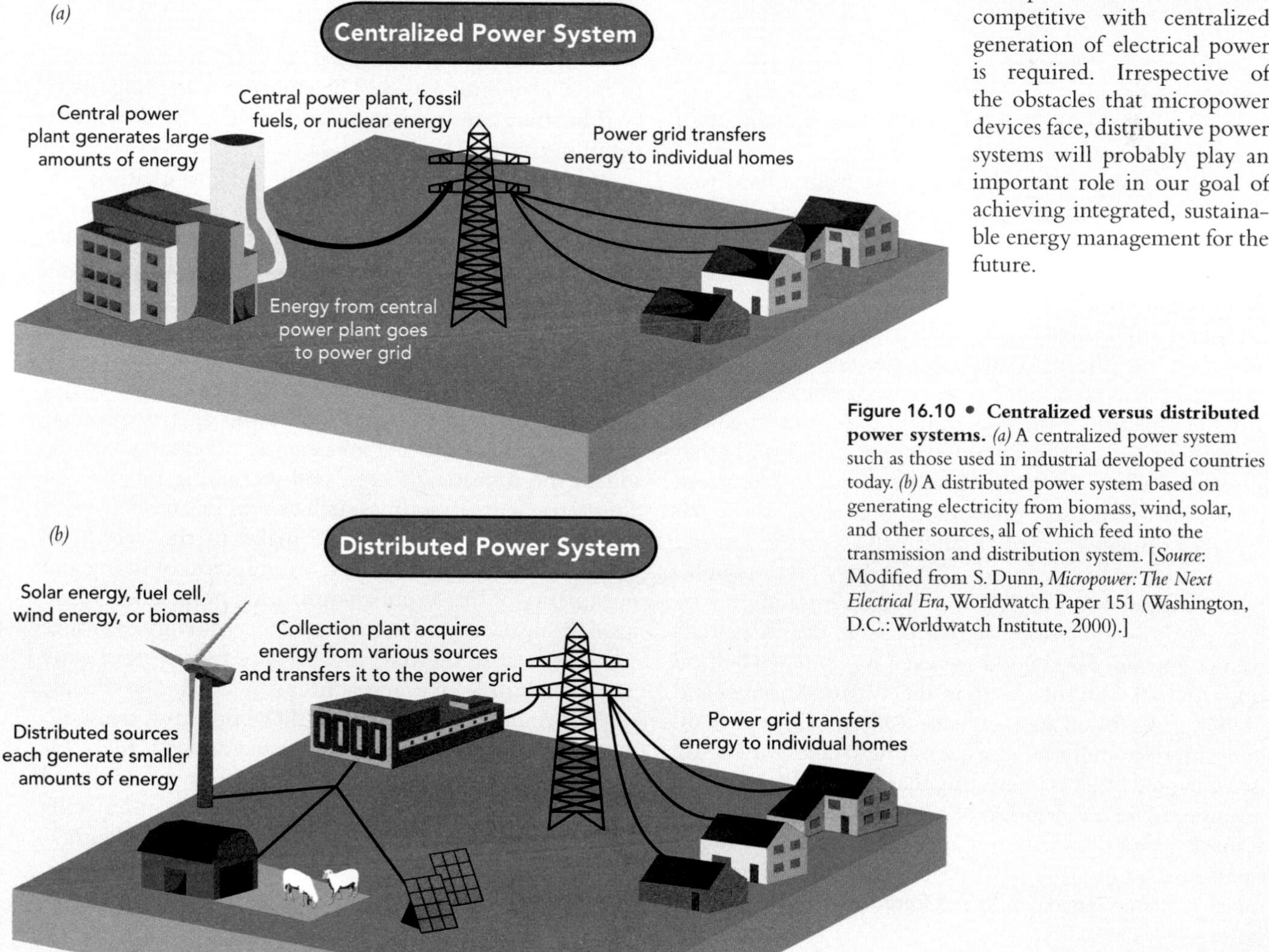

Figure 16.10 • **Centralized versus distributed power systems.** *(a)* A centralized power system such as those used in industrial developed countries today. *(b)* A distributed power system based on generating electricity from biomass, wind, solar, and other sources, all of which feed into the transmission and distribution system. [*Source:* Modified from S. Dunn, *Micropower: The Next Electrical Era*, Worldwatch Paper 151 (Washington, D.C.: Worldwatch Institute, 2000).]

CRITICAL THINKING ISSUE

Is There Enough Energy to Go Around?

The developing countries have most of the world's population (about 5 billion of 6 billion people) and are growing in population faster than developed countries. The average rate of energy use for individuals in developing countries is 1.0 kW, whereas that for persons in developed countries is 7.5 kW. If the current annual growth rate of 1.3% is maintained, the world's population will double in 54 years, to 12 billion people. More people will mean more energy use. In addition, people in developing countries will likely consume more energy per capita if they are to achieve a higher standard of living.

With a worldwide average energy-use rate of 2.6 kW per person, the 6 billion people on Earth use about 16 terawatt (TW) years annually (a terawatt is a trillion watts). A population of 12 billion with an average per capita energy use rate of 6.0 kW would use about five times as much as is presently used. Can the world support this much energy use?

The prominent energy policy analyst and educator John Holdren believes that a realistic goal is for annual per capita energy use to reach 3 kW, with the world population peaking at 10 billion individuals by the year 2100. If this goal is to be achieved, developing nations will be able to increase their populations by no more than 60% and their energy use by no more than 100%; developed nations can increase their population by only 10% and will have to reduce their energy use by 2% each year.

Critical Thinking Questions

1. What would the energy use rate be if Holdren's goals were realized? How much total energy would be required for all people on Earth to have a standard of living supported by 7.5 kW per person? How do these totals compare with the present energy-use rate worldwide?
2. In what specific ways could energy be used more efficiently in Canada? Compare your list with those of your classmates, and compile a class list.
3. In addition to increasing efficiency, what other changes in energy consumption might be required to provide an average energy-use rate of 7.5 kW per person or 3.0 kW per person in the future?
4. Would you consider Holdren's vision of the energy future an example of a hard or a soft path? Explain.
5. What are the implications for our personal and national ecological footprints, if Holdren's vision is realized? Discuss.

energy viewpoint, the next 20 to 30 years, as we move through the maximum production of petroleum, will be crucial to the industrialized world.

The transformation of energy use from a hard to a softer path has a long history, as seen in the example of the early Greek and Roman cultures. Canada and the United States experienced a shock in 1973 due to an oil shortage. Long lines at the gas pumps caused anxiety concerning our energy supply and the lifestyle that depends on abundant oil. The 1973 oil shock spawned new research and development of alternative energy sources. It was also the impetus for the government to provide financial incentives for the utilization of sunlight, wind, and other alternative energy sources. However, with the return of abundant, inexpensive oil in the 1980s, there was much less government support of alternative energy. During this period, China, with one-fifth of the world's population, continued to develop its emerging big industry by burning huge quantities of coal. Today, the industrialized countries of the world are even more dependent on imported oil than they were in the 1970s. Shortages and higher prices for oil are inevitable. One way to reduce oil consumption might be to establish new taxes on energy (see Chapter 17). It is interesting that rising oil prices have prompted a reconsideration of coal as an energy source in Canada. After decades of reliance on oil and natural gas, there is a return to coal-fired electricity generation in many parts of Canada and the United States, in response to these economic forces. The environmental impacts of coal burning are significant, and are discussed in more detail in chapters 17, 21, and 23.

Canada's economy is closely tied to that of the United States. Those ties make it difficult to change the current mix of energy sources or increase incentives for fuel efficiency, unless the U.S. takes similar measures. Energy choices also involve personal lifestyles, which are often slow to change. The National Energy Board of Canada predicts that Canada will continue to rely heavily on fossil fuels for at least the next 20 years and possibly into mid-century. The same agency predicts a growing focus on fuel and energy efficiency, including in the industrial sector, and increasing reliance on alternative energy sources such as wind power.[13]

The energy decisions we make in the very near future will greatly affect both our standard of living and our quality of life. From an optimistic point of view, we have the necessary information and technology to ensure a bright, warm, lighted, and moving future—but time may be running out, and action is needed now. People can continue to take things as they come and live with the results of our present dependence on fossil fuels. Or we can choose to build a sustainable energy future based on careful planning, innovative thinking, and a willingness to move from our dependency on petroleum.

Integrated, Sustainable Energy Management

The concept of **integrated energy management** recognizes that no single energy source can provide all the energy required by the various countries of the world.[14] A range of options that vary from region to region will have to be employed. Furthermore, the mix of technologies and sources of energy will involve both fossil fuels and alternative, renewable sources.

A basic goal of integrated energy management is to move toward **sustainable energy development** that is implemented at the local level. Sustainable energy development would have the following characteristics:

- It would provide reliable sources of energy.
- It would not cause destruction or serious harm to our global, regional, or local environments.
- It would help ensure that future generations inherit a high quality environment with a fair share of the Earth's resources.

To implement sustainable energy development, leaders in various regions of the world will need to implement energy plans based on local and regional conditions. The plans will integrate the sources of energy most appropriate for a particular region with potential for conservation and efficiency and with desired end uses for energy. Such plans will recognize that the preservation of resources can be profitable and that degradation of the environment and poor economic conditions go hand in hand. In other words, the degradation of air, water, and land resources results in a depletion of assets that ultimately will lower both the standard of living and the quality of life.

A good energy plan is part of an aggressive environmental policy with the goal of producing a quality environment for future generations. A good plan should do the following:

- Provide for sustainable energy development.
- Provide for aggressive energy efficiency and conservation.
- Provide for the diversity and integration of energy sources.
- Provide for a balance between economic health and environmental quality.
- Use second-law efficiencies as an energy policy tool (i.e., strive to produce a good balance between quality of energy source and end uses for that energy).

Such a plan recognizes that energy demands can be met in environmentally preferred ways. An important element of the plan involves the energy used for automobiles. This builds on policies of the past 30 years to develop hybrid engines that are part electric and part internal combustion and improve fuel technology to reduce both fuel consumption and emission of air pollutants. Finally, the plan should factor in the marketplace through pricing that reflects the economic cost of using the fuel as well as its cost to the environment. In summary, the plan should be an integrated energy management statement that moves toward sustainable development. Those who develop such plans recognize that a diversity of energy supplies will be necessary and that the key components are improvements in energy efficiency and conservation, and matching energy quality to end uses.[14]

The global pattern of ever-increasing energy consumption cannot be sustained without a new energy paradigm that includes changes in human values rather than a breakthrough in technology. Choosing to own lighter, more fuel-efficient automobiles and living in more energy-efficient homes is consistent with a sustainable energy system that focuses on how to provide and use energy to improve human welfare. A sustainable energy paradigm establishes and maintains multiple linkages among energy production, energy consumption, human well-being, and environmental quality.[15] It might also involve using more distributed production of energy (see A Closer Look 16.2).

Summary

- The first law of thermodynamics states that energy is neither created nor destroyed but is always conserved and is transformed from one kind to another. The first law is used to keep track of the quantity of energy.
- The second law of thermodynamics tells us that as energy is used, it always goes from a more usable (higher-quality) form to a less usable (lower-quality) form.
- Two fundamental types of energy efficiency are derived from the first and second laws of thermodynamics. In Canada and the United States today, first-law efficiencies average about 50%, which means that about 50% of the energy produced is returned to the environment as waste heat. Second-law efficiencies average 10–15%, so there is a high potential for saving energy through better matching of the quality of energy sources with their end uses.
- Energy conservation and improvements in energy efficiency can have significant effects on energy consumption. It takes three units of a fuel such as oil to produce one unit of electricity. As a result, each electricity unit conserved or saved through improved efficiency saves three units of fuel.
- There are arguments for both the hard and soft energy paths. The former has a long history of success and has produced the highest standard of living ever experienced. However, present sources of energy are causing serious environmental degradation and are not sustainable. Soft-path alternative energy sources are renewable, decentralized, diverse, and flexible; provide a better match between energy quality and end use; and emphasize second-law efficiencies.
- Sustainable, integrated energy management is needed to make the transition from fossil fuels to other energy sources. The goal is to provide reliable sources of energy that do not cause serious harm to the environment and ensure that future generations inherit a quality environment.

STUDY QUESTIONS

1. What evidence supports the notion that although present energy problems are not the first in human history, they are unique in other ways?
2. How do the terms energy, work, and power differ in meaning?
3. Compare and contrast potential advantages and disadvantages of a major shift from hard-path to soft-path energy development.
4. You have just purchased a 100-ha wooded island in the Strait of Georgia, British Columbia. Your house is uninsulated and built of raw timber. Although the island receives some wind, trees over 40 m tall block most of it. You have a diesel generator for electric power, and hot water is produced by an electric heater run by the generator. Oil and gas can be brought in by ship. What steps would you take in the next five years to reduce the cost of the energy you use with the least damage to the natural environment of the island?
5. How might better matching of end uses with potential sources yield improvements in energy efficiency?
6. What energy source is used to heat the building in which you live? Develop recommendations for home energy practices (for example, wall and roof insulation; door and window choices; heating and cooling temperatures) and energy sources that might lead to lower utility bills for your home.
7. How might plans using the concept of integrated energy management differ for the Victoria, B.C., area and the Montréal area? How might both of these plans differ from an energy plan for Mexico City, which is quickly becoming one of the largest urban areas in the world?
8. A recent energy analysis suggests that in the coming decades, energy sources might be natural gas (10%), solar power (30%), hydro-power (20%), wind power (20%), biomass (10%), and geothermal energy (10%). Do you think this is a likely scenario? What would be the major difficulties and points of resistance or controversy?

FURTHER READING

Berger, J. J. *Beating the Heat.* Berkeley, Calif.: Berkeley Hills Books, 2000. Excellent overview of all aspects of energy as they relate to global warming. Discusses how we can reduce or eliminate warming as an environmental threat.

Boyle, G., B. Everett, and J. Ramage. *Energy Systems and Sustainability.* Oxford, UK: Oxford University Press, 2003. An excellent summary of energy sources, uses, consumption, sustainability.

Fay, J. A. and D.S. Golomb. *Energy and the Environment.* New York: Oxford University Press, 2002. More quantitative treatment of basic energy principles.

Lovins, A. B. *Soft Energy Path: Towards a Durable Peace.* New York: Harper & Row, 1979. An interesting book that presents the argument for the soft path. Its message is more important today than when it was written.

United Nations Development Program. *Energy and the Challenge of Sustainability.* New York: United Nations Development Program, 2000. Good presentation of energy sources and sustainable development.

U.S.-Canada Power System Outage Task Force. *Final Report on the August 14, 2003 Blackout in the United States and Canada: Causes and Recommendations.* Washington: United States Department of Energy and Ottawa: Natural Resources Canada, 2004.

Wackernagel, Mathis and William Rees. *Our Ecological Footprint: Reducing Human Impact on the Earth.* Gabriola Island, B.C.: New Society Publishers, 1996.

chapter 17

Sport-Utility Vehicles (SUVs): A Fuelish Trend

Sport-utility vehicles have become vehicles of choice in Canada and the United States. They first appeared in the mid-1970s, and a million were produced in 1992; by 2002, more than four million SUVs had been sold, surpassing pickup trucks and vans and accounting for nearly one-half of the total automobiles sold. There is great variety among SUVs. They range from relatively small six-cylinder, two-wheel-drive vehicles that get about 10 km per litre, to super-SUVs that more closely resemble military vehicles, which they were modelled after. Environmentalists have labelled SUVs as harmful to the environment and SUV owners as being insensitive to environmental problems. In fact, SUVs have become a symbol and battleground over environmental issues. Why is there so much controversy about this type of vehicle?[1]

SUVs are marketed as providing their owners with a driving experience that allows them to enjoy nature by driving on dirt roads and off-road to view beautiful locations with their families. Actually, only a very small percentage of SUVs ever go off-road, and those that do may cause significant environmental damage to soils, plants, and animals. Most are used as family vehicles, with parents taking their children to school, doing their shopping, and taking trips. They have experienced tremendous popularity because they are large, are perceived as being safer (although some have a rollover problem, and their heavy weight can cause greater damage to vehicles they are involved in accidents with), and allow greater flexibility for number of passengers and baggage.

SUVs emit between 30% and 100% more carbon dioxide than standard automobiles. The automobile industry is capable of meeting tighter pollution restrictions. However, it is unlikely that automakers will do so voluntarily, and as a result, people are pushing for government to set the standards. Today, more than two-thirds of oil is used for vehicle gasoline. If we want the golden age of automobiles to last longer into the future without continuing to do serious harm to the environment, we must conserve gasoline and design engines to burn cleaner and get better fuel efficiency. Doing this will provide the necessary time to transition to a fuel economy not based on oil.[1, 2]

- *SUVs have been become a symbol for problems related to conservation of fossil fuels and reduction of the pollutants that enter our atmosphere. Fossil fuels are our primary energy source today, but the peak in oil production is likely to occur in the present century, and supplies will be much reduced in the future. In addition, the environmental impacts of fossil fuels have prompted a re-examination of other energy sources, and a reconsideration of our energy use overall.*

Energy Choices and the Environment

LEARNING OBJECTIVES

We rely almost completely on fossil fuels (oil, natural gas, and coal) for our energy needs. However, these are non-renewable resources, and their production and use have a variety of serious environmental impacts. After reading this chapter, you should understand:

- How oil, natural gas, and coal form.
- What the environmental effects are of producing and using oil, natural gas, and coal.
- What is nuclear fission and what are the basic components of a nuclear power plant.
- What is nuclear radiation and what are the three major types.
- How radioisotopes affect the environment and the major pathways of radioactive materials in the environment.
- What the future of nuclear power is likely to be.
- The advantages, disadvantages, and environmental impacts of developing different forms of alternative energy, including solar power, hydropower, wind power, biomass energy, and geothermal energy.
- What important policy issues will affect large-scale use of alternative energy sources.

Bumper-to-bumper traffic on a Canadian highway during rush hour. Large SUVs have been criticized for their high gas consumption and associated pollutant emissions.

Energy sources can be divided into two groups: **renewable energy** and **non-renewable energy**. Non-renewable alternative energy sources include fossil fuels, nuclear energy, and geothermal energy. Fossil fuels and nuclear energy are non-renewable because they require mineral fuels mined from the Earth. Geothermal energy is considered non-renewable for the most part because heat can be extracted from the Earth faster than it is naturally replenished from the Earth (output exceeds input; see Chapter 3). The renewable sources are solar energy, water (hydro) power, wind power, hydrogen produced from water, and energy derived from biomass (crops, wood, and so forth).

In this chapter, we review three categories of energy: fossil fuels, nuclear energy, and a range of alternative energy sources.

17.1 Fossil Fuels

Fossil fuels are forms of stored solar energy. Plants are solar energy collectors, because they can convert solar energy to chemical energy through photosynthesis (see Chapter 5). The main fossil fuels used today were created from incomplete biological decomposition of dead organic matter (mostly land and marine plants). This occurred when buried organic matter that was not completely oxidized was converted by chemical reactions over hundreds of millions of years to oil, natural gas, and coal. Biological and geologic processes in various parts of the geologic cycle produce the sedimentary rocks in which we find these fossil fuels.[3, 4]

The major fossil fuels—crude oil, natural gas, and coal—are our primary energy sources; on a worldwide basis, they provide approximately 90% of the energy consumed (Figure 17.1). In this chapter, we focus primarily on these major fossil fuels. We also briefly discuss two other fossil fuels, oil shale and tar sands, that may become increasingly important as oil, gas, and coal reserves are depleted.

In the following discussion, we will refer both to fossil fuel resources, those materials concentrated in a form that has the potential to be extracted, bought, and sold; and to fossil fuel reserves, the portion of a resource that is legally and economically viable to extract at the present time. The resource is the potential total, while the reserve is the total that is realistically available today.

Crude Oil and Natural Gas

Most geologists accept the hypothesis that **crude oil** (petroleum) and **natural gas** (a mixture of hydrocarbon and non-hydrocarbon gases dominated by methane) are derived from organic materials, mostly plants, that were buried with marine or lake sediments in what are known as depositional basins. Oil and gas are found primarily along geologically young tectonic belts at plate boundaries, where large depositional basins are more likely to occur (see Chapter 5). There are exceptions, such as in Texas, the Gulf of Mexico, and the North Sea, where oil in depositional basins far from active plate boundaries has been discovered.

The source material, or *source rock*, for oil and gas is fine-grained (less than 1/16 mm in diameter), organic-rich sediment buried to a depth of at least 500 m, where it is subjected to increased heat and pressure. The elevated temperature and pressure initiate the chemical transformation of the organic material in the sediment into oil and gas. The elevated pressure causes the sediment to be compressed; this, along with the elevated temperature in the source rock, initiates the upward migration of the oil and gas, which are relatively light, to a lower-pressure environment (known as the *reservoir rock*). The reservoir rock is coarser grained and relatively porous (has more and larger open spaces between the grains). Sandstone and porous limestone, which have a relatively high proportion (about 30%) of empty space in which to store oil and gas, are common reservoir rocks.

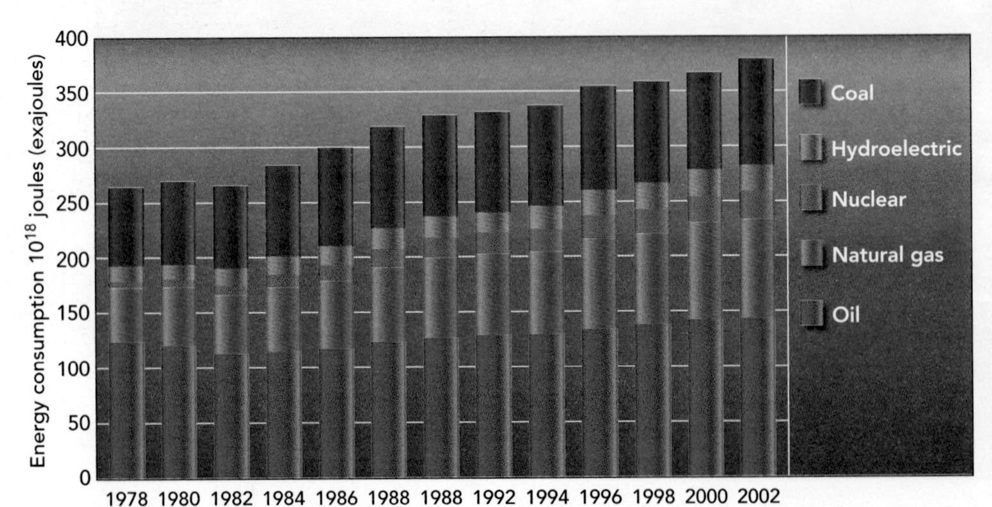

Figure 17.1 • World energy consumption by primary sources, 1978–2002. [*Source*: Modified after British Petroleum Company, BP Statistical Review of World Energy (London: British Petroleum Company, 2003).]

As mentioned, oil and gas are light; and if their upward mobility is not blocked, they will escape to the atmosphere. This explains why oil and gas are not generally found in geologically old rocks. Oil and gas in rocks older than about 0.5 billion years have had ample time to migrate to the surface, where they have either vaporized or eroded away.[4]

The oil and gas fields from which we extract resources are places where the natural upward migration of the oil and gas to the surface is interrupted or blocked by what is known as a *trap* (Figure 17.2). The rock that helps form the trap, known as the *cap rock*, is usually a very fine-grained sedimentary rock, such as shale, which is composed of silt and clay-sized particles. A favourable rock structure, such as an anticline (arch-shaped fold) or a fault (fracture in the rock along which displacement has occurred), is necessary to form traps, as shown in Figure 17.2. The important concept is that the combination of favourable rock structure and the presence of a cap rock allow deposits of oil and gas to accumulate in the geologic environment, where they are then discovered and extracted.[4]

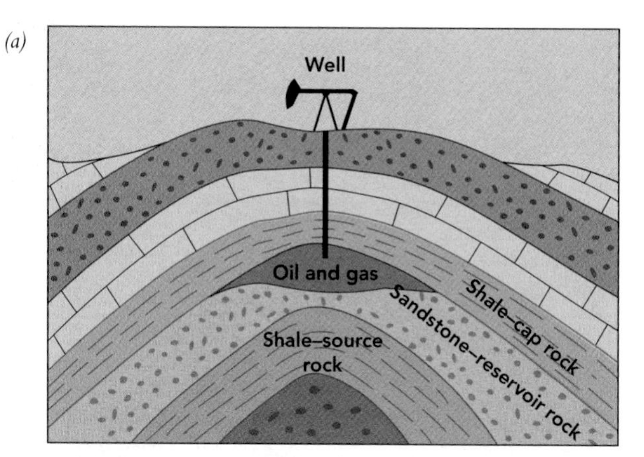

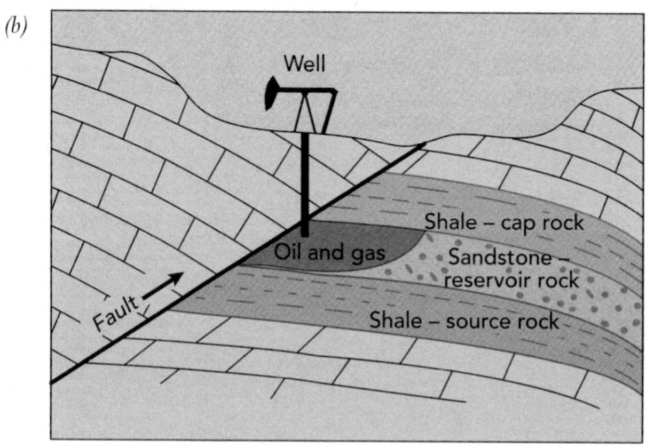

Figure 17.2 • Two types of oil and gas traps: *(a)* anticline and *(b)* fault.

Petroleum Production

Production wells in an oil field recover oil through both primary and enhanced methods. *Primary production* involves simply pumping the oil from wells, but this method can recover only about 25% of the petroleum in the reservoir. To increase the amount of oil recovered to about 60%, *enhanced* methods are used. In enhanced recovery, steam, water, or chemicals, such as carbon dioxide or nitrogen gas, are injected into the oil reservoir to push the oil toward the wells, where it can be more easily recovered by pumping.

Next to water, oil is the most abundant fluid in the upper part of the Earth's crust. Most of the known, proven oil reserves, however, are located in a few fields. Proven oil reserves are the part of the total resource that has been identified and can be extracted now at a profit. Of the total reserves, 65% is located in 1% of the fields, the largest of which are in the Middle East (Figure 17.3). Although new oil and gas fields have recently been and continue to be discovered in Alaska, Mexico, South America, and other areas of the world, the present known world reserves may be depleted in the next few decades.

The total resource always exceeds known reserves; it includes petroleum that cannot be extracted at a profit and petroleum that is suspected but not proved to be present. Several decades ago, the amount of oil that ultimately could be recovered was estimated to be about 1.6 trillion barrels. Today, that estimate is just over 3 trillion barrels.[5] The increases in proven oil reserves in the last few decades were due primarily to discoveries in the Middle East, Venezuela, Kazakhstan, and other areas.

Because so much of the world's oil is found in the Middle East, oil revenues have flowed into that area, resulting in huge trade imbalances. Figure 17.4 shows the major routes of trade for oil. Recent estimates of proven oil reserves suggest that, at present production rates, oil and natural gas will last only a few decades.[6, 7] The important question, however, is not how long oil is likely to last at present and future production rates but when we will we reach peak production. This is important because following peak production, less oil will be available. Forty percent of the world's population depends on oil as an energy source, and significant shortages will cause major problems, including price shocks.[9] World oil production is likely to peak between the years 2020 and 2050,[8] and estimates indicate that world oil production will be nearly exhausted by 2100.[9]

One way to make these issues more apparent to consumers is to raise the tax on gasoline. The number of motor vehicles in the world, most of which use petroleum as fuel, has increased about nine times since 1950. There are about 750 million motor vehicles in the world, over one-third of them in Canada and the United States. Each year, Canada uses over 2,306 petajoules (one petajoule equals 1×10^{15} joules) of energy

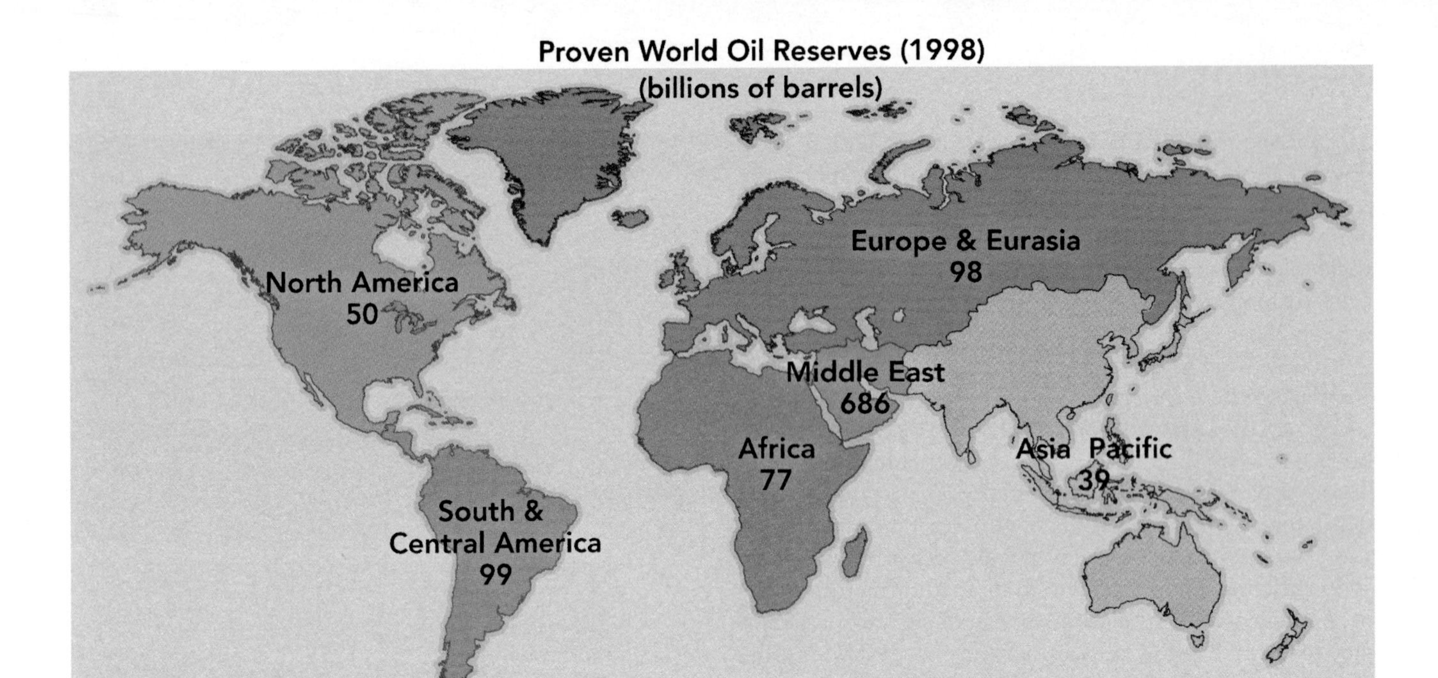

Figure 17.3 • Proven world oil reserves (billions of barrels) in 1998. The Middle East dominates with 65% of total reserves. [*Source*: Modified after British Petroleum Company, BP Statistical Review of World Energy (London: British Petroleum Company, 2003).]

Figure 17.4 • Major trade routes for the world's oil, emphasizing the countries that use Middle Eastern oil. Units are millions of metric tons. [*Source*: Modified after British Petroleum Company, BP Statistical Review of World Energy (London: British Petroleum Company, 2003).]

for transportation, roughly a third of all secondary energy use in Canada. In addition to the drain on a limited resource, the use of petroleum in transportation contributes significantly to air pollution (Chapter 21).

Some energy policymakers have suggested that increasing the tax on gasoline would encourage people to purchase more fuel-efficient vehicles and discourage unnecessary use of motor transportation, thereby lowering the demand for gasoline. In Canada, gasoline tax varies by province, but averages 40 to 50% of the price at the pump (Figure 17.5). By comparison, U.S. taxes are only

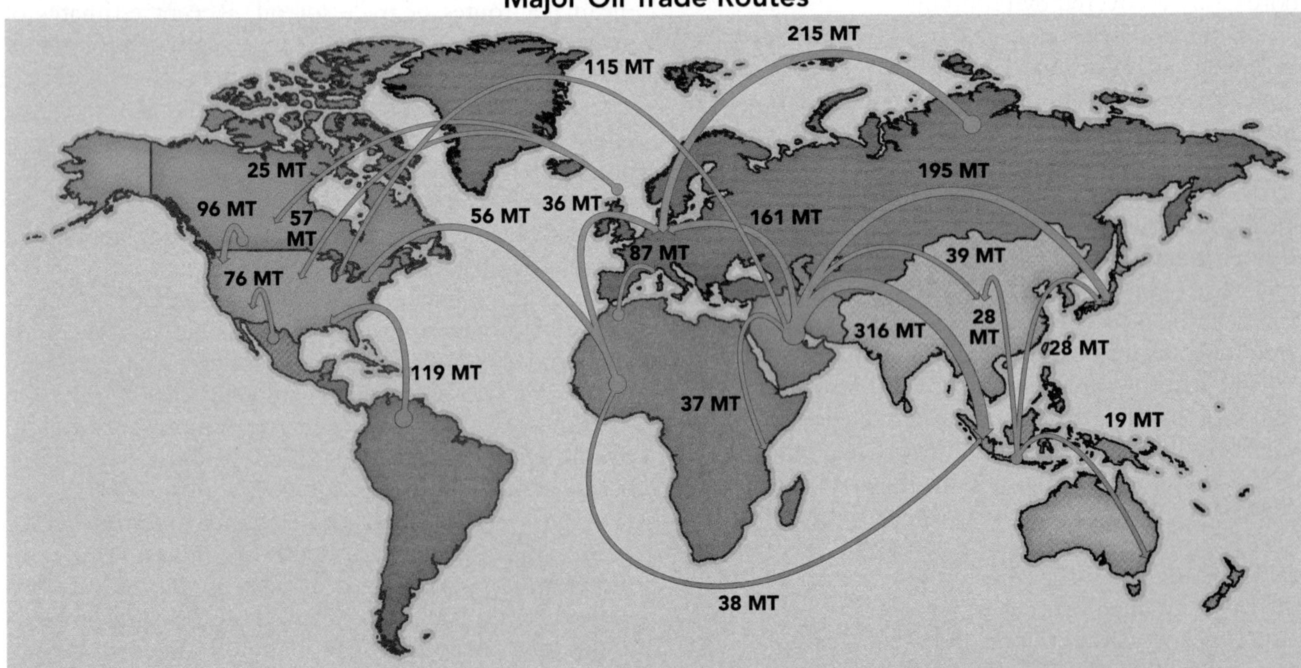

20 to 30% of the pump price—only 12 to 18% of those of other industrialized countries. Opponents, however, argue that higher taxes at the pump might have a greater impact on poorer families and would also unfairly punish those living in western regions, where driving distances are typically greater. As with many controversial issues, the facts themselves are in question. For example, do the poor spend proportionally more on transportation, and would they be disproportionately and adversely affected by a rise in gasoline taxes?

In the 2005 federal budget, the Canadian government proposed a fee on SUV purchases and other large vehicles, and a rebate system for people who buy smaller cars. This "feebate" is intended to encourage use of fuel-efficient vehicles, and will form part of Canada's response to Kyoto Protocol requirements to reduce greenhouse gases (Chapter 20). Ontario has had such a fee since 1989. In the Ontario system, vehicles with fuel consumption higher than 6 L/100 km pay tax on a sliding scale, with less efficient vehicles paying a higher fee. Vehicles with fuel consumption less than 6 L/100 km are eligible for rebates. Hybrid vehicles like the Toyota Prius are eligible for a rebate of all provincial sales tax, to a maximum of $1,000 per vehicle.

Natural Gas

We have only begun to seriously search for natural gas and to utilize this resource to its potential. One reason for this slow start is that natural gas is transported primarily by pipelines, and only in the last few decades have these been constructed in large numbers. In fact, until recently, natural gas found with petroleum was often simply burned off as waste; and in some cases, this practice continues.[10] The worldwide estimate of recoverable natural gas is about 155 trillion m^3, which at the current rate of world consumption will last approximately 70 years.[6] Large natural gas deposits were recently discovered off Nova Scotia and Newfoundland. Canadian gas is a major export to the United States, now supplying 13% of total U.S. gas demand (up to 50% in western regions). Canadian production of natural gas has been growing at the rate of about 3% a year, and is expected to continue at that rate for the foreseeable future.[3]

Natural gas that contains high concentrations of hydrogen sulphide (H_2S) is called sour gas. In Canada, about a quarter of natural gas production (and one-third of Alberta's production) is sour gas. Although sour gas extraction presents challenges because of the toxicity of hydrogen sulphide to humans and animals, its sulphur content can be recovered and sold for additional profit. Sulphur recovery from sour gas now plays an important part in the western Canadian economy.

Natural gas is a promising alternative to coal and oil. It is considered a clean fuel; burning it produces fewer pollutants than does burning other fossil fuels, so it causes fewer environmental problems. As a result, it is being considered as a possible transition fuel from oil and coal to alternative energy sources, such as solar power, wind power, and hydropower.

Figure 17.5 • **Gasoline taxes across Canada, 2004.** [*Source*: Petro-Canada.]

Base Price	Region	Federal Excise Tax	Provincial Tax			
40.0¢	Newfoundland	10.0¢	16.5¢	15%		
40.0¢	Prince Edward Island	10.0¢	17.0¢	7%		
40.0¢	Nova Scotia	10.0¢	15.5¢	15%		
40.0¢	New Brunswick	10.0¢	14.5¢	15%		
40.0¢	Quebec	10.0¢	15.2¢	7%	7%	
40.0¢	Ontario	10.0¢	14.7¢	7%		
40.0¢	Manitoba	10.0¢	11.5¢	7%		
40.0¢	Saskatchewan	10.0¢	15.0¢	7%		
40.0¢	Alberta	10.0¢	9.0¢	7%		
40.0¢	British Columbia	10.0¢	14.5¢	7%		
40.0¢	North West Territories	10.0¢	10.7¢	7%		
40.0¢	Nunavut	10.0¢	6.4¢	7%		
40.0¢	Yukon	10.0¢	6.2¢	7%		
40.0¢	Montreal	10.0¢	15.2¢	7%	7.5%	1.5¢
40.0¢	Vancouver	10.0¢	14.5¢	7%	6.0¢	

Base Price
Federal Excise Tax
Provincial Tax
GST/HST
Provincial Sales Tax
Transit Tax

In spite of the new discoveries and the construction of pipelines, long-term projections for a steady supply of natural gas are uncertain. The supply is finite, and at present rates of consumption, it is only a matter of time before the resources are depleted.

Coal-Bed Methane

The processes responsible for the formation of coal include the partial decomposition of plants buried by sediments that slowly convert the organic material to coal. This process also produces a lot of methane (natural gas) that is stored within the coal.[11] The methane is actually stored on the surfaces of the organic matter in the coal, and because there are many large internal surfaces in coal, the amount of methane for a given volume of rock is something like seven times more than could be stored in gas reservoirs associated with petroleum. The estimated amount of coal-bed methane in Canada is more than 5.4 trillion cubic metres, much of it thought to be recoverable with existing technology. These reserves extend right across the country, from Vancouver Island to Cape Breton. The British Columbia and Alberta reserves alone are thought to be sufficient to supply Canada's natural gas needs for decades to come.

Coal-bed methane is a promising energy source that comes at a time when we are importing vast amounts of energy and attempting to evaluate a transition from the fossil fuels to alternative fuels. However coal-bed methane (CBM) has been slow to gain acceptance in Canada. Although CBM now comprises about 8% of total gas production in the United States, Canada's current production is less than 0.2% of total domestic gas production. Several pilot projects are underway, for example at Corbett, 160 km northwest of Edmonton, Alberta, but in many regions, including British Columbia, there are no producing CBM wells, despite substantial known reserves.

The technology to recover coal-bed methane is a young one, but it is developing quickly. The big advantage of coal-bed methane wells is that they only need to be drilled to shallow depths (100 metres or so). Drilling can be done with conventional water-well technology, and the cost is about $100,000 per well compared to several million dollars for an oil well.There are several environmental concerns associated with coal-bed methane, however, including: (1) disposal of large volumes of water, which is produced when the methane is recovered; and (2) migration of methane, which may contaminate groundwater or migrate into residential areas.[12]

A major environmental benefit from burning coal-bed methane, as with other sources of methane, is that combustion produces much less carbon dioxide than does the burning of coal or petroleum. Furthermore, production of methane gas prior to mining coal reduces the amount of methane that would be released into the atmosphere. Both methane and carbon dioxide are greenhouse gases that contribute to global warming. However, because methane produces a lot less carbon dioxide, it is considered one of the main transitional fuels from fossil fuels to alternative energy sources.

There is also concern for the sustainability of water resources. Extraction of methane requires removal of vast amounts of water from groundwater aquifers. In some instances, springs have dried up following coal-bed methane extraction.[13] In other words, the "mining" of groundwater for coal-bed methane extraction will remove water that has perhaps taken hundreds of years to accumulate in the subsurface environment.

There is also concern for the migration of methane away from the well sites, possibly to nearby, more urban areas. The problem is that methane in its natural state is odourless, as compared to the smelly variety in homes, and it is explosive. Finally, coal-bed methane wells and related compressors and other needed equipment cause noise pollution. People living within a few hundred metres of coal-bed methane facilities have reported serious and distressing noise pollution.[13]

In summary, coal-bed methane is a tremendous source of energy. It is a relatively clean-burning fuel, but its extraction must be closely evaluated and studied to minimize environmental degradation.

Methane Hydrates

Beneath the sea floor, at depths of about 1,000 m, there exist deposits known as **methane hydrate**, a white, ice-like compound made up of molecules of methane gas (CH_4) molecular "cages" of frozen water. The methane has formed as a result of microbial digestion of organic matter in the sediments of the sea floor and has become trapped in these ice cages. Methane hydrates in the oceans were discovered over 30 years ago and are widespread in both the Pacific and Atlantic Oceans. Methane hydrates are also found on land; the first ones discovered were in permafrost areas of Siberia and North America, where they are known as marsh gas.[14]

Methane hydrates in the ocean are found in areas where deep, cold seawater provides high pressure and low temperatures. They are not stable at lower pressure and warmer temperatures. At a water depth of less than about 500 m, methane hydrates decompose rapidly and methane gas is freed from the ice cages to move up as a flow of methane bubbles (like rising helium balloons) to the surface and the atmosphere.

In 1998, researchers from Russia discovered the release of methane hydrates off the coast of Norway. During the release, scientists documented plumes of methane gas as tall as 500 m being emitted from methane

(a)

(b)

Figure 17.6 • **Drilling for oil** in *(a)* the Sahara Desert of Algeria and *(b)* under the ocean.

hydrate deposits on the sea floor. It appears that there have been large emissions of methane from the sea. The physical evidence includes fields of depressions, looking something like bomb craters that pockmark the bottom of the sea near methane hydrate deposits. Some of the craters are as large as approximately 30 m deep and 700 m in diameter, suggesting that they were produced by rapid if not explosive eruptions of methane. Canada is believed to have significant methane hydrate reserves in the Beaufort-Mackenzie region, although these, like others in the country, are not yet fully mapped.

Methane hydrates in the marine environment are a potential energy resource with approximately twice as much energy as all the known natural gas, oil, and coal deposits on Earth.[14] Methane hydrates look particularly attractive to countries such as Japan that rely exclusively on foreign oil and coal for fossil fuel needs. Unfortunately, mining methane hydrates will be a difficult task, at least for the near future. The hydrates are found along the lower parts of the continental slopes, where water depths are often greater than 1 km. The deposits themselves extend into the ocean floor sediments another few hundred metres. Most drilling rigs cannot operate safely at these depths, and development of a method to produce and transport the gas to land will be challenging.

Methane hydrates are famous for another reason: their possible link to the so-called Bermuda Triangle Mystery. The Bermuda Triangle is an area of ocean off the southeast coast of the United States, near the island of Bermuda, in which numerous small boats, ships, and even airplanes have disappeared over the years, seemingly without explanation. In the early 1980s, however, geochemist Richard McIver speculated that the Bermuda Triangle disappearances are in fact the result of ruptures in methane hydrate pockets, caused by undersea landslides. A rapid release of methane hydrate gas would shoot to the surface, making the water beneath it much less dense and quickly swallowing even the largest objects. Rock and sediment from the landslide would then quickly cover the craft, making it "disappear." The explanation seems reasonable: huge quantities of methane hydrate are known to exist in the area, and undersea landslides along the North American continental shelf are not uncommon.

Environmental Effects of Oil and Natural Gas

There is no escaping the fact that recovery, refining, and use of oil—and, to a lesser extent, natural gas—cause well-known, documented environmental problems, such as air and water pollution, acid rain, and global warming. Humans have gained many benefits from abundant, inexpensive energy, but at a price to the global environment and human health. The environmental effect of oil and natural gas include the following:

- Development of oil and gas fields involves drilling wells on land or beneath the sea floor (Figure 17.6). This involves use of land to construct roads, pads for wells, pipelines, and storage tanks; pollution of surface waters and groundwater from leaking equipment, pipes and tanks, and salty water brought to the surface with oil; accidental release of air pollutants, such as hydrocarbons and hydrogen sulphide (a toxic gas); land subsidence (sinking) as oil and gas are withdrawn, and loss or damage to fragile or pristine ecosystems (see the example of the Arctic National Wildlife Refuge in Alaska, discussed in Chapter 14).

- Oil production in the marine environment can result in oil seepage into the sea from normal operations or large spills from accidents, such as blowouts or pipe ruptures; release of drilling muds (heavy liquids injected into the bore hole during drilling to keep the hole open) containing heavy metals, such as barium, that may be toxic to marine life; and aesthetic degradation from the presence of offshore oil-drilling platforms, which some people think are unsightly.
- In the refining process, crude oil is heated so that its components can be separated and collected (this process is called fractional distillation). Other industrial processes are then used to make products such as gasoline and heating oil. These processes can leak, causing accidental release of gasoline and other products and substances from storage tanks and pipes, polluting soil and groundwater resources below the site. Massive groundwater cleaning projects have been required at several West Coast refineries.
- Delivery and use of oil and gas can cause extensive and significant environmental problems. Crude oil is mostly transported on land in pipelines or across the ocean by tankers, and both methods present the danger of oil spills, as occurred when a bullet from a high-powered rifle punctured the trans-Alaskan pipeline in 2001, causing a small but damaging oil spill. Additionally, strong earthquakes may pose a problem for pipelines in the future. However, proper engineering can minimize an earthquake hazard. Air pollution is perhaps the most familiar and serious environmental impact associated with the use (burning) of oil. Combustion of gasoline in automobiles produces pollutants that contribute to urban smog. The adverse effects of smog on vegetation and human health are well documented and are discussed in detail in Chapter 21.

Coal

Partially decomposed vegetation, when buried in a sedimentary environment, may be slowly transformed into the solid, brittle, carbonaceous rock we call **coal**. This process is shown in Figure 17.7. Coal is by far the world's most abundant fossil fuel, with total recoverable reserves of about 1,000 billion metric tons (Figure 17.8). The annual world consumption of coal is about 4 billion metric tons, sufficient for about 250 years at the current rate of use. However, if consumption increases in the coming decades, the resource will not last nearly so long.[15]

Coal is classified according to its energy and sulphur content as anthracite, bituminous, subbituminous, or lignite (see Table 17.1). Energy content is greatest in anthracite coal and lowest in lignite coal. Figure 17.8 shows that Canada is not a major coal producer relative to other countries, particularly Russia and the United States. Canada does however have extensive coal reserves, distributed from coastal British Columbia to Atlantic Canada and northward toward the Arctic. These reserves include all types of coal, from lignite to anthracite. Current recoverable reserves are estimated at about 6.5 billion tonnes (roughly 100 years' production), a small fraction of the total. The remainder is not currently considered recoverable, but if eventually accessed could extend Canada's coal production to approximately 1,000 years of production. The environmental impacts of coal mining[16] are discussed in more detail in Chapter 23.

The sulphur content of coal is important because low-sulphur coal emits less sulphur dioxide (SO_2) and as a result is more desirable as a fuel for power plants. Most of the low-sulphur coal used in Canada and the United States is relatively low-grade, low-energy lignite and subbituminous coal. Power plants using high-sulphur coal mined in their own region to lower its sulphur content before, during, or after combustion and thus avoid excessive air pollution. Although it is expensive, treating coal to reduce pollution may be more economical than transporting low-sulphur coal from distant regions.

Transporting coal from mining areas to large population centers where energy is needed is another significant environmental issue. Although coal can be converted at the production site to electricity, synthetic oil, or synthetic gas, these alternatives have their own problems. Power plants necessary to convert coal to electricity require water for cooling, and in semi-arid regions there may not be sufficient water. Furthermore, electricity transmission over long distances is inefficient and expensive (see Chapter 16). Converting coal to synthetic oil or

Table 17.1 • Typical Content of U.S. and Canadian Coal Resources

		Sulphur Content (%)		
Type of Coal	*Energy Content (millions of joules/kg)*	*Low (0–1)*	*Medium (1.1–3.0)*	*High (3 +)*
Anthracite	30–34	97.1	2.9	—
Bituminous coal	23–34	29.8	26.8	43.4
Subbituminous coal	16–23	99.6	0.4	—
Lignite	13–16	90.7	9.3	—

Note: Canada currently produces almost no anthracite and only a small amount of lignite.

Sources: Natural Resources Canada; U.S. Bureau of Mines Circular 8312, 1966; P. Averitt, "Coal," in D. A. Brobst and W. P. Pratt, eds., United States Mineral Resources, *U.S. Geological Survey, Professional Paper 820*, pp. 133–142.

gas also requires a tremendous amount of water; and the conversion process is expensive.[17]

Freight trains and coal-slurry pipelines (designed to transport pulverized coal mixed with water) are options to transport the coal itself over long distances. Trains are typically used and will continue to be used because they provide relatively low-cost transportation compared with the cost of constructing pipelines. The economic advantages of slurry pipelines are tenuous, especially in semi-arid regions, where large volumes of water to transport the slurry are difficult to obtain.[17]

Burning coal currently produces about 17% of the electricity produced in Canada, although usage varies widely by region. For example, Alberta and Nova Scotia generate about 80% of their electricity using coal, while 99% of Manitoba's electricity generation uses fuels other than coal.[18] Coal reserves in Canada approach 1 billion tonnes, roughly 1% of the world total, and we have enough coal to last at least several hundred years. However, serious concern has been raised about burning that coal. Giant coal-burning power plants in Canada and the United States are responsible for about 70% of the total emissions of sulphur dioxide, 30% of the nitrogen oxides, and 35% of the carbon dioxide. (The effects of these pollutants are discussed in Chapter 21.)

Emissions from coal-burning power plants are regulated under provincial legislation in Canada, and vary from region to region. In the 1970s, acid rain concerns prompted tougher emissions controls in most provinces, and by 1994, sulphur dioxide emissions from all sources were 54% lower than in 1980. Controls on coal-burning power plants were responsible for a large part of this reduction.

In 2002, U.S. President George Bush announced the introduction of a new Clean Coal Technology program, offering $330 million in matching funds for industrial research and development related to new-generation coal technology. Clean coal technologies include new ways of operating existing coal-burning plants to reduce emission of greenhouse gases and other pollutants, new technologies for burning coal and removing pollutants from air emissions, and energy-from-coal-waste initiatives. Examples of new and emerging clean coal strategies include:[19]

- Chemical and/or physical cleaning of coal prior to combustion.
- New boiler designs that permit a lower temperature of combustion, which reduces emissions of nitrogen oxides.
- Injection of material rich in calcium carbonate (such as pulverized limestone or lime) into the gases following burning of the coal. This practice, known as **scrubbing**, removes sulphur dioxides. In the scrubber, the carbonate reacts with sulphur dioxide, producing hydrated calcium sulphite as a sludge, which must be separately collected and disposed of.
- Conversion of coal at power plants into a gas (syngas, a methane-like gas) before burning. Syngas, although cleaner burning than coal, is still more polluting than natural gas.
- Consumer education about energy conservation and efficiency to reduce the demand for energy and thus the amount of coal burned and emissions released.

Figure 17.7 • Processes by which buried plant debris (peat) is transformed into coal.

The real shortages of oil and gas are still a few years away, but when they happen, they will put pressure on the coal industry to open more and larger mines. Increased use of coal will have significant environmental impacts including alteration of land in the mining process; emission of pollutants to air and water; creation and disposal of by-products such as ash, boiler slag (a rocklike cinder produced in the furnace), and calcium sulphite sludge produced through scrubbing. Although some wastes can be recovered or recycled, about 75% of the combustion products of burning coal, amounting to tens of millions of tonnes, end up on waste piles or in landfills.[20]

It seems unlikely that coal will be abandoned in the near future in Canada and the United States, because

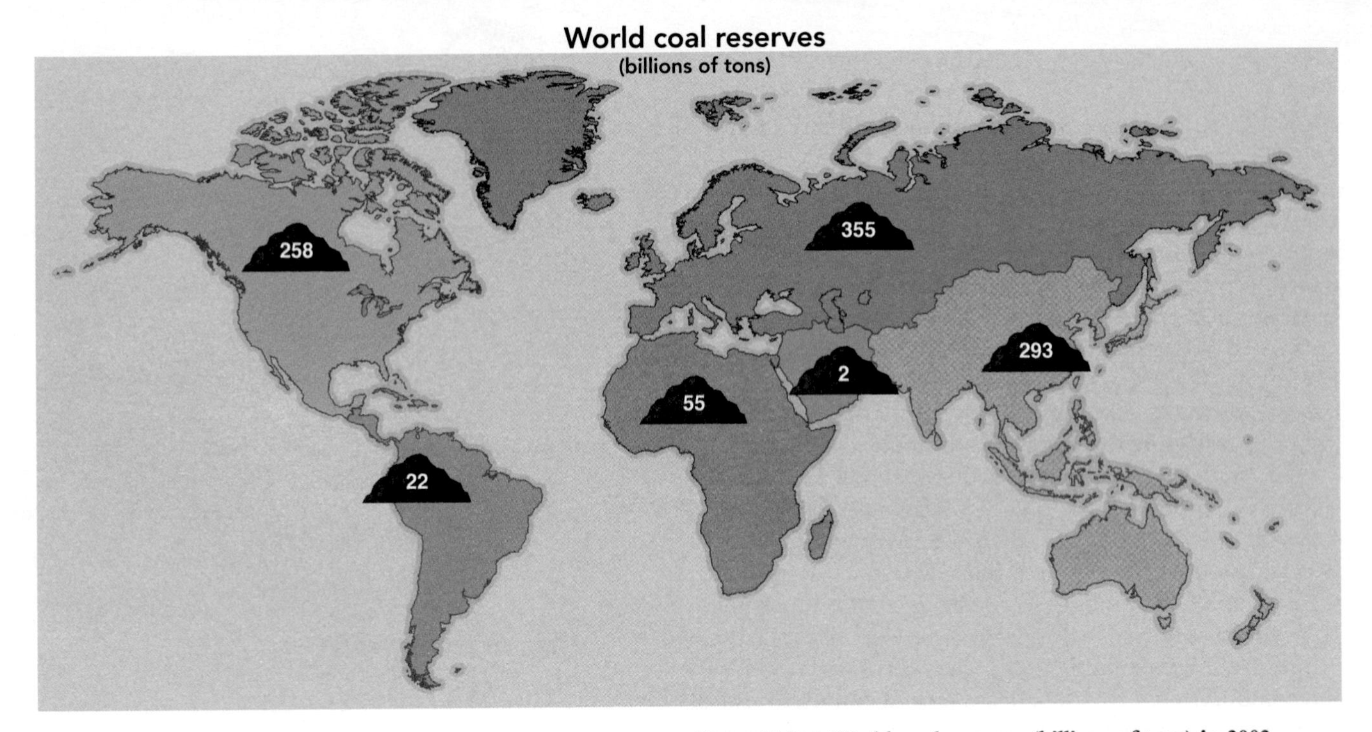

Figure 17.8 • **World coal reserves (billions of tons) in 2002.** [*Source*: Modified from British Petroleum Company, BP Statistical Review of World Energy (London: British Petroleum Company, 2003).]

we have so much of it and we have spent so much time and money developing coal resources over the past century. It has been suggested that we should now promote the use of natural gas in preference to coal because it burns so much cleaner. However, there is a concern that we might then become dependent on imports of natural gas. Regardless, it remains a fact that coal is the most polluting of all the fossil fuels.

Oil Shale and Tar Sands

Oil shale and tar sands play a minor role in today's mix of available fossil fuels. However, they may be more significant in the future when traditional oil from wells becomes scarce.[21]

Oil shale is a fine-grained sedimentary rock containing organic matter (kerogen). When heated to 500°C in a process known as destructive distillation, oil shale yields up to nearly 60 L of oil per tonne of shale. If not for the heating process, the oil would remain in the rock. The oil from shale is one of the so-called **synfuels** (from the words synthetic and fuel), which are liquid or gaseous fuels derived from solid fossil fuels. Total identified world oil shale resources are estimated to be equivalent to about 3 trillion bbl of oil. However, evaluation of the oil grade and the feasibility of economic recovery with today's technology is not complete.

The environmental impact of developing oil shale varies with the recovery technique used. Both surface and subsurface mining techniques have been considered.

Surface mining is attractive to developers because nearly 90% of the shale oil can be recovered, compared with less than 60% by underground mining. However, waste disposal will be a major problem with either surface or subsurface mining. Both require that oil shale be processed, or retorted (crushed and heated), at the surface. The volume of waste will exceed the original volume of shale mined by 20% to 30%, owing to the fact that crushed rock, because of added pore spaces, has more volume than the solid rock from which it came. The mines from which the shale is removed thus will not be able to accommodate all the waste, and its disposal will become a problem.[10]

In the 1970s, a tremendous rush to develop oil shale was expected. Interest in oil shale was heightened by the oil embargo in 1973 and by fear of continued shortages of crude oil. In the 1980s through the mid-1990s, though, plenty of cheap oil was available, and oil shale development was put on the back burner; it is much more expensive to extract a barrel of oil from oil shale than to pump it from a well. However, shortages of oil will occur in the future, and we will again turn to oil shale. This would result in significant environmental, social, and economic impacts in the oil shale areas resulting from rapid urbanization to house a large workforce, construction of industrial facilities, and an increased demand on water resources.

Tar sands are sedimentary rocks or sands impregnated with tar oil, asphalt, or bitumen. Petroleum cannot be recovered from tar sands by pumping wells or other usual commercial methods because the oil is too viscous (thick) to flow easily. Oil in tar sands is recovered by first mining the sands (which are very difficult to remove) and then washing the oil out with hot water.

Some 75% of the world's known tar sand deposits are in the Athabasca Tar Sands in Alberta. The total Canadian resource represents about 2 trillion bbl, but it is not known how much of this will eventually be recovered. Today's production of the Athabasca Tar Sands is several hundred thousand bbl of synthetic crude oil per day, about 10% of North America's production of oil.[22]

In Alberta, tar sand is mined in a large open-pit mine. The mining process is complicated by the fragile native vegetation, a water-saturated mat known as a muskeg swamp—a kind of wetland that is difficult to remove except when frozen. Restoration of this fragile, naturally frozen (permafrost) environment is difficult. In addition, there is a waste disposal problem because the mined sand material (as discussed for oil shale) occupies a greater volume than unmined material. The land surface after mining can be up to 20 m higher than the original ground surface.

17.2 Nuclear Energy

Nuclear energy is the energy of the atomic nucleus. Two nuclear processes can be used to release that energy to do work: fission and fusion. Nuclear **fission** is the splitting of atomic nuclei, and nuclear **fusion** is the fusing, or combining, of atomic nuclei. A by-product of both fission and fusion reactions is the release of enormous amounts of energy. (You may wish to review the discussion of matter and energy in A Closer Look 5.1.)

Nuclear energy for commercial use is produced by the splitting of atoms in **nuclear reactors**, which are devices that produce controlled nuclear fission. In Canada, there are 22 nuclear reactors, all of them using Canadian CANDU technology, in five nuclear power plants. Three of those plants (Pickering, Darlington, and Bruce) are in southern Ontario; a fourth is located in Gentilly, Québec, and the fifth is at Point Lepreau, New Brunswick.

Unlike the pressurized water reactors used in the United States, which employ uranium oxide as fuel, CANDU reactors use fuel pellets made from natural uranium.[23, 24] These technologies are discussed in more detail later in this chapter. Nuclear fusion is not yet used commercially, although it has been accomplished in experimental fusion reactors.

Fission Reactors

The first human-controlled nuclear fission, demonstrated in 1942 by Italian physicist Enrico Fermi at the University of Chicago, led to the development of nuclear energy to produce electricity. Today, in addition to power plants to supply electricity to homes and industry, nuclear reactors power submarines, aircraft carriers, and icebreaker ships.

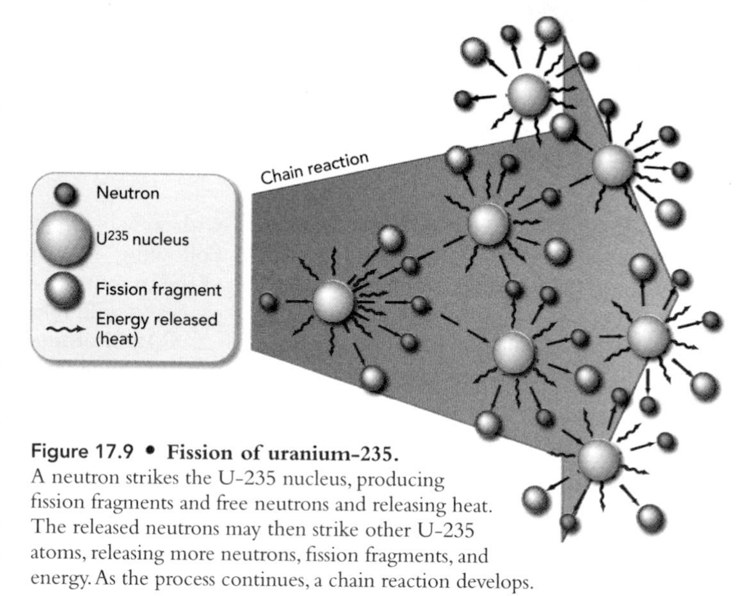

Figure 17.9 • Fission of uranium-235. A neutron strikes the U-235 nucleus, producing fission fragments and free neutrons and releasing heat. The released neutrons may then strike other U-235 atoms, releasing more neutrons, fission fragments, and energy. As the process continues, a chain reaction develops.

Russia is building ships that contain reactors to provide electricity to coastal towns, and the United States is designing reactors for deep-space missions.

Nuclear fission produces much more energy than other sources, such as burning fossil fuels. One kilogram of uranium fuel produces 1,000 times more energy than an equivalent amount of fossil fuel. This makes uranium an important source of energy in Canada and the rest of the world.

Three types, or isotopes, of uranium occur in nature: uranium-238, which accounts for approximately 99.3% of all natural uranium; uranium-235, which makes up about 0.7%; and uranium-234, which makes up about 0.005%. Uranium-235 and uranium-238 are two radioactive isotopes of uranium. However, uranium-235 is the only naturally occurring fissionable (or fissile) material; therefore, it is essential to the production of nuclear energy. Processing uranium (called enrichment) to increase the concentration of uranium-235 from 0.7% to about 3% produces enriched uranium, which is used as fuel for the fission reaction in light water reactors, such as are common in the United States. Canada's CANDU reactors use natural (unenriched) uranium as fuel, but require a more expensive moderator, "heavy" water (discussed below). Radiation is explained and related terms are defined in A Closer Look 17.1.

Fission reactors split uranium-235 by neutron bombardment (Figure 17.9). The reaction produces neutrons, fission fragments, and heat. The released neutrons strike other uranium-235 atoms, releasing more neutrons, fission products, and heat. The neutrons released are fast moving and must be slowed down slightly, or *moderated*, to increase the probability of fission. In light water reactors, the kind most commonly used in the United States, ordinary water is used as the moderator. As the process continues, a chain reaction develops as more and more uranium is split, releasing more neutrons and more heat.

A CLOSER LOOK 17.1

Radioactive Decay

To many people, radiation is a frightening subject shrouded in mystery. We cannot see it, taste it, smell it, or feel it, but we are told that radiation and fallout from atomic bomb testing can cause widespread human suffering. In this feature, we try to demystify some aspects of radiation by discussing the natural process of radiation, or radioactivity.

Radiation is a natural process that has been going on since the creation of the universe. Understanding it involves understanding the radioisotope, which is a form of a chemical element that spontaneously undergoes radioactive decay. During the decay process, the radioisotope changes from one isotope to another and emits one or more forms of radiation.

You may recall from Chapter 5 that isotopes are atoms of an element that have the same atomic number (the number of protons in the nucleus) but that vary in atomic mass number (the number of protons plus neutrons in the nucleus). For example, two isotopes of uranium are $^{235}U_{92}$ and $^{238}U_{92}$. The atomic number for both isotopes of uranium is 92; however, the atomic mass numbers are 235 and 238. The two different uranium isotopes may be written as uranium-235 and uranium-238 or ^{235}U and ^{238}U.

An important characteristic of a radioisotope is its half-life, the time required for one-half of a given amount of the isotope to decay to another form. Half-lives range from millions of years, for uranium-235 (700 million years) to a fraction of a second. Radioactive carbon-14 has a half-life of 5,730 years, which is in the intermediate range, and radon-222 has a relatively short half-life of 3.8 days. Polonium-218's half-life is about 3 minutes.

Table 17.2 illustrates the general pattern for decay in terms of the elapsed half-lives and the fraction remaining. When an element like polonium decays, where does it go? It has decayed to lead-214, another radioactive isotope, which has a half-life of about 27 minutes. The progression of changes associated with the decay process is often known as a radioactive decay chain. Now suppose we had started with 1 g uranium-235, with a half-life of 700 million years. Following 10 elapsed half-lives, 0.1% of the uranium would be left—but this process would take 7 billion years.

Radioisotopes with short half-lives initially undergo a more rapid rate of change (nuclear transformation) than do radioisotopes with long half-lives. Conversely, radioisotopes with long half-lives have a less intense and slower initial rate of nuclear transformation but may be hazardous for a much longer time.[28]

There are three major kinds of nuclear radiation: alpha particles, beta particles, and gamma rays. Alpha particles consist of two protons and two neutrons (a helium nucleus) and have the greatest mass of the three types of radiation (Figure 17.10*a*). As a result, they do not travel far, only about 5–8 cm in air, and about 0.005-0.008 cm in living tissue, before they stop. Because these distances are so short, to cause damage to living cells alpha particles must originate very close to the cells. Alpha particles can be stopped by a sheet or so of paper, and alpha-emitting isotopes stored in a container of any kind are relatively harmless.

Beta particles are electrons and have a mass of 1/1,840 of a proton. Beta decay occurs when one of the protons in the nucleus of an isotope spontaneously changes into a neutron or a neutron is transformed into a proton (Figure 17.10*b*). As a result of this process, another particle, known as a neutrino, is also ejected. A neutrino is a particle with no rest mass (the mass when the particle is at rest with respect to an observer).[29] Beta particles travel farther through air than the more massive alpha particles but are blocked by even moderate shielding, such as a thin sheet of aluminum foil or a block of wood. Beta radiation is intermediate in its toxicity, although most beta radiation is absorbed by the body when a beta emitter is ingested.

Table 17.2 • Generalized Pattern of Radioactive Decay

Elapsed Half-life	*Fraction Remaining*	*Percent Remaining*
0	—	100
1	1/2	50
2	1/4	25
3	1/8	13
4	1/16	6
5	1/32	3
6	1/64	1.5
7	1/128	0.8
8	1/256	0.4
9	1/512	0.2
10	1/1024	0.1

The third and most penetrating type of radiation comes from gamma decay. When gamma decay occurs, a gamma ray, a type of electromagnetic radiation, is emitted from the isotope. Gamma rays are similar to X-rays but are more energetic and penetrating; they travel the longest average distance of all types of radiation. Protection from gamma rays requires thick shielding, such as about a metre of concrete or several centimetres of lead. Gamma emitters are toxic and dangerous inside or outside the body, but when they are ingested, some of the radiation passes outside the body.

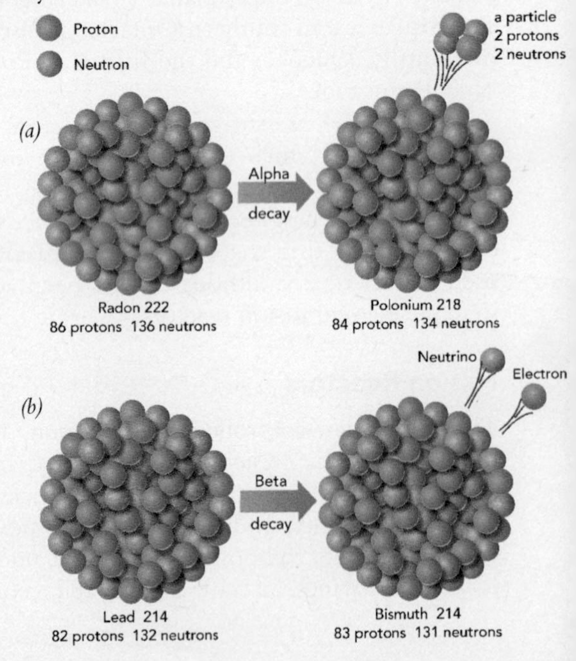

Figure 17.10 • Conceptual diagrams showing *(a)* alpha and *(b)* beta decay processes. [*Source*: D. J. Brenner, *Radon: Risk and Remedy* (New York: Freeman, 1989). Copyright 1989 by W. H. Freeman and Company. Reprinted with permission.]

A fourth type of radiation is neutron radiation, which occurs during the fission process and in some radioactive decay processes when neutrons are emitted from atomic nuclei. Neutron radiation also occurs naturally in the upper atmosphere, when cosmic radiation interacts with air. Neutron radiation is rare in the natural environment, but is a consideration in the decommissioning of large nuclear reactors.

Some radioisotopes, particularly those of very heavy elements such as uranium, undergo a series of radioactive decay steps (a decay chain) before finally reaching a stable nonradioactive isotope. For example, uranium decays through a series of steps to the stable nonradioactive isotope of lead. A decay chain for uranium-238 (with a half-life of 4.5 billion years) to stable lead-206 is shown in Figure 17.11. Also listed are the half-lives and types of radiation that occur during the transformations. Note that the simplified radioactive decay chain shown in Figure 17.11 involves 14 separate transformations and includes several environmentally important radioisotopes, including, in addition to uranium-238, radon-222, polonium-218, and lead-210. The decay from one radioisotope to another is often stated in terms of parent and daughter products. For example, uranium-238, with a half-life of 4.5 billion years, is the parent of daughter product thorium-234.

To sum up, when considering radioactive decay, two important facts to remember are: (1) the half-life and (2) the type of radiation emitted.

Radioactive Elements	Radiation Emitted			Half-Life		
	Alpha	Beta	Gamma	Minutes	Days	Years
Uranium-238 ↓	☢		☢			4.5 billion
Thorium-234 ↓		☢	☢		24.1	
Protactinium-234 ↓		☢	☢	1.2		
Uranium-234 ↓	☢		☢			247,000
Thorium-230 ↓	☢		☢			80,000
Radium-226 ↓	☢		☢			1,622
Radon-222 ↓	☢				3.8	
Polonium-218 ↓	☢	☢		3.0		
Lead-214 ↓		☢	☢	26.8		
Bismuth-214 ↓		☢	☢	19.7		
Polonium-214 ↓	☢			0.00016 (sec)		
Lead-210 ↓		☢	☢			22
Bismuth-210 ↓		☢			5.0	
Polonium-210 ↓	☢		☢		138.3	
Lead-206	None			Stable		

Figure 17.11 • Uranium-238 decay chain. [*Source*: Modified from F. Schroyer, ed., *Radioactive Waste*, 2nd printing (American Institute of Professional Geologists, 1985).]

CANDU reactors use "heavy" water—water enriched with deuterium isotopes of hydrogen, which have twice the mass of hydrogen.

Most reactors now in use consume more fissionable material than they produce and are known as burner reactors. The reactor is part of the nuclear steam supply system, which produces steam to run the turbine generators that produce the electricity.[25] Therefore, the reactor has the same function as the boiler that produces the heat in coal-burning or oil-burning power plants (Figure 17.12).

Figure 17.13 shows the main components of a CANDU reactor: the core (consisting of fuel and moderator), control rods, coolant, and reactor vessel. The core of the reactor is enclosed in the heavy, stainless steel reactor vessel; then, for extra safety and security, the entire reactor is contained in a reinforced concrete building.

In the reactor core, enriched uranium pellets (cylinders roughly 1.5 cm in diameter and 2.5 cm long) are packed in 50-cm hollow tubes and welded together in bundles weighing about 25 kg. A single fuel pellet is enough to provide an average family's annual energy requirements.

A minimum fuel concentration is necessary to keep the reactor *critical*—that is, to achieve a self-sustaining chain reaction. A stable fission chain reaction in the core is maintained by controlling the number of neutrons that cause fission. The control rods, which contain materials that capture neutrons, are used to regulate the chain reaction. As the control rods are moved out of the core, the chain reaction increases; as they are moved into the core, the reaction slows. Full insertion of the control rods into the core stops the fission reaction.[23]

The function of the coolant is to remove the heat produced by the fission reaction. This is an important point—the rate of generation of heat in the fuel *must match* the rate at which heat is carried away by the coolant. All major nuclear accidents have occurred when

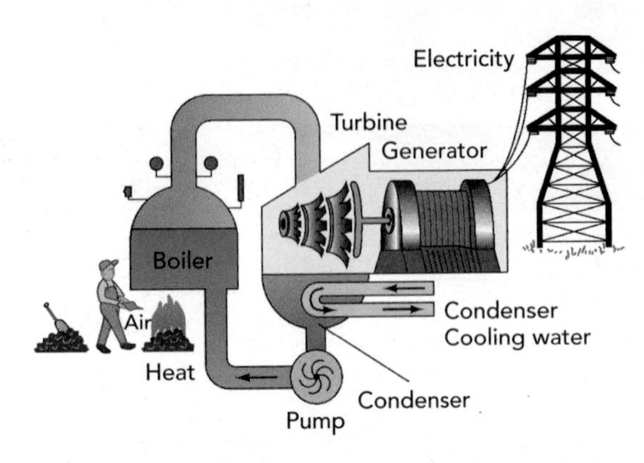

(a)

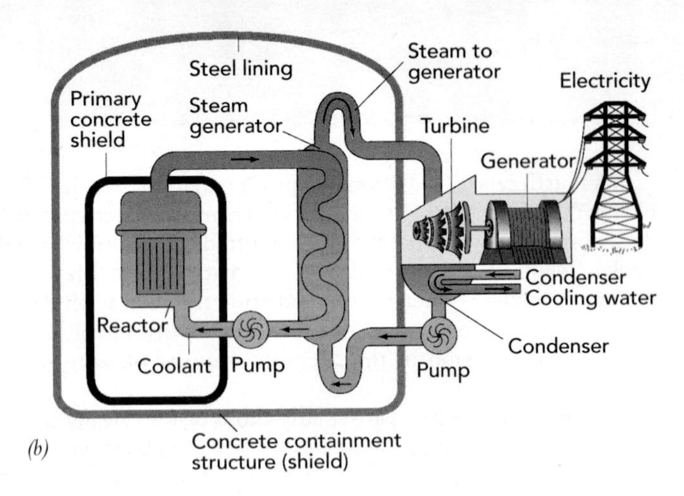

(b)

Figure 17.12 • Comparison of *(a)* a fossil fuel power plant and *(b)* a nuclear power plant with a boiling water reactor. Notice that the nuclear reactor has exactly the same function as the boiler in the fossil fuel power plant. The coal-burning plant *(a)* is Ratcliffe-on-Saw, located in Nottinghamshire, England, and the nuclear power station *(b)* is located in Leibstadt, Switzerland. (*Source:* American Nuclear Society, *Nuclear Power and the Environment*, 1973.)

something went wrong with the balance and heat built up in the reactor core.[24] A **meltdown** generally refers to a nuclear accident in which the nuclear fuel becomes so hot that it forms a molten mass that breaches the containment of the reactor and contaminates the outside environment with radioactivity.

Other parts of the nuclear steam supply system are the primary coolant loops and pumps, which circulate the coolant through the reactor, extracting heat produced by fission, and heat exchangers or steam generators, which use the fission-heated coolant to make steam (Figure 17.12*b*). In light water reactors, water is used as the coolant as well as the moderator.

Current thinking suggests that smaller, less complex nuclear reactors are the easiest to operate safely, because they require less extensive infrastructure. Small systems can also be designed with cooling systems that work by gravity and as a result are not as vulnerable to pump failure caused by power loss.[26] If a fuel assembly is small enough, it can generate useful power but cannot hold sufficient fuel to reach the temperatures necessary for a core meltdown.

Sustainability and Nuclear Power

Two things concern us about the sustainability of nuclear power: (1) nuclear power's role in creating alternative fuel supplies, and (2) the sustainability of nuclear fuel itself. In the first case, nuclear energy can be used to produce hydrogen from water or methane, to supply fuel cells in automobiles. Using hydrogen would ease the transition from our dependence on oil to a less environmentally damaging energy source (hydrogen). This is a central theme of sustainability that has the objective of meeting our energy needs in the future without harming the environment. The second aspect of sustainability with respect to nuclear energy has to do with nuclear fuel. This is especially important because uranium-fuelled nuclear power is a non-renewable resource.

Nuclear power plants are becoming safer and more economical. Although no new nuclear power plants have been built in over 20 years, existing plants provide an

Figure 17.13 • *(a)* Main components of a CANDU nuclear reactor. *(b)* Glowing spent fuel elements being stored in water at a nuclear power plant.

(a)

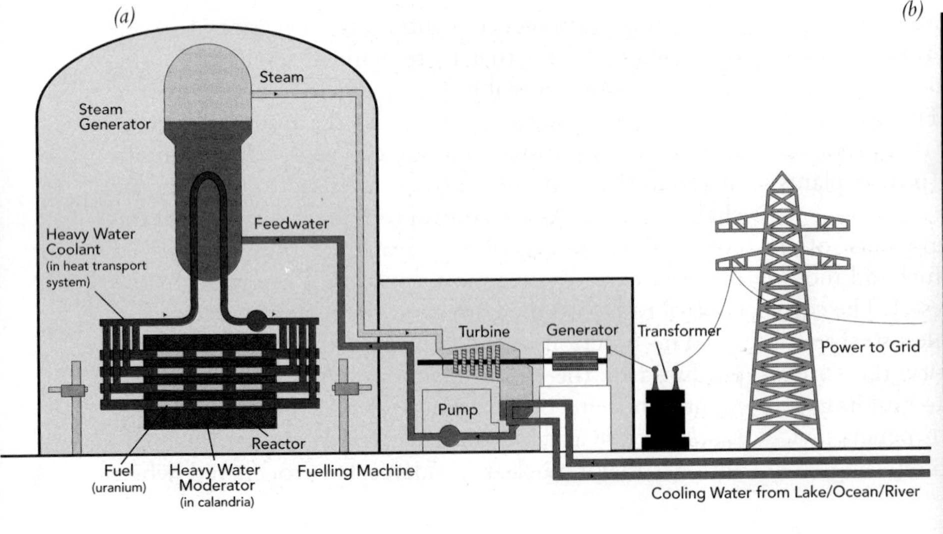

(b)

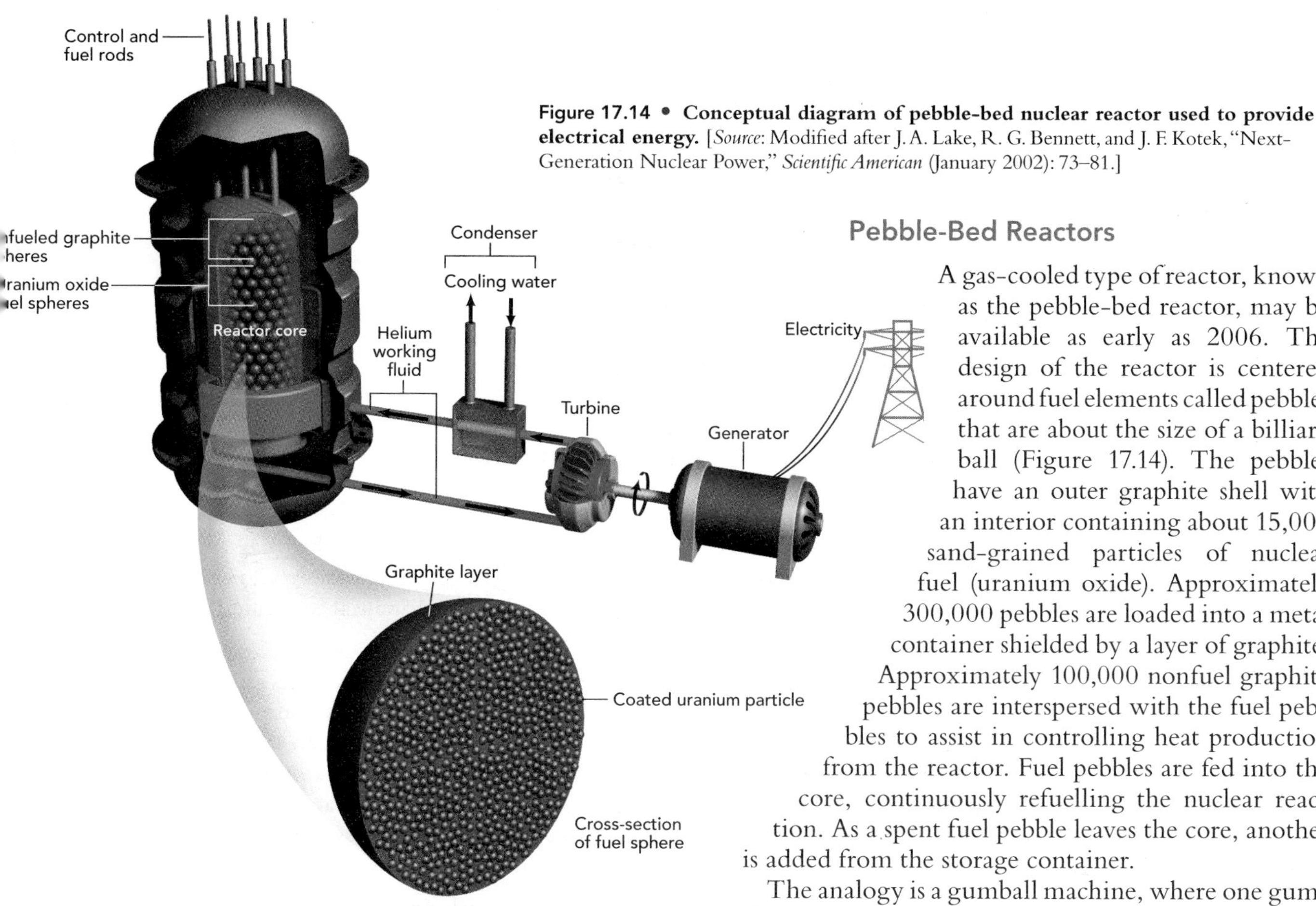

Figure 17.14 • Conceptual diagram of pebble-bed nuclear reactor used to provide electrical energy. [*Source*: Modified after J. A. Lake, R. G. Bennett, and J. F. Kotek, "Next-Generation Nuclear Power," *Scientific American* (January 2002): 73–81.]

increasing amount of electrical energy. Since the early 1990s, U.S. nuclear plants have added over 23,000 MW, equivalent to 23 large fossil fuel-burning power plants, and Canadian plants have also achieved modest increases. This increase is the result of more efficient use of existing nuclear power plants and has lowered the cost of energy production from nuclear plants.[27]

Light water reactors use uranium very inefficiently. Only about 1% of the uranium is used in the reactor; the other 99% ends up as waste. Those kinds of reactors are part of the nuclear waste problem and not a long-term solution to the energy problem. Advanced CANDU reactors have extended fuel lives that reduce spent fuel volume (and thus waste) by up to two-thirds, although long-term management of high-level and low-level nuclear waste remains a challenge in Canada.

One way for nuclear power to be sustainable for at least hundreds of years would be the use of a process known as *breeding*. **Breeder reactors** are designed to produce new nuclear fuel. They do this through a process in which they transform waste or lower-grade uranium into fissionable material. Breeder reactors, if constructed in sufficient numbers (several thousand), could supply about half the energy presently produced by fossil fuels for about 2,000 years.[24] Breeder reactors hold great promise for the future of nuclear power, provided that we can find ways of recycling and reprocessing their fuels.

Pebble-Bed Reactors

A gas-cooled type of reactor, known as the pebble-bed reactor, may be available as early as 2006. The design of the reactor is centered around fuel elements called pebbles that are about the size of a billiard ball (Figure 17.14). The pebbles have an outer graphite shell with an interior containing about 15,000 sand-grained particles of nuclear fuel (uranium oxide). Approximately 300,000 pebbles are loaded into a metal container shielded by a layer of graphite. Approximately 100,000 nonfuel graphite pebbles are interspersed with the fuel pebbles to assist in controlling heat production from the reactor. Fuel pebbles are fed into the core, continuously refuelling the nuclear reaction. As a spent fuel pebble leaves the core, another is added from the storage container.

The analogy is a gumball machine, where one gumball is removed from the dispenser and another takes it place. This is a safety feature of the reactor because the core at any time has just the right amount of fuel necessary for optimal production of energy. The pebble-bed reactors will probably be modular, inexpensive, and assembled from mass-produced parts, with each unit producing about 120 MW of power, or about one-tenth of that produced by a large centralized nuclear power plant. It is expected that pebble-bed reactors will compete economically with the new generation of natural gas power plants and be about 25% more efficient than present nuclear reactors.[27]

China, which will need an estimated 300 gigawatts of energy by 2050—almost 50 times the country's current production—is actively exploring the potential for pebble-bed reactors to meet its growing energy needs. Pebble-bed reactors are attractive because of their low cost and modular design, which allows several smaller facilities to be linked to a single control mechanism. They are also designed to be "meltdown-proof." However, some people believe that safe and responsible management of nuclear energy is easier when fuels and reactors are concentrated in a small number of locations, rather than distributed over many smaller, more accessible sites.

Fusion Reactors

In contrast to fission, which involves splitting heavy nuclei (such as uranium), fusion involves combining the nuclei of light elements (such as hydrogen) to form heavier ones (such as helium). As fusion occurs, heat energy is released (Figure 17.15). Nuclear fusion is the source of energy in our sun and other stars.

In a hypothetical fusion reactor, two isotopes of hydrogen—deuterium and tritium—are injected into the reactor chamber, where the necessary conditions for fusion are maintained. Products of the deuterium–tritium (DT) fusion include helium, producing 20% of the energy released, and neutrons, producing 80% of the energy released (Figure 17.16).[30]

Several conditions are necessary for fusion to take place. First, the temperature must be extremely high (approximately 100 million degrees Celsius for DT fusion). Second, the density of the fuel elements must be sufficiently high. At the temperature necessary for fusion, nearly all atoms are stripped of their electrons, forming a plasma. Plasma is an electrically neutral material consisting of positively charged nuclei (ions) and negatively charged electrons. Third, the plasma must be confined for a sufficient time to ensure that the energy released by the fusion reactions exceeds the energy supplied to maintain the plasma.[30, 31]

The potential energy available when and if fusion reactor power plants are developed is nearly inexhaustible. One gram of DT fuel (from a water and lithium fuel supply) has the energy equivalent of 45 barrels of oil. Deuterium can be extracted economically from ocean water, and tritium can be produced in a reaction with lithium in a fusion reactor. Lithium can be extracted economically from abundant mineral supplies.

Many problems remain to be solved before nuclear fusion can be used on a large scale. Research is still in the first stage, which involves basic physics, testing possible fuels (mostly DT), and magnetic confinement of plasma. Progress in fusion research has been steady in recent years, so there is optimism that useful power will eventually be produced from controlled fusion.[32]

Nuclear Energy and the Environment

The **nuclear fuel cycle** includes the processes involved in producing nuclear power from the mining and processing of uranium to controlled fission, the reprocessing of spent nuclear fuel, the decommissioning of power plants, and the disposal of radioactive waste. Throughout the cycle, radiation can enter and affect the environment (Figure 17.17). In order to understand the environmental effects of radiation, it is useful to be acquainted with the units used to measure radiation and

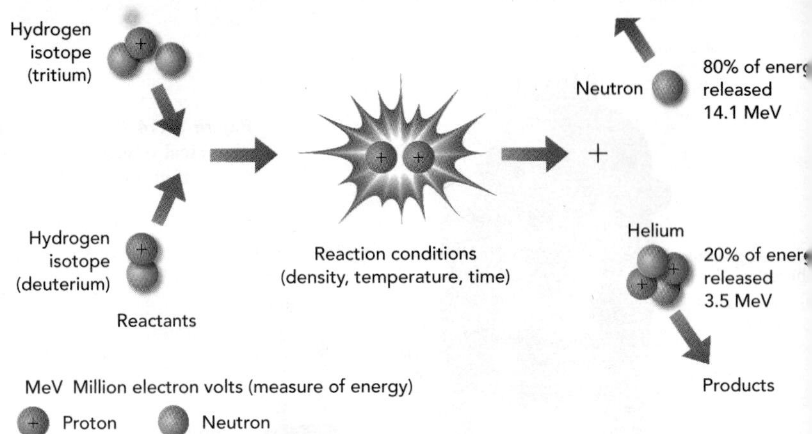

Figure 17.15 • Deuterium–tritium (DT) fusion reaction. [*Source*: Modified from U.S. Department of Energy, 1980.]

the amount or dose of radiation that may cause a health problem. These are explained in A Closer Look 17.2. Nuclear energy has the potential to create the following environmental effects:

- Uranium mines and mills produce radioactive waste material that can pollute the environment. Radioactive mine tailings have sometimes been used for foundation and building materials and have contaminated dwellings. (Tailings are materials that are removed by mining activity but are not processed and generally remain at the site, often unsupervised and open to the public.)
- Uranium-235 enrichment and fabrication of fuel assemblies also produce waste materials that must be carefully handled and disposed of.
- Site selection and construction of nuclear power plants, especially nuclear reactors, have been extremely controversial. The environmental review process is extensive and expensive, often centering on hazards related to the probability of events such as earthquakes.

Figure 17.16 • Experimental fusion nuclear reactor that magnetically confines plasma at very high temperatures.

- Waste disposal is a controversial part of the nuclear cycle because no one wants a nuclear waste disposal facility nearby. There is general public concern that hazardous nuclear waste cannot be adequately isolated from the environment for the long period of time (millions of years) that it remains hazardous.
- Nuclear power plants have a limited lifetime of several decades. Decommissioning (removal from service) or modernization of a plant is a controversial part of the cycle, involving the removal and disposal of contaminated machinery, with which we have little experience. Five Canadian reactors, at Pickering and Bruce in Ontario, are currently undergoing refurbishment, at significant cost. It is possible that the dismantling of decommissioned reactors may become one of the highest costs for the nuclear industry.[33]

In addition to the hazards of transporting and disposing of nuclear material, potential hazards are associated with supplying other nations with reactors. Terrorist activity and the possibility of irresponsible persons in governments add risks that are present in no other form of energy production. For example, Kazakhstan inherited a large nuclear weapons testing facility, covering hundreds of square kilometres, from the former Soviet Union. Several sites contain "hot spots" of plutonium in the soil that present a serious problem of toxic contamination. The facility also presents a security problem. There is some international concern that this plutonium could be collected and used by terrorists to produce dirty bombs (conventional explosives that disperse radioactive materials). There may even be enough plutonium to produce small nuclear bombs.[38] Nuclear energy may indeed be an answer to our energy problems, and perhaps someday it will provide unlimited cheap energy. However, with nuclear power comes responsibility.

Effects of Radioisotopes

As explained in A Closer Look 17.1, a **radioisotope** is an isotope of a chemical element that spontaneously undergoes radioactive decay. Radioisotopes affect the environment in two ways: by emitting radiation that affects other materials, and by entering the normal pathways of mineral cycling and ecological food chains.

The explosion of a nuclear atomic weapon does damage in both ways. At the time of the explosion, intense radiation of many kinds and energies is sent out, killing organisms directly. The explosion generates large amounts of radioactive isotopes, which are dispersed into the environment. Nuclear bombs exploded in the atmosphere produce a huge cloud that sends radioisotopes directly into the stratosphere, where the radioactive particles may be widely dispersed by winds. Atomic fallout—the deposit of these radioactive materials around the world—was an environmental problem in the 1950s and 1960s, when the United States, the former Soviet Union, China, France, and Great Britain were testing and exploding nuclear weapons in the atmosphere.

The pathways (Figure 17.21) of some of these isotopes illustrate the second way in which radioactive materials can be dangerous in the environment; they can enter ecological food chains. For example, cesium-137 was one of the radioisotopes emitted into the stratosphere by atomic explosions. It was deposited in relatively small concentrations but was widely dispersed in the Arctic region of North America, falling on reindeer moss, a

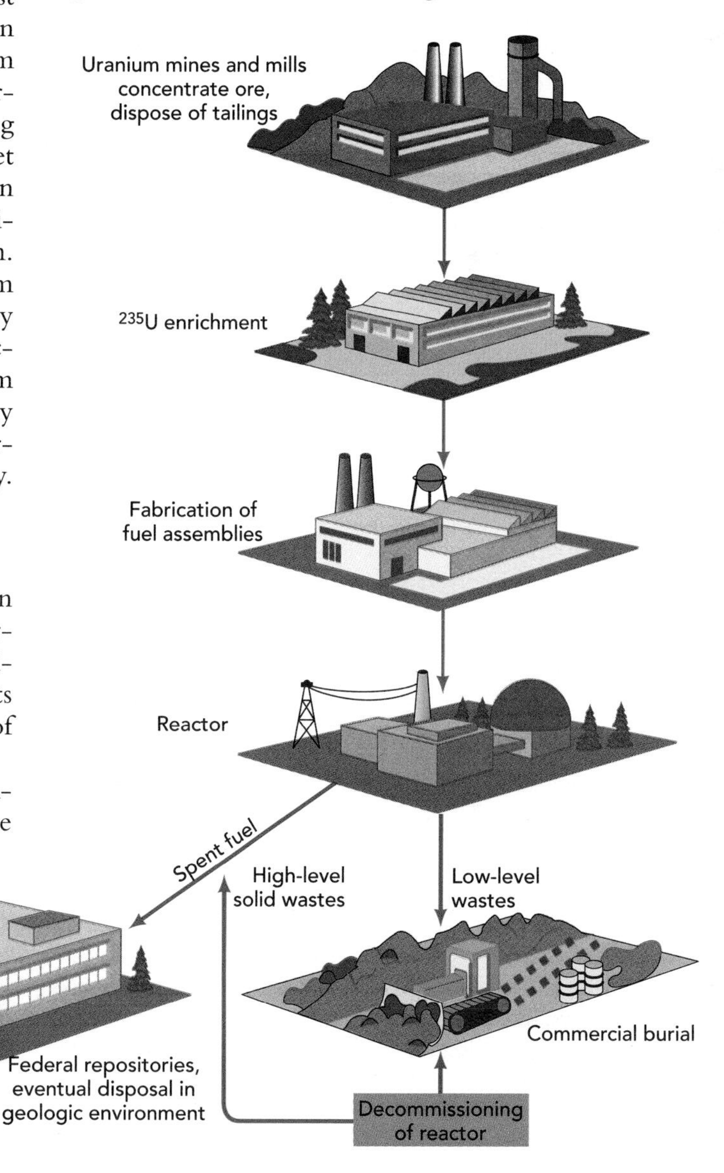

Figure 17.17 • Conceptual diagram showing the nuclear fuel cycle. Disposal of tailings, which because of their large volume may be more toxic than high-level waste, was treated casually in the past. [*Source*: Office of Industry Relations, The Nuclear Industry, 1974.]

A CLOSER LOOK 17.2

Radiation Units and Doses

The units used to measure radioactivity are complex and somewhat confusing. Nevertheless, a modest acquaintance with them is useful in understanding and talking about the effects of radiation on the environment.

A commonly used unit for radioactive decay is the curie (Ci), a unit of radioactivity equal to 37 billion nuclear transformations per second.[28]

The curie is named for Marie Curie and her husband, Pierre, who discovered radium in the 1890s. They also discovered polonium, which they named after Marie's homeland, Poland. The harmful effects of radiation were not known at that time, and both Marie Curie and her daughter died of radiation-induced cancer.[29] Her laboratory (Figure 17.18) is still contaminated today.

In the International System (SI) of measurement, the unit commonly used for radioactive decay is the becquerel (Bq), which is one radioactive decay per second. Units of measurement often used in discussions of radioactive isotopes, such as radon-222, are becquerels per cubic metre and picocuries per litre (pC/l). A picocurie is one-trillionth (10^{-12}) of a curie. Becquerels per cubic metre or picocuries per litre are therefore measures of the number of radioactive decays that occur each second in a cubic metre or litre of air.

When dealing with the environmental effects of radiation, we are most interested in the actual dose of radiation delivered by radioactivity. In the International System (SI), dose is commonly measured in grays (Gy) and sieverts (Sv). Grays are the units of the absorbed dose of radiation. Sieverts are units of equivalent dose, or effective equivalent dose.[28] The energy retained by living tissue that has been exposed to radiation is called the radiation absorbed dose. When very small doses of radioactivity are being considered, the millisievert (mSv)—that is, one-thousandth (0.001) of a sievert—is used.[28, 34, 35] For gamma rays, the unit commonly used is the coulomb per kilogram (C/kg).

People are exposed to a variety of sources of radiation from the sky, the air, and the food we eat (Figure 17.19). What is the background radiation received by people? This question is commonly asked by people concerned with radiation. The average Canadian receives about 2 to 4 mSv/yr. Of this, about 1 to 3 mSv/yr, or 50% to 75%, is natural. The differences are primarily due to elevation and geology. More cosmic radiation from outer space (which delivers about 0.3 to 1.3 mSv/yr) is received at higher elevations. Radiation from rocks and soils (such as granite and organic shales) containing radioactive minerals delivers about 0.3 to 1.2 mSv/yr. The amount of radiation delivered from rocks, soils, and water may be much larger in areas where radon gas (a naturally occurring radioactive gas) seeps into homes. As a result, mountain areas that also have an abundance of granitic rocks, such as British Columbia and Alberta, have greater background radiation than do areas that have a lot of limestone bedrock and are low in elevation, such as southern Ontario.[34]

The amount of radiation received by people from human sources is about 1.35 mSv/yr. Two sources include naturally occurring radioactive potassium-40 and carbon-14, which are present in our bodies and produce about 0.35 mSv/yr. Potassium is an important electrolyte in our blood, and one isotope of potassium (potassium-40) has a very long half-life. Although potassium-40 makes up only a very small percentage of the total potassium in our bodies, it is present in all of us, so we are all slightly radioactive. As a result, if you choose to share your life with another person, you are also exposed to a little bit more radiation.

Sources of low-level radiation from human processes include X-rays for medical and dental purposes, which may deliver an average of 0.8 to 0.9 mSv/yr; nuclear weapons testing, approximately 0.04 mSv/yr; the burning of fossil fuels such as coal, oil, and natural gas, 0.03 mSv/yr; and nuclear power plants (under normal operating conditions), 0.002 mSv/yr.[4,35,38]

A person's occupation and lifestyle can also affect the annual dose of radiation received. If you fly at high altitudes in jet aircraft, you receive an additional small dose of radiation—about 0.05 mSv for each flight across Canada. If you work at a nuclear power plant, you can receive up to about 3 mSv/yr. Living next door to a nuclear power plant adds 0.01 mSv/year, and sitting on a bench watching a truck carrying nuclear waste pass by would add 0.001 mSv to your annual exposure. Sources of radiation are summarized in Figure 17.20a, assuming an annual total of 3 mSv/yr.[36,37] The amount of radiation received at certain job sites, such as nuclear power plants and laboratories where X-rays are produced, is closely monitored. At such locations, personnel wear badges that indicate the dose of radiation received.

Figure 17.20*b* lists some of the common sources of radiation to which we are exposed. Notice that exposure to radon gas can equal what people were exposed to as a result of the Chernobyl nuclear power accident, which occurred in the Soviet Union in 1986. In other words, in some homes, people are exposed to about the same radiation as that experienced by the people evacuated from the Chernobyl area. (Radon gas is discussed in detail in Chapter 21.)

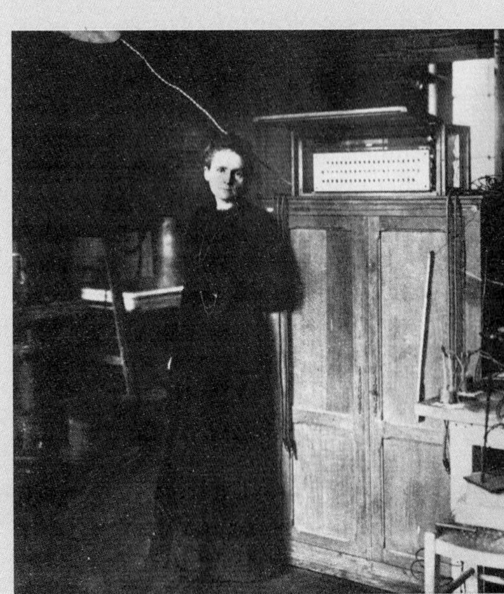

Figure 17.18 • Marie Curie in her laboratory.

From the sky
Outer space

About 8,000 cosmic rays pass through each of us every minute.

From the air
What we breathe

About 500 atoms of radioactive isotopes (such as radon, polonium, uranium, and lead) disintegrate in our lungs each minute, producing mostly alpha, beta radiation.

From our diet
What we eat

About 400 radioactive isotopes (mostly potassium and uranium) disintegrate inside us each minute.

Figure 17.19 • The major sources of natural radiation are the sky, the air we breathe, and the food we eat. [*Source*: Modified from National Radiological Protection Board, Living with Radiation, 3rd ed. (Reading, England: National Radiological Protection Board, 1986).]

lichen that is a primary winter food of the caribou. A strong seasonal trend in the levels of cesium-137 in caribou was discovered; the level was highest in the winter, when reindeer moss was the principal food, and lowest in the summer. Aboriginal peoples who obtained a high percentage of their protein from caribou ingested the radioisotope by eating the meat, and their bodies concentrated the cesium. The more that members of a group depended on caribou as their primary source of food, the higher was the level of the isotope in their bodies (Figure 17.22).

It is possible to predict the environmental pathways that radioisotopes will follow because we know the normal pathways of nonradioactive isotopes with the same chemical characteristics. Our knowledge of biomagnification and of large-scale air and oceanic movements that transport radioisotopes throughout the biosphere will also help us to understand the effects of radioisotopes.

Radiation Doses and Health

The most important question in studying radiation exposure in people involves determining the point at which the exposure or dose becomes a health hazard (see A Closer Look 17.2). Unfortunately, there are no simple answers to this seemingly simple question. We do know that a dose of about 5,000 mSv (5 sieverts) is considered lethal to 50% of people exposed to it. Exposure to 1,000–2,000 mSv is sufficient to cause health problems, including vomiting, fatigue, potential abortion of early pregnancies, and temporary sterility in males. At 500 mSv, physiological damage is recorded. The maximum allowed dose of radiation per year for workers in industry is 50 mSv, which is approximately 30 times the average natural background radiation.[28, 32] For the general public, Health Canada's maximum permissible annual drinking water dose (for infrequent exposure) is set at 0.1 mSv.

Most information concerning the effects of high doses of radiation on people comes from studies of the people who survived the atomic bomb detonations in Japan at the end of World War II. We also have information concerning people exposed to high levels of radiation in factories and uranium mines, and people treated with radiation therapy for disease.[39] Mine workers exposed to high levels of radiation have been shown to suffer a significantly higher rate of lung cancer than the general population, although there is a delay of 10 to 25 years between the time of exposure and the onset of disease. Hundreds of workers who painted watch dials with luminous paint containing radium, and licked their brushes to maintain a sharp point, later developed jaw disease, anemia or bone cancer.[29]

Radiation has a long history in the field of medicine. Drinking waters that contain radioactive materials goes back to Roman times. By 1899, the adverse effects of

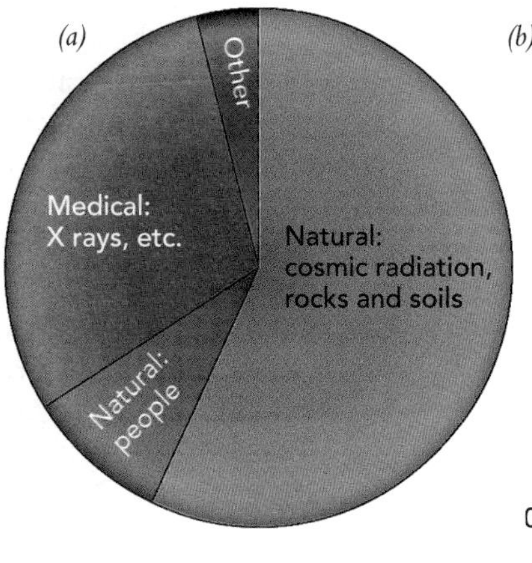

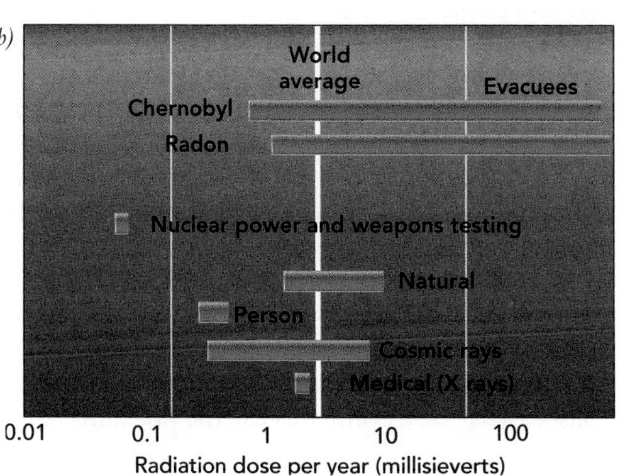

Figure 17.20 • *(a)* **Sources of radiation received by people**; assumes annual dose of 3.0 mSv/yr, with 66% natural and 33% medical and other (occupational, nuclear weapons testing, television, air travel, smoke detectors, etc.). [*Sources*: U.S. Department of Energy, 1999; New Encyclopedia Britannica, 1997. Radiation.V26, p. 487. *(b)* **Range in annual radiation dose to people from major sources.** [*Source*: Data in part from A.V. Nero, Jr., "Controlling Indoor Air Pollution," Scientific American 258(5) (1998) 42–48.]

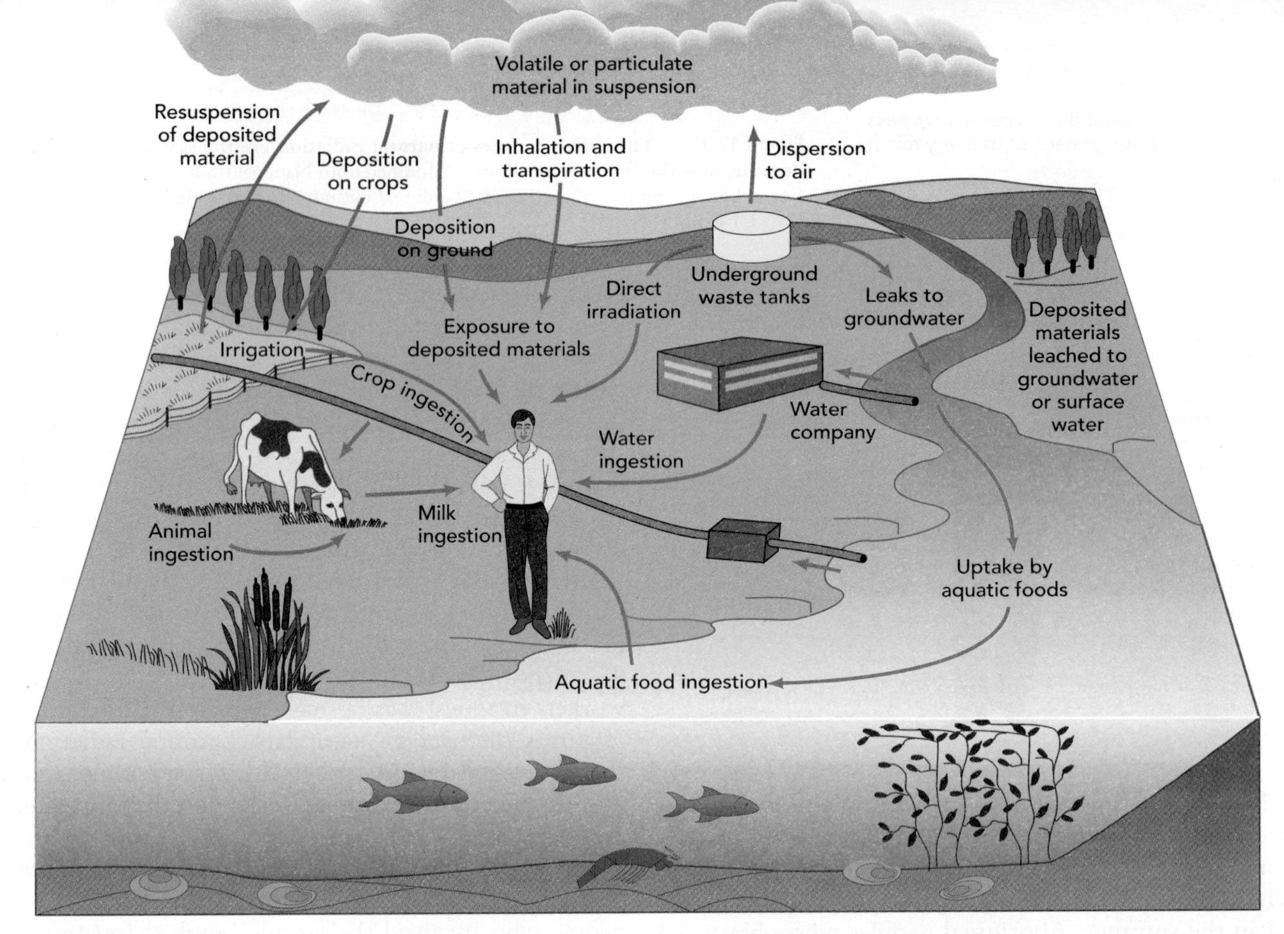

Figure 17.21 • How radioactive substances reach people. [*Source:* F. Schroyer, ed., Radioactive Waste, 2nd printing (American Institute of Professional Geologists, 1985).]

radiation had been studied and were well known; and in that year, the first lawsuit for malpractice in using X-rays was filed. Because science had shown that radiation could destroy human cells, however, it was a logical step to conclude that drinking water containing radioactive material such as radon might help fight diseases such as stomach cancer. In the early 1900s, it became popular to drink water containing radon, and the practice was supported by doctors, who stated that there were no known toxic effects. Although we now know that statement to be incorrect, radiotherapy, which uses radiation to kill cancer cells in humans, has been widely and successfully used for a number of years.[29]

Although there is a vigorous and ongoing debate about the nature and extent of the relationship between radiation exposure and cancer mortality, most scientists agree that radiation can cause cancer. Some believe that there is a linear relationship, such that any increase in radiation beyond the background level will produce an additional hazard. Others believe that the body is able to successfully handle and recover from low levels of exposure to radiation but that health effects (toxicity) become apparent beyond some threshold. The verdict is still out on this subject, but it seems prudent to take a conservative viewpoint—to use the precautionary principle—and accept that there may be a linear relationship. Unfortunately, long-term chronic health problems related to low-level exposure to radiation are neither well known nor well understood.

Nuclear Power Plant Accidents

Although the chance of a disastrous nuclear accident is estimated to be very low, the probability that an accident will occur increases with every reactor put into operation. One estimate by the U.S. Nuclear Regulatory Commission suggests that the probability of a large-scale core meltdown for a single reactor in any given year should be no greater than 0.01% (one chance in 10,000). However, if there were 1,500 nuclear reactors (about four times the present world total), a meltdown could, at the low annual probability of 0.01%, be expected every seven years. This is clearly an unacceptable risk.[24] Increasing safety by about 10 times would result in lower, more manageable risk; but the risk would still be appreciable because the potential consequences remain large.

Next, we discuss the two most well-known nuclear accidents, which occurred at the Three Mile Island and Chernobyl reactors. It is important to understand that these serious accidents resulted in part from human error.

Three Mile Island

One of the most dramatic events in the history of radiation pollution occurred on March 28, 1979, at the Three Mile Island nuclear power plant near Harrisburg, Pennsylvania. A valve malfunction, along with human errors (thought to be the major problem), resulted in a partial core meltdown. Intense radiation was released to the interior of the containment structure. Fortunately, the containment structure functioned as designed, and only a relatively small amount of radiation was released into the environment. Exposure from the radiation emitted into the atmosphere has been estimated at 1 mSv, which is low in terms of the amount of radiation required to cause acute toxic effects. Average exposure to radiation in the surrounding area is estimated to have been approximately 0.012 mSv, which is only about 1% of the natural background radiation received by people. However, radiation levels were much higher near the site. On the third day after the accident, 12 mSv/hour was measured at ground level near the site. By comparison, the average person receives about 2 mSv/year from natural radiation.

Because the long-term chronic effects of exposure to low levels of radiation are not well understood, the effects of Three Mile Island exposure, although apparently small, are difficult to estimate. However, the incident revealed many potential problems with the way in which society dealt with nuclear power. Historically, nuclear power had been relatively safe, and the state of Pennsylvania was unprepared to deal with the accident. For example, there was no state bureau for radiation help, and the state Department of Health did not have a single book on radiation medicine (the medical library had been dismantled two years earlier for budgetary reasons). One of the major impacts of the incident was fear; yet there was no state office of mental health, and no staff member from the Department of Health was allowed to sit in on important discussions following the accident.[40]

Chernobyl

Lack of preparedness to deal with a serious nuclear power plant accident was dramatically illustrated by events that began unfolding on the morning of April 28, 1986 (Figure 17.23). Workers at a nuclear power plant in Sweden, frantically searching for the source of elevated levels of radiation near their plant, concluded that it was not their installation that was leaking radiation. Rather, the radioactivity was coming from the Soviet Union by way of prevailing winds. Confronted, the Soviets announced that an accident had occurred at a nuclear power plant at Chernobyl two days earlier, on April 26. This was the first notice to the world of the worst accident in the history of nuclear power generation.

(a)

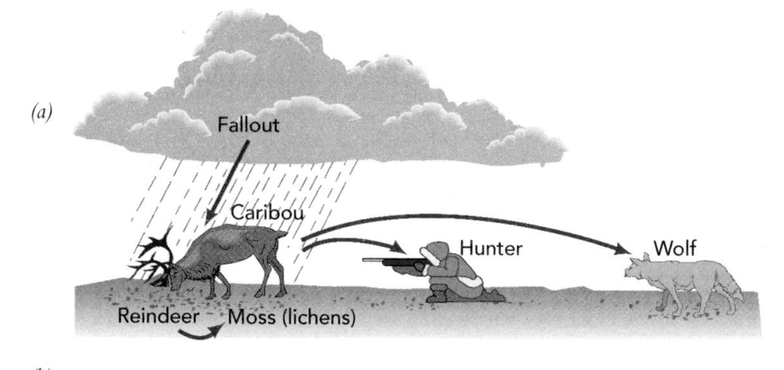

(b)

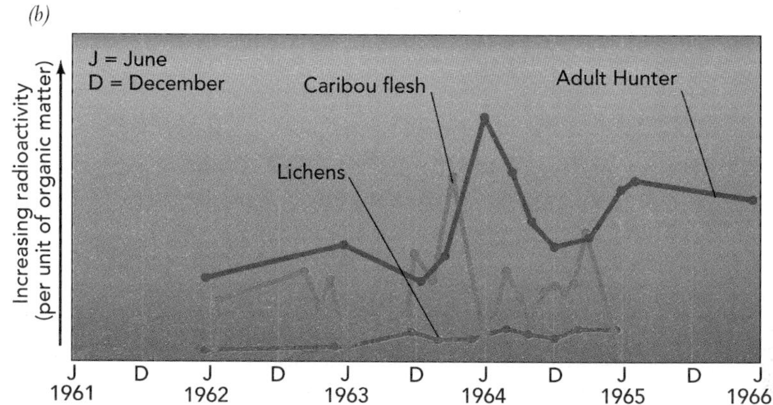

Figure 17.22 • Cesium-137 released into the atmosphere by atomic bomb tests was part of the fallout deposited onto soil and plants. *(a)* The cesium fell on lichens, which were eaten by caribou. The caribou were in turn eaten by First Nations people. *(b)* The cesium was concentrated by the food chain. Peaks in concentrations occurred first in the lichens, then in the caribou, and last in the people. [*Source*: *(b)* W. G. Hanson, "Cesium-137 in Alaskan Lichens, Caribou, and Eskimos," Modified from Health Physics 13 (1967): 383–389. Copyright 1967 by Pergamon Press. Reprinted with permission.]

Figure 17.23 • Chernobyl Nuclear Power Plant, just after the April 1986 disaster.

It is speculated that the system that supplied cooling waters for the Chernobyl reactor failed as a result of human error, causing the temperature of the reactor core to rise to over 3,000°C, melting the uranium fuel. Explosions removed the top of the building over the reactor, and the graphite surrounding the fuel rods used to moderate the nuclear reactions in the core ignited. The fires produced a cloud of radioactive particles that rose high into the atmosphere. There were 237 confirmed cases of acute radiation sickness, and 31 people died of radiation sickness.[41]

In the days following the accident, nearly 3 billion people in the Northern Hemisphere received varying amounts of radiation from Chernobyl. With the exception of the 30-km zone surrounding Chernobyl, the world human exposure was relatively small. Even in Europe, where exposure was highest, it was considerably less than the natural radiation received during one year.[41]

Approximately 115,000 people were evacuated from the 30-km radius around Chernobyl; and as many as 24,000 people were estimated to have received an average radiation dosage of 0.43 Sv (430 mSv). This group of people is being studied carefully. It was expected, based on results from Japanese bomb survivors, that approximately 122 spontaneous leukemias would likely occur during the period from 1986 through 1998.[41] Surprisingly, as of late 1998, there was no significant increase in the incidence of leukemia, even among the most highly exposed people, but an increase of the incidences of leukemia could still be expected in the future.[42]

Studies have found that the number of childhood thyroid cancer cases per year has however risen steadily in Belarus, Ukraine, and the Russian Federation (those most affected by Chernobyl) since the accident. Since the accident, a total of 1,036 thyroid cancer cases have been diagnosed in children under 15 years old. These cancer cases are believed to be linked to the released radiation from the accident, although other factors, such as environmental pollution, may also play a role. It is predicted that a few percent of the roughly 1 million children exposed to the radiation eventually will contract thyroid cancer.[43] Outside the 30-km zone, the increased risk of contracting cancer is very small and not likely to be detected from an ecological evaluation.[43] Nevertheless, cases of thyroid cancer are still appearing, and it is clear that final story of the world's most serious nuclear accident is still unfolding.[44] According to one estimate, Chernobyl will ultimately be responsible for approximately 16,000 deaths worldwide.[45]

Vegetation within 7 km of the power plant was either killed or severely damaged following the accident. Pine trees examined in 1990 around Chernobyl showed extensive tissue damage and still contained radioactivity. The distance between annual rings (a measure of tree growth) had decreased since 1986.[46] Animal populations seemed relatively unaffected, however. Scientists returning to the evacuated zone in the mid-1990s found, to their surprise, thriving and expanding animal populations. Species such as wild boar, moose, otters, waterfowl, and rodents seemed to be enjoying a population boom in the absence of humans. However, these animals may be paying a genetic price for living within the contaminated zone. Study of gene mutations in voles (small mammals resembling rats or mice) within the contaminated zone found gene mutation rates of over five mutations per animal, compared with a rate of only 0.4 per animal outside the zone. It is puzzling to scientists that the high mutation rate has not crippled the animal populations, but it appears so far that the benefit of excluding humans outweighs the negative factors associated with radioactive contamination.[45]

In the areas surrounding Chernobyl, radioactive materials continue to contaminate soils, vegetation, surface water, and groundwater, presenting a hazard to plants and animals. The evacuation zone may be uninhabitable for a very long time unless some way is found to remove the radioactivity.[41] Estimates of the total cost of the Chernobyl accident vary widely, but will probably exceed $200 billion.

Although Chernobyl is the most serious nuclear accident to date, it certainly was not the first, and it is not likely to be the last. At least 10 accidents have released radioactive particles in the last 34 years. Although the probability of a serious accident is very small at a particular site, the consequences may be great, perhaps resulting in an unacceptable risk to society. This is really not so much a scientific issue as a political one involving a question of values. Older CANDU reactors are now known to have a design flaw that allows sudden Loss of Coolant Accidents (LOCAs), leading to the accidental release of radionuclides into the environment. To date, such accidents have been minor, but some observers believe that the potential exists for more catastrophic coolant loss and associated heat increase in the reactor. Without adequate cooling, radioactive fuels can overheat quickly, releasing contaminated gases, and even melting. At minimum, an overheated CANDU reactor must be shut down, causing the temporary loss of power generation capacity for that facility and increasing the potential for brownouts.

Advocates of nuclear power have argued that nuclear power is safer than other sources of energy. They say that the number of additional deaths caused by air pollution resulting from burning fossil fuels is much greater than the number of lives lost through nuclear accidents. For example, the 16,000 deaths that might eventually be attributed to Chernobyl are fewer than the number of deaths caused each year by air pollution from burning coal.[26] Those arguing against nuclear power state that as long as people build nuclear power plants and manage them, there will be

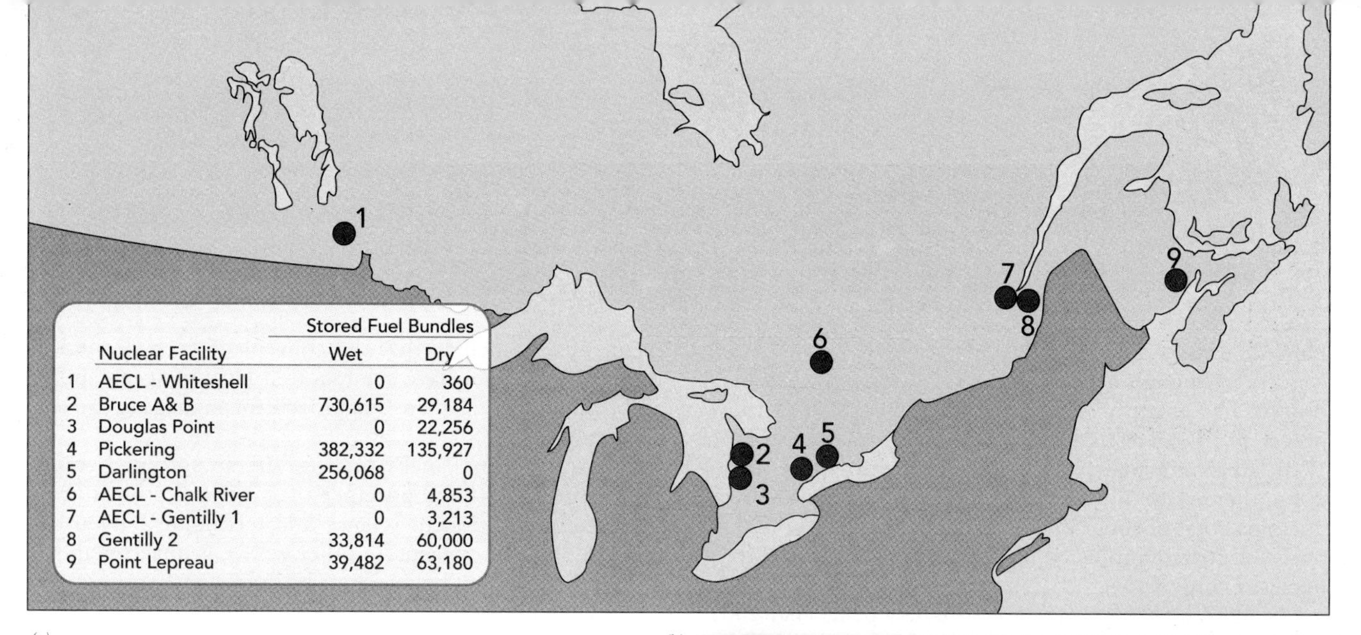

	Nuclear Facility	Stored Fuel Bundles Wet	Dry
1	AECL - Whiteshell	0	360
2	Bruce A& B	730,615	29,184
3	Douglas Point	0	22,256
4	Pickering	382,332	135,927
5	Darlington	256,068	0
6	AECL - Chalk River	0	4,853
7	AECL - Gentilly 1	0	3,213
8	Gentilly 2	33,814	60,000
9	Point Lepreau	39,482	63,180

(a)

(b)

Figure 17.24 • *(a)* **Location of Canadian used fuel waste disposal sites in December 2004.** *(b)* **Dry storage containers at the Pickering (Ontario) Used Fuel Dry Storage Facility.** [*Source*: Nuclear Waste Management Organization.]

the possibility of accidents. We can build nuclear reactors that are safer, but people will continue to make mistakes, and accidents will continue to occur.

Radioactive Waste Management

Examination of the nuclear fuel cycle (Figure 17.17) illustrates some of the sources of waste that must be disposed of as a result of using nuclear energy to produce electricity. Radioactive wastes are by-products that must be expected when electricity is produced at nuclear reactors; they may be grouped into three general categories: low-level waste, transuranic waste, and high-level waste. In addition, the tailings from uranium mines and mills must also be considered hazardous. In Ontario, Saskatchewan, and the western United States, tens of millions of metric tons of abandoned tailings will continue to produce radiation for at least 100,000 years, often in unsupervised locations open to the public.

Canada's nuclear industry, and its wastes, are regulated under three federal laws. The *Nuclear Safety and Control Act* (1997) replaced the earlier Atomic Energy Control Act of 1946. It provides new and explicit controls on the activities of the nuclear industry in Canada, and establishes the Canadian Nuclear Safety Commission (in place of the former Atomic Energy Control Board). A second important law is the *Nuclear Liability Act* (1970), which ensures that funds are available to compensate the victims of nuclear accidents in Canada. Finally, the *Nuclear Fuel Waste Act* (2002) established the Nuclear Waste Management Organization and oversees the management and disposal of nuclear waste in Canada.

Low-level radioactive waste contains sufficiently low concentrations or quantities of radioactivity that it does not present a significant environmental hazard if properly handled. Low-level waste includes a wide variety of items, such as residuals or solutions from chemical processing; solid or liquid plant waste, sludges, and acids; and slightly contaminated equipment, tools, plastic, glass, wood, and other materials.[47] In 1998 alone, Canada produced 4,300 cubic metres of low-level radioactive waste and one million tonnes of uranium mine and mill tailings. Total Canadian wastes are estimated at about 6,000 cubic metres of nuclear fuel waste (discussed below), 1.8 million cubic metres of low-level radioactive waste (typically stored on-site at nuclear plants), and 210 million tonnes of uranium mine and mill tailings.

In some countries, low-level waste has been buried in near-surface burial areas in which the hydrologic and geologic conditions were thought to severely limit the migration of radioactivity.[47] Studies have shown, however, that burial does not necessarily provide adequate protection for the environment. Of six original burial sites in the United States, three were forced to close prematurely due to unexpected leaks, financial problems, or loss of license; only two government site and one private facility remain in operation. Construction of new disposal sites

CRITICAL THINKING ISSUE

Public Perception of Nuclear Safety: Canada's Seaborn Panel

Nuclear energy is at a crossroads early in the twenty-first century. Although no new nuclear power plants had been ordered in Canada or the United States as of 2005, 31 new power plants were under construction worldwide (Table 17.3). Approximately 25% of these were in India. As of 2004, there were 440 operating nuclear power plants producing a total of about 362 GW of electricity. About 80% of the electricity produced in Lithuania and France is from nuclear energy.

In 1977, a federal government review panel chaired by University of Toronto geographer Ken Hare concluded that nuclear waste could best be stored in geological formations like the Canadian Shield. Over a decade later, in 1988, a federal environmental assessment of deep-rock nuclear waste disposal began, chaired by Blair Seaborn. This panel took 10 years to complete its work, conducting wide-ranging public consultations across Canada. In 1998, the panel concluded that while the concept of deep geological disposal had been adequately demonstrated, at least at a conceptual level, "from a social perspective, it has not." The panel went on to conclude that the concept of deep geological disposal "does not have the required level of acceptability to be adopted as Canada's approach for managing nuclear fuel wastes."

The Seaborn Panel's findings are important because they emphasize the importance of social and ethical perspectives in addition to science and technology. Even the best-engineered concept, in other words, is unacceptable if the public does not believe it to be safe.

Some countries, including China and Russia, are actively developing nuclear capacity. Figure 17.25a shows the result of U.S. public opinion polls from 1975 to 2003 in which people were asked to indicate whether they were favourably inclined to the idea of building new nuclear power plants in that country. The figure shows rising acceptance of nuclear power in that country.

Table 17.3 • World Nuclear Power Reactors under Construction (2002)

Country	*Units*	*Total MWe*
India	8	3,622
Ukraine	4	3,800
China	4	3,275
Russian Federation	3	2,825
Japan	3	3,696
Slovak Republic	2	776
Republic of Korea	2	1,920
Islamic Republic of Iran	2	2,111
Romania	1	655
Democratic People's Republic of Korea	1	1,040
Argentina	1	692
Total	31	24,412

Source: International Atomic Energy Agency. 2004. Power reactor information system.

Critical Thinking Questions

1. How might you interpret Figure 17.25*a*, which shows the percentage of people who favoured building more nuclear power plants from 1975 to 2002? What do you think the effects of nuclear accidents and rising oil prices have had on the number of people who favour building new power plants?
2. According to Figure 17.25*b*, approximately two-thirds of people in the United States today favour nuclear energy, with the remaining one-third opposing it. Try to construct arguments for those who would strongly favour nuclear power and contrast them with the arguments of those who would strongly oppose it. Do you think people in Canada share these views? Why or why not?
3. Why do you think there was a drop in the number of people who favour building more nuclear power plants from 2001 to 2002?
4. Putting together everything in this chapter and in Figure 17.25, what do you think the future of nuclear power will be in the world in the coming decades?

has been met with strong public opposition, and controversy remains as to whether low-level radioactive waste can be disposed of safely.[48]

Transuranic waste is composed of human-made radioactive elements heavier than uranium. It is produced in part by neutron bombardment of uranium in reactors and includes plutonium, americum, and einsteineum. Most transuranic waste is industrial trash, such as clothing, rags, tools, and equipment that has been contaminated. The waste is low-level in terms of intensity of radioactivity, but plutonium has a long half-life and requires isolation from the environment for about 250,000 years. Most transuranic waste is generated from the production of nuclear weapons and, more recently, from cleanup of former nuclear weapons facilities.

High-level radioactive waste consists of commercial and military spent nuclear fuel; uranium and plutonium derived from military reprocessing; and other radioactive nuclear weapons materials. It is extremely toxic, and a sense of urgency surrounds its disposal as the total volume

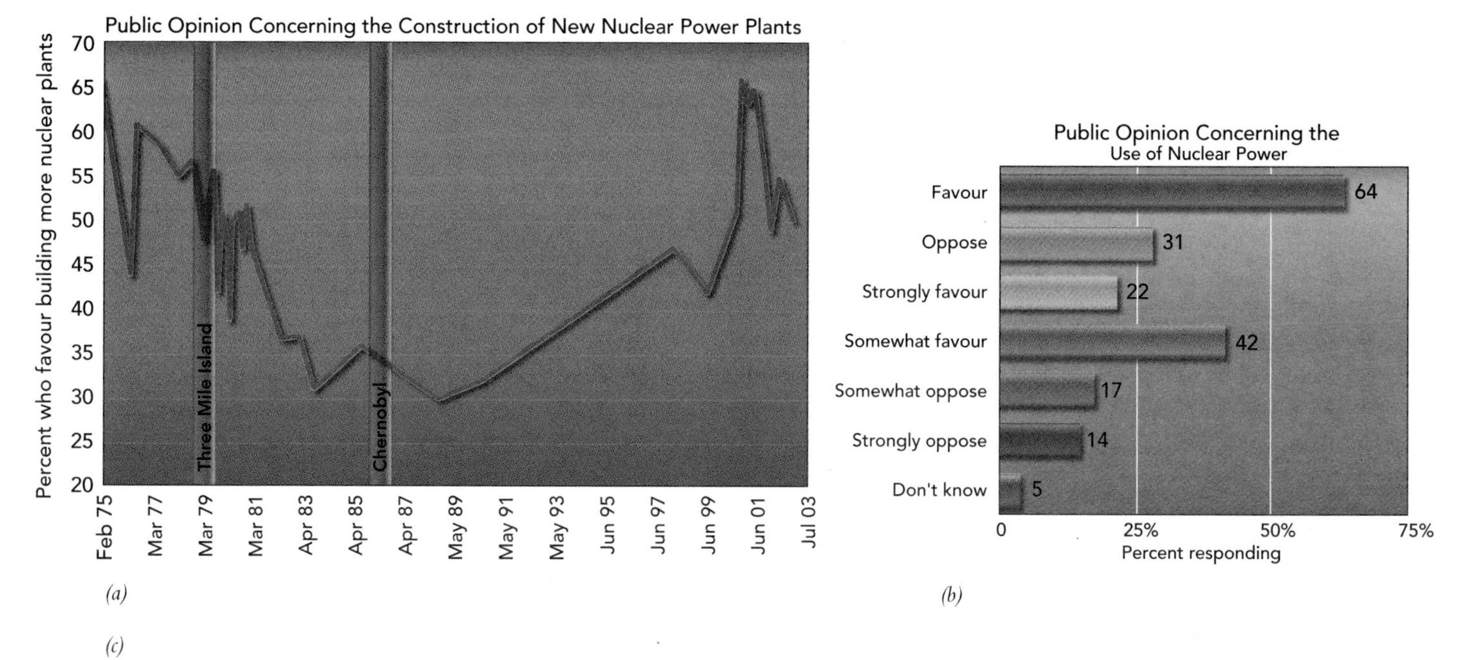

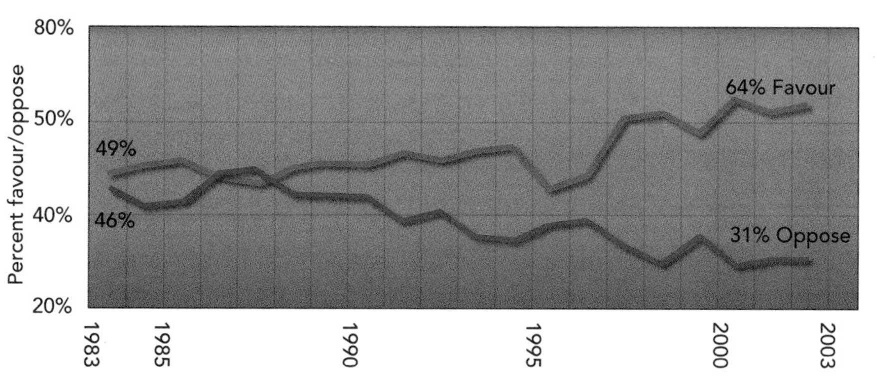

Figure 17.25 • Public opinion on nuclear power. *(a)* Public opinion concerning the construction of new nuclear power plants. [*Source*: E. A. Rosa and R. E. Dunlap, "Nuclear Power: Three Decades of Public Opinion," Public Opinion Quarterly 58 (1994): 295–325 (and www.nei.org).] *(b)* current distribution of public opinion on the use of nuclear power (2003 data); *(c)* long-term trends in public opinion about nuclear power [*Source*: Nuclear Energy Institure, 2003].

of spent fuel accumulates. Currently in Canada, used CANDU fuel bundles are removed from a power reactor after 12 to 18 months of use and stored in a pool of water at the reactor site. After about six years, the radioactivity and heat in the fuel bundles has decreased sufficiently for them to be handled safely. They are then removed from the pool and transferred to concrete storage containers on land, usually at the reactor site.[49, 50]

Used nuclear fuel is currently stored at seven locations in Canada: Whiteshell Laboratories, Manitoba; Bruce A&B, Pickering, and Darlington Power Generating Stations, Ontario; Chalk River Laboratories, Ontario; Gentilly-2 nuclear reactor, Trois-Rivières, Québec; and Point Lepreau Generating Station, St. John, New Brunswick (Figure 17.24).[51]

Despite 50 years of nuclear power generation, Canada does not have a secure long-term storage facility. Canada's Nuclear Waste Management Organization is currently working with governments and stakeholders to develop long-term secure storage solutions. Options under consideration include modular "casks" stored on the surface, secure disposal in shallow trenches, and 1,000 metre deep disposal in the granite rocks of the Canadian Shield. During the late 1970s, investigations of potential disposal sites in Northern Ontario met with enormous resistance from local residents. To date, the only feasibility test for deep-rock disposal has been conducted at Lac du Bonnet, Manitoba, near the Whiteshell Laboratories. At that site, a 500 metre shaft was sunk into the Lac du Bonnet batholith, a massive rock formation thought to offer the possibility of secure long-term storage. As a condition of leasing land for this purpose, the Manitoba government stipulated that no radioactive waste will actually be deposited in the shaft. A federal environmental assessment panel, called the Seaborn Panel after its chair, Blair Seaborn, spent more than 10 years investigating the merits of deep geologic disposal of high-level radioactive wastes. The panel concluded that although the concept appeared technologically sound, it did not have sufficient social support to be viable. (See the Critical Thinking Issue on page 326.)

In the United States, the Department of Energy specified through the *Nuclear Waste Policy Act* that high-level waste was to be disposed of underground in a deep, geologic waste repository. It was also specified that the Yucca Mountain site in Nevada was to be the only site evaluated. This does not mean that Yucca Mountain had been selected for actual disposal of nuclear waste, only that it is the only site being completely evaluated at this time. If the site is suitable, then it could accept high-level waste as early as 2010. As part of the assessment of that site, the Department of Energy evaluated the potential for volcanic eruptions, earthquakes, and other potential long-term changes in the storage environment, and is attempting to estimate the period of time over which wastes can be safely stored without migration to the external environment.[51] Extensive scientific evaluations of the Yucca Mountain site have been completed. However, use of this site is controversial and is generating considerable resistance from the state and people of Nevada as well as those scientists not confident in the plan.

Storage of high-level waste is at best a temporary solution, and serious problems with radioactive waste have occurred where it is being stored. Although improvements in storage tanks and other facilities will help, eventually some sort of disposal program must be initiated. Some scientists believe the geologic environment can best provide safe containment of high-level radioactive waste. Others disagree and have criticized proposals for long-term disposal of high-level radioactive waste underground.

One of the major questions concerning the disposal of high-level radioactive waste is this: How credible are long-range geologic predictions—those covering several thousand to a few million years?[52] Unfortunately, there is no easy answer to this question, because geologic processes vary over both time and space. Climates change over long periods of time, as do areas of erosion, deposition, and groundwater activity. For example, large earthquakes even thousands of kilometres from a site may permanently change groundwater levels. The earthquake record for most of North America extends back for only a few hundred years; therefore, estimates of future earthquake activity are tenuous at best.[52,53]

The bottom line is that geologists can suggest sites that have been relatively stable in the geologic past but cannot absolutely guarantee future stability. This means that policy-makers (not geologists) need to evaluate the uncertainty of predictions in light of pressing political, economic, and social concerns.[52] In the end, the geologic environment may be deemed suitable for safe containment of high-level radioactive waste, but care must be taken to ensure that the best possible decisions are made on this important and controversial issue.

The Future of Nuclear Energy

Nuclear energy as a power source for electricity is now being seriously evaluated. Advocates for nuclear power have argued that nuclear power is good for the environment for these reasons:

- It does not produce potential global warming through release of carbon dioxide (see Chapter 20).
- It does not cause air pollution or emit precursors (sulphates and nitrates) that cause acid rain (see Chapter 21).
- If breeder reactors are developed for commercial use, the amount of fuel available will be greatly increased.

Those in favour of nuclear power argue that it is safer than other means of generating power and that we should build many more nuclear power plants in the future. This argument is predicated on the understanding that such power plants would be considerably safer than those being used today. That is, if we standardize nuclear reactors and make them safer and smaller, nuclear power could provide much of our electricity in the future.[45] The safety of nuclear power under normal operating procedures is not disputed, although the possibility of accidents and the disposal of spent fuel are concerns. Since the Chernobyl accident, many European countries have been re-evaluating the use of nuclear power, and in most instances, the number of nuclear power plants being built has been significantly reduced. Indeed, in Germany, where about one-third of the country's electricity is produced by nuclear power, the decision was made to shut down all nuclear power plants in the next 25 years as they become obsolete. By contrast, France, with virtually no coal or oil resources, has chosen to focus on the development of nuclear power as a means of reducing its dependence on fossil fuels. Since the oil crisis of 1973, when oil prices quadrupled in France, the country has built 56 nuclear reactors, enough to supply its domestic energy needs and provide a surplus for sale to other countries. In contrast to public opinion in Canada and the United States (see Figure 17.5), nuclear energy is popular in France, and is seen as safe, clean, and economical.

The argument against reviving nuclear power is based on political and economic considerations as well as scientific uncertainty concerning safety issues. Those opposed to expanding nuclear power argue that converting from coal-burning plants to nuclear power plants for the purpose of reducing carbon dioxide emissions would require an enormous investment in nuclear power to make a real impact. This is true. Furthermore, critics say, given the fact that safer nuclear reactors are only just now being developed, there will be a time lag. As a result, nuclear power is not likely to have a real impact on environmental problems, such as air pollution, acid rain, and potential global warming, before at least 2050.[54] This point is debatable; the lag time could be shortened if research and development in advanced reactors become a higher priority and are better funded. Pressures associated with climate change (Chapter 20) and the environmental problems of fossil fuels suggest that nuclear power use may be increased in the future. Indeed, Ontario's recently announced plan to phase out coal-fired power generation has prompted renewed public discussion about the merits of nuclear power. However, the benefits of nuclear power must be balanced with the safety and waste disposal issues that have made nuclear energy an uncertain option for many people. The full impact of what began in 1942, when the atom was first split, is still to be determined.

17.3 Alternative Energy Sources

Renewable energy sources—sun, water, wind, and biomass—are often discussed as a group because they are all derived from the sun's energy. In other words, solar energy, broadly defined, comprises all the renewable energy sources, as shown in Figure 17.26. These energy sources are renewable, because they are regenerated (renewed) by the sun within a time period useful to humans.

Water and biomass, although renewable, are fundamentally different from solar and wind energy, which are there as long as the sun shines and the wind blows. Water power and biomass energy are not always automatically renewed by nature. They may be depleted if the environment necessary for their renewal is not maintained. For biomass to be renewable, both water and soil are necessary for plant growth. If either of these is depleted, then biomass production may decrease or even stop. Likewise, the water behind a dam depends on the climate to produce runoff. If the climate becomes arid, runoff may be reduced and the dam will store less water and produce less electricity. If people divert river water for crops upstream of a dam, there will be less water in the reservoir and hydropower will be reduced.

The total energy we may be able to extract from alternative energy sources is enormous (Table 17.4). For example, the estimated recoverable energy from solar energy is about 75 times the present annual human global energy consumption. The estimated recoverable energy from wind is comparable to current global energy consumption, as is the estimated recoverable energy from biomass.

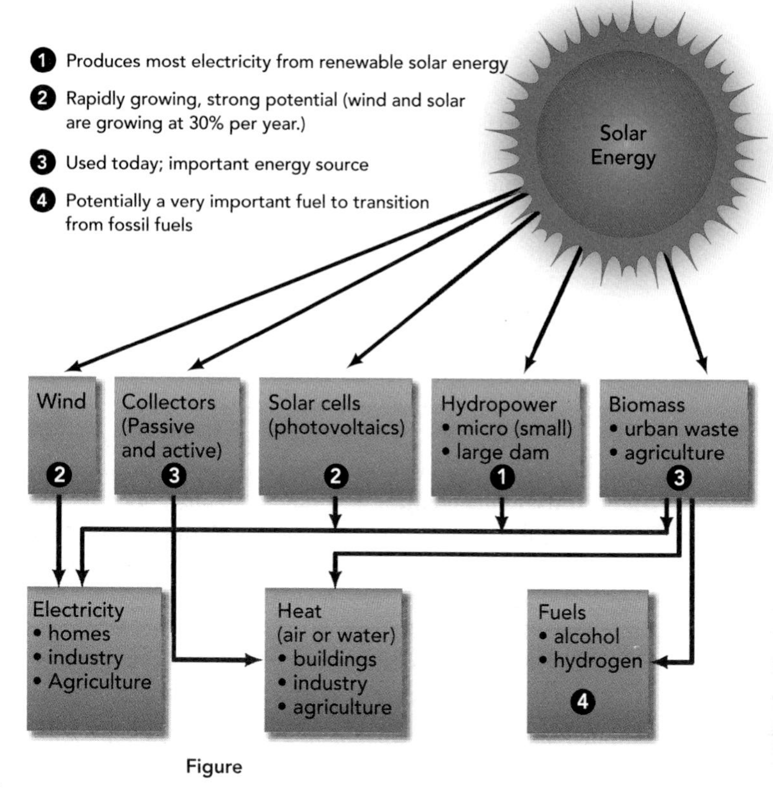

Figure

Table 17.4 • Global Resource-Based and Recoverable Energy for Selected Alternative Energy Sources

Source	*Resource Base (TW)**	*Recoverable Resource (TW)*
Solar	90,000	1,000
Wind	300–1,200	10
Water	10–30	2
Biomass	30	10
Geothermal	30	3

* 1 TW = 10^{12} W; global energy production consumption is about 13 TW. This is equivalent to an annual consumption of approximately 425 exajoules.

Source: Modified from T. Jackson, and R. Lofstedt, "Royal Commission on Environmental Pollution, Study on Energy and the Environment," 1998. Accessed November 29, 2000, at http://www.rcep.org.uk/studies/energy/98-6061/jackson.html.

It is true, however, that this energy may not necessarily always be available when we need it. Renewable energy sources, with the exception of biomass and water power (which can be stored), are intermittent, with daily and seasonal variations in supply. These variations will become a problem when renewable energy supplies make up about 40% of our total energy use. Beyond about 40%, the natural variation will periodically produce shortages in energy supply. As a result, the development of energy-storage systems to smooth out the variations in supply will be necessary.[55] For example, suppose you have solar collectors on your roof to produce electricity. At night, no energy is produced, but if you store some of the energy produced during the day in batteries, you can use stored energy at night. For large-scale production of renewable energy, large energy storage systems are necessary. Technology for battery storage and inverters (the devices that convert energy from sources like solar panels to the standard form of household electricity) is now well developed and low in cost. Ongoing research is also improving the performance of storage technologies for a range of climates, including Canada's cold winters, to ensure that battery performance remains high even when temperatures drop. Some private wind and solar energy producers can now store their surplus electricity and sell it back to the public system. This process, called **net metering** is now used in many locations in Canada, for example the Falls Brook Centre, a community education and demonstration centre in rural New Brunswick.

Renewable energy sources are not equally available in all locations. Therefore, matching energy production to appropriate sites is important. For example, in Canada, we might build solar energy power plants in the Prairie Provinces, where sunlight is often intense, and construct wind farms in the Prairie Provinces, coastal British Columbia, and Atlantic Canada , where the wind is strong and steady. Likewise, we could build biomass power plants in locations where forest and agriculture fuel resources are abundant.[56]

Generally speaking, renewable alternative energy sources are associated with minimal environmental degradation. In general, because no fuel is burned, these energy sources do not increase atmospheric carbon dioxide, which causes global warming. (An exception is the burning of biomass or its derivative, urban waste.) Renewable energy sources such as solar and wind power will not cause climate change or raise sea levels, increasing coastal erosion. Another advantage is that the construction lead time necessary to implement the technology for production of energy from renewable sources is often short compared with the construction of power plants that use fossil or nuclear fuels.

Alternative energy sources, particularly solar and wind, are growing at a tremendous rate. For the first time, it has become apparent that these energy sources may compete successfully with fossil fuels. Alternative renewable energy sources offer our best chance to develop a truly sustainable energy policy that will not harm the planet.[27,58] Canada has a very active national research and development program for alternative energy technologies; private companies such as Ballard Power Systems (Burnaby, B.C.) and Calgary's Vision Quest Windelectric are also working actively in this area.

This section has supplied a brief introduction to alternative energy sources. Next, we discuss individual sources and, where appropriate, the environmental advantages and disadvantages of each.

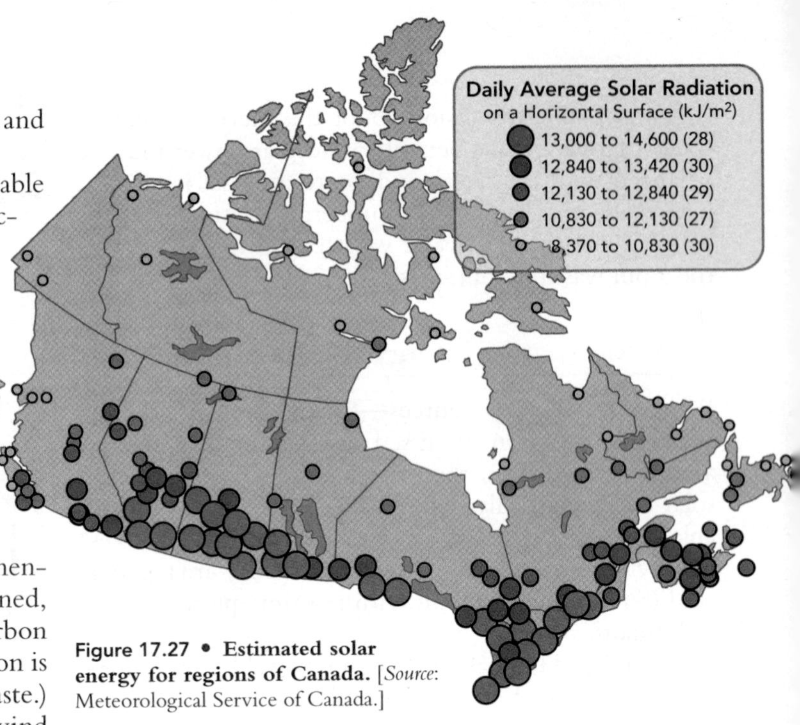

Figure 17.27 • Estimated solar energy for regions of Canada. [*Source*: Meteorological Service of Canada.]

(a)

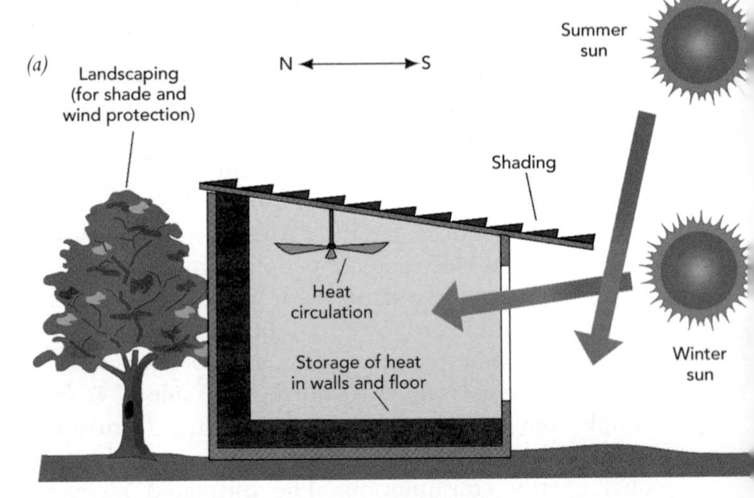

(b)

Figure 17.28 • *(a)* Essential elements of passive solar design. High summer sunlight is blocked by the overhang, but low winter sunlight enters the south-facing window. Other features are designed to facilitate the storage and circulation of passive solar heat. ***(b)* Design of this home utilizes passive solar energy.** Sunlight entears through the windows and strikes a specially designed masonry wall that is painted black. The masonry wall heats up and radiates this heat, warming the house during the day and into the evening. Another passive solar technique makes use of trees that lose their leaves during the winter, which are common in most of Canada. When planted on the sunny side of a building, these trees provide summer shade that cools the building. In the winter, with the leaves gone, winter sunlight enters the house. [*Source*: Moran, Morgan, and Wiersma, Introduction to Environmental Science (New York: Freeman, 1986). Copyright by W. H. Freeman & Company. Reprinted with permission.]

Solar Energy

The total amount of solar energy reaching the Earth's surface is tremendous. For example, on a global scale, 10 weeks of solar energy is roughly equivalent to the energy stored in all known reserves of coal, oil, and natural gas on Earth. Solar energy is absorbed at Earth's surface at an average rate of 120,000 TW (1 TW is 1012 W), which is about 10,000 times the total global demand for energy.[55] In Canada and the United States, on average, 13% of the sun's original energy entering the atmosphere arrives at the surface (equivalent to approximately 177 W/m^2 on a continuous basis). Solar energy is site specific, and detailed observation of a potential site is necessary to evaluate the daily and seasonal variability of its solar energy potential.[59]

Solar energy may be used through passive solar systems or active solar systems. **Passive solar energy systems** often involve architectural designs that enhance the absorption of solar energy by using and adjusting for natural changes that occur throughout the year without requiring mechanical power (Figure 17.28). During the last few thousand years, various societies have used passive solar energy (see Chapter 16). For example, Islamic architects have traditionally used passive solar energy in hot climates to cool buildings.[55]

Thousands of homes and other buildings in Canada now use passive solar systems for at least part of their energy needs.[56] Passive solar energy also provides natural lighting to buildings through windows and skylights. A special glazing on the glass transmits light and provides insulation. Use of passive solar energy features is most cost-effective when they are part of an new building design. However, in some cases, adding passive solar features to an existing building can also be cost-effective.

Active solar energy systems require mechanical power, usually electric pumps and other apparatus, to circulate air, water, or other fluids from solar collectors to a location where the heat is stored until used. We discuss solar collectors next and then describe several specific types of active solar energy systems.

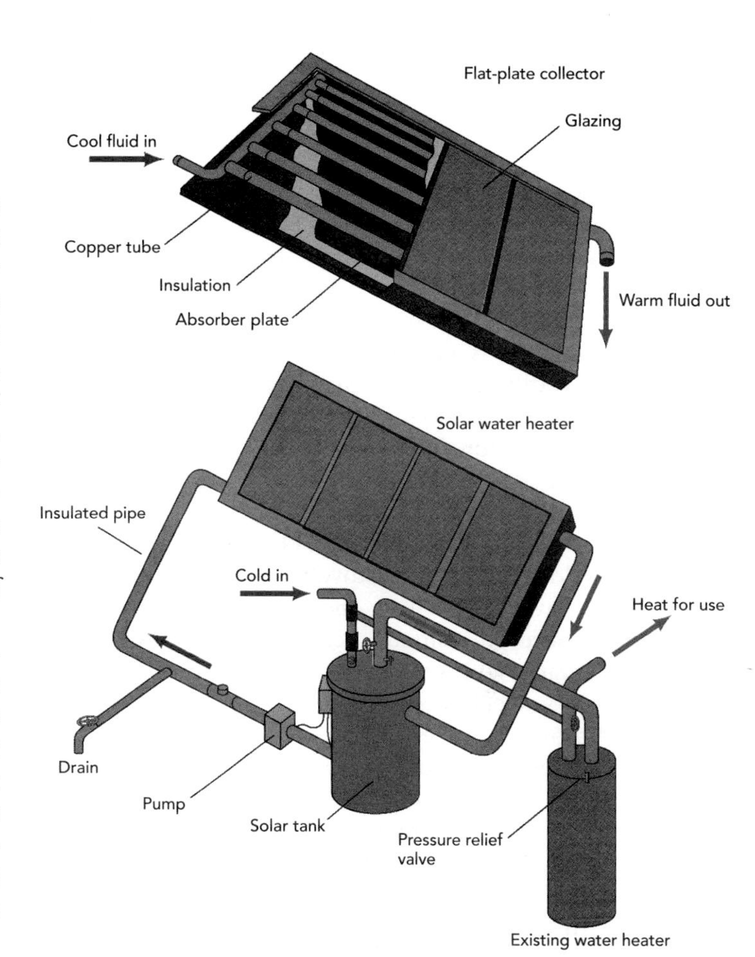

Figure 17.29 • **Detail of a flat-plate solar collector and pumped solar water heater.** [*Source*: Farallones Institute, The Integral Urban House (San Francisco: Sierra Club Books, 1979). Copyright 1979 by Sierra Club Books. Reprinted with permission.]

Figure 17.30 • **Conceptual diagram illustrating how solar cells work.**

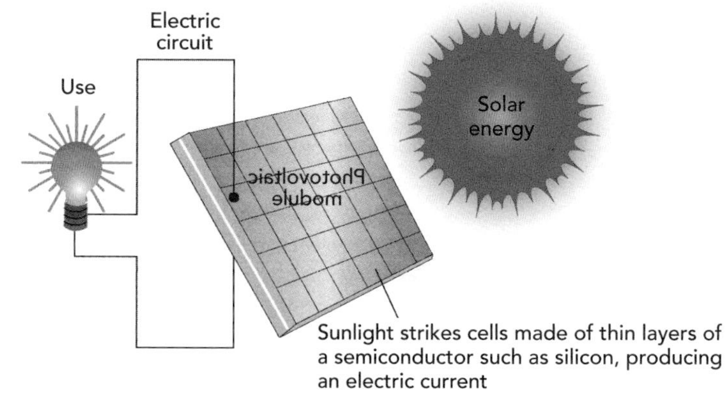

Solar Collectors

Solar collectors that provide space heating or, more commonly hot water, are usually flat panels consisting of a glass-covered plate over a black background where an absorbing fluid (water or a volatile liquid) is circulated through tubes (Figure 17.29). Solar radiation enters the glass and is absorbed by the black background. Heat is emitted from the black material, heating the fluid circulating in the tubes. If water is the absorbing fluid, it is heated to 38–93°C.[57] In cold climates, water is a poor absorbing fluid, because it can freeze. As a result, a liquid with a low freezing temperature is used.

A second type of solar collector is called the evacuated tube collector. Its design is similar to that of the flat-plate collector. The difference is that each tube, along with its absorbing fluid, passes through a larger tube that helps reduce heat loss.[55]

Photovoltaics

Photovoltaics is a technology that converts sunlight directly into electricity (Figure 17.30). The systems use special roof- or wall-mounted solar cells, also called photovoltaic cells, made of thin layers of semiconductors (silicon or other materials) and solid-state electronic components with few or no moving parts.[58,60] Solar cell technology is advancing rapidly, and first-law efficiency is as high as 10%. The cells are constructed in standardized modules, encapsulated in plastic or glass, which can be combined to produce systems of various sizes so that power output can be matched to the intended use.

Electricity is produced when sunlight strikes the cell. The different electronic properties of the layers cause electrons to flow out of the cell through electrical wires. Photovoltaic cells can also work with light energy other than the sun. For example, using the light of your desk lamp, a photovoltaic cell can produce electricity to light a small bulb in a clock or to power your hand-held calculator. Photovoltaics have been used to supply power to satellites and space vehicles, and as a power source in remote areas for electric equipment such as water-level sensors, meteorological stations, and emergency telephones (Figure 17.31).

Photovoltaics is emerging as a significant contributor to developing countries that do not have the financial ability to build large central power plants that burn fossil fuels. These countries are demonstrating that solar technology can be simple, relatively inexpensive, and capable of meeting energy needs for people in many places in the world by matching the demand with an appropriate supply, often at a cost of less than $400 per household.[61] About half a million homes, mostly in villages not linked to a countrywide electrical grid, now receive their electricity from photovoltaic cells.

In 1997, Germany installed photovoltaic cells on 10,000 roofs. The governments of India and the United States both announced that photovoltaic systems will be installed on a million roofs by 2010. Canada has about 10 MW of photovoltaic capacity installed, with plans to expand this capacity by at least 20% by 2008.

In the future, the alternative energy source for many people and communities might well be lightweight photovoltaics in which the solar cells are mounted together to form modules. The growth and technology changes in photovoltaics suggest that in the twenty-first century, solar energy is likely to become a multigigawatt-per-year industry that will provide a significant portion of the energy we use.[62]

Large-Scale Solar Systems

Large-scale solar energy systems have been installed in several locations in the world. Figure 17.32 shows the 10 MW solar power tower at Barstow, California, which collects heat from solar energy using an array of adjustable mirrors, and delivers that energy in the form of steam to turbines that produce electric power.[56]

Solar farms can also be linked into larger, modular systems. A system developed at Luz International is the most technically successful solar power experiment to date. The site, in the Mojave Desert, consists of nine solar farms. Each farm comprises a power plant surrounded by hundreds of solar collectors, and each produces 350 MW of electrical power, sufficient for about 550,000 people. The system uses solar collectors (curved mirrors) to heat synthetic oil that flows through heat exchangers, which

Figure 17.31 • Photovoltaic systems. *(a)* Panels of photovoltaic cells are being used here to power a small refrigerator to keep vaccines cool. The unit is designed to be carried by camels to remote areas in Chad. *(b)* Photovoltaics are used to power emergency telephones along a highway on the island of Tenerife in the Canary Islands.

(a)

(b)

Figure 17.32 • Solar power tower at Barstow, California. Sunlight is reflected and concentrated at the central collector, where the heat is used to produce steam to drive turbines and generate electric power.

in turn drive steam turbine generators. (See Figure 17.33.) The success of the system is based, in part, on the fact that it uses a natural gas burner backup system (75% solar and 25% natural gas), ensuring uninterrupted power generation both on cloudy days and during times of peak demand. Thus, the Luz system is really a combination of solar technology and conventional power generation.[63]

Shallow, salt-gradient **solar ponds** can be used to generate relatively low-temperature water to produce heat for commercial, industrial, and agricultural uses. The ponds are designed to collect incoming solar radiation, which produces a bottom-water temperature of about 70°C. The hot water is kept on the bottom by the addition of salt, which makes the water heavier. Circulation is restricted so that the dense bottom water does not mix with the water above. However, solar radiation penetrates to warm the lower water. The heat can then be extracted from the bottom and used.[64]

Finally, **ocean thermal conversion**, involves using part of the oceanic environment as a gigantic solar collector. The surface temperature of ocean water in the tropics is often about 28°C. However, at the bottom of the ocean, at a depth of about 600 m, the temperature may be 1–3°C. Equipment has been designed to take advantage of this temperature difference, either by using the seawater directly or by using a heat-exchange system in which a fluid such as ammonia or propane is vaporized by the warm water. The expanding vapour then propels a turbine and generates electricity. Following the generation of electricity, the vapour is cooled and condensed by the cold water. An experimental ocean thermal conversion installation in Hawaii needed large amounts of energy to pump water up from deep ocean regions, significantly increasing costs. Some of those costs were offset by aquaculture at the site, which employed nutrient rich deep ocean waters to grow salmon, oysters, abalone, and kelp.

The construction of large-scale ocean thermal plants depends on whether they can be built near potential markets (the people who use the energy) and whether they are economically viable.[59] Answers to these questions remain uncertain, and the future of ocean thermal plants is speculative.

Solar Energy and the Environment

The use of solar energy generally has a relatively low impact on the environment, but there are some environmental concerns. One disadvantage of solar energy is that, because it is not concentrated in space (like a thermal power plant), a large land area is required to generate a large amount of energy. This problem is negligible when solar collectors can be combined with existing structures, as with the addition of solar hot-water heaters on the roofs of existing houses. Highly centralized and high-technology solar energy units, such as solar power towers, have a greater impact on the land because they need considerable space. The impact of large centralized solar energy systems can be minimized by locating them in remote areas not used for other purposes and by making use of dispersed solar energy collectors on existing structures wherever possible.

Another concern involves the large variety of metals, glass, plastics, and fluids used in the manufacture and use of solar equipment. Some of these substances may cause environmental problems through production and by accidental release of toxic materials.

Hydrogen

Hydrogen, the fuel burned by our sun (producing solar energy), is the lightest, most abundant element in the universe. Hydrogen gas may be an important fuel of the future.[56]

Hydrogen is a high-quality fuel that can be easily used in the same way as fossil fuels, such as to power automobile and truck engines and to heat water and buildings. Hydrogen is used and stored in **fuel cells** (see A Closer Look 17.3), which are similar to batteries in that electrons flow between negative and positive poles. However, a fuel cell generates electricity rather than just storing it as in a battery. Hydrogen, like natural gas, can be transported in pipelines and stored in tanks; and it can be produced using solar and other renewable energy sources. It is a clean fuel; the combustion product of burning hydrogen is water, so it does not contribute to global warming, air pollution, or acid rain (see Chapters 20 and 21). Technological improvements in producing hydrogen

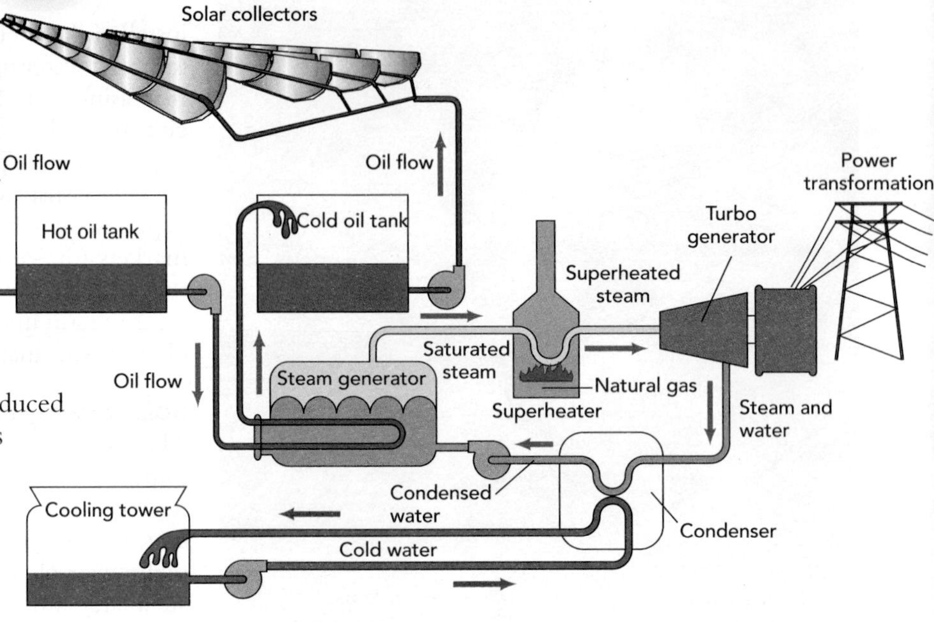

Figure 17.33 • Diagram illustrating how the Luz Solar Farm system works. [*Source*: Courtesy of Luz International.]

are certain, and the fuel price of hydrogen may be substantially reduced in the future.[64]

One way to produce hydrogen is to use an electric current to separate hydrogen (H) from water (H_2O), a process known as electrolysis. If the electricity for electrolysis is produced by solar or wind power, the hydrogen is produced in a clean, carbon-free process.[56] However, the most economical way to produce hydrogen is to use a thermal process in which steam is combined with natural gas (CH_4), removing the carbon (C) and leaving hydrogen (H). This process depends on a fossil fuel and produces some carbon dioxide (CO_2). Hydrogen can also be produced from gasification of biomass. Hydrogen is an ideal medium for the storage of energy, and could therefore help smooth out the natural variability of using renewable energy sources such as solar and wind power, providing electricity on cloudy days and at night or on days when the wind isn't blowing.[55]

The island nation of Iceland, with assistance from the European Union, is currently attempting to become the first hydrogen-based-energy economy. Although Iceland has enormous reserves of geothermal energy that can be used to produce hydrogen for fuel cells, it has no fossil fuels. The most important step will be to create the necessary infrastructure for storage, transport, and fuelling stations for hydrogen, which is as flammable as gasoline.[65]

Water Power

Water power is a form of stored solar energy, as mentioned earlier, because climate, which dictates the flow of water on Earth, is driven in part by differential solar heating of the atmosphere. Water power has been successfully harnessed since the time of the Roman Empire. Waterwheels that convert water power to mechanical energy were turning in western Europe in the seventeenth century; during the eighteenth and nineteenth centuries, large waterwheels provided energy to power grain mills, sawmills, and other machinery.

Today, hydroelectric power plants use the water stored behind dams. In Canada, hydroelectric plants generate over 60% of the total electricity produced in the nation, and about 14% of all electricity produced in the world. However, as with other energy sources, the distribution of hydropower varies across the country, from a high of more than 85% for Manitoba, Québec, Newfoundland and Labrador, and parts of British Columbia and the Yukon, to lows of 9.4% for Nova Scotia, 4.2% for Alberta, and 0% for Prince Edward Island. Figure 17.35*a* shows the major components of a hydroelectric power station.

Hydropower can also be used to store energy through the process of pump storage (Figure 17.35*b* and *c*). During times when demand for power is low (for example, at night in the summer), electricity produced from oil, coal, or nuclear plants in excess of the demand is used to pump water uphill to a higher reservoir (high pool). Then, during times when demand for electricity is high (for example, on hot summer days), the stored water flows back down to a low pool through generators to help provide energy. The advantage of pumped storage lies in the timing of energy production and use.

Small-Scale Systems

Small-scale hydropower systems, designed for individual homes, farms, or small industries, have a long history in Canada, especially in northern regions. Indeed, numerous First Nations communities in remote areas use small-scale hydropower as a principal energy source. These small systems, known as micro-hydropower systems, have power output of less than 100 kW.[68]

Numerous sites in many areas have the potential for producing small-scale electrical power. This is particularly true in mountainous areas, where potential energy from stream water is often available. Micro-hydropower development is site specific, depending on local regulations, economic situations, and hydrologic limitations. Hydropower can be used to generate either electrical power or mechanical power to run machinery; its use

A CLOSER LOOK 17.3

Fuel Cells—An Attractive Alternative

Power produced from burning fossil fuels, particularly coal and fuels used in internal combustion engines (cars, trucks, ships, and locomotives), is associated with serious environmental problems. As a result, we are searching for and developing environmentally benign technologies capable of generating power.[66] One promising technology uses fuel cells, which produce fewer pollutants, are relatively inexpensive, and have the potential to store and produce high-quality energy.

Fuel cells are highly efficient power-generating systems that produce electricity by combining fuel and oxygen in an electrochemical reaction. Hydrogen is the most common fuel type, although fuel cells that run on methanol, ethanol, and natural gas are available. Traditional generating technologies require combustion of fuel in order to convert the resultant heat into mechanical energy (to drive pistons or turbines), and this mechanical energy is then converted into electricity. With fuel cells, however, chemical energy is converted directly into electricity, thus increasing second-law efficiency (see Chapter 16) while reducing harmful emissions.

Basic components of a hydrogen-burning fuel cell are shown in Figure 17.34. Both hydrogen and oxygen are added to the fuel cell in an electrolyte solution. The reactants remain separated from one another, and a platinum membrane prevents electrons from flowing directly to the positive side of the fuel cell. The electrons are routed through an external circuit.[65, 66] The flow of electrons from the negative to the positive electrode is diverted along its path into an electrical motor, supplying current to keep the motor running. In order to maintain this reaction, hydrogen and oxygen are added as needed. When hydrogen is used in a fuel cell, the only waste products are oxygen and water. Using natural gas (CH_4) in fuel cells produces some pollutants, but the amount is only about 1% of what would be produced by burning fossil fuels in an internal combustion engine or a conventional power plant.[66]

Fuel cells are efficient and clean, and they can be arranged in a series to produce the appropriate amount of energy for a particular task. Additionally, the efficiency of a fuel cell is largely independent of its size and energy output. For these reasons, fuel cells are well suited to provide power for automobiles, homes, and large-scale power plants. They can also be used to store energy to be used as needed. Fuel cells are used in many locations. For example, they power buses in Vancouver and at Los Angeles International Airport, and provide heat and power at Vandenberg Air Force Base in California.[67] A Burnaby, British Columbia, company, Ballard Power Systems, is one of the world's leaders in developing and marketing fuel cell systems for transportation and power generation.

Figure 17.34 • Conceptual diagram showing how a fuel cell works and its application to power a vehicle.

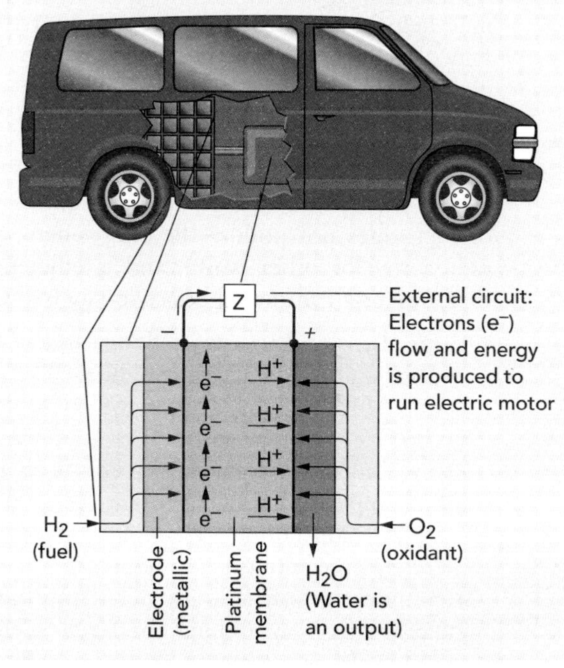

Figure 17.35 • *(a)* **Basic components of a hydroelectric power station.** *(b)* **A pumped storage system.** During light power load, water is pumped from low pool to high pool. *(c)* During peak power load, water flows from high pool to low pool through a generator. [*Source:* Modified from Council on Environmental Quality, Energy Alternatives: A Comparative Analysis (Norman: University of Oklahoma Science and Policy Program, 1975).]

(a)

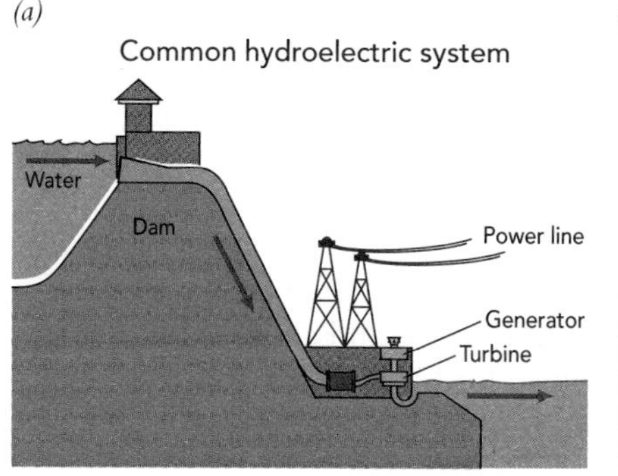

(b)

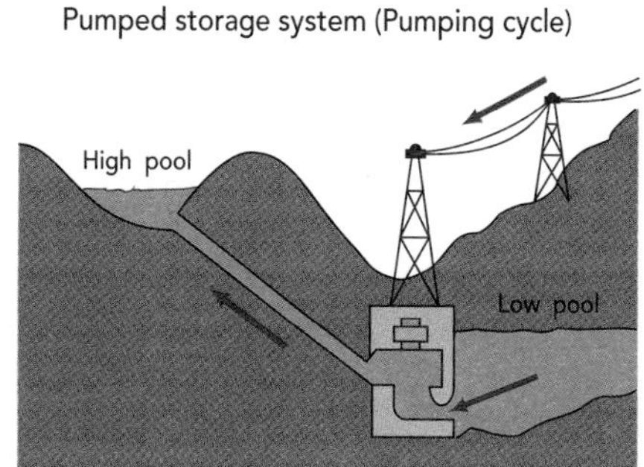

(c)

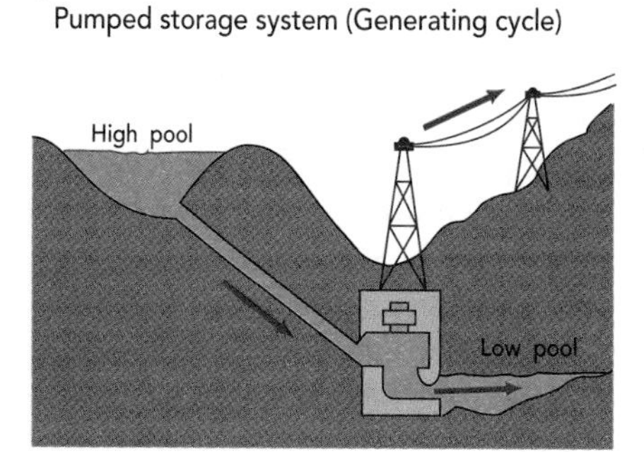

Figure 17.36 • **Tidal power station on the River Rance near Saint-Malo, France.**

may help reduce the high cost of importing energy. It may also help small operations become more independent of local utility providers.[68]

Water Power and the Environment

Water power is clean power; it requires no burning of fuel, does not pollute the atmosphere, produces no radioactive or other waste, and is efficient. However, there are environmental prices to pay (see Chapter 18):

- Large dams and reservoirs flood large tracts of urban and agricultural land that could have had other uses.
- Dams block the migration of some fish, such as salmon.
- Dams can cause adverse health effects on aquatic biota, for example with increased nitrogen gas concentrations in water falling over high dams, or high methyl mercury concentrations in reservoir waters (caused by a combination of naturally occurring inorganic mercury, low oxygen, and high bacterial levels in reservoir sediments).
- Dams trap sediment that would otherwise reach the sea and eventually replenish the beach sand.
- For a variety of reasons, many people do not want to turn wild rivers into a series of lakes.
- Reservoirs with large surface area increase evaporation of water compared to pre-dam conditions. In arid regions, evaporative loss of water from reservoirs is more significant than in more humid regions.

For all these reasons, and because many good sites for dams already have one, the likely growth of large-scale water power in the future (with the exception of a few areas, including Africa, South America, and China) appears limited. Indeed, in Canada and the United States, there is an emerging social movement to remove dams. Hundreds of dams, especially those with few useful functions, are being considered for removal; and a few have already been removed (see Chapter 18).

As mentioned, there does seem to be continued interest in micro-hydropower to supply either electricity or mechanical energy. However, small dams and reservoirs tend to fill more quickly with sediment than large reservoirs, rendering their useful life much shorter. In fact, many dams likely to be removed are small ones filled with sediment. Because micro-hydropower systems can adversely affect stream environments by blocking fish passage and changing downstream flow, careful consideration must be given to their construction. If the number of dams in a region is large, the total impact may be appreciable.

Tidal Power

The use of **tidal power**, a type of water power derived from ocean tides, can be traced back to tenth-century Britain, where tides were used to power coastal mills.[55] However, only in a few places with favourable topography—such as the north coast of France, the Bay of Fundy in Canada (Chapter 7), and the northeastern United States—are the tides sufficiently strong to produce commercial electricity. The tides in the Bay of Fundy have a maximum range of about 15 m. A minimum range of about 8 m is necessary for development of tidal power to be considered.

To harness tidal power, a dam is built across the entrance to a bay or estuary, creating a reservoir. As the tide rises (flood tide), water is initially prevented from entering the bay landward of the dam. Then, when there is sufficient water (from the ocean-side high tide) to run the turbines, the dam is opened, and water flows through it into the reservoir (the bay), turning the blades of the turbines and generating electricity. When the reservoir (the bay) is filled, the dam is closed, stopping the flow and holding the water in the reservoir. When the tide falls (ebb tide), the water level in the reservoir is higher than that in the ocean. The dam is then opened to run the turbines (which are reversible), and electric power is produced as the water is let out of the reservoir.

Figure 17.36 shows the La Rance tidal power plant on the north coast of France. La Rance, constructed in the 1960s, is the first and largest modern tidal power plant. The plant at capacity produces about 240,000 kW from 24 power units spread out across the dam. At the La Rance power plant, most electricity is produced from the ebb tide, which is easier to control.

Tidal power does have environmental impacts. The change in the hydrology of a bay or estuary caused by the dam can adversely affect the vegetation and wildlife. The dam restricts upstream and downstream passage of fish. Furthermore, the periodic rapid filling and emptying of the bay as the dam opens and closes with the tides rapidly changes habitats for birds and other organisms.

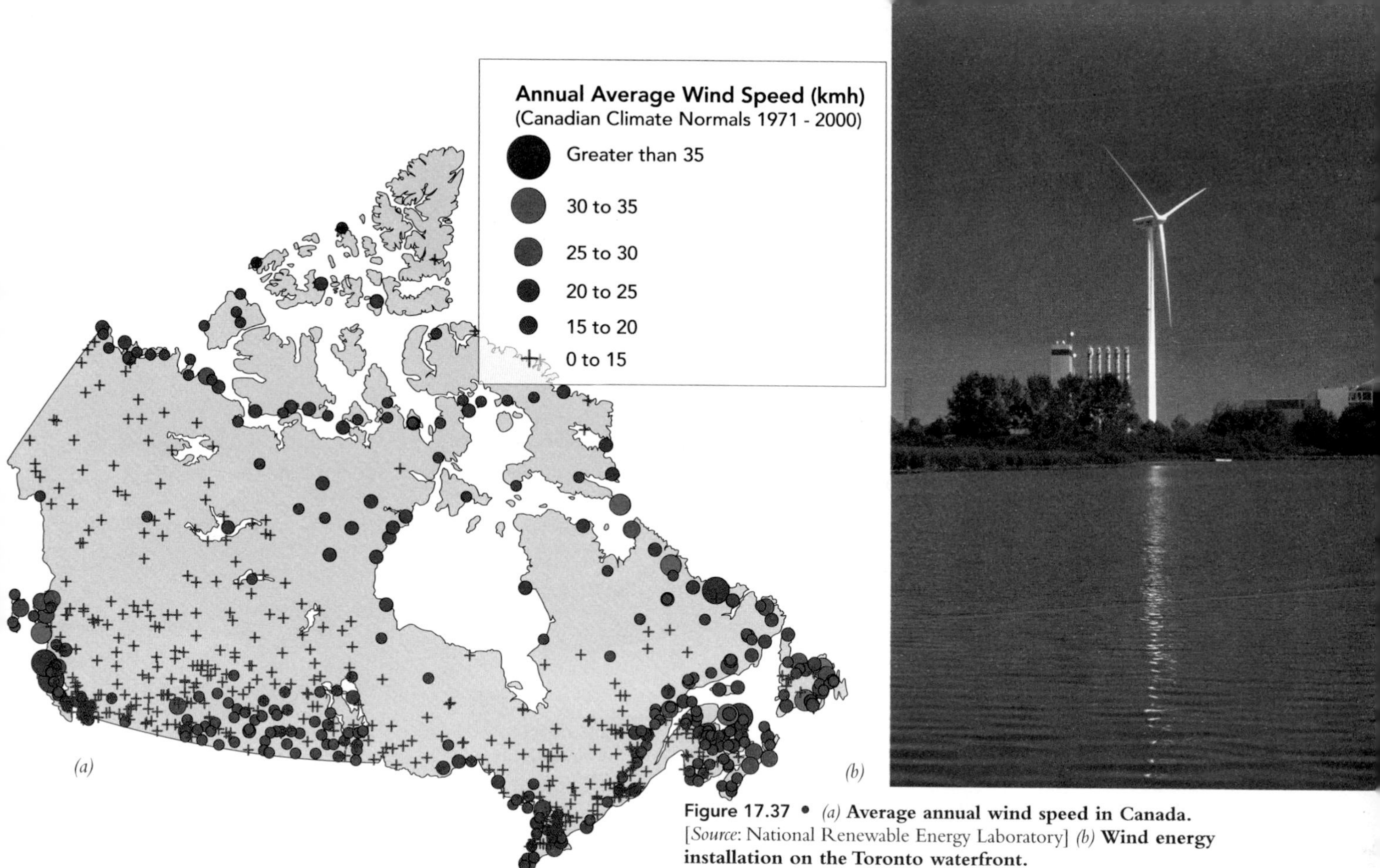

Figure 17.37 • *(a)* **Average annual wind speed in Canada.** [*Source*: National Renewable Energy Laboratory] *(b)* **Wind energy installation on the Toronto waterfront.**

Wind Power

Wind power, like solar power, has evolved over a long period of time, from early Chinese and Persian civilizations to the present. Wind has propelled ships and has driven windmills to grind grain and pump water. In the past, thousands of windmills in western Canada and the United States were used to pump water for ranches. More recently, wind has been used to generate electricity.

Basics of Wind Power

Winds are produced when differential heating of Earth's surface creates air masses with differing heat contents and densities. The potential for energy from the wind is large, and yet there are problems with its use because wind tends to be highly variable in time, place, and intensity.[69]

Wind prospecting has become an important endeavour. On a national scale, regions with the greatest potential for wind energy are coastal British Columbia, the coastal region of Atlantic Canada, the Gaspé Peninsula, and much of Manitoba, Saskatchewan, and Alberta. Other good sites include areas near the Great Lakes, such as the wind power system recently installed in Toronto, on the Lake Ontario shoreline (Figure 17.37*b*). A site with sustained wind velocity of about 5 m/sec or greater is considered a good prospect for wind energy development.[55] Wind speed, and thus wind energy potential, is strongly affected by local terrain. Good wind power locations can be found in most areas of Canada.

Even in a particular area, however, the direction, velocity, and duration of wind may be quite variable, depending on local topography temperature differences in the atmosphere.[69] For example, wind velocity often increases over hilltops, and wind may be funnelled through a mountain pass (Figure 17.38). The increase in wind velocity over a mountain is due to a vertical convergence of wind, whereas in a pass, the increase is partly due to a horizontal convergence. Because the shape of a mountain or a pass is often related to the local or regional geology, prospecting for wind energy is a geologic as well as a geographic and meteorologic problem.[70]

Significant improvements in the size of windmills and the amount of power they produce occurred from the late 1800s through approximately 1950, when many European countries became interested in large-scale generators driven by the wind. In Canada and the United States, thousands of small, wind-driven generators have been used on farms for many years. Today, many large utility companies are considering wind power in their long-range energy planning goals. Since 2000, Germany, with over 6,000 MW installed, has had the highest number of installed wind power systems in the world (Figure 17.39), and the European Union leads the world in using wind power. Where land for wind farms is scarce, wind turbines can be sited offshore.

The cost of producing electricity from wind must be competitive with other sources to be economically

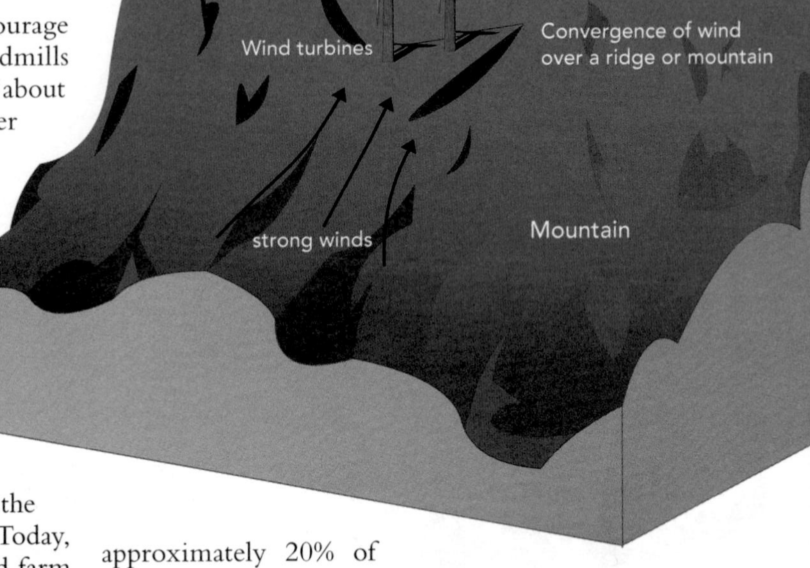

Figure 17.38 • Conceptual diagram showing how wind energy is concentrated by topography.

viable. California created tax incentives to encourage wind power. That state now has about 17,000 windmills installed, with a combined generating capacity of about 1,400 MW, about 60% of the U.S. wind power capacity of about 2,500 MW (Figure 17.37*a*). Clusters of windmills, called wind farms, are located in mountain passes, producing electricity that is transferred to utility lines. Wind power will likely be California's second least expensive source of power by 2010, second only to hydropower.

Wind power is rapidly gaining favour in Canada. In 1998, Calgary, Alberta initiated a program under which consumers can pay a modest surcharge to have their energy supplied by wind. In the pilot stage of the project, 3,000 households were given this option. Today, demand exceeds supply. The Cowley Ridge wind farm in southwestern Alberta, Canada's first commercial wind power installation, can supply only enough power for 7,000 homes, a fraction of those that would buy wind power if it were available.

Other major wind energy facilities are under construction across Canada. For example, Québec's Gaspé Peninsula, one of Canada's windiest regions, has emerged as one of the country's most important wind power sites, with 32% of installed capacity and one of the biggest wind farms in the world, the Parc Éolien du Renard, commissioned in 2003 with a capacity of 2.25 MW. A further 1,000 MW of wind energy is planned for the Gaspé by 2012.

Wind power potential in Canada far exceeds present demand, as is also the case in Britain and many other countries.[71] Consider the implications of wind power for nations such as China. China burns tremendous amounts of coal at a heavy environmental cost, including exposing millions of people to hazardous air pollution. In rural China, exposure to the smoke from burning coal in homes has increased the threat of lung cancer by a factor of nine or more. China could probably double its current capacity to generate electricity with wind alone![62]

Although wind now provides less than 1% of the world's demand for electricity, its growth rate suggests that it could be a major supplier of power in the relatively near future. One scenario suggests that wind power could supply 10% of the world's electricity in the coming decades and, in the long run, could provide more power than hydropower, which today supplies approximately 20% of the electricity in the world.

The wind energy industry has created thousands of jobs in recent years; it is also becoming a major investment opportunity. Technology is producing more efficient wind turbines, thereby reducing the price of wind power. Today, a large state-of-the-art wind turbine is about 100 m in diametre, as tall as a 30 storey building, and produces about 3 MW of electricity. One wind turbine provides enough power for several hundred homes.

In many regions, power generated from wind is already less expensive in some instances than electricity produced from coal-fired and natural gas power plants.[71] The cost of producing the electric energy is about $0.05 per kilowatt hour, which is competitive with electricity from burning natural gas.

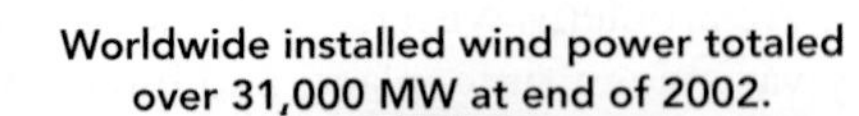

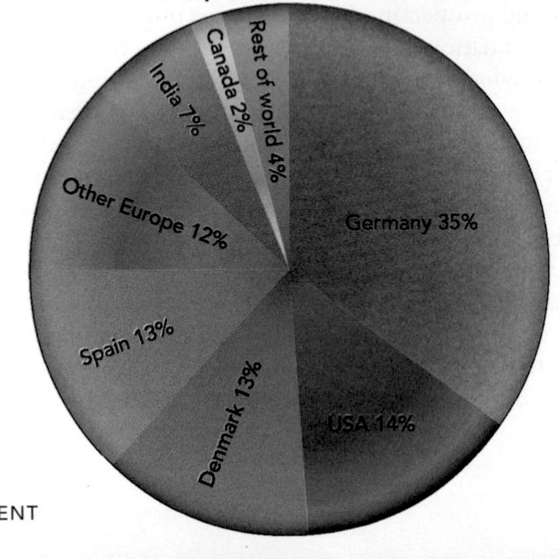

Figure 17.39 • Worldwide installed wind power totaled over 31,000 MW at the end of 2002. (*Source*: British Wind Energy Association, 2003.)

Wind Power and the Environment

The use of wind power will not solve all our energy problems, but as one of the major alternative energy sources, wind power can be used at particular sites to reduce dependence on fossil fuel and help achieve sustainability. Wind energy does have a few disadvantages, especially the use of large areas of land for roads, windmill pads, and similar equipment (and associated aesthetic impacts), and impacts on birds (birds of prey, such as hawks and falcons, are particularly vulnerable). However, everything considered, wind energy has a relatively low environmental impact, and its continued use is certain.

Biomass Energy

Biomass energy is energy recovered from biomass—organic matter, such as plant material and animal waste. Whereas the energy from most other renewable sources comes from physical processes, such as moving water and wind, the energy from biomass comes from chemical bonds formed through photosynthesis in living or once-living matter.[55] Biomass fuel is organic matter that can be burned directly or converted to a more convenient form and then burned. For example, we can burn wood in a stove or convert it into charcoal to use as a fuel. Energy from biomass is the oldest fuel used by humans. Our Pleistocene ancestors burned wood in caves to keep warm and cook food.

Biomass continued to be a major source of energy for human beings throughout most of the history of civilization. When North America was first settled, there was more wood fuel than could be used. The forests often were cleared for agriculture by girdling trees (cutting through the bark all the way around the base of a tree) to kill them and then burning the forests. During the mid-twentieth century, when coal, oil, and gas were plentiful, burning wood became old-fashioned and quaint. Burning wood was something done for pleasure in an open fireplace that conducted more heat up the chimney than it provided for space heating. Now, with other fuels reaching a limit in abundance and production, there is renewed interest in the use of natural organic materials for fuel.

More than 1 billion people in the world today still use wood as their primary source of energy for heat and cooking. In developing countries, biomass currently provides about 35% of the total energy supply.[55] Energy from biomass can take several routes: direct burning of biomass either to produce electricity or to heat water and air; heating of biomass to form a gaseous fuel (gasification); and distillation or processing of biomass to produce biofuels such as ethanol, methanol, or methane.[72] Energy crops such as sugarcane are grown to be fermented to produce ethanol, a fuel that can be used in automobiles. For example, Brazil produces about 12 billion litres of ethanol per year from sugarcane. This is equivalent to about 60% of the fuel used for the nation's automobiles. The wastes from ethanol distillation were originally dumped in rivers, causing water pollution. The wastes are now treated to produce biogas and liquid fertilizers (recycled to sugarcane fields).[55] In Hawaii, sugarcane waste is burned to produce electricity.

Other sources of biomass energy include cattle dung and dried peat, both of which are burned for cooking and heating in other parts of the world. In North America, the primary sources of biomass fuels are forest products, agricultural residues, energy crops, animal manure, and urban waste (Figure 17.40). The BIOCAP Canada Foundation estimates that Canadian forest resources have a biomass energy potential equal to 69 times the country's annual consumption of fossil fuels. Combined with agricultural resources and by-products from manufacturing operations, Canada's biomass energy potential has been estimated as 27% of the energy we currently derive from fossil fuels.[73]

Methane and biogas can also be produced from municipal solid waste in landfills and sewage at wastewater treatment plants. In rural China, about 5 million small facilities are used to treat sewage. The original purpose was to reduce disease, but their potential as an energy source was soon recognized and used.[55] Today, many facilities, including some in Canada, process municipal solid waste to generate electricity or to be used as a fuel. In Western Europe a number of countries use from one-third to one-half their municipal waste for energy production. With the end of inexpensive, available fossil fuels, additional energy recovery systems utilizing urban waste will no doubt emerge.[74]

Sweden is a leader in biomass energy, with 17% of its primary energy consumption from biomass, much of it from the country's pulp and paper industry.[55] In Canada and the United States, biomass energy is a small but growing renewable source of electric power. Sweden's example shows that we may be able to make much better use of this promising energy source.

Biomass Energy and the Environment

The use of biomass fuels can pollute the air and degrade the land. For most of us, the smell of smoke from a single campfire is part of a pleasant outdoor experience; but under certain weather conditions, wood smoke from many campfires or chimneys in narrow valleys can lead to air pollution. Combustion of biomass-derived fuels generally releases fewer pollutants such as sulphur dioxide and nitrogen oxides than combustion of coal and gasoline; however,[75] this is not always the case for burning urban waste. Although plastics, glass, and hazardous materials are removed before burning,

Build Your Environmental Skills 17.1

Multicriteria Evaluation of Alternative Energy Sources

The world is moving into a new era, one of transition from almost total dependence on fossil fuels to greater use of alternative renewable sources of energy. Although each of the alternatives offers a way out of the energy dilemma created by population growth and technological development, each has advantages and disadvantages. How can we evaluate the alternatives and select the right mix of energy sources for the coming decades? We can begin by comparing them on the basis of those characteristics most important to us: cost, jobs lost or gained, environmental impact, and potential for supplying energy. One technique for this is called multicriteria evaluation (MCE). In its simplest form, MCE is just a table showing alternatives down one axis, and the criteria we might use to judge each alternative down the other. The table below illustrates such a framework.

A Simple Multicriteria Procedure

1. Using what you have learned about alternative energy in this chapter and elsewhere, evaluate the environmental impacts of the energy sources listed in the accompanying table. Complete the last column of the table. You may wish to subdivide the column into advantages and disadvantages.
2. Using the numbers 1–10, where 10 represents the best and 1 the worst, assign a rating to each value in the table. For example, in the column for carbon reduction, you might assign a rating of 10 to wind because it results in 100% reduction of carbon emissions. Solar thermal energy would then receive a rating of 8.4. In rating environmental impact, you will have to use your judgment in assigning numerical values.
3. One way to evaluate the various alternatives would be to add up the rating scores for each energy source and see which ones received the highest score. However, you may feel that some of the characteristics are more important than others and therefore should be weighted more heavily. Assign a weight to each column of the table, taking into consideration the importance you believe each should have in decision making. For example, if you believe that costs are more important than land used, you will assign a higher value to costs. In order to be able to compare your evaluation with those of your classmates, use decimal fractions for the weights, such as 0.2. The total should add up to 1.0.
4. Now, for each energy source, multiply its rating in each column by the weight you have assigned to the column. What is the total weighted score for each energy source? What are the sources in order of score, from highest to lowest?
5. Based on this analysis, what policy and research recommendations would you make to the government concerning alternative sources of energy?

Energy Source	*Recoverable Resource*[a] *(exajoule/yr)*	*Costs in 1998 Cents*[b] *(per KWh)*		*Land Use*[c] *(m^2/GWh for 30 years)*	*Carbon Reduction (%)*	*Carbon Avoidance Cost*[d] *($/ton)*	*Number of Jobs*[e] *(per thousand GWh/yr)*	*Environmental Impact*
		1988	*2000*					
Wind	10–40	8	5	1,355	100	95	542	
Geothermal	Small	4	4	404	99	110	112	
Photovoltaic	35	30	10	3,237	100	819	—	
Solar thermal	65	8	6	3,561	84	180	248	
Biomass	13–26	5	NA	—	100[f]	125	—	
Combined-cycle coal	—	6[g]	—	3,642	10	954	116	
Nuclear	—	15[h]	—	—	86	535	100	

[a] Recoverable resource is a measure of how much of the energy can be captured or exploited. From M. Brower, *Cool Energy* (Washington, D.C.: Union of Concerned Scientists, 1990), p. 19.
[b] L. R. Brown, C. Flavin, and S. Postel, *Saving the Planet* (New York: Norton, 1991), p. 27.
[c] Ibid., p. 60.
[d] Based on comparison with existing coal-fired plants. From C. Flavin, "Slowing Global Warming," in *State of the World* (New York: W.W. Norton, 1990), p. 27.
[e] Brown et al., p. 62.
[f] Assumes that the amount of carbon dioxide released in combustion will be consumed by replanted vegetation.
[g] C. Flavin, "Building a Bridge to a Sustainable Future," in *State of the World* (New York: Norton, 1992), p. 35.
[h] A. K. Reddy and J. Goldenberg, "Energy for the Developing World," *Scientific American* 263 (3)(1990):116.

some inevitably slip through the sorting process and are burned, releasing air pollutants, including metals and organic compounds. Burning urban waste to recover energy is preferable to landfill disposal, but incineration of waste paper competes with recycling, which is preferable to disposal or burning.

The use of biomass as fuel can also place pressure on an already heavily used resource. A worldwide shortage of firewood is adversely affecting natural areas and endangered species. For example, the need for firewood has threatened the Gir Forest in India, the last remaining habitat of the Indian lion (not to be confused with the Indian tiger). The world's forests will also decrease if our need for forest products and forest biomass fuel exceeds the productivity of the forests.

If our forest and crop resources are managed properly (for sustainability), it may be possible to make biomass energy more attractive. Many millions of hectares of land that are unsuitable for food production could be used for growing biomass crops (trees and other plants) using short rotation times (time between harvests). However, forest plantations would have to be managed for sustainability, because deforestation accelerates the process of soil erosion (soils without vegetation cover erode more quickly). When fine-grained (silt and clay) particles from soil erosion enter streams and rivers, the water becomes muddy, and its quality is degraded.

Biomass energy in its various forms appears to have a future as an energy source.[55] It is abundant (at the global scale, roughly equivalent to total present energy consumption), easily stored, and renewable. However, questions remain about the amount of energy biomass can reasonably provide and sustainable extraction rates. Any use of biomass fuel must therefore be part of the general planning for all uses of the land's resources.[76]

Geothermal Energy

Geothermal energy is natural heat from the interior of the Earth that is converted to heat buildings and generate electricity. The idea of harnessing Earth's internal heat is not new. As early as 1904, geothermal power was used in Italy. Today, Earth's natural internal heat is being used to generate electricity in 21 countries, including Russia, Japan, New Zealand, Iceland, Mexico, Ethiopia, Guatemala, El Salvador, the Philippines, the United States, and Canada. Total worldwide production is approaching 9,000 MW (equivalent to nine large modern coal-burning or nuclear power plants)—double the amount in 1980. Some 40 million people today receive their electricity from geothermal energy at a cost competitive with that of other energy sources.[77] In El Salvador, geothermal energy supplies 30% of the total electric energy used, but worldwide it is only about 0.15% of the total energy supply.[55] Manitoba leads Canada in geothermal power, with 40% more installations in 2004 than in 2003 and 30% of the installations in Canada. This is in part the result of aggressive marketing efforts and financial incentives, aimed at reducing peak demands on Manitoba Hydro's generating stations by promoting alternative energy sources.

Geothermal energy may be considered a nonrenewable energy source when rates of extraction are greater than rates of natural replenishment. However, geothermal energy has its origin in the natural heat production within Earth, and only a small fraction of the vast total resource base is being utilized today. Although most geothermal energy production involves the tapping of high-heat sources, people are also using the low-temperature geothermal energy of groundwater in some applications.

Geothermal Systems

The average heat flow from the interior of the Earth is very low, about 0.06 W/m^2. This amount is trivial compared with the 177 W/m^2 from solar heat at the surface in Canada and the United States. However, in some areas, heat flow is sufficiently high to be useful for producing energy.[55] For the most part, areas of high-heat flow are associated with plate tectonic boundaries (see Chapter 5). Oceanic ridge systems (divergent plate boundaries) and areas where mountains are being uplifted and volcanic island arcs are forming (convergent plate boundaries) are areas where this natural heat flow is anomalously high. One such region is located in western North America, where recent tectonic and volcanic activity has occurred.[77, 78]

On the basis of geologic criteria, several types of hot geothermal systems (with temperatures greater than about 80°C, have been defined, and the resource base is larger than that of fossil fuels and nuclear energy combined. A common system for energy development is hydrothermal convection, characterized by the circulation of steam and/or hot water that transfers heat from depths to the surface. The Geysers Geothermal Field,

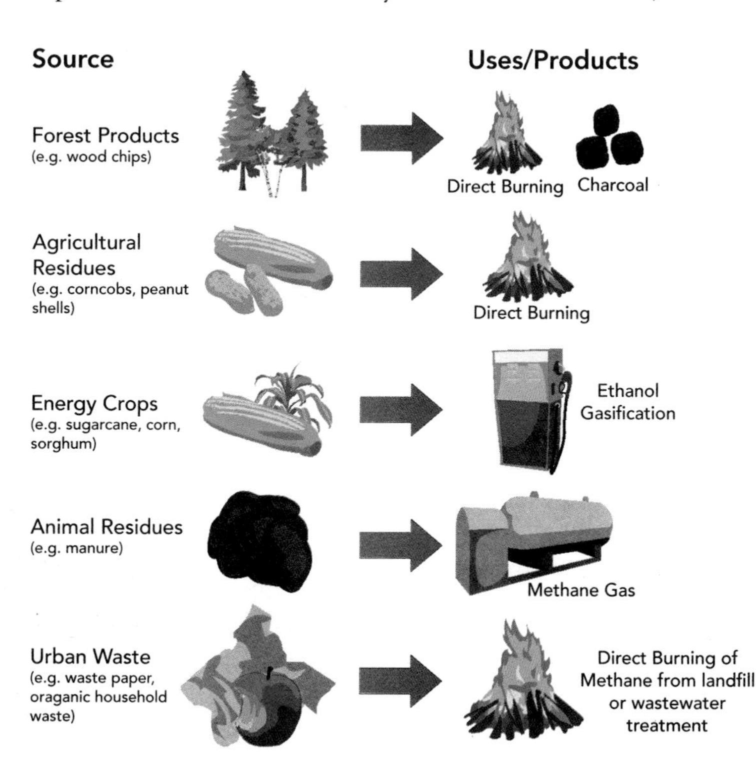

Figure 17.40 • Selected examples of biomass sources, uses, and products.

north of San Francisco, California, is the largest geothermal power operation in the world, producing about 2,000 MW of electrical energy (Figure 17.41).

It may come as a surprise to learn that relatively low-temperature groundwater can be considered a source of geothermal energy. It is geothermal because the normal internal heat flow from Earth keeps the temperature of groundwater at a depth of 100 m at about 13°C. Water at 13°C is cold for a shower; but compared with winter temperatures in much of Canada and the United States, it is warm and can help heat a house. Furthermore, compared with summer temperatures of 30–35°C, groundwater at 13°C is cool and can be used to provide air conditioning. In the summer, heat can be transferred from the warm air in a building to the cool groundwater. In the winter, when the outdoor temperature is below about 4°C, heat can be transferred from the groundwater to the air in the building, reducing the heating needed from other sources. Technology for such heat transfer is well known and available.

Geothermal Energy and the Environment

The environmental impact of geothermal energy may not be as extensive as that of other sources of energy, but it can be considerable. When geothermal energy is developed at a particular site, environmental problems include on-site noise, gas emissions, and disturbance of the land at drilling sites, disposal sites, roads, pipelines, and power plants. Development of geothermal energy does not require large-scale transportation of raw materials or refining of chemicals, as development of fossil fuels does. Furthermore, geothermal energy does not produce the atmospheric pollutants associated with burning fossil fuels or the radioactive waste associated with nuclear energy. However, geothermal development often does produce considerable thermal pollution from hot wastewaters, which may be saline or highly corrosive, producing disposal and treatment problems.

Geothermal power is not very popular in some locations among some people. For instance, geothermal energy has been produced for years on the island of Hawaii, where active volcanic processes provide abundant near-surface heat. There is controversy, however, over further exploration and development. Native Hawaiians and others have argued that the exploration and development of geothermal energy degrade the tropical forest as developers construct roads, build facilities, and drill wells. In addition, religious and cultural issues in Hawaii relate to the use of geothermal energy. For example, some people are offended by using the "breath and water of Pele" (the volcano goddess) to make electricity. This issue points out the importance of being sensitive to the values and cultures of people where development is planned.

17.6 Policy Issues

Renewable energy requires a broad energy policy that takes the long-term view. Policy issues and goals for alternative renewable energy include the following:[55]

- Encouraging and funding alternative energy research and development.
- Ensuring fair access to alternative energy sources.
- Developing a fiscal framework that takes into account the external costs of fossil fuels (air pollution and health costs) and the benefits of alternative energy sources (less environmental damage).
- Developing planning initiatives that promote environmentally preferred and socially acceptable alternative energy.
- Establishing industrial training programs to retrain employees in alternative energy industries, especially in rural agricultural regions and former coal mining communities.
- Encouraging funding institutions to support community investment in alternative energy.

These policy issues and goals are important in discussions concerning the future of alternative renewable energy sources. Before oil production declines, we need to rethink our entire energy planning strategy. In Canada, commitments to control greenhouse gas emissions under the Kyoto Protocol (Chapter 20) have proved to be a boon for the development of alternative energy sources, for instance through the creation of new financial incentives. (The "feebate" system described for hybrid automobiles is one example of this.) Phasing in alternative renewable energy in time to meet our

Figure 17.41 • Geyser Goeothermal Field, located north of San Francisco, California. The Geysers is the largest geothermal power operation in the world and produces energy directly from steam.

future energy demands is critical to avoiding a looming energy crisis in the years to come. However, obstacles remain. Experience elsewhere in the world suggests that a successful move to alternative energy sources will require three main policy actions: introduction of price subsidies for sales of non-traditional energy; grants and subsidies for the development of required infrastructure; and a tax system that preferentially rewards alternative energy producers.

Summary

- Fossil fuels are forms of stored solar energy. Most are created from the incomplete biological decomposition of dead organic material that is buried and converted by complex chemical reactions in the geologic cycle. Because fossil fuels are non-renewable, we will eventually have to develop other sources to meet our energy demands. We must decide when the transition to alternative fuels will occur and what the impacts of the transition will be.
- Environmental impacts related to oil and natural gas include those associated with exploration and development (damage to fragile ecosystems, water pollution, air pollution, and waste disposal); those associated with refining and processing (soil, water, and air pollution); and those associated with burning oil and gas for energy to power automobiles, produce electricity, run industrial machinery, heat homes, and so on (air pollution).
- Coal is a source of energy particularly damaging to the environment. The environmental impacts associated with mining, processing, transporting, and using coal are many. Burning coal can release air pollutants, including sulphur dioxide and carbon dioxide. Finally, burning coal produces a large volume of combustion products and by-products such as ash, slag, and calcium sulphite (from scrubbing).
- Nuclear fission is the process of splitting an atomic nucleus into smaller fragments. As fission occurs, energy is released. The major components of a fission reactor are the core, control rods, coolant, and reactor vessel.
- Nuclear radiation occurs when a radioisotope spontaneously undergoes radioactive decay and changes into another isotope. The three major types of nuclear radiation are alpha, beta, and gamma. Different types of radiation have different toxicities; and in terms of the health of humans and other organisms, it is important to know the type of radiation emitted and the half-life.
- The nuclear fuel cycle consists of mining and processing uranium, generating nuclear power through controlled fission, reprocessing spent fuel, disposal of nuclear waste, and decommissioning power plants. Each part of the cycle is associated with characteristic processes, all with different potential environmental problems.
- The dose response for radiation is fairly well established. We know the dose response for higher exposures, when illness or death occurs. However, there are vigorous debates concerning the health effects of low-level exposure to radiation and what relationships exist between exposure and cancer mortality. Most scientists believe that radiation can cause cancer. Ironically, radiation can be used to kill cancer cells, as in radiotherapy treatments.
- Nuclear power is again being seriously evaluated as an alternative to fossil fuels. On the one hand, it has advantages in that it emits no carbon dioxide, will not contribute to global warming or cause acid rain, and can be used to produce alternative fuels such as hydrogen. On the other hand, people are uncomfortable with nuclear power because of possible accidents and waste disposal problems.
- The use of renewable alternative energy sources, such as wind and solar energy, is growing rapidly. These energy sources do not cause air pollution, create health problems, or cause climate changes. They offer our best chance to replace fossil fuels and develop a sustainable energy policy.
- Solar energy, photovoltaics, and wind energy are particularly promising for Canada and other countries in the world. Other alternative energy sources, including geothermal energy, hydrogen gas used in fuel cells, and biomass energy, are not widely used but may become important in the future.
- Water power today provides over two-thirds of the total electricity produced in Canada. Most good sites for large dams have already been utilized. Water power is clean, but there is an environmental price to pay in terms of disturbance of ecosystems, sediment trapped in reservoirs, loss of wild rivers, and loss of productive land.
- Significant policy issues are related to the future of alternative renewable energy. These include financing of research and development, access to energy sources, education, and community involvement.

STUDY QUESTIONS

1. Compare the potential environmental consequences of burning oil, burning natural gas, and burning coal.
2. What actions can you take at a personal level to reduce consumption of fossil fuels?
3. What are some of the technical solutions to reducing air-pollutant emissions from burning coal? Which are best? Why?
4. What do you think about the idea of allowance trading as a potential solution to reducing pollution from burning coal?
5. If exposure to radiation is a natural phenomenon, why are we worried about it?
6. What is the normal background radiation that people receive? Why is it variable? How does it compare to radiation from human activities such as nuclear power generation and waste management?
7. What is the role of reactor coolant in a nuclear power plant? What is the role of the reactor moderator?
8. If carbon-14 has a half life of 5,730 years, how many years will it take until it has decayed to one-tenth of its original activity?
9. Suppose it is recommended that high-level nuclear waste be disposed of in the geologic environment of the region in which you live. How would you go about evaluating potential sites?
10. Are there good environmental reasons to develop and build new nuclear power plants? Discuss both sides of the issue.
11. What type of government incentives could be used to encourage use of alternative energy sources? Would their widespread use impact our economic and social environment?
12. Your town is near a large river that has a nearly constant water temperature of about 15°C. Could the water be used to cool buildings in the hot summers? How? What would be the environmental effects?
13. It is the year 2500, and natural oil and gas are rare curiosities that people see in museums. Given the technologies available today, what would be the most sensible fuel for airplanes? How would this fuel be produced to minimize adverse environmental effects?

FURTHER READING

Boyle, G., B. Everett, and J. Ramage. *Energy Systems and Sustainability*. Oxford (UK): Oxford University Press, 2003. See excellent discussion of fossil fuels.

Caleira, Ken, Atul Jain, and Martin Hoffert. "Climate Sensitivity Uncertainty and the Need for Energy without CO2 Emission." *Science* 299 (2003): 2052–2054. A good in-depth discussion of climate change, energy needs, and possibilities for alternative energy sources.

Fay, J. A., and D.S. Golomb. *Energy and the Environment*. New York. Oxford University Press, 2003. See Chapters 1–5 for fossil fuels.

Gross, R., M. Leach, and A. Bauen. "Progress in renewable energy." *Environment International* 29 2003 (10: 105-122.

Hellemans, Alexander. "Solar Homes for the Masses." *Science* 285 (1999):679. A look at the feasibility of solar homes with indications of the progress made to date.

Kaul, S. and R. Edinger. 2004. Efficiency versus cost of alternative fuels from renewable resources: outlining decision parameters. *Energy Policy* 32(7): 929-935. Discussion of the relative costs of energy production with renewable energy sources compared with traditional sources and how these technologies have and will continue to become more competitive with oil and coal.

Liu, P. I. *Introduction to Energy and the Environment*. New York: Van Nostrand Reinhold, 1993. A good summary of energy sources and issues, with discussions of the environmental effects of various energy sources.

Natural Resources Canada. *Renewable Energy Strategy: Creating a New Momentum*. Ottawa: Natural Resources Canada, Energy Resources Branch, 1996.

Nuclear Energy Agency (NEA) and Organization for Economic Co-Operation and Development (OECD). 1994. *Power Generation Choices: Costs, Risks, and Externalities*. Proceedings of an international symposium, Washington, D.C., September 23–24, 1993. NEA, OECD. Includes discussion of the economics of nuclear power versus other energy sources.

Nuclear Energy Agency (NEA) and Organization for Economic Co-Operation and Development (OECD). 1995. *Environmental and Ethical Aspects of Long-Lived Radioactive Waste Disposal*. Proceedings of an international workshop, Paris, September 1–2, 1994. NEA, OECD. Essays covering topics of environmental policies, ethical and environmental considerations, cost-benefit analysis, and disposal issues of long-lived radioactive waste.

Pimentel, D., M. Herz, M. Glickstein, M. Zimmerman, R. Allen, K. Becker, J. Evans, B. Hussain, R. Sarsfeld, A. Grosfeld, and T. Seidel. "Renewable energy: current and potential issues." *Bioscience* 52, no. 12 (2002): 1111-1120.

Sheffield, J. "The role of energy efficiency and renewable energies in the future and world energy market." *Renewable Energy* 10, no. 2 (1997):315–318.

Turner, John. "A Realizable Renewable Energy Future." *Science* 285 (1999):687–689. A good analytical treatment of the potential of alternative energy sources, particularly hydrogen.

Tammemagi, Hans and David Jackon. *Unlocking the Atom: The Canadian Guide to Nuclear Technology*. Hamilton, Ontario: McMaster University Press, 2002.

Wald, M. "Dismantling Nuclear Reactors," *Scientific American*, March 2003, pp. 60–69. This is an in-depth discussion of steps in dismantling a nuclear power plant and some of the unforeseen difficulties.

Wohlgemuth, N. and F. Missfeldt. "The Kyoto mechanisms and the prospects for renewable energy technologies." *Solar Energy* 69, no. 4 (2000): 305-314.

World Health Organization. *Health Consequences of the Chernobyl Accident*. Geneva, Switzerland: World Health Organization, 1995. A short book covering the accident, response, health consequences, findings, and proposed future work.

Young, J. P., and R. S. Yalow, eds. *Radiation and Public Perception: Benefits and Risks*. Washington, D.C.: American Chemical Society, 1995. A comprehensive look at public perception of radiation risks and health effects of radiation through experimentation, occupational exposure, atomic detonation, and nuclear reactor accidents.

Part III

Air, Water, and Waste

Waste or resource? This biofuels facility in Northern India uses rice husks to generate energy, transforming a waste material into a useful resource.

Photograph courtesy Isobel Heathcote

chapter 18

Removal of a Dam: A Québec Case Study

The idea that dams, unlike the pyramids of Egypt, are not a permanent feature of the human-made landscape is somewhat new to North Americans. The idea of actually removing a dam was, until recently, almost unheard of. Nevertheless, in the past few years hundreds of dams have been removed in Canada and the United States. A good example is the removal of the dam at Lac Édouard at La Mauricie National Park in Québec in 1996 (Figure 18.1*a*).

The Lac Édouard Dam was constructed in 1900 and was still in place when the area was designated a national park in 1971. Shortly after that, Parks Canada rebuilt the dam on its original foundations, in an effort to protect the lake's recreational beach. The new dam retained the water level of the original dam, at about 1.2 metres above the natural level.

Over the next 25 years, the rebuilt dam deteriorated steadily, and water levels began to fluctuate widely, adversely affecting fish communities in the lake and river. Park biologists were also concerned that the dam was damaging fisheries by blocking the fishes' access to upriver spawning grounds. In 1996, Parks Canada removed the dam entirely. In its place, they constructed three terraces of rock fill, so the site now has a more natural "stair step" appearance typical of natural streams. Park biologists also constructed new spawning habitat upstream and in the terraced steps. With the dam removed, a larger beach area is now possible. And although water levels in the lake now fluctuate slightly, they do so naturally and without the ecological impacts caused by the dam structure.[1]

- *Water is a critical, limited, renewable resource in many regions on Earth. This chapter discusses our water resources in terms of supply, use, management, and sustainability. It also addresses important environmental concerns related to water: wetlands, dams and reservoirs, channelization, and flooding. Near the end of this chapter we present a discussion of water use, management, and environmental concerns for the Fraser River and the complex social, economic, and biophysical aspects of that system.*

Water Supply, Use, and Management

LEARNING OBJECTIVES

Although water is one of the most abundant resources on Earth, many important issues and problems are involved in water management. After reading this chapter, you should understand:

- What a water budget is, and why it is useful in analyzing water supply problems and potential solutions.
- What groundwater is, and what environmental problems are associated with its use.
- How water can be conserved at home and in industrial and agricultural practice.
- Why sustainable water management will become more difficult as the demand for water increases.
- What the environmental impacts are of water projects such as dams, reservoirs, canals, and channelization.
- What a wetland is, how wetlands function, and why they are important.
- What hazards are presented by river flooding.
- Why we are facing a growing global water shortage linked to our food supply.

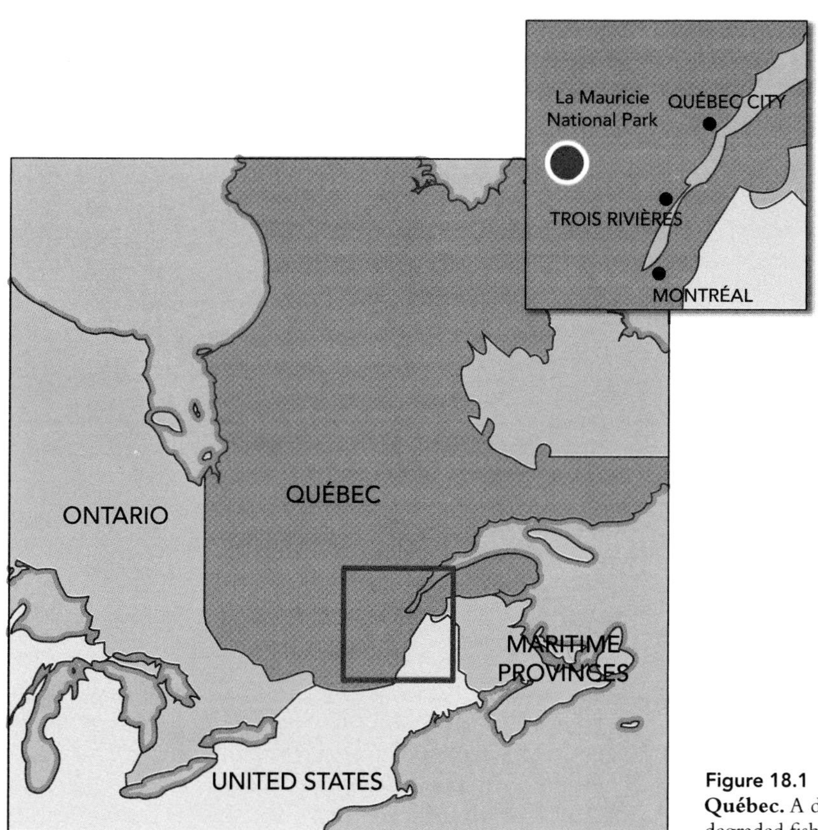

Figure 18.1 • **Lac Édouard, La Mauricie National Park, Québec.** A dam at Lac Édouard was removed in 1996 to restore degraded fish habitat. Location of La Mauricie National Park, Québec.

18.1 Water

To understand water as a necessity, as a resource, and as a factor in the pollution problem, we must understand its characteristics, its role in the biosphere, and its role in living things. Water is a unique liquid; without it, life as we know it is impossible. Consider the following:

- Compared with most other common liquids, water has a high capacity to absorb and store heat. The capacity of water to hold heat has important climatic significance. Solar energy warms the oceans of the world, storing huge amounts of heat. The heat can be transferred to the atmosphere to develop hurricanes and other storms. The heat in warm oceanic currents such as the Gulf Stream warms Great Britain and Western Europe, making these areas much more hospitable for humans than would otherwise be possible at such high latitudes.
- Water is the universal solvent. Because many natural waters are slightly acidic, they can dissolve a great variety of compounds, from simple salts to minerals, including sodium chloride (common table salt) and calcium carbonate (calcite) in limestone rock. Water also reacts with complex organic compounds, including many amino acids found in the human body.
- Compared with other common liquids, water has a high surface tension, a property that is extremely important in many physical and biological processes that involve moving water through, or storing water in, small openings or pore spaces.
- Among the common compounds, water is the only one whose solid form is lighter than its liquid form (it expands by about 8% when it freezes, becoming less dense). That is why ice floats. If ice were heavier than liquid water, it would sink to the bottom of the oceans, lakes, and rivers. If water froze from the bottom up, shallow seas, lakes, and rivers would freeze solid. All life in the water would die, because cells of living organisms are mostly water, and as water freezes and expands, cell membranes and walls rupture. If ice were heavier than water, the biosphere would be vastly different from what it is, and life, if it existed at all, would be greatly altered.[2]
- Sunlight penetrates water to variable depths, permitting photosynthetic organisms to live below the surface.

A Brief Global Perspective

The problem with the water supply is that there is a growing global water shortage linked to the food supply. We will return to this important concept at the end of the chapter, following a discussion of water use, supply, and management.

A review of the global hydrologic cycle, introduced in Chapter 5, is important here. The main process in the cycle is the global transfer of water from the atmosphere to the land and oceans and back to the atmosphere (Figure 18.2). Table 18.1 lists the relative amounts of water in the major storage compartments of the cycle. Notice that more than 97% of Earth's water is in the oceans; the next largest storage compartment, the ice caps and glaciers, accounts for another 2%. Together, these sources account for more than 99% of the total water, and both are generally unsuitable for human use because of salinity (seawater) and location (ice caps and glaciers). Only about 0.001% of the total water on Earth is in the atmosphere at any one time. However, this relatively small amount of water in the global water cycle, with an average atmosphere residence time of only about nine days, produces all our freshwater resources through the process of precipitation.

Water can be found in either liquid, solid, or gaseous form at a number of locations at or near Earth's surface. Depending on the specific location, the residence time may vary from a few days to many thousands of years (see Table 18.1). However, as mentioned, more than 99% of Earth's water in its natural state is unavailable or unsuitable for beneficial human use. Thus, the amount of water for which all the people, plants, and animals on Earth compete is much less than 1% of the total.

As the world's population and industrial production of goods increases, the use of water will also accelerate. The world per capita use of water in 1975 was about 700 m^3/year, or 2 m^3/day, and the total human use of water was about 3,850 km^3/year. Today, world use of water is about 6,000 km^3/yr, which is a significant fraction of the naturally available fresh water.

Compared with other resources, water is used in very large quantities. In recent years, the total mass (or weight) of water used on Earth per year has been approximately 1,000 times the world's total production of minerals, including petroleum, coal, metal ores, and nonmetals.[3] Because of its great abundance, water is generally a very inexpensive resource. However, in some parts of the southwestern United States, the cost of water has been kept artificially low as a result of government subsidies and programs.

Because the quantity and quality of water available at any particular time are highly variable, shortages of water have occurred and will probably continue to occur with increasing frequency. Such shortages can lead to serious economic disruption and human suffering.[4] In the Middle East and northern Africa, scarce water has resulted in harsh words and threats between countries. War over water is a possibility. In 1995, Canada Mortgage and Housing Corporation predicted that water use in Canada would exceed available supplies by the year 2020. Their current projections incorporate water efficiency measures, and indicate that available water resources

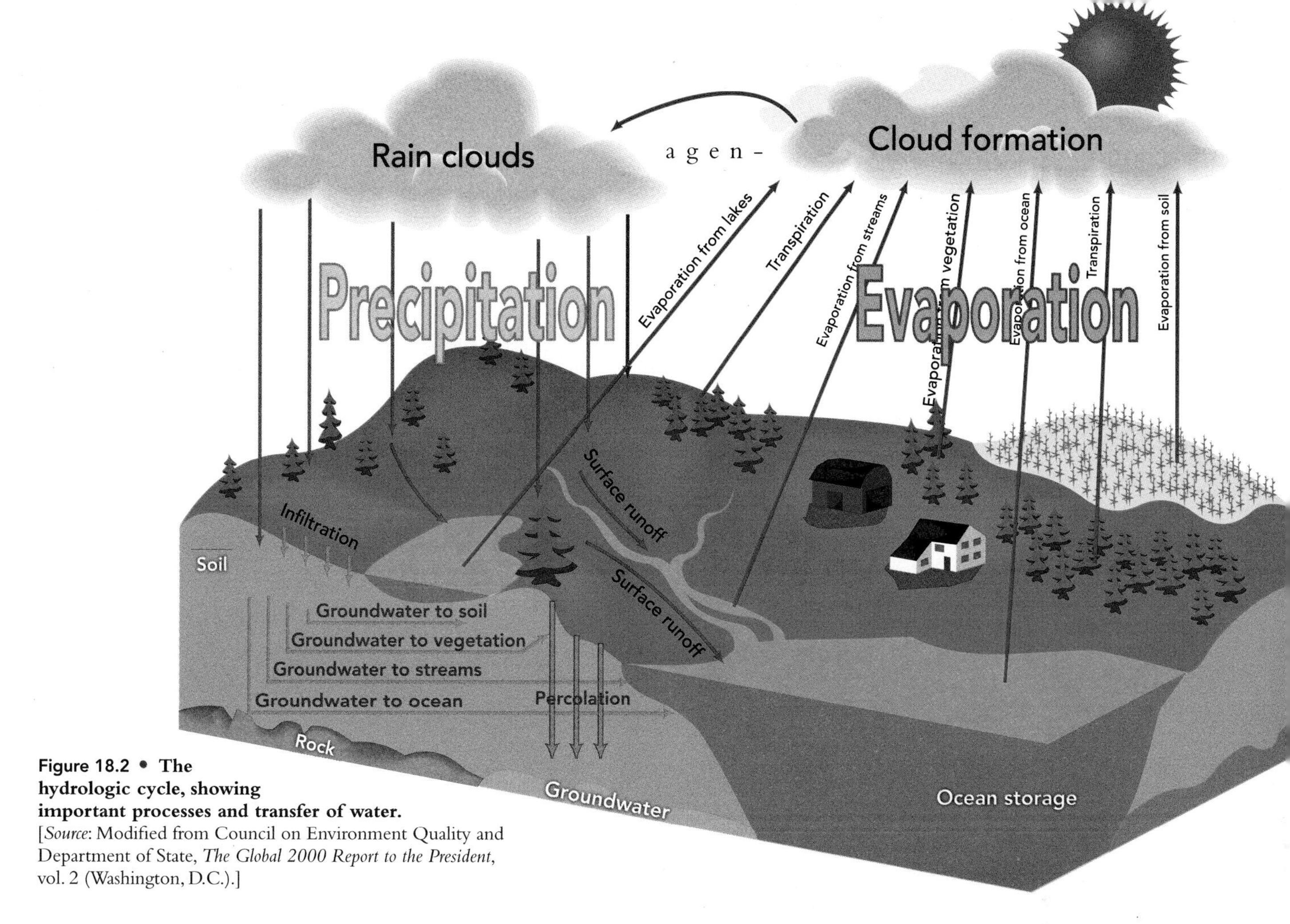

Figure 18.2 • The hydrologic cycle, showing important processes and transfer of water. [*Source*: Modified from Council on Environment Quality and Department of State, *The Global 2000 Report to the President*, vol. 2 (Washington, D.C.).]

should be sufficient for at least the next 50 years.[4] Water efficiency is therefore a critical component of water use forecasting. Therefore, an important question is, how can we best manage our water resources, use, and treatment to maintain adequate supplies?

Water shortages can also cause political conflict. Longstanding water allocation disputes exist in many places in the world, including the Middle East and along the U.S.–Mexico border. Canada and the United States established the International Joint Commission (IJC) under the Boundary Waters Treaty of 1909, with the express purpose of anticipating and resolving water disputes in the boundary waters. Under the auspices of the IJC and local agencies, the two countries have established and now operate a number of dams and other regulatory structures on binational river systems. Yet even this powerful and longstanding structure may not be enough. In the spring of 2005, the state of North Dakota expressed its intention of draining water from Devil's Lake into Manitoba's Red River and Lake Winnipeg, without the permission of the Canadian government or the involvement of the International Joint Commission. This announcement has caused great concern among observers, not least of which because of its implied threat to future U.S.–Canada relations involving water resources.

Table 18.1 • The World's Water Supply (Selected Examples)

Location	*Surface Area (km²)*	*Water Volume (km³)*	*Percentage of Total Water*	*Estimated Average Residence Time of Water*
Oceans	361,000,000	1,230,000,000	97.2	Thousands of years
Atmosphere	510,000,000	12,700	0.001	9 days
Rivers and streams	—	1,200	0.0001	2 weeks
Groundwater (shallow to depth of 0.8 km)	130,000,000	4,000,000	0.31	Hundreds to many thousands of years
Lakes (fresh water)	855,000	123,000	0.01	Tens of years
Ice caps and glaciers	28,200,000	28,600,000	2.15	Tens of thousands of years and longer

Source: U.S. Geological Survey.

Groundwater and Streams

Next, before moving onto issues of water supply and management, we introduce groundwater and surface water and the terms used in discussing them. An acquaintance with this terminology is important in understanding many environmental issues, problems, and solutions.

The term **groundwater** usually refers to the water below the water table, where saturated conditions exist. The upper surface of the groundwater is called the *water table*.

Rain that falls on the land evaporates, runs off the surface, or moves below the surface and is transported underground. Locations where surface waters move into, or infiltrate, the ground are known as *recharge zones*. Places where groundwater flows or seeps out at the surface, such as springs, are known as *discharge zones or discharge points*.

Water that moves into the ground from the surface first seeps through pore spaces (empty spaces between soil particles or rock fractures) in the soil and rock known as the *vadose zone*. This area is seldom saturated (not all pore spaces are filled with water). The water then enters the groundwater system, which is saturated (all pore spaces are filled with water).

An *aquifer* is an underground zone or body of earth material from which groundwater can be obtained (from a well) at a useful rate. Loose gravel and sand deposits with lots of pore space between grains and rocks or many open fractures generally make good aquifers. Groundwater in aquifers usually moves slowly at rates of centimetres or metres per day. When water is pumped from an aquifer, the water table is depressed around the well, forming a *cone of depression*. Figure 18.3 shows the major features of a groundwater and surface water system.

Streams may be classified as effluent or influent. In an **effluent stream**, the flow is maintained during the dry season by groundwater seepage into the stream channel from the subsurface. A stream that flows all year is called a perennial stream. Most perennial streams flow all year because they constantly receive groundwater to sustain flow. An **influent stream** is entirely above the water table and flows only in direct response to precipitation. Water from an influent stream seeps down into the subsurface. An influent stream is called an ephemeral stream, because it doesn't flow all year.

A given stream may have reaches (unspecified lengths of stream) that are perennial and other reaches that are

Figure 18.3 • Conceptual diagram illustrating some interactions between surface water and groundwater for a city in a semi-arid environment with adjacent agricultural land and reservoir. (1) Water pumped from wells lowers the groundwater level. (2) Urbanization increases runoff to streams. (3) Sewage treatment discharges nutrient-rich waters to stream, groundwater, and reservoir. (4) Agriculture uses irrigation waters from wells, and runoff to stream from fields contains nutrients from fertilizers. (5) Rural residences draw water from the reservoir and discharge nutrient-rich effluent to septic systems. (6) Reservoir waters infiltrate into groundwater systems.

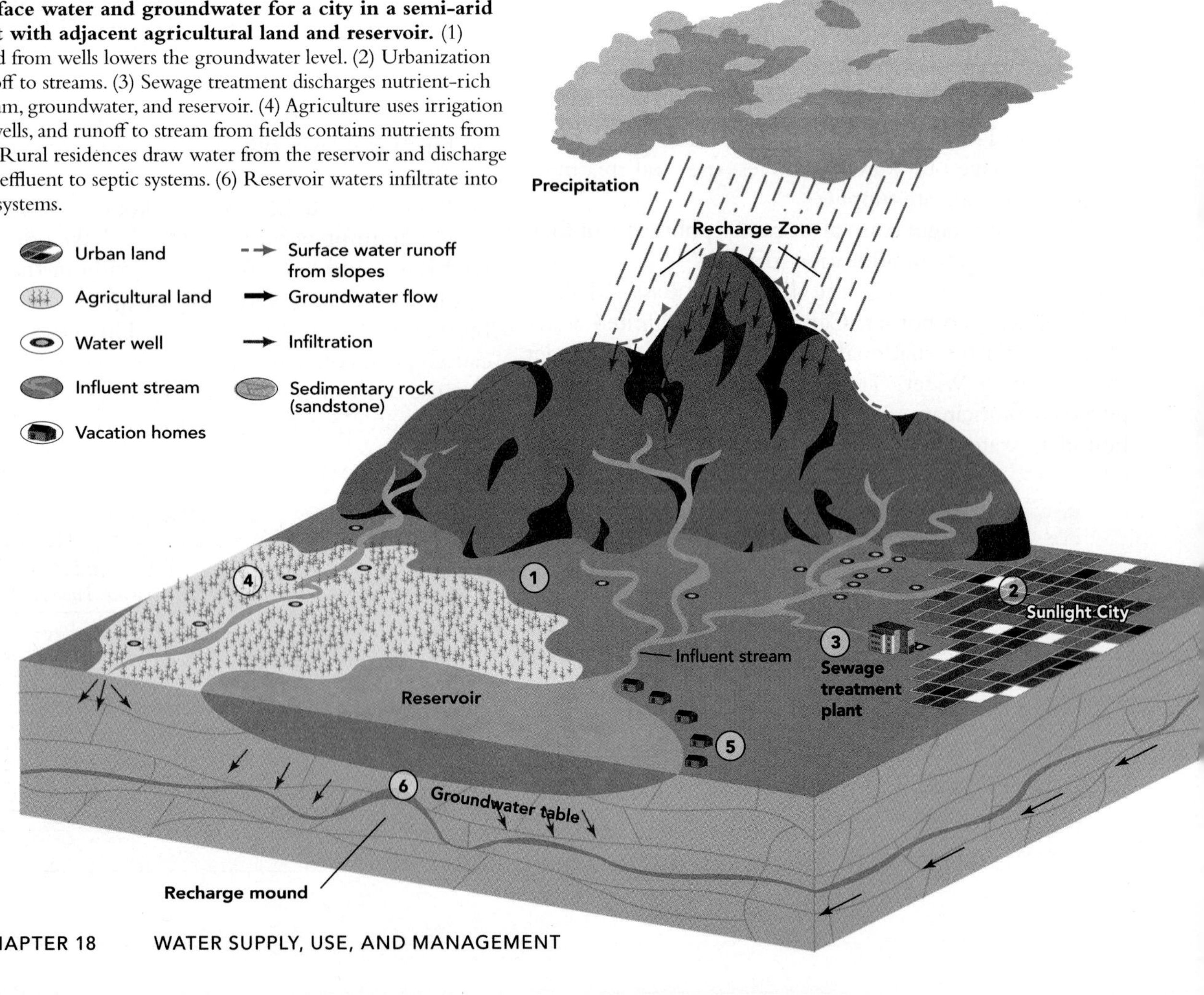

ephemeral. It may also have reaches, known as intermittent, that have a combination of influent and effluent flow varying with the time of year. For example, streams flowing from the mountains to the sea in southern California often have reaches in the mountain that are perennial, supporting populations of trout or endangered southern steelhead, and lower intermittent reaches that transition to ephemeral reaches. At the coast, these streams may receive fresh or salty groundwater and tidal flow from the ocean to become a perennial lagoon.

Interactions between Surface Water and Groundwater

Surface water and groundwater interact in many ways and should be considered part of the same resource. Nearly all natural surface water environments, such as rivers and lakes, as well as human-constructed water environments, such as reservoirs, have strong linkages with groundwater. For example, withdrawal of groundwater by pumping from wells may reduce stream flow, lower lake levels, or change the quality of surface water. Reduction of effluent stream flow by lowering the groundwater level may change a perennial stream that flows all year to an intermittent influent stream. Similarly, withdrawal of surface water by diversion from streams and rivers can deplete groundwater resources or change the quality of groundwater. Diverting surface waters that recharge groundwaters may result in an increase in concentrations of dissolved chemicals in the groundwater. This happens because dissolved chemicals present in the groundwater are not diluted by mixing with infiltrated surface water. Finally, groundwater pollution may result in pollution of surface water, and vice versa.[5]

Selected interactions between surface water and groundwater in a semi-arid urban and agricultural environment are shown in Figure 18.3. Urban and agricultural runoff increases the volume of water in the reservoir. Pumping groundwater for agricultural and urban uses lowers the groundwater level. The quality of surface water and groundwater is reduced by urban and agricultural runoff, which adds nutrients from fertilizers, oil from roads, and nutrients from treated wastewaters to streams and groundwater.

18.2 World Water Supply

The water supply at any particular point on the land surface depends on several factors in the hydrologic cycle, including the rates of precipitation, evaporation, transpiration (water in vapour form that directly enters the atmosphere from plants through pores in leaves and stems), stream flow, and subsurface flow. A useful concept in understanding water supply is the **water budget**, which is a model that balances the inputs, outputs, and storage of water in a system. Simple annual water budgets (precipitation − evaporation = runoff) for North America and other continents are shown on Table 18.2. The total average annual water yield (runoff) from Earth's rivers is approximately 47,000 km^3, but its distribution is far from uniform (see Table 18.2). Some runoff occurs in relatively uninhabited regions, such as Antarctica, which produces about 5% of Earth's total runoff. South America, which includes the relatively uninhabited Amazon basin, provides about one-fourth of Earth's total runoff. Total runoff in North America is about two-thirds that of South America. Unfortunately, much of the North American runoff occurs in sparsely settled or uninhabited regions, particularly in the northern parts of Canada and Alaska.

The daily global water budget is shown in Figure 18.4. Note that about half the precipitation that falls evaporates quickly or is transpired by vegetation. The remaining water enters the surface water or groundwater storage systems, flows to the oceans or across the nation's boundaries, is used by people, or evaporates from reservoirs. Owing to natural variations in precipitation that cause either floods or droughts, only a portion of this water can be developed for intensive uses.[4]

Precipitation and Runoff Patterns

In developing water budgets for water resources management, it is useful to consider annual precipitation and runoff patterns. Potential problems with water supply can be predicted in areas where average precipitation and runoff are relatively low, such as in the arid and semi-arid parts of Canadian prairies and the southwestern United

Table 18.2 • Annual Water Budgets for the Continents[a]

	Precipitation		*Evaporation*		*Runoff*
Continental	*mm/yr*	*km^3*	*mm/yr*	*km^3*	*km^3/yr*
North America	756	18,300	418	10,000	8,180
South America	1,600	28,400	910	16,200	12,200
Europe	790	8,290	507	5,320	2,970
Asia	740	32,200	416	18,100	14,100
Africa	740	22,300	587	17,700	4,600
Australia and Oceania	791	7,080	511	4,570	2,510
Antarctica	165	2,310	0	0	2,310
Earth (entire land area)	800	119,000	485	72,000	47,000[b]

[a] Precipitation − evaporation = runoff.
[b] Surface runoff is 44,800; groundwater runoff is 2,200.

Source: I. A. Shiklomanov, "World Fresh Water Resources," in P. H. Gleick, ed., *Water in Crisis* (New York: Oxford University Press, 1993), pp. 3–12.

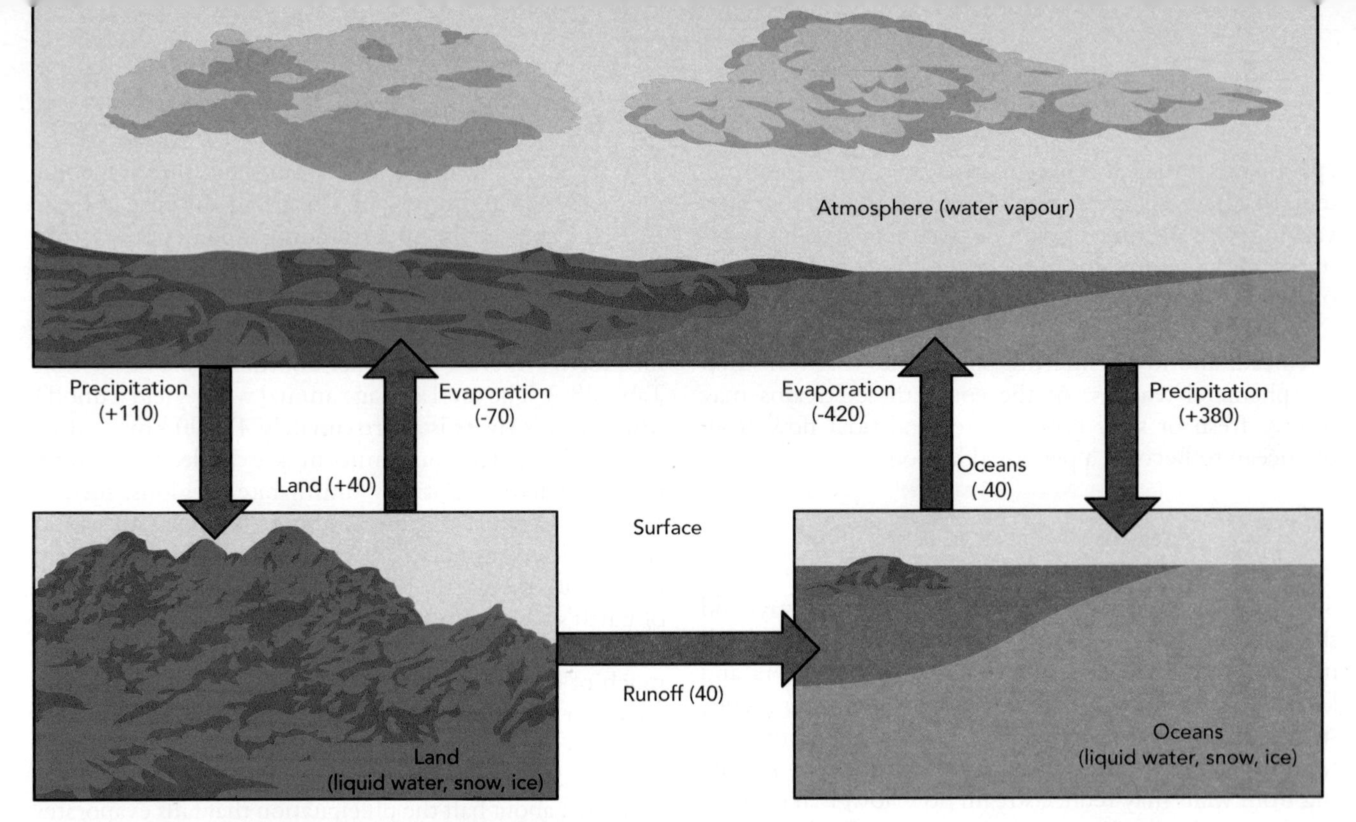

Figure 18.4 • Global water budget. Units are thousands of cubic kilometres per year.

States. Surface water supply can never be as high as the average annual runoff, because not all runoff can be successfully stored. Total storage of runoff is not possible because of evaporative losses from river channels, ponds, lakes, and reservoirs. As a result, shortages in the water supply are common in areas with natural low precipitation and runoff coupled with high evaporation. In these areas, strong conservation practices are necessary to help ensure an adequate supply of water.[4]

Droughts

Because there are large annual and regional variations in stream flow, even areas with high precipitation and runoff may periodically suffer from droughts. For example, the dry years near the end of the twentieth century in western Canada and the United States produced serious water shortages. Fortunately for the more humid eastern portions of the continent, stream flow there tends to vary less than in other regions, and drought is less likely.[4] (Serious droughts did however occur in southern Ontario during several years in the late 1990s, creating additional pressure on water resources in an area of fast-growing population.)

Groundwater Use and Problems

About one-third of the people in Canada use groundwater as a primary source for drinking water. It accounts for approximately 50% of Canada's total freshwater resources. Groundwater resources are especially important in the Prairie Provinces, where surface water is often in short supply. Manitoba obtains 20% of its water from groundwater, the highest proportion in the country except for Prince Edward Island, which is entirely reliant on groundwater supplies. Saskatchewan, at 9%, and Alberta (4%) extract less groundwater because of the abundance of surface waters arising in the Rocky Mountains. Almost all rural prairie residences obtain their drinking water from groundwater wells. Canada does not have a national inventory of groundwater resources. However, as in other regions of the world, not all groundwater is accessible for consumption. In some regions, the high cost of pumping limits the total amount of groundwater that can be economically recovered.[4]

In the driest parts of North America, groundwater withdrawal from wells can exceed natural inflow. In such cases of **overdraft**, we can think of water as a nonrenewable resource that is being mined. This can lead to a variety of problems, including damage to river ecosystems and land subsidence. Groundwater overdraft is a serious problem in the Texas–Oklahoma–High Plains area (which includes much of Kansas and Nebraska and parts of other states), as well as in California, Arizona, Nevada, New Mexico, and isolated areas of Louisiana, Mississippi, Arkansas, and the south Atlantic region.

The Ogallala aquifer, which is composed of water-bearing sands and gravels that underlie an area of about 400,000 km^2 from South Dakota into Texas, is the main groundwater resource through much of the southern United States. Although the aquifer holds a tremendous amount of groundwater, it is being used in some areas at a rate up to 20 times higher than the rate at which it is being naturally replaced. The water table in many parts of the aquifer has declined in recent years, causing yields from wells to decrease and energy costs for pumping the water to increase. There is concern that eventually a significant portion of land now being irrigated will be returned to dryland farming as the resource is used up.

Desalination as a Water Source

Seawater is about 3.5% salt; that means each cubic meter of seawater contains about 40 kg of salt. Desalination, a technology to remove salt from water, is being used at several hundred production plants around the world to produce water with reduced salt. The salt content of the water must be reduced to about 0.05% for the water to be used as a freshwater resource. Large desalination plants produce 20,000–30,000 m^3 of water per day.

The cost of desalinated water is about 10 times that paid for traditional water supplies in North America. Desalinated water has a *place value*, which means that the price increases quickly with the transport distance and the cost of moving water from the plant. Because the various processes that remove the salt require large amounts of energy, the cost of the water is also tied to ever-increasing energy costs. For these reasons, desalination will remain an expensive process used only when alternative water sources are not available.

Desalination also has environmental impacts. Discharge of very salty water from a desalination plant into another body of water, such as a bay, may locally increase salinity and kill some plants and animals intolerant to salt. The discharge from desalination plants may also cause wide fluctuations in salt content of local environments, which may damage ecosystems.

18.3 Water Use

In discussing water use, it is important to distinguish between off-stream and in-stream uses. **Off-stream use** refers to water removed from its source (such as a river or reservoir) for use. Much of this water is returned to the source after use; for example, the water used to cool industrial processes may go to cooling ponds and then be discharged to a river, lake, or reservoir. **Consumptive use** is an off-stream use in which water is consumed by plants and animals or used in industrial processes. The water enters human tissue or products or evaporates during use and is not returned to its source.[4]

In-stream use includes the use of rivers for navigation, hydroelectric power generation, fish and wildlife habitats, and recreation. These multiple uses usually create controversy, because each requires different conditions to prevent damage or detrimental effects. For example, fish and wildlife require certain water levels and flow rates for maximum biological productivity; these levels and rates will differ from those needed for hydroelectric power generation, which requires large fluctuations in discharges to match power needs. Similarly, in-stream uses of water for fish and wildlife will likely conflict with requirements for shipping and boating. Figure 18.5 demonstrates some of these conflicting demands on a graph that shows optimal discharge for various uses throughout the year. In-stream water use for navigation is optimal at a constant fairly high discharge. Some fish, however, prefer higher flows in the spring for spawning.

Another problem for in-stream use is how much water can be removed from a stream or river without damaging the stream's ecosystem. This is an issue in the Pacific Northwest, where fish, such as steelhead trout and salmon, are on the decline partly because diversions (removal of water for agricultural, urban, and other uses) have reduced stream flow to the extent that fish habitats are damaged.

The Aral Sea in Kazakhstan and Uzbekistan provides a wake-up call of environmental damage that can be caused by diversion of water for agricultural purposes. Diversion of water from the two rivers that flow into the Aral Sea has transformed one of the largest bodies of inland water in the world from a vibrant ecosystem into a dying sea. The present shoreline is surrounded by thousands of square kilometres of salt flats that formed as the surface area of the sea was reduced by about 40% in the past 40 years (Figure 18.6). The volume of the lake was reduced by more than 50%. The salt content of the water increased fish kills, such as sturgeon that are an important component of the economy. Dust raised by winds from the dry salt flats is producing a regional air pollution problem, and the climate in the region has changed as the moderating effect of the sea has been reduced. Winters have grown colder and summers warmer. Fishing centers such as Muynak in the south and Aralsk to the north that were once on the shore of the sea are now many kilometres inland. Loss of fishing with decline of tourism has damaged the local economy.

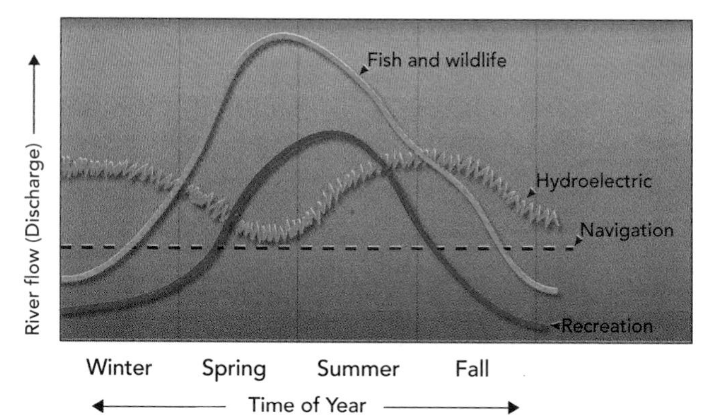

Figure 18.5 • **In-stream water uses and optimal discharges (volume of water flowing per second) for each use.** Discharge is the amount of water passing by a particular location and is measured in cubic meters per second. Obviously, all these needs cannot be met simultaneously.

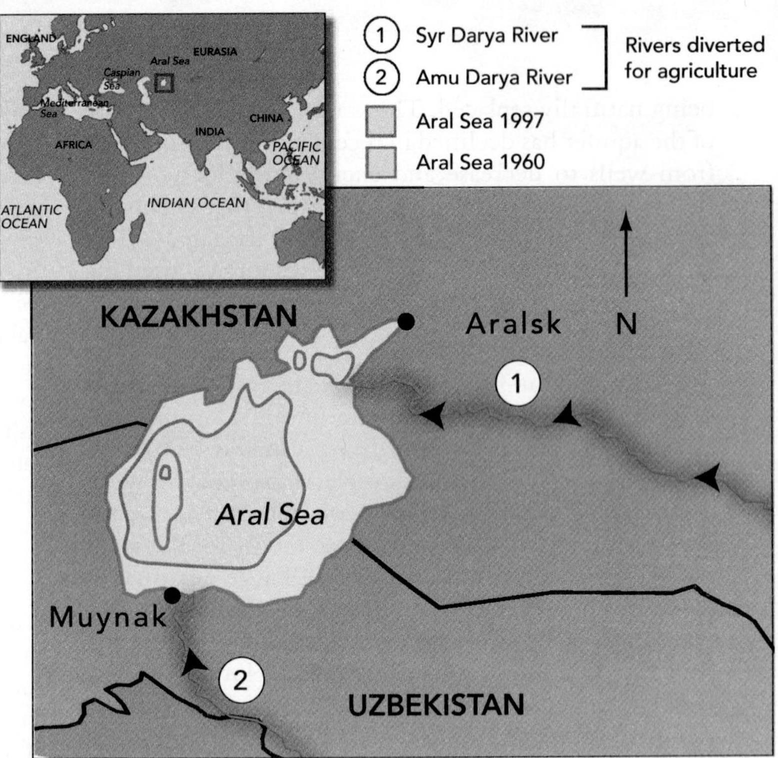

Figure 18.6 • The Aral Sea is drying up and dying as a result of diversion of water for agriculture. [Modified from original courtesy of Philip P. Micklin.]

Transport of Water

In many parts of the world, demands are being made on rivers to supply water to agricultural and urban areas. This is not a new trend. Ancient civilizations, including the Romans and First Nations, constructed canals and aqueducts to transport water from distant rivers to where it was needed. In our modern civilization, as in the past, water is often moved long distances from areas with abundant rainfall or snow to areas of high usage (usually agricultural areas). For instance, the Winnipeg Aqueduct carries almost 400 million litres of water daily 156 km from its source east of the city.

In a broader perspective, the cost of obtaining water for large urban centers from long distances, along with competition for available water from other sources and users, will eventually place an upper limit on the water supply of the city. As shortages develop, stronger conservation measures are implemented, and the cost of water increases. As with other resources, as the water supply is reduced and demand for water increases, so does its price. If the price becomes high enough, more expensive sources may be developed—for example, pumping from deeper wells or using desalination.

Some Trends in Water Use

Environment Canada estimates that about 89% of the country's municipal water is drawn from surface water; only 11% comes from groundwater. Data on allocation of water use, and especially on long-term trends, are not easily available for Canada. The United Nations Educational, Scientific, and Cultural Organization (UNESCO) reports that Canadian water withdrawals have risen dramatically from 1950 to the present, and are projected to continue increasing for at least the next 20 years (see Figure 18.7). Figure 18.7 also shows, however, that water consumption is much lower than withdrawals, and that it has levelled off since the early 1990s. This difference is explained by the huge proportion of water that is returned to lakes and rivers. For example, of the approximately 40.4 billion cubic metres of water withdrawn every year for thermal power generation, about 70% (28.3 billion cubic metres) is discharged again into surface water systems, and 29% (about 11.7 billion cubic metres) is recirculated. Only a little over 1% (508 million cubic metres) is actually consumed in the power generation process.[6] UNESCO estimates that Canada's total water consumption in 1950 was about 17% of total water withdrawals; by 2010, that proportion will likely be closer to 10%, reflecting efficiencies in water use.[6, 7]

Trends in freshwater withdrawals by water-use categories for Canada from 1972 to 1996 are shown in Figure 18.8. In summary:

1. The major uses of water are for the thermoelectric industry (63%) and manufacturing (20%); municipalities (8%), agriculture (6%) and mining (3%) are relatively minor water users.
2. The use of water for irrigation by agriculture has been fairly stable at about 4 billion cubic metres/year since the mid-1970s.
3. Water use by the thermoelectric industry has continued to rise, although use by other industries decreased slightly beginning in 1981.
4. Use of water for public and rural supplies continued to increase through the period from 1972 to 1996, presumably related to the increase in human population.[7] However, average daily per capita water use has declined slightly over the same period, to about 650 L/person/day, compared to a high of almost 700 L/person/day in 1989.[7] This suggests that improvements have been made in water management and water conservation.

18.4 Water Conservation

Water conservation is the careful use and protection of water resources. It involves both the quantity of water used and its quality. Conservation is an important component of sustainable water use. Because the field of water conservation is changing rapidly, it is expected that a

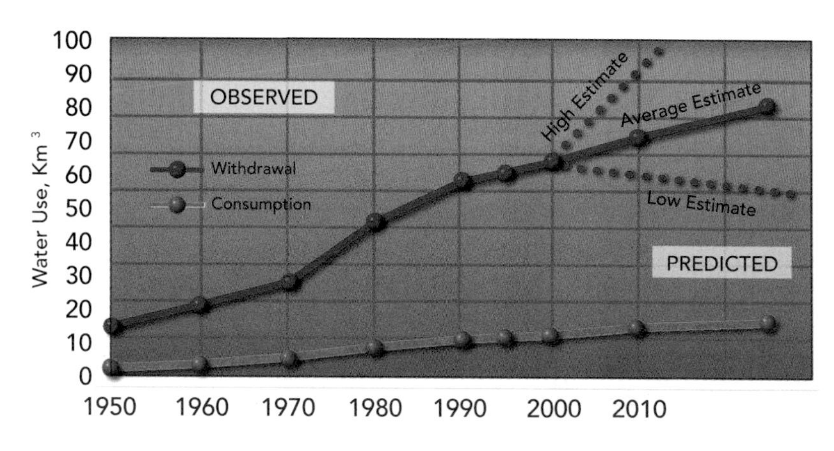

Figure 18.7 • Trends in past and projected Canadian fresh water withdrawals and consumption (1950–2010). [*Source*: UNESCO: Dynamics of Water Use in Canada 1998. http://webworld.unesco.org/water/ihp/db/shiklomanov/part'3/HTML/Tb_21'CN.html.]

number of innovations will reduce the total withdrawals of water for various purposes, even though consumption will continue to increase.[3,8]

Agricultural Use

Improved irrigation (Figure 18.9) could reduce agricultural withdrawals by between 20 and 30%. Because agriculture is a major water user in many parts of the world, this would be a tremendous savings. Suggestions for agricultural conservation include the following:

- Price agricultural water to encourage conservation (subsidizing water will encourage overuse).
- Use lined or covered canals that reduce seepage and evaporation.
- Use computer monitoring and schedule release of water for maximum efficiency.
- Integrate the use of surface water and groundwater to more effectively use the total resource. That is, irrigate with surplus surface water when it is abundant, and also use surplus surface water to recharge groundwater aquifers by applying the surface water to specially designed infiltration ponds or injection wells. When surface water is in short supply, use more groundwater.
- Irrigate at times when evaporation is minimal, such as at night or in the early morning.
- Use improved irrigation systems, such as sprinklers or drip irrigation, that more effectively apply water to crops.
- Improve land preparation for water application; that is, improve the soil to increase infiltration and minimize runoff. Where applicable, use mulch to help retain water around plants.
- Encourage the development of crops that require less water or are more salt tolerant so that less periodic flooding of irrigated land is necessary to remove accumulated salts in the soil.
- Pump water to controlled outlets for livestock watering, rather than using open ponds or free access to streams.

Domestic Use

Domestic use of water accounts for less than 10% of total national water withdrawals. However, because domestic water use is concentrated in urban areas, it may pose major local problems in areas where water is periodically or often in short supply. Most water in homes is used in the bathroom and for washing clothing and dishes. Water use for domestic purposes can be substantially reduced at a relatively small cost by implementing the following measures:

- In semi-arid regions, replace lawns with decorative gravels and native plants.
- Use more efficient bathroom fixtures, such as low-flow toilets that use 6 litres or less per flush rather than the standard 18 litres, and low-flow shower heads that deliver less but sufficient water.
- Turn off water when not absolutely needed for washing, brushing teeth, shaving, and so on.

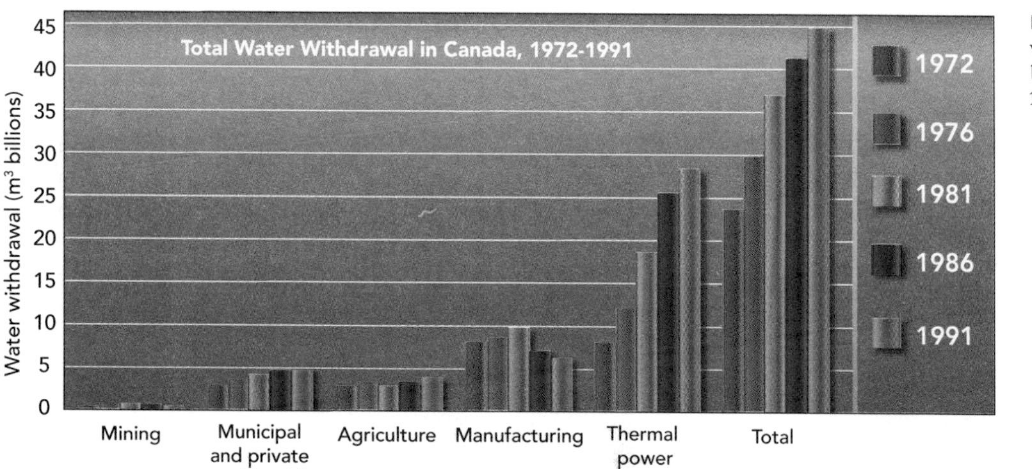

Figure 18.8 • Trends in Canadian water withdrawals (fresh and saline) by water-use category (1972–1991). [*Source*: Environment Canada.]

Agriculture, 1990

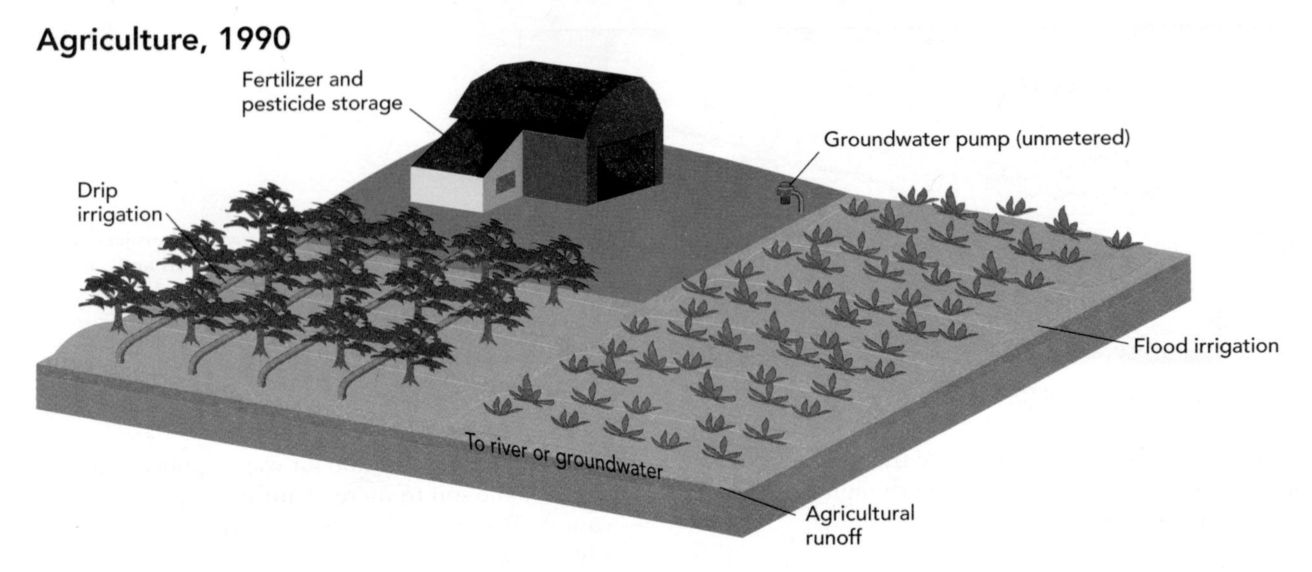

Agriculture, 2020

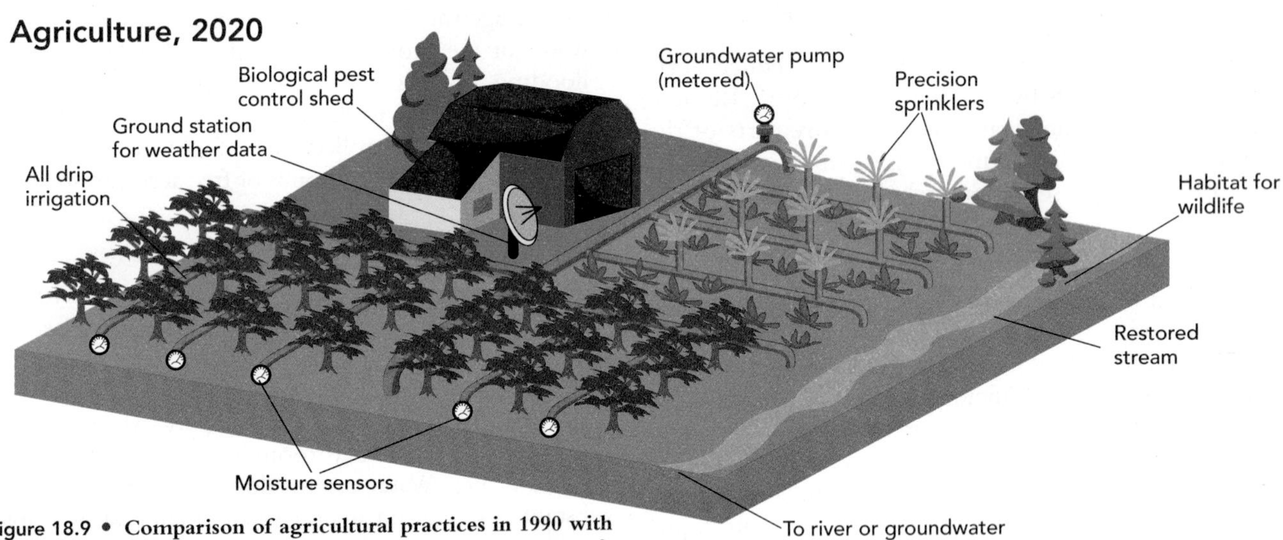

Figure 18.9 • **Comparison of agricultural practices in 1990 with what they might be by 2020.** The improvements call for a variety of agricultural procedures, from biological pest control to more efficient application of irrigation water to restoration of water resources and wildlife habitat. [*Source*: P. H. Gleick, P. Loh, S.V. Gomez, and J. Morrison, *California Water 2020, a Sustainable Vision* (Oakland, Calif.: Pacific Institute for Studies in Development, Environment and Security, 1995).]

- Flush the toilet only when really necessary.
- Fix all leaks quickly. Dripping pipes, faucets, toilets, or garden hoses waste water. A small drip can waste several litres per day; multiply this by millions of homes with a leak, and a large volume of water is lost.
- Purchase dishwashers and washing machines that minimize water consumption.
- Take a long bath rather than a long shower.
- Don't wash sidewalks and driveways with water (sweep them).
- Consider using grey water (from showers, bathtubs, sinks, and washing machines) to water vegetation. The grey water from washing machines is easiest to use, as it can be easily diverted before entering a drain.
- Water lawns and plants in the early morning, late afternoon, or at night to reduce evaporation.
- Use drip irrigation and place water-holding mulch around garden plants.
- Plant drought-resistant vegetation that requires less water.
- Learn how to read the water meter to monitor for unobserved leaks and record your conservation successes.

In addition, local water districts should encourage water pricing policies in which water is more expensive beyond some baseline amount determined by the number of people in a home and the size of the property.

Industry and Manufacturing Use

Water conservation measures taken by industry can be improved.[9] For instance, water removal for steam generation of electricity could be reduced 25% to 30% by using cooling towers that use less or no water. Manufacturing and industry could curb water withdrawals by increasing in-plant treatment and recycling water and by developing new equipment and processes that require less water.[4]

How Green is Your Campus?

Water Conservation

Many of the residential water conservation strategies discussed in this chapter are also applicable to larger institutions such as colleges and universities. How does your campus score on the following items?

- No lawn or garden watering, or watering only with grey or recycled water.
- Any lawn or garden watering conducted only in early morning, late afternoon, or evening to reduce evaporative loss.
- Gardens are mulched to retain water and reduce the need for irrigation.
- Low-flush toilets and showerheads in university residences and other buildings.
- Leaks and drips quickly detected and repaired.
- Water-efficient appliances in kitchens, laundries and residences.
- A campus-wide master plan for water conservation (*hint*: search your university's web site for this kind of information).

Perception and Water Use

How people perceive the water supply is important in determin ing how much water is used. Perception of water is based partly on its price and availability. If water is abundant and inexpensive, we don't think much about it. If water is scarce or expensive, it is another matter. Many places in Canada offer water at a fixed price, regardless of consumption. In cities with effective water conservation programs, such as Kitchener-Waterloo, Ontario, water meters are an essential part of the conservation strategy. Metering allows municipalities to charge consumers for the water they use. Some progressive municipalities also employ "increasing-block" pricing: water-efficient households pay a lower unit price for water, compared to users who exceed a target consumption level.

18.5 Sustainability and Water Management

Water is essential to sustain life and to maintain ecological systems necessary for the survival of humans. As a result, water plays important roles in ecosystem support, economic development, cultural values, and community well-being. Managing water use for sustainability is important in many ways.

Sustainable Water Use

From a water supply use and management perspective, **sustainable water** use can be defined as use of water resources by people in a way that allows society to develop and flourish into an indefinite future without degrading the various components of the hydrologic cycle or the ecological systems that depend on it.[10] Some general criteria for water use sustainability are as follows.[10]

- Develop water resources in sufficient volume to maintain human health and well-being.
- Provide sufficient water resources to guarantee the health and maintenance of ecosystems.
- Ensure minimum standards of water quality for the various users of water resources.
- Ensure that human actions do not damage or reduce long-term renewability of water resources.
- Promote the use of water-efficient technology and practice.
- Gradually eliminate water pricing policies that subsidize the inefficient use of water.

Groundwater Sustainability

The concept of sustainability by its very nature involves a long-term perspective. With groundwater resources, the length of time for effective management for sustainability is even longer than for other renewable resources. Surface waters, for example, may be replaced over a relatively short time. In contrast, groundwater development may take place over many years, at relatively slow rates. Effects of pumping groundwater at rates greater than natural replenishment rates may take years to be recognized. Similarly, effects of withdrawal of groundwater, such as drying up of springs or reduction of stream flow, may not be recognized until years after pumping begins. The long-term approach to sustainability with respect to groundwater often involves balancing withdrawals of groundwater resources with recharge of those resources, which is an important component of water management.[11]

Water Management

Management of water resources for water supply is a complex issue that will become more difficult as demand for water increases in the coming years. This difficulty will be especially apparent in semi-arid and arid parts of the world where water is or soon will be in short supply. Options for minimizing potential water supply problems include locating alternative water supplies and managing existing supplies better. In some areas, locating new supplies is unlikely, and serious consideration is being given to ideas as original as towing icebergs to coastal regions where fresh water is needed. It seems apparent that water will become much more expensive in the future; and if the price is right, many innovative programs are possible.

Some water-poor countries must rely mainly on rainwater for a water source. Bermuda once had a lens of fresh groundwater that was connected to the surrounding ocean at its margins. With time, that resource has become depleted, allowing saltwater to intrude on the freshwater supply. Today, houses in Bermuda are built with rainwater collection and storage systems (cisterns). Fresh water is used only in drinking, cooking and washing, with saltwater used for toilet flushing. Such systems would be possible in Canada, but have not been employed in most regions because of a perception of unlimited freshwater resources.

A method of water management utilized by a number of municipalities, mainly outside Canada, is known as the *variable-water-source approach*. For example, the city of Santa Barbara, California, has developed a variable-water-source approach that uses several interrelated measures to meet present and future water demands. The plan (Figure 18.10) includes importing water, developing new sources, using reclaimed water, and instituting a permanent conservation program. In essence, this seaside community has developed a master water plan.

A Master Plan for Water Management

Luna Leopold, a famous hydrologist, suggests that a new philosophy of water management is needed, one based on geologic, geographic, and climatic factors as well as on the traditional economic, social, and political factors. He argues that the management of water resources cannot be successful as long as it is naively perceived from an economic and political standpoint.

The essence of Leopold's water management philosophy is that surface water and groundwater are both subject to natural flux with time. In wet years, there is plenty of surface water, and the near-surface groundwater resources are replenished. During dry years, which must be expected even though they may not be accurately predicted, specific plans to supply water on an emergency basis to minimize hardships must be in place and ready to use.

For example, there are subsurface waters in some parts of the world that are too deep to be economically pumped from wells or that have marginal water quality. These waters may be isolated from the present hydrologic cycle and therefore not subject to natural recharge. Such water might be used when the need is great. However, advance planning to drill the wells and connect them to existing water lines is necessary if they are to be ready when the need arises.

Another possible emergency plan might involve the treatment of wastewater. Reuse of such water on a regular basis might be too expensive, but advance planning to reuse treated water during emergencies could be a wise decision.

Finally, we should develop plans to use surface water when available, and we should not be afraid to use groundwater as needed in dry years. During wet years, natural recharge as well as artificial recharge (pumping excess surface water into the ground) will replenish the groundwater resources. This water management plan recognizes that excesses and deficiencies in water are natural and can be planned for.[12]

Water Management and the Environment

Many agricultural and urban areas require water to be delivered from nearby (and, in some cases, not-so-nearby) sources. To deliver the water, a system is needed for water storage and routing by way of canals, pipelines, and aqueducts from reservoirs. As a result, dams

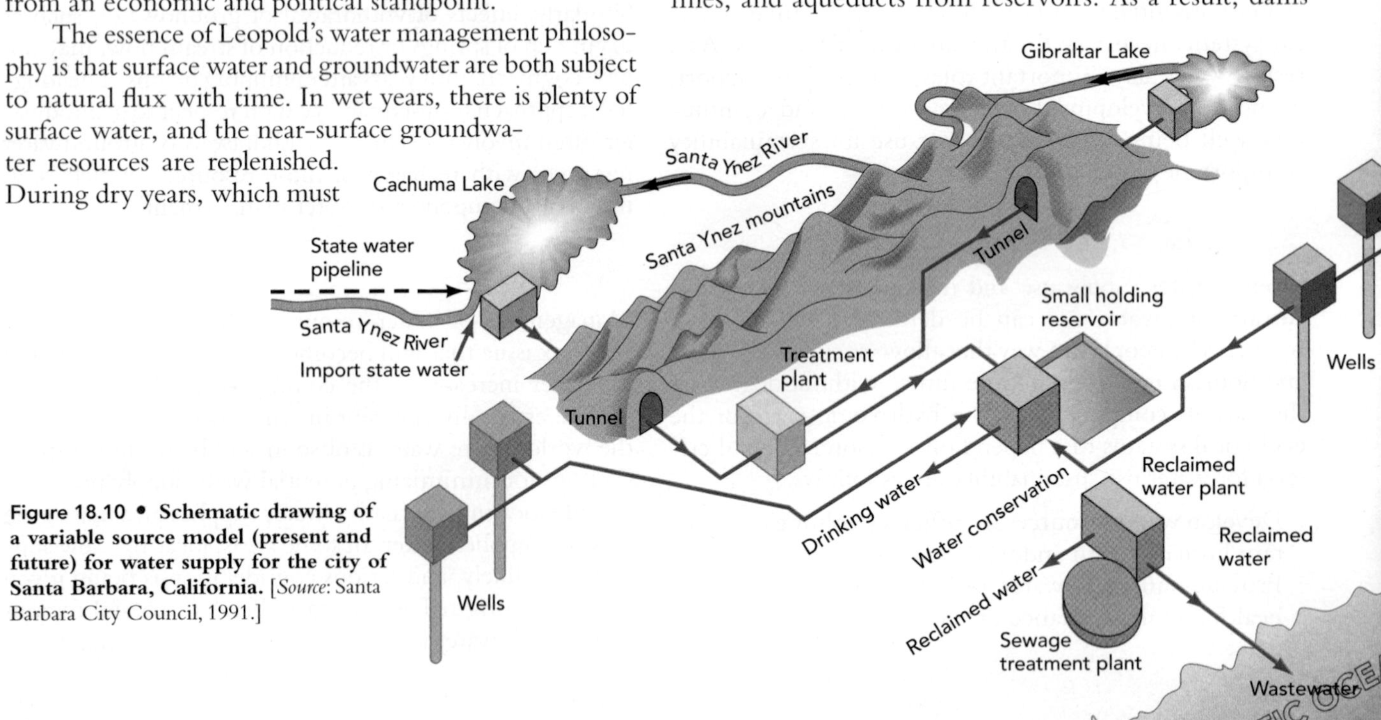

Figure 18.10 • Schematic drawing of a variable source model (present and future) for water supply for the city of Santa Barbara, California. [*Source*: Santa Barbara City Council, 1991.]

(a)

(b)

(c)

are built, wetlands may be modified, and rivers may be channelized to help control flooding. Often, a good deal of controversy surrounds water development.

The days of developing large projects in Canada and the United States without environmental and public review have passed. The resolution of development issues now involves input from a variety of government and public groups, which may have very different needs and concerns. These range from agricultural groups that see water development as critical for their livelihood, to groups primarily concerned with wildlife and wilderness preservation. It is a positive sign that the various parties with interests in water issues are encouraged—and, in some cases, required—to meet and communicate their desires and concerns. Next, we address the subjects of some of these concerns: wetlands, dams, channelization, and flooding.

18.6 Wetlands

Wetlands is a comprehensive term for landforms such as salt marshes, swamps, bogs, prairie potholes, and vernal pools (shallow depressions that seasonally hold water). Their common feature is that they are wet at least part of the year and as a result have a particular type of vegetation and soil. Figure 18.11 shows several types of wetlands.

Wetlands may be defined as areas that are inundated by water or where the land is saturated to a depth of a few centimetres for at least a few days per year. Three major components used to determine the presence of wetlands are: hydrology, or wetness; type of vegetation; and type of soil. Of these, hydrology is often the most difficult to define, because some freshwater wetlands may be wet for only a few days a year. The duration of inundation or saturation must be sufficient for the development of wetland soils, which are characterized by poor drainage and lack of oxygen, and for the growth of specially adapted vegetation.[13]

Natural Service Functions of Wetlands

Wetland ecosystems may serve a variety of natural service functions for other ecosystems and for people, including the following:

- Freshwater wetlands are a natural sponge for water. During high river flow they store water, reducing downstream flooding. Following a flood, they slowly release the stored water, nourishing low flows.
- Many freshwater wetlands are important as areas of groundwater recharge (water seeps into the ground from a prairie pothole, for instance) or discharge (water

Figure 18.11 • Several types of wetlands: *(a)* a prairie pothole ; *(b)* Oxbow Swamp, Ontario; *(c)* a marsh.

seeps out of the ground in a marsh that is fed by springs).

- Wetlands are one of the primary nursery grounds for fish, shellfish, aquatic birds, and other animals. It has been estimated that as many as 45% of endangered animals and 26% of endangered plants either live in wetlands or depend on them for their continued existence.[13]
- Wetlands are natural filters that help purify water; plants in wetlands trap sediment and toxins.
- Wetlands are often highly productive and are places where many nutrients and chemicals are naturally cycled.
- Coastal wetlands provide a buffer for inland areas from storms and high waves.
- Wetlands are an important storage site for organic carbon; carbon is stored in living plants, animals, and rich organic soils.
- Wetlands are aesthetically pleasing to people.

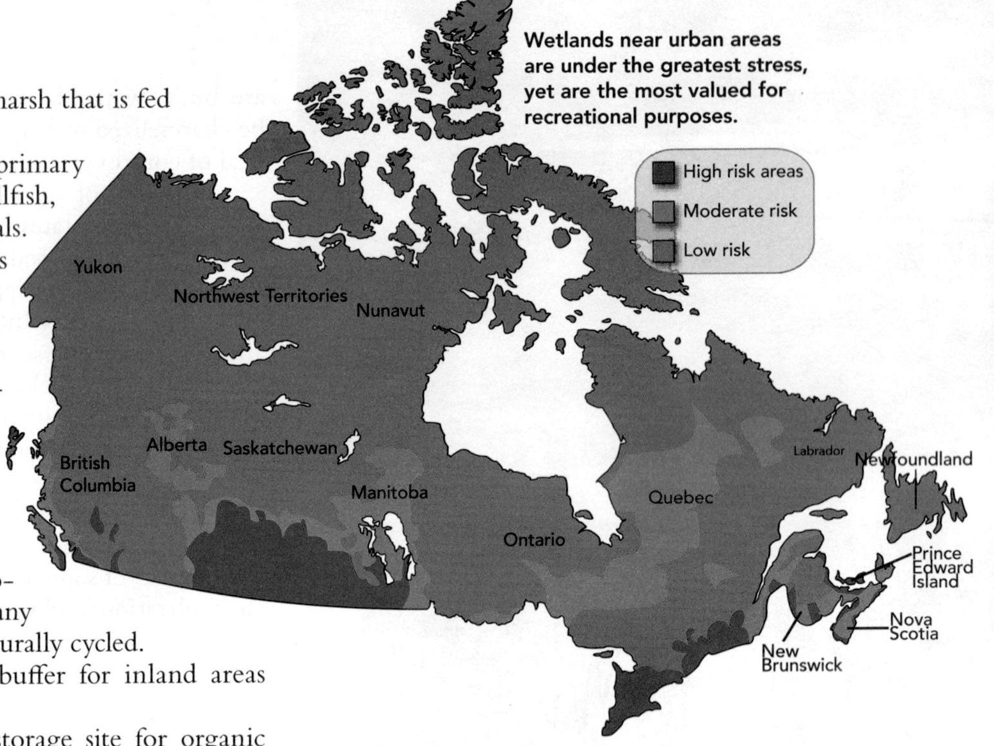

Figure 18.12 • Wetlands at risk in Canada [*Source*: Atlas of Canada.]

Freshwater wetlands are threatened in many areas. One percent of the nation's total wetlands are lost every two years, and freshwater wetlands account for 95% of this loss. Wetlands such as prairie potholes and vernal pools (temporary wetlands that form in low-lying areas after heavy rain or snowmelt) are particularly vulnerable because their hydrology is poorly understood and establishing their wetland status is more difficult.[14] Over the past 200 years, more than 50% of the wetlands in Canada have disappeared because they have been diked or drained for agricultural purposes or filled for urban or industrial development. Perhaps as much as three-quarters of the freshwater wetlands have disappeared in some regions, although most wetlands in the North remain intact. In those regions, muskeg—sphagnum peatland characterized by stunted black spruce and larch—is a common form of wetland, and one that is vulnerable to disturbance from oil, gas, and mineral exploration.

In the Peace-Athabasca delta region of northeastern Alberta, extensive wetland areas have been lost as the result of damming the Peace River in northeastern British Columbia. As downstream wetlands were deprived of natural river flows, wetland habitat, including nesting grounds for the endangered whooping crane, dried up, creating a patchwork of dry, fire-prone habitat that favoured large mammals like the bison over waterfowl. In an effort to restore the wetlands, dam managers periodically increased discharges from the dam, creating artificial floods. This solution has been less than ideal for local residents, and has not been fully successful in restoring the lost wetland habitat.

Although most coastal marshes are now protected in Canada and the United States, the extensive salt marshes at many major estuaries, where rivers entering the ocean widen and are influenced by tides, have been modified or lost. These include deltas and estuaries of major rivers such as the Fraser and the tributaries of the Bay of Fundy.[15] Modifications result not only from filling and diking but also from loss of water. Upstream changes can reduce the freshwater inflow, dramatically changing the hydrology of estuaries in terms of flow characteristics and water quality. Plant and animal communities in the deltas and estuaries also change as habitats for fish and wildfowl are eliminated.[15]

Most people agree that wetlands are valuable and productive lands for fish and wildlife. But wetlands are also valued as potential lands for agricultural activity, mineral exploitation, and building sites. Wetland management is drastically in need of new incentives for private land owners (who own tens of thousands of hectares of wetlands in Canada and the United States) to preserve wetlands rather than fill them in and develop the land.[14] Management strategies must also include careful planning to maintain the water quantity and quality necessary for wetlands to flourish or at least survive. Canada is a signatory to the Ramsar Convention on Wetlands, an international agreement focused on wetland conservation, and has a national Policy on Wetland Conservation to that end.[16] Debate continues, however, as to what constitutes a wetland and how property owners should be compensated for preserving wetlands.[13, 17]

Restoration of Wetlands

A related management issue is restoration of wetlands. A number of projects have attempted to restore wetlands, with varied success. The most important factor to be considered in most freshwater marsh restoration projects is the availability of water. If water is present, wetland soils and vegetation will likely develop. The restoration of salt marshes is more difficult because of the complex interactions among the hydrology, sediment supply, and vegetation that allow salt marshes to develop. Careful studies of relationships between the movement of sediment and the flow of water in salt marshes is providing information crucial to restoration, which makes successful re-establishment of salt marsh vegetation more likely. The restoration of wetlands has become an important topic in Canada because of the mitigation requirement related to environmental impact analysis, as set forth in the *Canadian Environmental Assessment Act* and parallel provincial legislation (see Chapter 26). According to this requirement, if wetlands are destroyed or damaged by a particular project, the developer must usually obtain or create additional wetlands at another site to compensate.[13] Unfortunately, the state of the art of restoration is not adequate to ensure that specific restoration projects will be successful.[18]

Constructing wetlands for the purpose of cleaning up agricultural runoff is an idea being attempted in areas with extensive agricultural runoff. Wetlands have the natural ability to remove excess nutrients, break down pollutants, and cleanse water. A series of wetlands are being created in many places in Canada, including Lunenburg County, Nova Scotia, to remove organic matter and nutrients (especially phosphorus) from household sewage and thus help restore the neighbouring LaHave River. (Phosphorus enrichment causes undesired changes in water quality and aquatic vegetation; see the discussion of eutrophication in the next chapter.) The human-made wetlands are designed to intercept and hold the nutrients so they do not enter and damage the river and its shellfish fishery.[19] In St. John, New Brunswick, constructed wetlands have been proposed to treat stormwater runoff entering the St. John River and harbour.

18.7 Dams and the Environment

Dams and their accompanying reservoirs generally are designed to be multifunctional structures. People who propose the construction of dams and reservoirs point out that reservoirs may be used for recreational activities and generating electricity as well providing flood control and ensuring a more stable water supply. However, it is often difficult to reconcile these various uses at a given site. For example, water demands for agriculture might be high during the summer, resulting in a drawdown of the reservoir and the production of extensive mudflats or an exposed bank area subject to erosion (Figure 18.13). Recreational users find the low water level and the mudflats aesthetically displeasing. Also, high water demand may cause quick changes in lake levels, which may interfere with wildlife (particularly fish) by damaging or limiting spawning opportunities. Another consideration is that dams and reservoirs tend to give a false sense of security to those living below these water retention structures. Dams may fail. Flooding may originate from tributary rivers that enter the main river below a dam; and dams cannot be guaranteed to protect people against floods larger than those for which they have been designed.

The environmental effects of dams are considerable and include the following:

- Loss of land, cultural resources, and biological resources in the reservoir area.
- Sediment stored behind the dam that would otherwise move downstream to coastal areas, where it would supply sand to beaches. The trapped sediment also reduces water storage capacity, limiting the life of the reservoir. Sediment-free water released from the dam then scours downstream river beds, removing fine sediments and altering aquatic habitat.
- Downstream changes in hydrology and in sediment transport that change the entire river environment and the organisms that live there.

Figure 18.13 • **Bank erosion along the shoreline of a reservoir following release of water, exposing bare banks.**

For a variety of reasons that include displacement of people, loss of land, loss of wildlife, and permanent, adverse changes to river ecology and hydrology, many people today are vehemently against turning remaining rivers into a series of reservoirs with dams. In Canada and the United States, a number of dams were recently removed, and others are being considered for removal as a result of the environmental damages they are causing. In contrast, China has the world's largest dam, as described in A Closer Look 18.1.

There is little doubt that if our present practices of water use continue we will need additional dams and reservoirs, and some existing dams will be heightened to increase water storage. However, there are few acceptable sites for new dams. Conflicts over the construction of additional dams and reservoirs are bound to occur. Water developers may view a valley dam site as a resource for water storage, whereas others may view it as a wilderness area and recreation site for future generations. The conflict is common, because good dam sites are often sites of high-quality scenic landscape.

Canals

Water from upstream reservoirs may be routed downstream by way of natural watercourses or canals and aqueducts. Canals are not hydrologically the same as creeks or rivers. They often have smooth, steep banks; and the water moves deceptively fast. Canals are a hazard attracting children to swim in them and animals to swim across them. Where they flow, drownings of people and animals are an ever-present threat.

The construction of canal systems, especially in developing countries, has led to serious, unanticipated environmental problems. For example, when the High Dam on the Nile River at Aswan, Egypt, was completed in 1964, a system of canals was built to convey the water to agricultural sites. The canals became infested with snails that carry the disease schistosomiasis (snail fever). This disease has always been a problem in Egypt, but the swift currents of the Nile floodwaters flushed the snails out each year. The tremendous expanse of waters in irrigation canals now provides happy homes for these snails. The disease is debilitating and so prevalent in parts of Egypt that virtually the entire population of some areas may be affected by it.

Removal of Dams

We opened this chapter with the story of the removal of the Lac Édouard dam in La Mauricie National Park, Québec. As mentioned, a number of Canadian and U.S. dams have been removed; and the largest project to date is the removal of the Matilija Dam on the Ventura River in southern California. The dam, completed by Ventura County in 1948, is about 190 m wide and 60 m high. The structure is in poor condition, with leaking, cracked concrete and a reservoir nearly filled with sediment. The dam serves no useful purpose and blocks endangered southern steelhead trout from their historic spawning grounds. The sediment trapped in the dam also reduces the natural nourishment of sand on beaches, increasing coastal erosion.

The trapped sediment however presents a problem in removing the dam. If released quickly, it could damage the downstream river environment, filling pools and killing river organisms such as fish, frogs, and salamanders. If the sediment can be slowly, more naturally released, the downstream damage can be minimized.

The removal process began with much fanfare in October 2000, when a 27-m section was removed from the top of the dam. The entire removal process may take years, after scientists have determined how to safely remove the sediment stored behind the dam. The cost of the dam in 1948 was about $300,000. The cost to remove the dam and sediment will be more than 10 times that amount.[22]

The perception concerning dams as permanent structures has clearly changed. What is learned from studying the removal of the Lac Édouard dam and the Matilija Dam in California will be useful in planning other dam removal projects. The studies will also provide important case histories to evaluate ecologic restoration of rivers following removal of dams.

18.8 Channelization and the Environment

Channelization of streams consists of straightening, deepening, widening, clearing, or lining existing stream channels. It is an engineering technique that has been used to control floods, improve drainage, control erosion, and improve navigation.[23] Of these objectives, flood control and drainage improvement are cited most often in channel improvement projects.

Thousands of kilometres of streams in Canada and the United States have been modified by channelization. The practice all too often has produced adverse environmental effects, including the following:

- Degradation of the stream's hydrologic qualities, turning a meandering stream with pools (deep, slow flow) and riffles (faster, shallow flow) into straight channels that are nearly all riffle flow, resulting in loss of important fish habitats.
- The removal of vegetation along the watercourse, which removes wildlife habitats and shading of the water.
- Downstream flooding where the channelized flow ends, because the channelized section has a larger channel area and carries a greater amount of floodwater

A CLOSER LOOK 18.1

Three Gorges Dam

In spite of known adverse effects of large dams, the world's largest is now being constructed in China. Three Gorges Dam on the Yangtze River (Figure 18.14) has drowned cities, farm fields, important archeological sites, and highly scenic gorges while displacing approximately two million people from their homes. In the river, habitat for endangered dolphins will likely be damaged. On land, habitats will be fragmented and isolated as mountaintops become islands in the reservoir.

The dam, which is approximately 185 m high and more than 1.6 km wide, produces a reservoir nearly 600 km long. Raw sewage and industrial pollutants that are discharged into the river are discharged into the reservoir, and there is concern that the reservoir will become seriously polluted. In addition, the Yangtze River has a high sediment load, and it is feared that the upstream end of the reservoir, where sediments will likely be deposited, will fill with sediment, damaging deep-water shipping harbours.

The dam may also produce a false sense of security in the people living in downstream cities. The presence of the dam may encourage further development in flood-prone areas, which will be damaged or lost if the dam and reservoir are unable to hold back floods in the future. If this happens, loss of property and life from flooding may be greater than if the dam had not been constructed. Contributing to this problem is the dam's location in a seismically active region where earthquakes and large landslides have been common in the past. Should the dam fail, then downstream cities such as Wushan, with a population of several million people, might be submerged with catastrophic loss of life.[20]

Figure 18.14 • Three Gorges on the Yangtze River is a landscape of high scenic value. Shown here is the Wu Gorge, near Wushan, one of the gorges flooded by the water in the reservoir.

A positive attribute of the giant dam and reservoir will be the capacity to produce about 18,000 MW of electricity, which is equivalent to about 18 large coal-burning power plants. As has been pointed out in earlier discussions, pollution from coal burning is a serious problem in China. Some opponents to the dam have pointed out, however, that a series of dams on tributaries to the Yangtze River could have produced similar electric power while not causing the anticipated environmental damage to the main river.[21]

than the downstream natural channel can carry without overbank flow or flooding.

- Damage to or loss of wetlands (because the water source is removed by the channelization, which often lowers the groundwater table and thus drains the wetlands).
- Aesthetic degradation (channelized streams are much less attractive than natural streams).

Figure 18.15 compares selected environmental features of natural and channelized streams.

A case study in problems with channelization involves Mink Creek, one of the tributaries draining into Dauphin Lake, Manitoba. In the early twentieth century, the lake supported a thriving sport and commercial walleye fishery. By the end of the century, the walleye catch had almost disappeared. The cause of the decline was traced to extensive habitat alteration, especially the degradation or disappearance of the stream's characteristic pool-and-riffle sequence, which provided important spawning habitat for the walleye. Creeks and agricultural drains emptying into Mink Creek had been straightened and channelized to improve drainage and reduce flooding. As a result, spring snowmelt flowed very quickly into the lake, over a much shorter period of time, than had been the case a hundred years earlier.[24]

Rehabilitation of Mink Creek began in 1985 and continues to this day. Early efforts involved experimental construction of pools and riffles in parts of the channelized stream. The changes created quiet sections where walleye eggs could settle and grow, rather than being carried into high siltation areas closer to the lake. Retention

General characteristics

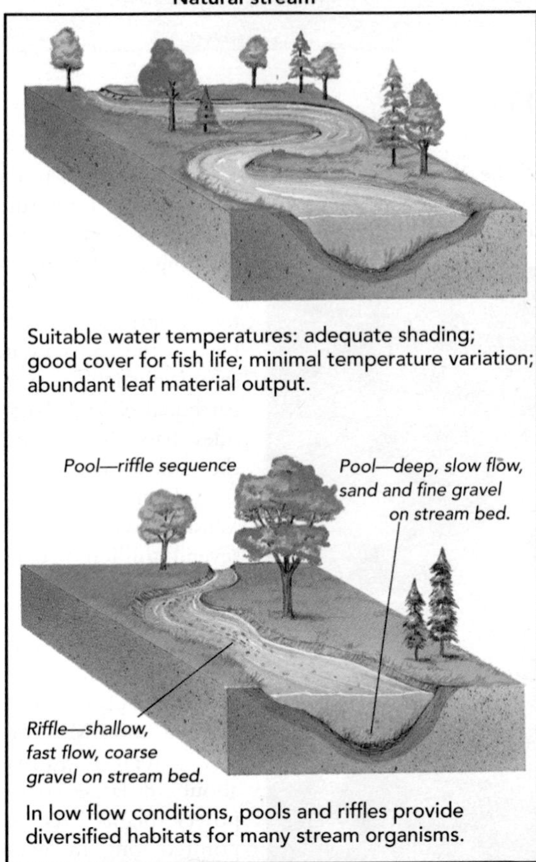

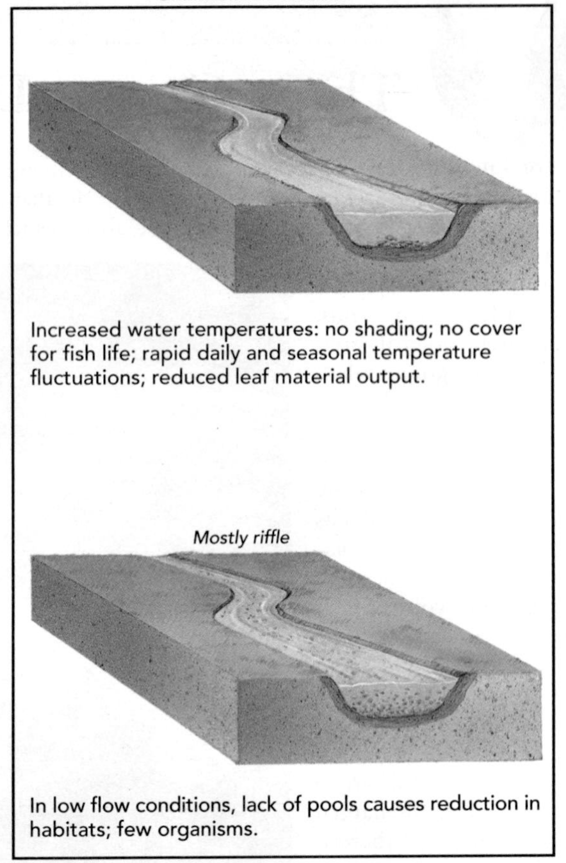

Figure 18.15 • A natural stream compared with a channelized stream in terms of general characteristics and pool environments. [*Source*: Modified from R.V. Corning, *Virginia Wildlife*, February 1975, pp. 6–8.]

of walleye eggs, and thus walleye spawning success, was much higher in the rehabilitated sections of the stream.

Not all channelization causes serious environmental degradation, and in many cases, drainage projects are beneficial. Moreover, channel design is being improved on the basis of past experience. Currently, more consideration is being given to the environmental aspects of channelization, and some projects are being designed with modified channels that behave more like natural streams than do straight ditches.

18.9 Flooding

River flooding is the most universally experienced natural hazard in the world today. A river and the flatland adjacent to it, known as the **floodplain** (Figure 18.17), together constitute a natural system. In most natural rivers, the water flows over the riverbanks and onto the floodplain every year or so. This natural process has many benefits for the environment:

- Water and nutrients are stored on the floodplain.
- Deposits on the floodplain contribute to the formation of nutrient-rich soils.
- Wetlands on the floodplain provide an important habitat for many birds, animals, plants, and other living things.
- The floodplain functions as a natural riparian (river-edge) greenbelt that is distinctly different from adjacent environments and provides environmental diversity.

Natural flooding is not a problem until people choose to build homes and other structures on floodplains. These structures are damaged when inundated by floodwaters. In Canada, about 100 people lose their lives in floods every year; and accompanying damages exceed $3 billion. In 1997, the "flood of the century" in the Red River Basin inundated 2,000 square kilometres in Manitoba, including almost 200,000 ha of agricultural land. At one point, the floodwaters created a "lake" 40 km wide, and water levels 12 m above normal (Figure 18.18). At least 22,000 people were evacuated from rural areas in Manitoba, and another 6,000 from Winnipeg. More than 2,500 homes were damaged by the floodwaters, 100 of them entirely destroyed. Total losses in Manitoba alone (the flood also affected many parts of neighbouring Minnesota and North Dakota) were over $150 million. These losses, although terrible, are relatively low compared with that in areas of the world that lack the sophisticated monitoring and warning systems Canada and the United States have for their rivers. For example, flooding associated

Figure 18.16 • **Restoring channelized streams.** *(a)* **A channelized urban stream.** *(b)* **A similar urban stream restored to its natural condition.**

with two cyclones that struck Bangladesh in 1970 and 1991 killed more than half a million people.

Urbanization and Flooding

A drainage basin is the land area that contributes water to a particular stream system. Many small drainage basins with areas less than 20 km^2 or so have been intensively urbanized. In these drainage basins, it may be hard to even find the streams, which have been forced to flow through underground pipes or concrete lined channels.

Urbanization has caused significant environmental impacts on these basins and streams. With urbanization, much of the land is paved over or covered with buildings and rendered impervious to water (pavement and buildings do not allow for infiltration). As a result, rainwater quickly runs off the artificial surfaces to drainage systems (storm sewers) and then to streams. Because both the amount of runoff and the speed at which it reaches streams increase as a result of urbanization, there is also an increase in the flood hazard. Thus, urban areas experience more frequent and larger floods than do natural systems of similar size.

Most major cities, including Halifax, Toronto, and Vancouver, have a single underground sewage system in older parts of the city. During times of no rain or light rain, this *combined sewer system* handles only sewage. But during periods of heavy rain, the runoff is mixed with the sewage and can exceed the capacity of sewage-treatment plants, causing sewage to be discharged downstream without sufficient treatment. In most cities that already have such systems, the expense of building a completely new and separate runoff system is very high, so other solutions, such as storage tunnels, must be found, at least in the short term.

Problems with flooding and overtaxing of storm sewage systems are made worse in many cities built on floodplains. Floodplains have often been chosen as sites for cities because the land is flat and easy to build on and river transportation is available nearby. More often than not, to protect against flooding, the river is channelized and levees are built along the shores. But as the citizens of Winnipeg learned in the heavy rains and floods of 1997, these levees have their limits. They concentrate the waters in a narrower channel, which increases the speed of water flow and the destructive force of the waters when they do rise above the levee. The levee is an example of the idea expressed earlier that the artificial structures of cities give us a sense of independence from the environment, whereas in fact these structures make us more dependent on the environment. An alternative is to allow for more

Figure 18.17 • **Floodplain of the Yamuna River, India.**

(a)

(b)

Figure 18.18 • *(a)* **Flood damage to farmlands and homes in Manitoba as a result of the 1997 Red River flood.** *(b)* **Destruction of homes and property in Cambridge, Ontario, during the 1974 flood of the Grand River.**

areas that have multiple uses and can flood without severe damage during times of heavy rains.

The flooding hazards associated with urbanization can be minimized in several ways. For example, runoff from urban areas during storms can be delayed by storing water in retention ponds, parking lots, and by other creative means. In their analysis of the 1997 Red River flood, the Canada-United States International Joint Commission concluded that widespread urbanization and construction of dikes had contributed to the magnitude of the flood. The Commission recommended a number of rehabilitation actions to increase natural water storage in the basin, for instance by the creation of new wetland areas.

Adjustments to the Flood Hazard

By far the most appropriate adjustment to flooding is land-use planning or floodplain zoning that restricts development and building on floodplains. This option is also the most desirable from an environmental perspective. Canada's federal Flood Damage Reduction Program (FDRP), a collaborative program undertaken with the support of the provinces, identifies, maps, and designates flood risk areas in Canada, and then applies policies to discourage development on flood plains. The program also includes flood forecasting and warning systems. By contrast, the flood insurance program in the United States has in a sense had the opposite effect; the program has been criticized as encouraging people to live on floodplains by allowing repeated insurance coverage in hazardous areas. The insurance program is being re-evaluated, and repeated coverage may be denied in favour of buyout programs for floodplain homes.

Getting the message across to people wishing to build on floodplains is not easy. For example, in Great Britain, thousands of homes and millions of people are threatened by flooding; and it is expected that by 2020, thousands of additional homes will be constructed on the nation's floodplains. The Environment Agency in Great Britain has often recommended that new buildings not be allowed on floodplains, but it does not have the authority to deny applications for development. That may change as a result of floods in the fall of 2000, which occurred following record rainfall. During the six-month period before the floods, which inundated more than 5,000 buildings (Figure 18.19), the agency advised that 200 applications for new buildings not be allowed. Local authorities in over 40% of the cases allowed the development to proceed. After the flood, the national government acknowledged that the Environment Agency and local authorities must now say no to future development on floodplains. It will be interesting to see the final policy decisions concerning floodplain development, as housing is expensive in Britain and there is pressure to build new homes, even in hazardous areas.

In Canada, many cities including Toronto and Vancouver are now mapping their floodplains, to provide a basis for limiting riparian development and protecting these important buffer zones for urban streams.

18.10 The Fraser River: Water Resources Management and the Environment

The history of the Fraser River emphasizes linkages among physical, biological, and social systems that are at the heart of environmental science.

The Fraser is the major river of British Columbia and extends from the Rocky Mountains southwest to Richmond, B.C., where it ends in the Pacific Ocean. With a total area of over 240,000 km^2 and a length of 1400 km, it is the fifth largest drainage basin in Canada. Its complex network of water systems, including lakes, rivers, marshes, bogs, swamps, and sloughs, is typical of major river basins in temperate regions.

Figure 18.19 • Floods in Great Britain in 2000. Shown here are people being rescued in Lewes, England, from floodwaters of the River Ouse, which inundated their homes.

From its headwaters on Mount Robson in the Rocky Mountains, the Fraser flows through alpine tundra down through forests and grasslands to the temperate rain forests and fertile valleys of the B.C. coast. Along its course, the Fraser supports one of the world's most famous salmon fisheries, with five species of salmon and 65 other fish species including giant sturgeon. The river is also a critical waterfowl breeding and overwintering area, and an important staging area on the Pacific flyway for migratory birds. At the mouth of the Fraser, a rich estuary supports hundreds of species of mammals, birds, reptiles and amphibians.

The huge Fraser basin is home to almost three million people, including 70,000 aboriginal people in 91 First Nations, representing eight language groups. Activities in the Fraser watershed, its delta, and the river itself contribute 80% of the province's economic output and 65% of total household income. Industries in the basin include half of all British Columbia's farms, almost 60% of its metal mine production, and two-thirds of its tourism revenue.[25]

Figure 18.20 • The Fraser River basin.

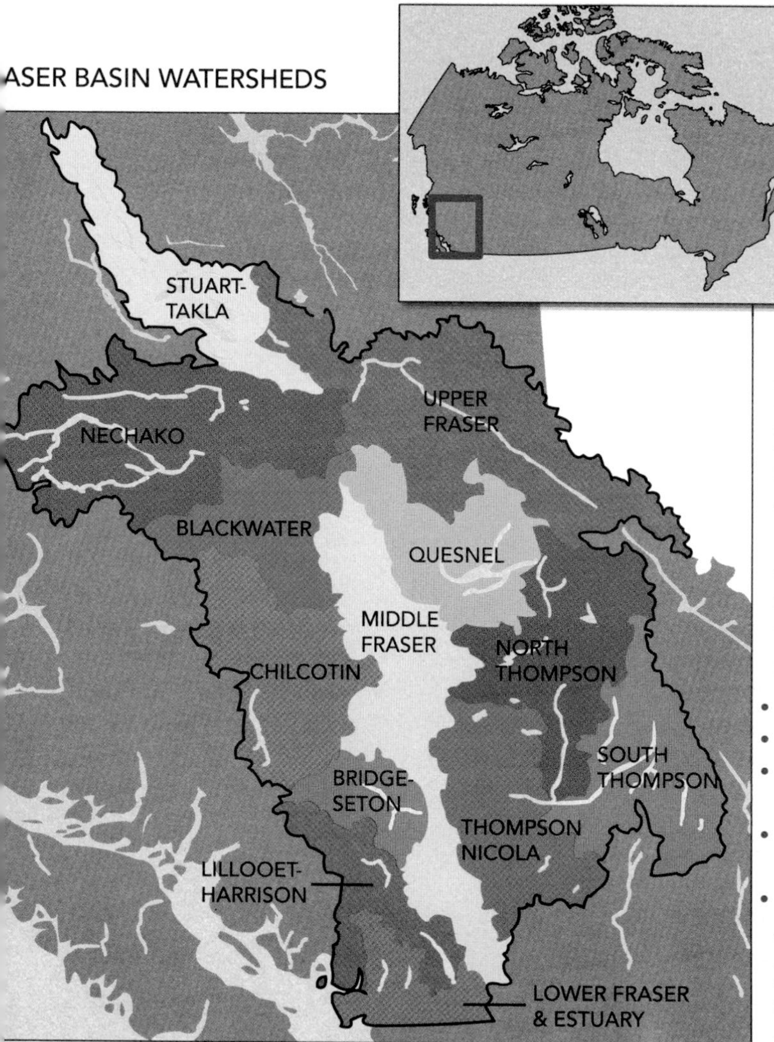

With such a diversity of peoples and uses, it is not surprising that conflict has arisen in the management of the Fraser's water resources. In 1997, the Fraser Basin Council was created as a framework for multidisciplinary, multistakeholder decision making for the Fraser. The Council's membership includes community groups, industrial representatives, First Nations, and four levels of government. Since its beginnings, the Fraser Basin Council has been an important impartial, non-political force in public education and conflict resolution. Their programs include activities as diverse as urban infrastructure planning, collaborative management of the river's important salmon fishery, and protection of montane habitats and water systems.[25,26,27,28]

Although its mandate is stewardship of the Fraser and its watershed ecosystem, the Fraser Basin Council recognizes that people are central to its success. In 2005, the Council's Board of Directors established the following five strategic priorities for the next five years:

- Strengthening Communities
- Fraser Fish and Fisheries Together
- Protecting People and Property from the Next Great Flood
- Measuring Progress Towards a Sustainable Fraser Basin, and
- Enhancing Aboriginal—Non-Aboriginal Collaboration

The complex issues of water management for the Fraser River illustrate major problems that are likely to be faced by other regions of the world in the coming years: How are water resources to be allocated?

Figure 18.21 • **Fraser River.**

Figure 18.22 • **First Nations woman drying salmon in Fraser River basin.**

How can we best control water quality? How can we protect river ecosystems? There are no easy answers to these questions.

18.11 Global Water Shortage Linked to Food Supply

As a capstone to this chapter, we present the hypothesis that we are facing a growing water shortage linked to our food supply. This is a potentially very serious problem! In the past few years, we have begun to realize that isolated water shortages are apparently indicators of a global pattern.[29,30,31] At numerous locations on Earth, both surface water and groundwater are being stressed and depleted:

- Groundwater in Canada, the United States, China, India, Pakistan, Mexico, and many other countries is being mined (used faster than it is being renewed) and so is being depleted.
- Large bodies of water—for example, the Aral Sea—are drying up (see Figure 18.6).
- Large rivers, including the Yellow in China, do not deliver any water to the ocean in some seasons or years. Others, such as the Nile in Africa, have had their flow to the ocean greatly reduced.

Water demand during the past half-century has tripled as human population has more than doubled. In the next half-century, the human population is expected to increase by 2–3 billion. There is growing concern that there won't be sufficient water to grow the food to feed the 8–9 billion people expected to inhabit the planet by the year 2050. Therefore, a food shortage linked to water resources seems a real possibility. The problem is that our increasing use of groundwater and surface water resources for irrigation has allowed for increased food production—mostly crops such as rice, corn, and soybeans. These same water resources are being depleted, and as water shortages for an agricultural region occur, food shortages may follow.

The solution to avoid food shortages resulting from water resource depletion is clear. We need to control human population growth, and conserve and sustain water resources. In this chapter, we have outlined a number of ways to conserve, manage, and sustain water. The good news is that a solution is possible—but it will take time, and we need to be proactive now before significant food shortages develop.

Figure 18.23 • **The huge Fraser River basin supports diverse urban, rural, recreational, and commercial water uses.**

Build Your Environmental Skills 18.2

How Much Water Do You Use?

This chapter discusses a number of options for water conservation. You may be amazed at how much water you use personally, in everyday activities such as washing and cooking. Log your water use for a week with the following table. Don't forget to include water used away from home.

	How Many Times Each Day											
Task	*S*	*M*	*T*	*W*	*Th*	*F*	*S*	*Total Times*	*Average*	*Actual*		*Total*
In the Bathroom (65%)												
Toilet Flushes											× 18 L	
Showers											× 100 L	
Baths											× 60 L	
Brushing Teeth											× 10 L	
Shaving											× 20 L	
In the Kitchen (10%)												
Cooking											× 20 L	
Dishes by hand											× 35 L	
Dishwasher											× 40 L	
In the Laundry Room (20%)												
Clothes washing											× 225 L	
Outdoors (5%)												
Car washing											× 400 L	
Lawn and garden watering											× 35 L/min.	
											Estimate:	
											Total Weekly:	

[*Source*: Adapted from Environment Canada: www.ec.gc.ca/water]

Summary

- Water is a liquid with unique characteristics that has made life on Earth possible.
- Although it is one of the most abundant and important renewable resources on Earth, more than 99% of Earth's water is unavailable or unsuitable for beneficial human use because of its salinity or location.
- The pattern of water supply and use on Earth at any particular point on the land surface involves interactions and linkages among the biological, hydrological, and rock cycles. To evaluate a region's water resources and use patterns, a water budget is developed to define the natural variability and availability of water.
- During the next several decades, it is expected that the total water withdrawn from streams and groundwater in Canada and the United States will decrease slightly, but the consumptive use will increase because of greater demands from a growing population and industry.
- Water withdrawn from streams competes with instream needs, such as maintaining fish and wildlife habitats and navigation, and may therefore cause conflicts.
- Groundwater use has resulted in a variety of environmental problems, including overdraft, loss of vegetation along watercourses, and land subsidence.
- Because agriculture is the biggest user of water, conservation of water in agriculture has the most significant effect on sustainable water use. However, it is also important to practice water conservation at the personal level in our homes and to price water to encourage conservation and sustainability.
- There is a need for a new philosophy in water resource management that considers sustainability and uses creative alternatives and variable sources. Development of a master plan involves inclusion of normal sources of surface water and groundwater, conservation programs, and use of reclaimed water.

CRITICAL THINKING ISSUE

How Wet Is a Wetland?

Areas where land meets water, whether fresh or salt are places where wetlands are found. Characteristically, wetlands are covered by surface water or have saturated soils. Landscape features such as swamps, bogs, marshes, potholes, and sloughs are wetlands. For most of our history, wetlands were considered wastelands and were destroyed by filling, draining, or polluting. According to the Atlas of Canada, about 14% of Canada's area is some form of wetland. Ontario, Manitoba, and the Northwest Territories have the largest proportion of wetland area nationally. Most of the pressure on Canada's wetlands is caused by agriculture, which accounts for 85% of wetland losses to date. For this reason, the most intense conflicts around wetland management occur in the most productive agricultural areas, such as southern Ontario. It is estimated that 20 million ha of wetland have been lost to agriculture since Europeans settled in Canada. This figure includes 65% of Atlantic Canada's coastal marshes, 70% of the wetlands in southern Ontario, 71% of those in the Prairie Provinces, and 80% of the Fraser River delta in British Columbia.

Today, however, it is generally recognized that wetlands have many values. They provide food, water, and cover for fish, shellfish, waterfowl, game animals, and many amphibians and reptiles. In addition, soil and plants in wetlands purify water by absorbing or destroying pollutants. Wetlands also help to recharge groundwater stores and, by holding water, control flooding and erosion. One-third of endangered and threatened species, two-thirds of the commercial saltwater fish and shellfish, one-third of the birds, and almost all the amphibians in North America depend on wetlands. Wetlands are among the most productive ecological communities in the world, many times more productive than a heavily fertilized cornfield.

Protection of these unique communities began with the goal of preserving wetlands used by wildlife, particularly ducks, but federal and provincial protection has since been broadened to include most of the remaining wetlands. Still, tens of thousands of hectares of wetlands are lost in Canada and the United States each year. Of particular concern are the freshwater wetlands, many of which are on private land. A recent policy of no net loss of wetlands has been praised by some environmental scientists but attacked by farmers and developers.

Particularly controversial are the small, seasonal potholes in the agricultural areas of the prairies and central Canada that may not appear wet to the casual observer. They do, however, provide habitats for many species, especially waterfowl. Critics of applying strict regulations to potholes say that if an area is not wet enough for a duck to land in and splash, it is not wet enough to qualify as a wetland.

Critical Thinking Questions

1. Some people have proposed defining wetlands so as to exclude many of the seasonal ones, over 5 million ha in Canada and the United States. What position would you take on such a proposal, and why? How would you reconcile the conflicting needs of the farmers and developers with the need to preserve wildlife habitats?
2. In what ways would you expect a substantial decrease in wetlands to affect populations of migratory birds?
3. In times of drought, what priority should be given to wetland protection compared with residential and industrial uses, and recreational uses such as golf courses and canoeing?

- Development of water supplies and facilities to more efficiently move water may cause considerable environmental degradation; construction of reservoirs and canals, and channelization of rivers should be considered carefully in light of potential environmental impacts.
- Wetlands serve a variety of functions at the ecosystem level that benefit other ecosystems and people.
- Flooding is perhaps the most universal natural hazard in the world; both the frequency and the severity of flooding of small streams are increased by urbanization. The preferable and most environmentally sound adjustment to flooding is land-use planning that avoids building on floodplains.
- The Fraser River in southwestern British Columbia is one Canada's largest and most diverse river systems. Understanding links among physical, biological, and social systems of the Fraser River is necessary to manage its water resources and ecosystems.
- We are facing a growing global water shortage linked to the food supply.

STUDY QUESTIONS

1. If water is one of our most abundant resources, why are we concerned about its availability in the future?
2. Which is more important from a national point of view, conservation of water use in agriculture or in urban areas? Why?
3. Distinguish between in-stream and off-stream uses of water. Why is in-stream use controversial?
4. What are some important environmental problems related to groundwater use?
5. How might your community better manage its water resources?
6. What are some of the major environmental impacts associated with the construction of dams and canals? How might these be minimized?
7. What are some of the environmental problems associated with channelizing rivers? How might the potential adverse effects of channelization be minimized?
8. How does urbanization affect flood hazard, and how can this hazard best be managed?
9. How can we reduce or eliminate the growing global water shortage? Do you believe the shortage is related to our food supply? Why? Why not?

FURTHER READING

Ashmore, P. and M. Church. *The impact of climate change in rivers and river processes in Canada.* Geological Survey of Canada Bulletin 555. Ottawa: Geological Survey of Canada, 2001.

Baxter, R. M. "Environmental effects of dams and impoundments." *Ann. Rev. Ecol. Syst.* 8 (1977): 255-283.

Gleick, P. H. "Global Freshwater Resources: Soft-Path Solutions for the 21st Century." *Science* 302 (2003): 1524–1528.

Gleick, P. H. *The World's Water 2000–2001.* Washington, D.C.: Island Press, (2000).

James, W. and J. Neimczynowicz, eds. *Water, Development and the Environment.* Florida: CRC Press, 1992. Covers problems with water supplies imposed by a growing population, including urban runoff, pollution and water quality, and management of water resources.

La Riviere, J. W. M. "Threats to the World's Water," *Scientific American* 261, no. 3 (1989):80–84. Summary of supply and demand for water and threats to continued supply.

Spulber, N., and A. Sabbaghi. *Economics of Water Resources: From Regulation to Privatization.* London: Kluwer Academic, 1994. Discussions of water supply and demand, pollution and its ecological consequences, and water on the open market.

Twort, A. C., F. M. Law, F. W. Crowley, and D. D. Ratnayaka. *Water Supply,* 4th ed. Edward Arnold, 1994. Good coverage of water topics from basic hydrology to water chemistry, its use, management, and treatment.

Wheeler, B. D., S. C. Shaw, W. J. Fojt, and R. A. Robertson. *Restoration of Temperate Wetlands.* New York: Wiley, 1995. Discussions of wetland restoration around the world.

chapter 19

Intensive Livestock Farming and Water Quality Impacts in North Carolina

Hurricane Floyd struck the Piedmont area of North Carolina in September 1999. The killer storm took a number of lives while flooding many homes and forcing some 48,000 people into emergency shelters. The storm had another, more unusual effect as well. Floodwaters containing thousands of dead pigs along with their feces and urine flowed through schools, churches, homes, and businesses. The stench was reported to be overwhelming, and the count of pig carcasses may have been as high as 30,000. The storm waters had overlapped and washed out over 38 pig lagoons with as much as 3.6 million cubic metres of liquid pig waste, which ended up in flooded creeks, rivers, and wetlands. In all, approximately 250 large commercial pig farms flooded out, drowning hogs whose floating carcasses had to be collected and disposed of (Figure 19.1).

North Carolina has a long history of hog production. The population of pigs grew from about 2 million in 1990 to nearly 10 million in 1997, when North Carolina became the second largest pig-farming state in the United States.[1] As the number of large commercial pig farms grew, the state allowed the hog farmers to build automated and very confining farms housing hundreds or thousands of pigs. There were no restrictions on farm location, and many farms were constructed on floodplains.

Each pig produces approximately 2 metric tonnes of waste per year. The North Carolina herd was producing approximately 20 million tonnes of waste a year, mostly manure and urine, which was flushed out of the pig barns and into open, unlined lagoons about the size of football fields. Favourable regulations, along with the availability of inexpensive waste disposal systems (the lagoons), were responsible for the tremendous growth of the pig population in North Carolina during the 1990s.

Following the hurricane, mobile incinerators were moved into the hog region to burn the carcasses; but there were so many dead pigs that hog farmers had to bury some animals in shallow pits. The pits were supposed to be at least 1 m deep and dry, but there wasn't always time to find dry ground; and for the most part, the pits were dug and filled on floodplains. As these pig carcasses rot, bacteria will leak into the groundwater and surface water for some appreciable time.

The pig farmers blamed the hurricane for the environmental catastrophe. However, it was clearly a human-induced disaster that had been expected. As early as 1995, a pig-waste-holding lagoon failed and sent approximately 1 million cubic metres of concentrated pig feces down the New River past the city of Jacksonville and into the New River Estuary. Adverse environmental effects of that spill on marine life lasted for approximately three months.

The lesson to be learned from this case is that we are vulnerable to environmental catastrophes caused by large-scale industrial agriculture. Economic growth and livestock production must be carefully planned to anticipate problems, and waste management facilities must be designed so as not to pollute local streams, rivers, and estuaries.[2]

- *The combination of extreme weather and intensive livestock farming produced a particularly visible and serious water pollution episode in a beautiful region with abundant natural resources. Other types of water pollution, such as waterborne disease in surface waters and pesticides in groundwater, are often much more difficult to identify without careful sampling and testing. This chapter discusses major categories of water pollution and traditional as well as innovative options for wastewater treatment.*

Water Pollution and Treatment

LEARNING OBJECTIVES

Degradation of our surface water and groundwater resources is a serious problem, the effects of which are not fully known. There are a number of steps we can take to treat water and to minimize pollution. After reading this chapter, you should understand:

- What constitutes water pollution and what are the major categories of pollutants.
- Why the lack of clean drinking water is the primary water pollution problem in many locations in the world.
- How point and nonpoint sources of water pollution differ.
- What biochemical oxygen demand is, and why it is important.
- What eutrophication is, why it is an ecosystem effect, and how human activity can cause cultural eutrophication.
- Why sediment pollution is a serious problem.
- What acid mine drainage is, and why it is a problem.
- How urban processes can cause shallow-aquifer pollution.
- What are the various methods of wastewater treatment, and why some are environmentally preferable.
- What are the environmental laws that protect water resources and ecosystems.

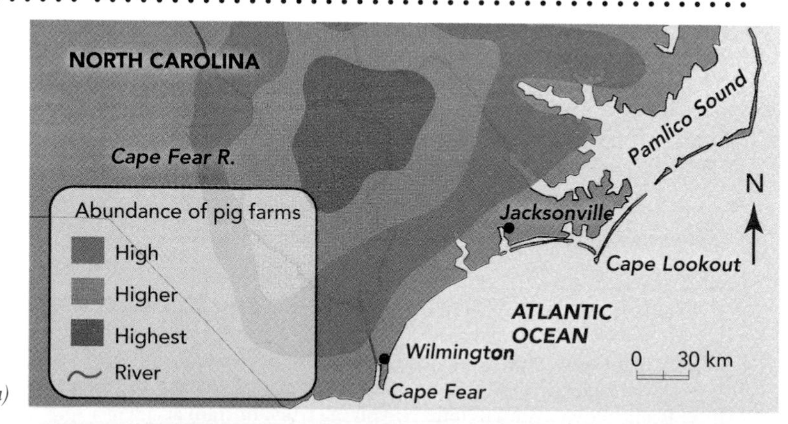

(a)

(b)

Figure 19.1 • North Carolina's Bay of Pigs. *(a)* Map of areas flooded by Hurricane Floyd in 1999 with relative abundance of pig farms. *(b)* Collecting dead pigs near Boulaville, North Carolina. The animals were drowned when floodwaters from the Cape Fear River inundated commercial pig farms.

19.1 Water Pollution

Water pollution refers to degradation of water quality. In defining pollution, we generally look at the intended use of the water, how far the water departs from the norm, and its effects on human and ecosystem health. From a public health or ecological view, a pollutant is any biological, physical, or chemical substance that, in identifiable excess, is harmful to other desirable living organisms. Water pollutants include metals, trace organic compounds, sediment, certain radioactive isotopes, heat, fecal coliform bacteria, phosphorus, nitrogen, sodium, and other useful (even necessary) elements, as well as certain pathogenic bacteria and viruses. In some instances, a material may be considered a pollutant to a particular segment of the population although it is not harmful to other segments. For example, excessive sodium as a salt is not generally harmful, but it may be harmful to people who must restrict salt intake for medical reasons.

Today, the primary water pollution problem in the world is the lack of clean, disease-free drinking water. In the past, epidemics (outbreaks) of waterborne diseases such as cholera have been responsible for the deaths of thousands of people in Canada and the United States. Fortunately, epidemics of such diseases have been largely eliminated in industrialized countries, as a result of treating drinking water prior to consumption. This certainly is not the case worldwide, however. Every year, several billion people are exposed to waterborne diseases. For example, an epidemic of cholera occurred in South America in the early 1990s, and outbreaks of waterborne diseases continue to be a threat, even in developed countries.

It is a fundamental principle that the quality of water determines its potential uses. The major uses for water today are power generation, agriculture, industrial processes, and domestic (household) supply. Water for domestic supply must be virtually free from constituents harmful to health, such as insecticides, pesticides, pathogens, and metals; it should taste good, should be odourless, and should not damage plumbing or household appliances. The quality of water required for industrial purposes varies widely depending on the process involved. Some processes may require distilled water; others need water that is not highly corrosive or that is free of particles that could clog or otherwise damage equipment. Since most vegetation is tolerant of a wide range of water quality, agricultural waters may vary widely in physical, chemical, and biological properties.[3]

Many different processes and materials may pollute surface water or groundwater. Some of these are listed in Table 19.1. All segments of society (urban, rural, industrial, and agricultural) may contribute to the problem of water pollution. Most of the sources result from runoff and pollutant leaks or seepage into surface water or groundwater. Pollutants are also transported by air and deposited in water bodies.

Increasing population often results in the introduction of more pollutants into the environment as well as demands on finite water resources.[4] As a result, sources of drinking water in some locations are expected to degrade in the near future. As many as one-quarter of drinking water systems in Canada and the United States have reported at least one violation of national health standards.[5]

Health Canada and the U.S. Environmental Protection Agency have set thresholds, or limits, on water pollution levels for some (but not all) pollutants. As a result of difficulties in determining effects of exposure to low levels of pollutants, maximum concentration

Table 19.1 • Some Sources and Processes of Water Pollution

Surface Water	*Groundwater*
• Urban runoff (oil, chemicals, metals, organic matter, etc.) (U, I, M) • Agricultural runoff (oil, metals, fertilizers, pesticides, etc.) (A) • Accidental chemical spills including oil (U, R, I, A, M) • Radioactive materials (often involving truck or train accidents) (I, M) • Runoff (solvents, chemicals, etc.) from industrial sites (factories, refineries, mines, etc.) (I, M) • Leaks from surface storage tanks or pipelines (gasoline, oil, etc.) (I, A, M) • Sediment from a variety of sources, including agricultural lands and construction sites (U, R, I, A, M) • Air fallout (particles, pesticides, metals, etc.) into rivers, lakes, oceans (U, R, I, A, M)	• Leaks from waste disposal sites (chemicals, radioactive materials, etc.) (I, M) • Leaks from buried tanks and pipes (gasoline, oil, etc.) (I, A, M) • Seepage from agricultural activities (nitrates, heavy metals, pesticides, herbicides, etc.) (A) • Saltwater intrusion into coastal aquifers (U, R, I, M) • Seepage from cesspools and septic systems (R) • Seepage from acid-rich water from mines (I) • Seepage from mine waste piles (I) • Seepage of pesticides, herbicide nutrients, and so on from urban areas (U) • Seepage from accidental spills (e.g., train or truck accidents) (I, M) • Inadvertent seepage of solvents and other chemicals including radioactive materials from industrial sites or small businesses (I, M)

Key: U = urban; R = rural; I = industrial; A = agricultural; M = military.

standards have been set for only a small fraction of the more than 700 identified drinking water contaminants. If the pollutant is present in a concentration greater than an established threshold, then the water is unsatisfactory for a particular use. A list of selected pollutants (contaminants) included in the national drinking water guidelines for Canada can be found in Table 19.2.

The following sections focus on several water pollutants to emphasize principles that apply to pollutants in general. (See Table 19.3 for categories and examples of water pollutants.) Other water pollutants are discussed elsewhere in this book (for example, metals, organic chemicals, and thermal pollution in Chapter 15, and radioactive materials in Chapter 17). Before proceeding to our discussion of pollutants, however, we first consider biochemical oxygen demand and dissolved oxygen. Dissolved oxygen is not a pollutant but rather is needed for healthy aquatic ecosystems.

Table 19.2 • Canadian Drinking Water Guidelines

Contaminant	*Maximum Acceptable Concentration (mg/l)*
Toxic metals	
Arsenic	0.025
Cadmium	0.005
Lead	0.010
Mercury	0.001
Selenium	0.01
Organic chemicals	
Methoxychlor	0.9
2,4-D	0.9
2,4,6-TCP	0.005
Volatile organic chemicals	
Benzene	0.005
Carbon tetrachloride	0.005
Trichloroethylene	0.05
Vinyl chloride	0.002
Microbiological organisms	
Escherichia coli (E.coli) bacteria	none detected

Source: Health Canada.

19.2 Biochemical Oxygen Demand (BOD)

Dead organic matter in streams decays. Bacteria that carry out this process require oxygen. If there is enough bacterial activity, the oxygen in the water available to fish and other organisms can be reduced to levels so low that they may die. A stream with low oxygen content is a poor environment for fish and most other organisms. A stream with an inadequate oxygen level is considered polluted for those organisms that require dissolved oxygen above the existing level.

The amount of oxygen required for biochemical decomposition processes is called the **biochemical**

Table 19.3 • Categories of Water Pollutants

Pollutant Category	*Examples of Sources*	*Comments*
Dead organic matter	Raw sewage, agricultural waste, urban garbage	Produces biochemical oxygen demand and diseases.
Pathogens	Human and animal excrement and urine	Examples: Recent cholera epidemics in South America and Africa; 1993 epidemic of *cryptosporidiosis* in Milwaukee, Wisconsin. See discussion of fecal coliform bacteria in Section 19.3.
Organic chemicals	Agricultural use of pesticides and herbicides (see Chapter 12); industrial processes that produce dioxin (Chapter 15)	Potential to cause significant ecological damage and human health problems. Many of these chemicals pose hazardous-waste problems (Chapter 24)
Nutrients	Phosphorus and nitrogen from agricultural and urban land use (fertilizers) and wastewater from sewage treatment	Major cause of artificial eutrophication. Nitrates in groundwater and surface waters can cause pollution and damage to ecosystems and people
Toxic metals	Agricultural, urban, and industrial use of mercury, lead, selenium, cadmium, and so on (Chapter 15)	Example: Mercury from industrial processes that is discharged into water (Chapter 15). Toxic metals can cause significant ecosystem damage and human health problems
Acids	Sulphuric acid (H_2SO_4) from coal and some metal mines; industrial processes that dispose of acids improperly	Acid mine drainage is a major water pollution problem in many mining areas, damaging ecosystems and spoiling water resources
Sediment	Runoff from construction sites, agricultural runoff, and natural erosion	Reduces water quality and results in loss of soil resources.
Heat (thermal pollution)	Warm to hot water from power plants and other industrial facilities	Causes ecosystem disruption (Chapter 15)
Radioactivity	Contamination by nuclear power industry, military, and natural sources (see Chapter 17)	Often related to storage of radioactive waste. Health effects vigorously debated (Chapters 15 and 17)

Figure 19.2 • Pollution control officer measuring oxygen content of the Grand River, Ontario.

oxygen demand (BOD). BOD is commonly used in water quality management (Figure 19.2). It measures the amount of oxygen consumed by micro-organisms as they break down organic matter within small water samples, which are analysed in a laboratory. The BOD is routinely measured as part of water quality testing, particularly at discharge points into surface water, such as at wastewater treatment plants. At treatment plants, the BOD of the incoming sewage water from sewer lines is measured, as is water from locations both upstream and downstream of the plant. This practice allows comparison of the upstream, or background BOD with the BOD of the water being discharged by the plant.

Dead organic matter—which produces BOD—is added to streams and rivers from natural sources (such as dead leaves from a forest) as well as from agricultural runoff and urban sewage. Approximately 33% of all BOD in streams results from agricultural activities. However, urban areas, particularly those with older, combined sewer systems (in which storm-water runoff and urban sewage share the same line), also considerably increase the BOD in streams. This results because during times of high flow, when sewage-treatment plants are unable to handle the total volume of water, raw sewage mixed with storm runoff overflows and is discharged untreated into streams and rivers.

When the BOD is high, as suggested earlier, the dissolved oxygen content of the water may become too low to support the life in the water. Typically, concern is raised (and in some cases a formal pollution alert issued) when the *dissolved oxygen content* of water is less than 5 mg/l. Figure 19.3 illustrates the effect of high BOD on dissolved oxygen content in a stream when raw sewage is introduced as a result of an accidental spill. Three zones are identified:

1. *A pollution zone*, where a high BOD exists. As decomposition of the waste occurs, oxygen is used by micro-organisms, and dissolved oxygen content of the water decreases.
2. *An active decomposition zone*, where the dissolved oxygen content reaches a minimum owing to rapid biochemical decomposition by micro-organisms as the organic waste is transported downstream.
3. *A recovery zone*, where the dissolved oxygen increases and the BOD is reduced, because most oxygen-demanding organic waste from the input of sewage has decomposed and the natural stream processes are replenishing the water with dissolved oxygen. For example, in quickly moving water, the water at the surface mixes with air, and oxygen enters the water.

All streams have some capability—assimilative capacity—to degrade organic waste. Problems result when the stream is overloaded with biochemical oxygen-demanding waste, exceeding its assimilative capacity and overpowering the stream's natural cleansing function.

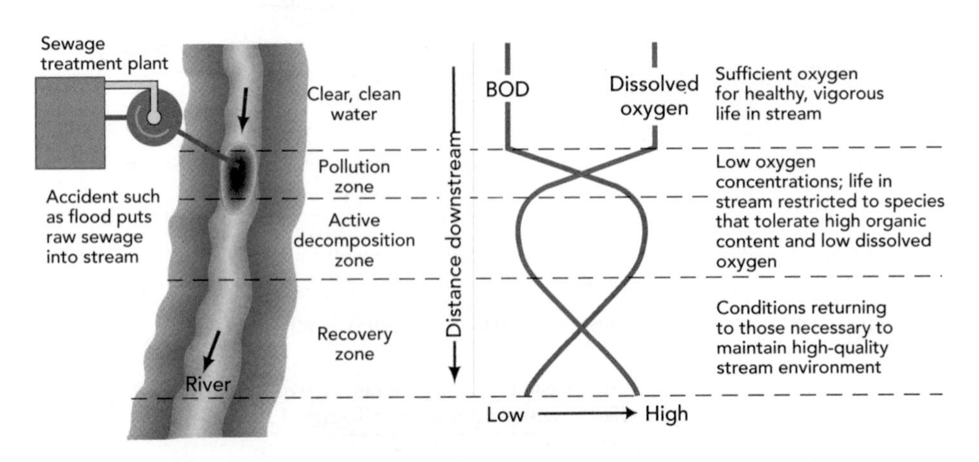

Figure 19.3 • Relationship between dissolved oxygen and biochemical oxygen demand (BOD) for a stream following the input of sewage.

19.3 Waterborne Disease

As mentioned earlier, the primary water pollution problem in the world today is the lack of clean, disease-free drinking water. Each year, particularly in less-developed countries, several billion people are exposed to waterborne diseases, the effects of which vary in severity from an upset stomach to death. As recently as the early 1990s, epidemics of cholera, a serious waterborne disease, caused widespread suffering and death in South America.

In Canada, we tend not to think much about waterborne illness. Although historically epidemics of waterborne disease killed thousands of people in cities such as Toronto and Halifax, public health programs have largely eliminated such epidemics by treating drinking water to remove disease-carrying micro-organisms and not allowing sewage to contaminate drinking water supplies. However, North America is not immune to **outbreaks**—sudden occurrences—of waterborne disease.

Fecal Coliform Bacteria

Since it is difficult to monitor disease-carrying organisms directly, the count of **fecal coliform bacteria** is often used as a measure and indicator of disease potential. The presence of fecal coliform bacteria in water indicates that fecal material from mammals or birds is present, so organisms that cause waterborne diseases may be present as well. Fecal coliform bacteria are usually (but not always) harmless bacteria that are normal constituents of human and animal intestines. They are present in all human and animal waste. The threshold used by Health Canada for swimming water is not more than 200 cells of fecal coliform bacteria per 100 ml of water; if fecal coliform is above the threshold level, the water is considered unfit for swimming. Water with any fecal coliform bacteria is unsuitable for drinking.

One type of fecal coliform bacteria, *Escherichia coli*, or *E. coli*, has been responsible for causing human illness and death. Outbreaks have resulted from eating contaminated meat and drinking contaminated juices or water. There have been outbreaks caused by contaminated meat at camps, and popular fast-food chains. It is clear that *E. coli* bacteria can be a real threat to human health and must be carefully regulated.[6]

Threat of waterborne diseases in nearshore coastal waters as well as lakes and rivers is responsible for thousands of warnings and beach closings per year in Canada and the United States. Notices are often posted that swimming may be hazardous to health (Figure 19.4). In most cases, the identified pollutant is fecal coliform bacteria, which may indicate the presence of a specific disease-causing virus, such as hepatitis. Pollutants to coastal waters have a variety of sources, including urban storm runoff, leaking sewer lines, sewage-treatment plant overflow or failure, and leaking sewage treatment facilities from individual homes (septic tanks; see Section 19.10). Coastal communities in many areas are facing potential loss of income from tourism resulting from beach closures. As a result, increased research and testing of nearshore coastal waters is becoming more routine. As sources of pollutants are identified, management plans are being designed and implemented to reduce the threat of pollution and future beach closures.

Outbreak in Walkerton, Ontario

One of the most serious outbreaks of *E. coli* bacterial infection in Canadian history developed in May 2000 in Walkerton, Ontario, a town with about 5,000 inhabitants. The strain of *E. coli* involved was a dangerous one found in the digestive system of cows. The likely cause of the contamination was cow manure washed into the water supply during heavy rains and flooding on May 12, 2000. The local public utility commission knew as early as May 18 that water from wells serving the town was contaminated with *E. coli*, but the commission did not report it immediately to health authorities. People were not advised to boil water (thus killing the bacteria) until it was too late to avoid the outbreak. By May 26, six people had died, more than 20 were in intensive care

Figure 19.4 • **"Stoop and scoop" ordinances requiring people to pick up pet feces are intended to reduce bacterial contamination of swimming beaches.**

units, and over 500 were ill with severe symptoms that included cramps, vomiting, and diarrhea. Older adults and young children are most vulnerable to the ravages of the disease, which can damage the kidneys. Two of the first deaths were a two-year-old baby and an 82-year-old woman. Finally, officials took over management of the water supply, and bottled water was distributed.

The town of Walkerton experienced a full-blown outbreak of a waterborne disease. Doctors warned that additional deaths were likely and that modern medicine could do little to treat the disease. The best advice doctors could give was to consume large amounts of safe water and avoid dehydration as the infection ran its course. Soon, people were asking why the outbreak had occurred. An investigation was launched by authorities, and one of the questions asked was why there had been a delay between identifying the potential problem and issuing a warning. If people had been notified earlier, much of the sickness might have been avoided. The outbreak might also have been detected earlier if the government of Ontario had not cut back on the level of testing of the public water supply (earlier regulations had required more testing). A major lesson learned from Walkerton is that we should remain vigilant in testing our water supplies and raising the alarm quickly if potential problems arise.[7]

Outbreaks of *Cryptosporidium* in Milwaukee, Wisconsin, and Canadian Cities

Major outbreaks of the waterborne disease *Cryptosporidiosis* have occurred in several major cities in recent years. The disease, which causes flulike symptoms, is a gastrointestinal illness carried by a micro-organism (a parasite) known as *Cryptosporidium*. Between March 11 and April 9, 1993, approximately 400,000 people of a total of 1.6 million people in a five-county area around Milwaukee, Wisconsin, became ill following exposure to *Cryptosporidium* in the drinking water. Most people who contracted the illness were sick for approximately nine days; but the disease can be fatal to people with depressed immune systems, such as cancer patients and AIDS patients. Approximately 100 people died. The parasite is resistant to chlorination and evidently passed through one of the city's water treatment plants. The source of the parasite remains unknown, but possible sources include cattle grazing along rivers that flow into Milwaukee Harbour, slaughterhouses, and human sewage. It is possible that runoff from spring rains and snowmelt transported the parasites to Lake Michigan, where they entered the intake to water treatment plants.[7–10] Similar but smaller outbreaks occurred in Kitchener-Waterloo (1993) and Collingwood (1996), Ontario, and in North Battleford, Saskatchewan (2001), affecting thousands of residents in the three communities.

These outbreaks were a wake-up call concerning the quality of our drinking water. Many other cities in Canada and the United States that utilize surface water resources are just as vulnerable as these.[7] In fact, recent tests suggest that *Cryptosporidium* is present in 65% to 97% of the surface waters in Canada and the United States.

The outbreaks happened even though the water treatment plants appeared to have met required operating and treatment standards. Concern was therefore raised as to whether waterborne parasites such as *Cryptosporidium* are being effectively removed by existing treatment processes. Although *Cryptosporidium* is very resistant to disinfectants, it may be removed by filtration. Upgrading water treatment plants is therefore a cost-effective mechanism for reducing the threat of waterborne disease. The price of inaction is very high. Considering the high costs of illness and death associated with drinking contaminated water, future investments in technology and facilities to treat water are an important government service that should be considered a bargain.[6]

19.4 Nutrients

Two important nutrients that cause water pollution problems are phosphorus and nitrogen, and both are released from sources related to land use. Forested land has the lowest concentrations of phosphorus and nitrogen in stream waters. In urban streams, concentrations of these nutrients are greater because of fertilizers, detergents, and products of sewage-treatment plants. Often, however, the highest concentrations of phosphorus and nitrogen

Figure 19.5 • An intensive swine operation in Thailand. High numbers of livestock in small areas have the potential to create both surface water and groundwater pollution because of runoff and infiltration of waste-contaminated water.

(a)

(b)

Figure 19.6 • *(a)* **Mats of dying green algae in a river undergoing eutrophication.** *(b)* **Cultural eutrophication in the Gulf of Mexico.**

are found in agricultural areas, where the sources are fertilized farm fields and feedlots (Figure 19.5). Over 90% of total nitrogen added to the environment by human activity comes from agriculture.

Eutrophication

Chapter 2 discussed how Canadian scientists used ecosystem-scale experiments in the Experimental Lakes Area to explore the causes of algae blooms. The phenomenon they investigated is called **eutrophication**—the process by which a body of water develops increasingly high concentrations of nutrients, such as nitrogen and phosphorus (in the forms of nitrates and phosphates). The nutrients cause an increase in the growth of aquatic plants in general, as well as production of photosynthetic blue-green bacteria and algae. Algae may form surface mats (Figure 19.6*a*), shading the water and reducing light to algae below the surface, greatly reducing photosynthesis. The bacteria and algae die; as they decompose, the BOD increases, oxygen in the water is consumed, and oxygen content is reduced. If the oxygen content is sufficiently lowered, other organisms, such as fish, will die.

The process of eutrophication of a lake is shown in Figure 19.7. A lake that has a natural high concentration of the chemical elements required for life is called a *eutrophic lake*. A lake with a relatively low concentration of chemical elements required by life is called an *oligotrophic lake*. Oligotrophic lakes have clear water that is pleasant for swimmers and boaters, with relatively low abundance of life. Eutrophic lakes have an abundance of life, often with mats of algae and bacteria and murky, unpleasant water.

When eutrophication is accelerated by human processes that add nutrients to a body of water, we say that **cultural eutrophication** is occurring.[11,12] Problems associated with the artificial eutrophication of bodies of water are not restricted to lakes, but can also affect

Figure 19.7 • The eutrophication of a lake. *(a)* In an oligotrophic, or low-nutrient, lake, the abundance of green algae is low, the water clear. *(b)* Phosphorus is added to streams and enters the lake. Algae growth is stimulated, and a dense layer is formed. *(c)* The algae layer becomes so dense that the algae at the bottom die. Bacteria feed on the dead algae and use up the oxygen. Finally, fish die from lack of oxygen.

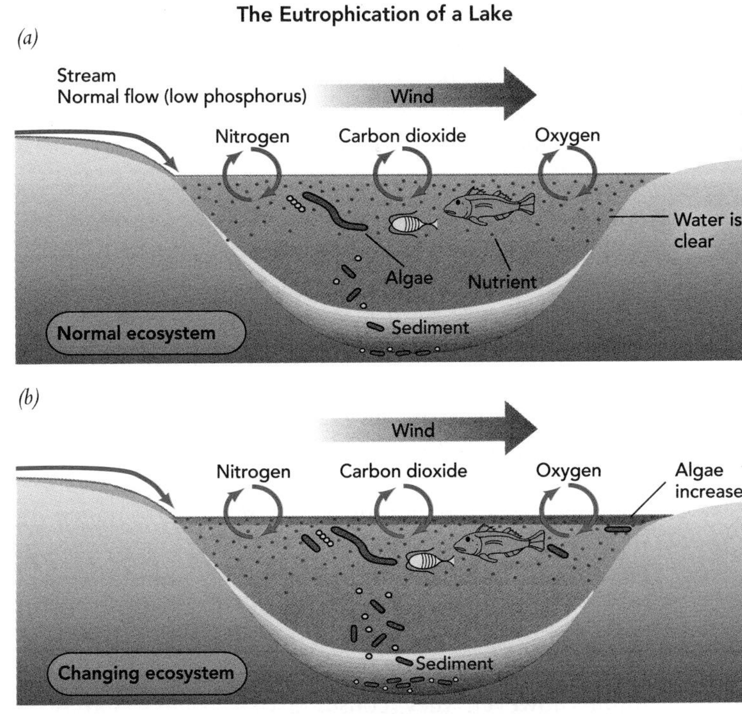

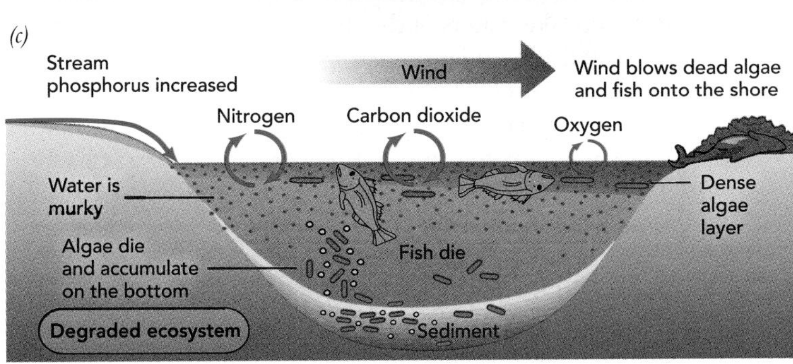

nearshore marine environments. In recent years concern has grown about the outflow of sewage from urban areas into tropical coastal waters and cultural eutrophication on coral reefs.[13,14] For example, parts of the famous Great Barrier Reef of Australia, as well as some reefs fringing the Hawaiian Islands, are being damaged by eutrophication.[15,16] The damage to corals occurs as nutrient input stimulates algal growth on the reef, which covers and smothers the coral. In Canada, eutrophication of most freshwater systems is related to elevated concentrations of phosphorus. Lake Erie provides a good example. During the 1960s and 1970s, Lake Erie was considered a dead ecosystem. Between 1948 and 1962, phosphorus levels in the lake had quintupled, largely as a result of human activities. Following the introduction of improved sewage treatment and low-phosphate cleaning products in the early 1970s, the lake's condition gradually improved. By the 1990s, the lake was considered an environmental success story—until the appearance in 2001 of a large "dead zone" in the lake's central basin. The causes of the dead zone are unclear. Phosphorus loadings to the lake are declining, but concentrations in the lake's waters continue to rise. Lake Erie is once again the subject of intense investigation, as researchers attempt to understand how phosphorus is cycling in the lake, and what role alien invasive species like zebra and quagga mussel may have had in that cycle.

Cultural Eutrophication in Lake Winnipeg

Each summer, thick algae blooms choke the north basin of Lake Winnipeg and force the closures of beaches in the lake's southern beaches. Blue-green algae growth is of particular concern, because of the toxins it produces. Almost every year in the past decade, blue-green algae mats have covered large portions—thousands of square kilometres—of the lake's north basin. At first, researchers hoped that the blooms were a fluke, encouraged by unusually low water levels and high temperatures in the late 1990s. But in 2004, temperatures were relatively cool, and the blue-green algae mats were still as "thick as paint."[13] As in other eutrophic systems, dense algae blooms produce abundant oxygen during the day. At night, however, they consume oxygen in respiration, starving lake waters of the dissolved oxygen necessary for healthy fish populations and other aquatic organisms.

The source of the problem is now believed to be nutrient enrichment, especially phosphorus, mainly from inadequately treated sewage, deteriorating septic systems, and agricultural runoff. For municipal systems, the costs of control are significant. It would take an estimated $300 million to upgrade the area's three sewage treatment plants for advanced phosphorus and nitrogen removal—and some estimates suggest that this would only reduce nutrient loadings by 6%. A public agency, the Lake Winnipeg Stewardship Board, has been established to oversee restoration activities. A partial reduction of nutrients (phosphorus and nitrogen) reaching Lake Winnipeg via the Red, Saskatchewan, Winnipeg, and Assiniboine Rivers may be accomplished by the following actions:[12]

- Modify agricultural practices to reduce the nutrients that enter the river from fertilizers by using fertilizers more effectively and efficiently.
- Restore and create buffer zones and river wetlands between farm fields and streams and rivers, particularly in areas known to contribute high amounts of nutrients. The wetland plants use the phosphorus and nitrogen, lowering the amount that enters the river.
- Implement nutrient reduction processes at wastewater treatment plants for towns, cities, and industrial facilities.
- Limit livestock access to streams, to reduce the likelihood that their wastes will enter the stream waters and contribute to nutrient enrichment.

Improving agricultural practices and upgrading sewage treatment plants could result in a major improvement in Lake Winnipeg's water quality. But those changes will not be simple to achieve. The bulk of the problem comes from agricultural lands in the lake's drainage basin, much of which lie within the United States. It is estimated that 30 to 40% of total nutrient loadings to Lake Winnipeg arise in the U.S., outside the jurisdiction of the Manitoba or Canadian governments.

There is no easy solution to cultural eutrophication in Lake Winnipeg. Clearly, though, a reduction in the amount of nutrient entering the lake is needed. Also needed is a better understanding of details of the phosphorus and nitrogen cycle within Lake Winnipeg and its tributary rivers. Developing this understanding will require monitoring of phosphorus and nitrogen, and development of mathematical models of sources, sinks, and rates of transfer. With an improved understanding of the phosphorus and nitrogen cycles, a management strategy to reduce or eliminate the blue-green algae blooms can be created and implemented.

The solution to artificial eutrophication is fairly straightforward and involves ensuring that high concentrations of nutrients from human sources do not enter lakes and other water bodies. This goal can be accomplished by using phosphate-free detergents, controlling nutrient-rich runoff from agricultural and urban lands, disposing of or reusing treated wastewater, and using more advanced water treatment methods, such as special filters and chemical treatments that remove more of the nutrients.

19.5 Oil

Oil discharged into surface water, usually in the ocean but also on land and in rivers, has caused major pollution problems. Several large oil spills from underwater oil drilling have occurred in recent years. However, although spills make headlines, normal shipping activities probably release more oil over a period of years than is released by the occasional spill. The cumulative impacts of these releases are not well known.

Exxon Valdez: Prince William Sound, Alaska

The best-known oil spills are caused by tanker accidents. On March 24, 1989, the supertanker *Exxon Valdez* ran aground on Bligh Reef south of Valdez in Prince William Sound, Alaska. Alaskan crude oil that had been delivered to Valdez through the trans-Alaskan pipeline poured out of ruptured tanks of the vessel at about 20,000 barrels per hour. The tanker was loaded with about 1.2 million barrels of oil, and about 250,000 barrels (about 44 million litres) entered the sound (Figure 19.8). The spill could have been larger than it was, but fortunately some of the oil in the tanker was off-loaded (pumped out) into another vessel. The *Exxon Valdez* spill produced an environmental shock that resulted in passage of the U.S. Oil Pollution Act of 1990 and in a renewed evaluation of cleanup technology.[17, 18]

The oil was spilled into what is considered one of the most pristine and ecologically rich marine environments of the world.[17] Many species of fish, birds, and marine mammals are present in the sound. The effects of the spill included the death of 13% of harbour seals, 28% of sea otters, and 100,000–645,000 seabirds.[18]

Figure 19.8 • *Exxon Valdez* tanker accident in Prince William Sound (1989). Oil is being offloaded from the leaking *Exxon Valdez* (left) to the smaller *Exxon Baton Rouge* (right).

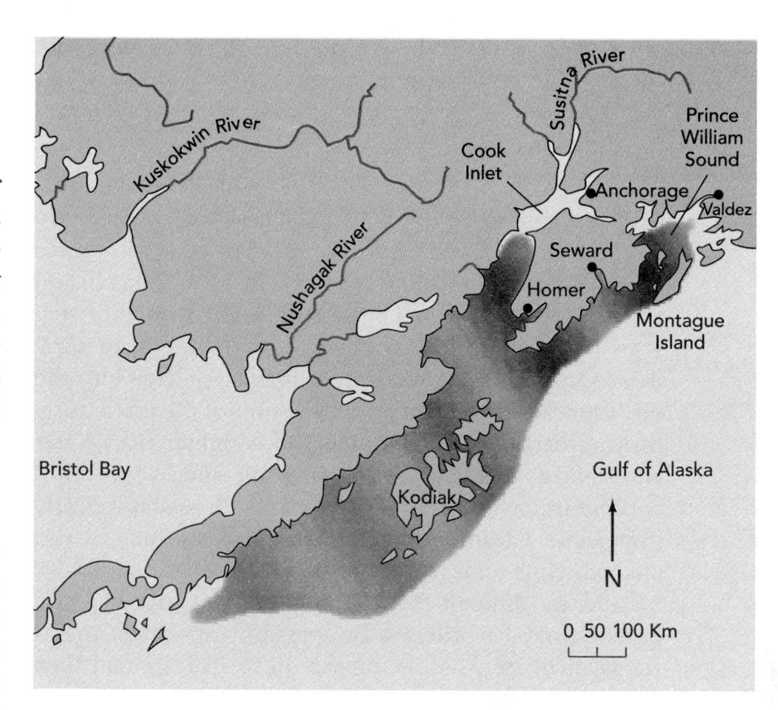

Figure 19.9 • Extent of Alaskan oil spill of 1989. [*Source:* Alaska Department of Fish and Game. 1989 *Alaska Fish and Game* 19(4), Special Issues.]

Within three days of the spill, winds began blowing the slick beyond any hope of containment. Of the 44 million litres of spilled oil, about 20% evaporated and 50% was deposited on the shoreline. Only 14% was collected by skimming and waste recovery. The extent of oil sheens, tar balls (formed by sticky, less volatile components in the oil), and mousse (a thick, weathered patch of oil with the consistency of soft pudding) is shown in Figure 19.9.

Before the *Exxon Valdez* spill, it was generally believed that the oil industry was capable of dealing with oil spills. However, to date more than U.S.$3 billion has been spent to clean the spill, and few people are satisfied with the results. Some scientists argue that the recovery might have been faster if some of the cleanup methods, such as spraying rocks and beaches with high-pressure hot water, had not been used. They argue that the coastal organisms that live under the rocks and that had survived the initial impact of the spill were killed by the high pressure and heat.[18] There is no doubt that the cleanup work posed enormous problems. Photographs and videotapes of workers attempting to clean individual pebbles on beaches are a vivid reminder of the difficulty and virtual futility of achieving an effective cleanup after an event of this magnitude. In addition, the oil spill disrupted the lives of the people who live and work in the vicinity of Prince William Sound.

The long-term effects of large oil spills are uncertain. We know that the effects can last several decades; toxic levels of oil have been identified in salt marshes 20 years following a spill.[18]

The *Exxon Valdez* spill demonstrated that the technology for dealing with oil spills is inadequate. The first and most important step is to avoid large spills; a primary method for doing so is to use supertankers with double hulls designed to minimize the release of oil on collision and rupture of tanks. The second most important step is to pump the oil out of a tanker as soon as an accident occurs, thereby preventing further spillage into the sea. Once a spill occurs, the collection of oil at sea using floating barriers and skimmers (oil is lighter than water and so floats on water) is a worthwhile endeavour; but if conditions include high winds and rough seas, it is nearly impossible. Cleaning oil from birds and mammals is also a worthwhile endeavour, but because the animals ingest oil and are difficult to clean, mortality is high. Oil on beaches may be collected by spreading absorbent material, such as straw, allowing the oil to soak in, and then collecting and disposing of the oily straw.

Jessica: Galápagos Islands

Another shock occurred on January 22, 2001, when a small tanker, the *Jessica*, made a navigational error off the coast of Ecuador near the Galápagos Islands and ran aground, spilling light diesel oil into the ocean. Although the spill (over 400,000 litres) was small compared with the *Exxon Valdez* spill in Alaska, there was grave concern, and Ecuador declared a state of emergency. The Galápagos are environmental treasures and icons of the environment, where Charles Darwin worked in developing his theory of evolution of the species. The United States responded quickly with a Coast Guard ship designed to pump oil from a damaged tanker. Some of the oil was washed ashore onto a small island, injuring birds, seals, and other marine life. The oil slick spread over 3,000 km^2 in the first week but was fortunately carried by currents and trade winds away from the nearby Galápagos. The spill was yet another warning regarding the use of better-built tankers with double hulls and, where possible, the routing of tankers away from ecological areas of particular concern.

19.6 Sediment

Sediment consisting of rock and mineral fragments ranging from gravel particles greater than 2 mm in diameter to finer sand, silt, clay, and even finer colloidal particles can produce a *sediment pollution* problem. In fact, by volume and mass, sediment is our greatest water pollutant. In many areas, it chokes streams; fills lakes, reservoirs, ponds, canals, drainage ditches, and harbours; buries vegetation; and generally creates a nuisance that is difficult to remove. Sediment pollution is a twofold problem: It results from erosion, which depletes a land resource (soil) at its site of origin (Figure 19.10), and it reduces the quality of the water resource it enters.[19]

Many human activities affect the pattern, amount, and intensity of surface water runoff, erosion, and sedimentation. Streams in naturally forested or wooded areas may be nearly stable; that is, there is relatively little erosion or sedimentation. However, converting forested land to agriculture generally increases the runoff and sediment yield or erosion of the land. Application of soil conservation procedures to farmland can minimize but not eliminate soil loss. The change from agricultural, forested, or rural land to highly urbanized land has even more dramatic effects. Large quantities of sediment may be produced during the construction phase of urbanization. Fortunately, sediment production and soil erosion can be minimized by on-site erosion control measures, which are now required under various federal, provincial, and municipal laws.

Reducing sediment pollution in an urbanizing area through control measures is demonstrated by a study in a rapidly growing area north of Toronto.[20] The study showed that 100 times more sediment can be released during construction than under normal urban conditions, and that this quantity can cause significant damage in receiving lakes and streams. The results also showed that these sediments contained other pollutants, including metals and pesticides washed off city lawns and streets. The study recommended changes in existing stormwater pond design, to trap the smaller sediment sizes that tended to clog downstream river habitat. Other recommendations included the need to tailor the development to the natural topography, expose a minimum amount of land, provide temporary protection for exposed soil, minimize surface runoff from critical areas, and trap eroded sediment on the construction site.[20]

Figure 19.10 • Erosion on a residential construction site. The erosion produces gullies and removes vegetation. The sediment may be transported off-site to degrade streams, rivers, ponds, and lakes.

19.7 Acid Mine Drainage

The term **acid mine drainage** refers to water with a high concentration of sulphuric acid (H_2SO_4) that drains from mines—mostly coal mines but also some metal mines (copper, lead, and zinc) in sulphide rocks, which contain the mineral iron sulphide, also known as fool's gold or pyrite (FeS_2). When the pyrite, which may be finely disseminated in the rock and coal, comes into contact with oxygen and water, it weathers, producing sulphuric acid. The acid is produced when surface water or shallow groundwater runs through or moves into and out of mines or tailings (Figure 19.11). If the acid-rich water runs off to a natural stream, pond, or lake, significant pollution and ecological damage may result. The acidic water is toxic to the plants and animals of an aquatic ecosystem; it damages biological productivity, and fish and other aquatic life may die. Acid-rich water can also seep into and pollute groundwater.

Acid mine drainage is a significant water pollution problem in many parts of Canada, especially British Columbia, Ontario, and Québec. The total impact is significant because thousands of kilometres of streams have been damaged.

Even abandoned mines can cause serious problems. For example, subsurface mining in the Canadian Shield area of northern Ontario for sulphide deposits containing copper, nickel, and zinc began in the late nineteenth century and ended in some areas in the 1970s and 1980s. When the mines were operating, they were kept dry by pumping out the groundwater that seeped in. However, since the termination of mining, some of them have flooded and overflowed into nearby creeks, polluting the creeks with acid-rich water. Natural Resources Canada has established the Mine Environment Neutral Drainage Program to assist mining companies in developing sustainable solutions for acid mine drainage.

Figure 19.11 • **Section through a tailings deposit showing the impact of acid mine drainage.** In the upper layers, sulphur-rich iron compounds have reacted with oxygen to create light-coloured ferric (oxidized iron) compounds and sulphuric acid. Deeper layers are protected from oxygen and remain in the ferrous (reduced) state, as indicated by their dark blue colour.

19.8 Trace Inorganic and Organic Pollutants

Our society uses many inorganic and organic substances, and traces of these are found in most lakes and streams. Metals such as iron, copper, nickel, zinc, cadmium, chromium, lead, and mercury occur naturally in many areas of Canada, and are commonly used in industry. Metals are also present in automobile, truck, and vehicle exhaust, and are deposited on the road and highway surfaces where they can be washed off by rainwater. Especially in northern Canada, impoundments associated with hydroelectric projects can be a major source of mercury to water and air, because of the high mercury content of local rocks. As we demonstrated with the example of mercury pollution in Minamata, Japan, and the English-Wabigoon River system of Ontario and Manitoba (Chapter 15), metals like mercury can impair nervous system and enzyme function in humans and other animals.

Industrial organic compounds, pesticides, and herbicides are also of concern in surface and groundwaters. Because their concentrations are often very low, for many years scientists assumed they could not be important in human or ecosystem health. Chapter 15 described our growing understanding of the toxicology of those compounds. Many are almost undetectable in water, but may have a high affinity for particulate matter or fats and oils. When these compounds are persistent and bioaccumulative, they are of special concern because of their potential to biomagnify in food chains.

Metals and trace organic compounds are among the water pollution problems that have plagued Hamilton Harbour, Ontario, the recipient of wastewater discharges from two steel mills and other industrial and municipal operations. Restoration of Hamilton Harbour is described in more detail in A Closer Look 19.1.

19.9 Surface Water Pollution

Pollution of surface waters occurs when too much of an undesirable or harmful substance flows into a body of water, exceeding the natural ability of the water body to remove the undesirable material, dilute it to a harmless concentration, or convert it to a harmless form.

Water pollutants, like other pollutants, are categorized as being emitted from point or nonpoint sources (see Chapter 15). **Point sources** are distinct and confined, such as pipes from industrial or municipal sites that empty into streams or rivers (Figure 19.12). In general, point source pollutants from industries are controlled through on-site treatment or disposal and are regulated by permit. Municipal point sources are also regulated by permit. In older cities in eastern Canada and many parts of the United States, most point sources are outflows from storm sewers and combined sewer systems. As mentioned earlier, such systems combine storm-water flow with municipal wastewater. During heavy rains, urban storm runoff may exceed the capacity of the sewer system, causing it to overflow and deliver pollutants to nearby surface waters.

Nonpoint sources, such as runoff, are diffused and intermittent and are influenced by factors such as land use, climate, hydrology, topography, native vegetation, and geology. Common urban nonpoint sources include runoff from streets or fields; such runoff contains all sorts of pollutants, from metals to chemicals and sediment. Rural sources of nonpoint pollution are generally associated with agriculture, mining, or forestry. Nonpoint sources are difficult to monitor and control.

From an environmental view, two approaches to dealing with surface water pollution are to reduce the sources and to treat the water to remove pollutants or convert them to forms that can be disposed of safely. Which of the options is used depends on the specific circumstances of the pollution problem. Reduction at the source is the most environmentally preferable way of dealing with pollutants. For example, air-cooling towers, rather than water-cooling towers, may be utilized to dispose of waste heat from power plants, thereby avoiding thermal pollution of water. The second method—water treatment—is used for a variety of pollution problems. Water treatments include chlorination to kill micro-organisms such as harmful bacteria, and filtering to remove heavy metals.

There is a growing list of success stories in the treatment of water pollution. One of the most notable is the cleanup of the Thames River in Great Britain. For centuries, London's sewage had been dumped into that river, and there were few fish to be found downstream in the estuary. In recent decades, however, improvement in water treatment has led to the return of a number of species of fish, some not seen in the river in centuries.

Many large cities in Canada, such as Montréal, Ottawa, Toronto, Winnipeg, and Calgary, grew on the banks of rivers; but the rivers were often nearly destroyed with pollution and concrete. Today, there are grassroots movements all across the country to restore urban rivers and adjacent lands as greenbelts, parks, or other environmentally sensitive developments.[21]

19.10 Groundwater Pollution

Approximately one-third of all people in Canada today depend on groundwater as their source of drinking water. (Water for domestic use in Canada is discussed in A Closer Look 19.2). People have long believed that groundwater is in general pure and safe to drink. In fact, however, groundwater can be easily polluted by any one of several sources (see Table 19.1); and the pollutants, even though they are very toxic, may be difficult to recognize. (Groundwater processes were discussed in Section 18.1.)

In Canada today, only a small portion of the groundwater is known to be seriously contaminated; but as mentioned earlier, the problem may become worse as human population pressure on water resources increases. Already, the extent of the problem is growing as the testing of groundwater becomes more common. For example, the city of Elmira, Ontario, has long been threatened by polluted groundwater that is slowly migrating toward its wells.

Figure 19.12 • **This pipe is a point source of chemical pollution from an industrial site entering a river in England.**

Water for Domestic Use: How Safe Is It?

Water for domestic use in Canada is drawn from surface waters and groundwater. Although some groundwater sources have high water quality and need little or no treatment, most sources are treated to conform to national drinking water guidelines (revisit Table 19.3).

Before treatment, water is usually stored in reservoirs or special ponds. Storage allows for solids, such as fine sediment and organic matter, to settle out, improving the clarity of water. The water is then passed through a water plant, where it is filtered and disinfected before it is distributed to individual homes. Once in people's homes, water may be further treated. For example, many people run their tap water through readily available charcoal filters before using it for drinking and cooking.

Some people prefer not to drink tap water. Instead, they use bottled water for personal consumption; as a result, production of bottled water has become a multi-billion-dollar industry.[6] Some people prefer not to drink water that contains chlorine or that passes through metal pipes. Furthermore, water supplies vary in clarity, hardness (concentration of calcium and magnesium), and taste; and the water available locally may not be to some people's liking. A common complaint about tap water is a chlorine taste, which may occur with chlorine concentrations as low as 0.2–0.4 mg/l. People may also fear contamination by minute concentrations of pollutants. The drinking water in Canada is some of the safest in the world; there is no doubt that disinfection of water with chlorine has nearly eliminated waterborne diseases, such as typhoid and cholera, which previously caused widespread suffering and death in the developed world and still do in many parts of the world. However, we need to know much more about the long-term effects of exposure to low concentrations of toxins in our drinking water. How safe is the water in Canada? It's much safer than it was 100 years ago, but low-level contamination (below what is thought dangerous) of organic chemicals and metals is a concern that requires continued research and evaluation.

It is estimated that a majority of the known waste disposal sites in Canada and the United States are producing plumes of hazardous chemicals that are migrating into groundwater resources. Many of the chemicals are toxic or are suspected carcinogens. Thus, it appears that we have inadvertently been conducting a large-scale experiment of how chronic low-level exposure to potentially harmful chemicals affects people. The final results of the experiment will not be known for many years.[22] Preliminary results suggest that we had better act now before a hidden time bomb of health problems explodes.

The hazard presented by a particular groundwater pollutant depends on several factors, including the concentration or toxicity of the pollutant in the environment and the degree of exposure of people or other organisms to the pollutants.[23] (See the section on risk assessment in Chapter 15.)

Principles of Groundwater Pollution: An Example

Some general principles of groundwater pollution are illustrated by an example. Pollution from leaking buried gasoline tanks belonging to automobile service stations is a widespread environmental problem that no one thought very much about until only a few years ago. Underground tanks are now strictly regulated. Many thousands of old, leaking tanks have been removed, and the surrounding soil and groundwater have been treated to remove the gasoline. Cleanup can be a very expensive process, involving removal and disposal of soil (as a hazardous waste) and treatment of the water using a process known as vapour extraction (Figure 19.14). Treatment may also be accomplished under the ground by micro-organisms that consume the gasoline. This is known as **bioremediation** and is much less expensive than removal, disposal, and vapour extraction.

Pollution from leaking buried gasoline tanks emphasizes some important points about groundwater pollutants:

- Some pollutants, such as gasoline, are lighter than water and thus float on the groundwater.
- Some pollutants have multiple phases: liquid, vapour, and dissolved. Dissolved phases chemically combine with the groundwater (e.g., salt dissolves into water).
- Some pollutants are heavier than water and sink or move downward through groundwater. Examples of sinkers include some particulates and cleaning solvents. Pollutants that sink may become concentrated deep in groundwater aquifers.
- The method used to treat or eliminate a water pollutant must take into account the physical and chemical properties of the pollutant and how these interact with surface water or groundwater. For example, the extraction well for removing gasoline from a groundwater resource (Figure 19.14) takes advantage of the fact that gasoline floats on water.
- Since cleanup or treatment of water pollutants in groundwater is very expensive, and undetected or untreated pollutants may cause environmental damage, the emphasis should be placed on preventing pollutants from entering groundwater in the first place.

A CLOSER LOOK 19.2

Hamilton Harbour, Ontario: Cleaning Up a Pollution Hot Spot

Sprawled around a natural embayment at the western tip of Lake Ontario, the city of Hamilton owes much of its history and economic development to its lakefront location. A natural shipping centre and rail hub, Hamilton grew quickly as its excellent links to land and water transportation attracted workers and goods from all parts of the province. In 1854, the new Great Western Railway had located its maintenance yards in Hamilton, prompting a demand for high-quality steel nearby. By the beginning of the twentieth century, Steel City, as it was nicknamed, was well established as a major centre of heavy industry, including two steel companies, the Steel Company of Canada (Stelco) and the Dominion Foundries and Steel Company (Dofasco).

As in the case of the Sydney tar pits, discussed later in this chapter, early waste management practices in the steel industry often had devastating impacts on the environment. In Hamilton Harbour, those impacts occurred mainly in the water and sediments of the harbour itself. In 1985, the International Joint Commission designated Hamilton Harbour as one of 43 "Areas of Concern" in the Great Lakes, requiring that governments in co-operation with stakeholders develop a formal Remedial Action Plan that would lead to cleanup.

The harbour has a number of serious water quality concerns, including low dissolved oxygen levels, high concentrations of polynuclear aromatic hydrocarbons (PAHs), coal tar, metals, and other toxics, and serious bacteriological contamination. A particular focus of remediation efforts has been Randle Reef, a portion of the harbour near the steel mills. Randle Reef is so highly contaminated, with such a complex mixture of chemicals, that it is often cited as one of the most challenging sediment remediation sites in the Great Lakes.

The Hamilton Harbour Remedial Action Plan is community-based, and decisions about remedial actions are proposed and discussed by a multistakeholder advisory group. This group, the Public Advisory Group, recommended a three-part strategy for Randle Reef. About 130,000 cubic metres of the most contaminated sediments would be covered in place in a 9.5 ha dyked disposal facility. Approximately 500,000 cubic metres, of less contaminated sediment would be moved from adjacent areas and disposed of in the same containment facility. When complete, the facility would create new naturalized shoreline areas and port lands.

Three levels of government, one university (McMaster University in Hamilton), industrial partners, two conservation authorities, the Royal Botanical Garden, and public interest groups collaborate in the funding, planning, and management of the Hamilton Harbour RAP. This multistakeholder approach has had many benefits for the city and its water resources. Not least of these has been the building of trust and collaborative working relationships among the city's many diverse interests.[32]

Figure 19.13 • Hamilton Harbour, at the west end of Lake Ontario, has received discharges of industrial and municipal wastewaters for over a hundred years. It is currently rated among the most polluted water systems in the Great Lakes.

Another form of groundwater pollution results from overpumping of aquifers near the ocean. As groundwater resources shrink, salt water, found below the freshwater, can rise closer to the surface[24], thereby contaminating the water resource by a process called saltwater intrusion (Figure 19.15).

Groundwater pollution differs in several ways from surface water pollution. Groundwater often lacks oxygen, a situation that kills aerobic types of micro-organisms (which require oxygen-rich environments) but may provide a happy home for anaerobic varieties (which live in oxygen-deficient environments). The breakdown of pollutants that occurs in the soil and in material a meter or so below the surface does not occur readily in groundwater. Furthermore, the channels through which groundwater moves are often very small and variable. Thus, the rate of movement is low in most cases, and the opportunity for dispersion and dilution of pollutants is limited.

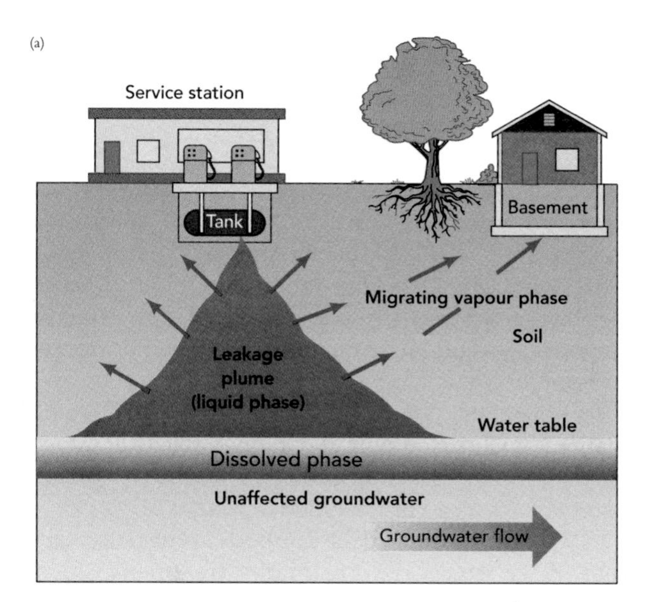

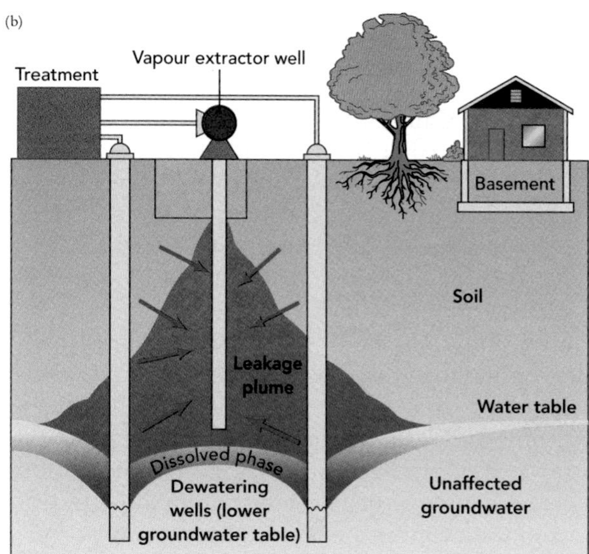

Figure 19.14 • Managing leaking underground storage tanks. Diagram illustrating *(a)* leak from a buried gasoline tank and *(b)* possible remediation using vapor extractor system. Notice that the liquid gasoline and the vapor from the gasoline are above the water table; a small amount dissolves into the water. All three phases of the pollutant (liquid, vapor, and dissolved) float on the denser groundwater. The extraction well takes advantage of this situation. The function of the dewatering wells is to pull the pollutants in where the extraction is most effective. [*Source*: Courtesy of the University of California Santa Barbara Vadose Zone Laboratory and David Springer.]

Sydney Tar Ponds

The case of Sydney, Nova Scotia's tar ponds illustrates how careless industrial practices can cause groundwater pollution problems and affect public water supplies. Eighty years ago, the 33 ha ponds were constructed as a means of disposing of various wastes from a steel mill. Today, they hold over 700,000 tonnes of sediment contaminated with polychlorinated biphenyls (PCBs), polynuclear aromatic hydrocarbons (PAHs) and other substances from the steel making process. Considered by many to be Canada's most contaminated site, the tar ponds contain 20 times more toxic waste than all of New York's Love Canal (Chapter 24).

Like the Love Canal, the Sydney tar pond site is bordered by ordinary residential development, including homes, schools, and playgrounds. Groundwater concentrations in the area and in local streams exceed drinking water guidelines for numerous substances, including barium, PCBs, and PAHs. It has taken decades to assess the extent of surface and groundwater contamination, and to plan the systematic removal of the waste.

In 2004, the governments of Canada and Nova Scotia announced $400 million in funding for remediation of the Sydney tar ponds. The cleanup of the site itself is expected to take 10 years. In the process, all of the contaminated sediments will be removed and burned, and other materials like ovens and structures will be treated on site or removed to safe disposal facilities. Eventually, residential buildings will be demolished and, like the Love Canal, the site will be converted to parkland with walking trails. The fate of contaminated groundwater—difficult to map and even harder to clean—is less certain.

19.11 Wastewater Treatment

Water used for industrial and municipal purposes is often degraded during use by the addition of suspended solids, salts, nutrients, bacteria, and oxygen-demand-

Figure 19.15 • How saltwater intrusion might occur. The upper drawing *(a)* shows the groundwater system near the coast under natural conditions, and the lower drawing *(b)* shows a well with both a cone of depression and a cone of ascension. If pumping is intensive, the cone of ascension may be drawn upward, delivering salt water to the well.

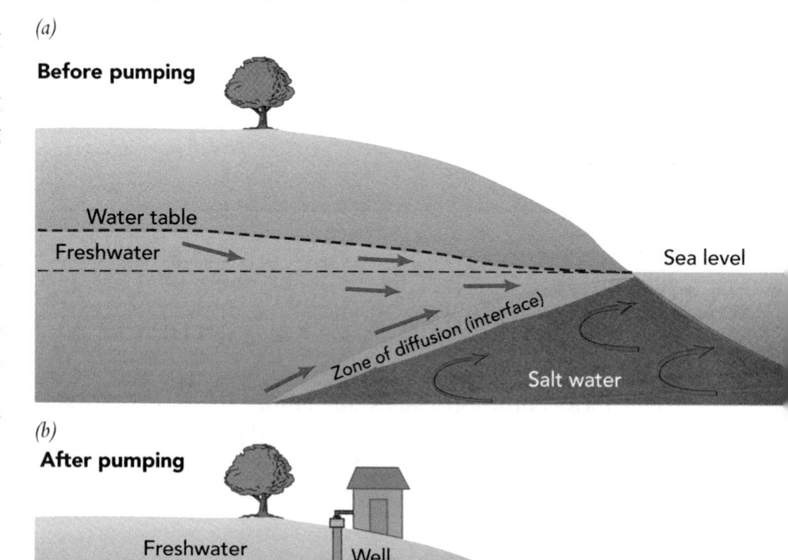

ing material. In Canada, by law, these waters must be treated before being released back into the environment, although the required level of treatment varies across the country. **Wastewater treatment**, or sewage treatment, costs billions of dollars per year in Canada and the United States, and the cost continues to increase. Wastewater treatment will continue to be big business.

Conventional methods of wastewater treatment include septic-tank disposal systems in rural areas and centralized wastewater treatment plants in cities. Recent, innovative approaches include the application of wastewater to the land and wastewater renovation and reuse. We discuss the conventional methods in this section and some newer methods in later sections.

Septic-Tank Disposal Systems

In many rural areas, including those on the developing fringe of cities, no central sewage systems or wastewater treatment facilities are available. As a result, individual septic-tank disposal systems, not connected to sewer systems, continue to be an important method of sewage disposal in rural areas as well as outlying areas of cities. However, not all land is suitable for the installation of a septic-tank disposal system, so the law requires an evaluation of each site before a permit can be issued. An alert buyer should ensure that the site is satisfactory for septic-tank disposal before purchasing property in a rural setting or on the fringe of an urban area where such a system is necessary.

The basic parts of a septic-tank disposal system are shown in Figure 19.16. The sewer line from the house leads to an underground septic tank in the yard. The tank is designed to separate solids from liquid, digest (biochemically change) and store organic matter through a period of detention, and allow the clarified liquid to discharge into the drain field (absorption field) from a system of piping through which the treated sewage seeps into the surrounding soil. As the wastewater moves through the soil, the natural processes of oxidation and filtering treat it. By the time the water reaches any freshwater supply, it should be safe for other uses.

Sewage absorption fields may fail for several reasons. The most common causes are failure to pump out the septic tank when it is full of solids, and poor soil drainage, which allows the effluent to rise to the surface in wet weather. When a septic-tank absorption field does fail, pollution of groundwater and surface water may result. Solutions to septic-tank system problems include siting septic tanks on well-drained soils, making sure systems are of adequate size, and practicing proper maintenance.

Wastewater Treatment Plants

In urban areas, wastewater treatment occurs at specially designed plants that accept municipal sewage from homes, businesses, and industrial sites. The raw sewage is delivered to the plant through a network of sewer pipes. Following treatment, the wastewater is discharged into the surface water environment (river, lake, or ocean) or, in some limited cases, used for another purpose, such as crop irrigation. The main purpose of standard treatment plants is to break down and reduce the BOD and kill bacteria. A simplified diagram of a wastewater treatment plant is shown in Figure 19.17.

Wastewater treatment methods are usually divided into three categories: **primary treatment**, **secondary treatment**, and **advanced treatment**. Wastewater treatment is controlled by the provinces in Canada, and therefore varies from region to region. Most provinces require both primary and secondary treatment for municipal sewage treatment plants, although some cities (Halifax is one example) have no centralized sewage treatment at all. In other areas, only primary treatment may be required. Where secondary treatment is not sufficient to protect the quality of the surface water into which the treated water is discharged—for example, a river with endangered fish species that must be protected—advanced treatment may be required.[25]

Primary Treatment

Incoming raw sewage enters the plant from the municipal sewer line and passes first through a series of screens to remove large floating organic material. The sewage then enters the grit chamber, where sand, small stones, and grit are removed and disposed of. In the next stage, the sewage enters the primary sedimentation tank, where particulate matter settles out to form a sludge. Sometimes, chemicals are used to help the settling process. The sludge, sometimes called biosolids, is removed and transported to the digester for further processing. Primary treatment

Figure 19.16 • Septic-tank sewage disposal system and location of the absorption field with respect to the house and well. [*Source*: Based on Indiana State Board of Health.]

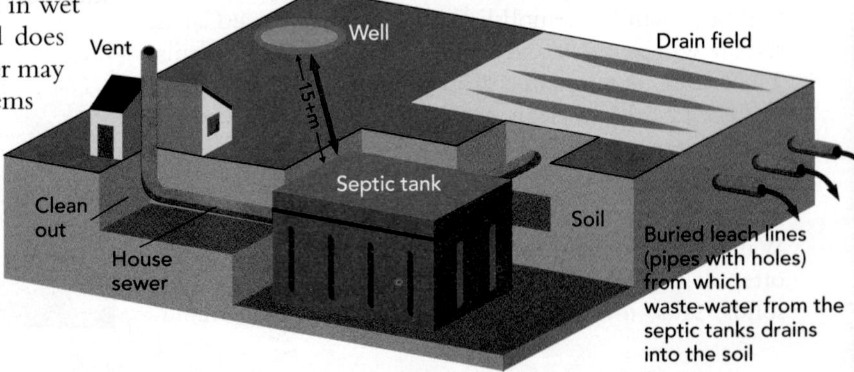

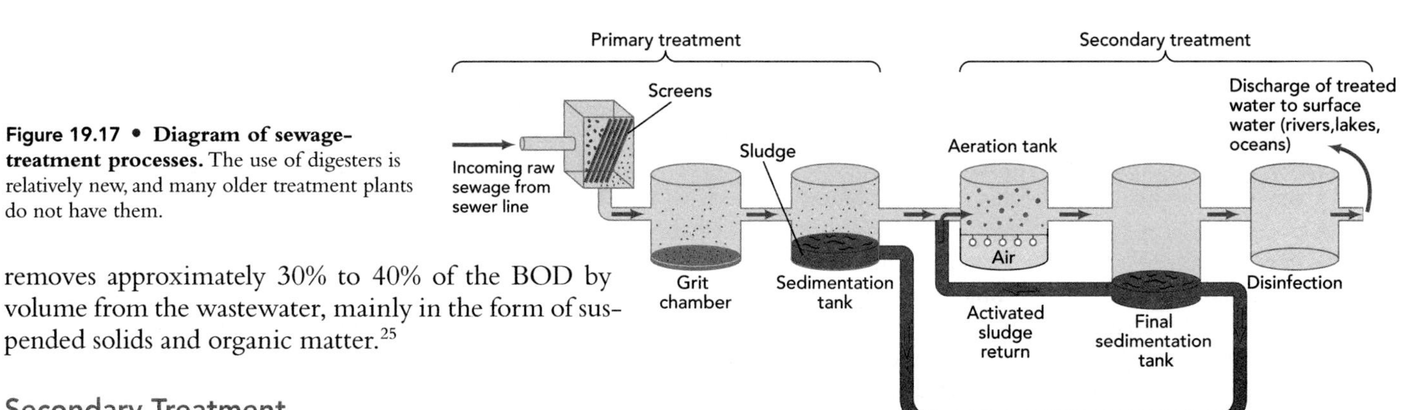

Figure 19.17 • Diagram of sewage-treatment processes. The use of digesters is relatively new, and many older treatment plants do not have them.

removes approximately 30% to 40% of the BOD by volume from the wastewater, mainly in the form of suspended solids and organic matter.[25]

Secondary Treatment

There are several methods of secondary treatment. What is described here is known as activated sludge, the most common treatment. In this procedure, the wastewater from the primary sedimentation tank enters the aeration tank (Figure 19.17), where the wastewater is mixed with air (which is pumped in) and some of the sludge from the final sedimentation tank. The sludge contains aerobic bacteria that consume organic material (BOD) in the waste. The wastewater then enters the final sedimentation tank, where sludge settles out. Some of this activated sludge, which is rich in bacteria, is recycled and mixed again in the aeration tank with air and new, incoming wastewater acting as a starter; and the bacteria are used again and again. Most of the sludge from the final sedimentation tank, however, is transported to the sludge digester. There, along with sludge from the primary sedimentation tank, it is treated by anaerobic bacteria, which further degrade the sludge by microbial digestion.

Methane gas (CH_4) is a product of the anaerobic digestion and may be used at the plant as a fuel to run equipment or heat and cool buildings. In some cases it is burned off. Wastewater from the final sedimentation tank is then disinfected, usually by chlorination, to eliminate disease-causing organisms. The treated wastewater is discharged into a river, lake, or ocean or, in some limited cases, used to irrigate farmland. Secondary treatment removes about 90% of the BOD that entered the plant in the sewage.[25]

The sludge from the digester is dried and disposed of in a landfill or applied to improve soil. In some instances, treatment plants in urban and industrial areas contain many pollutants, such as metals, that are not removed in the treatment process. Sludge from these plants is too polluted to use to improve the soil, and the sludge must be disposed of. Some communities, however, require industries to pretreat sewage to remove heavy metals before the sewage is sent to the treatment plant; in these instances, the sludge can be more safely used for soil improvement.

Advanced Wastewater Treatment

Primary and secondary treatment does not remove all pollutants from incoming sewage. Some additional pollutants can be removed by adding more treatment steps. For example, nutrients such as phosphates and nitrates; organic chemicals; and metals can be removed by specifically designed treatments such as sand filters, carbon filters, and chemicals applied to assist in the removal process.[25] Treated water is then discharged into surface water or may be used for irrigation of agricultural lands or municipal properties, such as golf courses, city parks, and grounds surrounding wastewater treatment plants.

Advanced wastewater treatment is used when it is particularly important to maintain good water quality. For example, if a treatment plant discharges treated wastewater into a river and there is concern that nutrients remaining after secondary treatment may cause damage to the river ecosystem (eutrophication), advanced treatment may be used to reduce the nutrients.

Chlorine Treatment

As mentioned, chlorine is frequently used to disinfect water as part of wastewater treatment. (Disinfection using ultraviolet light is another technique used in some systems.) Chlorine treatment is very effective in killing the pathogens that historically caused outbreaks of serious waterborne diseases, which killed many thousands of people. However, a recently discovered potential is that chlorine treatment produces minute quantities of chemical by-products, some of which have been identified as potentially hazardous to humans and other animals. For example, a recent study in Britain revealed that in some rivers, male fish sampled downstream from wastewater treatment plants had testes containing both eggs and sperm. This is likely related to the concentration of sewage effluent and the treatment method used.[26] Evidence also suggests that these by-products in the water may pose risk of cancer and other human health effects. The degree of risks is controversial and is currently being debated.

The Sierra Legal Defence Fund publishes a National Sewage Report Card for Canada. The most recent edition, published in 2004, evaluates sewage treatment practices in 22 Canadian cities. Of those, 14 had improved their treatment levels over the previous five years, three had deteriorated, and four showed no change. The city of Victoria was suspended from the assessment because it remains the only city in Canada to discharge all of its sewage without treatment, despite a decade of warnings from Sierra Legal Defence. By contrast, Calgary received the group's only A+ rating, reflecting its high level of treatment, extensive toxicity testing, and use of ultraviolet disinfection in place of chlorine.[27]

19.12 Land Application of Wastewater

The practice of applying wastewater to the land results from the fundamental belief that waste is simply a resource out of place. Land application of untreated human waste was practiced for hundreds if not thousands of years before the development of wastewater treatment plants, which have sanitized the process through the reduction of BOD and the use of chlorination. Today, land application of biosolids, the solid residues of the wastewater treatment process, is common in Canada and the United States. Biosolids can, however, include metals and trace organic compounds attached to sediment particles, and can pose a hazard to human and ecosystem health. For that reason, most provinces have established guidelines for land application of sewage biosolids.

The Wastewater Renovation and Conservation Cycle

The ideal land application system is sometimes called the wastewater renovation and conservation cycle and is shown schematically in Figure 19.18. The major steps in the cycle are the following:

1. Return of treated wastewater (following primary treatment) to crops via a sprinkler or other irrigation system.
2. Renovation, or natural purification by slow percolation of the wastewater into the soil, to eventually recharge the groundwater resource with clean water (an advanced form of treatment).
3. Reuse of the treated water, which is pumped out of the ground for municipal, industrial, institutional, or agricultural purposes.

Recycling wastewater is now being practiced at many sites in Canada. Approximately 3% of the wastewater in British Columbia is now reused; in Alberta, treated wastewater is widely used for irrigation.

Wastewater treatment technology is rapidly evolving. An important question being asked is: Can we develop environmentally preferred, economically viable wastewater treatment plants that are fundamentally different from those in use today? An idea for such a plant, called a *resource recovery* wastewater treatment plant, is shown in Figure 19.19.[28] The term resource recovery here refers to the production of resources, including methane gas (which can be burned as a fuel), as well as ornamental plants and flowers that have commercial value.

The processes in the resource recovery treatment plant are as follows: First, the wastewater is run through filters to remove large objects. Second, the water undergoes anaerobic processing (this process produces the methane gas). Third, the nutrient-rich water flows over an inclined surface containing plants (the plants use the nutrients and further purify the water). This process can often clean the water to the same standards obtained from secondary treatment in conventional wastewater treatment plants. If further purification is necessary, other living plants may be used to process the water before being discharged into the environment.

Wastewater treatment that utilizes the resource recovery concept is in the experimental stage in small pilot plants. This technology must overcome several problems before it is likely to be used more widely. First, there has been a tremendous investment in traditional wastewater treatment plants, and engineers and other

Figure 19.18 • The wastewater renovation and conservation cycle. [*Source*: R. R. Parizek, L. T. Kardos, W. E. Sopper, E. A. Myers, D. E. Davis, M. A. Farrel, and J. B. Nesbitt, "Pennsylvania State Studies: Waste Water Renovation and Conservation," *University Studies* 23. Copyright 1967 by the Pennsylvania State University. Reproduced by permission of the Pennsylvania State University Press.]

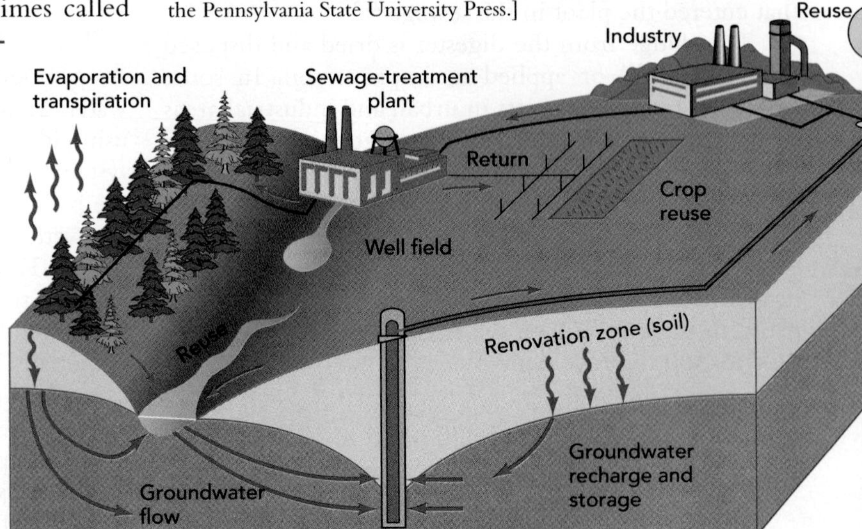

technicians are familiar with how to build and operate them. Second, economic incentives to provide for new technologies are not sufficient. Third (and perhaps most significant), there are not sufficient personnel trained to design and operate new types of wastewater treatment plants. This may be changing, however, because more universities are developing environmental engineering programs that take a broader view of technological development and applications.[28]

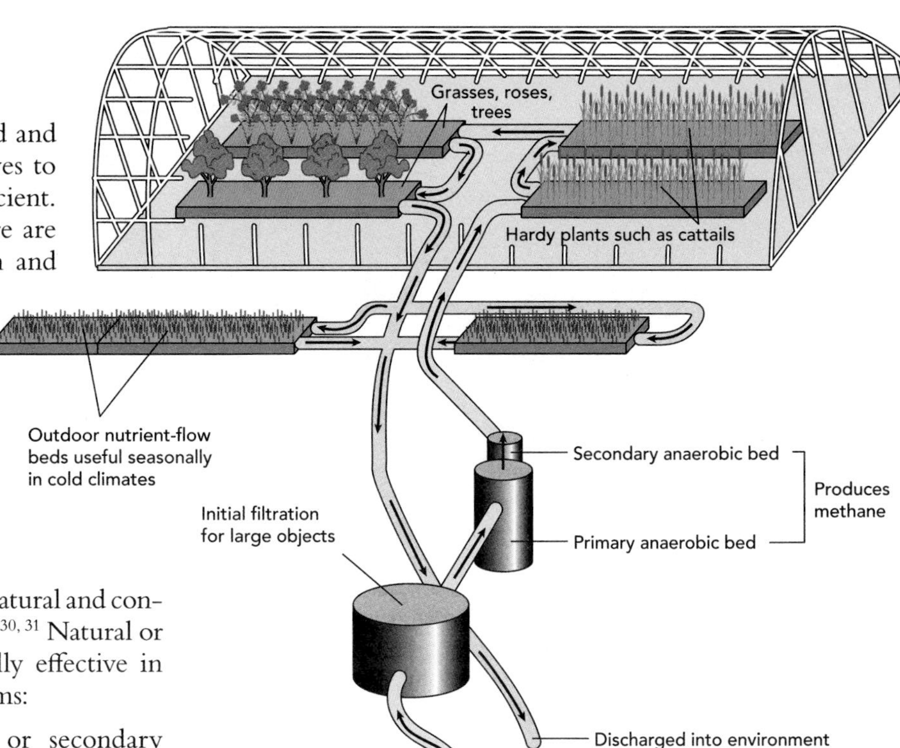

Figure 19.19 • Components of a resource recovery wastewater treatment plant. For this model, two resources are recovered: methane, which can be burned to produce energy from the anaerobic beds, and ornamental plants, which can be sold. [*Source*: Based on W. J. Jewell, "Resource-Recovery Wastewater Treatment," *American Scientist* (1994) 82:366–375.]

Wastewater and Wetlands

Wastewater is being applied successfully to natural and constructed wetlands at a variety of locations.[29, 30, 31] Natural or human-constructed wetlands are potentially effective in treating the following water quality problems:

- Municipal wastewater from primary or secondary treatment plants (BOD, pathogens, phosphorus, nitrate, suspended solids, and metals).
- Storm water runoff (metals, nitrate, BOD, pesticides, oils).
- Industrial wastewater (metals, acids, oils, solvents).
- Agricultural wastewate r and runoff (BOD, nitrate, pesticides, suspended solids).
- Mining waters (metals, acidic water, sulphates).
- Groundwater seeping from landfills (BOD, metals, oils, pesticides).

Treatment of wastewater through wetland systems is particularly attractive to communities that find it difficult to purchase traditional wastewater treatment plants.[29] For example, the city of Dawson Creek, British Columbia, makes use of a wetland as part of its wastewater treatment system. Raw sewage from the city's 11,000 residents is pumped to two small anaerobic (oxygen-free) lagoons, and from there, to larger aerated cells. The final stage in the treatment system is a wetland site that provides extensive wildlife habitat and a valuable aesthetic resource for the community.

19.13 Water Reuse

Water reuse can be inadvertent, indirect, or direct. *Inadvertent water reuse* results when water is withdrawn, treated, used, treated, and returned to the environment, followed by further withdrawals and use. Inadvertent water use is very common and a fact of life for millions of people who live along large rivers. Many sewage treatment plants are located along rivers and discharge treated water into the rivers. Downstream, other communities withdraw, treat, and consume the water.

Several risks are associated with inadvertent reuse:

1. Inadequate treatment facilities may deliver contaminated or poor-quality water to downstream users.
2. Since the fate of all disease-causing viruses during and after treatment is not completely known, the environmental health hazards of treated water remain uncertain.
3. Every year, new potentially hazardous chemicals are introduced into the environment. Harmful chemicals are often difficult to detect in the water; if they are ingested in low concentrations over many years, their effects on humans may be difficult to evaluate.[29]

Indirect water reuse is a planned endeavour. An example is the wastewater renovation and conservation cycle previously discussed and illustrated in Figure 19.18. Similar plans have been used in many places in the world, where several thousand cubic metres of treated wastewater per day have been applied to surface recharge areas. The treated water eventually enters groundwater storage to be reused for agricultural and municipal purposes. Effluent reuse is common in the Canadian prairies, where a total of over 10,000 ha of agricultural lands are currently irrigated with treated wastewater. In those regions, water reuse is seen as a low-cost solution for municipal disposal of effluents. If such practices are to be sustainable, it is

Figure 19.20 • Constructed wetland used for the treatment of mining wastes.

Figure 19.21 • Water reuse at a Las Vegas, Nevada, resort hotel.

important that they be carefully monitored for quality and impacts on surface and groundwater systems.

Direct water reuse refers to use of treated wastewater that is piped directly from a treatment plant to the next user. In most cases, the water is used in industry, in agricultural activity, or to irrigate golf courses, institutional grounds (such as university campuses), and parks. Direct water reuse is growing rapidly. Direct reuse of water by factories for industrial processes is the norm. In Las Vegas, Nevada, new resort hotels that use a great deal of water for fountains, rivers, canals, and lakes are required to treat wastewater and reuse it (Figure 19.21). Very little direct reuse of water is planned for human consumption (except in emergencies) because of perceived risks and negative cultural attitudes toward using treated wastewater.

19.14 Water Pollution and Environmental Law

Environmental law, the branch of law dealing with conservation and use of natural resources and control of pollution, is very important as we debate environmental issues and make decisions about how best to protect our environment (see Chapter 26). In Canada, laws at the federal, provincial, and local levels address these issues.

As discussed in Chapter 16, responsibility for environmental matters is not clearly assigned under Canada's Constitution. Federal responsibility for water management relates to the protection of fisheries, under the *Fisheries Act*, one of Canada's oldest statutes. Originally enacted in 1868, the *Fisheries Act* is intended to protect Canadian fisheries and the streams, rivers, and lakes in which they live. The *Canadian Environmental Protection Act* limits the amount of pollution that can be emitted from specific industries, including the pulp and paper industry. Other federal legislation important for water includes the *Canadian Environmental Assessment Act*, which requires evaluation of major public projects such as dams, and the *Pest Control Products Act*, which regulates the introduction and use of pesticides.

As we have seen, under Canada's Constitution the provinces were granted control of natural resources, and the business activities that process them. This has meant that most of the responsibility for regulating water use and water pollution has fallen to the provinces. Therefore, most provinces have statutes regulating pollution discharges to water, and provincial environmental assessment legislation that binds public projects undertaken by the province or municipalities.

Municipalities have limited influence over water use and water quality. They do, however, control water pricing and local water conservation by-laws, operate sewage and water treatment plants, and oversee land use planning. The last can have an important impact on the quality and flow of urban lakes and streams, by controlling the percentage of impervious cover in a drainage basin.

Table 19.4 summarizes jurisdictional responsibilities for water in Canada and lists some important federal laws relating to water use. Each of these major pieces of legislation has had a significant impact on water quality issues.

Table 19.4 • Responsibility for Water Management in Canada

Level of Government	*Water Management Responsibilities*
Federal	- More than 20 federal statutes and at least 10 agencies may be involved. - Many responsibilities are shared with provinces, often through federal-provincial agreements - *Canadian Environmental Protection Act* and *Fisheries Act* are key statutes for management of pollution discharges to water. - Establishes national guidelines for water use, including drinking water, but these are not legally binding - Key role in technical support and research
Provincial	- Under the Canadian Constitution, the provinces have the principal authority for managing natural resources, including water - Establish environmental statutes and regulations for the allocation of water, public water supplies (including drinking water quality), wastewater treatment and discharge of treated effluents, and water quality and flow monitoring.
Municipal	- Responsible for local management and delivery of water and water services, including sewer and water systems, water treatment, sewage treatment, testing and analysis of water supplies, water pricing, water metering, and related matters - Local public health units have a central role in monitoring of public health risks related to water, for instance in drinking water supplies and bathing beaches; can for instance issue "boil water" orders and close swimming beaches if conditions are judged unsafe.

Summary

- The primary water pollution problem in the world today is the lack of disease-free drinking water.
- Water pollution is degradation of quality that renders water unusable for its intended purpose.
- Major categories of water pollutants include disease-causing organisms, dead organic material, metals, organic chemicals, acids, sediment, heat, and radioactivity.
- Sources of pollutants may be point sources, such as pipes that discharge into a body of water, or nonpoint sources, such as runoff, which are diffused and intermittent.
- Eutrophication is a natural or human-induced increase in the concentration in water of nutrients required for living things, such as phosphorus and nitrogen. A high concentration of such nutrients may cause a population explosion of photosynthetic bacteria. As the bacteria die and decay, the concentration of dissolved oxygen in the water is lowered, leading to the death of fish.
- Sediment pollution is a twofold problem: soil is lost through erosion, and water quality is reduced when sediment enters a body of water.
- Acid mine drainage is a serious water pollution problem that results when water and oxygen react with sulphide minerals, often associated with coal or metal sulphide deposits, forming sulphuric acid. Acidic water draining from mines or tailings pollutes streams and other bodies of water, damaging aquatic ecosystems and degrading water quality.
- Urban processes—for example, waste disposal in landfills, application of fertilizers, and dumping chemicals such as motor oil and paint—can contribute to shallow-aquifer contamination. Overpumping aquifers near the ocean may cause salt water, found below the freshwater, to rise closer to the surface, thereby contaminating the water resource by a process called saltwater intrusion.
- Wastewater treatment at conventional treatment plants includes primary, secondary, and occasionally advanced treatment. In some locations, natural ecosystems, such as wetlands and soils, are being used as part of the treatment process.
- Water reuse is the norm for millions of people living along rivers where series of sewage-treatment plants discharge treated wastewater back into the river water. People who withdraw river water downstream are reusing some of the treated wastewater.
- Industrial reuse of water is the norm for many factories.
- Deliberate use of treated wastewater for irrigating agricultural lands, parks, golf courses, etc., is growing rapidly as demand for water increases.
- Cleanup and treatment of water pollution for both surface water and groundwater resources are expensive and may not be completely successful. Furthermore, environmental damage may result before a pollution problem is identified and treated. Therefore, we should continue to focus on preventing pollutants from entering water, which is a goal of much water quality legislation.

CRITICAL THINKING ISSUE

How Can Polluted Waters Be Restored?

The Illinois River begins in the northeast part of that state and flows west and south, draining parts of Indiana and Wisconsin (see Figure 19.22). From Lake Michigan, which is connected to the river by a canal at Chicago, to the river's confluence with the Mississippi is a distance of 526 km. The surrounding floodplains, once a mixture of prairie and oak–hickory forest, are now primarily used for raising crops. Formerly, the river was highly productive, especially in the lower 320 km; it produced 10% of the U.S. freshwater fish catch in 1908 (11 million kg; 200 kg/ha). By the 1970s, the same stretch of river produced a mere 0.32% of the total freshwater fish harvest (4.5 kg/ha). Two major factors are responsible for the change in the productivity of the Illinois River: diversion of Chicago's sewage from Lake Michigan to the river, and agriculture. A brief history of events related to water quality in the Illinois River is given in Table 19.5.

Critical Thinking Questions

1. Develop a hypothesis to explain why the fish population peaked in 1908, after the construction of the Chicago Sanitary and Ship Canal, and declined after that. Your hypothesis should also be able to explain the recovery of the fish in the 1920s and 1930s, and the causes of the environmental problems of the 1940s and 1950s. Design a controlled experiment to test your hypothesis.
2. Why did water quality show some improvement by 1990, although the Tunnel and Reservoir Plan (TARP) was not yet completed? (Hint: See Table 19.5.)
3. The most important variables that affect the life of a river or stream are energy source (the amount of organic material entering the stream from sources outside it), water quality, habitat quality, water flow, and interactions among living things. In the case of the Illinois River, which variables are affected by human activities? For each variable, cite examples of specific activities, their environmental effects, and what could be done to further improve water quality in the river.
4. There is a conflict between managing the Illinois River for waterfowl and managing it for fish. Why is this so? How could the conflict be resolved?

Figure 19.22 • **Illinois River Watershed.**

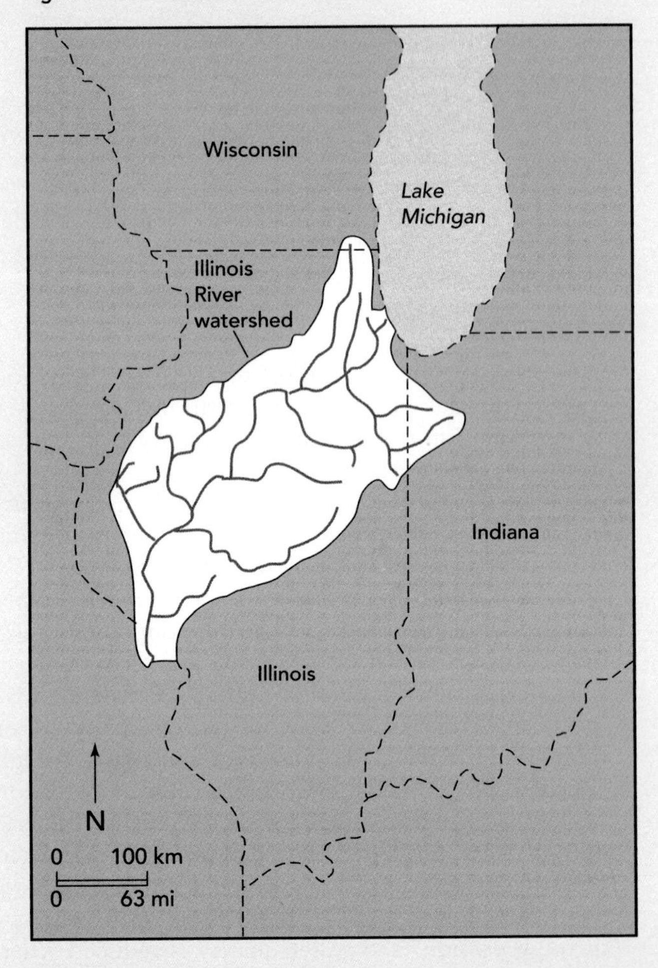

Table 19.5 • Illinois River Water Quality History

Year	*Critical Event*	*Environmental Impact*
1854–1855	After heavy rains, untreated sewage from Chicago entered Lake Michigan and then the city's drinking water	Cholera and typhoid epidemic in Chicago
1900	Chicago Sanitary and Ship Canal built to convey sewage away from Lake Michigan and into Illinois River	Waste entered Illinois River; commercial fish yield from river reached peak in 1908; by 1920, fish populations in river had declined
1920–1940	Most cities along river built sewage treatment plants	Some recovery in fish population
1940–1960	Rapid population growth in Chicago and other cities along the river; increase in agricultural area in backwaters and lakes of the river	Lower oxygen levels in river; further declines in fish populations; sport fish and ducks declined
1977	Construction of Chicago Tunnel and Reservoir Plan (TARP) to capture and treat sewage overflows initiated	Some improvement in water quality by 1990, but no change in turbidity or total phosphorus; sodium increased

STUDY QUESTIONS

1. Do you think that outbreaks of waterborne diseases will be more common or less common in the future? Why? Where are outbreaks most likely to occur?
2. What was learned from the *Exxon Valdez* oil spill that might help reduce the number of future spills and their environmental impact?
3. What is meant by the term water pollution, and what are several major processes that contribute to water pollution?
4. Compare and contrast point and nonpoint sources of water pollution. Which is easier to treat, and why?
5. What is the twofold effect of sediment pollution?
6. In the summer, you buy a house with a septic system that appears to function properly. In the winter, effluent discharges at the land surface. What could be the environmental cause of the problem? How could the problem be alleviated?
7. Describe the major steps in wastewater treatment (primary, secondary, advanced). Can natural ecosystems perform any of these functions? Which ones?
8. In a city along an ocean coast, rare water birds inhabit a pond that is part of a sewage-treatment plant. How could this have happened? Is the water in the sewage pond polluted? Consider this question from the birds' and from your point of view.
9. How does water that drains from mines become contaminated with sulphuric acid? Why is this an important environmental problem?
10. What is eutrophication, and why is it an ecosystem effect?
11. How safe do you believe the drinking water is in your home? How did you reach your conclusion? Are you worried about low-level contamination by toxins in your water? What could the sources of contamination be?

FURTHER READING

Borner, H., ed. *Pesticides in Ground and Surface Water.* Vol. 9 of *Chemistry of Plant Protection.* New York: Springer-Verlag, 1994. Essays on the fate and effects of pesticides in surface water and groundwater, including methods to minimize water pollution from pesticides.

Dunne, T. and L. B. Leopold. *Water and Environmental Planning.* San Francisco: W. H. Freeman, 1978. A great summary and detailed examination of water resources and problems.

Hester, R. E. and R. M. Harrison, eds. *Agricultural Chemicals and the Environment.* Cambridge: Royal Society of Chemistry, Information Services, 1996. A good source for information about the impact of agriculture on the environment, including eutrophication and the impact of chemicals on water quality.

Manahan, S. E. *Environmental Chemistry.* Chelsea, Mich.: Lewis, 1991. A detailed primer on the chemical processes pertinent to a broad array of environmental problems, including water pollution and treatment.

Newman, M. C. *Quantitative Methods in Aquatic Ecotoxicology.* Chelsea, Mich.: Lewis, 1995. Up-to-date text on fate, effects, and measurement of pollutants in aquatic ecosystems.

O'Connor, D. *Report of the Walkerton Inquiry: A Strategy for Safe Drinking Water.* Parts One and Two. Toronto: Ministry of the Attorney General, 2002.

Rao, S. S., ed. *Particulate Matter and Aquatic Contaminants.* Chelsea, Mich.: Lewis, 1993. Coverage of the biological, microbiological, and ecotoxicological principles associated with interaction between suspended particulate matter and contaminants in aquatic environments.

chapter 20

Climate Change and the Polar Bears of Hudson Bay

Although scientific uncertainties remain, it is now apparent that human-induced warming is occurring in many parts of the world.[1] The questions have been reduced to how much, how fast, and where. The story of the plight of polar bears in the western Hudson Bay begins our exploration of climate change. Polar bears are the largest carnivore in North America; they can reach 2.5 m in length and weigh over 700 kg. The polar bears of western Hudson Bay are under threat, and the early breaking up of sea ice each spring is thought to be the problem. During the past 40 years, sea ice, on a global basis, has thinned by as much as 40% and decreased in extent by about 10%, presumably in response to climate change. Some of the most significant impacts of climate change on wildlife are occurring in the Arctic, because temperature changes are more dramatic there than at lower latitudes.

For polar bears in the western Hudson Bay, sea ice is a critical habitat for hunting seals. In the spring, polar bears prey on seals, especially on very young seals, providing the opportunity for the bears to fatten up before the annual melting of the sea ice. After the ice melts, the bears move to land, where they go on a fast that may last for months. In particular, pregnant female bears fast for up to eight months and need a significant reserve of fat to carry, care for, and feed their cubs until they can again return to the ice to feed.

Since 1981, when studies on polar bears in western Hudson Bay began, the bears there have weighed less than average and have given birth to fewer cubs. Biologists have established a link between the decline in polar bears and the early breakup of sea ice. If the trend continues, there will be a continued drop in the bear population. Another consequence is that bears will be forced onto land earlier, where dangerous contact with people is more likely.[2, 3]

The situation in eastern Hudson Bay at present is different, because sea ice is more permanent there, and some polar bear populations live their entire lives on sea ice. These bears are less likely to be adversely affected by climate change. However, massive melting of sea ice would be a serious problem for all polar bears, including those of eastern Hudson Bay.[2]

- *Our story of polar bears in the Hudson Bay suggests that global climate warming can cause serious problems in the biosphere. This chapter addresses global issues related to weather and climate change, with an emphasis on the role of humans in climate change.*

Polar bear moving through thin ice while hunting for seals in Hudson Bay.

The Atmosphere, Climate, and Climate Change

LEARNING OBJECTIVES

Earth's atmosphere is a dynamic system that is changing continuously while undergoing complex physical and chemical processes. After reading this chapter, you should understand:

- What the basic composition and structure of the atmosphere are.
- How the processes of atmospheric circulation, climate, and microclimate work.
- What the four major processes that remove materials from the atmosphere are.
- How the climate has changed during the last million years.
- What the science behind human-induced climate change is.
- How human activity has resulted in increased emissions of greenhouse gases.
- How positive- and negative-feedback cycles in the atmosphere might affect global temperature change.
- What effects climate change might have, and how we can adjust to those changes.

20.1 The Atmosphere

The **atmosphere** is the thin layer of gases that envelops Earth. We begin the chapter by examining basic features of atmospheric composition, structure, and processes.

Composition of the Atmosphere

The atmosphere is composed of gas molecules held close to Earth's surface by a balance between gravitation and thermal movement of air molecules (90% of the mass of the atmosphere is in the first 12 km above the surface of the Earth). Major gases in the atmosphere are nitrogen (78%), oxygen (21%), argon (0.9%), and carbon dioxide (0.03%). The atmosphere also contains trace amounts of numerous elements and compounds, including methane, ozone, hydrogen sulphide, carbon monoxide, oxides of nitrogen and sulphur, hydrocarbons, chlorofluorocarbons (CFCs), and various particulates or aerosols (small particles). Water vapour is also present in the lower few kilometres of atmosphere.

The atmosphere is a dynamic system, changing continuously. Physical movement of air masses, each with a different temperature, pressure, moisture, and aerosol content, produces weather and climate. A vast, chemically active system, the atmosphere is fuelled by sunlight, high-energy compounds (for example, oxygen, methane, and carbon dioxide) emitted by living things, and human industrial and agricultural activities. Many complex chemical reactions take place in the atmosphere, changing from day to night and with the chemical elements available.

The main chemical components in Earth's atmosphere either are produced primarily by biological activity or are greatly affected by it. The atmosphere has been greatly modulated by life during the last 3.5 billion years, but for the most part in ways and rates that have produced an atmosphere whose makeup is relatively constant and essential for our survival.

Ever since life began on Earth, the atmosphere has been an important resource for chemical elements and a medium for the deposition of wastes. The earliest bacteria that carried out photosynthesis released oxygen as a waste product into the atmosphere. The long-term increase in atmospheric oxygen, in turn, made possible the development and survival of life-forms that required high rates of metabolism and rapid use of energy. For our biological ancestors and for us, oxygen became a necessary resource for respiration, the cellular process by which we burn our internal biological fuels and provide the energy to sustain our life processes.

Structure of the Atmosphere

The atmosphere is made up of several layers, as illustrated in Figure 20.1. The lower part of the atmosphere (lower 10 to 12 km) is known as the **troposphere**, and it is here that weather occurs. In the troposphere, the temperature of the atmosphere decreases systematically with elevation (from about 17°C at the surface to −60°C at 12 km above Earth's surface, the top of the troposphere) at a global average rate of decrease of approximately 6.5°C/km. At the top of the troposphere, the **tropopause** (about 12 to 20 km above sea level), with a constant temperature of about −60°C, produces a lid, or cold trap, on the troposphere. The cold trap causes condensation of atmospheric water vapour. Therefore, there is very little water vapour in the **stratosphere**, which lies above the troposphere and is where the atmosphere warms with increasing altitude. Condensation of water in the troposphere produces clouds. The role of clouds, including how they develop and move, is an important area of research to understand better the global processes that operate in the atmosphere. Above the stratosphere is the **mesosphere**, in which meteors and rock fragments entering the atmosphere burn up. At the top of the mesosphere is the **thermosphere**, the

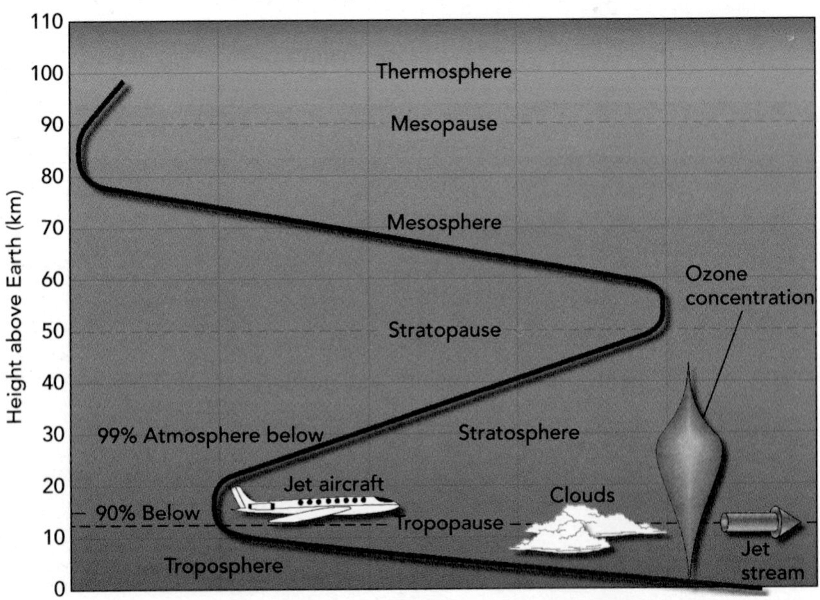

Figure 20.1 • The structure of the atmosphere showing temperature profile and ozone layer of the atmosphere to an altitude of 110 km. Note that 99% of the atmosphere (by mass) is below 30 km, the ozone layer is thickest at about 25–30 km, and the weather occurs below about 11 km—about the elevation of the jet stream. [*Source*: Modified from A. C. Duxbury and A. B. Duxbury, *An Introduction to the World's Oceans*, © 1997. Wm. C. Brown Publishers, 5th ed.].

layer in which aurora borealis (northern lights) occurs, and in which the space shuttle orbits. At the top of the atmosphere is a very thin layer, the **exosphere**, in which the atmosphere merges into space.

Figure 20.1 also shows the **stratospheric ozone layer**, which extends from the tropopause to an elevation of approximately 40 km, with a maximum concentration of ozone above the equator at about 25 to 30 km. Stratospheric ozone (O^3) protects life in the lower atmosphere from receiving harmful doses of ultraviolet radiation and is discussed in detail in Chapter 22.

Atmospheric Processes: Temperature, Pressure, and Global Zones of High and Low Pressure

Two important measurable quantities in the atmosphere are pressure and temperature. *Pressure* is force per unit area. Atmospheric pressure is the weight of overlying atmosphere (air) per unit area; it decreases as altitude increases because there is less weight from overlying air. At sea level, atmospheric pressure is 105 N/m^2 (newtons per square metre). *Temperature* refers to relative hotness or coldness of materials, such as air, water, soil, and living organisms, and is measured with a thermometer. In a quantitative sense, temperature is a measure of thermal energy, which is the kinetic energy of the motion of atoms and molecules in a substance. Figure 20.2 shows the three common temperature scales. This book uses the Kelvin (K) and Celsius (°C) scales.

In addition to pressure and temperature, air is characterized by its water vapour content. In the lower atmosphere, water vapour content varies from approximately 1% to 4% by volume. The amount of water vapour present in the atmosphere at a particular location depends on many factors, including air temperature, air pressure, and availability of water vapour from processes such as evaporation from water bodies and soil, and transpiration from vegetation (loss of water from plants to the air). When air holds the maximum amount of water that it can, given its particular temperature, it is said to be *saturated*, which means that no more water vapour can be added to the air. The term *relative humidity* is a measure of how close the air is to saturation. For example, a relative humidity of 100% means that the air is completely saturated; a relative humidity of 5% means that the air is very dry. At high relative humidity, people and other animals may feel uncomfortable and perceive the air as sticky. In desert regions with low relative humidity, the air is dry; even with a warm temperature, to many people the air feels more comfortable than air with a lower temperature but a higher relative humidity.[4]

Winds, cloud movement, and transitions from stormy to clear skies show us that the atmosphere changes rapidly and continuously. Heat energy is transferred to the air from the land and from bodies of water by the evaporation of water. As warm air rises, it cools; winds are produced as cooler surface air is drawn in to replace rising warm air. The lower atmosphere is said to be unstable because it tends to circulate and mix, particularly in the lowest 4 km or so.

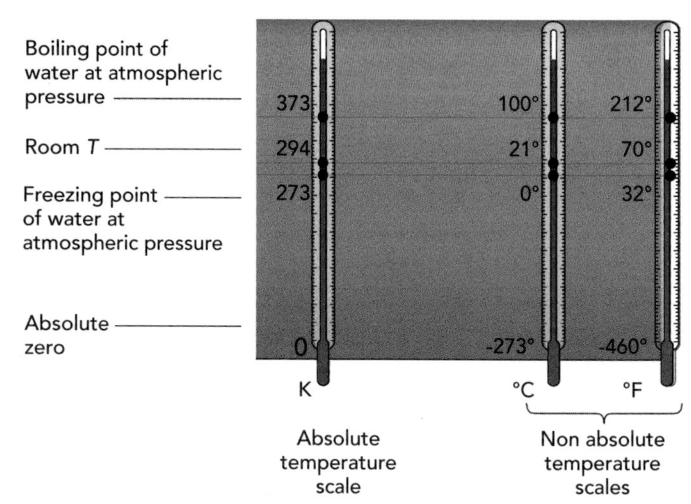

Figure 20.2 • Temperature scales. °C = Celsius; °F = Fahrenheit; K = kelvin. °C = 5/9 (°F –32); °C = K –273.

On a global scale, atmospheric circulation results primarily from Earth's rotation and the differential heating of Earth's surface and atmosphere. These processes produce global patterns that include prevailing winds and latitudinal belts of low and high air pressure from the equator to the poles (Figure 20.3). In general, belts of low air pressure develop both at the equator and at 50° to 60° north and south latitude as a result of rising columns of air, producing precipitation. Belts of high pressure resulting from descending air develop at 25° to 30° north and south latitude, producing arid conditions.

The latitudinal belts have names, such as the "doldrums," regions at the equator with little air movement; "trade winds," northeast and southeast winds important in the early days of international trade, when sailing ships moved the world's goods; and "horse latitudes," two belts centered about 30° north and south of the equator with descending air and high pressure. Earth's major deserts occur in the horse latitudes as a result of pervasive high pressure and low precipitation sandwiched between the equatorial and midlatitudinal zones of low pressure with higher precipitation.

Processes that Remove Materials from the Atmosphere

Understanding processes that remove materials from the atmosphere is important in solving atmospheric pollution problems. Four processes are responsible for removing human-induced particles and chemicals from the atmosphere:

- *Sedimentation.* Particles heavier than air settle out as a result of gravitational attraction to Earth. For example, particulates from volcanic eruptions or burning coal will settle out over time as a dry deposition.
- *Rain out.* Precipitation (rain, ice, or snow) can physically and chemically flush material from the atmosphere. For example, raindrops form by condensation of water on small particles in the atmosphere, bringing the particles to Earth with the raindrops. Carbon dioxide combines with water in the atmosphere to form weak carbonic acid by the following chemical reaction:

$$CO_2 + H_2O \rightarrow H_2CO_3$$

- This process effectively removes some carbon dioxide from the atmosphere and explains why natural rainfall is slightly acidic (see Chapter 21).
- *Oxidation.* Oxidation is a reaction in which oxygen is chemically combined with another substance. For example, sulphur dioxide in the atmosphere oxidizes easily to sulphur trioxide (SO^3), which may dissolve in water, forming sulphuric acid and producing acid rain (see Chapter 21).
- *Photodissociation.* Solar radiation (light) can break down chemical bonds in a chemical process known as photodissociation. For example, ozone (O^3) in the atmosphere may break down to O^2 as a result of photodissociation.

With our basic discussion of the composition of the atmosphere and some of the important atmospheric processes behind us, we next discuss climate and climatic change.

20.2 Climate

Climate refers to the representative or characteristic atmospheric conditions for a region on Earth. The term climate refers to these conditions over long time periods, such as seasons, years, or decades, whereas weather refers to shorter periods of time, such as hours, days, or weeks. When we say it is hot and humid in Toronto today or raining in Victoria, we are speaking of weather. When we say Halifax has cool, wet winters and warm, dry summers, we are referring to the Halifax climate. Climate depends in part on precipitation and temperature, both of which show tremendous variability on a global scale. However, because the climate of a particular location may depend on extreme or infrequent conditions, climate is more than just the average temperature and precipitation of a region.

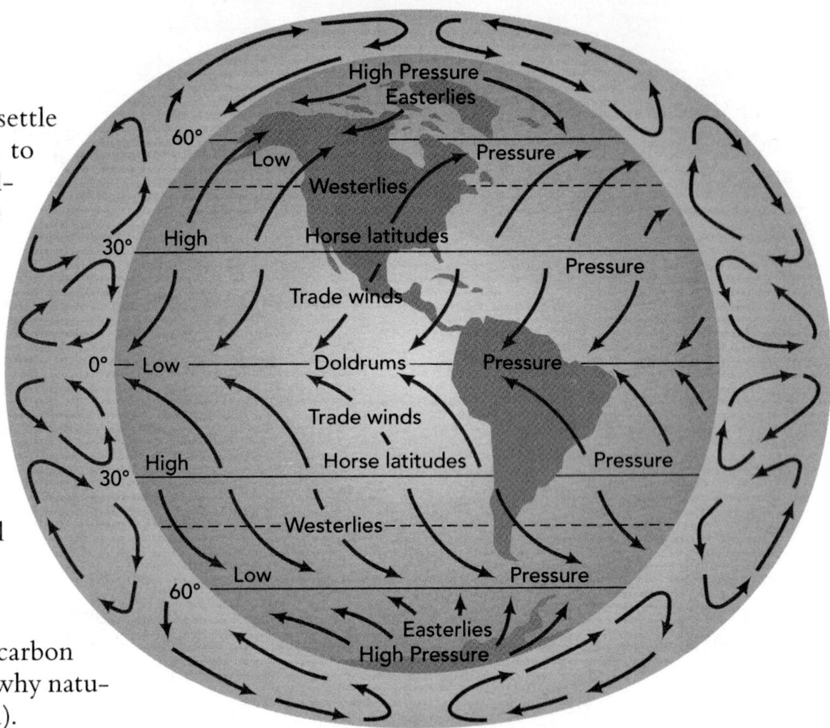

Figure 20.3 • Generalized circulation of the atmosphere. [*Source*: Modfied from Williamson, *Fundamentals of Air Pollution*, ©1973. Figure 5.5. Reprinted with permission of Addison-Wesley, Reading, Mass.]

The simplest classification of climate is by latitude—tropical, subtropical, midaltitudinal (continental), subarctic (continental), and arctic. Several other categories are used as well, including humid continental, Mediterranean, monsoon, desert, and tropical wet–dry (Figure 20.4). Although detailed discussion of climatic types is beyond the scope of this book, it is important to recognize the significance of potential climatic variability in determining what kinds of organisms live where. Recall from the discussion of biogeography in Chapter 8 that similar climates produce similar kinds of ecosystems. This concept is important and useful for environmental science. Knowing the climate, we can predict a great deal about what kinds of life we will find in an area and what kinds could survive there if introduced.

On a regional scale, air masses that cross oceans and continents can have a profound influence on seasonal patterns of precipitation and temperature. On a local scale, climatic conditions can also vary considerably and produce a local effect referred to as a **microclimate**. Microclimate can vary even from one side of a small rock to another or from one side of a tree to another. Organisms often take advantage of these different conditions.

Urban areas produce a characteristic microclimate with important environmental consequences. The very presence of a city affects the local climate, and as the city changes, so does its climate. For example, in the middle of the eighteenth century, according to the writer Peter

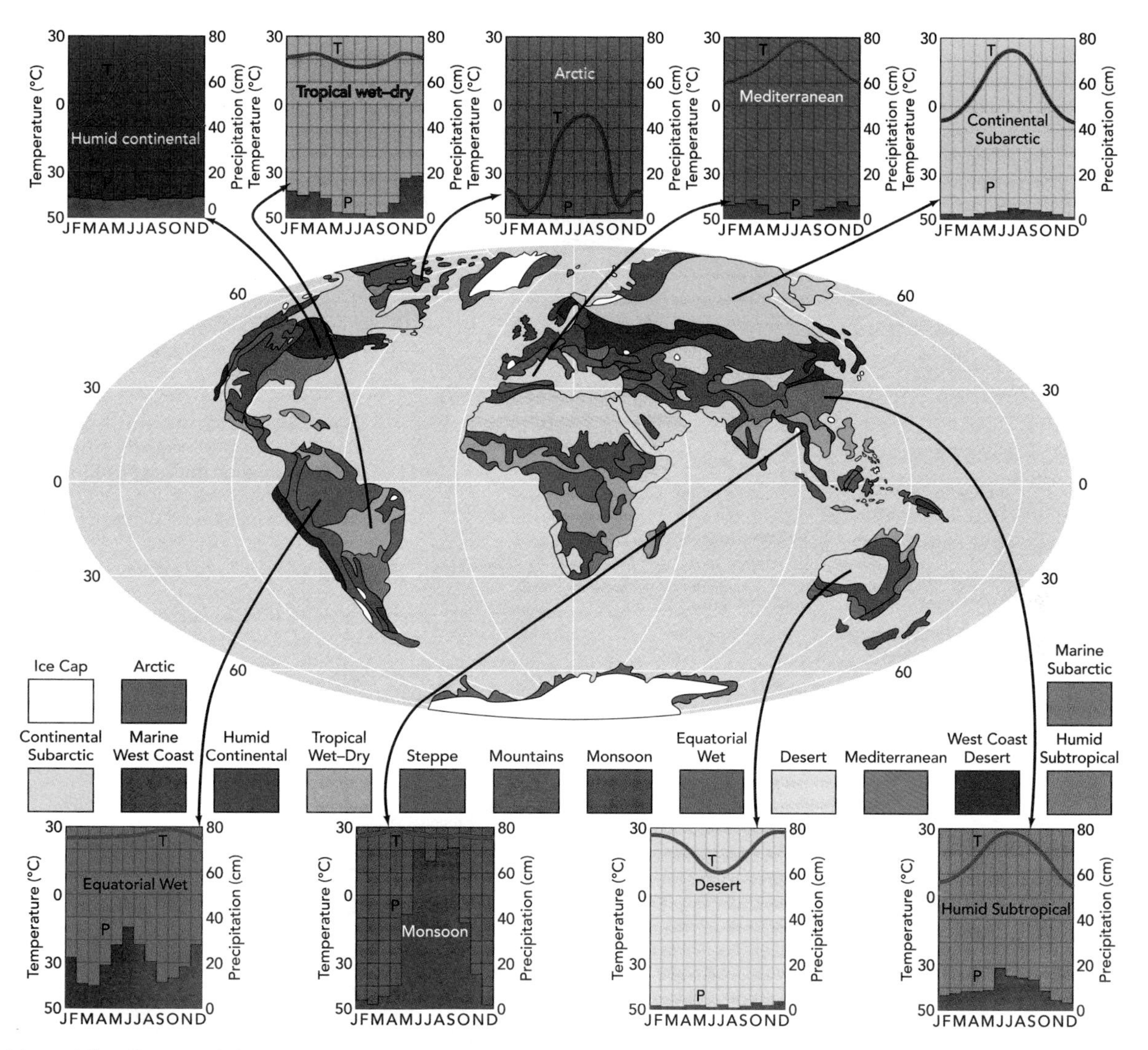

Figure 20.4 • **The climates of the world and some of the major climate types in terms of characteristic precipitation and temperature conditions.** [*Source*: Modified from W. M. Marsh and J. Dozier, *Landscape*, © 1981, John Wiley & Sons, reprinted with permission of John Wiley & Sons, Inc.]

Kam, who was traveling in North America, Manhattan Island was "generally reckoned very healthy," perhaps because of its nearness to the ocean and its relatively unobstructed ocean breezes. Today, air pollution and the effects of tall buildings on air flow lead the average visitor to Manhattan to a quite different conclusion.

Although air quality in urban areas is in part a function of the amount of pollutants present or produced, it is also affected by the city's ability to ventilate and flush out pollutants. The amount of ventilation depends on several aspects of the urban microclimate. In terms of climate, cities in midlatitude regions tend to differ from surrounding rural areas in several ways:[5]

1. They are warmer (0.5–2°C) because of heat from burning fuel for transportation, industry, and homes.
2. They are less humid (6–10%) because of less evaporation of water from standing water and soil.
3. They have up to 10 times as many dust particles, due to pollution, forming the **urban dust dome**.
4. They have 30–100% more fog and 5–10% more precipitation because of the presence of small particles in the city air.

The abundance of particulates above a city reduces incoming solar radiation by as much as 30% and thus cools the city, but the cooling effects of particles are relatively small compared with the processes that heat the city.[5] The combination of lingering air and abundant particles and other pollutants produces a heat island effect (see Chapter 26). The effect causes a circulation pattern in which cooler, often cleaner air moves from rural or suburban areas toward the inner city, where heat rises and then flows out to rural areas, completing the pattern.

Climate Change

Another important aspect of climate is **climate change**. The mean annual temperature of the Earth has swung up and down by several degrees Celsius over the past million years (Figure 20.5). Times of high temperature reflect relatively ice-free periods (interglacial periods)

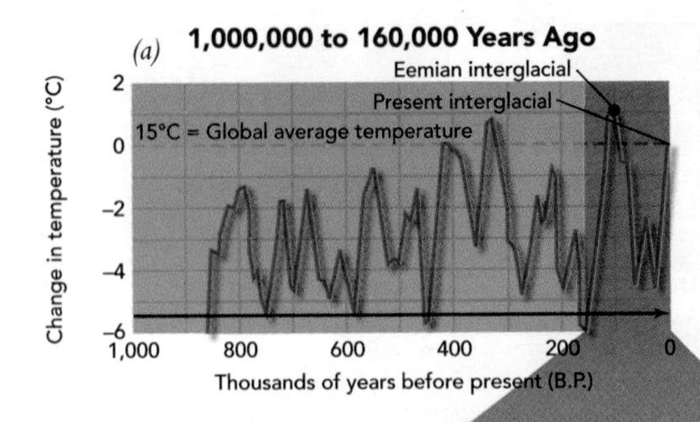

- The 1990s were the warmest decade in the 142 years since temperatures have been recorded, and in the last 1,000 years, according to geologic data.
- Warming since the mid-1970s has been approximately three times as rapid as in the preceding 100 years.
  - The 10 warmest years have all occurred since 1990 and the five warmest since 1997.

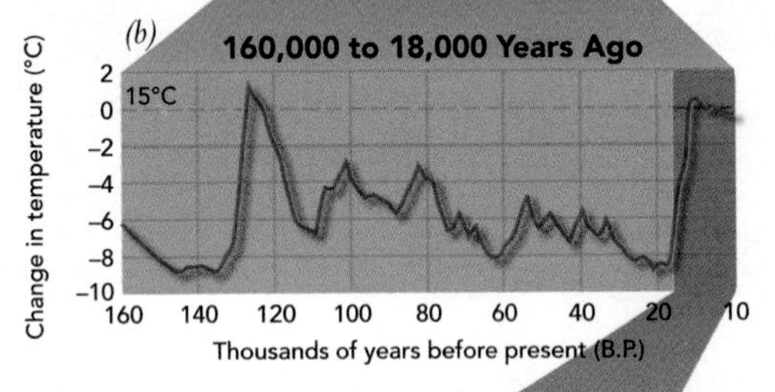

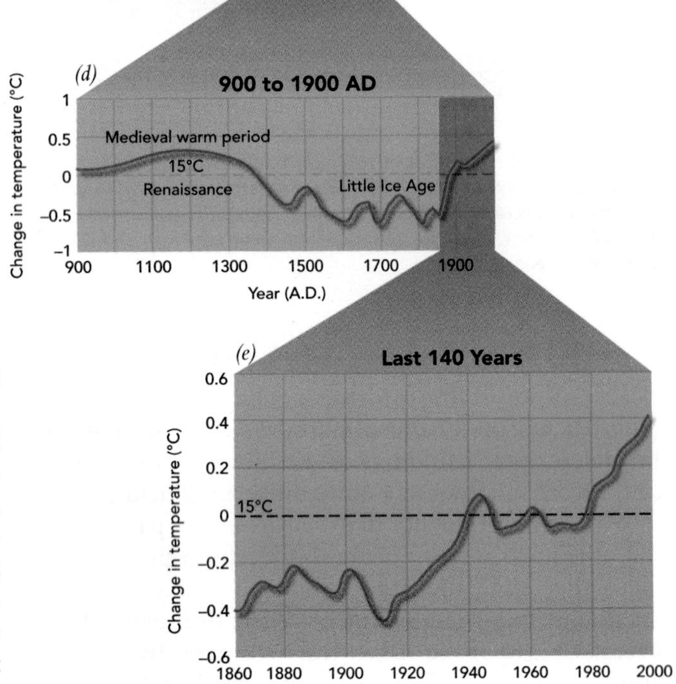

Figure 20.5 • Changes in Earth's temperature over varying time periods during the past million years. [*Sources*: Modified from UCAR/DIES, "Science Capsule, Changes in the Temperature of the Earth," *Earth Quest 5*, no. 1 (Spring 1991); Houghton, J. T., G. L. Jenkins, and J. J. Ephranns, eds. *Climate Change, the Science of Climate Change* (Cambridge: Cambridge University Press, 1996); U.K. Meteorological Office, *Climate Change and Its Impacts: A Global Perspective* 1997.]

over much of the planet; times of low temperature reflect the glacial events (Figure 20.5*a*, *b*). It is not yet clear whether our current warm climate marks the end of the ice ages or is merely an interglacial period with another glacial age due.

Global climate also changes in shorter time frames. For example, continental glaciation ended about 12,500 years ago with very rapid warming, perhaps over a period as brief as a few decades.[6] This was followed by a short global cooling about 11,500 years ago (Figure 20.5*c*). Climatic change over the last 18,000 years reflects several warming and cooling trends that have greatly affected people. For example, during a major warming trend from 1100 to 1300 (medieval warm period, Figure 20.5*c, d*), the Vikings colonized Iceland, Greenland, and North America. When glaciers advanced during the cold period starting around 1400 (Little Ice Age), the Viking settlements in North America and parts of Greenland were abandoned.

Starting in approximately 1850, a warming trend became apparent. It lasted until the 1940s, when temperatures began again to cool. Figure 20.5*e* shows change over the last 140 years. On this scale, the 1940s warming event is clearer. As you can see, it was followed by a levelling off of temperature in the 1950s and then a further drop during the 1960s. After that time, temperature increased steadily through the 1990s. What is evident from the record of the last 100 years is that global mean annual temperature has increased by approximately 0.6° C. (Figure 20.5*e*). This period includes the warmest years of the twentieth century[7, 8], known as the "late-twentieth-century increase in global temperature."[9]

- The warmest year on record was 1998, with 2002 and 2003 tied for second.
- In much of North America, 2003 was cooler and wetter than average in much of the eastern part of the continent, and warmer and drier in much of the western

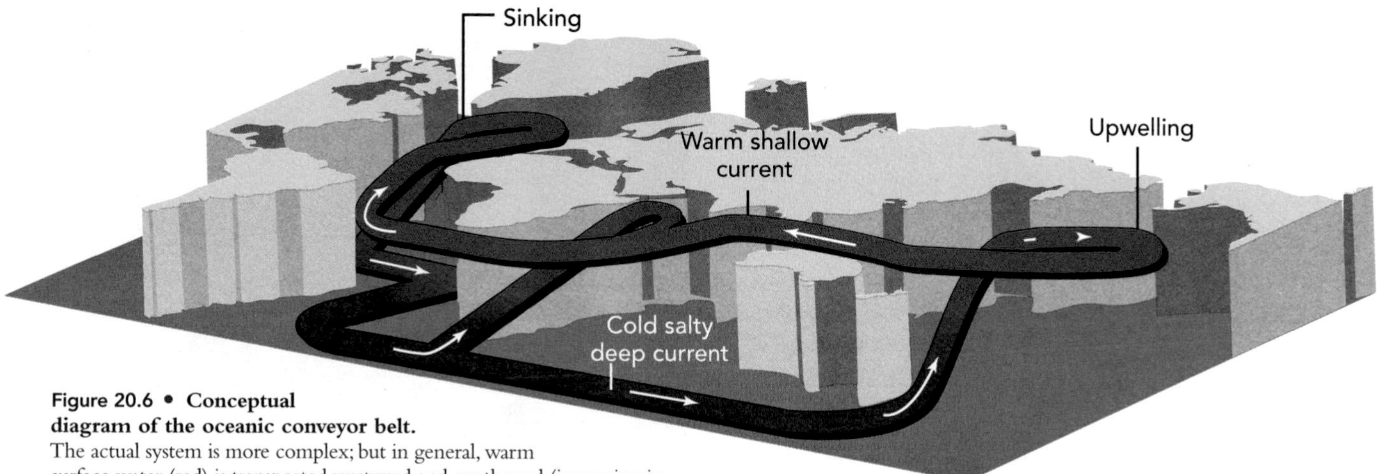

Figure 20.6 • Conceptual diagram of the oceanic conveyor belt. The actual system is more complex; but in general, warm surface water (red) is transported westward and northward (increasing in salinity because of evaporation) to near Greenland, where it cools from contact with cold Canadian air. As the water increases in density, it sinks to the bottom and flows south, then east to the Pacific, then north where upwelling occurs in the north Pacific. The masses of sinking and upwelling waters balance, and the total flow rate is about 20 million m^3/sec. The heat released to the atmosphere from the warm water keeps northern Europe 5°C to 10°C warmer than if the oceanic conveyor belt were not present. [*Source*: Modified from W. Broker, "Will Our Ride into the Greenhouse Future Be a Smooth One?" *Geology Today* 7, no. 5 (1997): 2–6.]

part. The western part of the continent was warmer than average; New Mexico had its warmest year on record, and Alaska and the Yukon were warmer than average in all four seasons. In 2003, Europe experienced summer heat waves, with the warmest seasonal temperatures ever recorded in Spain, France, Switzerland, and Germany. Approximately 15,000 people died in heat waves in France during the summer.
- Warm conditions with drought contributed to severe wildfires in Australia, southern California, and British Columbia.

Of course, a year or two of high temperatures with drought, heat waves, and wildfires is not by itself an indication of longer-term climate change. The persistent trend of increasing temperatures over three decades is more compelling evidence that climate change is real and happening.

The question that begs to be asked is: Why does climate change occur? Examination of Figure 20.5*a* suggests there are cycles of change about 100,000 years long separated by shorter cycles of 20,000 to 40,000 years in duration. Milutin Milankovitch first identified these cycles in the 1920s as a hypothesis to explain climate change. Milankovitch realized that the spinning Earth is like a wobbling top unable to keep a constant position in relation to the sun, and that position determines (in part) the amount of sunlight reaching and warming Earth. He discovered that variations in Earth's orbit around the sun follow a cycle of approximately 100,000 years, which correlates with the major glacial and interglacial periods of Figure 20.5*a*. Cycles of approximately 40,000 and 20,000 years are the result of changes in the tilt and wobble of the Earth's axis. Milankovitch cycles are consistent with most of the long-term cycles we see in the climate. However, the cycles are not sufficient, by themselves, to produce the large-scale climatic variations in the geologic record. Therefore, the Milankovitch cycles can be looked at as natural mechanisms that, along with other processes, may produce climatic change.

Shorter cycles have also been suggested. In fact, one study suggests that during the past 4,000 years there has been a cycle of about 1,500 years that may help explain the medieval warm period followed by the Little Ice Age, as well as the present warming trend, which is predicted to continue naturally until approximately 2400.[10] If this is correct, then any warming caused by human activity would be superimposed on a system that is already slowly warming.

Our climate system may be inherently unstable and capable of changing quickly from one state (cold) to another (hot) in as short a time as a few decades.[11] Part of what may drive the climate system and its changes is the "ocean conveyor belt"—a global circulation of ocean waters characterized by strong northward movement of upper warm waters of the Gulf Stream in the Atlantic Ocean. These waters are approximately 12–13°C when they arrive near Greenland and are cooled in the North Atlantic to a temperature of 2–4°C (Figure 20.6).[11] As the water is cooled, it becomes more salty and so increases in density, causing it to sink to the bottom. The cold, deep current flows southward, then eastward, and finally northward in the Pacific Ocean. Upwelling in the north Pacific starts the warm shallow current again. The flow in this conveyor belt current is huge (20 million m^3/sec), about equal to 100 Amazon Rivers. The amount of warm water and heat released at the surface is sufficient to keep northern Europe 5–10°C warmer than it would be if the conveyor belt were not present. If the system were to shut down, global cooling could result, and northern Europe would become much less habitable. If this were to happen in the future when there are a few more billion people to feed, a global catastrophe might result.[11]

Some scientific uncertainties remain regarding the human role in the observed warming of Earth's climate (see the Critical Thinking Issue on page 429). The science of climate change is discussed in detail following consideration of the tools used to study global change.

20.3 Earth System Science and Global Change

Until very recently, it was generally thought that human activity could only cause local or, at most, regional environmental changes. We now know otherwise! The main goal of the emerging science known as **Earth system science** is to obtain a fundamental understanding of how our planet works as a system. From a pragmatic point of view, the research priorities of Earth system science and global change can be summarized as follows:[12]

- Establishment of worldwide measurement stations to better understand physical, hydrologic, chemical, and biological processes that are significant in the evolution of Earth on a variety of time scales.
- Documentation of global changes, especially those that occur in a time period of several decades, that are of particular interest to the human environment.
- Development of quantitative models useful in the prediction of future global change.
- Provide information needed by decision makers at regional and global levels.

The major tools for studying global change are:

- Evaluation of the geologic record
- Monitoring
- Mathematical models

Geologic Record

The sediments deposited on floodplains, in bogs, and on lake and ocean bottoms can be read just as the pages of a history book can. Organic material, such as skeletal material, shells, pollen, bits of wood, leaves, and other plant parts, is often deposited with sediments and can yield valuable information concerning Earth's history (see Chapter 2). In addition, organic material can be dated to provide the necessary chronology to establish past changes. Finally, both sediments and organic material can be used to evaluate the past climate—what lived where, what kinds of changes have occurred, and how extensive the changes were.

One interesting use of the geologic record has been the examination of glacial ice. The process of transformation of snow to glacial ice involves recrystallization and an increase in the density of the ice. The process also traps air bubbles, which can be analysed to provide information concerning the concentration of carbon dioxide in the atmosphere at the time the ice formed. Thus, glacial ice can be thought of as a time capsule that stores information about the atmosphere in the past. To study the ice, scientists extract long cores of glacial ice (Figure 20.7) and carefully sample trapped air bubbles. This method has been used to analyse the carbon dioxide content of the atmosphere up to 160,000 years ago. Figure 20.8 shows the record from 1500 to 2000.[13] Ice cores have also yielded data on variability of solar radiation through measurement of accumulation of cosmogenic isotopes in the ice.[8] (Cosmogenic isotopes are rare

Figure 20.7 • **A scientist examines a glacial ice core stored in a freezer.**

Figure 20.8 • **Average concentration of atmospheric carbon dioxide, 1500–2000.** [*Source*: Modified from W. M. Post, T. Peng, W. R. Emanuel, A. W. King, V. H. Dale, and D. L. De Angelis, *American Scientist* 78, no. 4 (1990): 310–326. By permission of *American Scientist*, Journal of Sigma Xi, The Scientific Research Society.]

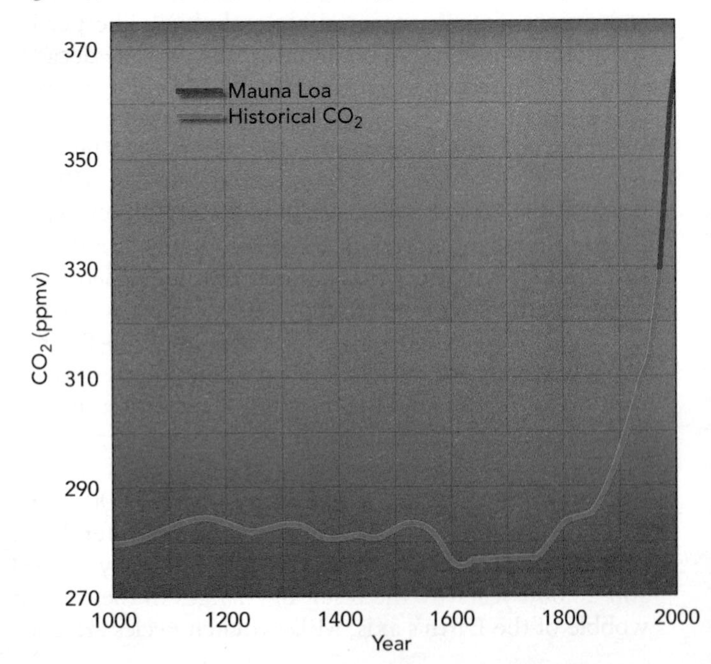

A CLOSER LOOK 20.1

Monitoring of Atmospheric Carbon Dioxide Concentrations

The activities of humans and other living things affect characteristics of Earth's surface, waters, and atmosphere, even in areas removed from such activities. For example, air pollutants and other artifacts of human society are found in glacial ice, and pesticides are found in lakes far from agricultural activity.

The U.S. National Oceanic and Atmospheric Administration (NOAA) coordinates an international network of air sampling stations, the Carbon Cycle Greenhouse Gases (CCGG) group, begun in 1967. Air samples are collected approximately weekly from dozens of sites representing every continent. One of the longest-operating of these sites is the summit of Mauna Loa, Hawaii. Mauna Loa is one of the largest active volcanoes and the highest mountain in the world, based on elevation change from base to top. Air samples are taken there because it is located far away from direct effects of human life and other biological activity. The data from Mauna Loa and other CCGG sites has helped show another dimension of how life and human activity affect the atmosphere.

One substance of interest in the Mauna Loa samples is carbon dioxide. Carbon dioxide is taken up by green plants during photosynthesis and released in respiration from all oxygen-breathing organisms; thus, a measure of carbon dioxide in the atmosphere is analogous to a measure of the breathing in and out of all life on Earth.

The Mauna Loa measurements of carbon dioxide are truly remarkable. Two important aspects of the record are shown in Figure 20.9: a strong upward trend over the years and an annual cycle, which is obvious and regular. Carbon dioxide concentration reaches a peak in winter and a trough in summer. The annual curve is a measure of life activities in the entire Northern Hemisphere. In summer, green plants are most active, and the total amount of photosynthesis exceeds the total amount of respiration. As a result, carbon dioxide is removed from the atmosphere in summer. In winter, photosynthesis decreases and becomes less than total respiration, so the carbon dioxide concentration in the atmosphere increases.

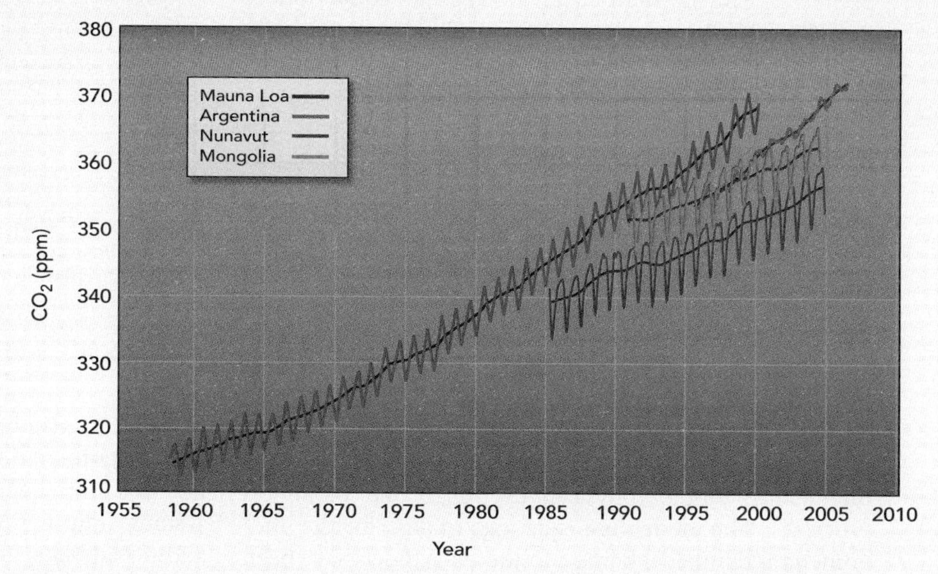

Figure 20.9 • **Monthly average carbon dioxide concentration and long-term trend for world locations.** [*Source*: NOAA: http://www.cmdl.noaa.gov/ccgg/iadv]

Figure 20.9 demonstrates that similar observations have been made Nunavut, South America, and other places in all parts of the world. In each case, the same trends are observed: a strong upward trend in concentration of carbon dioxide in the atmosphere and an annual cycle. The annual cycle from Antarctica is smaller in amplitude than the Mauna Loa cycle because of the relatively smaller land area in the Southern Hemisphere and the smaller amount of vegetation.

The upward trend, or increase, in carbon dioxide concentration in the atmosphere shown in Figure 20.9 is believed to be due to the addition of carbon dioxide from burning fossil fuels and other human activities, such as cutting forests and burning wood (burning releases the stored carbon in wood, which combines with oxygen, producing carbon dioxide). The Mauna Loa data provided some of the first evidence that directly indicated that life touches the entire Earth and that human activities have begun to affect the atmosphere of our planet.

The data in Figure 20.9 clearly illustrate that changes in atmospheric carbon dioxide are global in nature, underscoring the need for a global network of monitoring stations and the benefits of long-term data collection. Funding for long-term projects is often difficult to maintain because funding agencies may prefer to sponsor new projects rather than long-term monitoring. Nevertheless, understanding global change depends on collection and maintenance of supportive data. To that end, the NOAA CO_2 measurement project is unique, and its effectiveness is a tribute to the people who initiated it and nurtured it for so many years.

isotopes created in space or Earth's atmosphere when cosmic radiation interacts with atomic nuclei.)

Real-Time Monitoring

Monitoring can be defined as the regular collection of data for specific purposes. For example, we monitor rainfall and the flow of water in rivers to evaluate water resources or flood hazards. We collect the data to provide baseline conditions from which to evaluate changes in the future. Similarly, samples of atmospheric gases, particulates, and chemicals are helping establish trends in the composition of the atmosphere (see A Closer Look 20.1). Finally, measurements of the temperature, composition, and chemistry of ocean waters can be used to help evaluate changes in the marine environment.

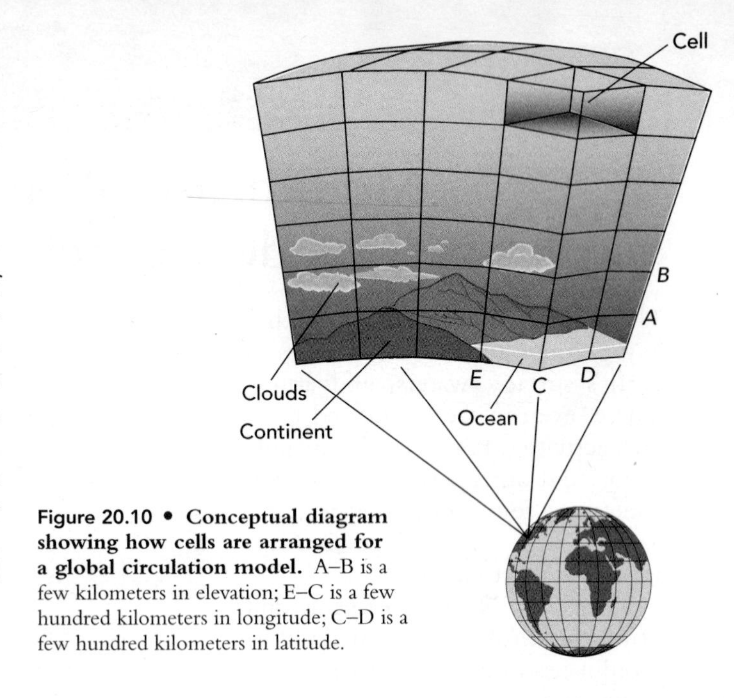

Figure 20.10 • Conceptual diagram showing how cells are arranged for a global circulation model. A–B is a few kilometers in elevation; E–C is a few hundred kilometers in longitude; C–D is a few hundred kilometers in latitude.

Mathematical Models

Mathematical models are attempts to represent, through numerical means, real-world phenomena, linkages, and interactions among physical, chemical, and biological processes. The models use equations to describe the phenomena and linkages being considered. Models have been developed to predict the flow of surface water and groundwater, ocean circulation, and atmospheric circulation. In the area of global change, the models that have gained the most attention are the **global circulation models (GCMs)**. The objective of GCMs is to predict atmospheric changes on a global scale.[14]

The variables utilized in GCMs include temperature, relative humidity, and wind conditions. Values for many of these variables are estimated for the past based on data derived from such sources as tree-ring records. Annual growth rings of trees can be evaluated to provide a time scale (called *dendrochronology*). Growth rings can also be helpful in inferring climatic information, such as precipitation and runoff, which are useful in calibrating results from mathematical models. To organize the data necessary for the calculations in GCMs, the surface of Earth is divided into large cells measuring several degrees of latitude and longitude. The typical cell is about the size of Japan or Great Britain, or about twice the size of the island of Newfoundland. Several layers of data are necessary. Most GCMs use 6 to 20 levels of vertical data collected throughout the lower atmosphere (Figure 20.10). Mathematical relationships are used in the calculations to predict future atmospheric circulation. The GCMs are complex and require supercomputers for their operation.

Results from these models are relatively crude and so may not accurately predict future conditions.[14, 15] Furthermore, it is difficult to estimate interactions of other factors, such as cloud cover, which may significantly affect atmospheric energy relations. As a result, GCMs can only be considered a first approach to solving complex atmospheric problems. In spite of their limitations, mathematical models are providing information necessary for evaluating global change and Earth as a system. The GCMs are also helping to pinpoint what additional data are necessary to better predict future change. The models do predict, in a relative sense, areas or regions that are likely to be wetter or drier if certain changes in the atmosphere occur. Thus, predictions of future global change from mathematical models are being taken seriously, and their importance as a tool for studying global change will continue to increase.

Our discussion of the tools used to study global change leads us to consider human-induced climate change. The major question is: How much has human activity contributed to climate change?

20.4 Climate Change: Earth's Energy Balance and the Greenhouse Effect

Climate change is defined as a natural or human-induced increase in the average global temperature of the atmosphere near the Earth's surface. The temperature at or near the surface of the Earth is determined by four main factors:[14]

- The amount of sunlight Earth receives.
- The amount of sunlight Earth reflects.
- Retention of heat by the atmosphere.
- Evaporation and condensation of water vapour.

Electromagnetic Radiation and Earth's Energy Balance

In order to understand climate change, it is necessary to have a modest acquaintance with electromagnetic radiation and Earth's energy balance. Our Earth is part of a planetary system receiving energy from the sun. This energy undergoes changes; affects life, oceans, atmosphere, and climate; and is eventually emitted as heat back into the depths of space. In this system, Earth is an intermediate between the *source* (the sun) and the *sink* (space).

Nearly all the energy available at Earth's surface comes from the sun (Figure 20.11), with small additional amounts coming from human activities, geothermal energy (from the interior of Earth), and tides. Since energy is conserved, the amount emitted to space matches that received from the sun plus the other small contributions. The matching of input of energy from the sun with output from the Earth defines Earth's energy balance. Although Earth intercepts only a very tiny fraction of the total energy emitted by the sun, solar energy sustains life on Earth and greatly influences climate and weather.

Energy is emitted from the sun in the form of electromagnetic radiation (EMR). Different forms of electromagnetic energy can be distinguished by their wavelengths. (See Chapters 15 and 17 and A Closer Look 17.1 for additional discussion on the importance of electromagnetic radiation.) The collection of all possible wavelengths of electromagnetic energy, considered a continuous range, is known as the **electromagnetic spectrum** (Figure 20.12). The electromagnetic spectrum is one of the most important phenomena in the physical sciences and is fundamental to understanding many environmental topics. Gamma rays, X-rays, ultraviolet light, visible light, infrared radiation, television waves, radio waves, and radar are all different types of electromagnetic radiation. The relatively long

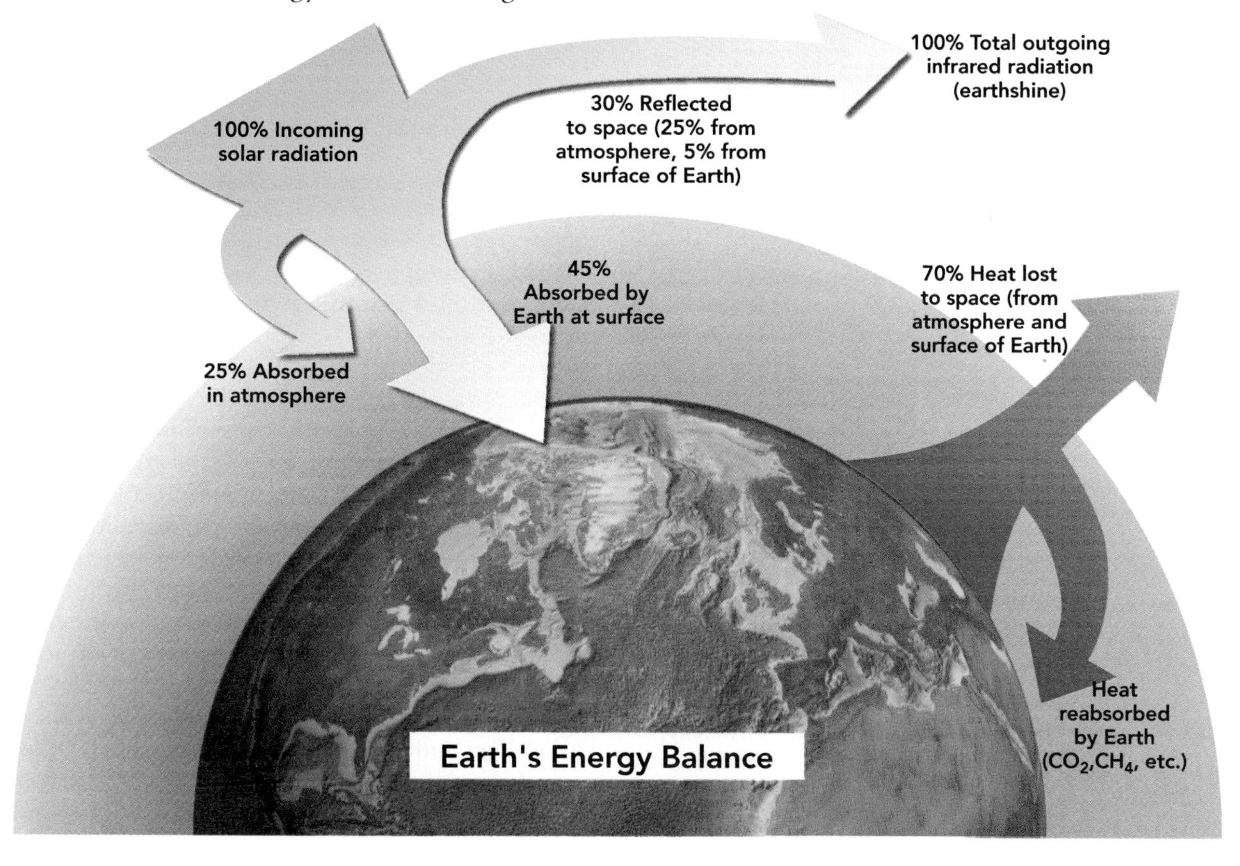

Figure 20.11 • **Conceptual diagram showing Earth's energy balance.** Incoming solar radiation amounts to approximately 5.5 million exajoules, and that is balanced by a similar amount leaving Earth. A very small component of heat (1,000 exajoules) entering the system is generated within the Earth; it represents only about 0.002% of the budget and so is neglected in the values presented here. Approximately 30% of the total incoming solar radiation (1.7 million exajoules) is reflected immediately in the atmosphere, mostly by clouds and at the surface of the Earth. Of the remaining 70% of incoming solar radiation, approximately 25% is absorbed in the upper atmosphere, and 45% reaches the Earth and is absorbed at the surface. As the energy from the sun is absorbed by plants, soils, rocks, water, and other surface materials, it is transformed into heat, radiates as infrared radiation, and is eventually lost to space from Earth's upper atmosphere. Part of the infrared radiation thus released is reabsorbed by compounds, including carbon dioxide and methane, known as greenhouse gases (discussed later in the chapter). This energy, too, is ultimately lost to space as heat, completing the energy balance of Earth. About two-thirds of the incoming solar radiation is available to drive processes at or near the surface of the Earth, including the hydrologic cycle, wind, and so forth. [*Source*: Modified from N. L. Pruitt, L. S. Underwood, and W. Surver. *Bio Inquiry* (New York: Wiley, 2002).]

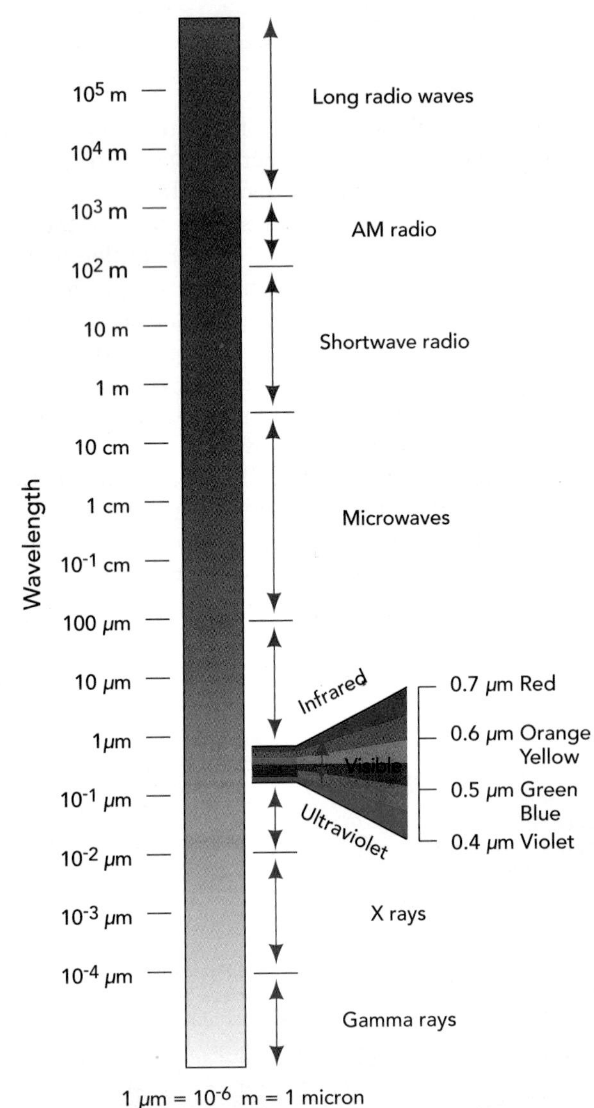

Figure 20.12 • Types of electromagnetic radiation (EMR). Notice that the spectrum of wavelengths from radio waves to gamma rays is over nine orders of magnitude (1 $\mu m = 10^{-6}$ m = 1 micron).

wavelengths (greater than 1 m in the electromagnetic spectrum) include radio waves, and the shortest wavelengths are those of gamma rays and X-rays. The electromagnetic radiation to which our eyes are sensitive, *visible electromagnetic radiation*, is a very small fraction of the total spectrum, as you can see in Figure 20.12. Other types of electromagnetic radiation with environmental significance include radar, microwaves, and infrared radiation.

The amount of energy per unit time radiated from a body such as the sun or Earth varies with the fourth power of the absolute temperature of the body. Thus, if a body's temperature doubles, the energy radiated increases by 24, or 16 times. This phenomenon explains why the sun, with a temperature of 5,800°C, radiates a tremendously greater amount of energy than does the surface of Earth, which radiates at an average temperature of 15°C. Figure 20.13 illustrates this and another important point: The sun emits strongly in the visible region (relatively short-wave radiation of about 0.4–0.7 μm), whereas Earth emits in the infrared region (relatively long-wave radiation of about 10 μm). The hotter an object is, the more rapidly it radiates heat and the shorter the wavelength of its predominant radiation. Earth's surface and the surfaces of animals, plants, clouds, water, and rocks are cool enough to radiate heat predominantly in the infrared wavelength, which is invisible to us.

Energy from the sun travels to Earth at the speed of light through the vacuum of space. The Earth and lower atmosphere absorb less than half the solar energy that reaches our atmosphere. The rest is either absorbed by the upper atmosphere or reflected back into space. Much radiation that is harmful to living organisms, such as X-rays and ultraviolet radiation, is filtered out by the upper atmosphere. The stratospheric ozone layer, extending from about 15 to 40 km above Earth's surface (Figure 20.1), is particularly important in absorbing ultraviolet radiation from the sun and protecting living organisms at Earth's surface. The depletion of the ozone layer is discussed in Chapter 22.

The Greenhouse Effect

Sunlight that reaches Earth warms both the atmosphere and the surface. Earth's surface and atmospheric system then re-radiate heat as infrared radiation.[14] Certain gases in Earth's atmosphere absorb and re-emit this radiation. Some of it returns to the Earth's surface, making Earth warmer than it otherwise would be. In trapping heat, the gases act a little like the panes of glass in a greenhouse (although the process by which the heat is trapped is not the same as in a greenhouse). Accordingly, the effect is called the **greenhouse effect**, and the gases—which include water vapour, carbon dioxide, methane, and chlorofluorocarbons (CFCs)—are called **greenhouse gases**.

It is important to understand that the greenhouse effect is a natural phenomenon that has been occurring for millions of years on Earth as well as on other planets in our solar system. After learning how the greenhouse effect works, you will be on your way to understanding how the effect moderates the temperature of the lower atmosphere and helps keep Earth habitable for life. You will also understand how human activity is interacting with the greenhouse effect to cause human-induced climate change.

Most natural greenhouse warming is due to water in the atmosphere. On a global level, water vapour and small particles of water in the atmosphere produce about 85% and 12%, respectively, of our total greenhouse warming. The greenhouse gases with which we are concerned are those that result in part from *anthropogenic* processes—that is, from human activities. These include carbon dioxide, CFCs, methane, nitrous

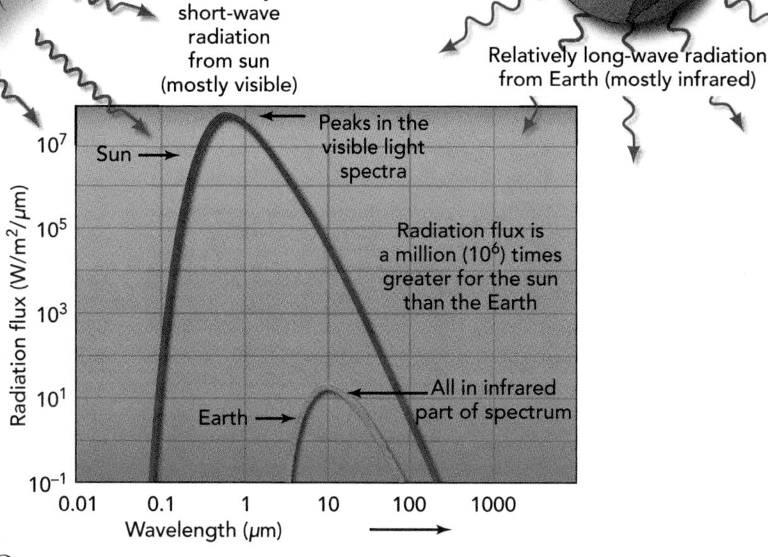

Figure 20.13 • Emission of energy from the sun compared with that from Earth. Notice that solar emissions have a relatively short wavelength, whereas those from Earth have a relatively long wavelength. [*Sources*: Modified from W. M. Marsh and J. Dozier, Landscape (Reading, Mass.: 1981); and L. R. Kump, J. F. Kasting, and R. G. Crane, *The Earth System* (Upper Saddle River, N.J.: Prentice Hall, 1999).]

oxides, and ozone, all of which have increased significantly in the atmosphere in recent years. It has been hypothesized that Earth is warming because of the increases in these greenhouse gases of anthropogenic origin.

The following discussion focuses on the anthropogenic greenhouse effect as it relates to three factors:

- Burning fossil fuels, which in recent years has added about 5.5 GtC (gigatons, or billions of metric tonnes, of carbon) per year to the atmosphere. The carbon combines with oxygen to form CO_2. The Intergovernmental Panel on Climate Change (IPCC) reports that these emissions appear to be rising steadily.[1]
- Deforestation, by burning trees, increases the concentration of atmospheric CO_2, adding 1.6 GtC per year. Burning trees releases carbon stored in the wood that combines with oxygen to form CO_2.
- Human activities that emit other greenhouse gases, such as CFCs, ozone, methane, and nitrous oxides.

How the Greenhouse Effect Works

A conceptual diagram showing some important aspects of the greenhouse effect is presented in Figure 20.14. The arrows labelled "energy input" represent the energy from the sun absorbed at or near the surface of Earth. The arrows labelled "energy output" represent energy emitted from the upper atmosphere and the surface of the Earth, which balances the input, consistent with Earth's energy balance. The highly contorted lines near the surface of Earth represent the absorption of infrared radiation (IR) occurring there and producing the 15°C near-surface temperature. Following many scatterings and absorptions and re-emissions, the infrared radiation emitted from levels near the troposphere corresponds to a temperature of approximately –18°C.

The one output arrow that goes directly through Earth's atmosphere represents radiation emitted through what is called the *atmospheric window* (Figure 20.15). The atmospheric window, centered on a wavelength of 10 μm, represents a region of wavelengths (8–12 μm) where outgoing radiation from Earth is not absorbed well by natural greenhouse gases (water vapour and carbon dioxide). Anthropogenic CFCs do absorb in this region, however; and CFCs significantly contribute to the greenhouse effect in this way.

Let us look more closely at the relation of the greenhouse effect to the Earth's energy balance, which was shown in Figure 20.11 in a simplistic way. The figure showed that, of the incoming solar radiation, approximately 30% is reflected back to space from the atmosphere as short-wave solar radiation, while 70% is absorbed by the Earth's surface and atmosphere. The 70% that is absorbed is eventually re-emitted as infrared radiation (IR) into space. Thus, the sum of the reflected solar radiation and the outgoing infrared radiation balances with the energy arriving from the sun.

This simple balance becomes much more complicated when we consider exchanges of IR within the atmospheric and Earth surface systems. In some instances, these internal radiation fluxes may have magnitudes greater than the amount of energy entering the Earth's atmospheric system from the sun, as shown in Figure 20.16. A major contributor to the fluxes is the greenhouse effect.

At first glance, you might think it would be impossible to have internal radiation fluxes greater than the total amount of incoming solar radiation (shown as 100 units in Figure 20.16). However, this apparent contradiction is possible because the infrared radiation bounces around many times in the atmosphere, resulting in high internal fluxes. For example, in terms of the figure, the amount of IR absorbed at the surface of the Earth from

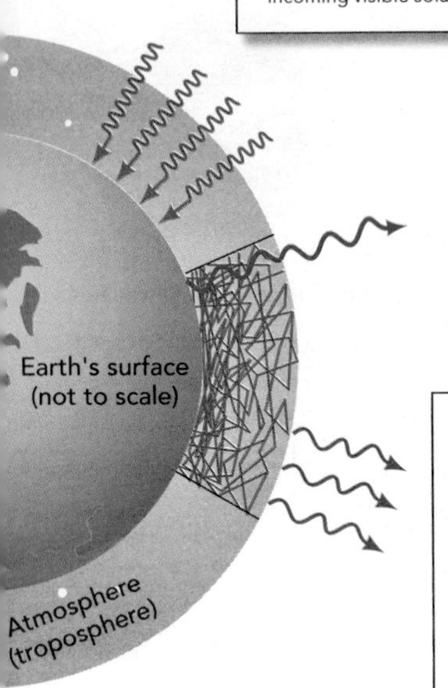

Energy input
Close to a third of the energy that descends on the Earth from the sun is reflected (scattered) back into space. The bulk of the remaining incoming visible solar radiation is absorbed by the Earth's surface.

Energy output
The atmosphere transmits outgoing infrared radiation from the surface (about 8% of the total outgoing radiation) at wavelengths between 8 and 13 microns and corresponds to a surface temperature of 15°C. This radiation appears in the atmospheric window, where the natural greenhouse gases do not absorb very well. However, the anthropogenic chlorofluorocarbons do absorb well in this wavelength region.

Most of the outgoing radiation after many scatterings, absorptions, and re-emissions (about 92% of the total outgoing radiation) is emitted from levels near the top of the atmosphere (troposphere) and corresponds to a temperature of –18°C. Most of this radiation originates at the Earth's surface, and the bulk of it is absorbed by greenhouse gases at heights on the order of 100 m. By various atmospheric energy exchange mechanisms, this radiation diffuses to the top of the troposphere, where it is finally emitted to outer space.

Figure 20.14 • **Conceptual diagram showing the greenhouse effect.** Incoming visible solar radiation is absorbed by the Earth's surface, to be re-emitted in the infrared region of the electromagnetic spectrum. Most of this re-emitted infrared radiation is absorbed by the atmosphere, maintaining the greenhouse effect. [*Source*: Developed by M. S. Manalis and E. A. Keller, 1990.]

the greenhouse effect is approximately 88 units, which is about twice the amount of short-wave solar radiation (45 units) absorbed by the Earth's surface. In spite of the large internal fluxes, the overall energy balance remains the same. At the top of the atmosphere, the net downward solar radiation (70 units) balances the outgoing IR from the top of the atmosphere (70 units).

The important point in taking a more detailed look at Earth's energy balance is recognizing the strength of the greenhouse effect. For example, notice in the figure that, of the 104 units of IR emitted by the surface of the Earth, only four go directly to the upper atmosphere and are emitted. The remainder are reabsorbed and re-emitted by greenhouse gases. Of these, 88 units are directed downward to Earth and 66 units upward to the upper atmosphere.

All this may sound somewhat complicated, but if you read and study carefully the points mentioned in Figure 20.16 and work through the balances of the various parts of the energy fluxes, you will gain a deeper understanding of why the greenhouse effect is so important. The greenhouse effect keeps the lower atmosphere of the Earth approximately 33°C warmer than it would otherwise be. In addition, the greenhouse effect provides other important service functions. For example, the strong downward emission of IR from the atmosphere that results from the greenhouse effect keeps variations in surface temperature between day and night relatively small. Without this effect, the land surface would cool much more rapidly at night and warm much more quickly during the day. Thus, the greenhouse effect not only maintains our relatively comfortable warm surface temperatures but also helps to limit temperature swings from day to night over the land.[3] Therefore, it is not the greenhouse effect itself that causes concern for humans and other life forms, but rather climate change caused by anthropogenic greenhouse gases.

Changes in Greenhouse Gases

The major anthropogenic greenhouse gases are listed in Table 20.1. The table also lists the recent rate of increase for each gas and its relative contribution to the anthropogenic greenhouse effect. Greenhouse gases vary in their radiative properties. Carbon dioxide has the greatest radiative forcing (greenhouse) potential, followed by methane and nitrous oxide.

Carbon Dioxide

Approximately 200 billion metric tonnes of carbon in the form of carbon dioxide enter and leave Earth's atmosphere each year as a result of a number of biological and physical processes. Not surprisingly, carbon dioxide has received a lot of attention with respect to climate change; 50–60% of the anthropogenic greenhouse effect is attributed to this gas (Table 20.1).

In order to evaluate recent changes in atmospheric carbon dioxide, we need to take a broad view of Earth's history. As explained earlier, we can find out about ancient air by sampling the air bubbles trapped in glacial ice. Measurements of carbon dioxide trapped in air bubbles in the Antarctic ice sheet suggest that during the 160,000 years prior to the Industrial Revolution, the atmospheric concentration of carbon dioxide varied from approximately 200 to 300 ppm.[13] The highest level or concentration of carbon dioxide in the atmosphere

Figure 20.15 • **Absorption spectra of greenhouse gases, water vapor, and carbon dioxide in the long-wave infrared radiation region.** The atmospheric window is a region where neither water vapor nor carbon dioxide absorbs, but where CFCs do absorb. [*Source*: Modified from T. G. Spiro and W. M. Stigliani, *Environmental Science in Perspective* (Albany: State University of New York Press, 1980).]

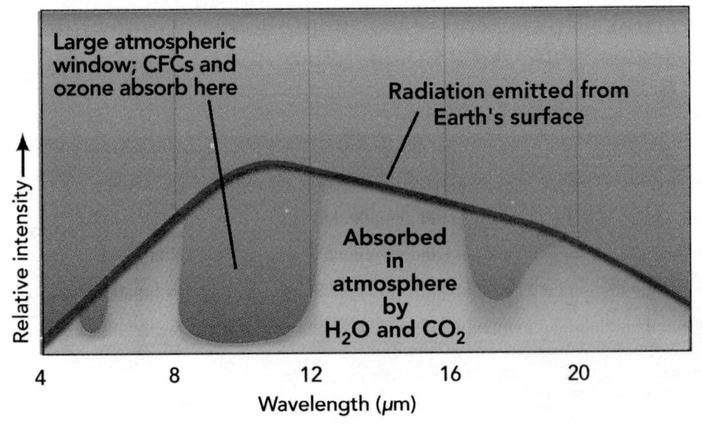

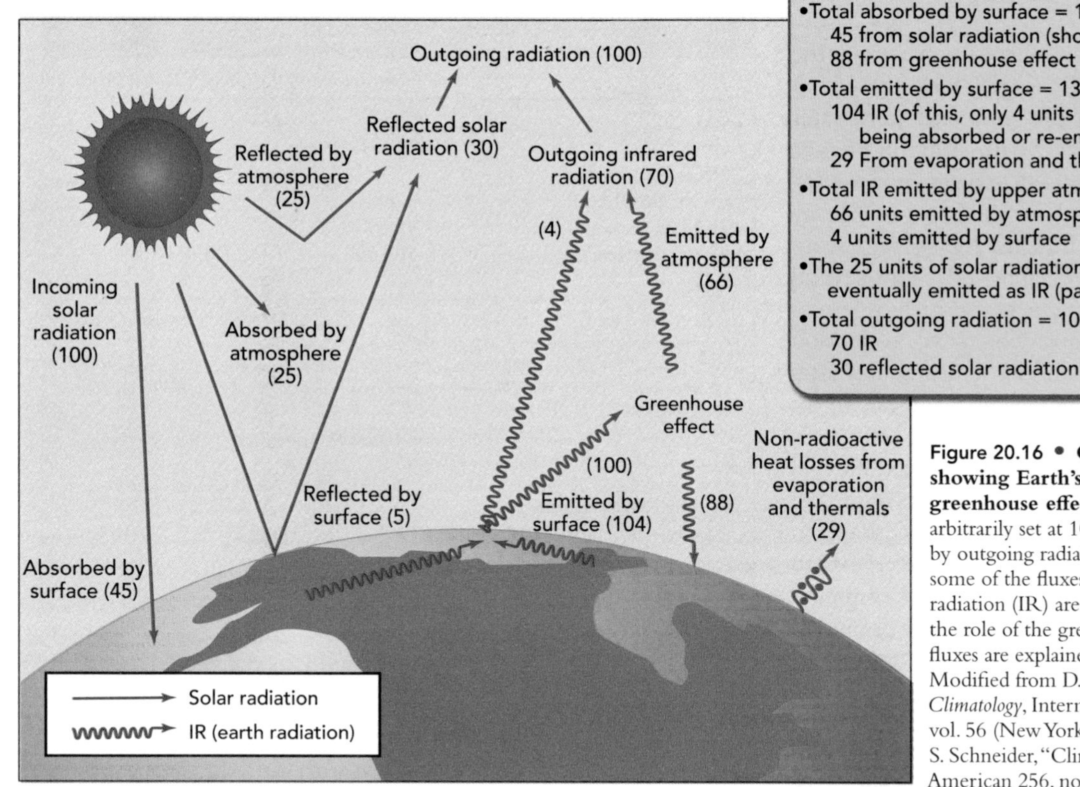

Figure 20.16 • Conceptual diagram showing Earth's energy balance and the greenhouse effect. Incoming solar radiation is arbitrarily set at 100 units, and this is balanced by outgoing radiation of 100 units. Notice that some of the fluxes (rates of transfer) of infrared radiation (IR) are greater than 100, reflecting the role of the greenhouse effect. Some of these fluxes are explained in the diagram. [*Source:* Modified from D. L. Hartmann, *Global Physical Climatology*, International Geophysics Series, vol. 56 (New York: Academic Press, 1994); and S. Schneider, "Climate Modeling," Scientific American 256, no. 5 (1987):72–80.]

other than at present occurred during the major interglacial period about 125,000 years ago.

About 140 years ago, at the beginning of the Industrial Revolution, the atmospheric concentration of carbon dioxide was approximately 280 ppm, a level that had apparently remained constant over the previous 700 years.[14] Since the beginning of the Industrial Revolution, the concentration of carbon dioxide in the atmosphere has grown exponentially. Currently, the rate of increase is about 0.5% per year. If growth continues at this rate, we will see a doubling of the concentration before the end of the twenty-second century. Today, the concentration of carbon dioxide in the atmosphere is about 370 ppm, and it is predicted that the level may rise to approximately 450 ppm by the year 2050, more than 1.5 times the pre-industrial level.[16] (Data before the mid-twentieth century are from measurements of air trapped in glacial ice. The remaining data are from direct measurement from the monitoring station at Mauna Loa, Hawaii, discussed in A Closer Look 20.1.) It is interesting to note that the rate of increase of carbon emissions (not carbon dioxide) from anthropogenic processes has been approximately 4.3% per year since the Industrial Revolution began, more than 8 times the 0.5% per year rate of increase in the concentration of carbon dioxide in the atmosphere.

The high rate of carbon dioxide emissions and the high growth rate in emissions would seem to suggest that the increase in carbon dioxide in the atmosphere is a direct result of the anthropogenic input of carbon. Establishing this seemingly simple relationship as a fact has been difficult, however, because the global carbon cycle is complex; all the linkages and flows of carbon from various sources to sinks are not yet well understood. What is apparent is that if all the carbon dioxide produced by human activities remained in the atmosphere, the concentration of that gas should be even higher than it is today. Therefore, we must hypothesize that there are sinks for carbon dioxide in the oceans or on the land that are not identified or well understood (see Chapter 5).

In spite of all these cautions, it is clear that carbon dioxide concentrations in the atmosphere have increased significantly since the Industrial Revolution. Furthermore, it is a reasonable assumption that these increases will continue to contribute to climate change via the greenhouse effect. Assuming that approximately 50–60% of the anthropogenic greenhouse effect is due to carbon dioxide, we can conclude that the remaining greenhouse gases must account for approximately 40–50% of the effect.

Methane

The concentration of methane (CH_4) in the atmosphere has more than doubled in the past 200 years, and it is thought to contribute approximately 12–20% of the anthropogenic greenhouse effect.[17] As with carbon dioxide, there are important uncertainties in our understanding of the sources and sinks of methane in the atmosphere.

Trace Gases	Relative Contribution (%)	Growth Rate (%/yr)
CH_4	12[a]–20[b]	0.4[c]
CFC	15[a]–25[b]	5
N_2O	5[d]	0.2
O_3 (troposphere)	8[d]	0.5
Total of these gases	40–50	
Contribution of CO_2	**50–60**	**0.3[e]–0.5[d,f]**

Table 20.1 • Relative Contribution of Trace Gases to the Anthropogenic Greenhouse Effect

[a] W. A. Nierenberg, "Atmospheric CO2: Causes, Effects, and Options," *Chemical Engineering* Progress 85, no. 8 (August 1989): 27.

[b] J. Hansen, A. Lacis, and M. Prathe r, "Greenhouse Effect of Chlorofluorocarbons and Other Trace Gases," *Journal of Geophysical Research* 94 (November 20, 1989): 16, 417.

[c] Over the past 200 yrs.

[d] H. Rodhe, "A Comparison of the Contribution of Various Gases to the Greenhouse Effect," *Science* 248 (1990):1218, Table 2.

[e] W. W. Kellogg, "Economic and Political Implications of Climate Change," paper presented at Conference on Technology-based Confidence Building: Energy and Environment, University of California, Los Alamos National Laboratory, July 9–14, 1989.

[f] H. Abelson, "Uncertainties about Global Warming," *Science* 247 (March 30, 1990):1529.

Natural environments release methane into the atmosphere. Major contributors are termites, which produce methane as they process wood; freshwater wetlands and hydroelectric dam impoundments, where decomposing plants in oxygen-poor environments produce and release methane as a decay product; seepage from oil fields; and seepage from methane hydrates (see Chapter 17). The several anthropogenic sources of methane include emissions from landfills, burning of biomass, production of coal and natural gas, and agricultural activities such as the cultivation of rice and the raising of cattle. (Methane is released by anaerobic activity in flooded lands where rice is grown, and cattle expel methane gas as part of their digestive processes.)

Rates of increase in emissions of methane in recent years have been variable. In the 1980s to 1992, leaks in production and transport of methane pushed the global rate of increase to about 1% per year.[18, 19] In the last decade of the twentieth century the rate of increase decreased, but variability of emissions continued. For example, emissions of methane were anomalously high in 1998. The increase most likely resulted from unusually warm, wet weather in 1998 that caused an increase in methane released from warming wetlands along with large forest fires that also released methane, as the trees burned.

Chlorofluorocarbons

Chlorofluorocarbons, or CFCs, are inert, stable compounds that have been or are being used in spray cans as aerosol propellants and in refrigeration units. Use of CFCs as propellants was banned in Canada and the United States in 1978. (CFCs are explicitly designated as toxic substances under the *Canadian Environmental Protection Act*.) Although CFCs are no longer used in spray cans in many countries, however, they have not yet been banned worldwide.

Deliberate release and accidental leaks of CFCs into the atmosphere have been considerable. The rate of increase of CFCs in the atmosphere in the recent past was about 5% per year. It has been estimated that approximately 15–25% of the anthropogenic greenhouse effect may be related to CFCs.[20]

In 1987, 24 countries signed a treaty, the Montreal Protocol, agreeing to reduce and eventually eliminate the production of CFCs and to accelerate the development of alternative chemicals. As a result of this treaty, production of CFCs was nearly phased out by 2000. However, not all countries signed the treaty, and illegal production and use of CFCs continues in some countries. If CFCs had not been regulated by the Montreal Protocol, by the early 1990s they would have become the major contributor to the anthropogenic greenhouse effect.[14] Reduced emissions are evidently responsible for the recent decrease in the concentration of atmospheric CFCs.

The potential climate change from CFCs is considerable, because they absorb in the atmospheric window, as explained earlier; and each CFC molecule may absorb hundreds or even thousands of times more infrared radiation emitted from Earth than is absorbed by a molecule of carbon dioxide. Furthermore, because CFCs are highly stable, their residence time in the atmosphere is long. Even if production of these chemicals is drastically reduced or eliminated within the next few years, their concentrations in the atmosphere will remain significant (although reduced from today's concentrations) for many years, perhaps for as long as a century.[15, 20] (CFCs are discussed in more detail in Chapter 22, which examines stratospheric ozone depletion.)

Nitrous Oxide

Nitrous oxide (N_2O) is increasing in the atmosphere and is probably contributing as much as 5% of the anthropogenic greenhouse effect.[16] Anthropogenic sources of nitrous oxide include agricultural activities (application of fertilizers) and burning fossil fuels. Reductions in the use of fertilizers and burning fossil fuels would reduce emissions of nitrous oxide. However, this gas also has a long residence time; even if emissions were stabilized or reduced, elevated concentrations of nitrous oxide would persist for at least several decades.[17]

In summary, as indicated in Table 20.1, carbon dioxide contributes between 50% and 60% of the anthropogenic greenhouse effect. The rest of the human-made effect comes from trace gases, the most important of which are the CFCs and methane. These trace gases contribute between 27% and 45% of the anthropogenic greenhouse effect, and they have accumulated in the atmosphere at much faster rates than carbon dioxide.

20.5 Science of Climate Change

As we suggested earlier, there is good reason to argue that increases in carbon dioxide and other greenhouse gases are related to an increase in the mean global temperature of Earth.[10, 20] Over the past 160,000 years, a strong correlation has existed between the concentration of atmospheric CO_2 and global temperature (Figure 20.17). When CO_2 has been high, temperature has also been high; conversely, low concentrations of CO_2 have correlated with a low global temperature.

Global models of the climate suggest that warming as a result of anthropogenic increases in greenhouse gases will occur (some scientists, however, find faults with the models). According to the models, the average global temperature will rise 1.5–6.0°C by 2100. In the most optimistic case, if there are large reductions in emissions of greenhouse gases, the warming may be closer to 1°C.[1]

In order to understand global climate change, we need to consider both the positive and negative feedbacks that occur on Earth, on the sun, and in the atmospheric system. We also need to consider the major variables that may cause (force) a global change in temperature, including solar emissions, volcanic eruption, El Niño warming, and anthropogenic input of particulates and greenhouse gases. The process of changing global temperature is called forcing.

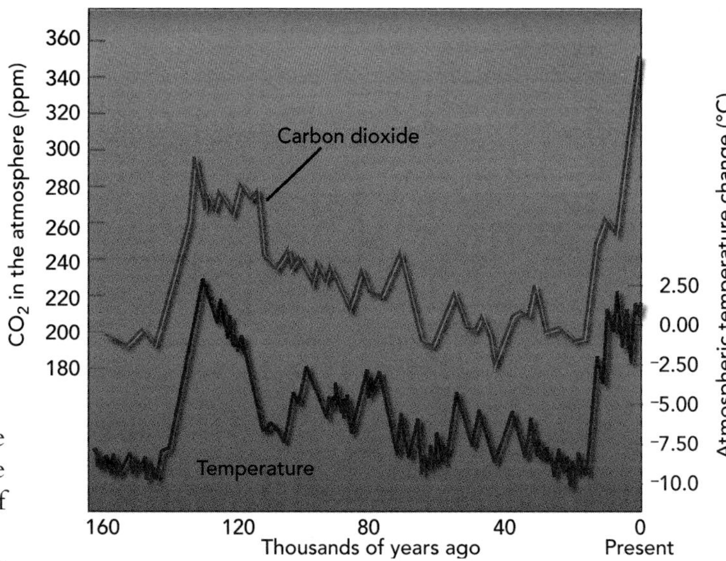

Figure 20.17 • Inferred concentration of atmospheric carbon dioxide and temperature change for the past 160,000 years. The relation is based on evidence from Antarctica and indicates a high correlation between temperature and CO2. [*Source*: Adapted from S. H. Schneider "The Changing Climate," © September 1989, *Scientific American*, Inc. All rights reserved. vol. 261, p. 74.]

Negative and Positive Feedbacks

Greenhouse warming is very complex. The warming effect initiates both negative- and positive-feedback loops that can offset any temperature increase or enhance it. Negative-feedback loops are self-regulating and help stabilize global temperature change in response to a warming circumstance. Positive feedbacks are self-enhancing; thus, a process that causes an increase in global temperature leads to further increases in temperature. We first discussed positive and negative feedbacks with respect to Earth systems and changes in Chapter 3.

Several of the potential negative and positive feedbacks concerning climate change are shown in Figure 20.18.[21] It is important to remember that if the negative feedbacks are strong and persistent, climate change may not occur. In contrast, if the negative-feedback systems are weak relative to positive feedback, warming is likely to occur more readily.

Both negative- and positive-feedback processes occur simultaneously in the atmosphere. Which are more important? The nod goes to positive feedbacks, because the mean global temperature is increasing. A great deal of research is currently being carried out to better understand negative-feedback processes associated with clouds and their water vapour. Many discussions on the greenhouse effect state that if Earth's atmosphere did not trap heat, our planet would be approximately 33°C cooler at the surface; and as a result, all water would be frozen. However, since water vapour is the major greenhouse gas in the atmosphere, no greenhouse effect implies no (or very little) water vapour in the atmosphere. This implies no clouds, which would lead to a substantial reduction in the atmospheric reflection of incoming sunlight, which would result in warmer surface temperatures on Earth. The dual role of atmospheric water vapour as both a negative and a positive feedback with respect to climate change is extremely important to understanding possible climatic modifications created by an anthropogenic greenhouse effect.

Solar Forcing

The sun is responsible for heating the Earth; therefore, in evaluating climatic change, we must consider solar variation a possible cause. As mentioned earlier, the Milankovich cycles are consistent with most of the long-term variability of climate that has occurred during the past several hundred thousand years of Earth's history. However, the Milankovich cycles by themselves are not sufficient to produce the magnitude of observed climate changes.

When we examine the history of climate during the past thousand years, the variability of solar energy plays a role. Isotopes from the outer layer of the sun are carried to Earth by the solar wind and deposited in

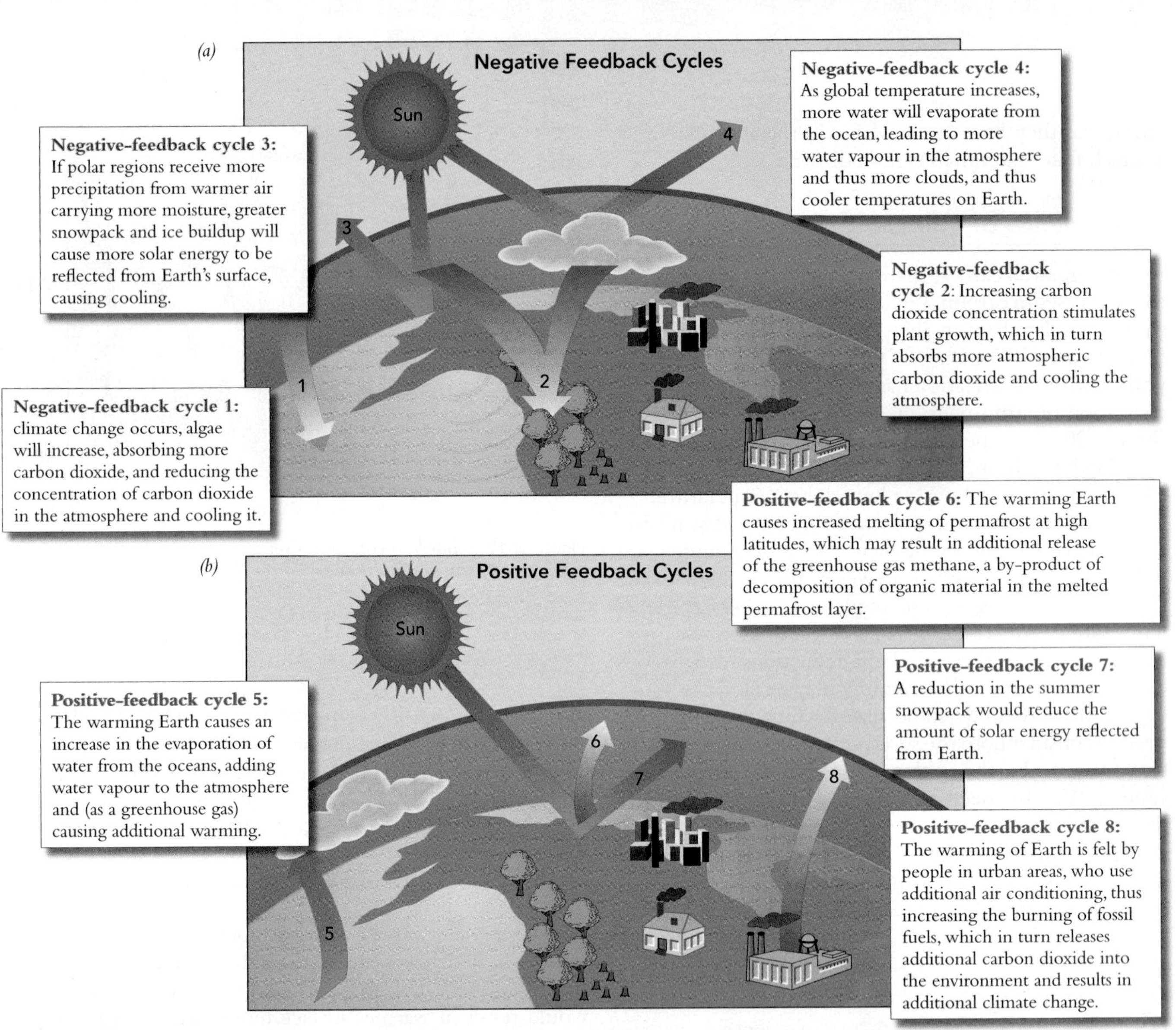

Figure 20.18 • *(a)* **Negative-feedback cycles and** *(b)* **positive-feedback cycles associated with the greenhouse effect.** [*Source*: Modified from J. R. Luoma, "Gazing into Our Greenhouse Future," *Audubon* 93, no. 2 (1991):57.]

glacial ice. Analysis of such isotopes in ice cores from glaciers reveals that during the medieval warm period, from approximately 1100 to 1300, the amount of solar energy reaching Earth was relatively high, comparable to the warming we see today. Evaluation of the data also suggests that minimum solar activity occurred during the fourteenth century, coincident with the beginning of the Little Ice Age (Figure 20.5). Thus, it appears that the variability of solar input of energy to the Earth can indeed explain some of the climatic variability we have experienced in the past thousand years. The effect, however, is relatively small.[9]

Aerosols and Volcanic Forcing

An aerosol is a particle with a diameter less than 10 μm. These particles tend to remain in the atmosphere for a long time, because of the effects of collisions with air molecules dominate over gravity. In contrast, bigger particles (with diameters greater than 10 μm) drop out of the atmosphere faster because of gravity. Emissions of aerosols to the atmosphere by human processes have increased significantly since the Industrial Revolution, and volcanic eruptions have periodically released aerosols as well. Recent research has indicated that aerosols emitted from coal (sulphates) and volcanic eruptions may contribute to global cooling, because sulphates act as seeding for clouds. The aerosol particles provide surfaces on which water can condense; clouds form and reflect incoming solar energy. The aerosol particles also provide a "dust veil" that reflects a significant amount of sunlight.

In the early 1990s, net cooling owing to sulphate aerosols from burning fossil fuels offset some of the climate change expected from the anthropogenic greenhouse effect.[22, 23] However, emissions of the sulphate SO_2 (sulphur dioxide) from human activity are decreasing because of programs to reduce air pollution.

The effect of reduction of SO_2 emissions will mean a strengthening of the greenhouse effect from CO_2.

Reflection of solar radiation from air pollution particles has apparently reduced incoming solar radiation by as much as 10% in some regions, offsetting up to 50% of expected warming due to increases in greenhouse gases. The process of the reduction of incoming solar radiation by reflection from particles and their interaction with water vapour in the atmosphere (especially clouds) is known as **global dimming**.

Tremendous explosions from Mount Pinatubo in the Philippines in 1991 sent volcanic ash to elevations of 30 km into the stratosphere (Figure 20.19). The aerosol cloud of ash, with 20 million tonnes of sulphur dioxide, remained in the atmosphere circling Earth for several years. The particles of ash and sulphur dioxide scattered incoming solar radiation, resulting in a slight cooling of the global climate during 1991 and 1992. Calculations suggest that aerosol additions to the atmosphere from the Mount Pinatubo eruption counterbalanced the warming effects of greenhouse gas additions through 1992. However, by 1994, most aerosols from the eruption had fallen out of the atmosphere, and global temperatures had returned to previous higher levels. Volcanic climate forcing from pulses of volcanic eruptions is also believed to have significantly contributed to the cooling associated with the Little Ice Age, from about 1450 to 1850 (see Figure 20.6).[9]

El Niño

Another natural perturbation that affects global climate is the occurrence of El Niño events (see A Closer Look 20.2). During an El Niño, the normal conditions of equatorial upwelling of deep oceanic waters in the eastern Pacific are diminished or eliminated. Upwellings release carbon dioxide to the atmosphere as carbon dioxide–rich deep water reaches the surface. El Niño events thus reduce the amount of oceanic carbon dioxide outgassing, influencing the global carbon dioxide cycle. Climatic models generally predict that as the Earth warms, El Niño events will become more common.[1]

The 1982–1983 El Niño event was particularly strong. It is thought to have produced climatic events such as floods and droughts that killed several thousand people and caused billions of dollars in damage to crops, structures, and other facilities. El Niño events may change the patterns of the upper troposphere (jet streams) to produce wetter winters and larger storms in parts of North America. El Niño conditions developed in 1991–1993 and again in 1997–1998; although they were not as strong as the 1982–1983 event, they contributed to a slight climate change.

Recognition of El Niño events is important for understanding processes that could affect the global climate. The events also provide data for global models that predict change. Volcanic eruptions and El Niño events are relatively short term, and therefore provide important information from which models can be calibrated and tested.

Methane Forcing

Methane (CH_4), or natural gas, is a strong greenhouse gas contributing as much as 20% of the anthropogenic greenhouse effect. There may also be a periodic natural methane forcing that causes rapid warming to end a glacial period. (See the discussion of methane hydrates in Chapter 17.)

One mechanism that releases large amounts of methane is linked to lowering of sea level. Remember from Chapter 17 that methane hydrates remain stable in deep (high-pressure) cold water. During glacial times, sea level drops because of the huge volume of water stored in glaciers. The lower sea level causes water pressure at the bottom of the sea to be reduced. Under these conditions, methane hydrates could become unstable and be released into the water. The gas, once released from the ocean bottom, would rise and enter the atmosphere. A second process favouring the release of methane from hydrates is the warming of ocean water. Warming of deep water occurs at the beginning of interglacial periods when sea level is low.[25]

If sufficient methane is released, either by lowering of the sea level or warming of the water, the gas may cause climate change, contributing to ending the glacial

Figure 20.19 • **Eruption column from Mount Pinatubo in the Philippines during 1991 eruptions.** Such eruptions injected vast amounts of dust and sulfur dioxide as high as 30 km into the atmosphere.

A CLOSER LOOK 20.2

El Niño

El Niño became a household word during the winter of 1997–1998, when it was blamed for everything from tornadoes and thunderstorms in Florida to catastrophic fires in Indonesia (see Chapter 21). The term El Niño means "little boy" in Spanish and refers in particular to the Christ Child, because the event often begins off the coast of South America near Christmas time.

El Niño events disrupt the ocean-atmosphere system in the tropical Pacific. They are in part responsible for weather phenomena that can cause billions of dollars in property damage and the loss of thousands of human lives. Occurring at intervals of two to seven years, El Niño events typically last for 12 to 18 months. They start with the weakening of east-to-west trade winds and the warming of eastern Pacific Ocean waters. As a result, tropical rainfall shifts from Indonesia to South America, as shown in Figure 20.20.

Let's look at this process in more detail. Under non–El Niño conditions, trade winds blow west across the tropical Pacific. The warm surface water in the western Pacific tends to pile up, so that the sea surface can be as much as 0.5 m higher at Indonesia than at Peru. In contrast, during El Niño, the trade winds weaken and may even reverse. As a result, the eastern equatorial Pacific Ocean becomes unusually warm, and the westward moving equatorial ocean current weakens or reverses. The rise in temperature of sea surface waters off the South American coast inhibits the upwelling of nutrient-rich cold water from deeper levels; the upwelling normally supports a diverse marine ecosystem and major fisheries. Rainfall follows warm water eastward during El Niño years, so there are high rates of precipitation and flooding in Peru, while droughts and fires are commonly observed in Australia and Indonesia. Warm ocean water provides an atmospheric heat source, so El Niño changes global atmospheric circulation, which causes changes in weather in regions that are far removed from the tropical Pacific.

It is important to remember that El Niño events are natural phenomena and part of a dynamic system involving the coupling of Earth's atmosphere and ocean. El Niño events can alternate with contrasting phenomena called La Niña events. La Niña events are characterized by unusually cool ocean water temperatures, and the effects are opposite those of El Niño. For example, during a 1998–1999 La Niña, winter temperatures were cooler than normal in coastal British Columbia and the U.S. Northwest. A La Niña event does not necessarily follow every El Niño event; but together, they constitute a natural cycle of ocean-atmospheric disruption with global consequences.[24]

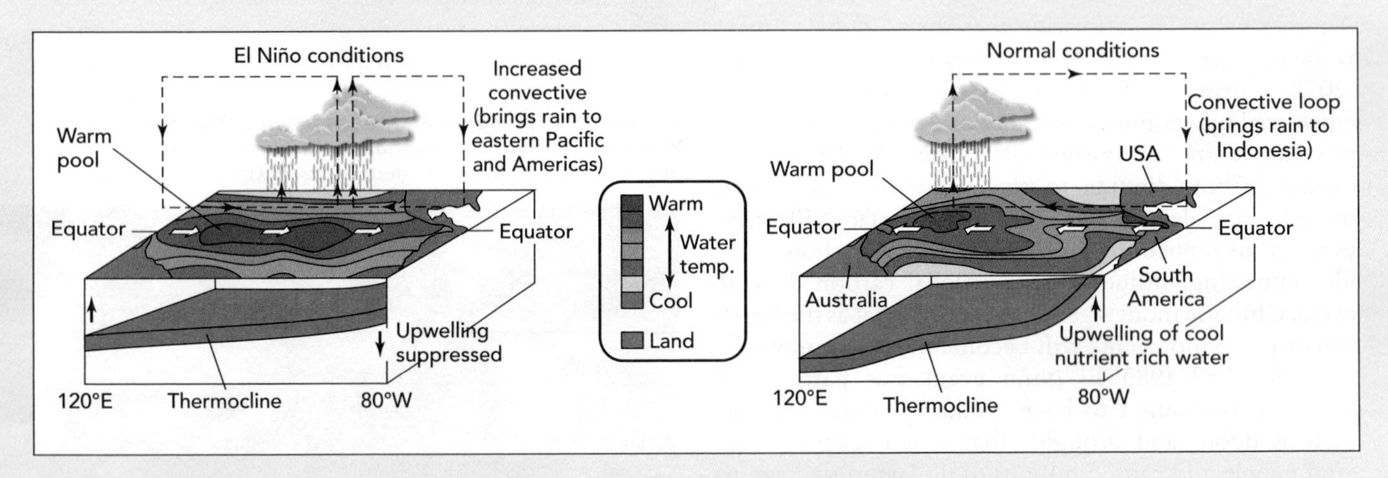

Figure 20.20 • Conceptual diagram comparing selected normal and El Niño conditions. (*Source*: NOAA.)

period. For example, release of methane may have been responsible for a rapid increase in the global temperature that occurred over a relatively short period at the end of the last ice age about 15,000 years ago.

Anthropogenic Forcing from Greenhouse Gases

Until recently, there was controversy concerning whether the anthropogenic component of climate change was significant or not. The question now seems to have been answered, and we can state that there is a significant human footprint on observed climate change. A recent study evaluated climate change over the past thousand years, allowing late-twentieth-century warming to be placed within a historical context. The study used a mathematical model to remove major natural forcing mechanisms, including solar radiation and volcanic eruptions, so that climatic forcing by greenhouse gases could be directly estimated. The research established that the natural variability in the climate system in the past thousand years is far less than what occurred at the end of the twentieth century. That is, present warming far exceeds natural variability and closely agrees with the warming predicted by global circulation models that take account of increasing greenhouse gases.[9] Results of the modeling are shown in Figure 20.21. Notice that the late-twentieth-century rise in temperature far exceeds the zone of maximum variability that could result from forcing mechanisms such as solar variability, volcanic activity, and global dimming.

It appears that the major scientific issues concerning climate change have been resolved. Significant climate change as a result of human activity is occurring. We now discuss some potential effects of and possible adjustments to climate change. What we ultimately decide to do about climate change will reflect our values.

20.6 Potential Effects of Climate Change

If we continue emitting large amounts of carbon dioxide into the atmosphere, it is estimated that by 2030 the concentration of carbon dioxide in the atmosphere will have doubled from pre–Industrial Revolution concentrations. The average global temperature (according to mathematical models) will have risen approximately 1°–2°C, with significantly greater temperature change at the polar regions.[16] Climate change causes greater temperature increases at polar regions in part because of a positive-feedback mechanism; sea ice melts from warming, and water reflects much less light than white ice, resulting in enhanced warming. Solar energy that would have been reflected by sea ice is absorbed by the ice-free water. This mechanism is part of what is termed **polar amplification**.

Figure 20.21 • Global temperature change during the past 1,000 years. Solar and volcanic forcing have been removed, leaving anthropogenic forcing from greenhouse gases. [*Source*: Modified from T. J. Crowley, 2000. "Causes of Climate Change over the Past 1000 Years," *Science* 289 (2000):270–277.]

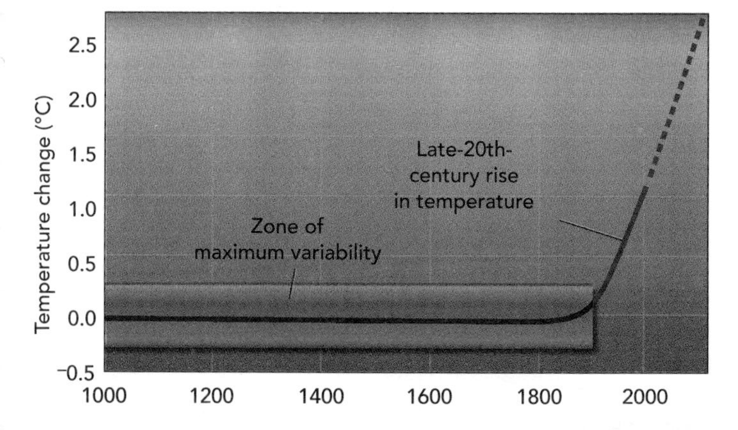

Specific effects of climate change are difficult to predict from global models. However, it is expected that warming will likely have the following consequences:[1]

- As a result of regional changes in climate, semi-arid land areas will become drier, while other regions will become wetter.
- World distribution of biomes will likely change. Some, such as alpine tundra, may be lost, while others, such as midlatitude deserts, may expand.
- Agricultural production in some regions will decrease.
- The incidence of diseases such as malaria and dengue fever will increase in tropical countries.
- Rising sea levels and accompanying coastal erosion will threaten low-lying islands and will displace tens of millions of people worldwide.
- The biosphere will change as a result of damage to ecosystems, from tropical coral reefs to birds and bears in the Arctic.

Next, we briefly discuss selected aspects of these consequences.

Changes in Climate

Various estimates have been made of what changes in annual temperature and precipitation are likely to occur as a result of climate change. Figure 20.22 shows changes expected to occur by 2050 assuming concentrations of greenhouse gases increasing at about 1% per year.[1, 26] In central North America (the grain-growing region), warming is expected to vary from approximately 2° to 4°C, with a small increase in precipitation. As a result, soil moisture may decrease in the summer by as much as 20%. Clearly, this could have a significant effect on the grain-growing areas of Canada and the United States.

Changes in relative runoff of water linked to changes in temperature and precipitation for 2080 are shown in Figure 20.23. Lower runoff is projected for much of Mexico, South America, southern Europe, India, southern Africa, and Australia. It is important to keep in mind that these projections are based on global circulation models that are controversial and subject to variability. Nevertheless, most of the models predict changes in the directions indicated, and as a result are being taken seriously by both scientists and policymakers.

As already suggested, rising global temperatures are expected to have a significant influence on patterns of rainfall, soil moisture, and other climatic factors related to agricultural productivity. Studies using global circu-

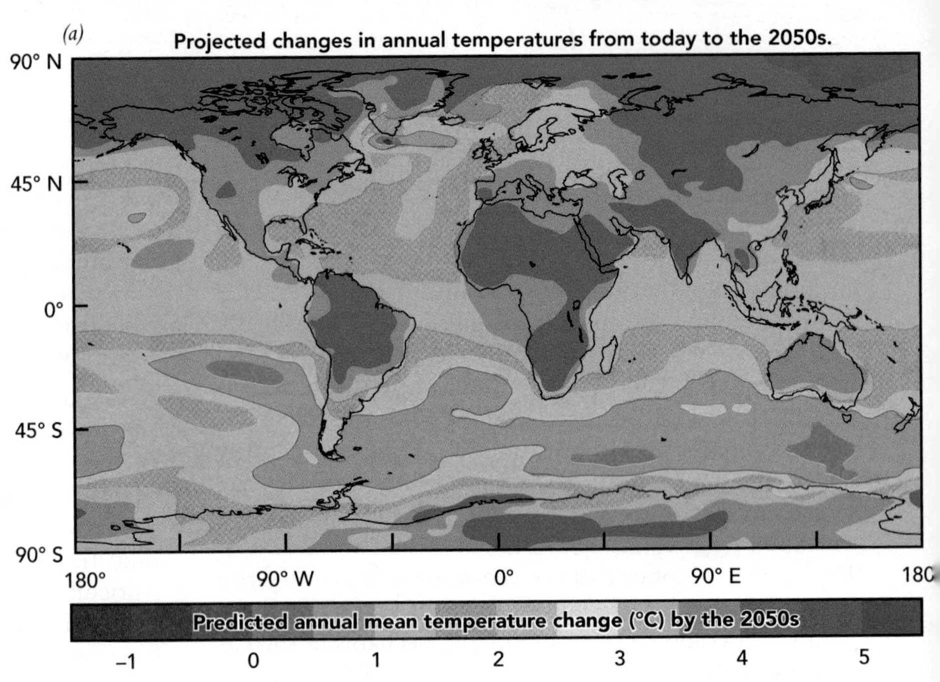

Figure 20.22 • *(a)* Projected changes in annual temperatures from today to the 2050s. Notice that changes are greatest at the polar regions. *(b)* Projected changes in annual precipitation from today to the 2050s. Highest increases are near the equator. Modeling that predicted these changes assumes an increase in greenhouse gas of about 1% per year. [*Source*: Adapted from Met Office, Hadley Center for Climate Prediction and Research, in R.T. Watson, presentation at the Sixth Conference of the Parties of the United Nations Framework Convention on Climate Change, Intergovernmental Panel on Climate Change, November 13, 2000. Accessed December 1, 2000, www.ipcc.ch/press/sp-COPG.htm.]

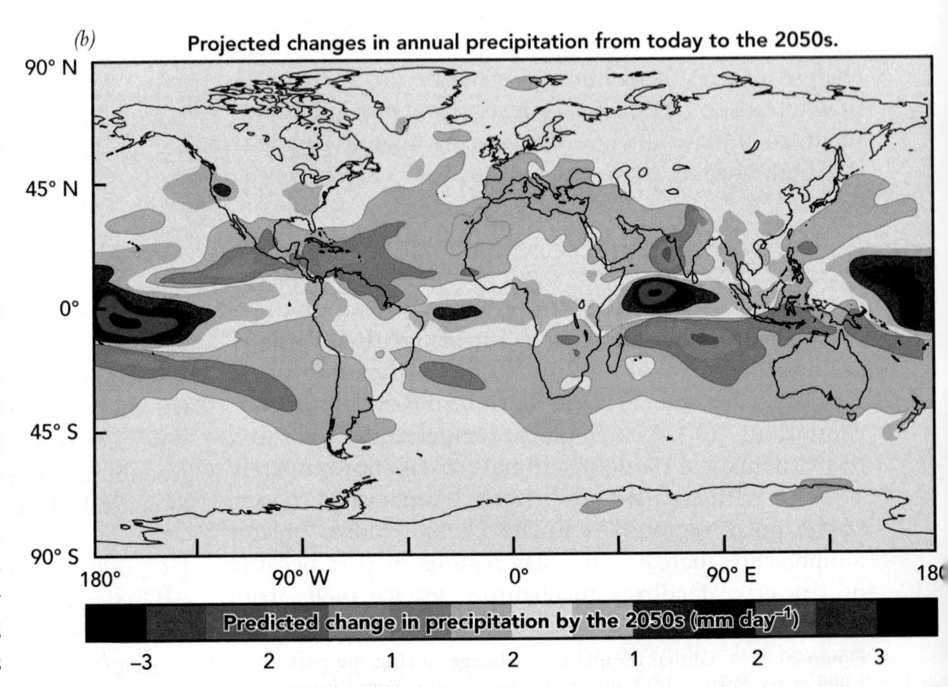

lation models to predict patterns for the Northern Hemisphere suggest that some of the more northern areas, such as Canada and Russia, may become more productive. Although climate change might move North America's prime farming climate north from the midwestern United States to the region of Saskatchewan, the U.S. loss would not simply be translated into a gain for Canada. Saskatchewan would have the optimum climate for growing, but the Canadian soils are somewhat thinner and less fertile than those in the U.S. Midwest. Therefore, a climate shift could have serious negative effects on midlatitude food production. In addition, lands in the southern part of the Northern Hemisphere may become more arid; and as a result, soil moisture relationships will change.

People are anxious when uncertainty exists and they are particularly anxious when that uncertainty involves their food supply. There is real concern that hydrologic changes associated with climatic change resulting from climate change may seriously affect the global food supply. Today, 800 million people are malnourished. As the human population increases, so will demand for food. During the twenty-first century, agricultural production in many subtropical and tropical regions (especially in South America and Africa) will likely decrease (Figure 20.24) if the mean annual temperature increases by more than 2°C. Unfortunately, millions of the poorest people on Earth live in the tropics and subtropics, where risk of hunger is the greatest.[1]

There is also concern that climate change will alter normal weather and climatic patterns, including the frequency or intensity of violent storms. This possibility may be more important than changes in climate. The hypothesis is that warming ocean waters could feed more energy into high-magnitude storms, such as cyclones and hurricanes, causing a significant increase in their frequency or intensity. Approximately half of Earth's human population live in coastal areas. Potential problems are exacerbated by the fact that many of these areas are low lying and are experiencing rapid population growth. In addition, greenhouse warming is expected to result in wetter winters, hotter and drier summers, an increased frequency of large storm events, and an increased possibility of droughts in the northern temperate latitudes.[27]

In summary, climate change could affect people and ecosystems on Earth in various ways, often for the worse. These changes may affect hydrology, crop production,

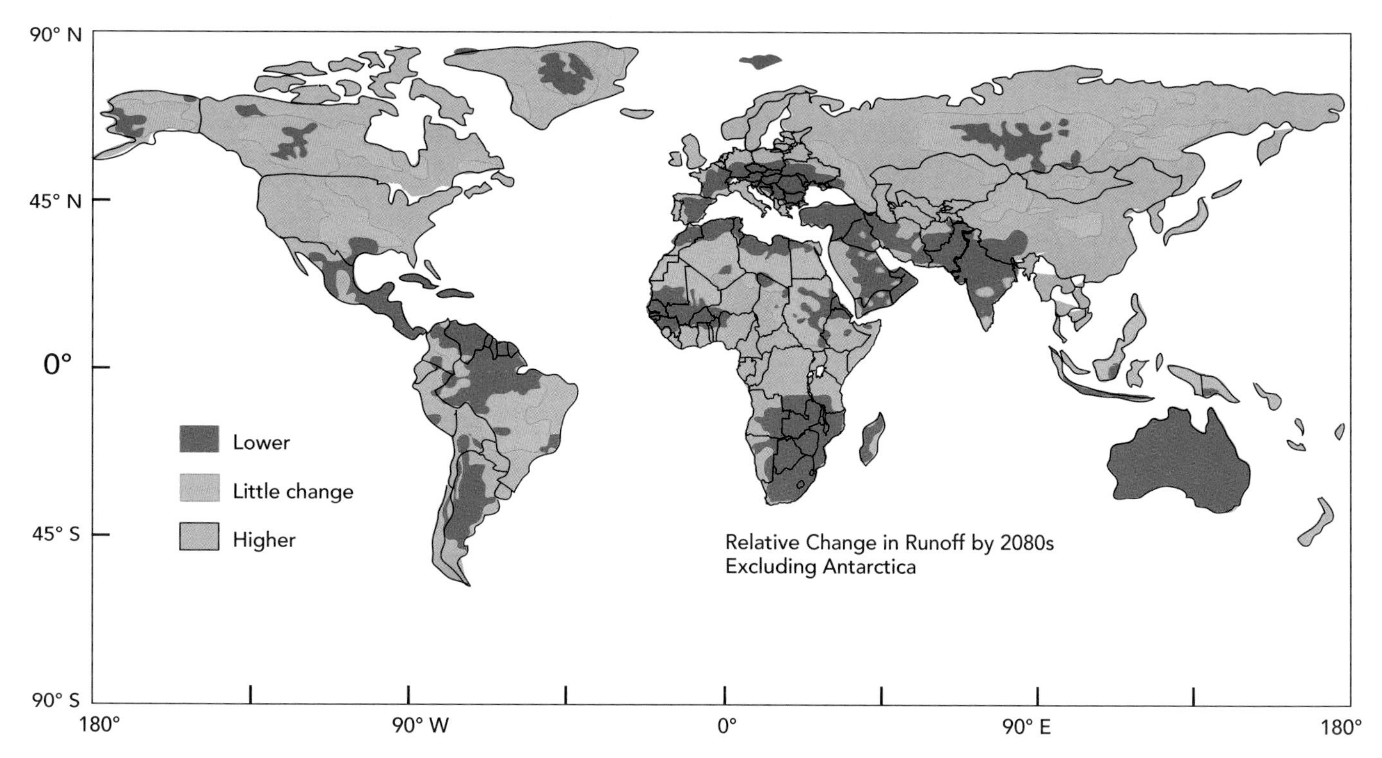

Figure 20.23 • **Predicted relative change in runoff of water by the 2080s linked to changes in temperature and precipitation** (Figure 20.22a, b). [*Source:* Adapted from University of Southampton, in R. T. Watson, presentation at the Sixth Conference of the Parties of the United Nations Framework Convention on Climate Change, Intergovernmental Panel on Climate Change, November 13, 2000. Accessed December 1, 2000, at www.ipcc.ch/press/sp-COPG.htm.]

forestry, and human health. Midlatitude climate zones could shift northward by as much as 550 km over the next century. At this rapid rate, some tree species may be nearly eliminated. Furthermore, the expected expansion of tropical climate zones will lead to an increase in tropical diseases such as malaria, dengue fever, yellow fever, and viral encephalitis.[28]

Rise in Sea Level

A rise in sea level is a potentially serious problem related to climate change. As mentioned, about half of the people on Earth live in the coastal zone, and about 50 million people per year experience flooding due to storm surges. As sea level rises and population increases, the number of people vulnerable to coastal flooding will increase. For example, two cyclones that hit highly populated Bangladesh in the last 25 years killed more than 400,000 people and caused over $1.6 billion in property damage (see Chapter 4). The double impact of rising sea level and more frequent and powerful cyclones and other tropical disturbances (owing to warmer oceans, as discussed earlier) would have devastating effects on people in developing countries.

Although a precise estimate of the total potential rise in sea level is not possible at this time, there is a consensus that the level of the sea will in fact rise. In fact, sea level along much of the North American coast is already rising at a rate of 1 to 2 mm/yr.[29] The cause for the rise is thought to be twofold: thermal expansion of warming ocean water (the primary cause) and melting of glacial ice (a secondary cause).

Various models predict that sea level may rise anywhere from 20 cm to approximately 2 m in the next century; the most likely rise is probably 20–40 cm. One estimate is that sea level will likely rise 15 cm by 2050 and 34 cm by 2100. When other factors—such as land subsidence and compaction, groundwater depletion, and natural climate variation—are considered, some coastal regions could experience a sea level rise of 45–55 cm by 2100.[29]

Such a change would have significant environmental impacts. It could cause increased coastal erosion on open beaches of 50–100 m, making buildings and other structures in the coastal zone more vulnerable to damage from waves generated by high-magnitude storms. It could also cause a landward migration of estuaries and salt marshes, lead to loss of coastal wetlands (see Chapter 18), and put additional pressure on human structures in the coastal zone.[16] The people and wetlands at risk from a 44-cm rise in sea level are shown by region on Figure 20.25. Finally, groundwater supplies for coastal communities may be threatened by saltwater intrusion should sea levels rise (see Chapter 19).

A rise in sea level of approximately 1 m would have even more serious consequences, threatening the existence of some low-lying islands. People would have to significantly alter the coastal environment to protect investments, and communities would be forced to choose either to make very heavy expenditures to

Figure 20.24 • Predicted changes in average yield of wheat, corn, and rice in the 2050s linked to changes in temperature and precipitation (Figure 20.22*a, b*). [*Sources*: Jackson Institute, University College, London; Goddard Institute for Space Studies; International Institute for Applied Systems Analysis; and in R. T. Watson, presentation at the Sixth Conference of the Parties of the United Nations Framework Convention on Climate Change, Intergovernmental Panel on Climate Change, November 13, 2000. Accessed December 1, 2000, at www.ipcc.ch/press/sp-COPG.htm.]

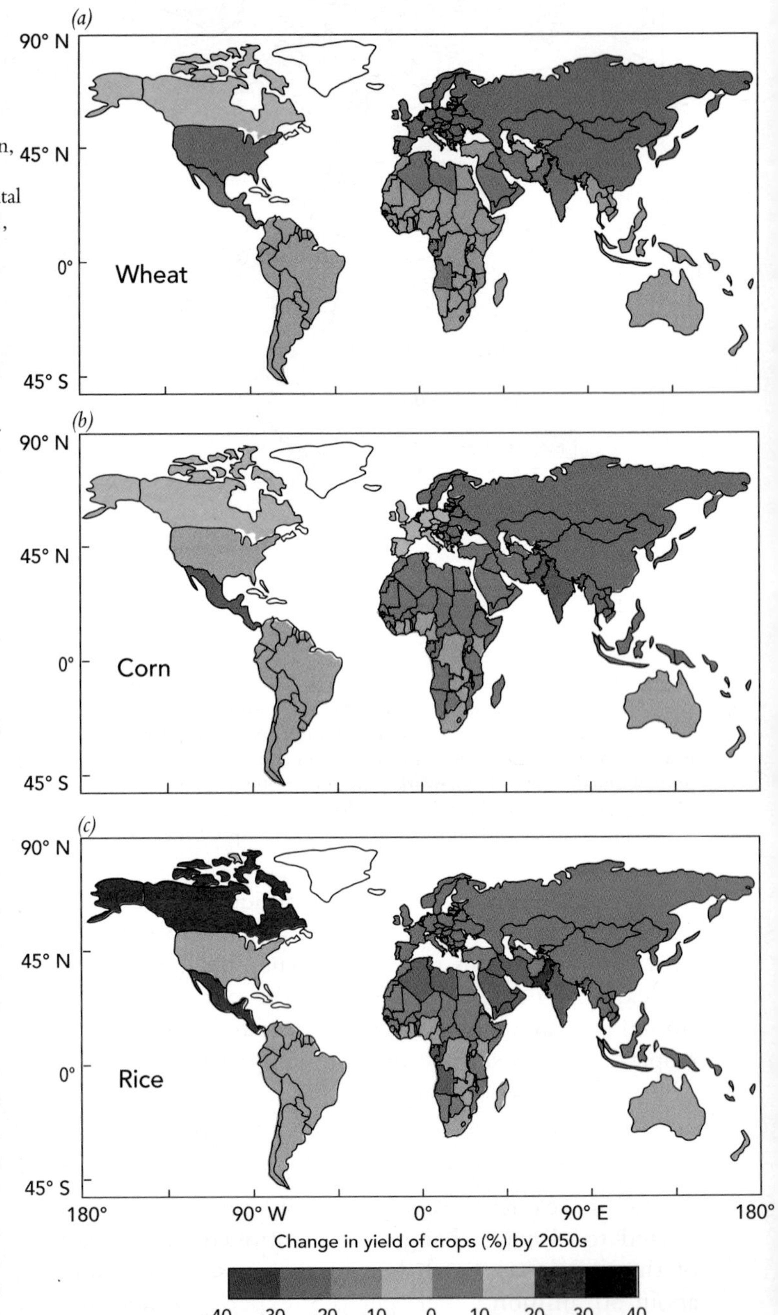

control coastal erosion or to allow considerable loss of property.[16]

It seems inevitable that a rise in sea level will increase coastal erosion and lead to further investments to protect cities in the coastal zone. Construction of seawalls, dikes, and other erosion-controlling structures will become more common as coastal erosion threatens urban property. In more rural areas, where development is set well back from the coastal zone, the most likely response to rising sea level will be to adjust to the erosion that occurs. Coastal erosion is a difficult problem that is very expensive to mitigate. In many cases, it is best to allow erosion to take place naturally where feasible and only defend against coastal erosion where absolutely necessary.

Changes in Biosphere

A growing body of evidence indicates that climate change is probably initiating a number of changes in the biosphere, sometimes threatening ecological systems and people. These changes include shifts in the ranges of plants and animals. Some such shifts are summarized in Figure 20.26.

One change mentioned on Figure 20.26 is a shift in the range of mosquitoes that carry diseases including malaria and dengue fever. Malaria, which causes chills and fever, is particularly worrisome, as it kills about 3,000 people per day, mostly children. Projected climate change will increase the land area where the disease can be transmitted. Today, that area contains 45% of the world's population. With climate change, malaria could threaten 60% of the world's population.[30, 31]

The possible relationship between climate change and the emergence of the West Nile virus in 1999 is an example of linkages between physical, biological, and social systems—the essence of environmental science. How the West Nile virus arrived in North America is not known, but its spread from mosquitoes to birds to people was enhanced by a warm winter followed by a dry spring and a hot, wet summer; these conditions are symptoms of climate change. This is how it works:[31]

- During a mild winter, more mosquitoes survive in sewers as well as in still water in a variety of locations, including ponds, waste cans, and abandoned tires.
- During a dry spring, surface water sites, such as small ponds, decrease in size, concentrating both birds and mosquitoes at the water sites.
- Mosquitoes infected with the virus bite uninfected birds, passing the virus on to them.
- Infected birds are bitten by uninfected mosquitoes, passing the virus to more mosquitoes.
- Hot summer months, with their warm air and heavy rains, cause the mosquito population to mature and grow rapidly. More mosquitoes become infected and pass on the virus both to birds and, eventually, to people.

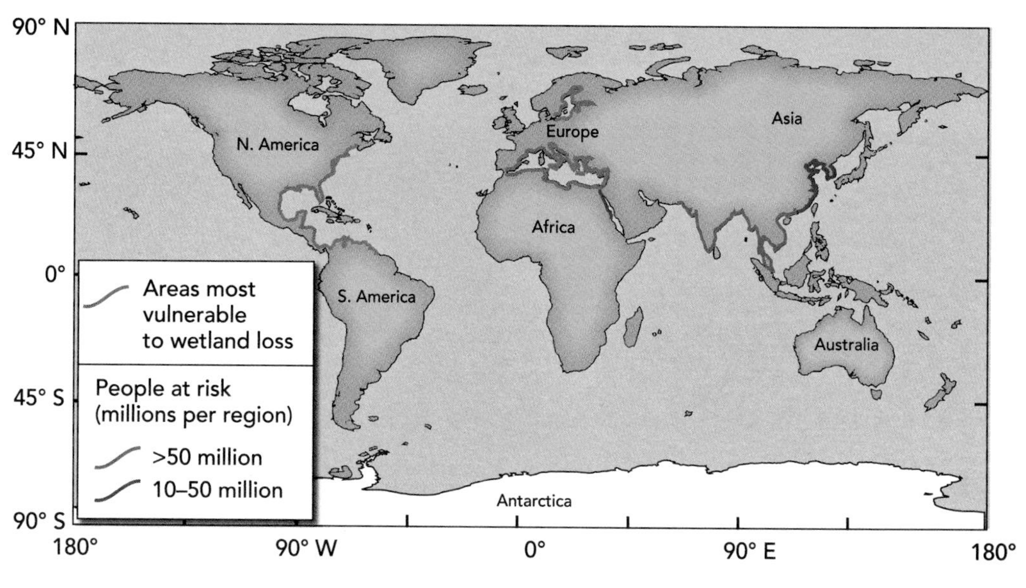

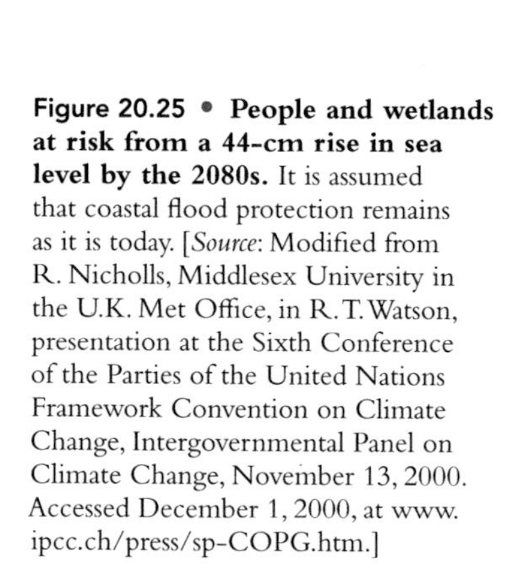

Figure 20.25 • People and wetlands at risk from a 44-cm rise in sea level by the 2080s. It is assumed that coastal flood protection remains as it is today. [*Source*: Modified from R. Nicholls, Middlesex University in the U.K. Met Office, in R. T. Watson, presentation at the Sixth Conference of the Parties of the United Nations Framework Convention on Climate Change, Intergovernmental Panel on Climate Change, November 13, 2000. Accessed December 1, 2000, at www.ipcc.ch/press/sp-COPG.htm.]

Thus, the unusual weather, possibly from climate change, results in an increase in the mosquito population. Once the virus appears, a strong positive-feedback cycle between birds and mosquitoes (Figure 20.27) increases the number of infected mosquitoes, eventually producing a health problem for people.

What is happening to some fish species is also of concern. The alewife is a saltwater species that has found its way up the St. Lawrence Seaway and other rivers into freshwater lakes and streams. Already under added physiological stress from the low-salinity environment, the alewife needs very specific temperature conditions to survive and spawn successfully. Recent alewife die-offs in the Great Lakes have been linked to water temperature changes associated with climate change.

As another example, as early as 1989, Edith's checkerspot butterflies (*Euphydryas editha*; Figure 20.28) in some high meadows of the Sequoia National Forest

Figure 20.26 • Selected examples of how rising global temperatures are shifting the range of plants and animals in North America. (*Sources*: Modified from S. Levy, "Wildlife on the Hot Seat," *National Wildlife* 38, no. 5(2000):20–27; and N. Holmes, "Has Anyone Checked the Weather [Map]?" *Amicus Journal* 21, no. 4(2000):50–51.)

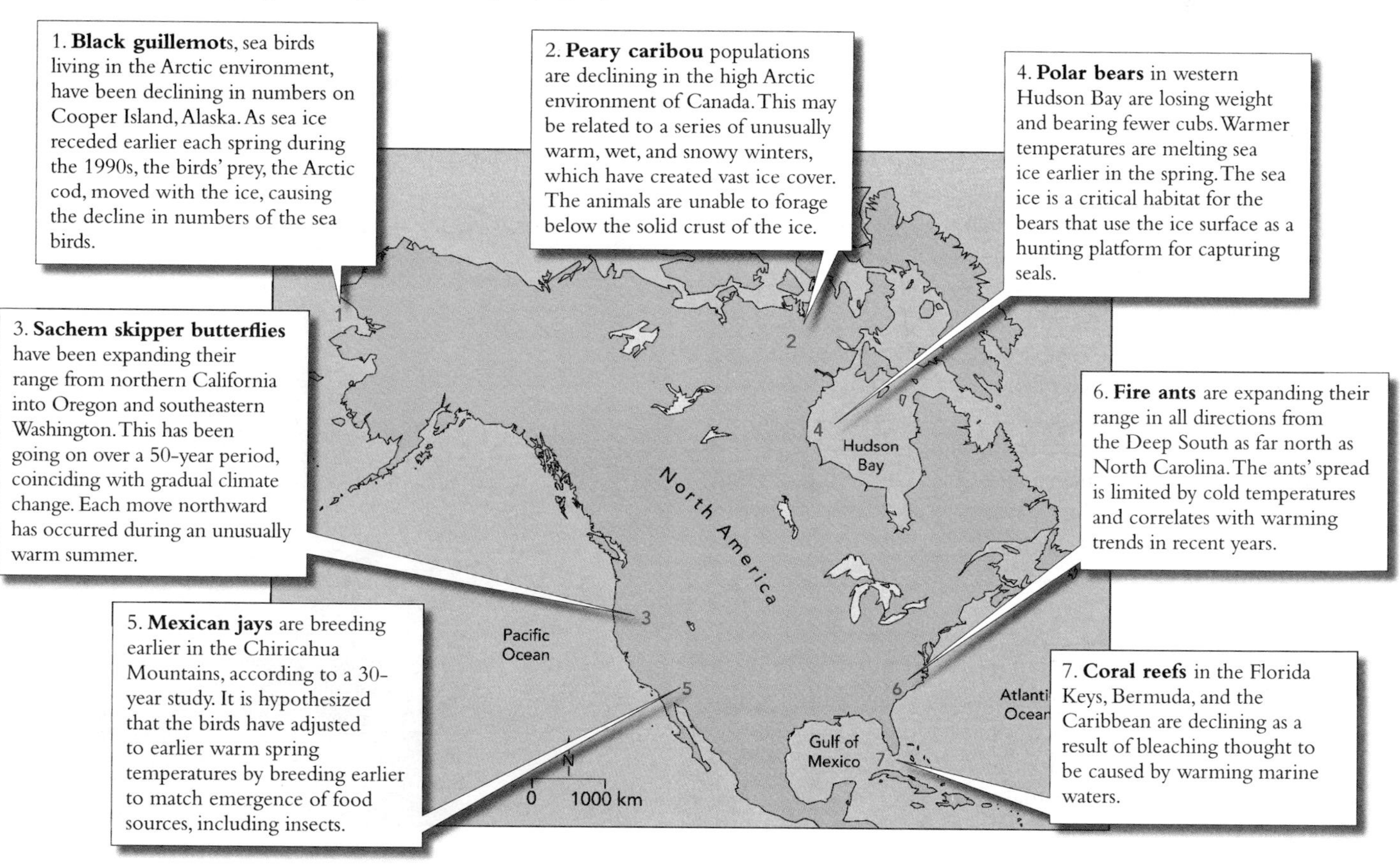

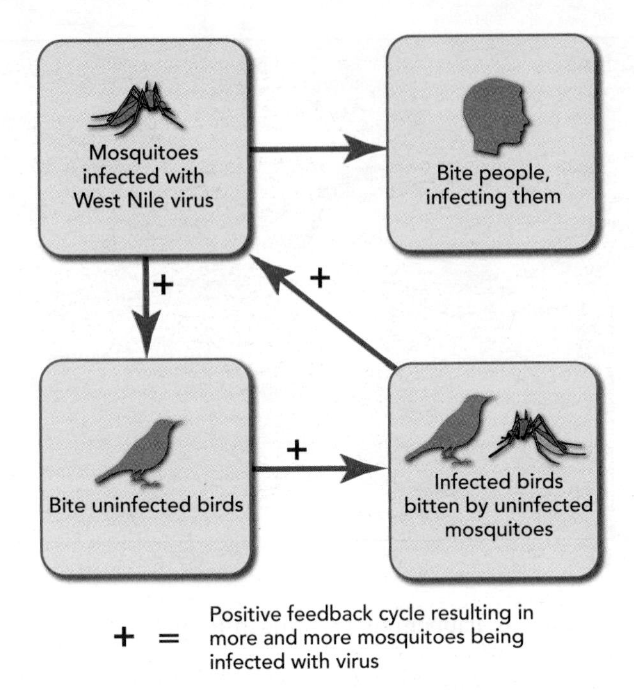

Figure 20.27 • **As a result of a positive-feedback cycle, more and more mosquitoes are infected with West Nile virus, eventually infecting people.**

in California died, not from poisoning or disease, but because the snowpack melted early and the warm temperatures caused the butterflies to emerge before the nectar-rich plants they feed on were available.

As a final example of ecological effects of climate change, consider the declining numbers of black guillemots in Canada's high arctic. Increases in temperature in the 1990s caused the sea ice to recede farther from locations such as Lancaster Sound, off Baffin Island, and Cooper Island, Alaska, each spring. The recession of sea ice often occurs before the black guillemot chicks were mature enough to survive on their own. (Figure 20.29). Parent birds feed on Arctic cod found under the sea ice, then return to the nest to feed their chicks. The distance from feeding grounds to nest must be less than about 30 km; but in recent years, the ice in the spring has been receding as much as 250 km from the island before the chicks are able to leave the nests. As a result, the black guillemots lose an important source of food. Evidence from Cooper Island shows declining numbers of the bird; their future will clearly depend on future springtime weather. Too warm, and the birds may disappear. Too cold, and there may be too few snow-free days for breeding; in this case, too, they will disappear. Thus, the survival of black guillemots in Canada's north is precarious, given our changing global climate.[2]

An increase in the risk of species extinction is a potential change to the biosphere resulting from climate change. This results because warming may cause a shifting of climate that impacts the distribution and abundance of a species. As climate changes, a species may attempt to move to a climatically suitable area. Some will be able to do so, but others will not. Furthermore, efforts of some species to relocate may be restricted by other environmental disruptions, such as habitat loss or fragmentation of habitat from land-use change.[32]

What is apparent from our examples is that the ecological consequences of climate change are not the concern of a distant future. Changes in the biosphere and ecological communities as a result of warming are happening now. As individual species in an ecosystem are affected, changes will occur in other species. This is the principle of environmental unity discussed in Chapter 3.

20.7 Adjustments to Potential Climate Change

Before we consider possible adjustments to climate change, we need to recognize that the problem is complex. We learned in Chapter 3 that solving complex environmental problems is often difficult due to exponential growth, lag times, and the possibility of irreversible change. We now return to that concept when considering climate change. Peak emissions of CO_2 will

Figure 20.28 • **Edith's checkerspot butterflies** in some locations are emerging before the plants they feed on bloom and produce the nectar necessary for their survival.

Figure 20.29 • **Black guillemots** in northern regions experienced a reduction in population in the 1990s as a result of receding sea ice in the spring.

likely occur in the next 50 to 100 years (Figure 20.30). Response to the emission of carbon dioxide up to and beyond the peak will include exponential growth of the concentration of carbon dioxide in the atmosphere as well as exponential growth in global temperature and rise in sea level. Following lag time of hundreds to a thousand years or more, these will eventually stabilize. Some irreversible environmental consequences might include extinction of some species.

Debate continues over some details of climatic change related to human activities that release greenhouse gases into the atmosphere. Figure 20.31 summarizes the framework for human-induced climate change with potential adjustments. Scientists do agree that we need to develop better models and gain a better understanding of both climatic change and variability and short-term trends in weather. At this point, two important statements can be made concerning climate change and climate change:

- Global temperature in the last few decades has increased approximately 0.5°C.
- The bulk of the evidence suggests that massive carbon dioxide emissions into the atmosphere from use of fossil fuels are causing some significant portion of the observed climate change. Furthermore, computer models predict that climate change will continue to occur as a result of increased atmospheric levels of greenhouse gases.

Given these statements, should we spend large sums of money to reduce CO_2 and other greenhouse gas emissions? There is no easy answer to this question, and the answer will not come entirely from scientific evaluation. Figure 20.32 shows cost estimates to stabilize atmospheric concentration of CO_2 at several levels. For example, at 550 ppm (double the pre-industrial level) the cost is estimated to be about $40 billion per year over a century.

In terms of science, the question of whether human-induced warming is occurring is solved: a significant part of climate change is human-induced. We now must evaluate potential consequences and risks to society and the biosphere. Policy-makers must take action to address risks posed by climate change. They must do this even though significant scientific uncertainties remain concerning whether climate change and associated climate-induced changes can be reversed.[1]

What should we do about potential climate change? There are two basic adjustments:

- Adapt; that is, do nothing to combat it, and live with future global climatic change.
- Work to mitigate the situation (reduce its severity) by reducing emissions of greenhouse gases and enhancing biospheric sinks for CO_2.

Living with Global Change

Considering our present understanding of global climate and recent trends in energy use and deforestation, a likely adjustment will be learning to live with the changes. As discussed earlier, these include a warmer climate, more variability in weather patterns, changes in the biosphere, and a higher sea level. If the changes are relatively slow and total mean climate change is less than 2°C, then learning to live with new conditions may be feasible. In some

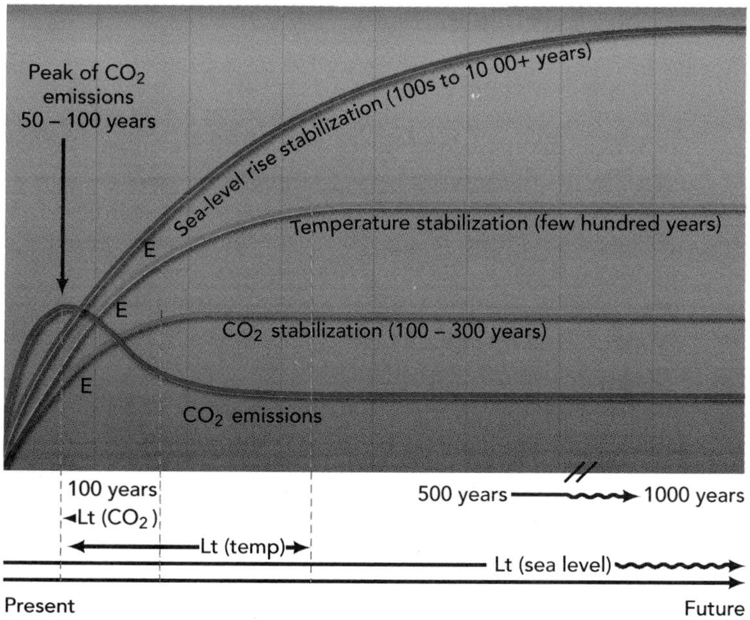

Figure 20.30 • **Exponential growth (E) with lag times of atmospheric concentration (Lt) following peak in CO_2 emissions for stabilization of CO_2, global temperature and sea level rise are hundreds to more than 1000 years.** [*Source:* Modified after Intergovernmental Panel on Climate Change 2001 at www.ipcc.ch/press/sp-COPG.htm.].

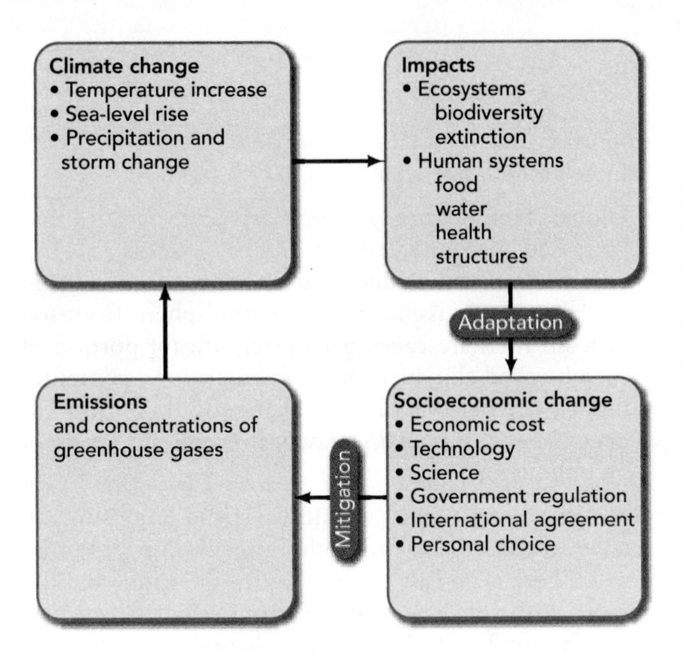

Figure 20.31 • Framework for human induced climate change. [*Source*: Modified after Intergovernmental Panel on Climate Change 2001.]

cases, the changes will offer opportunities. However, overall, this may not be the best adjustment. Surprises and unexpected problems will emerge. A 2°C change is at the low end of predicted climate change; at increases greater than 2°C, potential consequences to humans increase significantly. Investigation of sea surface temperature over the past several hundred thousand years from the geologic record suggests that warming will more likely be about 5°C.[33] Finally, although crop yields will vary on a regional and local level, problems in food production will occur in tropical and subtropical regions, where many millions of people live.[1]

In summary, one possible adjustment to climate change will be to live with it. However, if we are prudent, we will plan instead to reduce emissions of carbon dioxide and other greenhouse gases. By doing so, we may be able to limit warming to less than 2°C.

Mitigating Climate change

The attempt to control emissions of greenhouse gases, particularly carbon dioxide, was initiated when a major scientific conference on the issue of climate change was held in 1988 in Toronto, Canada. At that meeting, scientists recommended a 20% reduction in carbon dioxide emissions by 2005. The meeting was a catalyst for scientists and other concerned people to work with politicians to initiate international agreements for reductions in emissions of greenhouse gases. Although at that time many uncertainties remained concerning climate change, the prevailing attitude was that it was advisable to be conservative and reduce emissions before problems became apparent.

In 1992, at the Earth Summit in Rio de Janeiro, Brazil, a general blueprint for the reduction of global emissions was suggested. Some in the United States, however, objected that the reductions in CO_2 emissions would be too costly. Furthermore, agreements from the Earth Summit did not include legally binding limits. Following the meetings in Rio de Janeiro, governments worked to strengthen a climate-control treaty that included specific limits on the amounts of greenhouse gases that could be admitted into the atmosphere by each industrialized country.

Legally binding emission limits were discussed in Kyoto, Japan, in December 1997, but specific aspects of the agreement divided the delegates. Canada ratified the Kyoto protocol on December 17, 2002, and therefore committed to a six percent reduction in emissions below 1990 levels, to be achieved by 2012. Although we do not yet have legally binding requirements for emissions reductions, in November 2002 the Canadian federal government established a Climate Change Plan for Canada. The Plan takes a three-part approach to emissions reductions, including voluntary covenants with major emitters, establishment of a domestic emissions trading system, and various other measures. In March 2004, the Canadian government announced its One-Tonne Challenge (see Build Your Environmental Skills 20.1), aimed at reducing the 25% of the country's greenhouse gas emissions that arise from private residences and automobiles.

The United States has refused to ratify the Kyoto agreement, but eventually agreed to cut emissions to about 7% below 1990 levels. However, that was far below the reductions suggested by scientists, who recommended reductions of 60–80% below 1990 levels. In fact, following the conference, it was realized that emissions of carbon dioxide in 2010 would likely be about 30% above the 1990 emissions.

Since that time, the science question has been answered, and the human component of climate change has increased, as evidenced by a series of very warm years and biosphere changes. In spite of this, some sectors of the fossil fuels industry continue to resist lowering emissions, arguing that the economic cost would be too great.

Every time an environmental solution to a problem has been suggested, there have been those who said it could not be done, yet it has been proved time and time again that it can be done. It appears that advances in alternative renewable energy are sufficient that society could in fact meet goals for reductions of carbon dioxide emissions without jeopardizing our economic future. An energy scenario that emphasizes sustainable alternative renewable energy sources is shown in Figure 20.33. This scenario (prepared by a major oil company) predicts that total energy consumed will triple by 2060. All of the growth in energy use after about March 2020 is assumed to be from alternative sources, which under this scenario would supply about 60% of the total global demand. The energy scenario of Figure 20.33 continues the present growth path and is not sustainable. Nevertheless, the scenario is significant in recognizing a path toward more sustainable energy development. To have alternative energy resources supplying twice the total energy consumed today in the next 60 years seems overly optimistic but perhaps possible with significant energy policy reforms, education, and enhanced research and development in alternative energy.

How can carbon dioxide emissions be reduced? Energy planning that relies heavily on energy conservation and efficiency, along with use of alternative energy sources has the potential to reduce emissions of carbon dioxide. Increased use of nuclear power would also assist in reducing carbon dioxide emissions. Changing the balance of fossil fuels to burn more natural gas would also be helpful, because natural gas releases less carbon than other fossil fuels (Table 20.2).[34] Other strategies to reduce emissions of carbon dioxide include the use of mass transit and thereby decrease the use of automobiles; providing greater economic incentives to energy-efficient technology; requiring higher fuel-economy standards for cars, trucks, and buses; and requiring higher standards of energy efficiency.

Burning forests to convert land to agricultural uses accounts for about 20% of anthropogenic emissions of carbon dioxide into the atmosphere. Management plans with the objective of minimizing burning and protection of the world's forests would help reduce the threat of climate change. Planting more trees (reforestation) is also a potential strategy: reforestation would increase biospheric sinks for carbon dioxide. Other natural sinks for carbon dioxide—such as soils, forests, and grasslands—might be enhanced and managed better to sequester more carbon dioxide than they currently do.[35]

Geologic (rock) sequestration of carbon is another possible mitigation measure to reduce the amount of carbon dioxide that would otherwise enter the atmosphere. The general principle of geologic sequestration of carbon is fairly straightforward. The idea is to capture carbon dioxide from power plants and industrial smokestacks and inject the carbon into deep subsurface geologic reservoirs. Geologic environments suitable for carbon sequestration are sedimentary rocks that contain salt water and sedimentary rocks that are the sites of depleted oil and gas fields. Sedimentary rock environments, which are widespread at numerous locations on Earth, have large capacity or potential to sequester as much as 1,000 gigatons of carbon. In order to mitigate the adverse effects of carbon dioxide emissions that result in climate change, we need to sequester at least 2 gigatons of carbon per year.[36]

The process of placing carbon dioxide in the geologic environment involves compressing the gas and changing it to a mixture of both liquid and gas and then injecting it deep underground. Individual injection projects can sequester approximately 1 million tonnes of carbon dioxide per year.

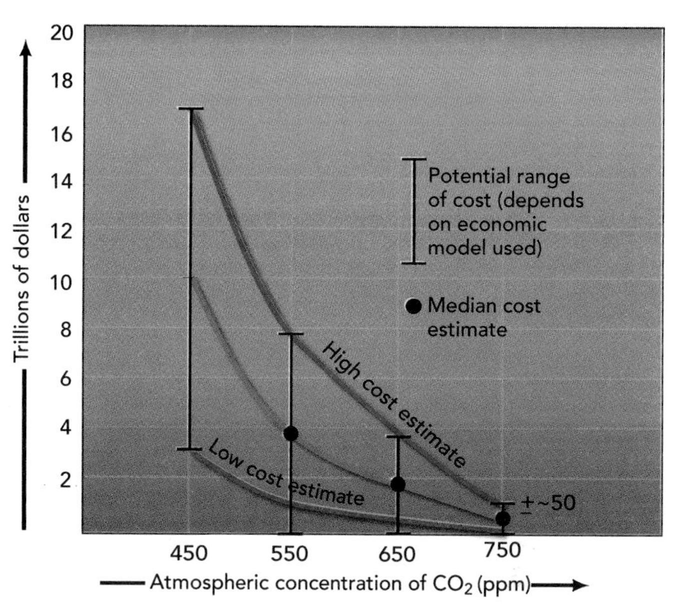

Figure 20.32 • **Cost of stabilizing CO_2 concentrations to the year 2100.** Cost at 550ppm is at the concentration of approximately 2 times pre industry levels. [*Source:* Modified after Intergovernmental Panel on Climate Change 2001.]

Figure 20.33 • **Global energy consumption, 1860–2000, predicted to 2060, based on a model of movement toward sustainable energy development.** [*Source*: Modified from Shell International, in R. T. Watson, presentation at the Sixth Conference of the Parties of the United Nations Framework Convention on Climate Change, Intergovernmental Panel on Climate Change, November 13, 2000. Accessed December 1, 2000, at www.ipcc.ch/press/sp-COPG.htm.]

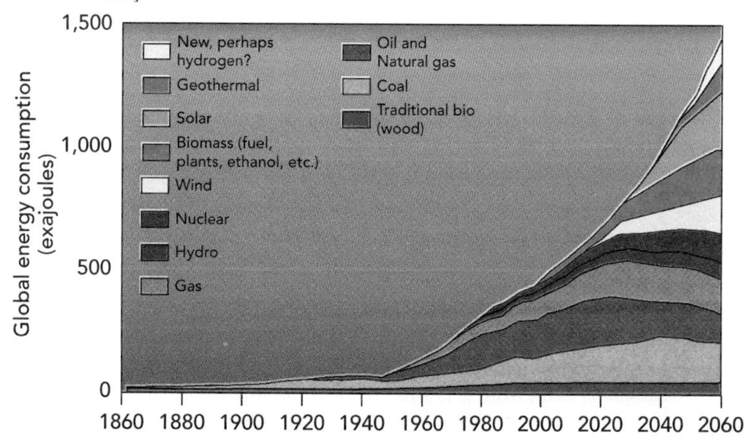

Build Your Environmental Skills 20.1

Canada's One-Tonne Challenge: How do you measure up?

Personal actions such as washing clothes, lighting homes, and driving cars account for 25% of Canada's greenhouse gas emissions. In March 2004, the Canadian government announced its One-Tonne Challenge. The Challenge asks each Canadian to reduce their personal contributions to greenhouse gases by one tonne. Review the following table of actions suggested by the One-Tonne Challenge, and see where you may have opportunities to reduce your personal greenhouse gas emissions. How many of the 32 suggested actions have you adopted?

Now visit the One-Tonne Challenge website and use the online calculator (http://www.climatechange.gc.ca/onetonne/calculator/english/) to estimate your personal greenhouse gas reductions. (Reductions must be calculated separately for each person and household, because they can vary widely depending on type of home, family size, fuel used, and location in the country.)

Table 20.3 • Options for reducing personal greenhouse gas emissions

Home Heating and Cooling	*Water and Waste*	*On the Road*
Install a ceiling fan to circulate air	Wash clothes in cold water	Don't buy more car than you need
Install an energy-efficient furnace	Don't run the tap	Don't idle your car engine
Install storm windows	Use low-flow showerheads	Drive 10% less; walk, cycle or jog more
Keep blinds closed during summer days	Insulate hot water pipes	Drive at the posted speed limit
Keep curtains open during winter days	Take a quick shower	Give up the second vehicle
Keep furnace well maintained	Use an energy efficient water heater	Use a block heater in very cold weather
Lower your thermostat in winter	Compost kitchen waste	Use your car's air conditioner sparingly
Set air conditioner at 24°	Avoid over-packaged goods	Keep your vehicle well maintained
Seal and insulate warm-air ducts	Reuse and recycle materials where possible	Measure tire pressure once a month
Upgrade home insulation		Remove roof racks when not in use
Use caulking and weather strips		Use ethanol-blended gasoline
Use energy efficient light bulbs		

Source: Canada's One-Tonne Challenge; http://www.climatechange.gc.ca/onetonne/

A carbon-sequestration project is under way in Norway beneath the North Sea. The carbon dioxide from a large natural gas production facility is injected approximately 1,000 metres deep into sedimentary rocks below a natural gas field. The project, started in 1996, injects about 1 million tonnes of carbon dioxide every year. It is estimated that the entire reservoir can hold up to about 600 billion tonnes of carbon. To put this number in perspective, it is about the amount of carbon dioxide that is likely to be produced from all of Europe's fossil fuel plants in the next several hundred years.[36, 37] The cost of sequestering carbon beneath the North Sea is expensive, but it saves the company from paying carbon dioxide taxes for emissions into the atmosphere.

A pilot project to demonstrate the potential of sequestering carbon into oil reservoirs has been initiated in Weyburn, Saskatchewan. In Joffre, Alberta, a chemical company extracts carbon dioxide from an ethylene manufacturing operation and sells it to an oil company for use in enhanced oil recovery.

Table 20.2 • Comparison of Common Fossil Fuels in Terms of Share of World Energy Supply and Percentage of Total Carbon Released

Fossil Fuel	*World Energy Supply (%)**	*Carbon Emitted as % of Total Carbon from Fossil Fuels*	*Comparison*
Oil	32	43	Releases 23% less carbon per unit of energy than coal
Natural gas	22	21	Releases 28% less carbon per unit of energy than oil or gas
Coal	21	36	Releases much more carbon per unit of energy than oil or gas

* In total, fossil fuels comprise 75% of world energy supply.
Source: Data from J. Dunn, "Decarbonizing the Energy Economy," in World Watch Institute State of the World 2001 (New York: W. W. Norton, 2001), pp. 83–102.

CRITICAL THINKING ISSUE

The Climate Change Controversy

Climate change is not without controversy. Several issues are debated:

1. Are observed climate fluctuations part of a normal cycle of variation or somehow abnormal?
2. Are human activities, especially the burning of fossil fuels, responsible for the changes we observe?
3. What changes can we expect in the future, and what should we do about them?

Although the debate is couched in scientific terms, at its heart are economics (Chapter 25). Some people argue that the economic impacts of reducing fossil fuel use would be devastating to national economies. Others argue that immediate action to control greenhouse gases is essential if we are to avoid much worse impacts in the future. The question is whether such actions will result in any measurable improvement in global temperature.

Most scientists now agree that the evidence of global climate change is persuasive. According to a 2002 report from the Intergovernmental Panel on Climate Change (IPCC), the bulk of the scientific evidence suggests that the warming is due in part to human activities. As a consequence of projected emissions of carbon dioxide (CO_2), mean surface temperatures are expected to increase by 1.5–6.0°C by 2100. The IPCC's evidence continues to accumulate. In its most recent report, the IPCC states that, "The Earth's climate system has demonstrably changed on both global and regional scales since the pre-industrial era, with some of these changes attributable to human activities." The IPCC's cautious statement reveals the crux of the scientific debate. Most of the temperature change we now observe occurred before the bulk of modern carbon dioxide emissions.

Supporters of the notion that human activities are causing climate change—including the IPCC, Environment Canada, and the U.S. National Academy of Science—argue that global warming has accelerated over the last century, and that natural climate variability patterns cannot explain the phenomena we now observe. They warn of more, and worse, climate effects in the future, and advocate for actions to prevent or reduce those changes.

Opponents of global climate change theories argue that the scientific basis for these conclusions is weak and contradictory, that the models used to predict climate charge are inadequately tested, and that a correlation between rising carbon dioxide levels and rising temperatures does not prove that the one caused the other. Some believe that sunspots or urban "heat islands" (Chapter 26) have caused the warming trends we observe in some areas. This group includes a number of prominent academics including S. Fred Singer, a retired professor of atmospheric physics at the University of Virginia; Richard Lindzen, a professor of meteorology at the Massachusetts Institute of Technology; and Ross McKitrick, a professor of environmental economics at the University of Guelph, Ontario.

Critical Thinking Questions

1. What position do you take on the climate change controversy, and why? How would you respond to the objections of opponents of your views?
2. In what ways would you expect climate change to affect the area in which you live? What industries in your region would be most at risk of impact from climate change, or from measures to mitigate climate change? Describe the kinds of impacts that you might expect.
3. In your view, how urgent is the problem of climate change, compared to other environmental issues such as air and water pollution, waste management, and urban sprawl?

Summary

- The atmosphere, a layer of gases that envelops Earth, is a dynamic system that is constantly changing. A great number of complex chemical reactions take place in the atmosphere, and atmospheric circulation takes place on a variety of scales, producing the world's weather and climates.
- Nearly all the compounds found in the atmosphere either are produced primarily by biological activity or are greatly affected by life.
- The four main processes that remove particles and pollutants from the atmosphere are sedimentation, rain out, oxidation, and photodissociation.
- On a local scale, climatic conditions vary considerably and may produce microclimates. For example, the presence of a city significantly affects local climate, producing an urban microclimate. Thus, some midlatitude cities are cloudier, warmer, rainier, and less humid than the surrounding areas.
- Major climatic changes have occurred during the past 2 million years, with periodic appearances and retreats of glaciers. During the past 1,500 years, several warming and cooling trends have affected people. During the past 100 years, the mean global annual temperature has apparently increased by about 0.5°C.
- Water vapour and several other gases, including carbon dioxide, tend to warm Earth's atmosphere through the greenhouse effect. The vast majority of the greenhouse effect is produced by water vapour, which is a natural constituent of the atmosphere. Carbon dioxide and other greenhouse gases also occur naturally in the

atmosphere. However, especially since the Industrial Revolution, human activity has added substantial amounts of carbon dioxide to the atmosphere, along with such greenhouse gases as methane and CFCs.

- The mean global temperature of Earth is rising. Climatic models suggest that when carbon dioxide has doubled from its pre-industrial levels in the next few decades, the mean global temperature may rise by 1–2°C. Total warming during the twenty-first century may range from 1.5°C to 6.0°C.
- Many complex positive-feedback and negative-feedback cycles affect the atmosphere. Nevertheless, it appears that increases in the emission of greenhouse gases from human activity are a significant cause of late-twentieth-century and present climate change. Natural cycles, solar forcing, aerosols, volcanic eruptions, and El Niño events also affect the temperature of Earth.
- Major effects of climate change include: (1) changes in climatic patterns and frequency and intensity of storms, (2) a rise in sea level, and (3) changes in the biosphere.
- Changes in climatic patterns and storms are worrisome because they may adversely affect agriculture and thus food supply. A rise in sea level is a potentially serious problem because so many people live in or near coastal areas and coastal erosion is a difficult problem. As sea level rises and more heat energy is fed into the atmosphere, cyclones and tropical disturbances will pose a greater hazard for people living in vulnerable areas.
- Changes in the biosphere include shifts in where specific plants and animals live, bleaching of coral, and spreading of diseases such as malaria to higher elevations.
- Adjustments to climate change include learning to live with the changes and attempting to mitigate warming through reduction of emissions of greenhouse gases. It seems likely that the adjustment chosen will be to learn to live with change. A danger of this path is that climatic change may be rapid, resulting in problems that will be difficult to address.
- Because a significant component of climate change appears to result from anthropogenic processes, it is prudent to reduce emissions of greenhouse gases. With control of CO_2 emissions, climate change may be limited to less than 2°C, which would cause far less harm than would a larger increase. Carbon dioxide emissions can be reduced through energy conservation, sequestration of carbon in the biosphere and lithosphere, and the use of alternative energy sources.

STUDY QUESTIONS

1. What is the composition of Earth's atmosphere and how has life affected the atmosphere during the past several billion years?
2. How do midlatitude cities differ from surrounding rural areas in climate?
3. What is the greenhouse effect? What is its importance to global climate?
4. What is an anthropogenic greenhouse gas? Discuss the various anthropogenic greenhouse gases in terms of their potential to cause climate change.
5. What are some of the major negative-feedback cycles and positive-feedback cycles that might increase or decrease climate change?
6. In terms of the effects of climate change, do you think that a change in climate patterns and storm frequency and intensity is likely to be more serious than a global rise in sea level? Illustrate your answer with specific problems and areas where the problems are likely to occur.
7. How would you refute or defend the statement that the best adjustment to climate change is to do little or nothing and learn to live with change?

FURTHER READING

Anthes, A. R. *Meteorology*, 6th ed. New York: Macmillan 1992. A short text providing a good overview of basic meteorology and atmospheric processes.

Fay, J. A., and Golumb. *Energy and Environment*. Oxford University Press, 2002. See Chapter 10 on climate change.

Goodess, C. M., J. P. Palutikof, and T. D. Davies. *The Nature and Causes of Climate Change: Assessing the Long-Term Future*. Lewis, 1992. A good text discussing short- and long-term climate changes and anthropogenic versus natural forcing in climate change.

IPCC. *The Intergovernmental Panel on Climate Change Scientific Assessment*. New York: Oxford University Press, 2001. A detailed scientific review and assessment of climate change.

Mohnen, V. A., W. Goldstein, and C. W. Wang. "The Conflict over Global Warming," Global and Environmental Change 1 no. 2 (1991):109–123. A paper providing detailed information on the global warming controversy that existed at the time.

Organisation for Economic Co-operation and Development, International Energy Agency. *The Economics of Climate Change*. Proceedings of an OECD/IEA Conference, OECD, IEA, 1994. These proceedings discuss the economics of climate change, differences of opinion on the matter, methods to link economic studies and climate change policy, and directions of international policy to deal with climate change.

Titus, J. G. and V. K. Narayanan. *The Probability of Sea Level Rise*. Washington, D.C.: U.S. Environmental Protection Agency, 1995. A close look at future sea-level rise due to climate change using mathematical models to predict changes.

chapter 21

London Smog and Indonesian Fires

In December 1952, air in London became stagnant, and cloud cover blocked incoming solar radiation. The temperature dropped rapidly. At noon it was about −1°C, and humidity climbed to 80%. Thick fog developed; cold and dampness increased demand for home heating. Because the primary fuel used in homes was coal, emissions of ash, sulphur oxides, and soot increased quickly. Stagnant air was filled with pollutants from home heating fuels and automobile exhaust. At the height of the crisis, visibility was so reduced that automobiles used headlights at midday. Between December 4 and 10, about 4,000 Londoners died from the pollution. The siege of smog ended when the weather changed and the air pollution dispersed. Environment, not human activities, finally solved this problem. What went wrong?

During the London smog crisis, stagnant weather conditions, combined with emissions from homes burning coal and from cars burning gasoline, exceeded the atmosphere's ability to remove or transform pollutants. As a result, sulphur dioxide remained in the air and fog became acidic, adversely affecting people and other organisms. Human health effects were especially destructive because small acid droplets became fixed on larger particulates, which were easily drawn into the lungs.

The 1952 London smog crisis was a landmark event. Finally, the natural abilities of the atmosphere to serve as a sink for removal of wastes had been exceeded by human activities. The crisis was due, in part, to a reinforcing feedback situation. Burning fossil fuels added particulates to the air, increasing the formation of fog and decreasing visibility and light transmission. The dense, smoggy layer increased the dampness and cold, which accelerated use of home-heating fuels. As the weather and pollution worsened, more people burned fuel, which further worsened weather and pollution. London had been known for its fogs; long before 1952, what had been little known was the role of coal burning in exacerbating the fog conditions. The good news is that, since 1952, London fogs associated with air pollution have been greatly reduced because coal has been replaced by cleaner natural gas as the primary home-heating fuel.

Forty-five years after the London smog crisis and following the most severe drought in 50 years, a serious air pollution event in Indonesia dwarfed the London event. As a result of a strong El Niño (see Chapter 20), huge fires ravaged Indonesia for months (Figure 21.1), producing a thick, toxic haze of smoke. What had gone wrong in this instance?

Slash-and-burn practices have been part of farming in tropical rain forests for centuries. Each family burns a few hectares of rainforest, plants it, and after harvest moves on to a new area. Fire is the preferred method of land clearing because clearing land with bulldozers is too expensive. It is generally thought that slash-and-burn farming has not been particularly disturbing to the ecology of rain forests, as long as it is not too widespread and fires are carefully controlled.[1] Unfortunately, in 1997 a combination of events related to El Niño and out-of-control burns resulted in what some consider one of the world's greatest environmental disasters.

Monsoon rains were late that year, and people took advantage of dry weather to burn more forest. Although much of the blame was placed on small farmers, agriculture on an industrial scale was also responsible for clear-cut logging and burning. At least 20,000 hectares of land burned during the 1997 dry season. Vast amounts of smoke and particulate matter entered the atmosphere, causing a severe pollution problem. At one point, people were so desperate for relief from the dense, blinding haze that they attempted to use surgical masks as filters, but this provided little protection. Some 20 million Indonesians were treated for a variety of illnesses directly caused or aggravated by the fires. Smoke was so dense that a passenger airline crashed in Sumatra, killing 234 people. The Air Quality Index (AQI)—a measure of air quality—was at about 800 because of the fire. An AQI of 500 is considered very hazardous, and it is recommended that all people remain indoors (the AQI is discussed later in this chapter).

Near the end of summer, haze began to drift across the South China Sea to Malaysia and Singapore. Many of the fires were extinguished with the arrival of monsoons, but when the rains stopped in February 1998, more fires followed. There is worry that annual cycles of fire have begun that will have cumulative adverse environmental impacts.

Air Pollution

LEARNING OBJECTIVES

The atmosphere has always been a sink—a place for deposition and storage—for gaseous and particulate wastes. When the amount of waste entering the atmosphere in an area exceeds the ability of the atmosphere to disperse or breakdown the pollutants, problems result. After reading this chapter, you should understand:

- Why human activities that pollute the air, combined with meteorological conditions, may exceed the natural abilities of the atmosphere to remove wastes.
- What the major categories and sources of air pollutants are.
- What acid precipitation is, how it is produced, what its environmental impacts are, and how they might be minimized.
- What methods are useful in the collection, capture, and retention of pollutants before they enter the atmosphere.
- Why determining the economics of air pollution is controversial and difficult.
- Why indoor air pollutants cause some of our most serious environmental health problems.
- What the major indoor air pollutants are, and where they come from.
- Why concentration of pollutants found in the indoor environment may be much greater than concentrations of the same pollutants generally found outdoors.
- What the major strategies are to control and minimize indoor air pollution.

The damage from the fires is not limited to humans. Wildlife—including endangered species, such as rhinoceros—and wilderness reserves have also suffered damage.[1]

- *The London smog of 1952 and the Indonesian fires of 1997–1998 are examples of serious, human-induced air pollution. This chapter discusses the major air pollutants, urban air, acid precipitation, and control of air pollution.*

Figure 21.1 • **Image of air pollution (smoke from forest fires) over Indonesia and Indian Ocean on October 22, 1997.** White is smoke near fires. Red, yellow, and green represent decreasing amounts of lower atmosphere pollution being transported westward.

21.1 A Brief History of Air Pollution

As the fastest-moving fluid medium in the environment, the atmosphere has always been one of the most convenient places to dispose of unwanted materials. Ever since we first used fire, the atmosphere has been a sink for waste disposal.

People have long recognized the existence of atmospheric pollutants, both natural and human-induced. In 1550, Leonardo da Vinci wrote that a blue haze formed from materials emitted into the atmosphere from trees. He had observed a natural photochemical smog, resulting from hydrocarbons given off by living trees, whose cause is still not completely understood. This haze gave rise to the name Smoky Mountains for the range in the southeastern United States. The phenomenon of acid precipitation was first described in the seventeenth century; and by the eighteenth century it was known that smog and acid rain damaged plants in London.

Beginning with the Industrial Revolution in the eighteenth century, air pollution became more noticeable. The word smog was introduced by a physician at a public health conference in 1905 to denote poor air quality resulting from a mixture of smoke and fog.

An air pollution event in Donora, Pennsylvania, in 1948, remains one of the worst industrial air pollution incidents in recent history, causing 20 deaths and 5,000 illnesses. What was called the "Donora fog" involved pollutants from the Donora Zinc Works metal-smelting plant and other sources. Pollutants, including sulphur dioxide, carbon monoxide, and heavy metals, were trapped by weather conditions in a narrow valley. The event lasted about three days until pollutants were washed out and dispersed by rainstorms.

The Donora event was followed in 1952 by the London smog crisis described in the opening case study. Following the Donora and London events, regulations to control air quality began. Today, in Canada, the United States and other countries, legislation to reduce emission of air pollutants has been successful, but more needs to be done. Chronic exposure to high levels of air pollutants continues to contribute to illnesses that kill people around the world.

What are the chances that another killing smog will occur somewhere in the world? Unfortunately, the chances are all too good, given the tremendous amount of air pollution in some large cities. For example, Beijing might be a candidate; the city uses an immense amount of coal, and coughing is so pervasive among its residents that they often refer to it as the "Beijing cough." Another likely candidate is Mexico City, which has one of the worst air pollution problems anywhere in the world today.

Figure 21.2 • This steel mill in Beijing, China is a major source of air pollution.

21.2 Stationary and Mobile Sources of Air Pollution

What are the sources of air pollution? The two major categories are stationary sources and mobile sources. **Stationary sources** are those that have a relatively fixed location. These include point sources, fugitive sources, and area sources.

- *Point sources*, as discussed in Chapter 15, emit pollutants from one or more controllable sites, such as smokestacks of power plants (Figure 21.2).
- *Fugitive sources* generate air pollutants from open areas exposed to wind processes. Examples include burning for agricultural purposes (Figure 21.3), as well as dirt roads, construction sites, farmlands, storage piles, surface mines, and other exposed areas from which particulates may be removed by wind.
- *Area sources*, also discussed in Chapter 15, are well-defined areas within which are several sources of air pollutants—for example, small urban communities, areas of intense industrialization within urban complexes, and agricultural areas sprayed with herbicides and pesticides.

Mobile sources of air pollutants move from place to place while emitting pollutants. These include automobiles, trucks, buses, aircraft, ships, and trains (Figure 21.4).[2]

21.3 General Effects of Air Pollution

Air pollution affects many aspects of our environment: its visual qualities, vegetation, animals, soils, water quality, natural and artificial structures, and human health. Air pollutants affect visual resources by discolouring the atmosphere and by reducing visual range and atmospheric clarity so that the visual contrast of distant objects is decreased. We cannot see as far in polluted air, and what we do see has less colour contrast. These effects were once

Figure 21.3 • **Fugitive source of air pollution.** Burning sugarcane fields, Maui, Hawaii.

limited to cities, but they now extend to some wide-open spaces. For example, the British Columbia government has recently expressed concern about episodes of poor visibility, related to air pollution, in the Cowichan Valley area of Vancouver Island, an area famous for its mountains, old growth forests, and lovely rivers.[2]

The effects of air pollution on vegetation are numerous. They include damage to leaf tissue, needles, and fruit; reduction in growth rates or suppression of growth; increased susceptibility to a variety of diseases, pests, and adverse weather; and disruption of reproductive processes.[2]

Air pollution is a significant factor in the human death rate for many large cities. For example, it has been estimated that in Athens, Greece, the number of deaths is several times higher on days when the air is heavily polluted; and in Hungary, where air pollution has been a serious problem in recent years, it may contribute to as many as 1 in 17 deaths. Canada and the United States are certainly not immune to health problems related to air pollution. Areas with recurring air quality problems include southern Ontario (especially Windsor, Kitchener-Waterloo, and Hamilton), Montréal, parts of Nova Scotia, and the lower Fraser River Valley of British Columbia. In these areas, millions of people are exposed to unhealthy air. It is estimated that as many as 175 million people live in areas of Canada and the United States where exposure to air pollution contributes to lung disease, which causes more than 310,000 deaths per year. Air pollution in Canada and the United States is directly responsible for annual health costs of about $55 billion.[3] In China, whose large cities have serious air pollution problems, mostly from burning coal, the health cost is now about $50 billion per year and may increase to about $100 billion per year by 2020 (Figure 21.5).

Air pollutants can affect human health in several ways (Figure 21.6). The effects on an individual depend on the dose or concentration (see the discussion of dose response in Chapter 15) and other factors, including individual susceptibility. Some of the primary effects of air pollutants include toxic poisoning, cancer, birth defects, eye irritation, and irritation of the respiratory

Figure 21.4 • **Exhaust from older diesel buses such as this one in Thailand are a major source of air pollution.**

Figure 21.5 • **Annual cost of air pollution to human health in China.** Note the growing costs associated with premature deaths and chronic bronchitis. [*Source*: Modified from World Bank, *Clear Water, Blue Skies; China's Environment in the New Century* (World Bank, 1997).]

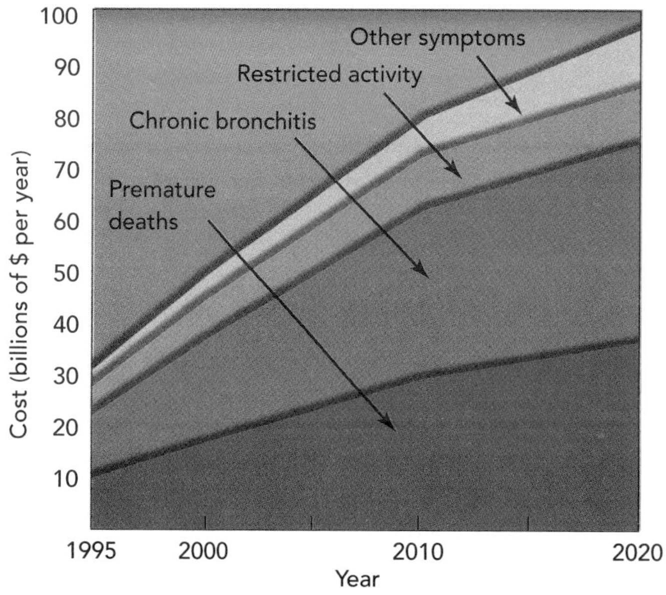

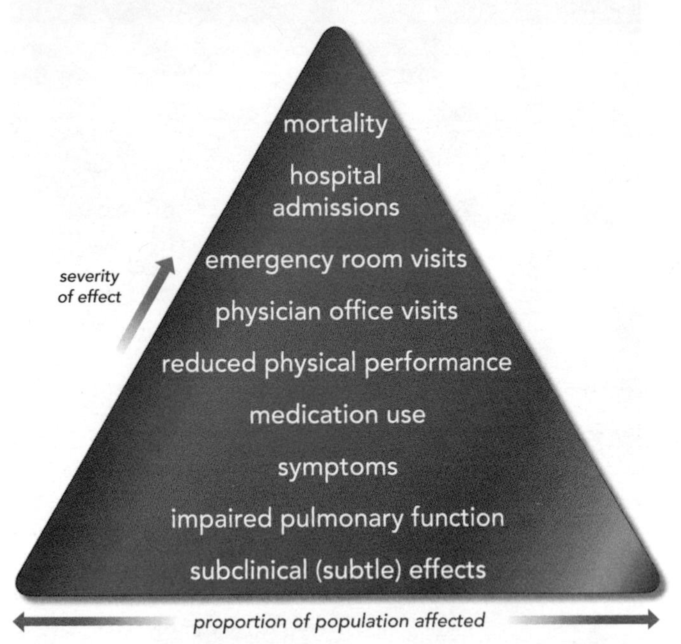

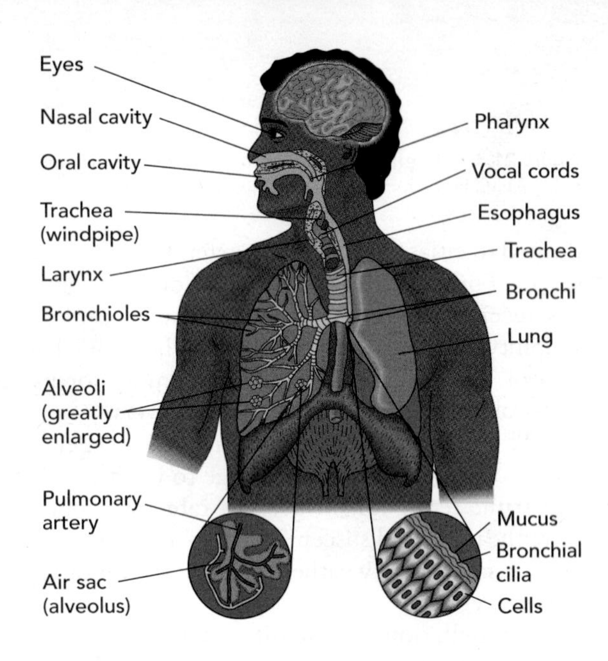

Figure 21.6 • *(a)* **The health effects pyramid.** The worst effects occur in only a few people, but many more have less severe symptoms. [*Source*: Health Canada http://www.hc-sc.gc.ca/hecs-sesc/air_quality/health_effects.htm] *(b)* **Some of the parts of the human body (brain, cardiovascular system, and pulmonary system) that can be damaged by common air pollutants.** The most severe health risks from normal exposures are related to particulates. Other substances of concern include carbon monoxide, photochemical oxidants, sulphur dioxide, and nitrogen oxides. Toxic chemicals and tobacco smoke also can cause chronic or acute health problems.

system; increased susceptibility to viral infections, causing pneumonia and bronchitis; an increased susceptibility to heart disease; and aggravation of chronic diseases, such as asthma and emphysema. People suffering from respiratory diseases are the most likely to be affected by air pollutants. Healthy people tend to acclimate to pollutants in a relatively short period of time. However, this is a physiological tolerance; as explained in Chapter 15, it does not mean that the pollutants are doing no harm.

Many air pollutants have *synergistic effects* (in which the combined effects are greater than the sum of the separate effects). For example, sulphate and nitrate may attach to small particles in the air, facilitating their inhalation deep into lung tissue. There, they may do greater damage to the lungs than a combination of the two pollutants would be expected to do based on their separate effects. This phenomenon has obvious health consequences; consider joggers breathing deeply of particulates as they run along the streets of a city.

The effects of air pollutants on vertebrate animals in general include impairment of the respiratory system; damage to eyes, teeth, and bones; increased susceptibility to disease, parasites, and other stress-related environmental hazards; decreased availability of food sources (such as vegetation affected by air pollutants); and reduced ability for successful reproduction.[2]

Air pollution can also degrade soil and water resources when pollutants from the air are deposited. Soils and water may become toxic from the deposition of various pollutants. Soils may also be leached of nutrients by pollutants that form acids. The effects of air pollution on human-made structures include discoloration, erosion, and decomposition of building materials. These effects are described when we explore the topic of acid precipitation later in this chapter. Effects of major air pollutants discussed in this chapter are listed in Table 21.1.

21.4 Primary and Secondary Pollutants, Natural and Human

The major air pollutants occur either in gaseous forms or as particulate matter (PM). Particulate matter pollutants (PM 10 or PM 2.5) are particles of solid or liquid substances less than 10 μm or 2.5 mm in diameter and may be organic or inorganic. The gaseous pollutants include sulphur dioxide (SO_2), nitrogen oxides (NO_X), carbon monoxide (CO), ozone (O_3); and volatile organic compounds (VOC) such as hydrocarbons, hydrogen sulphide (H_2S), and hydrogen fluoride (HF).

Air pollutants can be classified as primary or secondary. **Primary pollutants** are those emitted directly into the air. They include particulates, sulphur dioxide, carbon monoxide, nitrogen oxides, and hydrocarbons. **Secondary pollutants** are produced through reactions between primary pollutants and normal atmospheric compounds. For example, ozone forms over urban areas through reactions of primary pollutants, sunlight, and natural atmospheric gases. Thus, ozone is a secondary pollutant.

The primary pollutants that account for nearly all air pollution problems are carbon monoxide (58%), volatile organic compounds (11%), nitrogen oxides (15%), sulphur oxides (13%), and particulates (3%). In Canada and the United States today, over 150 million

Table 21.1 • Effects of Major Air Pollutants on People, Plants, and Materials

Pollutant	*Effects on People*[a]	*Effects on Plants*[a]	*Effects on Materials*[a]
Ozone (O_3)	Strong irritant, aggravates asthma; injures respiratory cells, decreased elasticity of lung tissue, coughing, chest discomfort; eye irritation	Flecking, stippling, spotting, and/or bleaching of plant leaves and stems; oldest leaves are most sensitive; tips of conifer needles become brown and die; lower yields and damage to crops including lettuce, grapes, and corn	Cracks rubber; reduces durability and appearance of paint, causes fabric dyes to fade
Sulphur dioxide (SO_2)	Increase in chronic respiratory disease; shortness of breath; narrowing of airways for people with asthma	Bleaching of leaves; decay and death of tissue; younger leaves are most sensitive; sensitive crops and trees include alfalfa, barley, cotton, white pine, white birch, and trembling aspen; if oxidized to sulphuric acid, causes damage associated with acid precipitation	If oxidized to sulphuric acid, damages buildings and monuments, corrodes metal; causes paper to become brittle; turns leather to red-brown dust; SO_2 fades dyes of fabrics, damages paint
Nitrogen oxides (NO_x)[b]	A mostly nonirritating gas; may aggravate respiratory infections and symptoms (sore throat, cough, nasal congestion, fever) and increase risk of chest cold, bronchitis, and pneumonia in children	No perceptible effects on many plants, but may suppress growth for some, and may be beneficial at low concentrations; if oxidized to nitric acid, causes damage associated with acid precipitation	Causes fading of textile dyes; if oxidized to nitric acid, may damage buildings and monuments
Carbon monoxide (CO)	Reduces the ability of the circulatory system to transport oxygen; causes headache, fatigue, nausea; impairs performance of tasks that require concentration; reduces endurance; may be lethal, causing asphyxiation	None perceptible	None perceptible
Particulate matter (PM 2.5, PM 10)	Increased chronic and acute respiratory diseases; depending on chemical composition of particulates, may irritate tissue of throat, nose, lungs, and eyes	Depending on chemical composition of particles, may damage trees and crops; dry deposition of SO_2, when oxidized, is a form of acid precipitation	Contributes to and may accelerate corrosion of metal; may contaminate electrical contacts; damages paint appearance and durability; fades textile dyes

[a] Effects depend on dose (concentration of pollutant and time of exposure) and susceptibility of people, plants, and materials to a particular pollutant. For example, older people, children, and those with chronic lung diseases are more susceptible to O_3, SO_2, and NO_x

[b] In NO_x, x refers to the number of oxygen atoms in the gas molecules, as in NO (nitric oxide) and NO_2 (nitrogen dioxide).

Sources: Modified from U.S. Environmental Protection Agency; R. W. Bunbel, D. L. Fox, D. B. Turner, and A. C. Stern. *Fundamentals of Air Pollution*, 3rd ed. (San Diego: Academic Press, 1994); and T. Godish. *Air Quality*, 3rd ed. (Boca Raton: Lewis Publishers, 1997).

metric tonnes of these materials enter the atmosphere from human-related processes. If these pollutants were uniformly distributed in the atmosphere, the concentration would be only a few parts per million by weight. Unfortunately, pollutants are not uniformly distributed but tend to be released, produced, and concentrated locally or regionally (e.g., in large cities).

In addition to pollutants from human sources, our atmosphere contains many pollutants of natural origin. Examples of natural emissions of air pollutants include the following:

- Release of sulphur dioxide from volcanic eruptions. For example, volcanic activity on the island of Hawaii emits SO_2 and other pollutants, which react in the atmosphere to produce volcanic smog called "vog." The smog can present a health hazard to people and cause local acid precipitation.
- Release of hydrogen sulphide from geysers and hot springs and from biological decay in bogs and marshes.
- Release of ozone in the lower atmosphere as a result of unstable meteorological conditions, such as violent thunderstorms.
- Emission of a variety of particles from wildfires and windstorms.[2]
- Natural hydrocarbon seeps; for example, the La Brea Tar Pits in Los Angeles.

The data in Table 21.2 suggest that, with the exception of sulphur and nitrogen oxides, natural emissions of air pollutants exceed human-produced emissions. Nevertheless, it is the human component that is most abundant in urban areas and that leads to the most severe problems for human health.

21.5 Major Air Pollutants: Some Details

According to a 2000 report by the Commission for Environmental Cooperation, Ontario ranked third from the bottom among Canadian and U.S. jurisdictions in its air pollution emissions. Even more disturbing, Canada's emissions of toxic air pollutants rose 7%

from 1998 to 2000, while U.S. emissions declined by 8% over the same period. Air pollution emissions vary widely across Canada (see Figure 21.7) and appear to be improving in some regions. Everywhere in Canada and the United States, automobiles are a leading cause of air pollution, accounting for about 40% of all nitrogen oxides and 25% of carbon dioxide emissions.

The Canadian government has expressed concern about rising air pollution, and especially a trend to more driving, in larger vehicles, nationwide. In February 2001, Environment Canada announced a 10-year Plan of Action for Cleaner Vehicles, Engines and Fuels. The plan includes reductions in the sulphur content of diesel fuel by 2006, and incentives for manufacturers to introduce low-emission vehicles.

Because air pollution crosses borders, Canada and the U.S. share an interest in reducing air pollution. In 1991, the two countries signed the Canada-U.S. Air Quality Agreement to reduce air pollution emissions; in 2000, ground-level ozone was added to the agreement. Three binational programs have their basis in this initiative: the Great Lakes Basin Airshed Management Framework, the Georgia Basin/Puget Sound International Airshed Strategy, and a joint technical study on the feasibility of emissions trading for nitrogen and sulphur oxides.

We now consider some major air pollutants in more detail. Summaries of these sources are shown in Table 21.2.

Sulphur Dioxide

Sulphur dioxide (SO_2) is a colourless and odourless gas normally present at Earth's surface at low concentrations. A significant feature of SO_2 is that once it is emitted into the atmosphere, it can be converted through complex oxidation reactions into fine particulate sulphate (SO_4) and removed from the atmosphere by wet or dry deposition. The major anthropogenic source of sulphur dioxide is the burning of fossil fuels, mostly coal in power plants (see Table 21.2). Another major source comprises a variety of industrial processes, ranging from petroleum refining to the production of paper, cement, and aluminum.[2, 4]

Adverse effects associated with sulphur dioxide depend on the dose or concentration present (see Chapter 15) and include injury or death to animals and plants, as well as corrosion of paint and metals. Crops such as alfalfa, cotton, and barley are especially susceptible. Sulphur dioxide is capable of causing severe damage to the lungs of human and other animals, particularly in the sulphate form. It is also an important precursor to acid precipitation.[2, 4] (See A Closer Look 21.1.)

Environment Canada reports that concentrations of SO_2 declined 61% on average between 1974 and 1992 (Figure 21.7). Canada's total emissions averaged about 2.79 metric tonnes per 1000 sq. km (19th in the world); U.S. emissions were much lower at about 1.68 metric tonnes per 1000 sq. km (38th in the world). Emissions peaked in the early 1970s and since have been reduced by about 50%, as a result of effective emission controls.

Nitrogen Oxides

Although nitrogen oxides (NOx) occur in many forms in the atmosphere, they are emitted largely in two forms: nitric oxide (NO) and nitrogen dioxide (NO_2); and only these two forms are subject to emission regulations. The more important of the two is NO_2, a yellow-brown to reddish-brown gas. A major concern with nitrogen dioxide is that it may be converted by complex

Figure 21.7 • Air quality trends in Canada. Percent of regulatory standard for total suspended particulates, carbon monoxide, sulfur dioxide, nitrogen dioxide, lead and ground-level ozone (*Source*: Environment Canada accessed March 22, 2005 at www.ec.gc.ca/pdb/pa/trmap_e.gif.)

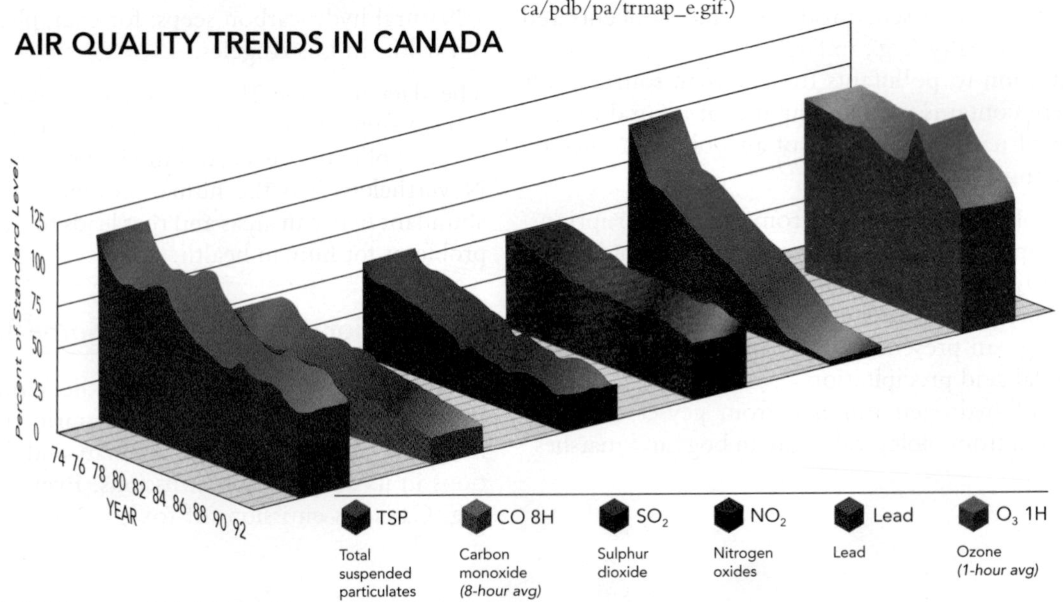

Table 21.2 • Major Natural and Human-Produced Components of Air Pollutants

Air Pollutants	Emissions (% of total) Natural	Human-Produced	Major Sources of Human-Produced Components	Percent
Particulates	85	15	Fugitive (mostly dust)	85
			Industrial processes	7
			Combustion of fuels (stationary sources)	8
Sulphur oxides (SO_X)	50	50	Combustion of fuels (stationary sources, mostly coal)	84
			Industrial processes	9
Carbon monoxide (CO)	91	9	Transportation (automobiles)	54
Nitrogen dioxide (NO_2)		Nearly all	Transportation (mostly automobiles)	37
			Combustion of fuels (stationary sources, mostly natural gas and coal)	38
Ozone (O_3)	A secondary pollutant derived from reactions with sunlight, NO_2, and oxygen (O_2)		Concentration present depends on reactions in lower atmosphere involving hydrocarbons and thus automobile exhaust	
Hydrocarbons (HC)	84	16	Transportation (automobiles)	27
			Industrial processes	7

reactions in the atmosphere to an ion, NO_3^{2-}, within small water particulates, impairing visibility. Both NO and NO_2 are major contributors to the development of smog, and NO_2 is a major contributor to acid precipitation (see A Closer Look 21.1). Nearly all NO_2 is emitted from anthropogenic sources. The two main sources are automobiles and power plants that burn fossil fuels.[2]

The environmental effects of nitrogen oxides on humans are variable but include irritation of eyes, nose, throat, and lungs and increased susceptibility to viral infections, including influenza (which can cause bronchitis and pneumonia).[2] Nitrogen oxides may suppress plant growth. When the oxides are converted to their nitrate form in the atmosphere, they impair visibility. However, when nitrate is deposited on the soil, it can promote plant growth through nitrogen fertilization.

Environment Canada reports an average reduction of 38% in nitrogen dioxide concentrations in Canadian cities between 1974 and 1992. In 2000, Canadian emission rates of NOx were about 1.15 metric tonnes/1000 sq. km (17th in the world); the equivalent U.S. emissions were 1.29 metric tonnes/1000 sq. km (13th in the world). Emissions are primarily from combustion of fuels in power plants and vehicles. They have been reduced by about 17% since 1970.

Carbon Monoxide

Carbon monoxide (CO) is a colourless, odourless gas that even at very low concentrations is extremely toxic to humans and other animals. The high toxicity results from a physiological effect—carbon monoxide and hemoglobin in blood have a strong natural attraction for one another. The hemoglobin in our blood will take up carbon monoxide nearly 250 times more rapidly than it will oxygen. Therefore, if there is any carbon monoxide in the vicinity, a person will take it in very readily, with potentially dire effects. Many people have been accidentally asphyxiated by carbon monoxide produced from incomplete combustion of fuels in campers, tents, and houses. The effects depend on the dose or concentration of exposure and range from dizziness and headaches to death. Carbon monoxide is particularly hazardous to people with known heart disease, anemia, or respiratory disease. In addition, it may cause birth defects, including mental retardation, and impaired fetus growth.[2] Finally, the effects of carbon monoxide tend to be worse at higher altitudes, where oxygen levels are naturally lower. Detectors (similar to smoke detectors) are now in common use to warn people if CO in a building becomes concentrated at a potentially harmful level.

Approximately 90% of the carbon monoxide in the atmosphere comes from natural sources. The other 10% comes mainly from fires, automobiles, and other sources of incomplete burning of organic compounds. Concentrations of carbon monoxide can build up and cause serious health effects in a localized area.

Environment Canada reports a 70% reduction in carbon monoxide concentrations in Canadian cities between 1974 and 1972. In 1995, total Canadian emissions of CO were estimated at 52.5 million metric tonnes; equivalent U.S. emissions totalled 90.2 million metric tonnes, roughly half the levels of the early 1970s. Most emissions are through vehicle tailpipes. Reductions have resulted largely from cleaner-burning automobile engines.

Ozone and Other Photochemical Oxidants

Photochemical oxidants result from atmospheric interactions of nitrogen dioxide and sunlight. The most common photochemical oxidant is ozone (O_3), a colourless gas with a slightly sweet odour. In addition to ozone, a number of photochemical oxidants known as PANs (peroxyacyl nitrates) occur with photochemical smog (discussed later in the chapter).

Ozone is a form of oxygen in which three atoms of oxygen occur together rather than the normal two. Ozone is relatively unstable and releases its third oxygen atom readily, so it oxidizes or burns things more readily and at lower concentrations than does normal oxygen. Ozone is sometimes used to sterilize; for example, bubbling ozone gas through water is a method used to purify water. The ozone is toxic to and kills bacteria and other organisms in the water. When it is released into the air or produced in the air, ozone may injure living things.

Ozone is very active chemically, and it has a short average lifetime in the air. Because of the effect of sunlight on normal oxygen, ozone forms a natural layer high in the atmosphere (stratosphere). This ozone layer protects us from harmful ultraviolet radiation from the sun. Thus, although ozone is considered a pollutant in the lower atmosphere when concentrations are above the National Ambient Air Quality Standard threshold, it is beneficial in the stratosphere. The important topic of stratospheric ozone depletion is discussed in Chapter 22.

Ozone is a secondary pollutant produced on bright, sunny days in areas where there is much primary pollution. The major sources of the chemicals that produce ozone, as well as other oxidants, are automobiles, fossil fuel burning, and industrial processes that produce nitrogen dioxide. Because of the nature of its formation, ozone is difficult to regulate. It is the pollutant whose health standard is most frequently exceeded in urban areas of Canada and the United States.[5, 6] The adverse environmental effects of ozone and other oxidants, like those of other pollutants, depend in part on the dose or concentration of exposure and include damage to plants and animals as well as to materials such as rubber, paint, and textiles.

Effects of ozone on plants can be subtle. At very low concentrations, ozone can reduce growth rates while not producing any visible injury. At higher concentrations, ozone kills leaf tissue and, if pollutant levels remain high, whole plants. The death of white pine trees along highways in Eastern North America is believed to be due in part to ozone pollution. Ozone's effect on animals, including people, involves various kinds of damage, especially to the eyes and the respiratory system. Many millions of Americans are often exposed to ozone levels that damage cell walls in lungs and airways. This causes tissue to redden and swell, inducing cellular fluids to seep into the lungs. Eventually, the lungs decrease in elasticity and are more susceptible to bacterial infection; and scars and lesions may form in the airways. Even young, healthy people may not be able to breathe normally, and on especially polluted days, breathing may be shallow and painful.[6] In Canada, the national maximum acceptable objective of 82 parts per billion was exceeded in five to six days per year in the early 1980s. By the mid-1990s, that number had dropped to one or two days a year.

Volatile Organic Compounds

Volatile organic compounds (VOCs) include a variety of organic compounds used as solvents in industrial processes such as dry cleaning, degreasing, and graphic arts. Hydrocarbons—compounds composed of hydrogen and carbon—comprise one group of VOCs. Thousands of hydrocarbon compounds exist, including natural gas, methane (CH_4) butane (C_4H_{10}) and propane (C_3H_8). Analysis of urban air has identified many hydrocarbons, some of which react with sunlight to produce photochemical smog. Potential adverse effects of hydrocarbons are numerous. Many are toxic to plants and animals, and some may be converted to harmful compounds through complex chemical changes that occur in the atmosphere.

On a global basis, only about 15% of hydrocarbon emissions (primary pollutants) are anthropogenic. In Canada and the United States, however, nearly half of hydrocarbons entering the atmosphere are emitted from anthropogenic sources. The largest human source of hydrocarbons in Canada and the United States is automobiles. Anthropogenic sources are particularly abundant in urban regions. However, in some southeastern U.S. cities such as Atlanta, Georgia, natural emissions probably exceed those from automobiles and other human sources.[3]

In 1995, Canadian emission of VOCs was estimated at about 4.6 million metric tonnes, a little less than a quarter of the U.S. figure of 19.5 metric tonnes despite Canada'. Like sulphur dioxide and nitrogen oxide emissions, VOCs peaked in the early 1970s. As noted, a major source of hydrocarbons (VOC) is automobiles. Effective government-mandated emission controls for automobiles are thus responsible for the 50% lower emissions.

Particulate Matter (PM 10 and PM 2.5)

Particulate matter (PM 10) is made up of particles less than 10 μm in diameter. The term is used for varying mixtures of particles suspended in the air we breathe. Particles are present everywhere, but high concentrations and/or specific types of particles have been found to present a serious danger to human health. Asbestos, for example, is particularly dangerous.

Farming adds considerable particulate matter to the atmosphere, as do windstorms in areas with little vegetation and volcanic eruptions. Nearly all industrial processes, as well as burning fossil fuels, release particulates

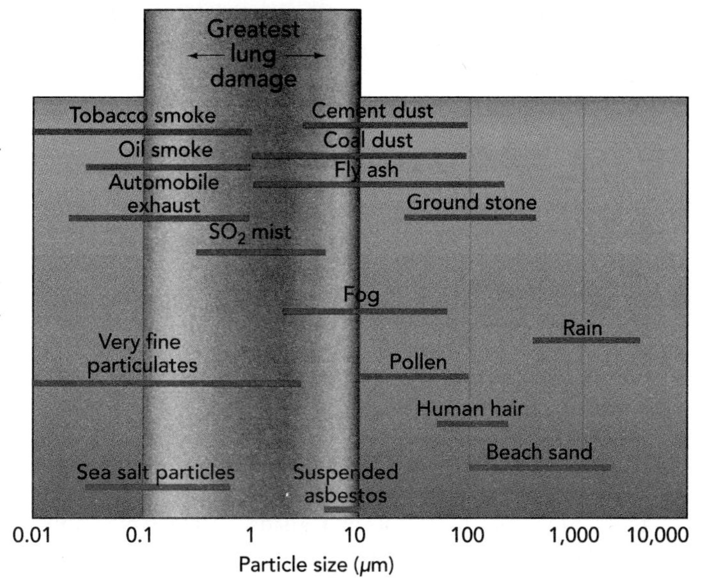

Figure 21.8 • **Size of selected particulates.** Shaded area shows size range that produces the greatest lung damage. [*Source*: Modified from Fig. 7–8, p. 244 from *Chemistry, Man and Environmental Change: An Integrated Approach*, by J. Calvin Giddings. Copyright © 1973 by J. Calvin Giddings. Reprinted by permission of HarperCollins Publishers, Inc.]

into the atmosphere. Much particulate matter is easily visible as smoke, soot, or dust; other particulate matter is not easily visible. Particulates include materials such as airborne asbestos particles and small particles of heavy metals, such as arsenic, copper, lead, and zinc, which are usually emitted from industrial facilities such as smelters.

Of particular concern are very fine particle pollutants (PM 2.5) less than 2.5 μm in diameter (2.5 millionths of a meter; Figure 21.8). For comparison, the diameter of human hair is about 60 μm to 150 μm. Fine particles are easily inhaled into the lungs, where they can be absorbed into the bloodstream or remain embedded for a long period of time. Among the most significant fine particulate pollutants are sulphates and nitrates. These are mostly secondary pollutants produced in the atmosphere through chemical reactions between normal atmospheric constituents and sulphur dioxide and nitrogen oxides. These reactions are important in the formation of sulphuric and nitric acids in the atmosphere and are further discussed when we consider acid precipitation.[2]

When measured, particulate matter is often referred to as *total suspended particulates* (TSPs). Values for TSPs tend to be much higher in large cities in developing countries such as Mexico, China, and India than in developed countries such as Japan, Canada, and the United States (Figure 21.9).

Particulates affect human health, ecosystems, and the biosphere. In Canada and the United States, particulate air pollution contributes to the death of an estimated 75,000 people annually.[7] Recent studies estimate that 2–9% of human mortality in cities is associated with particulate pollution; risk of mortality is about 15–25% higher in cities with the highest levels of fine particulate pollution.[8] As mentioned, particulates that enter lungs may lodge there, with chronic effects on respiration. Particulates are linked to both lung cancer and bronchitis. Particulate matter is especially hazardous to the elderly and to individuals with respiratory problems, such as asthma. There is a direct relationship between particulate pollution and increased hospital admissions for respiratory distress.

Dust raised by road building and ploughing not only makes breathing more difficult for humans and other animals, but also can be deposited on surfaces of green plants, where it may interfere with absorption of carbon dioxide and oxygen and release of water (transpiration). On a larger scale, particulates associated with large construction projects, such as housing developments, shopping centres, and industrial parks, may injure or kill plants and animals and damage surrounding areas, changing species composition, altering food chains, and thereby affecting ecosystems. Modern industrial processes have greatly increased total suspended particulates in Earth's atmosphere. Particulates block sunlight and may cause changes in climate. Such changes have lasting effects on the biosphere.

Environment Canada reports a 54% reduction in total suspended particulate concentrations in Canadian cities between 1974 and 1992. In 1995, total Canadian anthropogenic emissions of suspended particulates in Canada were approximately 15.8 million tonnes.

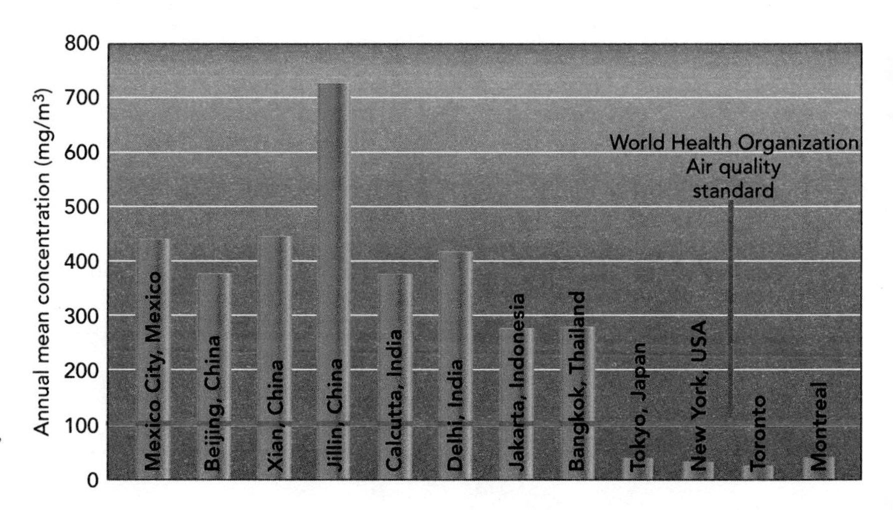

Figure 21.9 • **Total suspended particulates (TSP) for several large cities in developing countries (orange) and developed countries (green).** The value of 100 μg/m3 is the air quality standard set by the World Health Organization. [*Source*: Modified from R. T. Watson, Intergovernment Panel on Climate Change, presentation at the Sixth Conference of Parties to the United Nations Framework Convention on Climate Change, November 13, 2000, Figure 20.]

A CLOSER LOOK 21.1

Acid Precipitation

Acid precipitation—acidic rain and snow—is one of the most prominent environmental issues of the last 30 years. Although it has likely been around since the beginning of the Industrial Revolution, acid precipitation is now known to be a global problem affecting all major coal-using countries, including much of Eastern Europe, Scandinavia, Great Britain, and China. Most of eastern Canada and the U.S. Northeast are affected, as well as parts of western British Columbia.

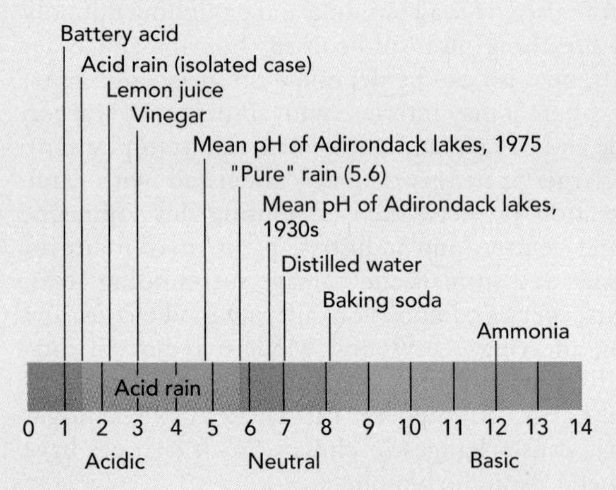

Figure 21.10 • The pH scale. [*Source*: Modified from U.S. Environmental Protection Agency, 1980.]

Many people are surprised to learn that all rainfall is slightly acidic; water reacts with atmospheric carbon dioxide to produce weak carbonic acid. Thus, pure rainfall has a pH of about 5.6, where 1 is highly acidic and 7 is neutral (see Figure 21.10). Because the pH scale is logarithmic, a pH value of 3 is 10 times more acidic than a pH value of 4 and 100 times more acidic than a pH value of 5. Automobile battery acid has a pH value of 1.

Acid precipitation includes both wet (rain, snow, fog) and dry (particulate) acidic depositions. The depositions occur near and downwind of areas where the burning of fossil fuels creates major emissions of sulphur dioxide (SO_2) and nitrogen oxides (NO_X) (Figure 21.7). Emissions of SO_2 peaked in the 1970s at about 30 million metric tonnes per year and had declined to about 18 million metric tonnes per year by 1998. Nitrogen oxides levelled off at about 23 million metric tonnes per year in the mid-1980s. Although these oxides are the primary contributors to acid precipitation, other acids are also involved. An example is hydrochloric acid, which is emitted from coal-fired power plants.

In the atmosphere, sulphur dioxide and nitrogen oxides are transformed by reactions with oxygen and water vapour to sulphuric and nitric acids. These acids may travel long distances with prevailing winds to be deposited as acid precipitation (Figure 21.11). Sulphur dioxide is emitted primarily from stationary sources, such as power plants that burn fossil fuels, whereas nitrogen oxides are emitted from both stationary sources and transport-related sources, such as automobiles.

In some areas, such as Sudbury, Ontario, stationary sources have attempted to reduce the local effects of emissions by constructing taller emission stacks. Taller stacks have reduced local concentrations of air pollutants but increased regional effects by spreading pollution more widely. Many Canadian areas affected by acid precipitation can trace the source of those pollutants to areas thousands of kilometres to the south and west. About one-third of the total amount deposited over eastern Canada and the northeastern United States originates from sources farther away than 500 km.[9]

Figure 21.11 • Conceptual diagram showing selected aspects of acid precipitation formation and paths.

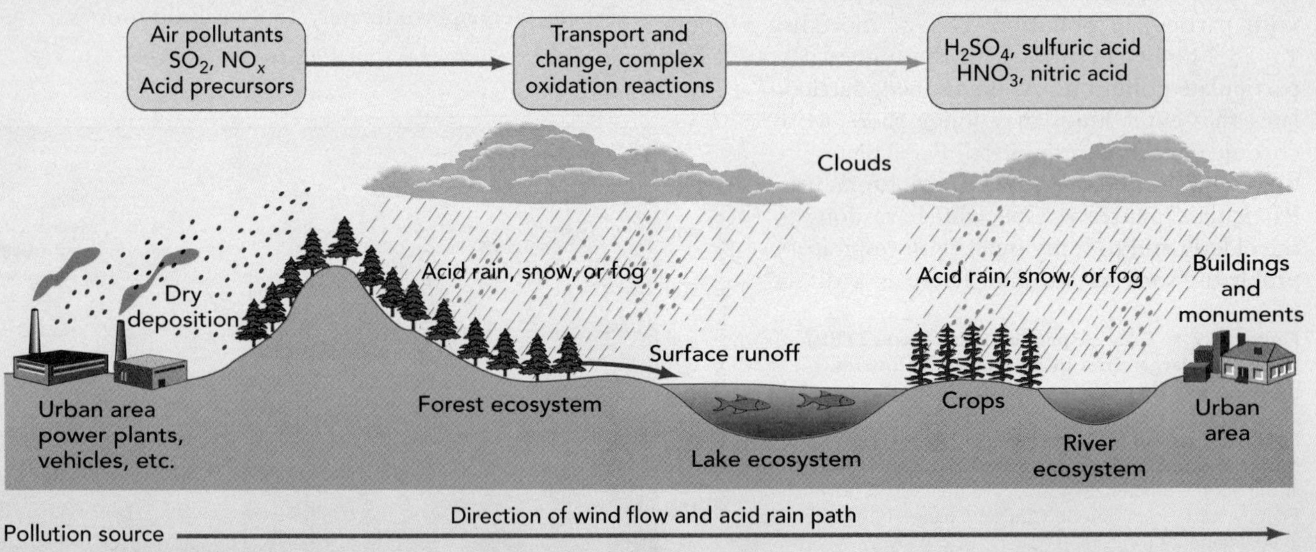

Environmental Impacts of Acid Precipitation

Geology and climatic patterns as well as types of vegetation and soil composition affect the potential impact of acid precipitation (Figure 21.12). Sensitive areas are those in which bedrock or soil cannot buffer (neutralize) acid input. Areas with bedrock containing calcium carbonate ($CaCO_3$) have high natural buffering capacity, because the calcium carbonate reacts with hydrogen in the acid, removing hydrogen ions from solution and raising the pH. Areas less likely to suffer damage from acid precipitation are those in which bedrock and soils contain limestone or other carbonate material. In contrast, areas with abundant granitic rocks and areas in which soils have little buffering action—including much of the Canadian Shield—are sensitive to acid precipitation.[10,11]

It has long been suspected that acid precipitation, whether snow, rain, fog, or dry deposition, adversely affects ecosystems. Scientists have now identified acid precipitation and other air pollution as the cause of death for thousands of acres of evergreen trees in Bavaria, New England, and eastern Canada. Forest soils and agricultural land exposed to acid precipitation may lose their fertility, because acid water moving through them can leach nutrients and releases elements that are toxic to plants. This results in weakening of plants, so that they become more susceptible to disease, drought, and consumption by herbivores, with associated alteration of habitat and food availability for consumers, and reduced crop yields in agricultural areas.

Acid precipitation is also known to have caused major declines in freshwater fish populations in Canada, Scandinavia, and other northern countries. (Figure 21.13) Acidic water encourages the dissolution of chemical elements like phosphorus that are necessary for life in the lake. Once in solution, the necessary elements are lost from the lake with the water outflow. With few nutrients, primary production in the lake is reduced, and higher trophic levels have little on which to feed.

Canadian scientists added sulphuric acid to a lake in the Experimental Lakes Area of northwest Ontario (Chapter 2) over a period of years and observed the effects. When the experiment started, the pH of the lake was 6.8. The following year, owing to addition of the acid, the pH dropped to 6.1. The initial drop in pH was not harmful to the lake; but as more and more acid was added, the pH dropped first to 5.8, then to 5.6, then to 5.4, and finally, five years after the project started, to 5.1. The problems started when the pH was lowered to 5.8. Some species disappeared, and others experienced reproductive failure. At a pH of 5.6, the death rate among lake trout embryos increased. When the pH was lowered to 5.4, lake trout reproduction failed.[10]

Thousands of rivers and lakes in Canada and the United States located in areas sensitive to acid precipitation are currently in various stages of acidification. In Nova Scotia, at least a dozen rivers have water so acidic part of the year that they no longer support healthy populations of Atlantic salmon.

Acid precipitation damages not only forests and lakes but also many building materials, including steel, galvanized steel, paint, plastics, cement, masonry, and several types of rock, especially limestone, sandstone, and marble (Figure 21.14). Classical buildings on the Acropolis in Athens and in other cities show chemical weathering that accelerated in the twentieth century as a result of air pollution. The problem has grown to such an extent that restoration costs for buildings, statues and other monuments now approach billions of dollars a year. Particularly important statues in Greece and other areas have been removed and placed in protective glass containers, with replicas standing in their former outdoor locations for tourists to view.[12]

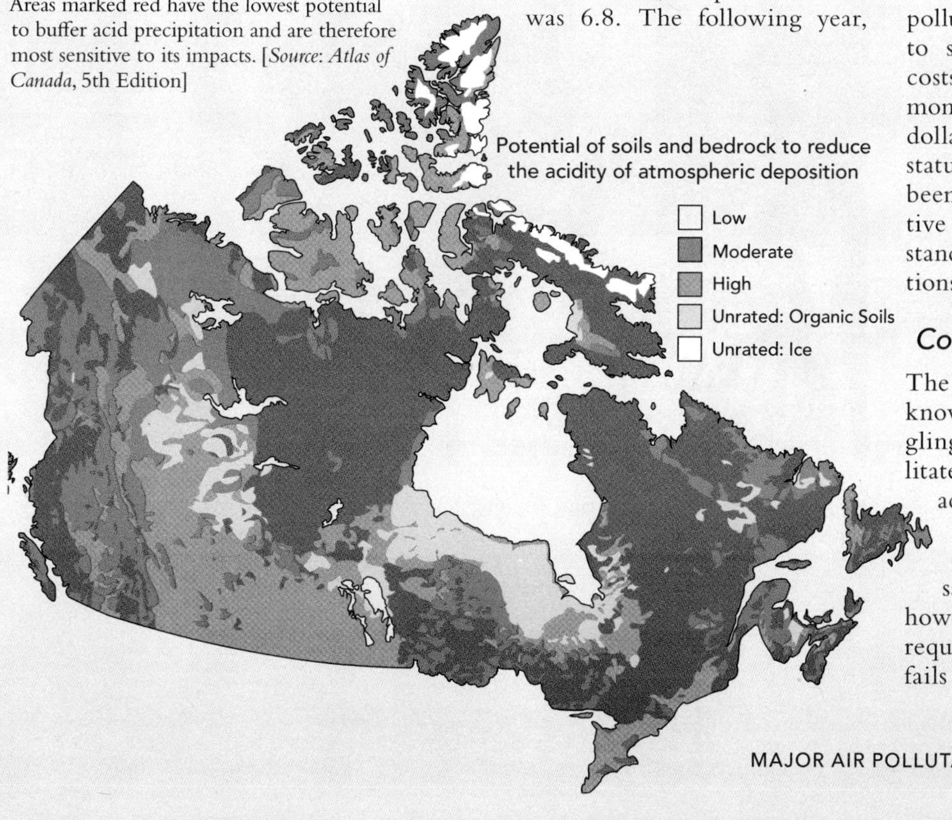

Figure 21.12 • The potential of Canadian soils to reduce acid rain. Areas marked red have the lowest potential to buffer acid precipitation and are therefore most sensitive to its impacts. [*Source*: *Atlas of Canada*, 5th Edition]

Control of Acid Precipitation

The cause of acid precipitation is known; it is the solution we are struggling with. One option is to rehabilitate acidified lakes by the periodic addition of lime, as has been done in Ontario, Sweden, and New York State. This solution is not satisfactory over a long period, however, because it is expensive and requires a continuing effort; it also fails to address the root cause of the problem. The only long-term solution for acid precipitation is to ensure that emissions of acid-forming components to the atmosphere are minimized.

From an environmental point of view, the best strategy is increasing energy efficiency and conservation measures that result in burning less coal and using non-polluting alternative energy sources. Another strategy is to utilize pollution abatement technology at power plants to lower emissions of air pollutants.[13] Such technology is readily available but expensive, and adds cost in producing energy. Nevertheless, sulphur dioxide emissions have been reduced about 50% since 1970. This is a big improvement that will reduce acid precipitation.

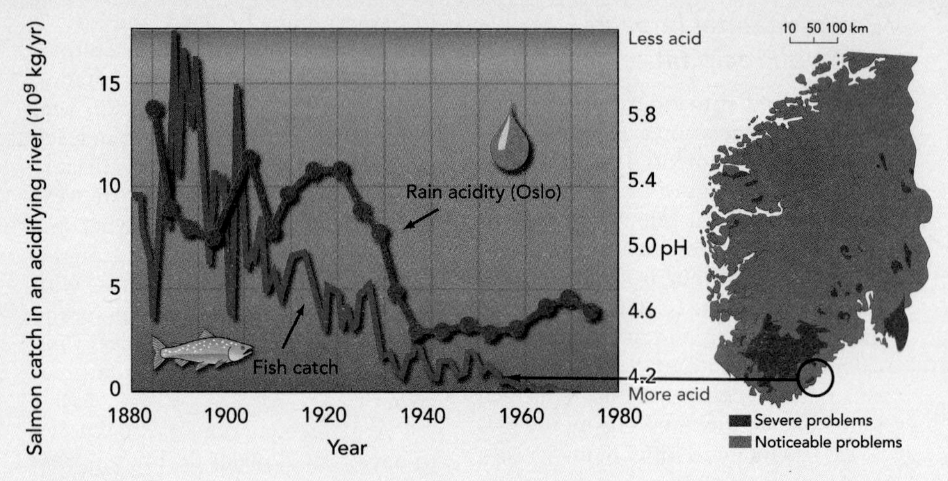

Figure 21.13 • Acid Rain in Norway. In Norway, many lakes in the south have severe problems with acid precipitation. The precipitation has become more acidic over the last 20 years, as measured at Oslo and at five other sites in southern Norway. At the same time, the salmon catch in the Tovdalselva River of southern Norway has decreased dramatically. [*Sources*: Adapted from I. P. Muniz, H. Leivestad, E. Gjessing, E. Joranger, and D. Svalastog, Acid Precipitation: Effects on Forest and Fish, SNSF Project, IR 13/75 (As, Norway: Government of Norway, 1976); S. Oden, "The Acidity Problem—An Outline of Concepts," *Water, Air, and Soil Pollution* 6 (1976):137–166; E. Snekvik, Norwegian Directorate for Game and Freshwater Fish, unpublished report, 1970; and R. F. Wright, T. Dale, E. G. Gjessing, G. R. Hendrey, A. Henriksen, M. Johannessen, and I. P. Muniz, "Impact of Acid Precipitation on Freshwater Ecosystems in Norway," *Water, Air, and Soil Pollution* 6 (1976):438–499.]

Figure 21.14 • Damage to a statue resulting from acid deposition (left) and the same statue following restoration (right).

Hydrogen Sulphide

Hydrogen sulphide (H_2S) is a highly toxic corrosive gas easily identified by its rotten egg door. Hydrogen sulphide is produced from natural sources such as geysers, swamps, and bogs and from human sources such as industrial plants that produce petroleum or that smelt metals. Potential effects of hydrogen sulphide include functional damage to plants and health problems ranging from toxicity to death for humans and other animals.[4]

Hydrogen Fluoride

Hydrogen fluoride (HF) is a gaseous pollutant released by some industrial activities such as production of aluminum, coal gasification, and burning of coal in power plants. Hydrogen fluoride is extremely toxic. Even a small concentration (as low as 1 ppb) of HF may cause problems for plants and animals. HF is potentially dangerous to grazing animals because some forage plants can become toxic when exposed to this gas.[2]

Other Hazardous Gases

It is a rare month when the newspapers do not carry a story of a truck or train accident that releases toxic chemicals in a gaseous form into the atmosphere. People are often evacuated until the leak is stopped or the gas has dispersed to a non-toxic level. Chlorine gases and a variety of other materials used in chemical and agricultural processes may be involved.

Another source of gaseous air pollution is sewage treatment plants. Urban areas deliver a variety of organic chemicals, including paint thinner, industrial solvents, chloroform, and methyl chloride, to treatment plants through sewers. These materials are not usually removed in treatment plants; in fact, the treatment processes facilitate the evaporation of the chemicals into the atmosphere, where they may be inhaled. Many of the chemicals are toxic or are suspected of being carcinogens.

Some chemicals are so toxic that extreme care must be taken to ensure they do not enter the environment. This was demonstrated in an extreme event on December 3, 1984, when a toxic liquid from a pesticide plant leaked, vaporized, and formed a deadly gas cloud that settled over a 64-km^2 area of Bhopal, India. The gas leak lasted less than an hour; yet more than 2,000 people were killed and more than 15,000 were injured. The colourless gas that resulted from the leak was methyl isocyanate, which causes severe irritation (burns on contact) to eyes, nose, throat, and lungs. Breathing the gas in concentrations of only a few parts per million (ppm) causes violent coughing, swelling of the lungs, bleeding, and death. Less exposure can cause a variety of problems, including loss of sight. The Bhopal accident has implications related to our key theme "an urban world." Clearly, chemicals that can cause catastrophic injuries and death should not be stored close to large population centres. In addition, chemical plants need to have reliable accident-prevention equipment and personnel trained to control and prevent potential problems.

Lead

Lead is an important constituent of automobile batteries and many other industrial products. When lead is added to gasoline, it helps protect engines and promotes more even fuel consumption. Lead in gasoline (which is still used in some countries) is emitted into the air with the exhaust. By this process, lead has been spread widely around the world and has reached high levels in soils and waters along roadways.

Once released, lead can be transported through the air as particulates to be taken up by plants through the soil or deposited directly on plant leaves. Thus, it enters terrestrial food chains. When lead is carried by streams and rivers, deposited in quiet waters, or transported to oceans or lakes, it is taken up by aquatic organisms and enters aquatic food chains.

Lead reaches Greenland as airborne particulates and via seawater and is stored in glacial ice. The concentration of lead in Greenland glaciers was essentially zero in A.D. 800 and reached measurable levels with the beginning of the Industrial Revolution in the mid-eighteenth

Figure 21.15 • Vancouver, British Columbia: haze resulting from particulate (PM 10) pollution. Sources of particulates include construction sites and dirt roads (60% of total particulates) and natural and other sources (40%).

century. The lead content of the glacial ice increased steadily from 1750 until about 1950, when the rate of lead accumulation began to increase rapidly. This sudden upsurge reflects rapid growth in the use of lead additives in gasoline. The accumulation of lead in the Greenland ice illustrates that our use of heavy metals in the twentieth century reached the point of affecting the entire biosphere.

Lead has now been removed from nearly all gasoline in Canada, the United States, and much of Europe. In Canada, mean air concentration of lead has dropped by about 98% since the early 1970s (Figure 21.7). The reduction and eventual elimination of lead in gasoline is a good start in reducing levels of anthropogenic lead in the biosphere.

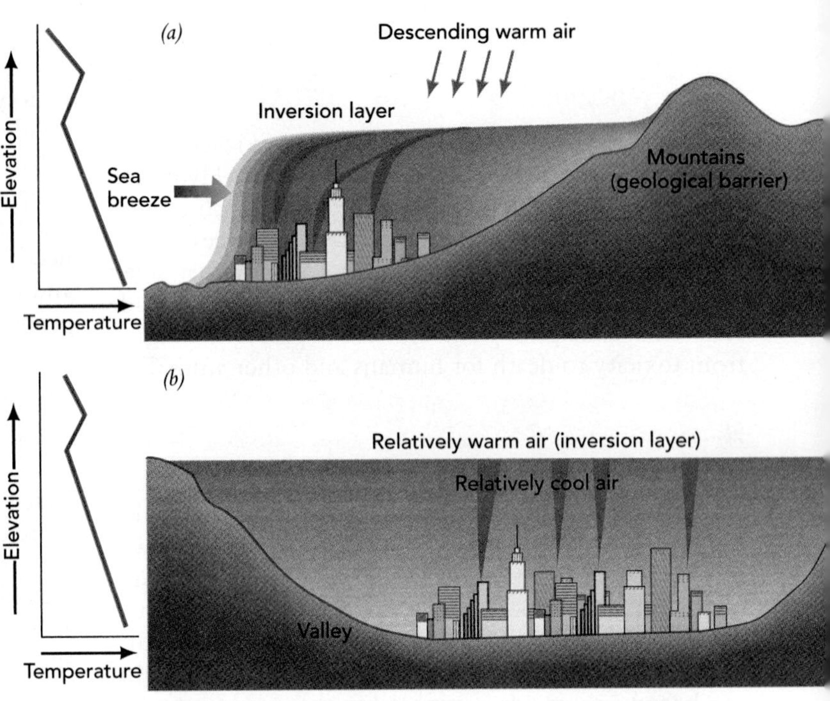

Figure 21.16 • Two causes for the development of atmospheric inversion, which may aggravate air pollution problems. *(a)* Descending warm air forms a semipermanent inversion layer. Polluted air blown inland from the ocean moves up canyons but is blocked by the mountains and trapped. The air pollution that develops occurs primarily during the summer and fall. This scenario is typical of smog in Los Angeles, California. *(b)* Relatively cool air is overlain by warm air, for instance when cloud cover develops over an urban area. Incoming solar radiation is blocked by the clouds, which reflect and absorb some of the solar energy and are warmed. On the ground, or near Earth's surface, the air cools. Fog may form when moist air cools. The air is cold, and people burn more fuel for heat, so more pollutants enter and build up in the atmosphere. It was this mechanism that caused the deadly 1952 London smog.

21.6 Variability of Air Pollution

Pollution problems vary in different regions of the world. There is great variance even within Canada. For example, as we will see, in many cities, nitrogen oxides and hydrocarbons are particularly troublesome because they combine in the presence of sunlight to form photochemical smog. Most of the nitrogen oxides and hydrocarbons are emitted from automobiles, a collection of mobile sources. In other regions, such as the Great Lakes region, air quality problems also result from emissions of sulphur dioxide and particulates from industry and from coal-burning power plants, which are point sources.

Air pollution also varies with the time of year. For example, smog is usually a problem mostly in the summer months when there is a lot of sunlight; and particulates are a problem in dry months when wildfires are likely and during months when the wind blows across bare soil.

Haze from Afar

Air quality concerns are not restricted to urban areas. It is logical to assume that air in the Arctic environments of Canada's far north would be pristine in quality, except perhaps near areas where petroleum is being vigorously developed. However, ongoing studies suggest that even these remote areas have air pollution problems that originate from sources in Eastern Europe and Eurasia.

It is suspected that pollutants from burning fossil fuels in Eurasia are transported via the jet stream, moving at speeds that may exceed 400 km/hr, northeast from Eurasia over the North Pole and eventually to the Canadian Arctic and Alaska. There, the air mass slows, stagnates, and produces what is known as the Arctic haze. Concentrations of air pollutants, which include oxides of sulphur and nitrogen, are high enough that the air quality is being compared with that of some eastern cities, such as Toronto or Boston. Air quality problems in remote areas have significance as we try to understand air pollution at the global level.[14]

Another global event occurred in the spring of 2001, when a white haze consisting of dust from Mongolia and industrial particulate pollutants arrived in North America. The haze could be seen from Canada to Mexico. The particulates were close enough to the ground to cause respiratory problems for people. In much of the United States, pollution levels from the haze alone were as high as two-thirds of federal health limits. The haze demonstrates what was formerly believed—that pollution from Asia is carried by winds across the Pacific Ocean.

21.7 Urban Air Pollution

Wherever many sources emit air pollutants over a wide area (whether automobile emissions in Toronto or smoke from wood-burning stoves in Nova Scotia), air pollution can develop. Whether air pollution does develop depends on topography and meteorological

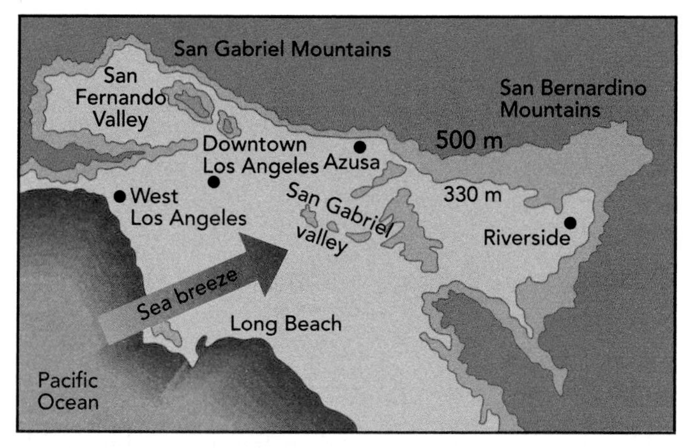

Figure 21.17 • Part of southern California showing the Los Angeles basin (south coast air basin). [*Source*: Modified from S. J. Williamson, Fundamentals of Air Pollution, © 1973, by Addison-Wesley, Reading, Mass.]

conditions, because these factors determine the rate at which pollutants are transported away from their sources and converted to harmless compounds in the air. When the rate of production exceeds the rate of degradation and of transport, dangerous conditions can develop, as illustrated in the case study that opened this chapter.

Influences of Meteorology and Topography

Meteorological conditions can determine whether air pollution is a nuisance or a major health problem. The primary adverse effects of air pollution are damage to green plants and aggravation of chronic illnesses in people; most of these effects are due to relatively low-level concentrations of pollutants over a long period of time. Periods of pollution generally do not directly cause large numbers of deaths. However, as with the London and Pennsylvania cases described earlier, serious pollution events (disasters) can develop over a period of days and lead to increases in illnesses and deaths.

In the lower atmosphere, restricted circulation associated with inversion layers may lead to pollution events. An **atmospheric inversion** occurs when warmer air is found above cooler air, and it poses a particular problem when there is a stagnant air mass. Figure 21.16 shows two types of atmospheric inversion that may contribute to air pollution problems.

Cities situated in a valley or topographic bowl surrounded by mountains are more susceptible to smog problems than are cities in open plains. Surrounding mountains and the occurrence of temperature inversions prevent the pollutants from being transported by winds and weather systems. The production of air pollution is particularly well documented for Los Angeles, which has mountains surrounding part of the urban area and lies within a region where the air lingers, allowing pollutants to build up (Figure 21.17).[15-17]

Potential for Urban Air Pollution

We have seen that topographical and meteorological conditions are important in the development of air pollution. More specifically, the potential for air pollution in urban areas is determined by the following factors:

1. The rate of emission of pollutants per unit area.
2. The downwind distance that a mass of air moves through an urban area.
3. The average speed of the wind.
4. The elevation to which potential pollutants can be thoroughly mixed by naturally moving air in the lower atmosphere (Figure 21.18).[15]

The concentration of pollutants in the air is directly proportional to the first two factors. That is, as either the emission rate or downwind travel distance increases, so will the concentration of pollutants in the air. A good example is provided by the Los Angeles basin (see Figure 21.21). If there is a wind from the ocean, as is generally the case, coastal areas such as West Los Angeles will experience much less air pollution than more inland areas.

Assuming a constant rate of emission of air pollutants, the air mass will collect more and more pollutants as it moves through the urban area. The inversion layer acts as a lid for the pollutants; however, near a geological barrier, such as a mountain, there may be a chimney effect, in which the pollutants spill over the top of the mountain (see Figures 21.17 and 21.18). This effect has been noticed in the Los Angeles basin, where pollutants may climb several thousand metres,

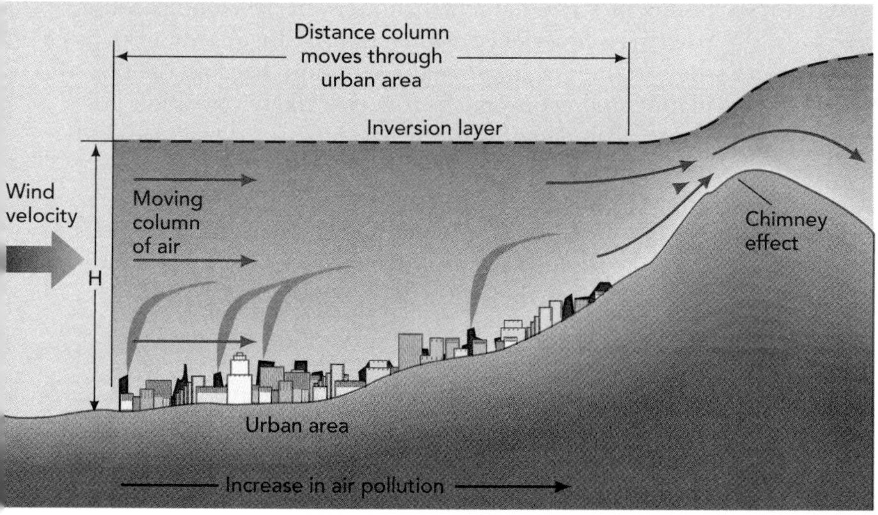

Figure 21.18 • Environmental influences on pollution levels. The higher the wind velocity and the thicker the mixing layer (shown here as H), the less the air pollution. The greater the emission rate and the longer the downwind length of the city, the greater the air pollution. The chimney effect allows polluted air to move over a mountain and down into an adjacent valley.

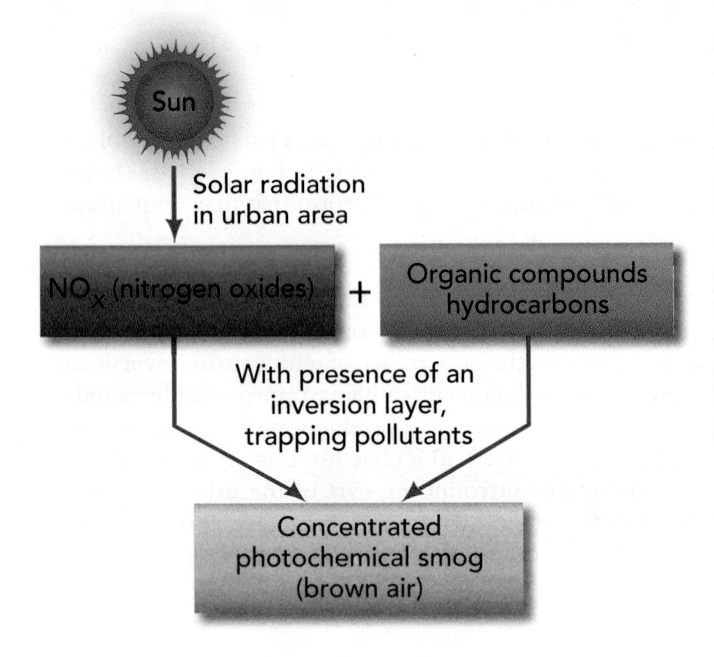

Figure 21.19 • How photochemical smog is produced.

damaging mountain pine trees and other vegetation and spoiling the air of mountain valleys.

City air pollution decreases with increases in the third and fourth factors, which are meteorological: the wind velocity and the height of mixing. The stronger the wind and the higher the mixing layer, the lower the pollution.

Smog

Smog is a general term first used in 1905 for a mixture of smoke and fog that produced unhealthy urban air. It is the most recognized term for urban air pollution. There are two major types of smog: **photochemical smog,** which is sometimes called *L.A.-type smog or brown air;* and **sulphurous smog**, which is sometimes referred to as *London-type smog, gray air, or industrial smog.*

Solar radiation is particularly important in the formation of photochemical smog (Figure 21.19). The reactions that occur in the development of photochemical smog are complex and involve both nitrogen oxides (NO_x) and organic compounds (hydrocarbons).

Figure 21.20 • Development of photochemical smog over an urban area on a typical warm day.

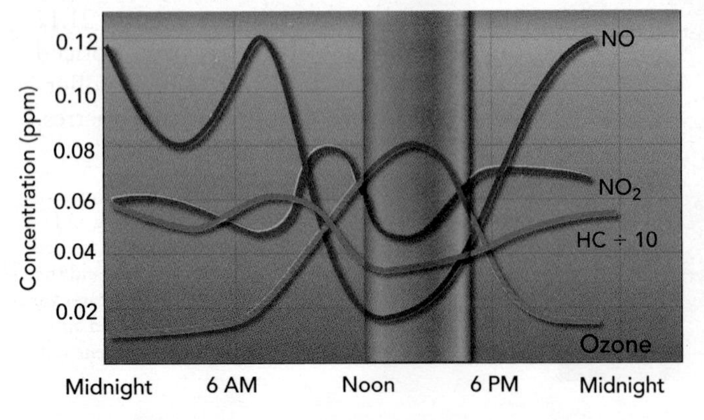

The development of photochemical smog is directly related to automobile use. Figure 21.20 shows a characteristic pattern in terms of how the nitrogen oxides, hydrocarbons, and oxidants (mostly ozone) vary through a typically smoggy day. Early in the morning, when commuter traffic begins to build up, the concentrations of nitrogen oxide (NO) and hydrocarbons begin to increase. At the same time, the amount of nitrogen dioxide, NO_2, may decrease, because sunlight breaks it down to produce NO plus atomic oxygen (NO + O). The atomic oxygen (O) is then free to combine with molecular oxygen (O_2) to form ozone (O_3). As a result, the concentration of ozone also increases after sunrise. Shortly thereafter, oxidized hydrocarbons react with NO to increase the concentration of NO_2 by midmorning. This reaction causes the NO concentration to decrease and allows ozone to build up, producing a midday peak in ozone and a minimum in NO. As the smog develops, visibility may be greatly reduced (Figure 21.21) as light is scattered by the pollutants.

Sulphurous smog is produced primarily by the burning of coal or oil at large power plants. Sulphur oxides and particulates combine under certain meteorological conditions to produce a concentrated sulphurous smog (Figure 21.22).

Future Trends for Urban Areas

What does the future hold for urban areas with respect to air pollution? The optimistic view is that urban air quality will continue to improve as it has in the past 35 years because we know so much about the sources of air pollution and have developed effective ways to reduce it. The pessimistic view is that in spite of this knowledge, population pressures and economics will dictate what happens in many parts of the world, and the result will be poorer air quality (more air pollution) in many locations.

The actual situation in the twenty-first century is likely to be a mixture of the optimistic and pessimistic points of view. Large urban areas in developing countries may experience a reduction in air quality even as they attempt to improve the situation, because the population and economic factors will likely outweigh pollution abatement. Large urban areas in developed and more affluent countries may well continue to experience improved air quality in coming years (see Figure 21.7).

Air pollution abatement in Canadian and U.S. cities will take massive efforts much different from past strategies, including the following features:[16]

- Strategies to discourage automobile use and reduce the number of cars.
- Stricter emission controls for automobiles.

(a)

(b)

Figure 21.21 • The city of Los Angeles, on *(a)* a clear day and *(b)* a smoggy day.

- A requirement for a certain number of zero-pollutant automobiles (electric cars) and hybrid cars with fuel cell and gasoline engines.
- A requirement for more gasoline to be reformulated to burn cleaner.
- Improvements in public transportation and incentives for people to use it.
- Mandatory carpooling.
- Increased controls on industrial and household activities known to contribute to air pollution.

At the household level, for example, common materials such as paints and solvents will be reformulated so that their fumes will cause less air pollution. Eventually, certain equipment, such as gasoline-powered lawn mowers that contribute to air pollution, may be banned.

As we saw earlier, there are encouraging signs of improvement in air quality. Peak levels of key air pollutants have dropped to half their 1970s values, despite huge growth in the population and number of vehicles. Nevertheless, even if all the aforementioned controls in urban areas are implemented, air quality will continue to be a significant problem in coming decades, particularly if the urban population continues to increase.

Many large and not-so-large Canadian cities have poor air quality a significant part of the year. Based on the criteria of 30 days per year of unhealthy air resulting from ozone pollution, millions of Canadians live in cities where hazardous air pollution exists. Table 21.3 lists the 24 most polluted cities with respect to ozone and particulate matter in Canada. These include Kitchener, Toronto, Windsor, and Hamilton, Ontario. The Windsor-Detroit corridor is one of the most industrialized regions in Canada and has among the worst air pollution. That area has been the focus of numerous binational (Canada-United States) studies examining the sources of and solutions for air pollution, and the links between air pollution and public health. In contrast, some of the cities with the cleanest air in Canada include St. John's, Newfoundland and Victoria, British

(a)

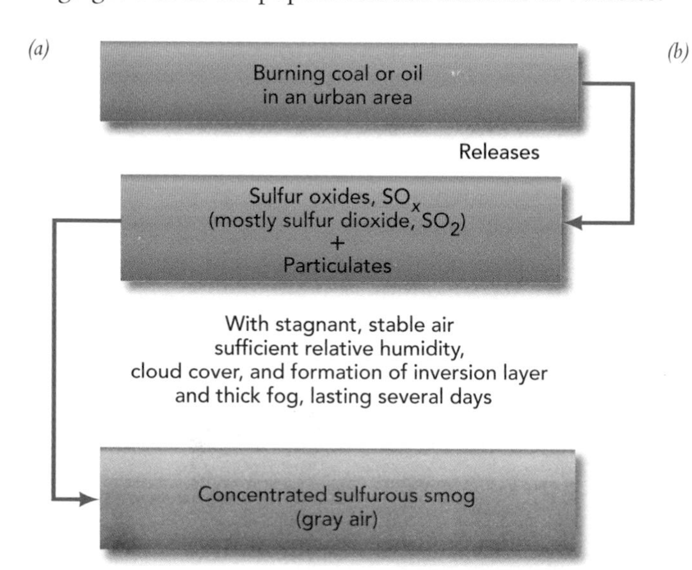

(b)

Figure 21.22 • *(a)* **How concentrated sulphurous smog and smoke might develop.** *(b)* Sulphurous smoke emitted from a smelter.

Columbia.[6] However, few regions are free from ozone pollution and associated adverse health.

Developing Countries

As mentioned earlier, cities in less developed countries with burgeoning populations are particularly susceptible to air pollution now and will be in the future (see Figures 21.5 and 21.9). They often do not have the financial base necessary to fight air pollution; they are more concerned with finding ways to house and feed their growing populations.

A good example is Mexico City. With a population of about 25 million, Mexico City is one of the four largest urban areas in the world. Cars, buses, industry, and power plants in the city emit hundreds of thousands of metric tonnes of pollutants into the atmosphere each year. The city is at an élevation of about 2,255 m in a natural basin surrounded by mountains, a perfect situation for a severe air pollution problem. It is becoming a rare day in Mexico City when the mountains can be seen, and physicians report a steady increase in respiratory diseases. Headaches, irritated eyes, and sore throats are common when the pollution settles in. Doctors advise parents to take their children out of the city permanently. The people in Mexico City do not need to be told they have an air pollution problem; it is all too apparent. However, developing a successful strategy to improve the quality of the air is difficult.

A major source of pollutants in Mexico City is motor vehicles. There are some 50,000 buses and taxis and several million automobiles in the city. Most are old and in poor running condition, and they pump immense amounts of pollutants into the atmosphere. Another major source of air pollution is leaks of liquefied petroleum gas (LPG, a hydrocarbon) used in homes for cooking and heating water. The leaking LPGs produce atmospheric precursors to the formation of ozone, a major component of urban photochemical smog, and LPG leaks in Mexico City may be responsible for a significant portion of the city's ozone pollution.[18]

In an attempt to reduce air pollution in the urban area, officials shut down a large oil refinery. For almost 60 years, the refinery had emitted nearly 90,000 metric tonnes of air pollutants into the atmosphere annually. Thousands of other industrial plants have been ordered to relocate. Although these measures will help the air quality of the urban area, industrial facilities are not the primary source of pollutants. Air pollution will continue to be a serious problem for many years if the city is unable to control increases in population; the use of buses, taxis, and automobiles; and leaks of liquefied petroleum gas. Indeed, Mexico City may eventually experience a pollution event of catastrophic proportions.

In summary, future trends in urban air problems and solutions will include a mixture of success stories and potential or actual tragedies. What is apparent is that urban air pollution is important to people, and ambitious air pollution control plans are being drawn up in many urban areas. Whether these plans are put into action will depend on numerous factors: global, regional, and local economies (reducing air pollution is expensive); population growth (more people means more air pollution); international co-operation (air pollutants travel across international borders); and the priority given to pollution abatement relative to other environmental concerns such as sanitation and clean water. With these thoughts in mind, we turn to a discussion of how to reduce air pollution.

21.8 Pollution Control

For both stationary and mobile sources of air pollutants, the most reasonable strategies for control have been to reduce, collect, capture, or retain the pollutants before they enter the atmosphere. From an environmental viewpoint, the reduction of emissions through energy efficiency and conservation measures (for example,

Table 21.3 • Canadian Cities Ranked by Air Pollution Indicators

City	*Province*	*Rank by Amount of Ground-Level Ozone*	*Rank by Amount of Particulate Matter*
Kitchener	ON	1	5
Toronto	ON	2	7
Windsor	ON	3	3
Simcoe	ON	4	11
Hamilton	ON	5	10
Egbert	QC	6	4
Montréal	QC	7	1
St. Catharines	ON	8	6
Kejimkujik	NS	9	7
Oshawa	ON	10	22
Halifax	NS	11	13
Québec	QC	12	18
St. John	NB	13	7
Vancouver	BC	14	16
Calgary	AB	15	11
Kamloops	BC	16	1
Prince George	BC	17	22
Winnipeg	MB	18	15
Chilliwack	BC	19	14
Kelowna	BC	20	21
Nanaimo	BC	21	18
Victoria	BC	22	18
St. John's	NF	23	17
Ottawa	ON	24	24

*Source: Canadian Geographic.*2000 (May-June). "Blowin' in the Wind: Canada's Smoggiest Places."

burning less fuel) is the preferred strategy, with clear advantages over all other approaches (see Chapters 16 and 17). Here, we discuss pollution control for selected air pollutants.

Pollution Control: Particulates

Particulates emitted from fugitive, point, or area stationary sources are much easier to control than are the very small particulates of primary or secondary origin released from mobile sources, such as automobiles. As we learn more about these very small particles, we will have to devise new methods to control them.

A variety of settling chambers or collectors are used to control emissions of coarse particulates from power plants and industrial sites (point or area sources) by providing a mechanism that causes particles in gases to settle out in a location where they can be collected for disposal in landfills. In recent decades, tremendous gains have been made in the control of particulates, such as ash, from power plants and industry. Cities are now much cleaner. Particulate pollutants no longer pose serious health risks in these cities. In contrast, such risks have plagued parts of Eastern Europe in recent years.

Particulates from fugitive sources (such as waste piles) must be controlled on-site so that the wind does not blow them into the atmosphere. Methods include protecting open areas, controlling dust, and reducing the effects of wind. For example, waste piles can be covered by plastic or other material, and soil piles can be vegetated to inhibit wind erosion; water or a combination of water and chemicals can be spread to hold dust down; and structures or vegetation can be positioned to lessen wind velocity near the ground, thus retarding wind erosion of particles.

Pollution Control: Automobiles

Control of pollutants such as carbon monoxide, nitrogen oxides, and hydrocarbons in urban areas is best achieved through pollution control measures for automobiles. Control of these materials will also limit ozone formation in the lower atmosphere, since ozone forms through reactions with nitrogen oxides and hydrocarbons in the presence of sunlight.

Nitrogen oxides from automobile exhausts are controlled by recirculating exhaust gas, diluting the air-to-fuel mixture being burned in the engine. Dilution reduces the temperature of combustion and decreases the oxygen concentration in the burning mixture, resulting in the production of fewer nitrogen oxides. Unfortunately, the same process increases hydrocarbon emissions. Nevertheless, exhaust recirculation to reduce nitrogen oxide emissions has been common practice in Canada and the United States for more than 20 years.[19]

The most common device to reduce carbon monoxide and hydrocarbon emissions from automobiles is the exhaust system's catalytic converter. In the converter, oxygen from outside air is introduced and exhaust gases from the engine are passed over a catalyst, typically platinum or palladium. Two important chemical reactions occur: (1) carbon monoxide is converted to carbon dioxide; and (2) hydrocarbons are converted to carbon dioxide and water.

As government regulations controlling emissions became stronger, it became difficult to meet new standards without the aid of computer-controlled engine systems. Computer-controlled fuel injection began to replace carburetors in the 1980s and has resulted in lower fuel consumption and lower exhaust emissions.[19]

It has been argued that recent regulatory initiatives in Canada and the United States have not been effective in reducing pollutants. Pollutants may be relatively low when a car is new, but many people do not take care of their automobiles well enough to ensure that the emission control devices continue to work. Some people even disconnect smog control devices. Evidence suggests that these devices tend to become less efficient with each year following purchase.

It has been suggested that effluent fees replace emission controls as the primary method of regulating air pollution from automobiles.[20] Under this scheme, vehicles would be tested each year for emission control, and fees would be assessed on the basis of test results. Fees would provide an incentive to purchase automobiles that pollute less, and annual inspections would ensure that pollution control devices were properly maintained. Although there is considerable controversy regarding enforced pollution inspections, such inspections are common in a number of areas and are expected to increase as air pollution abatement becomes essential.

Another approach to reducing urban air pollution from vehicles involves various measures aimed at reducing the number and type of cars on roads. Other measures include developing cleaner automobile fuels through use of fuel additives and reformulation; requiring new cars to use less fuel; and encouraging the use of cars with electric engines and hybrid cars that have both an electric engine and an internal combustion engine.

Pollution Control: Sulphur Dioxide

Sulphur dioxide emissions can be reduced through abatement measures performed before, during, or after combustion. Technology to clean up coal so it will burn more cleanly is already available. Although the cost of removing sulphur makes fuel more expensive, the expense must be balanced against the long-term consequences of burning sulphur-rich coal.

Changing from high-sulphur coal to low-sulphur coal seems an obvious solution to reducing sulphur dioxide emissions. In some regions, this change will work. In others, coal must be transported over long distances; and use of low-sulphur coal is a solution only in cases where it is economically feasible.

Another possibility is cleaning up relatively high-sulphur coal by washing it to remove sulphur. In this process, finely ground coal is washed with water. Iron sulphide (mineral pyrite) settles out because of its relatively high density. Although the washing process is effective in removing nonorganic sulphur from minerals such as pyrite (FeS_2), it is ineffective for removing organic sulphur bound up with carbonaceous material. Cleanup by washing is therefore limited, and it is also expensive.

Another option is **coal gasification**, which converts coal that is relatively high in sulphur to a gas in order to remove the sulphur. The gas obtained from coal is quite clean and can be transported relatively easily, augmenting supplies of natural gas. The synthetic gas produced from coal is still fairly expensive compared with gas from other sources, but its price may become more competitive in the future.

Sulphur oxide emissions from stationary sources, such as power plants, can be reduced by removing the oxides from the gases in the stack before they reach the atmosphere. Perhaps the most highly developed technology for the cleaning of gases in tall stacks is flue gas desulphurization, or **scrubbing** (Figure 21.23). The technology to scrub sulphur dioxide and other pollutants at power plants was developed in the 1970s. However, it was not initially implemented in Canada or the United States because regulators chose to allow plants to disperse pollutants using tall smokestacks rather than using scrubbing to remove them. This increased the regional acid precipitation problem.

Scrubbing occurs after coal is burned. The SO_2-rich gases are treated with a slurry (a watery mixture) of lime (calcium oxide, CaO) or limestone (calcium carbonate, $CaCO_3$). The sulphur oxides react with the calcium to form calcium sulphite, which is collected and then disposed of, usually in a landfill.

In 1980, a German company purchased coal-scrubbing technology and improved on it as part of efforts to reduce air pollution and acid precipitation. Rather than disposing of the calcium sulphite-rich sludge formed during the scrubbing process, the company further processes it to produce building materials (gypsum, $CaSO_4 \cdot 2H_2O$, which is sheet rock or wallboard), which are sold worldwide.

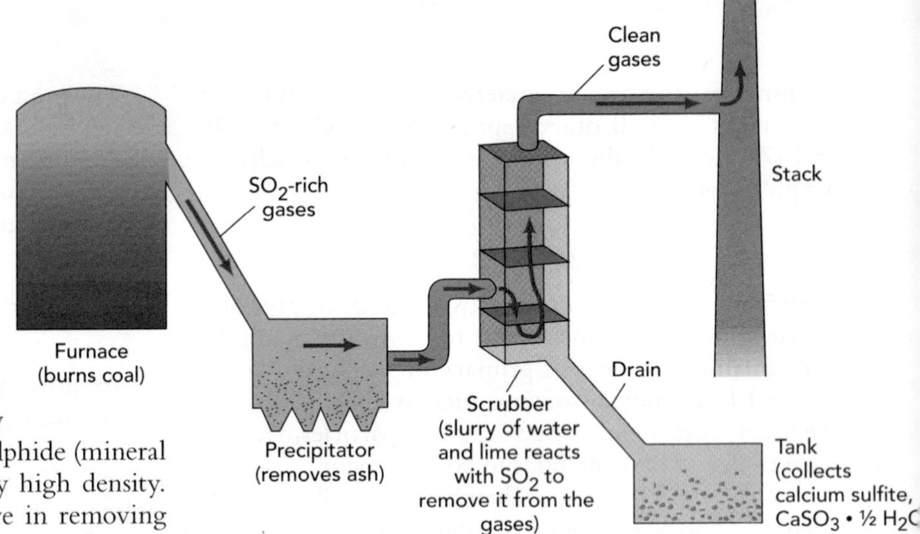

Figure 21.23 • Scrubber used to remove sulfur oxides from the gases emitted by tall stacks.

Another innovative approach to removing sulphur has been taken at a large coal-burning power plant near Mannheim, Germany. The smoke from combustion is cooled, and then treated with liquid ammonia (NH_3), which reacts with the sulphur to produce ammonium sulphate. In this process, the sulphate-contaminated smoke is cooled in a heat-exchange process by outgoing clean smoke to a temperature that allows the chemical reaction between the sulphur-rich smoke and ammonia to take place. The cooled, cleaned, outgoing smoke is then heated by dirty smoke in the same sort of heat-exchange process to force it out the vent. Waste heat from the cooling towers is used to heat nearby buildings, and the plant sells the ammonium sulphate in a solid granular form to farmers to use as fertilizer.

Thus, Germany, in response to tough pollution control regulations, has substantially reduced its sulphur dioxide emissions (as well as many other pollutant emissions) and has boosted its economy in the process.

21.9 Air Pollution Legislation and Standards

As we mentioned in earlier chapters of this book, Canada's Constitution does not differentiate clearly between federal and provincial responsibilities for managing the environment. In general, air pollution control is a provincial responsibility in Canada, because it relates to management of business activities and natural resources. Some pollutants, such as CFCs, are controlled under federal legislation because they are thought to be of national concern.

On May 27, 2000, Canada's Ministers of Health and Environment jointly announced their decision to designate particulate matter less than 10 microns (PM 10) as toxic under the *Canadian Environmental Protection Act*.

Under this provision, industrial emitters of particulate matter must set reduction targets and timetables to meet those targets. In June 2000, the Canadian federal, provincial and territorial governments ratified a national standard for particulate matter less than 2.5 microns (PM 2.5). Under this agreement, provinces and territories will establish regulatory requirements for the new lower limit, while the federal government will work on controls relating to the design and sale of cleaner vehicles and fuels. As we have seen, air pollution in urban areas is commonly associated with automobile exhaust. Strategies like the new PM 2.5 standard are intended to reduce the occurrence of urban smog. Expected impacts of the legislation include increases in the cost of automobile fuels and in the price of new automobiles.

Ambient Air Quality Standards

Air quality standards are important because they are tied to emission standards that attempt to control air pollution. Many countries have developed air quality standards, including France, Japan, Israel, Italy, Germany, Norway, the United States, and Canada. All of Canada's provinces also have such standards, although allowable emissions vary somewhat from region to region. Typical provincial standard values are given in Table 21.4. Under a Canada-Wide Accord on Environmental Harmonization, the federal and provincial ministers of the environment (with the exception of Québec) agreed to establish Canada-wide standards (CWS) for particulate matter and ozone. The current CWS for ozone is 65 ppm. All participating jurisdictions are expected to meet the CWS by 2010 and begin to report on achievement of target levels by 2011.

The CWS for PM 2.5 is 30 micrograms per cubic metre. As for the ozone standard, participating jurisdictions are expected to meet the CWS by 2010 and begin reporting on achievement by 2011.

Implementing the new standards will take years; ways will have to be found to enforce them. Nevertheless, it is expected that the new standards will eventually be achieved.

Air Quality Index

The Air Quality Index (AQI), familiar from television weather reports, is essentially a communications tool used to convert the concentration of several pollutants into a single indicator. AQIs are issued by federal, provincial, and local (including health unit) agencies, but the form and basis for these messages varies from region to region. Typically, a numerical value (often between 1 and 100) and a rating such as "good," "fair," or "poor" are reported. AQIs are usually based on ambient concentrations of sulphur dioxide, nitrogen oxides, ozone, carbon monoxide, PM 10 and PM 2.5. The higher the number reported, the poorer the air quality. Ontario's AQI system, based on six pollutants and a 100-point scale, has five categories:

Very Good (0–15)
Good (16–31)
Moderate (32–49)
Poor (50–99)
Very Poor (100+)

Most Canadian cities have air quality in the "Very Good" to "Good" range, although poor quality does occur, especially in hot summer weather when smog

Table 21.4 • Canadian Ambient Air Quality Standards

Pollutant	*Standard Value*		*Standard Type*
Carbon monoxide (CO)			
8-hour average	15 ppm	(10 $\mu g/m^3$)	Typical provincial standard; no federal equivalent
Nitrogen dioxide (NO_2)			
Annual arithmetic mean	0.100 ppm	(100 $\mu g/m^3$)	Typical provincial standard; no federal equivalent
Ozone (O_3)			
1-hour average	0.82 ppm	(235 $\mu g/m^3$)	Federal guideline; standard in several provinces
Lead (Pb)			
24-hour average		2 $\mu g/m^3$	Typical provincial standard; no federal equivalent.
Total Suspended Particulates			
Annual arithmetic mean		70 $\mu g/m^3$	Federal guideline; standard in several provinces
Particulate (PM 10) *10 micrometres or less*			
24-hour average		50 $\mu g/m^3$	Typical provincial standard; no federal equivalent
Particulate (PM 2.5) *2.5 micrometres or less*			
24-hour average		25 $\mu g/m^3$	Typical provincial standard; no federal equivalent
Sulphur Dioxide (SO_2)			
Annual arithmetic mean		60 $\mu g/m^3$	Typical provincial standard; no federal equivalent
24-hour average		300 $\mu g/m^3$	Typical provincial standard; no federal equivalent
1-hour average		900 $\mu g/m^3$	Typical provincial standard; no federal equivalent

Source: Environment Canada.

production is likeliest. AQI values greater than 100 are generally recorded for a particular city only a few times per year. In comparison, for large urban cities outside of Canada with dense human populations and numerous uncontrolled sources of pollution, AQIs greater than 200 are frequent.

During a pollution episode, hourly ozone levels are reported. In very poor air quality conditions, an air pollution alert is issued and people are requested to remain indoors and minimize physical exertion. Driving private automobiles and operating gasoline-fuelled lawnmowers may be discouraged, and industry may be required to reduce emissions to a minimum during the episode. Recall that the AQI was 800 during the Indonesian fires of 1997–1998!

In 2002, the Canadian federal government announced the development of a consistent health-based AQI for Canada. Still under discussion, the new index will be based on health research and associated public consultations conducted by Health Canada.

21.10 Cost of Air Pollution Control

The cost of air pollution control varies tremendously from one industry to another. For example, consider the incremental control costs (costs to remove an additional unit of pollution) for utilities burning fossil fuels and for an aluminum plant. The cost for incremental control in a fossil fuel–burning utility is a few hundred dollars per additional tonne of particulates removed. For the aluminum plant, the cost to remove an additional ton of particulates may be as much as several thousand dollars.[22,23] Some economists would argue that it is wise to increase the standards for utilities and relax or at least not increase them for aluminum plants. This practice would lead to more cost-efficient pollution control while maintaining good air quality. However, the geographic distribution of various facilities will obviously determine the trade-offs possible.[24]

Another economic consideration is that, as the degree of control of a pollutant increases, a point is reached at which the cost of incremental control is very high in relation to the additional benefits of the increased control. Because of this and other economic factors, it has been argued that enforcing fees or taxes for emitting pollutants might make more economic sense than attempting to evaluate the uncertain costs and benefits associated with enforcement of standards. Another approach is to issue vouchers to allow businesses to emit a certain total amount of pollution in a region. These vouchers are bought and sold on the open market. All these economic alternatives are controversial and may be objectionable to people who believe that polluters should not be allowed to buy their way out of doing what is socially responsible (that is, not polluting our atmosphere).

Economic analysis of air pollution is not simple. There are many variables, some of which are hard to quantify. We do know the following:

- With increasing air pollution controls, the capital cost for technology to control air pollution increases.
- As the controls for air pollution increase, the loss from pollution damages decreases.
- The total cost of air pollution is the cost of pollution control plus the environmental damages of the pollution.

Although the cost of pollution-abatement technology is fairly well known, it is difficult to adequately determine the loss from pollution damages, particularly when considering health problems an damage to vegetation, including food crops. For example, exposure to air pollution may cause or aggravate chronic respiratory diseases in human beings, with a very high cost. A recent study by the Ontario Medical Association estimated that the annual cost associated with air pollution in Ontario is about $10 billion in economic damages, with loss of life and pain and suffering accounting for between $4.1 and $4.8 billion of that total.[25] The annual health care costs associated with air pollution are estimated at $600 million, with a similar amount attributed to lost productivity. These figures are shocking enough, but the authors of the report remind us that unless we reduce our urban air pollution significantly, those costs will continue to rise over the foreseeable future.

How do we determine the real and total benefits and costs of controlling or reducing air pollution? As we have seen, there are no easy answers to this question. In spite of our inability to determine all benefits and costs, it seems worthwhile to reduce the air pollution level below some particular regulatory standard. However, as discussed, it is also a good idea to consider alternatives, such as charging fees or taxes for emissions. If such charges are determined carefully and emissions are closely monitored, the charges should provide an incentive for the installation of control measures. The end result would be better air quality.[26,27]

21.11 Indoor Air Pollution

The potential sources of indoor air pollution are incredibly varied (Figure 21.24). Two common indoor air pollutants are shown in Figure 21.25. Indoor air pollutants can arise from both human activities and natural processes. In recent years the public has been made aware of several of these sources, described in the following list:

- Environmental tobacco smoke (second-hand smoke) is the most hazardous common indoor air pollutant,

Figure 21.24 • Some potential sources of indoor air pollution.

1. Heating, ventilation, and air-conditioning systems of buildings may be sources of indoor air pollutants, including moulds and bacteria, if filters and equipment are not maintained properly. Gas and oil furnaces release carbon monoxide, nitrogen dioxide, and particles.
2. Restrooms may be sources of a variety of indoor air pollutants, including second-hand smoke, moulds, and fungi resulting from humid conditions.
3. Furniture and carpets in buildings often contain toxic chemicals (formaldehyde, organic solvents, asbestos), which may be released over time in buildings.
4. Coffee machines, fax machines, computers, and printers can release particles and chemicals, including ozone (O_3), which is highly oxidizing.
5. Pesticides can contaminate buildings with cancer-causing chemicals.
6. Fresh air intake that is poorly located—as, for example, above a loading dock or first-floor restaurant exhaust fan—can bring in air pollutants.
7. People who smoke inside buildings, perhaps in restaurants, and people who smoke outside buildings, particularly near open or revolving doors, may cause pollution as the environmental tobacco smoke (second-hand smoke) is drawn into and up through the building by the chimney effect.
8. Remodelling, painting, and other such activities often bring a variety of chemicals and materials into a building. Fumes from such activities may enter the building's heating, ventilation, and air-conditioning system, causing widespread pollution.
9. A variety of cleaning products and solvents used in offices and other parts of buildings contain harmful chemicals, whose fumes may circulate throughout a building.

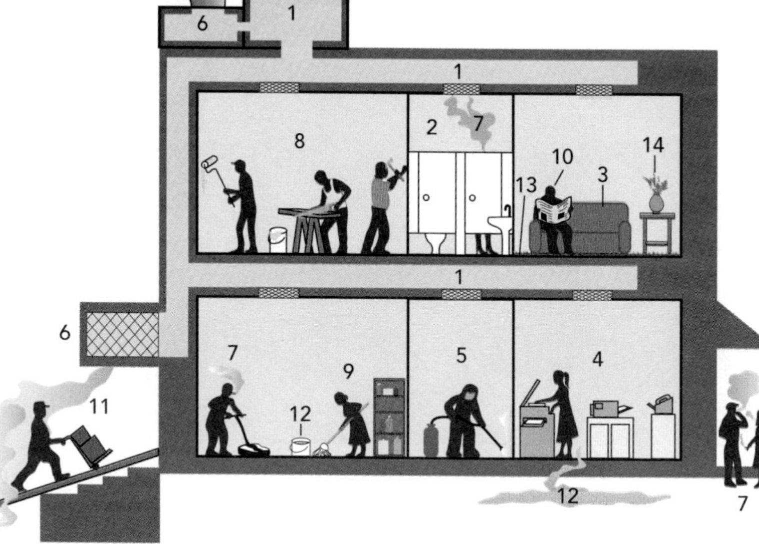

10. People can increase carbon dioxide levels; they can emit bioeffluents and spread bacterial and viral contaminants.
11. Loading docks can be sources of organics from garbage containers, of particulates, and of carbon monoxide from vehicles.
12. Radon gas can seep into a building from soil; rising damp (water), which facilitates the growth of moulds, can enter foundations and rise up walls.
13. Dust mites and moulds can live in carpets and other indoor places.
14. Pollen can come from inside and outside sources.

associated with over 40,000 deaths per year (mostly heart disease and lung cancer) in Canada and the United States.

- *Legionella pneumophila*, a bacterium that normally lives in pond water, causes a type of pneumonia called Legionnaires' disease when inhaled. This disease is usually spread through air-conditioning equipment, which harbours the disease-causing bacteria in pools of stagnant water in air ducts and filters. Bacteria are transported through a building as a bacterial aerosol when heating or cooling units are in use. However, spread of the disease is not limited to this pathway. One epidemic occurred in a hospital as a result of contamination from an adjacent construction site. A recent outbreak of Legionnaires' disease caused by bacteria in the air-conditioning system of an aquarium killed four people.[28]
- Some moulds (fungal growths) in buildings release toxic spores. When inhaled over a period of time, the spores can cause chronic inflammation and scarring of lungs, as well as hypersensitivity pneumonitis and pulmonary fibrosis. These medical conditions are painful, disabling, and can even cause death.[28] It is believed that moulds may be responsible for up to half of all health complaints resulting from indoor air environments.
- Radon gas seeps up naturally from soils and rocks below buildings and is thought to be the second most common cause of lung cancer.
- Pesticides that are deliberately or inadvertently applied in buildings to control ants, flies, fleas, moths, and rodents are toxic to people as well.
- Some varieties of asbestos, used as an insulating material and fireproofing material in homes, schools, and offices, are known to cause a particular type of lung cancer (see Chapter 15).
- Formaldehyde (a volatile organic compound with a chemical formula of CH_2O) is used in some foam insulation materials, as a binder in particleboard and wood paneling, and in many other materials found in homes and offices. These materials can emit formaldehyde as a gas into buildings. Some mobile homes have been found to have high concentrations of formaldehyde because products containing the chemical are used in their construction (wood paneling, for example).
- Dust mites and pollen irritate the respiratory system, nose, eyes, and skin of people who are sensitive to them.

Common indoor air pollutants and guidelines for allowable exposure are listed in Table 21.5. Many of the products and processes used in our homes and workplaces are sources of pollution. Furthermore, common indoor air pollutants are often highly concentrated compared

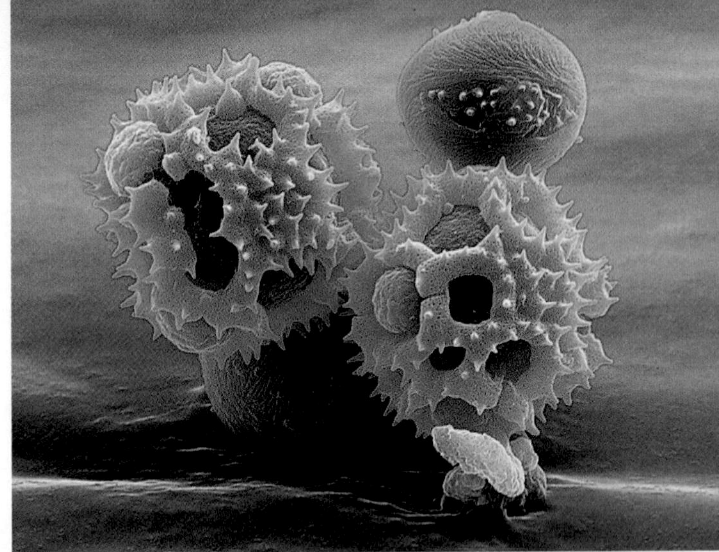

Figure 21.25 • **Two common indoor air pollutants.** *(a)* This dust mite (magnified about 140 times) is an eight-legged relative of spiders. It feeds on human skin in household dust. It lives in materials such as fabrics on furniture. Dead dust mites and their excrement can produce allergic reactions and asthma attacks in some people. *(b)* Microscopic pollen grains that in large amounts may be visible as a brown or yellow powder. The pollen here are dandelion and horse chestnut.

with outdoor levels. For example, carbon monoxide, particulates, nitrogen dioxide, radon, and carbon dioxide are generally found in much higher concentrations indoors than outdoors. This important concept is shown in more detail in Figure 21.26, which provides a comparison of indoor with outdoor pollutants.

Why are concentrations of indoor air pollutants generally greater than those found outdoors? One obvious reason is that there are so many potential indoor sources of pollutants. Another reason is somewhat ironic: the effectiveness of the steps we have taken to conserve energy in homes and other buildings has led to the trapping of pollutants inside.

Two of the best ways to conserve energy in homes and other buildings are to increase the insulation and to decrease the infiltration of outside air. Constructing our buildings with windows that do not open and applying extensive caulking and weather-stripping does reduce energy consumption, but it also tends to affect the air quality of the building by reducing natural ventilation. An important function of ventilation is that it replaces the indoor air with outdoor air in which the concentrations of pollutants are generally much lower. With less natural ventilation, we must depend more on the ventilation systems that are part of heating and air-conditioning systems.

Heating, Ventilation, and Air-Conditioning Systems

Heating, ventilation, and air-conditioning systems are designed to provide a comfortable indoor environment for people. Design of these systems depends on a number of variables, including the activity of people in the building, air temperature and humidity, and air quality. The interaction among these factors determines whether people are comfortable indoors. If the heating, ventilation, and air-conditioning system is designed correctly and functions properly, it will provide thermal comfort for people inhabiting the building. It will also provide the necessary ventilation (utilizing outdoor air) and remove common air pollutants via exhaust fans and filters.[29]

Personal comfort levels in terms of temperature and humidity vary depending on age, physiology, and level of activity. Furthermore, different portions of buildings may have different temperatures and air quality because of their location in relation to heat sources, cold surfaces, and large windows. Humidity should be carefully controlled. High humidity may facilitate the growth of adverse mildews or moulds, whereas low humidity may be a source of discomfort to some people.[29]

Regardless of the type of heating, ventilation, and air-conditioning system used in a home or other building, the effectiveness of that unit depends on the proper design of the equipment relative to the building, on proper installation, and on correct maintenance and operating procedures.[29] Indoor air pollution may result if any one of these factors concentrates pollutants from the many possible sources. If filters become plugged or contaminated with fungi, bacteria, or other potentially infectious agents, serious problems can result. In addition, as we see later in this chapter, ventilation systems are not generally designed to reduce some types of indoor pollution.[29,30]

Pathways, Processes, and Driving Forces

Many air pollutants originate within buildings and may be concentrated there because of lack of proper ventilation with the outside atmosphere. Other air pollutants may enter a building by infiltration, either through cracks and other openings in the foundations and walls or by way of ventilation systems.

Table 21.5 • Sources, Concentrations, Occurrences, and Possible Health Effects of Indoor Air Pollutants

Pollutant	*Source*	*Guidelines (Dose or Concentrations)*	*Possible Health Effects*
Asbestos	Fireproofing; insulation, vinyl floor, and cement products; vehicle brake linings	0.2 fibres/mL for fibres larger than 5 μm	Skin irritation, lung cancer
Biological aerosols/ micro-organisms	Infectious agents, bacteria in heating, ventilation, and air-conditioning systems; allergens	None available	Diseases, weakened immunity
Carbon dioxide	Motor vehicles, gas appliances, smoking	1,000 ppm	Dizziness, headaches, nausea
Carbon monoxide	Motor vehicles, kerosene and gas space heaters, gas and wood stoves, fireplaces; smoking	10,000 $\mu g/m^3$ for 8 hours; 40,000 $\mu g/m^3$ for 1 hour	Dizziness, headaches, nausea, death
Formaldehyde	Foam insulation; plywood, particleboard, ceiling tile, paneling, and other construction materials	120 $\mu g/m^3$	Skin irritant, carcinogen
Inhalable particulates	Smoking, fireplaces, dust, combustion sources (wildfires, burning trash, etc.)	55–110 $\mu g/m^3$ annual; 350 $\mu g/m^3$ for 1 hour	Respiratory and mucous irritant, carcinogen
Inorganic particulates			
Nitrates	Outdoor air	None available	
Sulphates	Outdoor air	4 $\mu g/m^3$ annual; 12 $\mu g/m^3$ for 24 hours	
Metal particulates			Toxic, carcinogen
Arsenic	Smoking, pesticides, rodent poisons	None available	
Cadmium	Smoking, fungicides	2 $\mu g/m^3$ for 24 hours	
Lead	Automobile exhaust	1.5 $\mu g/m^3$ for 3 months	
Mercury	Old fungicides; fossil fuel combustion	2 $\mu g/m^3$ for 24 hours	
Nitrogen dioxide	Gas and kerosene space heaters, gas stoves, vehicular exhaust	100 $\mu g/m^3$ annual	Respiratory and mucous irritant
Ozone	Photocopying machines, electrostatic air cleaners, outdoor air	235 $\mu g/m^3$ for 1 hour	Respiratory irritant, causes fatigue
Pesticides and other semivolatile organics	Sprays and strips, outdoor air	5 $\mu g/m^3$ for chlordane	Possible carcinogens
Radon	Soil gas that enters buildings, construction materials, groundwater	4 pCi/L	Lung cancer
Sulphur dioxide	Coal and oil combustion, kerosene space heaters, outside air	80 $\mu g/m^3$ annual; 365 $\mu g/m^3$ for 24 hours	Respiratory and mucous irritant
Volatile organics	Smoking, cooking, solvents, paints, varnishes, cleaning sprays, carpets, furniture, draperies, clothing	None available	Possible carcinogens

Sources: N. L. Nagda, H. E. Rector, and M. D. Koontz, 1987; M. C. Baechler et al., 1991; E. J. Bardana Jr. and A. Montaro (eds.), 1997; M. Meeker, 1996; D. W. Moffatt, 1997.

The driving forces that control or modify the flow of air in buildings result from a variety of processes related to both natural forces and human activity. Both natural and human processes in buildings create differential pressures that move air and contaminants from one area to another. Areas of high pressure may develop on the windward side of a building, whereas pressure is lower on the leeward, or protected, side. As a result, air is drawn into a building from the windward side. Opening and closing doors produces pressure differentials that induce air to move within buildings, and wind can affect the movement of air in a building, particularly if the structure is leaky.[29]

A **chimney effect** (or **stack effect**) occurs when there is a temperature differential between the indoor and outdoor environments. Warm air rises within a building. If the indoor air is warmer than that found outdoors, as the warmer air rises in the building to the upper levels, it is replaced in the lower portion of the building by outdoor air drawn in through a variety of openings, such as windows, doors, or cracks in the foundations and walls. Facilities such as elevator shafts and stairwells provide corridors through which air

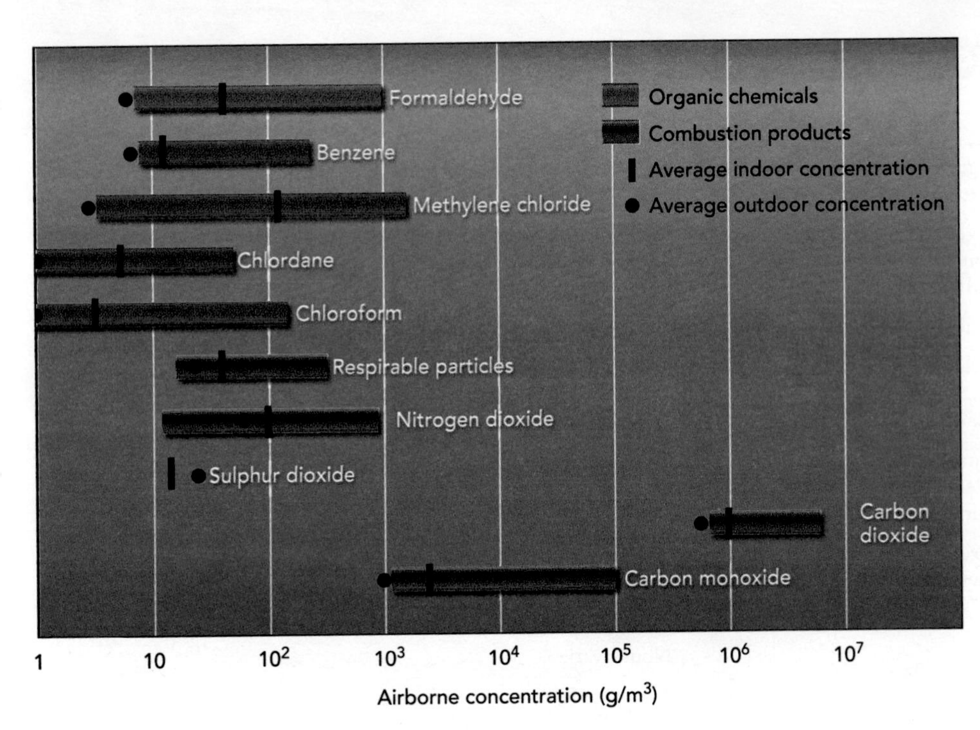

Figure 21.26 • Concentrations of common indoor air pollutants compared with outdoor concentrations plotted on a log scale, that is, $10^2=100$; $10^3=1{,}000$, $10^4=10{,}000$, etc. [*Source*: A.V. Nero, Jr., "Controlling Indoor Air Pollution," *Scientific American*, 258, no. 5 (1998): 42–48.]

can move from one floor to another.[29] Environmental tobacco smoke, or second-hand smoke, may also be drawn into a building by the chimney effect if smokers move outside near revolving or open doors to smoke.[28]

Because air is such a fluid medium, the possible interactions between the driving forces and the building are complex and the distribution of potential air contaminants and pollutants is extensive. One outcome of this situation is that people in various parts of a building may complain about the air quality even if they are separated by considerable distances from each other and from potential sources of pollution.[29]

Susceptibility to Indoor Air Pollution

People have varying sensitivity to air pollutants. One person may be adversely affected by a particular pollutant, whereas others in nearby areas may seem to be unaffected. Sometimes the problem is a matter of concentration rather than sensitivity; the person most affected by a particular indoor air pollutant might be experiencing the greatest exposure. A person's susceptibility to a particular air pollutant also depends on genetic factors, lifestyle, and age (see Chapter 15).

As a result of these individual differences, the response of one person bothered by a particular pollutant may differ from the response of another affected individual, making it difficult to evaluate indoor pollution problems.

Nevertheless, older people with impaired health and children (because of their activity level and developing lungs) are generally more sensitive to air pollutants.[30] People suffering from chronic lung or respiratory diseases, such as chronic bronchitis, allergies, or asthma, are especially likely to be adversely affected by poor indoor air quality. Another group more strongly affected comprises individuals who have suppressed immune systems owing to disease or medical treatment, such as chemotherapy or radiation therapy.[29] Some people, when exposed to chemicals, develop what is known as multiple chemical sensitivity (MCS), a controversial disease in which people are allergic to any material or product containing human-produced chemicals. Some sufferers of MCS are so sensitive that they effectively have to live in a "plastic bubble" that keeps out chemicals.[28]

A great variety of symptoms can result from exposure to indoor air pollutants (Table 21.6). Some chemical pollutants can cause nosebleeds, chronic sinus infections, headaches, and irritation of the skin or eyes, nose, and throat. More serious problems include loss of balance and memory, chronic fatigue, difficulty in speaking, and allergic reactions, including asthma. For example, chlorine tablets, which are often used in swimming pools and hot tubs, are a tremendous irritant if dust from the tablets is inhaled; shortness of breath and coughing result. Other pollutants cause dizziness or nausea. Exposure to carbon monoxide results in shortness of breath at low concentrations. At high concentrations, extreme toxicity and death can result. Tissues sensitive to carbon monoxide include the brain, heart, and muscles.[31]

The symptoms just described may have a quick onset after exposure. Other pollutants, including radon, asbestos, and chemicals such as benzene, may have long-term chronic health effects, including diseases such as cancer. Because of long lag times between exposure and disease, it may be difficult to establish relationships between a particular indoor air environment and disease in an individual.

Sick Buildings

An entire building can be considered sick because of environmental problems. There are two types of sick buildings:

- Buildings with identifiable problems, such as occurrences of toxic moulds or bacteria known to cause disease. The diseases are known as building-related illnesses (BRI).
- Buildings with sick building syndrome (SBS), where the symptoms people report cannot be traced to any one known cause.

Sick building syndrome is a condition associated with an indoor environment that appears to be unhealthy. A number of people in such a building report adverse health effects that they believe are related to the amount of time they spend in the building. The range of complaints may vary from funny odours to more serious symptoms, such as headaches, dizziness, nausea, and so forth. In addition, a number of people in the building may be sick; and a group of people may have contracted a disease, such as cancer.

In many cases, it is difficult to establish what may be causing a particular sick building syndrome. Sometimes, the problem has been found to be related to poor management practices and worker morale rather than exposure to toxins in the building. When the occupants of a building report adverse health effects and a study follows, often the cause is not detected. A number of things may be happening:[29]

- The complaints result from the combined effects of a number of contaminants present in the building.
- Environmental stress from a source other than air quality—such as noise, high or low humidity, poor lighting, or overheating—is responsible.
- Employment-related stress—such as stress resulting from poor relations between labour and management, poor morale, or overcrowding—may be leading to the symptoms reported.
- Other unknown factors may be responsible. For example, pollutants or toxins may be present but not identified.

Of course, sick building syndrome may be the combined effect of various aspects of some or all of these factors. As noted, one common aspect of sick building syndrome is that often no one specific disease or cause is easily identified.[29]

We have reviewed some basics concerning sources, processes, and effects of indoor air pollution. Next, we consider two selected pollutants: environmental tobacco smoke and radon gas.

Environmental Tobacco Smoke

Environmental tobacco smoke (ETS), also known as second-hand smoke, comes from two sources: smoke exhaled by smokers and smoke emitted from burning tobacco in cigarettes, cigars, or pipes. People who are exposed to ETS are sometimes referred to as passive smokers.[32]

ETS is the best known of the hazardous indoor air pollutants. It is hazardous for the following reasons:[32, 33]

- Tobacco smoke contains several thousand chemicals, many of which are irritants. Examples include NO_X,

Table 21.6 • Some Symptoms of Indoor Air Pollution

Symptoms	*ETS*[a]	*Combustion Products*[b]	*Biologic Pollutants*[c]	*VOCs*[d]	*Metals*[e]	*SBS*[f]
Respiratory						
Inflammation of mucous membranes of the nose, nasal congestion	Yes	Yes	Yes	Yes	No	Yes
Nosebleed	No	No	No	Yes	No	Yes
Cough	Yes	Yes	Yes	Yes	No	Yes
Wheezing, worsening asthma	Yes	Yes	No	Yes	No	Yes
Laboured breathing	Yes	No	Yes	No	No	Yes
Severe lung disease	Yes	Yes	Yes	No	No	Yes
Other						
Irritation of mucous membranes of eyes	Yes	Yes	Yes	Yes	No	Yes
Headache or dizziness	Yes	Yes	Yes	Yes	Yes	Yes
Lethargy, fatigue, malaise	No	Yes	Yes	Yes	Yes	Yes
Nausea, vomiting, anorexia	No	Yes	Yes	Yes	Yes	No
Cognitive impairment, personality change	No	Yes	No	Yes	Yes	Yes
Rashes	No	No	Yes	Yes	Yes	No
Fever, chills	No	No	Yes	No	Yes	No
Abnormal heartbeat	Yes	Yes	No	No	Yes	No
Retinal hemorrhage	No	Yes	No	No	No	No
Muscle pain, cramps	No	No	No	Yes	No	Yes
Hearing loss	No	No	No	Yes	No	No

[a] Environmental tobacco smoke.
[b] Combustion products include particles, NO_x, CO, and CO_2.
[c] Biologic pollutants include moulds, dust mites, pollen, bacteria, and viruses.
[d] Volatile organic compounds, including formaldehyde and solvents.
[e] Metals include lead and mercury.
[f] Sick building syndrome.

Source: Modified from American Lung Association, Environmental Protection Agency, and American Medical Association, "Indoor Air Pollution—An Introduction for Health Professionals," 523–217/81322 (Washington, D.C.: GPO, 1994).

CO, hydrogen cyanide, and about 40 carcinogenic chemicals.
- Studies of non-smoking workers exposed to ETS found that they have reduced airway functions comparable to what would be caused by smoking up to 10 cigarettes per day. They suffer more illnesses, such as coughs, eye irritation, and colds, and lose more work time than those not exposed to ETS.
- In Canada, about 300 deaths from lung cancer and about 4,000 deaths from heart disease are thought to be associated with ETS each year.

ETS exposure depends on a variety of factors, including the number of people smoking in a room, the size of the room, and the rate of ventilation. Separating smokers from non-smokers reduces but does not eliminate exposure to ETS. Thus, many provincial and local governments have banned smoking in restaurants, bars, and public buildings to protect citizens from ETS.

Smokers must realize that they are harming not only themselves but also others, including their families, their friends, and the general public, by polluting the air that others breathe. Indeed, many people consider being exposed to ETS as a transgression of their right to breathe clean air in their homes, their workplaces, and public buildings.

The number of smokers in Canada has declined, but there are still about 7 million. The rate is higher in the developing world, where health warnings are few or non-existent. Smoking is extremely addictive, because tobacco contains nicotine, a highly addictive substance. Nevertheless, education and social pressure have worked to some extent to influence some thoughtful people to quit smoking and encourage others to quit.

Radon Gas

It has become apparent within the past few decades that radon gas may constitute a significant environmental health problem in Canada and the United States.[11, 12] An interesting aspect of the radon gas hazard is that it comes from natural processes rather than human activities.

Radon is a naturally occurring radioactive gas that is colourless, odourless, and tasteless. It is a member of the naturally occurring radioactive decay chain from radiogenic uranium to stable lead (Figure 21.27). Radon-222, which has a half-life of 3.8 days, is the product of radioactive decay of radium-226. Radon decays with emission of an alpha particle to polonium-218, which has a half-life of approximately three minutes. (See Chapter 17 for a discussion of half-lives.)

Radon was discovered in 1900 by Ernest Dorn, a German chemist, and was considered a health boon in the early years of the twentieth century. Products such as chocolate candies, bread, toothpaste, and even contraceptive jelly containing radium were readily available to consumers.[11] It wasn't until 1984, when a nuclear power plant worker set off radiation alarms on his way into work, that radon was recognized as a potential health risk. The worker's home had a radiation level of 3,200 pCi/L, 800 times higher than the action level of 4 pCi/L set by regulatory agencies.[11, 13, 15]

Radon gas enters homes and other buildings in three main ways (Figure 21.28):

1. It migrates up from soil and rock into basements and lower floors.
2. Dissolved in groundwater, it is pumped into wells and then into homes.
3. Radon-contaminated materials, such as building blocks, are used in construction.

Most Canadian provinces have areas with elevated concentrations of radon gas. Particularly high levels have been observed in the Northwest Territories and in parts of Saskatchewan and Ontario with past or present uranium mining activities. The Radiation Safety Institute of Canada estimates that every home in Canada contains

Figure 21.27 • Simplified diagram showing the radioactive decay chain for radon. Not all isotopes are shown. Half-lives and types of decay are shown for some. Radon is a gas and can move up from rock and soil into the air. The isotopes of polonium and lead are particles that may be suspended in air or attach to dust particles and move with air currents. They may also settle out of air.

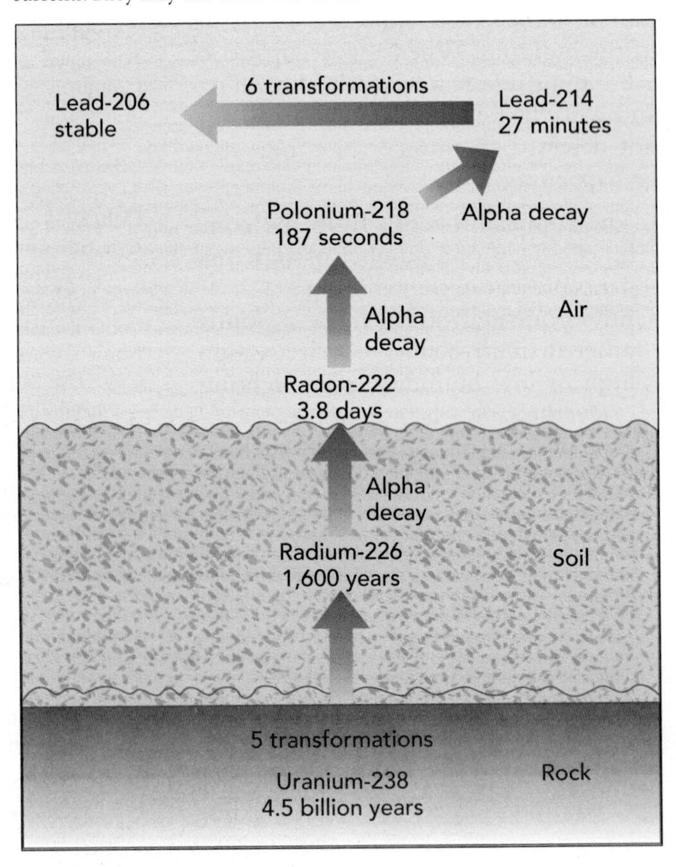

some radon, but only about one-tenth of one percent have dangerous concentrations. Once identified, radon contamination is often easily fixed, often simply by locating and sealing entry points. Additional ventilation, using fans and other devices, may also be necessary.[34, 35]

Control of Indoor Air Pollution

In terms of the workplace, there are strong financial incentives to provide workers with a clean air environment. As much as $250 billion per year might be saved by decreasing illnesses and increasing productivity through improving the work environment.[25] A good starting point would be passing environmental legislation requiring minimum indoor air quality standards. This should include increasing the inflow of fresh air through ventilation. In Europe, systems of filters and pumps in many office buildings circulate air three times as frequently as is typical in buildings in Canada. Many building codes in Europe require that workers have access to fresh air (windows) and natural light. Unfortunately for Canadian workers, no similar codes exist in Canada, and many buildings use central air-conditioning with windows permanently sealed.[28]

You might think that heating, ventilating, and air-conditioning systems, when operating properly and well maintained, will ensure good indoor air quality. Unfortunately, these systems are not designed to maintain all aspects of air quality. For example, commonly used ventilation systems do not generally reduce radon gas. Ventilation is one control strategy when faced with high concentrations of any indoor air pollutant, including radon. Other strategies, shown in Table 21.7, include source removal, source modification, and air cleaning.[31] These strategies do not constitute a complete list, and some combination of them may be the best approach.

Figure 21.28 • How radon can enter homes. (1) Radon in groundwater enters a well and goes to a house, where it is used for water supply, dishwashing, showers, and other purposes. (2) Radon gas in rocks and soil migrates into a basement through cracks in the foundation and pores in construction. (3) Radon gas is emitted from construction materials used in building a house. (*Source*: Environmental Protection Agency.)

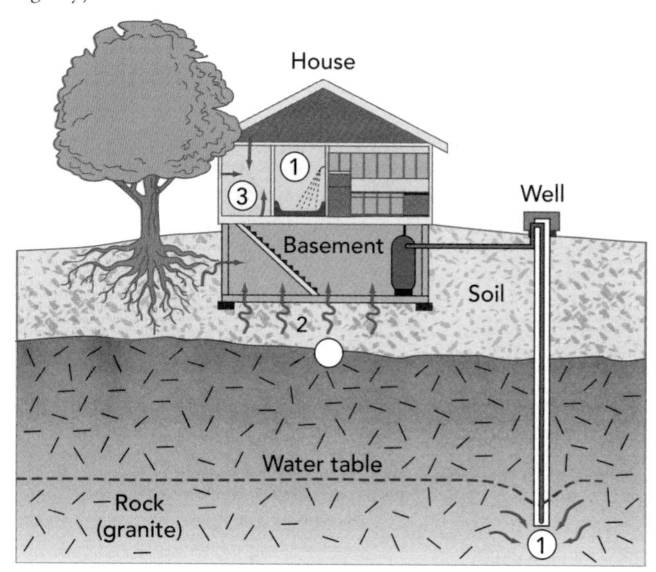

Table 21.7 • Strategies to Control Indoor Air Pollution

Ventilation: General ventilation system; spot (zone or localized) ventilation (exhaust fans, etc.)

Source removal: Material or product substitution; restrictions on source use (e.g., establishment of smoking areas; restrictions on sale of particular items and on activities that cause indoor air pollution)

Source modification: Change in combustion design (e.g., maximize efficiency of a gas stove); material substitution (use materials that don't cause air pollution); reduction in emission rates by intervention of barriers (e.g., apply coatings over lead paint or asbestos)

Air cleaning (pollutant removal): Particle filtering; gas and vapour removal; passive scavenging or absorption

Education: Consumer information on products and materials; public information on health, productivity, and nuisance effects; resolution of legal rights and liabilities of consumer, tenant, manufacturer, and so on, related to indoor air quality

Source: Modified from Committee on Indoor Pollutants, *Indoor Pollutants* (Washington, D.C.: 1981), p. 489.

One of the principal means for controlling the quality of indoor air is by dilution with fresh outdoor air via a ventilating air-conditioning system and windows that can be opened. Outside air is brought in and mixed with air in the return flow system from the building; the air is filtered, heated or cooled, and supplied to the building. Various types of air-cleaning systems for residential and non-residential buildings are available to reduce potential pollutants, such as particles, vapours, and gases. These systems can be installed as part of the heating, ventilation, and air-conditioning system or as stand-alone appliances.[31]

Education also plays an important role in understanding and developing strategies for reducing indoor air pollution problems; it empowers people with knowledge necessary to make intelligent decisions. At one level, this may involve deciding not to install unvented or poorly vented appliances. A surprising (and tragic) number of people are killed each year by carbon monoxide poisoning resulting from poor ventilation in homes, campers, and tents. At other levels, educated people are more aware of their legal rights with respect to product liability and safety. Furthermore, education provides people with the information necessary to make decisions concerning exposure to chemicals, such as paints and solvents, and strategies to avoid potentially hazardous conditions in the home and workplace.[31]

CRITICAL THINKING ISSUE

Where Does Arctic Haze Come From, and How Does It Affect the Environment?

A dark gray haze, full of industrial pollutants, hovers over the ground and extends to an altitude of 8 km. It is not Los Angeles but the frozen Arctic, thousands of miles from heavy industry. The polluted air mass includes the entire atmosphere above the Arctic Circle, as well as lobes extending into Eurasia and North America. The total area of polluted air mass, about as large as the African continent, is present during winter months and disappears in summer.

Scientists describe this haze as an aerosol—that is, microscopic particles dispersed in a gas, smoke, or fog. They have found dust from Mongolia and sea salt in the haze; but of greater concern are the pollutants, mainly sulphates, carbon soot (dark carbon), organic compounds, and toxic metals, including mercury, lead, and vanadium. The gaseous atmosphere itself contains elevated levels of carbon dioxide, methane, and carbon monoxide, as well as chemicals destructive of the ozone layer.

Like detectives searching for fingerprints, scientists used ratios of six elements (arsenic, antimony, zinc, indium, manganese, and vanadium) to a seventh, selenium, to track down sources of Arctic haze. Emissions from burning fossil fuels differ in amounts of these elements, depending on the type of fuel used. The ratios are so characteristic that scientists can tell whether pollution is from hard or soft coal. The ratios found in the haze were compared with ratios characteristic of types of fuel known to be used in various regions of the world. For example, manganese was present in greater amounts and vanadium in lesser amounts in the haze than in emissions typical of North America.

Based on the ratios and on knowledge of air circulation patterns, scientists identified Eastern Europe and Russia as major sources of Arctic air pollution, with a significant but lesser contribution from the United Kingdom and Western Europe. Two major routes (see Figure 21.29) are (1) from Eastern Europe and Russia across the Taimyr Peninsula and the North Pole to the Alaskan Arctic, and (2) from the United Kingdom and Western Europe across Scandinavia to the Norwegian Arctic. (A different study argued that none of the pollution comes from Western Europe or from China. Rather, it comes from Eastern Europe and Russia, with some possibly originating within the Arctic Circle itself.)

Major atmospheric forces driving pollution along these routes begin with large temperature differences between the equator and the poles in winter, creating strong air currents from 0° to 90° latitude. The air currents are also propelled by seasonal lows in the North Atlantic and highs on the Eurasian continent. Once masses of pollutant-laden air reach the dry, stable air of Arctic winter, they form layers, which remain relatively intact. In spring, when the northward flow of air diminishes, the haze disperses and is carried to higher levels in the atmosphere and back to the midlatitudes.

Now that scientists have a better idea of the sources of Arctic haze, they are beginning to speculate on the effects it has on the ecology of the Arctic and on global climate. Furthermore, they are calling for research into environmental impacts of the haze and for co-operation among countries in the Northern Hemisphere to reduce the amount of haze. In the meantime, humans are producing enough toxic emissions to carry as far as 10,000 km and produce pollution levels at the North Pole as great as those found in medium-sized industrial cities.

Critical Thinking Questions

1. Although laden with pollutants, Arctic haze drops less of its aerosol on the ground than would happen in other areas of the world. What is it in the Arctic environment that could account for this?
2. Elements such as mercury and lead are present in small amounts in Arctic haze (the levels are often comparable to those near industrial sites and 10–20 times those in Antarctica). If the amounts are so low, why are scientists concerned about their effects? (You may wish to review Chapter 15.

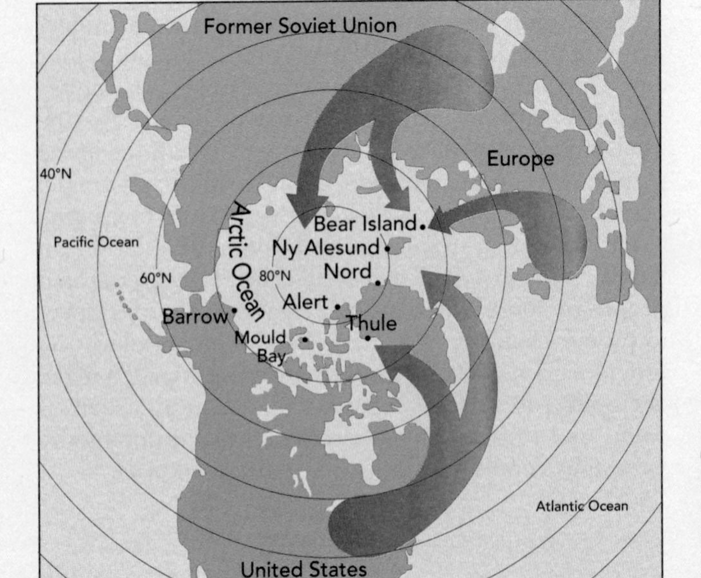

Figure 21.29 • Two airstreams out of the former Soviet Union (top arrows) carry most of the pollution that becomes Arctic haze. According to some findings, Europe is also an important source of smog (arrow at right). But the northeastern United States (bottom arrows) is not a big contributor to the haze, because pollutants are often washed out by storms before they reach the Arctic. The dots represent air-sampling stations.

3. Sulphates in haze react with water to form acids. Scientists predict that the Arctic will be sensitive to additional acids. Why might this be so?
4. What are possible effects of dark carbon particles such as soot on the Arctic climate? Make a diagram. Put the term "dark carbon particles" in the centre of a sheet of paper. Write as many direct consequences of dark carbon as you can around the term, and connect the consequences to the term with arrows. Now, what would be the secondary consequences of these direct effects? Place them around the edge of the diagram, connecting them by arrows to the relevant direct effect. If you can think of third- and fourth-order effects, place these on the diagram in a similar manner.

Summary

- Every year, almost 200 million metric tonnes of primary pollutants enter the atmosphere above Canada and the United States from processes related to human activity. Considering the enormous volume of the atmosphere, this is a relatively small amount of material. If it were distributed uniformly, there would be little problem with air pollution. Unfortunately, the pollutants generally are not evenly distributed but are concentrated in urban areas or in other areas where the air naturally lingers.
- The two major types of pollution sources are stationary and mobile. Stationary sources have a relatively fixed position and include point sources, area sources, and fugitive sources.
- There are two main groups of air pollutants: primary and secondary. Primary pollutants are those emitted directly into the air: particulates, sulphur dioxide, carbon monoxide, nitrogen oxides, and hydrocarbons. Secondary pollutants are those produced through reactions between primary pollutants and other atmospheric compounds. A good example of a secondary pollutant is ozone, which forms over urban areas through photochemical reactions between primary pollutants and natural atmospheric gases.
- The effects of the major air pollutants are considerable. They include effects on the visual quality of the environment, vegetation, animals, soil, water quality, natural and artificial structures, and human health.
- The combustion of large quantities of fossil fuels results in the emission of sulphur and nitrogen oxides into the atmosphere, creating acid precipitation. Environmental degradations associated with acid precipitation include loss of fish and other life in lakes, damage to trees and other plants, leaching of nutrients from soils, and damage to stone statues and buildings in urban areas.
- There are two major types of smog: photochemical and sulphurous. Each type of smog brings particular environmental problems that vary with geographic region, time of year, and local urban conditions.
- Meteorological conditions greatly affect whether polluted air is a problem in an urban area. In particular, restricted circulation in the lower atmosphere associated with temperature inversion layers may lead to pollution events.
- From an environmental viewpoint, the preferred method of reducing the emission of air pollutants produced from burning fossil fuels is to practice energy efficiency and conservation so that smaller amounts of fossil fuels are burned. Another option is to increase the use of alternative energy sources, such as solar and wind power that do not emit air pollutants.
- Methods to control air pollution are tailored to specific sources and types of pollutants. These methods vary from settling chambers for particulates to scrubbers that use lime to remove sulphur before it enters the atmosphere. In urban regions, emission control efforts centre on automobiles, buses, and other vehicles, because they account for most of the pollutants that enter the urban atmosphere.
- Emissions of air pollutants in Canada and the United States are decreasing. In developing countries, air pollution in large urban centres is often a serious problem. Air quality in Canadian urban areas is usually reported to the public in terms of whether the quality is good, moderate, poor, or very poor. These levels are defined in terms of the Air Quality Index (AQI).
- The relationships between emission control and environmental cost are complex. The minimum total cost is a compromise between capital costs to control pollutants and losses or damages resulting from pollution. If additional controls are necessary to lower the pollution to a more acceptable level, additional costs are incurred. Beyond a certain level of pollution abatement, these costs can increase rapidly.
- Indoor air pollution has been with us for thousands of years, since people first built structures and burned fuel indoors. It is one of our most serious environmental health problems.
- Sources of indoor air pollution are extremely varied. They may be associated with the materials with which we build our buildings, the furnishings we put in them, and the types of equipment we use for heating and cooling, as well as natural processes that allow gases to seep into buildings.

- Concentrations of indoor air pollutants are generally greater than concentrations of the same pollutants found outdoors. The common method for controlling indoor air pollution is ventilation. However, natural ventilation has been reduced through tighter construction of buildings, and commonly used ventilation systems are not generally designed to reduce certain types of indoor air pollutants. In addition, these systems require careful maintenance.
- A variety of pathways, processes, and driving forces affect the air quality of a building. The most common natural processes involve the differential pressure produced by wind and the chimney, or stack, effect, which occurs when there is a temperature differential between indoor and outdoor environments.
- Indoor air pollution has different effects on different people, and some groups of people are particularly susceptible to air pollution problems. Often, the symptoms reported by people working in a building vary; and some symptoms may result from factors other than air pollution.
- There are two basic types of sick buildings: those with an identifiable problem, such as mould or bacteria, and those with sick building syndrome, in which people's symptoms can't be traced to any one cause.
- Environmental tobacco smoke, or second-hand smoke, is the most hazardous indoor air pollutant.
- Radon gas that seeps into homes is thought to be a serious environmental health hazard in Canada today. Studies have suggested that exposure to elevated concentrations of radon is associated with an increased risk of lung cancer.
- Control of indoor air pollution involves several strategies, including ventilation, source removal, source modification, and installation of air-cleaning equipment, as well as education.

STUDY QUESTIONS

1. Compare and contrast the London 1952 fog event with smog problems in the Los Angeles basin.
2. Why do we have air pollution problems when the amount of pollution emitted into the air is a very small fraction of the total material in the atmosphere?
3. What is the difference between point and nonpoint sources of air pollution? Which type is easier to manage?
4. What are the differences between primary and secondary pollutants?
5. Carefully examine Figure 21.19, which shows a column of air moving through an urban area, and Figure 21.20, which shows relative concentrations of pollutants that develop on a typical warm day in Los Angeles. What linkages between the information in these two figures might be important in trying to identify and learn more about potential air pollution in an area?
6. Why is acid deposition a major environmental problem, and how can it be minimized?
7. Why will air pollution-abatement strategies in developed countries probably be much different in terms of methods, process, and results from air pollution-abatement strategies in developing countries?
8. Why is it so difficult to establish national air quality standards?
9. In a highly technological society, is it possible to have 100% clean air? Is it feasible or likely?
10. How good are the air quality standards being used by your province? How might their usefulness be evaluated? Do you think the standards will change in the future? If so, what are the likely changes?
11. What are some of the common sources of air pollutants where you live, work, or attend classes?
12. Develop a research plan to complete an audit of the indoor air quality in your local library. How might that research plan differ from a similar audit for the science buildings on your campus?
13. What do you think about the concept of sick building syndrome? If you were working for a large corporation and a number of your employees stated that they were becoming sick and listed a series of symptoms and problems, how would you react? What could you do? Play the role of the administrator, and develop a plan to look at the potential problem.
14. Some people argue that the potential hazard from radon gas in homes is much less than suggested by regulatory authorities. Do you agree or disagree? How might potential differences of opinion ultimately be answered?
15. Develop a plan to study the potential radon hazard in your community. Where would you start? How would you gather data, and so on? If your community has undergone extensive testing already, review the results and decide if further testing is necessary.

FURTHER READING

Boubel, R. W., D. L. Fox, D. B. Turner, and A. C. Stern.*Fundamentals of Air Pollution*, 3rd ed. New York: Academic, 1994. A thorough book covering the sources, mechanisms, effects, and control of air pollution.

Brenner, D. J. *Radon: Risk and Remedy*. New York: Freeman, 1989. A wonderful book concerning the hazard of radon gas. It covers everything from the history of the problem to what was happening in 1989, as well as solutions, and is highly recommended.

Brooks, B. O., and W. F. Davis. *Understanding Indoor Air Quality*. Ann Arbor, Mich.: CRC Press, 1992. A comprehensive evaluation of indoor air pollution. It discusses most of the sources of indoor air pollutants, as well as health effects and controls.

Health Canada. *Exposure Guidelines for Residential Indoor Air Quality: A Report of the Federal-Provincial Advisory Committee on Environmental and Occupational Health*. Environmental Health Directorate, Health Protection Branch. Ottawa: Health Canada, 1995.

Hewitt, D. N., W. T. Sturges, and NOAA, eds. *Global Atmospheric Chemical Change*. New York: Chapman & Hall, 1995. A book describing aspects of global air pollution, including chemical changes in the atmosphere, climate change, acid deposition, and other anthropogenic pollutants.

Kay, J. G., G. E. Keller, and J. F. Miller. *Indoor Air Pollution*. Chelsea, Mich.: Lewis, 1991. Essays on problems with biological and nonbiological air pollution in the indoor environment.

Marconi, M., B. Seifert, and T. Lindvall. *Indoor Air Quality: A Comprehensive Reference Book*. New York: Elsevier, 1995. A thorough text covering all aspects of indoor air pollution and air quality.

Rose, J., ed. *Acid Rain: Current Situation and Remedies*. Philadelphia: Gordon and Breach Science, 1994. Essays covering issues and consequences of acid precipitation in Europe and the United States.

Stieb, D. M., L. D. Pengelly, N. Arron, S. M. Taylor and M. E. Raizenne. "Health effects of air pollution in Canada: Expert panel findings for the Canadian Smog Advisory Program." *Can Respir.* J. 2(3): 156-160, 1995.

Stone, R. "Air Pollution: Counting the Cost of London's Killer Smog," *Science* 298 (2002): 2106–2107. An in-depth look at events involved in the opening case study.

Wang, L. "Paving out Pollution," *Scientific American* (February 2002), p. 20. Discussion of an innovative approach to reducing air pollution.

chapter 22

Our Skin Cancer Epidemic

People in Canada and the United States are experiencing an epidemic of skin cancer. A worldwide increase in skin cancer has occurred since early 1970. The Canadian Cancer Society reported that deaths from melanoma, one form of skin cancer, rose 41% in men and 23% in women between 1988 and 2003—the highest rate of increase for any cancer in Canada. In parts of the United States, the increase since 1970 is estimated at about 90%. It is suspected that the increase in skin cancer rates is linked to ozone depletion.[1] Ozone depletion and, in particular, development of the "ozone hole" (discussed later in this chapter) have been most dramatic since about 1975, and this correlates with the increased incidence of skin cancers.

As ozone depletion occurs, ultraviolet B (UVB) radiation, a normal part of the solar radiation that reaches Earth's surface, increases. UVB causes damage and mutation to DNA in skin cells, which may initiate cancer. It is believed that a 1% decrease in ozone causes an increase of UVB radiation of about 1–2%. For each 1% increase in UVB radiation, it is projected that skin cancer will increase about 2%.

Since 1970, ozone depletion in the atmosphere above Canada and the United States has been about 10%. Under a worst-case scenario, this 10% decrease in ozone could cause an increase in skin cancer of 20–40% over the same period. However, the observed increase has been 90%! Thus, either we are incorrectly evaluating the effects of ozone depletion and increased UVB radiation on skin cancers; or diagnosis of skin cancer has improved, so more are found; or other factors are affecting the incidence of skin cancers. Indeed, because disease seldom involves a one cause–one effect relationship, the epidemic of skin cancers is probably due to multiple interrelated causes, including the following:

- People are living longer, and cancers are a disease of aging.
- Cancer seems to be more a problem of urbanized and industrial societies.
- People in Canada and the United States today are more affluent than in the past and spend more time outdoors exposed to cancer-inducing UVB radiation.

Older people, who for years worked outside as fishermen, farmers, lifeguards, or in construction, as well as those who sunbathed regularly, are at greater risk of contracting skin cancer. Sunburns that produce blistering and severe sunburns in childhood are thought to be especially hazardous, increasing the risk of melanoma in later life.[1] The risk to an individual depends on two factors: the extent of exposure to UVB and the amount of protective melanin pigment in the skin. Melanin pigment absorbs UVB and provides a natural protection. Thus, fair-skinned people who have little melanin compared with darker-skinned people have a higher incidence of skin cancer.

Applying sunscreen. Sunscreen reduces exposure to UVB, which is known to cause skin cancer.

Ozone Depletion

LEARNING OBJECTIVES

Ozone depletion in the stratosphere is recognized as a major environmental problem with serious effects. After reading this chapter, you should understand:

- What ozone is and how ozone is naturally formed and destroyed in the stratosphere.
- What the so-called ozone shield is, and why it is important.
- How chemical and physical processes and reactions link emissions of chlorofluorocarbons (CFCs) to stratospheric ozone depletion.
- What role polar stratospheric clouds play in ozone depletion.
- Why ozone depletion is a long-term problem.
- What the environmental effects of ozone depletion are, and what options are available to minimize ozone depletion.
- Why international co-operation, including significant economic aid from wealthy to less wealthy nations, is necessary to encourage future reduction or elimination of emissions of ozone-depleting chemicals into the atmosphere.

- *Ozone depletion is a serious global environmental problem. We understand what causes ozone depletion and have implemented appropriate policies to solve the depletion problem.*[2] *The science of ozone depletion and the policies to reduce it are the two main subjects of this chapter.*

22.1 Ozone

The air we breathe at sea level is composed of approximately 21% diatomic oxygen (O_2), which is two oxygen atoms bonded together. Ozone (O_3) is a triatomic form of oxygen, in which three atoms of oxygen are bonded. Ozone is a strong oxidant and chemically reacts with many materials in the atmosphere. In the lower atmosphere, ozone is a pollutant produced by photochemical reactions involving sunlight, nitrogen oxides, hydrocarbons, and diatomic oxygen.

Figure 22.1 shows the structure of the atmosphere and concentrations of ozone. The highest concentrations are in the stratosphere, ranging from about 15 km to 40 km in altitude. Approximately 90% of the ozone in the atmosphere is found in the stratosphere, where peak concentrations are about 400 ppb. The altitude of peak concentration varies from about 30 km near the equator to about 15 km in polar regions.[3]

Ultraviolet Radiation and Ozone

The ozone layer in the stratosphere is often called the ozone shield, because it absorbs most of the potentially hazardous ultraviolet radiation that enters Earth's atmosphere from the sun. Figure 22.2 shows part of the electromagnetic spectrum, discussed in Chapter 20. Ultraviolet radiation consists of wavelengths between 0.1 and 0.4 μm and is subdivided into ultraviolet A (UVA), ultraviolet B (UVB), and ultraviolet C (UVC). Ultraviolet radiation with a wavelength of less than about 0.3 μm is potentially very hazardous to life. If much of this radiation reached Earth's surface, it would injure or kill most living things.

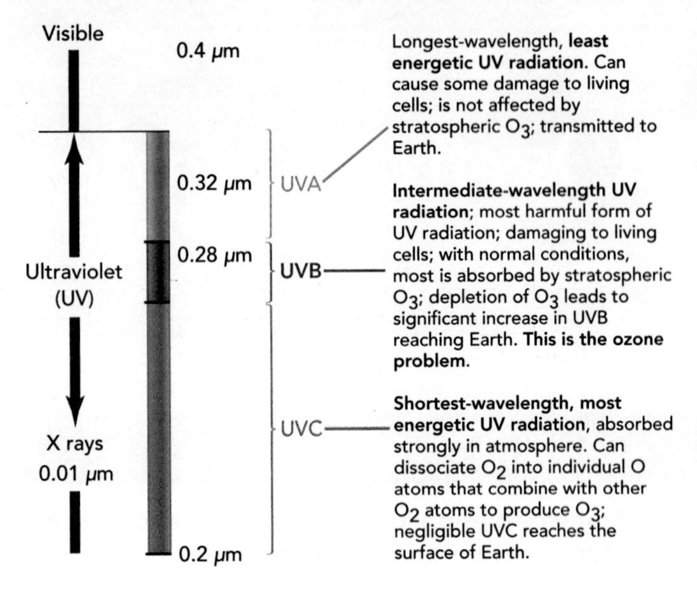

Figure 22.2 • Part of the electromagnetic spectrum showing ultraviolet radiation with wavelengths between 0.01 and 0.4 μm.

Ultraviolet C (UVC) has the shortest wavelength and is the most energetic of the types of ultraviolet radiation. It has sufficient energy to break down diatomic oxygen (O_2) in the stratosphere into two oxygen atoms. Each of these oxygen atoms may combine with an O_2 molecule to create ozone. Ultraviolet C is strongly absorbed in the stratosphere, and negligible amounts reach the surface of Earth.[3, 4]

Ultraviolet A (UVA) radiation has the longest wavelength and the least energy of the three types of ultraviolet radiation. UVA can cause some damage to

Figure 22.1 • *(a)* Structure of the atmosphere and ozone concentration. *(b)* Reduction of the most potentially biologically damaging ultraviolet radiation by ozone in the stratosphere.

[*Source*: Ozone concentrations modified from R. T. Watson, "Atmospheric Ozone," in J. G. Titus, ed., *Effects of Change in Stratospheric Ozone and Global Climate,* vol. 1, Overview, p. 70 (U.S. Environmental Protection Agency).]

(a) Structure of the atmosphere and ozone concentration.

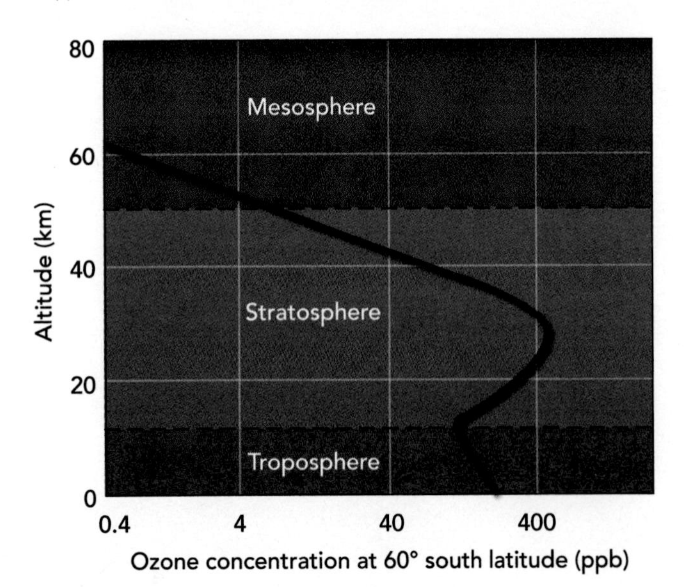

(b) Reduction of the most potentially biologically damaging ultraviolet radiation by ozone in the stratosphere.

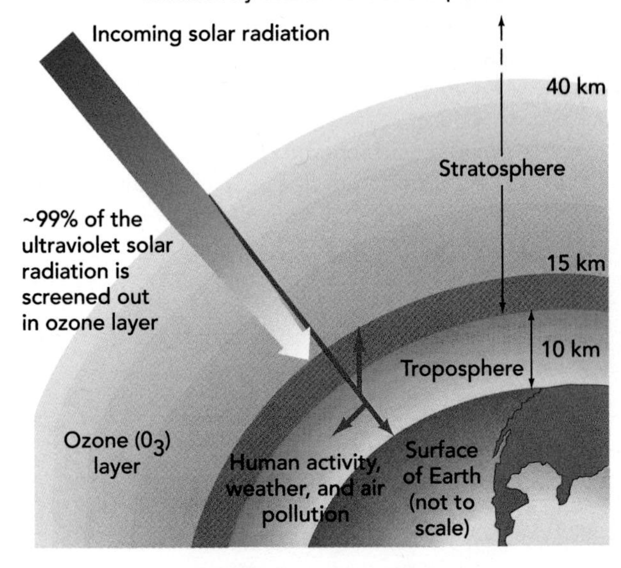

living cells, is not affected by stratospheric ozone, and is transmitted to the surface of Earth.[3]

Ultraviolet B (UVB) radiation is energetic and is strongly absorbed by stratospheric ozone. Ozone is the only known gas that absorbs UVB. As a result, depletion of ozone in the stratosphere results in an increase in the UVB that reaches the surface of the Earth. Because UVB radiation is known to be hazardous to living things,[3, 4] this increase in UVB is the hazard we are talking about when we discuss the ozone problem.

Processes that produce ozone in the stratosphere are illustrated in Figure 22.3. The first process, or step, in ozone production occurs when intense ultraviolet radiation (UVC) breaks apart an oxygen molecule (O_2) through the process of photodissociation into two oxygen atoms. These atoms then react with another oxygen molecule to form two ozone molecules. Ozone, once produced, may absorb UVC radiation, which breaks the ozone molecule into an oxygen molecule and an oxygen atom. This is followed by the recombination of the oxygen atom with another oxygen molecule to re-form into ozone. As part of this process, UVC radiation is converted to heat energy in the stratosphere.[5] Natural conditions that prevail in the stratosphere result in a dynamic balance between the creation and destruction of ozone.

In summary, approximately 99% of all ultraviolet solar radiation (all UVC and most UVB) is absorbed or screened out in the ozone layer. The absorption of ultraviolet radiation by ozone is a natural service function of the ozone shield and protects us from the potentially harmful effects of ultraviolet radiation.

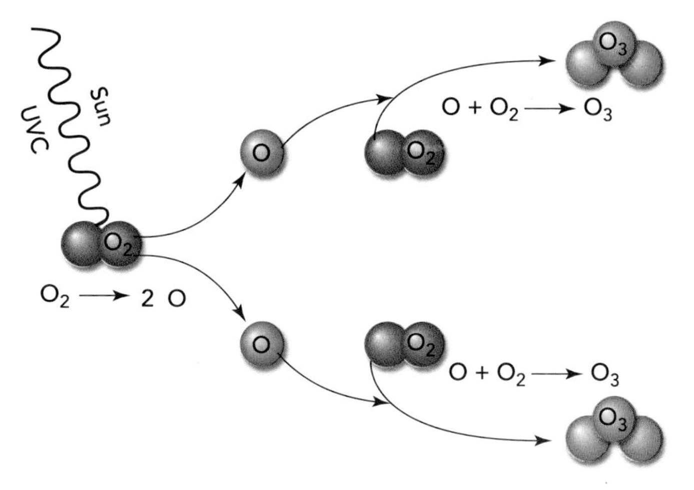

Figure 22.3 • Production of ozone (O_3) in the stratosphere. Photo dissociation of the oxygen molecule (O_2) yields two atoms of oxygen. Each combines with an oxygen molecule to form ozone (O_3). [*Source*: Modified from NASA-GSFC, "Stratospheric Ozone," accessed August 22, 2000 at http://see.gsfc.nasa.gov.]

Measurement of Stratospheric Ozone

Scientists first measured the concentration of atmospheric ozone in the 1920s, from the ground, using an instrument known as a Dobson ultraviolet spectrometer. The **Dobson unit (DU)** is still commonly used to measure the concentration of ozone; 1 DU is equivalent to a concentration of 1 ppb O_3. Today, we have a record of ozone concentrations from more than 30 locations around the world over about 30 years. Most of the measurement stations are in the middle latitudes, and the accuracy of the data varies with different levels of quality control.[3] Satellite measurements of concentrations of atmospheric ozone began in 1970 and continue today.

Ground-based measurements first identified ozone depletion over the Antarctic. Members of the British Antarctic Survey began to measure ozone in 1957 and in 1985 published the first data that suggested significant ozone depletion over Antarctica. The data are taken during October of each year—the Antarctic spring—and show that the concentrations of ozone hovered around 300 DU from 1957 to about 1970, and then declined to about 200 DU by 1983. They then sharply dropped to approximately 150 DU by 1986. Since then, the variability of the minimum ozone concentration has been considerable, with a high of about 175 DU in 1989 and a low of about 90 DU in 1995 (Figure 22.4). By 2002, the value was about 140 DU, but dropped again to about 100 DU in 2003. In spite of the variations, the direction of change, with minor exceptions (1989 and 2002), is clear; ozone concentrations in the stratosphere during the Antarctic spring have been decreasing since the mid-1970s.[6-9]

Satellite measurements of ozone recorded before 1985 had also indicated a significant reduction in ozone concentration; however, the values were so low that they were not believed. After the 1985 announcement of the decrease in ozone over Antarctica, the satellite

Figure 22.4 • Average Antarctic minimum ozone concentration, 1980 to 2003. Values in the 1970s were about 300 DU. Modified from NASA 2003. [Antarctic ozone hole accessed 3/24/04 at http://jwoky.gsfc.nasa.gov/multi/min_ozone.gif.]

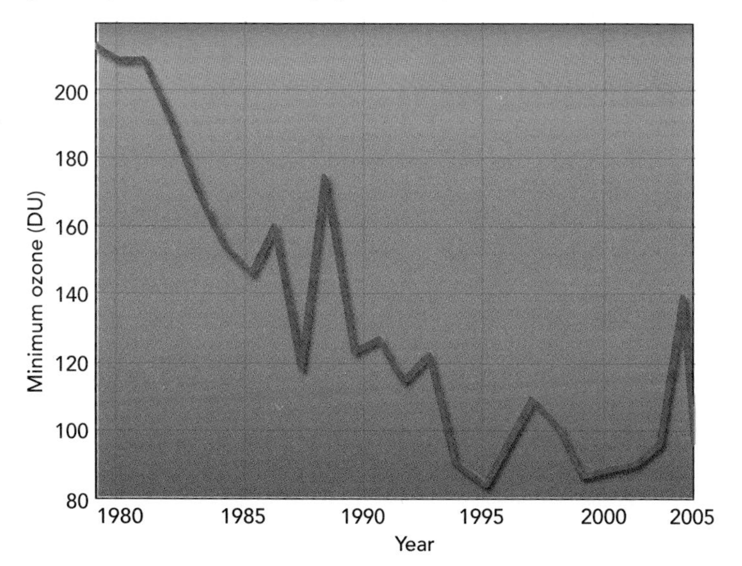

measurements were re-evaluated and found to confirm the observations reported by the British Antarctic Survey. This depletion in ozone was dubbed the ozone hole. However, there is no actual hole in the ozone shield where all the ozone is depleted; rather, the term describes a relative depletion in the concentration of ozone that occurs during the Antarctic spring.

22.2 Ozone Depletion and CFCs

The hypothesis that ozone in the stratosphere is being depleted by the presence of **chlorofluorocarbons (CFCs)** was first suggested in 1974 by Mario Molina and F. Sherwood Rowland.[10] This hypothesis, based for the most part on physical and chemical properties of CFCs and knowledge about atmospheric conditions, was immediately controversial. The idea received a tremendous amount of exposure, both in newspapers and on television, and was vigorously debated by scientists, companies producing CFCs, and other interested parties.

The public became concerned because everyday products, such as shaving cream, hair spray, deodorants, paints, and insecticides, were packaged in spray cans that carried CFCs as a propellant. The idea that these products could be responsible for threatening their health and the well-being of the environment captured people's imagination. Many people responded by writing to their elected representatives and making individual decisions to purchase fewer products containing CFCs.[11]

The major features of the Molina and Rowland hypothesis are as follows:[3]

- The CFCs emitted in the lower atmosphere by human activity are extremely stable. They are unreactive in the lower atmosphere and therefore have a very long residence time (about 100 years). Another way of stating this is to say that no significant tropospheric sinks for CFCs are known. A possible exception is soils, which evidently do remove an unknown amount of CFCs from the atmosphere at Earth's surface.[12]
- Because CFCs have a long residence time in the lower atmosphere and because the lower atmosphere is very fluid, with abundant mixing, the CFCs eventually (by the process of dispersion) wander upward and enter the stratosphere. Once they have reached altitudes above most of the stratospheric ozone, they may be destroyed by the highly energetic solar ultraviolet radiation. This process releases chlorine, a highly reactive atom.
- The reactive chlorine released may then enter into reactions that deplete ozone in the stratosphere.
- The result of the depletion of ozone is an increase in the amount of UVB radiation that reaches Earth's surface. Ultraviolet B is a cause of human skin cancers and is also thought to be harmful to the human immune system.

Emissions and Uses of Ozone-Depleting Chemicals

Emissions of the commonly used chemicals related to ozone depletion are shown in Table 22.1. These emissions are thought to destroy stratospheric ozone amount to approximately 1.5 million metric tonnes per year, with CFCs accounting for approximately 60% of the total emissions.

Table 22.1 also shows the approximate atmospheric lifetimes of these chemicals, which vary from about five years to more than 100 years. Because CFCs, in particular, have such long lifetimes in the atmosphere, they will be with us for many years.

Table 22.1 • Emissions of Some Chemicals Associated with Stratospheric Ozone Depletion in 1989, with Growth Rate of CFCs for 1998 and Substitutes for CFCs

Chemical	*Emissions (thousand of tonnes)*	*Atmospheric Lifetime[a] (years)*	*Applications*	*Annual Growth Rate to 1989 (%)*	*Share of Contribution to Depletion (%)*	*Annual Growth Rate, 1998 (%)*
CFC-12	454	139	Air-conditioning, refrigeration, aerosols, foams	5	45	0.5
CFC-11	262	76	Foams, aerosols, refrigeration	5	26	−0.1
CFC-113	152	92	Solvents	10	12	0.0
Carbon tetrachloride	73	35	Solvents	1	8	
Methyl chloroform	522	5	Solvents	7	5	
Halon 1301	3	65	Fire extinguishers	n.a.	4	
Halon 1211	79	11	Refrigeration, foams	11	0	
HCFC-22		12	Substitute for CFC			
HCFC-123		2	Substitute for CFC			
HCFC-124		6	Substitute for CFC			

a. Sources for atmospheric lifetimes of substances include C. P. Shea, "Mending the Earth's Shield," *World Watch* (January–February 1989):27–34; NASA-GSFC, "Stratospheric Ozone," accessed August 22, 2000 at http://see.gsfc.nasa.gov; and U.S. Environmental Protection Agency, "Ozone-Depleting Substances," accessed March 1, 2001 at http://www.epa.gov/ozone/ods/html.

Various uses of the chemicals are also shown in the table. CFCs have been used as aerosol propellants in spray cans, as a working gas in refrigeration and air-conditioning units, and in the foam-blowing process for the production of Styrofoam. A variety of cleaning solvents, such as carbon tetrachloride and methyl chloroform, contain chlorine and thus destroy ozone, as does halon, which contains bromine (another chemical like chlorine) that is used in fire extinguishers.[3, 11]

One of the first restrictions on CFCs included its use as a propellant gas for spray cans. This practice was banned in the late 1970s in a number of countries, setting a trend that has continued, with the result that CFCs as aerosol propellants are no longer a problem.[3] In contrast, the use of CFCs as a refrigerant has increased dramatically in recent years (Figure 22.5), especially in developing countries, such as China.

Simplified Stratospheric Chlorine Chemistry

CFCs are considered responsible for most of the ozone depletion observed by scientists. Let us look more closely at how this effect occurs.

Earlier, we noted that there are no tropospheric sinks for CFCs. That is, the processes that remove most chemicals in the lower atmosphere—destruction by sunlight, rain-out, and oxidation—do not break down CFCs, because CFCs are transparent to sunlight, are essentially insoluble, and are non-reactive in the oxygen-rich lower atmosphere.[13] Indeed, the fact that CFCs are non-reactive in the lower atmosphere was one reason they were attractive for use as propellants.

When CFCs wander to the upper part of the stratosphere, however, reactions do occur. Highly energetic ultraviolet radiation (UVC) splits up the CFC, releasing chlorine. When this happens, the following two reactions can take place:[13]

$$(1)\ Cl + O_3 \rightarrow ClO + O_2$$
$$(2)\ ClO + O \rightarrow Cl + O_2$$

These two equations define a chemical cycle that can deplete ozone (Figure 22.6). In the first reaction, chlorine combines with ozone to produce chlorine monoxide, which in the second reaction combines with monatomic oxygen to produce chlorine again. The chlorine can then enter another reaction with ozone and cause additional ozone depletion. This series of reactions is known as a *catalytic chain reaction*. Because the chlorine is not removed but reappears as a product of the second reaction, the process may be repeated over and over again. It has been estimated that each chlorine atom may destroy approximately 100,000 molecules of ozone over a period of one or two years before the chlorine is finally removed from the stratosphere through other chemical reactions and rain-out.[13] The significance of these reactions is apparent

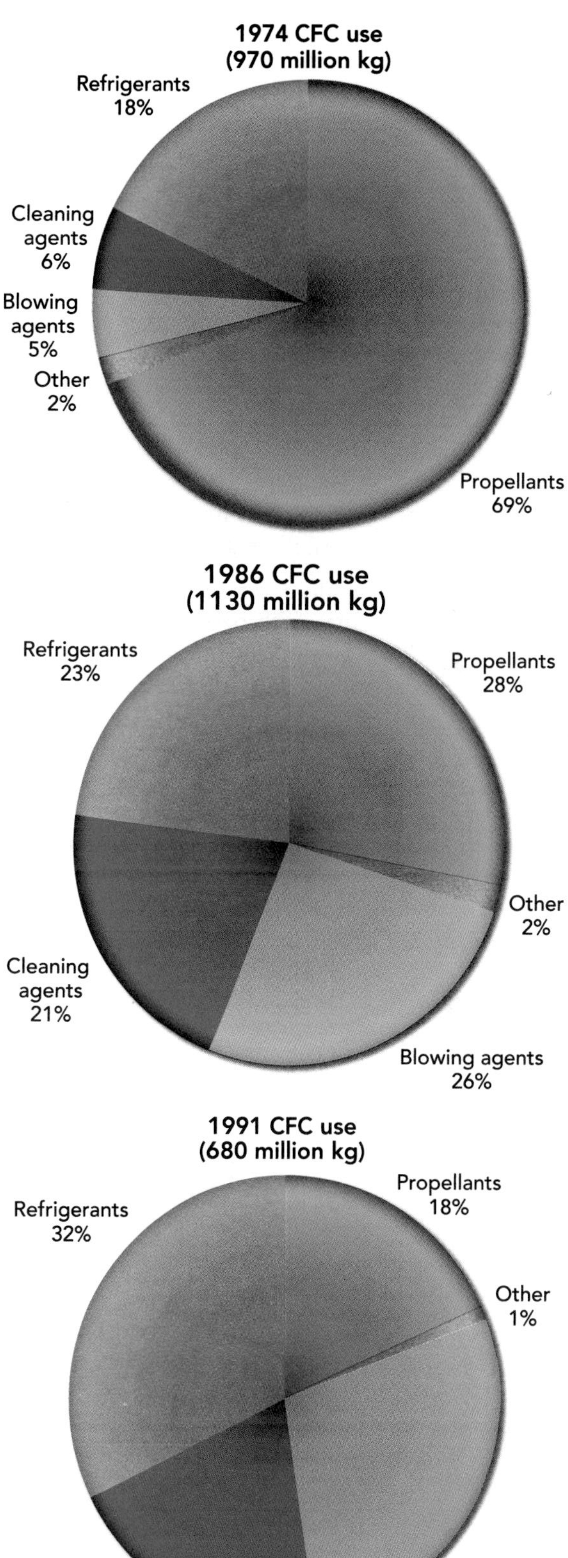

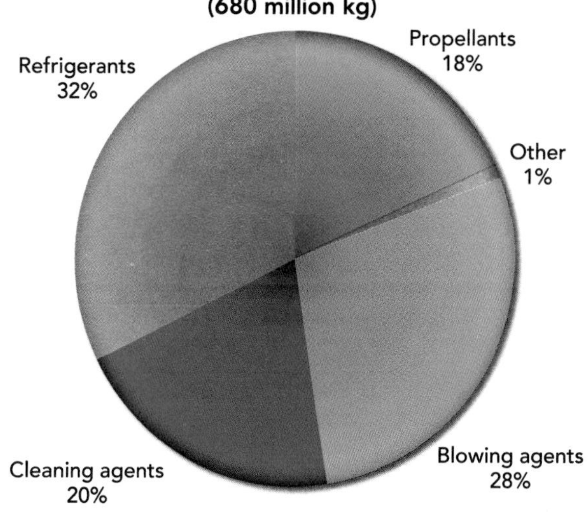

Figure 22.5 • Uses of CFC and amounts emitted in *(a)* **1974;** *(b)* **1986; and** *(c)* **1991.** Use of CFCs as propellants in spray cans decreased dramatically, while use of CFCs as refrigerants increased. Total emissions in 1991 are far below those in 1986 and less than those in 1974. [*Source*: Modified from NASA-GSFC, "Stratospheric Ozone," accessed August 22, 2000 at http://see.gsfc.nasa.gov.]

when we realize how many metric tonnes of CFCs have been emitted into the atmosphere (see Table 22.1).

It should be noted that what actually happens chemically in the stratosphere is considerably more complex than the two equations shown here. The atmosphere is essentially a chemical soup in which a variety of processes related to aerosols and clouds take place (some of these are addressed in the discussion of the ozone hole). Nevertheless, these equations show us the basic chemical chain reaction that occurs in the stratosphere to deplete ozone.

The catalytic chain reaction just described can be interrupted through storage of chlorine in other compounds in the stratosphere. Two possibilities are as follows:

1. Ultraviolet light breaks down CFCs to release chlorine, which combines with ozone to form chlorine monoxide (ClO), as already described. This is the first reaction discussed. The chlorine monoxide may then react with nitrogen dioxide (NO_2) to form a chlorine nitrate ($ClONO_2$). If this reaction occurs, ozone depletion is minimal. The chlorine nitrate, however, is only a temporary reservoir for chlorine. The compound may be destroyed, and the chlorine released again.
2. Chlorine released from CFCs may combine with methane (CH_4) to form hydrochloric acid (HCl). The hydrochloric acid may then diffuse downward. If it enters the troposphere, rain may remove it, thus removing the chlorine from the ozone-destroying chain reaction. This is the ultimate end for most chlorine atoms in the stratosphere. However, while the hydrochloric acid molecule is in the stratosphere, it may be destroyed by incoming solar radiation, releasing the chlorine for additional ozone depletion.

It has been estimated that the chlorine chain reaction that destroys ozone may be interrupted by the processes just described as many as 200 times while a chlorine atom is in the stratosphere.[3, 14]

In part as a result of the ozone-depletion reactions, concentrations of ozone have declined in both northern and southern temperate latitudes. Figure 22.7 shows the estimated decline in stratospheric ozone from 1970 to 2002 by latitude in the Southern Hemisphere. Similar declines are thought to be occurring in the Northern Hemisphere. Massive destruction of ozone identified in the Antarctic constitutes the ozone hole, which continues to be a source of concern.[3]

In discussing the global distribution of ozone, it is important to remember that, for the Southern Hemisphere, under natural conditions, the highest concentration of ozone is found in the polar regions (about 60° south latitude) and the lowest near the equator. (See Figure 22.7 for 1970, before massive ozone destruction occurred.) At first, this may seem strange, because ozone is produced in the stratosphere by solar energy, and more solar energy is found near the equator. Much of the world's ozone is produced near the equator, but the ozone in the stratosphere moves from the equator toward the poles with global air circulation patterns, which are not well understood.[8]

22.3 The Antarctic Ozone Hole

Since the Antarctic ozone hole was first reported in 1985, it has captured the interest of many people around the world. Every year since then, ozone depletion has been observed in the Antarctic in October, the spring season there.

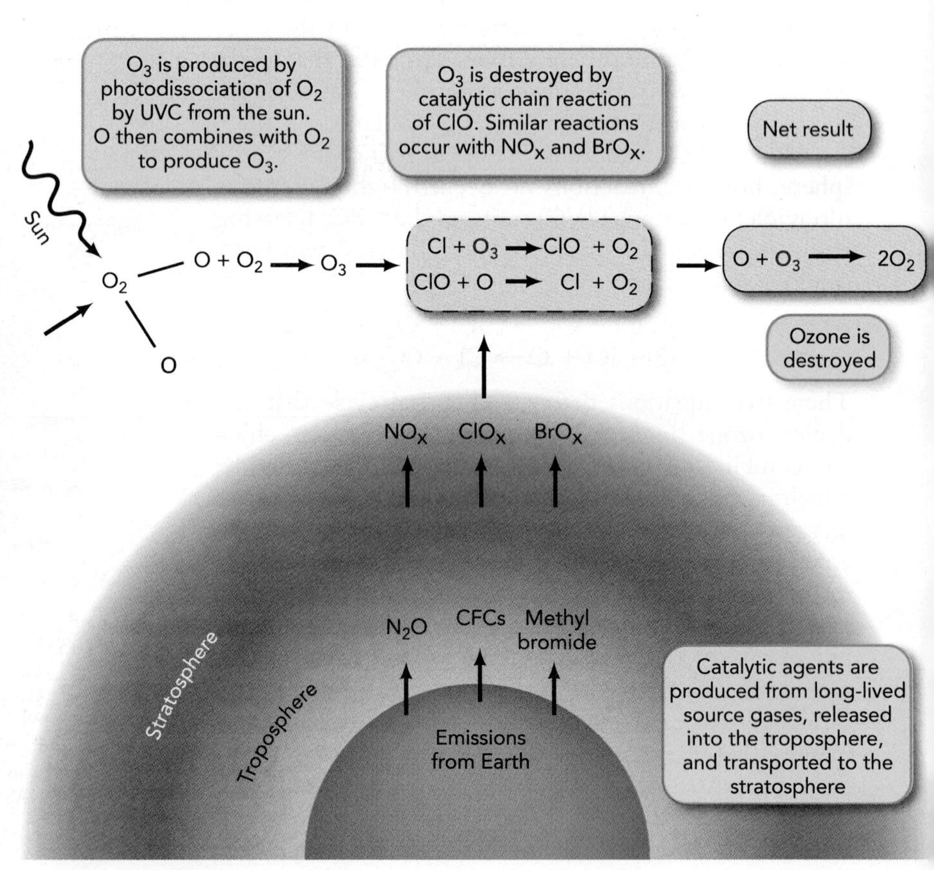

Figure 22.6 • Processes of natural formation of ozone and destruction by CFCs, N_2O, and methylbromide. [*Source*: Modified from NASA-GSFC, "Stratospheric Ozone," accessed August 22, 2000 at http://see.gsfc.nasa.gov.]

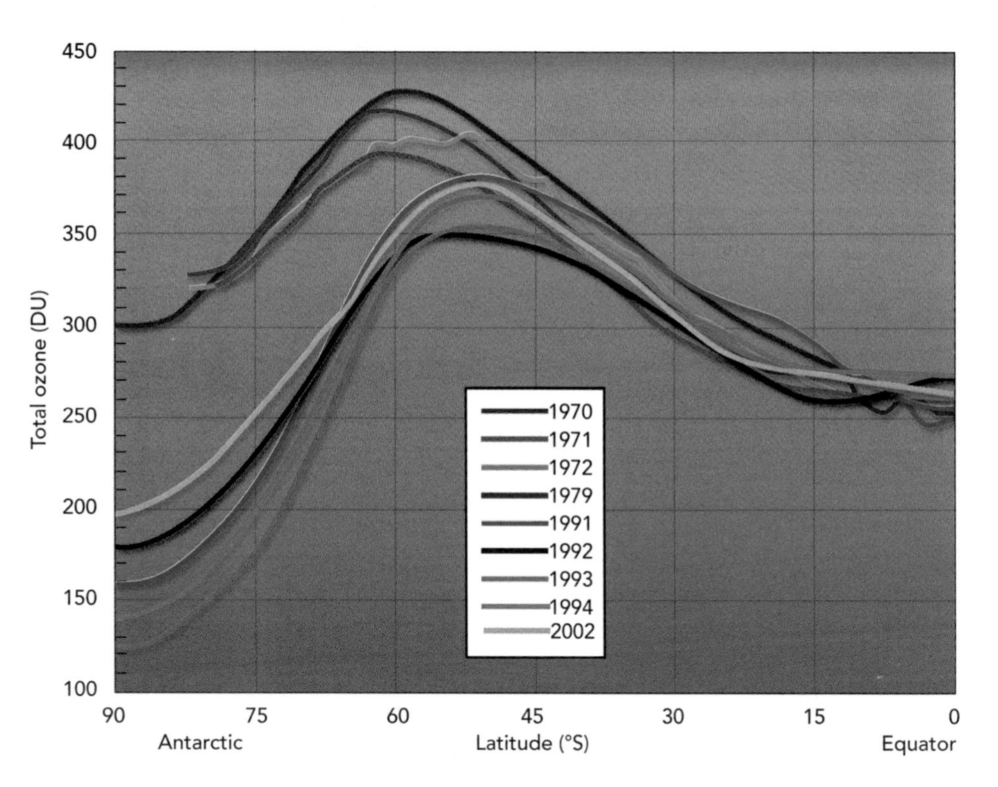

Figure 22.7 • Average change in concentration of stratospheric ozone, 1970–2002 (October mean total), by latitude. Notice that ozone reduction varies most in the Antarctic, whereas concentrations at the equator are nearly constant. [*Source*: Modified from NASA-GSFC, "Stratospheric Ozone," accessed October 2, 2003 at http://www.ccpp.edu.edu/SEES.]

The thickness of the ozone layer above the Antarctic during the springtime has been declining since the mid-1970s, and the geographic area covered by the ozone hole continues to increase.[15] The ozone hole has increased in size from a million or so square kilometres in the late 1970s and early 1980s to about 25 million square kilometres today, exceeding the area of North America (Figure 22.8).

Development of the ozone hole from the 1970s to the early to middle 1990s is shown in Figure 22.9. Also shown are the percent differences between the two time periods, emphasizing the development of the massive ozone hole in the early to middle 1990s.[5]

Polar Stratospheric Clouds

The minimum concentration of ozone in the Antarctic since 1980 has varied from about 50% to 70% of that in the 1970s (300 DU) (refer to Figure 22.4). Lesser depletion in some years is thought to be related to fewer polar stratospheric clouds over the Antarctic. By contrast, in years when the ozone depletion was greater, the regions with the most ozone depletion were in the lower stratosphere at an altitude between 14 and 24 km, where polar stratospheric clouds are present. What is the significance of these clouds?

Polar stratospheric clouds have been observed for at least the past hundred years at altitudes of approximately 20 km above the polar regions. The clouds are approximately 10–100 km in length and several kilometres thick.[14] They have an eerie beauty and an iridescent glow, with a color reminiscent of mother-of-pearl (Figure 22.10).[14]

Polar stratospheric clouds form during the polar winter (called the polar night because of the lack of sunlight, which results from the tilt of Earth's axis). During the polar winter, the Antarctic air mass is isolated from the rest of the atmosphere and circulates about the pole in what is known as the Antarctic **polar vortex**. The vortex, which rotates counter-clockwise because of the rotation of Earth in the Southern Hemisphere, forms as the isolated air mass cools, condenses, and descends.[7] The cooling occurs because the isolated air mass continues to lose heat through radiation, and no more heat is supplied because of the lack of sunlight.

Clouds are formed in the vortex when the air mass reaches a temperature between 195 K and 190 K (–78° to –83°C). At these very low temperatures, small sulphuric

Figure 22.8 • Size of the Antarctic ozone hole, 1979–2001 (September 5–October 25 average). The size of the hole started increasing rapidly in the 1980s, leveled off in the early to middle 1990s, and then increased again. [*Source*: Modified from EPA, "The Size and Depth of the Ozone Hole," accessed October 2, 2003 at http://www.epa.gov/ozone/science/hole/size.html.]

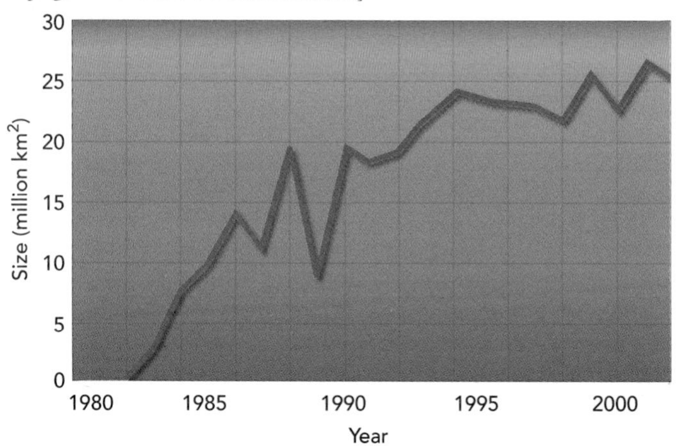

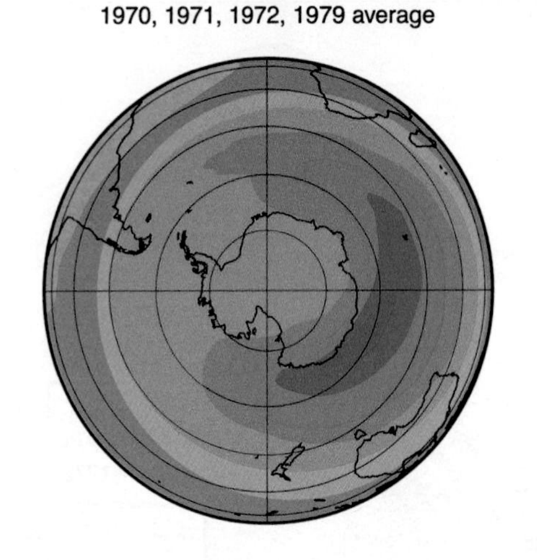

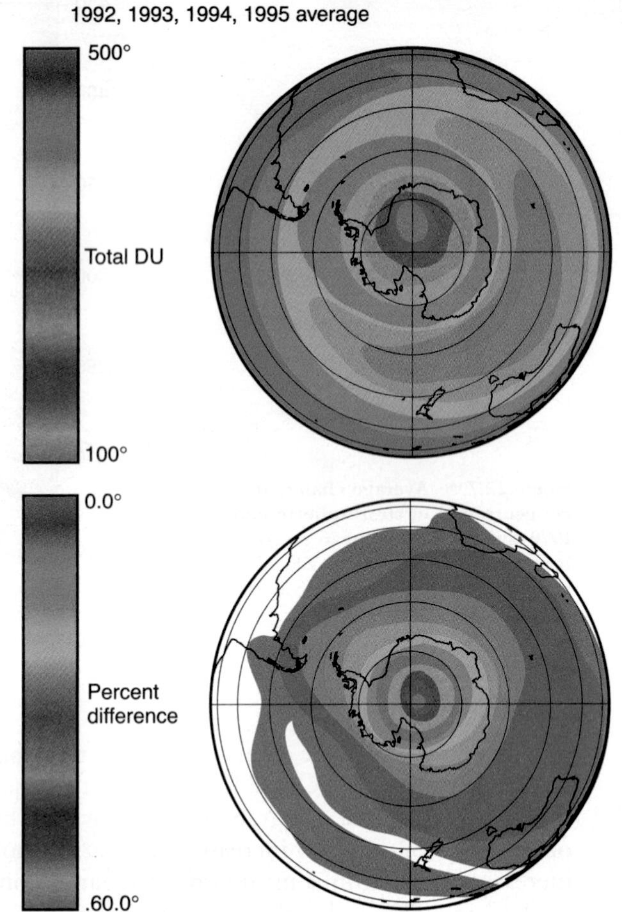

Figure 22.9 • Average development of the ozone hole in the 1970s compared with the early to middle 1990s, with percent differences between those two periods. [*Source*: Modified from NASA-GSFC, "Stratospheric Ozone," accessed August 22, 2000 at http://see.gsfc.nasa.gov.]

acid particles (approximately 0.1 μm) are frozen and serve as seed particles for nitric acid (HNO_3). These clouds are called Type I polar stratospheric clouds.

If temperatures drop below 190 K (–83°C), water vapour condenses around some of the earlier-formed Type I cloud particles, forming Type II polar stratospheric clouds, which contain larger particles. Type II polar stratospheric clouds have a mother-of-pearl color visible in polar areas.

During the formation of polar stratospheric clouds, nearly all the nitrogen oxides in the air mass are held in the clouds as nitric acid. The nitric acid particles grow large enough to fall out by gravitational settling from the stratosphere. This phenomenon has the important result of leaving very little nitrogen oxide in the atmosphere in the vicinity of the clouds.[3,7,14] This process facilitates ozone-depleting reactions, which ultimately may reduce stratospheric ozone in the polar vortex by as much as 1% to 2% per day in the early spring, when sunlight returns to the polar region.

An idealized diagram of the polar vortex that forms over Antarctica is shown in Figure 22.11*a*. Ozone-depleting reactions that occur within the vortex are illustrated in Figure 22.11*b*. As this figure shows, in the dark Antarctic winter, almost all the available nitrogen oxides are tied up on the edges of particles in the polar stratospheric clouds or have settled out. Hydrochloric acid and chlorine nitrate (the two important sinks of chlorine) act on particles of polar stratospheric clouds to form dimolecular chlorine (Cl_2) and nitric acid through the following reaction:[16]

$$HCl + ClONO_2 \rightarrow Cl_2 + HNO_3$$

In the spring, when sunlight returns and breaks apart chlorine (Cl_2), the ozone-depleting reactions discussed earlier occur. Nitrogen oxides are absent from the Antarctic stratosphere in the spring; thus, the chlorine cannot be sequestered to form chlorine nitrate, one of its major sinks. Therefore, chlorine is free to destroy ozone. In the early spring of the Antarctic, these ozone-depleting reactions can be rapid, producing

Figure 22.10 • Polar stratospheric clouds, February 12, 1989, photographed from an aircraft cruising at an altitude of approximately 12 km in the polar region north of Stavanger, Norway. The red haze and thin orange or brown layers at lower altitudes are Type I clouds. The red coloring is probably due to scattering from nitric acid particles. The higher white clouds are Type II polar stratospheric clouds, which consist mostly of water molecules frozen as ice.

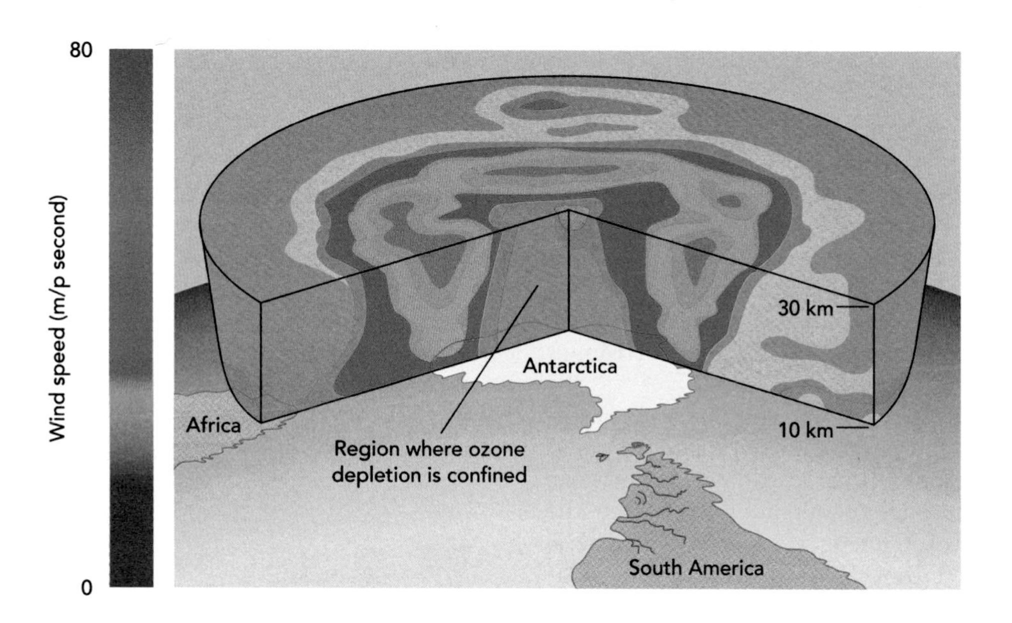

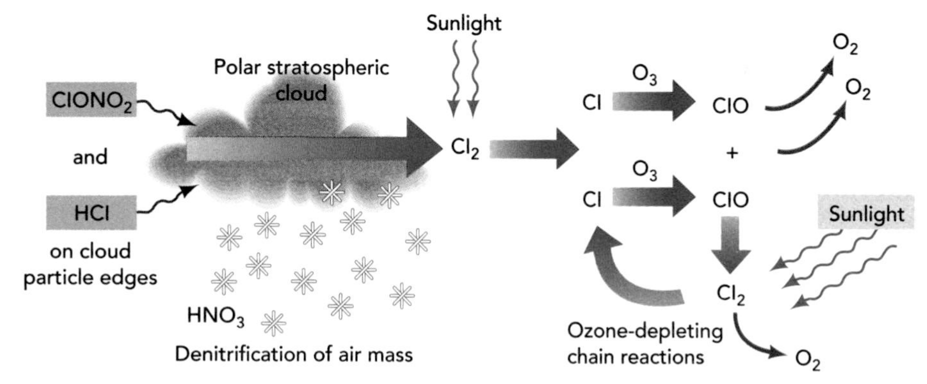

Figure 22.11 • ***(a)* Conceptual diagram of the Antarctic polar vortex and *(b)* the role of polar stratospheric clouds in the ozone depletion chain reaction.** [*Source*: Based on O. B. Toon and R. P. Turco, "Polar Stratospheric Clouds and Ozone Depletion," *Scientific American*, 264, no. 6, (1991):68–74.]

the 70% reduction in ozone observed in 1995. Ozone depletion in the Antarctic vortex ceases later in spring as the environment warms and the polar stratospheric clouds disappear, releasing nitrogen back into the atmosphere, where it can combine with chlorine and thus be removed from ozone-depleting reactions. Stratospheric ozone concentrations then increase as ozone-rich air masses again migrate to the polar region.

An Arctic Ozone Hole?

A polar vortex also forms over the North Pole area, but it is generally weaker than the Antarctic polar vortex and does not last as long. Nevertheless, ozone depletion occurs over the North Pole and it is speculated that if the vortex persists for a month or more, ozone losses in the affected air mass may be as high as 30% to 40%.

A major concern regarding the northern polar vortex is that, as it breaks up, it sends ozone-deficient air masses southward, where they may drift over populated areas of Europe and North America. In January 1992, for example, satellite data indicated a large air mass containing high levels of chlorine monoxide (ClO) stretching from Great Britain eastward over northern Europe.[17] (ClO is sometimes referred to as the "smoking gun" in the ozone problem because it plays a major part in ozone depletion—recall the earlier discussion of ozone reactions.) In contrast, the Antarctic polar vortex tends to remain more stationary, although in 1987 an ozone-depleted air mass that formed over Antarctica in October drifted northward over Australia and New Zealand in December, resulting in record low concentrations of stratospheric ozone in that region.[3]

In summary, although not as severe as the ozone depletion over the Antarctic, ozone depletion over the Arctic each winter is troublesome. Since the Arctic polar vortex is relatively weak, warmer air from midlatitudes is usually able to dissipate the vortex before ozone depletion becomes severe. However, in 1995, ozone levels were as much as 40% below normal. Scientists studying the developing ozone hole speculated that the unusually cold Arctic winter of 1995 triggered the record losses of ozone, leading to an ozone hole similar to the one that forms over the Antarctic.[18]

22.4 Tropical and Midlatitude Ozone Depletion

It has been firmly established that ozone depletion in polar regions occurs as a result of reactions that take place on particles in polar stratospheric clouds. Ice particles also

A CLOSER LOOK 22.1

How Did Mount Pinatubo's Eruption Affect Ozone Levels?

The June 1991 eruption of Mount Pinatubo in the Philippines was the largest volcanic eruption of the twentieth century. The eruption injected approximately 20 million metric tonnes of sulphur dioxide into the atmosphere, creating a stratospheric aerosol cloud larger than all the polar stratospheric clouds combined.[19]

It was hypothesized that the sulphur-based aerosol cloud contained particles that facilitated ozone-depleting reactions. Figure 22.12*a* shows the sulphur-dioxide–rich cloud (orange band) that circled Earth in the tropics after the Mount Pinatubo eruption. Figure 22.12*b* shows a corresponding band of air relatively low in ozone (violet band). The correlation between the two suggests that they might be related. However, such a correlation cannot be used to establish a cause-and-effect relationship without further supporting evidence.[19] In fact, additional data and analysis from the Pinatubo event suggest that anomalously low values of tropical ozone (Figure 22.12*b*) were probably due to effects of the eruptions on physical atmospheric circulation rather than chemical reactions that produce loss of ozone (J. W. Waters, 1994, person communication). That is, the eruptions may have caused air at lower elevations and low ozone concentrations to heat and rise to higher elevations, where higher values of ozone are normally present.

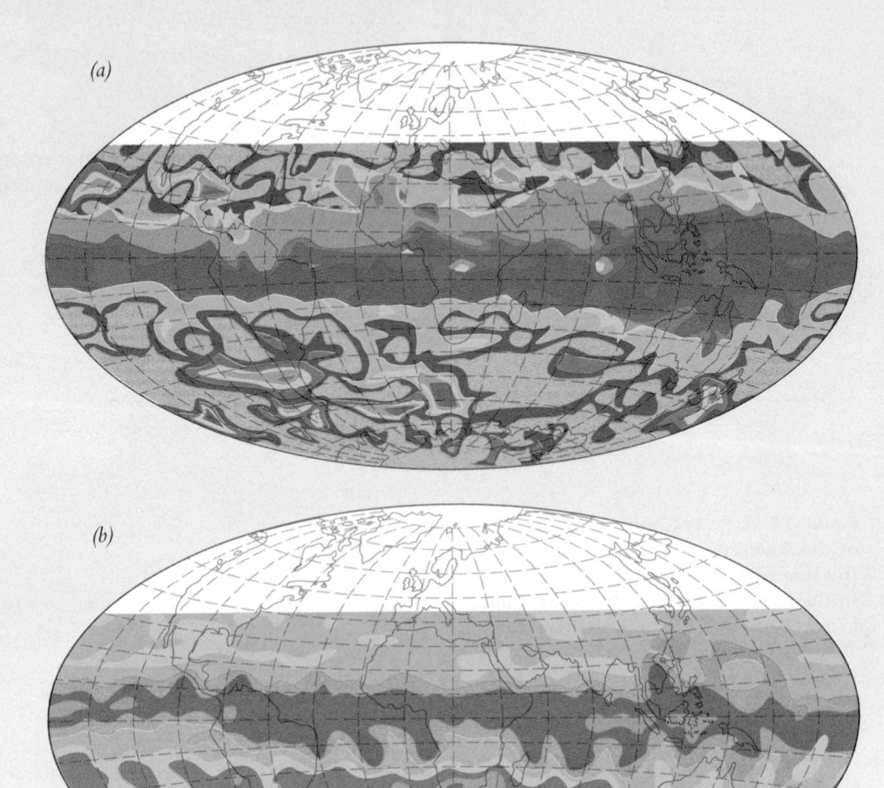

Figure 22.12 • *(a)* **Tropical belt of sulfur-dioxide–rich aerosol cloud (orange) and** *(b)* **corresponding belt of air with lower concentration of ozone (violet).** [*Source*: J. B. Waters, Jet Propulsion Laboratory, in J. Horgan, "Volcanic Disruption," *Scientific American* 266, no. 3 (1992):28, 29.]

occur in the stratosphere over the tropics; and at times, sulphuric acid aerosols are abundant in the stratosphere because of injection of sulphur by volcanic eruptions. That these particles may cause ozone depletion is only a hypothesis; there is no substantial evidence. (See A Closer Look 22.1.) Nevertheless, following the 1982 eruption of the volcano El Chichón in Mexico, a 10% reduction in ozone in the Northern Hemisphere was measured.[19]

At the South Pole, stratospheric ozone was lost at a rate of about 35% per year from the 1970s to the 1990s (see Figure 22.4).[20] However, evidence also suggests an increase in ozone depletion at middle latitudes over areas including Canada, the United States, and Europe. In summary, although we know the most information about ozone depletion in polar regions (particularly Antarctica), depletion of ozone is a global concern, from the poles to the tropics.

22.5 The Future of Ozone Depletion

A troubling aspect of ozone depletion is that if the manufacture, use, and emission of all ozone-depleting chemicals were to stop today, the problem would not go away, because millions of metric tonnes of those chemicals are now in the lower atmosphere, working their way up to the stratosphere. As shown in Table 22.1, several CFCs have atmospheric lifetimes of 75 to 140 years. Thus, about 35% of the CFC-12 molecules in the atmosphere are expected to still be there in 2100, and approximately 15% will be there in 2200.[3] In addition, approximately 10% to 15% of the CFC molecules manufactured in recent years have not yet been admitted to the atmosphere, because they remain tied up in foam insulation, air-conditioning units, and refrigerators.[3] Nevertheless, indicators suggest that growth in the concentrations of CFCs has been reduced and in some cases reversed.

Figure 22.13 shows concentrations of CFC-11 and CFC-12 from 1977 to 1998. CFC-11 concentrations reached a peak in about 1992 and then levelled off. CFC-12 concentrations peaked in 1995 and then levelled off. The growth in concentrations of CFC-12, which accounts for nearly 50% of ozone depletion (see Table 22.1), was about 5% per year from 1978 to 1995. It was about 0.5% per year by 1998 (see Figure 22.13).

Environmental Effects

Ozone depletion has several serious potential environmental effects, including damage to Earth's food chains on land and in the oceans and damage to human health, including increases in all types of skin cancers, cataracts, and suppression of immune systems.[21, 22] As mentioned earlier, a 1% decrease in ozone can cause an increase of 1–2% in UVB radiation and an increase of 2% in the incidence of skin cancer.[22]

Figure 22.13 • Concentrations of CFC-11 and CFC-12 from 1977 to 1998. [*Source*: Data from Mauna Loa, Hawaii, modified from NOAA Climate Monitoring and Diagnostic Lab, accessed August 22, 2000 at http://www.cmdl.noaa.gov.]

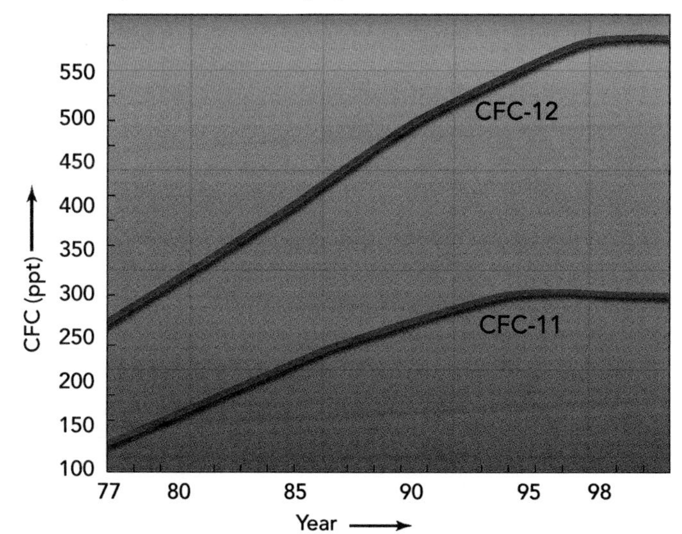

There has been speculation for some time that ozone depletion might lead to a reduction of primary productivity in the world's oceans. A loss of productivity of the phytoplankton (microscopic marine algae and photosynthetic bacteria that float near the surface of the ocean) would have a negative impact on a variety of other marine organisms, because phytoplankton are at the base of the food chain. Because ozone over the Antarctic area has been depleted by as much as 70% in recent years, more UVB radiation is reaching the surface of the ocean there. A study of the Antarctic waters beneath the mass of ozone-depleted air suggests that a reduction of primary productivity of at least 6–12% is associated with ozone depletion.[4] This disruption of the food chain might eventually affect people who consume marine resources, because fewer resources would be available for harvest. Also, because plankton is a sink for atmospheric carbon dioxide, their disruption might increase the concentration of CO_2 in the atmosphere, thereby increasing global warming.

If ozone depletion becomes more widespread and affects major food crops (such as beans, wheat, rice, and corn), serious social disruption could occur. Even a small decrease in food production might have large social and political consequences around the world. A loss of 10–15% production could be catastrophic.

The range of human health effects of ozone depletion is being vigorously researched and debated. There is general agreement that the effects will be negative and will result in increases in a variety of diseases, perhaps at epidemic levels compared with what would otherwise be expected. As mentioned, one of the most serious hazards anticipated is an increase in skin cancers of all types, including the often-fatal melanoma.

In 1987, low ozone measurements over Antarctica in October were followed by low readings over Australia and New Zealand in December,[23] when a mass of ozone-depleted air separated from the Antarctic vortex as it broke up in late spring. There is fear that large ozone holes will increasingly threaten populated areas in midlatitude regions.

Already, around the world, the incidence of skin cancers has increased (see the case study at the beginning of this chapter). For many years, a suntan was considered a healthy look, and people deliberately exposed their bodies to sunlight. Today, sunblock lotions and hats are replacing tanning oils for health-conscious people. Newspapers and televisions weather networks are providing their readers with the Ultraviolet (UV) Index (Table 22.2). The index predicts UV intensity levels on a scale from 1 to 11+. Some news agencies also use the index to recommend the level of sunblock. It is speculated that the incidence of skin cancer caused by ozone depletion will increase until about

2060 and will then decline as the ozone shield recovers as a result of controls on CFC emissions.[20]

Individuals can reduce their risk of skin cancer and other skin damage from UV exposure by following a few simple precautions:

- Limit exposure to the sun between the hours of 10 a.m. and 4 p.m. That is, avoid the hours of intense solar radiation.
- When possible, remain in the shade.
- Use a sunscreen with an SPF of at least 30. (Remember that protection from sunscreen with any SPF is diminished with time of exposure.)
- Wear a wide-brimmed hat and, where possible, tightly woven full-length clothing (for example, loose-fitting light cotton).
- Wear UV-protective sunglasses.
- Avoid tanning salons and sunlamps.
- Consult the UV Index before going out (see Table 22.2).

Following these precautions may help ensure that your skin will remain healthy, with fewer wrinkles into old age.

In addition to harming skin, ultraviolet radiation can damage eyes, causing cataracts, an eye disease in which the lens becomes opaque and vision is impaired. People are now more often choosing eyeglasses that block ultraviolet radiation. It is believed that an increased exposure to ultraviolet radiation may also damage or reduce the efficiency of the human immune system.[21] In turn, decreases in the effectiveness of the human immune system would result in higher numbers of a variety of diseases. Finally, a variety of environmental pollutants in the air and water could have synergistic effects, increasing potential health risks of exposure to ultraviolet radiation.

Management Issues

A key issue in management of ozone depletion is whether the depletion is natural or human-induced. If stratospheric ozone depletion were a natural process, then we would not expect to see the continuous, dramatic reductions since the mid-1970s, when the real impact of production of CFCs began. World production of CFCs increased significantly from 1970 to 1994. The most dramatic declines in ozone in Antarctica have occurred since 1980. Thus, it appears that chlorine from CFCs is the "smoking gun" mentioned earlier. Supporting this hypothesis, an investigation determined that the concentration of stratospheric chlorine (which is responsible for ozone destruction) is more than five times what could be expected from natural emissions from oceans or other natural processes. The authors concluded that, beyond a reasonable doubt, CFCs are responsible for ozone depletion in the stratosphere.[24] We turn next to a discussion of management issues and strategies to deal with CFC-induced ozone depletion.

The Montreal Protocol

A diplomatic achievement of monumental proportions was completed with the signing of the Montreal Protocol in September 1987; 27 nations signed the agreement originally, and 119 additional nations signed later. The protocol outlined a plan for the eventual reduction of global emissions of CFCs to 50% of 1986 emissions. The protocol originally called for elimination of the production of CFCs by 1999. Because of scientific evidence that stratospheric ozone depletion was occurring faster than predicted, the timetable for elimination of CFC production was shortened. Most industrialized countries, including Canada and the United States, had stopped production by the end of 1995; the deadline for developing countries is the end of 2005. An eventual phase-out of all CFC consumption is part of the Montreal Protocol.

Assessment of the effects of the protocol, along with effects of other agreements and amendments made in London (1990) and Copenhagen (1992), suggests that stratospheric concentrations of ozone-depleting substances (CFCs) will return to pre-1980 levels by about 2050. Already, the rate of increase of CFC emissions has been reduced (see Table 22.1 and Figure 22.13). Nevertheless, as already mentioned, because of the long residence times of CFCs in the stratosphere, ozone depletion from CFCs is expected to continue for many years to come.[25, 26] That's the bad news. The good news is that ozone levels in the stratosphere will slowly increase in the next few decades.

Unfortunately, not all industrial nations are responding to the urgency of this issue, partly because of the economic gap between wealthy and poorer nations. For example, China and India chose not to participate in the protocol. Their refusal was probably related to the substantial investments they are making in refrigeration and the fact that replacement chemicals currently considered

Table 22.2 • Ultraviolet (UV) Index for Human Exposure

Exposure Category	*UV Index*	*Comment*
Low	< 2	Sunblock recommended for all exposure
Moderate	3 to 5	Sunburn can occur quickly
High	6 to 7	Potentially hazardous
Very high	8 to 10	Potentially very hazardous
Extreme	11+	Potentially very hazardous

Source: Modified after World Health Organization. Accessed March 20, 2005 at www.who.int/uv/publications/en/GlobalUVI.pdf. Beginning in February 2004, Canada officially adopted the World Health Organization UV index.

Note: At moderate exposure to UV, sunburn can occur quickly; at high exposure, fair-skinned people may burn in 10 minutes or less of exposure.

for CFCs are approximately six times as expensive. The poorer nations will probably not participate in the reduction of CFCs unless they are assisted by the wealthier countries.[27] The Montreal Protocol stipulates a transfer of both technology and funds from industrial to developing countries to speed up CFC phase-out. We return to this issue in the discussion of substitutes for CFCs.

A black-market trade in CFCs has hindered the attempt to eliminate their production and consumption. Illegal smuggling of CFCs into Europe from Russia in recent years has equalled as much as 10% of the total CFCs legally allowed in those countries. Both Canada and the United States have also received illegal imports of CFCs.[28]

Collection and Reuse of CFCs

One way to lower emissions of CFCs into the atmosphere is to develop ways to collect and reuse them. Every year, approximately 50 million refrigerators are discarded worldwide, and each one contains approximately 1.2 kg of CFCs, mostly tied up in the foam plastic insulation. Methods have been developed to liberate and collect these CFCs when refrigerators are recycled. The CFCs used as coolant gases can also be collected. One company in Germany recycles approximately 6,000 refrigerators a month.[29] The same techniques can be used to recover the CFCs in air conditioners used in automobiles and homes.

Substitutes for CFCs

Of primary importance in developing substitutes for CFCs is finding substitutes that are both safe and effective. Two substitutes for CFCs being experimented with today are **hydrofluorocarbons (HFCs)** and **hydrochlorofluorocarbons (HCFCs)**. These chemicals are controversial and more expensive than CFCs but do have advantages.

The advantage of HFCs is that they do not contain chlorine. They do contain fluorine, though; and when fluorine atoms are released into the stratosphere, they participate in reactions similar to those of chlorine. Thus, they can cause ozone depletion. However, ozone depletion is not thought to be a significant problem with fluorine, because it is approximately 1,000 times less efficient in those reactions.[3, 27] Additionally, some blends of HFCs have no ozone-depleting potential.[30]

HCFCs contain an atom of hydrogen in place of a chlorine atom. They can be broken down in the lower atmosphere, in which case they do not inject chlorine into the stratosphere; however, they can cause ozone depletion if they do reach the stratosphere before being broken down. Although their atmospheric lifetime is much shorter than those of the CFCs (see Table 22.1), when HCFCs are used in tremendous quantities, they do cause ozone depletion.[3] HCFCs are at best a transitional chemical to be used until substitutes that do not cause ozone depletion are readily available. HCFCs will be phased out by 2030.

Another option being considered is the use of helium as the working gas in refrigerators. At present, however, the world supply of helium is very limited. Furthermore, helium is expected to be too expensive for home refrigeration because of the cost of making the equipment. Nevertheless, research in this area continues, and the price may come down, making its use more feasible.

The hydrocarbon propane, a colourless gas easily separated from crude oil and natural gas, is a common fuel that can be easily liquefied and stored in containers. It is also an inexpensive substitute for CFCs that has not received nearly as much attention as have HFCs and HCFCs. The cost of propane is about 10% that of CFCs and less than 2% that of HFCs and HCFCs. A refrigerator in a London laboratory was inexpensively converted to use propane as the coolant, and oil refineries have used propane as a coolant in industrial processes.[31] Why, then, hasn't propane been more seriously considered as a replacement for CFCs? There are two main reasons:[31]

- The chemical companies that manufacture CFCs and have patented HFCs and HCFCs are not in the refrigeration industry, and their focus is on their products rather than on finding an inexpensive substitute.
- Propane is thought to be dangerous in refrigerators and air conditioners because it is potentially explosive.

For most uses in household refrigerators and air conditioners, the concern about the danger of propane is overstated. The modern refrigerator would use about 100 g of propane in a sealed system, the equivalent of the amount of propane in a hand-held cigarette lighter. Because the cooling system is sealed, it is unlikely that a house fire could generate sufficient heat to rupture the system. If it did, the resulting risk should be weighed against the risk of CFCs in the same situation; combustion products associated with a fire burning CFCs includes toxic nerve gas. In contrast, in commercial refrigeration systems that use large amounts of coolant, the use of propane would be a potential hazard, and careful engineering would be required to ensure safety.

The use of propane as a refrigeration coolant is being tested in several countries, including Canada and the United States. The results will be important, especially to developing countries, where most CFC use is for refrigeration and air-conditioning. Developing countries tend not to have large industrial uses for CFCs. For example, in India, about 75% of the CFCs are for refrigerators and air conditioners, a pattern characteristic of developing countries. To reduce or eliminate CFC use by replacement with HCFCs and HFCs, developing countries

CRITICAL THINKING ISSUE

Human-Made Chemicals and the Ozone Hole: Why Was There Controversy?

By 1993, scientists had accumulated enough evidence to support earlier predictions that stratospheric ozone was being depleted over the Antarctic. Most of them blamed the damage on organic chlorine compounds (those containing both carbon and chlorine) manufactured by humans, such as CFCs. But consensus among most of the scientists in the field did not prevent a continuing storm of controversy over these findings. Critics included meteorologists, scientists from other fields, amateur scientists, journalists, talk-show hosts, and authors of non-technical books on the environment. The critics charged that natural sources of chlorine, not those generated by humans, were responsible for ozone depletion and that the environmental and health threats of ozone depletion were greatly exaggerated.

Accusations levelled against scientists, NASA officials, and some industrialists included claims that the ozone scare was a sham, a scam, a hoax, or a conspiracy. Scientific uncertainties, such as those concerning the causes of ozone depletion in the Northern Hemisphere and whether depletion was allowing more ultraviolet radiation to reach Earth's surface, were emphasized.

In his 1993 presidential address to the American Association for the Advancement of Science, F. Sherwood Rowland, principal architect of the hypothesis that CFCs damage the ozone layer, cited poor communication between scientists and non-scientists as the cause of the controversy. His colleague, Mario Molina, stated that the arguments put forth by critics about natural causes of ozone depletion had been tested; no results had been found to support those arguments in the 20 years since the hypothesis that CFCs were responsible for ozone depletion had been proposed.

Others felt that the controversy fed on lack of understanding of the nature of science, particularly discomfort with uncertainty, on the part of most non-scientists. In making this point, science fiction writer Frederic Pohl quoted the late Nobel Prize physicist Richard Feynman: "Scientific knowledge is a body of statements of varying degrees of certainty—some most unsure, some nearly sure, but none absolutely certain."

Critical Thinking Questions

1. Table 22.3 summarizes the major points of argument about whether ozone depletion is primarily a result of natural or of human-made chemicals. Evaluate the arguments in the table as best you can in light of what you have learned about ozone in this chapter. Identify each statement as (a) a hypothesis, (b) an inference, (c) a fact confirmed by observation or experiment (fact), (d) pseudoscience, or (e) an absolute certainty. If you feel the information is incomplete, what additional observations or experiments would you suggest? Based on your analysis, what is your position on the question of whether natural or human-made chemicals are depleting the ozone layer?
2. Some critics have pointed out that ultraviolet radiation increases by 50 times (5000%) in going from pole to equator. The expected increase in ultraviolet radiation in the mid-latitudes, they say, is like moving from sea level to an elevation of 500 m What is your reaction to this criticism?
3. In February 1992, NASA scientists reported unusually high levels of chlorine over the Northern Hemisphere and warned that this might lead to significant ozone loss over heavily populated areas. Congress reacted quickly to enact legislation to speed up discontinuing use of CFCs. When the ozone loss over the Northern Hemisphere was much less than the scientists had expected, they were attacked by some people as being unnecessarily alarmist and using scare tactics to obtain additional funding. Do you think these attacks were justified? Would it have been better if the scientists had presented a much more conservative scenario?
4. F. Sherwood Rowland identified two sources of public misunderstanding about scientific issues: poor communication about science and widespread scientific illiteracy. Do you agree or disagree with Rowland? Explain your position. If you agree, what remedies do you suggest?

would need massive financial aid. They could do it on their own if an inexpensive, safe alternative were available. Research is being conducted to determine if propane or some other chemical could be that alternative.[31]

Short-Term Adaptation to Ozone Depletion

As we have seen, there is good news in the ozone depletion story. Concentrations of CFCs in the upper atmosphere where ozone depletion occurs have apparently peaked. Depletion of stratospheric ozone will be a story of gradual recovery by the mid-twenty-first century.[20, 28] Recovery will take place as a result of restrictions in the production of ozone-depleting chemicals, especially CFCs.

Given the nature of the ozone-depletion problem and the atmospheric lifetimes of the chemicals that produce the depletion, the major short-term adaptation by people will be learning to live with higher levels of exposure to ultraviolet radiation. In the long term, achievement of sustainability with respect to stratospheric ozone will require management of human-produced ozone-depleting chemicals.

Table 22.3 • Major Arguments Regarding Sources of Ozone Depletion
Natural sources of chlorine are so large that CFCs are insignificant by comparison.
1. CFCs are heavier than air and would not reach the stratosphere in significant amounts.
2. No measurements have shown CFCs to be present in the stratosphere.
3. Volcanoes, which produce 20 times as much chlorine as do CFCs, caused the Antarctic ozone hole. Mount Erebus in Antarctica has been erupting since 1973 and emits more than 1,000 tonnes of chlorine a day.
4. Evaporation of sodium chloride from ocean water is a source of stratospheric chlorine.
5. Volcanoes release hydrogen chloride, which has increased in the stratosphere in the last 10 years.
6. Burning releases methyl chloride.
Humans are producing chemicals that significantly deplete stratospheric ozone in the Antarctic and possibly in the Northern Hemisphere.
1. Within the atmosphere, large masses of air are constantly mixing.
2. CFCs are found in stratospheric samples.
3. Volcanic activity has been around for a long time, and recent eruptions have not been unusual. Mount Erebus does not erupt forcefully enough to propel chlorine into the stratosphere in significant amounts. Furthermore, it produces only 15,000 tonnes of chorine a year.
4. Sodium chloride, unlike CFCs, is water soluble and is washed out of the atmosphere when it rains. No sodium is found in the lower stratosphere.
5. Both hydrogen chloride and hydrogen fluoride, also water soluble, are increasing, which is what would be predicted if CFCs were reaching the stratosphere. There are almost no natural sources of hydrogen fluoride, and increases in it would not be consistent with a volcanic source for chlorine. Sulphuric acid aerosols from volcanoes that do eject into the stratosphere promote destruction of ozone.

Summary

- Concentration of atmospheric ozone has been measured for more than 70 years. In the last decade, measurements have been taken from instruments mounted on satellites. Evaluation of the available data shows a clear trend: ozone concentrations in the stratosphere have been decreasing since the mid-1970s.
- In 1974, Mario Molina and F. Sherwood Rowland advanced the hypothesis that stratospheric ozone might be depleted as a result of CFC emissions into the lower atmosphere. Major features of the hypothesis are: CFCs are extremely stable and have a long residence time in the atmosphere; eventually, the CFCs reach the stratosphere, where they may be destroyed by highly energetic solar ultraviolet radiation, releasing chlorine; the chlorine may then enter into a catalytic chain reaction that depletes ozone in the stratosphere. An environmentally significant result of the depletion is that more ultraviolet radiation reaches the lower atmosphere, where it can damage living cells.
- The Antarctic ozone hole was first reported in 1985 and since has captured the imagination of people around the world. Of particular importance to understanding the ozone hole are the complex reactions that occur in the polar vortex and the development of polar stratospheric clouds. Reactions in the clouds tend to denitrify the air mass in the vortex; and during the polar spring, chlorine is released to react in the catalytic ozone depletion cycle. The reactions can be very rapid, producing the observed 70% reduction in stratospheric ozone in only a few weeks.
- Tropical and midlatitude ozone depletion is also hypothesized. Polar stratospheric clouds may be present above the tropics; in addition, belts of sulphur dioxide–rich aerosol clouds may result from volcanic eruptions. These clouds may be related to processes that denitrify the atmosphere and facilitate ozone depletion.
- Millions of tonnes of chemicals with the potential to deplete stratospheric ozone are now in the lower atmosphere and working their way to the stratosphere. As a result, if all production, use, and emissions of these chemicals were stopped today, the problem would continue. The good news is that the concentrations of CFCs in the atmosphere have apparently peaked and are now static or in slow decline.
- Potential environmental effects related to ozone depletion include damage to Earth's food chain, both on land and in the ocean, and human health effects, including increases in skin cancers, cataracts, and suppression of the immune system.
- Many nations around the world have agreed to the Montreal Protocol, which will reduce global emissions of CFCs to 50% of the 1986 levels. The agreement called for elimination of production of the chemicals by 1996 for industrialized nations and by

2006 for developing nations. A serious hurdle to compliance for some nations relates to the economic fact that most chemical replacements for CFCs are more expensive than CFCs. Therefore, it appears that financial aid will be required if the less wealthy nations are to eliminate CFC use.

- Potential management strategies for the ozone depletion problem, along with eliminating production of CFCs include: (1) collecting and reusing CFCs, and (2) using substitutes for CFCs.
- Given the lifetimes of the ozone-depleting chemicals, people will have to continue to live with higher levels of exposure to ultraviolet radiation over the next few decades. Banning chemicals that can deplete stratospheric ozone is a step in the right direction and will ultimately result in the reduction of atmospheric ozone depletion.

STUDY QUESTIONS

1. Do you think that a gap ever existed between people's understanding of the ozone problem and their taking action to find a solution to the problem? Compare the situation with the global warming problem.
2. Given that primary productivity is reduced in the Antarctic by ozone depletion, how could you test the hypothesis that penguins might be adversely affected?
3. Consider the possibility of building a small chamber to test the hypothesis that polar stratospheric clouds facilitate ozone depletion. What might be some of the problems in attempting to do this? What are alternative approaches to studying the role of stratospheric clouds?
4. Suppose that next year all our understanding of ozone depletion is changed by the discovery that concentrations of stratospheric ozone have natural cycles and that the lower concentrations in recent years have resulted from natural rather than human-induced processes. How would you put all the information in this chapter into perspective? Would you think that science had let you down?
5. What types of economic and political changes will be necessary to encourage less developed countries to support plans to eliminate chemicals responsible for ozone depletion? Do you believe that richer countries have an obligation to help poorer ones?
6. Do you agree or disagree that most of the people in the world will adjust to ozone depletion by doing little to decrease their personal exposure to ultraviolet radiation?
7. The Montreal Protocol (originally signed in 1987) placed limits on the production of ozone-depleting substances, but peak concentrations of those chemicals in the atmosphere are only now being reached. Suggest some explanations for this phenomenon.

FURTHER READING

Christie, M. *The Ozone Layer: A Philosophy of Science Perspective*. Cambridge: Cambridge University Press, 2000. A complete look at the history of the ozone hole from the first discovery of its existence to more recent studies of the hole over Antarctica.

Hamill, P. and O. B. Toon. "Polar Stratospheric Clouds and the Ozone Hole," *Physics Today* 44, no 12. (1991):34–42, 1991. A good review of the ozone problem and important chemical and physical processes related to ozone depletion.

Jones, R. R. and T. Wigley. *Ozone Depletion: Health and Environmental Consequences*. New York: Wiley, 1989. Proceedings of the International Conference on the Health and Environmental Consequences of Stratospheric Ozone Depletion, held at the Royal Institute of British Architects, London, November 28–29, 1988.

Litfin, K. T. *Ozone Discourses: Science and Politics in Global Environmental Cooperation*. New York: Columbia University Press, 1994. Discussions on the importance of science and scientific discourse in shaping world politics and policy with regard to ozone depletion.

Makhijani, A. and K. R. Gurney. *Mending the Ozone Hole*. Cambridge, Mass.: MIT Press, 1995. A good overview of causes and consequences of stratospheric ozone depletion, sources of and alternatives to ozone-depleting chemicals, and policy recommendations to reverse ozone damage.

Reid, S. *Ozone and Climate Change: A Beginner's Guide*. Amsterdam: Gordon & Breach Science Publishers, 2000, 210 p. A good look at the science behind the ozone hole and future predictions written for a general audience to make the science understandable.

Rowland, F. S. "Stratospheric Ozone Depletion by Chlorofluoro-carbons." *AMBIO* 19, no. 6–7 (1990):281–292, 1990. An excellent summary of stratospheric ozone depletion that discusses some of the major issues.

Shea, C. P. "Mending the Earth's Shield," *World Watch* 2, no1(1989):28–34, 1989. An article focusing on solutions to the ozone problem. It is primarily concerned with control strategies to stop ozone-depleting chemicals from being emitted into the atmosphere.

Toon, O. B. and R. P. Turco. "Polar Stratospheric Clouds and Ozone Depletion," *Scientific American* 246(6):68–74, 1991. An article that provides valuable information concerning polar stratospheric clouds and their importance in ozone depletion. It offers a good explanation of the formation of polar stratospheric clouds and the chemistry that occurs there.

chapter 23

Canada's Butchart Gardens: From Eyesore to Eden

It's often been stated that one person at the right time and in the right place has the opportunity to make a real difference. Jenny Butchart in the early 1900s had a vision to transform her husband's exhausted mine pit into a garden. Today, we would use the term mine reclamation for such a project. The story that followed was one of transformation. Gardens, which began as a limestone quarry that was an eyesore, became a tourist attraction visited by about a million people each year—Butchart Gardens, Vancouver Island, British Columbia. The Butchart Garden story offers an early illustration of the principles of environmental restoration and sequential land use.

The story is closely related to the cement industry. The process of producing Portland cement was developed in 1824 in England. The procedure includes fine grinding of an exact mixture of limestone and shale that has been heated in a rotary kiln to near-fusion temperature. This material is then finely ground again, placed in sacks, and sold to consumers as Portland cement. Adding aggregate such as gravel to the cement makes concrete, which is an integral part of construction in our urban environment.

Robert Butchart was educated in the process of making Portland cement while on honeymoon in England in 1884. In 1902, he learned of the availability of limestone located approximately 20 km north of the city of Victoria on Vancouver Island. He determined that the site was ideal for the establishment of a cement plant and started the Cod Inlet Cement Plant there in 1904. The limestone used to produce the Portland cement was exhausted by 1908, and what was left behind was a large, unsightly open excavation about 20 m deep. The walls of the quarry were nearly vertical gray limestone.

That's when Jenny Butchart entered the picture. She had little experience with gardens but was fascinated with the site and the idea of transforming the quarry into a sunken garden. Her husband supported this endeavour, and with the help of many workmen, she brought in massive amounts of topsoil by

Butchart Gardens, Vancouver Island, Canada. The site of the gardens was once a limestone quarry.

Minerals and the Environment

LEARNING OBJECTIVES

Modern society depends on the availability of mineral resources, which can be considered a non-renewable heritage from the geologic past. After reading this chapter, you should understand:

- The standard of living in modern society is related in part to the availability of natural resources.
- What processes are responsible for the distribution of mineral deposits.
- What the differences are between mineral resources and reserves.
- What factors control the environmental impact of mineral exploitation.
- How wastes generated from the use of mineral resources affect the environment.
- What the social impacts are from mineral exploitation.
- How sustainability may be linked to use of non-renewable minerals.

horse and cart to form garden beds on the floor of the quarry. The garden, which was completed in 1921, soon became a tourist attraction. Today, the garden is much larger and includes fountains, rose gardens, lakes, Japanese gardens, and other attractions in addition to the original sunken garden. The Butchart Gardens case history illustrates the power of the vision of one person to transform the landscape from a degraded site to a beautiful and esthetically pleasing place.[1]

- *Butchart Gardens is a unique instance of mine reclamation; but each renovation is somewhat unique, based on local physical, hydrological, and biological conditions. This chapter discusses the origin of mineral deposits as well as environmental consequences of mineral development.*

23.1 The Importance of Minerals to Society

Modern society depends on the availability of mineral resources.[2] Many mineral products are found in a typical Canadian home (see Table 23.1). Consider your breakfast this morning. You probably drank from a glass made primarily of sand, ate food from dishes made from clay, flavoured your food with salt mined from Earth, ate fruit grown with the aid of fertilizers such as potassium carbonate (potash) and phosphorus, and used utensils made from stainless steel, which comes from processing iron ore and other minerals.

Minerals are so important to people that the standard of living increases with the availability of minerals in useful forms. The availability of mineral resources is one measure of the wealth of a society. Those who have been successful in locating and extracting or importing and using minerals have grown and prospered. Without mineral resources to grow food, construct buildings and roads, and manufacture everything from computers to televisions to automobiles, modern technological civilization as we know it would not be possible. In maintaining our standard of living in North America, every person requires about 10 tonnes of nonfuel minerals per year.[3]

Minerals can be considered our non-renewable heritage from the geologic past. Although new deposits are still forming from present Earth processes, these processes are producing new mineral deposits too slowly to be of use to us today. Because mineral deposits are generally located in small, hidden areas, they must be discovered. Unfortunately, most of the easy-to-find deposits have been exploited; if modern civilization were to vanish, our descendants would have a harder time discovering rich mineral deposits than we did. It is interesting to speculate that they might mine landfills for metals thrown away by our civilization. Unlike biological resources, minerals cannot be easily managed to produce a sustained yield; the supply is finite. Recycling and conservation will help, but eventually the supply will be exhausted.

23.2 How Mineral Deposits Are Formed

Metals in mineral form are generally extracted from naturally occurring, anomalously high concentrations of Earth materials. When metals are concentrated in anomalously high amounts by geologic processes, ore deposits are formed. The discovery of natural ore deposits allowed early peoples to exploit copper, tin, gold, silver, and other metals while slowly developing skills in working with metals.

The origin and distribution of mineral resources is intimately related to the history of the biosphere and to the entire geologic cycle (see Chapter 5). Nearly all aspects and processes of the geologic cycle are involved to some extent in producing local concentrations of useful materials. In this section, we look first at the distribution of mineral resources on Earth and then describe processes that form mineral deposits.

Distribution of Mineral Resources

Earth's outer layer, or crust, is silica rich, made up mostly of rock-forming minerals containing silicon, oxygen, and a few other elements. The elements are not evenly distributed in the crust: nine elements account for about 99% of the crust by weight (oxygen, 45.2%; silicon, 27.2%; aluminum, 8.0%; iron, 5.8%; calcium, 5.1%; magnesium, 2.8%; sodium, 2.3%; potassium, 1.7%; and titanium, 0.9%). In general, remaining elements are found in trace concentrations.

The ocean, covering nearly 71% of Earth, is another reservoir for many chemicals other than water.

Table 23.1 • Mineral Products in a Typical Canadian Home

Building materials: sand, gravel, stone, brick (clay), cement, steel, aluminum, asphalt, glass, petroleum products
Plumbing and wiring materials: iron and steel, copper, brass, lead, cement, asbestos, glass, tile, plastic
Insulating materials: fibreglass, gypsum (plaster and wallboard)
Paint and wallpaper: mineral pigments (such as iron, zinc, and titanium) and fillers (such as talc and asbestos)
Plastic floor tiles, other plastics: mineral fillers and pigments, petroleum products
Appliances: iron, copper, and many rare metals
Furniture: synthetic fibres made from minerals (principally coal and petroleum products); steel springs
Clothing: natural fibres grown with mineral fertilizers; synthetic fibres made from minerals (principally coal and petroleum products)
Food: grown with mineral fertilizers; processed and packaged by machines made of metals
Drugs and cosmetics: suspensions, tablets, soaps and shampoos containing minerals and petroleum products
Other items: windows, screens, light bulbs, porcelain fixtures, china utensils, jewellery made from mineral products

Source: U.S. Geological Survey, *Professional Paper* 940, 1975.

Most elements in the ocean have been weathered from crustal rocks on the land and transported to the oceans by rivers. Other elements are transported to the ocean by wind or glaciers. Ocean water contains about 3.5% dissolved solids, mostly chlorine (55.1% by mass) and sodium (30.9%). Each cubic kilometre of ocean water contains about 2.0 metric tonnes of zinc, 2.0 metric tonnes of copper, 0.8 metric tonne of tin, 0.3 metric tonne of silver, and 0.01 metric tonne of gold. These concentrations are low compared with those in the crust, where corresponding values (in metric tonnes/km^3) are: zinc, 170,000; copper, 86,000; tin, 5,700; silver, 160; and gold. After rich crustal ore deposits are depleted, we will be more likely to extract metals from lower-grade ore deposits or even from common rock than from ocean water; however, if mineral extraction technology becomes more efficient, this prognosis could change.

We have mentioned that minerals we mine occur in deposits—with anomalously high local concentrations. Why is this so? Planetary scientists now believe that Earth, like the other planets in the solar system, formed by condensation of matter surrounding the sun. Gravitational attraction brought together the matter dispersed around the forming sun. As the mass of the proto-Earth increased, the material condensed and was heated by the process. The heat was sufficient to produce a molten liquid core, consisting primarily of iron and other heavy metals, which sank toward the center. The crust formed from generally lighter elements and is a mixture of many different elements. The elements in the crust are not uniformly distributed because geologic processes and some biological processes selectively dissolve, transport, and deposit elements and minerals. We discuss these processes next.

Plate Boundaries

Plate tectonics is responsible for the formation of some mineral deposits. According to the theory of plate tectonics (see Chapter 5), the continents (which are crustal rocks and part of the lithosphere) are composed mostly of relatively light rocks. As the tectonic plates of the lithosphere slowly move across Earth's surface, so do the continents. Metallic ores are thought to be deposited in the crust both where the tectonic plates separate, or diverge, and where they come together, or converge.

At divergent plate boundaries, cold ocean water comes in contact with hot molten rock. The heated water is lighter and more active chemically. It rises through fractured rocks and leaches metals from them. The metals are carried in solution and deposited as metal sulphides (see Chapter 5) when the water cools.

At convergent plate boundaries, rocks saturated with seawater are forced together, heated, and subjected to intense pressure, which causes partial melting. The combination of heat, pressure, and partial melting mobilizes metals in the molten rocks, or magma. Most major mercury deposits, for example, are associated with the volcanic regions that occur close to convergent plate boundaries. Geologists believe that the mercury is distilled out of the tectonic plate as the plate moves downward; as the plate cools, the mercury migrates upward and is deposited at shallower depths, where the temperature is lower.

Figure 23.1 • **Diamond mine near Kimberley, South Africa.** This is the largest hand-dug excavation in the world.

Igneous Processes

Igneous processes are related to molten rock material known as magma. Ore deposits may form when magma cools. As the molten rock cools, heavier minerals that crystallize (solidify) early may slowly sink or settle toward the bottom of the magma, whereas lighter minerals that crystallize later are left at the top. Deposits of an ore of chromium, called chromite, are thought to be formed in this way. When magma containing small amounts of carbon is deeply buried and subjected to very high pressure during slow cooling (crystallization), diamonds (which are pure carbon) may be produced (Figure 23.1).[4]

Hot water moving within the crust is perhaps the source of most ore deposits. It is speculated that circulating water, including that released during metamorphism and mineralization, is heated and enriched with minerals on contact with deeply buried rocks. This water then moves up or laterally to other, cooler rocks, often in the vicinity of joints, faults, and fractures, where the cooled water deposits the dissolved minerals.[5] Minerals of economic value are usually the product of multiple steps, beginning with cooling magma and proceeding through secondary (often hydrothermal) accumulation and concentration stages.

Sedimentary Processes

Sedimentary processes relate to the transport of sediments by wind, water, and glaciers. These processes often concentrate materials in amounts sufficient for extraction.

As sediments are transported, running water and wind help segregate the sediments by size, shape, and density. This sorting function is useful to people. The best sand, or sand and gravel deposits for construction purposes, for example, are those in which the finer materials have been removed by water or wind. Sand dunes, beach deposits, and deposits in stream channels are good examples. The sand and gravel industry amounts to several billion dollars annually, and in terms of the total volume of materials mined, it is one of the largest nonfuel mineral industries in Canada.[3]

Stream processes transport and sort all types of materials according to size and density. Therefore, if the bedrock in a river basin contains heavy metals, such as gold, streams draining the basin may concentrate the metals in areas where there is less water turbulence or velocity. These concentrations, called placer deposits, are often found in open crevices or fractures at the bottoms of pools, on the inside curves of bends, or on riffles, where shallow water flows over rocks. Placer mining of gold (which was known as a poor man's method because a miner needed only a shovel, a pan, and a strong back to work the stream-side claim) played an important role in the settling of the Yukon, Alaska, the British Columbia interior, and California.

Rivers and streams that empty into oceans and lakes carry tremendous quantities of dissolved material derived from the weathering of rocks. Over geologic time, a shallow marine basin may be isolated by tectonic activity that uplifts its boundaries. In other cases, climatic variations, such as the ice ages, produce large inland lakes with no outlets. These basins and lakes eventually dry up. As evaporation progresses, the dissolved materials precipitate (drop out of solution), forming a wide variety of compounds, minerals, and rocks that have important commercial value. Most of these *evaporates* (deposits originating by evaporation) can be grouped into one of three types:[6]

- *Marine evaporates* (solids)—potassium and sodium salts, gypsum, and anhydrite.
- *Nonmarine evaporates* (solids)—sodium and calcium carbonate, sulphate, borate, nitrate, and limited iodine and strontium compounds.
- *Brines* (liquids derived from wells, thermal springs, inland salt lakes, and seawaters)—bromine, iodine, calcium chloride, and magnesium.

Heavy metals (such as copper, lead, and zinc) associated with brines and sediments are important resources that may be exploited in the future. Such deposits are found in many locations, including the Red Sea and the Salton Sea, a California water body formed in 1905 from overflows of fresh water from the Colorado River into a basin called the Salton Sink. Because it lacks an outflow and is located in an area of high evaporation, the Salton Sea has become progressively more saline over the last 100 years. Evaporite minerals are widely used in industrial and agricultural activities, and their annual value is over $1 billion.[6] Evaporite and brine resources in Canada and the United States are substantial, ensuring an ample supply for many years. Mesozoic shales of marine origin underlie much of the Canadian prairie region. Their rich salt content has resulted in saline lakes and groundwater in some areas, and caused the salinization of prairie soils.

Biological Processes

Most minerals are formed by inorganic processes, and indeed conventional definitions of the term "mineral" require inorganic origin. However, some materials formed by biological processes are now recognized as **organic minerals**. Many are formed under conditions of the biosphere that have been greatly altered by life. Examples include phosphates (discussed in Chapter 5) and iron ore deposits.

The major iron ore deposits exist in sedimentary rocks that were formed more than 2 billion years ago.[7] There are several types of iron deposits. Grey beds, an important type, contain unoxidized iron. Red beds contain oxidized iron (the red colour is the colour of iron oxide). Grey beds formed when there was little oxygen in the atmosphere, and red beds formed when there was relatively more oxygen. Although the processes are not completely understood, it appears that major deposits of iron stopped forming when the atmospheric concentration of oxygen reached its present level.[8]

Organisms are able to form many kinds of minerals, such as the calcium minerals in shells and bones. Some of these minerals cannot be formed inorganically in the biosphere. Thirty-one different biologically produced minerals have been identified. Minerals of biological origin contribute significantly to sedimentary deposits.[9]

Weathering Processes

Weathering, the chemical and mechanical decomposition of rock, concentrates some minerals in the soil. When insoluble ore deposits, such as native gold, are weathered from rocks, they may accumulate in the soil unless removed by erosion. Accumulation occurs most readily when the parent rock is relatively soluble, as is limestone. Intensive weathering of certain soils derived from aluminum-rich igneous rocks may concentrate oxides of aluminum and iron. (The more soluble elements, such as silica, calcium, and sodium, are selectively removed by soil and biological processes.) If sufficiently concentrated, residual aluminum oxide forms an ore of aluminum known as bauxite. Important nickel and cobalt deposits are also found in soils developed from iron- and magnesium-rich igneous rocks.

Weathering produces sulphide ore deposits from low-grade primary ore through *secondary enrichment* processes. Near the surface, primary ore containing minerals such as iron, copper, and silver sulphides is in contact with slightly acidic soil water in an oxygen-rich environment. As the sulphides are oxidized, they dissolve, forming solutions that are rich in sulphuric acid as well as silver and copper sulphate. These solutions migrate downward, producing a leached zone devoid of ore minerals (Figure 23.2). Below the leached zone and above the groundwater table, oxidation continues, and sulphate solutions continue their downward migration. Below the water table, if oxygen is no longer available, the solutions are precipitated as sulphides, enriching the metal content of the primary ore by as much as 10 times. In this way, low-grade primary ore is rendered more valuable, and high-grade primary ore is made even more attractive.[10, 11]

Figure 23.2 • **Typical zones that form during secondary enrichment processes.** Sulphide ore minerals in the primary ore vein are oxidized and altered and then are leached from the oxidized zone by descending groundwater and redeposited in the enriched zone. The iron oxide cap is generally a reddish color and may be helpful in locating ore deposits that have been enriched. [*Source*: Modified from R. J. Foster, *General Geology*, 4th ed. (Columbus, Ohio: Charles E. Merrill, 1983).]

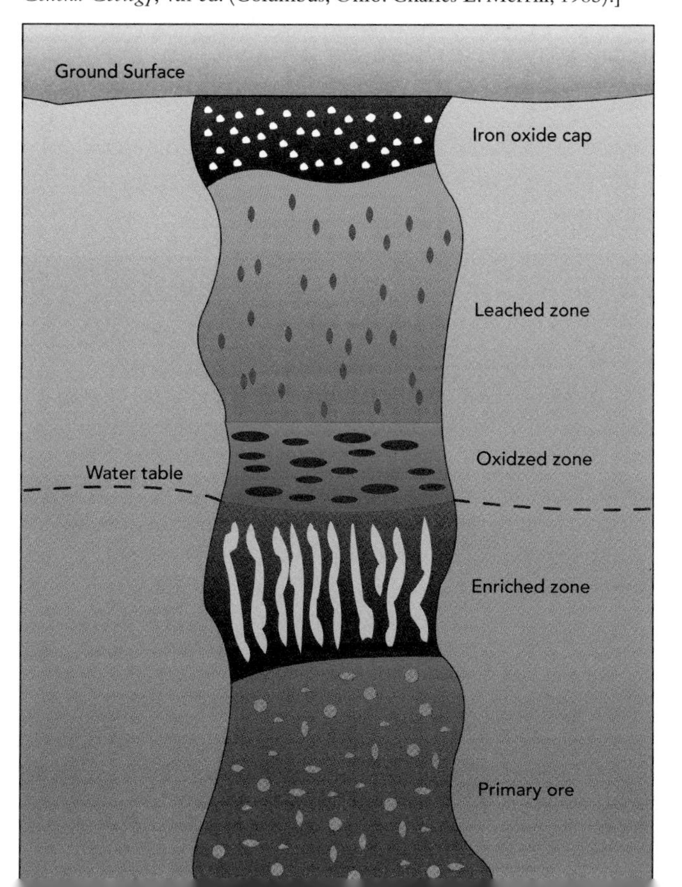

For example, secondary enrichment of a disseminated copper deposit at Miami, Arizona, increased the grade of the ore from less than 1% copper in the primary ore to as much as 5% in some areas.[10]

23.3 Resources and Reserves

We can classify minerals as resources or reserves. **Mineral resources** are broadly defined as elements, chemical compounds, minerals, or rocks concentrated in a form that can be extracted to obtain a usable commodity—something that can be bought and sold. It is assumed that a resource can be extracted economically or at least has the potential for economic extraction. A **reserve** is that portion of a resource that is identified and from which usable materials can be legally and economically extracted *at the time of evaluation* (Figure 23.3).

Whether a mineral deposit is classified as part of the resource base or as a reserve may be a question of economics. For example, if an important metal becomes scarce, the price may rise, which would encourage exploration and extraction (mining). As a result of the price increase, previously uneconomic deposits (part of the resource base before the scarcity and price rise) may become profitable, and those deposits would be reclassified as reserves.

The main point here is that *resources are not reserves*. An analogy from a student's personal finances may help clarify this point. A student's reserves are liquid assets, such as money in the bank, whereas the student's resources include the total income the student can expect to earn during his or her lifetime. This distinction is often critical to the student in school, because resources that may become available in the future cannot be used to pay this month's bills.[12]

Regardless of potential problems, it is important for planning purposes to estimate future resources. This task requires continual reassessment of all components of a total resource through consideration of new technology, the probability of geologic discovery, and shifts in economic and political conditions.[2]

Silver provides an illustration of some important points about resources and reserves. Based on geochemical estimates of the concentration of silver in rocks, Earth's crust (to a depth of 1 km) contains almost 2×10^{12} metric tonnes of silver, an amount much larger than annual world use, which is approximately 10,000 metric tonnes. If this silver existed as pure metal concentrated in one large mine, it would represent a supply sufficient for several hundred million years at current levels of use. Most of the silver, however, exists in extremely low concentrations, too low to be extracted economically with current technology. The known reserves of silver, reflecting the amount we could obtain immediately with known techniques, are about 200,000 metric

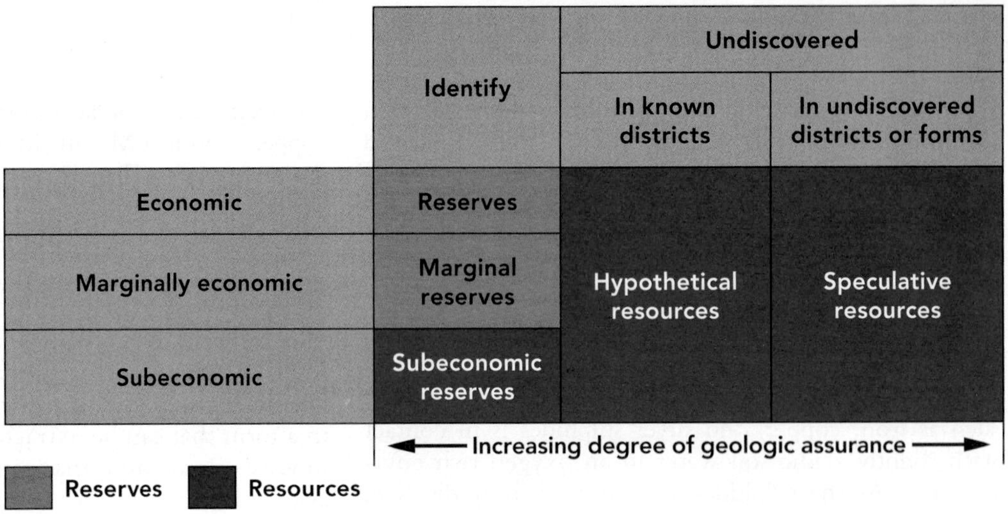

	Identify	Undiscovered	
		In known districts	In undiscovered districts or forms
Economic	Reserves	Hypothetical resources	Speculative resources
Marginally economic	Marginal reserves		
Subeconomic	Subeconomic reserves		

← Increasing degree of geologic assurance →

Reserves   Resources

Figure 23.3 • A system for the classification of mineral resources. [*Source*: Modified from *Principles of a Resource Reserve Classification for Minerals*, U.S. Geological Survey Circular 831, 1980.]

tonnes, or a 20-year supply at current use levels and no new reserves.

The problem with silver, as with all mineral resources, is not with its total abundance but with its concentration and relative ease of extraction. When an atom of silver is used, it is not destroyed but remains an atom of silver. It is simply dispersed and may become unavailable. In theory, given enough energy, all mineral resources could be recycled, but this is not possible in practice. Consider lead, which is mined from minerals in which it is concentrated. The lead that was used in gasoline for many years is now scattered along highways across the world and deposited in low concentrations in forests, fields, and salt marshes close to these highways. Recovery of this lead is, for all practical purposes, impossible.

23.4 Classification, Availability, and Use of Mineral Resources

Earth's mineral resources can be divided into several broad categories, depending on our use of them: elements for metal production and technology; building materials; minerals for the chemical industry; and minerals for agriculture. Metallic minerals can be further classified according to their abundance. The abundant metals include iron, aluminum, chromium, manganese, titanium, and magnesium. Scarce metals include copper, lead, zinc, tin, gold, silver, platinum, uranium, mercury, and molybdenum.

Some mineral resources, such as salt (sodium chloride), are necessary for life. Aboriginal peoples traveled long distances to obtain salt when it was not locally available. Other mineral resources are desired or considered necessary to maintain a particular level of technology.

When we think about mineral resources, we usually think of the metals; but, with the exception of iron, the predominant mineral resources are not metallic. Consider the annual world consumption of a few selected elements. Sodium and iron are used at a rate of approximately 100–1,000 million metric tonnes per year. Nitrogen, sulphur, potassium, and calcium are used at a rate of approximately 10–100 million metric tonnes per year, primarily as soil conditioners or fertilizers. Elements such as zinc, copper, aluminum, and lead have annual world consumption rates of about 3–10 million metric tonnes, and gold and silver have annual consumption rates of 10,000 metric tonnes or less. Of the metallic minerals, iron makes up 95% of all the metals consumed; and nickel, chromium, cobalt, and manganese are used mainly in alloys of iron (as in stainless steel). Thus, with the exception of iron, the non-metallic minerals are consumed at much greater rates than are elements used for their metallic properties.

Availability of Mineral Resources

The basic issue with mineral resources is not actual exhaustion or extinction but the cost of maintaining an adequate stock within an economy through mining and recycling. At some point, the costs of mining exceed the worth of material. When the availability of a particular mineral becomes a limitation, there are four possible solutions:

1. Find more sources.
2. Recycle and reuse what has already been obtained.
3. Reduce consumption.
4. Find a substitute.

Which choice or combination of choices is made depends on social, economic, and environmental factors.

The availability of a mineral resource in a certain form, in a certain concentration, and in a certain amount is a geologic issue determined by Earth's history. What is considered a resource and at what point a resource becomes limited are ultimately social questions. Before metals were discovered, they could not be considered resources. Before smelting was invented, the only metal ores were those in which the metals appeared in their pure form. For example, originally gold was obtained as a pure, or native, metal. Many gold mines are now deep beneath the surface, and the recovery process involves reducing tons of rock to ounces of gold. Although Nunavut possesses valuable mineral deposits, including gold, some are in remote locations too costly to develop. Prevailing prices would have to rise substantially for those deposits to be considered resources.

Mineral resources are limited, which raises important questions. How long will a particular resource last? How much short-term or long-term environmental deterioration are we willing to accept to ensure that resources are developed in a particular area? How can we make the best use of available resources? These questions have no easy answers. We are now struggling with ways to better estimate the quality and quantity of resources.

Mineral Consumption

We can use a particular mineral resource in several ways: rapid consumption, consumption with conservation, or consumption and conservation with recycling. Which option is selected depends in part on economic, political, and social criteria. Figure 23.4 shows the hypothetical depletion curves corresponding to these three options. Historically, with the exception of precious metals, rapid consumption has dominated most resource utilization. However, as the supply of resources becomes short, increased conservation and recycling are expected. Certainly, the trend toward recycling is well established for metals such as copper, lead, and aluminum.

From a global viewpoint, limits on our mineral resources and reserves threaten our affluence. As the world population and the desire for a higher standard of living increase, the demand for mineral resources expands at a faster and faster rate. Today the more developed countries consume a disproportionate amount of the mineral resources extracted. For example, the United States, western Europe, Japan, and more recently China, with its burgeoning capitalist economy, collectively use most of the aluminum, copper, and nickel that is extracted from the Earth.[5] Predicted increases in the world use of iron, copper, and lead, when linked with expected population increases, suggest that the rate of production of these metals will have to increase by several times if the world per capita consumption rate is to rise to the level of consumption in developed countries today. Such an increase is very unlikely; affluent countries will thus have to find substitutes for some minerals or use a smaller proportion of the world's annual production. This situation parallels the Malthusian predictions discussed in Chapter 4: It is impossible in the long run to support an ever-increasing population on a finite resource base.

Figure 23.4 • **Several hypothetical resource depletion curves.**

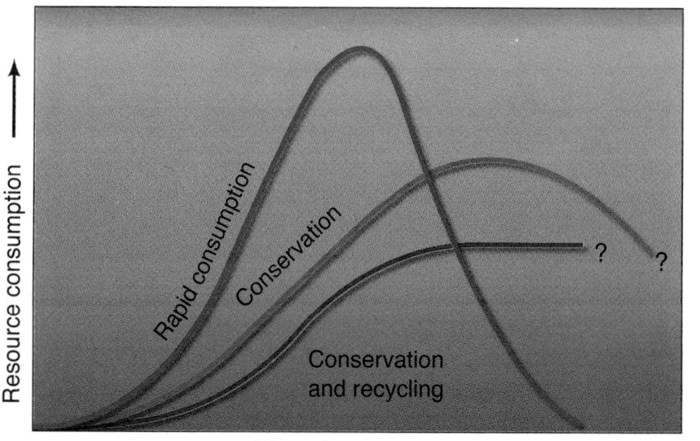

Canada's Supply of Mineral Resources

More than 60 different minerals are mined in Canada, including base metals (such as copper, nickel, and lead); ferrous minerals including iron; coal; diamonds; gold; silver; uranium; salt; various other metals, like tungsten; and so-called industrial minerals like sand, gravel, and natural cement rock. Of these, Canada's most important minerals in terms of dollar value are gold ($2.3 billion production value in 2002); nickel ($1.9 billion); potash ($1.6 billion); copper ($1.4 billion); and cement ($1.4 billion). There are operating mines in every province and territory of Canada except for Prince Edward Island. Nevertheless, domestic supplies of many mineral resources in Canada are insufficient for current use and must be supplemented by imports from other nations.

The fact that Canada—along with many other countries—depends on a steady supply of imports to meet the mineral demand of industries does not necessarily mean that the imported minerals do not exist within the country in quantities that could be mined. Rather, it suggests that there are economic, political, or environmental reasons that make it easier, more practical, or more desirable to import the material. This situation has resulted in political alliances that otherwise would be unlikely. Industrial countries often need minerals from countries with whose policies they do not necessarily agree; as a result, they make political concessions on human rights and other issues that they would not otherwise make.

Mining has been an important part of Canada's economy for almost 200 years. In 2003, Canada produced a total of $20.2 billion in minerals, in about 190 major metal, non-metal and coal mines, 3000 stone quarries, sand, and gravel pits, and about 50 nonferrous smelters, refineries, and steel mills. Major Canadian mining operations are shown in Figure 23.5. About half a million Canadians are employed in mining, refining, smelting, or other mineral-related activities.

23.5 Environmental Impacts of Mineral Development

The impact of mineral exploitation on the environment depends on such factors as ore quality, mining procedures, local hydrologic conditions, climate, rock types, size of operation, topography, and many more interrelated factors. The impact varies with the stage of development of the resource. For example, the exploration and testing stages involve considerably less impact than do the mining and processing stages. In addition, our use of mineral resources has a significant social impact.

Exploration activities for mineral deposits vary from collection and analysis of remote-sensing data gathered from airplanes or satellites, to fieldwork involving surface mapping and drilling. Generally, exploration has a minimal impact on the environment, provided that care is taken in sensitive areas, such as arid lands, marshes, and areas underlain by permafrost. Some arid lands are covered by a thin layer of pebbles over fine silt several centimetres thick. The layer of pebbles, called desert pavement, protects the finer material from wind erosion. When road building or other activities disturb the desert pavement, the fine silts may be eroded, impairing physical, chemical, and biological properties of the soil and possibly scarring the land for many years. Similarly, marshes and other wetlands, such as the northern tundra, are very sensitive to even seemingly small disturbances such as vehicular traffic.

The mining and processing of mineral resources generally have a considerable impact on land, water, air, and biological resources. Furthermore, as it becomes necessary to use ores of lower and lower grades, negative effects on the environment tend to become greater problems. For example, there is concern about asbestos fibres in the drinking water (from Lake Superior) of Duluth, Minnesota, as a result of the disposal of iron mining waste (tailings) from mining low-grade iron ore.

A major practical issue is whether surface or subsurface mines should be developed in an area. Surface mining is cheaper but has more direct environmental effects. The trend in recent years has been away from subsurface mining and toward large, open-pit (surface) mines, such as the proposed Voisey's Bay nickel mine in Labrador (Figure 23.6). The Voisey's Bay deposit is among the largest in the world, and is located in one of the world's last great wild areas. Proposed mining activities include both an open-pit mine and a subsurface mine, and extensive infrastructure including roads, docks, fuel storage, air strips, and residential buildings. The project has come under considerable criticism from local First Nations communities and environmental groups, both because of its size and because of the fragility of the natural environment.

Out of a total land area of about 1 billion ha, less than 0.03% (an area less than half the size of Prince Edward Island) is used for mining in Canada. Even though the impact of these operations is a local phenomenon, numerous local occurrences will eventually constitute a larger problem. Environmental degradation tends to

Figure 23.5 • *(a)* **Location of mining sites in Canada.** *(b)* **The Cantung tungsten mine, Northwest Territories.**

(a)

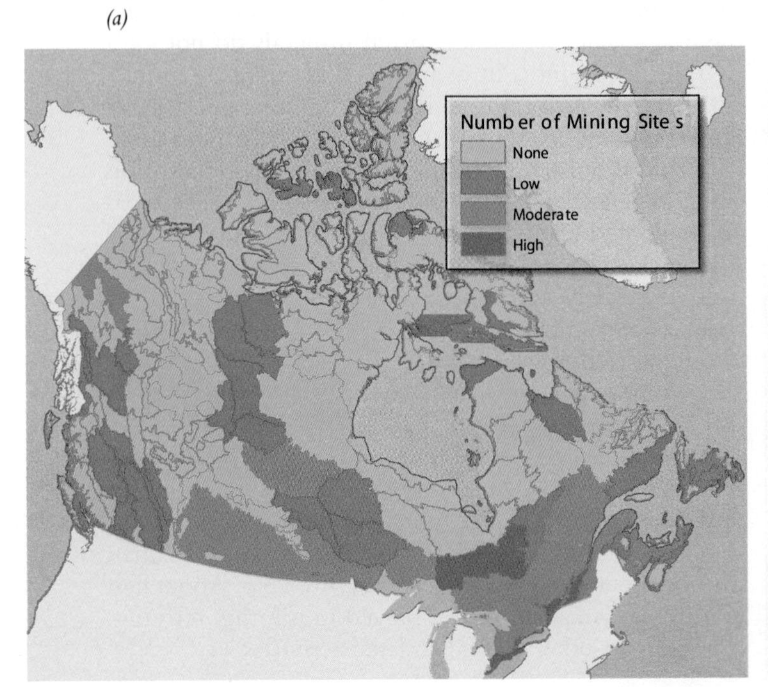

(b)

(a)

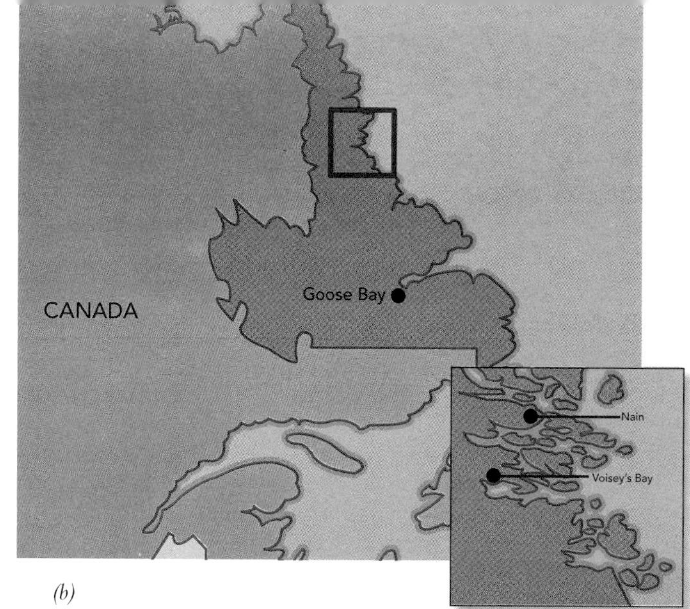

(b)

Figure 23.6 • *(a)* **Aerial photograph of Discovery Hill, the centre of the Voisey's Bay nickel deposit in coastal Labrador.** *(b)* **Map of coastal Labrador showing location of the proposed Voisey's Bay mine complex and associated infrastructure.**

extend beyond the land actually being mined. Large mining operations disturb the land by directly removing material in some areas and dumping waste in others, thus changing topography. At the very least, these actions produce severe aesthetic degradation. In addition, dust at mines may affect air resources, even though care is often taken to reduce dust production by sprinkling water on roads and on other sites that generate dust.

Water Resources Impacts of Mineral Development

A potential problem associated with mineral resource development is the possible release of harmful trace elements to the environment. Water resources are particularly vulnerable to such degradation, even if drainage is controlled and sediment pollution is reduced. Surface drainage is often altered at mine sites, and runoff from precipitation may infiltrate waste material, leaching out trace elements and minerals. The effects of mine wastes and trace elements (cadmium, cobalt, copper, lead, molybdenum, and others) on non-human biota are still poorly understood. Specially constructed ponds to collect such runoff help, but cannot be expected to eliminate all problems.

Of particular concern are tailings ponds. Tailings are the slurry of finely ground rock mixed with water, which are formed in the process of mining and ore concentration. Tailings ponds (Figure 23.7) are used for the disposal of many kinds of rock tailings, including base metals, uranium, and gold. Abandoned tailings ponds are often unsupervised and are open to the public. Sustained exposure to tailings can cause a variety of human and ecosystem health impacts, depending on the nature of the materials present. For example, some gold operations use mercury or cyanide to enhance the extraction of gold, and these materials can be present in gold mine and mill wastes. Similarly, arsenic is often present in the same rock as gold, and can be released in the mining and milling process. Chemicals added in the mining and refining process, including oils, solvents, and surfactants can also foul surface and groundwater systems near mines.

Acid mine drainage (Chapter 19) occurs when surface water (H_2O) reacts chemically with sulphide minerals, such as pyrite (FeS_2), to produce sulphuric acid (H_2SO_4). The acid then pollutes streams and groundwater resources. Acid water also drains from underground mines and roads cut in areas where coal and pyrite are abundant, but the problem of acid mine drainage is magnified when large areas of disturbed material remain exposed to surface waters. Acid mine drainage can be minimized from active mines by channelling surface runoff or groundwater before it enters a mined area and diverting it around the potentially polluting materials. However, diversion is not feasible in heavily mined regions where spoil banks from unreclaimed mines may cover hundreds of square kilometres. In these areas, acid mine drainage will remain a long-term problem. Acid mine drainage is a particular problem in sulphide mining areas, including parts of the Canadian Shield. Acid drainage from the abandoned Britannia Mine, near Britannia Beach, British Columbia, has contaminated Howe Sound for decades.

In arid and semiarid regions, water problems associated with mining are not as pronounced as in wetter regions, but the land may be more sensitive to activities related to mining, such as exploration and road building. In some arid areas, the land is so sensitive that tire tracks can remain for years. To complicate matters, soils are often thin, water is scarce, and reclamation work is difficult.

Groundwater may also be polluted by mining operations when waste comes into contact with slow-moving subsurface water. Surface water infiltration or groundwater movement causes leaching of sulphide minerals that may pollute groundwater and eventually seep into

Figure 23.7 • A modern tailings pond. Few older tailings facilities are designed with this level of care. Poorly managed tailings can cause serious impacts in downstream ecosystems.

streams to pollute surface water. Groundwater problems are particularly troublesome because reclamation of polluted groundwater is very difficult and expensive.

Land Impacts of Mineral Development

Physical changes in the land, soil, water, and air associated with mining directly and indirectly affect the biological environment. Plants and animals killed by mining activity or contact with toxic soil or water are examples of direct impacts. Indirect impacts include changes in nutrient cycling, total biomass, species diversity, and ecosystem stability owing to alterations in groundwater or surface water availability or quality. Periodic or accidental discharge of low-grade pollutants through failure of barriers, ponds, or water diversions or through breach of barriers during floods, earthquakes, or volcanic eruptions also may damage local ecological systems to some extent.

One of the main ways in which mining alters the land is through surface mining techniques such as strip mining and open pit mining, in which the overlying layer of soil and rock—often massive quantities of material—is stripped off to reach the underlying mineral. The target mineral and any waste rock are removed in layers, creating a series of steps or benches at the edge of the mine. Several benches may be in operation at the same time, giving the mine a stair-step appearance.

The practice of surface mining started in the late nineteenth century and has steadily increased because it tends to be cheaper and easier than underground mining. Surface mining is used extensively in coal mining, but also in "hard rock" mining for base and precious metals. It is likely that more and larger surface mines will be developed as the demand for coal increases.

The impact of large surface mines varies from region to region, depending on topography, climate, and reclamation practices. Surface mining has the potential to pollute or damage water, land, and biologic resources. However, good reclamation practices can minimize the damage (Figure 23.8). Reclamation practices required by law necessarily vary by site. Some of the principles of reclamation are illustrated in the case history of a modern coal mine in Fording River, British Columbia (see A Closer Look 23.1).

In 2002, there were 20 surface coal mines in operation in Canada, and dozens of metal and asbestos mines planned or in operation (like the Voisey's Bay nickel mine). In Canada and the United States, thousands of square kilometres of land have been disturbed by surface mining, and only about half this land has been reclaimed. Reclamation is the process of restoring and improving disturbed land, often by re-forming the surface and replanting vegetation (see Chapter 10). Unreclaimed coal dumps from open-pit mines are numerous and continue to cause environmental problems. Because little reclamation occurred before about 1960, and mining started much earlier, abandoned mines are common. Figure 23.9 shows an abandoned mine site before and after reclamation.

The Canadian Mining Regulations (under the federal *Territorial Lands Act* of 1985) do not explicitly require reclamation of former mine sites, although other legislation provides tax advantages for such activities. Mining activities on public lands are subject to environmental assessment under provincial legislation (if provincial lands are involved) or the *Canadian Environmental Protection Act* (for federal lands, including national parks and First Nations territory). Environmental assessment normally encompasses not only the construction and operation phases of a mine, but also its decommissioning and the restoration of affected lands. Reclamation includes disposing of wastes, contouring the land, and replanting vegetation. Reclamation is often difficult, and it is rarely completely successful. In fact, some environmentalists argue that success stories with reclamation are the exception and that surface mining should not be allowed in some fragile environments where reclamation cannot be guaranteed.

Canada currently has hundreds of underground mines, many of them abandoned. Underground mining is a dangerous profession; there are always hazards of collapse, explosion, and fire. Underground miners are at increased risk of respiratory illnesses such as black lung disease, which is related to exposure to coal dust and has killed or disabled many miners.

Land subsidence can occur over underground mines. Vertical subsidence occurs when the ground above mine tunnels collapses, often resulting in a crater-shaped pit at

A CLOSER LOOK 23.1

The Fording River Mine

The Fording River Mine on Eagle Mountain, northeast of Elkford, British Columbia, is a good example of a new generation of large open pit coal mines. The mine began operation in 1971, and currently produces about 10.5 million tonnes of coal every year. Its reserves are sufficient to continue mining for at least another 20 years.

The mine uses conventional open pit mining technology to access its 18 coal seams. The seams are separated by layers of rock, called overburden (rocks without coal), and there is additional overburden above the top seam of coal. The depth of the overburden varies from 0 to about 50 m. Blasting dislodges waste rock and exposes the coal seam; waste rock is then removed to a storage pile. Over 80 million cubic metres of waste rock and coal is blasted and hauled each year. Once removed from the seam, bulk coal is crushed, screened, washed, and finally dried in a special furnace (which burns coal as its fuel source). The finished "clean coal" is stored in a silo for later loading into rail cars.

The habitat surrounding Fording River is typical of montane systems elsewhere in the Rocky Mountains (see Chapter 8). The forest is a mixture of spruce and pine interspersed with subalpine meadows. Large mammals such as cougar, bear, and elk are common. The Fording River mine works actively to reclaim waste rock and open pit sites as mining progresses. Important components of this reclamation include reshaping and replanting waste rock dumps to encourage revegetation. About 40,000 tree seedlings are planted every year, most of them started in the mine's own greenhouse. Every year, about 40 ha are revegetated in this way. To date, a total of 600 ha of disturbed land has been reclaimed, recreating forest and wildlife forage areas and restoring local fisheries affected by the mine.

Fording River has been a leader in Canada in finding and using the best methods for mine reclamation in mountainous regions, and for the success of their efforts to date. In recognition of these efforts, Fording River has received 23 environmental awards since 1978, most recently the 2003 Citation for Outstanding Achievement for Reclamation at a Coal Mine at the 28th annual British Columbia Mine Reclamation Symposium.

Figure 23.8 • The Fording River coal mine. Resloping waste rock dumps during restoration following mining. Topsoil is spread prior to planting of vegetation.

the surface. Subsidence of underground mines has been a problem in many parts of Canada, including abandoned base metal mines in northern Ontario and Québec (Figure 23.10). For example, underground mines near Timmins, Ontario, often used loose sand to backfill mined-out sections. As early as the 1920s, large holes would appear in public areas near the town as this sand washed out or compacted. As mines were abandoned, water gradually seeped into the workings, washing out the loose sand and removing what little structural support remained. By the late 1990s, water levels had risen to the point that large mine areas were flooded. Large subsidence craters began to appear on public and private lands, including residential areas and a golf course.

(a)

(b)

Figure 23.9 • Mine reclamation *(a)* gold tailings before reclamations *(b)* a similar deposit after reclamation.

Air Impacts of Mineral Development

Smelting—the process of refining metals by heating and separating them—uses coal-burning furnaces. Acid precipitation (Chapter 21), caused by sulphur and nitrogen air emissions from smelting activities, has caused widespread damage to aquatic ecosystems in eastern Canada, for example near Sudbury, Ontario. Smelter emissions can also carry metals into the air. The Trail Smelter, in Trail, British Columbia, has been cited as the source of metal contamination in the Columbia River, Washington.

Mining also produces voluminous amounts of dust that settles on towns and fields, polluting the land and causing or facilitating lung diseases including asthma. Protests and complaints by communities in the path of mining that formerly were ignored are now receiving more attention by regulators. As people become better educated about mining laws, they are more effective in confronting mining companies to get to them reduce potential adverse consequences of mining—however, much more needs to be done.

Finally, coal fires in underground mines may be either naturally caused or deliberately set. The fires may belch smoke and hazardous fumes, causing people exposed to them to suffer from a variety of respiratory diseases. Such fires may be difficult to extinguish. An underground coal fire in Australia's Mount Wingen, between Brisbane and Sydney, is thought to have been burning for at least 6,000 years. The fire is of natural origin, and has given rise to the area's nickname of Burning Mountain.

23.6 Social Impacts of Mineral Development

Social impacts associated with large-scale mining result from the rapid influx of workers into areas unprepared for growth. Stress is placed on local services such as water supplies, sewage and solid-waste disposal systems, schools, and housing. Land use shifts from open range, forest, and agriculture to urban patterns. The additional people also increase the stress on nearby recreation and wilderness areas, some of which may be in a fragile ecological balance. Construction activity and urbanization affect local streams through sediment pollution, reduced water quality, and increased runoff. Air quality is reduced as a result of more vehicles, dust from construction, and generation of power.

Adverse social impacts also occur when mines are closed; towns surrounding large mines come to depend on the income of employed miners. Mine closures produced many ghost towns in the western United States and parts of Canada. Today, the price of coal and other minerals directly affects the livelihood of many small towns. This relationship is especially evident in Atlantic Canada and the Appalachian Mountain region of the United States, where closures of coal mines have taken a toll in past decades. These mine closings were partly the result of lower prices for coal and partly the result of rising mining costs. In recent years, coal prices have once again begun to rise, revitalizing coal communities in some areas.

One of the reasons mining costs are rising is the increased level of environmental regulation of the mining industry. Of course, regulations have also helped make mining safer and have facilitated land reclamation. Some miners, however, believe the regulations are not flexible enough, and there is some truth to their arguments. For example, some mined areas might be reclaimed for use as farmland now that the original hills have been levelled. Regulations, however, may require the restoration of the land to its original hilly state, even though hills make inferior farmland.

23.7 Minimizing the Impact of Mineral Development

Minimizing the environmental impact of mineral development requires consideration of the entire cycle of mineral resources shown in Figure 23.11. Inspection of this diagram reveals that many components of the cycle are related to the generation of waste material. In fact, the major environmental impacts of mineral resource utilization are related to waste products, which cause pollution of natural ecosystems, human health impairment, and undesirable aesthetic changes. Waste also depletes nonrenewable mineral resources and, when simply disposed of, provides no offsetting benefits for human society.

Minimization of environmental effects associated with mineral development can take several paths:

- *Environmental regulations at the federal and provincial levels.* Regulations such as the (federal) Canadian Mining Regulations address topics such as sediment, air, and water pollution resulting from all aspects of the mineral cycle. They may also address reclamation of land used for mining. Today in Canada, approximately 50% of Canada's mines have undergone some form of reclamation activity.
- *On-site and off-site treatment of waste.* Minimizing on-site and off-site problems by controlling sediment, water, and air pollution through good engineering and conservation practices is an important goal. Of particular interest is the development of biotechnological processes such as biooxidation, bioleaching, and biosorption, as well as genetic engineering of microbes. These practices have enormous potential for both extracting metals and minimizing environmental degradation. For example, engineered (constructed) wetlands are being used at several sites (Figure 19.21); acid-tolerant plants in the wetlands remove metals from mine wastewaters and neutralize acids by biological activity.[13] Reclamation can occur during mine operation or following closure. At the abandoned Island Copper Mine on Vancouver Island, for example, sulphate-reducing bacteria are used to produce bicarbonate, which neutralizes acid mine drainage, and hydrogen sulphide, which helps transform dissolved metals into harmless insoluble compounds (Figure 23.12)[14].
- *Practicing the three R's of waste management.* That is, reducing the amount of waste produced; reusing materials in the waste stream as much as feasible; and maximizing recycling opportunities.

Figure 23.10 • **Subsidence at a base metal mine near Timmins, Ontario**

Let's look at this third option in greater detail. Wastes from some parts of the mineral cycle may themselves be referred to as ores, because they contain materials that might be recycled and used again to provide energy or useful products.[15,16] The notion of reusing waste materials is not new. Metals such as iron, aluminum, copper, and lead have been recycled for many years and are still being recycled today. For example, the metal from almost all of the millions of automobiles discarded annually in Canada and the United States is recycled.[16, 17]

The total value of recycled metals is about $50 billion. Of this, iron and steel are approximately 90% by weight and 40% by total value of recycled metals. Iron and steel are recycled in such large volumes for three reasons.[18] First, the market for iron and steel is huge, and as a result, there is a large scrap collection and processing industry. Second, an enormous economic burden would result from failure to recycle. Third, failing to recycle would create significant environmental impacts related to disposal of over 50 million tonnes of iron and steel.

It is estimated that each metric tonne of recycled steel saves 1,136 kg of iron ore, 455 kg of coal, and 18 kg of limestone. In addition, only one-third as much energy is required to produce steel from recycled scrap as from native ore.[17]

Other metals that are recycled in large quantities include lead (63%), aluminum (38%), and copper (36%).[3] Recycling aluminum (for instance, in soft drink cans) reduces our need to import raw aluminum ore and saves approximately 95% of the energy required to produce new aluminum from bauxite.[17]

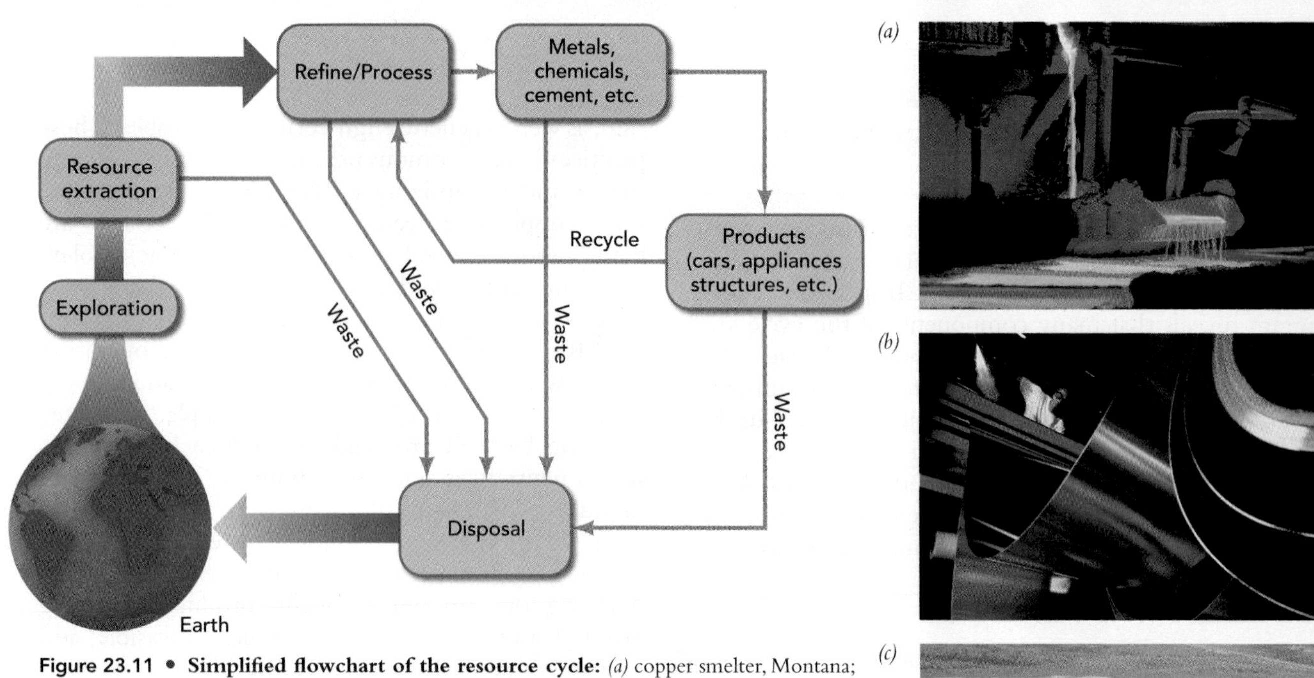

Figure 23.11 • Simplified flowchart of the resource cycle: *(a)* copper smelter, Montana; *(b)* sheets of copper for industrial use; *(c)* disposal of mining waste into a tailings pond from a gold mine.

23.7 Minerals and Sustainability

Simultaneously considering sustainable development and mineral exploitation and use is problematic. This is because non-renewable mineral resources are consumed over time and sustainability is a long-term concept that includes finding ways to provide future generations a fair share of Earth's resources. Recently, it was argued that given human ingenuity and sufficient lead-time, we can find solutions for sustainable development that incorporate non-renewable mineral resources.

Human ingenuity is important because often it is not the mineral we need so much as what we use the mineral for. For example, we mine copper and use it to transmit electricity in wires or electronic pulses in telephone wires. It is not the copper itself we desire but the properties of copper that allow these transmissions. We can use fibreglass cables in telephone wires, eliminating the need for copper. Digital cameras have eliminated the need for film development that uses silver. The message is that it is possible to compensate for a non-renewable mineral by finding new ways to do things. We are also learning that we can use raw mineral materials more efficiently. For example, when the Eiffel Tower was constructed in the late 1800s, 8,000 metric tonnes of steel was used. Today, the tower could be constructed with a quarter of that amount of steel.[19]

Finding substitutes or ways to use non-renewable resources more efficiently generally requires several decades of research and development. A measure of the time available for finding the solutions to depletion of non-renewable reserves is the R-to-C ratio, where R is the known reserves (e.g., hundreds of thousands of tons of a metal) and C is the rate of consumption (e.g., thou-

Figure 23.12 • Section through a tailings deposit showing action of sulphur-reducing bacteria. Thin black layers show where sulphur-reducing bacteria have converted the oxygen-rich orange-coloured tailings into darker-coloured sulphide compounds.

CRITICAL THINKING ISSUE

Will Mining with Microbes Help the Environment?

Mining is an ancient technology first practiced at least 6,500 years ago. Modern mining methods are more technologically sophisticated but use the same basic processes (digging and smelting) to isolate valuable metals. To be economic, these methods have traditionally required high-grade ore and cheap sources of energy as well as an acceptance that mining to a lesser or greater extent would damage the environment. Although these conditions have prevailed for most of human history, they are changing. Earlier exploitation of mineral resources is pushing the mining industry toward lower-grade ores; non-renewable energy sources are expensive and disappearing; and concern over degradation of the environment and health threats to humans and other species is growing. Demand for minerals, however, is increasing, because of both population growth and technological development.

As an example, the average grade of copper ore has dropped from 6% to 0.6% over the last century, making copper mining more energy intensive and more wasteful. Five metric tonnes of coal are required to produce 1 metric tonne of copper, and every kilogram of copper produced represents 89 kg of waste. Open-pit mining of copper causes acids and heavy metals, such as arsenic, to contaminate surface water and groundwater. Smelting produces sulphur dioxide and other gaseous compounds, as well as particles, which contribute to air pollution.

Microscopic organisms produced by biotechnology offer an entirely new approach to mining. By 1989, more than 30% of copper mined in the United States depended on a biochemical process that begins with a microbe, *Acidithiobacillus ferrooxidans*. (This species is, however, also known to cause acid mine drainage, by facilitating the oxidative dissolution of pyrite and other sulphide minerals, and may thus also cause unintended adverse effects.) Biological processes have also been used in mining uranium and gold. Research is under way to use microbes to remove sulphur from coal and cyanide from mining waste. The union of biological processes and mining is called biohydrometallurgy.

In the future, it may be possible to use microbes on ores without removing them from Earth. Metallurgists envision drilling wells into the ore and fracturing it, then injecting bacteria into the wells and fractures. The ore could be removed by flooding the wells with water, removing the ore, and recycling the water. Biotechnologists hope to use genetic engineering to develop bacteria to mine specific metals when no naturally occurring bacteria exist to extract them.

The disadvantages of biohydrometallurgy are that it is slow, requiring decades rather than years, and that methods for breaking ores into small enough particles for efficient extraction are not yet available. Already, however, biological methods are economically feasible for low-grade ores that elude conventional methods. Further technological innovations may make them competitive in more situations.

Critical Thinking Questions

1. What are the environmental advantages of biohydrometallurgy over conventional methods? What are the possible disadvantages?
2. How would you assess the possible dangers from genetically engineered organisms developed for mining compared with the dangers from organisms engineered for use in agriculture and medicine?
3. Some experts believe that without economic pressure (e.g., the decline in high-grade ore, increased energy requirements for extracting metal, governmental regulations on clean air and water, economic recessions), the mining industry would not continue to explore biochemical methods. How could the government encourage the industry to devote more research and development efforts to expanding biochemical mining? Develop a proposal for a government policy.

In summary, we may approach sustainable development and use of non-renewable mineral resources by finding ways to more wisely use resources, developing more efficient ways of mining resources, more efficiently using available resources, and applying human ingenuity to find substitutes for a particular function that a non-renewable mineral resource is used.

Summary

- Mineral resources are usually extracted from naturally occurring, anomalously high concentrations of Earth materials. Natural deposits allowed early peoples to exploit minerals while slowly developing technological skills.
- The origin and distribution of mineral resources is intimately related to the history of the biosphere and the geologic cycle. Nearly all aspects and processes of the geologic cycle are involved to some extent in producing local concentrations of useful materials.
- Mineral resources are not mineral reserves. Unless discovered and developed, resources cannot be used to address present shortages.
- The availability of mineral resources is one measure of the wealth of a society. Modern technological civilization would not be possible without the exploitation of mineral resources. However, it is important to recognize that mineral deposits are not infinite and that we cannot maintain exponential population growth on a finite resource base.
- Canada, the United States, and many other affluent nations rely on imports for their supplies of many mineral resources. As other nations industrialize and develop, imports may be more difficult to obtain, and affluent countries may have to find substitutes for some minerals or use a smaller portion of the world's annual production.
- The environmental impact of mineral exploitation depends on many factors, including mining procedures, local hydrologic conditions, climate, rock types, size of operation, topography, and other factors.
- The mining and processing of mineral resources greatly affect the land, water, air, and biological resources and create social impacts as a result of the increased demand for housing and services in mining areas.
- Because the demand for mineral resources will increase in the future, we must strive to minimize both on-site and off-site problems by controlling sediment, water, and air pollution through good engineering and conservation practices.
- Sustainable development and use of non-renewable resources are not necessarily incompatible. Reducing consumption, reusing, recycling, and finding substitutes are environmentally preferable ways to delay or alleviate possible crises caused by the convergence of a rapidly rising population and a limited resource base.

STUDY QUESTIONS

1. What is the difference between a resource and a reserve?
2. Under what circumstances might sewage sludge be considered a mineral resource?
3. If surface mines and quarries cover only a small portion of the land surface of Canada and the United States, why is there so much environmental concern about them?
4. Which biological processes can influence mineral deposits?
5. An environmentalist claims that the oceans can provide all our mineral resources with no negative environmental effects. Do you agree or disagree? Explain the reasons for your answer.
6. What factors determine the availability of a mineral resource?
7. Utilizing a mineral resource involves four general phases: (a) exploration, (b) recovery, (c) consumption, and (d) disposal of waste. Which phase do you think has the greatest environmental effect?
8. How can use of non-renewable mineral resources be compatible with sustainable development?

FURTHER READING

Brookins, D. G. *Mineral and Energy Resources*. Columbus, Ohio: Charles E. Merrill, 1990. A good summary of mineral resources.

Douglas, R.J.W. (ed.) *Geology and Economic Minerals of Canada*. Geological Survey of Canada, 1970.

Fulton, R. J. (ed.). *Quaternary Geology of Canada and Greenland*. Geological Survey of Canada. Geology of Canada Series No. 1. Ottawa: Geological Survey of Canada, 1989.

Kesler, S. F. *Mineral Resources*, Economics and the Environment. Upper Saddle River, NJ: Prentice Hall, 1994.. A good book about mineral resources.

Muise, D. A. *Coal Mining in Canada*: A historical and comparative overview. Ottawa: National Museum of Science and Technology, 1996.

chapter 24

Toronto's Garbage Crisis

Canada's largest city has run out of space to put its garbage. The city's Keele Valley Landfill closed on December 31, 2002, leaving the city with few waste management options. Beginning in January 2003, 130 truckloads of Toronto's garbage are shipped daily to commercial landfills north of Detroit, Michigan. Over a million tonnes of Toronto's garbage travel Ontario's Highway 401 each year, adding to the traffic and air pollution on that already-congested route.

The Michigan solution isn't new. As early as 1995, tipping fees at the Keele Valley site had reached $150/tonne, far higher than rates available through private landfill operators Michigan sites. Soon, much of Toronto's waste was being shipped to the U.S. by private waste haulers whose traffic had been an important revenue stream for Keele Valley. In an attempt to regain customers, Keele Valley cut its tipping fees to $50. And with disposal costs so low, Toronto had no reason to reduce its waste stream. Even though Keele Valley was known to have limited capacity, Toronto seemed unable to come up with a reasonable waste management solution.

The Keele Valley closure prompted reconsideration of another option that had first been proposed in 1989: to dispose of Toronto's garbage in an abandoned, flooded, open-pit mine near Kirkland Lake, Ontario. The mine project had long been controversial among local residents. The proposal called for 1.3 million tonnes of mixed solid waste, including wastes from homes, farms, construction, and even biomedical operations, deposited in the mine over a 20-year period. Estimates suggested that over 300 million litres of leachate would have to be pumped and treated each year—a total of almost 100 billion litres over the life of the project. Treated effluent was to be discharged into Lake Temiscaming and the Ottawa River, which is shared between Ontario and Québec. Federal lands, including First Nations reserves, would potentially be affected. The project's many critics pointed out that while Toronto would benefit from the Adams Mine disposal site, it would bear none of its costs, some of which could be significant. Rather, people far from Toronto, some of them in another province, would have to contend with the air and water pollution, increased truck traffic, dust, and noise associated with the site. Worse, many argued that approving the Adams Mine project essentially gave Toronto permission to continue its wasteful practices for another 20 years.[1]

And it is Toronto's wasteful practices that are at the heart of the problem. At present, only 30% of Toronto's waste stream is recycled, far below levels elsewhere in the world. Some estimates suggest that as much as 80 to 90% of municipal solid waste can be diverted through intensive recycling and composting programs. Toronto has argued that the costs of centralized composting are significant, especially for apartment complexes.

In the summer of 2004, the Ontario government passed legislation preventing the Adams Mine from ever becoming a solid waste disposal site, and banning the

Figure 24.1 • Toronto's Keele Valley landfill, which closed in 2002, was one of Canada's largest waste disposal sites.

Waste Management

LEARNING OBJECTIVES

The waste management concept of "dilute and disperse" (e.g., dumping waste into a river) is a holdover from the frontier days, when people mistakenly believed land and water were limitless resources. The next attempt was to "concentrate and contain" waste in disposal sites—a practice that also proved to pollute land, air, and water resources. The current focus is on managing materials to eliminate waste. Finally, we are getting it right! After reading this chapter, you should understand:

- What the advantages and disadvantages are of each of the major methods that constitute integrated waste management.
- How the physical and hydrologic conditions at a site affect its suitability as a landfill.
- What multiple barriers for landfills are, and how landfill sites can be monitored.
- Why management of hazardous chemical waste is one of our most serious environmental concerns.
- What the various methods are of managing hazardous chemical waste.
- What the major pathways are by which hazardous waste from a disposal site can enter the environment.
- What problems are related to lake and ocean dumping, and why these problems may persist for some time.

development of future dumps in bodies of water larger than a hectare. The government has also proposed a goal of diverting 60% of the province's wastes by 2008. That will require some big changes in Toronto's garbage practices.[2]

Some of those changes are already happening. Following a successful pilot project, the city hopes to extend collection of compostable materials to the whole city in 2005, and collection of household hazardous waste is under discussion (a Toxics Taxi service is currently available for larger quantities of such wastes).

As Toronto struggles with its garbage crisis, it must also manage the legacy of past disposal practices. Keele Valley will require systematic reclamation and maintenance for many years. Like Marie Curtis Park and Riverdale Park in Toronto, both of which are former dump sites, it may eventually become a public park. However, it would have been preferable to preserve the land in its original condition, or at least to extend the life of that site by thoughtful waste management practices. The need to find new landfill sites and reclaim old ones emphasizes the failure of our past waste management policy, based on a throwaway mentality.

- *Toronto's garbage crisis represents a tremendous financial burden to Canadian society and reminds us that we have failed in the past 50 years to move from a throwaway, waste-oriented society to a society that sustains natural resources through improved materials management. We are now moving in that direction by producing less waste and recycling more. With this in mind, we introduce in this chapter concepts of waste management applied to urban waste, hazardous chemical waste, and waste in the freshwater and marine environment.*

24.1 Early Concepts of Waste Disposal

During the first century of the Industrial Revolution, the volume of waste produced in Canada and the United States was relatively small, and waste could be managed through the concept of "dilute and disperse." Factories were located near rivers because the water provided a number of benefits, including easy transport of materials by boat, sufficient water for processing and cooling, and easy disposal of waste into the river. With few factories and a sparse population, dilute and disperse was sufficient to remove the waste from the immediate environment.[3]

As industrial and urban areas expanded, the concept of dilute and disperse became inadequate, and a new concept, known as "concentrate and contain," came into use. It has become apparent, however, that containment was, and is, not always achieved. Containers, whether simple trenches excavated in the ground or metal drums and tanks, may leak or break and allow waste to escape. Health hazards resulting from past waste disposal practices have led to the present situation, in which many people have little confidence in government or industry to preserve and protect public health.[4]

The problem of waste disposal in Toronto, presented in the opening case study, is not unique. In Canada, as well as in many other parts of the world, people are facing a serious solid-waste disposal problem. The problem exists because we are producing a great deal of waste and the acceptable space for permanent disposal is limited. It has been estimated that within the next few years approximately half the cities in Canada and the United States may run out of landfill space. Toronto, for example, is essentially out of landfill space now and is bargaining with other jurisdictions on a monthly or yearly basis to dispose of its trash.

To say we are actually running out of space for landfills isn't altogether accurate. Land used for landfills is minute compared to the land area of Canada. Rather, existing sites are being filled, and it is difficult to site new landfills. After all, no one wants to live near a waste disposal site, be it a sanitary landfill for municipal waste, an incinerator that burns urban waste, or a hazardous-waste disposal operation for chemical materials. This attitude is widely known as NIMBY ("not in my backyard"). Although it is often criticized as a self-centred attitude, NIMBY plays an important role in making the environmental impacts of waste management transparent, and forcing society to confront our growing waste management challenge.

Another major limiting factor is the cost of disposal. A decade or so ago, the cost of disposal of 1 metric tonne of urban refuse was approximately $10 to $20. Today, the average cost is about $50; and some cities, such as London, Ontario, are considering fees as high as $150 per metric tonne for waste disposal.[4,5] These costs are only a small part of waste management expenditures. Disposal or treatment of liquid and solid waste costs over $20 billion in Canada and the United States every year. It is one of our most costly environmental expenditures.[6]

24.2 Modern Trends

The environmentally correct concept with respect to waste management is to consider wastes as resources out of place. Although we may not soon be able to reuse and recycle all waste, it seems apparent that the increasing cost of raw materials, energy, transportation, and land will make it financially feasible to reuse and recycle more resources. Moving toward this objective is moving toward an environmental view that there is no such thing as waste. Under this concept, waste would not exist because it would not be produced or, if produced, would be a resource to be used again. This concept is referred to as the "zero waste" movement.

Zero waste is the essence of what is known as industrial ecology, the study of relationships among industrial systems and their links to natural systems. Under the principles of industrial ecology, our industrial society would function much as a natural ecosystem functions. Waste from one part of the system would be a resource for another part (see A Closer Look 24.1).[7]

The concept of zero waste production was until recently considered unreasonable in the waste management arena. The concept, however, is catching on. The city of Canberra, Australia may be the first community to propose a plan to have zero waste, a goal it hopes to reach by 2010. Thousands of kilometres away, in the Netherlands, a national waste reduction goal of 70–90% has been set. How this is to be accomplished is not entirely clear; but a large part of the planning involves taxation of waste in all its various forms, from emissions from smokestacks to solids delivered to landfills. Already in the Netherlands, discharges of heavy metals to waterways have been nearly eliminated by implementation of pollution taxes. At the household level, the government is considering programs—known as "pay as you throw"—in which people are charged by the volume of waste they produce. As disposal of waste, including household waste, is taxed, people produce less waste.[8]

Of particular importance to waste management is the growing awareness that many of our waste management programs involve moving waste from one site to another and not really managing it. For example, waste from urban areas may be placed in landfills; but eventually these landfills may cause new problems by producing methane gas or noxious liquids that leak

SCIENCE AND VALUES

A CLOSER LOOK 24.1

Industrial Ecology

As noted in the text, industrial ecology can be defined as the study of relationships among industrial systems and their linkages to natural systems. Industrial ecology and sustainable development are often linked. In part, this results from the analogy drawn between industrial ecology and natural ecology, with its focus on ecosystems. Just as sustaining life on Earth is a function of ecosystems, sustainability in our industrial society is a function of industrial ecology.[7]

To understand the concept of waste according to industrial ecology, consider waste in natural ecosystems. What might be thought of as waste in one part of an ecosystem is often a resource for another part. For example, the elephant eats and produces waste that becomes a resource for the dung beetle.

Now consider an example of industrial ecology. Suppose that there exists a coal-burning power plant that produces electricity for a town. Waste from producing the power includes ash from the coal, exhaust heat, and products from combustion that might go out a smokestack, including carbon dioxide, sulphur dioxide, and heat. For our hypothetical power plant and surrounding town, let's assume that the waste heat is used to warm homes and provide heat for industrial activities. Sulphur dioxide is removed from the system by scrubbing to produce gypsum, which is the major constituent of wallboard used in construction. Carbon dioxide is sequestered and used in local greenhouses along with waste heat to force and prolong the growing cycle of plants. We are left with the ash from the coal, which is used for road surfacing.

This hypothetical scenario is being partly played out at several power plants today. It is expected that, as principles of industrial ecology are better understood and applied, we will see more applications in a variety of industries as well as in agriculture.

As we attempt to produce sustainable economic development on local, regional, and global scales, however, we need to acknowledge that industrial ecology based on application of science and technology alone is not adequate to achieve sustainability of global systems. We also need to clarify our values. Science can provide potential solutions to problems we face; however, which solutions we choose will reflect our values. That is one of the key themes of this book.

from the site and contaminate the surrounding areas. Managed properly, however, methane produced from landfills is a resource that can be burned as a fuel (an example of industrial ecology).

Concepts involving our use of materials and the waste we produce are changing. Previous notions of waste disposal are no longer acceptable, and we are rethinking how we deal with materials, with the objective of eliminating the concept of waste entirely. In this way, we can reduce the consumption of virgin materials, which depletes our environment, and can live within our environment more sustainably.[8]

24.3 Integrated Waste Management

The dominant concept today in managing waste is known as integrated waste management (IWM), which is best defined as a set of management alternatives that includes reuse, source reduction, recycling, composting, landfill, and incineration.[4]

The three often-cited R's of IWM are **reduce, reuse, and recycle**. The ultimate objective of the three R's is to reduce the amount of urban and other waste that must be disposed of in landfills, incinerators, and other waste management facilities. Study of the waste stream (the waste produced) in areas that utilize IWM technology suggests that the amount (by weight) of urban refuse disposed of in landfills or incinerated can be reduced by at least 50% and perhaps as much as 80%. A 50% reduction by weight of urban waste could be facilitated by:[4]

- Reduced consumption and better design of packaging to reduce waste, an element of source reduction (10% reduction).
- Large-scale composting programs (10% reduction).
- Establishment of recycling programs (30% reduction).

The three R's are listed in order of importance. Our first step must be to reduce. Wasteful consumption lies at the heart of our solid waste management problem. Our throwaway society has become accustomed to highly packaged food and consumer products, often wrapped in several layers of packaging. If we use less, or buy in larger quantities with less packaging, we can reduce our personal waste generation significantly. The second R, recycling, is also a major player in the reduction of the urban waste stream. Can recycling in fact reduce the waste stream by 50%? Recent work suggests that the 50% goal is reasonable. In fact, it has been reached in some parts of Canada. The potential upper limit for recycling is considerably higher. It is estimated that as much as 80–90% of the waste stream might be recovered through what is known as intensive recycling.[9] Guelph, Ontario, was the first Canadian community to introduce an integrated two-stream waste separation system. Over 70% of the city's waste is diverted from landfills to composting and recycling (Figure 24.2). Many more

Figure 24.2 • **Guelph, Ontario's wet-dry recycling facility has a goal of 70% diversion of waste from municipal landfills.**

communities have partial recycling programs, which target a specified number of materials, such as glass, aluminum cans, plastic, organic material, and newsprint. Partial recycling can provide a significant reduction, and in many places it is approaching or even exceeding 50%.[10, 11] Recent legislation in Europe and the United States requires a high proportion of recycled fibre in some paper products, a trend that will increase incentives to recycle paper.

Public Support for Recycling

An encouraging sign associated with public support for the environment is an increase in the willingness of industry and business to support recycling on a variety of scales. For example, fast-food restaurants are using less packaging for their products and providing on-site bins for recycling used paper and plastic. Grocery stores are encouraging the plastic and paper bag recycling by providing bins for their collection Some food stores offer inexpensive canvas shopping bags to people who prefer them to disposable plastic and paper bags. Companies are redesigning products so that they can be more easily disassembled after use and the various parts recycled. As this concept catches on, small appliances, such as electric frying pans and toasters, may be recycled rather than ending up in landfills. The automobile industry is also responding by designing automobiles with coded parts so that they can be more easily disassembled (by professional recyclers) and recycled, rather than left to become rusting eyesores in junkyards.

On the consumer front, people are now more likely to purchase products that can be recycled or that come in containers that are more easily recycled or composted. Many consumers have purchased small home appliances that crush bottles and aluminum cans, reducing their volume and facilitating recycling. The entire area is rapidly changing, and innovations and opportunities will undoubtedly continue.

Markets for Recycled Products

As with many other environmental solutions, implementing the IWM concept successfully can be a complex undertaking. In some communities where recycling has been successfully implemented, it has resulted in glutted markets for the recycled products, which has sometimes required temporary stockpiling or suspension of recycling of some items. It is apparent that if recycling is to be successful, markets and processing facilities will also have to be developed to ensure that recycling is a sound financial venture as well as an important part of IWM.

Recycling of Human Waste

The use of human waste or "night soil" on croplands is an ancient practice. In Asia, recycling of human waste has a long history. Chinese agriculture was sustained for thousands of years through collection of human waste, which was spread over agricultural fields. The practice also spread, and by the early twentieth century, land application of sewage was one of the primary methods of disposal in many metropolitan areas in countries including Canada, Mexico, Australia, and the United States.[12] These early uses of human waste for agriculture occasionally spread infectious diseases through agents, including bacteria, viruses, and parasites, applied to crops along with the waste. Today, with the globalization of agriculture, we still see occasional warnings about fruits and vegetables contaminated in this way.

One of the major problems of recycling human waste is that, along with human waste, thousands of chemicals and metals flow through our modern waste stream. Even garden waste that is composted may contain harmful chemicals such as pesticides.

Metals, petroleum products, industrial solvents, and pesticides may end up in our wastewater collection systems and sewage treatment plants. Because many toxic materials are likely to be present with the waste, we must be very skeptical of utilizing sewage sludge for land application. Of course, the contents of sewage sludge vary from place to place and even from day to day. Nevertheless, studies have shown that high levels of toxic chemicals may be present in the sludge of cities or towns with industries that use toxic materials.[12] The good news is that fewer toxic materials end up at sewage-treatment plants than several decades ago because many industries are now pre-treating their waste to remove materials that previously contaminated wastewaters. In Canada, discharges to municipal sewer systems are controlled through municipal sewer-use bylaws, which vary considerably from region to region. Federal and provincial legislation aimed at reducing the concentrations of pollutants in industrial wastewaters has also made a difference. For example, Ontario's Municipal-Industrial Strategy for Abatement regulations have the goal of virtually eliminating toxic substances in industrial wastewater effluents.

Discussions involving various federal, local, and other government agencies, as well as industries, have addressed the question of how much toxic material in the waste stream constitutes a problem. This is really not the correct question to ask, however. The goal should be that sewage sludge contains no toxic materials at all. A problem is that the sewer lines from urban homes are the same ones utilized by industry. As a result, conventional waste disposal technology is unlikely to produce a form of sludge that is safe for us and other living things. A possible solution is to separate urban waste from industrial waste. A second possibility is to pretreat waste from industrial sources to remove hazardous components before they enter the wastewater stream. Many industries today pretreat waste, as mentioned earlier. Finally, some communities are considering smaller wastewater treatment facilities intended to treat waste from homes; local farms would use the recycled waste. In the future, as the cost of oil, which is necessary to produce fertilizers, continues to rise, the age-old practice of recycling human waste may again be financially viable and necessary in many more places than today.[12]

24.4 Materials Management

The recycling option of IWM has been seriously attempted for over two decades. Recycling has been responsible for generating entire systems of waste management that have produced tens of thousands of jobs in Canada and the United States and reduced the amount of urban waste sent from homes to landfills from 20 to 40% in the 1980s to about 65% today. Many firms have combined waste reduction with recycling to reduce by 50% to 90% the waste they deliver to landfills. However, in spite of this success, IWM has been criticized for overemphasizing recycling and failing to effectively advance policies to prevent waste production.

Futuristic waste management has the goal of zero production of waste, consistent with the ideals of industrial ecology. This visionary goal will require more sustainable use of materials combined with resource conservation in what is being termed **materials management**. It is believed that the goal could be pursued in the following ways:[13]

- Eliminate subsidies for extraction of raw materials such as minerals, oil, and timber.
- Establish "green building" incentives that encourage the use of recycled-content materials and products in new construction.
- Assess financial penalties for production that uses negative materials management practices.
- Provide financial incentives for industrial practices and products that benefit the environment by enhancing sustainability (e.g., reducing waste production and using recycled materials).
- Increase the number of new jobs in the technology of reuse and recycling of resources.

Germany's Green Dot program has been a world leader in materials management. With a large population and small land area, Germany faced a major waste management crisis. In the mid-1990s, the country passed legislation requiring 80% recycling of all packaging. Manufacturers that choose not to comply must pay a surcharge of 30%, which is passed on to the consumer in the form of higher prices. Participating manufacturers pay a fee based on the size and type of packaging, allowing them to display the green dot symbol on their products. The fees pay for the collection and return of packaging to its producers. Green Dot products are offered at a lower price than non-conforming products, creating an incentive for consumers to "buy green." The Green Dot system has had a major influence on the development of refillable and reusable packaging in Germany and elsewhere in the world.

Materials management in Canada today is beginning to have effects on where industries are located. For example, approximately 50% of the steel produced in the country now comes from scrap. As a result, new steel mills are no longer located where resources such as coal and iron ore are close by. New steel mills are located in a variety of places, from British Columbia to Atlantic Canada; and their resource is the local supply of scrap steel. Because they are starting with scrap metal, the new industrial facilities use far less energy and cause much less pollution than older steel mills that produce their product from iron ore.[14] Similarly, paper recycling is changing where new paper mills are constructed. In the past, mills were built near forested areas where timber necessary for paper production was being logged. As a result of the recycling of voluminous amounts of paper, mills are now being built near cities where supplies of recycled paper exist. For example, New Jersey has 13 paper mills utilizing recycled paper and eight steel "mini-mills" producing steel from scrap metal. What is remarkable is that New Jersey has little forested land and no iron mines! Resources for the paper and steel mills come from materials already in use and exemplify the power of materials management.[14]

Although materials management is providing alternatives to waste disposal, we still need to deal with the solid waste we produce in both urban and rural areas. We now discuss solid-waste management in more detail.

24.5 Solid-Waste Management

Solid-waste management continues to be a problem in most parts of the world, including Canada. In many developing countries, waste management practices are inadequate. These practices, which include poorly controlled open dumps and illegal roadside dumping, can spoil scenic resources, pollute soil and water resources, and produce potential health hazards.

Illegal dumping is a social problem as much as a physical one, because many people are simply disposing of their waste as inexpensively and as quickly as possible. They may not see dumping their garbage or trash as an environmental problem. If nothing else, this is a tremendous waste of resources; much of what is dumped could be recycled or reused. In areas where illegal dumping has been reduced, the keys have been awareness, education, and alternatives. Education programs teach people about the environmental problems of unsafe, unsanitary waste dumping. Funding is provided for trash cleanup and inexpensive trash collection and recycling at sites of origin.

The next section discusses the composition of solid waste in Canada. Then, in the remainder of this section, we describe specific disposal methods: on-site disposal, composting, incineration, open dumps, and sanitary landfills.

Composition of Solid Waste

In 2000, Canada generated an estimated 31.4 million tonnes of municipal solid waste, less than a quarter of which was recycled. The general composition of solid waste likely to end up at a disposal site in Canada is shown in Figure 24.3. It is no surprise that paper (28%) and food and garden wastes (33%) are by far the most abundant of these solid wastes. However, this is only an average content, and considerable variation can be expected based on factors such as land use, economic base, industrial activity, climate, and season of the year.

Infectious wastes from hospitals and clinics sometimes make their way to disposal sites, where they can create potential health problems if they were not properly sterilized before disposal. Some hospitals have facilities to incinerate such wastes, and incineration is probably the surest way to manage infectious medical waste. In addition, in urban areas, a large amount of toxic material may end up at disposal sites; as a result, many older urban landfills are now being considered hazardous-waste sites that will require costly cleanup.

People have many misconceptions about our waste stream.[15] There has been much publicity concerning urban waste associated with fast-food packaging, polystyrene foam, and disposable diapers. Therefore, many people assume that these products make up a large percentage of the total waste stream and are responsible for the rapidly filling landfills. However, excavations into modern landfills using archeological tools have cleared up some misconceptions concerning these items. We now know that fast-food packaging accounts for only about 0.25% of the average landfill; disposable diapers, approximately 0.8%; and polystyrene products, about 0.9%.[16] Paper, as mentioned, is one of the major constituents in landfills (perhaps as much as 30% by volume). The largest single item is newsprint, which accounts for as much as 18% by volume.[15] Newsprint is one of the major items targeted for recycling because big environmental dividends can be expected. However (and this is a value judgment), the need to deal with the major contributors does not mean that we need not reduce our use of disposable diapers, polystyrene, and other paper products. In addition to creating a need for disposal, these products are made from resources that might be better managed. An emerging problem is so-called "e-waste"—discarded electronic products such as television sets, hair dryers, and computer products. Several provinces, including Alberta and Ontario, are developing policies for the recycling and disposal of e-waste.

On-Site Disposal

We turn now to specific methods of solid-waste disposal. A common on-site disposal method in urban areas is the mechanical grinding of kitchen food waste. Garbage-disposal devices are installed in the wastewater pipe system at the kitchen sink, and the garbage is

Figure 24.3 • Composition of Canadian urban solid waste (by weight). [*Source*: Fraser Institute.]

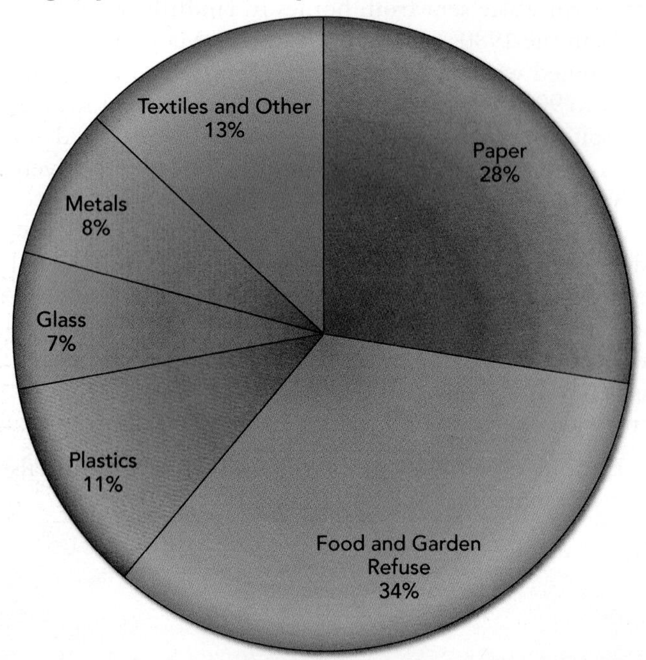

ground and flushed into the sewer system. This effectively reduces the amount of handling and quickly removes food waste. Final disposal is transferred to sewage-treatment plants, where solids remaining as sewage sludge still must be disposed of.[17]

Composting

Composting is a biochemical process in which organic materials such as lawn clippings and kitchen scraps decompose to a rich, soil-like material. The process involves rapid partial decomposition of moist solid organic waste by aerobic organisms. Although simple backyard compost piles may come to mind, as a waste management option, large-scale composting is generally carried out in the controlled environment of mechanical digesters.[17] This technique is popular in Europe and Asia, where intense farming creates a demand for the compost.[17] A major drawback of composting is the necessity of separating organic material from other waste. Therefore, it is probably economically advantageous only where organic material is collected separately from other waste. Composting plant debris previously treated with herbicides may produce a compost toxic to some plants. Nevertheless, composting is an important component of IWM, and its contribution will undoubtedly grow in the future.

Incineration

In **incineration**, combustible waste is burned at temperatures high enough (900°–1,000°C) to consume all combustible material, leaving only ash and non-combustibles to dispose of in a landfill. Under ideal conditions, incineration may reduce the volume of waste by 75–95%.[17] In practice, however, the actual decrease in volume is closer to 50% because of maintenance problems as well as waste supply problems. Besides reducing a large volume of combustible waste to a much smaller volume of ash, incineration has another advantage in that the process of incineration can be used to supplement other fuels and generate electrical power.

Incineration of urban waste is not necessarily a clean process. Incineration may produce air pollution and toxic ash. For example, incineration in Canada and the United States apparently is a significant source of environmental dioxin, a carcinogenic toxin (see Chapter 15).[18] Smokestacks from incinerators may emit oxides of nitrogen and sulphur, which lead to acid rain; heavy metals such as lead, cadmium, and mercury; and carbon dioxide, which is related to global warming.

In modern incineration facilities, smokestacks are fitted with special devices to trap pollutants; but the process of pollutant abatement is expensive. Furthermore, the plants themselves are expensive, and government subsidization may be needed to aid in their establishment. Recent evaluation of the urban waste stream suggests that with an investment of under $10 billion, a sufficient number of incinerators could be constructed in Canada and the United States to burn approximately 25% of the solid waste that is generated. However, a similar investment in source reduction, recycling, and composting could result in diversion from landfills of as much as 75% of Canada's urban waste stream.[9]

The economic viability of incinerators depends on revenue from the sale of the energy produced by burning the waste. As recycling and composting are increased, they will compete with incineration for their portion of the waste stream, and sufficient waste (fuel) to generate a profit from incineration may not be available. The main conclusion that can be drawn based on IWM principles is that a combination of reusing, recycling, and composting could reduce the volume of waste requiring disposal at a landfill by at least as much as incineration.[9]

Open Dumps

In the past, solid waste was often disposed of in open dumps, where the refuse was piled up without being covered or otherwise protected. Thousands of open dumps have been closed in recent years, and new open dumps are banned in Canada, the United States, and many other countries. Nevertheless, many are still being used worldwide (Figure 24.4).

Dumps are located wherever land is available, without regard to safety, health hazards, or aesthetic degradation. Common sites are abandoned mines and quarries, where gravel and stone have been removed (sometimes by ancient civilizations); natural low areas, such as swamps or

Figure 24.4 • **Open garbage dump in Kanpur, India.**

floodplains; and hillside areas above or below towns. The waste is often piled as high as equipment allows. In some instances, the refuse is ignited and allowed to burn. In others, the refuse is periodically levelled and compacted.[17]

As a general rule, open dumps create a nuisance by being unsightly, providing breeding grounds for pests, creating a health hazard, polluting the air, and sometimes polluting groundwater and surface water. Fortunately, open dumps are giving way to the better-planned and -managed sanitary landfills. In some less developed countries, however, open dumps provide an important livelihood for "garbage pickers," who comb wastes for materials that can be recovered and sold. The photograph at the beginning of this chapter shows a paper recycling facility in India that uses mainly scrap paper collected by garbage pickers. Although lucrative, garbage picking carries significant health risks. Some systems, such as Mumbai, India, are encouraging garbage pickers to go door-to-door, collecting wastes in special containers. In addition to traditional waste sales and recycling, some garbage pickers now compost organic waste, and use those materials to maintain nurseries in which plants are grown for later sale.

Sanitary Landfills

A sanitary landfill is designed to concentrate and contain refuse without creating a nuisance or hazard to public health or safety. The idea is to confine the waste to the smallest practical area, reduce it to the smallest practical volume, and cover it with a layer of compacted soil at the end of each day of operation or more frequently if necessary. Covering the waste is what makes the landfill sanitary. The compacted layer restricts (but does not eliminate) continued access to the waste by insects, rodents, and other animals, such as seagulls. It also isolates the refuse, minimizing the amount of surface water seeping into and gas escaping from the waste.[19]

Leachate

The most significant hazard from a sanitary landfill is pollution of groundwater or surface water. If waste buried in a landfill comes into contact with water percolating down from the surface or with groundwater moving laterally through the refuse, leachate—noxious, mineralized liquid capable of transporting bacterial pollutants—is produced.[20] Landfills dating from the 1930s and 1940s have been observed to produce subsurface leachate trails (plumes) several hundred metres wide that have migrated kilometres from the disposal site. The nature and strength of the leachate produced at a disposal site depends on the composition of the waste, the amount of water that infiltrates or moves through the waste, and the length of time that infiltrated water is in contact with the refuse.[17]

Site Selection

The siting of a sanitary landfill is very important. A number of factors must be taken into consideration, including topography, location of the groundwater table, amount of precipitation, type of soil and rock, and location of the disposal zone in the surface water and groundwater flow system. A favourable combination of climatic, hydrologic, and geologic conditions helps to ensure reasonable safety in containing the waste and its leachate.[21]

The best sites are in arid regions, where disposal conditions are relatively safe because little leachate is produced. In a humid environment, some leachate is always produced; therefore, an acceptable level of leachate production must be established to determine the most favourable sites in such environments. What is acceptable varies with local water use, regulations, and the ability of the natural hydrologic system to disperse, dilute, and otherwise degrade the leachate to harmless levels.

Elements of the most desirable site in a humid climate with moderate to abundant precipitation are shown in Figure 24.5. The waste is buried above the water table in relatively impermeable clay and silt soils, material through which water cannot easily move. Any leachate produced remains in the vicinity of the site and degrades by natural filtering action and chemical reactions between clay and leachate.[22, 23]

The siting of waste disposal facilities also involves important social considerations. Often, planners choose sites where they expect local resistance to be minimal or where they perceive land to have little value. Waste disposal facilities are frequently located in areas where residents tend to have low socioeconomic status or belong to a particular racial or ethnic group. The study of social issues in siting waste facilities, chemical plants,

Figure 24.5 • **The most desirable landfill site is a humid environment.** Waste is buried above the water table in a relatively impermeable environment. [*Source*: W. J. Schneider, Hydraulic Implications of Solid-Waste Disposal, U.S. Geological Survey Circular 601F, 1970.]

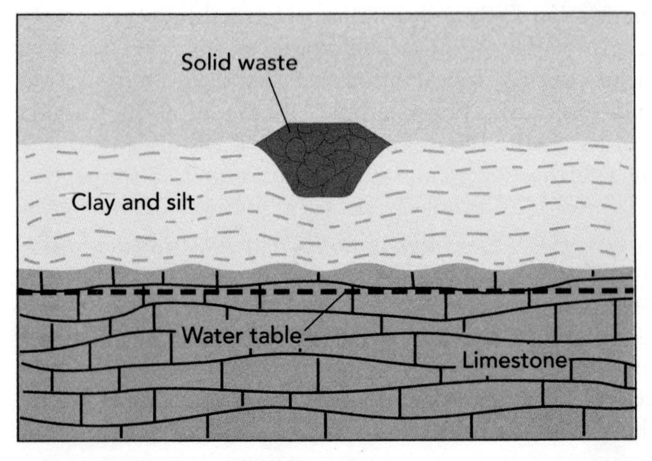

A CLOSER LOOK 24.2

Environmental Justice: Demographics of Hazardous Waste

An emerging field in the social sciences known as environmental justice focuses on investigating social issues related to the siting of facilities that many people find objectionable on environmental grounds. These facilities include sanitary landfills and other waste facilities, as well as chemical plants; and people's concerns include risk to nearby residents in case of accidental spills, fires, explosions, or illegal discharge of waste or chemicals. In particular, environmental justice addresses the potential for inequitable impacts from the environmental hazards related to the waste facility and chemical industries.[24]

Many Canadian First Nations have been affected by improper waste disposal activities. One example is the mercury-contaminated fish in the English-Wabigoon River system (Chapter 15). In that case, the source of the problem was a pulp and paper mill in Dryden, Ontario, but the First Nations communities who consumed the fish bore the health and economic costs of the pollution. Health Canada reports that 22% of people in aboriginal communities have methyl mercury levels above the level considered safe.

These patterns are evident in other low-income areas. A recent analysis reports that the Afro-Canadian community of Africville, now a part of Halifax, Nova Scotia, was at one time surrounded by a variety of heavy industries, including, "three systems of railway tracks; an open city dump; disposal pits for Halifax toxic waste; a hospital for infectious diseases; a stone and coal crushing plant; a toxic waste dump; a bone-meal plant; a cotton factory; a rolling mill/nail factory; a slaughterhouse; sewage disposal units; a prison; and a port facility for handling coal."[25] As this example shows, industrial land use in these areas is often much higher than the county average, while property values are well below the average. Evidence from this and other studies suggests that race is a more important factor than income in locations of waste facilities. These findings suggest that people fighting for environmental justice have real cause for concern.

and other such facilities is an emerging field known as environmental justice (See A Closer Look 24.2).[24]

Monitoring Pollution in Sanitary Landfills

Once a site is chosen for a sanitary landfill and before filling starts, monitoring the movement of groundwater should begin. Monitoring is accomplished by periodically taking samples of water and gas from specially designed monitoring wells. Monitoring the movement of leachate and gases should continue as long as there is any possibility of pollution. This procedure is particularly important after the site is completely filled and a final, permanent cover material is in place. Continued monitoring is necessary because a certain amount of settlement always occurs after a landfill is completed; and if small depressions form, surface water may collect, infiltrate, and produce leachate. Monitoring and proper maintenance of an abandoned landfill reduces its pollution potential.[19]

How Pollutants Can Enter the Environment from Sanitary Landfills

Pollutants from a solid-waste disposal site can enter the environment by as many as eight paths (Figure 24.6):[26]

1. Methane, ammonia, hydrogen sulphide, carbon dioxide, and nitrogen gases can be produced from compounds in the soil and the waste and can enter the atmosphere.
2. Heavy metals, such as lead, chromium, and iron, can be retained in the soil.
3. Soluble materials, such as chloride, nitrate, and sulphate can readily pass through the waste and soil to the groundwater system.
4. Overland runoff can pick up leachate and transport it into streams and rivers.
5. Some plants (including crops) growing in the disposal area can selectively take up heavy metals and other toxic materials. These materials are passed up the food chain as people and animals eat the plants.
6. If plant residue from crops left in fields contains toxic substances, these substances return to the soil.
7. Streams and rivers may become contaminated by waste from groundwater seeping into the channel (3) or by surface runoff (4).
8. Toxic materials can be transported to other areas by the wind, either attached to dust particles or as gases.

Modern sanitary landfills are engineered to include multiple barriers: clay, plastic and geotextile liners to limit the movement of leachate; surface and subsurface drainage to collect leachate; systems to collect methane gas produced as waste decomposes (the Vancouver Landfill installed such a system in 2003); and groundwater monitoring to detect leaks of leachate below and adjacent to the landfill. A thorough monitoring program considers all eight possible paths by which pollutants enter the environment. In practice, however, seldom does monitoring include all

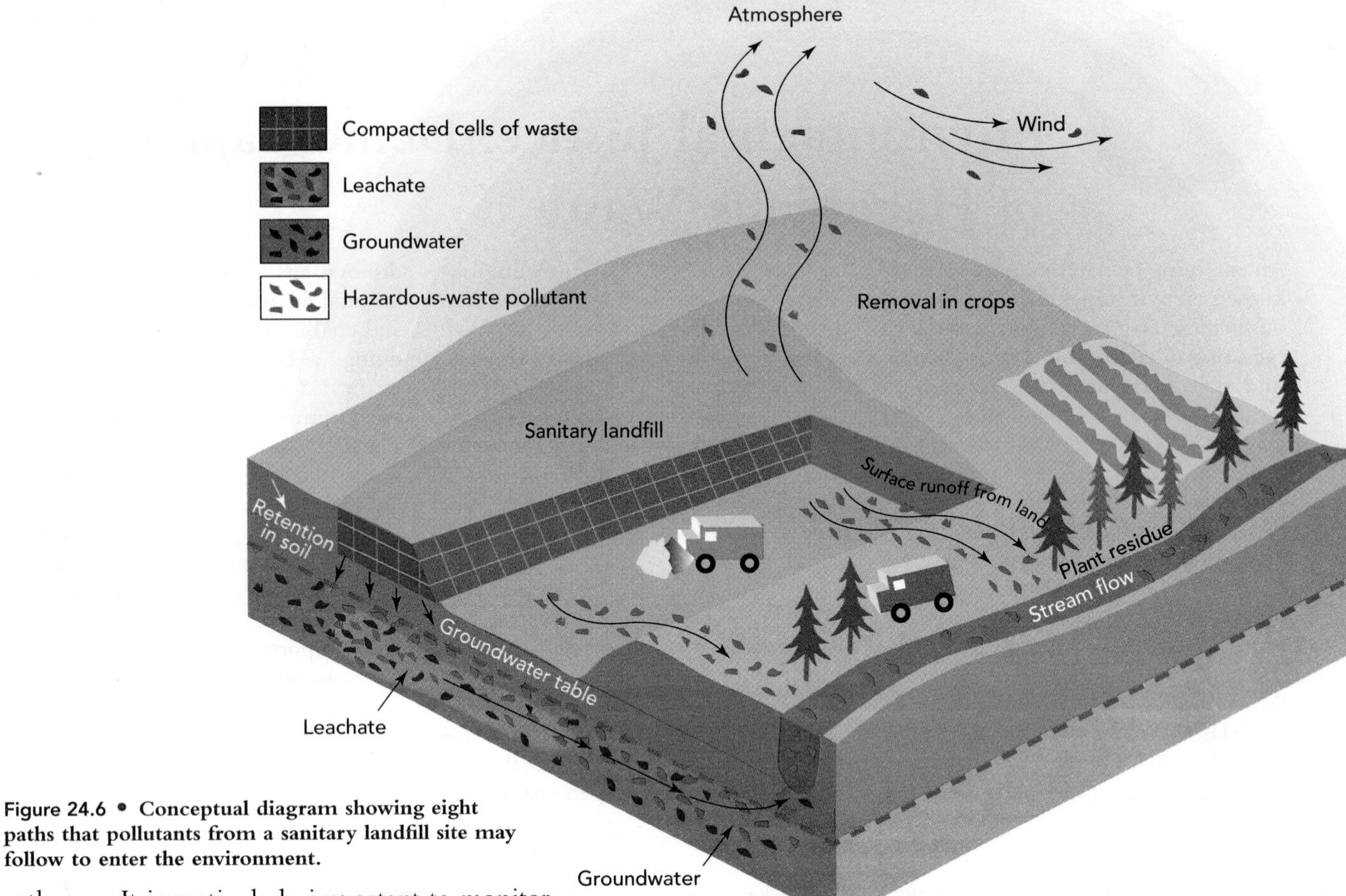

Figure 24.6 • Conceptual diagram showing eight paths that pollutants from a sanitary landfill site may follow to enter the environment.

pathways. It is particularly important to monitor the zone above the water table to identify potential pollution problems before they reach and contaminate groundwater resources, where correction would be very expensive. Figure 24.7 shows a conceptual diagram of a landfill that utilizes the multiple-barrier approach and a photograph of a landfill site under construction.

In Canada, waste management is a provincial responsibility but is often delegated to local municipalities. Typical waste management regulations require that:

- Landfills may not be sited on floodplains, wetlands, earthquake zones, unstable land, or near airports (birds at sites are a hazard to aircraft).
- Landfills must have liners and multiple barriers between stored wastes and the natural environment.
- Landfills must have a leachate collection system.
- Landfill operators must monitor groundwater for many specified toxic chemicals.
- Landfill operators must meet financial assurance criteria to ensure that monitoring continues for 30 years after closure of the landfill.

Two other points deserve mention. First, in addition to the costs of construction, operation, maintenance, and monitoring, landfills have a hidden cost. They are perceived to render land unusable for other purposes, and so in a sense waste the potential value of the land on which they are situated. The second point is that our society tends to think of landfills as the final disposal sites for waste. In fact, in the future landfills may become an important source of raw materials, to be "mined" using

Table 24.1 • Products and the Potentially Hazardous Waste They Generate

Products We Use	*Potentially Hazardous Waste*
Leather	Heavy metals, organic solvents
Medicines	Organic solvents and residues, heavy metals (e.g., mercury and zinc)
Metals	Heavy metals, fluorides, cyanides, acid and alkaline cleaners, solvents, pigments
Oil, gasoline, and other petroleum products	Oil, phenols and other organic compounds, heavy metals, ammonia salts, acids
Paints	Heavy metals, pigments, solvents, organic residues
Pesticides	Organic chlorine compounds, organic phosphate compounds
Plastics	Organic chlorine compounds
Textiles	Heavy metals, dyes, organic chlorine compounds, solvents

Source: Environment Canada; U.S. Environmental Protection Agency.

technology that has yet to be developed. As we site and design future landfills, we should therefore be thinking in terms of long-term storage, rather than secure disposal.

24.6 Hazardous Waste

So far in this chapter, we have discussed integrated waste management and materials management for the everyday waste stream from homes and businesses. We now consider the important topic of hazardous waste.

Creation of new chemical compounds has proliferated in recent years. In Canada and the United States, approximately 1,000 new chemicals are marketed each year, and about 70,000 chemicals are currently on the market. Although many of these chemicals have been beneficial to people, approximately 35,000 chemicals currently in use are classified as definitely or potentially hazardous to the health of people or ecosystems (see Table 24.1).

Canada currently produces about 6 million tonnes of hazardous chemical waste per year, referred to more commonly as hazardous waste. About 68% of this is generated in Ontario; Québec accounts for another 15%. About half of the total by weight is generated by chemical products industries, with the electronics industry and petroleum and coal products industries each contributing about 10%.[27, 28] Overall, about 55% of hazardous waste is destined for recycling.

Another source of hazardous chemicals are buildings destroyed by fires or hurricanes. Chemicals such as paints, solvents, and pesticides that were stored in the buildings may be released into the environment when debris from damaged buildings is burned or buried. As a result, collection of potentially hazardous chemicals following natural disasters is an important goal in managing hazardous materials.

In the recent past, much of the total volume of hazardous waste produced in Canada was indiscriminately dumped.[28] Some hazardous waste was illegally dumped on public or private lands, a practice called "midnight dumping." Buried drums of hazardous waste from illegal dumping have been discovered at hundreds of sites by contractors constructing buildings and roads. Cleanup of the waste has been costly and has delayed projects (see A Closer Look 24.3).[27]

Canada does not have a national inventory of abandoned waste sites. In the United States, there are an estimated 32,000–50,000, often now abandoned, waste disposal sites where past dumping was totally unregulated. Of these, probably 1,200–2,000 contain sufficient hazardous waste to be a threat to public health and the environment. For this reason, many scientists believe management of hazardous chemical materials may be the most serious environmental problem in Canada or the United States.

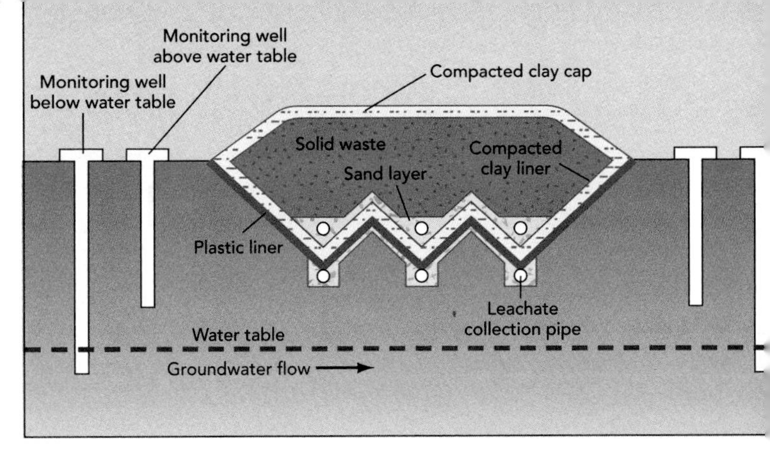

(a)

(b)

Figure 24.7 • *(a)* Conceptual diagram of a solid-waste facility (sanitary landfill) illustrating multiple-barrier design, monitoring system, and leachate collection system. *(b)* The Vancouver, British Columbia landfill is a modern sanitary landfill engineered with a compacted clay liner and methane gas collection and utilization system.

Uncontrolled dumping of chemical waste has polluted soil and groundwater in several ways:

- Chemical waste may be stored in barrels, either stacked on the ground or buried. The barrels eventually corrode and leak, polluting surface water, soil, and groundwater.
- When liquid chemical waste is dumped into an unlined lagoon, contaminated water may percolate through soil and rock to the groundwater table.
- Liquid chemical waste may be illegally dumped in deserted fields or even along roads.

Some sites pose particular dangers. The floodplain of a river, for example, is not an acceptable site for storing hazardous chemical waste. Yet that is exactly what occurred at a site on the floodplain of the River Severn near a village in one of the most scenic areas of England. Several fires at the site in 1999 were followed by a large fire of unknown

A CLOSER LOOK 24.3

Love Canal

The story of Love Canal has become a well-known hazardous-waste horror story. In 1976, in a residential area near Niagara Falls, New York, trees and gardens began to die (Figure 24.8). Rubber on running shoes and bicycle tires disintegrated. Puddles of toxic substances began to ooze through the soil. A swimming pool popped from its foundation and floated in a bath of chemicals.

The story of Love Canal started in 1892 when a canal 8 km long was excavated by William Love as part of the development of an industrial park. The development didn't need the canal when inexpensive electricity arrived, and it was never completed. The canal remained unused for decades and became a dump for wastes. From 1920 to 1952, approximately 20,000 metric tonnes of more than 80 chemicals were dumped in the canal. In 1953, the Hooker Chemical Company—which produced the insecticide DDT as well as a herbicide and chlorinated solvents and which had dumped chemicals into the canal—was pressured to donate the land to the city of Niagara Falls for $1. The city knew that chemical wastes were buried there, but no one expected any problems.[27] Eventually, several hundred homes and an elementary school were built on and near the site. For years, everything seemed fine. Then, in 1976–1977, heavy rains and snows triggered the events listed in the opening of this story, making Love Canal a household word.

A study of the site identified many substances suspected of being carcinogens, including benzene, dioxin, dichlorethylene, and chloroform. Although officials admitted that little was known about the impact of these chemicals, there was grave concern for people living in the area. Eventually, concern centred on alleged high rates of miscarriages, blood and liver abnormalities, birth defects, and chromosome damage. Although a study by New York health authorities suggested that no chemically caused health effects had been absolutely established,[29-31] the decision was made to clean up the site.

Cleanup of Love Canal demonstrates the technology for hazardous-waste treatment. The objective was to contain waste, stop migration of wastes through the groundwater flow system, and remove and treat contaminated soil and sediment in streambeds and sewers. To minimize further contamination, the area was covered with 1 m of compacted clay and a polyethylene plastic cover to reduce infiltration of surface water. Water is inhibited from entering the site by specially designed barriers. These procedures greatly reduce subsurface seepage of water, and water that does seep out is collected and treated.[29-32]

By 1990, $275 million had been spent for cleanup and relocation projects at Love Canal. Homes in an adjacent area were purchased, and about 200 homes and a school had to be destroyed by the government. The U.S. Environmental Protection Agency eventually declared the area clean, and about 280 remaining homes were marketed for resale.

As the result of a court-mediated settlement in 1994, the company responsible for the chemical waste paid the state of New York $98 million and took responsibility for $25 million for treatment operations.[33] In 1995, the company agreed to pay an additional $129 million to the federal government as reimbursement for its costs.[34]

The story of Love Canal raises many questions. What went wrong? How can we avoid such disasters in the future? How many other Love Canals are there across the country, each a potential time bomb waiting to explode?[29, 30]

Figure 24.8 • Aerial infrared photograph of the Love Canal area in New York state. Healthy vegetation is bright red. The canal, running from the upper left to the lower right, is a scar on the landscape where hazardous chemical waste rose to the surface, making Love Canal a household name for hazardous-waste problems.

origin on October 30, 2000. In that fire, approximately 200 tonnes of chemicals, including industrial solvents (xylene and toluene), cleaning solvents (methylene chloride), and various insecticides and pesticides, produced a fireball that rose into the night sky (Figure 24.9). The fire occurred during a rainstorm with wind gusts of hurricane strength. Toxic smoke and ash spread to nearby farmlands and villages, necessitating evacuation. People exposed to the smoke complained of a variety of symptoms, including headaches, stomach aches and vomiting, sore throats, coughs, and difficulty in breathing. Then, a few days later, on November 3, the site flooded. The floodwaters interfered with cleanup of the site after the fire and increased the risk of contamination of downstream areas by way of waterborne transport of hazardous chemical wastes. In one small village, contaminated floodwaters apparently inundated farm fields, gardens, and even homes.[35] Of course, the solution to this problem is to clean up the site and move waste storage facilities to a safer location. A similar fire in the town of St. Basile-le-Grand, outside Montréal, required decontamination of 25,000 square metres of surfaces, and 3 million litres of water contaminated with PCBs, dioxins and furans (PCB concentrations in rainwater were found to exceed 12,000 :g/L!). Three tonnes of PCBs were eventually recovered from the site in decontamination procedures.

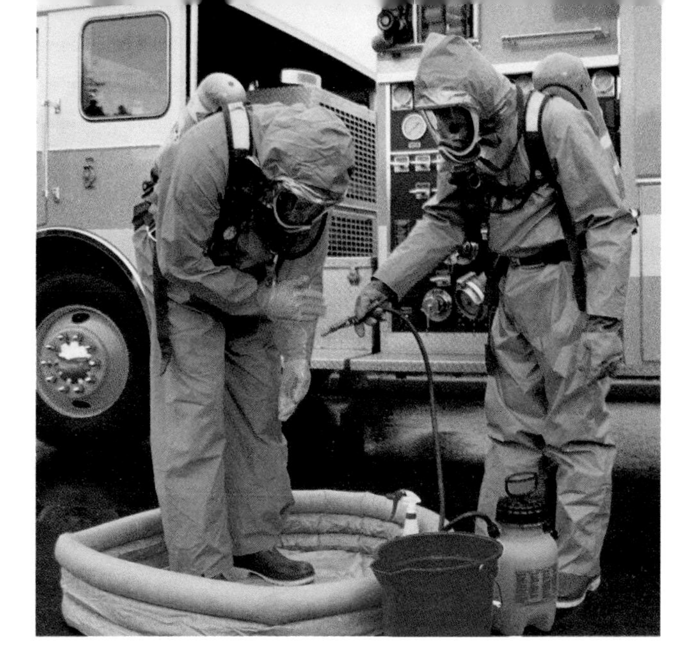

Figure 24.9 • **A hazardous materials cleanup crew washes down their protective suits.** Fires and spills can release quantities of hazardous materials into the air, water, and soil.

24.7 Hazardous-Waste Legislation

The recognition in the 1970s that hazardous waste was a danger to people and the environment and that the waste was not being properly managed led to important federal legislation in Canada. Although individual provinces have their own laws regulating hazardous waste, the Canadian federal government has enacted two laws that control movement of toxic materials inside and outside the country. Major changes to the federal regulatory regime were introduced as part of revisions to the *Canadian Environmental Protection Act* (CEPA) in 2000. The new changes affect the generators, treaters, and shippers of hazardous waste and other dangerous materials. These regulations replace an earlier statute, the *Transportation of Dangerous Goods Act* (1992), which was administered by Transport Canada.

Interprovincial Movement of Hazardous Waste Regulations (1999)

The goal of the interprovincial regulations under the *Canadian Environmental Protection Act* is to ensure that all hazardous materials are properly tracked and reported from "cradle to grave." At the heart of the regulations is identification of hazardous wastes and their life cycles. The idea was to issue guidelines and responsibilities to those who manufactured, transported, and disposed of hazardous waste. This is known as "cradle to grave" management. Regulations require stringent record keeping and reporting to verify that wastes do not present a public nuisance or a health problem.

The earlier *Transportation of Dangerous Goods Act* (TDGA) had established a system of hazardous waste classification, manifests, and associated tracking. The new regulations became necessary when the TDGA was revised in 2002, and its new regulations no longer required hazardous waste manifests or tracking.

Under the new CEPA regulations, a waste is hazardous if its concentration, volume, or infectious nature may contribute to serious disease or death or if it poses a significant hazard to people and the environment as a result of improper management (storage, transport, or disposal).[27] The act classifies hazardous wastes in several categories: materials highly toxic to people and other living things; wastes that may ignite when exposed to air; extremely corrosive wastes; and reactive unstable wastes that are explosive or generate toxic gases or fumes when mixed with water.

Export and Import of Hazardous Waste Regulations (1999)

The Export and Import of Hazardous Wastes Regulations, also under CEPA, establish a permit system for the export, import, or transit of hazardous wastes in Canada. All categories of hazardous waste are covered by the regulations, whether they are intended for disposal or recycling. People or companies who wish to import or export hazardous wastes must obtain a permit from the Transboundary Movement Branch of Environment Canada, showing the proposed route of the waste through Canada. Waste cannot be transported without a permit. The regulations also establish a waste manifest system parallel to that used for the interprovincial movement of hazardous wastes.

Canada does not have an equivalent to the United States' *Comprehensive Environmental Response, Compensation, and Liability Act* (CERCLA), sometimes called the Superfund Act. CERCLA defined policies and procedures for release of hazardous substances into the environment (for example, landfill regulations). It also mandated development of a list of the sites where hazardous substances were likely to or already had produced the most serious environmental problems and established a revolving fund (Superfund) to clean up the worst abandoned hazardous-waste sites. In 1984 and 1986, CERCLA was strengthened by amendments that made the following changes:

- Improved and tightened standards for disposal and cleanup of hazardous waste (e.g., requiring double liners, monitoring landfills).
- Banned land disposal of certain hazardous chemicals, including dioxins, polychlorinated biphenyls (PCBs), and most solvents.
- Initiated a timetable for phasing out disposal of all untreated liquid hazardous waste in landfills or surface impoundments.
- Increased the size of the Superfund. The fund was allocated about $8.5 billion in 1986. Congress approved another $5.1 billion for fiscal year 1998, which almost doubled the Superfund budget.[36]

Superfund has experienced significant management problems, and cleanup efforts are far behind schedule. Unfortunately, the funds available are not sufficient to pay for decontamination of all targeted sites. Furthermore, there is concern that present technology is not sufficient to treat all abandoned waste disposal sites; it may be necessary to simply try to confine waste to those sites until better disposal methods are developed. It seems apparent that abandoned disposal sites are likely to persist as problems for some time to come.

Environmental Liability

Concerns about the cost and health implications of cleaning up abandoned waste sites have changed the ways in which real estate business is conducted. For example, under some provincial legislation, property owners may be held liable for costly cleanup of hazardous waste present on their property even if they did not directly cause the problem. As a result, banks and other lending institutions might be held liable for release of hazardous materials by their tenants. Abandonment of contaminated property is a growing problem in Canada and the United States. The Insurance Bureau of Canada is considering establishing a separate class of insurance for environmental liability. At present, however, it is very difficult to obtain insurance against environmental liability, including that created by the ownership or cleanup of hazardous waste sites.

24.8 Hazardous-Waste Management: Land Disposal

Management of hazardous chemical waste involves several options, including recycling, on-site processing to recover by-products with commercial value, microbial breakdown, chemical stabilization, high-temperature decomposition, incineration, and disposal by secure landfill or deep-well injection. A number of technological advances have been made in toxic-waste management; and as land disposal becomes more expensive, a recent trend toward on-site treatment is likely to continue. However, on-site treatment will not eliminate all hazardous chemical waste; disposal of some waste will remain necessary.

Table 24.2 compares hazardous-waste reduction technologies for treatment and disposal. Notice that all available technologies cause some environmental disruption. There is no simple solution for all waste management issues. In this section, we consider land disposal of hazardous waste; in the following section we discuss alternative approaches.

Secure Landfill

A secure landfill for hazardous waste is designed to confine the waste to a particular location, control the leachate that drains from the waste, collect and treat the leachate, and detect possible leaks. This type of landfill is similar to the modern sanitary landfill; it is an extension of the sanitary landfill for urban waste. Because in recent years it has become apparent that urban waste contains much hazardous material, the design of sanitary landfills and that of secure landfills for hazardous waste have converged to some extent.

Design of a secure landfill is shown in Figure 24.10. A dike and liner (made of clay or other impervious material, such as plastic) confine waste, and a system of internal drains concentrates leachate in a collection basin, from which it is pumped and transported to a wastewater treatment plant. Modern facilities include multiple barriers consisting of several impermeable layers and filters. The function of impervious liners is to ensure that leachate does not contaminate soil and groundwater resources. However, this type of waste disposal procedure, like the sanitary landfill from which it evolved, must have several monitoring wells to alert personnel if leachates leak out of the system and threaten water resources.

It has recently been argued that there is no such thing as a really secure landfill, implying that they all leak to

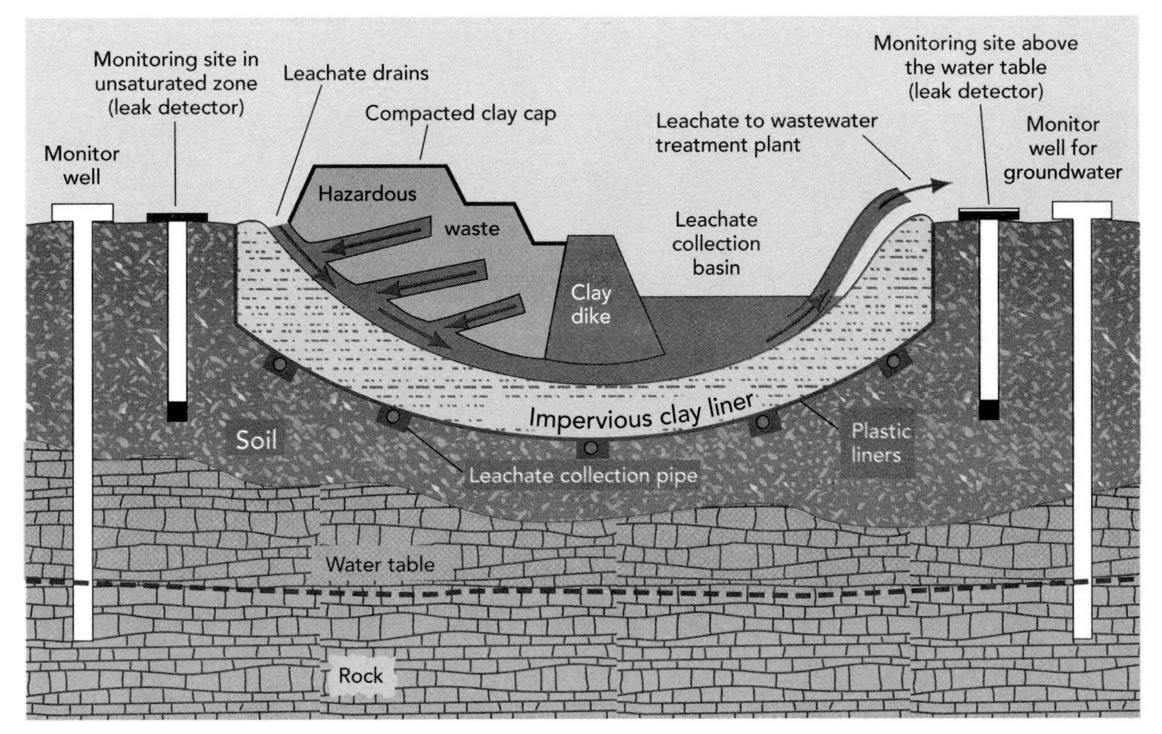

Figure 24.10 • A secure landfill for hazardous chemical waste. The impervious liners, systems of drains, and leak detectors are integral parts of the system to ensure that leachate does not escape from the disposal site. Monitoring in the unsaturated zone is important and involves periodic collection of soil water.

some extent. Some estimates suggest that even a well-designed and engineered landfill may only be secure for about 100 years. Impervious plastic liners, filters, and clay layers can fail, even with several backups; and drains can become clogged and cause overflow. Animals, such as gophers, ground squirrels, groundhogs, and muskrats, can chew through plastic liners, and some may burrow through clay liners, thus promoting or accelerating leaks. Yet careful siting and engineering can minimize problems. As with sanitary landfills, preferable sites are those with good natural barriers to migration of leachate: thick clay deposits, an arid climate, or a deep water table. Nevertheless, land disposal should be used only for specific chemicals compatible with and suitable for the method.

Land Application: Microbial Breakdown

Intentional application of waste materials to the surface soil is referred to as land application, land spreading, or land farming. We discussed land application of human waste earlier in the chapter. Land application of waste may be a desirable method of treatment for certain biodegradable industrial waste, such as oily petroleum waste and some organic chemical-plant wastes. A good indicator of the usefulness of land application for a particular waste is the biopersistence of the waste—how long the material remains in the biosphere—the greater the biopersistence, the less suitable the waste for land application procedures. Land application is not an effective treatment or disposal method for inorganic substances such as salts and heavy metals.[37]

Land application of biodegradable waste works because, when such materials are added to the soil, they are attacked by microflora (bacteria, moulds, yeasts, and other organisms) that decompose the waste material in a process known as microbial breakdown. The soil thus can be thought of as a microbial farm that constantly recycles organic and inorganic matter by breaking it down into more fundamental forms useful to other living things in the soil. Because the upper soil zone contains the largest microbial populations, land application is restricted to the uppermost 15–20 cm of the soil.[38] However, the effectiveness of biological systems for waste breakdown varies widely with climate and soil chemistry, and is still poorly understood in cold northern climates such as Canada's.

Surface Impoundment

Both natural topographic depressions and human-made excavations have been used to hold hazardous liquid waste in a method known as surface impoundment. The depressions or excavations are primarily formed of soil or other surface materials but may be lined with manufactured materials such as plastic. Examples include aeration pits and lagoons at hazardous-waste facilities.

Surface impoundments have been criticized because they are especially prone to seepage, resulting in pollution of soil and groundwater. Evaporation from surface impoundments can also produce an air pollution problem. This type of storage or disposal system for hazardous waste is controversial, and many sites have been closed.

Table 24.2 • Comparison of Hazard Reduction Technologies

	Disposal			*Treatment*		
Parameter Compared	*Landfills and Impoundments*	*Injection Wells*	*Incineration and Other Thermal Destruction*	*High-temperature Decomposition*[a]	*Chemical Stabilization*	*Microbial Breakdown*
Effectiveness: how well it contains or destroys hazardous characteristics	Low for volatiles, high for unsoluble solids	High, for waste compatible with the disposal environment	High	High for many chemicals	High for many metals	High for many metals and some organic waste such as oil
Reliability issues	Siting, construction, and operation. Uncertainties: long-term integrity and cover	Site history and geology, well depth, construction, and operation	Monitoring uncertainties with respect to high degree of DRE: surrogate measures, PICs, incinerability[b]	Mobile units; on-site treatment avoids hauling risks Operational simplicity	Some inorganics still soluble Uncertain leachate production	Monitoring uncertainties during construction and operation
Environment media most affected	Surface water and groundwater	Surface water and groundwater	Air	Air	Groundwater	Soil, groundwater
Least compatible wastes[c]	Highly toxic, persistent chemicals	Reactive; corrosive; highly toxic, mobile, and persistent	Highly toxic organics, high heavy-metal concentration	Some inorganics	Organics	Highly toxic persistent chemicals
Relative costs	Low to moderate	Low	Moderate to high	Moderate to high	Moderate	Moderate
Resource recovery potential	None	None	Energy and some acids	Energy and some metals	Possible building material	Some metals

a Molten salt, high-temperature fluid well, and plasma arc treatments.
b DRE = destruction and removal efficiency; PIC = product of incomplete combustion.
c Wastes for which this method may be less effective for reducing exposure, relative to other technologies. Wastes listed do not necessarily denote common usage.
Source: Modified after Council on Environmental Quality, 1983.

Deep-Well Disposal

Deep-well disposal, another controversial method of waste disposal, involves injecting waste into deep wells. A deep well must penetrate to a depth below and completely isolated from all freshwater aquifers, thereby assuring that waste will not contaminate or pollute existing or potential water supplies. Typically, the waste is injected into a permeable rock layer several thousand metres below the surface in geologic basins capped by relatively impervious, fracture-resistant rock such as shale or salt deposits.[3]

Deep-well injection of oil-field brine (salt water) has been important in the control of water pollution in oil fields for many years. Huge quantities of liquid waste (brine) pumped up with oil have been injected back into the rock.[39]

Deep-well disposal of industrial wastes should not be viewed as a quick and easy solution to industrial waste problems.[40] Even where geologic conditions are favourable for deep-well disposal, there are a limited number of suitable sites; and within these sites, there is limited space for disposal of waste. Finally, disposal wells must be carefully monitored by additional wells, known as monitoring wells, to determine if the waste is remaining in the disposal site.

Summary of Land Disposal Methods

Direct land disposal of hazardous waste is often not the best initial alternative. There is consensus that even with extensive safeguards and state-of-the-art designs, land disposal alternatives cannot guarantee that the waste is contained and will not cause environmental disruption in the future. This concern holds true for all land disposal facilities, including landfills, surface impoundments, land application, and injection wells. Pollution of air, land, surface water, and groundwater may result from failure of a land disposal site to contain hazardous waste. Pollution of groundwater is perhaps the most significant risk, because groundwater provides a convenient route for pollutants to reach humans and other living things. Figure 24.11 shows some of the paths that pollutants may take from land disposal sites to contaminate the environment. These paths include leakage and runoff to surface water or groundwater from improperly designed or maintained landfills; seepage, runoff, or air emissions from unlined lagoons; percolation and seepage from failure of surface land application of waste to soils; leaks in pipes or other equipment associated with deep-well injection; and leaks from buried drums, tanks, or other containers.

24.9 Alternatives to Land Disposal of Hazardous Waste

Our methods for handling hazardous chemical waste should be multifaceted. In addition to the disposal methods just discussed, chemical waste management should include such processes as source reduction, recycling and resource recovery, treatment, and incineration.[41] Recently, it has been argued that these alternatives to land disposal are not being utilized to their full potential; that is, the volume of waste could be reduced and the remaining waste could be recycled or treated in some form prior to land disposal of the residues of the treatment processes.[41] Advantages to source reduction, recycling, treatment, and incineration include the following:

- Useful chemicals can be reclaimed and reused.
- Treatment of wastes may make them less toxic and therefore less likely to cause problems in landfills.
- The actual waste that must eventually be disposed of is reduced to a much smaller volume.
- Because a reduced volume of waste is finally disposed of, there is less stress on the dwindling capacity of waste disposal sites.

Although some of the following techniques have been discussed as part of IWM, the techniques have special implications and complications where hazardous wastes are concerned.

Green Chemistry

Green chemistry is an emerging environmental philosophy based on the principles that chemical processes can be designed to be benign, energy efficient, renewable (through reuse of feedstocks), and safely degraded. The Canadian Green Chemistry Network is a non-profit foundation whose mandate is to promote research and increase public and industrial awareness of green chemistry.

Source Reduction

The object of source reduction in hazardous-waste management is to reduce the amount of hazardous waste generated by manufacturing or other processes. For example, changes in the chemical processes involved, equipment used, raw materials used, or maintenance measures may successfully reduce the amount or toxicity of the hazardous waste produced.[41]

Recycling and Resource Recovery

Hazardous chemical waste may contain materials that can be recovered for future use. For example, acids and solvents collect contaminants when they are used in manufacturing processes. These acids and solvents can be processed

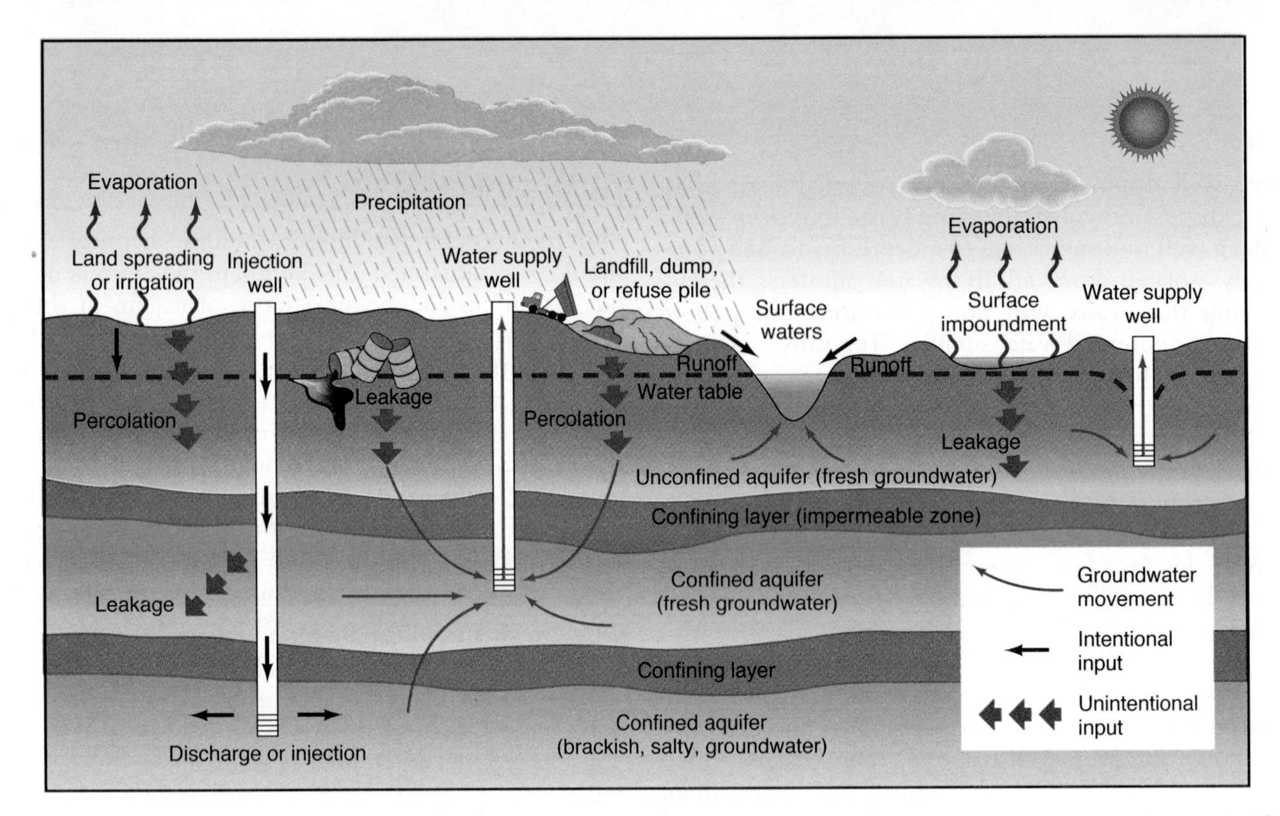

Figure 24.11 • Examples of how land disposal and treatment of hazardous waste may contaminate the environment. [*Source*: Modified from C. B. Cox, The Buried Threat, California Senate Office of Research, No. 115-5, 1985.]

to remove the contaminants and can then be reused in the same or other different manufacturing processes.[41]

Treatment

Hazardous chemical waste can be treated by a variety of processes to change the physical or chemical composition of the waste and so to reduce its toxic or hazardous characteristics. For example, acids can be neutralized, heavy metals can be separated from liquid waste, and hazardous chemical compounds can be broken up through oxidation.[41]

Incineration

Hazardous chemical waste can be successfully destroyed by high-temperature incineration. Incineration is considered a waste treatment rather than a disposal method because the process produces an ash residue, which must then be disposed of in a landfill operation. Hazardous waste has also been incinerated offshore on ships, creating potential air pollution and ash disposal problems for the marine environment.

The technology used in incineration and other high-temperature decomposition or destruction is changing rapidly. Figure 24.12 shows a diagram of one type of high-temperature incineration system that can be used to burn toxic waste. Waste (as liquid, solid, or sludge) enters the rotating combustion chamber, where it is rolled and burned. Ash from this burning process is collected in a water tank; the remaining gaseous materials move into a secondary combustion chamber, where the process is repeated. Finally, the remaining gases and particulates move through a scrubber system that eliminates surviving particulates and acid-forming components. Carbon dioxide, water, and air are then emitted from the stack. As shown in Figure 24.12, ash particulates and wastewater are produced at various stages of the incineration process; and these must be treated or disposed of in a landfill.

More advanced techniques for the incineration and thermal decomposition of waste are being developed. One of these utilizes a molten salt bed that should be useful in destroying certain organic materials. Other incineration techniques include liquid-injection incineration on land or sea and multiple-hearth furnaces. Which incineration method is used for a particular waste depends on the nature and composition of the waste and the temperature necessary to destroy the hazardous components. For example, the generalized incineration system shown in Figure 24.12 could be used to destroy PCBs.

In contrast to many European nations, where incineration is the preferred waste disposal option, Canada incinerates less than 10% of its municipal waste. Incineration of hospital wastes is common, however, and (because of the high proportion of plastic in hospital wastes) potentially a significant source of chlorinated compounds in air.

24.10 Disposal in Rivers, Lakes, and Oceans

Waste management issues are not restricted to the land. Some waste ends up in the freshwater systems and the world's oceans.

Water covers more than 70% of Earth's surface. Rivers, lakes, and oceans play a part in maintaining our global environment and are of major importance in the cycling of carbon dioxide, which helps regulate the global climate. They are also important in cycling many chemical elements important to life, such as nitrogen and phosphorus, and are a valuable resource to people because they provide necessities such as foods and minerals.It seems reasonable that such an important resource would receive preferential treatment, and yet rivers, lakes, and oceans have long been dumping grounds for waste. Part VI of the *Canadian Environmental Protection Act* (The Ocean Dumping Control Act) regulates the transfer, disposal and incineration of most materials at sea. This is Canada's response to the 1972 Convention on the Prevention of Marine Pollution by Dumping of Wastes and Other Matters, an international treaty signed by 80 countries. Ocean dumping is not a large problem in Canada. The Canadian Institute for Environmental Law and Policy estimates that 90% of all material dumped at sea is ocean sediment, much of it contaminated by land-based activities. CEPA establishes a permit system for ocean dumping (see the example of ocean disposal of the ship Chaudière in Chapter 10). The system has however been criticized for failing to recover the full costs of monitoring, enforcement and cleanup from dumpers.

The law bans ocean dumping of specified radiological, chemical, and biological warfare agents, various types of industrial waste, and high-level radioactive waste.

The types of wastes dumped in the oceans include the following:[42]

- Dredge spoils—solid materials, such as sand, silt, clay, rocks, and pollutants, deposited in the ocean from industrial and municipal discharges and removed from water bodies, generally to improve navigation.
- Industrial waste—acids, refinery wastes, paper mill wastes, pesticide wastes, and assorted liquid wastes.
- Sewage sludge—solid material (sludge) that remains after municipal wastewater treatment.
- Construction and demolition debris—cinder block, plaster, dirt, stone, and tile.
- Solid waste—plastic, refuse, garbage, explosives, radioactive waste, and untreated urban sewage.

Dredge spoils, construction debris, and similar materials are also deposited in fresh water systems, and indeed are a continuing concern in major systems like the Great Lakes. Disposal of wastes in rivers, lakes, and oceans contributes to the larger problem of freshwater and marine pollution. Locations in the oceans of the world that are accumulating pollution continuously, that have intermittent pollution problems, or that have potential for pollution from ships in the major shipping lanes are shown on Figure 24.13. Notice that the areas

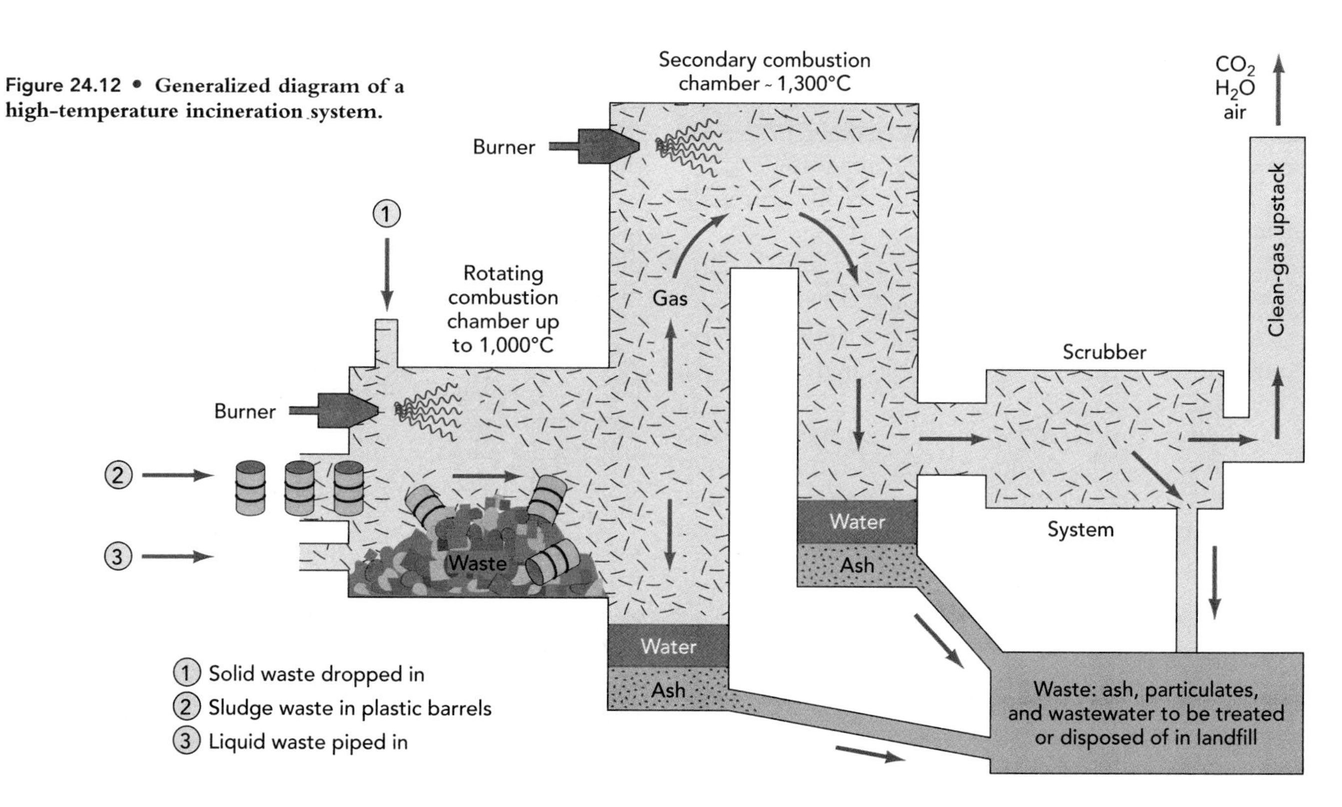

Figure 24.12 • Generalized diagram of a high-temperature incineration system.

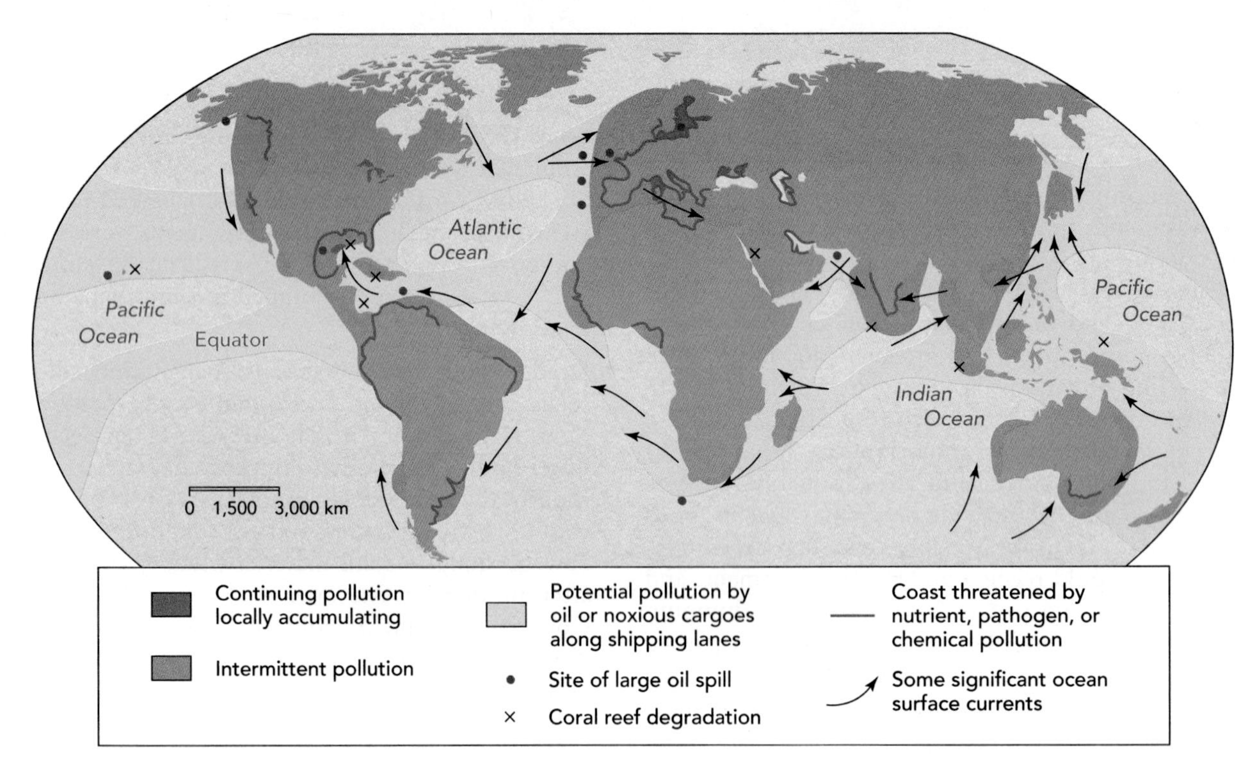

Figure 24.13 • Ocean pollution of the world. Notice that the areas of continuing and locally accumulating pollution, as well as the areas with intermittent pollution, are in nearshore environments. [*Source*: Modified from the Council on Environmental Quality, Environmental Trends, 1981, with additional data from A. P. McGinn, "Safeguarding the Health of the Oceans," WorldWatch Paper 145 (Washington, D.C.: WorldWatch Institute, 1999), pp. 22–23.]

with continual or intermittent pollution are located near shore. Unfortunately, these are also areas of high productivity with valuable fisheries. Shellfish have been found to contain organisms that produce diseases, such as polio and hepatitis. In Canada and the United States, at least 20% of the nation's commercial shellfish beds have been closed (mostly temporarily) because of pollution. Lifeless zones in the marine environment have been created. Heavy kills of fish and other organisms have occurred, and profound changes in marine ecosystems have taken place (see Chapter 19).[42, 43]

The effects of freshwater and marine dumping include:

- Death or retarded growth, vitality, and reproductivity of freshwater and marine organisms.
- Reduction in the dissolved oxygen content necessary for life because of increased biochemical oxygen demand.
- Eutrophication caused by nutrient-rich waste in rivers and lakes, and the shallow waters of estuaries, bays, and parts of the continental shelf, resulting in depletion of oxygen and subsequent killing of algae, which may wash up and pollute coastal areas.
- Habitat change caused by waste disposal practices that subtly or drastically change entire freshwater or marine ecosystems.[42]

The European marine waters are in particular trouble, in part as a result of urban and agricultural pollutants that have raised concentrations of nutrients in seawater. Blooms (heavy, sudden growth) of toxic algae are becoming more common. In 1988, in the waterway connecting the North Sea to the Baltic Sea, a bloom was responsible for killing nearly all marine life to a depth of about 15 m. It is believed that urban waste and agricultural runoff contributed to the toxic bloom.

Although the oceans and major lakes such as the Great Lakes are vast, they are basically giant sinks for materials from continents; and parts of their environment are extremely fragile.[43] One area of concern is the microlayer, the upper 3 mm of a water body. The base of freshwater and marine food chains consists of planktonic life abundant in the microlayer. The young of certain fish and shellfish also reside in the upper few millimetres of water in the early stages of their life. Unfortunately, the upper few millimetres of lakes and oceans also tend to concentrate pollutants, such as toxic chemicals and heavy metals. One study reported that the concentrations of heavy metals—including zinc, lead, and copper—in the microlayer are from 10 to 1,000 times higher than in the deeper waters. There is fear that disproportionate pollution of the microlayer will have especially serious effects on marine organisms.[43] There is also concern that some ecosystems in the oceans, such as coral reefs, estuaries and salt marshes, and mangrove swamps, are threatened by ocean pollution.

Marine pollution can have major impacts on people and society. Contaminated marine organisms may transmit toxic elements or diseases to people who eat them. When solid waste, oil, and other materials pollute beaches and harbours there may be damage to marine life as well as a loss of visual appeal and other amenities. Economic loss is considerable. Loss of shellfish from pollution in Canada and the United States, for example, amounts to many millions of dollars per year. In addition, a great deal of money is spent cleaning up solid waste, liquid waste, and other pollutants in coastal areas.[41]

24.11 Pollution Prevention

As we saw at the beginning of the chapter, approaches to waste management are changing. During the first several decades of environmental concern and management (the 1970s and 1980s), most industrialized countries approached waste management through government regulations and control of waste. Control was accomplished through chemical, physical, or biological treatment and collection (for eventual disposal), transformation, or destruction of pollutants after they had been generated. This was thought to be the most cost-effective approach to waste management.

With the 1990s came a growing emphasis on pollution prevention, which involves identifying ways to prevent the generation of waste rather than finding ways to dispose of it. This approach, which is part of materials management, reduces the need for management of waste, because less waste is produced. This area has prompted considerable research in recent years. For example, Dalhousie University's Eco-Efficiency Centre works in co-operation with the Nova Scotia of Department of Environment and Labour to produce a pollution prevention guide for Nova Scotia industries.

Approaches to pollution prevention include the following:[44]

- Purchasing the proper amount of raw materials so that no excess remains to be disposed of.
- Exercising better control of materials used in manufacturing processes so that less waste is produced.
- Substituting non-toxic chemicals for hazardous or toxic materials currently in use.
- Improving engineering and design of manufacturing processes so that less waste is produced.

These approaches are often called P-2 approaches, for "pollution prevention." Probably the best way to illustrate the P-2 process is through case histories.[44]

The Canadian Centre for Pollution Prevention works with industries, municipalities, and governments to develop preventive approaches. One of their flagship initiatives is the Green Dry Cleaners Program. Dry cleaners are a particular challenge for Canada and the United States, because there are so many of them and because cost-effective alternatives to toxic chemicals have been hard to find. There are an estimated 2,500 dry cleaners in Ontario alone, and 90% of them use a toxic chemical called perchloroethylene ("perc"), a designated toxic substance under CEPA. In the mid-1990s, dry cleaners used about half the perchloroethylene in Canada.

In 1995, the Canadian Centre for Pollution Prevention signed a Memorandum of Understanding with a number of dry cleaners' associations. The principal goal of the agreement has been to eliminate the use of perchlorethylene in the dry cleaning process. Other activities focus on improvements in operating practices, modernizing equipment, and collecting and reusing plastic bags and hangers. Today, a large proportion of Canadian dry cleaners employ "green dry cleaning" and have reduced the costs and hazards of using a toxic chemical.

As another example, a Wisconsin firm that produced cheese was faced with disposal of about 8,000 litres a day of a salty solution generated as part of the cheese-making process. Initially, the firm spread the salty solution on nearby agricultural lands—common practice for firms that could not discharge wastewater into publicly owned treatment plants. This method of waste disposal, if the solution was applied incorrectly, caused the level of salts in the soil to rise so much that crops were damaged. As a result, the Department of Natural Resources in Wisconsin placed limitations on the discharge of salt to the land.

The cheese-manufacturing firm decided to modify its cheese-making processes to recover salt from the solution and reuse it in production. This involved developing a recovery process using an evaporator. The recovery process reduced the salty waste by about 75% and at the same time reduced the amount of the salt the company needed to purchase by 50%. The operating and maintenance costs for recovery were approximately three cents per pound of salt recovered, and the time necessary to reclaim the extra cost of the new equipment was only two months. The firm saved thousands of dollars a year by purchasing less salt.

As a final example, a large international chemical company, 3M, initiated a pollution prevention program and was able to stop the release of a billion pounds of toxic chemicals and save the company $500 million in the process. These case histories suggest that often rather minor changes can result in large reductions of waste produced. Thousands of similar cases exist today as we move from the era of recognizing environmental problems and regulating them at a national level to providing economic incentives and new technology to better manage materials.[44]

CRITICAL THINKING ISSUE

Can We Make Recycling a Financially Viable Industry?

There is tremendous public support for recycling in Canada today. Many people understand that management of our waste has many advantages to society as a whole and the environment in particular. People like the notion of recycling because they correctly assume they are helping conserve resources such as forests that make up much of the planet's nonurban environment. Large cities across North America have initiated recycling programs, but there is continued concern that recycling is not yet "cost-effective." However, there are success stories, such as the integrated wet-dry waste system in Guelph, Ontario. Other cities have implemented a "pay-as-you-throw-away" approach; businesses and individuals are charged for disposal of garbage but not for materials that are recycled. Waste materials from Canadian and U.S. cities are shipped as far away as China and the Philippines to be recycled into usable products; organic waste is sent to agricultural areas; metals such as aluminum are sent to other locations in Canada, where they are recycled. New York, on the other hand, recycles about 20% of its waste and in July of 2002 suspended all recycling of glass and plastic. By doing so, the city expected to save more than $40 million a year. This saving occurs because recycling metal, glass, and plastic costs approximately $100 more per tonne than sending it to out-of-state landfills for disposal. The people of New York and their city officials are not against recycling, but the economics, to them in 2002, simply did not add up when the city was cash-poor. Having to decide to cut social programs or to cut recycling, they decided that recycling of plastic and glass was simply not cost-effective. In an about turn, New York then reinstated recycling of glass and plastic in 2003–2004.

To understand some of the issues concerning recycling and its cost, consider the following points:

- The average cost of disposal at a landfill is about $50/metric tonne in Canada, and even at a higher price of about $80/tonne, it may exceed the cost of recycling.
- Landfill fees in Europe range from $200–$300/tonne.
- Europe has been more successful in recycling, in part because countries such as Germany make manufacturers responsible for the disposal cost of packaging and industrial goods they produce.
- In Canada and the United States, packaging accounts for approximately one-third of the entire waste produced by manufacturing.
- The cost to cities such as Toronto, which must export their waste out of the province, is steadily rising and is expected to exceed the cost of recycling within about 10 years.
- Placing a 10-cent refundable deposit on all beverage containers except milk would greatly increase the number of such containers recycled. For example, those states with a deposit system have an average recycling rate of about 70–95% of bottles and cans, whereas those states that do not have a refundable deposit system average less than 30%.
- When people have to pay for the trash that has to be disposed of at a landfill, but are not charged for materials—such as paper, plastic, glass, and metals—that are recycled, the success of recycling is greatly enhanced.
- Companies that make beverages are not particularly excited about a proposal that would require a refundable deposit for containers. They claim that the additional costs would be several billion dollars but do agree that recovery rates would be higher and that this would help provide a more steady supply of recycled metal, such as aluminum, as well as plastic.
- Education is a big issue with recycling. Many people still don't know which items are in fact recyclable and which are not. Furthermore, they don't understand that mixing of recyclable and non-recyclable items in their waste results in much higher cost of separation at recycling centers.
- Global markets for recyclable materials such as paper and metals have potential for expansion, particularly for large urban areas on the seacoast where the shipping of materials is economically viable. Recycling in Canada and the United States today is a $16 billion industry, and if it is done right, it generates new jobs and revenue for participating communities.

Critical Thinking Questions

1. What can be done to assist recycling industries to become more cost-effective?
2. What are some of the indirect benefits to society and the environment from recycling?
3. Defend or criticize the contention that if we really want to do something to improve the environment through reduction of our waste, we have to move beyond evaluating benefits of recycling based simply on the fact that it may cost more than dumping waste at a landfill.
4. What are the recycling efforts in your community and university or college, and how could improvements be made?

Summary

- The history of waste disposal practices since the Industrial Revolution has progressed from the practice of dilution and dispersion to the concept of integrated waste management (IWM), which emphasizes the three R's of reducing waste, reusing materials, and recycling.
- The emerging concept of industrial ecology has as a goal a system in which the concept of waste doesn't exist, because waste from one part of the system would be a resource for another part.
- The most common method for disposal of solid waste is the sanitary landfill. However, around many large cities, space for landfills is hard to find; and few people wish to live near any waste disposal operation.
- Physical and hydrologic conditions of a site greatly affect its suitability as a landfill. These include landform, topography, rock and soil type, depth to groundwater, and amount of precipitation.
- Hazardous chemical waste is one of the most serious environmental problems in Canada and the United States. Hundreds or even thousands of uncontrolled disposal sites could be time bombs that will eventually cause serious public health problems. We know that we will continue to produce some hazardous chemical waste. Therefore, it is imperative to develop and use safe disposal methods.
- Management of hazardous chemical wastes involves several options, including on-site processing to recover by-products with commercial value, microbial breakdown, chemical stabilization, incineration, and disposal by secure landfill or deep-well injection.
- Dumping of wastes in rivers, lakes and oceans is a significant source of water pollution. The most seriously affected areas are near shore, where valuable fisheries often exist.
- Pollution prevention (P-2)—identifying and using ways to prevent the generation of waste—is an important emerging area of materials management.

STUDY QUESTIONS

1. Have you ever contributed to the hazardous-waste problem through disposal methods used in your home, school laboratory, or other location? How big a problem do you think such events are? For example, how bad is it to dump paint thinner down a drain?
2. Why is it so difficult to ensure safe land disposal of hazardous waste?
3. Would you approve the siting of a waste disposal facility in your part of town? If not, why not, and where do you think such facilities should be sited?
4. Why might there be a trend toward on-site disposal rather than land disposal of hazardous waste? Consider physical, biological, social, legal, and economic aspects of the question.
5. Is government doing enough to clean up abandoned hazardous waste dumps? Do private citizens have a role in choosing where cleanup funds should be allocated?
6. Do you think we should collect household waste and burn it in special incinerators to make electrical energy? What problems and what advantages do you see for this method compared with other disposal options?
7. Many jobs will be available in the next few years in the field of hazardous-waste monitoring and disposal. Would you take such a job? If not, why not? If so, do you feel secure that your health would not be jeopardized?
8. Should companies that dumped hazardous waste years ago when the problem was not understood or recognized be held liable today for health-related problems to which their dumping may have contributed?
9. Suppose you found that the home you had been living in for 15 years was located over a buried waste disposal site. What would you do? What kinds of studies could be done to evaluate the potential problem?
10. What are you personally doing in terms of the Three R's? List the activities that are currently part of your lifestyle. What more could you be doing? What would persuade you to make those changes?

FURTHER READING

Allenby, B. R. *Industrial Ecology: Policy Framework and Implementation.* Upper Saddle River, N.J.: Prentice Hall, 1999. Primer on industrial ecology.

Ashley, S. "It's Not Easy Being Green." *Scientific American* (April 2002), pp. 32–34. A look at the economics of developing biodegradable products and a little of the chemistry involved.

Canadian Council of Ministers of the Environment. Guidance document on the management of contaminated sites in Canada. Winnipeg: Canadian Council of Ministers of the Environment, 1997.

Great Britain Department of the Environment. *Landfill Design, Construction and Operational Practice.* London: Her Majesty's Stationery Office, 1995.

Rhyner, C. R., L. J. Schwartz, R. B. Wenger, and M. G. Kohrell. *Waste Management and Resource Recovery.* Boca Raton, Fla.: CRC, Lewis, 1995. Discussions of the archeology of waste, waste generation, source reduction and recycling, wastewater treatment, incineration and energy recovery, hazardous waste, and costs of waste systems and facilities.

Watts, R. J. *Hazardous Wastes.* New York: John Wiley, 1998. A to Z of hazardous wastes.

Part IV
Managing Sustainably

Panoramic view of London, England.

Photo courtesy Will Gorlitz

chapter 25

The Economics of Softwood Lumber

What price should a Canadian company pay for the right to harvest trees? Most of Canada's harvestable forests are located on crown lands, so it seems reasonable that logging companies should pay a fee, called stumpage, to cut trees on those lands. But should the fee be large or small?

This seemingly simple question has been the basis of a dispute between Canada and the United States that dates back hundreds of years. Softwood lumber—easily cut lumber, usually from evergreen trees like spruce and pine (Figure 25.1)—constitutes a large portion of Canada's exports to the United States, worth almost $7 billion in 2003 alone. But the two countries disagree on what constitutes fair and unfair trading in this important commodity. At the centre of the dispute are stumpage fees. Most Canadians feel that stumpage fees should be low, reflecting the value of the resource in the Canadian economy. Many Americans, however, see low fees as creating hidden—and, under existing trade agreements like the North American Free Trade Agreement, illegal—subsidies that give Canadian companies an unfair advantage in the marketplace.

The softwood lumber trade between Canada and the United States operates under a formal agreement that expired in March 2001. In 2001, the Bush administration implemented new surcharges on exports of Canadian lumber to the U.S. These charges originally imposed an 18.8% tax on exported lumber that was deemed to have been unfairly subsidized. A further duty of 8.4% was later added for lumber that was considered to have been "dumped"— exported and sold below cost in the U.S. These duties were applied only to lumber from public lands. In Atlantic Canada, most lumber comes from private lands, and was therefore not subject to the surcharges.

When the softwood lumber agreement was signed in 1996, Atlantic Canada produced about 5% of Canada's lumber. By 2001, production in Nova Scotia and New Brunswick had increased 62%. In British Columbia, 15,000 forestry workers were laid off. The United States continued to press for higher stumpage fees; Canada continued to refuse.

Subsequent attempts to reach a compromise failed, and eventually the dispute went to NAFTA for adjudication. The NAFTA panel's decision, which is legally binding and must be acted upon within 60 days, ruled that the U.S. duties were too high and ordered the U.S. to review its position. In December, 2003, a World Trade Organization panel similarly concluded that the U.S. duties were too harsh, and that although low stumpage fees are beneficial to Canadian companies, they are not low enough to constitute an unfair subsidy.

The Canada-U.S. softwood lumber dispute is estimated to have cost over $1.5 billion and tens of thousands of jobs, mainly in western Canada. A tentative settlement in December 2003 eliminates the hefty U.S. duty on Canadian lumber, but caps Canadian exports at 31.5% of the U.S. market, well below current levels. Canada is seeking the return of just over half of the duties already paid, amounting to at least $1 billion (U.S. companies would keep the rest).

Opinion on the issue is divided on both sides of the border. Everyone wants free trade in softwood lumber, but people disagree about how best to achieve that goal. U.S. markets need Canada's high quality, reliable lumber supply. Canadians need jobs.

In April 2005, Canada's Minister of International Trade announced $20 million in assistance for the softwood lumber industry, to offset the legal costs the Canadian industry has incurred in defending its interests. Despite the NAFTA and WTO rulings, the U.S. continues to impose duties on Canadian lumber, amounting to over $4.5 billion since 2001.[1]

Environmental Policy: Economics and Law

LEARNING OBJECTIVES

Other chapters in this text have explained the causes of environmental problems and discussed technical solutions. The scientific solutions, however, are only part of the answer. This chapter introduces some basic concepts of environmental economics and shows how they have been applied in the analysis of environmental issues. After reading this chapter, you should understand:

- When and how it is possible to put a dollar value on environment.
- What the "tragedy of the commons" is, and how it leads to an overexploitation of resources.
- How the perceived future value of an environmental benefit affects our willingness to pay for it now.
- What externalities are, and why it is important to evaluate them in determining the costs of actions that affect the environment.
- What factors may be involved in determining a level of acceptable environmental risk and risk to human life.
- Why it is difficult, yet important, to evaluate environmental intangibles, such as landscape beauty.

- *The case study of Canada's softwood lumber dispute with the United States illustrates the importance of conflicts over values, including economic ones. In this chapter we will see that economic analysis can help us understand how to sustain renewable resources.*

Figure 25.1 • **A clear-cut forest, British Columbia.**

25.1 The Economic Importance of the Environment

In 2000, Canada spent just over $3 billion to deal with pollution, about 2% of the nation's gross national product. Although costly, cleaning our environment has economic benefits. Populations subject to high levels of certain pollutants (people in inner cities, for example) have lower average life expectancies and higher incidences of certain diseases.[2, 3] Air pollution in Canadian cities is a contributor to thousands of deaths annually,[4] and 2–9% of total mortality in cities is associated with particulate air pollution.[5] Ontario's Anti-Smog Action Plan is expected to reduce the health and economic costs of air pollution by over a billion dollars. And these are from a single province.

Environmental decision making often involves analysis of tangible and intangible factors. A mudslide that results from altering the slope of land is an example of a tangible factor; the beauty of the slope before the mudslide and its ugliness afterward is an example of an intangible factor. Of the two, the intangibles are obviously more difficult to deal with because they are hard to measure and to value economically. Nonetheless, evaluation of intangibles is becoming more important. One task of **environmental economics** is to develop methods for evaluating intangibles that provide good guidelines, are easy to understand, and are quantitatively credible. Not an easy goal!

25.2 The Environment as a Commons

Often people who use a natural resource do not act in a way that maintains that resource and its environment in a renewable state—they do not seek sustainability. At first glance this seems puzzling. Why do people not act in their own best interest? Economic analysis suggests that the profit motive, by itself, will not always lead a person to act in the best interests of the environment. Here, we give two reasons why this may be so.

The first reason has to do with what the ecologist Garrett Hardin called "the tragedy of the commons."[6] When a resource is shared, an individual's personal share of profit from exploitation of the resource is usually greater than that individual's share of the resulting loss.

The second has to do with the low growth rate, and therefore the low productivity, of a resource. A **commons** is land (or another resource) owned publicly with public access for private uses. The term commons originated from land owned publicly and set aside in English towns where all the farmers of the town could graze their cattle. The practice of sharing the grazing area worked as long as the number of cattle was low enough to prevent overgrazing. It would seem that

Figure 25.2 • **Softwood lumber used in residential construction.**

people of goodwill would understand the limits of a commons. But take a dispassionate view and think about the benefits and costs to each farmer as if it were a game. Phrased simply, each farmer tries to maximize personal gain and must periodically consider whether to add more cattle to the herd on the commons. The addition of one cow has both a positive and a negative value. The positive value is the benefit when the herder sells that cow. The negative value is the additional grazing by the cow. The benefit to an individual of selling a cow for personal profit is greater than that individual's share of the loss in the degradation of the commons. The short-term successful game plan, therefore, is always to add another cow.

So an individual will act to increase use of the common resource. Eventually the common grazing land is so crowded with cattle that none can get adequate food and the pasture is destroyed. In the short run, everyone seems to gain, but in the long run, everyone loses. This applies generally: Complete freedom of action in a commons inevitably brings ruin to all. The implication seems clear: Without some management or control, all natural resources treated like a commons will inevitably be destroyed.

How can we deal with the tragedy of the commons? It is an unsolved puzzle. As several scientists wrote recently, "no single broad type of ownership—government, private or community—uniformly succeeds or fails to halt major resource deterioration." In trying to solve this puzzle, economic analysis can be helpful.

There are many examples of commons, both past and present. In Canada, 94% (393 million hectares) of forests are on publicly owned lands. Resources in international regions, such as ocean fisheries away from coastlines and the deep-ocean seabed, where valuable mineral deposits lie, are international commons not controlled by any single nation (Figure 25.3). Most of the continent of Antarctica is a commons, although there are some national territorial claims there, and international

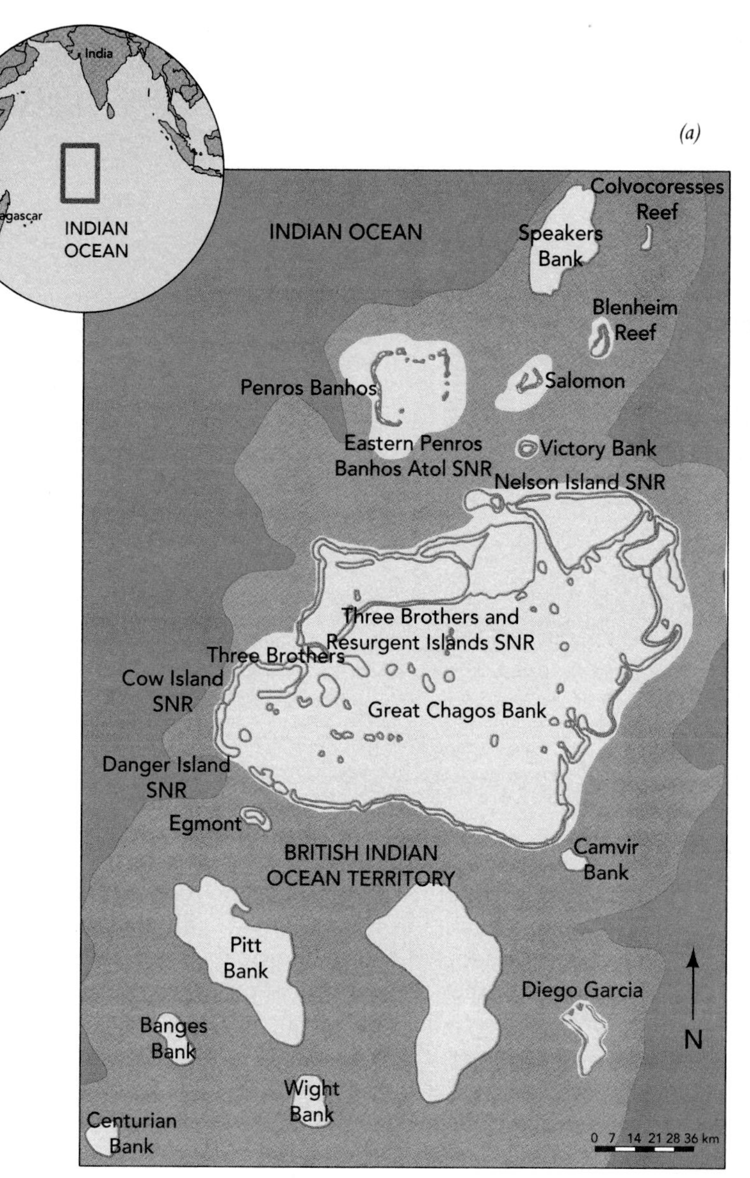

(b) Fish in the waters of the Chagos coral reefs.

(c) A red-footed booby on a palm tree in the Chagos atolls.

Figure 25.3 • British Indian Ocean Territory: A Kind of Commons. *(a)* This large region, called Chagos Archipelago, contains the Great Chagos Bank, the largest atoll structure in the world, covering 13,000 square kilometres. Among other uses, it functions as a major tuna fishery and is a global commons in this way. *(b)* It is also home to many rare species, such as beautiful water birds *(c)* and serves as a biodiversity commons as well. Fish in the waters of the Chagos coral reefs.

negotiations have continued for years about conserving Antarctica and possible use of its resources.

The atmosphere, too, is a commons, both nationally and internationally. Consider the possibility of global warming. Individuals, corporations, public utilities, motor vehicles, and nations add carbon dioxide to the air by burning fossil fuels. Just as Garrett Hardin suggested, people tend to respond by benefiting themselves (by burning more fossil fuel) rather than by benefiting the commons (burning less fossil fuel). The picture here is quite mixed, however, with much ongoing effort to bring co-operation to this common issue.

In the nineteenth century, burning wood in fireplaces was the major source of heating in Canada and the United States (and fuelwood is still the major source of heat in many nations). Until the 1980s, a wood fire in a fireplace or woodstove was considered a simple good, providing warmth and beauty; people enjoyed sitting around a fire and watching the flames. This activity must have a long history in human societies. But in the 1980s, with increases in populations and recreational houses in regions such as British Columbia and northern Ontario, home burning of wood began to pollute air locally. Especially in valley towns surrounded by mountains, the air became fouled, visibility declined, and there was a potential for effects on human health and environmental conditions. As a result, some communities restrict or prohibit the use of fireplaces and woodstoves. The local air is a commons, and its overuse required a societal change.

Recreation is a problem of the commons—overcrowding of national parks, wilderness areas, and other nature-recreation areas. An example is Algonquin Provincial Park in northern Ontario. The park, located on the edge of the boreal forest biome (see Chapter 8) of North America, contains many lakes and islands and is an excellent place for fishing, hiking, canoeing, and viewing wildlife. Before the area became a provincial park, it was used for motorboating, snowmobiling, trapping, and hunting; a number of people in the region made their living from tourism based on these kinds of recreation. Some environmental groups argue that Algonquin Provincial Park is ecologically fragile and needs to be legally designated as a wilderness area in order to protect it from overuse and from the adverse effects of motorized vehicles. Others argue that the park can withstand a moderate level of hunting and motorized transportation, and that these uses should be allowed.

At the heart of this conflict is the problem of the commons, which in this case can be phrased as: What is the appropriate public use of public lands? Should all public lands be open to all public uses? Should some public lands be protected from people? At present, Canada has a policy of different uses for different lands. In general, national parks are open to the public for many kinds of recreation, whereas designated wildernesses have restricted visitorship and kinds of uses.

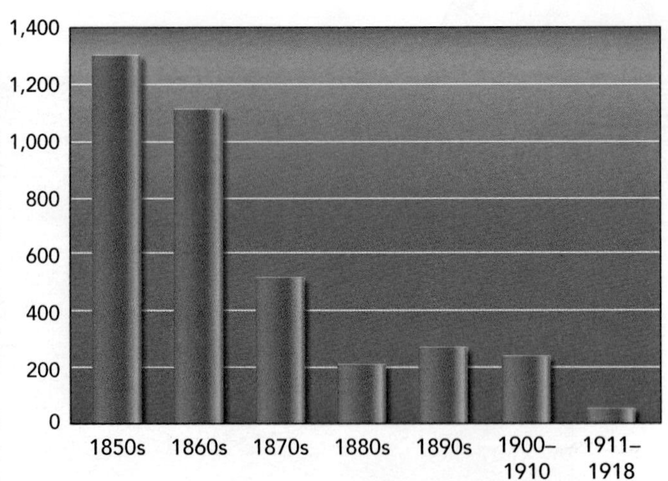

Figure 25.4 • Bowhead Whales Killed by Pelagic Whalers (Caught and Killed) from 1849 to 1914. The fact that the number killed declined rapidly, although hunting continued, indicates that the population rapidly decreased—it was unable to reproduce at a rate that could replace the large catches in the first two decades. [*Source*: Redrawn from J. R. Bockstoce, and D. B. Botkin, *The Historical Status and Reduction of the Western Arctic Bowhead Whale (Balaena mysticetus) Population by the Pelagic Whaling Industry, 1848–1914.* Final report to the U.S. National Marine Fisheries Service by the Old Dartmouth Historical Society, 1980.]

25.3 Low Growth Rate and Therefore Low Profit as a Factor in Exploitation

Recall that the second reason individuals tend to overexploit natural resources held in common is the low growth rate of the resource.[7] For example, one way to view whales economically is to consider them solely in terms of whale oil (see Chapter 14). Whale oil, a marketable product, can be thought of as the capital investment of the industry. How can whalers get the best return on their capital? (Here we need to remember that whale populations, like other populations, increase only if there are more births than deaths.) We will examine two approaches: resource sustainability and maximum profit. If whalers adopt a simple, one-factor resource sustainability policy, they will harvest only the net biological productivity each year and thus maintain the total abundance of whales at the current level; that is, they will stay in the whaling business indefinitely. In contrast, if they adopt a simple approach to maximizing immediate profit, they will harvest all the whales now, sell the oil, get out of the whaling business, and invest the profits.

Suppose they adopt the first policy. What is the maximum gain they can expect? Whales, like other large, long-lived creatures, reproduce slowly; typically, a calf is born every three or four years (Figure 25.4). The total net growth of a whale population is unlikely to be more than 5% per year. If all the oil in the whales in the oceans today represented a value of $100 million, then the most the whalers could expect to take in each year would be 5% of this amount, or $5 million. Meanwhile, they would have to pay the cost of upkeep on ships and other equipment, interest on loans, and employee salaries—all of which would decrease profit. If whalers adopted the second policy and harvested all the whales, then they can invest the money from the oil. Although investment income varies, even a conservative investment of $100 million would very likely yield more than 5%, especially when this income could be received without the cost of paying a crew, maintaining the ships, buying fuel, marketing the oil, and so on.

It is quite reasonable and practical, if one considers only direct profit, to adopt the second policy: Harvest all the whales, invest the money, and relax. Whales simply are not a highly profitable long-term investment under the resource sustainability policy. Note that this discussion suggests that whaling on the open seas can be also viewed as a problem of a commons, complicated by low growth rate.

It is no wonder that there are fewer and fewer whaling companies and that companies have left the whaling business when their ships became old and inefficient. Few nations support whaling; those that do have stayed with whaling for cultural reasons. For example, whaling is important to some First Nations cultures, and some harvest of bowheads takes place in Alaska; whale meat is a traditional Japanese food, and whale harvest is maintained for this reason.

Another factor to consider in resource use is the relative scarcity of a necessary resource, which affects its value and therefore its price. For example, if a whaler lived on an isolated island where whales were the only food and had no communication with other people, then his primary interest in whales would be in his remaining alive. He could not choose to sell off all whales to maximize profit, as he would have no one to sell them to. It would seem to make sense that he would harvest in a way that would maintain the population of whales. Or the whaler might estimate that his own life expectancy was only 10 years and that to avoid starvation he had to consume the whales beyond their ability to reproduce. He might try to harvest the whales so that they would become extinct at the same time that he would die. "You can't take it with you" would be his attitude.

If ships began to land regularly at this island, he could trade and begin to benefit from some of the future value of whales. If ocean property rights existed so that he could "own" the whales that lived within a certain distance of his island, then he might consider the economic value of owning this right to the whales. He could sell rights to future whalers, or mortgage against them, and thus reap the benefits during his lifetime for whales that could be caught after his death. Causing the extinction of whales would not be necessary.

From harvesting whales, we see that we must think beyond the immediate, direct economic advantages of whaling. Policies that seem ethically good may not be the most profitable for an individual. Economic analysis clarifies how an environmental resource is used, what is perceived as its intrinsic values, and therefore its price. This brings us to the question of externalities.

25.4 Externalties

One gap in our thinking about whales, an environmental economist would say, is that we must be concerned with externalities in whaling. An **externality, also called an indirect cost**, is an effect not normally accounted for in the cost–revenue analysis of producers and often not recognized by them as part of their costs and benefits.[7] Simply put, externalities are costs or benefits that don't show up in the price tag.[8] In the case of whaling, externalities include the loss of revenue to tourist boats used to view whales and the loss of an ecological role played by whales in marine ecosystems. Classically, economists agree that the only way for a consumer to make a rational decision is by comparing the true costs against the benefits the consumer seeks. If the true costs are not revealed, then the price will be wrong, and purchasers cannot act rationally.

Air and water pollution provide other good examples of externalities. Consider production of nickel from ore at the Sudbury, Ontario, smelters, which has serious environmental effects, as discussed in Chapter 15. Traditionally, the economic costs associated with the production of commercially usable nickel from an ore are the **direct costs**—that is, those borne by the producer and passed directly on to the user or purchaser. In this case, direct costs include the costs of purchasing the ore, buying energy to run the smelter, building the plant, and paying employees. Meanwhile, externalities include costs associated with degradation of the environment from emissions from the plant. For example, prior to implementation of pollution control, the Sudbury smelter destroyed vegetation over a wide area, which led to an increase in erosion. Although air emissions from smelters have been substantially reduced and restoration efforts have initiated a slow recovery of the area, pollution remains a problem, and total recovery of the local ecosystem may take a century or more.[9] There are costs associated with the value of trees and soil, with restoration of vegetation and land to a productive state.

Problem number one: What is the true cost of clean air over Sudbury? Economists say that there is plenty of disagreement about such a price but that everyone agrees that it is larger than zero. But in spite of this, clean air and water get traded and dealt with in today's world as if their value was zero. How do we get the value of clean air and water and other environmental benefits to be recognized socially as greater than zero? In some cases, the dollar value can be determined. Water resources for power or other uses may be evaluated by the amount of flow of the rivers and the quantity of water storage in rivers and lakes; forest resources may be evaluated by the number, type, and sizes of trees and their subsequent yield of lumber; and mineral resources may be evaluated by the estimated number of metric tons of economically valuable mineral material at particular locations. Quantitative evaluation of the tangible natural resources—such as air, water, forests, and minerals—prior to development or management of a particular area is now standard procedure.

Problem number two: Who should bear the burden of these costs? Some suggest that environmental and ecological costs should be included in costs of production through taxation or fees. The expense would be borne by the corporation that benefits directly from the sale of the resource (nickel in the case of Sudbury) or would be passed on, in increased sale prices, to users (purchasers) of nickel. Others suggest that these costs should be shared by the entire society and paid for by general taxation (such as a sales tax or income tax). Stated simply, the question is whether it is better to finance pollution control using tax dollars or a "polluter pays" approach. Today, economists generally agree that the "polluter pays" approach provides much stronger incentives for cost-effective pollution reduction.

25.5 Natural Capital, Environmental Intangibles, and Ecosystem Services

Public Service Functions of Nature

A complicating factor in our perception of maintaining clean air and water is that ecosystems do some of this without our help—and before the Industrial Revolution did much of it for us. Forests absorb particulates, salt marshes convert toxic compounds to non-toxic forms, wetlands and organic soils treat sewage (see Chapter 13). These are called the public service functions of nature. For example, it is estimated that bees pollinate over $20 billion worth of crops in Canada and the United States. The cost of pollinating these crops by hand would be exorbitant, and a pollutant that eliminated bees would have large indirect economic consequences. We rarely think of this benefit of bees. Recently, however, an outbreak of bee parasites in the United States reduced the abundance of bees, bringing this once-intangible factor to public attention (Figure 25.5).

As another example, bacteria fix nitrogen in the oceans, lakes, rivers, and soils. The cost of replacing this function in terms of production and transport of artificially produced nitrogen fertilizers would be immense, but again we rarely think about this activity of bacteria. Bacteria also clean water in the soil by decomposing toxic chemicals.

The atmosphere performs a public service by acting as a large disposal site for toxic gases. For instance, carbon monoxide is eventually converted to non-toxic carbon dioxide either by inorganic chemical reactions or by bacteria.

Figure 25.5 • Public service functions of living things. Wild creatures and natural ecosystems perform public service functions for us—carrying out tasks important for our survival that would be extremely expensive for us to accomplish by ourselves. For example, bees pollinate millions of flowers important for food production, timber supply, and aesthetics.

Only when our environment loses a public service function do we usually begin to recognize its economic benefits. Then, what had been accepted as an economic externality (indirect cost) suddenly may become a direct cost.

Estimates have been attempted to determine the dollar value of public service functions. At this time, we have to consider these estimates only rough approximations, as the value is difficult to measure. Public service functions of living things that benefit human beings and other forms of life have been estimated to provide between $3 trillion and $33 trillion per year.[10, 11] Economists refer to the ecological systems that provide these benefits as natural capital.

Valuing The Beauty of Nature

The beauty of nature—often referred to by the more general term, *landscape aesthetics*—is another important environmental intangible, one that has probably been important to people as long as our species has existed and certainly has been important since people have written, because the beauty of nature is a continuous theme in literature and art. Once again, as with forests cleaning the air, we face the difficult question: How do we arrive at a price for the beauty of nature? The problem is even more complicated because among the kinds of scenery we enjoy are many modified by people (Figure 25.6).

One of the perplexing problems associated with aesthetic evaluation is personal preference. One person may appreciate a high mountain meadow far removed from civilization. A second person prefers visiting with others on a patio at a trailhead lodge. A third person may prefer to visit a city park. A fourth may prefer the austere beauty of a desert. If we are going to consider aesthetic factors in environmental analysis, we must develop a method of aesthetic evaluation that allows for individual differences—another yet unsolved topic.

Some philosophers suggest that there are specific characteristics of landscape beauty and that we can use these to help us set the value of intangibles. Some suggest that the three key elements of landscape beauty are coherence, complexity, and mystery—mystery in the form of something seen in part but not completely, or not completely explained. Other philosophers suggest that the primary aesthetic qualities are unity, vividness, and variety.[12] Unity refers to the quality or wholeness of the perceived landscape—not as an assemblage but as a single, harmonious unit. Vividness refers to that quality of landscape that makes a scene visually striking; it is

Figure 25.6 • A landscape extensively modified by humans. An ornamental garden in the centre of a city.

related to intensity, novelty, and clarity. People differ in what they believe are the key qualities of landscape beauty, but, once again, almost everyone would agree that the value is greater than zero.

25.6 How Is the Future Valued?

The discussion about whaling—explaining why whalers may not find it valuable to conserve whales—reminds us of the old saying, "A bird in the hand is worth two in the bush." In economic terms, a profit now is worth much more than a profit in the future. This brings up another economic concept important to environmental issues—the future value compared with the present value of anything.

As an example, suppose you are dying of thirst in a desert and meet two people; one offers to sell you a glass of water now, and the other offers to sell you a glass of water if you can be at the well tomorrow. How much is each glass worth? If you believe you will die today without water, the glass of water today is worth all your money, and the glass tomorrow is worth nothing. If you believe you can live another day without water, but will die in two days, you might place more value on tomorrow's glass than today's.

In practice, things are rarely so simple and distinct. We know we are mortal, so we tend to value personal wealth and goods more if they are available now than if they are promised in the future. This evaluation is made more complex, however, because we are accustomed to thinking of the future—to planning a nest egg for retirement or for our children. Indeed, many people today argue that we have a debt to future generations and must leave the environment in at least as good a condition as we found it; these people would argue that the future environment is not to be valued less than the present one.

Since the future existence of whales and other endangered species has value to those interested in biological conservation, the question arises: Can we place an economic (quantitative dollar) value on the future existence of anything (Figure 25.7)? The future value depends on how long a time period you are talking about. For example, the future times associated with some important global environmental topics, such as stratospheric ozone depletion and global warming, extend longer than a century. This is so because chlorofluorocarbons (CFCs) have such a long residence time in the atmosphere (see Chapter 22) and because of the time necessary to realize potential benefits from changing energy policy to offset global climate change.

Another aspect of future versus present value is that spending on the environment can be viewed as diverting resources from alternative forms of productive investment that will be of benefit to future generations. (This assumes that spending on the environment is not itself a productive investment.) For example, in the twentieth century, asbestos was used as an insulator in buildings, around hot-water pipes and similar devices. Then it was discovered that asbestos can cause cancer. Insulation containing asbestos was found in some schools. As long as the asbestos is well contained within an external covering, it does not pose an immediate threat. In some cases, people

Figure 25.7 • Economic value as a function of time—a way of comparing the value of having something now with the value of having it in the future. A negative value means that there is more value attached to having something in the present than having it in the future. A positive value means that there is more value attached to having something in the future than having it today.

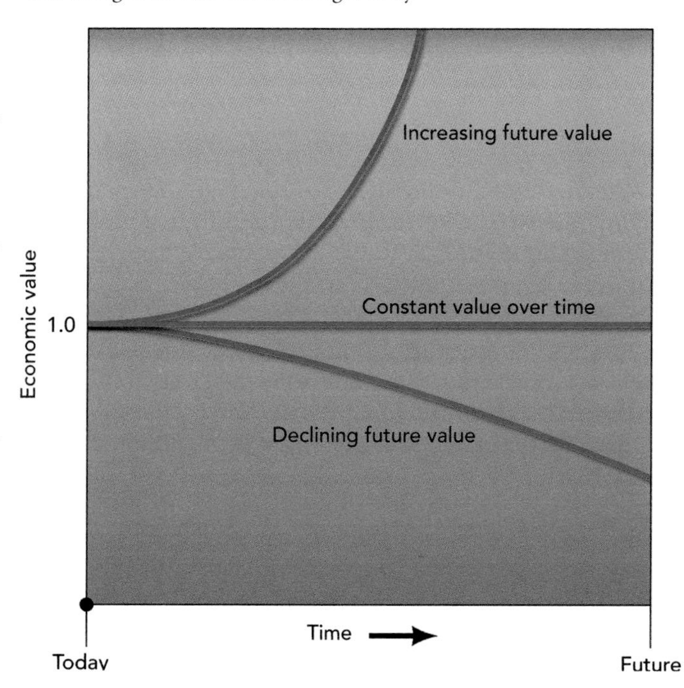

decided to spend the money now to remove the asbestos in order to reduce future risk to the students. Another approach is to leave the well-protected asbestos intact and spend the money that it would have cost to remove the asbestos on improving other educational facilities in a school, only acting to deal with the asbestos when it begins to enter the air in the school.

A further complicating issue is that as we get wealthier, the value we place on many environmental assets (like wilderness areas) increases dramatically. Thus, if society continues to grow in wealth over the next century as it has over the past century, the environment will be worth far more to our great-grandchildren than to our great-grandparents, at least in terms of willingness to pay to protect it. The implication—which complicates this topic even more—is that conserving resources and environment for the future is tantamount to taking from the poor today and giving to the rich in the future. If history is a guide, North Americans in the twenty-first century will be far better off than North Americans at the end of the twentieth century. To what extent should we ask the average person today to sacrifice now for much richer great-great grandchildren? How can we know the future usefulness of today's sacrifices? Put another way, what would you have liked your ancestors in 1900 to have sacrificed for our benefit today? Should they have increased research and development on electric transportation? Should they have saved more tall grass prairie or restricted whaling?

Economists observe that it is an open question whether something promised in the future can have *more* value than it does today. Future economic value is difficult enough to predict because it is affected by how future consumers view consumption. But if, in addition, something has greater value in the future than it does today, then that leads to an impossible mathematical situation: In the very long run, the future value will become infinite, which of course is impossible. So in terms of the future, the basic issues are: (1) We are so much richer and better off than our ancestors that their sacrificing for us might have been inappropriate. (2) Even if they had wanted to sacrifice, how would they have known what sacrifices would be important to us?

As a general rule, one answer to these thorny questions about future value is: Do not throw away or destroy something that cannot be replaced if you are not sure of its future value. For example, if we do not fully understand the value of the wild relatives of potatoes that grow in Peru but do know that their genetic diversity might be helpful in developing future strains of potatoes, then we ought to preserve those wild strains (Figure 25.8).

Figure 25.8 • **Wild wheat.** Wild relatives of cultivated plants can be important sources of genetic diversity for the development of future strains.

25.7 Risk–Benefit Analysis

Death is the fate of all individuals, and every activity in life involves some risk of injury or death. How, then, do we place a value on saving a life by reducing the level of a pollutant? This question raises another important area of environmental economics—**risk–benefit analysis**, in which the riskiness of a present action in terms of its possible outcomes is weighed against the benefit, or value, of the action. Here, too, difficulties arise.

Acceptability of Risks and Costs

With some activities, the relative risk is clear. It is much more dangerous to stand in the middle of a busy highway than to stand on the sidewalk. Hang gliding has a much higher mortality rate than hiking. The effects of pollutants are often more subtle, so the risks are harder to pinpoint and quantify. Table 25.1 gives the risk associated with a variety of activities and some forms of pollution. The table shows the lifetime risk of death from each cause. In looking at the table, remember that, since the ultimate fate of everyone is death, the total lifetime risk of death from all causes must be 100%. So if you are going to die of something and you smoke a pack of cigarettes a day, you have 8 chances in 100 that your death will be a result of smoking. At the same time, your risk of death from driving an automobile is 1 in 100. Risk tells you the chance of an event, but not its timing. So you might smoke all you want and die from the automobile risk first.

One of the striking things about Table 25.1 is that death from outdoor environmental pollution is comparatively low—even compared to the risks from drowning or dying in a fire. This suggests that the primary reason we value lowering air pollution is an improvement in the quality of our lives, rather than an increase in the time we are alive. Considering the great interest people now show in air pollution, quality of life is much more important than is generally recognized. We are willing to spend money on improving that quality, not just the length of our lives.

Another striking observation in this table is that natural *indoor* air pollution is much more deadly than most outdoor air pollution—unless, of course, you live at a toxic waste facility (see Chapter 21).

Societies differ in socially, psychologically, and ethically acceptable levels of risk for any cause of death or injury. It is commonly believed that future discoveries will help to decrease various risks, perhaps eventually allowing us to approach a zero-risk environment. But complete elimination of risk is generally either technologically impossible or prohibitively expensive. We can make some generalizations about the acceptability of various risks. One factor is the number of people affected. Risks that affect a small population (such as employees at nuclear power plants) are usually more acceptable than those that involve all members of a society (such as risk from radioactive fallout).

In addition, novel risks appear to be less acceptable than long-established or natural risks, and society tends to be willing to pay more to reduce such risks. For example, France spends approximately $1 million to reduce the likelihood of one air-traffic death but only $30,000 for the same reduction in automobile deaths.[13] Some argue that the greater safety of commercial air travel compared with automobile travel is in part a function of the relatively novel fear of flying compared with the more ordinary fear of death from a road accident. That is, because the risk is newer to us and thus less acceptable, we are willing to spend more per life to reduce the risk from flying than to reduce the risk from driving.

People's willingness to pay for reducing a risk also varies with how essential and desirable the activity is. For example, many people accept much higher risks for athletic or recreational activities than they would for transportation- or employment-related activities (see Table 25.1). The risks associated with playing a sport or using transportation are assumed to be inherent in the activity. The risks to human health from pollution may be widespread and linked to a large number of deaths. Although risks from pollution are often unavoidable and unseen, people want a lesser risk from pollution than from, say, driving a car or playing a sport.

In an ethical sense, it is impossible to put a value on a human life. However, it is possible to determine how much people are willing to pay for a certain reduction in risk or a certain probability of an increase in longevity. For example, a study by the Rand Corporation considered measures that would save the lives of heart-attack victims, including increasing ambulance services and initiating pre-treatment screening programs. According to the study, which identified the likely cost per life saved and the willingness of people to pay, people were willing to pay approximately $32,000 per life saved, or $1,600 per year of longevity.[13] Although information is incomplete, it is possible to estimate the cost of extending lives in terms of the dollars per person per year for various actions (Figure 25.9 and Table 25.1). For example, on the basis of direct effects on human health, it costs more to increase longevity through a reduction in air pollution than to directly reduce deaths through the addition of a coronary ambulance system. Such a comparison is useful as a basis for decision-making. Clearly, though, when a society chooses to reduce air pollution, many factors beyond the direct, measurable health benefits are considered. Pollution directly affects more than just our health, and ecological and aesthetic damage can also indirectly affect human health (see Section 25.4). We might want to choose a slightly higher risk of death in a more pleasant environment (spend money to clean up the air instead of to increase ambulance services) rather than increase the chances of living longer in a poor environment (spend the money on reducing heart attacks).

Such comparisons may make you feel uncomfortable. But like it or not, we cannot avoid making choices of this kind. The issue boils down to whether we should improve the quality of life for the living or extend life expectancy regardless of the quality of life.[14]

The degree of risk is an important concept in our legal processes. For example, the *Canadian Environmental Protection Act* states that no one may manufacture a new chemical substance or process a chemical substance for a new use without obtaining a clearance from Environment Canada. The act establishes procedures to estimate the hazard to the environment and to human health of any new chemical before it becomes widespread. Environment Canada examines the data provided and judges the degree of risk associated with all aspects of the production of the new chemical or process, including extraction of raw materials, manufacturing, distribution, processing, use, and disposal. The chemical can be banned or restricted

Table 25.1 • Risk of Death from Various Activities

Activity	*Result*	*Risk of Death (per lifetime)*	*Lifetime Risk of Death (%)*	*Comment*
Cigarette smoking (pack a day)	All causes: cancer, effect on heart, lungs, etc.	8 in 100	8.00000%	
Air in the home containing radon	Cancer	1 in 100	1.00000%	Naturally occurring
Automobile driving		1 in 100	1.00000%	
Death from a fall		4 in 1,000	0.40000%	
Drowning		3 in 1,000	0.30000%	
Fire		3 in 1,000	0.30000%	
Artificial chemicals in the home	Cancer	2 in 1,000	0.20000%	Paints, cleaning agents, pesticides
Sunlight exposure	Melanoma	2 in 1,000	0.20000%	Of those exposed to sunlight
Electrocution		4 in 10,000	0.04000%	
Air outdoors in an industrial area		1 in 10,000	0.01000%	
Artificial chemicals in water		1 in 100,000	0.00100%	
Artificial chemicals in foods		less than 1 in 100,000	0.00100%	
Airplane passenger (commercial airline)		less than 1 in 1,000,000	0.00010%	

Source: From *Guide to Environmental Risk* (1991), U.S. EPA Region 5 Publication Number 905/91/017.

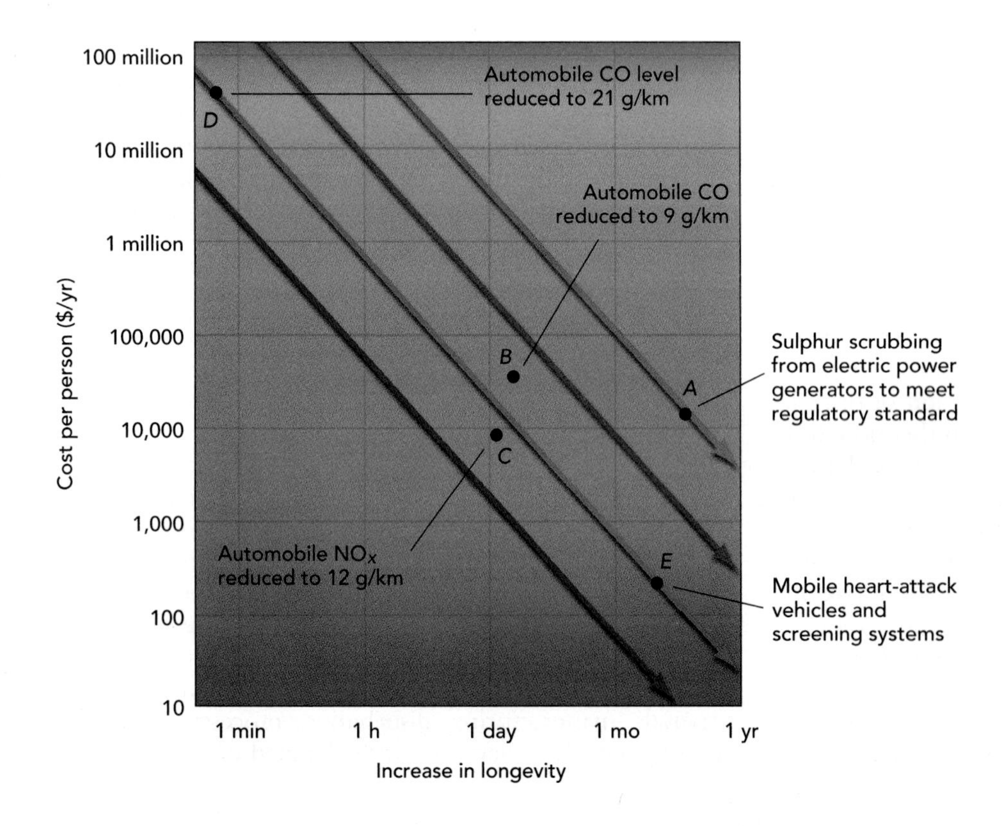

Figure 25.9 • **One way to rank the effectiveness of various efforts to reduce pollutants is to estimate the cost of extending a life in dollars per year.** This graph shows that reducing sulphur emissions from power plants to a regulatory target level (A) would extend a human life 1 year at a cost of about $10,000. Similar restrictions applied to automobile emissions (B, C) would increase lifetimes by 1 day. More stringent automobile controls would be much more expensive (D), and mobile units and screening programs for heart problems would be much cheaper (E). This graph represents only one step in an environmental analysis. (*Source*: Based on R. Wilson, "Risk/Benefit Analysis for Toxic Chemicals," *Ecotoxicology and Environmental Safety* 4 [1980]:370–383.)

Risk–Benefit Analysis and DDT

A review of the history of the use of DDT illustrates the difficulty of completely eliminating a pollution risk. As noted in Chapter 12, DDT was first applied widely in the 1940s to control the spread of diseases such as malaria via insects and was subsequently used to control crop pests. At first, tests of the safety of DDT focused on human health effects, which were believed to be small. In the late 1950s, however, DDT was discovered in the livers of sharks. Because DDT had not been used in the ocean, at first it was believed that this observation was simply a measurement error or the result of an unknown dumping of DDT directly into the ocean.

Gradually, and to their surprise, scientists began to understand that DDT had spread from farmland through surface water runoff and through the air into the ocean and that the chemical had become a worldwide contaminant. DDT was found in the tissues of penguins in the Antarctic and seals in the Pribilof Islands of the Bering Sea. By 1970, it had become clear that DDT was everywhere and was a global environmental problem.[16] Although DDT is banned for use in Canada and the United States, U.S. corporations continue to be among the largest producers of this chemical, which is shipped to other nations. DDT is still used widely elsewhere, especially in developing nations of the tropics.

Complete elimination of DDT residues from all the environments of the world—and from all the sharks, penguins, seals, and birds—no longer seems feasible. Aside from the difficulty of eliminating pollutants, another lesson of this example is that what had seemed to be an economic externality of DDT (its indirect ecological effects on birds and ocean animals) became a major societal issue.

in either manufacturing or use if the evidence suggests that it will pose an unreasonable risk of injury to human health or to the environment.

But what is unreasonable?[15] This question brings us back to Table 25.1 and makes us realize that "unreasonable" involves judgments about the quality of life as well as the risk of death. The level of acceptable pollution (and thus risk) is a social–economic–environmental trade-off. Moreover, the level of acceptable risk changes over time in society, depending on changes in scientific knowledge, comparison with risks from other causes, the expense of decreasing the risk, and the social and psychological acceptability of the risk.

For example, when DDT was first used, no one understood the subtle environmental transport and ecological effects of the chemical; scientific observations revealed these effects later (see A Closer Look 25.1). At that time, there was relatively little concern with environmental issues, and society was not yet willing to pay for many of these costs. Now, people widely agree that the environment is a major concern, and there is less willingness to accept indirect environmental effects. What had been considered externalities to the use of DDT have become internal cost factors.

As explained in Chapter 21, the total cost of pollution is the sum of the costs to control pollution and the loss from pollution damages. In some cases, these two factors have opposite trends in terms of economic cost (as one goes down, the other goes up), and their intersection point is the minimum total cost, as shown by point *A* in Figure 25.9. In other cases, total costs may stabilize or even decline as companies utilizing pollution control become more efficient and external costs in the form of environmental damage are minimized. If the minimum total cost involves a pollution level that is too high a risk, then additional control may add considerable expense.

The risk associated with a pollutant can be determined by the present levels of exposure and predicted future trends. These trends depend on the production and origin of the pollutant, pathways it follows through the environment, and changes it undergoes along these pathways. Dose-response curves establish risk to a population from a particular level of pollutant (see Chapter 15). Relative risks of different pollutants can be determined by comparing their current levels and their dose-response curves.

The previous discussion suggests that, when adequate data are available, it is possible to take scientific and technological steps to estimate the level of risk and from this to estimate the cost of reducing risk and compare the cost with the benefit. However, what constitutes an acceptable risk is more than a scientific or technical issue. Acceptability of a risk involves ethical and psychological attitudes of individuals and society. We must therefore ask several questions. What risk from a particular pollutant is acceptable to us? How much is a given reduction in risk from that pollutant worth to us? How much will each of us, as individuals or collectively as a society, be willing to pay for a given reduction in that risk? The answers depend not only on facts but also on societal and personal values. What must also be factored into the equation is that costs associated with cleanup of polluted areas and with restoration programs can be minimized or even eliminated if a recognized pollutant is controlled initially. The total cost of pollution control need not increase indefinitely.

Although pollution control may involve many dollars, the average cost per family in Canada is low, especially compared with other costs. It is estimated that the per-family pollution control cost is between $30 and $60 per year for a family with a median income. In addition to the low cost per family, pollution control has many benefits whose quantitative value can be estimated. For example, air quality standards are estimated to reduce the risk of asthma by 3% and the risk to locally exposed adults of chronic bronchitis and emphysema by 10% to 15%. Estimates of the total cost of the direct and indirect effects on human health from stationary sources of air pollution are $250 per family per year. Air pollution contributes to inflation by reducing the number of productive workdays, reducing work efficiency, adding to direct expenditures for health treatments, and necessitating repair of non-human environmental damage. On this basis, air pollution control appears to be cost-effective; in fact, it has economic benefits.[17]

25.8 Global Issues: Who Bears the Costs?

Global environmental problems make us more aware of the public service functions of the environment of our planet and of life around us as well as raising new economic questions. An important case in point is the possibility of global warming. The problem is that our technological society is adding carbon dioxide and other greenhouse gases to the atmosphere that have the potential to warm the climate. (See the discussion of global climate change in Chapter 20.)[16,18] The direct solution is to decrease the release of these gases, but to do so would require a worldwide decrease in the burning of fossil fuels. Although most of the production of greenhouse gases today is from the industrial nations, in the future developing nations, especially China and India, will contribute large quantities of these gases.

The economist Ralph D'Arge points out an economic problem arising from this global issue: The less-developed countries did not share in the economic benefits of the burning of fossil fuels during the first two centuries of the Industrial Revolution, but they are sharing in the disadvantages of this activity.[19] Now the industrialized nations are suggesting that all nations, including the less-developed ones, restrict their use of fossil fuels and participate in future disadvantages without obtaining the benefits of cheap energy. Developing nations tend to think that industrial nations, which enjoyed the past benefits, should accept most of the future costs. At the same time, why shouldn't the developing nations proceed to develop and burn fossil fuels?

This perspective, limited to what benefits individual nations, may be too restricted in light of the global environmental effects of our technological civilization. It may be necessary to reduce the total production of greenhouse gases. If so, then the economic question is: Who pays, and how? At present, this is an unresolved issue in environmental economics. One suggestion is that the developed nations pay for the reduction in greenhouse emissions of the less developed nations. Another suggestion is that the developed countries share their technology with developing countries, thus helping the developing countries to reduce both local and global pollution. These issues were a major concern at the 1992 Earth Summit in Rio de Janeiro.[20]

25.9 How Do We Achieve a Goal? Environmental Policy Instruments

How does a society achieve an environmental goal, such as preservation and use of a resource or reduction of a pollutant? Any society has several methods to achieve such goals. Means to implement a society's policies are known among economists as **policy instruments** (see Table 25.2). These include moral suasion (which politicians call "jawboning," i.e., persuading people by talk, publicity, and social pressure); direct controls, which include laws; market processes, which affect the price of goods and include taxation of various kinds, subsidies, licenses, and deposits; and government investments, which include research and education. Society also has administrative mechanisms to ensure that the policy instruments chosen actually function.

Marginal Costs and the Control of Pollutants

How clean is clean? When have we done enough to think that the environment is "good" and that we have achieved a reasonable balance between benefits and costs? In deciding this, we have to look at the cost of each additional step we take, given the previous steps. This leads us to the concept of **marginal costs**. In controlling pollutants, marginal cost is the cost to reduce one additional unit of pollutant. By analogy, with conservation of an endangered species or a rare habitat or ecosystem, the marginal cost would be the cost of adding one more unit (however we might define a unit) to those already conserved.

With pollution control, the marginal cost often increases rapidly as the percentage of reduction increases. For example, the marginal cost of reducing the biological oxygen demand in wastewater from petroleum refining increases exponentially. When 20% of the pollutants have been removed, the cost of removing an additional kilogram is five cents. When 80% of the pollutants have been removed, it costs 49 cents to remove an additional

Table 25.2 • Performance of Various Policy Instruments

Policy Instrument	Reliability	Permanence	Adaptability to Growth	Resistance to Inflation	Incentive for Improved Effort	Economy	Feasibility without Metering	Non-interference in Private Decisions	Political Attraction	
									Actual	*Potential*
Moral suasion	Good[a]	Poor	Good[b]	Good[b]	Fair	Poor[c]	Excellent	Excellent	Excellent	—
Direct controls By quota	Fair	Poor	Fair	Excellent	Poor	Poor	Poor	Poor	Excellent	—
By specification of technique	Fair	Poor	Good[b]	Good[b]	Poor	Poor	Excellent	Poor	Excellent	—
Fees	Excellent	Excellent	Fair	Fair	Excellent	Excellent	Poor	Excellent	Poor	Good
Sale of permits or licenses	Excellent	Excellent	Excellent	Excellent	Excellent	Excellent[d]	Poor	Excellent	Poor	Good
Subsidies										
Per-unit reduction	Fair[e]	Good	Fair	Fair	Excellent	Good	Poor	Excellent	Good	—
For equipment purchase	Fair	Good	Fair	Fair	Excellent	Good	Poor	Excellent	Good	—
Government investment										
Facilities	Excellent	?	?	?	—	?	Excellent	—	Good	—
Regeneration	Fair	Good	?	—	—	Poor	Excellent	Poor	Good	—
Technical support	Fair	Poor	Excellent	Excellent	—	Fair	Excellent	Fair	Poor	—
Research	Fair	Poor	Excellent	Excellent	—	Fair	Excellent	Good	Good	—
Education	Poor	Poor	Excellent	Excellent	—	Good	Excellent	Good	Poor	—

[a] For short periods of time when urgency of appeal is made very clear.

[b] Baumol and Oates's judgment.

[c] Induces contributions from decision makers who are most cooperative, not necessarily from those able to do the job most effectively.

[d] Tends to allocate reduction quotas among firms in a cost-minimizing manner, but if the number of emissions permitted is too small it will force the community to devote an excessive quantity of resources to environmental protection.

[e] Tends to allocate reduction quotas among firms in cost-minimizing manner but introduces inefficiency into the environmental protection process by attracting more polluting firms into the subsidized industry, so that aggregate response is questionable.

Source: Adapted from W. J. Baumol and W. E. Oates, *Economics, Environmental Policy and the Quality of Life.* (Englewood Cliffs, N.J.: Prentice-Hall, 1979).

kilogram. Extrapolating from these results, it would cost an infinite amount to remove all the pollution.

Three common methods of direct control of pollution are shown in Table 25.2: (1) setting maximum levels of pollution emission, (2) requiring specific procedures and processes that reduce pollution, and (3) charging fees for pollution emission. In the first case, a regulatory body could set a maximum level for the amount of sulphur emitted from the smokestack of an industry. In the second, it could restrict the kind of fuel the industry could use. Many areas have chosen the latter method by prohibiting the burning of high-sulphur coal.

The problem with the first approach—controlling emissions—is that careful monitoring is required indefinitely to make certain the allowable levels are not exceeded. Such monitoring may be costly and difficult to carry out. The disadvantages of the second approach—requiring specific procedures—are that the required methodology may impose a severe financial burden on the producer of the pollutant, restrict the kinds of production methods open to an industry, and become technologically obsolete. (See the discussion of air pollution and laws regulating it in Chapter 21.) However, after an initial expenditure on procedures to reduce pollution, production efficiency may be increased and other costs such as monitoring and waste disposal can be reduced. For example, Japan, whose economy is one of the most energy efficient in the world, has reduced its air pollution more than any other industrial nation. In recent years, Japanese industry has used only 5 megajoules to produce $1 of gross domestic product (GDP), while U.S. industry required 12 megajoules.

Although Canada and the United States have emphasized the use of direct regulation to control pollution, other countries have been successful in controlling pollution by charging effluent fees. For example, charges for effluents into the Ruhr River in Germany are assessed on the basis of both the concentration of pollutant and the total quantity of polluted water emitted into the river. In response, plants have introduced water recirculation and internal treatment to reduce emissions.[21]

Studies of the uses of different policy instruments for environmental matters have resulted in some ability to evaluate their relative success (see Table 25.2). For example, moral suasion is reliable but not very permanent in its effect. Sale of licenses or permits has been found to be among the more successful option.

In every environmental matter, there is a desire on the one hand to maintain individual freedom of choice and on the other to achieve a specific social goal. In ocean fishing, for example, how does a society allow every individual to choose whether or not to fish and yet prevent everyone from fishing at the same time and bringing fish species to extinction? This interplay between private good and public good is at the heart of environmental issues. Some argue that the market itself will provide the proper control. For example, it can be argued that people will stop fishing when there is no longer a profit to be made. We have already seen, however, that this may not be so. Two factors interfere with this argument: (1) By the time the reduced fish population results in no economic gain for fishers, it may be too late to avoid eventual extinction. (2) Even when it is not possible to make a sustained annual profit, there may be an advantage in harvesting the entire resource and getting out of the business (see A Closer Look 25.2).

25.10 Environment and Law: The Policy Process in Canada

The legal system of most of Canada and the United States has historical origins in the British common law system—that is, laws derived from custom, judgement, and decrees of the courts, rather than from legislation. (In Canada, the exception is Québec, whose legal system is based on the French Napoleonic Code, a highly prescriptive system that relies much more on written law and less on custom and precedent.) The Canadian legal system preserved and strengthened British law to protect the individual from society. Individual freedom—nearly unlimited discretion to use property as the owner pleased—was given high priority, and the powers of the federal government were strictly limited.

But there is a caveat: *When individual behaviour infringed on the property or well-being of others, the common law provided protection through doctrines prohibiting trespass and nuisance.* For example, if you own land that is damaged by erosion or flooding caused by improper management of your neighbour's land, then you have recourse under common law. If the harm was more widely spread through the community, then only the government has the authority to take action—for instance, to limit certain air and water pollution to abate a public nuisance.

The common law provides another doctrine, that of public trust, which both grants and limits the authority of government over certain natural areas of special character. Under this doctrine, navigable and tidal waters were entrusted to the government to hold in trust for public use. For such resources, the government has the strong responsibility of a trustee to provide protection and is not permitted to transfer such properties into private ownership.

We can view the history of federal legislation affecting land and natural resources as occurring in three stages. In the first stage, the goal for public lands was to convert them to private uses. During this phase, the Canadian government passed laws that were not intended to address

A CLOSER LOOK 25.2

Making Policy Work: Fishing Resources and Policy Instruments

Ocean fishing illustrates different ways of making a policy work, referred to as policy instruments. The oceans outside of national territorial waters are commons, and thus the fish and mammals that live in them are common resources. What is a common resource may change over time, however. The move by many nations to define international waters as beginning 325 km from their coasts has turned some fisheries from completely open common resources to national resources open only to domestic fishers.

In fisheries, there are four main management options:[22]

1. Establish total-catch quotas for the entire fishery and allow anybody to fish until the total is reached.
2. Issue a restricted number of licenses but allow each licensed fisherman to catch many fish.
3. Tax the catch (the fish brought in) or the effort (the cost of ships, fuel, and other essential items).
4. Allocate fishing rights—that is, assign each fisherman a transferable and salable quota.

With total-catch quotas, the fishery is closed when the quota is reached. Whales, Pacific halibut, tropical tuna, and anchovies have been regulated in this way. When the practice was used in Alaska, all of the halibut were caught in a few days, with the result that restaurants no longer had halibut available for most of the year. This undesirable result led to a change in policy; the total-catch approach was replaced by the sale of licenses.

Although regulating the total catch can be done in a way that helps the fish, it tends to increase the number of fishers and the capacity of vessels, and the end result is a hardship on fishers. Recent economic analysis suggests that taxes that take into account the cost of externalities can work to the best advantage of fishers and fish. Similar results are achieved by allocating a transferable and salable quota to each fisherman.

Determining which management method achieves the best use of a desirable environmental resource is not simple. The answer varies with the specific attributes of both the resource and the users. The tools of economics can be used to determine the methods that will work best within a given social framework.

environmental issues but did affect land, water, minerals, and living resources—and thereby had large effects on the environment. A good example is the federal *Fisheries Act*, first enacted in 1868 to protect the economic resource of the Canadian fisheries from careless disposal of wastes. The nineteenth century also saw the enactment of early public health legislation, regulating the provision of water supplies and the disposal of sewage.

The second stage began in the first half of the twentieth century, when governments began to pass laws that conserved public lands for recreation, scenic beauty, and historic preservation. Late in the nineteenth century, Canadians and Americans came to believe that their nations' grand scenery should be protected, and that public lands provided benefits, some directly economic, such as providing tourism benefits or rangelands for private ranching.

Federal laws created Banff and Jasper National Parks in the second half of the nineteenth century in response to Canadians' growing interest in their scenic resources. By 1920, there were five national parks in the Rocky and Selkirk Mountains, and St. Lawrence Islands National Park (1904) and Point Pelee National Park (1918) in Ontario. A formal plan for managing Canada's National Park System was created in the 1970s. Today, the system consists of 39 national parks.

Also in the second stage, interest in forest management began to increase, and along with it the establishment of forestry courses at the Universities of New Brunswick (1908) and Toronto (1907). The focus of these early forestry programs was on production of useful products rather than environmental stewardship.

The third stage began in the 1970s, with growing awareness of the environmental consequences of human activities. The 1970s were the period when most federal and provincial environmental legislation was introduced, and the structure of modern Canadian environmental agencies began to take shape.

Government regulation of land and resources has also given rise to controversy: How far should the government be allowed to go to protect what appears to be the public good against what have traditionally been private rights and interests? Part of achieving a sustainable future in Canada will be finding a balance among these uses, as well as a balance between the amount of land that should be public and the amount of land that need not be.

In 1867, the British Parliament passed the *British North America Act*, creating the Dominion of Canada and giving the new country some powers of self-government and a constitutional framework. As we have seen in earlier chapters, that document, now called the *Constitution Act 1982*, divides jurisdiction (law-making power) between the federal and provincial governments.

The Making of Laws in Canada

Today, Canada continues to base its national legal system on British common law, although over time many statutes have been written down, to clarify and sometimes alter common law precedents. Statutes must be drafted, approved by elected governments, and enforced by public agencies. In most democratic countries, including Canada, proposed statutes undergo extensive public scrutiny and a multi-step, multi-party approval process.

Statutes can be divided into two types. **Criminal statutes** prohibit the kinds of actions that society considers wholly wrong and unacceptable, such as theft, rape, and murder. **Public welfare statutes** address more routine matters, for instance the administration of universities, the conduct of professional engineers, and the protection of public health. Environmental legislation is this type. Statutes may have **regulations** attached to them. Regulations are the detailed rules that specify matters such as allowable pollutant discharge levels, the names of industries covered by the law, the classification of wastes, and similar information. For example, the *Canadian Environmental Protection Act* (a statute) has as its general goal "pollution prevention and the protection of the environment and human health in order to contribute to sustainable development". Associated with the Act are a number of regulations that articulate rules governing specific activities (e.g., the export and import of hazardous wastes, discussed in Chapter 24) and industries (e.g., the pulp and paper industry).

In Canada, the federal law-making (legislative) branch of government is the House of Commons and Senate, a **bicameral** (two-house) system. Provincial legislatures and city councils have parallel law-making powers at the provincial and municipal levels of government, respectively. Members of Parliament are elected representatives of the general public, normally are members of a political party, and serve for a limited term; senators are appointed by the government in power, with no fixed term.

A new law, or bill, may be proposed by any member of the public, by elected officials, or by bureaucrats, but must be presented to the House of Commons by a sponsoring Member of Parliament. Most new bills are introduced by the political party currently in power.

New bills pass through three readings in the House of Commons, and must be approved by a majority of members at each reading. The first reading is essentially the introduction of the proposal to the House. After the second reading, Members of Parliament may debate the contents of the bill, propose amendments, or reject all or part of it. As part of this process, a new law is usually gazetted—that is, published in draft form for scrutiny by the legal profession—before it receives final reading. The third reading presents the bill, with any revisions, to the House for a final vote. If approved, the bill then passes to the Senate for endorsement, and finally to the Governor General of Canada, who is the Queen's representative in Canada, for royal assent—formal approval on behalf of the Queen. At this point, the bill is law, but it is not binding until it is "proclaimed in force." Proclamation in force normally occurs at the same time as royal assent, but may be delayed for months or even years if the bill is considered to impose high costs or other burdens on the regulated parties.

The Canadian law-making process is designed with abundant opportunity for multi-party and public scrutiny of proposed legislation. In theory, therefore, a law has the support of a majority of Canadian citizens, as represented by their elected representatives in the House of Commons.

25.11 Who Stands for Nature? The Jumbo Glacier and Mineral King Resorts

Planning for recreational activities on natural lands (including national forests and parks) is controversial. At the heart of the controversy are two different moral positions, both of which have wide support in Canada. On one side, some argue that public land must be open to public use, and therefore the resources within those lands should be available to citizens and corporations for economic benefit. On the other side are those who argue that public lands should serve the needs of society first and individuals second, and that public lands can and must provide for land uses not possible on private lands.

A classic example of this controversy took place in the 1990s over a plan by Glacier Resorts Ltd., a private developer, to build a year-round ski resort at the foot of Jumbo Mountain and Jumbo Glacier, near Invermere, British Columbia (Figure 25.10). The location is the site of a former sawmill, and has been logged at intervals over the past 80 years. In addition, a major mine operated at the valley's entrance from the 1930s to the 1990s. Although the site is not designated as a park or wilderness reserve, it is near a number of existing and proposed protected areas. The resort project will build ski lifts to nearby glaciers up to 3,400 m, and a 5,500-bed resort with accommodations for an additional 750 staff. It is expected that the development will bring an additional 2,000 to 3,000 visitors to the area in high season.[23]

Opponents of the project argue that although the site is not pristine, it is wild and a valuable ecosystem in its present state. They argue that the ecological impacts of the project on the region's many mammals, birds and fish cannot be mitigated. In particular, they are concerned that noise, traffic, and habitat degradation may put Canada's

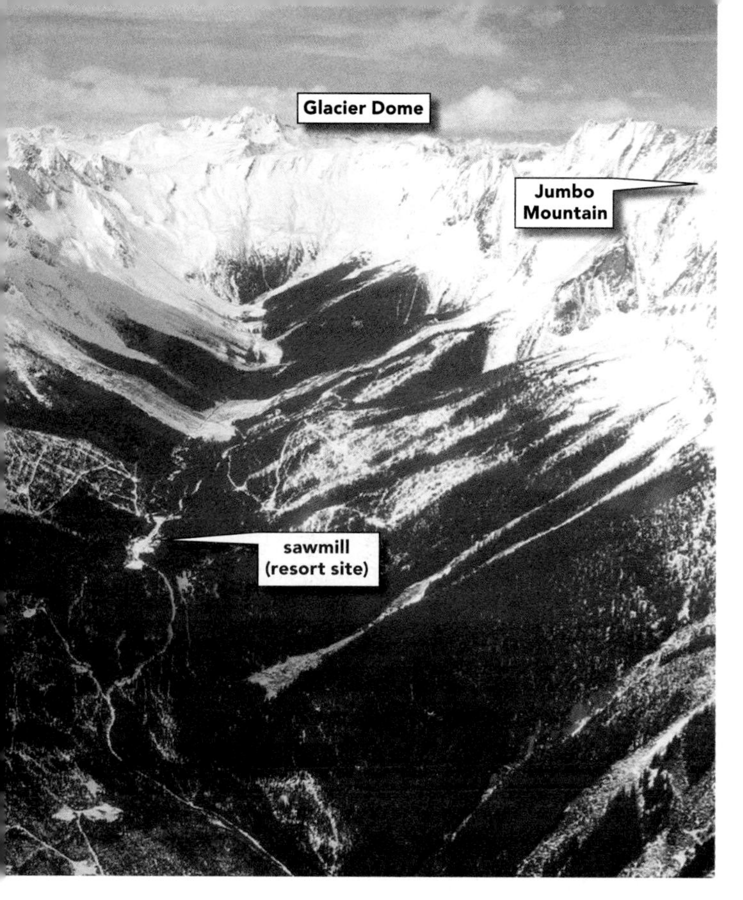

Figure 25.10 • Proposed location of the Jumbo Glacier Resort.

southern grizzly population at risk of extirpation. The proponent's environmental consultants, who have studied the project in depth, disagree and have indicated that with proper planning, impacts can be minimized and mitigated. In October 2004, after more than nine years of review in the British Columbia Environmental Assessment Office, the Jumbo Glacier Resort project was granted an environmental assessment certificate and allowed to proceed with its planning.

The outcome of the development will not be known with certainty until it is built and operating. But the case brought up a curious question: *If the project does in fact have an adverse impact on the environment, who will have been wronged?*

In a recent U.S. case, the Sierra Club opposed the Disney Corporation's proposal to build a ski area on federal land in the Mineral King Valley of California's Sierra Nevada mountains. The California courts decided that neither the Sierra Club nor local government would be directly harmed by the building of the Mineral King project. When the case was appealed to the U.S. Supreme Court, the court confirmed the lower court's ruling, stating the Sierra Club did not have a sufficient "personal stake in the outcome of the controversy" to bring the case to court. But in a dissenting statement, Justice William O. Douglas addressed the question of legal standing. (Standing is a legal term relating here to the right to bring suit.) Douglas proposed the establishment of a new federal rule that would allow "environmental issues to be litigated before federal agencies or federal courts in the name of the inanimate object about to be despoiled, defaced, or invaded by roads and bulldozers and where injury is the subject of public outrage." In other words, trees would have legal standing.

While trees did not achieve legal standing in that case, it was a landmark in that legal rights and ethical values were explicitly discussed for natural systems. This subject in ethics is still a lively, controversial one. *Should our ethical values be extended to non-human, biological communities and even to the life-support system of Earth?* What position you take will depend in part on your understanding of the characteristics of wilderness, natural systems, and other environmental factors and features, and in part on your values.

25.12 Environment and Law II: How You Can Be an Actor in Legal Processes

Environmental Litigation

The cases of the Jumbo Glacier and Mineral King Resorts bring up the question: What is the role of our legal system—laws, courts, judges, lawyers—in achieving environmental goals? The current answer is that environmental groups working through the courts have been a powerful force in shaping the direction of environmental quality control since the early 1970s. Their influence arose in part because the courts, appearing to respond to the national sense of environmental crisis of that time, took a more activist stance and were less willing to defer to the judgment of government agencies. At the same time, citizens were granted unprecedented access to the courts and, through them, to environmental policy.

Citizen Actions

Even without specific legislative authorization for citizens' suits, courts have allowed citizen actions in environmental cases as part of a trend to liberalize standing requirements.[12]

In the 1980s, a new type of environmentalism (which some people would label radical) arose, based in part on the premise that when it comes to the defense of wilderness there can be no compromise. Methods used by these new environmentalists have included holding sit-ins to block roads that provide access to forest areas where mining or timber harvesting is scheduled to take place; sitting in trees to block timber harvesting; implanting large steel spikes in trees to discourage timber harvesting; and sabotaging equipment, such as bulldozers (a practice known as "ecotage").

Figure 25.11 • Protesters at Clayoquot Sound, summer 1993. More than 850 people were arrested and over 12,000 demonstrated in this protest against logging of old-growth forests.

Civil disobedience and ecotage have undoubtedly been responsible for millions of dollars' worth of damage to a variety of industrial activities related to the use of natural resources in wilderness areas. One result of civil disobedience by some environmental groups is that other environmental groups, such as the Sierra Club, are now considered moderate in their approach to protecting the environment.

There is no doubt, however, that civil disobedience has been successful in defending the environment in some instances. For example, in the summer of 1993, more than 850 people were arrested and over 12,000 demonstrated in opposition to logging operations in old growth areas of Clayoquot Sound, British Columbia (Figure 25.11). In Robert F. Kennedy Jr.'s words, the protesters were "asserting public ownership over resources which, under Canadian law, are owned by all Canadians, but, in historical practice, were treated as the personal fiefdoms of a few giant timber companies."

Environmentalists are now relying more on the law when arguing for ecosystem protection. The *Species at Risk Act* has been used as a tool in attempts to halt activities such as timber harvesting and development. Although rarely is the presence of an endangered species responsible for stopping a proposed development, those species are increasingly being used as a weapon in attempts to save remaining portions of relatively undisturbed ecosystems.

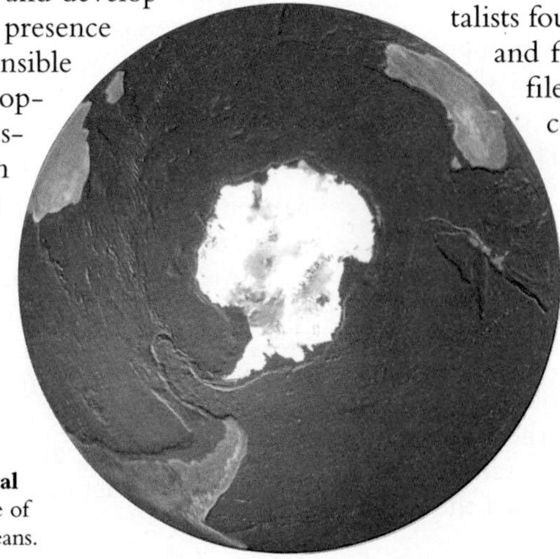

Figure 25.12 • International agreements determine environmental practices in Antarctica. Satellite image of Antarctica and surrounding southern oceans.

Mediation

The expense and delay of litigation have led people to seek other ways to resolve disputes. In environmental conflicts, an alternative that has recently received considerable attention is mediation, a negotiation process between the adversaries guided by a neutral facilitator. The task of the mediator is to clarify the issues in contention, help each party understand the position and the needs of the other parties, and attempt to arrive at a compromise whereby each party gains enough to prefer a settlement to the risks and costs of litigation. Often, citizens' suits or the possibility that a suit might be filed gives an environmental group a place at the table in the mediation process. Litigation, which may delay a project for years, becomes something that can be bargained away in return for concessions of decreased environmental impact (mitigation) from a developer. There have been some successes with mediation where bargaining positions were approximately equal and where participants who truly represented the conflicting interests could be identified and persuaded to devote the considerable effort that the process demands. The *Canadian Environmental Assessment Act* formally designated mediation as an alternative to the environmental assessment hearing process. In some U.S. states, mediation is required by law as an alternative or as a precedent to litigation in the highly contentious siting of waste treatment facilities.

A classic example of a situation in which mediation could have saved millions of dollars in legal expenses and years of litigation is the Storm King Mountain case. The case involved a conflict between a utility company and conservationists. In 1962, the Consolidated Edison Company of New York announced plans for a new hydroelectric project in the Hudson River highlands, an area considered to have many unique aesthetic values as well as thriving fisheries. The utility company argued that it needed the new facility, and the environmentalists fought to preserve the beautiful landscape and fisheries resource. The first lawsuit was filed in 1965, and after 16 years of intense courtroom battles, the litigation ended in 1981. Incredibly, the paper trail exceeded 20,000 pages. After millions of dollars had been spent, the various parties finally managed to get together and settle the case with the assistance of an outside mediator. If they had been able to sit down and talk about the issues at an early stage, mediation

CRITICAL THINKING ISSUE

Fisheries: How Can They Be Made Sustainable?

Both overfishing and pollution have been blamed for the alarming decline in groundfish (cod, haddock, flounder, redfish, pollack, hakes) in the north Atlantic Ocean. Most scientists and fisheries managers have focused on overfishing, but attempts to regulate fishing have generated bitter disputes with fishers, many of whom contend that restrictions on fishing make them scapegoats for pollution problems. The controversy has become a classic battle between short-term economic interests and long-term environmental concerns.

In 1977, in response to overfishing in Canadian waters by foreign factory ships, the Canadian government extended the nation's coastal waters from 19 to 322 km. Within the zone, Canada imposed strict fishing controls, but the government could not control fishing of important stocks of cod, flounder, and redfish that extended beyond the boundary. A decision about Canadian and American fishing rights in the Georges Bank, the most prolific area in the North Atlantic, in favour of Canada in 1984 by the International Court of Justice in The Hague intensified competition for remaining fishing waters among U.S. fishers. Overfishing knows no boundaries; however; in 1992, Canada was forced to suspend all cod fishing to save the stock from complete annihilation. Portions of the Georges Bank are now closed indefinitely to fishing for some fish species, including yellowtail, cod, and haddock. Landings of yellowtail in 1993 were 3,800 metric tonnes, the lowest on record and a mere 6% of the historical maximum in 1969.

A 1992 study concluded that key characteristics in declining flounder populations were consistent with overfishing rather than pollution. Furthermore, if pollution had been the cause, the skate and dogfish populations would have been expected to decline as well. However, S. J. Correia, the author of the study, cautioned that the issue should not be seen as an either/or question and that "overfishing may currently mask any pollution-induced population growth constraints."

Can the Atlantic fishery be made sustainable? If so, how? Some agencies, like Fisheries and Oceans Canada and the U.S. National Marine Fisheries Service, advocate a system of individual transferable quotas (ITQs) by which permits are issued to boat owners to allow them to harvest a fixed amount of fish each year. They can lease, sell, or bequeath the permits to others. Although ITQs have been successful in several cases, some small operators fear that large corporations will buy up permits and dominate the industry.

Other approaches focus on promoting the domestic fishery. At least 3,000 people in Atlantic Canada have suffered severe income losses as a direct result of the closure of the fishery. To encourage domestic fishers, Fisheries and Oceans Canada provided loan guarantees for replacing older vessels and equipment with newer boats with high-tech equipment for locating fish. During this same period, demand for fish increased as people became more concerned about cholesterol levels in red meat. Consequently, the number of fishing boats, the number of days at sea, and fishing efficiency increased sharply. As a result, 50–60% of the populations of some species were landed each year—a significant impact on species whose population dynamics are not well understood.

Critical Thinking Questions

1. Using the example of the North Atlantic fishery, what are some arguments for and against the proposition that all natural resources treated like commons will inevitably be destroyed unless controls are instituted?
2. Which measures described attempt to convert the fishing industry from a commons system to private ownership? How might these measures help prevent overfishing? Is it right to institute private ownership of public resources?
3. What approach to future value (approximately) does each of the following people assume for fish?
   Fisher: If you don't get it now, someone else will.
   Fisheries manager: By sacrificing now, we can do something to protect fish stocks.
4. Develop a list of the environmental and economic advantages and disadvantages of ITQs. Would you support instituting ITQs for the North Atlantic fishery? Explain.
5. Do you think it possible to reconcile economic and environmental interests in the case of the North Atlantic fishing industry? If so, how? If not, why not?

might have settled the issue much sooner and at much less cost to the individual parties and to society.[13] The Storm King Mountain case is often cited as a major victory for environmentalists, but the cost was great to both sides.

International Environmental Law and Diplomacy

Legal issues involving the environment are difficult enough within a nation. They become extremely difficult in international situations. International law is different in basic concept from domestic law because there is no world government with enforcement authority over nations. As a result, international law must depend on the agreement of the parties concerned to bind them to behaviour that many residents of a particular nation may oppose. Certain issues of multinational concern are addressed by a collection of policies, agreements, and treaties that are loosely called international environmental law. There have been encouraging developments in this area, such as agreements to reduce air pollutants that destroy stratospheric ozone (the Montreal Protocol of 1987 see Chapter 22).

Antarctica provides a positive example of using international law to protect the environment. Antarctica, a continent of 14 million km^2, was first visited by a Russian ship in 1820, and people soon recognized that the continent contained unique landscapes and life forms (Figure 25.12). By 1960, a number of countries had claimed parts of Antarctica to exploit mineral and fossil fuel resources. Then, in 1961, an international treaty was established designating Antarctica a scientific sanctuary. Thirty years later, in 1991, a major environmental agreement, the Protocol of Madrid, was reached, protecting Antarctica, including islands and sea south of 60° latitude. The continent was designated "nuclear-free," and access to resources was restricted. This was the first step in conserving Antarctica from territorial claims and establishing the "White Continent" as a heritage for all people of Earth.

Other environmental problems that have been addressed at the international level include persistent organic pollutants (POPs) such as dioxins, DDT, and other pesticides. Following several years of negotiations in South Africa and Sweden, 127 nations adopted a treaty in May 2001 to greatly reduce or eliminate the use of toxic chemicals known to contribute to cancer and harm the environment.

Summary

- An economic analysis can help us understand why environmental resources have been poorly conserved in the past and how we might more effectively achieve conservation in the future.
- Economic analysis is applied to two different kinds of environmental issues: the use of desirable resources (fish in the ocean, oil in the ground, forests on the land) and the minimization of undesirable pollution.
- Resources may be common property or privately controlled. The kind of ownership affects the methods available to achieve an environmental goal. There is a tendency to overexploit a common property resource and to harvest to extinction nonessential resources whose innate growth rate is low, as suggested in Hardin's "tragedy of the commons."
- Future worth compared with present worth can be an important determinant of the level of exploitation.
- The relation between risk and benefit affects our willingness to pay for an environmental good.
- Evaluation of environmental intangibles, such as landscape aesthetics and scenic resources, is becoming more common in environmental analysis. When quantitative, such evaluation balances the more traditional economic evaluation and helps separate facts from emotion in complex environmental problems.
- Societal methods to achieve an environmental goal include the categories of moral suasion, direct controls through laws (statutes and regulations), market processes, and government investment. Many kinds of controls have been applied to the use of desirable resources and the control of pollution.

STUDY QUESTIONS

1. What is meant by the term the tragedy of the commons? Which of the following are the result of this tragedy:
   a. The fate of the Vancouver Island marmot
   b. The fate of the right whale
   c. The high price of walnut wood used in furniture
2. What is meant by risk–benefit analysis?
3. Cherry and walnut are valuable woods used to make fine furniture. Basing your decision on the information in the following table, which would you invest in? (*Hint*: Refer to the discussion of whales in this chapter.)
   a. A cherry plantation
   b. A walnut plantation
   c. A mixed stand of both species
   d. An unmanaged woodland where you see some cherry and walnut growing

Species	*Longevity*	*Maximum Size*	*Maximum Value*
Walnut	400 years	1 m	$15,000/tree
Cherry	100 years	1 m	$10,000/tree

4. Flying over a major city, you see smog below. Your neighbour in the next seat says, "That smog looks bad, but eliminating it would save only a few lives. Doing that isn't worth the cost. We should spend the money on other things, like new hospitals." Do you agree or disagree? Give your reasons.

5. Which of the following are intangible resources? Which are tangible?
   a. The view of Mount Robson in British Columbia
   b. Owning property with a view of Mount Robson
   c. Porpoises in the ocean
   d. Tuna fish in the ocean
   e. Clean air
   f. Owning property outside the smog area of Toronto
6. What kind of future value is implied by the statement "Extinction is forever"? Discuss how we might approach providing an economic analysis for extinction. (See Chapter 14 for additional information about extinction.)
7. Which of the following can be thought of as commons in the sense meant by Garrett Hardin? Explain.
   a. Tuna fisheries in the open ocean
   b. Trout in artificial freshwater ponds
   c. Grizzly bears in Banff National Park
   d. A view of Riding Mountain National Park in Manitoba.
   e. Air over Riding Mountain National Park in Manitoba.

FURTHER READING

Daly, H. E., and J. C. Farley. *Ecological Economics: Principles and Applications.* Island Press, Washington, D.C, 2004. Discusses an interdisciplinary approach to the economics of environment.

Goodstein, E. S. *Economics and the Environment*, 3rd ed. New York: Wiley, 2000. 2000.

Hardin, G. "Tragedy of the Commons." *Science* 162 (1968):1243–1248. One of the most cited papers in both science and social science, this classic work outlines the differences between individual interest and the common good.

Hodge, I. *Environmental Economics: Individual Incentives and Public Choice.* New York: St. Martin's Press, 1995.

National Academy of Public Administration. *The Environment Goes to Market: The Implementation of Economic Incentives for Pollution Control.* Washington, D.C.: National Academy Press, 1994. Description and analysis of a number of implementation issues as evidenced in four case studies: air pollution trading, pollution charges, solid waste recycling, and a deposit-refund program.

Schnaiberg, A., and K. A. Gould. *Environment and Society: The Enduring Conflict.* New York: St. Martin's Press, 1994. An examination of several myths related to economics and environmental problems.

Tietenberg, T. H. *Environmental and Natural Resource Economics*, 3rd ed. New York: HarperCollins, 1992. A book that provides a broad range of national and international examples of the application of economic principles as they relate to environmental issues and natural-resource management.

Tisdell, C. A. *Economics of Environmental Conservation: Economics of Environmental and Ecological Management.* New York: Elsevier Science, 1991. A book focusing on the ecological dimensions of environmental economics, concentrating on living or biological resources and their life-support systems.

Tisdell, C. A. *Environmental Economics: Policies for Environmental Management and Sustainable Development.* Aldershot, Hants, England: Edward Elgar, 1993. Includes discussions of externalities, cost–benefit analysis, sustainable development and economic activity, ecological economics, and global resource conservation.

chapter 26

Craik, Saskatchewan: Sustainability Begins at Home

Can a town of 400 people make a difference in the global environment? Craik, Saskatchewan, thinks so. Four years ago, the town's mayor and reeve met with Dr. Lynn Oliphant, a retired professor from the University of Saskatchewan (and co-founder of Earth Day), to explore the possibility of creating an ecologically friendly community. Dr. Oliphant's original idea was to create small groups of 8 to 10 people living in low-consumption, low-waste rural developments. Craik agreed to donate 52 hectares to the project, but they had bigger ideas. They created a five-year plan and began by building a 552 square metre, energy-efficient Eco-centre, using mainly recycled materials. The straw-bale walls and cellulose-filled roof provide excellent insulation against the prairie winters—roughly two and a half times the typical insulation in a Canadian home. South-facing windows and field stone walls tap and store the sun's heat. A system of pipes embedded in the concrete floor carries ethanol pumped from three-metre deep trenches, bringing gentle warmth in winter and cooling in summer. Five solar panels on the roof capture solar energy. A metal roof collects rainwater in a 17,500-litre cistern, for treatment on site. Composting toilets virtually eliminate the use of water for flushing. All of the building's lights are energy efficient, including the exit signs, which use 70% less energy than normal. Even the building's restaurant is energy-efficient. A brick oven (made from 3,000 bricks from an old school house) can be heated and then turned off, allowing normal baking for up to four hours using its residual heat. Oh, and the restaurant serves organic foods.[1,2]

Craik's Eco-centre and Eco-village are impressive achievements, but what is more remarkable is the unintended impact the projects have had on rural Saskatchewan and across Canada. The Eco-centre's unusual design and building materials attracted a lot of attention in the area. The original idea was to build a facility that would demonstrate simple technologies rural residents could use in their own homes. But as word about Craik's projects spread, people from elsewhere in Canada began to visit the site, many interested in joining the Eco-village.

Craik residents are proud of their achievement. Glenn Hymers, Chair of the project's Steering

Figure 26.1 • Craik's Eco-centre. *(a)* Under construction; and *(b)* the completed structure.

(a)

(b)

Imagine a Sustainable Future

LEARNING OBJECTIVES

- Putting it all together.

Committee, believes that the projects have fostered a better sense of community and co-operation. Display gardens will demonstrate how landscaping can be used as windbreaks and noise barriers, wildlife habitat, and carbon sequestration. New businesses and tourism activities are also emerging. A Vancouver business, Hemptown, liked Craik's philosophy so much that it will soon open a new hemp mill outside Craik, providing local jobs and markets for crops. By 2008, the town hopes to have produced an ecologically sound village with full-scale production and marketing of food. An energy-efficient marina, wind power system, and new waste management strategy are all in the planning stage.

- *Craik's example illustrates some major themes of environmental science. People and environment are intimately connected, and changes in one lead to changes in the other. We can make simple choices that replace our use of natural resources and increase the sustainability of our communities. Environmental issues involve values and attitudes as well as scientific understanding. And local changes can have regional, national, and international impact—even for a very small town.*

26.1 Imagine

Imagine a future in which we use our environment wisely—an "ecotopia." The learning objective of this chapter is for you to work out, to the best that present information allows, what you think is possible and desirable for this future world, focusing primarily on Canada, and also to describe how this might be accomplished. This may seem an empty academic exercise, but unless we have an idea of what we want, we have no direction in which to seek our future. Ideas are powerful, as history has proved. Wars have been fought over ideas. Ideas led Europeans to the New World and forged the current Canadian culture. So what seems simply an academic exercise could be a powerful force for the future. We, the authors, will give you our best try at what an ecotopia might be, but this is only to provide a starting point for you, who will be part of a future attempt to create ecotopia for an even more distant future.

What would that ecotopia be like? Here are a few of its qualities:

- Since human population growth is the underlying environmental problem, our ecotopia would have to include a human population that had stabilized or even perhaps declined through a systematic decrease in the global birth rate.
- All living resources would be sustainable, as would harvests of those resources.
- Everyone would have opportunities for recreation and for the enjoyment of nature, and pollution would be minimized.
- The risk of extinction of many species would be minimized.
- There would be enough functioning ecosystems to handle the "public service functions" that ecosystems perform for us and other species.
- There would be enough wilderness and other kinds of natural or naturalistic areas to provide some level of recreation.
- An ecotopia would sustain representatives of all natural ecosystems in their dynamic ecological states.

Developing a Sustainable Future

In Chapter 1, we discussed the concept of our ecological footprint, pointing out that developed nations like Canada and the United States have a huge ecological footprint relative to most other countries in the world. In large part, this is because of our excessive per capita consumption of natural resources, and the wastes that we generate in using and disposing of those materials. Stating that we wish to develop a sustainable future acknowledges that our present practices are not sustainable. Indeed, continuing in our present paths of high resource consumption and associated waste generation will not lead to sustainability. What is needed is development of new concepts that will mould industrial, social, and environmental interests into an integrated, harmonious system. In other words, we need to develop a new paradigm—a new pattern—as an alternative to our present model for running society and creating wealth.[3] The new paradigm might be described in the following ways:[4]

- *Evolutionary rather than revolutionary.* Development of a sustainable future will require an evolution in our values that involves our lifestyles as well as social, economic, and environmental justice. It will not be accomplished overnight, in a single radical change in our behaviour.
- *Inclusive, not exclusive.* All peoples on Earth must be included. This means bringing the people of the world to a higher standard of living in a sustainable way that will not compromise our environment. It may also mean changes in the standard of living for those who live in industrialized nations.
- *Proactive, not reactive.* We must plan for change and for events such as human population problems, resource shortages, and natural hazards rather than waiting for them to surprise us and then reacting.
- *Attracting, not attacking.* People must be attracted to the new paradigm because it is the best path for our species and those species which inhabit earth with us for the future. Those who speak for our environment should not take a hostile stand but, through sound, scientific argument and appropriate values, should attract people to the path of sustainability.
- *Assisting the disadvantaged, not taking advantage.* This involves issues of environmental justice. All people should have the right to live and work in a safe, clean environment.

The consensus among environmentalists today is that the best future would have no endangered species. We would rephrase this to say that any species of interest to us would have less than an X probability of going extinct during the next Y number of years—you pick the numbers. Since there are about 1.5 million named species and we would like none of them to go extinct during a reasonable time, we could say that the chance should be less than one in 1.5 million that any species would go extinct within, let's say, a human generation—30 years. We can then turn to our mathematical experts to calculate the probability of extinction for any species for any year.

Starting with our present world, we cannot abruptly halt the momentum of population growth, but we can plan for the smallest expected increase, about 3 billion

A CLOSER LOOK 26.1

An Environmental History of Cities

The Rise of Towns

The first cities emerged on the landscape thousands of years ago during the New Stone Age with the development of agriculture, which provided food in sufficient quantity to maintain a city.[8] In this first stage, the number of people per square kilometre in cities was much higher than in the surrounding countryside, but the density was still too low to cause rapid, serious disturbance to the land. In fact, the waste from city dwellers and their animals was an important fertilizer for the surrounding farmlands. In this stage, the city's size was restricted by the primitive transportation methods for bringing food and necessary resources into the city and removing waste. Because of such limitations, no European medieval town served only by land transportation had a population greater than 15,000.[8]

The Urban Centre

In the second stage, more efficient transportation made possible the development of much larger urban centers. Boats, barges, canals, and wharves, as well as roads, horses, carriages, and carts, made it possible for cities to be located farther from agricultural areas. Rome, originally dependent on local produce, became a city fed by granaries in Africa and the Near East.

The population of a city is limited by how far a person can travel in one day to and from work and by how many people can be packed into an area (density). In the second stage, the internal size of a city was limited by pedestrian travel. A worker had to be able to walk to work, do a day's work, and walk home the same day. The density of people per square kilometre was limited by architectural techniques and primitive waste disposal. These cities never exceeded a population of 1 million, and only a few approached this size, most notably Rome and some cities in China.

The Industrial Metropolis

The Industrial Revolution allowed greater modification of the environment than had been possible before. In the Victorian era, two technological advances that had significant effects on the city environment were improved sanitation, which led to the control of many diseases, and improved transportation (Figure 26.2).

Efficient transportation makes a larger city possible. Workers can live farther from their place of work and commerce, and communication can extend over larger areas. Air travel has freed cities even more from the traditional limitation of situation. We now have thriving urban areas where previously transportation was poor: in the north (Whitehorse, Yukon) and on islands (Newfoundland). These changes increase urban dwellers' sense of separateness from their natural environment.

Subways and commuter trains have also allowed the development of suburbs. In some cities, however, the negative effects of urban sprawl have led many people back to the urban centres or to smaller, satellite cities surrounding the central city. The drawbacks of suburban commuting and the destruction of the landscape in suburbs have brought new appeal to the city centre.

The Centre of Civilization

We are at the beginning of a new stage in the development of cities. With modern telecommunications, people can work at home or at distant locations. Perhaps, as telecommunication frees us from the necessity for certain kinds of commercial travel and related activities, the city can become a cleaner, more pleasing centre of civilization.

An optimistic future for cities requires a continued abundance of energy and material resources, which are certainly not guaranteed, and wise use of these resources. If energy resources are rapidly depleted, mass transit may fail, fewer people will be able to live in suburbs, and the cities will become more crowded. Reliance on coal and wood will increase air pollution. The continued destruction of the land within and near cities could compound transportation problems, making local production of food impossible. The future of our cities depends on our ability to plan and to conserve and use our resources wisely.

Figure 26.2 • **Early water and sewer infrastructure** ensured a reliable water supply and effective sewage disposal, and contributed to the creation of livable environments in cities. Shown here: an artist's reconstruction of a Roman drain and sewer system. (*Source*: Modified from David Macaulay, 1975, *City A Story of Roman Planning and Construction*. New York: Houghton Mifflin.)

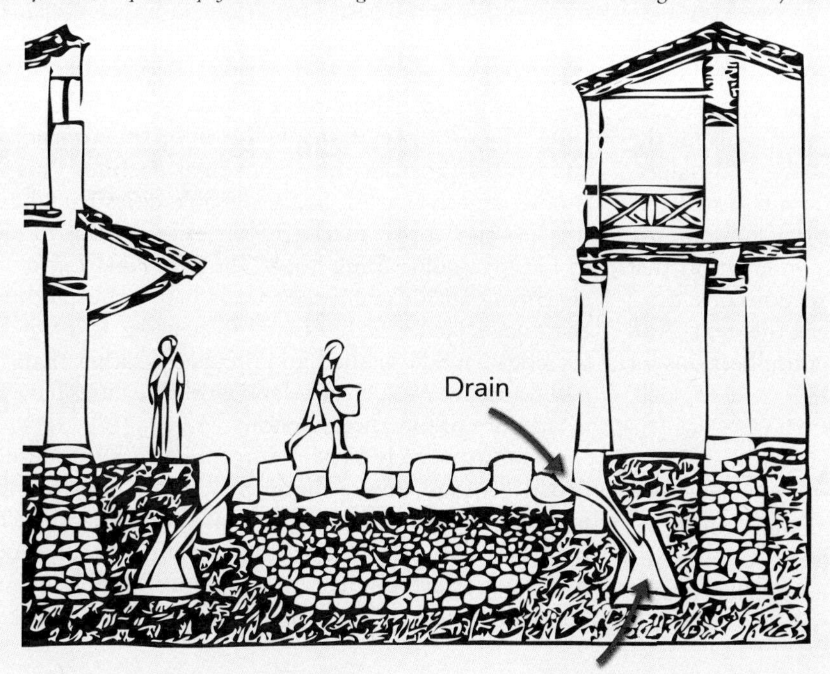

more than are on Earth today. In this ideal world, we would hope that each person would have an opportunity for a decent quality of life. In terms of our connections to nature, this would mean including at least an adequate diet, shelter, sources of energy, water, minerals, daily surroundings that are pleasing, and access to nature as appropriate for an individual's culture and interests.

Just to keep things simple, and knowing that many people have less than they need, we estimate that we will have to double production of all living resources that we harvest; in particular, we will have to double food production, timber production, edible fish production, and energy production, while reducing pollution by half. That's challenging enough for starters.

There is more to specify, but let's just use these goals for now—they're enough to think about. Can nature do it on its own? If not, to what extent do we have to participate? How would this Earth have to be organized in space and varied in time for us to have any hope of achieving our goals? What social mechanisms might help us achieve them? This raises another question: How well can we hope to forecast the future and achieve sustainability with our present knowledge? In our pursuit of ecotopia, we will accept also the oft-repeated saying that "those who ignore the past are destined to repeat it." This means that the past must be a lesson for the future.

Recent thinking about the environment has focused on the big picture: What is necessary at a national scale, or at some landscape scale, to achieve our goals? Once organized entirely around political regions, our society is gradually thinking more and more in terms of planning around large watersheds. This regional approach may help us move closer to the dream of our ecotopia.

Scientific studies of ecosystems and landscapes also lead to speculation about the best way to conserve biological resources. Some argue that nature can be saved only in the large, for instance through projects like the Yellowstone-to-Yukon initiative described in Chapter 13. As we have seen throughout this book, proposals for the environment of the future involve science and values, and people and nature. So what do you want? A vast area of Canada returned to what might be self-functioning ecosystems? Or some open system of conservation that integrates people and allows for more freedom of action? The choices lie with your generation and the next. And tests of their validity are also yours. The implications for environment and for people are huge.

26.2 Planning for Sustainable Cities

In the past, the emphasis of environmental action has most often been on wilderness, wildlife, endangered species, and the impact of pollution on natural landscapes outside cities. Yet as we saw in Chapter 4, worldwide we are becoming an increasingly urbanized species. In Canada, about 80% of the population live in urban areas and about 20% live in rural areas. Today, approximately 45% of the world's population—2.75 billion people—live in cities.[5] It is projected that 62% of the population, 6.5 billion people, will live in cities by the year 2025.[6] Economic development leads to urbanization; 75% of people in developed countries live in cities, but only 38% of the people in the poorest developing countries are city dwellers.[5]

Megacities—huge metropolitan areas with more than 8 million residents—are cropping up more and more. In 1950, the world had only two: New York City and nearby urban New Jersey (12.2 million residents) and greater London (12.4 million). By 1975, Mexico City, Los Angeles, Tokyo, Shanghai, and São Paulo, Brazil, had joined this list. In 1995, there were 23 such areas, 17 of them in the developing world. At present, Los Angeles and New York City are among the 10 largest megacities.[6, 7] As mentioned in Chapter 1, Canada currently has no megacities. With about 4.2 million people, the Greater Toronto Area (GTA) is Canada's largest city, but still small compared to the world's megacities.

It is estimated that by 2015 the world will have 36 megacities, 23 of them in Asia. In the future, most people will live in cities. In most nations, most urban residents will live in the country's single largest city. In the future, for most people, living in an environment of good quality will mean living in a city that is managed carefully to maintain that environmental quality.

In the development of the modern environmental movement in the 1960s and 1970s, it was fashionable to consider everything about cities bad and everything about wilderness good. Cities were thought of as polluted, lacking in wildlife and native plants, dirty, and artificial—and therefore bad. Wilderness was thought of as unpolluted, clean, full of wildlife and native plants, and natural—and therefore good. Although it was fashionable to disdain cities, the majority of people live in urban environments and have suffered directly from their decline. This view is now seen as flawed. In fact, livable cities have been at the heart of many successful societies. In the foreword to her groundbreaking book *The Death and Life of Great American Cities*, the eminent Canadian author and urban critic Jane Jacobs wrote that, "Whenever and wherever societies have flourished and prospered rather than stagnated and decayed, creative and workable cities have been at the core of the phenomenon."

Comparatively little public concern has focused on urban ecology; many urban people see environmental issues as outside their realm. But the reality is just the opposite: City dwellers are at the center of some of the most important environmental issues. People are realizing that city and nature are inextricably connected.

A CLOSER LOOK 26.2

A Brief History of City Planning

Two dominant themes in formal city planning have been planning for defence and planning for beauty. Ancient Roman cities were typically designed along simple geometric patterns that had both practical and aesthetic benefits. The symmetry of the design was considered beautiful but also provided a useful layout for streets.

During the height of Islamic culture, in the first millennium A.D., Islamic cities typically contained beautiful gardens, often within the grounds of royalty. One of the most famous urban gardens in the world is the garden of the Alhambra, a palace in Granada, Spain (Figure 26.3). The garden was created when this city was a Moorish capital, and it was maintained after Islamic control of Granada ended in 1492. Today, as a tourist attraction that receives 2 million visitors a year, the Alhambra garden demonstrates the economic benefits of aesthetic considerations in city planning. It also illustrates that making a beautiful park a specific focus in a city benefits the city environment by providing relief from the city itself.

After the demise of the Roman Empire in Western Europe, the earliest planned towns and cities in Europe were walled fortress cities designed for defence. But even in these, city planners considered the aesthetics of the town. In the fifteenth century, one such planner, Leon Battista Alberti, argued that large and important towns should have broad and straight streets; smaller, less fortified towns should have winding streets to increase their beauty. He also advocated the inclusion of town squares and recreational areas, which continue to be important considerations in city planning.[23] One of the most successful of these walled cities is Carcasonne, in southern France, now the third most visited tourist site in that country. Today, walled cities have become major tourist attractions, again illustrating the economic benefits of good aesthetic planning in urban development.

The usefulness of walled cities essentially ended with the invention of gunpowder. The Renaissance sparked an interest in the ideal city, which in turn led to the development of the park city. A preference for gardens and parks, emphasizing recreation, developed in Western civilization in the seventeenth and eighteenth centuries. It characterized the plan of Versailles, France, with its famous formal parks of many sizes and tree-lined walks, and also the work of the Englishman Capability Brown, who designed parks in England and was one of the founders of the English school of landscape design, which emphasized naturalistic gardens.

Figure 26.3 • Planned beauty. The Alhambra gardens of Granada, Spain, illustrate how vegetation can be used to create beauty within a city.

Fortunately, we are experiencing a rebirth of interest in urban environments and in urban ecology. Environment Canada has added several urban monitoring sites, including the Metropolitan Toronto Zoo and London, Ontario's Upper Thames Conservation Authority, to their Ecological Monitoring and Assessment Network (EMAN). In the following section, we examine major aspects of urban ecology.

26.3 Site and Situation: The Location of Cities

Cities are not located at random but develop mainly because of local conditions and regional benefits. In most cases, they grow up at crucial transportation locations (an aspect of what is called the city's situation) and can be readily defended, with good building locations, water supplies, and access to resources (qualities related to what is called site). The primary exceptions are cities that have been located primarily for political reasons. For example, Ottawa was a fur and timber town in the mid-1850s, when squabbles among Québec City, Toronto, Montreal, and Kingston over the location of a new national capital caused Queen Victoria to choose a compromise location. Although pleasantly positioned on a large river, Ottawa's site was distant from all major trading centres and lacked a deepwater harbour.

The location of a city is influenced by the two factors just mentioned: site, which is the summation of all the environmental features of that location, and situation, which is the placement of the city with respect to other areas. A good site includes a geologic substrate suitable for buildings, such as a firm rock

(a)

(b)

Figure 26.4 • The importance of a city's site. *(a)* The geologic, topographic, and hydrologic conditions of a city's site greatly influence how successful the city can be. If these conditions, known collectively as the city's site, are poor, much time and effort are necessary to create a livable environment. Bangkok, an important location (called the situation) for a city, sits where the water table is at or near the surface, at the mouth of the Chao Phraya River, where wetlands provide a poor base for building, afford a breeding ground for mosquitoes, and make construction difficult. *(b)* In contrast, St. John's, Newfoundland has a strong bedrock base for buildings and a soil that is sufficiently above the water table so that flooding and mosquitoes are much less a problem.

base and well-drained soils that are above the water table; nearby supplies of drinkable water; good nearby lands suitable for agriculture; and forests (Figure 26.4). It is also easier to build a city where the climate is benign—meaning that it does not suffer extremes of temperature and rainfall and is not subject to frequent storms.

The environmental situation strongly affects the development and importance of a city, particularly with regard to transportation and defence. Most early cities, including all-important cities in the ancient Roman Empire, were located on or near waterways. Other crucial transportation points, such as markets, river crossings, and forts, have also influenced the location of cities. Newcastle, England, and Budapest, Hungary, are located at the lowest bridging points on their rivers. Other cities, such as Geneva, are located where a river enters or leaves a major lake. Some well-known cities are located at the confluence of major rivers: Calgary lies at the confluence of the Bow and Elbow Rivers; Montréal, Manaus (Brazil), Koblenz (Germany), and Khartoum (Sudan) are all located at the confluence of several rivers. Many famous cities are located at crucial defensive locations, such as on or adjacent to easily defended rock outcrops. Examples include Edinburgh, Athens, and Salzburg. Other cities are situated on peninsulas—for example, Monaco and Istanbul (Figure 26.5).

Cities are often founded close to a mineral resource, such as salt (Salzburg, Austria), metals (Kalgoorlie, Australia), or medicated waters and thermal springs (Spa, Belgium; Bath, Great Britain; Vichy, France; and Banff, Alberta). A successful city can grow and spread over surrounding terrain, so its original purpose may be obscured to a resident. Its original market or fort may have evolved into a square or a historical curiosity. In most cases, though, cities originated where the situation provided a natural meeting point for people.

An ideal location for a city has both a good site and a good situation, but such a place is difficult to find. Paris is perhaps one of the best examples of a perfect location for a city—one with both a good site and good situation. Paris began on an island more than 2,000 years ago, the situation providing a natural moat for defence and waterways for transportation. Surrounding countryside, a fertile lowland called the Paris basin, affords good local agricultural land and other natural resources.

Site Modification

The environment provides the site, but technology and environmental change can alter a site for better or worse. People can improve the site of a city and have done so when the situation of the city made it important and when its citizens could afford large projects. An excellent situation can sometimes compensate for a poor site. However, improvements are almost always required to the site so the city can persist.

For example, Bangkok (Figure 26.4) has a good situation but a poor site. An important transportation center at the mouth of the Chao Phraya River, it lies on low mudflats of the delta, which are unstable and provide poor substrate for construction. Backwaters and swamps offer little as a local resource for agriculture but provide breeding habitats for mosquitoes. Fishing in the Gulf of Thailand is an advantage, however, and the situation of Bangkok has proved to be important for transportation and as a link between coastal and inland regions. Similarly, Venice, Italy, was founded by people escaping invader hordes at the end of the Roman Empire. Its location in marshland along the Adriatic Sea was easily defended and well suited for trade and transportation. Like Bangkok, however, Venice's site presented environmental challenges that had to be solved—using poles driven into the mud, on which buildings could be constructed—before the city could develop. More than 1,000 years later, Venice still stands on that foundation.

However, changes in a site over time can also have adverse effects on a city. For example, Bruges, Belgium, developed as an important centre for commerce in the thirteenth century because its harbour on the English Channel permitted trade with England and other European nations. By the fifteenth century, however, the harbour had seriously silted in, and the limited technology of the time did not make dredging possible. This problem, combined with political events, led to a decline in the importance of Bruges—a decline from which it never recovered. Nevertheless, today, Bruges still lives, a beautiful city with many fine examples of medieval architecture. Ironically, the fact that these buildings were never replaced with new ones makes Bruges a popular tourist destination. Ghent, Belgium, and Ravenna, Italy, are other examples of cities whose harbours silted in.[8] As human effects on the environment bring about global change, there may be rapid, serious changes in the sites of many cities. For instance, if global warming occurs and sea levels rise, many coastal cities will be subject to flooding.

Figure 26.5 • **The city of Istanbul, Turkey**, is located on a peninsula of land jutting into the Bosphorus strait, which separates Europe from Aisa.

City Planning and the Environment

Many cities in history grew without any conscious plan. **City planning**—formal, conscious planning for new cities—occurred in many ancient civilizations, including those of Rome, China, Egypt, and Latin America. Beauty has always been important in the design of cities, as for instance in the case of Ottawa. Since 1899, Ottawa's planning has been overseen by the National Capital Commission and its predecessors. These agencies were specifically charged with "beautifying" Ottawa through a system of scenic driveways and bridges, and the creation of urban parks. The Todd Plan of 1903 was the first to envision Ottawa as more than buildings. It set out a comprehensive vision of a beautiful city region, defined by a system of parks. This design has made Ottawa one of the most pleasant cities in Canada.

Parks have become more and more important in cities. Vancouver's Stanley Park, one of the first large public parks in Canada, was planned and built in the nineteenth century on a 400 ha peninsula formerly used for logging. In a visionary move, Vancouver's City Council established an autonomous Board of Parks and Recreation to oversee planning and operation of the park. The Board remains the only elected body of its kind in Canada. The park's designers took site and situation into account and attempted to blend improvements to the site with the aesthetic qualities of the city.

An extension of the park idea was the **garden city**, a term coined in 1902 by Ebenezer Howard. Howard believed that city and countryside should be planned together. A garden city was one that was surrounded by a **greenbelt**. The idea was to locate garden cities in a set connected by greenbelts, forming a system of countryside and urban landscapes. The idea caught on, and garden cities were planned and developed in Great Britain and the United States. Letchworth, Welwyn Garden City and Hampstead Garden Suburb, England, are examples of Howard's garden city concept. In Canada, Howard's ideas prompted the development of neighbourhoods like Rosedale in Toronto and Westmount in Montréal.

26.4 The City as an Ecosystem

Cities are ecological systems of a special kind. Like any other life-supporting system, a city must maintain a flow of energy, provide necessary material resources, and have ways of removing wastes. These ecosystem functions are maintained in a city by transportation and communication with outlying areas. A city is not a self-contained

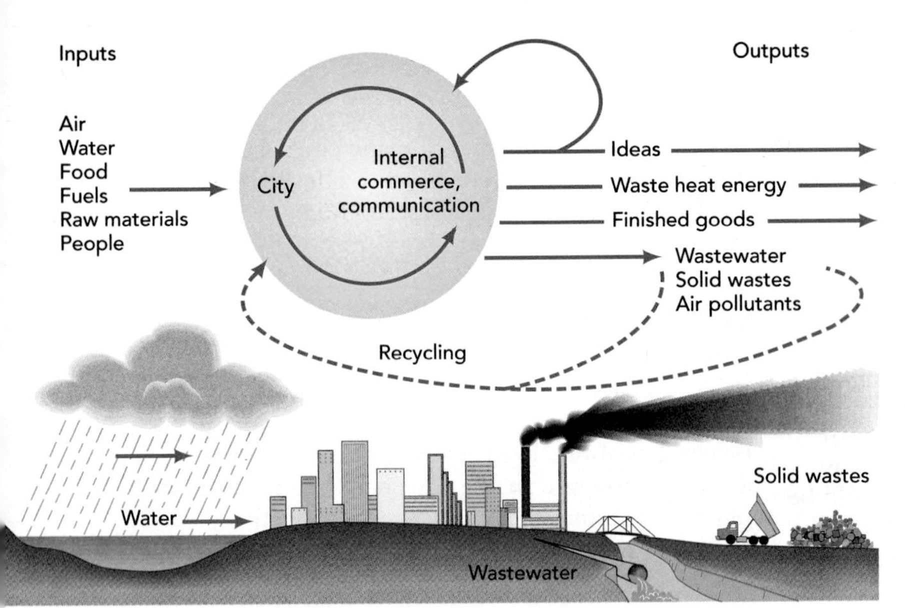

Figure 26.6 • The city as a system with flows of energy and materials. A city must function as part of a city–countryside ecosystem, with an input of energy and materials, internal cycling, and an output of waste heat energy and material wastes. As in any natural ecosystem, recycling of materials can reduce the need for input and the net output of wastes.

ecosystem; it depends on other cities and rural areas. A city takes in raw materials from the surrounding countryside: food, water, wood, energy, and mineral ores, everything that a human society uses. In turn, the city produces and exports material goods and, if it is a truly great city, exports ideas, innovations, inventions, arts, and the spirit of civilization. A city cannot exist without countryside to support it. This is a reflection of the city's ecological footprint—the area of land necessary to produce the resources used by the city, and assimilate the city's wastes (Figure 26.6).[9] The average city resident in an industrial nation annually uses (directly or indirectly) about 208,000 kg of water, 660 kg of food, and 3,146 kg of fossil fuels and produces 1,660,000 kg of sewage, 660 kg of solid wastes, and 200 kg of air pollutants. For this reason, cities are sometimes said to cast an urban shadow over surrounding lands, because of the influence of development, transportation, resource use and waste generation, recreational uses, and other pressures from the urban population on outlying areas. It therefore follows that if the environment of a city declines, almost certainly the environment of its surroundings will also decline. The reverse is also true: If the environment around a city declines, the city itself will be threatened. Some people suggest, for example, that the ancient Native American settlement in Chaco Canyon, Arizona, declined after the environment surrounding it experienced environmental change, for instance a decline in rainfall, depleted forest resources from overharvesting, or lost soil fertility from poor farming practices.[10]

A city changes the landscape; and because it does, it also changes the relationship between biological and physical aspects of the environment. Many of these changes were discussed in earlier chapters as aspects of pollution, water management, or climate. We next describe how a city can fit within, use, and avoid destroying the ecological systems on which it depends and how the city itself can serve human needs and desires as well as environmental functions.

Like any ecological and environmental system, a city has an "*energy budget*." The city exchanges energy with its environment in the following ways: (1) absorption and reflection of solar energy, (2) evaporation of water, (3) conduction of air, (4) winds (air convection), (5) transport of fuels into the city and burning of fuels by people within the city, and (6) convection of water (subsurface and surface stream flow). These affect the climate within the city, and the city may affect the climate in the nearby surroundings, a possible landscape effect.

Cities also affect the local climate; as the city changes, so does its climate (see Chapter 20). Cities are generally less windy than nonurban areas because buildings and other structures obstruct the flow of air. But city buildings also channel the wind, sometimes creating local wind tunnels with high wind speeds. The flow of wind around one building is influenced by nearby buildings, and the total wind flow through a city

Figure 26.7 • A typical urban heat island profile. The graph shows temperature changes correlated with the density of development and trees. [*Source:* Modified from Andrasko and Huang, in H. Akbari et al., *Cooling Our Communities: A Guidebook on Tree Planting and Light-Colored Surfacing* (Washington, D.C.: U.S. EPA Office of Policy Analysis, 1992).]

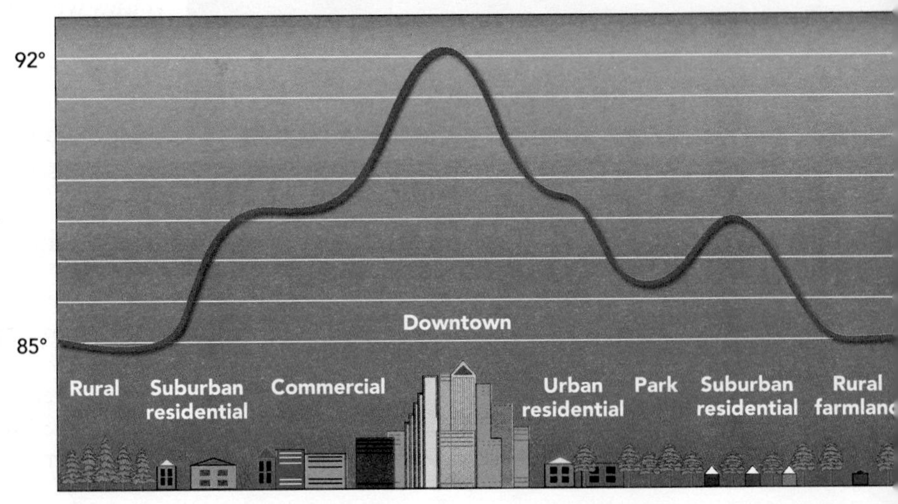

Figure 26.8 • **Planned for better drainage.** This parking lot incorporates wetland drainage and soils, which temporarily absorb runoff.

is the result of the relationships among all the buildings. Thus, plans for a new building must take into account its location among other buildings as well as its shape. Recall that a city typically receives less sunlight than the countryside because of the higher levels of particulates in the atmosphere over cities.[11] Despite this, cities are warmer than surrounding areas (a city is a *heat island*), for two reasons. One is increased heat production from burning fossil fuels and other industrial and residential activities. The other is a lower rate of heat loss, partly because buildings and paving materials act as solar collectors (Figure 26.7).[12]

Cities also affect the *water cycle*, in turn affecting soils and, consequently, plants and animals in the city. Because paved city streets and city buildings prevent water infiltration, most rain runs off into storm sewers. Street pavements and sidewalks prevent water in the soil from evaporating to the atmosphere, a process that cools natural ecosystems; without this evaporative cooling, cities tend to be hotter than their surroundings. Chances of flooding increase both within the city and downstream outside the city.

New, ecological methods of managing storm water can alleviate some of these problems by controlling the speed and quality of water running off pavements and into streams (Figure 26.8). The University of Victoria has recently completed an innovative integrated stormwater management plan. For the Engineering and Computer Science Building, for example, the plan calls for rooftop storage of stormwater, and porous pavement and grassed swales to encourage natural infiltration. Whereas a traditional design would have incorporated more impervious surfaces and directed all rainfall to storm sewers, the innovative design is aimed at storing and infiltrating more stormwater on site. (The university is also one of the first in Canada to use solar energy to heat water for its aquatic centre.)

Cities have long used their rivers as places to dump wastes rather than as an aesthetic or recreational resource. Edmonton, Alberta, on the North Saskatchewan River, illustrates the traditional treatment. The North Saskatchewan's floodplain provided a convenient transportation corridor, so the south shore was dominated by

Figure 26.9 • **A city and its river.** *(a)* One of Europe's oldest cities, Heidelberg, Germany is closely linked to the Neckar River. *(b)* Ottawa's Rideau Canal is an integrating feature for urban recreation and aesthetics.

(a)

(b)

railroads, a railway station, and a hotel, while the downtown area developed on the north shore. In Edmonton's early years, the river had little place in the city as a source of recreation and relief for its citizens or in the conservation of nature. Contrast Edmonton with Ottawa, discussed earlier. In Ottawa, the Rideau River has always been an integral part of city planning and remains an important recreational corridor, incorporating public buildings, shops, and pleasant walkways and pocket parks that lead into the city's downtown(Figure 26.9*b*).

In urban environments, long stretches of major rivers are channelized. Recall from Chapter 18 that channelization has two negative effects. First, when the river is maintained in an artificial channel, its sediment load is not deposited on the land and the land's fertility is not renewed. Second, this sediment load passes downriver and is lost at the mouth, where it causes siltation and may fill in important harbours and damage cities at the ocean side. It is important to note that these negative effects are an indirect result of building cities on floodplains without proper planning and then trying to prevent the inevitable floods.

A city has a great impact on *soils*. Cement, asphalt, or stone cover most of a city's soil; the soil no longer has its natural cover of vegetation, and the natural exchange of gases between the soil and air is greatly reduced. Because they are no longer replenished by vegetation growth, these soils lose organic matter and soil organisms die from lack of food and oxygen. In addition, the process of construction and the weight of the buildings compact the soil, which restricts water flow. City soils, then, are more likely to be compacted, waterlogged (or dry), impervious to water flow, and lacking in organic matter than they otherwise would be.

A kind of soil important in cities is the soil that occurs on human-made lands, which are lands created from fill, sometimes as waste dumps of all kinds, sometimes directly to create more land for construction. The soils of **made lands** are different from those of the original landscape. They may be made of all kinds of trash, from newspapers to bathtubs, and may contain some toxic materials. The fill material is unconsolidated, meaning that it is loose material without rock structure. Thus, it is not well suited to be a foundation for buildings. Fill material is particularly vulnerable to the shaking from earthquakes. In such events, the fill can act somewhat like a liquid and amplify the effects of the earthquake on buildings. However, some made lands have been turned into well-used parks. For example, several marina parks in Toronto, Ontario, are built on "lakefill" created from solid wastes, dredged lake sediments and soils from construction excavations. These parks extend into Lake Ontario, providing public access to beautiful scenery. (See Chapter 24 for more information about solid-waste disposal.)

In a city, everything is concentrated, including *pollutants*. City dwellers are exposed to more kinds of toxic chemicals in higher concentrations and to more human-produced noise, heat, and particulates than their rural neighbours (see Chapter 15). This environment makes life riskier. Human lives are shortened by an average of one to two years in the most polluted cities. In Chapter 21, we discussed the impact of Windsor, Ontario's air quality on human health in that city. Sources of air pollutants in the city include motor vehicles, stationary power sources, home heating, and industries.[13] Although it is impossible to eliminate exposure to pollutants in a city, it is possible to reduce the exposure through careful design, planning, and development. For example, air pollution impacts can be reduced by placing houses and recreational areas away from roads and by developing a buffer zone using trees that are resistant to the pollutant and that absorbed pollutants.

Vegetation such as trees, shrubs, and flowers improves the beauty of a city. Plants fill different needs in different locations.[14] Trees provide shade, which reduces the need for air conditioning and makes travel much more pleasant in hot weather. In parks, vegetation provides places for quiet contemplation; trees and shrubs can block some of the city sounds, and their complex shapes and structures create a sense of solitude. Plants also provide habitats for wildlife such as birds and squirrels, which many urban residents consider pleasant additions to a city (Figure 26.10). In many cities, trees are now considered an essential element of the urban visual scene, and major cities have large tree-planting programs. For example, the Waterfront Regeneration

Figure 26.10 • **Paris was one of the first modern cities to use trees along streets to provide beauty and shade, as shown in this picture of the famous Champs Elysées.**

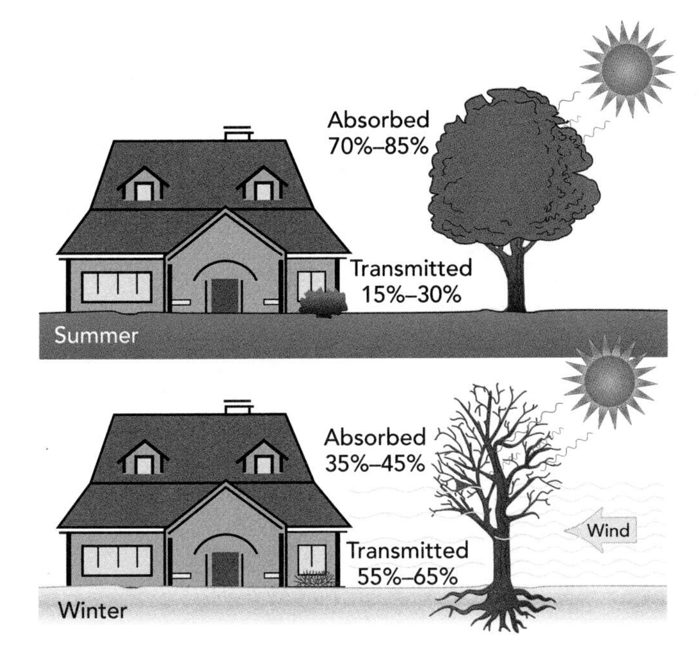

Figure 26.11 • Trees cool homes. Trees can improve the microclimate near a house, protecting the house from winter winds and providing shade in the summer while allowing sunlight through in the winter. [*Source*: Modified from J. Huang and S. Winnett, in H. Akbari et al., *Cooling Our Communities: A Guidebook on Tree Planting and Light-Colored Surfacing* (Washington, D.C.: U.S. EPA Office of Policy Analysis, 1992).]

Trust in Toronto planted over 300,000 native trees and shrubs in 2001.[15]

Trees can also soften the effects of climate near houses. In colder climates, rows of conifers planted to the north of a house can protect it from winter winds. Deciduous trees to the south can provide shade in the summer, reducing requirements for air conditioning, yet allowing sunlight to warm the house in the winter (Figure 26.11).[16] The lifetime of trees in a city is usually shorter than in their natural woodland habitats unless they are given considerable care. Care must also be taken to create diverse urban planting. Reliance on one or a few species results in ecologically fragile urban planting, as we learned when Dutch elm disease spread throughout eastern and central Canada and the United States, destroying urban elms.[17]

Cities, of course, have many recently disturbed areas, including abandoned lots and corridors between lanes in boulevards and highways. Disturbed areas provide habitat for early successional plants, including many that we call "weeds," which are often introduced (exotic) plants, such as European mustard and dandelions. Therefore, wild plants that do particularly well in cities are those characteristic of disturbed areas and of early stages in ecological succession (see Chapter 10).

With the exception of some birds and small, docile mammals such as squirrels, most forms of *wildlife* in cities are considered pests. Vancouver, for example, has a large population of urban coyotes, and many cities have abundant raccoon, that play havoc with unattended garbage and backyard bird feeders. But there is much more wildlife in cities, a great deal of it unnoticed. For example, foxes in London, England, feed on garbage and roadkill (animals run over by motor vehicles); shy and nocturnal, they are seen by few Londoners.[18] Their food chain includes pets and other animals, fruits and vegetables, earthworms and insects, and scavenged items.

There is growing recognition that urban areas can be modified to provide more habitats for wildlife that people can enjoy. This can be an important method of biological conservation.[19, 20] For example, chimney swifts once lived in hollow trees but are now common in chimneys and other vertical shafts, where they glue their nests to the walls with saliva. Urban zoos play an important role in the conservation of endangered species, and the importance of parks and zoos will increase as truly wild areas shrink. Finally, cities that are seaports often have many species of marine wildlife at their doorsteps. Halifax Harbour's waters include sharks, cod, redfish, halibut, monkfish, mackerel, squid, lobster, and dozens of other species of fish and shellfish.[18] Cities can even be home to rare or endangered species. Peregrine falcons, once endangered as a result of organic chemical pollution, have been reintroduced into cities. (Figure 26.12).[3] Urban drainage structures can also be designed as wildlife habitats if they are planned to maintain or create stream and marsh habitats with meandering waterways and storage areas that do not interfere with city processes. Such areas can become habitats for fish and mammals (Figure 26.13).[21]

Animal pests, including cockroaches, fleas, termites, rats, and pigeons are familiar to urban dwellers. Before modern sanitation and medicine, diseases carried by animals played a major role in limiting human population density in cities. Bubonic plague is spread by fleas found on rodents; mice and rats in cities promoted the spread

Figure 26.12 • Peregrine falcon nest on top of an office building

Figure 26.13 • How water drainage systems in a city can be modified to provide wildlife habitat. In the community on the right, concrete-lined ditches result in rapid runoff and have little value to fish and wildlife. In the community on the left, the natural stream and marsh were preserved; water is retained between rains and an excellent habitat is provided.

of the Black Death (see Chapter 1). Bubonic plague continues to be a health threat in cities. The World Health Organization reports several thousand cases a year.[22] Poor sanitation and high population densities of people and rodents set up a situation where the disease can strike. We can best control pests by recognizing how they fit their natural ecosystem and identifying their natural controlling factors. One of the keys to controlling pests is to eliminate their habitats. For example, the best way to control rats is to reduce the amount of open garbage and eliminate areas to hide and nest.

26.5 How Much Green Space is Enough? Landscape Management in Canada

Even if most of us will live in cities in the future, our world will need green space—natural areas in and around cities, and in wild, remote areas. How much green space is enough?

In Canada, the landscape perspective for environmental management has been actively promoted by Environment Canada, Canada's Biodiversity Convention Office, the National Round Table on the Environment and Economy, and similar bodies. Tackling environmental issues from a landscape perspective does however require different kinds of government structures and decision processes than might have been the case in the past. It is not enough to manage some pieces of the land independent of others; successful landscape management requires linkage of cities to outlying agricultural and other "working" lands, and to protected areas more distant from cities. The governments of Canada and other countries are only just beginning to explore these important concepts.

As we have seen in previous chapters, scientific studies of ecosystems and landscapes also lead to speculation about the best way to conserve biological resources. Some argue that only very large-scale action will suffice. As an extreme example, in the United States, a group called the Wildlands Project has argued that big predators, referred to as "umbrella species," are keys to ecosystems and that landscapes must be large enough to protect these predators' huge home ranges. They believe that even the largest national parks, such as Yellowstone, are not big enough, and much larger areas

Figure 26.14 • Wildlands Project diagram of land divisions. Leaders of the Wildlands Project propose that large areas of the continent be managed around the needs of big predators and that we replan our landscapes to provide a combination of core areas in which no human activities would take place, corridors, and inner and outer buffers in which human activity would be restricted.

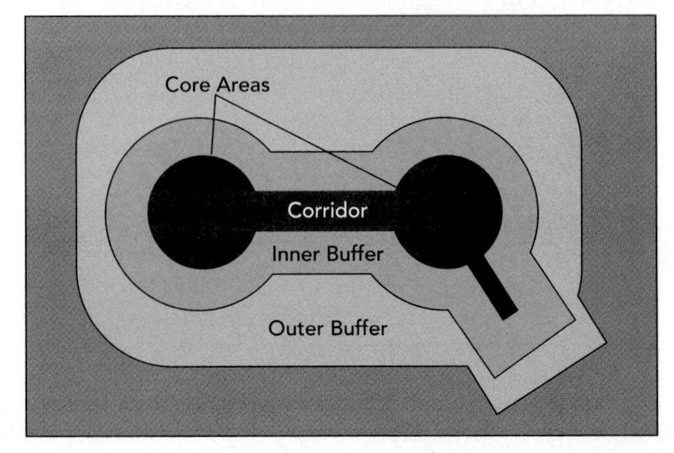

of North America need "rewilding"[24] (Figure 26.14). Critics of the project say that, while some ecological research suggests that large predators may be important, what controls populations in all ecosystems is far from understood. Similarly, the idea of keystone species, central to the rationale of the Wildlands Projects, lacks an adequate scientific base.

More realistic landscape-scale conservation proposals include the Yellowstone-to-Yukon and Algonquin-to-Adirondack conservation initiatives described in Chapter 13. These proposals would link existing protected areas in a system of core areas and connecting corridors. The Yellowstone-to-Yukon initiative is supported by 180 organizations and individuals on both sides of the Canada-U.S. border. (Figure 26.15.) Its value therefore lies not only in its landscape-scale vision, but also in the co-operative management network that is at its core. Farther to the east, the Algonquin-to-Adirondack Conservation Association, an independent non-governmental organization, aims to "restore, enhance, and maintain ecological connectivity, ecosystem function, and native biodiversity, while respecting sustainable human land uses" in a large corridor of land between Ontario's Algonquin Provincial Park and New York's Adirondack Park. They plan to work with local landowners to restore natural landscape connections that have been severed by human activities.

Proposals for the environment of the future thus involve science and values, and people and nature. So what do you want? A vast area of Canada returned to what might be self-functioning ecosystems? Or some open system of conservation that integrates people and allows for more freedom of action? The choices lie with your generation and the next. And tests of their validity are also yours. The implications for environment and for people are huge.

How then should we proceed? We next discuss processes of environmental decision making, and how they can be used to plan a sustainable future.

26.6 The Process of Planning a Future

Our society has formal planning processes for land use. These processes have two qualities: a set of rules (laws, regulations, etc.) requiring forms to be filled out and certain procedures followed; and an imaginative attempt to use land and resources in ways that are beautiful, economically beneficial, and sustainable. All human civilizations plan the development and use of land and resources in one way or another—through customs or by fiat of a ruling monarch, if not by democratic processes. For thousands of years, experts have created formal plans for cities and for important buildings and other architectural structures, such as bridges. In some cases,

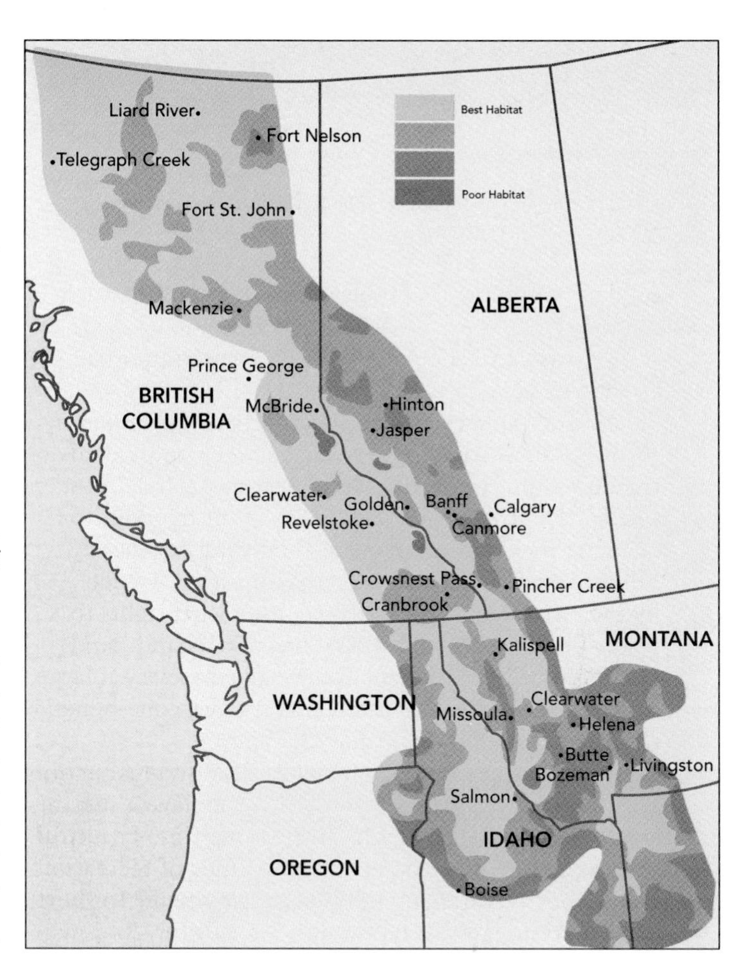

Figure 26.15 • **Example of Yellowstone-to-Yukon studies of grizzly bear habitat.**

land development has proceeded uncontrolled, without an explicit or apparent plan, but as a result of need and custom. For example, cities develop at important transportation centers and where local resources can support a high density of people. In medieval Europe, bridges and other transportation aids developed in response to local needs. People arriving at a river would pay the farmer whose land lay along the river to row them across. Sometimes this would become more profitable than farming or an important addition to the farmer's income. Eventually he might build a toll bridge.[25]

Issues of environmental planning and review are closely related to how land is used—what proportion is allocated to people, to farms, to industry, and to wild areas. Land use in Canada is dominated by agriculture and forestry; only a small portion of land (about 0.2%) is urban. However, rural lands are being converted to non-agricultural uses at about 9,000 km^2 per year. About half the conversion is for wilderness areas, parks, recreational areas, and wildlife refuges. The other half is for urban development, transportation networks, and other facilities. On a national scale, there is relatively little conversion of rural lands to urban uses. But in rapidly growing urban areas, increasing urbanization may be viewed as destroying agricultural land and exacerbating

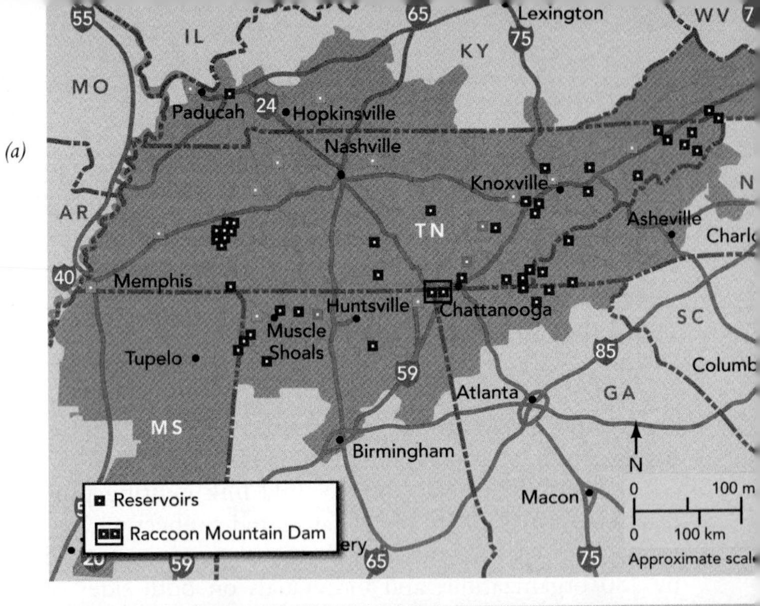

Figure 26.16 • The Tennessee Valley Authority *(a)* A map showing the region encompassed by the TVA (darker area) and one of the major impoundments, The Raccoon Mountain Pumped Storage facility; *(b)* a large reservoir created by one of the TVA dams.

urban environmental problems, and urbanization in remote areas with high scenic and recreational value may be viewed as potentially damaging to important ecosystems.

In a democracy, planning with the environment in mind leads to a tug-of-war between individual freedoms and the welfare of society as a whole. On one hand, citizens of a democracy want freedom to do what they want, wherever they want, especially on land that, in Western civilizations, is "owned" by the citizens or where citizens have legal rights to water or other resources. On the other hand, land and resource development and use affect society at large, and in either direct or indirect ways everyone benefits or suffers from a specific development.

How can we balance freedom of individual action with effects on society? How can we achieve a sustainable use of resources that is at the same time beautiful, spiritually fulfilling, open to many kinds of recreation, and supportive of many kinds of employment? In short, the questions are: Who speaks for nature? And who legally represents the environment? The landowner? Society at large? At this time, we have no definitive answers. Planning is a social experiment in which we all participate. Planning occurs at every level of activity, from a garden to a house, a neighbourhood, a city park and its surroundings, a village town or city, a county, a state, and a nation. However, the history of our laws provides insight into our modern dilemma.

Regional Planning

It is increasingly clear that planning for a sustainable future will require a regional perspective, not just isolated local concerns. The next sections of this chapter give examples of these experiments at various spatial scales. Their successes and failures may be helpful guides to us in developing paths to our ecotopia.

One kind of important experiment of the twentieth century was regional planning. One of the best-known regional plans began in 1933, when U.S. President Franklin D. Roosevelt proposed the establishment of the Tennessee Valley Authority (TVA), a semi-independent authority responsible for promoting economic growth and social well-being for the people throughout parts of seven states, which were economically depressed at the time the authority was established.[26] The TVA was and is one of the most ambitious examples of regional planning in the twentieth century (Figure 26.16). It is characterized by multi-dimensional and multi-level planning to manage land and water resources. The authority is involved in the production and regulation of electrical power, flood control, navigation, and outdoor recreation. Regional and watershed-scale environmental planning was also one of the driving forces in the creation of the Fraser Basin Council (Chapter 18), Ontario's system of Conservation Authorities (which oversee resource management within a particular watershed), and even the Canada-United States International Joint Commission, which was created to prevent and resolve disputes about shared water resources. Other examples of regional land use planning include the Yellowstone-to-Yukon and Adirondacks-to-Algonquin initiatives described earlier, and comprehensive planning for Canada's huge tracts of publicly owned land.

Government officials and scientists involved in developing plans for public lands are faced with complex land-use problems at a variety of levels. Often, the problem is so complex that no easy answers can be found. Nonetheless, because action or inaction today can have serious consequences tomorrow, it is best to have at least some plans to protect and preserve a quality environment for future generations.

CRITICAL THINKING ISSUE

How Can Urban Sprawl Be Controlled?

As the world becomes increasingly urbanized, individual cities are growing in area as well as population. Residential areas and shopping centers move into undeveloped land near cities, impinging on natural areas and creating a chaotic, unplanned human environment. Millions of hectares of rural land were developed in Canada and the United States between 1970 and the present, a process that is continuing at the rate of 160,000 ha per year. Urban sprawl has become a serious concern in communities all across Canada and the United States.

Many Canadian cities are attempting to limit sprawl by imposing urban growth boundaries. In Ontario, most municipalities have Official Plans that designate the location and type of future development. These plans can be used to control the future shape of cities, although they are seldom used for that purpose because they are easily, and frequently, modified to accommodate development pressures. The larger problem may be encouraging cities to work together to create a regional development plan. In February 2005, the Ontario government announced their intention to designate a Greenbelt Area in southern Ontario's Golden Horseshoe, the densely populated region around Toronto along the north shore of Lake Ontario. Although controversial, the proposal is important as a major step toward regional coordination of land use planning.

The city of Boulder, Colorado, has been in the forefront of sprawl control since 1959, when it created the "blue line"—a line at an elevation of 1,761 m (the city itself is at 1,606 m) above which it would not extend city water or sewer services. Boulder's citizens felt, however, that the blue line was insufficient to control development and maintain the city's scenic beauty in the face of rapid population growth. To prevent uncontrolled development in the area between the city and the blue line, Boulder systematically began to establish a greenbelt around the city, and limited the building of new homes to 2% per year.

The benefits of controlled-growth initiatives often include a defined urban–rural boundary; rational, planned development; protection of sensitive environmental areas and scenic vistas; and large areas of open space within and around the city for recreation. But residential growth controls can also have implications for traffic and air pollution in transportation corridors, as workers are displaced to "bedroom communities" on the outskirts of cities. Toronto's Golden Horseshoe region now encompasses bedroom communities more than 100 km to the east, west, and north of the city, while its major feeder highways, Highways 400, 401, and 427, are jammed with traffic at almost any time of the day.

Critical Thinking Questions

1. Is a city an open or a closed system (Chapter 3)? Use examples from the case of Toronto or Boulder to support your answer.
2. As a city takes steps to limit growth, it becomes an even more desirable place to live, which subjects it to even greater growth pressures. What ways can you suggest to avoid such a positive-feedback loop?
3. One response to urban sprawl is "smart growth" planning, now underway in most regions of Canada and the United States. In part, smart growth encourages higher residential density within cities. How do you think people living in those cities will accept this approach? What are the advantages and disadvantages of increasing density?

Activities on government lands can be more easily regulated than those occurring elsewhere. However, park management may be difficult if goals are not clear and natural processes not understood.

26.7 Environmental Impact Assessment

Environmental impact assessment is the process of determining the probable effects of human use of the land. Such analysis originated in the United States following the passage of the *National Environmental Protection Act* (NEPA) in 1969. The United States Congress had become frustrated by repeated delays of public projects when the environmental and economic impacts of those projects emerged only after construction had begun. Congress established provisions under NEPA that require environmental impact statements (EIS) by federal agencies before they undertake actions with the potential to damage the environment, including direct, indirect, and cumulative effects.Canada's first environmental assessment legislation was introduced in Ontario in 1973.

Environmental assessment law requires that each major public project that might affect the quality of the human environment be preceded by an evaluation of the project and its potential impact on the environment. The evaluation would be presented in an environmental impact statement. All provinces and the Canadian federal government now have environmental assessment legislation in place.

In the early years of environmental assessment, the process of environmental impact analysis was criticized because it produced a tremendous volume of paperwork that tended to obscure the important issues. In response to that criticism, most jurisdictions have introduced requirements for scoping; most also now record the decision in a formal document. Scoping involves early identification of important environmental issues that require detailed evaluation. Citizen groups as well as federal, provincial, and local agencies are asked to participate in the scoping process by identifying issues and alternatives that they believe should be addressed as part of the environmental analysis.

Environmental impact assessments typically incorporate a concise statement by the proponent (the agency planning the proposed project) that outlines the alternatives that were considered and discusses which alternatives are environmentally preferable. The document also explains the rationale for choosing a particular alternative and describes how environmental degradation might be minimized or avoided for the alternative chosen.

Mitigation identifies ways to avoid, lessen, or compensate for anticipated adverse environmental impacts resulting from a particular project. For example, if a project in the coastal zone will damage wetlands, a possible mitigation might be the creation or enhancement of wetlands at another site. Mitigation is fast becoming a common requirement of many environmental impact statements and is useful in many instances, but it should not be used routinely to sidestep the adverse environmental impacts of a project. Sometimes there is no way to mitigate a particular environmental problem. Indeed, current research suggests that some mitigation strategies cannot adequately meet the productive capacity of some ecosystems (or their parts) being replaced.[27]

26.8 Development Status, Conflict and the Environment

Achieving sustainability in the world today has strong political and economic components, but it also has an environmental component. Over 1 billion people on Earth today live in poverty with little hope for the future. In some large urban regions, tens of millions of people exist in crowded, unsanitary living conditions, with unsafe drinking water and inadequate sewage disposal. In the countryside, rural people in many developing countries are being terrorized and displaced by armed conflicts over the control of valuable resources, such as oil, diamonds, and timber. Examples include oil in Nigeria, Sudan, and Colombia; diamonds in Sierra Leone, Angola, and the Democratic Republic of the Congo; and timber in Cambodia, Indonesia, and Borneo.[28] Recent terrorist attacks, such as the September 11, 2001, attacks on the Pentagon in Washington and New York's World Trade Towers, have increased our awareness of threats to the security of our water and food.

The 1992 Rio Earth Summit on Sustainable Development had the objective of addressing global environmental problems of both developed and developing countries, with an emphasis on solving conflicts between economic interests and environmental concerns. In many countries today, the gap between the rich and poor is even wider than it was in the early 1990s. As a result, political, social, and economic security remains threatened and serious environmental damage from overpopulation and resource exploitation continues. The environment remains inadequately funded. Worldwatch Institute reported in 2002 that the United Nations' annual budget for the environment is about $100 million, while the governments of the world are spending $2 billion per day for military purposes.[28]

26.9 Challenge to Students of the Environment

To end this book on an optimistic note—and there are reasons to be optimistic—we note that the Earth Summit on Sustainable Development, held in the summer of 2002 in Johannesburg, South Africa, had the following objectives:

- To continue to work toward environmental and social justice for all the people in the world.
- To enhance the development of sustainability.
- To minimize local, regional, and global environmental degradation resulting from overpopulation, deforestation, mining, agriculture, and pollution of the land, water, and air—in other words, to begin to shrink the size of our personal and national ecological footprints.
- To develop and support international agreements to control global warming and pollutants, and to foster environmental and social justice.

Solving our environmental problems will help build a more secure and sustainable future. This is becoming your charge and responsibility, as students of the environment and our future leaders, as you graduate from colleges and universities. This transfer of knowledge and leadership is a major reason why we wrote this book.

How Green is Your Campus?

Making a Commitment to Sustainability

Few Canadian universities or colleges have a comprehensive environmental planning strategy. The University of Victoria is an exception, and recently created the position of Sustainability Coordinator within its Facilities Management group. UVic's sustainability coordinator has responsibility for the following areas. Which of these kinds of activities does your university have in place? Which office at your university oversees environmental matters such as energy use, water use, and waste management? Think about how you and other concerned members of your university community could encourage your university's administration to begin thinking and acting more sustainably.

Natural Areas Management
- Restoration projects
- Natural areas inventory
- Hands-on learning
- Integrated pest management

Energy Management:
- Building retrofits
- Energy conservation
- Renewable energy applications
- Recycling

Water Management
- Stormwater management
- Water conservation
- Water reuse applications

Waste Management
- Recycling
- Composting
- Waste prevention education

Transportation Demand
- Management
- Campus infrastructure
- Sustainable transportation programs

Planning and Projects
- Green renovations
- Low impact development
- Campus master plan implementation

Summary

- The path toward sustainable development involves an evolution of our values to include all people within a framework that is economically, socially, and environmentally just and requires that we consider what a desirable environment in the future might be. This means reducing our personal and national ecological footprints.
- As an urban society, we must recognize the city's relation to the environment. A city influences and is influenced by its environment and is an environment itself.
- Like any other life-supporting system, a city must maintain a flow of energy, provide necessary material resources, and have ways of removing wastes. These functions are accomplished through transportation and communication with outlying areas.
- As the human population continues to increase, we can envision two futures: one in which people are dispersed widely throughout the countryside and cities are abandoned except by the poor; and another in which cities attract most of the human population, freeing much landscape for conservation of nature, production of natural resources, and public service functions of ecosystems.
- International environmental law is proving useful in addressing several important environmental problems, including preservation of resources and pollution abatement. Mediation is a method of conflict or dispute resolution that seeks to avoid lengthy and expensive litigation by arriving at a compromise whereby each party gains enough to prefer a settlement to litigation.
- Land-use planning includes definition of goals and objectives; collection and analysis of data; development of alternatives; formulation of plans; and review, implementation, and revision of plans.
- Global security, sustainability, and environment are linked in complex ways. Solving environmental problems will improve both sustainability and security.

CRITICAL THINKING ISSUE

Euro Tunnel: England to France: Tunnel Rubble to a Nature Reserve

The tunnel across the Dover Straits of the North Sea from England to France—sometimes called the Chunnel—is one of the world's most amazing engineering feats, long proposed, often thought impossible, now a reality. A tunnel under the English Channel had been considered for almost 200 years, and a false start had been made in 1974. Trains travel between London and Paris, going through the tunnel, which is approximately 50 km long, about 38 km of it under the sea (Figure 26.17). Serious construction of the tunnel started in 1984, and it was completed in 1994. Nearly 14,000 technicians, engineers, and other workers were involved in the project, which eventually cost approximately $15 billion. The project is actually two tunnels built close together: an eastbound and a westbound tunnel, each with a diameter of approximately 7.6 m.

Extensive surface and subsurface geologic and hydrologic evaluations completed before construction determined that the best route for the tunnel lay within a clay-rich chalk (a type of limestone) sandwiched between an overlying chalk, which forms the famous White Cliffs of Dover, and underlying clay stones (Figure 26.18). The undersea sections of the tunnel are contained within the clay-rich chalk beds for over 90% of their total length. The primary reason the clay-rich chalk was chosen was that the clay allows very little seepage of water. On completion, the tunnel was lined with concrete. Today, the train ride under the sea takes about 20 minutes at speeds up to 160 km/h.

As might be expected, such a large undertaking moved a great deal of rock. The British share of the tunnel rubble, or spoil, was nearly 9 million m3 of mostly clay-rich chalk, and disposing of such a large quantity of rubble posed a problem. Ultimately, about 5 million m3 of the material was dumped at the base of the White Cliffs of Dover at a location known as Samphire Hoe, approximately 3 km west of Dover. It was necessary first to construct a seawall nearly 2 km long, behind which the tunnel spoil was dumped. This increased the size of England by approximately 36 ha (90 acres). Buildings were removed, and the spoil was levelled, creating a nature reserve seaward of the Dover cliffs. The nature reserve has walking trails, and plants have colonized the new ground, providing habitat for butterflies, birds, and other wildlife. The project has produced a truly unique area.

The natural landscape in the area consists of tall white chalk cliffs, locally known as the Shakespeare Cliffs because the famous playwright mentioned them in King Lear. The cliffs comprise a classic seacliff and beach environment that are eroding at a rate of approximately 0.1 m per year. In the area of Samphire Hoe, however, this dynamic environment will be changed as seacliff erosion at the base ceases.

On the one hand, those who constructed and use the nature reserve view it very positively, for these reasons:

1. It has added new land and new environments to England.
2. It has provided new habitats for a variety of wildlife.
3. It provides an offshore view of the White Cliffs of Dover.
4. It offers a quiet area for people to visit in our busy world.

On the other hand, one may question the advisability of dumping the tunnel spoil in this location, for the following reasons:

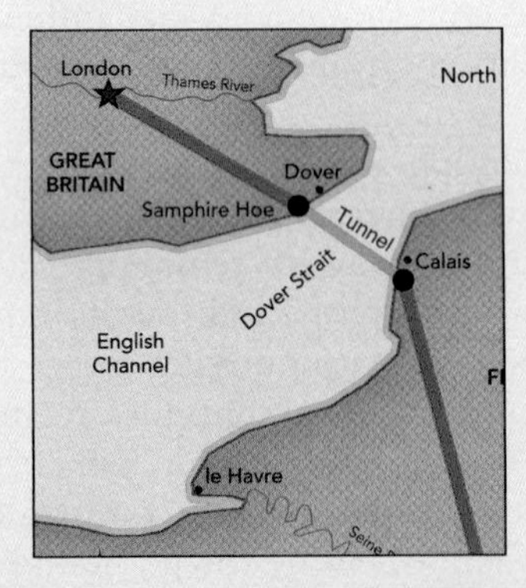

Figure 26.17 • The location of the tunnel beneath the English Channel from England to France.

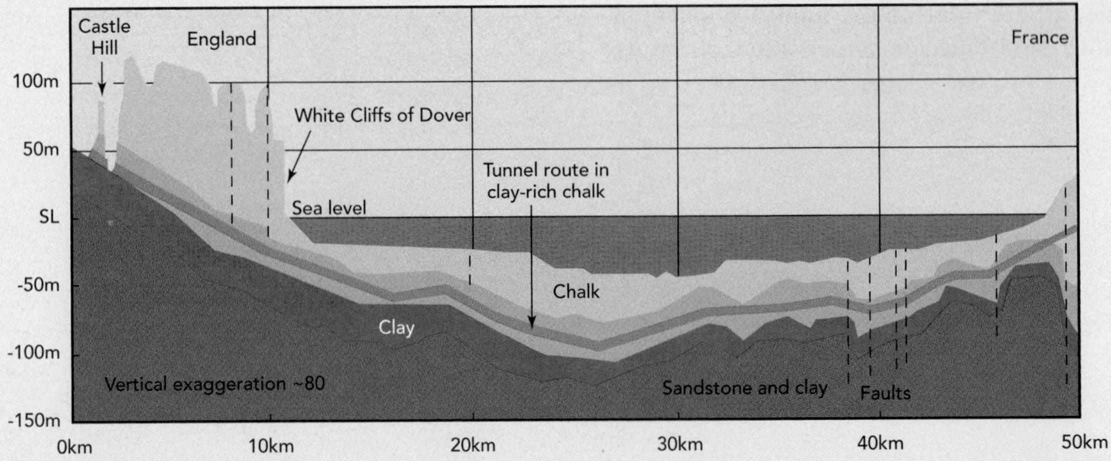

Figure 26.18 • Generalized geology along the tunnel.

1. The White Cliffs of Dover comprise a unique environment that will be changed as a result of the addition of the tunnel rubble.
2. There is loss of beach environment, which is a valued resource.
3. Construction of a seawall and protrusion of the land into the sea will change local coastal processes as the adjacent areas continue to erode. Eventually the site may be turned into a point, with increased wave activity and potential for erosion and destruction of the wall.
4. Some people find seawalls offensive from an aesthetic viewpoint; and while a nature reserve is a contribution, other sites for a reserve might have been preferable.

Critical Thinking Questions

1. Consider the project of creating the nature reserve carefully and write a few statements about potential environmental impact analyses that could have been done. Include (1) a statement of the purpose of the project; (2) a description of the environment before the project; (3) possible alternatives to the project, including the tunnel itself; and (4) the environmental consequences of each of the alternatives. Finally, defend either the project or one of your alternatives.
2. Do you think environmental analyses would have been improved if a scoping process had been completed prior to beginning the project? Explain your view.

STUDY QUESTIONS

1. How does the environment influence the location of cities?
2. You are the manager of High Park in Toronto and receive the following offers and requests. Which would you approve? Explain your reasons.
   a. A gift of $1 million to plant trees from all of eastern Canada.
   b. A gift of $1 million to set aside half the park to be forever untouched, thus producing an urban wilderness.
   c. A gift of the construction of an asphalt jogging track and a gym for physical fitness. The donor says that lack of physical fitness is a major urban health problem.
   d. A request to install a skating rink with artificially made ice. Facilities include an elegant restaurant with many views of the park.
3. Your province asks you to locate and plan a new town. The purpose of the town is to house people who will work at a wind farm—a large area of many windmills, all linked to produce electricity. You must first locate the site for the wind farm and then plan the town. How would you proceed? What factors would you take into account?
4. Visit your town centre. What changes, if any, would make better use of the environmental location? How could the area be made more livable?
5. It is popular to suggest that in the information age people can work at home and live in the suburbs and the countryside, so that cities are no longer necessary. List five arguments for and five arguments against this point of view.
6. In your view, just how big should a wilderness area be? Justify your answer.
7. The famous ecologist Garrett Hardin argued that designated wilderness areas should not have provisions for people with handicaps, even though he himself was handicapped—confined to a wheelchair. He believed that wilderness should be truly natural in the ultimate sense, that is, without any trace of civilization. Argue for or against Garrett Hardin's position. In your argument, consider the People and Nature theme of this book.
8. How can we balance freedom of individual action with the need to sustain our environment?
9. Should trees—and other non-human organisms—have legal standing? Explain your position on this topic.
10. Since there are no international "laws" that are binding in the same way that laws govern people within a nation, what can be done to achieve a sustainable environment for world fisheries or other international resources?

FURTHER READING

Akbari, H., S. Davis, S. Dorsano, S. J. Huang, and S. Winnett. *Cooling Our Cities: A Guidebook on Tree Planting and Light-Colored Surfacing.* Washington, D.C.: U.S. EPA Office of Policy Analysis, 1992. One of the most complete reviews of research on the uses of trees in cities.

Arnold, F. S. *Economic Analysis of Environmental Policy and Regulation.* New York: Wiley, 1995. A wide variety of practical applications of economics to environmental regulatory and policy analysis.

Beveridge, C. E., and P. Rocheleau. *Frederick Law Olmsted: Designing the American Landscape.* New York: Rizzoli International, 1995.

Bornkamm, R., J. A. Lee, and M. R. D. Seaward. eds. *Urban Ecology.* Oxford: Blackwell Scientific, 1982. A collection of papers presented at the Second European Ecological Symposium, held in Berlin in 1980. The papers cover a broad range of subjects, from the constitution of urban ecosystems to the impact of human activity in urban areas to the application of ecological knowledge in urban environments.

Bregman, J. I., and K. M. Mackenthum. *Environmental Impact Statements.* Boca Raton, Fla.: Lewis, 1992. An up-to-date examination of the many scientific and technical aspects of EIS preparation.

Burton, J. A. *Worlds Apart: Nature in the City.* Garden City, N.Y.: Doubleday, 1977. An intriguing description of wildlife, including foxes, inhabiting temperate-zone cities, little known to the people in the cities.

Butti, K., and J. Perlin. *The Golden Thread: 2500 Years of Solar Architecture and Technology.* New York: Cheshire, 1980. A history and outline of the use of solar energy from ancient Greece to modern times.

Cole, R. J. "Green Buildings: In Transit to a Sustainable World." *Canadian Architect* 41 no. 7 (1996):12-13.

Cronon, W. *Changes in the Land: Indians, Colonists, and the Ecology of New England.* New York: Hill and Wang, 1983. A description of the changes that occurred in New England's ecosystem following the arrival of European settlers.

Gilbert, O. L. *The Ecology of Urban Habitats.* London: Chapman & Hall, 1991.

Glacken, C. L. *Traces on the Rhodian Shore: Nature and Culture in Western Thought from Ancient Times to the End of the Eighteenth Century.* Berkeley: University of California Press, 1967. The best history of the idea of nature in Western civilization.

Henderson, L. J. *The Fitness of the Environment.* Boston: Beacon, 1966. A unique book, originally published in 1913, discussing the curious features of the universe and the Earth that make both suitable to life as we know it.

Hengeveld, H., and C. De Vocht, eds. *Role of Water in Urban Ecology.* Volume 1. *Developments in Landscape Management and Urban Planning.* New York: Elsevier Scientific, 1982. A publication that addresses the role of water in urban ecology. The contents are from the Proceedings of the Second International Environmental Symposium, held in Amsterdam in 1979.

Howard, Ebenezer. *Garden Cities of Tomorrow.* Cambridge, Mass.: MIT Press, 1965 (reprint). A classic work of the nineteenth century that has influenced modern city design. It presents a methodology for designing cities with the inclusion of parks, parkways, and private gardens.

Jacobs, Jane. *The Death and Life of Great American Cities.* New York: Vintage Books, 1961. A groundbreaking book in its day, analyzing the problems and needs of modern cities.

Jacobs, Jane. *Edge of Empire: Postcolonialism and the City.* London: Routledge, 1996. An examination of the colonial roots of three cities in the industrialized world.

Jain, R. K., L. V. Urban, G. S. Stacey, and H. E. Balbach. *Environmental Assessment.* New York: McGraw-Hill, 1993. A good book covering the importance of, and steps involved in, investigating environmental considerations concerning a proposed action.

McHarg, I. L. *Design with Nature.* Garden City, N.Y.: Doubleday, 1971. Examines case studies and presents a methodology for suburban regions based on an understanding of ecological processes.

Nilon, C. H., A. R. Berkowitz, and K. S. Hollweg. "Understanding *Urban Ecosystems*: A New Frontier for Science and Education." Urban Ecosystems 3 no. 1 (1999):3–5.

Noss, Reed F., Allen Y. Cooperrider (contributor), and Rodger Schlickeisen. *Saving Nature's Legacy.* Washington, D.C.: Island Press, 1994.

Rosenzweig, Michael, L. *Win-Win Ecology: How The Earth's Species Can Survive in the Midst of Human Enterprise.* New York: Oxford University Press, 2003. The author argues that ecological science leads to a belief that people and nature are not in opposition to one another.

Schaefer, V., H. Rudd and J. Vala. *Urban Biodiversity: Exploring Natural Habitat and its Value in Cities.* New Westminster, BC: Douglas College Centre for Environmental Studies and Urban Ecology, 2002.

Sears, P. *Deserts on the March.* Norman: University of Oklahoma Press, 1935.

Spirn, A. W. *The Granite Garden: Urban Nature and Human Design.* New York: Basic Books, 1984. Comprehensive strategies for designing cities in concert with natural processes, focusing on the look and shape of the city rather than on economics or public policy.

Teubner, G., L. Farmer, and D. Murphy, eds. *Environmental Law and Ecological Responsibility: The Concept and Practice of Ecological Self-Organization.* New York: Wiley, 1994. Argues that the old-style command-and-control regulation of business has reached its limits and that a new, externally induced, internal self-organizing process.

APPENDIX

A Prefixes and Multiplication Factors

Number	*10^x, Power of 10*	*Prefix*	*Symbol*
1,000,000,000,000,000,000	10^{18}	exa	E
1,000,000,000,000,000	10^{15}	peta	P
1,000,000,000,000	10^{12}	tera	T
1,000,000,000	10^{9}	giga	G
1,000,000	10^{6}	mega	M
10,000	10^{4}	myria	
1,000	10^{3}	kilo	k
100	10^{2}	hecto	h
10	10^{1}	deca	da
0.1	10^{-1}	deci	d
0.01	10^{-2}	centi	c
0.001	10^{-3}	milli	m
0.000 001	10^{-6}	micro	μ
0.000 000 001	10^{-9}	nano	n
0.000 000 000 001	10^{-12}	pico	p
0.000 000 000 000 001	10^{-15}	femto	f
0.000 000 000 000 000 001	10^{-18}	atto	a

B Periodic Table

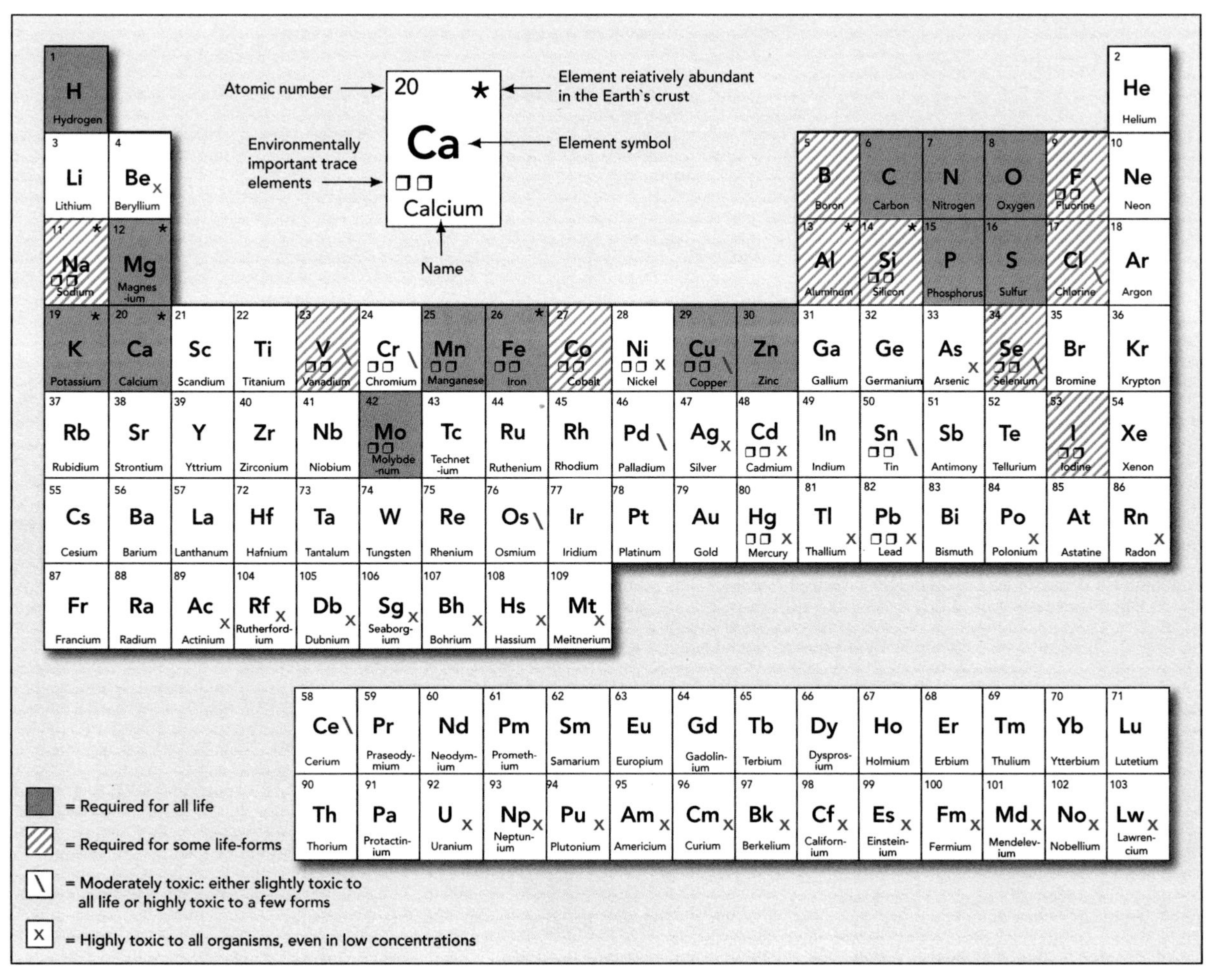

C Geologic Time Scale and Biologic Evolution

Era	*Approximate Age in Millions of Years Before Present*	*Period*	*Epoch*	*Life Form*
	Less than 0.01		Recent (Holocene)	
	0.01–2	Quaternary	Pleistocene	Humans
	2			
	2–5		Pliocene	
	5–23		Miocene	
Cenozoic	23–35	Tertiary	Oligocene	
	35–56		Eocene	Mammals
	56–65		Paleocene	
	65			
	65–146	Cretaceous		
Mesozoic	146–208	Jurassic		Flying reptiles, birds
	208–245	Triassic		Dinosaurs
	245			
	245–290	Permian		Reptiles
	290–363	Carboniferous		Insects
Paleozoic	363–417	Devonian		Amphibians
	417–443	Silurian		Land plants
	443–495	Ordovician		Fish
	495–545	Cambrian		
	545			
	700			Multicelled organisms
	3,400			One-celled organisms
	4,000	Approximate age of oldest rocks discovered on Earth		
Precambrian				
	4,600	Approximate age of the Earth and meteorites		

GLOSSARY

A

Abortion rate The estimated number of abortions per 1,000 women aged 15 to 44 in a given year. (Ages 15 to 44 are taken to be the limits of ages during which women can have babies. This, of course, is an approximation, made for convenience.)

Abortion ratio The estimated number of abortions per 1,000 live births in a given year.

Acid mine drainage Does not refer to an acid mine but to acidic water that drains from mining areas (mostly coal but also metal mines). The acidic water may enter surface water resources, causing environmental damage.

Acid precipitation Rain, fog, and snow made artificially acid by pollutants, particularly sulphur and nitrogen oxides. (Natural rainwater is slightly acid owing to the effect of carbon dioxide dissolved in the water.)

Active solar energy systems Direct use of solar energy that requires mechanical power; usually consists of pumps and other machinery to circulate air, water, or other fluids from solar collectors to a heat sink where the heat may be stored.

Adaptive radiation The process that occurs when a species enters a new habitat that has unoccupied niches and evolves into a group of new species, each adapted to one of these niches.

Advanced wastewater treatment Treatment of wastewater beyond primary and secondary procedures. May include sand filters, carbon filters, or application of chemicals to assist in removing potential pollutants such as nutrients from the wastewater stream.

Aerobic Characterized by the presence of free oxygen.

Aesthetic justification for the conservation of nature An argument for the conservation of nature on the grounds that nature is beautiful and that beauty is important and valuable to people.

Age dependency ratio The ratio of dependent-age people (those unable to work) to working-age people. It is customary to define working-age people as those aged 15 to 65.

Age structure (of a population) A population divided into groups by age. Sometimes the groups represent the actual number of each age in the population; sometimes the groups represent the percentage or proportion of the population of each age.

Agroecosystem An ecosystem created by agriculture. Typically, it has low genetic, species, and habitat diversity.

Air quality standards Levels of air pollutants that delineate acceptable levels of pollution over a particular time period. Valuable because they are often tied to emission standards that attempt to control air pollution.

Allowance trading Approach to managing coal resources and reducing pollution through buying, selling, and trading of allowances to emit pollutants from burning coal. The idea is to control pollution by controlling the number of allowances issued. The notion of allowance trading is now gaining acceptance in other areas of air pollution control (e.g., greenhouse gas emissions) and in the management of nutrients on farmland.

Alpha particles One of the major types of nuclear radiation, consisting of two protons and two neutrons (a helium nucleus).

Alternative energy Renewable and non-renewable energy resources that are alternatives to fossil fuels.

Ammonification The conversion of nitrates and nitrites to ammonia by bacterial action.

Anaerobic Characterized by the absence of free oxygen.

Aquaculture Production of food from aquatic habitats.

Aquifer An underground zone or body of earth material from which groundwater can be obtained from a well at a useful rate.

Area sources Sometimes also called nonpoint sources. These are diffused sources of pollution such as urban runoff or automobile exhaust. These sources include emissions that may be over a broad area or even over an entire region. They are often difficult to isolate and correct because of the widely dispersed nature of the emissions.

Asbestos A term for several minerals that have the form of small, elongated particles. Some types of particles are believed to be carcinogenic or to carry with them carcinogenic materials. White, or chrysotile, asbestos is the most common form in Canada.

Assimilative capacity The capacity of a stream to degrade organic waste.

Atmosphere Layer of gases surrounding Earth.

Atmospheric inversion A condition in which warmer air is found above cooler air, restricting air circulation; often associated in urban areas with a pollution event.

Autotroph An organism that produces its own food from inorganic compounds and a source of energy. There are photoautotrophs (photosynthetic plants) and chemical autotrophs.

Average residence time A measure of the time it takes for a given part of the total pool or reservoir of a particular material in a system to be cycled through the system. When the size of the pool and rate of throughput are constant, average residence time is the ratio of the total size of the pool or reservoir to the average rate of transfer through the pool.

B

Balance of nature An environmental myth that states that the natural environment, when not influenced by human activity, will reach a constant status, unchanging over time, referred to as a balance or equilibrium state.

Barrier island An island separated from the mainland by a salt marsh. It generally consists of a multiple system of beach ridges and is separated from other barrier islands by inlets that allow the exchange of seawater with lagoon water.

Becquerel The unit commonly used for radioactive decay in the International System (SI) of measurement.

Beta particles One of the three major kinds of nuclear radiation; electrons that are emitted when one of the protons or neutrons in the nucleus of an isotope spontaneously changes.

Bicameral system A two-house system of government. Canada's government is bicameral, comprising the House of Commons and the Senate.

Bioaccumulation The accumulation of a substance in the tissue of a single organism. Different from biomagnification, which refers to the tendency of some chemicals to increase in concentration in each trophic level.

Biochemical oxygen demand (BOD) A measure of the amount of oxygen necessary to decompose organic material in a unit volume of water. As the amount of organic waste in water increases, more oxygen is used, resulting in a higher BOD.

Biogeochemical cycle The cycling of a chemical element through the biosphere; its pathways, storage locations, and chemical forms in living things, the atmosphere, oceans, sediments, and lithosphere.

Biogeography The large-scale geographic pattern in the distribution of species, and the causes and history of this distribution.

Biohydromettalurgy Combining biological and mining processes, usually involving microbes to help extract valuable metals such as gold from the ground. May also be used to remove pollutants from mining waste.

Biological control A set of methods to control pest organisms by using natural ecological interactions, including predation, parasitism, and competition. Part of integrated pest management.

Biological diversity Used loosely to mean the variety of life on the Earth, but scientifically, typically used to consist of three components: (1) genetic diversity (the total number of genetic characteristics); (2) species diversity; and (3) habitat or ecosystem diversity (the number of kinds of habitats or ecosystems in a given unit area). Species diversity includes three concepts: species richness, evenness, and dominance.

Biological evolution The change in inherited characteristics of a population from generation to generation, which can result in new species.

Biological production The capture of usable energy from the environment to produce organic compounds in which that energy is stored.

Biomagnification Also called biological concentration or bioconcentration. The tendency for some substances to concentrate with each trophic level. Organisms preferentially store certain chemicals and excrete others. When this occurs consistently among organisms, the stored chemicals increase as a percentage of the body weight as the material is transferred along a food chain or trophic level. For example, the concentration of DDT is greater in herbivores than in plants and greater in plants than in the non-living environment.

Biomass The amount of living material, or the amount of organic material contained in living organisms, both as live and dead material, as in the leaves (live) and stem wood (dead) of trees.

Biomass energy The energy that may be recovered from biomass, which is organic material such as plants and animal waste.

Biomass fuel A new name for the oldest fuel used by humans. Organic matter, such as plant material and animal waste that can be used as a fuel.

Biome A kind of ecosystem. The rain forest is an example of a biome; rain forests occur in many parts of the world but are not all connected with each other.

Bioremediation A method of treating groundwater pollution problems that utilizes micro-organisms in the ground to consume or break down pollutants.

Biosphere The word has several meanings. One is that part of a planet where life exists. On Earth it extends from the depths of the oceans to the summit of mountains, but most life exists within a few metres of the surface. A second meaning is the planetary system that includes and sustains life, and therefore is made up of the atmosphere, oceans, soils, upper bedrock, and all life.

Biota All the organisms of all species living in an area or region up to and including the biosphere, as in "the biota of the Mojave Desert" or "the biota in that aquarium."

Biotic province A geographical region inhabited by life forms (species, families, orders), of common ancestry, bounded by barriers that prevent the spread of the distinctive kinds of life to other regions and the immigration of foreign species.

Birth rate The rate at which births occur in a population, measured either as the number of individuals born per unit of time or as the percentage of births per unit of time compared with the total population.

Black lung disease Often called coal miner disease because it is caused by years of inhaling coal dust, resulting in damage to lungs.

Body burden The amount of concentration of a toxic chemical in an individual.

Breeder reactor A type of nuclear reactor that utilizes between 40% and 70% of its nuclear fuel and converts fertile nuclei to fissile nuclei faster than the rate of fission. Thus, breeder reactors actually produce nuclear fuels.

Brines With respect to mineral resources, refers to waters with a high salinity that contain useful materials such as bromine, iodine, calcium chloride, and magnesium.

Buffers Materials (chemicals) that have the ability to neutralize acids. Examples include the calcium carbonate that is present in many soils and rocks. These materials may lessen potential adverse effects of acid rain.

Burner reactors A type of nuclear reactor that consumes more fissionable material than it produces.

C

Capillary action The rise of water along narrow passages, facilitated and caused by surface tension.

Carbon cycle Biogeochemical cycle of carbon. Carbon combines with and is chemically and biologically linked with the cycles of oxygen and hydrogen that form the major compounds of life.

Carbon monoxide (CO) Colourless, odourless gas that at very low concentrations is extremely toxic to humans and animals.

Carbon-silicate cycle A complex biogeochemical cycle over time scales as long as one-half billion years. Included in this cycle are major geologic processes, such as weathering, transport by ground and surface waters, erosion, and deposition of crustal rocks. The carbonate-silicate cycle is believed to provide important negative feedback mechanisms that control the temperature of the atmosphere.

Carcinogen Any material that is known to produce cancer in humans or other animals.

Carnivores Organisms that feed on other live organisms; usually applied to animals that eat other animals.

Carrying capacity The maximum abundance of a population or species that can be maintained by a habitat or ecosystem without degrading the ability of that habitat or ecosystem to maintain that abundance in the future.

Cash crops Crops grown to be traded in a market.

Catastrophe A situation or event that causes significant damage to people and property, such that recovery and/or rehabilitation is a long and involved process. Examples of natural catastrophes include hurricanes, volcanic eruptions, large wildfires, and floods.

Catch per unit effort The number of animals caught per unit of effort, such as the number of fish caught by a fishing ship per day. It is used to estimate the population abundance of a species.

Certification of forestry The formal process of comparing the actual forestry practices of specific corporations or government agencies with practices that are believed to be consistent with sustainability. Operations found to be using "sustainable" practices receive formal certification to that effect. Forest certification programs are of questionable value, given that the long-term impacts of so-called sustainable practices are still largely unknown.

Channelization An engineering technique that consists of straightening, deepening, widening, clearing, or lining existing stream channels. The purpose is to control floods, improve drainage, control erosion, or improve navigation. It is a very controversial practice that may have significant environmental impacts.

Chaparral A dense scrubland found in areas with Mediterranean climate (a long warm, dry season and a cooler rainy season).

Chemical reaction The process in which compounds and elements undergo a chemical change to become a new substance or substances.

Chemoautotrophs Autotrophic bacteria that can derive energy from chemical reactions of simple inorganic compounds.

Chemosynthesis Synthesis of organic compounds by energy derived from chemical reactions.

Chimney (or stack) effect Process whereby warmer air rises in buildings to upper levels and is replaced in the lower portion of the building by outdoor air drawn through a variety of openings, such as windows, doors, or cracks in the foundations and walls.

Chlorofluorocarbons (CFCs) Highly stable compounds that have been or are being used in spray cans as aerosol propellants and in refrigeration units (the gas that is compressed and expanded in a cooling unit). Emissions of chlorofluorocarbons have been associated with potential global warming and stratospheric ozone depletion.

Chronic disease A disease that is persistent in a population, typically occurring in a relatively small but constant percentage of the population.

Chronic hunger A condition in which there is enough food available per person to stay alive, but not enough to lead a satisfactory and productive life.

Chronic patchiness A situation where ecological succession does not occur. One species may replace another, or an individual of the first species may replace it, and no overall general temporal pattern is established. Characteristic of harsh environments such as deserts.

City planning Formal, conscious planning for new cities before they are built.

Clay May refer to a mineral family or to a very fine-grained sediment. It is associated with many environmental problems, such as shrinking and swelling of soils and sediment pollution.

Clear-cutting In timber harvesting, the practice of cutting all trees in a stand at the same time.

Climate The representative or characteristic conditions of the atmosphere at particular places on Earth. Climate refers to the average or expected conditions over long periods; weather refers to the particular conditions at one time in one place.

Climate change Natural or human-induced change in global mean annual temperature of the atmosphere near the Earth's surface and other aspects of climate over periods of time ranging from decades to hundreds of years to several million years.

Climax stage (or ecological succession) The final stage of ecological succession and therefore an ecological community that continues to reproduce itself over time.

Closed system A type of system in which there are definite boundaries to factors such as mass and energy such that exchange of these factors with other systems does not occur.

Closed-canopy forest Forests in which the leaves of adjacent trees overlap or touch, so that the trees form essentially continuous cover.

Coal Solid, brittle carbonaceous rock that is one of the world's most abundant fossil fuels. It is classified according to energy content as well as carbon and sulphur content.

Coal gasification Process that converts coal that is relatively high in sulphur to a gas in order to remove the sulphur.

Co-dominants In forestry, fairly common species that share the canopy or top part of the forest.

Cogeneration The capture and use of waste heat; for example, using waste heat from a power plant to heat adjacent factories and other buildings.

Cohort All the individuals in a population born during the same time period. Thus, all the people born during the year 1980 represent the world human cohort for that year.

Common law Law derived from custom, judgement, or decrees of courts rather than from legislation.

Commons Land or other resources that belong to the public, not to individuals. Historically, a part of English towns where all the farmers could graze their cattle.

Community, ecological A group of populations of different species living in the same local area and interacting with one another. A community is the living portion of an ecosystem.

Community effect (community-level effect) When the interaction between two species leads to changes in the presence or absence of other species or in a large change in abundance of other species, then a community effect is said to have occurred.

Competition The situation that exists when different individuals, populations, or species compete for the same resource(s) and the presence of one has a detrimental effect on the other. Sheep and cows eating grass in the same field are competitors.

Competitive exclusion principle The idea that two populations of different species with exactly the same requirements cannot persist indefinitely in the same habitat—one will always win out and the other will become extinct.

Composting Biochemical process in which organic materials, such as lawn clippings and kitchen scraps are decomposed to a rich, soil-like material.

Comprehensive plan Official plan adopted by local government formally stating general and long-range policies concerning future development.

Cone of depression A cone-shaped depression in the water table around a well caused by withdrawal by pumping of water at rates greater than the rates at which the water can be replenished by natural groundwater flow.

Conservation With respect to resources such as energy, refers to changing our patterns of use or simply getting by with less. In a pragmatic sense the term means adjusting our needs to minimize the use of a particular resource, such as energy.

Consumptive use A type of off-stream water use. This water is consumed by plants and animals or in industrial processes or evaporates during use. It is not returned to its source.

Contamination Presence of undesirable material that makes something unfit for a particular use.

Continental drift The movement of continents in response to seafloor spreading. The most recent episode of continental drift started about 200 million years ago with the breakup of the supercontinent Pangaea.

Continental shelf Relatively shallow ocean area between the shoreline and the continental slope that extends to approximately a 200 m water depth surrounding a continent.

Contour ploughing Ploughing land along topographic contours, as much in a horizontal plane as possible, thereby decreasing the erosion rate.

Controlled experiment A controlled experiment is designed to test the effects of independent variables on a dependent variable by changing only one independent variable at a time. For each variable tested, there are two set-ups (an experiment and a control) that are identical except for the independent variable being tested. Any difference in the outcome (dependent variable) between the experiment and the control can then be attributed to the effects of the independent variable tested.

Convection The transfer of heat involving the movement of particles; for example, the boiling water in which hot water rises to the surface and displaces cooler water, which moves toward the bottom.

Convergent evolution The process by which species evolve in different places or different times and, although they have different genetic heritages, develop similar external forms and structures as a result of adaptation to similar environments. The similarity in the shapes of sharks and porpoises is an example of convergent evolution.

Convergent plate boundary Boundary between two lithosphere plates in which one plate descends below the other (subduction).

Cosmopolitan species A species with a broad distribution, occurring wherever in the world the environment is appropriate.

Creative justification for the conservation of nature An argument for the conservation of nature on the grounds that people often find sources of artistic and scientific creativity in their contacts with the unspoiled natural world.

Criminal statutes Laws that prohibit the kinds of actions that society considers wholly wrong and unacceptable, such as theft, rape, and murder.

Crop rotation A series of different crops planted successively in the same field, with the field occasionally left fallow, or grown with a cover crop.

Crude oil Naturally occurring petroleum, normally pumped from wells in oil fields. Refinement of crude oil produces most of the petroleum products we use today.

Cultural eutrophication Human-induced eutrophication that involves nutrients such as nitrates or phosphates that cause a rapid increase in the rate of plant growth in ponds, lakes, rivers or the ocean.

Curie Commonly used unit to measure radioactive decay; the amount of radioactivity from 1 gram of radium 226 that undergoes about 37 billion nuclear transformations per second.

D

Death rate The rate at which deaths occur in a population, measured either as the number of individuals dying per unit time or as the percentage of a population dying per unit time.

Decomposers Organisms that feeds on dead organic matter.

Deductive reasoning Drawing a conclusion from initial definitions and assumptions by means of logical reasoning.

Deep-well disposal Method of disposal of hazardous liquid waste that involves pumping the waste deep into the ground below and completely isolated from all freshwater aquifers. A controversial method of waste disposal that is being carefully evaluated.

Demand for food The amount of food that would be bought at a given price if it were available.

Demand-based agriculture Agriculture with production determined by economic demand and limited by that demand rather than by resources.

Demographic transition The pattern of change in birth and death rates as a country is transformed from undeveloped to developed. There are three stages: (1) in an undeveloped country, birth and death rates are high and the growth rate low; (2) the death rate decreases, but the birthrate remains high and the growth rate is high; (3) the birthrate drops toward the death rate and the growth rate therefore also decreases.

Demography The study of populations, especially their patterns in space and time.

Dependent variable *See* Variable, dependent

Denitrification The conversion of nitrate to molecular nitrogen by the action of bacteria, an important step in the nitrogen cycle.

Density-dependent population effects Factors whose effects on a population change with population density.

Density-independent population effects Changes in the size of a population due to factors that are independent of the population size. For example, a storm that knocks down all trees in a forest, no matter how many there are is a density-independent population effect.

Desalination The removal of salts from seawater or brackish water so that the water can be used for purposes such as agriculture, industrial processes, or human consumption.

Desertification The process of creating a desert where there was not one before.

Dioxin An organic compound composed of oxygen, hydrogen, carbon, and chlorine. About 75 types are known. Dioxin is not normally manufactured intentionally but is a by-product resulting from chemical reactions in the production of other materials, such as herbicides. Known to be extremely toxic to mammals, its effects on the human body are being intensively studied and evaluated.

Direct cost A cost that is borne by the producer and passed on directly to the consumer. For example, the price of steel includes direct costs such as the cost of iron, the cost of energy used to make steel, and the cost of buildings and labour associated with steel manufacturing.

Disprovability The idea that a statement can be said to be scientific if someone can clearly state a method or test by which it might be disproved.

Divergent evolution Organisms with the same ancestral genetic heritage migrate to different habitats and evolve into species with different external forms and structures, but typically continue to use the same kind of habitats. The ostrich and the emu are believed to be examples of divergent evolution.

Divergent plate boundary Boundary between lithospheric plates characterized by the production of new lithosphere; found along oceanic ridges.

Diversity, genetic The total number of genetic characteristics, sometimes of a specific species, subspecies, or group of species.

Diversity, habitat The number of kinds of habitats in a given unit area.

Diversity, species Used loosely to mean the variety of species in an area or on the Earth. Technically, it is composed of three components: species richness (the total number of species); species evenness (the relative abundance of species); and species dominance (the most abundant species).

Dobson unit Commonly used to measure the concentration of ozone. One Dobson unit is equivalent to a concentration of 1 ppb ozone.

Dominant species Generally, the species that are most abundant in an area, ecological community, or ecosystem.

Dominants In forestry, the tallest, most numerous, and most vigorous trees in a forest community.

Dose dependency Dependence on the dose or concentration of a substance for its effects on a particular organism.

Dose response The principle that the effect of a certain chemical on an individual depends on the dose or concentration of that chemical.

Doubling time The time necessary for a quantity of whatever is being measured to double.

Drainage basin The area that contributes surface water to a particular stream network.

Drip irrigation Irrigation by the application of water to the soil from tubes that drip water slowly, greatly reducing the loss of water from direct evaporation and increasing yield.

E

Early successional species Species that occur only or primarily during early stages of succession.

Earth system science The Science of the Earth as a system. It includes understanding of processes and linkages between the lithosphere, hydrosphere, biosphere, and atmosphere.

Ecological community This term has two meanings. (1) A conceptual or functional meaning: a set of interacting species that occur in the same place (sometimes extended to mean a set that interacts in a way to sustain life). (2) An operational meaning: a set of species found in an area, whether or not they are interacting.

Ecological economics Study and evaluation of relations between humans and the economy with emphasis on long-term health of ecosystems and sustainability.

Ecological gradient A change in the relative abundance of a species or group of species along a line or over an area.

Ecological island An area that is biologically isolated so that a species occurring within the area cannot mix (or only rarely mixes) with any other population of the same species.

Ecological justification for the conservation of nature An argument for the conservation of nature on the grounds that a species, an ecological community, an ecosystem, or the Earth's biosphere provides specific functions necessary to the persistence of our life or of benefit to life. The ability of trees in forests to remove carbon dioxide produced in burning fossil fuels is such a public benefit and an argument for maintaining large areas of forests.

Ecological niche The general concept is that the niche is a species' "profession"—what it does to make a living. The term is also used to refer to a set of environmental conditions within which a species is able to persist.

Ecological resilience The ability of an ecosystem to withstand disturbance because of the presence of complex, interactive processes and structures.

Ecological succession The process of the development of an ecological community or ecosystem, usually viewed as a series of stages-early, middle, late, mature (or climax), and sometimes postclimax. Primary succession is an original establishment; secondary succession is a reestablishment.

Ecology The science of the study of the relationships between living things and their environment.

Ecosystem An ecological community and its local, non-biological community. An ecosystem is the minimum system that includes and sustains life. It must include at least an autotroph, a decomposer, a liquid medium, a source and sink of energy, and all the chemical elements required by the autotroph and the decomposer.

Ecosystem effect Effects that result from interactions among different species, effects of species on chemical elements in their environment, and conditions of the environment.

Ecosystem energy flow The flow of energy through an ecosystem—from the external environment through a series of organisms and back to the external environment.

Ecotourism Tourism based on an interest in observation of nature.

ED-50 The effective dose, or dose that causes an effect in 50% of the population on exposure to a particular toxicant. It is related to the onset of specific symptoms, such as loss of hearing, nausea, or slurred speech.

Edge effect An effect that occurs following the forming of an ecological island; in the early phases the species diversity along the edge is greater than in the interior. Species escape from the cut area and seek refuge in the border of the forest, where some may last only a short time.

Efficiency The ratio of output to input. With machines, usually the ratio of work or power produced to the energy or power used to operate or fuel them. With living things, efficiency may be defined as either the useful work done or the energy stored in a useful form compared with the energy taken in.

Efficiency improvements With respect to energy, refers to designing equipment that will yield more energy output from a given amount of energy input.

Effluent Any material that flows outward from something. Examples include wastewater from hydroelectric plants and water discharged into streams from waste-disposal sites.

Effluent stream Type of stream where flow is maintained during the dry season by groundwater seepage into the channel.

El Niño Natural perturbation of the physical earth system that effects global climate. Characterized by development of warm oceanic waters in the eastern part of the tropical Pacific Ocean, a weakening or reversal of the trade winds, and a weakening or even reversal of the equatorial ocean currents. Reoccurs periodically and effects the atmosphere and global temperature by pumping heat into the atmosphere.

Electromagnetic fields (EMF) Magnetic and electrical fields produced naturally by our planet and also by appliances such as toasters, electric blankets, and computers. Currently, there is controversy concerning potential adverse health effects related to exposure to EMF in the workplace and home from such artificial sources as power lines and appliances.

Electromagnetic spectrum All the possible wavelengths of electromagnetic energy, considered as a continuous range. The spectrum includes long wavelength (used in radio transmission), infrared, visible, ultraviolet, X- rays, and gamma rays.

Endangered species A species that faces threats that might lead to its extinction in a short time.

Endemic Referring to all factors confined to a given region, such as an island or a country; the whooping crane is endemic to North America.

Endemic species A species that is native to a particular area.

Energy An abstract concept referring to the ability or capacity to do work.

Energy efficiency Refers to both first-law efficiency and second-law efficiency, where first-law efficiency is the ratio of the actual amount of energy delivered to the amount of energy supplied to meet a particular need, and the second-law efficiency is the ratio of the maximum available work needed to perform a particular task to the actual work used to perform that task.

Energy flow The movement of energy through an ecosystem from the external environment through a series of organisms and back to the external environment. It is one of the fundamental processes common to all ecosystems.

Entropy A measure in a system of the amount of energy that is unavailable for useful work. As the disorder of a system increases, the entropy in a system also increases.

Environment All factors (living and non-living) that actually affect an individual organism or population at any point in the life cycle. Environment is also sometimes used to denote a certain set of circumstances surrounding a particular occurrence (environments of deposition, for example).

Environmental audit Process of determining the past history of a particular site, with special reference to the existence of toxic materials or waste.

Environmental economics The study of relationships of the importance of the environment to the economy, including the impact of environment as a result of economic activity, regulation of the economy and economic processes with the objective of balancing environmental and economic goals of society, and development of economic policy to minimize environmental degradation and finding solutions to environmental problems.

Environmental ethics A school, or theory, in philosophy that deals with the ethical value of the environment, including especially the rights of non-human objects and systems in the environment, for example, trees and ecosystems.

Environmental geology The application of geologic information to environmental problems.

Environmental impact The effects of some action on the environment, particularly action by human beings.

Environmental impact assessment The formal process of evaluating and comparing the effect of a proposed project and its alternatives; required under Canadian federal and provincial laws and those of at least 100 other countries. Normally, only public projects such as dams and highways are subject to environmental assessment legislation. Private projects such as golf courses and shopping malls are not usually required to undergo environmental assessment, although such analysis may be required by regulatory agencies if the project is very large in scale or is expected to have significant environmental impacts.

Environmental impact report (EIR) Similar to the environmental impact statement (EIS), a report describing potential environmental impacts resulting from a particular project, often at the state level.

Environmental impact statement (EIS) A written statement that assesses and explores possible impacts associated with a particular project that may affect the human environment. Environmental justice The principle of dealing with environmental problems in such a way as to not discriminate against people based upon socio-economic status, race, or ethnic group.

Environmental law A field of law concerning the conservation and use of natural resources and the control of pollution.

Environmental tobacco smoke Commonly called second-hand smoke from people smoking tobacco.

Environmental connectivity A principle of environmental sciences that states that everything affects everything else, meaning that a particular course of action leads to an entire potential string of events. Another way of stating this idea is that you can't only do one thing.

Environmentalism A social, political, and ethical movement concerned with protecting the environment and using its resources wisely.

Epidemic disease A disease that appears occasionally in the population, affects a large percentage of it, and declines or almost disappears for a while only to reappear later.

Equilibrium A point of rest. At equilibrium, a system remains in a single, fixed condition and is said to be in equilibrium. Compare with steady state.

Eukaryote organism An organism whose cells have nuclei and organelles that distinguish it from the prokaryotes. The eukaryotes include flowering plants, animals, and many single-cell organisms.

Eutrophic Referring to bodies of water having an abundance of the chemical elements required for life.

Eutrophication Increase in the concentration of chemical elements required for living things (for example, phosphorus). Increased nutrient loading may lead to a population explosion of photosynthetic algae and blue-green bacteria that become so thick that light cannot penetrate the water. Bacteria deprived of light beneath the surface die; as they decompose, dissolved oxygen in the lake is lowered and eventually a fish kill may result. Eutrophication of lakes caused by human-induced processes, such as nutrient-rich sewage water entering a body of water, is called cultural eutrophication.

Even-aged stands A forest of trees that began growth from seeds and roots planted in or about the same year.

Evolution biological The change in inherited characteristics of a population from generation to generation, sometimes resulting in a new species or populations.

Evolution, non-biological Outside the realm of biology, the term evolution is used broadly to mean the history and development of something.

Exosphere The top-most layer of the atmosphere, where the atmosphere merges into space.

Exotic species Species introduced into a new area, usually assumed to occur as a result of human action.

Experimental errors There are two kinds of experimental errors, random and systematic. Random errors are those due to chance events, such as air currents pushing on a scale and altering a measurement of weight. In contrast, a miscalibration of an instrument would lead to a systematic error. Human errors can be either random or systematic.

Explanation A reason or description that is consistent with currently accepted hypotheses. Sometimes an explanation takes the form or a computer, mathematical, or physical model of an ecological system.

Exponential growth Growth in which the rate of increase is a constant percentage of the current size; that is, the growth occurs at a constant rate per time period.

Externality In economics, an effect not normally accounted for in the cost-revenue analysis of producers.

Extinction Disappearance of a life form from existence; usually applied to a species.

Extirpation Extinction of a species over a geographic region; local extinction.

F

Facilitation During succession, one species prepares the way for the next (and may even be necessary for the occurrence of the next).

Fact Something that is known based on actual experience and observation.

Fall line The point on a river where there is an abrupt drop in elevation of the land and where numerous waterfalls occur. The line in eastern North America is located where streams pass from harder to softer rocks.

Fallow A field allowed to grow with a cover crop without harvesting for at least one season.

Fecal coliform bacteria Standard measure of microbial pollution and an indicator of disease potential for a water source.

Feedback A kind of system response that occurs when output of the system also serves as input leading to changes in the system.

First law of thermodynamics The principle that energy may not be created or destroyed but is always conserved.

First-law efficiency The ratio of the actual amount of energy delivered where it is needed to the amount of energy supplied in order to meet that need; expressed as a percentage.

Fission The splitting of an atom into smaller fragments with the release of energy.

Flooding The natural process whereby waters emerge from their stream channel to cover part of the floodplain. Natural flooding is not a problem until people choose to build homes and other structures on floodplains.

Floodplain Flat topography adjacent to a stream in a river valley that has been produced by the combination of overbank flow and lateral migration of meander bends.

Fluidized-bed combustion A process used during the combustion of coal to eliminate sulphur oxides. Involves mixing finely ground limestone with coal and burning it in suspension.

Food chain The linkage of who feeds on whom.

Food-chain concentration *See* Biomagnification.

Food web A network of who feeds on whom or a diagram showing who feeds on whom. It is synonymous with food chain.

Force A push or pull that affects motion. The product of mass and acceleration of a material.

Forcing With respect to global change, processes capable of changing global temperature, such as changes in solar energy emitted from the sun, or volcanic activity.

Fossil fuels Forms of stored solar energy created from incomplete biological decomposition of dead organic matter. Includes coal, crude oil, and natural gas.

Fuel cell A device that produces electricity directly from a chemical reaction in a specially designed cell. In the simplest case the cell uses hydrogen as a fuel, to which an oxidant is supplied. The hydrogen is combined with oxygen as if the hydrogen were burned, but the reactants are separated by an electrolyte solution that facilitates the migration of ions and the release of electrons (which may be tapped as electricity, energy source).

Fugitive sources Type of stationary air pollution sources that generate pollutants from open areas exposed to wind processes.

Fusion, nuclear Combining of light elements to form heavier elements with the release of energy.

G

Gaia hypothesis The Gaia hypothesis states that the surface environment of the Earth, with respect to such factors as the atmospheric composition of reactive gases (for example, oxygen, carbon dioxide, and methane), the acidity-alkalinity of waters, and the surface temperature are actively regulated by the sensing, growth, metabolism and other activities of the biota. Interaction between the physical and biological system on the Earth's surface has led to a planetwide physiology which began more than 3 billion years ago and the evolution of which can be detected in the fossil record.

Gamma rays One of the three major kinds of nuclear radiation. A type of electromagnetic radiation emitted from the isotope similar to X-rays but more energetic and penetrating.

Garden city Land use planning that considers a city and countryside together.

Gene A single unit of genetic information comprised of a complex segment of the four DNA base pair compounds.

Genetic drift Changes in the frequency of a gene in a population as a result of chance rather than of mutation, selection, or migration.

Genetic risk Used in discussions of endangered species to mean detrimental change in genetic characteristics not caused by external environmental changes. Genetic changes can occur in small populations from such causes as reduced genetic variation, genetic drift, and mutation.

Genetically modified crops Crop species modified by genetic engineering to produce higher crop yields and increase resistance to drought, cold, heat, toxins, plant pests, and disease.

Genetically modified organisms Also known as genetic engineering, refers to the altering of genes or genetic material with the objectives of producing new organisms or organisms with desired characteristics, or to eliminate undesirable characteristics in organisms.

Geochemical cycles The pathways of chemical elements in geologic processes, including the chemistry of the lithosphere, atmosphere, and hydrosphere.

Geographic Information System (GIS) Technology capable of storing, retrieving, transferring, and displaying environmental data.

Geologic cycle The formation and destruction of earth materials and the processes responsible for these events. The geologic cycle includes the following subcycles: hydrologic, tectonic, rock, and geochemical.

Geometric growth *See* Exponential growth.

Geopressurized systems Geothermal system that exists when the normal heat flow from the Earth is trapped by impermeable clay layers that act as an effective insulator.

Geothermal energy The useful conversion of natural heat from the interior of the Earth.

Global circulation models (GCM) A type of mathematical model used to evaluate global change, particularly related to climatic change. GCMs are very complex and require supercomputers for their operation.

Global dimming The process of the reduction of incoming solar radiation by reflection from suspended particles in the atmosphere and their interaction with water vapour (especially clouds).

Global extinction Disappearance or extinction of a species everywhere.

Global forecasting Process of predicting or forecasting future change in environmental areas such as world population, natural resource utilization, and environmental degradation.

Gravel Unconsolidated, generally rounded fragment of rocks and minerals greater than 2 mm in diameter.

Green plans Long-term strategies for identifying and solving global and regional environmental problems. The philosophical heart of green plants is sustainability.

Green revolution Name attached to post-World War II agricultural programs that have led to the development of new strains of crops with higher yield, better resistance to disease, or better ability to grow under poor conditions.

Greenbelt The idea of locating garden cities in a set connected by undeveloped areas, forming a system of countryside and urban landscapes.

Greenhouse effect Process of trapping heat in the atmosphere. Water vapour and several other gases warm the Earth's atmosphere because they absorb and remit radiation; that is, they trap some of the heat radiating from the Earth's atmospheric system.

Greenhouse gases The suite of gases that have a greenhouse effect, such as carbon dioxide, methane, and water vapour.

Gross production (biology) Production before respiration losses are subtracted.

Groundwater Water found beneath the Earth's surface within the zone of saturation, below the water table.

Growth efficiency Gross production efficiency (P/C), or ratio of the material produced (P = net production) by an organism or population to the material ingested or consumed (C).

Growth rate The net increase in some factor per unit time. In ecology, the growth rate of a population is sometimes measured as the increase in numbers of individuals or biomass per unit time and sometimes as a percentage increase in numbers or biomass per unit time.

H

Habitat Where an individual, population, or species exists or can exist. For example, the habitat of the Douglas Fir is the temperate rainforest of North America.

Half-life The time required for half the amount of a substance to disappear; the average time required for one-half of a radioisotope to be transformed to some other isotope; the time required for one-half of a toxic chemical to be converted to some other form.

Hard path Energy policy based on the emphasis of energy quantity generally produced from large centralized power plants.

Hazardous waste Waste that is classified as definitely or potentially hazardous to the health of people. Examples include toxic or flammable liquids and a variety of metals, pesticides, and solvents.

Heat energy Energy of the random motion of atoms and molecules.

Heat island Warmer air of a city than surrounding areas as a result of increased heat production and decreased rate of heat loss caused by the abundance of building and paving materials, which act as solar collectors.

Heat island effect Urban areas are several degrees warmer than their surrounding areas. During relatively calm periods there is an upward flow of air over heavily developed areas accompanied by a downward flow over nearby greenbelts. This produces an air-temperature profile that delineates the heat island.

Heat pumps Devices that transfer heat from one material to another, such as from groundwater to the air in a building.

Herbivore An organism that feeds on an autotroph.

Heterotrophs Organisms that cannot make their own food from inorganic chemicals and a source of energy and therefore live by feeding on other organisms.

High-level radioactive waste Extremely toxic nuclear waste, such as spent fuel elements from commercial reactors. A sense of urgency surrounds determining how we may eventually dispose of this waste material.

Historical range of variation The known range of an environmental variable, such as the abundance of a species or the depth of a lake, over some past time interval.

Homeostasis The ability of a cell or organism to maintain a constant environment. Results from negative feedback, resulting in a state of dynamic equilibrium.

Hormonally active agents (HAAs) Chemicals in the environment able to cause reproductive and developmental abnormalities in animals, including humans.

Hot igneous systems A geothermal system that involves hot, dry rocks with or without the presence of near-surface molten rock.

Human carrying capacity Theoretical estimates of the number of humans who could inhabit the earth at the same time.

Human demography The study of human population characteristics, such as age structure, demographic transition, total fertility, human population and environment relationships, death-rate factors, and standard of living.

Hutchinsonian niche The idea of a measured niche, the set of environmental conditions within which a species is able to persist.

Hydrocarbons Compounds containing only carbon and hydrogen are a large group of organic compounds, including petroleum products, such as crude oil and natural gas.

Hydrochlorofluorocarbons Also known as HCFCs, which are a group of chemicals containing hydrogen, chlorine, fluorine, and carbon, produced as a potential substitute for chlorofluorocarbons (CFCs).

Hydrofluorocarbons Chemicals containing hydrogen, fluorine, and carbons, produced as potential substitutes for chlorofluorocarbons (CFCs).

Hydrologic cycle Circulation of water from the oceans to the atmosphere and back to the oceans by way of evaporation, runoff from streams and rivers, and groundwater flow.

Hydrology The study of surface and subsurface water.

Hydroponics The practice of growing plants in a fertilized water solution on a completely artificial substrate in an artificial environment such as a greenhouse.

Hydrothermal convection systems A type of geothermal energy characterized by circulation of steam and/or hot water that transfers to the surface.

Hypothesis In science, an explanation set forth in a manner that can be tested and is capable of being disproved. A tested hypothesis is accepted until and unless it has been disproved.

I

Igneous rocks Rocks formed from the solidification of magma. They are extrusive if they crystallize on the surface of the Earth and intrusive if they crystallize beneath the surface.

Incineration Combustion of waste at high temperature, consuming materials and leaving only ash and noncombustibles to dispose of in a landfill.

Independent variable *See* Variable, independent

Indicator species Species with known tolerance levels, whose presence or absence can provide insight into abiotic ecosystem conditions.

Indirect costs *See* Costs, indirect

Inductive reasoning Drawing a general conclusion from a limited set of specific observations.

Industrial ecology Process of designing industrial systems to behave more like ecosystems where waste from one part of the system is a resource for another part.

Inference (1) A conclusion derived by logical reasoning from premises and/or evidence (observations or facts), or (2) a conclusion, based on evidence, arrived at by insight or analogy, rather than derived solely by logical processes.

Influent stream Type of stream that is everywhere above the groundwater table and flows in direct response to precipitation. Water from the channel moves down to the water table, forming a recharge mound.

Inspirational justification for the conservation of nature An argument for the conservation of nature on the grounds that direct experience of nature is an aid to spiritual or mental well-being.

In-stream use A type of water use that includes navigation, generation of hydroelectric power, fish and wildlife habitat, and recreation.

Integrated energy management Use of a range of energy options that vary from region to region, including a mix of technology and sources of energy.

Integrated pest management Control of agricultural pests using several methods together, including biological and chemical agents. A goal is to minimize the use of artificial chemicals; another goal is to prevent or slow the buildup of resistance by pests to chemical pesticides.

Integrated waste management (IWM) Set of management alternatives including reuse, source reduction, recycling, composting, landfill, and incineration.

Interference During succession, one species prevents the entrance of later successional species into an ecosystem. For example, some grasses produce such dense and thick mats that seeds of trees cannot reach the soil to germinate. As long as these grasses persist, the trees that characterize later stages of succession cannot enter the ecosystem.

Intermediate species In forestry, species that form a layer of growth below the dominant species.

Island arc A curved group of volcanic islands associated with a deep oceanic trench and subduction zone (convergent plate boundary).

Island biogeography A theory relating island size and distance from mainland to the number of species present. The theory states that islands have fewer species than continents, and the smaller the island, and the farther away an island is from a continent, the fewer the species present.

Isotope Atoms of an element that have the same atomic number (the number of protons in the nucleus of the atom) but vary in atomic mass number (the number of protons plus neutrons in the nucleus of an atom).

J

Jurisdiction The power of a government to make laws. Jurisdiction is normally stipulated in a country's constitution. In Canada's Constitution Act of 1982, certain powers such as interprovincial transportation and trade, and the power to make criminal law, were granted to the federal government. Under the Constitution, the provinces were granted more local powers, such as education, regulation of local business, and the rights to natural resources within the province. The provinces were also granted the right to create municipalities and delegate power to them.

K

Keystone species A species, such as the sea otter, that has a large effect on its community or ecosystem so that its removal or addition to the community leads to major changes in the abundances of many or all other species.

Kinetic energy The energy of motion. For example, the energy in a moving car that results from the mass of the car travelling at a particular velocity.

Kwashiorkor Lack of sufficient protein in the diet, which leads to a failure of neural development in infants and therefore to learning disabilities.

L

Landscape A set of ecosystems connected across a geographic area.

Landscape perspective The concept that effective management and conservation recognizes that ecosystems, populations, and species are interconnected across large geographic areas.

Land application Method of disposal of hazardous waste that involves intentional application of waste material to surface soil. Useful for certain biodegradable industrial waste, such as oil and petroleum waste, and some organic chemical waste.

Land ethic A set of ethical principles that affirm the right of all resources, including plants, animals, and earth materials, to continued existence and, at least in some locations, to continued existence in a natural state.

Land-use planning Complex process involving development of a land-use plan to include a statement of land-use issues, goals, and objectives; a summary of data collection and analysis; a land-classification map; and a report that describes and indicates appropriate development in areas of special environmental concern.

Late successional species Species that occur only or primarily in, or are dominant in late stages in succession.

Law of the minimum (Liebig's law of the minimum) The concept that the growth or survival of a population is directly related to the single life requirement that is in least supply (rather than due to a combination of factors).

LD-50 A crude approximation of a chemical toxicity defined as the dose at which 50% of the population dies on exposure.

Leachate Noxious, mineralized liquid capable of transporting bacterial pollutants. Produced when water infiltrates through waste material and becomes contaminated and polluted.

Leaching Water infiltration from the surface, dissolving soil materials as part of chemical weathering processes and transporting the dissolved materials laterally or downward.

Lead A heavy metal that is an important constituent of automobile batteries and other industrial products. A toxic metal capable of causing environmental disruption and producing a health problem to people and other living organisms.

Liebig's law of the minimum *See* Law of the minimum.

Life expectancy The estimated average number of years (or other time period used as a measure) that an individual of a specific age can expect to live.

Limiting factor The single requirement for growth available in the least supply in comparison to the need of an organism. Originally applied to crops but now often applied to any species.

Lithosphere Outer layer of Earth, approximately 100 km thick, of which the plates that contain the ocean basins and the continents are composed.

Littoral drift Movement caused by wave motion in nearshore and beach environment.

Local extinction The disappearance of a species from part of its range, but continued persistence elsewhere; also called extirpation.

Logistic carrying capacity In terms of the logistic curve, the population size at which births equals deaths and there is no net change in the population.

Logistic equation The equation that results in a logistic growth curve, that is, the growth rate $dN/dt = rN [(K-N)/N]$, where r is the intrinsic rate of increase, K is the carrying capacity, and N is the population size.

Logistic growth curve The S-shaped growth curve that is generated by the logistic growth equation. In the logistic, a small population grows rapidly, but the growth rate slows down, and the population eventually reaches a constant size.

Low-level radioactive waste Waste materials that contain sufficiently low concentrations or quantities of radioactivity so as not to present a significant environment hazard if properly handled.

Luz solar electric generating system Solar energy farms comprising a power plant surrounded by hundreds of solar collectors (curved mirrors) that heat a synthetic oil, which flows through heat exchangers to drive steam turbine generators.

M

Macronutrients Elements required in large amounts by living things. These include the big six—carbon, hydrogen, oxygen, nitrogen, phosphorus, and sulphur.

Made lands Man-made areas created artificially with fill, sometimes as waste dumps of all kinds and sometimes directly, to make more land available for construction.

Magma A naturally occurring silica melt, a good deal of which is a liquid state.

Malnourishment The lack of specific components of food, such as proteins, vitamins, or essential chemical elements.

Manipulated variable *See* Variable, independent

Marasmus Progressive emaciation caused by a lack of protein and calories.

Marginal cost In environmental economics, the cost to reduce one additional unit of a type of degradation, for example, pollution.

Marginal land An area of the Earth with minimal rainfall or otherwise limited severely by some necessary factor, so that it is a poor place for agriculture and easily degraded by agriculture. Typically, these lands are easily converted to deserts when used for light grazing and crop production.

Mariculture Production of food from marine habitats.

Marine evaporites With respect to mineral resources, refers to materials such as potassium and sodium salts resulting from the evaporation of marine waters.

Materials management Term related to waste management consistent with the ideals of industrial ecology that makes better use of materials, leading to more sustainable use of resources.

Matter Anything that occupies space and has mass. It is the substance of which physical objects are composed.

Maximum lifetime Genetically determined maximum possible age to which an individual of a species can live.

Maximum sustainable population The population size that has the maximum production (net increase in number of individuals or net change in biomass) that allows the population to be sustained indefinitely without decreasing the ability of the population to provide the same level of production.

Maximum sustainable yield (MSY) The maximum usable production of a biological resource that can be obtained in a specified time period.

Mediation Negotiation process between adversaries, guided by a neutral facilitator.

Megacities Urban areas with at least eight million inhabitants.

Meltdown Refers to a nuclear accident in which the nuclear fuel forms a molten mass that breaches the containment of the reactor, contaminating the outside environment with radioactivity.

Mesosphere The portion of the atmosphere above the stratosphere, where meteors and rock fragments entering the atmosphere burn up.

Metals Refers to a number of metals, including lead, mercury, arsenic, and silver (among others) that have a relatively high atomic number (the number of protons in the nucleus of an atom). They are often toxic at relatively low concentrations, causing a variety of environmental problems. (The term "heavy metals" is sometimes used, but has no basis in science.)

Methane (CH_4) Molecule of carbon and hydrogen, which is a naturally occurring gas in the atmosphere. One of the so-called greenhouse gases.

Methane hydrate A white ice-like compound made up of molecules of methane gas trapped in "cages" of frozen water in the sediments of the deep seafloor.

Microclimate The climate of a very small local area. For example, the climate under a tree, near the ground within a forest, or near the surface of streets in a city.

Micronutrients Chemical elements required in very small amounts by at least some forms of life. Boron, copper, and molybdenum are examples of micronutrients.

Micropower The production of electricity using smaller distributed systems rather than relying on large central power plants.

Migration The movement of an individual, population, or species from one habitat to another or more simply from one geographic area to another.

Migration corridor Designated passageways among parks or preserves allowing migration of many life forms among several of these areas.

Mineral Naturally occurring inorganic material with a definite internal structure and physical and chemical properties that vary within prescribed limits.

Mineral resources Elements, chemical compounds, minerals, or rocks concentrated in a form that can be extracted to obtain a usable commodity.

Minimum viable population The minimum number of individuals that have a reasonable chance of persisting for a specified time period.

Missing carbon sink Substantial amounts of carbon dioxide released into the atmosphere but apparently not reabsorbed and thus remaining unaccounted for.

Mitigated negative declaration Special type of negative declaration that suggests that the adverse environmental aspects of a particular action may be mitigated through modification of the project in such a way as to reduce the impacts to near insignificance.

Mitigation Process that identifies actions to avoid, lessen, or compensate for anticipated adverse environmental impacts.

Mobile sources Sources of air pollutants that move from place to place, for example, automobiles, trucks, buses, and trains.

Model A deliberately simplified explanation, often physical, mathematical, pictorial, or computer-simulated, of complex phenomena or processes.

Monitoring Process of collecting data on a regular basis at specific sites to provide a database from which to evaluate change. For example, collection of water samples from beneath a landfill to provide early warning should a pollution problem arise.

Monoculture (Agriculture) The planting of large areas with a single species or even a single strain or subspecies in farming.

Moral justification for the conservation of nature An argument for the conservation of nature on the grounds that aspects of the environment have a right to exist, independent of human desires, and that it is our moral obligation to allow them to continue or to help them persist.

Multiple use Literally, using the land for more than one purpose at the same time. For example, forestland can be used to produce commercial timber but at the same time serve as wildlife habitat and land for recreation. Usually multiple use requires compromises and trade-offs, such as striking a balance between cutting timber for the most efficient production of trees at a level that facilitates other uses.

Mutation Stated most simply, a chemical change in a DNA molecule. It means that the DNA carries a different message than it did before, and this change can affect the expressed characteristics when cells or individual organisms reproduce.

Mutualism *See* Symbiosis.

N

Natural catastrophe Sudden catastrophic change in the environment, not the result of human actions.

Natural gas Naturally occurring gaseous hydrocarbon (predominantly methane) generally produced in association with crude oil or from gas wells; an important efficient and clean-burning fuel commonly used in homes and industry.

Natural selection A process by which organisms whose biological characteristics better fit them to the environment are better represented by descendants in future generations than those whose characteristics are less fit for the environment.

Nature preserve An area set aside with the primary purpose of conserving some biological resource.

Negative feedback A type of feedback that occurs when the system's response is in the opposite direction of the output. Thus, negative feedback is self-regulating.

Net growth efficiency Net production efficiency (P/A), or ratio of the material produced (P) to the material assimilated (A) by an organism. The material assimilated is less than the material consumed, because some food taken is egested as waste (discharged) and never used by an organism.

Net metering The process in which private wind and solar energy producers can store their surplus electricity and sell it back to the public system.

Net production (biology) The production that remains after utilization. In a population, net production is sometimes measured as the net change in the numbers of individuals. It is also measured as the net change in biomass or in stored energy. In terms of energy, it is equal to the gross production minus the energy used in respiration.

New forestry The name for a new variety of timber harvesting practices to increase the likelihood of sustainability, including recognition of the dynamic characteristics of forests and of the need for management within an ecosystem context.

Niche (1) The "profession," or role, of an organism or species; or (2) all the environmental conditions under which the individual or species can persist. The fundamental niche is all the conditions under which a species can persist in the absence of competition; the realized niche is the set of conditions as they occur in the real world with competitors.

Nitrogen cycle A complex biogeochemical cycle responsible for moving important nitrogen components through the biosphere and other Earth systems. This is an extremely important cycle because all living things require nitrogen.

Nitrogen fixation The process by which atmospheric nitrogen is converted to ammonia, nitrate ion, or amino acids. Micro-organisms perform most of the conversion, but a small amount is also converted by lightning.

Nitrogen oxides Occur in several forms (NO, NO_2, and NO_3). Most important as an air pollutant is nitrogen dioxide, which is a visible yellow brown to reddish brown gas. It is a precursor of acid rain and produced through the burning of fossil fuels.

Noise pollution A type of pollution characterized by unwanted or potentially damaging sound.

Nonmarine evaporites With respect to mineral resources, refers to useful deposits of materials such as sodium and calcium bicarbonate, sulphate, borate, or nitrate produced by evaporation of surficial waters on the land, as differentiated from marine waters in the oceans.

Nonpoint sources Sources of pollutants that are diffused and intermittent and are influenced by factors such as land use, climate, hydrology, topography, native vegetation, and geology.

Non-renewable energy Alternative energy sources, including nuclear and geothermal, that are dependent on fuels or a resource that may be used up much faster than it is replenished by natural processes.

Non-renewable resource A resource that is cycled so slowly by natural Earth processes that once used, it is essentially not going to be made available within any useful time framework.

No-till agriculture Combination of farming practices that includes not ploughing the land and using herbicides to keep down weeds.

Nuclear cycle The series of processes that begins with the mining of uranium to be processed and used in nuclear reactors and ends with the disposal of radioactive waste.

Nuclear energy The energy of the atomic nucleus that, when released, may be used to do work. Controlled nuclear fission reactions take place within commercial nuclear reactors to produce energy.

Nuclear fuel cycle Processes involved with producing nuclear power from the mining and processing of uranium to control fission, reprocessing of spent nuclear fuel, decommissioning of power plants, and disposal of radioactive waste.

Nuclear reactors Devices that produce controlled nuclear fission, generally for the production of electric energy.

O

Obligate symbionts A symbiotic relationship between two organisms in which neither by themselves can exist without the other.

Observations Information obtained through one or more of the five senses or through instruments that extend the senses. For example, some remote sensing instruments measure infrared intensity, which we do not see, and convert the measurement into colours, which we do see.

Ocean thermal conversion Direct utilization of solar energy using part of a natural oceanic environment as a gigantic solar collector.

Off-site effect An environmental effect occurring away from the location of the causal factors.

Off-stream use Type of water use where water is removed from its source for a particular use.

Oil shale A fine-grained sedimentary rock containing organic material known as kerogen. On distillation, yields significant amounts of hydrocarbons including oil.

Old growth forest A non-technical term often used to mean a virgin forest (one never cut) but also used to mean a forest that has been undisturbed for a long, but usually unspecified, time.

Oligotrophic Referring to bodies of waters having a low concentration of the chemical elements required for life.

Omnivores Organisms that eat both plants and animals.

On-site effect An environmental effect occurring at the location of the causal factors.

Open dumps Area where solid waste is disposed of by simply dumping it. These dumps often cause severe environmental problems, such as water pollution, and create a health hazard. They are illegal in Canada and in many other countries around the world.

Open system A type of system in which exchanges of mass or energy occur with other systems.

Open woodlands Areas in which trees are a dominant vegetation form but the leaves of adjacent trees generally do not touch or overlap, so that there are gaps in the canopy. Typically, grasses or shrubs grow in these gaps among the trees.

Operational definitions Definitions that tell you what you need to look for or do in order to carry out an operation, such as measuring, constructing, or manipulating.

Optimal carrying capacity This term has several meanings, but the major idea is the maximum abundance of a population or species that can persist in an ecosystem without degrading the ability of the ecosystem to maintain: (1) that population or species; (2) all necessary ecosystem processes; and (3) the other species found in that ecosystem.

Optimum sustainable population (OSP) The population level that results in an optimum sustainable yield; or the population level that is in some way best for the population, its ecological community, its ecosystem, or the biosphere.

Optimum sustainable yield (OSY) The largest yield of a renewable resource achievable over a long time period without decreasing the ability of the population or its environment to support the continuation of this level of yield.

Ore deposits Earth materials in which metals are concentrated in high concentrations, sufficient to be mined.

Organic compound A compound of carbon; originally used to refer to the compounds found in and formed by living things.

Organic farming Farming that is more "natural" in the sense of not using artificial pesticides and, more recently, not using genetically modified crops. In recent years governments have begun to set up legal criteria for what constitutes organic farming.

Organism A living being, such as an individual plant or animal.

Outbreaks An epidemic or sudden occurrence of a disease in a population, normally affecting many individuals.

Overdraft Groundwater withdrawal when the amount pumped from wells exceeds the natural rate of replenishment.

Overgrazing When the carrying capacity of land for a herbivore, such as cattle or deer, is exceeded.

Overshoot and collapse This occurs when growth in one part of a system over time exceeds carrying capacity, resulting in sudden decline in one or both parts of the system.

Ozone (O_3) Form of oxygen in which three atoms of oxygen occur together. Is chemically active and has a short average lifetime in the atmosphere. Forms a natural layer high in the atmosphere (stratosphere) that protects us from harmful ultraviolet radiation from the sun.

Ozone shield Stratospheric ozone layer that absorbs ultraviolet radiation.

P

Parasitism *See* Predation-parasitism.

Particulate matter Small particles of solid or liquid substances that are released into the atmosphere by many activities, including farming, volcanic eruption, and burning fossil fuels. Particulates affect human health, ecosystems, and the biosphere.

Passive solar energy system Direct use of solar energy through architectural design to enhance or take advantage of natural changes in solar energy that occur throughout the year without requiring mechanical power.

Pasture Land ploughed and planted to provide forage for domestic herbivorous animals.

Pebble A rock fragment between 4 and 64 mm in diameter.

Pedology The study of soils.

Pelagic whaling Practice of whalers taking to the open seas and searching for whales from ships that remained at sea for long periods.

Per capita availability The amount of a resource available per person.

Per capita demand The economic demand per person.

Per capita food production The amount of food produced per person.

Permafrost Permanently frozen ground.

Persistent organic pollutants (POPs) Synthetic carbon-based compounds, often containing chlorine that do not easily break down in the environment. Many were introduced decades before their harmful effects were fully understood and are now banned or restricted.

Pesticides, broad spectrum Pesticides that kill a wide variety of organisms. Arsenic, one of the first elements used as a pesticide, is toxic to many life forms, including people.

Phosphorus cycle Major biogeochemical cycle involving the movement of phosphorus throughout the biosphere and lithosphere. This cycle is important because phosphorus is an essential element for life and often is a limiting nutrient for plant growth.

Photochemical oxidants Result from atmospheric interactions of nitrogen dioxide and sunlight. Most common is ozone (O_3).

Photochemical smog Sometimes called L.A.-type smog or brown air. Directly related to automobile use and solar radiation. Reactions that occur in the development of the smog are complex and involve both nitrogen oxides and hydrocarbons in the presence of sunlight.

Photosynthesis Synthesis of sugars from carbon dioxide and water by living organisms using light as energy. Oxygen is given off as a by-product.

Photovoltaics Technology that converts sunlight directly into electricity using a solid semiconductor material.

Physiographic province A region characterized by a particular assemblage of landforms, climate, and geomorphic history.

Pioneer species Species found in early stages of succession.

Placer deposit A type of ore deposit found in material transported and deposited by agents such as running water, ice, or wind; for example, gold and diamonds found in stream deposits.

Plantations Managed forest similar to mechanized farming, in which a single species is planted in straight rows and harvested in a specified rotation period of time.

Plate tectonics A model of global tectonics that suggests that the outer layer of Earth, known as the lithosphere, is composed of several large plates that move relative to one another. Continents and oceans basins are passive riders on these plates.

Point sources Sources of pollution such as smokestacks, pipes, or accidental spills that are readily identified and stationary. They are often thought to be easier to recognize and control than are area sources. This is true only in a general sense, as some very large point sources emit tremendous amounts of pollutants into the environment.

Polar amplification Processes in which global warming causes greater temperature increases at polar regions.

Polar stratospheric clouds Clouds that form in the stratosphere during the polar winter.

Polar vortex Arctic air masses that in the winter become isolated from the rest of the atmosphere and circulate about the pole. The vortex rotates counter-clockwise because of the rotation of the Earth in the Southern Hemisphere.

Policy instruments The means to implement a society's policies. Include moral suasion (jawboning—persuading people by talk, publicity, and social pressure); direct controls, including regulations; and market processes affecting the price of goods and processes, such as subsidies, licenses, and deposits.

Pollutant In general terms, any factor that has a harmful effect on living things or their environment.

Pollution The process by which something becomes impure, defiled, dirty, or otherwise unclean.

Pollution prevention Identifying ways to avoid the generation of waste rather than finding ways to dispose of it.

Pool (geology) Common bed form produced by scour in meandering and straight channels.

Population A group of individuals of the same species living in the same area or interbreeding and sharing genetic information.

Population age structure The number of individuals or proportion of the population in each age class.

Population dynamics The study of changes in population sizes and the causes of these changes.

Population momentum or lag effect The continued growth of a population that occurs after replacement level fertility is reached.

Population regulation *See* Density-dependent population effects and Density-independent population effects.

Population risk Used in discussions of endangered species to mean random variation in population rates—birthrates and death rates—possibly causing species in low abundance to become extinct.

Positive feedback A type of feedback that occurs when an increase in output leads to a further increase in output. This is sometimes known as a vicious cycle, since the more you have the more you get.

Potential energy Energy that is stored. Examples include the gravitational energy of water behind a dam; chemical energy in coal, fuel oil, and gasoline; and nuclear energy (in the forces that hold atoms together).

Power The time rate of doing work.

Precautionary principle The idea that in spite of the fact that full scientific certainty is often not available to prove cause and effect, we should still take cost-effective precautions to solve environmental problems when there exists a threat of potential serious and/or irreversible environmental damage.

Predation-parasitism Interaction between individuals of two species in which the outcome benefits one and is detrimental to the other.

Predator An organism that feeds on other, live organisms, usually of other species. The term is usually applied to animals that feed on other animals, but sometimes it is used to mean herbivore.

Premises In science, initial definitions and assumptions.

Primary pollutants Air pollutants emitted directly into the atmosphere. Included are particulates, sulphur oxides, carbon monoxide, nitrogen oxides, and hydrocarbons.

Primary production *See* Production, primary

Primary succession The initial establishment and development of an ecosystem.

Primary treatment (of wastewater) Removal of large particles and organic materials from wastewater through screening.

Probability The relative probability that an event will occur.

Production, ecological The amount of increase in organic matter, usually measured per unit area of land surface or unit volume of water, as in grams per square metre (g/m^2). Production is divided into primary (that of autotrophs) and secondary (that of heterotrophs). It is also divided into net (that which remains stored after use) and gross (that added before any use).

Production, primary The production by autotrophs.

Production, secondary The production by heterotrophs.

Productivity, ecological The rate of production; that is, the amount of increase in organic matter per unit time (for example, grams per metre squared per year).

Prokaryote A kind of organism that lacks a true cell nucleus and has other cellular characteristics that distinguish it from the eukaryotes. Bacteria are prokaryotes.

Pseudoscientific Ideas which are claimed to have scientific validity but are inherently untestable and/or lack empirical support and/or were arrived at through faulty reasoning or poor scientific methodology.

Public service functions Functions performed by ecosystems for the betterment of life and human existence. For example, the cleansing of the air by trees and removal of pollutants from water by infiltration through the soil.

Public trust Grants and limits the authority of government over certain natural areas of special character.

Public welfare statutes Laws that address routine civil matters, such as the administration of universities, the conduct of professional engineers, and the protection of public health. Environmental legislation is of this type. (*See* also Criminal statutes.)

Q

Qualitative data Data that are distinguished by qualities or attributes that cannot or are not expressed as quantities. For example, blue and red are qualitative data about the electromagnetic spectrum.

Quantitative data Data that are expressed as numbers or numerical measurements. For example, the wavelengths of specific colours of blue and red light (460 and 650 nanometers, respectively) are quantitative data about the electromagnetic spectrum.

R

R-to-C ratio A measure of the time available for finding the solutions to depletion of non-renewable reserves, where R is the known reserves (for example, hundreds of thousands of tons of a metal) and C is the rate of consumption (for example, thousands of tons per year used by people).

Radiation absorbed dose Energy retained by living tissue that has been exposed to radiation.

Radioactive decay A process of decay of radioisotopes that change from one isotope to another and emit one or more forms of radiation.

Radioactive waste Type of waste produced in the nuclear fuel cycle; generally classified as high level or low level.

Radioisotope A form of a chemical element that spontaneously undergoes radioactive decay.

Radon Naturally occurring radioactive gas. Radon is colourless, odourless, and tasteless and must be identified through proper testing.

Rangeland Land used for grazing.

Range of tolerance The set of abiotic conditions in which an organism is able to grow and reproduce.

Rare species Species with a small total population size, or restricted to a small area, but not necessarily declining or in danger of extinction.

Realms (Ecological) Major biogeographic regions of Earth that are based upon fundamental features of the plants and animals found in those regions.

Recreational justification for the conservation of nature An argument for the conservation of nature on the grounds that direct experience of nature is inherently enjoyable and the benefits derived from it are important and valuable to people.

Recycle Integral part of waste management that attempts to identify resources in the waste stream that may be collected and reused.

Reduce With respect to waste management, refers to practices that will reduce the amount of waste we produce.

Reduce, reuse, and recycle The three Rs of integrated waste management.

Regulations The detailed rules attached to a statute (law). Whereas the statute sets out the general intent and administration of a law, a regulation specifies detailed operational considerations such as allowable pollutant discharge levels, the names of industries covered by the law, the classification of wastes, and similar information. Regulations are more easily amended than are statutes.

Renewable energy Alternative energy sources, such as solar, water, wind, and biomass, that are more or less continuously made available in a time framework useful to people.

Renewable resource A resource, such as timber, water, or air, that is naturally recycled or recycled by artificial processes within a time framework useful for people.

Replacement level fertility Fertility rate required for the population to remain a constant size.

Representative natural areas Parks or preserves set aside to represent presettlement conditions of a specific ecosystem type.

Reserves Known and identified deposits of earth materials from which useful materials can be extracted profitably with existing technology and under present economic and legal conditions.

Resilience *See* Ecological resilience.

Resource-based agriculture Agricultural practices that rely on extensive use of resources, so that production is limited by the availability of resources.

Resources Reserves plus other deposits of useful earth materials that may eventually become available.

Respiration The complex series of chemical reactions in organisms that make energy available for use. Water, carbon dioxide, and energy are the products of respiration.

Responding variable *See* Variable, dependent

Restoration ecology The field within the science of ecology with the goal to return damaged ecosystems to ones that are functional, sustainable, and more natural.

Reuse With respect to waste management, refers to finding ways to reuse production and materials so they need not be disposed of.

Riffle A section of stream channel characterized at low flow by fast, shallow flow. Generally contains relatively coarse bed-load particles.

Risk assessment The process of determining potential adverse environmental health effects to people following exposure to pollutants and other toxic materials. Generally includes the four steps of identification of the hazard, dose-response assessment, exposure assessment, and risk characterization.

Risk-benefit analysis In environmental economics, the riskiness of the future that influences the value we place on things in the present.

Rock (engineering) Any earth material that has to be blasted in order to be removed.

Rock (geologic) An aggregate of a mineral or minerals.

Rock cycle A group of processes that produce igneous, metamorphic, and sedimentary rocks.

Rotation time Time between cuts of a stand or area of forest.

Rule of climatic similarity Similar environments lead to the evolution of organisms similar in form and function (but not necessarily in genetic heritage or internal makeup) and to similar ecosystems.

Ruminants Animals having a four-chambered stomach within which bacteria convert the woody tissue of plants to proteins and fats that, in turn, are digested by the animal. Cows, camels, and giraffes are ruminants; horses, pigs, and elephants are not.

S

Sand Grains of sediment between 1/16 and 2 mm in diameter; often sediment composed of quartz particles of this size.

Sand dune A ridge or hill of sand formed by wind action.

Sanitary landfill A method of disposal of solid waste without creating a nuisance or hazard to public health or safety. Sanitary landfills are highly engineered structures with multiple barriers and collection systems to minimize environmental problems.

Savanna An area with trees scattered widely among dense grasses.

Scientific method Actually a set of methods which are the systematic methods by which scientists investigate natural phenomena, including gathering data, formulating and testing hypotheses, and developing scientific theories and laws.

Scientific theory A grand scheme that relates and explains many observations and is supported by a great deal of evidence, in contrast to a guess, a hypothesis, a prediction, a notion, or a belief.

Scoping The process of early identification of important environmental issues that require detailed evaluation.

Scrubbing A process of removing sulphur from gases emitted from power plants burning coal. The gases are treated with a slurry of lime and limestone, and the sulphur oxides react with the calcium to form insoluble calcium sulphides and sulphates that are collected and disposed of.

Second growth Forest that has been logged and regrown.

Second law of thermodynamics A fundamental principle of energy that states that energy always tends to go from a more usable (higher quality) form to a less usable (lower quality) form. When we say that energy is converted to a less useful form we mean that entropy (a measure of the energy unavailable to do useful work) of the system has increased.

Secondary enrichment A weathering process of sulphide ore deposits that may concentrate the desired minerals.

Secondary pollutants Air pollutants produced through reactions between primary pollutants and normal atmospheric compounds. An example is ozone that forms over urban areas through reactions of primary pollutants, sunlight, and natural atmospheric gases.

Secondary production *See* Production, secondary

Secondary succession The re-establishment of an ecosystem where there are remnants of a previous biological community.

Secondary treatment (of wastewater) Use of biological processes to degrade wastewater in a treatment facility.

Second-law efficiency The ratio of the minimum available work needed to perform a particular task to the actual work used to perform that task. Reported as a percentage.

Secure landfill A type of landfill designed specifically for hazardous waste. Similar to a modern sanitary landfill in that it includes multiple barriers and collection systems to ensure that leachate does not contaminate soil and other resources.

Sediment pollution By volume and mass, sediment is our greatest water pollutant. It may choke streams, fill reservoirs, bury vegetation, and generally create a nuisance that is difficult to remove.

Seed-tree cutting A logging method in which mature trees with good genetic characteristics and high seed production are preserved to promote regeneration of the forest. It is an alternative to clear-cutting.

Seismic Referring to vibrations in the Earth produced by earthquakes.

Selective cutting In timber harvesting, the practice of cutting some, but not all, trees, leaving some on the site. There are many kinds of selective cutting. Sometimes the biggest trees with the largest market value are cut, and smaller trees are left to be cut later. Sometimes the best trees are left to provide seed for future generations. Sometimes trees are left for wildlife habitat and recreation.

Shelterwood cutting A logging method in which dead and less desirable trees are cut first; mature trees are cut later. This ensures that young, vigorous trees will always be left in the forest. It is an alternative to clear-cutting.

Sick building syndrome Condition associated with a particular indoor environment that appears to be unhealthy to the human occupants.

Silicate minerals The most important group of rock-forming minerals.

Silt Sediment between 1/16 and 1/256 mm in diameter.

Silviculture The practice of growing trees and managing forests, traditionally with an emphasis on the production of timber for commercial sale.

Sinkhole A surface depression formed by the solution of limestone or the collapse over a subterranean void such as a cave.

Site (in relation to cities) Environmental features of a location that influences the placement of a city. For example, Bangkok, Thailand, is built on low-lying mud, which forms a poor site, while St. John's, Newfoundland is built on an island of strong bedrock, an excellent site.

Site quality Used by foresters to mean an estimator of the maximum timber crop the land can produce in a given time.

Situation (in relation to cities) The relative geographic location of a site that makes it a good location for a city. For example, Bangkok, Thailand, has a good situation because it is located at the mouth of the Chao Phraya River and is therefore a natural transportation junction.

Smog A term first used in 1905 for a mixture of smoke and fog that produced unhealthy urban air. There are several types of fog, including photochemical smog and sulphurous smog.

Soft path Energy policy that relies heavily on renewable energy resources as well as other sources that are diverse, flexible, and matched to the end-use needs.

Soil The top layer of a land surface where the rocks have been weathered to small particles. Soils are made up of inorganic particles of many sizes, from small clay particles to large sand grains. Many soils also include dead organic material.

Soil (in engineering) Earth material that can be removed without blasting.

Soil (in soil science) Earth material modified by biological, chemical, and physical processes such that the material will support rooted plants.

Soil fertility The capacity of a soil to supply the nutrients and physical properties necessary for plant growth.

Soil horizon A layer in soil (A, B, C) that differs from another layer in chemical, physical, and biological properties.

Solar cell (photovoltaic) Device that directly converts light into electricity.

Solar collector Device for collecting and storing solar energy. For example, home water heating is done by flat panels consisting of a glass cover plate over a black background on which water is circulated through tubes. Short-wave solar radiation enters the glass and is absorbed by the black background. As long-wave radiation is emitted from the black material, it cannot escape through the glass, so the water in the circulating tubes is heated, typically to temperatures of 38° to 93°C.

Solar energy Collecting and using energy from the sun directly.

Solar pond Shallow pond filled with water and used to generate relatively low-temperature water.

Solar power tower A system of collecting solar energy that delivers the energy to a central location where the energy is used to produce electric power.

Source reduction Process of waste management, the object of which is to reduce the amounts of materials that must be handled in the waste stream.

Species A group of individuals capable of interbreeding.

Stable equilibrium A condition in which a system will remain if undisturbed and to which it will return when displaced.

Stack effect *See* Chimney effect.

Stand An informal term used by foresters to refer to a group of trees.

Stationary sources Air pollution sources that have a relatively fixed location, including point sources, fugitive sources, and area sources.

Steady state When input equals output in a system, there is no net change and the system is said to be in a steady state. A bathtub with water flowing in and out at the same rate maintains the same water level and is in a steady state. Compare with equilibrium.

Stratosphere The part of the atmosphere above the troposphere; the portion of the atmosphere that warms with increasing altitude.

Stratospheric ozone layer A protective layer extending from the troposphere to an altitude of about 40 km, with a maximum concentration above the equator at about 25 to 30 km. Stratospheric ozone protects life in the lower atmosphere from receiving harmful doses of ultraviolet radiation. *See also* Ozone shield.

Stress Force per unit area. May be compression, tension, or shear.

Strip cutting In timber harvesting, the practice of cutting narrow rows of forest, leaving wooded corridors.

Strip mining Surface mining in which the overlying layer of rock and soil is stripped off to reach the resource. Large strip mines are some of the largest excavations caused by people in the world.

Subduction A process in which one lithospheric plate descends beneath another.

Subsidence A sinking, settling, or otherwise lowering of parts of the crust of the Earth.

Subsistence crops Crops used directly for food by a farmer or sold locally where the food is used directly.

Succession The process of establishment and development of an ecosystem.

Sulphur dioxide (SO_2) Colourless and odourless gas normally present at the Earth's surface in low concentrations. An important precursor to acid rain. Major anthropogenic source is burning fossil fuels.

Sulphurous smog Produced primarily by burning coal or oil at large power plants. Sulphur oxides and particulates combine under certain meteorological conditions to produce a concentrated form of this smog.

Suppressed In forestry, tree species growing in the understory, beneath the dominant and intermediate species.

Surface impoundment Method of disposal of some liquid hazardous waste. This method is controversial and many sites have been closed.

Sustainability Management of natural resources and the environment with the goals of allowing the harvest of resources to remain at or above some specified level, and the ecosystem to retain its functions and structure.

Sustainable development The ability of a society to continue to develop its economy and social institutions and also maintain its environment for an indefinite time.

Sustainable ecosystem An ecosystem that is subject to some human use, but at a level that leads to no loss of species or of necessary ecosystem functions.

Sustainable energy development A type of energy management that provides for reliable sources of energy while not causing environmental degradation and ensuring that future generations will have a fair share of the Earth's resources.

Sustainable forestry Effort to manage a forest so that a resource in it can be harvested at a rate that does not decrease the ability of the forest ecosystem to continue to provide that same rate of harvest indefinitely.

Sustainable resource harvest An amount of a resource that can be harvested at regular intervals indefinitely.

Sustainable water use Use of water resources that does not harm the environment and provides for the existence of high-quality water for future generations.

Symbiont Each partner in symbiosis.

Symbiosis An interaction between individuals of two different species that benefits both. For example, lichens contain an alga and a fungus that require each other to persist. Sometimes this term is used broadly, so that domestic corn and people could be said to have a symbiotic relationship—domestic corn cannot reproduce without the aid of people, and some peoples survive because they have corn to eat.

Symbiotic Relationships that exist between different organisms that are mutually beneficial.

Synergism Co-operative action of different substances such that the combined effect is greater than the sum of the effects taken separately.

Synergistic effect When the change in availability of one resource affects the response of an organism to some other resource.

Synfuels Synthetic fuels, which may be liquid or gaseous, derived from solid fuels, such as oil from kerogen in oil shale or oil and gas from coal.

Synthetic organic compounds Compounds of carbon produced synthetically by human industrial processes, as for example pesticides and herbicides.

System A set of components that are linked and interact to produce a whole. For example, the river as a system is composed of sediment, water, bank, vegetation, fish, and other living things that all together produce the river.

T

Taiga Forest of cold climates of high latitudes and high altitudes, also known as boreal forest.

Tar sands Sedimentary rocks or sands impregnated with tar oil, asphalt, or bitumen.

Taxa Categories that identify groups of living organisms based upon evolutionary relationships or similarity of characters.

Taxon A grouping of organisms according to evolutionary relationships.

TD-50 The toxic dose defined as the dose that is toxic to 50% of a population exposed to the toxin.

Tectonic cycle The processes that change Earth's crust, producing external forms such as ocean basins, continents, and mountains.

Terminator gene A genetically modified crop, which has a gene to cause the plant to become sterile after the first year.

Tertiary treatment (of wastewater) Advanced form of wastewater treatment involving chemical treatment or advanced filtration. An example is chlorination of water.

Theories Scientific models that offer broad, fundamental explanations of related phenomena and are supported by consistent and extensive evidence.

Thermal (heat energy) The energy of the random motion of atoms and molecules.

Thermal pollution A type of pollution that occurs when heat is released into water or air and produces undesirable effects on the environment.

Thermodynamic system Formed by an energy source, ecosystem, and energy sink, where the ecosystem is said to be an intermediate system between the energy source and the energy sink.

Thermodynamics, first law of *See* First Law of Thermodynamics.

Thermodynamics, second law of *See* Second Law of Thermodynamics.

Thermosphere The portion of the atmosphere above the mesosphere; the layer in which aurora borealis (northern lights) occurs, and in which the space shuttle orbits.

Thinning The timber-harvesting practice of selectively removing only smaller or poorly formed trees.

Threatened species Species experiencing a decline in the number of individuals to the degree that a concern is raised about the possibility of extinction of that species.

Threshold A point in the operation of a system at which a change occurs. With respect to toxicology, it is a level below which effects are not observable and above which effects become apparent.

Tidal power Form of water utilizing ocean tides in places where favourable topography allows for construction of a power plant.

Time series The set of estimates of a population over a number of years.

Tolerance The ability to withstand stress resulting from exposure to a pollutant or harmful condition.

Total fertility rate (TFR) Average number of children expected to be born to a woman during her lifetime. (Usually defined as the number born to a woman between the ages 15 to 44, taken conventionally as the lower and upper limit of reproductive ages for women.)

Toxic Harmful, deadly, or poisonous.

Toxicology The science concerned with study of poisons (or toxins) and their effects on living organisms. The subject also includes the clinical, industrial, economic, and legal problems associated with toxic materials.

Transuranic waste Radioactive waste consisting of human-made radioactive elements heavier than uranium. Includes clothing, rags, tools, and equipment that have been contaminated.

Trophic level In an ecological community, all the organisms that are the same number of food-chain steps from the primary source of energy. For example, in a grassland, the green grasses are on the first trophic level, grasshoppers are on the second, birds that feed on grasshoppers are on the third, and so forth.

Trophic level efficiency The ratio of the biological production of one trophic level to the biological production of the next lower trophic level.

Tropopause The top portion of the troposphere, approximately 12 to 20 km above sea level. The tropopause has a constant temperature of about -60°C, and produces a lid or cold trap on the troposphere.

Troposphere The lower part of the atmosphere (lower 10 to 12 km).

Tundra The treeless land area in alpine and arctic areas characterized by plants of low stature and including bare areas without any plants and covered areas with lichens, mosses, grasses, sedges, and small flowering plants, including low shrubs.

U

Ubiquitous species Species that are found almost anywhere on the Earth.

Ultraviolet A The longest wavelength of ultraviolet radiation (0.32-0.4 micrometres), not affected by stratospheric ozone, and is transmitted to the surface of the earth.

Ultraviolet B Intermediate-wavelength radiation that is the ozone problem. Wavelengths are from approximately 0.28-0.32 micrometres and is the most harmful of the ultraviolet radiation types. Most of this radiation is absorbed by stratospheric ozone, and depletion of ozone has led to increase in ultraviolet B radiation reaching the earth.

Ultraviolet C Shortest wavelength of the ultraviolet radiation with wavelengths from approximately 0.2-0.28 micrometres. Is the most energetic of the ultraviolet radiation and is absorbed strongly in the atmosphere. Only a negligible amount of Ultraviolet C reaches the surface of the earth.

Undernourishment The lack of sufficient calories in available food, so that one has little or no ability to move or work.

Uneven-aged stands Forest area with at least three distinct age classes.

Unified soil classification system A classification of soils, widely used in engineering practice, based on the amount of coarse particles, fine particles, or organic material.

Uniformitarianism The principle that processes that operate today operated in the past. Therefore, observations of processes today can explain events that occurred in the past and leave evidence, for example, in the fossil record or in geologic formations.

Urban dust dome Polluted urban air produced by the combination of lingering air and abundance of particulates and other pollutants in the urban air mass.

Urban forestry The practice and profession of planting and maintaining trees in cities, including trees in parks and other public areas.

Urban shadow The impact of a city on surrounding lands, through the influence of development, transportation, resource use and waste generation, recreational uses, and other pressures from the urban population on outlying areas.

Utilitarian justification for the conservation of nature An argument for the conservation of nature on the grounds that the environment, an ecosystem, habitat, or species, provides individuals with direct economic benefit or is directly necessary to their survival.

Utility Value or worth in economic terms.

UVA Least energetic form of ultraviolet radiation. It is capable of causing some damage to living cells, is not affected by stratospheric ozone, and is therefore transmitted to the Earth.

UVB Intermediate-wavelength ultraviolet radiation, damaging to living cells. Most is absorbed by stratospheric ozone, and therefore depletion of ozone leads to significant increase of this radiation. This is the ozone problem.

UVC Shortest wavelength and most energetic of the ultraviolet radiation. It is strongly absorbed in the atmosphere and negligible amounts reach the surface of the Earth.

V

Vadose zone Zone or layer above the water table where water may be stored as it moves laterally or down to the zone of saturation. Part of the vadose zone may be saturated part of the time.

Variable A quantity or condition that may vary.

Variable, dependent A variable that changes in response to changes in an independent variable; a variable taken as the outcome of one or more other variabl es.

Variable, independent In an experiment, the variable that is manipulated by the investigator. In an observational study, the variable that is believed by the investigator to affect an outcome, or dependent, variable.

Variable, manipulated *See* Variable, independent.

Variable, responding *See* Variable, dependent.

Vulnerable species Another term for threatened species (species experiencing a decline in the number of individuals).

W

Waldsterben German phenomenon of forest death as the result of acid rain, ozone, and other air pollutants.

Wallace's realms Six biotic provinces, or biogeographic regions, divided on the basis of fundamental inherited features of the animals found in those areas, suggested by A. R. Wallace (1876). His realms are Nearctic (North America), Neotropical (Central and South America), Palearctic (Europe, northern Asia, and northern Africa), Ethiopian (central and southern Africa), Oriental (the Indian subcontinent and Malaysia), and Australian.

Wastewater renovation and conservation cycle Practice of applying wastewater to the land. In some systems, treated wastewater is applied to agricultural crops, and as the water infiltrates through the soil layer it is naturally purified. Reuse of the water is by pumping it out of the ground for municipal or agricultural uses.

Wastewater treatment Process of treating wastewater (primarily sewage) in specially designed plants that accept municipal wastewater. Generally divided into three categories: primary treatment, secondary treatment, and advanced wastewater treatment.

Water budget Inputs and outputs of water for a particular system (a drainage basin, region, continent, or the entire Earth).

Water conservation Practices designed to reduce the amount of water we use.

Water power Alternative energy source derived from flowing water. One of the world's oldest and most common energy sources. Sources vary in size from microhydropower systems to large reservoirs and dams.

Water reuse The use of wastewater following some sort of treatment. Water reuse may be inadvertent, indirect, or direct.

Water table The surface that divides the zone of aeration from the zone of saturation, the surface below which all the pore space in rocks is saturated with water.

Watershed An area of land that forms the drainage of a stream or river. If a drop of rain falls anywhere within a watershed to become surface runoff it can flow out only through the same stream.

Weathering Changes that take place in rocks and minerals at or near the surface of Earth in response to physical, chemical, and biological changes; the physical, chemical, and biological breakdown of rocks and minerals.

Wetlands Comprehensive term for landforms such as salt marshes, swamps, bogs, prairie potholes, and vernal pools. Their common feature is that they are wet at least part of the year and as a result have a particular type of vegetation and soil. Wetlands form important habitats for many species of plants and animals, while serving a variety of natural service functions for other ecosystems and people.

Wilderness A term that is defined differently by different people. To some people, "wilderness" is an area undisturbed by people. To others, it is simply a natural area. The International Union for the Conservation of Nature defines wilderness as "a large area of unmodified or slightly modified land, and/or sea, retaining its natural character and influence, without permanent or significant habitation."

Wind power Alternative energy source that has been used by people for centuries. More recently, thousands of windmills have been installed to produce electric energy.

Work (physics) Force times the distance through which it acts when work is done we say energy is expended.

Z

Zero population growth A population in which the number of births equals the number of deaths so that there is no net change in the size of the population.

Zone of aceration The zone or layer above the water table in which some water may be suspended or moving in a downward migration toward the water table or laterally toward a discharge point is often called the vadose zone.

Zone of saturation Zone or layer below the water table in which all the pore space of rock or soil is saturated.

Zooplankton Small aquatic invertebrates that live in the sunlit waters of streams, lakes, and oceans and feed on algae and other invertebrate animals.

NOTES

Chapter 1 Notes

1. Quinby, P.A. An Overview of Ancient Forest Ecology in the Lake Temagami Site Region. Forest Landscape Baseline No. 11. Toronto: Ancient Forest Exploration and Research. Available online at http://www.ancient forest.org/flb11.html.
2. Clark, T.P. and A.H. Perera. 1995. An Overview of Ecology of Red and White Pine Old-Growth Forests in Ontario. Ontario Forest Research Institute, Ministry of Natural Resources, Sault Ste. Marie, Ontario.
3. Quinby, Peter A. 1991. Self-replacement in old-growth white pine forests of Temagami, Ontario. *Forest Ecology and Management.* 41: 95-109.
4. Everett, G. D. 1961. One man's family. *Population Bulletin* 17:153–169.
5. Ehrlich, P. R., A. H. Ehrlich, and P. H. Holdren. 1977. *Ecoscience: Population, Resources, Environment*, 3rd ed., San Francisco: Freeman.
6. Rees, William. 1996. Revisiting carrying capacity: area-based indicators of sustainability. *Population and Environment: A Journal of Interdisciplinary Studies* 17(3). Available online at http://dieoff.org/page110.htm.
7. Wackernagel, Mathis, Larry Onisto, Alejandro Callejas Linares, Ina Susana López Falfán, Jesus Méndez García, Ana Isabel Suárez Guerrero, Ma. Guadalupe Suárez Guerrero. 1997. Ecological Footprints of Nations. Rio +5 Forum Report Commissioned by The Earth Council, Costa Rica. Available online at http://www.ecouncil.ac.cr/rio/focus/report/english/footprint/.
8. Deevey, E. S. 1960. The human population. *Scientific American* 203:194–204.
9. Keyfitz, N. 1989. The growing human population. *Scientific American* 261:118–126.
10. Gottfield, R. S. 1983. *The black death: Natural and human disaster in medieval Europe.* New York: Free Press.
11. Field, J. O., ed. 1983. *The challenge of famine: Recent experience, lessons learned.* Hartford, Conn.: Kumarian Press.
12. Glantz, M. H., ed. 1987. *Drought and hunger in Africa: Denying famine a future.* Cambridge: Cambridge University Press.
13. Seavoy, R. E. 1989. *Famine in East Africa: Food production and food politics.* Westport, Conn.: Greenwood.
14. United Nations Food and Agricultural Organization. 2002. *Global Information and Early Warning System* (GIEWS), "Food Supply Situation and Crop Prospects in Sub-Saharan Africa." N. Y., N. Y.
15. Gower, B. S. 1992. What do we owe future generations? In D. E. Cooper and J. A. Palmers, eds., *The environment in question: Ethics and global issues*, pp. 1–12. New York: Routledge.
16. Haub, C., and D. Cornelius. 2000. *World population data sheet.* Washington, D.C.: Population Reference Bureau.
17. World Resources Institute. 1998. *Teacher's guide to world resources: Exploring sustainable communities.* Washington, D.C.: World Resources Institute. (http://www.igc.org/wri/wr-98-99/citygrow.htm)
18. World Resources Institute. 1999. *Urban growth.* Washington, D.C.: World Resources Institute.
19. Margulis, L., and J. E. Lovelock. 1989. *Urban growth.* Washington, D.C.: World Resources Institute.
20. Riddell, Brian E. and Art F. Tautz. 2004. State of Pacific salmon and their habitats: Canada and the United States. Chapter 7 in Patricia Gallauger and Laurie Wood (eds.). 2004. Proceedings, the World Summit on Salmon. June 10-13, 2003. Simon Fraser University Continuing Studies in Science. Vancouver, B.C.: Simon Fraser University.
21. Botkin, D. B., M. Caswell, J. E. Estes, and A. Orio, eds. 1989. *Changing the global environment: Perspectives on human involvement.* New York: Academic Press.
22. Botkin, D. B., and C. E. Beveridge. 1997. Cities as environments. *Urban Ecosystems* 1(1):3–20.
23. Botkin. Daniel B. 1990. *Discordant Harmonies: A New Ecology for the 21st Century.* New York: Oxford University Press.
24. Nash, R. F. 1988. *The rights of nature: A history of environmental ethics.* Madison: University of Wisconsin Press.
25. Bryant, D., A. Burke, J. McManus, and M. Spalding, 1998. *Reefs at risk: A map-based indicator of threats to the world's coral reefs.* Washington, D.C.: World Resources Institute.
26. Coles, S. L., and L. Ruddy. 1995. Comparison of water quality and reef coral mortality and growth in southeastern Kaneohe Bay, Oahu, Hawaii, 1990 to 1992, with conditions before sewage diversion. *Pacific Science* 49(3): 247–265.
27. Hinrichsen, D. 1997. Coral reefs in crisis. *Bioscience* 47(9): 554–558.

Chapter 1 Critical Thinking Issue Notes

Bryant, Dirk, Auretta Burke, John McManus, and Mark Spalding. 1998. *Reefs at risk: A map-based indicator of threats to the world's coral reefs.* Washington, D.C.: World Resources Institute.

Coles, S. L., and L. Ruddy. 1995. Comparison of water quality and reef coral mortality and growth in southeastern Kaneohe Bay, Oahu, Hawaii, 1990 to 1992, with conditions before sewage diversion, *Pacific Science* 49(3): 247–265.

Hinrichsen, D. 1997 (October). Coral Reefs in Crisis. *Bioscience* 47(9): 554–558.

Jameson, S. C., J. W. McManus, and M. D. Spalding. 1995 (May). *State of the reefs: Regional and global perspectives.* International Coral Reef Initiative Executive Secretariat (Background Paper).

Chapter 2 Notes

1. Schindler, D. W. and E. J. Fee. 1974. The Experimental Lakes Area: whole-lake experiments in eutrophication. *J. Fish. Res. Board Can.* 31: 937-953.
2. Botkin, D. B. 2001. *No man's garden: Thoreau and a new vision for civilization and nature.* Washington, D.C.: Island Press.
3. Taylor, F. S. 1949. *Science and scientific thought.* New York: Norton.
4. Schmidt, W. E. 1991 (September 10). "Jovial con men" take credit (?) for crop circles. *New York Times*, p. B1.
5. Tuohy, W. 1991 (September 10). "Crop circles" their prank, 2 Britons say, *Los Angeles Times*, p. A14.
6. This information about crop circles is from Crop Circle News, http://cropcirclenews.com.

7. Gibbs, A., and A. E. Lawson. 1992. The nature of scientific thinking as reflected by the work of biologists and by biology textbooks. *The American Biology Teacher* 54:137–152.
8. Pease, C. M., and J. J. Bull. 1992. Is science logical? *Bioscience* 42:293–298.
9. Handa, I. T., R. Harmsen, and R. L. Jefferies. 2000. Assisted revegetation trials in degraded salt-marshes. *Journal of Applied Ecology* 37: 944-958.
10. Handa, I. T., R. Harmsen, and R. L. Jefferies. 2002. Patterns of vegetation change and the recovery potential of degraded areas in a coastal marsh system of the Hudson Bay lowlands. *Journal of Ecology* 90: 86-99.
11. Lerner, L. S., and W. J. Bennetta. 1988 (April). The treatment of theory in textbooks. *The Science Teacher*, pp. 37-41.
12. Kuhn, T. S. 1970. *The structure of scientific revolutions*. Chicago: University of Chicago Press.
13. Trefil, J. S. 1978. A consumer's guide to pseudoscience. *Saturday Review* 4:16–21.
14. Hastings and Hastings. 1992. Telepathy. American Institute of Public Opinion poll from June 1990.
15. Bryant, A. A., H. M. Schwantje, and N. I. de With. 1997. Disease and unsuccessful reintroduction of Vancouver Island marmots (*Marmota vancouverensis*). In: Armitage, K. B. and V. U. Rumianstev. (eds.) *Holarctic Marmots as a Factor of Biodiversity*. Proceedings of the 3rd International Conference on Marmots (Cheboksary, Chuvash Republic, Russia, 25-30 August 1997). Moscow: ABF Publishing House.
16. Feynman, R. P. 1998. The meaning of it all: Thoughts of a citizen-scientist. Reading, Mass.: Addison-Wesley, Perseus Books, p. 8. (From lectures at the University of Washington in 1963.)
17. Tyser, R. W., and Cerbin, W. H. 1991. Critical thinking exercises for introductory biology courses. *Bioscience* 41:41–46.

Chapter 2 Critical Thinking Issue References

Ford, R. 1998 (March). "Critically Evaluating Scientific Claims in the Popular Press." *The American Biology Teacher*, vol. 60, no. 3, pp. 174–180.

Marshall, E. 1998 (May 15). "The Power of the Front Page of the New York Times." *Science*, vol. 280, pp. 996–997.

Wilson, E.O. 1998. Scientists, scholars, knaves and fools. *American Scientist* 86:6-7.

Baines, C.J. 1998. Junk science must be stopped: misinformation and hysteria are driving public health policy in ways that harm us all. *National Post* November 23, 1998.

Chapter 3 Notes

1. Western, D., and C. Van Prat. 1973. Cyclical changes in habitat and climate of an East African ecosystem. *Nature* 241(549):104–106.
2. Dunne, T., and L. B. Leopold. 1978. *Water in environmental planning*. San Francisco: Freeman.
3. Bartlett, A. A. 1980. Forgotten fundamentals of the energy crisis. *Journal of Geological Education* 28:4–35.
4. Meadows, D. H., D. L. Meadows, and J. Randers. 1992. *Beyond the limits: Confronting global collapse; envisioning a sustainable future*. Post Mills, Vt.: Chelsea Green Publishers.
5. Wootton, J. T., M. S. Parker, and M. E. Power. 1996. Effects of disturbances on river food webs. *Science* 273:1558–1561.
6. Leach, M. K., and T. J. Givnich. 1996. Ecological determinants of species loss in remnant prairies. *Science* 273:1555–1558.
7. Lovelock, J. 1995. *The ages of Gaia: A biography of our living earth*, rev. ed. New York: Norton.
8. Gardner, G. T., and P. C. Stern. 2002. *Environmental problems and human behavior*, 2nd ed. Boston: Pearson Custom Publishing.
9. Foster, K. R., P. Vecchia, and M. H. Repacholi. 2000. Science and the Precautionary Principle. *Science* 288: 979-981.

Chapter 3 Critical Thinking Issue References

Barlow, C. 1993. *From Gaia to selfish genes*. Cambridge: MIT Press.

Kirchner, J. W. 1989. The Gaia hypothesis: Can it be tested? *Reviews of Geophysics* 27:223–235.

Lovelock, J. E. 1995. *Gaia: A new look at life on Earth*. New York: Oxford University Press.

Lyman, F. 1989. What hath Gaia wrought? *Technology Review* 92(5):55–61.

Resnik, D. B. 1992. Gaia: From fanciful notion to research program. *Perspectives in Biology and Medicine* 35(4):572–582.

Schneider, S. H. 1990. Debating Gaia. *Environment* 32(4):4–9, 29–32.

Chapter 3 Build Your Environmental Skills 3.1 Reference

Bartlett, A. A. 1993. The arithmetic of growth: Methods of calculation. *Population and Environment* 14(4):359–387.

Chapter 4 Notes

1. Population Reference Bureau, *2002 World Population Data Sheet*. The World Bank Data and Statistics, website http://devdata.worldbank.org/externalCPProfile.asp?SelectedCountry=BGD&CCODE=BGD&CNAME=Bangladesh&PTYPE=CP. Last accessed 4/28/04.
2. Keyfitz, N. 1992. Completing the worldwide demographic transition: The relevance of past experience. *AMBIO* 21:26-30.
3. Graunt, J. [1662] 1973. *Natural and political observations made upon the bill of mortality*. London, 662.
4. Dumond, D. E. 1975. The limitation of human population: A natural history. *Science* 187:713–721.
5. Zero Population Growth. 2000. *U.S. population*. Washington, D.C.: Zero Population Growth.
6. World Bank. 1984. *World development report 1984*. New York: Oxford University Press.
7. Erhlich, P. R. 1971. *The population bomb*, rev. ed. New York: Ballantine.
8. Malthus, T. R. [1803] 1992. *An essay on the principle of population*. Selected and introduced by Donald Winch. Cambridge, England: Cambridge University Press.
9. Statistics Canada (www.statcan.ca) and United Nations Statistics Division (http://unstats.un.org/unsd/demographic/sconcerns/popchar/popchar2.htm).
10. Xinhua News Agency, China's cross-border tourism prospers in 2002, December 31, 2002. From the Population Reference Bureau Web site available at http://www.prb.org/Template.cfm?Section=PRB&template=/ContentManagement/ContentDisplay.cfm&ContentID=8661.

11. U. S. Centers for Disease Control website available at http://www.cdc.gov/ncidod/sars/factsheet.htm and http://www.cdc.gov/ncidod/dvbid/westnile/qa/overview.htm.
12. Population Reference Bureau Web site, available at http://www.prb.org/Template.cfm?Section=PRB&template=/ContentManagement/ContentDisplay.cfm&ContentID=8661.
13. Joint United Nations Programme on HIV/AIDS. 1999. *AIDS epidemic update*. Geneva: Switzerland.
14. Population Division, United Nations Department of Economic and Social Affairs, 1998. *World population growing despite AIDS spread.* United Nations, N.Y. accessed at www.un.org/esa/population/publications/AID impact. Last accessed 4/28/04.
15. World Bank. 1985. *World development report*. New York: Oxford University Press.
16. Central Intelligence Agency. 1999. *The world factbook*. Washington, D.C.: CIA.
17. World Bank. 2000. *World development indicators 2000*. Washington, D.C.: World Bank.
18. World Bank. 1992. *World development report. The relevance of past experience.* Washington, D.C.: World Bank.
19. Guz, D., and J. Hobcraft. 1991. Breastfeeding and fertility: A comparative analysis. *Population Studies* 45:91–108.
20. Fathalla, M. F. 1992. Family planning: Future needs. *AMBIO* 21:84–87.
21. Alan Guttmacher Institute. *Sharing responsibility: Women, society and abortion worldwide*. New York: AGI, 1999.
22. Haupt, A., and T. T. Kane. 1978. *The Population Reference Bureau's population handbook*. Washington, D.C.: Population Reference Bureau.
23. Planned Parenthood Federation of America, Public Policy Division. 1997 (June). *International family planning: The need for services*. Planned Parenthood Federation of America. N. Y.
24. Xinhua News Agency March 13, 2002 untitled, available at http://www.16da.org.cn/english/archiveen/28691.htm.

Chapter 5 Notes

1. State of the Canadian Cryosphere (http://www.socc.ca/glaciers/glaciers_future_e.cfm) accessed March 28, 2005.
2. Henderson, L. J. [1913] 1966. *The fitness of the environment*. Boston: Beacon.
3. Van Koevering, T. E., and N. J. Sell. 1986. *Energy: A conceptual approach*. Englewood Cliffs, N.J.: Prentice Hall, p. 271.
4. Isacks, B., J. Oliver, and L. Sykes. 1968. Seismology and the new global tectonics. *Journal of Geophysical Research* 73:5855–5899.
5. Dewey, J. F. 1972. Plate tectonics. *Scientific American* 22:56–68.
6. Botkin, D. B. 1990. *Discordant harmonies: A new ecology for the 21st century*. New York: Oxford University Press.
7. Ehrlich, P. R., A. H. Ehrlich, and J. P. Holdren. 1970. *Ecoscience: Population, resources, environment*. San Francisco: W. H. Freeman, p. 1051.
8. Post, W. M., T. Peng, W. R. Emanuel, A. W. King, V. H. Dale, and D. L. De Angelis. 1990. The global carbon cycle. *American Scientist* 78:310–326.
9. Keeling, C. D., T. P. Whorf, M. Wahlen, and J. van der Plicht. 1995. Interannual extremes in the rate of rise of atmospheric carbon dioxide since 1980. *Nature* 375:666–670.
10. Hudson, R. J. M., S. A. Gherini, and R. A. Goldstein. 1994. Modeling the global carbon cycle: Nitrogen fertilization of the terrestrial biosphere and the "missing" CO2 sink. *Global Biogeochemical Cycles* 8:307–333.
11. Woods Hole. 2000. The missing carbon sink. http://www.whrc.org/science/carbon/missingc.htm. Accessed August 18, 2003.
12. Houghton, R. 2003. Why are estimates of the global carbon balance so different? *Global Change Biology* 9:500–509.
13. Houghton, R. 2003. Revised estimates of the annual net flux of carbon to the atmosphere from changes in land use and land management 1850–2000. *Tellus* 55 B:378–390.
14. Herring, D., and R. Kannenberg. 2000. The mystery of the missing carbon. http://earthobservatory.nasa.gov/cgi-bin/printall?/study/BOREAS/missing_carbon.html. Accessed July 5, 2000.
15. Chameides, W. L., and E. M. Perdue. 1997. *Biogeochemical cycles*. New York: Oxford University Press.
16. Agren, G. I., and E. Bosatta. 1996. *Theoretical ecosystem ecology*. New York: Cambridge University Press.
17. Kasting, J. F., O. B. Toon, and J. B. Pollack. 1988. How climate evolved on the terrestrial planets. *Scientific American* 258:90–97.
18. Berner, R. A. 1999. A new look at the long-term carbon cycle. *GSA Today* 9(11):2–6.
19. Carter, L. J. 1980. Phosphate: Debate over an essential resource. *Science* 209:4454.

Chapter 5 Critical Thinking Issue References

Asner, G. P., T. R. Seastedt, and A. R. Townsend. 1997 (April). The decoupling of terrestrial carbon and nitrogen cycles. *Bioscience* 47 (4):226–234.

Hellemans, A. 1998 (February 13). Global nitrogen overload problem grows critical. *Science*, 279:988–989.

Smil, V. 1997 (July). Global populations and the nitrogen cycle. *Scientific American*, 76–81.

Vitousek, P. M., J. Aber, R. W. Howarth, G. E. Likens, P. A. Matson, D. W. Schindler, W. H. Schlesinger, and G. D. Tilman. 1997. Human alteration of the global nitrogen cycle: Causes and consequences. *Issues in Ecology*. http://esa.sdsc.edu/

Chapter 6 Notes

1. Line, L. 1996 (April 16). Ticks and moths, not just oaks, linked to acorns, *New York Times*.
2. Ostfield, R. S., C. G. Jones, and J. O. Wolff. 1996 (May). Of mice and mast: Ecological connections in eastern deciduous forests. *BioScience* 46(5):323–330.
3. Morowitz, H. J. 1979. *Energy flow in biology*. Woodbridge: Conn.: Oxbow Press.
4. Canadian Species at Risk: Banff Springs Snail. http://www.speciesatrisk.gc.ca/search/speciesDetails_e.cfm?SpeciesID=311. Accessed March 5, 2005
5. Parks Canada. The Endangered Banff Springs Snail: The Most At-Risk Species in the Park. http://www.pc.gc.ca/regional/sourcesthermales-hotsprings/natcul/natcul3_E.asp. Accessed March 5, 2005.

6. Lavigne, D. M., W. Barchard, S. Innes, and N. A. Oritsland. 1976. *Pinniped bioenergetics.* ACMRR/MM/SC/12. Rome: United Nations Food and Agriculture Organization.
7. Servheen, Gregg and L. Jack Lyon. 1989. Habitat use by woodland caribou in the Selkirk Mountains. *Journal of Wildlife Management.* 53(1): 230-237.
8. Gray, D. R. 1999. Updated Status Report on the Woodland Caribou (caribou des bois) *Rangifer tarandus dawsoni and Rangifer tarandus caribou in Canada.* Committee on the Status of Endangered Wildlife in Canada. 38 p.
9. Miller, Frank L. 1987. Caribou. In: Chapman, Joseph A.; Feldhamer, George A., eds. *Mammals of North America.* Baltimore, MD: Johns Hopkins University Press: 923-959.
10. Bradshaw, C. J. A., D. M. Hebert, A. B. Rippin, and S. Boutin. 1995. Winter peatland habitat selection by Woodland Caribou in northeastern Alberta. *Can. J. Zool.* 73:1567-1574.
11. Bradshaw, C. J. A., S. Boutin, and D. M. Hebert. 1997. Effects of petroleum exploration on Woodland Caribou in northeastern Alberta. *Journal of Wildlife Management* 61:1127-1133.
12. Paine, R. T. 1969. A note on trophic complexity and community stability. *American Naturalist* 100:65–75.
13. Bradbury, R. H., J. D. Van Der Laan, and D. F. Green. 1996. The idea of complexity in ecology. *Senckenbergiana marit.* 27(3/6): 89-96.

Chapter 7 Notes

1. Lotze, Heike and Inka Milewski. 2002. *Two Hundred Years of Ecosystem and Food Web Changes in the Quoddy Region, Outer Bay of Fundy.* Fredericton, N.B.: Conservation Council of New Brunswick.
2. Cicero, *The Nature of the Gods* (44 B.C.).
3. World Health Organization. 1998. *Malaria.* Fact Sheet No. 94. WHO.
4. United Nations World Health Organization. 2003. http://www.who.int/mediacentre/releases/2003/pr33/en/.
5. World Health Organization. 2000. *Overcoming antimicrobial resistance: World health report on infectious diseases 2000.* WHO.
6. James A. A. 1992. Mosquito molecular genetics: The hands that feed bite back. Science 257:37–38; Kolata, G. 1984. The search for a malaria vaccine. *Science* 226:679–682; Miller, L. H. 1992. The challenge of malaria. Science 257:36–37; World Health Organization. 1999. (June). Using malaria information. News Release No. 59. WHO; World Health Organization. 1999 (July). Sequencing the *Anopheles gambiae* genome. News Release No. 60. WHO.
7. Hutchinson, G. E. 1965. *The ecological theater and the evolutionary play.* New Haven, Conn.: Yale University Press.
8. Woese, C. R., O. Kandler, and M. L. Wheelis. 1990. Towards a natural system of organisms: Proposals for the domains Archaea, Bacteria, and Eucharya, *Proceedings of the National Academy of Sciences* (USA) 87:4576–4579.
9. Mather, J. R., and G. A. Yoskioka. 1968. The role of climate in the distribution of vegetation. *Annals of the Association of American Geography* 58:29–41.
10. Hardin, G. 1960. The competitive exclusion principle. *Science* 131:1292–1297.
11. Rogers, C. 1996. Red squirrel: *Sciurus vulgaris.* The Wild Screen Trust.
12. Schoener, T. W. 1983. Field experiments in interspecific competition. *American Naturalist* 1222:240–285.
13. MacArthur, Robert. 1958. Population ecology of some warblers of northeastern coniferous forests. *Ecology* 39: 599-619.
14. Hutchinson, G. E. 1958. Concluding remarks. Cold Spring *Harbor Symposium in Quantitative Biology* 22:415–427.
15. Miller, R. S. 1967. Pattern and process in competition. *Advances in Ecological Research* 4:1–74.
16. Botkin, D. B. 1985. The need for a science of the biosphere. *Interdisciplinary Science Reviews* 10:267–278.
17. Forman, Richard T. T. 1995. *Land mosaics: The ecology of landscapes and regions.* New York: Cambridge University Press.

Chapter 8 Notes

1. Beck, K. 1994. Natural control for the purple loosestrife. *North Coast Newsletter,* Ohio Lake Erie Commission, 1–2.
2. Carroll, D. 1994. Subduing purple loosestrife. *The Conservationist* 49:6–9.
3. Malecki, R. A., B. Blossey, S. D. Hight, D. Schroeder, L. T. Kok, and J. R. Coulson. 1993. Biological control of purple loosestrife. *BioScience* 43:680–686.
4. Wallace, A. R. 1896. *The geographical distribution of animals.* Vol. 1. New York: Hafner. Reprint 1962.
5. Pielou, E. C. 1979. *Biogeography.* New York: Wiley.
6. Good, R. 1974. *The geography of the flowering plants,* 4th ed. London: Longman Group.
7. Takhtadzhian, A. L. 1986. *Floristic regions of the world.* Berkeley: University of California Press.
8. Udvardy, M. 1975. *A classification of the biogeographical provinces of the world.* IUCN Occasional Paper 18. Morges, Switzerland: IUCN.
9. Lentine, J. W. 1973. Plates and provinces, a theoretical history of environmental discontinuity. In N. F. Hughes, ed., *Organisms and continents through time,* pp. 79–92. Special Papers in Paleontology 12.
10. Hallam, A. 1975. Alfred Wegener and the hypothesis of continental drift. *Scientific American* 232:88–97.
11. Hurley, P. M. 1968. The confirmation of continental drift. *Scientific American* 218:52–64.
12. Mather, J. R., and G. A. Yoshioka. 1968. The role of climate in the distribution of vegetation. *Ann. Ass. American Geography* 58:29–41.
13. Prentice, I. C., W. Cramer, S. P. Harrison, R. Leemans, R. A. Monserud, and A. M. Solomon. 1992. A global biome model based on plant physiology and dominance, soil properties and climate. *Journal of Biogeography* 19:117–134.
14. Waring, R. H., and J. F. Franklin. 1979. Evergreeen coniferous forests of the Pacific Northwest. *Science* 204:1380–1386.
15. North, M. P., J. F. Franklin, A. B. Carey, E. D. Forsman, and T. Hamer. 1999. Forest stand structure of the northern spotted owl's foraging habitat. *Forest Science* 45:520–527.
16. Botkin, D. B. 2004. *Our natural history: The lessons of Lewis and Clark.* New York: Oxford University Press.
17. Shapiro, Beth, Alexei J. Drummond, Andrew Rambaut, Michael C. Wilson, Paul E. Matheus, Andrei V. Sher, Oliver G. Pybus, M. Thomas P. Gilbert, Ian Barnes, Jonas Binladen, Eske Willerslev, Anders J. Hansen, Gennady F. Baryshnikov, James A. Burns, Sergei Davydov, Jonathan C. Driver, Duane G. Froese, C. Richard Harington, Grant Keddie, Pavel Kosintsev, Michael L. Kunz, Larry D. Martin, Robert O.

Stephenson, John Storer, Richard Tedford, Sergei Zimov, and Alan Cooper. 2004. Rise and fall of the Beringian steppe bison. *Science* 306 (26 November 2004): 1561-1565.

18. Darwin, C. R. 1859. *The origin of species by means of natural selection or the preservation of favored races in the struggle for life.* London: Murray.
19. Grant, P. R. 1986. *Ecology and evolution of Darwin's finches.* Princeton, N.J.: Princeton University Press.
20. Cox, C. B., I. N. Healey, and P. D. Moore. 1973. *Biogeography.* New York: Halsted.
21. MacArthur, R. H., and E. O. Wilson. 1967. *The theory of island biogeography.* Princeton, N.J.: Princeton University Press.
22. Atlas of Canada: http://atlas.gc.ca/site/english/maps/archives/4thedition/environment/climate/047_48. Accessed March 12, 2005.
23. Tallis, J. H. 1991. *Plant community history.* London: Chapman & Hall.
24. Missouri Botanic Garden, Botany in North America: In Honor of the XVI International Botanical Congress, August 1–7, 1999, St. Louis, MO.

Chapter 9 Notes

1. Rackham, O. 1986. *The history of the countryside.* London: Dent & Sons, p. 63.
2. Perlin, J. 1989. *A forest journey: The role of wood in the development of civilization.* New York: Norton.
3. Schrödinger, E. 1942. *What is life?* Cambridge: Cambridge University Press.
4. Zhaohua, Zhu, Cai Mantang, Warg Shiji and Jiang Youxu (eds) with Cherla B. Sastry and A. N. Rao. 1991. *Agroforestry Systems in China.* Singapore: International Development Research Centre (Canada).
5. Morowitz, H. J. 1979. *Energy flow in biology.* Woodbridge, Conn.: Oxbow Press.
6. Slobodkin, L. B. 1960. Ecological energy relations at the population level. *American Naturalist* 95:213–236.
7. Peterson, R.O. 1995. *The wolves of Isle Royale: A broken balance.* Minocqua, Wis.: Willow Creek Press.
8. Jordan, J. D., D. B. Botkin, and M. I. Wolf, 1971. *Biomass dynamics in a moose population.* Ecology 52:147–152.
9. Kozlovsky, D. G. 1968. A critical evaluation of the trophic level concept: I. Ecological efficiencies. *Ecology* 49:147–160.
10. Schaefer, M., 1991. Secondary production and decomposition. In E. Rohrig and B. Ulrich, eds., *Temperate deciduous forests. Ecosystems of the world*, vol. 7. Amsterdam: Elsevier.
11. Golley, F. B. 1989. Energy dynamics of a food chain of an old-field community. *Ecol. Monographs* 30:187–291.
12. Bagley, P. B. 1989. Aquatic environments in the Amazon basin, with an analysis of carbon sources, fish production and yield. In D. P. Dodge, ed., *Proc. Int. Large Rivers Symp. Can. Spec. Publ. Fish. Aquat. Sci.* 106:385–398.
13. Gaill, F., B. Shillito, F. Menard, G. Goffinet, and J. Childress. 1997. Rate and process of tube production by the deep sea hydrothermal vent tubeworm *Riftia pachyptila. Marine Ecology Progress Series* 148:135–143.
14. Wills, J. 1996. Upwelling. FMF Glossary. First Millennial Foundation.

Chapter 10 Notes

1. Rodger, Lindsay. 1998. *Tallgrass Communities of Ontario: A Recovery Plan.* Report for World Wildlife Fund and Ontario Ministry of Natural Resources. Toronto: World Wildlife Fund. (http://www.tallgrassontario.org/Publications/TallgrassRecoveryPlan.pdf). See also Pinery Provincial Park controlled burn program: http://www.ontarioparks.com/english/parkzine/art-01-04-2003.html. Accessed March 25, 2005.
2. Forman, R. T. T., 1995. *Landscape mosaics.* Cambridge: Harvard University Press.
3. Houseal, G., and D. Smith. 2000. Source-identified seed: The Iowa roadside experience. *Ecological Restoration* 18(3):173–183.
4. Hall, F. G., D. B. Botkin, D. E. Strebel, K. D. Woods, and S. J. Goetz. 1991. Large-scale patterns in forest succession as determined by remote sensing. *Ecology* 72:628–640.
5. Gorham, E., P. M. Vitousek, and W. A. Reiners. 1979. The regulation of chemical budgets over the course of terrestrial ecosystem succession. *Ann, Rev. Ecol. Syst.* 10:53–84.
6. Vitousek, P. M., and L. R. Walker, 1987. Colonization, succession and resource availability: Ecosystem-level interactions. In A. J. Gray, M. J. Crawley, and P. J. Edwards, eds. *Colonization, succession and stability*, pp. 207–223. British Ecol. Soc. 26th Symp. Oxford: Blackwell Scientific Publications.
7. Connell, J. H., and R. O. Slatyer. 1977. Mechanism of succession in natural communities and their role in community stability and organization. *American Naturalist* 111:1119–1144.
8. Pickett, S. T. A., S. L. Collins, and J. J. Armesto. 1987. Models, mechanisms and pathways of succession. *Botanical Review* 53:335–371.
9. Gomez-Pompa, A., and C. Vazquez-Yanes. 1981. Successional studies of a rain forest in Mexico. In D. C. West, H. H. Shugart, and D. B. Botkin, eds., *Forest succession: Concepts and application*, pp. 246–266. New York: Springer-Verlag.
10. MacMahon, J. A. 1981. Successional processes: Comparison among biomes with special reference to probable roles of and influences on animals. In D. C. West, H. H. Shugart, and D. B. Botkin, eds., *Forest succession: Concepts and application*, pp. 277–304. New York: Springer-Verlag.
11. Walthern, P. 1986. Restoring derelict lands in Great Britain. In G. Orians, ed., *Ecological knowledge and environmental problem-solving: Concepts and case studies*, pp. 248–274. Washington, D. C.: National Academy Press.

Chapter 10 Critical Thinking Issue References

Arena, Paul T., Lance K. B. Jordan, David S. Gilliam, Robin L. Sherman, Kenneth Banks, and Richard E. Spieler. 2000. Shipwrecks as artificial reefs: a comparison of fish assemblage structure on ships and their surrounding natural reefs areas offshore southeast Florida: preliminary results. Proceedings, 53rd Annual Gulf and Caribbean Fisheries Institute Meeting, Biloxi, Mississippi, USA.

Environment Canada's Disposal at Sea regulations under the Canadian Environmental Protection Act: http://www.pyr.ec.gc.ca/disposal_at_sea/index_e.htm.

Georgia Strait Alliance. Why Sinking Ships as Artificial Reefs Makes No Sense. http://www.georgiastrait.org/Articles/art1.php. Accessed March 28, 2005.

Chapter 11 Notes

1. Haub, Carl, and Diana Cornelius. 2000. World population data sheet. Washington, D.C.: Population Reference Bureau.
2. Brown, L. R. 1995. *Who will feed China? Wakeup call for a small planet.* New York: Norton.
3. World Resources Institute. 2000. Facts and figures: Country environmental data. Washington, D.C.: World Resources Institute.
4. A food crisis–or a blip? 1996 (February). *World Press Review* 43(2):34.
5. Cohen, J. E. 1995 (November/December). How many people can the earth support? The Sciences, pp. 18–23.
6. Holmes, B. 1993 (February 8). Feeding a world of 10 billion. *U. S. News & World Report* 114(5):55.
7. Livernash, R. 1995 (July/August). The future of populous economies: China and India shape their destinies. *Environment* 37(6):6–11, 25–34.
8. Ryan, M., and C. Flavin. 1995. Facing China's limits. *State of the world,* 1995. New York: Norton.
9. Tyler, P. E. 1996 (May 23). China's fickle rivers: Dry farms, needy industry bring a water crisis. *New York Times.*
10. Hawthorne, P. 1998 (April 3). Rebirth. *Time 100/Africa* 151(15), Time 100/Leaders and Revolutionaries.
11. United Nations Food and Agriculture Organization. 1998. FAOSTAT database. Rome: UNFAO.
12. Biological Resources Division USGS. 1998. Historical interrelationships between population settlement and farmland in the conterminous United States, 1790–1992. Land use history of North America. USGS.
13. Field, J. O., ed. 1993. *The challenge of famine: Recent experience, lessons learned.* Hartford, Conn.: Kumarian Press.
14. World Food Programme. 1998. *Tackling hunger in a world full of food: Tasks ahead for food aid.*
15. Raven, P. H., R. F. Evert, and S. E. Eichhorn. 1999. *Biology of plants.* New York: W. H. Freeman/Worth.
16. Statistics are from the UN FAO Web site FAOSTAT 2003.
17. United Nations Food and Agriculture Organization. 2000 (April). Food emergencies persist in 34 countries throughout the world. *Food Outlook 2,* p. 4. Rome: UNFAO.
18. United Nations Food and Agriculture Organization. 1998 (September). Global information and early warning system on food and agriculture. *Global Watch, Food Outlook* Rome: UNFAO.
19. Bardach, J. E. 1968. Aquaculture. *Science* 161:1098–1106.
20. Smil, V. 1984. *The bad Earth: Environmental degradation in China.* Armonk, N.Y.: M. E. Sharpe.
21. Himnan, W. 1984. New crops for arid lands. *Science* 225:1445–1256.
22. Flannery, K. V. 1965. The ecology of early food production in Mesopotamia. *Science* 147:1247–1256.
23. Murdock, W. M. 1980. *The poverty of nations: The political economy of hunger and population.* Baltimore: Johns Hopkins University Press.
24. Selecting new crops using strategic marketing management. McConnell, Chai. 1995. *The Sixth Conference of the Australasian Council on Tree and Nut Production.* Available at http://www.newcrops.uq.edu.au/acotanc/papers/mcconnel.htm
25. Macey, Anne. 2004. "Certified Organic": *The Status of the Canadian Organic Market in 2003.* Report for Agriculture and Agri-Food Canada.
26. Agriculture and Agri-Food Canada. "Canada's Organic Industry": http://ats-sea.agr.gc.ca/supply/3313_e.pdf. Accessed March 15, 2005.
27. Statistics Canada. 2001 Census of Agriculture: http://www.statcan.ca/english/agcensus2001/.
28. Statistics Canada: "Net Cash Receipts for Canadian Agriculture": http://www.statcan.ca/english/Pgdb/economind.htm#agri
29. Borland, N. E. 1983. Contributions of conventional plant breeding to food production. *Science* 147:689–693.
30. PEW Biotechnology Initiative, Pew Charitable Trusts. Web site http://pewagbiotech.org/resources/factsheets/display.php3?FactsheetID=2
31. Smith, J. B., and D. Tirpak. 1989. *The potential effects of global climate change on the United States.* Report to Congress. Washington, D.C.: U.S. Environmental Protection Agency, EPA-230-05-89-050.
32. Botkin, D. B., R. A. Nisbet, and T. E. Reynolds. 1989. Effects of climate change on forests of the Great Lake states. In J. B. Smith and D. A. Tirpka, eds., *The potential effects of climate change of the United States,* pp. 2.1–2.31. Washington, D.C.: U.S. Environmental Protection Agency, EPA-302-05-89-0.
33. World Watch Institute. 1995. Rising food prices threaten political stability. *World Watch Magazine.* Washington, D.C.: World Watch Institute.
34. Brown, L. R., M. Renner, and C. Flavin. 1998. *Vital signs:* 1998. New York: W. W. Norton.
35. Food and Agricultural Organization of the United Nations (FAO). 1996. *Food for all.* Rome: FAO.

Chapter 12 Notes

1. Atlantic Coastal Action Plan (ACAP) case study of Bedeque Bay: http://www.atl.ec.gc.ca/community/cap_case_study,_bedeque_bay_pei/1.html. Accessed March 18, 2005.
2. *Encyclopedia Britannica* online, 2003, http://www.britannica.com/eb/article?eu=61977&tocid=0&query=plow&ct=.
3. World Resources Institute. *Annual report 1992–93.* Washington, D.C.
4. Vitousek, P. M. 1987. Personal communication.
5. Pimentel, D. E. C. Terhune, R. Dyson-Hudson, S. Rochereau, R. Samis, E. A. Smith, D. Deanma, D. Reifschneider, and M. Shepard, 1976. Land degradation: Effects on food and energy resources. *Science* 194:149–155.
6. Pimentel, D., and E. L. Skidmore. 1999. Rates of soil erosion. *Science* 286:1477–1478.
7. Rees, H. W., T. L. Chow, P. J. Loro, J. Lavoie, J. O. Monteith, and A. Blaauw. 2002. Hay mulching to reduce runoff and soil loss under intensive potato production in northwestern New Brunswick, Canada. *Can. J. Soil Sci.* 82: 249-258.
8. Chow, T. L., H. W. Rees, and J. Monteith. 2000. Seasonal distribution of runoff and soil loss under four tillage treatments in the upper St. John River valley, New Brunswick, Canada. *Can. J. Soil Sci.* 80: 649-660.
9. Lashof, J. C., ed. 1979. *Pest management strategies in crop protection.* Vol. 1. Washington, D.C.: Office of Technology Assessment, U. S. Congress.

10. Baldwin, F. L., and P. W. Santelmann. 1980. Weed science in integrated pest management. *BioScience* 30:675–678.
11. From UNEP Global Programme of Action for the Protection of Marine Environment from Lands-Based Activities, 2003, Web site: http://pops.gpa.unep.org/11aldi.htm
12. Michigan State University Web site, 2003, http://www.msue.msu.edu/vanburen/ofm.htm
13. Barfield, C. S., and J. L. Stimac, 1980. Pest management: An entomological perspective. *BioScience* 30:683–688.
14. May, R. M. 1985. Evolution of pesticide resistance. Nature 315:12–13.
15. Botkin, D. B. 1999. *Passage of discovery.* Putnam Books.
16. Buschbacher, R. J. 1986. Tropical deforestation and pasture development.
17. Grainger, A. 1982. *Desertification: How people make deserts, how people can stop and why they don't*, 2nd ed. London: Russell Press, Ltd.
18. United Nations. 1978. *United Nations conference on desertifiction: Roundup plan of action and resolutions.* New York: United Nations.
19. U.N. Food and Agricultural Organization. 1998. *The United Nations convention to combat desertification: An Explanatory leaflet.* Food and Agricultural Organization of the United Nations.
20. U.N. Food and Agricultural Organization. 1998. *What is desertification?* Food and Agricultural Organization of the United Nations.
21. Sheridan, D. 1981. *Desertification of the United States.* Washington, D.C.: Council on Environmental Quality.
22. Sears, P. B. 1959. *Deserts on the March*, 3rd ed. (revised). Norman, Ok: University of Oklahoma Press.

Chapter 13 Notes

1. Carolinian Canada's Big Picture Program: http://www.carolinian.org/ConservationPrograms_BigPicture.htm. Accesssed March 28, 2005.
2. Catfish Creek Slope and Floodplain Forest (Carolinian Canada protected site): http://www.carolinian.org/CarolinianSites_CatfishCreek.htm. Accessed March 28, 2005.
3. Botkin, D. B. 1990. *Discordant harmonies: A new ecology for the 21st century.* New York: Oxford University Press.
4. Burley, Jeffery. 1994. *World forestry: the professional scientific challenges.* The 1994 Leslie L. Schaeffer Lecture in Forest Science, April 6, 1994, University of British Columbia. http://www.forestry.ubc.ca/schaffer/burley.html, accessed April 24, 2005.
5. Likens, G. E., F. H. Borman, R. S. Pierce, J. S. Eaton, and N. M. Johnson. 1977. *The biogeochemistry of a forested ecosystem.* New York: Springer-Verlag.
6. The Hubbard Brook ecosystem continues to be one of the most active and long-term ecosystem studies in North America. An example of a recent publication is: Bailey, S. W., D. C. Buso, and G. E. Likens. 2003. Implications of sodium mass balance for interpreting the calcium cycle of a forested ecosystem. *Ecology* 84(2):471–484.
7. Stelfox, J.B. (ed.). 1995. *Relationships between stand age, stand structure, and biodiversity in aspen mixedwood forests in Alberta.* Joint publication of the Alberta Environmental Centre (AECV95-R1), Vegreville, AB, and the Canadian Forest Services (Project No. 0001A), Edmonton, AB.
8. Fredriksen, R. L. 1971. Comparative chemical water quality—natural and disturbed streams following logging and slash burning. In *Forest Land Use and Stream Environment*, pp. 125–137. Corvallis: Oregon State University.
9. Hagner, Stig. 1999. *Forest Management in Temperate and Boreal Forests: Current Practices and the Scope for Implementing Sustainable Forest Management.* United Nations Food and Agriculture Organization Working Paper FAO/FPIRS/03, prepared for the World Bank Forest Policy Implementation Review and Strategy. Rome: UN Food and Agriculture Organization.
10. Sedjo, R. A. 1983. The comparative economics of plantation forestry: A global assessment. Unpublished research paper, Johns Hopkins University, Baltimore.
11. Kimmins, H. 1995. *Proceedings of the conference on certification of sustainable forestry practices.* Malaysia.
12. Jenkins, Michael B. 1999. *The business of sustainable forestry.* Washington, D.C.: Island Press.
13. United Nations Food and Agriculture Organization. 2001. Rome: UN FAO. Available at ftp://ftp.fao.org/docrep/fao/003/y0900e/y0900e02.pdf.
14. *Atlas of Canada*: landcover in Canada. http://atlas.gc.ca/site/english/maps/environment/forest/forestcanada/landcover. Accessed March 14, 2005.
15. Canada's National Forest Inventory: http://nfi.cfs.nrcan.gc.ca/canfi/index_e.html,. Accessed March 12, 2005.
16. World Resources Institute. *Disappearing land: Soil degradation.* Washington, D.C.: WRI.
17. World Resources Institute. 1993. *World resources* 1992–93. New York: Oxford University Press.
18. Manandhar, A. 1997. *Solar cookers as a means for reducing deforestation in Nepal.* Nepal: Center for Rural Technology.
19. Council on Environmental Quality and U.S. Department of State. 1981. *The global 2000 report to the president: Entering the twenty-first century.* Washington, D.C.: Council on Environmental Quality.
20. Botkin, D. B., and L. Simpson. 1990. The first statistically valid estimate of biomass for a large region. *Biogeochemistry* 9:161–174.
21. Perlin, J. 1989. *A forest journey: The role of wood in the development of civilization.* New York: Norton.
22. World Resources Institute. 1999. *Deforestation: The global assault continues. Global trends, resources at risk, world resources* 1998–1999. Washington, D.C.: WRI.
23. Bryant, D., D. Nielson, and L. Tangley. 1997. *The last frontier forests: Ecosystems and economies on the edge.* Washington, D.C.: WRI.
24. Society of American Foresters, 2003. Available at http://www.safnet.org/archive/tropical1100.cfm Analysis of more than 300 satellite images shows that the rate of deforestation in tropical countries was 10% percent less in the past 10 years than in the 1980s. Of the 300 images, half show a reduced rate of deforestation, while 20% percent indicate an increase.
25. United Nations Food and Agriculture Organization. 1999. *State of the world's forests 1999.* Rome: UNFAO.
26. Botkin, D. B., R. A. Nisbet, and T. E. Reynales. 1989. Effects of climate change on forests of the Great Lakes states. In J. B. Smith and D. A. Tirpak, eds., *The potential effects of global climate change on the United States*, pp. 2.1–2.31. Washington, D.C.: United States Environmental Protection Agency, EPA-203-05-89-0; Zabinski, C., and M. B. Davis.

1989. Hard times ahead for Great Lakes forests: A climate threshold model predicts responses to CO2-induced climate change. In J. B. Smith and D. A. Tirpak, eds., The potential effects of global climate change on the United States, pp. 5.1–5.10. Washington, D.C.: United States Environmental Protection Agency, EPA-203-05-89-0.

27. Botkin, D. B., D. A. Woodby, and R. A. Nisbet. 1991. Kirtland's warbler habitats: A possible early indicator of climatic warming. *Biological Conservation* 56:63–78.
28. Yosemite National Park Web site, available at http://www.nps.gov/yose/nature/history.htm
29. Quotations from Alfred Runte cited by the Wilderness Society on its Website, available at http://www.wilderness.org/NewsRoom/Statement/20031216.cfm.
30. Costa Rica's TravelNet. National parks of Costa Rica. 1999. Costa Rica: Costa Rica's TravelNet.
31. Kenyaweb. 1998. National parks and reserves. Kenya: Kenyaweb.
32. Federation des Parcs naturels regionaux de France. 1999. List of regional nature parks in France. Paris: Federation des Parcs naturels regionaux de France.
33. World Conservation Monitoring Centre. 1996. Protected areas information: 1996 global protected areas summary statistics. Cambridge, England: WCMC.
34. Atlas of Canada: Protected Areas. http://atlas.gc.ca/site/english/maps/environment/ecology/protecting/protectedareas. Accesssed March 19, 2005.
35. Nash, R. 1978. International concepts of wilderness preservation. In J. C. Hendee, G. H. Stankey, and R. C. Lucas, eds., *Wilderness management*, pp. 43–59. United States Forest Service Misc. Pub. No. 1365.
36. National Park Service Web site, available at http://www.wilderness.net/index.cfm?fuse=NWPS&sec=fastfacts.
37. The Wilderness Society. 2000. *The future of wilderness: The national wilderness preservation system should triple in size.* Washington, D.C.: The Wilderness Society.
38. Hendee, J. C., G. H. Stankey, and R. C. Lucas. 1978. *Wilderness management.* United States Forest Service Misc. Pub. No. 1365.

Chapter 14 Notes

1. Aleutian goose removed from Oregon endangered species list: http://www.reporter-times.com/?module=displaystory&story_id=24464&format=html.
2. Snow goose population threatens arctic tundra habitat: http://www.mhhe.com/biosci/pae/es_map/articles/article_40.mhtml
3.. Data on the Hawaiian goose are from the following web site: http://www.aloha-hawaii.com/hawaii/nene/ (accessed March 15, 2005).
4. Botkin, D. B., and L. M. Talbot, 1992. Biological diversity and forests. In N. Sharma, ed., *Contemporary issues in forest management: Policy implications.* Washington, D.C.: World Bank.
5. World Resources Institute 1993. *A short list of plant-based medicinal drugs.* Washington, D.C.: World Resources Institute.
6. Principe, P. P. 1989. The economic significance of plants and their constituents as drugs. In H. Wagner, H. H. Hikino, and N. R. Farnsworth, eds., *Economic and medicinal plant research.* Vol. 3, pp. 1–17. New York: Academic Press.
7. Myers, N. 1983. *A wealth of wild species.* Boulder, Colo.: Westview Press.
8. U.S. Congress Office of Technology Assessment. 1987. *Technologies to maintain biological diversity.* Washington, D.C.: U.S. Government Printing Office, OTA-330, p. 45.
9. Naess A. 1989. *Ecology, community, and lifestyle.* Cambridge, England: Cambridge University Press. Naess does admit to the need to eat. He writes, "It is against my intuition of unity to say 'I can kill you because I am more valuable' but not against the intuition to say 'I will kill you because I am hungry.' In the latter case, there would be an implicit regret: 'Sorry, I am now going to kill you because I am hungry.' In short I find obviously right, but often difficult to justify, different sorts of behavior with different sorts of living beings. But this does not imply that we classify some as intrinsically more valuable than others." (p. 168).
10. Botkin, D. B. 2004. *Our natural history: The lessons of Lewis and Clark.* New York: Oxford University Press.
11. Haines, F. 1970. *The buffalo.* Thomas Y. Crowell, N.Y. See also: Sandoz, M. 1954. *The buffalo hunters.* University of Nebraska Press.
12. *Tom Stehn's whooping crane report* Aransas National Wildlife Refuge- December 10, 2003. Available at http://www.birdrockport.com/tom_stehn_whooping_crane_report.htm. See also Binkley, Clark S. and Richard S. Miller. 1988. Recovery of the Whooping Crane *Grus Americana.* Biological Conservation 45 (1988): 11-20.
13. Whooping Crane Conservation Association. 2003. Available at http://whoopingcrane.com/wccaflockstatus.htm.
14. Friends of the Earth. 1979. *The whaling question: The inquiry by Sir Sidney Frost of Australia.* San Francisco: Friends of the Earth.
15. Bockstoce, J. R., and D. B. Botkin. 1980. *The historical status and reduction of the western Arctic bowhead whale* (Balaena mysticetus) *population by the pelagic whaling industry,* 1848–1914. New Bedford, Conn.: Old Dartmouth Historical Society.
16. United Nations Food and Agriculture Organization. 1978. *Mammals in the seas.* Report of the FAO Advisory Committee on Marine Resources Research, Working Party on Marine Mammals. FAO Fisheries Series 5, vol. 1. Rome: UNFAO.
17. International Whaling Commission. 2003. Table of estimates of whale abundances, available at http://www.iwcoffice.org/Estimate.htm.
18. World Wildlife Fund. 2000. *Gray whales.* Washington, D.C.: WWF.
19. Perry, M. 1996. Climate change biggest risk to whales, says IWC. Sydney, Australia: Reuters.
20. UN FAO Statistics http://apps.fao.org/lim500/nph-wrap.pl?FishCatch&Domain=FishCatch
21. World Resources Institute. 1997. Water and fisheries, *World resources: A guide to the global environment. Washington,* D.C.: WRI.
22. Cushing, D. 1975. *Fisheries resources of the sea and their management.* London: Oxford University Press.
23. NOAA. 2003. World Fisheries. Available at http://www.st.nmfs.gov/st1/fus/current/04_world2002.pdf.
24. Bell, F. W. 1978. *Food from the sea: The economics and politics of ocean fisheries.* Boulder, Colo.: Westview Press.

25. Myers, A., and B. Worm. 2003 (May 15). Rapid worldwide depletion of predatory fish, communities, *Nature*.
26. UN FAO Statistics http://apps.fao.org/lim500/nph-wrap.pl?FishCatch&Domain=FishCatch
27. http://www.eubusiness.com/afp/03120412436.eq98k2mi
28. Martin, P. S. 1963. *The last 10,000 years*. Tucson: University of Arizona Press.
29. Dennis, B., P. L. Munholland, and J. M. Scott. 1991. Estimation of growth and extinction parameters for endangered species. *Ecological Monographs* 61:115–143.
30. Norman A. Wood, quoted in Mayfield, p. 23.
31. Information about the Kirtland's warbler and its habitat is from Byelich et al., p. 12; Mayfield, H. 1969. *The Kirtland's warbler*. Bloomfield Hills, Mich.: Cranbrook Institute of Science, pp. 24–25.
32. Ehrlich, Paul and Anne Ahrlich. 1981. *Extinction: the causes and consequences of the disappearance of species*. New York: Random House.
33. Cobhentz, B. E. 1990. Exotic organisms: a dilemma for conservation biology. *Conservation Biology* 4:261–265.
34. The discussion of the Kirtland's warbler is taken directly from Botkin, D. B. 1990. *Discordant harmonies: A new ecology for the 21st century*. New York: Oxford University Press.
35. Heinselman, M. F. 1985. Fire and succession in the conifer forests of northern North America. In D.C. West, H.H. Shugart, and D.B. Botkin, eds., *Forest succession*. New York: Springer-Verlag.
36. Botkin, D. B. 1977. Strategies for the reintroduction of species into damaged ecosystems. In J. Cairns, ed., *Recovery and restoration of damaged ecosystems*. Charlottesville: University of Virginia Press, pp. 241–260.

Chapter 14 Critical Thinking Issue References

Lohr, Christine, Warren B. Ballard, and Alistair Vath. 1996. Attitudes toward gray wolf reintroduction to New Brunswick. *Wildlife Society Bulletin* 24(3): 414-420.

Hopsack, D. A. 1996. Biological potential for eastern timber wolf re-establishment in the Adirondack Park. In *Wolves of America Conference Proceedings*, November 14–16, 1996. Albany, N.Y., and Washington, D.C.: Defenders of Wildlife.

Stevens, W. K. 1997 (March 3). Wolves may reintroduce themselves to East. *New York Times*.

Chapter 15 Notes

1. Committee on Hormonally Active Agents in the Environment, National Research Council, National Academy of Sciences. 1999. *Hormonally active agents in the environment*. Washington, D.C.: National Academy Press.
2. Krimsky, S. 2001. Hormone disrupters: A clue to understanding the environmental cause of disease. *Environment* 43(5): 22–31.
3. Royte, E. 2003. Transsexual frogs. *Discover* 24(2): 26–53.
4. Hayes, T., K. Haston, M. Tsui, A. Hong, C. Haeffele, and A. Vock. 2002. Feminization of male frogs in the wild. *Nature* 419: 495–496.
5. Hopps, H. C. 1971. Geographic pathology and the medical implications of environmental geochemistry. In H. L. Cannon and H. C. Hopps, eds., *Environmental geochemistry in health*, pp. 1–11. Geological Society of America Memoir 123. Boulder, Colo.: Geological Society of America.
6. Warren, H. V., and R. E. DeLavault. 1967. A geologist looks at pollution: Mineral variety. *Western Mines* 40:23–32.
7. Evans, W. 1996. Lake Nyos. Knowledge of the fount and the cause of disaster. *Science* 379:21.
8. Krajick, K. 2003. Efforts to tame second African killer lake begin. *Science* 379:21.
9. Gunn, J., ed. 1995. *Restoration and recovery of an industrial region: Progress in restoring the smelter-damaged landscape near Sudbury*, Canada. New York: Springer-Verlag.
10. Blumenthal, D. S., and Ruttenber. 1995. *Introduction to environmental health*, 2nd ed. New York: Springer.
11. U.S. Geological Survey. 1995. *Mercury contamination of aquatic ecosystems*. USGS FS 216-95.
12. McGinn, A. P. 2000 (April 1). POPs culture. *World Watch*, pp. 26–36.
13. Ehrlich, P. R., A. H. Ehrlich, and J. P. Holdren. 1970. *Ecoscience: Population, resources, environment*. San Francisco: Freeman.
14. Waldbott, G. L. 1978. *Health effects of environmental pollutants*, 2nd ed. Saint Louis: Moseby.
15. Carlson, E. A. 1983. International symposium on herbicides in the Vietnam War: An appraisal. *BioScience* 33:507–512.
16. Cleverly, D., J. Schaum, D. Winters, and G. Schweer. 1999. Inventory of sources and releases of dioxin-like compounds in the United States. Paper presented at the 19th International Symposium on Halogenated Environmental Organic Pollutants and POPs, September 12–17, Venice, Italy. Short paper in *Organohalogen Compounds* 41:467–472.
17. Grady, D. 1983 (May). The dioxin dilemma. *Discover*, pp. 78–83.
18. Roberts, L. 1991. Dioxin risks revisited. *Science* 251:624–626.
19. Kaiser, J. 2000. Just how bad is Dioxin? *Science* 5473: 1941–1944.
20. Johnson, J. 1995. SAB advisory panel rejects dioxin risk characterization. *Environmental Science & Technology* 29:302A.
21. Thomas, V. M., and T. G. Spiro. 1996. The U.S. dioxin inventory: Are there missing sources? *Environmental Science & Technology* 30:82A–85A.
22. U. S. Environmental Protection Agency. 1994 (June). Estimating exposure to dioxin-like compounds. Review draft. Office of Research and Development, EPA/600/6-88/005Ca-c.
23. U.S. Environmental Protection Agency. 1994. Health assessment document for 2,3,7,8-tetrachlorodibenzo-p-dioxin (TCDD) and related compounds. Vols. I–III. External review draft. Washington, D.C.: EPA.
24. Johnson, J. 1995. Dioxin risk assessment stalls: EPA to create new review panel. *Environmental Science & Technology* 29:492A.
25. Chanlett, E. T. 1979. *Environmental protection*, 2nd ed. New York: McGraw-Hill.
26. Ross, M. 1990. Hazards associated with asbestos minerals. In B. R. Doe, ed., *Proceedings of a U.S. Geological Survey Workshop on Environmental Geochemistry*, pp. 175–176. U.S. Geological Survey Circular 1033.
27. Pool, R. 1990. Is there an EMF–cancer connection? *Science* 249:1096–1098.
28. Linet, M. S., E. E. Hatch, R. A. Kleinerman, L. L. Robison, W. T. Kaune, D. R. Friedman, R. K. Severson, C. M. Haines, C. T. Hartsock, S. Niwa, S. Wacholder, and R. E.

Tarone. 1997. Residential exposure to magnetic fields and acute lymphoblastic leukemia in children. *New England Journal of Medicine* 337(1):1–7.

29. Kheifets, L. I., E. S. Gilbert, S. S. Sussman, P. Guaenel, S. D. Sahl, D. A. Savitz, and G. Thaeriault, G. 1999. Comparative analyses of the studies of magnetic fields and cancer in electric utility workers: Studies from France, Canada, and the United States. *Occupational and Environmental Medicine* 56(8):567–574.
30. Francis, B. M., 1994. *Toxic substances in the environment.* New York: John Wiley & Sons.
31. Poisons and poisoning. 1997. *Encyclopedia Britannica.* Vol. 25, p. 913. Chicago: Encyclopedia Britannica.
32. Air Risk Information Support Center (Air RISC), U.S. Environmental Protection Agency. 1989. *Glossary of terms related to health exposure and risk assessment.* EPA/450/3-88/016. Research Triangle Park, N.C.
33. Foster, K. R., P. Vecchia, and M. H. Repacholi. 2000. Science and the precautionary principle. *Science* 288: 979–981.
34. Easton, T. A. and T. D. Goldfarb, eds. 2003. Is the precautionary principle a sound basis for international policy?, In *Taking sides, environmental issues,* 10th ed., (pp.76–101. Guilford, Conn.) McGraw-Hill/Dushkin.

Chapter 15 Critical Thinking Issue References

Needleman, H. L., J. A. Riess, M. J. Tobin, G. E. Biesecker, and J. B. Greenhouse. 1996. Bone lead levels and delinquent behavior. *Journal of the American Medical Association* 275:363–369.

Centers for Disease Control. 1991. *Preventing lead poisoning in young children.* Atlanta: Public Health Service, Centers for Disease Control.

Goyer, R. A. 1991. Toxic effects of metals. In M. O. Amdur, J. Doull, and C. D. Klaassen, eds., *Toxicology,* pp. 623–680. New York: Pergamon.

Bylinsky, G. 1972. Metallic nemesis. In B. Hafen, ed., *Man, health and environment,* pp. 174–185. Minneapolis: Burgess.

Hong, S., J. Candelone, C. C. Patterson, and C. F. Boutron. 1994. Greenland ice evidence of hemispheric lead pollution two millennia ago by Greek and Roman civilizations. *Science* 265:1841–1843.

Chapter 16 Notes

1. U.S.-Canada Power System Outage Task Force. *Final Report on the August 14, 2003 Blackout in the United States and Canada: Causes and Recommendations.* Washington: United States Department of Energy and Ottawa: Natural Resources Canada, 2004.
2. Feynman, R. P., R. B. Leighton, and M. Sands. 1964. *The Feynman lectures on physics.* Reading, Mass.: Addison-Wesley.
3. Ehrlich, P. R., A. H. Ehrlich, and J. P. Holdren. 1970. *Ecoscience: Population, resources, environment.* San Francisco: W. H. Freeman.
4. Butti, K., and J. Perlin. 1980. *A golden thread: 2500 years of solar architecture and technology.* Palo Alto, Calif.: Cheshire Books.
5. Ménard, Marinka. 2005. Canada, a Big Energy Consumer: A Regional Perspective. Statistics Canada, Manufacturing, Construction and Energy Division. Available online at http://www.statcan.ca:8096/bsolc/english/bsolc?catno=11-621-M2005023.
6. Darmstadter, J., H. H. Landsberg, H. C. Morton, and M. J. Coda. 1983. *Energy today and tomorrow: Living with uncertainty.* Englewood Cliffs, N.J.: Prentice-Hall.
7. Strickland, Catherine and John Nyboer. 2002. *Cogeneration Potential in Canada: Phase 2.* Report for Natural Resources Canada, April 2002.
8. Olkowski, H., B. Olkowski, and T. Javits. (Farallones Institute). 1979. *The integral urban house: Self reliant living in the city.* San Francisco: Sierra Club Books.
9. Flavin, C. 1984. *Electricity's future: The shift to efficiency and small-scale power.* Worldwatch Paper 61. Washington, D.C.: Worldwatch Institute.
10. Berger, J. J. 2000. *Beating the heat.* Berkeley, Calif.: Berkeley Hills Books.
11. Lovins, A. B. 1979. *Soft energy paths: Towards a durable peace.* New York: Harper & Row.
12. *Atlas of Canada* (1997 data): http://atlas.gc.ca/site/english/maps/economic/generatingstations/utilitybytech/
13. National Energy Board of Canada. 2003. *Canada's Energy Future: Scenarios for Supply and Demand to 2025.* Ottawa: National Energy Board.
14. Brown, L. R., C. Flavin, and S. Postel (Worldwatch Institute). 1991. *Saving the planet: How to shape an environmentally sustainable global economy.* New York: W. W. Norton.
15. Flavin, C., and S. Dunn. 1999. Reinventing the energy system. In L. R. Brown et al., eds., *State of the world 1999: A Worldwatch Institute report on progress toward a sustainable society.* New York: W. W. Norton.
16. Dunn, S. 2000. *Micropower, the next electrical era.* Worldwatch Paper 151. Washington, D.C.: Worldwatch Institute.

Chapter 16 Critical Thinking Issue References

Ehrlich, P. R., and A. H. Ehrlich, 1991. *Healing the planet.* Reading, Mass.: Addison-Wesley.

Fickett, A. P. 1990. Efficient use of electricity. *Scientific American* 263(3):65–74.

Holdren, J. P. 1990. Energy in transition. *Scientific American* 263(3):157–163.

Lean, G. 1990. *Atlas of the environment.* New York: Prentice-Hall.

U.S. Census Bureau. 1998. World population and growth rates. http://www.census.gov/ipc/www/world.html.

Chapter 17 Notes

1. Sierra Club, 2004. Driving up the heat: SUVs and global warming—assessed 3/29/04 http://www.sierra club.org.
2. Union of Concerned Scientists, 2002. News. Sport utility vehicles. Positions. http://www.ucsusa.org.
3. Van Koevering, T. E., and N. J. Sell. 1986. *Energy: A conceptual approach.* Englewood Cliffs, N.J.: Prentice-Hall.
4. McCulloh, T. H. 1973. In D. A. Brobst and W. P. Pratt, eds., *Oil and gas in United States mineral resources,* pp. 477–496. U.S. Geological Survey Professional Paper 820.
5. Maugeri, L. 2004. Oil: Never cry wolf—when the petroleum age is far from over. *Science* 304:1114–1115.
6. British Petroleum Company. 1999. *B.P. statistical review of world energy.* London: British Petroleum Company.
7. Kerr, R. A. 2000. USGS optimistic on world oil prospects. *Science* 289:237.
8. Youngquist, W. 1998. Spending our great inheritance. Then what? *Geotimes* 43(7):24–27.

9. Edwards, J. D. 1997. Crude oil and alternative energy production forecast for the twenty-first century: The end of the hydrocarbon era. *American Association of Petroleum Geologists Bulletin* 81(8):1292–1305.
10. Darmstadter, J., H. H. Landsberg, H. C. Morton, and M. J. Coda. 1983. *Energy today and tomorrow: Living with uncertainty.* Englewood Cliffs, N.J.: Prentice-Hall.
11. Nuccio, V. 1997. Coal-bed methane—an untapped energy resource and an environmental concern. US Geological Survey Fact Sheet. FS-019-97.
12. Nuccio, V. 2000. *Coal-bed methane: Potential environmental concerns.* US Geological Survey. USGS Fact Sheet. FS-123-00.
13. Wood, T. 2003. Prosperity's brutal price. *Los Angeles Times Magazine.* February 2, 2003.
14. Suess, E., G. Bohrmann, J. Greinert, and E. Lauch. 1999. Flammable ice. *Scientific American* 28(5):76–83.
15. Rahn, P. H. 1996. *Engineering geology: An environmental approach*, 2nd ed. New York: Elsevier.
16. U.S. Environmental Protection Agency. 1973. *Processes, procedures and methods to control pollution from mining activities.* EPA-430/9-73-001. Washington, D.C.: U.S. Environmental Protection Agency.
17. Council on Environmental Quality. 1978. *Progress in environmental quality.* Washington, D.C.: Council on Environmental Quality.
18. Stone, Kevin. 2003. "Coal". In: *Minerals Yearbook 2003.* Ottawa: Natural Resources Canada.
19. Corcoran, E. 1991. Cleaning up coal. *Scientific American* 264:106–116.
20. Energy Information Administration. 1995 (February). Coal data: A reference. Washington, D.C.: U.S. Department of Energy.
21. Knapp, D. H. 1995. Non-OPEC oil supply continues to grow. *Oil & Gas Journal* 93:35–45. Paris: International Energy Agency.
22. Peterson, G. 2003. New statute for Canadian Oil Sands. *Geotimes* 48(3) 7.
23. Garland, William. How and why CANDU is designed the way it is: Introduction. Available online at http://canteach.candu.org/library/20000101.pdf,
24. Till, C. E. 1989. Advanced reactor development. *Ann. Nucl. Energy* 16(6):301–305.
25. Churchill, A. A. 1993 (July). Review of WEC Commission: Energy for tomorrow's world. *World Energy Council Journal*, pp. 19–22.
26. Cohen, B. L. 1990. *The nuclear energy option: An alternative for the 90s.* New York: Plenum.
27. Lake, J. A, R. G. Bennett, and J. F. Kotek. 2002 (January). Next-generation nuclear power. *Scientific American*, pp. 73–81.
28. Ehrlich, P. R., A. H. Ehrlich, and J. P. Holdren. 1970. *Ecoscience: Population, resources, environment.* San Francisco: Freeman.
29. Brenner, D. J. 1989. *Radon: Risk and remedy.* New York: Freeman.
30. U.S. Department of Energy. 1980. *Magnetic fusion energy.* DOE/ER-0059. Washington, D.C.: U.S. Department of Energy.
31. U.S. Department of Energy. 1979. *Environmental development plan, magnetic fusion.* DOE/EDP-0052. Washington, D.C.: U.S. Department of Energy.
32. Cordey, J. G., R. J. Goldston, and R. R. Parker. 1992. Progress toward a Tokamak fusion reactor. *Physics Today* 45:22–30.
33. Greenberg, P. A. 1993. Dreams die hard. *Sierra* 78:78.
34. Van Koevering, T. E., and N. J. Sell. 1986. *Energy: A conceptual approach.* Englewood Cliffs, N.J.: Prentice-Hall.
35. Waldbott, G. L. 1978. *Health effects of environmental pollutants.* 2nd ed. Saint Louis: C. V. Moseby.
36. *New Encyclopedia Britannica.* 1997. Radiation. V26. p. 487.
37. U.S. Department of Energy. 1999. Radiation (in) waste isolation pilot plant. 1999. Carlsbad, New Mexico. Accessed at www.wipp.carlsbad.nm.us.
38. Stone, R. 2003. Plutonium fields forever. *Science* 300:1220–1224.
39. University of Maine and Maine Department of Human Services. 1983 (February). Radon in water and air. *Resource Highlights.*
40. MacLeod, G. K. 1981. Some public health lessons from Three Mile Island: A case study in chaos. *AMBIO* 10:18–23.
41. Anspaugh, L. R., R. J. Catlin, and M. Goldman. 1988. The global impact of the Chernobyl reactor accident. *Science* 242:1513–1518.
42. Nuclear Energy Agency. 2002. Chernobyl Assessment of Radiological and Health Impacts: 2002 Update of Chernobyl: Ten Years On.
43. Balter, M. 1995. Chernobyl's thyroid cancer toll. *Science* 270:1758.
44. Fletcher, M. 2000 (November 14). The last days of Chernobyl. *Times 2* (London), pp. 3–5.
45. Williams, N. 1995. Chernobyl: Life abounds without people. *Science* 269:304.
46. Skuterud, L., N. I. Goltsova, R. Naeumann, T. Sikkeland, and T. Lindmo. 1994. Histological changes in *Pinus sylvestris* L. in the proximal-zone around the Chernobyl power plant. *The Science of the Total Environment* 157:387–397.
47. Weisman, J. 1996. Study inflames Ward Valley controversy. *Science* 271:1488.
48. Atomic Energy of Canada Limited. Waste Storage fact sheet. http://www.aecl.ca/index.asp?menuid=500&miid=545&layid=3&csid=302. Accessed March 29, 2005.
49. Nuclear Waste Management Organization. Centralized Extended Storage Fact Sheet. http://www.nwmo.ca/default.aspx?DN=897,177,20,1,Documents. Accessed March 29, 2005.
50. Nuclear Waste Management Organization. How Nuclear Fuel Waste is Managed in Canada. Fact Sheet: http://www.nwmo.ca/adx/asp/adxGetMedia.asp?DocID=177,20,1,Documents&MediaID=459&Filename=Fact_Sheet+_3_Managed.pdf. Accessed March 29, 2005.
51. Hanks, T. C., I. J. Winograd, R. E. Anderson, T. E. Reilly, and E. P. Weeks. 1999. *Yucca Mountain as a radioactive-waste repository.* U.S. Geological Survey Circular 1184
52. Bredehoeft, J. D., A. W. England, D. B. Stewart, J. J. Trask, and I. J. Winograd. 1978. *Geologic disposal of high-level radioactive wastes—Earth science perspectives.* U.S. Geological Survey Circular 779. Arlington, Va.: U.S. Department of the Interior.

53. Timmerman, Peter. 2003. Ethics of High Level Nuclear Fuel Waste Disposal in Canada: Background Paper. Nuclear Waste Management Organization: http://www.nwmo.ca/adx/asp/adxGetMedia.asp?DocID=277,276,274,20,1,Documents&MediaID=1076&Filename=21_NWMO_background_paper.pdf
54. Flavin, C. 1991. The case against reviving nuclear power. In L. R. Brown, ed., *The Worldwatch reader*, pp. 205–220. New York: Norton.
55. Jackson, T., and R. Lofstedt. 1998. Royal commission on environmental pollution. Study on energy and the environment. Accessed November 29, 2000, at http://www.rcep.org.uk/studies/energy/98-6061/jackson.html.
56. Berger, J. J. 2000. *Beating the heat. Berkeley*, Calif.: Berkeley Hills Books.
57. Miller, E. W. 1993. *Energy and American society: A reference handbook*. Santa Barbara, Calif.: ABC-CLIO.
58. Flavin, C., and S. Dunn. 1999. Reinventing the energy system. In L. R. Browne et al., *State of the world 1999: A Worldwatch Institute report on progress toward a sustainable society*. New York: W. W. Norton.
59. Eaton, W. W. 1978. Solar energy. In L. C. Ruedisili and M. W. Firebaugh, eds., *Perspectives on energy*, 2nd ed., pp. 418–436. New York: Oxford University Press.
60. Brown, L. R. 1999 (March–April). Crossing the threshold. *Worldwatch*, pp. 12–22.
61. Mayur, R., and B. Daviss. 1998 (October). The technology of hope. *The Futurist*, pp. 46–51.
62. Quinn, R. 1997 (March). Sunlight brightens our energy future. *The World and I*, pp. 156–163.
63. Johnson, J. T. 1990 (May). The hot path to solar electricity. *Popular Science*, pp. 82–85.
64. Schatz solar hydrogen project. N.D. Pamphlet. Arcata, Calif.: Humboldt State University.
65. Piore, A. 2002 (April 15). Hot springs eternal: hydrogen power. *Newsweek*, pp. 32H.
66. Kartha, S., and P. Grimes. 1994. Fuel cells: energy conversion for the next century. *Physics Today* 47:54–61.
67. Haggin, J. 1995. Fuel-cell development reaches demonstration stage. *Chemical & Engineering News* 73:28–30.
68. Alward, R., S. Eisenbart, and J. Volkman. 1979. *Micro-hydro power: Reviewing an old concept*. Butte, Mont.: National Center for Appropriate Technology, U.S. Department of Energy.
69. Nova Scotia Department of Mines and Energy. 1981. *Wind power*.
70. Demeo, E. M., and P. Steitz. 1990. The U.S. electric utility industry's activities in solar and wind energy. In K. W. Böer, ed., *Advances in solar energy*, Vol. 6, pp. 1–218. New York: American Solar Energy Society.
71. De Miguel Ichaso, A. 2000 (August). Wind power development in Spain, the model of Navarra. *DEWI Magazine* 17:49–54.
72. U.S. Congress, Office of Technology Assessment. 1993. *Potential environmental impacts of bioenergy crop production*. Background paper. Washington, D.C.: U.S. Government Printing Office.
73. Natural Resources Canada. 2004 *Improving Energy Performance in Canada: Report to Parliament under the Energy Efficiency Act*. 2002-2003 annual report. Ottawa: Natural Resources Canada
74. Council on Environmental Quality. 1979. *Environmental quality*.
75. Wihersaari, M. 1996. Energy consumption and greenhouse gas emissions from biomass production chains. *Energy Conversion and Management* 37:1217.
76. Sterzinger, G. 1995. Making biomass energy a contender. *Technology Review* 98:34–40.
77. Wright, P. 2000. Geothermal energy. *Geotimes* 45(7):16–18.
78. Duffield, W. A., J. H. Sass, and M. L. Sorey. 1994. *Tapping the Earth's natural heat*. U.S. Geological Survey Circular 1125.

Chapter 17 Critical Thinking Issue References

Bleviss, D. L. 1988. *The new oil crisis and fuel economy technologies*. New York: Quorum Books.

Bleviss, D. L., and P. Walzer. 1990. Energy for motor vehicles. *Scientific American* 263(3):103–109.

Corson, W. H., ed. 1990. *The global ecology handbook*. Boston: Beacon.

Driving down the deficit. 1993. *U.S. News & World Report* 114(2):58–60.

Energy Information Administration. 1996 (January). *Monthly energy review*. Washington, D.C.: U.S. Department of Energy.

Greenwald, J. 1993. Why not a gas tax? *Time* 141(7):25–27.

Miller, E. W., and R. M. Miller. 1993. *Energy and American society. A reference handbook*. Santa Barbara, Calif.: ABC-CLIO.

Nadis, S., and J. J. MacKenzie. 1993. *Car trouble*. Boston: Beacon.

Ahearne, J. F. 1993. The future of nuclear power. *American Scientist* 81(1):24–35.

Fox, M. R. 1987. Perspectives in risk: Compared to what? *Vital Speeches of the Day* 53(23):730–732.

Greenberg, P. A. 1993. Dreams die hard. *Sierra* 78(6):78.

Rosa, E. A., and R. E. Dunlap. 1994. Nuclear power: three decades of public opinion. *Public Opinion Quarterly* 58:295–325.

Canadian Environmental Assessment Agency (CEAA). 1998. *Nuclear Fuel Waste Management and Disposal Concept* (Seaborn Report). Report of the Nuclear Fuel Waste Management and Disposal Concept Environmental Assessment Panel. B. Seaborn (chairman). Canada: CEAA.

Chapter 18 Notes

1. Parks Canada. 1997. *State of the Parks 1997 Report*. Ottawa: Parks Canada.
2. Henderson, L. J. 1913. *The fitness of the environment: An inquiry into the biological significance of the properties of matter*. New York: Macmillan.
3. Council on Environmental Quality and U.S. Department of State. 1980. *The global 2000 report to the president: Entering the twenty-first century*, Vol. 2. Washington, D.C.
4. Water Resources Council. 1978. *The nation's water resources, 1975–2000*, Vol. 1. Washington, D.C.
5. Winter, T. C., J. W. Harvey, O. L. Franke, and W. M. Alley. 1998. *Groundwater and surface water: A single resource*. U.S. Geological Survey Circular 1139.
6. Environment Canada. Fact sheet on Water Use in Canada, 1996: http://www.ec.gc.ca/water/images/manage/use/a4fle.htm. Accessed April 29, 2005.
7. United Nations Educational, Scientific, and Cultural Organization: Dynamics of Water Use in Canada, 1998: http://webworld.unesco.org/water/ihp/db/shikloma

nov/part'3/HTML/Tb_21'CN.html. Accessed April 29, 2005. This paper summarizes estimates of current and forecasted Canadian water withdrawals and consumption based on the work of numerous authors.

8. U. S. General Accounting Office. 2003. *Freshwater supply: States' view of how federal agencies could help them meet the challenges of expected shortages.* Report GAO-03-514.
9. Alexander, G. 1984 (February/March). Making do with less. *National Wildlife*, special report, pp. 11–13.
10. Gleick, P. H., P. Loh, S. V. Gomez, and J. Morrison. 1995. *California water 2020, a sustainable vision.* Oakland, Calif.: Pacific Institute for Studies in Development, Environment and Security.
11. Alley, W. M., T. E. Reilly, and O. L. Franke. 1999. *Sustainability of ground-water resources.* U.S. Geological Survey Circular 1186.
12. Leopold, L. B. 1977. A reverence for rivers. *Geology* 5:429–430.
13. Holloway, M. 1991. High and dry. *Scientific American* 265:16–20.
14. Levinson, M. 1984 (February/March). Nurseries of life. *National Wildlife*, special report, pp. 18–21.
15. Nichols, F. H., J. E. Cloern, S. N. Luoma, and D. H. Peterson. 1986. The modification of an estuary. *Science* 231:567–573.
16. The Canadian federal government policy on wetland conservation: http://www.ramsar.org/wurc/wurc_policy_canada.htm. Accessed June 16, 2005.
17. Hileman, B. 1995. Rewrite of Clean Water Act draws praise, fire. *Chemical & Engineering News* 73:8.
18. Kaiser, J. 2001. Wetlands restoration: Recreated wetlands no match for original. *Science* 293;25a.
19. United States Environmental Protection Agency. 2000. Constructed Wetlands Treatment of Municipal Wastewaters. Report no. EPA/625/R-99/010. Available online at http://www.epa.gov/owow/wetlands/pdf/Design_Manual2000.pdf
20. Pearce, M. 1995 (January). The biggest dam in the world. *New Scientist*, pp. 25–29.
21. Zich, R. 1997. China's three gorges: Before the flood. *National Geographic* 192(3):2–33.
22. Booth, W. 2000 (December 12). Restoring rivers—at a high price. *Washington Post*, p. A3.
23. U.S. Congress. 1973. *Stream channelization: What federally financed draglines and bulldozers do to our nation's streams.* House Committee Report No. 93-530. Washington, D.C.: U.S. Government Printing Office.
24. Agriculture and Agri-Food Canada. Case Study of Mink Creek rehabilitation: http://res2.agr.gc.ca/publications/hw/07b1_e.htm. Accessed March 29, 2005.
25. Boeckh, I., V. S. Christie, and A. H. J. Dorcey. 1991. Human settlement and development in the Fraser River Basin. In: Dorcey, A. H. J. and J. R. Griggs (eds.) 1991. *Water in Sustainable Development: Exploring Our Common Future in the Fraser River Basin.* Westwater Research Centre, University of British Columbia. Vancouver, B.C.: University of British Columbia.
26. Dale, N.G. 1991. The quest for consensus on sustainable development in the use and management of Fraser River salmon. In: Dorcey, A. H. J. (ed.) 1991. *Perspectives on Sustainable Development in Water Management: Towards Agreement in the Lower Fraser River Basin.* Westwater Research Centre, University of British Columbia. Vancouver, B.C.: University of British Columbia.
27. Dorcey, A.H.J. 1997. Collaborating towards sustainability together: the Fraser Basin Management Board and program. In: Shrubsole, D. and Mitchell, B. (eds.) *Practising Sustainable Water Management: Canadian and International Experiences.* Cambridge, Ontario: Canadian Water Resources Association.
28. Blomquist, W., K. S. Calbick and Ariel Dinar. 2005. *Institutional and Policy Analysis of River Basin Management: the Fraser River Basin, Canada.* World Bank Working Paper No. number 3525. Available online at http://ideas.repec.org/p/wbk/wbrwps/3525.html#provider.
29. Heathcote, Isobel W. 1998. *Integrated Watershed Management: Principles and Practice.* New York: John Wiley and Sons.
30. Ballard, S. C., D. D. Michael, M. A. Chartook, M. R. Clines, C. E. Dunn, C. M. Hock, G. D. Miller, L. B. Parker, D. A. Penn, and G. W. Tauxe. 1982. *Water and western energy: Impacts, issues, and choices.* Boulder, Colo.: Westview Press.
31. Brown, L. R. 2003. *Plan B: Rescuing a planet under stress and a civilization in trouble.* New York: Norton.

Chapter 18 Critical Thinking Issues References

Baldwin, M. F. 1987. Wetlands: Fortifying federal and regional cooperation. *Environment* 29(7):16–20, 39.

Leidy, R. A., P. L. Fiedler, and E. R. Micheli. 1992. Is wetter better? *Bioscience* 40(9):58–61, 65.

National Institute for Urban Wildlife. *Wetlands conservation and use* (issue pak). Columbia, Md.: National Institute for Urban Wildlife.

Stevens, W. K. 1990 (March 13). Efforts to halt wetland loss are shifting to inland areas. *New York Times*, p. C1.

World Resources Institute. 1992. *The 1992 information please environmental almanac.* Boston: Houghton Mifflin.

Chapter 19 Notes

1. Mallin, M. A. 2000. Impacts of industrial animal production on rivers and estuaries. *American Scientist* 88(1): 26–37.
2. Bowie, P. 2000. No act of God. *The Amicus Journal* 21(4):16–21.
3. Groundwater: Issues and answers 1985. Arvada, Colo.: American Institute of Professional Geologists.
4. Gleick, P. H. 1993. An introduction to global fresh water issues. In P. H. Gleick, ed., *Water in crisis*, pp. 3–12. New York. Oxford University Press.
5. Hileman, B. 1995. Pollution tracked in surface- and groundwater. *Chemical & Engineering News* 73:5.
6. O'Connor, Dennis. 2002. *Report of the Walkerton Inquiry.* Parts 1 and 2. Toronto, Ontario: Ministry of the Attorney General.
7. Smith, R. A. 1994. Water quality and health. *Geotimes* 39:19–21.
8. MacKenzie, W. R., et al. 1994. A massive outbreak in Milwaukee of Cryptosporidium infection transmitted through the public water supply. *The New England Journal of Medicine* 331:161–167.
9. Centers for Disease Control and Environmental Protection Agency. 1995. Assessing the public health threat associated with waterborne Cryptosporidiosis: Report of a workshop. *Journal of Environmental Health* 58:31.
10. Kluger, J. 1998. Anatomy of an outbreak. *Time* 152(5):56–62.

11. Maugh, T. H. 1979. Restoring damaged lakes. *Science* 203:425–427.
12. Mitsch, W. J., J. W. Day, Jr., J. W. Gilliam, P. M. Groffman, D. L. Hey, G. W. Randall, and N. Wang. 1999. Reducing Nutrient Loads, Especially Nitrate—Nitrogen, to Surface Water, Ground Water, and the Gulf of Mexico: Topic 5 Report for the Integrated Assessment on Hypoxia in the Gulf of Mexico. NOAA Coastal Ocean Program Decision Analysis Series No. 19. NOAA Coastal Ocean Program, Silver Spring, MD. 111 pp.
13. Hinga, K. R. 1989. Alteration of phosphorus dynamics during experimental eutrophication of enclosed marine ecosystems. *Marine Pollution Bulletin* 20:624–628.
14. Richmond, R. H. 1993. Coral reefs: Present problems and future concerns resulting from anthopogenic disturbance. *American Zoologist* 33:524–536.
15. Bell, P. R. 1991. Status of eutrophication in the Great Barrier Reef Lagoon. *Marine Pollution Bulletin* 23:89–93.
16. Hunter, C. L., and C. W. Evans. 1995. Coral reefs in Kaneohe Bay, Hawaii: Two centuries of Western influence and two decades of data. *Bulletin of Marine Science* 57:499.
17. Department of Alaska Fish and Game 1918. *Alaska Fish and Game* 21(4), Special Issue.
18. Holway, M. 1991. Soiled shores. *Scientific American* 265:102–106
19. Robinson, A. R. 1973. Sediment, our greatest pollutant? In R. W. Tank, ed., *Focus on environmental geology*, pp. 186–192. New York: Oxford University Press.
20. Yorke, T. H. 1975. Effects of sediment control on sediment transport in the northwest branch, Anacostia River basin, Montgomery County, Maryland. *Journal of Research* 3:487–494.
21. Poole, W. 1996. Rivers run through them. *Land and People* 8:16–21.
22. Carey, J. 1984 (February/March). Is it safe to drink? *National Wildlife*, Special Report, pp. 19–21.
23. Pye, U. I., and R. Patrick. 1983. Groundwater contamination in the United States. *Science* 221:713–718.
24. Foxworthy, G. L. 1978. Nassau County, Long Island, New York—Water problems in humid county. In G. D. Robinson and A. M. Spieker, eds., *Nature to be commanded*, pp. 55–68. U.S. Geological Survey Professional Paper 950. Washington, D.C.: U.S. Government Printing Office.
25. Van der Leeden, F., F. L. Troise, and D. K. Todd. 1990. *The water encyclopedia*, 2nd ed. Chelsea, Mich.: Lewis Publishers.
26. Jobling, S., M. Nolan, C. R. Tyler, G. Brighty, and J. P. Sumpter. 1998. Widespread sexual disturbance in wild fish. *Environmental Science and Technology* 32(17):2498–2506.
27. Sierra Legal Defence Fund. 2004. The National Sewage Report Card III: http://www.sierralegal.org/reports/sewage_report_card_III.pdf..
28. Jewell, W. J. 1994. Resource-recovery wastewater treatment. *American Scientist* 82:366–375.
29. Task Force on Water Reuse. 1989. Water reuse: Manual of practice SM-3. Alexandria, Va.: Water Pollution Control Federation.
30. Kadlec, R. H., and R. L. Knight. 1996. *Treatment wetlands*. New York: Lewis Publishers.
31. Breaux, A. M., and J. W. Day, Jr. 1994. Policy considerations for wetland wastewater treatment in the coastal zone: A case study for Louisiana. *Coastal Management* (22):285–307.
32. Hamilton Harbour Remedial Action Plan fact sheet: http://www.on.ec.gc.ca/water/raps/hamilton/intro_e.html.

Chapter 19 Build Your Environmental Skills References

Allan, J. D., and A. S. Flecker. 1993. Biodiversity conservation in running waters. *BioScience* 43(1):32–43.

Armour, C. L., D. A. Duff, and W. Elmore. 1991. The effects of livestock grazing on riparian and stream ecosystems. *Fisheries* 16(1):7–11.

Karr, J. R., L. A. Toth, and D. R. Dudley. 1985. Fish communities of midwestern rivers: A history of degradation. *BioScience* 35(2):90–95.

Sparks, R. 1992. The Illinois River floodplain ecosystem. In National Research Council, *Restoration of aquatic ecosystems*, pp. 412–432. Washington, D.C.: National Academy Press.

Stevens, W. K. 1993 (January 26). River life through U.S. broadly degraded. *New York Times*, pp. B5, B8.

Chapter 20 Notes

1. Watson, R. T. 2000. Presentation at the Sixth Conference of the Parties of the United Nations Framework Convention on Climate Change, Intergovernment Panel on Climate Change, November 13, 2000. Accessed December 1, 2000, at www.IPCC.ch/press/sp-COPG.htm.
2. Levy, S. 2000. Wildlife on the hot seat. *National Wildlife* 38(5):20–27.
3. Hartmann, D. L. 1994. *Global physical climatology*. International Geophysics Series, vol. 56. New York: Academic Press.
4. Anthes, R. A. 1992. *Meteorology*. 6th ed. New York: Macmillan.
5. Detwyler, T. R., and M. G. Marcus. 1972. *Urbanization and the environment: The physical geography of the city*. Belmont, Calif.: Duxbury Press.
6. Steig, E. J., E. J. Brook, J. W. C. White, C. M. Sucher, M. L. Bender, S. J. Lehman, D. L. Morse, E. D. Waddington, and G. D. Clow. 1998. Synchronous climate changes in Antarctica and the North Atlantic. *Science* 282:92–94.
7. Union of Concerned Scientists. 1989. *The greenhouse effect*. Cambridge, Mass.
8. Kerr, R. A. 1996. 1995 the warmest year? Yes and no. *Science* 271:137–138.
9. Crowley, T. J. 2000. Causes of climate change over the past 1000 years. *Science* 289:270–277.
10. Campbell, I. D., C. Campbell, N. J. Apps, N. W. Rutter, and A. B. G. Bush. 1998. Late Holocene approximately 1500 year climatic periodicities and their implications. *Geology* 26(5):471–473.
11. Broecker, W. 1997. Will our ride into the greenhouse future be a smooth one? *GSA Today*, 7(5):2–6.
12. Earth System Sciences Committee. 1988. *Earth system science: A preview*. Boulder, Colo.: University Corporation for Atmospheric Research.
13. Post, W. M., T. Peng, W. R. Emanuel, A. W. King, V. H. Dale, and D. L. De Angelis. 1990. The global carbon cycle. *American Scientist* 78:310–326.
14. Moss, M. E., and H. F. Lins. 1989. *Water resources in the twenty-first century: A study of the implications of climate uncertainty*. U.S. Geological Survey Circular 1030. Washington, D.C.: U.S. Department of the Interior.

15. Hansen, J., A. Lacis, and M. Prather. 1989. Greenhouse effect of chlorofluorocarbons and other trace gases. *Journal of Geophysical Research* 94(D13):16417–16421.
16. Titus, J. G., S. P. Leatherman, C. H. Everts, Moffatt and Nichol Engineers, D. L. Kriebel, and R. G. Dean. 1985. *Potential impacts of sea level rise on the beach at Ocean City, Maryland.* Washington, D.C.: U.S. Environmental Protection Agency, Office of Policy Planning and Evaluation.
17. Rodhe, H. 1990. A comparison of the contribution of various gases to the greenhouse effect. *Science* 248:1217–1219.
18. Kerr, R. A. 1994. Methane increases put on pause. *Science* 263:751.
19. Dlugokencky, E. J., L. P. Steele, P. M. Lang, and K. A. Masarie. 1994. The growth rate and distribution of atmospheric methane. *Journal of Geophysical Research* 99(D8):17021–17043.
20. Council on Environmental Quality. 1990. *Environmental trends 1989.* Washington, D.C.
21. Luoma, J. R., and D. Hiser. 1991. *Gazing into our greenhouse future.* Audubon 93(2).
22. Charlson, R. J., S. E. Schwartz, J. M. Hales, R. D. Cess, J. A. J. Coakley, J. E. Hansen, and D. J. Hofmann. 1992. Climate forcing by anthropogenic aerosols. *Science* 255:423–430.
23. Kerr, R. A. 1995. Study unveils climate cooling caused by pollutant haze. *Science* 268:802.
24. NOAA. 1998. What is an El Niño? Accessed October 2, 1998 at http://www.elnino.noaa.gov.
25. Kennett, J. P., K. G. Cannariato, I. L. Hendy, and R. J. Behl. 2000. Carbon isotopic evidence for methane hydrate instability during Quaternary interstadials. *Science* 288:128–133.
26. Mohnen, V. A., W. Goldstein, and W. Wang. 1991. The conflict over global warming. *Global Environmental Change* 1:109–123.
27. Kerr, R. A. 1995. U.S. climate tilts toward the greenhouse. *Science* 268:363–364.
28. Kerr, R. A. 1995. Greenhouse report foresees growing global stress. *Science* 270:731.
29. Titus, J. G., and V. K. Narayanan. 1995. *The probability of sea level rise.* Washington, D.C.: U.S. Environmental Protection Agency.
30. Holmes, N. 2000. Has anyone checked the weather (map)? *Amicus Journal* 21(4):50–51.
31. Epstein, P. R. 2000. Is global warming harmful to health? *Scientific American* 283(2):50–57.
32. Thomas, C. D., et. al. 2004. Extinction risk of climate change. *Nature* 427:145–148.
33. Lea, D. W. 2004. The 100,000-yr cycle in tropical SST, greenhouse forcing, and climate sensitivity. *Journal of Climate* 17 (11) 2170–2179.
34. Dunn, S. 2001. Decarbonizing the energy economy. In *WorldWatch Institute state of the world 2001.* New York: W. W. Norton, pp. 83–102.
35. Rice, C. W. 2002. Storing carbon in soil: Why and how. *Geotimes* 47(1):14–17.
36. Friedman, S. J. 2003. Storing carbon in Earth. *Geotimes* 48(3):16–20.
37. Bartlett, K. 2003. Demonstrating carbon sequestration. *Geotimes* 48(3):22–23.

A Closer Look 20.1 References

Edahl, C. A., and C. D. Keeling. 1973. *Carbon and the biosphere.* Oak Ridge, Tenn.: Technical Information Service.

Keeling, C. D., T. P. Worf, and J. Van der Plicht. 1995. *Nature* 375:666–670.

Chapter 20 Critical Thinking Issue Reference

King, D. A. 2004. Climate change science: Adapt, mitigate, or ignore? *Science* 303:176–177.

Intergovernmental Panel on Climate Change. 2001. *Climate Change 2001: Synthesis Report.* Cambridge, UK: Cambridge University Press.

Oreskes, N. 2004. Beyond the ivory tower: the scientific consensus on climate change. *Science* 306:1686.

Chapter 21 Notes

1. Simons, L. M. 1998. Plague of fire. *National Geographic* 194(2):100–119.
2. Luginaah, Isaac N., Karen Y. Fung, Fevin M. Gorey, Greg Webster, and Chris Wills. 2004. Association of ambient air pollution with respiratory hospitalization in a government designated "Area of Concern": the case of Windsor, Ontario. Environmental Health Perspectives Online: http://dx.doi.org/. 14 December 2004.
3. American Lung Association. 1998. American Lung Association outdoor fact sheet. Accessed September 18, 1998 at http://www.lungusa.org.
4. Godish, T. 1991. *Air quality,* 2nd ed. Chelsea, Mich.: Lewis Publishers.
5. Seitz, F., and C. Plepys. 1995. Monitoring air quality in healthy people 2000. *Healthy people 2000: Statistical notes no. 9.* Atlanta: Centers for Disease Control and Prevention, National Center for Health Statistics.
6. American Lung Association, 2001. *State of the Air 2000.*
7. Moore, C. 1995. Poisons in the air. *International Wildlife* 25:38–45.
8. Pope, C. A., III, D. V. Bates, and M. E. Raizenne. 1995. Health effects of particulate air pollution: Time for reassessment? *Environmental Health Perspectives* 103:472–480.
9. Office of Technology Assessment. 1984. Balancing the risks. *Weatherwise* 37:241–249.
10. Canadian Department of the Environment. 1984. *The acid rain story.* Ottawa: Minister of Supply and Services.
11. Lippmann, M., and R. B. Schlesinger. 1979. *Chemical contamination in the human environment.* New York: Oxford University Press.
12. Winkler, E. M. 1998 (September). The complexity of urban stone decay. *Geotimes*, pp. 25–29.
13. How many more lakes have to die? 1981. *Canada Today* 12 (2).
14. Tyson, P. 1990. Hazing the Arctic. *Earthwatch* 10:23–29.
15. Pittock, A. B., L. A. Frakes, D. Jenssen, J. A. Peterson, and J. W. Zillman, eds. 1978. *Climatic change and variability: A southern perspective.* New York: Cambridge University Press.
16. Brown, L. R., ed. 1991. *The Worldwatch reader on global environmental issues.* New York: Norton.
17. Lents, J. M., and W. J. Kelly. 1993. Clearing the air in Los Angeles. *Scientific American* 269:32–39.

18. Blake, D. R., and F. S. Rowland. 1995. Urban leakage of liquefied petroleum gas and its impact on Mexico City air quality. *Science* 269:953.
19. Pountain, D. 1993 (May). Complexity on wheels. *Byte*, pp. 213–220.
20. Stern, A. C., R. T. Boubel, D. B. Turner, and D. L. Fox. 1984. *Fundamentals of air pollution*, 2nd ed. Orlando, Fla.: Academic Press.
21. Moore, C. 1995. Green revolution in the making. *Sierra* 80:50.
22. Kolstad, C. D. 2000. *Environmental economics*. New York: Oxford University Press.
23. Molnia, B. F. 1991. Washington report. *GSA Today* 1:33.
24. Crandall, R. W. 1983. *Controlling industrial pollution: The economics and politics of clean air.* Washington, D.C.: Brookings Institution.
25. Hall, J. V., A. M. Winer, M. T. Kleinman, F. W. Lurmann, V. Brajer, and S. D. Colome. 1992. Valuing the health benefits of clean air. *Science* 255:812–816.
26. Krupnick, A. J., and P. R. Portney. 1991. Controlling urban air pollution: A benefits-cost assessment. *Science* 252:522–528.
27. Lipfert, F. W., S. C. Morris, R. M. Friedman, and J. M. Lents. 1991. Air pollution benefit-cost assessment. *Science* 253:606.
28. Conlin, M. 2000 (June 5). Is your office killing you? *Business Week*, pp. 114–124.
29. U.S. Environmental Protection Agency. 1991. *Building air quality: A guide for building owners and facility managers.* EPA/400/1-91/033, DHHS (NIOSH) Pub. No. 91–114. Washington, D.C.: Environmental Protection Agency.
30. Zummo, S. M., and M. H. Karol. 1996. Indoor air pollution: Acute adverse health effects and host susceptibility. *Environmental Health* 58:25–29.
31. Committee on Indoor Air Pollution. 1981. *Indoor pollutants.* Washington, D.C.: National Academy Press.
32. Godish, T. 1997. *Air quality*, 3rd. ed. Boca Raton, Fla.: Lewis Publishers.
33. O'Reilly, J. T., P. Hagan, R. Gots, and A. Hedge. 1998. *Keeping buildings healthy.* New York: Wiley.
34. U.S. Environmental Protection Agency. 1986. *Radon reduction techniques for detached houses: Technical guidance.* EPA 625/5-86-019. Research Triangle Park, N.C.: Air and Energy Engineering Research Laboratory, Office of Research and Development, U.S. Environmental Protection Agency.
35. Osborne, M. C. 1988. *Radon-resistant residential new construction.* EPA 600/8-88/087. Research Triangle Park, N.C.: Air and Energy Engineering Research Laboratory, Office of Research and Development, U.S. Environmental Protection Agency.

Chapter 21 Critical Thinking Issue References

Khalil, M., and R. A. Rasmussen 1993. Arctic haze—patterns and relationships to regional signatures of trace gases. *Global Biogeochemical Cycles* 7(1):27–36.

Rahn, K. A. 1984. Who's polluting the Arctic? *Natural History* 93(5):31–38.

Shaw, G. E. 1995. The Arctic haze phenomenon. *Bulletin of the American Meteorological Society* 76(12):2403–2413.

Soroos, M. S. 1992. The odyssey of Arctic haze. *Environment* 34(10):6–27.

Young, O. R. 1990. Global commons: The Arctic in world affairs. *Technology Review* 93:52–61.

Chapter 22 Notes

1. Kane, R. P. 1998. Ozone depletion, UV-B changes, and increased cancer incidence. *International Journal.*
2. Brown, L. R., C. Flavin, and S. Postel. 1991. *Saving the planet: How to shape an environmentally sustainable global economy.* New York: Norton.
3. Rowland, F. S. 1990. Stratospheric ozone depletion of chlorofluorocarbons. *AMBIO* 19:281–292.
4. Smith, R. C., B. B. Prezelin, K. S. Baker, R. R. Bidigare, N. P. Boucher, T. Coley, D. Karentz, S. Macintyre, H. A. Matlick, D. Menzies, M. Ondrusek, Z. Wan, and K. J. Waters. 1992. Ozone depletion: Ultraviolet saturation and phytoplankton biology in Antarctic waters. *Science* 255:952–959.
5. NASA-GSFC. 2000. Stratospheric ozone. Accessed August 22, 2000 at http://see.gsfc.nasa.gov.
6. Environmental Protection Agency. 1995. *Protection of the ozone layer.* EPA 230-N-95-00. Washington, D.C.: U.S. EPA Office of Policy, Planning, and Evaluation and Office of Air and Radiation.
7. NASA. 2004. Average Antarctic miminum ozone concentration. Accessed 3/24/04 at http://jwocky.gsfc.nasa./gov/multi/min_ozone/gif.
8. Hamill, P., and O. B. Toon. 1991. Polar stratospheric clouds and the ozone hole. *Physics Today* 44:34–42.
9. Stolarski, R. S. 1988. The Antarctic ozone hole. *Scientific American* 258:30–36.
10. Molina, M. J., and F. S. Rowland. 1974. Stratospheric sink for chlorofluoromethanes: Chlorine-atom catalyzed distribution of ozone. *Nature* 249:810–812.
11. Brouder, P. 1986 (June). Annals of chemistry in the face of doubt. New Yorker, pp. 20–87.
12. Khalil, M. A. K., and R. A. Rasmussen. 1989. The potential of soils as a sink of chlorofluorocarbons and other man-made chlorocarbons. *Geophysical Research Letters* 16:679–682.
13. Rowland, F. S. 1989. Chlorofluorocarbons and the depletion of stratospheric ozone. *American Scientist* 77:36–45.
14. Toon, O. B., and R. P. Turco. 1991. Polar stratospheric clouds and ozone depletion. *Scientific American* 264:68–74.
15. Kerr, R. A. 1994. Antarctic ozone hole fails to recover. *Science* 266:217.
16. Webster, C. R., R. D. May, D. W. Toohey, L. M. Avallone, J. G. Anderson, P. Newman, L. Lait, M. Schoeberl, J. W. Elkins, and K. R. Chay. 1993. Chlorine chemistry on polar stratospheric cloud particles in the Arctic winter. *Science* 261:1130–1134.
17. Kerr, R. A. 1992. New assaults seen on Earth's ozone shield. *Science* 255:797–798.
18. Zurer, P. 1995. Record low ozone levels observed over Arctic. *Chemical & Engineering News* 73:8.
19. Horgan, J. 1992. Volcanic disruption. *Scientific American* 266:28–29.
20. Cutter Information Corp. 1996. Reports discuss present and future state of ozone layer. *Global Environmental Change Report* V, VIII, 21, no. 22, pp. 1–3. Dunster, B.C., Canada: Cutter Information Corp.
21. Shea, C. P. 1989. Mending the Earth's shield. *World Watch* 2:28–34.

22. Kerr, J. B., and C. T. McElroy. 1993. Evidence for large upward trends of ultraviolet-B radiation linked to ozone depletion. *Science* 262:1032–1034.
23. Atkinson, R. J., W. A. Matthews, P. A. Newman, and R. A. Plumb. 1989. Evidence of the mid-latitude impact of Antarctic ozone depletion. *Nature* 340:290–294.
24. Russell, J. M., M. Luo, R. J. Cicerone, and L. E. Deaver. 1996. Satellite confirmation of dominance of chlorofluorocarbons in the global stratospheric chlorine budget. *Nature* 379:526.
25. Showstack, R. 1998. Ozone layer is on slow road to recovery, new science assessment indicates. *Eos* 79(27):317–318.
26. Spurgeon, D. 1998. Surprising success of the Montreal protocol. *Nature* 389(6648):219.
27. Makhijani, A., and A. Bickel, 1990. Still working on the ozone hole. *Technology Review* 93:52–59.
28. Brown, L. R., N. Lenssen, and H. Kane. 1995. CFC production plummeting. In Worldwatch Institute, *Vital Signs* 1995. New York: Norton.
29. Shea, C. P. 1991. Disarming refrigerators. *World Watch* 4:36.
30. U.S. Environmental Protection Agency. 2003. Ozone depletion: accessed October 8, 2003 at http://www.epa.gov/ozone/index.html.
31. MacKenzie, D. 1990. Cheaper alternatives for CFCs. *New Scientist* 126:39–40.

Chapter 22 Critical Thinking Issue References

Brasseur, G., and C. Branier. 1992. Mount Pinatubo: Aerosols, chlorofluorocarbons, and ozone depletion. *Science* 257:1239–1241.

Kerr, R. A. 1993. Ozone takes a nose dive after the eruption of Mt. Pinatubo. *Science* 260:490–491.

Pohl, F., and J. P. Hogan. 1993. Ozone politics: They call this science? *Omni* 15(8):34–42, 91.

Rowland, F. S. 1993. President's lecture: The need for scientific communication with the public. *Science* 260:1571–1576.

Russel, J. M., M. Luo, R. J. Cicerone, and L. E. Deaver. 1996. Satellite confirmation of dominance of chlorofluorocarbons in the global stratospheric chlorine budget. *Nature* 379(6565):526.

Tabazadeh, A., and R. P. Turco. 1993. Stratospheric chlorine injection by volcanic eruptions: HCI scavenging and implications for ozone. *Science* 260:1082–1085.

Taubes, G. 1993 (June 11). The ozone backlash. *Science* 260, pp. 1580–1583.

Chapter 23 Notes

1. *The Butchart Gardens*. 1998. Victoria, B.C.: The Butchart Gardens, Ltd.
2. McKelvey, V. E. 1973. Mineral resource estimates and public policy. In D. A. Brobst and W. P. Pratt, eds., *United States mineral resources*, pp. 9–19. U.S. Geological Survey Professional Paper 820.
3. U.S. Department of the Interior, Bureau of Mines. 1993. *Mineral commodity summaries*, 1993. I 28.149:993. Washington, D.C.: U.S. Department of the Interior.
4. Meyer, H. O. A. 1985. Genesis of diamond: A mantle saga. *American Mineralogist* 70:344–355.
5. Kesler, S. F. 1994. *Mineral resources, economics, and the environment*. New York: Macmillan.
6. Smith, G. I., C. L. Jones, W. C. Culbertson, G. E. Erickson, and J. R. Dyni. 1973. Evaporites and brines. In D. A. Brobst and W. P. Pratt, eds., *United States mineral resources*, pp. 197–216. U.S. Geological Survey Professional Paper 820.
7. Awramik, S. A. 1981. The pre-Phanerozoic biosphere—three billion years of crises and opportunities. In M. H. Nitecki, ed., *Biotic crises in ecological and evolutionary time*, pp. 83–102. Spring Systematics Symposium. New York: Academic Press.
8. Margulis, L., and J. E. Lovelock. 1974. Biological modulation of the Earth's atmosphere. *Icarus* 21:471–489.
9. Lowenstam, H. A. 1981. Minerals formed by organisms. *Science* 211:1126–1130.
10. Bateman, A. M. 1950. *Economic mineral deposits*, 2nd ed. New York: Wiley.
11. Park, C. F., Jr., and R. A. MacDiarmid. 1970. *Ore deposits*, 2nd ed. San Francisco: Freeman.
12. Brobst, D. A., W. P. Pratt, and V. E. McKelvey. 1973. Summary of United States mineral resources. U.S. Geological Survey Circular 682.
13. Jeffers, T. H. 1991 (June). Using microorganisms to recover metals. *Minerals Today*. Washington, D.C.: U.S. Department of Interior, Bureau of Mines, pp. 14–18.
14. Island Copper Mine remediation program fact sheet: http://www.microbialtech.com/coppermine.html. Accessed March 29, 2005.
15. Sullivan, P. M., M. H. Stanczyk, and M. J. Spendbue. 1973. *Resource recovery from raw urban refuse*. U.S. Bureau of Mines Report of Investigations 7760.
16. Davis, F. F. 1972 (May). Urban ore. *California Geology*, pp. 99–112.
17. U.S. Geological Survey. 2000. *Minerals yearbook 1998—Recycling metals*. Accessed August 21, 2000 at http://minerals.usgs.gov.
18. Brown, L., N. Lenssen, and H. Kane. 1995. Steel recycling rising. In *Vital Signs 1995*. Worldwatch Institute.
19. Wellmar, F. W., and M. Kosinowoski. 2003. Sustainable development and the use of non-renewable sources. *Geotimes* 48(12): 14–17.

Chapter 23 Critical Thinking Issue Reference

Debus, K. H. 1990 (August/September). Mining with microbes. *Technology Review* 93:50.

Chapter 24 Notes

1. City of Toronto backgrounder on the Keele Valley Landfill situation: http://wx.toronto.ca/inter/it/newsrel.nsf/0/65b63ef988a3391b85256df60045ca0e?OpenDocument; accessed April 28, 2005.
2. City of Vaughan news release: Maple Valley Parks and Open Space Concept/Strategy Plan. http://www.city.vaughan.on.ca/newscentre/projects/city_programs02-06.cfm; accessed April 28, 2005.
3. Galley, J. E. 1968. Economic and industrial potential of geologic basins and reservoir strata. In J. E. Galley, ed., *Subsurface disposal in geologic basins: A study of reservoir strata*, pp. 1–19. American Association of Petroleum Geologists Memoir 10. Tulsa, Okla.: American Association of Petroleum Geologists.
4. Relis, P., and A. Dominski. 1987. *Beyond the crisis: Integrated waste management*. Santa Barbara, Calif.: Community Environmental Council.

5. Repa, D. W., and A. Blakey. 1996. Municipal solid waste disposal trends: 1996 update. *Waste Age* 27:42–54.
6. Council on Environmental Quality. 1973. *Environmental quality*—1973. Washington, D.C.: U.S. Government Printing Office.
7. Allenby, B. R. 1999. *Industrial ecology: Policy framework and implementation*. Upper Saddle River, N.J.: Prentice-Hall.
8. Garner, G., and P. Sampat. 1999 (May). Making things last: Reinventing of material culture. *The Futurist*, pp. 24–28.
9. Relis, P., and H. Levenson. *Discarding solid waste as we know it: Managing materials in the 21st century*. Santa Barbara, Calif.: Community Environmental Council.
10. Young, J. E. 1991. Reducing waste-saving materials. In L. R. Brown, ed., *State of the world*, 1991, pp. 39–55. New York: W. W. Norton.
11. Steuteville, R. 1995. The state of garbage in America: Part I. *BioCycle* 36:54.
12. Gardner, G. 1998 (January–February). Fertile ground or toxic legacy? *World Watch*, pp. 28–34.
13. McGreery, P. 1995. Going for the goals: Will states hit the wall? *Waste Age* 26:68–76.
14. Brown, L. R. 1999 (March–April). Crossing the threshold. *World Watch*, pp. 12–22.
15. Brown, Jeremy, Kenneth Green, Steve Hansen, and Liv Fredricksen. 2004. *Environmental Indicators*.6th Edition. Vancouver, B.C.: Fraser Institute. Available online at http://www.fraserinstitute.ca/admin/books/chapterfiles/2EnvInd2004complete.pdf#.
16. Rathje, W. L. 1991. Once and future landfills. *National Geographic* 179(5):116–134.
17. Schneider, W. J. 1970. Hydrologic implications of solid-waste disposal. 135(22). U.S. Geological Survey Circular 601F. Washington, D.C.: U.S. Geological Survey.
18. Thomas, V. M., and T. G. Spiro. 1996. The U.S. dioxin inventory: Are there missing sources? *Environmental Science & Technology* 30:82A–85A.
19. Turk, L. J. 1970. Disposal of solid wastes—acceptable practice or geological nightmare? In *Environmental Geology*, pp. 1–42. Washington, D.C.: American Geological Institute Short Course, American Geological Institute.
20. Hughes, G. M. 1972. Hydrologic considerations in the siting and design of landfills. *Environmental Geology Notes*, no. 51. Urbana: Illinois State Geological Survey.
21. Bergstrom, R. E. 1968. Disposal of wastes: Scientific and administrative considerations. *Environmental Geology* Notes, no. 20. Urbana: Illinois State Geological Survey.
22. Cartwright, K., and Sherman, F. B. 1969. Evaluating sanitary landfill sites in Illinois. *Environmental Geology Notes*, no. 27. Urbana: Illinois State Geological Survey.
23. Rahn, P. H. 1996. *Engineering geology*, 2nd ed. Upper Saddle River, N.J.: Prentice-Hall.
24. Bullard, R. D. 1990. *Dumping in Dixie: Race, class and environmental quality*. Boulder, CO: Westview Press.
25. Allen, Denise. 2001. Lessons from Africville. Shunpiking 6 September 2001. Available online at http://www.shunpiking.com/bhs/Lessons%20from%20Africville.htm/
26. Walker, W. H. 1974 Monitoring toxic chamical pollution from land disposal sites in humid regions. *Ground Water* 12: 213–218.
27. Watts, R. J. 1998. *Hazardous wastes*. New York: John Wiley & Sons.
28. Wilkes, A. S. 1980. Everybody's problem: Hazardous waste. SW-826. Washington, D.C.: U.S. Environmental Protection Agency, Office of Water and Waste Management.
29. Elliot, J. 1980. Lessons from Love Canal. *Journal of the American Medical Association* 240:2033–2034, 2040.
30. Kufs, C., and C. Twedwell. 1980. Cleaning up hazardous landfills. *Geotimes* 25:18–19.
31. Albeson, P. H. 1983. Waste management. *Science* 220:1003.
32. New York State Department of Environmental Conservation. 1994. *Remedial chronology: The Love Canal hazardous waste site*. New York State.
33. Kirschner, E. 1994. Love Canal settlement: OxyChem to pay New York state $98 million. *Chemical & Engineering News* 72:4–5.
34. Westervelt, R. 1996. Love Canal: OxyChem settles federal claims. *Chemical Week* 158:9.
35. Whittell, G. 2000 (November 29). Poison in paradise. (London) *Times* 2, p. 4.
36. U.S. Environmental Protection Agency. 2003. Key Dates in Superfund. Accessed November 6, 2003 at http://www.epa.gove/superfund/action/law/keydates.htm.
37. Bedient, P. B., H. S. Rifai, and C. J. Newell. 1994. *Ground water contamination*. Englewood Cliffs, N.J.: Prentice-Hall.
38. Huddleston, R. L. 1979. Solid-waste disposal: Land farming. *Chemical Engineering* 86:119–124.
39. McKenzie, G. D., and W. A. Pettyjohn. 1975. Subsurface waste management. In G. D. McKenzie and R. O. Utgard, eds., *Man and his physical environment: Readings in environmental geology*, 2nd ed., pp. 150–156. Minneapolis: Burgess Publishing.
40. National Research Council, Committee on Geological Sciences. 1972. *The earth and human affairs*. San Francisco: Canfield Press.
41. Cox, C. 1985. The buried threat: *Getting away from land disposal of hazardous waste*. No. 115-5. California Senate Office of Research.
42. Environment Canada's web site on solid and hazardous waste management in Canada: http://www.ec.gc.ca/wastes_e.html. Accessed March 29, 2005.
43. Lenssen, N. 1989 (July–August). The ocean blues. *World Watch*, pp. 26–35.
44. U.S. Environmental Protection Agency. 2000. Forward pollution protection: The future look of environmental protection. Accessed August 12, 2000, at http://www.epa.gov/p2/p2case. htm#num4.

Chapter 24 Critical Thinking Issue References

Schueller, G. H. 2002. Is recycling on the skids? *Wasting Away on Earth* 24(3), pp. 21–23.

Chapter 25 Notes

1. There is a wealth of information about the softwood lumber dispute on the Internet, and that information is regularly updated as negotiations proceed. Two good sources are the Canadian federal government's International Trade Canada site (http://www.dfait-mae ci.gc.ca/eicb/softwood/menu-en.asp), and the British Columbia government's Ministry of Forests and Range softwood lumber site (http://www.for.gov.bc.ca/HET/

Softwood/). Both sites contain extensive current information and links to numerous other relevant sources.

2. Ontario Medical Association. 2000. "Illness Costs of Air Pollution. Phase II: Estimating Health and Economic Damages." Available online at http://www.oma.org/phealth/report/techrep.pdf.
3. Ontario Medical Association. 1998. "The Health Effects of Ground-Level Ozone." Position paper, available online at http://www.oma.org/phealth/ground.htm.
4. Moore, C. E. 1995. Poisons in the air. *International Wildlife* 25:38–45.
5. Pope, C. A., III, D. V. Bates, and M. E. Raizenne. 1995. Health effects of particulate air pollution: Time for reassessment: *Environmental Health Perspectives* 103:472–480.
6. Hardin, G. 1968. The tragedy of the commons. *Science* 162:1243–1248.
7. Clark, C. W. 1973. The economics of overexploitation. *Science* 181:630–634.
8. Freudenburg, W. R. 2004. Personal communication.
9. Gunn, J. M., ed. 1995. *Restoration and recovery of an industrial region: Progress in restoring the smelter-damaged landscape near Sudbury, Canada.* New York: Springer-Verlag.
10. Costanza, R., et. al. 1997. The value of the world's ecosystem services and natural capital. *Nature* 387:253–260.
11. James, A., K. T. Gaston, and A. Blamford. 2001. Can we afford to conserve biodiversity? *BioScience* 51(1):43–52.
12. Litton, R. B. 1972. Aesthetic dimensions of the landscape. In J. V. Krutilla, ed., *Natural environments.* Baltimore, Johns Hopkins University Press.
13. Schwing, R. C. 1979. Longevity and benefits and costs of reducing various risks: *Technological Forecasting and Social Change* 13:333–345.
14. Gori, G. B. 1980. The regulation of carcinogenic hazards. *Science* 208:256–261.
15. Cairns, J. Jr., 1980. Estimating hazard. *BioScience* 20:101–107.
16. James, A., K. T. Gaston, and A. Blamford. 2001. Can we afford to conserve biodiversity? *BioScience* 51(1):43–52.
17. Ostro, B. D. 1980. Air pollution, public health, and inflation. *Environmental Health Perspectives* 345:185–189.
18. Office of Technology Assessment. 1991. *Changing by degrees: Steps to reduce greenhouse gases.* Washington, D.C.: U.S. Superintendent of Documents.
19. D'Arge, R. 1989. Ethical and economic systems for managing the global commons. In D. B. Botkin, M. Caswell, J. E. Estes, and A. Orio, eds., *Changing the global environment: Perspectives on human involvement*, pp. 327–337. New York: Academic Press.
20. Rogers, A. 1993. The Earth summit: A planetary reckoning. Global View Press.
21. Baumol, W. J., and W. E. Oates, 1979. *Economics, environmental policy, and the quality of life.* Englewood Cliffs, N.J.: Prentice-Hall.
22. Clark, C. W. 1981. Economics of fishery management. In T. L. Vincent and J. M. Skowronski, eds., *Renewable resource management: Lecture notes in biomathematics*, pp. 9–111. New York: Springer-Verlag.
23. The Jumbo Glacier Resort home page (http://www.jumboglacierresort.com) presents the perspective of the proponents of this project; an opposing view is provided by the Jumbo Creek Conservation Society (http://www.jumbowild.com/).

Chapter 25 Critical Thinking Issue References

Anthony, V. C. 1993. The state of groundfish resources off the northeastern United States. *Fisheries* 18 (3):12–17.

Correia, S. J. 1992 (3rd quarter). Flounder population declines: Overfishing or pollution! *Division of Marine Fisheries News.* Boston: Massachusetts Division of Marine Fisheries.

How to fish. 1988 (December 10). *The Economist* 309 (7580):93–96.

Keen, E. A. 1991. Ownership and productivity of marine fisheries resources. *Fisheries* 16:18–22.

Lawren, B. 1992. Net loss. *National Wildlife* 30(6):47—52.

Leal, D. R. 1992 (July 30). Using property rights to regulate fish harvest. *Christian Science Monitor* 84:18.

National Marine Fisheries Service. 1995. *Status of the fishery resources off the northeastern United States for 1994.* Woods Hole, Mass.: U.S. Department of Commerce, NOAA, NMFS Northeast Fisheries Science Center.

Pierce, D. 1992 (2nd quarter). New England council to cut fishing effort in half over next 5 years. *Division of Marine Fisheries News.* Boston: Massachusetts Division *of Marine Fisheries.*

Satchell, M. 1992. *The rape of the oceans.* U.S. News and World Report 112(24):64–75.

Chapter 25 Build Your Environmental Skills References

Estrin, David and John Swaigen. 1993. Environment on Trial.3rd Edition. Toronto, Ontario: Emond Montgomery Publishers.

Greenbaum, Allan, Alex Wellington, and Ellen Baar. 1995. *Social Conflict and Environmental Law: Ethics, Economics and Equity.* North York, Ontario: Captus Press.

Saxe, Dianne. 1990. *Environmental Offences: Corporate Responsibility and Executive Liability.* Aurora, Ontario: Canada Law Book.

Chapter 26 Notes

1. Craik Sustainable Living Project: www.craikecovillage.ca; accessed May 14, 2005
2. Ruddy, Jenn. 2004. Craik Eco-village: a model for sustainable living. The Commonwealth 64(5) (December 2004). Available online at www.saskndp.com/cw/64.5/craikecovillage.html.
3. Hawken, P., A. Lovins, and L. H. Lovins. 1999. *Natural capitalism.* Boston: Little, Brown.
4. Hubbard, B. M. 1998. *Conscious evolution.* Novato, Calif.: New World Library.
5. Haub, C., and D. Cornelius. 2000. *World population data sheet.* Washington, D.C.: Population Reference Bureau.
6. Population Reference Bureau. 1998. *World and regional population.* Washington, D.C.: Population Reference Bureau.
7. World Resources Institute. 1999. *Urban growth.* Washington, D.C.: World Resources Institute.
8. Leibbrand, K. 1970. *Transportation and town planning.* Translated by N. Seymer. Cambridge, Mass.: MIT Press.
9. Mumford, L. 1972. The natural history of urbanization. In R. L. Smith, ed., *The ecology of man: An ecosystem approach*, pp. 140–152. New York: Harper & Row.
10. Cronon, W. 1991. *Nature's metropolis: Chicago and the great West.* New York: Norton.

11. Detwyler, T. R., and M. G. Marcus, eds. 1972. *Urbanization and the environment: The physical geography of the city.* North Scituate, Mass.: Duxbury Press.
12. Butti, K., and J. Perlin. 1980. *A golden thread: 2500 years of solar architecture and technology.* New York: Cheshire.
13. Ford, A. B., and O. Bialik. 1980. Air pollution and urban factors in relation to cancer mortality. *Archives of Environmental Health* 35:350–359.
14. Moll, G. 1989. Designing the ecological city. *American Forests* 85:61–64.
15. Nadel, I. B., C. H. Oberlander, and L. R. Bohm. 1977. *Trees in the city.* New York: Pergamon.
16. Moll, G., P. Rodbell, B. Skiera, J. Urban, G. Mann, and R. Harris. 1991 (April/May). Planting new life in the city. *Urban Forests,* pp. 10–20. Washington, D.C.: American Forestry Association.
17. Dreistadt, S. H., D. L. Dahlsten, and G. W. Frankie. 1990. Urban forests and insect ecology. *BioScience* 40:192–198.
18. Burton, J. A. 1977. *Worlds apart: Nature in the city.* Garden City, N.Y.: Doubleday.
19. Leedly, D. L., and L. W. Adams. 1984. *A guide to urban wildlife management.* Columbia, Md.: National Institute for Urban Wildlife.
20. Tylka, D. 1987. Critters in the city. *American Forests* 93:61–64.
21. Adams, L. W., and L. E. Dove. 1989. *Wildlife reserves and corridors in the urban environment.* Columbia, Md.: National Institute for Urban Wildlife.
22. http://www.responsiblewildlifemanagement.org/bubo nic_plague.htm
23. Reps, J. W. 1965. *The making of urban America: A history of city planning in the United States,* 2nd ed. Princeton, N.J.: Princeton University Press.
24. Soulé, M., and R. Noss. 1998. Rewilding and biodiversity: complementary goals for continental conservation. *Wild Earth* 8(3):18-28.
25. Jusserand, J. 1897. *English wayfaring life in the Middles Ages* (XIVth Century). London: T. Fisher Unwin.
26. Steiner, F. 1983. Regional planning: Historic and contemporary examples. *Landscape Planning* 10:297–315.
27. Jones, N.E., W.M. Tonn, G.J. Scrimgeour, and C. Katopodis. 2003. Productive capacity of an artificial stream in the Canadian Arctic: assessing the effectiveness of fish habitat compensation. *Canadian Journal of Fisheries and Aquatic Science* 60:849-863.
28. Renner, M. 2002. Breaking the link between resources and repression. In *Worldwatch Institute state of the world* 2002. New York: Norton.

Chapter 26 Critical Thinking Issue References

Fautley, C. 2000 (October 15). Geology of the channel tunnel. *London Times*: Available at www.geologyshop.co.uk/chtung. htm.

PHOTO CREDITS

Part I Opener: Carole Ann Lacroix, Herbarium Curator, University of Guelph.

Chapter 1 Fig. 1.1: Ian Mackenzie. Fig. 1.4b: Corbis-Bettman. Fig. 1.5 a: Peter Turnley/Corbis Images. Fig. 1.5 b: Viviane Moos/SABA. Fig. 1.6: C. Mayhew & R. Simmon (NASA/GSFC), NOAA/ NGDC, DMSP Digital Archive. Fig. 1.7: Isobel W. Heathcote. Fig. 1.9 a: Prof. Randall Schaetzi. Fig. 1.9b: © J.P. Ferrero/Jacana/Photo Researchers. Fig. 1.10: Daniel Botkin. Fig. 1.11: Bart Hawkins. Fig. 1.12a: Florida Keys National Marine Sanctuary. Fig. 1.12b: OAR/National Undersea Research Program (NURP); University of North Carolina at Wilmington.

Chapter 2 Fig. 2.1: Dr. Karen Scott. Fig. 2.4a: Gideon Mendel// Magnum Photos, Inc. Fig. 2.5: Reprinted with permission from Weir, et al., SCIENCE 297:981 (2002) AAAS. Fig. 2.6: Dietmar Nill/Nature Picture Library. Fig. 2.12: Andrew Bryant. Fig. 2.14a: Raymond Gehman/Corbis Images. Fig. 2.14b: Renee Lynn/Photo Researchers.

Chapter 3 Fig. 3.1: Gerard D. Hertel, West Chester University, www.forestryimages.org. Fig. 3.2: M. Renaudeau/HOA-QUI/ Alpha Press. Fig. 3.3: Elspeth Evans. Fig. 3.4: Isobel W. Heathcote. Fig. 3.8: Isobel W. Heathcote. Fig. 3.9: Tom Till/DRK Photo. Fig. 3.12 a and b: Isobel W. Heathcote.

Chapter 4 Fig. 4.1: World Food Programme.

Chapter 5 Opener: Luke Copland. Fig. 5.1a: © Parks Canada, J. Butterill, 11. Fig. 5.1b: Frank Granshaw, Artemis Science. Fig. 5.8: © Parks Canada, W. Lynch, 19. Fig. 5.9: Manfred Gottschalk/Tom Stack & Associates.

Chapter 6 Fig. 6.1a: Bill Ivy/Ivy Images. Fig. 6.1b: O. Spielman/ CNRI/Phototake. Fig. 6.1c: Stephen J. Krasemann/Photo Researchers. Fig. 6.1d: USDA APHIS PPQ Archives, www.forestryimages.org. Fig. 6.1e: Alvin E. Staffan/Photo Researchers. Fig. 6.2a: © Photodisk/Getty Images. Fig. 6.2 b: Courtesy of Parks Canada; Douglas Driediger, artist Fig. 6.7a: © Parks Canada, W. Lynch, 11. Fig 6.7b: © Her Majesty the Queen in Right of Canada. Fig. 6.8a: Photographer's Choice/Getty Images. Fig. 6.8b: Riding Mountain National Park of Canada. Fig. 6.9: © Parks Canada, J. Butterill, 128. Fig. 6.10: Courtesy USDA Forest Service.

Chapter 7 Fig. 7.1: Nova Scotia Tourism, Culture and Heritage. Fig. 7.2: Purchased with funds donated by Blount Canada Ltd., with assistance from The Canada Council, 1983 Macdonald Stewart Art Centre Collection - MS983.001. Fig. 7.3a: Oliver Meckes/Photo Researchers. Fig. 7.3b: Oliver Meckes/Photo Researchers. Fig. 7.3c: Adam Jones/Photo Researchers. Fig. 7.4: Phanie/Photo Researchers, Inc. Fig. 7.5a: Zoë Belk. Fig. 7.5b: Geoff Scott, University of Winnipeg. Fig. 7.6: © Sinclair Stammers/Science Photo Library/Photo Researchers. Fig. 7.7a: © Parks Canada, W. Lynch, 23. Fig. 7.10a: Dr. Patricia Schultz/ Peter Arnold, Inc. Fig. 7.10b: Kari Lounatmaa/Photo Researchers.

Chapter 8 Fig 8.1a: Stephen G. Maka/DRK Photo. Fig 8.1b: Stephan C. White/Kansas Department of Agriculture Fig 8.5a: Stephen J. Krasemann/DRK Photo. Fig 8.5b: Kim Heacox/DRK Photo. Fig 8.5c: William M. Ciesla, Forest Health Management International, www.forestryimages.org. Fig 8.7a: Ferrero/Labat/ Auscape International Pty. Ltd. Fig 8.7b: Toni Angermayer/Photo Researchers. Fig 8.7c: Hawkindale Emu Farms. Fig 8.9: Courtesy NASA. Fig 8.11: © Parks Canada, W. Lynch, 22. Fig 8.12: Isobel W. Heathcote. Fig 8.13: David McEown. Fig 8. 14 Geoff Scott, University of Winnipeg. Fig 8.15 Isobel W. Heathcote. Fig 8.16 Reid Kreutzwiser. Fig 8.17 Isobel W. Heathcote. Fig 8.18 Isobel W. Heathcote. Fig 8.19 Elspeth Evans. Fig 8.24 F. Denhz/BIOS/ Peter Arnold, Inc.

Chapter 9 Fig. 9.1 Isobel W. Heathcote. Fig. 9.2 Heather Angel. Fig 9.3 Mona Lisa Production/Photo Researchers, Inc. Fig. 9.4a: Dr. Jianwu Wang, South China Agricultural University. Fig. 9.4b: Joyce Photographics/Photo Researchers, Inc. Fig. 9.6a: Daniel Botkin. Fig. 9.6b: Daniel Botkin. Fig. 9.6c: Daniel Botkin. Fig. 9.6d: Daniel Botkin. Fig. 9.9: © Al Giddings/Al Giddings Images.com. Fig. 9.10: Isobel W. Heathcote.

Part II Opener: Isobel W. Heathcote.

Chapter 10 Fig. 10.1a: © T. Crabe/Pinery Provincial Park Fig. 10.1b: © T. Crabe/Pinery Provincial Park. Fig. 10.2: Courtesy NASA Goddard Space Flight Center and the authors of Hall, F.G., D.B. Botkin, D.E. Strbel, K.D. Woods and S.J. Goetz, 1991, Large Scale Patterns in Forest Succession As Determined by Remote Sensing, Ecology, 72: 628-640. Fig. 10.4a: Masha Nordbye/Bruce Coleman, Inc. Fig. 10.4b: © Grant Heilman Photography. Fig. 10.5a: Isobel W. Heathcote. Fig. 10.5b: Isobel W. Heathcote. Fig. 10.5c: Isobel W. Heathcote. Fig. 10.6a: © Barret&MacKay Photo/bmpstock.com. Fig. 10.7: Marcy Sangret, Corporation of Delta. Fig. 10.11: David Tomlinsin/Windrush Photos.

Chapter 11 Fig. 11. 1b: Stone/Getty Images Fig. 11.3: Staffan Widstrand/Corbis Images Fig. 11.5: Mark A. Ernste/UNEP GRID, Sioux Falls (http://na.unep.net). Fig. 11.6: © AP/ Wide World Photos. Fig. 11.8a: United States Department of Agriculture Natural Resources Conservation Service. Fig. 11.8b: United States Department of Agriculture Natural Resources Conservation Service. Fig. 11.8c: United States Department of Agriculture Natural Resources Conservation Service. Fig. 11.12: Rossi/Tips Images. Fig. 11.13: Blue Revolution Consulting Group. Fig. 11.17: J. Victolero/International Rice Research Institute. Fig. 11.21: Pallava Bagla/Corbis Images.

Chapter 12 Fig. 12.1: John Sylvester Photography. Fig. 12.2a: Corbis-Bettmann. Fig. 12.2b: U.S. Dept. of Agriculture Photography Ctr. Fig. 12.3: Bettmann/Corbis Images. Fig. 12.4a: Thomas Siccama, Yale School of Forestry. Fig. 12.4b: Thomas Siccama, Yale School of Forestry. Fig. 12.5a: USDA Natural Resources Conservation Service. Fig. 12.5b: USDA Natural Resources Conservation Service. Fig. 12.8: Larry A. Hull. Fig. 12.9a: David Epstein, Howard Russell, Integrated Pest Management Program, Michigan State University. Fig. 12.9b: ©Remi Coutin/OPIE. Fig. 12.11a: Simko/Visuals Unlimited. Fig. 12.11b: USDA Natural Resources Conservation Service. Fig. 12.11c: Bob Gurr/DRK Photo. Fig. 12.12: Isobel W. Heathcote. Fig. 12.13: USDA Natural Resources Conservation Service. Fig. 12.15a: Bill Bachman/Photo Researchers. Fig. 12.15b: USDA Natural Resources Conservation Service.

Chapter 13 Fig. 13.1: Tony Difazio, Catfish Creek Conservation Authority. Fig. 13.3: Isobel W. Heathcote. Fig. 13.4: David Muench Photography. Fig. 13.8: Earthroots. Fig. 13.10: Zoë Belk Fig. 13.14a: Steve McCurry//Magnum Photos, Inc. Fig. 13.14b: Isobel W. Heathcote. Fig. 13.15: NASA/Science Source/Photo Researchers. Fig. 13.17: Parks Canada. Fig. 13.18: U.S. Fish and Wildlife Service/Liaison Agency, Inc./Getty Images. Fig. 13.19:

U.S. Fish and Wildlife Service/Liaison Agency, Inc./Getty Images.

Chapter 14 Fig. 14.1: Stephen J. Krasemann/DRK Photo. Fig. 14.2: W. Grenfell/Visuals Unlimited. Fig. 14.5: © Parks Canada, W. Lynch, 01 Fig. 14.6a: Tom Bledsoe/DRK Photo. Fig. 14.6b: Smithsonian American Art Museum, Washington, D.C./Art Resource, N.Y. 11832-33. Fig. 14.7a: Ken Lucas/ Visuals Unlimited. Fig. 14.8a: Richard Ellis/Photo Researchers. Fig. 14.8b: Marc Epstein/DRK Photo. Fig. 14.9a: Francis Gohier/ Photo Researchers. Fig. 14.9b: Seapics.com. Fig. 14.13a: Tom McHugh/Photo Researchers, Inc. Fig. 14.16: John Cunningham/ Visuals Unlimited. Fig. 14. 17a: Ron Austing Photography. Fig. 14.17b: Ross Frid/Visuals Unlimited. Fig. 14.18a: Maurice Jassak. Fig. 14.18b: © Parks Canada, T. Grant photographer, 77. Fig. 14.18c: Dianne Blell/Peter Arnold, Inc. Fig. 14.19: U.S. Fish & Wildlife Service. Fig. 14.20a: © Parks Canada, J. Pleau phtographer, 140.

Chapter 15 Opener: Stephen Dalton/Photo Researchers. Fig. 15.2a: T. Orban/Corbis Sygma. Fig. 15.2b: David & Peter Turnley/Corbis Images. Fig. 15.3: Prof. Ed Keller. Fig. 15.4a: J.K. Enright. Fig. 15.4b: © Mike Grandmaison Photography. Fig. 15.7: O. Franken/Corbis/Sygma. Fig. 15.8c: Martin Bond/Photo Researchers. Fig. 15.9: Isobel W. Heathcote. Fig. 15.15: Elspeth Evans

Chapter 16 Opener a and b: C. Elvidge, U.S. Air Force, DMSP). Fig. 16.3: Keith Sarver, www.keithsarver.com. Fig. 16.6: Studio MMA Atelier d'architecture. Fig. 16.7: Reproduced with the permission of Toyota Canada Inc., all rights reserved. Fig. 16.8: http://www.nrcan.gc.ca/es/etb/cetc/cetc01/images/cars_rtf7_full.jpg, Natural Resources Canada. Reproduced with the permission of the Minister of Public Works and Government Services Canada, 2005.

Chapter 17 Opener: Isobel W. Heathcote. Fig. 17.6a: George Hunter/Stone/Getty Images. Fig. 17.6b: Petro-Canada. Fig. 17.13b: Roger Ressmeyer/Starlight/Corbis Images. Fig. 17.16: Courtesy the Princeton Plasma Physics Laboratory. Fig. 17.18: Corbis-Bettmann. Fig. 17.23: Novosti/Photo Researchers, Inc. Fig. 17.24a: Courtesy of Nuclear Waste Management Organization. Fig. 17.24b: Courtesy of Ontario Power Generation. Fig. 17.28: Tom Bean. Fig. 17.31a: H. Gruyaeart//Magnum Photos, Inc. Fig. 17.31b: Prof. Ed Keller. Fig. 17.32: T.J. Florian/ Rainbow. Fig. 17.36: C.Delis/Explorer. Fig. 17.37b: Courtesy of Ontario Power Generation. Fig. 17.41: Courtesy Pacific Gas & Electric Company.

Part III Opener: Isobel W. Heathcote.

Chapter 18 Fig. 18.1: © Parks Canada, J. leau photographer, 2002. Fig. 18.11a: © O. Bierwagen, Ivy Images. Fig. 18.11b: Tony Difazio, Catfish Creek Conservation Authority. Fig. 18.11c: Isobel W. Heathcote. Fig. 18.13: Courtesy Ed Keller. Fig. 18.14: Liu Liqun/Corbis Images. Fig. 18.16a: Peter Boyer, International Joint Commission. Fig. 18.16b: Isobel W. Heathcote. Fig. 18.17: Isobel W. Heathcote. Fig. 18.18a: http://gsc.nrcan.gc.ca/floods/redriver/images/view4_100.jpg, G.R. Brooks, Natural Resources Canada. Reproduced with the permission of the Minister of Public Works and Government Services Canada, 2005. Fig. 18.18b: Grand River Conservation Authority. Fig. 18.19: © AP/Wide World Photos. Fig. 18.21: Fraser Basin Council. Fig. 18.22: Fraser Basin Council. Fig. 18.23: Fraser Basin Council.

Chapter 19 Fig. 19.1: © AP/Wide World Photos. Fig. 19.2: Grand River Conservation Authority. Fig. 19.4: Andrew McGlinchey. Fig. 19.5: Isobel W. Heathcote. Fig. 19.6a: Grand River Conservation Authority. Fig. 19.6b: A. Savtchenko, (NASA/GSFC). Fig. 19.8: Courtesy of Exxon Valdez Oil Spill Trustee Council. Fig. 19.10: Isobel W. Heathcote. Fig. 19.11: Ward Chesworth. Fig. 19.12: David Woodfall/DRK Photo. Fig. 19.13: National Water Research Institute, Environment Canada. Fig. 19.20: Dr. Hans Schreier. Fig. 19.21: Prof. Ed Keller.

Chapter 20 Opener: Johnny Johnson/DRK Photo. Fig. 20.7: Roger Ressmeyer/Corbis Images. Fig. 20.19: USGS/Cascades Volcano Observatory. Fig. 20.28: George D. Lepp/Corbis. Fig. 20.29: Uwe Walz/Corbis Images.

Chapter 21 Fig. 21.1: Courtesy NASA Fig. 21.2: F. Hoffman/ The Image Works. Fig. 21.3: Courtesy Ed Keller. Fig. 21.4: Jules Bucher/Photo Researchers. Fig. 21.14a and b: Don & Pat Valenti. Fig. 21.15: Big Stock Photography. Fig. 21.21a and b: Jim Mendenhall. Fig. 21.22b: Isobel W. Heathcote. Fig. 21.25a: Andrew Syred/Photo Researchers. Fig. 21.25b: Oliver Meckes/ Photo Researchers.

Chapter 22 Opener: Edward Belk. Fig. 22.10: Courtesy NASA.

Chapter 23 Opener: Courtesy Ed Keller. Fig. 23.1: Helen Thompson/Maxximages.com. Fig. 23.5a: This map was taken from the Atlas of Canada http://atlas.gc.ca/© 2005. Produced under license from Her Majesty the Queen in Right of Canada, with permission of Natural Resources Canada. Fig. 23.5b: Photo by David Sinclair. Used with permission of the Minister of Public Works and Government Services Canada, 2005 and Courtesy of Natural Resources Canada, Geological Survey of Canada. Fig. 23.6a: Isobel W. Heathcote. Fig. 23.7: Ed Keller. Fig. 23.8: Courtesy Elk Valley Coal Corporation. Fig. 23.9a: Danielle Fortin, University of Ottawa. Fig. 23.9b: Danielle Fortin, University of Ottawa. Fig. 23.10: Ministry of Northern Development and Mines, © Queen's Printer, 2003. Reproduced with permission. Fig. 23.11a: Craig Aurness/Corbis Images. Fig. 23.11b: © R. Maissonneuve/Publicphoto/Photo Researchers Fig. 23.11c: Marli Miller. Fig. 23.12: Danielle Fortin, University of Ottawa.

Chapter 24 Fig. 24.1: Courtesy Conestoga-Rovers & Associates. Fig. 24.2: Courtesy of the City of Guelph, Environment and Transportation Group. Fig. 24.4: Isobel W. Heathcote. Fig. 24.7b: Colin Jewell. Fig. 24.8: Courtesy NY State Dept. of Environmental Conservation. Fig. 24.9: © Pete Saloutos/ CORBIS.

Part IV Opener: Will Gorlitz.

Chapter 25 Fig. 25.1: © Greenpeace/Barclay. Fig. 25.2: Isobel W. Heathcote. Fig. 25.3b: © Mark Spalding, PhD. Fig. 25.3c: © Mark Spalding, PhD. Fig. 25.5: C. Bradley Simmons/Bruce Coleman, Inc. Fig. 25.6: Isobel W. Heathcote. Fig. 25.8: © Fred Voetsch. Fig. 25.10: Pheidias Project Management Corp. Fig. 25.11: © Greenpeace/Vinai. Fig. 25.12a: Tom Van Sant/ Photo Researchers.

Chapter 26 Fig. 26.1a and b: Linda McMillan. Fig. 26.3: Patrick Ward/Corbis Images. Fig. 26.4a: Isobel W. Heathcote. Fig. 26.4b Isobel W. Heathcote. Fig. 26.5: © Danny Lehman/CORBIS. Fig. 26.8: Isobel W. Heathcote. Fig. 26.9a: Isobel W. Heathcote. Fig. 26.9b: National Capital Commission. Fig. 26.10: Robert Holmes/Corbis Images. Fig. 26.12: Ralph Ginzburg/Peter Arnold, Inc. Fig. 26.16b: Tennessee Valley Authority.

INDEX

C

D

F

G

H

I

J

K

L

M

N

O

P

T

U

V

W

Y

Z